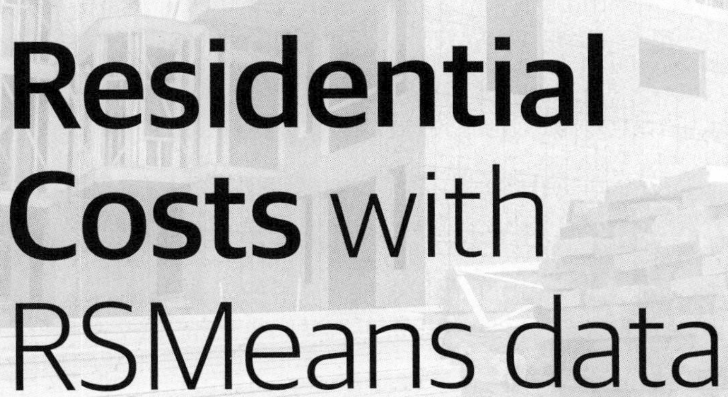

Residential Costs with RSMeans data

Thomas Lane, Senior Editor

2019

38th annual edition

Chief Data Officer
Noam Reininger

Engineering Director
Bob Mewis *(1, 3, 4, 5, 11, 12)*

Contributing Editors
Brian Adams *(21, 22)*
Christopher Babbitt
Sam Babbitt
Michelle Curran
Matthew Doheny *(8, 9, 10)*
John Gomes *(13, 41)*
Derrick Hale, PE *(2, 31, 32, 33, 34, 35, 44, 46)*
Michael Henry
Joseph Kelble *(14, 23, 25,)*
Charles Kibbee
Gerard Lafond, PE
Thomas Lane *(6, 7)*
Jake MacDonald
Elisa Mello
Michael Ouillette *(26, 27, 28, 48)*
Gabe Sirota
Matthew Sorrentino
Kevin Souza
David Yazbek

Product Manager
Andrea Sillah

Production Manager
Debbie Panarelli

Production
Jonathan Forgit
Mary Lou Geary
Sharon Larsen
Sheryl Rose

Data Quality Manager
Joseph Ingargiola

Technical Support
Kedar Gaikwad
Todd Klapprodt
John Liu

Cover Design
Blaire Collins

Data Analytics
Tim Duggan
Todd Glowac
Matthew Kelliher-Gibson

**Numbers in italics are the divisional responsibilities for each editor. Please contact the designated editor directly with any questions.*

RSMeans data from Gordian
Construction Publishers & Consultants
1099 Hingham Street, Suite 201
Rockland, MA 02370
United States of America
1.800.448.8182
RSMeans.com

Printed in the United States of America
ISSN 0896-8624
ISBN 978-1-946872-66-1

0179

$266.99 per copy (in United States)
Price is subject to change without prior notice.

Related Data and Services

Our engineers recommend the following products and services to complement *Residential Costs with RSMeans data:*

Annual Cost Data Books

2019 Square Foot Costs with RSMeans data
2019 CPG Residential Repair and Remodeling Costs with RSMeans data

Reference Books

Estimating Building Costs
RSMeans Estimating Handbook
Green Building: Project Planning & Estimating
How to Estimate with RSMeans data
Plan Reading & Material Takeoff
Project Scheduling & Management for Construction
Universal Design Ideas for Style, Comfort & Safety

Seminars and In-House Training

Unit Price Estimating
Training for our online estimating solution
Practical Project Management for Construction Professionals
Scheduling with MSProject for Construction Professionals
Mechanical & Electrical Estimating

RSMeans data Online

For access to the latest cost data, an intuitive search, and an easy-to-use estimate builder, take advantage of the time savings available from our online application. To learn more visit: RSMeans.com/2019online.

Enterprise Solutions

Building owners, facility managers, building product manufacturers, and attorneys across the public and private sectors engage with RSMeans data Enterprise to solve unique challenges where trusted construction cost data is critical. To learn more visit: RSMeans.com/Enterprise.

Custom Built Data Sets

Building and Space Models: Quickly plan construction costs across multiple locations based on geography, project size, building system component, product options, and other variables for precise budgeting and cost control.

Predictive Analytics: Accurately plan future builds with custom graphical interactive dashboards, negotiate future costs of tenant build-outs, and identify and compare national account pricing.

Consulting

Building Product Manufacturing Analytics: Validate your claims and assist with new product launches.

Third-Party Legal Resources: Used in cases of construction cost or estimate disputes, construction product failure vs. installation failure, eminent domain, class action construction product liability, and more.

API

For resellers or internal application integration, RSMeans data is offered via API. Deliver Unit, Assembly, and Square Foot Model data within your interface. To learn more about how you can provide your customers with the latest in localized construction cost data visit: RSMeans.com/API.

Table of Contents

Foreword

The Value of RSMeans data from Gordian

Since 1942, RSMeans data has been the industry-standard materials, labor, and equipment cost information database for contractors, facility owners and managers, architects, engineers, and anyone else that requires the latest localized construction cost information. More than 75 years later, the objective remains the same: to provide facility and construction professionals with the most current and comprehensive construction cost database possible.

With the constant influx of new construction methods and materials, in addition to ever-changing labor and material costs, last year's cost data is not reliable for today's designs, estimates, or budgets. Gordian's cost engineers apply real-world construction experience to identify and quantify new building products and methodologies, adjust productivity rates, and adjust costs to local market conditions across the nation. This adds up to more than 22,000 hours in cost research annually. This unparalleled construction cost expertise is why so many facility and construction professionals rely on RSMeans data year over year.

About Gordian

Gordian originated in the spirit of innovation and a strong commitment to helping clients reach and exceed their construction goals. In 1982, Gordian's chairman and founder, Harry H. Mellon, created Job Order Contracting while serving as chief engineer at the Supreme Headquarters Allied Powers Europe. Job Order Contracting is a unique indefinite delivery/indefinite quantity (IDIQ) process, which enables facility owners to complete a substantial number of repair, maintenance, and construction projects with a single, competitively awarded contract. Realizing facility and infrastructure owners across various industries could greatly benefit from the time and cost saving advantages of this innovative construction procurement solution, he established Gordian in 1990.

Continuing the commitment to providing the most relevant and accurate facility and construction data, software, and expertise in the industry, Gordian enhanced the fortitude of its data with the acquisition of RSMeans in 2014. And in an effort to expand its facility management capabilities, Gordian acquired Sightlines, the leading provider of facilities benchmarking data and analysis, in 2015.

Our Offerings

Gordian is the leader in facility and construction cost data, software, and expertise for all phases of the building life cycle. From planning to design, procurement, construction, and operations, Gordian's solutions help clients maximize efficiency, optimize cost savings, and increase building quality with its highly specialized data engineers, software, and unique proprietary data sets.

Our Commitment

At Gordian, we do more than talk about the quality of our data and the usefulness of its application. We stand behind all of our RSMeans data—from historical cost indexes to construction materials and techniques—to craft current costs and predict future trends. If you have any questions about our products or services, please call us toll-free at 800.448.8182 or visit our website at gordian.com.

How the Cost Data Is Built: An Overview

Unit Prices*

All cost data have been divided into 50 divisions according to the MasterFormat® system of classification and numbering.

Assemblies*

The cost data in this section have been organized in an "Assemblies" format. These assemblies are the functional elements of a building and are arranged according to the 7 elements of the UNIFORMAT II classification system. For a complete explanation of a typical "Assembly", see "RSMeans data: Assemblies—How They Work."

*Residential Models**

Model buildings for four classes of construction—economy, average, custom, and luxury—are developed and shown with complete costs per square foot.

*Commercial/Industrial/Institutional Models**

This section contains complete costs for 77 typical model buildings expressed as costs per square foot.

*Green Commercial/Industrial/Institutional Models**

This section contains complete costs for 25 green model buildings expressed as costs per square foot.

*References**

This section includes information on Equipment Rental Costs, Crew Listings, Historical Cost Indexes, City Cost Indexes, Location Factors, Reference Tables, and Change Orders, as well as a listing of abbreviations.

- **Equipment Rental Costs:** Included are the average costs to rent and operate hundreds of pieces of construction equipment.
- **Crew Listings**: This section lists all the crews referenced in the cost data. A crew is composed of more than one trade classification and/or the addition of power equipment to any trade classification. Power equipment is included in the cost of the crew. Costs are shown both with bare labor rates and with the installing contractor's overhead and profit added. For each, the total crew cost per eight-hour day and the composite cost per labor-hour are listed.

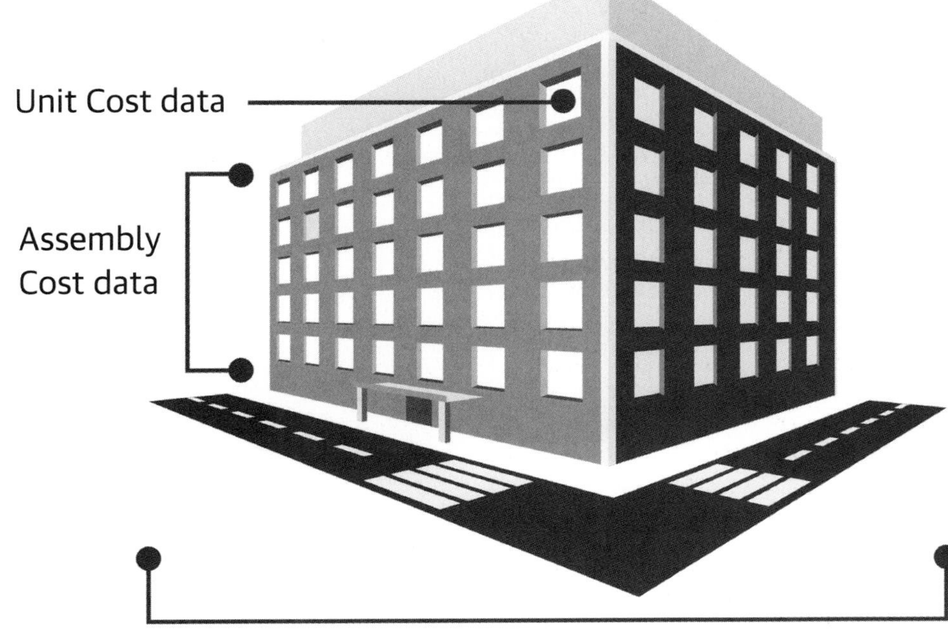

- **Historical Cost Indexes**: These indexes provide you with data to adjust construction costs over time.
- **City Cost Indexes**: All costs in this data set are U.S. national averages. Costs vary by region. You can adjust for this by CSI Division to over 730 cities in 900+ 3-digit zip codes throughout the U.S. and Canada by using this data.
- **Location Factors**: You can adjust total project costs to over 730 cities in 900+ 3-digit zip codes throughout the U.S. and Canada by using the weighted number, which applies across all divisions.
- **Reference Tables**: At the beginning of selected major classifications in the Unit Prices are reference numbers indicators. These numbers refer you to related information in the Reference Section. In this section, you'll find reference tables, explanations, and estimating information that support how we develop the unit price data, technical data, and estimating procedures.
- **Change Orders**: This section includes information on the factors that influence the pricing of change orders.
- **Abbreviations**: A listing of abbreviations used throughout this information, along with the terms they represent, is included.

Index (printed versions only)

A comprehensive listing of all terms and subjects will help you quickly find what you need when you are not sure where it occurs in MasterFormat®.

Conclusion

This information is designed to be as comprehensive and easy to use as possible.

The Construction Specifications Institute (CSI) and Construction Specifications Canada (CSC) have produced the 2016 edition of MasterFormat®, a system of titles and numbers used extensively to organize construction information.

All unit prices in the RSMeans cost data are now arranged in the 50-division MasterFormat® 2016 system.

* Not all information is available in all data sets

Note: The material prices in RSMeans cost data are "contractor's prices." They are the prices that contractors can expect to pay at the lumberyards, suppliers'/distributors' warehouses, etc. Small orders of specialty items would be higher than the costs shown, while very large orders, such as truckload lots, would be less. The variation would depend on the size, timing, and negotiating power of the contractor. The labor costs are primarily for new construction or major renovation rather than repairs or minor alterations. With reasonable exercise of judgment, the figures can be used for any building work.

Estimating with RSMeans data: Unit Prices

Following these steps will allow you to complete an accurate estimate using RSMeans data Unit Prices.

1. Scope Out the Project

- Think through the project and identify the CSI divisions needed in your estimate.
- Identify the individual work tasks that will need to be covered in your estimate.
- The Unit Price data have been divided into 50 divisions according to CSI MasterFormat® 2016.
- In printed versions, the Unit Price Section Table of Contents on page 1 may also be helpful when scoping out your project.
- Experienced estimators find it helpful to begin with Division 2 and continue through completion. Division 1 can be estimated after the full project scope is known.

2. Quantify

- Determine the number of units required for each work task that you identified.
- Experienced estimators include an allowance for waste in their quantities. (Waste is not included in our Unit Price line items unless otherwise stated.)

3. Price the Quantities

- Use the search tools available to locate individual Unit Price line items for your estimate.
- Reference Numbers indicated within a Unit Price section refer to additional information that you may find useful.
- The crew indicates who is performing the work for that task. Crew codes are expanded in the Crew Listings in the Reference Section to include all trades and equipment that comprise the crew.
- The Daily Output is the amount of work the crew is expected to complete in one day.
- The Labor-Hours value is the amount of time it will take for the crew to install one unit of work.
- The abbreviated Unit designation indicates the unit of measure upon which the crew, productivity, and prices are based.
- Bare Costs are shown for materials, labor, and equipment needed to complete the Unit Price line item. Bare costs do not include waste, project overhead, payroll insurance, payroll taxes, main office overhead, or profit.
- The Total Incl O&P cost is the billing rate or invoice amount of the installing contractor or subcontractor who performs the work for the Unit Price line item.

4. Multiply

- Multiply the total number of units needed for your project by the Total Incl O&P cost for each Unit Price line item.
- Be careful that your take off unit of measure matches the unit of measure in the Unit column.
- The price you calculate is an estimate for a completed item of work.
- Keep scoping individual tasks, determining the number of units required for those tasks, matching each task with individual Unit Price line items, and multiplying quantities by Total Incl O&P costs.
- An estimate completed in this manner is priced as if a subcontractor, or set of subcontractors, is performing the work. The estimate does not yet include Project Overhead or Estimate Summary components such as general contractor markups on subcontracted work, general contractor office overhead and profit, contingency, and location factors.

5. Project Overhead

- Include project overhead items from Division 1-General Requirements.
- These items are needed to make the job run. They are typically, but not always, provided by the general contractor. Items include, but are not limited to, field personnel, insurance, performance bond, permits, testing, temporary utilities, field office and storage facilities, temporary scaffolding and platforms, equipment mobilization and demobilization, temporary roads and sidewalks, winter protection, temporary barricades and fencing, temporary security, temporary signs, field engineering and layout, final cleaning, and commissioning.
- Each item should be quantified and matched to individual Unit Price line items in Division 1, then priced and added to your estimate.
- An alternate method of estimating project overhead costs is to apply a percentage of the total project cost—usually 5% to 15% with an average of 10% (see General Conditions).
- Include other project related expenses in your estimate such as:
  - Rented equipment not itemized in the Crew Listings
  - Rubbish handling throughout the project (see section 02 41 19.19)

6. Estimate Summary

- Include sales tax as required by laws of your state or county.
- Include the general contractor's markup on self-performed work, usually 5% to 15% with an average of 10%.
- Include the general contractor's markup on subcontracted work, usually 5% to 15% with an average of 10%.
- Include the general contractor's main office overhead and profit:
  - RSMeans data provides general guidelines on the general contractor's main office overhead (see section 01 31 13.60 and Reference Number R013113-50).
  - Markups will depend on the size of the general contractor's operations, projected annual revenue, the level of risk, and the level of competition in the local area and for this project in particular.
- Include a contingency, usually 3% to 5%, if appropriate.
- Adjust your estimate to the project's location by using the City Cost Indexes or the Location Factors in the Reference Section:
  - Look at the rules in "How to Use the City Cost Indexes" to see how to apply the Indexes for your location.
  - When the proper Index or Factor has been identified for the project's location, convert it to a multiplier by dividing it by 100, then multiply that multiplier by your estimated total cost. The original estimated total cost will now be adjusted up or down from the national average to a total that is appropriate for your location.

Editors' Note:
We urge you to spend time reading and understanding the supporting material. An accurate estimate requires experience, knowledge, and careful calculation. The more you know about how we at RSMeans developed the data, the more accurate your estimate will be. In addition, it is important to take into consideration the reference material such as Equipment Listings, Crew Listings, City Cost Indexes, Location Factors, and Reference Tables.

How to Use the Cost Data: The Details

What's Behind the Numbers? The Development of Cost Data

RSMeans data engineers continually monitor developments in the construction industry in order to ensure reliable, thorough, and up-to-date cost information. While overall construction costs may vary relative to general economic conditions, price fluctuations within the industry are dependent upon many factors. Individual price variations may, in fact, be opposite to overall economic trends. Therefore, costs are constantly tracked and complete updates are performed yearly. Also, new items are frequently added in response to changes in materials and methods.

Costs in U.S. Dollars

All costs represent U.S. national averages and are given in U.S. dollars. The City Cost Index (CCI) with RSMeans data can be used to adjust costs to a particular location. The CCI for Canada can be used to adjust U.S. national averages to local costs in Canadian dollars. No exchange rate conversion is necessary because it has already been factored in.

G The processes or products identified by the green symbol in our publications have been determined to be environmentally responsible and/or resource-efficient solely by RSMeans data engineering staff. The inclusion of the green symbol does not represent compliance with any specific industry association or standard.

Material Costs

RSMeans data engineers contact manufacturers, dealers, distributors, and contractors all across the U.S. and Canada to determine national average material costs. If you have access to current material costs for your specific location, you may wish to make adjustments to reflect differences from the national average. Included within material costs are fasteners for a normal installation. RSMeans data engineers use manufacturers' recommendations, written specifications, and/or standard construction practices for the sizing and spacing of fasteners. Adjustments to material costs may be required for your specific application or location. The manufacturer's warranty is assumed. Extended warranties are not included in the material costs. **Material costs do not include sales tax.**

Labor Costs

Labor costs are based upon a mathematical average of trade-specific wages in 30 major U.S. cities. The type of wage (union, open shop, or residential) is identified on the inside back cover of printed publications or selected by the estimator when using the electronic products. Markups for the wages can also be found on the inside back cover of printed publications and/or under the labor references found in the electronic products.

- If wage rates in your area vary from those used, or if rate increases are expected within a given year, labor costs should be adjusted accordingly.

Labor costs reflect productivity based on actual working conditions. In addition to actual installation, these figures include time spent during a normal weekday on tasks, such as material receiving and handling, mobilization at the site, site movement, breaks, and cleanup.

Productivity data is developed over an extended period so as not to be influenced by abnormal variations and reflects a typical average.

Equipment Costs

Equipment costs include not only rental but also operating costs for equipment under normal use. The operating costs include parts and labor for routine servicing, such as the repair and replacement of pumps, filters, and worn lines. Normal operating expendables, such as fuel, lubricants, tires, and electricity (where applicable), are also included. Extraordinary operating expendables with highly variable wear patterns, such as diamond bits and blades, are excluded. These costs are included under materials. Equipment rental rates are obtained from industry sources throughout North America—contractors, suppliers, dealers, manufacturers, and distributors.

Rental rates can also be treated as reimbursement costs for contractor-owned equipment. Owned equipment costs include depreciation, loan payments, interest, taxes, insurance, storage, and major repairs.

Equipment costs do not include operators' wages.

Equipment Cost/Day—The cost of equipment required for each crew is included in the Crew Listings in the Reference Section (small tools that are considered essential everyday tools are not listed out separately). The Crew Listings itemize specialized tools and heavy equipment along with labor trades. The daily cost of itemized equipment included in a crew is based on dividing the weekly bare rental rate by 5 (number of working days per week), then adding the hourly operating cost times 8 (the number of hours per day). This Equipment Cost/Day is shown in the last column of the Equipment Rental Costs in the Reference Section.

Mobilization, Demobilization—The cost to move construction equipment from an equipment yard or rental company to the job site and back again is not included in equipment costs. Mobilization (to the site) and demobilization (from the site) costs can be found in the Unit Price Section. If a piece of equipment is already at the job site, it is not appropriate to utilize mobilization or demobilization costs again in an estimate.

Overhead and Profit

Total Cost including O&P for the installing contractor is shown in the last column of the Unit Price and/or Assemblies. This figure is the sum of the bare material cost plus 10% for profit, the bare labor cost plus total overhead and profit, and the bare equipment cost plus 10% for profit. Details for the calculation of overhead and profit on labor are shown on the inside back cover of the printed product and in the Reference Section of the electronic product.

General Conditions

Cost data in this data set are presented in two ways: Bare Costs and Total Cost including O&P (Overhead and Profit). General Conditions, or General Requirements, of the contract should also be added to the Total Cost including O&P when applicable. Costs for General Conditions are listed in Division 1 of the Unit Price Section and in the Reference Section.

General Conditions for the installing contractor may range from 0% to 10% of the Total Cost including O&P. For the general or prime contractor, costs for General Conditions may range from 5% to 15% of the Total Cost including O&P, with a figure of 10% as the most typical allowance. If applicable, the Assemblies and Models sections use costs that include the installing contractor's overhead and profit (O&P).

Factors Affecting Costs

Costs can vary depending upon a number of variables. Here's a listing of some factors that affect costs and points to consider.

Quality—The prices for materials and the workmanship upon which productivity is based represent sound construction work. They are also in line with industry standard and manufacturer specifications and are frequently used by federal, state, and local governments.

Overtime—We have made no allowance for overtime. If you anticipate premium time or work beyond normal working hours, be sure to make an appropriate adjustment to your labor costs.

Productivity—The productivity, daily output, and labor-hour figures for each line item are based on an eight-hour work day in daylight hours in moderate temperatures and up to a 14' working height unless otherwise indicated. For work that extends beyond normal work hours or is performed under adverse conditions, productivity may decrease.

Size of Project—The size, scope of work, and type of construction project will have a significant impact on cost. Economies of scale can reduce costs for large projects. Unit costs can often run higher for small projects.

Location—Material prices are for metropolitan areas. However, in dense urban areas, traffic and site storage limitations may increase costs. Beyond a 20-mile radius of metropolitan areas, extra trucking or transportation charges may also increase the material costs slightly. On the other hand, lower wage rates may be in effect. Be sure to consider both of these factors when preparing an estimate, particularly if the job site is located in a central city or remote rural location. In addition, highly specialized subcontract items may require travel and per-diem expenses for mechanics.

Other Factors—

- season of year
- contractor management
- weather conditions
- local union restrictions
- building code requirements
- availability of:
  - adequate energy
  - skilled labor
  - building materials
- owner's special requirements/restrictions
- safety requirements
- environmental considerations
- access

Unpredictable Factors—General business conditions influence "in-place" costs of all items. Substitute materials and construction methods may have to be employed. These may affect the installed cost and/or life cycle costs. Such factors may be difficult to evaluate and cannot necessarily be predicted on the basis of the job's location in a particular section of the country. Thus, where these factors apply, you may find significant but unavoidable cost variations for which you will have to apply a measure of judgment to your estimate.

Rounding of Costs

In printed publications only, all unit prices in excess of $5.00 have been rounded to make them easier to use and still maintain adequate precision of the results.

How Subcontracted Items Affect Costs

A considerable portion of all large construction jobs is usually subcontracted. In fact, the percentage done by subcontractors is constantly increasing and may run over 90%. Since the workers employed by these companies do nothing else but install their particular products, they soon become experts in that line. As a result, installation by these firms is accomplished so efficiently that the total in-place cost, even with the general contractor's overhead and profit, is no more, and often less, than if the principal contractor had handled the installation. Companies that deal with construction specialties are anxious to have their products perform well and, consequently, the installation will be the best possible.

Contingencies

The allowance for contingencies generally provides for unforeseen construction difficulties. On alterations or repair jobs, 20% is not too much. If drawings are final and only field contingencies are being considered, 2% or 3% is probably sufficient and often nothing needs to be added. Contractually, changes in plans will be covered by extras. The contractor should consider inflationary price trends and possible material shortages during the course of the job. These escalation factors are dependent upon both economic conditions and the anticipated time between the estimate and actual construction. If drawings are not complete or approved, or a budget cost is wanted, it is wise to add 5% to 10%. Contingencies, then, are a matter of judgment.

Important Estimating Considerations

The productivity, or daily output, of each craftsman or crew assumes a well-managed job where tradesmen with the proper tools and equipment, along with the appropriate construction materials, are present. Included are daily set-up and cleanup time, break time, and plan layout time. Unless otherwise indicated, time for material movement on site (for items that can be transported by hand) of up to 200' into the building and to the first or second floor is also included. If material has to be transported by other means, over greater distances, or to higher floors, an additional allowance should be considered by the estimator.

While horizontal movement is typically a sole function of distances, vertical transport introduces other variables that can significantly impact productivity. In an occupied building, the use of elevators (assuming access, size, and required protective measures are acceptable) must be understood at the time of the estimate. For new construction, hoist wait and cycle times can easily be 15 minutes and may result in scheduled access extending beyond the normal work day. Finally, all vertical transport will impose strict weight limits likely to preclude the use of any motorized material handling.

The productivity, or daily output, also assumes installation that meets manufacturer/designer/ standard specifications. A time allowance for quality control checks, minor adjustments, and any task required to ensure proper function or operation is also included. For items that require connections to services, time is included for positioning, leveling, securing the unit, and making all the necessary connections (and start up where applicable) to ensure a complete installation. Estimating of the services themselves (electrical, plumbing, water, steam, hydraulics, dust collection, etc.) is separate.

In some cases, the estimator must consider the use of a crane and an appropriate crew for the installation of large or heavy items. For those situations where a crane is not included in the assigned crew and as part of the line item cost, then equipment rental costs, mobilization and demobilization costs, and operator and support personnel costs must be considered.

Labor-Hours

The labor-hours expressed in this publication are derived by dividing the total daily labor-hours for the crew by the daily output. Based on average installation time and the assumptions listed above, the labor-hours include: direct labor, indirect labor, and nonproductive time. A typical day for a craftsman might include but is not limited to:

- Direct Work
  - Measuring and layout
  - Preparing materials
  - Actual installation
  - Quality assurance/quality control
- Indirect Work
  - Reading plans or specifications
  - Preparing space
  - Receiving materials
  - Material movement
  - Giving or receiving instruction
  - Miscellaneous
- Non-Work
  - Chatting
  - Personal issues
  - Breaks
  - Interruptions (i.e., sickness, weather, material or equipment shortages, etc.)

If any of the items for a typical day do not apply to the particular work or project situation, the estimator should make any necessary adjustments.

Final Checklist

Estimating can be a straightforward process provided you remember the basics. Here's a checklist of some of the steps you should remember to complete before finalizing your estimate.

Did you remember to:

- factor in the City Cost Index for your locale?
- take into consideration which items have been marked up and by how much?
- mark up the entire estimate sufficiently for your purposes?
- read the background information on techniques and technical matters that could impact your project time span and cost?
- include all components of your project in the final estimate?
- double check your figures for accuracy?
- call RSMeans data engineers if you have any questions about your estimate or the data you've used? Remember, Gordian stands behind all of our products, including our extensive RSMeans data solutions. If you have any questions about your estimate, about the costs you've used from our data, or even about the technical aspects of the job that may affect your estimate, feel free to call the Gordian RSMeans editors at 1.800.448.8182.

Square Foot Cost Section

Table of Contents

Introduction to the Square Foot Cost Section

The Square Foot Cost Section contains costs per square foot for four classes of construction in seven building types. Costs are listed for various exterior wall materials which are typical of the class and building type. There are cost tables for wings and ells with modification tables to adjust the base cost of each class of building. Non-standard items can easily be added to the standard structures.

Cost estimating for a residence is a three-step process:

1. Identification
2. Listing dimensions
3. Calculations

Guidelines and a sample cost estimating procedure are shown on the following pages.

Identification

To properly identify a residential building, the class of construction, type, and exterior wall material must be determined. The "Building Classes" information has drawings and guidelines for determining the class of construction. There are also detailed specifications and additional drawings at the beginning of each set of tables to further aid in proper building class and type identification.

Sketches for eight types of residential buildings and their configurations follow. Definitions of living area are next to each sketch. Sketches and definitions of garage types follow.

Living Area

Base cost tables are prepared as costs per square foot of living area. The living area of a residence is that area which is suitable and normally designed for full time living. It does not include basement recreation rooms or finished attics, although these areas are often considered full time living areas by the owners.

Living area is calculated from the exterior dimensions without the need to adjust for exterior wall thickness. When calculating the living area of a 1-1/2 story, two story, three story, or tri-level residence, overhangs and other differences in size and shape between floors must be considered.

Only the floor area with a ceiling height of seven feet or more in a 1-1/2 story residence is considered living area. In bi-levels and tri-levels, the areas that are below grade are considered living areas, even when these areas may not be completely finished.

Base Tables and Modifications

Base cost tables show the base cost per square foot without a basement, with one full bath and one full kitchen for economy and average homes, and an additional half bath for custom and luxury models. Adjustments for finished and unfinished basements are part of the base cost tables. Adjustments for multi-family residences, additional bathrooms, townhouses, alternative roofs, and air conditioning and heating systems are listed in Modifications, Adjustments, and Alternatives tables below the base cost tables.

Costs for other modifications, adjustments, and alternatives, including garages, breezeways, and site improvements, follow the base tables.

Listing of Dimensions

To use this section, only the dimensions used to calculate the horizontal area of the building and additions, modifications, adjustments, and alternatives are needed. The dimensions, normally the length and width, can come from drawings or field measurements. For ease in calculation, consider measuring in tenths of feet, i.e., 9'-6" = 9.5 ft. and 9'-4" = 9.3 ft.

In all cases, make a sketch of the building. Any protrusions or other variations in shape should be noted on the sketch with dimensions.

Calculations

The calculations portion of the estimate is a two-step activity:

1. The selection of appropriate costs from the tables
2. Computations

Selection of Appropriate Costs

To select the appropriate cost from the base tables, the following information is needed:

1. Class of construction
   - Economy
   - Average
   - Custom
   - Luxury
2. Type of residence
   - 1 story
   - 1-1/2 story
   - 2 story
   - 2-1/2 story
   - 3 story
   - Bi-level
   - Tri-level
3. Occupancy
   - One family
   - Two family
   - Three family
4. Building configuration
   - Detached
   - Town/Rowhouse
   - Semi-detached
5. Exterior wall construction
   - Wood frame
   - Brick veneer
   - Solid masonry
6. Living areas

Modifications are classified by class, type, and size.

Computations

The computation process should take the following sequence:

1. Multiply the base cost by the area.
2. Add or subtract the modifications, adjustments, and alternatives.
3. Apply the location modifier.

When selecting costs, interpolate or use the cost that most nearly matches the structure under study. This applies to size, exterior wall construction, and class.

How to Use the Residential Square Foot Cost Section

The following is a detailed explanation of a sample entry in the Residential Square Foot Cost Section. Each bold number below corresponds to the item being described in the following list with the appropriate component of the sample entry in parentheses. Prices listed are costs that include overhead and profit of the installing contractor. Total model costs include an additional markup for the general contractor's overhead and profit, as well as fees specific to the class of construction.

RESIDENTIAL | Average 1 | 2 Story 2

- **Simple design from standard plans**
- **Single family — 1 full bath, 1 kitchen**
- **No basement** 3
- **Asphalt shingles on roof**
- **Hot air heat**
- **Gypsum wallboard interior finishes**
- **Materials and workmanship are average**

Note: The illustration shown may contain some optional components (for example: garages and/or fireplaces) whose costs are shown in the modifications, adjustments, & alternatives below or at the end of the square foot section.

Base cost per square foot of living area

Exterior Wall 4	Living Area 5										
	1000	1200	1400	1600	1800	2000	2200	2600	3000	3400	3800
Wood Siding - Wood Frame	153.05	138.30	131.10	126.15	121.05	115.85	112.15	105.50	99.10	96.00	93.35
Brick Veneer - Wood Frame	159.40	144.20	136.65	131.45	126.00	120.60	116.70	109.60	102.90	99.65	96.80
Stucco on Wood Frame	147.55	133.25	126.35	121.60	116.80	111.70	108.30	101.90	95.80	92.95	90.35
Solid Masonry	173.85	157.60	149.15	143.45	137.25	131.45	127.00	118.95	111.55	107.85	104.55
Finished Basement, Add 7	23.65	23.30	22.55	22.00	21.45	21.05	20.60	19.90	19.30	18.90	18.60
Unfinished Basement, Add	9.40	8.75	8.25	7.95	7.60	7.30	7.10	6.60	6.30	6.05	5.85

(6 marks the 2000 sq. ft. Wood Siding value; 8 marks the basement rows.)

Modifications

Add to the total cost

Upgrade Kitchen Cabinets 9	$ + 5907
Solid Surface Countertops (Included)	
Full Bath - including plumbing, wall and floor finishes	+ 8111
Half Bath - including plumbing, wall and floor finishes	+ 4789
One Car Attached Garage	+ 15,383
One Car Detached Garage	+ 20,232
Fireplace & Chimney	+ 7845

Adjustments

For multi family - add to total cost

Additional Kitchen	$ + 10,268
Additional Bath	+ 8111
Additional Entry & Exit	+ 1804
Separate Heating	+ 1748
Separate Electric	+ 1868

For Townhouse/Rowhouse - Multiply cost per square foot by

Inner Unit	10 .90
End Unit	.95

Alternatives

Add to or deduct from the cost per square foot of living area

Cedar Shake Roof	+ 1.75
Clay Tile Roof	+ 3.45
Slate Roof	+ 3.90
Upgrade Walls to Skim Coat Plaster 11	+ .63
Upgrade Ceilings to Textured Finish	+ .61
Air Conditioning, in Heating Ductwork	+ 3.10
In Separate Ductwork	+ 5.96
Heating Systems, Hot Water	+ 1.38
Heat Pump	+ 1.45
Electric Heat	– .97
Not Heated	– 2.90

Additional upgrades or components

Kitchen Cabinets & Countertops	Page 93
Bathroom Vanities	94
Fireplaces & Chimneys	94
Windows, Skylights & Dormers	94
Appliances	95
Breezeways & Porches	95
Finished Attic	95
Garages	96
Site Improvements	96
Wings & Ells 12	56

1 Class of Construction (Average)

The class of construction depends upon the design and specifications of the plan. The four classes are economy, average, custom, and luxury.

2 Type of Residence (2 Story)

The building type describes the number of stories or levels in the model. The seven building types are 1 story, 1-1/2 story, 2 story, 2-1/2 story, 3 story, bi-level, and tri-level.

3 Specification Highlights (Hot Air Heat)

These specifications include information concerning the components of the model, including the number of baths, roofing types, HVAC systems, materials, and workmanship. If the components listed are not appropriate, modifications can be made by consulting the information shown below or in the Assemblies section.

4 Exterior Wall System (Wood Siding-Wood Frame)

This section includes the types of exterior wall systems and the structural frames used. The exterior wall systems shown are typical of the class of construction and the building type shown.

5 Living Areas (2,000 S.F.)

The living area is that area of the residence which is suitable and normally designed for full time living. It does not include basement recreation rooms or finished attics. Living area is calculated from the exterior dimensions without the need to adjust for exterior wall thickness. When calculating the living area of a 1-1/2 story, 2 story, 3 story, or tri-level residence, overhangs and other differences in size and shape between floors must be considered. Only the floor area with a ceiling height of seven feet or more in a 1-1/2 story residence is considered living area. In bi-levels and tri-levels, the areas that are below grade are considered living areas, even when these areas may not be completely finished. A range of various living areas for the residential model is shown to aid in the selection of values from the matrix.

6 Base Costs per Square Foot of Living Area ($115.85)

Base cost tables show the cost per square foot of living area without a basement, with one full bath and one full kitchen for economy and average homes, and an additional half bath for custom and luxury models. When selecting costs, interpolate or use the cost that most nearly matches the residence under consideration for size, exterior wall system, and class of construction. Prices listed are costs that include overhead and profit of the installing contractor, a general contractor markup, and an allowance for plans that vary by class of construction. For additional information on contractor overhead and architectural fees, see the Reference Section.

7 Basement Types (Finished)

The two types of basements are finished and unfinished. The specifications and components for both are shown under Building Classes in the Introduction to this section.

8 Additional Costs for Basements ($21.05 or $7.30)

These values indicate the additional cost per square foot of living area for either a finished or an unfinished basement.

9 Modifications and Adjustments (Upgrade Kitchen Cabinets $5,907)

Modifications and Adjustments are costs added to or subtracted from the total cost of the residence. The total cost of the residence is equal to the cost per square foot of living area times the living area. Typical modifications and adjustments include kitchens, baths, garages, and fireplaces.

10 Multiplier for Townhouse/Rowhouse (Inner Unit 0.90)

The multipliers shown adjust the base costs per square foot of living area for the common wall condition encountered in townhouses or rowhouses.

11 Alternatives (Skim Coat Plaster $0.63)

Alternatives are costs added to or subtracted from the base cost per square foot of living area. Typical alternatives include variations in kitchens, baths, roofing, and air conditioning and heating systems.

12 Additional Upgrades or Components (Wings & Ells)

Costs for additional upgrades or components, including wings or ells, breezeways, porches, finished attics, and site improvements, are shown at the end of each quality section and at the end of the Square Foot Section.

How to Use the Residential Square Foot Cost Section (Continued)

Average 2 Story

❶ **Living Area - 2000 S.F.**
Perimeter - 135 L.F.

❷

		Cost Per Square Foot Of Living Area			% of Total
		Mat.	Inst.	Total	(rounded)
1 Site Work	Site preparation for slab; 4' deep trench excavation for foundation wall.		1.33	1.33	1.1%
2 Foundation	Continuous reinforced concrete footing, 10" deep x 20" wide; dampproofed and insulated 8" thick reinforced concrete block foundation wall, 4' deep; trowel finished 4" thick concrete slab on 4" crushed stone base and polyethylene vapor barrier.	3.67	4.79	8.46	7.3%
3 Framing ❸	Exterior walls - 2" x 6" wood studs, 16" O.C.; 1/2" sheathing; gable end roof framing, 2" x 10" rafters, 16" O.C. with 1/2" plywood sheathing; 2" x 10" floor joists, 16" O.C. with bridging and 5/8" subflooring; 2" x 4" interior partitions.	7.75 ❹	10.14 ❺	17.89 ❻	15.4% ❼
4 Exterior Walls	Beveled wood siding and housewrap on insulated wood frame walls; R38 attic insulation; double hung wood windows; flush solid core doors, frame and hardware, painted finish; aluminum storm and screen doors.	15.45	6.13	21.58	18.6%
5 Roofing	25 year asphalt roof shingles; #15 felt building paper; aluminum gutters, downspouts, drip edge and flashings.	1.17	1.42	2.59	2.2%
6 Interiors	Walls & ceilings, 1/2" taped & finished gypsum wallboard, primed & painted with 2 coats of finish paint; birch faced hollow core interior doors, frames & hardware, painted finish; medium weight carpeting with pad, 40%; sheet vinyl, 15%; oak hardwood, 40%; ceramic tile, 5%; hardwood tread stairway.	15.75	15.36	31.11	26.9%
7 Specialties	Average grade kitchen cabinets and countertop; stainless steel kitchen sink; 40 gallon electric water heater.	3.93	1.22	5.15	4.4%
8 Mechanical	Three fixture bathroom: bathtub, water closet, vanity and sink; gas fired hot air heating system.	3.49	3.40	6.89	5.9%
9 Electrical	200 amp electric service; wiring, duplex & GFI receptacles, wall switches, door bell, appliance circuits, fans and communications cabling; average grade lighting fixtures	1.55	2.44	3.99	3.4%
10 Overhead	Contractor's overhead and profit and plans. ❽	8.99	7.87	16.86	14.6%
	Total	61.75	54.10	**115.85** ❾	

1 Specifications

The parameters for an example dwelling from the previous pages are listed here. Included are the square foot dimensions of the proposed building. Living Area takes into account the number of floors and other factors needed to define a building's total square footage. Perimeter dimensions are defined in terms of linear feet.

2 Building Type

This is a sketch of a cross section view through the dwelling. It is shown to help define the living area for the building type. For more information, see the Building Types in the Introduction.

3 Components (3 Framing)

This page contains the ten components needed to develop the complete square foot cost of the typical dwelling specified. All components are defined with a description of the materials and/or task involved. Use cost figures from each component to estimate the cost per square foot of that section of the project. The components listed here are typical of all sizes of residences from the facing page. Specific quantities of components required would vary with the size of the dwelling and the exterior wall system.

4 Materials (7.75)

This column gives the amount needed to develop the cost of materials. The figures given here are not bare costs. Ten percent has been added to bare material cost for profit.

5 Installation (10.14)

Installation includes labor and equipment costs. The labor rates included here incorporate the total overhead and profit costs for the installing contractor. The average mark-up used to create these figures is 72% over and above bare labor costs. The equipment rates include 10% for profit.

6 Total (17.89)

This column lists the sum of two figures. Use this total to determine the sum of material cost plus installation cost. The result is a convenient total cost for each of the ten components.

7 % of Total (rounded) (15.4%)

This column represents the percent of the total cost for this component group.

8 Overhead

The costs in components 1 through 9 include overhead and profit for the installing contractor. Item 10 is overhead and profit for the general contractor. This is typically a percentage mark-up of all other costs. The amount depends on the size and type of dwelling, building class, and economic conditions. An allowance for plans or design has been included where appropriate.

9 Bottom Line Total (115.85)

This figure is the complete square foot cost for the construction project and equals the sum of total material and total labor costs. To determine total project cost, multiply the bottom line total by the living area.

Building Classes

Economy Class

An economy class residence is usually built from stock plans. The materials and workmanship are sufficient to satisfy building codes. Low construction cost is more important than distinctive features. The overall shape of the foundation and structure is seldom other than square or rectangular.

An unfinished basement includes a 7' high, 8" thick foundation wall composed of either concrete block or cast-in-place concrete.

Included in the finished basement cost are inexpensive paneling or drywall as the interior finish on the foundation walls, a low cost sponge backed carpeting adhered to the concrete floor, a drywall ceiling, and overhead lighting.

Average Class

An average class residence is a simple design and built from standard plans. Materials and workmanship are average but often exceed minimum building codes. There are frequently special features that give the residence some distinctive characteristics.

An unfinished basement includes a 7'-6" high, 8" thick foundation wall composed of either cast-in-place concrete or concrete block.

Included in the finished basement are plywood paneling or drywall on furring that is fastened to the foundation walls, sponge backed carpeting adhered to the concrete floor, a suspended ceiling, overhead lighting, and heating.

Custom Class

A custom class residence is usually built from plans and specifications with enough features to give the building a distinct design. Materials and workmanship are generally above average with obvious attention given to construction details. Construction normally exceeds building code requirements.

An unfinished basement includes a 7'-6" high, 10" thick cast-in-place concrete foundation wall or a 7'-6" high, 12" thick concrete block foundation wall.

A finished basement includes painted drywall on insulated 2" × 4" wood furring as the interior finish to the concrete walls, a suspended ceiling, carpeting adhered to the concrete floor, overhead lighting, and heating.

Luxury Class

A luxury class residence is built from an architect's plan for a specific owner. It is unique both in design and workmanship. There are many special features, and construction usually exceeds all building codes. It is obvious that primary attention is placed on the owner's comfort and pleasure. Construction is supervised by an architect.

An unfinished basement includes an 8' high, 12" thick foundation wall that is composed of cast-in-place concrete or concrete block.

A finished basement includes painted drywall on 2" × 4" wood furring as the interior finish, a suspended ceiling, tackless carpet on wood subfloor with sleepers, overhead lighting, and heating.

Configurations

Detached House

This category of residence is a free-standing separate building with or without an attached garage. It has four complete walls.

Semi-Detached House

This category of residence has two side-by-side living units. The common wall is fireproof. Semi-detached residences can be treated as a rowhouse with two end units. Semi-detached residences can be any of the building types.

Town/Rowhouse

This category of residence has a number of attached units made up of inner units and end units. The units are joined by common walls. The inner units have only two exterior walls. The common walls are fireproof. The end units have three walls and a common wall. Town/rowhouses can be any of the building types.

Building Types

One Story

This is an example of a one-story dwelling. The living area of this type of residence is confined to the ground floor. The headroom in the attic is usually too low for use as a living area.

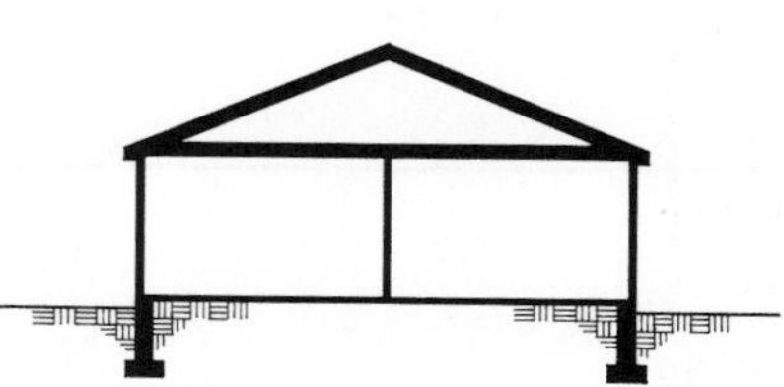

One-and-one-half Story

The living area on the upper level of this type of residence is 50% to 90% of the ground floor. This is made possible by a combination of this design's high-peaked roof and/or dormers. Only the upper level area with a ceiling height of seven feet or more is considered living area. The living area of this residence is the sum of the ground floor area plus the area on the second level with a ceiling height of seven feet or more.

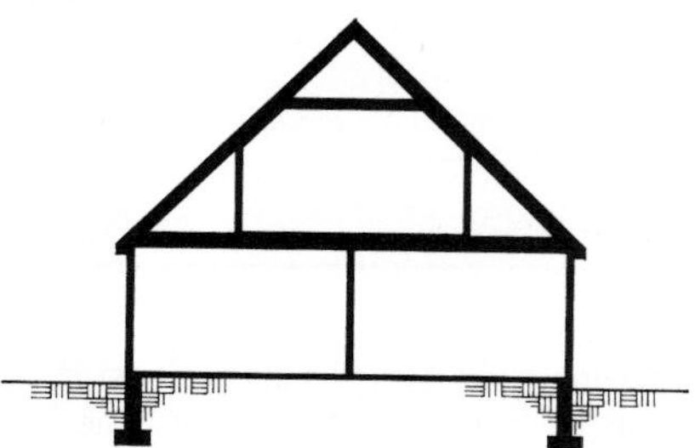

One Story with Finished Attic

The main living area in this type of residence is the ground floor. The upper level or attic area has sufficient headroom for use as a living area. This is made possible by a high-peaked roof. The living area in the attic is less than 50% of the ground floor. The living area of this type of residence is the ground floor area only. The finished attic is considered an adjustment.

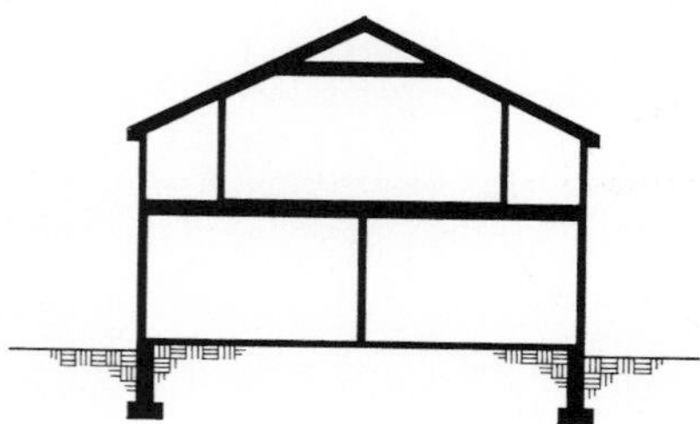

Two Story

This type of residence has a second floor or upper level area which is equal or nearly equal to the ground floor area. The upper level of this type of residence can range from 90% to 110% of the ground floor area, depending on setbacks or overhangs. The living area is the sum of the ground floor area and the upper level floor area.

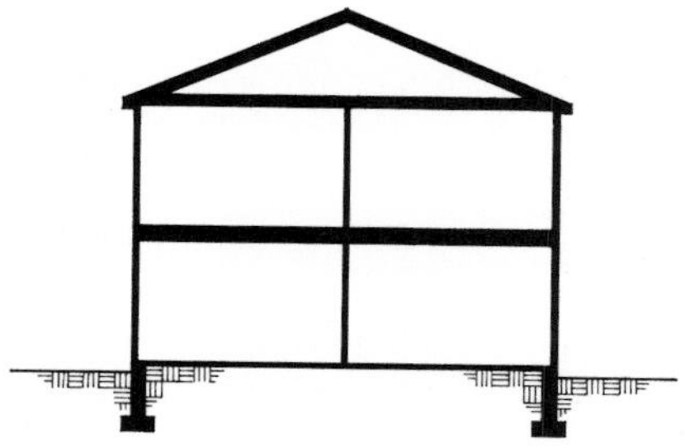

Two-and-one-half Story

This type of residence has two levels of equal or nearly equal area and a third level which has a living area that is 50% to 90% of the ground floor. This is made possible by a high-peaked roof, extended wall heights, and/or dormers. Only the upper level area with a ceiling height of seven feet or more is considered living area. The living area of this residence is the sum of the ground floor area, the second floor area, and the area on the third level with a ceiling height of seven feet or more.

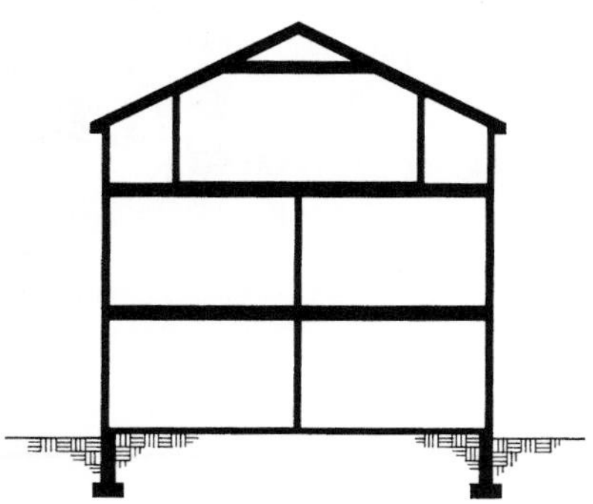

Bi-level

This type of residence has two living areas, one above the other. One area is about four feet below grade and the second is about four feet above grade. Both areas are equal in size. The lower level in this type of residence is designed and built to serve as a living area and not as a basement. Both levels have full ceiling heights. The living area is the sum of the lower level area and the upper level area.

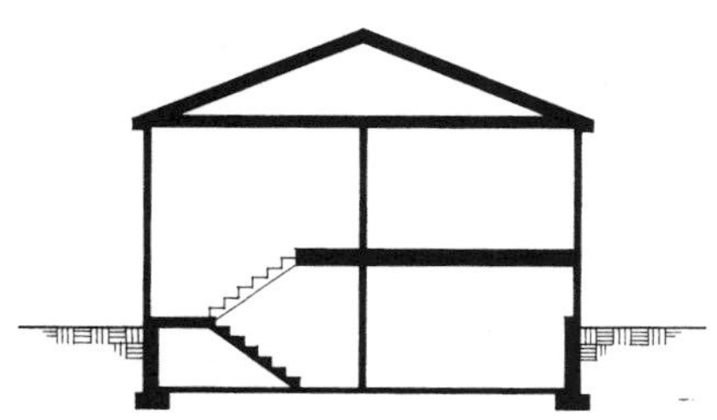

Three Story

This type of residence has three levels which are equal or nearly equal. As in the two story residence, the second and third floor areas may vary slightly depending on setbacks or overhangs. The living area is the sum of the ground floor area and the two upper level floor areas.

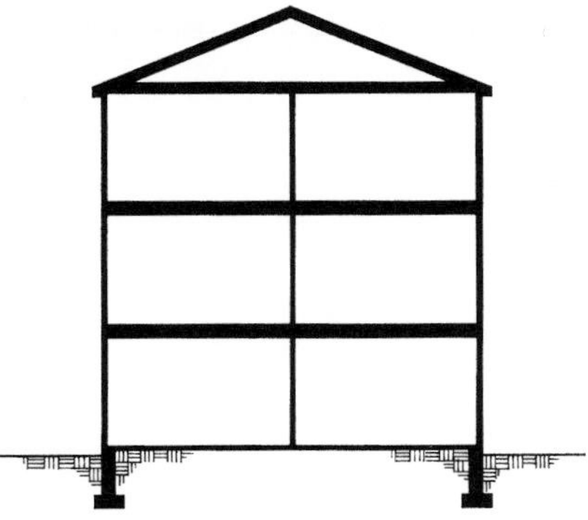

Tri-level

This type of residence has three levels of living area: one at grade level, one about four feet below grade, and one about four feet above grade. All levels are designed to serve as living areas. All levels have full ceiling heights. The living area is a sum of the areas of each of the three levels.

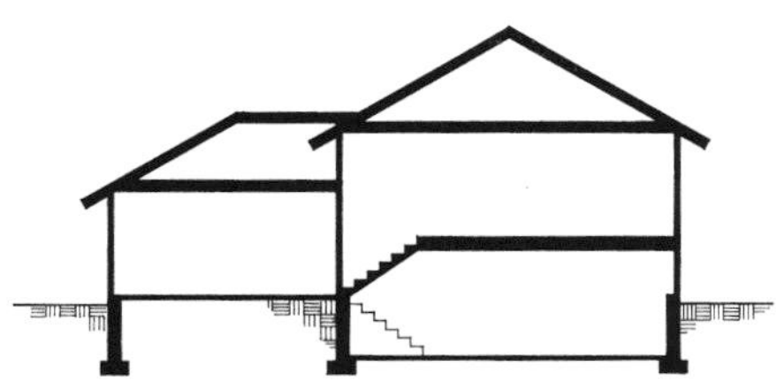

Garage Types

Attached Garage

Shares a common wall with the dwelling. Access is typically through a door between the dwelling and garage.

Basement Garage

Constructed under the roof of the dwelling but below the living area.

Built-In Garage

Constructed under the second floor living space and above the basement level of the dwelling. Reduces gross square feet of the living area.

Detached Garage

Constructed apart from the main dwelling. Shares no common area or wall with the dwelling.

Building Components

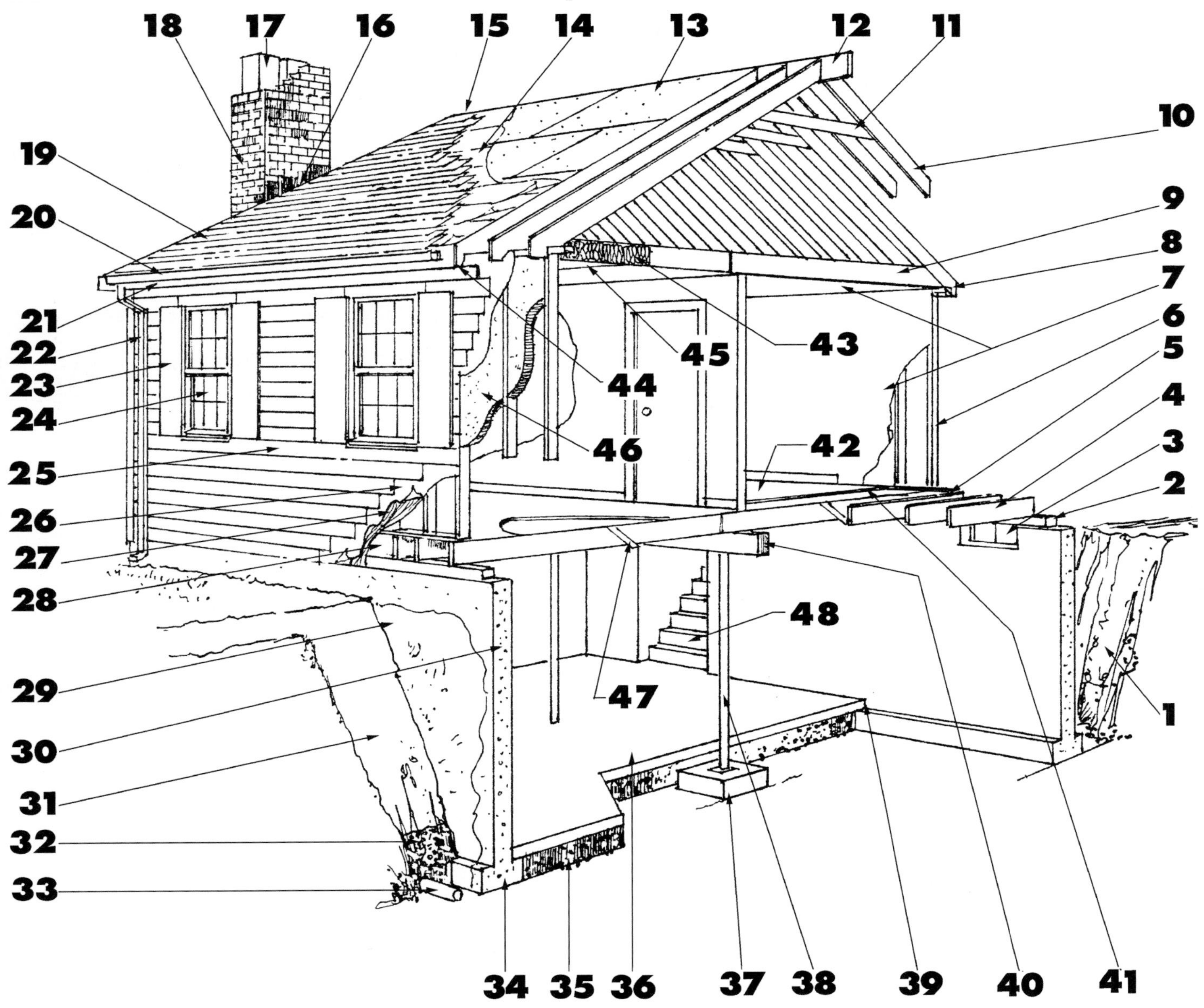

1. Excavation
2. Sill Plate
3. Basement Window
4. Floor Joist
5. Shoe Plate
6. Studs
7. Drywall
8. Plate
9. Ceiling Joists
10. Rafters
11. Collar Ties
12. Ridge Board
13. Roof Sheathing
14. Roof Felt
15. Roof Shingles
16. Flashing
17. Flue Lining
18. Chimney
19. Roof Shingles
20. Gutter
21. Fascia
22. Downspout
23. Shutter
24. Window
25. Wall Shingles
26. Weather Barrier
27. Wall Sheathing
28. Fire Stop
29. Dampproofing
30. Foundation Wall
31. Backfill
32. Drainage Stone
33. Drainage Tile
34. Wall Footing
35. Gravel
36. Concrete Slab
37. Column Footing
38. Pipe Column
39. Expansion Joint
40. Girder
41. Sub-floor
42. Finish Floor
43. Attic Insulation
44. Soffit
45. Ceiling Strapping
46. Wall Insulation
47. Cross Bridging
48. Bulkhead Stairs

Exterior Wall Construction

Typical Frame Construction

Typical wood frame construction consists of wood studs with insulation between them. A typical exterior surface is made up of sheathing, building paper, and exterior siding consisting of wood, vinyl, aluminum, or stucco over the wood sheathing.

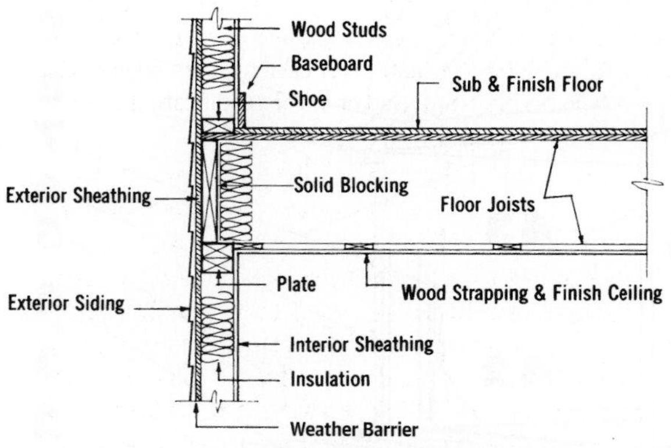

Brick Veneer

Typical brick veneer construction consists of wood studs with insulation between them. A typical exterior surface is sheathing, building paper, and an exterior of brick tied to the sheathing with metal strips.

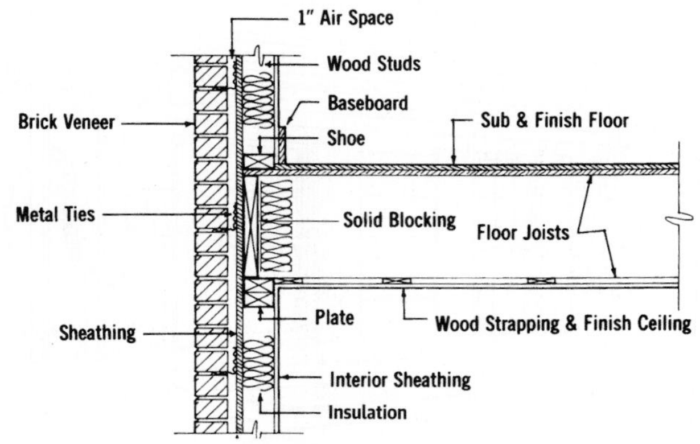

Stone

Typical solid masonry construction consists of a stone or block wall covered on the exterior with brick, stone, or other masonry.

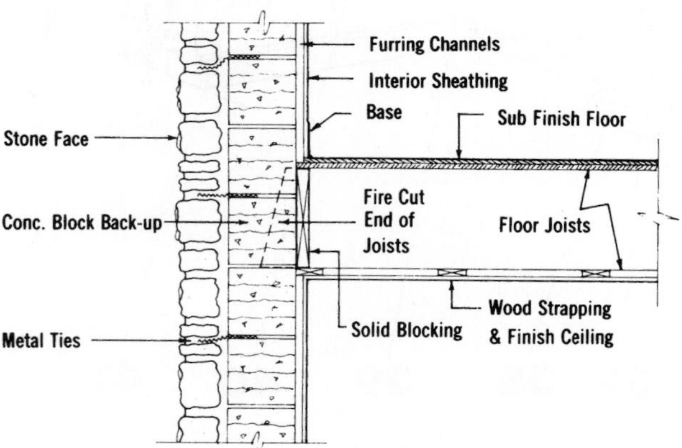

Residential Cost Estimate Worksheet

Worksheet Instructions

The residential cost estimate worksheet can be used as an outline for developing a residential construction or replacement cost. It is also useful for insurance appraisals. The design of the worksheet helps eliminate errors and omissions. To use the worksheet, follow the example below.

1. Fill out the owner's name, residence address, the estimator or appraiser's name, some type of project identifying number or code, and the date.
2. Determine from the plans, specifications, owner's description, photographs, or any other means possible the class of construction. The models in this data set use economy, average, custom, and luxury as classes. Fill in the appropriate box.
3. Fill in the appropriate box for the residence type, configuration, occupancy, and exterior wall. If you require clarification, the pages preceding this worksheet describe each of these.
4. Next, the living area of the residence must be established. The heated or air conditioned space of the residence, not including the basement, should be measured. It is easiest to break the structure up into separate components as shown in the example: the main house (A), a one-and-one-half story wing (B), and a one-story wing (C). The breezeway (D), garage (E), and open covered porch (F) will be treated differently. Data entry blocks for the living area are included on the worksheet for your use. Keep each level of each component separate, and fill out the blocks as shown.
5. By using the information on the worksheet, find the model, wing, or ell in the following square foot cost pages that best matches the class, type, exterior finish, and size of the residence being estimated. Use the Modifications, Adjustments, and Alternatives to determine the adjusted cost per square foot of living area for each component.
6. For each component, multiply the cost per square foot by the living area square footage. If the residence is a town/rowhouse, a multiplier should be applied based upon the configuration.
7. The second page of the residential cost estimate worksheet has space for the additional components of a house. The cost for additional bathrooms, finished attic space, breezeways, porches, fireplaces, appliance or cabinet upgrades, and garages should be added on this page. The information for each of these components is found with the model being used or in the Modifications, Adjustments, and Alternatives.
8. Add the total from page one of the estimate worksheet and the items listed on page two. The sum is the adjusted total building cost.
9. Depending on the use of the final estimated cost, one of the remaining two boxes should be filled out. Any additional items or exclusions should be added or subtracted at this time. The data contained in this data set are a national average. Construction costs are different throughout the country. To allow for this difference, a location factor based upon the first three digits of the residence's zip code must be applied. The location factor is a multiplier that increases or decreases the adjusted total building cost. Find the appropriate location factor and calculate the local cost. If depreciation is a concern, a dollar figure should be subtracted at this point.
10. No residence will match a model exactly. Many differences will be found. At this level of estimating, a variation of +/- 10% should be expected.

Adjustments Instructions

No residence matches a model exactly in shape, material, or specifications. The common differences are:

1. Two or more exterior wall systems:
   - Partial basement
   - Partly finished basement
2. Specifications or features that are between two classes
3. Crawl space instead of a basement

Examples

Below are quick examples. See pages 17–19 for complete examples of cost adjustments for these differences:

1. Residence "A" is an average one-story structure with 1,600 S.F. of living area and no basement. Three walls are wood siding on wood frame, and the fourth wall is brick veneer on wood frame. The brick veneer wall is 35% of the exterior wall area.

   Use page 38 to calculate the Base Cost per S.F. of Living Area. Wood Siding for 1,600 S.F. = $119.40 per S.F. and Brick Veneer for 1,600 S.F. = $123.25 per S.F.

   0.65 ($119.40) + 0.35 ($123.25) = $120.75 per S.F. of Living Area.

2a. Residence "B" is the same as Residence "A" but it has an unfinished basement under 50% of the building. To adjust the $118.10 per S.F. of living area for this partial basement, use page 38.

   $120.75 + 0.5 ($11.90) = $126.70 per S.F. of Living Area.

2b. Residence "C" is the same as Residence "A" but it has a full basement under the entire building. 640 S.F. or 40% of the basement area is finished.

   Using Page 38:

   $120.75 + 0.40 ($33.90) + 0.60 ($11.90) =
   $141.45 per S.F. of Living Area.

3. When specifications or features of a building are between classes, estimate the percent deviation, and use two tables to calculate the cost per S.F.

   A two-story residence with wood siding and 1,800 S.F. of living area has features 30% better than Average, but 70% less than Custom.

   From pages 42 and 64:

Custom 1,800 S.F. Base Cost	=	**$155.25 per S.F.**
Average 1,800 S.F. Base Cost	=	**$121.05 per S.F.**
DIFFERENCE	=	**$34.20 per S.F.**
Cost is $121.05 + 0.30 ($34.20)	=	**$131.31 per S.F. of Living Area.**

4. To add the cost of a crawl space, use the cost of an unfinished basement as a maximum. For specific costs of components to be added or deducted, such as vapor barrier, underdrain, and floor, see the "Assemblies" section (pages 97 to 281).

Model Residence Example

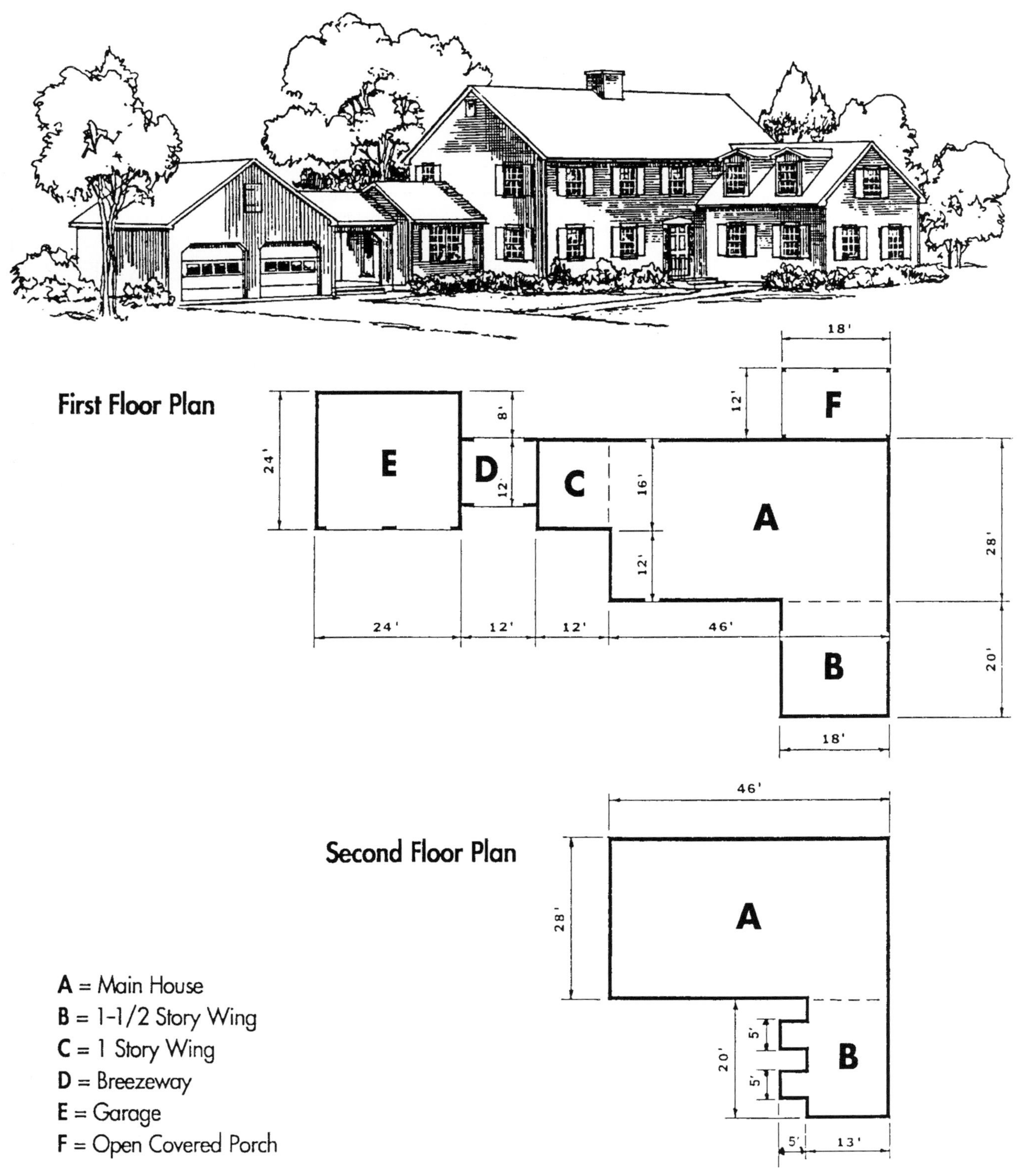

RESIDENTIAL
COST ESTIMATE

OWNERS NAME:	**Albert Westenberg**	APPRAISER:	**Nicole Wojtowicz**
RESIDENCE ADDRESS:	**300 Sygiel Road**	PROJECT:	**# 55**
CITY, STATE, ZIP CODE:	**Three Rivers, MA 01080**	DATE:	**Jan. 1, 2019**

CLASS OF CONSTRUCTION	RESIDENCE TYPE	CONFIGURATION	EXTERIOR WALL SYSTEM
☐ ECONOMY	☐ 1 STORY	☑ DETACHED	☑ WOOD SIDING - WOOD FRAME
☑ AVERAGE	☐ 11/2 STORY	☐ TOWN/ROW HOUSE	☐ BRICK VENEER - WOOD FRAME
☐ CUSTOM	☑ 2 STORY	☐ SEMI-DETACHED	☐ STUCCO ON WOOD FRAME
☐ LUXURY	☐ 2 1/2 STORY		☐ PAINTED CONCRETE BLOCK
	☐ 3 STORY	OCCUPANCY	☐ SOLID MASONRY (AVERAGE & CUSTOM)
	☐ BI-LEVEL	☑ ONE FAMILY	☐ STONE VENEER - WOOD FRAME
	☐ TRI-LEVEL	☐ TWO FAMILY	☐ SOLID BRICK (LUXURY)
		☐ THREE FAMILY	☐ SOLID STONE (LUXURY)
		☐ OTHER	

* LIVING AREA (Main Building)		* LIVING AREA (Wing or Ell) (B)		* LIVING AREA (Wing or Ell) (C)	
First Level	1288 S.F.	First Level	360 S.F.	First Level	192 S.F.
Second level	1288 S.F.	Second level	310 S.F.	Second level	S.F.
Third Level	S.F.	Third Level	S.F.	Third Level	S.F.
Total	2576 S.F.	Total	670 S.F.	Total	192 S.F.

* Basement Area is not part of living area.

MAIN BUILDING				COSTS PER S.F. LIVING AREA	
Cost per Square Foot of Living Area, from Page	42			$	105.50
Basement Addition:	____ % Finished,	100	% Unfinished	+	6.60
Roof Cover Adjustment:	Cedar Shake	Type, Page 42	(Add or Deduct)	( )	1.75
Central Air Conditioning:	☐ Separate Ducts	☑ Heating Ducts, Page	42	+	3.08
Heating System Adjustment:	____	Type, Page ____	(Add or Deduct)	( )	———
Main Building: Adjusted Cost per S.F. of Living Area				$	# 116.93

MAIN BUILDING TOTAL COST	$ 116.93 /S.F.	x	2,576 S.F.	x	———	=	$ 301,212
	Cost per S.F. Living Area		Living Area		Town/Row House Multiplier (Use 1 for Detached)		TOTAL COST

~~WING~~ OR ELL (B) 1 - 1/2 STORY				COSTS PER S.F. LIVING AREA	
Cost per Square Foot of Living Area, from Page	56			$	100.50
Basement Addition:	100 % Finished,	____	% Unfinished	+	29.45
Roof Cover Adjustment:	____	Type, Page ____	(Add or Deduct)	( )	——
Central Air Conditioning:	☐ Separate Ducts	☑ Heating Ducts, Page	42	+	3.08
Heating System Adjustment:		Type, Page ____	(Add or Deduct)	( )	——
~~Wing~~ or Ell (B): Adjusted Cost per S.F. of Living Area				$	# 133.03

WING OR ELL (B) TOTAL COST	$ 133.03 /S.F.	x	670 S.F.	=	$ 89,130
	Cost per S.F. Living Area		Living Area		TOTAL COST

WING OR ~~ELL~~ (C) 1 STORY				COSTS PER S.F. LIVING AREA	
Cost per Square Foot of Living Area, from Page	# 56	(WOOD SIDING)		$	# 153.45
Basement Addition:	____ % Finished,	____	% Unfinished	+	——
Roof Cover Adjustment:	-	Type, Page #	(Add or Deduct)	( )	——
Central Air Conditioning:	☐ Separate Ducts	☐ Heating Ducts, Page	____	+	——
Heating System Adjustment:	____	Type, Page ____	(Add or Deduct)	( )	——
Wing or Ell (C) Adjusted Cost per S.F. of Living Area				$	# 153.45

WING OR ELL (C) TOTAL COST	$ 153.45 /S.F.	x	192 S.F.	=	$ 29,462
	Cost per S.F. Living Area		Living Area		TOTAL COST

TOTAL THIS PAGE **419,804**

RESIDENTIAL
COST ESTIMATE

		QUANTITY		UNIT COST	
Total Page 1					$ **419,804**
Additional Bathrooms: **2** Full, **1** Half **2 @ 8111 1 @ 4789**					+ **21,011**
Finished Attic: **N/A** Ft. x ______ Ft.			S.F.		
Breezeway: ☑ Open ☐ closed **12** Ft. x **12** Ft.		**144**	S.F.	**41.11**	+ **5,920**
Covered Porch: ☑ Open ☐ Enclosed **18** Ft. x **12** Ft.		**216**	S.F.	**38.02**	+ **8,212**
Fireplace: ☑ Interior Chimney ☐ Exterior Chimney ☑ No. of Flues (**2**) ☑ Additional Fireplaces **1 - 2nd Story**					+ **13,150**
Appliances:					+ —
Kitchen Cabinets Adjustments:	(±)				—
☑ Garage ☐ Carport: **2** Car(s) Description **Wood, Attached**	(±)				**26,080**
Miscellaneous:					+

ADJUSTED TOTAL BUILDING COST $ **494,177**

REPLACEMENT COST		
ADJUSTED TOTAL BUILDING COST	$	**494,177**
Site Improvements		
(A) Paving & Sidewalks	$	
(B) Landscaping	$	
(C) Fences	$	
(D) Swimming Pools	$	
(E) Miscellaneous	$	
TOTAL	$	**494,177**
Location Factor	x	**1.060**
Location Replacement Cost	$	**523,828**
Depreciation	-$	**52,380**
LOCAL DEPRECIATED COST	$	**471,448**

INSURANCE COST		
ADJUSTED TOTAL BUILDING COST	$	
Insurance Exclusions		
(A) Footings, sitework, Underground Piping	-$	
(B) Architects Fees	-$	
Total Building Cost Less Exclusion	$	
Location Factor	x	
LOCAL INSURABLE REPLACEMENT COST	$	

SKETCH AND ADDITIONAL CALCULATIONS

Economy Class

1 Story

1-1/2 Story

2 Story

Bi-Level

Tri-Level

- **Mass produced from stock plans**
- **Single family — 1 full bath, 1 kitchen**
- **No basement**
- **Asphalt shingles on roof**
- **Hot air heat**
- **Gypsum wallboard interior finishes**
- **Materials and workmanship are sufficient to meet codes**

Note: The illustration shown may contain some optional components (for example: garages and/or fireplaces) whose costs are shown in the modifications, adjustments, & alternatives below or at the end of the square foot section.

Base cost per square foot of living area

Exterior Wall	Living Area 600	800	1000	1200	1400	1600	1800	2000	2400	2800	3200
Wood Siding - Wood Frame	150.15	135.50	124.30	115.45	107.65	102.85	100.20	96.90	90.30	85.50	82.20
Brick Veneer - Wood Frame	155.60	140.35	128.75	119.45	111.35	106.25	103.50	100.00	93.10	88.10	84.55
Stucco on Wood Frame	141.20	127.45	117.05	**108.95**	101.75	97.25	94.85	91.85	85.70	81.25	78.25
Painted Concrete Block	146.20	131.95	121.10	112.55	105.05	100.35	97.85	94.70	88.25	83.60	80.40
Finished Basement, Add	33.70	31.70	30.25	29.00	28.00	27.30	26.90	26.35	25.60	24.95	24.45
Unfinished Basement, Add	15.20	13.60	12.45	11.45	10.60	10.05	9.75	9.35	8.70	8.20	7.85

Modifications

Add to the total cost

Upgrade Kitchen Cabinets	$ + 1255
Solid Surface Countertops	+ 888
Full Bath - including plumbing, wall and floor finishes	+ 6489
Half Bath - including plumbing, wall and floor finishes	+ 3831
One Car Attached Garage	+ 14,279
One Car Detached Garage	+ 18,467
Fireplace & Chimney	+ 6742

Adjustments

For multi family - add to total cost

Additional Kitchen	$ + 5585
Additional Bath	+ 6489
Additional Entry & Exit	+ 1804
Separate Heating	+ 1748
Separate Electric	+ 1089

For Townhouse/Rowhouse - Multiply cost per square foot by

Inner Unit	.95
End Unit	.97

Alternatives

Add to or deduct from the cost per square foot of living area

Composition Roll Roofing	− 1.15
Cedar Shake Roof	+ 4.15
Upgrade Walls and Ceilings to Skim Coat Plaster	+ .86
Upgrade Ceilings to Textured Finish	+ .61
Air Conditioning, in Heating Ductwork	+ 4.95
In Separate Ductwork	+ 7.23
Heating Systems, Hot Water	+ 1.46
Heat Pump	+ 1.20
Electric Heat	− 2.24
Not Heated	− 3.72

Additional upgrades or components

Kitchen Cabinets & Countertops	Page 93
Bathroom Vanities	94
Fireplaces & Chimneys	94
Windows, Skylights & Dormers	94
Appliances	95
Breezeways & Porches	95
Finished Attic	95
Garages	96
Site Improvements	96
Wings & Ells	34

Important: See the Reference Section for Location Factors (to adjust for your city) and Estimating Forms.

Economy 1 Story

Living Area - 1200 S.F.
Perimeter - 146 L.F.

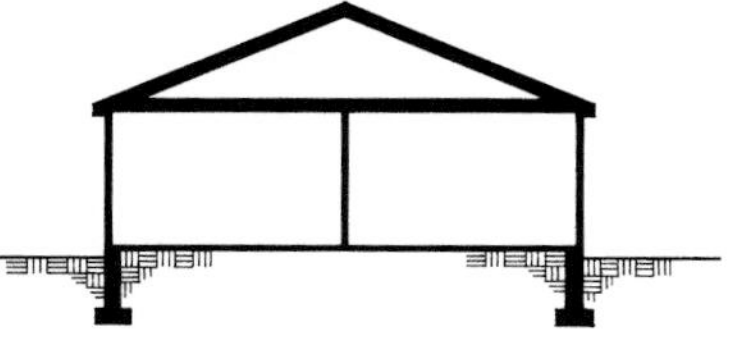

		Cost Per Square Foot Of Living Area			% of Total
		Mat.	Inst.	Total	(rounded)
1 Site Work	Site preparation for slab; 4′ deep trench excavation for foundation wall.		2.15	2.15	2.0%
2 Foundation	Continuous reinforced concrete footing, 10″ deep x 20″ wide; dampproofed and insulated 8″ thick reinforced concrete block foundation wall, 4′ deep; trowel finished 4″ thick concrete slab on 4″ crushed stone base and polyethylene vapor barrier.	6.64	8.58	15.22	14.0%
3 Framing	Exterior walls - 2″ x 4″ wood studs, 16″ O.C.; 1/2″ sheathing; wood truss roof frame, 24″ O.C. with 1/2″ plywood sheating; 2″ x 4″ interior partitions.	5.79	7.77	13.56	12.4%
4 Exterior Walls	Metal lath reinforced stucco exterior on insulated wood frame walls; R38 attic insulation; sliding wood windows; flush solid core doors, frame and hardware, painted finish; aluminum storm and screen doors.	8.61	7.91	16.52	15.2%
5 Roofing	25 year asphalt roof shingles; #15 felt building paper; aluminum gutters, downspouts, drip edge and flashings.	2.27	2.76	5.03	4.6%
6 Interiors	Walls and ceilings, 1/2″ taped and finished gypsum wallboard, primed and painted with 2 coats of finish paint; hollow core wood interior doors, frames and hardware, painted finish; lightweight carpeting with pad, 80%; sheet vinyl flooring, 20%.	12.29	13.38	25.67	23.6%
7 Specialties	Economy grade kitchen cabinets and countertops; stainless steel kitchen sink; 30 gallon electric water heater.	3.64	1.13	4.77	4.4%
8 Mechanical	Three fixture bathroom: bathtub, water closet and wall hung lavatory; gas fired hot air heating system.	4.63	3.96	8.59	7.9%
9 Electrical	100 amp electric service; wiring, duplex & GFI receptacles, wall switches, door bell, appliance circuits and communications cabling; economy grade lighting fixtures.	1.07	2.14	3.21	2.9%
10 Overhead	Contractor's overhead and profit.	6.76	7.47	14.23	13.1%
	Total	51.70	57.25	**108.95**	

- **Mass produced from stock plans**
- **Single family — 1 full bath, 1 kitchen**
- **No basement**
- **Asphalt shingles on roof**
- **Hot air heat**
- **Gypsum wallboard interior finishes**
- **Materials and workmanship are sufficient to meet codes**

Note: The illustration shown may contain some optional components (for example: garages and/or fireplaces) whose costs are shown in the modifications, adjustments, & alternatives below or at the end of the square foot section.

Base cost per square foot of living area

Exterior Wall	Living Area										
	600	800	1000	1200	1400	1600	1800	2000	2400	2800	3200
Wood Siding - Wood Frame	176.75	146.60	131.60	124.45	119.30	111.30	107.40	103.40	95.15	92.00	88.45
Brick Veneer - Wood Frame	184.65	152.35	136.90	129.40	123.95	115.55	111.40	107.30	98.60	95.20	91.45
Stucco on Wood Frame	163.95	137.30	123.00	116.35	111.60	104.30	100.75	97.15	89.60	86.70	83.60
Painted Concrete Block	170.85	142.35	127.60	120.70	115.70	108.05	104.30	100.50	92.60	89.50	86.20
Finished Basement, Add	26.00	22.00	21.00	20.25	19.75	18.95	18.50	18.15	17.30	16.95	16.50
Unfinished Basement, Add	13.35	10.25	9.45	8.85	8.40	7.75	7.45	7.10	6.45	6.20	5.80

Modifications

Add to the total cost

Upgrade Kitchen Cabinets	$ + 1255
Solid Surface Countertops	+ 888
Full Bath - including plumbing, wall and floor finishes	+ 6489
Half Bath - including plumbing, wall and floor finishes	+ 3831
One Car Attached Garage	+ 14,279
One Car Detached Garage	+ 18,467
Fireplace & Chimney	+ 6742

Adjustments

For multi family - add to total cost

Additional Kitchen	$ + 5585
Additional Bath	+ 6489
Additional Entry & Exit	+ 1804
Separate Heating	+ 1748
Separate Electric	+ 1089

For Townhouse/Rowhouse - Multiply cost per square foot by

Inner Unit	.95
End Unit	.97

Alternatives

Add to or deduct from the cost per square foot of living area

Composition Roll Roofing	– .75
Cedar Shake Roof	+ 3
Upgrade Walls and Ceilings to Skim Coat Plaster	+ .86
Upgrade Ceilings to Textured Finish	+ .61
Air Conditioning, in Heating Ductwork	+ 3.69
In Separate Ductwork	+ 6.35
Heating Systems, Hot Water	+ 1.38
Heat Pump	+ 1.32
Electric Heat	– 1.78
Not Heated	– 3.43

Additional upgrades or components

Kitchen Cabinets & Countertops	Page 93
Bathroom Vanities	94
Fireplaces & Chimneys	94
Windows, Skylights & Dormers	94
Appliances	95
Breezeways & Porches	95
Finished Attic	95
Garages	96
Site Improvements	96
Wings & Ells	34

Important: See the Reference Section for Location Factors (to adjust for your city) and Estimating Forms.

Economy 1-1/2 Story

Living Area - 1600 S.F.
Perimeter - 135 L.F.

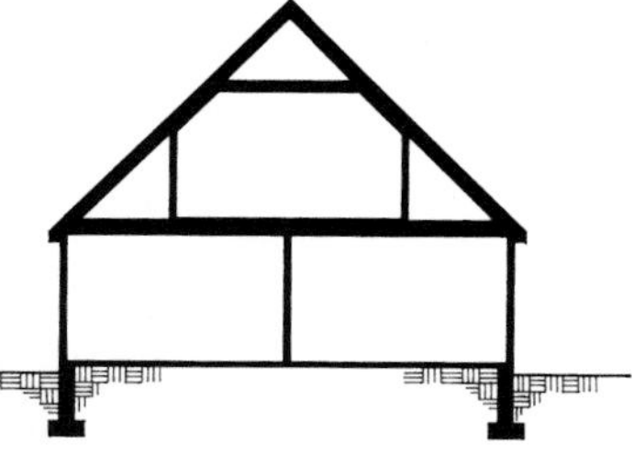

			Cost Per Square Foot Of Living Area			% of Total
			Mat.	Inst.	Total	(rounded)
1	**Site Work**	Site preparation for slab; 4′ deep trench excavation for foundation wall.		1.61	1.61	1.4%
2	**Foundation**	Continuous reinforced concrete footing, 10″ deep x 20″ wide; dampproofed and insulated 8″ thick reinforced concrete block foundation wall, 4′ deep; trowel finished 4″ thick concrete slab on 4″ crushed stone base and polyethylene vapor barrier.	4.44	5.82	10.26	9.2%
3	**Framing**	Exterior walls - 2″ x 4″ wood studs, 16″ O.C.; 1/2″ sheathing; gable end roof framing, steep pitch 2″ x 8″ rafters, 16″ O.C. with 1/2″ plywood sheathing; 2″ x 8″ floor joists, 16″ O.C. with bridging and 5/8″ subflooring; 2″ x 4″ interior partitions.	6.87	10.26	17.13	15.4%
4	**Exterior Walls**	Beveled wood siding and housewrap on insulated wood frame walls; R38 attic insulation; sliding wood windows; flush solid core doors, frame and hardware, painted finish; aluminum storm and screen doors.	15.86	6.05	21.91	19.7%
5	**Roofing**	25 year asphalt roof shingles; #15 felt building paper; aluminum gutters, downspouts, drip edge and flashings.	1.65	2.00	3.65	3.3%
6	**Interiors**	Walls and ceilings, 1/2″ taped and finished gypsum wallboard, primed and painted with 2 coats of finish paint; hollow core wood interior doors, frames and hardware, painted finish; lightweight carpeting with pad, 80%; sheet vinyl flooring, 20%; hardwood tread stairway.	14.09	14.27	28.36	25.5%
7	**Specialties**	Economy grade kitchen cabinets and countertops; stainless steel kitchen sink; 30 gallon electric water heater.	2.73	.85	3.58	3.2%
8	**Mechanical**	Three fixture bathroom: bathtub, water closet and wall hung lavatory; gas fired hot air heating system.	3.73	3.56	7.29	6.5%
9	**Electrical**	100 amp electric service; wiring, duplex & GFI receptacles, wall switches, door bell, appliance circuits and communications cabling; economy grade lighting fixtures.	1.01	1.97	2.98	2.7%
10	**Overhead**	Contractor's overhead and profit.	7.57	6.96	14.53	13.1%
		Total	57.95	53.35	**111.30**	

- **Mass produced from stock plans**
- **Single family — 1 full bath, 1 kitchen**
- **No basement**
- **Asphalt shingles on roof**
- **Hot air heat**
- **Gypsum wallboard interior finishes**
- **Materials and workmanship are sufficient to meet codes**

Note: The illustration shown may contain some optional components (for example: garages and/or fireplaces) whose costs are shown in the modifications, adjustments, & alternatives below or at the end of the square foot section.

Base cost per square foot of living area

	Living Area										
Exterior Wall	1000	1200	1400	1600	1800	2000	2200	2600	3000	3400	3800
Wood Siding - Wood Frame	134.50	121.80	115.75	111.55	107.20	102.55	99.40	93.45	87.80	85.15	82.75
Brick Veneer - Wood Frame	140.10	127.00	120.65	116.20	111.55	106.80	103.40	97.15	91.10	88.30	85.85
Stucco on Wood Frame	125.20	113.20	107.70	103.90	99.95	95.60	92.80	87.50	82.30	79.90	77.80
Painted Concrete Block	130.65	118.25	112.35	108.30	104.25	99.65	96.65	90.95	85.50	82.95	80.70
Finished Basement, Add	17.60	16.85	16.30	15.90	15.45	15.15	14.85	14.25	13.85	13.55	13.30
Unfinished Basement, Add	8.20	7.60	7.10	6.80	6.45	6.25	5.95	5.50	5.15	4.95	4.75

Modifications

Add to the total cost

Upgrade Kitchen Cabinets	$ + 1255
Solid Surface Countertops	+ 888
Full Bath - including plumbing, wall and floor finishes	+ 6489
Half Bath - including plumbing, wall and floor finishes	+ 3831
One Car Attached Garage	+ 14,279
One Car Detached Garage	+ 18,467
Fireplace & Chimney	+ 7450

Adjustments

For multi family - add to total cost

Additional Kitchen	$ + 5585
Additional Bath	+ 6489
Additional Entry & Exit	+ 1804
Separate Heating	+ 1748
Separate Electric	+ 1089

For Townhouse/Rowhouse - Multiply cost per square foot by

Inner Unit	.93
End Unit	.96

Alternatives

Add to or deduct from the cost per square foot of living area

Composition Roll Roofing	− .55
Cedar Shake Roof	+ 2.10
Upgrade Walls and Ceilings to Skim Coat Plaster	+ .87
Upgrade Ceilings to Textured Finish	+ .61
Air Conditioning, in Heating Ductwork	+ 3.01
In Separate Ductwork	+ 5.83
Heating Systems, Hot Water	+ 1.34
Heat Pump	+ 1.39
Electric Heat	− 1.56
Not Heated	− 3.24

Additional upgrades or components

Kitchen Cabinets & Countertops	Page 93
Bathroom Vanities	94
Fireplaces & Chimneys	94
Windows, Skylights & Dormers	94
Appliances	95
Breezeways & Porches	95
Finished Attic	95
Garages	96
Site Improvements	96
Wings & Ells	34

Important: See the Reference Section for Location Factors (to adjust for your city) and Estimating Forms.

Economy 2 Story

Living Area - 2000 S.F.
Perimeter - 135 L.F.

			Cost Per Square Foot Of Living Area			% of Total
			Mat.	Inst.	Total	(rounded)
1	Site Work	Site preparation for slab; 4′ deep trench excavation for foundation wall.		1.29	1.29	1.3%
2	Foundation	Continuous reinforced concrete footing, 10″ deep x 20″ wide; dampproofed and insulated 8″ thick reinforced concrete block foundation wall, 4′ deep; trowel finished 4″ thick concrete slab on 4″ crushed stone base and polyethylene vapor barrier.	3.57	4.66	8.23	8.0%
3	Framing	Exterior walls - 2″ x 4″ wood studs, 16″ O.C.; 1/2″ sheathing; wood truss roof frame, 24″ O.C. with 1/2″ plywood sheating; 2″ x 8″ floor joists, 16″ O.C. with bridging and 5/8″ subflooring; 2″ x 4″ interior partitions.	6.24	9.05	15.29	14.9%
4	Exterior Walls	Beveled wood siding and housewrap on insulated wood frame walls; R38 attic insulation; sliding wood windows; flush solid core doors, frame and hardware, painted finish; aluminum storm and screen doors.	15.69	5.99	21.68	21.1%
5	Roofing	25 year asphalt roof shingles; #15 felt building paper; aluminum gutters, downspouts, drip edge and flashings.	1.14	1.38	2.52	2.5%
6	Interiors	Walls and ceilings, 1/2″ taped and finished gypsum wallboard, primed and painted with 2 coats of finish paint; hollow core wood interior doors, frames and hardware, painted finish; lightweight carpeting with pad, 80%; sheet vinyl flooring, 20%; hardwood tread stairway.	13.77	14.19	27.96	27.3%
7	Specialties	Economy grade kitchen cabinets and countertops; stainless steel kitchen sink; 30 gallon electric water heater.	2.19	.68	2.87	2.8%
8	Mechanical	Three fixture bathroom: bathtub, water closet and wall hung lavatory; gas fired hot air heating system.	3.17	3.32	6.49	6.3%
9	Electrical	100 amp electric service; wiring, duplex & GFI receptacles, wall switches, door bell, appliance circuits and communications cabling; economy grade lighting fixtures.	.98	1.87	2.85	2.8%
10	Overhead	Contractor's overhead and profit.	7.00	6.37	13.37	13.0%
		Total	53.75	48.80	**102.55**	

- **Mass produced from stock plans**
- **Single family — 1 full bath, 1 kitchen**
- **No basement**
- **Asphalt shingles on roof**
- **Hot air heat**
- **Gypsum wallboard interior finishes**
- **Materials and workmanship are sufficient to meet codes**

Note: The illustration shown may contain some optional components (for example: garages and/or fireplaces) whose costs are shown in the modifications, adjustments, & alternatives below or at the end of the square foot section.

Base cost per square foot of living area

Exterior Wall	Living Area 1000	1200	1400	1600	1800	2000	2200	2600	3000	3400	3800
Wood Siding - Wood Frame	124.50	112.55	107.00	103.25	99.40	95.05	92.30	87.00	81.80	79.45	77.40
Brick Veneer - Wood Frame	128.80	116.55	110.75	106.80	102.75	98.25	95.35	89.80	84.35	81.85	79.75
Stucco on Wood Frame	117.45	106.00	100.95	97.40	93.95	89.75	87.25	82.45	77.60	75.50	73.60
Painted Concrete Block	121.50	109.75	104.45	100.75	97.10	92.80	90.15	85.10	80.00	77.75	75.80
Finished Basement, Add	17.60	16.85	16.30	15.90	15.45	15.15	14.85	14.25	13.85	13.55	13.30
Unfinished Basement, Add	8.20	7.60	7.10	6.80	6.45	6.25	5.95	5.50	5.15	4.95	4.75

Modifications

Add to the total cost

Upgrade Kitchen Cabinets	$ + 1255
Solid Surface Countertops	+ 888
Full Bath - including plumbing, wall and floor finishes	+ 6489
Half Bath - including plumbing, wall and floor finishes	+ 3831
One Car Attached Garage	+ 14,279
One Car Detached Garage	+ 18,467
Fireplace & Chimney	+ 6742

Adjustments

For multi family - add to total cost

Additional Kitchen	$ + 5585
Additional Bath	+ 6489
Additional Entry & Exit	+ 1804
Separate Heating	+ 1748
Separate Electric	+ 1089

For Townhouse/Rowhouse - Multiply cost per square foot by

Inner Unit	.94
End Unit	.97

Alternatives

Add to or deduct from the cost per square foot of living area

Composition Roll Roofing	– .55
Cedar Shake Roof	+ 2.10
Upgrade Walls and Ceilings to Skim Coat Plaster	+ .83
Upgrade Ceilings to Textured Finish	+ .61
Air Conditioning, in Heating Ductwork	+ 3.01
In Separate Ductwork	+ 5.83
Heating Systems, Hot Water	+ 1.34
Heat Pump	+ 1.39
Electric Heat	– 1.56
Not Heated	– 3.24

Additional upgrades or components

Kitchen Cabinets & Countertops	Page 93
Bathroom Vanities	94
Fireplaces & Chimneys	94
Windows, Skylights & Dormers	94
Appliances	95
Breezeways & Porches	95
Finished Attic	95
Garages	96
Site Improvements	96
Wings & Ells	34

Important: See the Reference Section for Location Factors (to adjust for your city) and Estimating Forms.

Economy Bi-Level

Living Area - 2000 S.F.
Perimeter - 135 L.F.

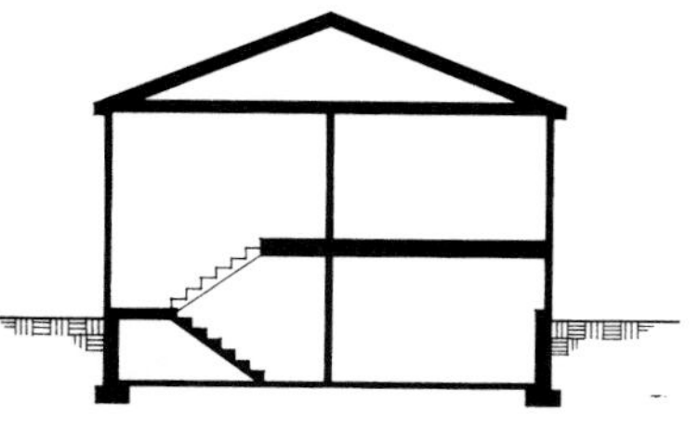

		Cost Per Square Foot Of Living Area			% of Total
		Mat.	Inst.	Total	(rounded)
1 Site Work	Site preparation for slab; 4' deep trench excavation for foundation wall.		1.29	1.29	1.4%
2 Foundation	Continuous reinforced concrete footing, 10" deep x 20" wide; dampproofed and insulated 8" thick reinforced concrete block foundation wall, 4' deep; trowel finished 4" thick concrete slab on 4" crushed stone base and polyethylene vapor barrier.	3.57	4.66	8.23	8.7%
3 Framing	Exterior walls - 2" x 4" wood studs, 16" O.C.; 1/2" sheathing; wood truss roof frame, 24" O.C. with 1/2" plywood sheating; 2" x 8" floor joists, 16" O.C. with bridging and 5/8" subflooring; 2" x 4" interior partitions.	5.78	8.38	14.16	14.9%
4 Exterior Walls	Beveled wood siding and housewrap on insulated wood frame walls; R38 attic insulation; sliding wood windows; flush solid core doors, frame and hardware, painted finish; aluminum storm and screen doors.	12.30	4.71	17.01	17.9%
5 Roofing	25 year asphalt roof shingles; #15 felt building paper; aluminum gutters, downspouts, drip edge and flashings.	1.14	1.38	2.52	2.7%
6 Interiors	Walls and ceilings, 1/2" taped and finished gypsum wallboard, primed and painted with 2 coats of finish paint; hollow core wood interior doors, frames and hardware, painted finish; lightweight carpeting with pad, 80%; sheet vinyl flooring, 20%; hardwood tread stairway.	13.47	13.74	27.21	28.6%
7 Specialties	Economy grade kitchen cabinets and countertops; stainless steel kitchen sink; 30 gallon electric water heater.	2.19	.68	2.87	3.0%
8 Mechanical	Three fixture bathroom: bathtub, water closet and wall hung lavatory; gas fired hot air heating system.	3.17	3.32	6.49	6.8%
9 Electrical	100 amp electric service; wiring, duplex & GFI receptacles, wall switches, door bell, appliance circuits and communications cabling; economy grade lighting fixtures.	.98	1.87	2.85	3.0%
10 Overhead	Contractor's overhead and profit.	6.40	6.02	12.42	13.1%
	Total	49.00	46.05	**95.05**	

- **Mass produced from stock plans**
- **Single family — 1 full bath, 1 kitchen**
- **No basement**
- **Asphalt shingles on roof**
- **Hot air heat**
- **Gypsum wallboard interior finishes**
- **Materials and workmanship are sufficient to meet codes**

Note: The illustration shown may contain some optional components (for example: garages and/or fireplaces) whose costs are shown in the modifications, adjustments, & alternatives below or at the end of the square foot section.

Base cost per square foot of living area

Exterior Wall	Living Area 1200	1500	1800	2000	2200	2400	2800	3200	3600	4000	4400
Wood Siding - Wood Frame	115.75	106.15	99.05	96.20	92.05	88.75	86.10	82.45	78.40	76.95	73.75
Brick Veneer - Wood Frame	119.75	109.80	102.30	99.30	95.00	91.50	88.75	84.90	80.65	79.15	75.85
Stucco on Wood Frame	109.20	100.20	93.65	91.05	87.20	84.15	81.65	78.40	74.60	73.25	70.30
Solid Masonry	112.95	103.55	96.70	93.95	89.95	86.75	84.15	80.65	76.75	75.35	72.25
Finished Basement, Add*	21.15	20.20	19.35	18.95	18.60	18.20	17.95	17.45	17.05	16.90	16.55
Unfinished Basement, Add*	9.05	8.30	7.65	7.35	7.00	6.70	6.45	6.10	5.80	5.65	5.40

*Basement under middle level only.

Modifications

Add to the total cost

Upgrade Kitchen Cabinets	$ + 1255
Solid Surface Countertops	+ 888
Full Bath - including plumbing, wall and floor finishes	+ 6489
Half Bath - including plumbing, wall and floor finishes	+ 3831
One Car Attached Garage	+ 14,279
One Car Detached Garage	+ 18,467
Fireplace & Chimney	+ 6742

Adjustments

For multi family - add to total cost

Additional Kitchen	$ + 5585
Additional Bath	+ 6489
Additional Entry & Exit	+ 1804
Separate Heating	+ 1748
Separate Electric	+ 1089

For Townhouse/Rowhouse - Multiply cost per square foot by

Inner Unit	.93
End Unit	.96

Alternatives

Add to or deduct from the cost per square foot of living area

Composition Roll Roofing	– .75
Cedar Shake Roof	+ 3
Upgrade Walls and Ceilings to Skim Coat Plaster	+ .75
Upgrade Ceilings to Textured Finish	+ .61
Air Conditioning, in Heating Ductwork	+ 2.57
In Separate Ductwork	+ 5.41
Heating Systems, Hot Water	+ 1.29
Heat Pump	+ 1.45
Electric Heat	– 1.34
Not Heated	– 3.14

Additional upgrades or components

Kitchen Cabinets & Countertops	Page 93
Bathroom Vanities	94
Fireplaces & Chimneys	94
Windows, Skylights & Dormers	94
Appliances	95
Breezeways & Porches	95
Finished Attic	95
Garages	96
Site Improvements	96
Wings & Ells	34

Important: See the Reference Section for Location Factors (to adjust for your city) and Estimating Forms.

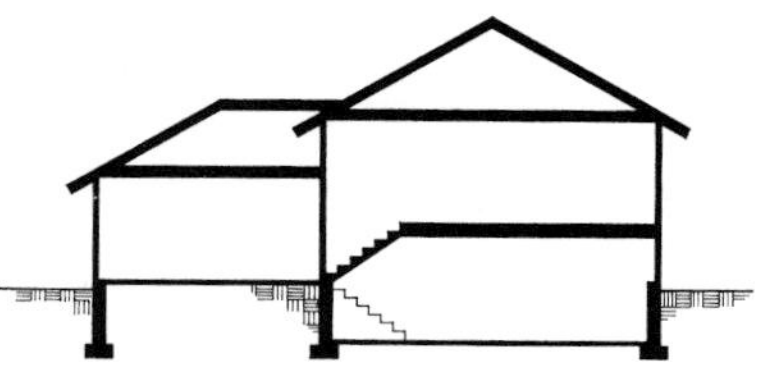

Economy Tri-Level

Living Area - 2400 S.F.
Perimeter - 163 L.F.

			Cost Per Square Foot Of Living Area		% of Total
		Mat.	Inst.	Total	(rounded)
1 Site Work	Site preparation for slab; 4′ deep trench excavation for foundation wall.		1.07	1.07	1.2%
2 Foundation	Continuous reinforced concrete footing, 10″ deep x 20″ wide; dampproofed and insulated 8″ thick reinforced concrete block foundation wall, 4′ deep; trowel finished 4″ thick concrete slab on 4″ crushed stone base and polyethylene vapor barrier.	4.00	5.05	9.05	10.2%
3 Framing	Exterior walls - 2″ x 4″ wood studs, 16″ O.C.; 1/2″ sheathing; wood truss roof frame, 24″ O.C. with 1/2″ plywood sheating; 2″ x 8″ floor joists, 16″ O.C. with bridging and 5/8″ subflooring; 2″ x 4″ interior partitions.	5.34	7.51	12.85	14.5%
4 Exterior Walls	Beveled wood siding and housewrap on insulated wood frame walls; R38 attic insulation; sliding wood windows; flush solid core doors, frame and hardware, painted finish; aluminum storm and screen doors.	10.83	4.10	14.93	16.8%
5 Roofing	25 year asphalt roof shingles; #15 felt building paper; aluminum gutters, downspouts, drip edge and flashings.	1.51	1.84	3.35	3.8%
6 Interiors	Walls and ceilings, 1/2″ taped and finished gypsum wallboard, primed and painted with 2 coats of finish paint; hollow core wood interior doors, frames and hardware, painted finish; lightweight carpeting with pad, 80%; sheet vinyl flooring, 20%; hardwood tread stairway.	12.36	12.42	24.78	27.9%
7 Specialties	Economy grade kitchen cabinets and countertops; stainless steel kitchen sink; 30 gallon electric water heater.	1.82	.57	2.39	2.7%
8 Mechanical	Three fixture bathroom: bathtub, water closet and wall hung lavatory; gas fired hot air heating system.	2.82	3.16	5.98	6.7%
9 Electrical	100 amp electric service; wiring, duplex & GFI receptacles, wall switches, door bell, appliance circuits and communications cabling; economy grade lighting fixtures.	.96	1.80	2.76	3.1%
10 Overhead	Contractor's overhead and profit.	5.96	5.63	11.59	13.1%
	Total	45.60	43.15	**88.75**	

1 Story — Base cost per square foot of living area

Exterior Wall	Living Area 50	100	200	300	400	500	600	700
Wood Siding - Wood Frame	200.25	152.15	131.30	108.10	101.25	97.15	94.45	94.95
Brick Veneer - Wood Frame	212.90	161.15	138.80	113.05	105.75	101.35	98.45	98.80
Stucco on Wood Frame	179.65	137.45	119.05	99.90	93.90	90.25	87.90	88.60
Painted Concrete Block	192.45	146.55	126.60	104.95	98.45	94.50	91.95	92.55
Finished Basement, Add	51.70	42.15	38.15	31.50	30.15	29.40	28.80	28.45
Unfinished Basement, Add	29.05	21.55	18.45	13.25	12.25	11.60	11.20	10.90

1-1/2 Story — Base cost per square foot of living area

Exterior Wall	Living Area 100	200	300	400	500	600	700	800
Wood Siding - Wood Frame	158.70	127.60	108.40	96.15	90.35	87.45	83.95	82.95
Brick Veneer - Wood Frame	170.00	136.60	115.90	102.05	95.80	92.60	88.75	87.75
Stucco on Wood Frame	140.30	112.85	96.15	86.65	81.55	79.15	76.05	75.15
Painted Concrete Block	151.70	122.00	103.70	92.50	87.00	84.30	80.95	80.00
Finished Basement, Add	34.55	30.50	27.85	24.95	24.15	23.60	23.10	23.00
Unfinished Basement, Add	17.90	14.80	12.70	10.40	9.75	9.35	8.95	8.95

2 Story — Base cost per square foot of living area

Exterior Wall	Living Area 100	200	400	600	800	1000	1200	1400
Wood Siding - Wood Frame	162.85	120.70	102.35	83.30	77.30	73.65	71.30	72.05
Brick Veneer - Wood Frame	175.40	129.70	109.85	88.25	81.75	77.85	75.30	75.90
Stucco on Wood Frame	142.25	105.95	90.05	75.10	69.90	66.75	64.75	65.75
Painted Concrete Block	154.95	115.10	97.70	80.15	74.45	71.05	68.80	69.65
Finished Basement, Add	25.90	21.10	19.10	15.80	15.10	14.70	14.45	14.25
Unfinished Basement, Add	14.55	10.80	9.25	6.65	6.15	5.85	5.60	5.45

Base costs do not include bathroom or kitchen facilities. Use Modifications/Adjustments/Alternatives on pages 93–96 where appropriate.

Important: See the Reference Section for Location Factors (to adjust for your city) and Estimating Forms.

1 Story

1-1/2 Story

2 Story

2-1/2 Story

Bi-Level

Tri-Level

- **Simple design from standard plans**
- **Single family — 1 full bath, 1 kitchen**
- **No basement**
- **Asphalt shingles on roof**
- **Hot air heat**
- **Gypsum wallboard interior finishes**
- **Materials and workmanship are average**

Note: The illustration shown may contain some optional components (for example: garages and/or fireplaces) whose costs are shown in the modifications, adjustments, & alternatives below or at the end of the square foot section.

Base cost per square foot of living area

Exterior Wall	Living Area 600	800	1000	1200	1400	1600	1800	2000	2400	2800	3200
Wood Siding - Wood Frame	176.20	158.05	144.55	134.10	125.20	119.40	116.15	112.30	104.70	99.20	95.40
Brick Veneer - Wood Frame	182.30	163.50	149.55	138.60	129.30	123.25	119.85	115.80	107.95	102.20	98.10
Stucco on Wood Frame	170.70	153.15	140.20	130.15	121.60	115.95	112.85	109.20	101.95	96.65	93.05
Solid Masonry	195.30	175.15	160.05	148.10	137.90	131.35	127.60	123.15	114.65	108.40	103.90
Finished Basement, Add	41.15	39.75	37.90	36.20	34.80	33.90	33.40	32.60	31.55	30.80	30.10
Unfinished Basement, Add	17.15	15.55	14.35	13.30	12.45	11.90	11.55	11.10	10.50	10.05	9.60

Modifications

Add to the total cost

Upgrade Kitchen Cabinets	$ + 5907
Solid Surface Countertops (Included)	
Full Bath - including plumbing, wall and floor finishes	+ 8111
Half Bath - including plumbing, wall and floor finishes	+ 4789
One Car Attached Garage	+ 15,383
One Car Detached Garage	+ 20,232
Fireplace & Chimney	+ 7100

Adjustments

For multi family - add to total cost

Additional Kitchen	$ + 10,268
Additional Bath	+ 8111
Additional Entry & Exit	+ 1804
Separate Heating	+ 1748
Separate Electric	+ 1868

For Townhouse/Rowhouse - Multiply cost per square foot by

Inner Unit	.92
End Unit	.96

Alternatives

Add to or deduct from the cost per square foot of living area

Cedar Shake Roof	+ 3.55
Clay Tile Roof	+ 6.90
Slate Roof	+ 7.80
Upgrade Walls to Skim Coat Plaster	+ .53
Upgrade Ceilings to Textured Finish	+ .61
Air Conditioning, in Heating Ductwork	+ 5.11
In Separate Ductwork	+ 7.50
Heating Systems, Hot Water	+ 1.48
Heat Pump	+ 1.25
Electric Heat	− 1.24
Not Heated	− 3.12

Additional upgrades or components

Kitchen Cabinets & Countertops	Page 93
Bathroom Vanities	94
Fireplaces & Chimneys	94
Windows, Skylights & Dormers	94
Appliances	95
Breezeways & Porches	95
Finished Attic	95
Garages	96
Site Improvements	96
Wings & Ells	56

Important: See the Reference Section for Location Factors (to adjust for your city) and Estimating Forms.

Average 1 Story

Living Area - 1600 S.F.
Perimeter - 163 L.F.

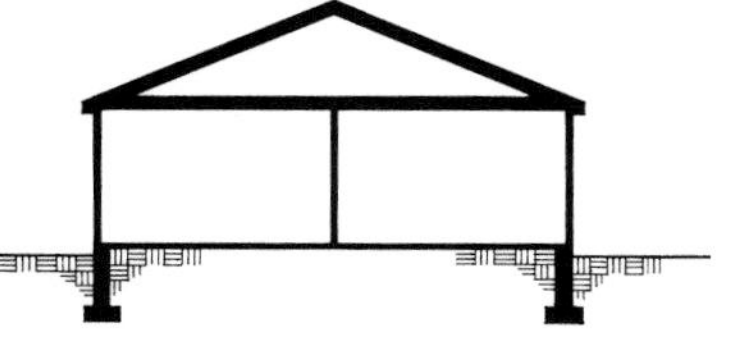

		Cost Per Square Foot Of Living Area			% of Total
		Mat.	Inst.	Total	(rounded)
1 Site Work	Site preparation for slab; 4′ deep trench excavation for foundation wall.		1.66	1.66	1.4%
2 Foundation	Continuous reinforced concrete footing, 10″ deep x 20″ wide; damproofed and insulated 8″ thick reinforced concrete block foundation wall, 4′ deep; trowel finished 4″ thick concrete slab on 4″ crushed stone base and polyethylene vapor barrier.	6.16	7.80	13.96	11.7%
3 Framing	Exterior walls - 2″ x 6″ wood studs, 16″ O.C.; 1/2″ sheathing; gable end roof framing, 2″ x 10″ rafters, 16″ O.C. with 1/2″ plywood sheathing; 2″ x 4″ interior partitions.	7.40	9.82	17.22	14.4%
4 Exterior Walls	Beveled wood siding and housewrap on insulated wood frame walls; R38 attic insulation; double hung wood windows; flush solid core doors, frame and hardware, painted finish; aluminum storm and screen doors.	13.49	5.35	18.84	15.8%
5 Roofing	25 year asphalt roof shingles; #15 felt building paper; aluminum gutters, downspouts, drip edge and flashings.	2.33	2.84	5.17	4.3%
6 Interiors	Walls & ceilings, 1/2″ taped & finished gypsum wallboard, primed & painted with 2 coats of finish paint; birch faced hollow core interior doors, frames & hardware, painted finish; medium weight carpeting with pad, 40%; sheet vinyl, 15%; oak hardwood, 40%; ceramic tile, 5%	13.37	13.43	26.80	22.4%
7 Specialties	Average grade kitchen cabinets and countertop; stainless steel kitchen sink; 40 gallon electric water heater.	4.88	1.53	6.41	5.4%
8 Mechanical	Three fixture bathroom: bathtub, water closet, vanity and sink; gas fired hot air heating system.	4.14	3.66	7.80	6.5%
9 Electrical	200 amp electric service; wiring, duplex & GFI receptacles, wall switches, door bell, appliance circuits, fans and communications cabling; average grade lighting fixtures.	1.61	2.60	4.21	3.5%
10 Overhead	Contractor's overhead and profit and plans.	9.07	8.26	17.33	14.5%
	Total	62.45	56.95	**119.40**	

- **Simple design from standard plans**
- **Single family — 1 full bath, 1 kitchen**
- **No basement**
- **Asphalt shingles on roof**
- **Hot air heat**
- **Gypsum wallboard interior finishes**
- **Materials and workmanship are average**

Note: The illustration shown may contain some optional components (for example: garages and/or fireplaces) whose costs are shown in the modifications, adjustments, & alternatives below or at the end of the square foot section.

Base cost per square foot of living area

Exterior Wall	Living Area										
	600	800	1000	1200	1400	1600	1800	2000	2400	2800	3200
Wood Siding - Wood Frame	201.10	166.65	149.10	140.40	134.25	125.20	120.60	116.05	106.70	103.00	98.90
Brick Veneer - Wood Frame	209.95	173.10	155.00	146.00	139.55	129.95	125.15	120.35	110.50	106.60	102.25
Stucco on Wood Frame	193.20	160.90	143.80	135.45	129.55	120.85	116.55	112.15	103.25	99.70	95.90
Solid Masonry	227.75	186.05	167.00	157.25	150.25	139.65	134.35	129.10	118.20	113.85	109.00
Finished Basement, Add	33.65	29.30	28.05	27.05	26.35	25.30	24.70	24.20	23.10	22.65	22.00
Unfinished Basement, Add	14.85	11.65	10.80	10.20	9.75	9.10	8.75	8.40	7.75	7.50	7.05

Modifications

Add to the total cost

Upgrade Kitchen Cabinets	$ + 5907
Solid Surface Countertops (Included)	
Full Bath - including plumbing, wall and floor finishes	+ 8111
Half Bath - including plumbing, wall and floor finishes	+ 4789
One Car Attached Garage	+ 15,383
One Car Detached Garage	+ 20,232
Fireplace & Chimney	+ 7100

Adjustments

For multi family - add to total cost

Additional Kitchen	$ + 10,268
Additional Bath	+ 8111
Additional Entry & Exit	+ 1804
Separate Heating	+ 1748
Separate Electric	+ 1868

For Townhouse/Rowhouse - Multiply cost per square foot by

Inner Unit	.92
End Unit	.96

Alternatives

Add to or deduct from the cost per square foot of living area

Cedar Shake Roof	+ 2.55
Clay Tile Roof	+ 4.95
Slate Roof	+ 5.65
Upgrade Walls to Skim Coat Plaster	+ .61
Upgrade Ceilings to Textured Finish	+ .61
Air Conditioning, in Heating Ductwork	+ 3.88
In Separate Ductwork	+ 6.56
Heating Systems, Hot Water	+ 1.40
Heat Pump	+ 1.37
Electric Heat	– 1.12
Not Heated	– 3

Additional upgrades or components

Kitchen Cabinets & Countertops	Page 93
Bathroom Vanities	94
Fireplaces & Chimneys	94
Windows, Skylights & Dormers	94
Appliances	95
Breezeways & Porches	95
Finished Attic	95
Garages	96
Site Improvements	96
Wings & Ells	56

Important: See the Reference Section for Location Factors (to adjust for your city) and Estimating Forms.

Average 1-1/2 Story

Living Area - 1800 S.F.
Perimeter - 144 L.F.

		Cost Per Square Foot Of Living Area			% of Total
		Mat.	Inst.	Total	(rounded)
1 Site Work	Site preparation for slab; 4′ deep trench excavation for foundation wall.		1.47	1.47	1.2%
2 Foundation	Continuous reinforced concrete footing, 10″ deep x 20″ wide; dampproofed and insulated 8″ thick reinforced concrete block foundation wall, 4′ deep; trowel finished 4″ thick concrete slab on 4″ crushed stone base and polyethylene vapor barrier.	4.41	5.74	10.15	8.4%
3 Framing	Exterior walls - 2″ x 6″ wood studs, 16″ O.C.; 1/2″ sheathing; gable end roof framing, steep pitch 2″ x 10″ rafters, 16″ O.C. with 1/2″ plywood sheathing; 2″ x 10″ floor joists, 16″ O.C. with bridging and 5/8″ subflooring; 2″ x 4″ interior partitions.	8.18	10.71	18.89	15.7%
4 Exterior Walls	Beveled wood siding and housewrap on insulated wood frame walls; R38 attic insulation; double hung wood windows; flush solid core doors, frame and hardware, painted finish; aluminum storm and screen doors.	14.83	5.84	20.67	17.1%
5 Roofing	25 year asphalt roof shingles; #15 felt building paper; aluminum gutters, downspouts, drip edge and flashings.	1.70	2.06	3.76	3.1%
6 Interiors	Walls & ceilings, 1/2″ taped & finished gypsum wallboard, primed & painted with 2 coats of finish paint; birch faced hollow core interior doors, frames & hardware, painted finish; medium weight carpeting with pad, 40%; sheet vinyl, 15%; oak hardwood, 40%; ceramic tile, 5%; hardwood tread stairway.	15.82	15.26	31.08	25.8%
7 Specialties	Average grade kitchen cabinets and countertop; stainless steel kitchen sink; 40 gallon electric water heater.	4.35	1.35	5.70	4.7%
8 Mechanical	Three fixture bathroom: bathtub, water closet, vanity and sink; gas fired hot air heating system.	3.78	3.50	7.28	6.0%
9 Electrical	200 amp electric service; wiring, duplex & GFI receptacles, wall switches, door bell, appliance circuits, fans and communications cabling; average grade lighting fixtures.	1.57	2.51	4.08	3.4%
10 Overhead	Contractor's overhead and profit and plans.	9.31	8.21	17.52	14.5%
	Total	63.95	56.65	**120.60**	

- **Simple design from standard plans**
- **Single family — 1 full bath, 1 kitchen**
- **No basement**
- **Asphalt shingles on roof**
- **Hot air heat**
- **Gypsum wallboard interior finishes**
- **Materials and workmanship are average**

Note: The illustration shown may contain some optional components (for example: garages and/or fireplaces) whose costs are shown in the modifications, adjustments, & alternatives below or at the end of the square foot section.

Base cost per square foot of living area

Exterior Wall	Living Area 1000	1200	1400	1600	1800	2000	2200	2600	3000	3400	3800
Wood Siding - Wood Frame	153.05	138.30	131.10	126.15	121.05	115.85	112.15	105.50	99.10	96.00	93.35
Brick Veneer - Wood Frame	159.40	144.20	136.65	131.45	126.00	120.60	116.70	109.60	102.90	99.65	96.80
Stucco on Wood Frame	147.55	133.25	126.35	121.60	116.80	111.70	108.30	101.90	95.80	92.95	90.35
Solid Masonry	173.85	157.60	149.15	143.45	137.25	131.45	127.00	118.95	111.55	107.85	104.55
Finished Basement, Add	23.65	23.30	22.55	22.00	21.45	21.05	20.60	19.90	19.30	18.90	18.60
Unfinished Basement, Add	9.40	8.75	8.25	7.95	7.60	7.30	7.10	6.60	6.30	6.05	5.85

Modifications

Add to the total cost

Upgrade Kitchen Cabinets	$ + 5907
Solid Surface Countertops (Included)	
Full Bath - including plumbing, wall and floor finishes	+ 8111
Half Bath - including plumbing, wall and floor finishes	+ 4789
One Car Attached Garage	+ 15,383
One Car Detached Garage	+ 20,232
Fireplace & Chimney	+ 7845

Adjustments

For multi family - add to total cost

Additional Kitchen	$ + 10,268
Additional Bath	+ 8111
Additional Entry & Exit	+ 1804
Separate Heating	+ 1748
Separate Electric	+ 1868

For Townhouse/Rowhouse - Multiply cost per square foot by

Inner Unit	.90
End Unit	.95

Alternatives

Add to or deduct from the cost per square foot of living area

Cedar Shake Roof	+ 1.75
Clay Tile Roof	+ 3.45
Slate Roof	+ 3.90
Upgrade Walls to Skim Coat Plaster	+ .63
Upgrade Ceilings to Textured Finish	+ .61
Air Conditioning, in Heating Ductwork	+ 3.10
In Separate Ductwork	+ 5.96
Heating Systems, Hot Water	+ 1.38
Heat Pump	+ 1.45
Electric Heat	– .97
Not Heated	– 2.90

Additional upgrades or components

Kitchen Cabinets & Countertops	Page 93
Bathroom Vanities	94
Fireplaces & Chimneys	94
Windows, Skylights & Dormers	94
Appliances	95
Breezeways & Porches	95
Finished Attic	95
Garages	96
Site Improvements	96
Wings & Ells	56

Important: See the Reference Section for Location Factors (to adjust for your city) and Estimating Forms.

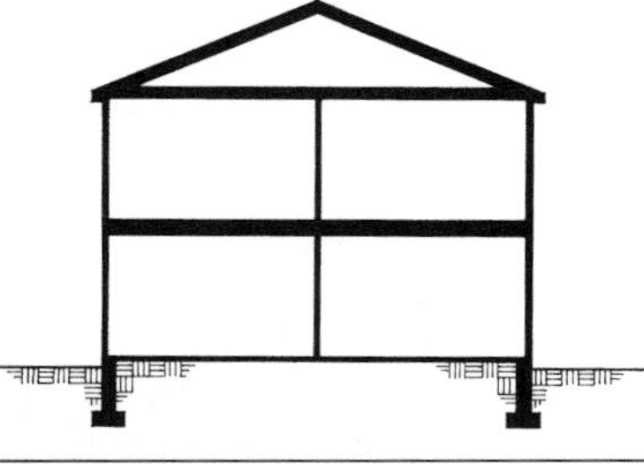

Average 2 Story

Living Area - 2000 S.F.
Perimeter - 135 L.F.

		Cost Per Square Foot Of Living Area			% of Total
		Mat.	Inst.	Total	(rounded)
1 Site Work	Site preparation for slab; 4′ deep trench excavation for foundation wall.		1.33	1.33	1.1%
2 Foundation	Continuous reinforced concrete footing, 10″ deep x 20″ wide; dampproofed and insulated 8″ thick reinforced concrete block foundation wall, 4′ deep; trowel finished 4″ thick concrete slab on 4″ crushed stone base and polyethylene vapor barrier.	3.67	4.79	8.46	7.3%
3 Framing	Exterior walls - 2″ x 6″ wood studs, 16″ O.C.; 1/2″ sheathing; gable end roof framing, 2″ x 10″ rafters, 16″ O.C. with 1/2″ plywood sheathing; 2″ x 10″ floor joists, 16″ O.C. with bridging and 5/8″ subflooring; 2″ x 4″ interior partitions.	7.75	10.14	17.89	15.4%
4 Exterior Walls	Beveled wood siding and housewrap on insulated wood frame walls; R38 attic insulation; double hung wood windows; flush solid core doors, frame and hardware, painted finish; aluminum storm and screen doors.	15.45	6.13	21.58	18.6%
5 Roofing	25 year asphalt roof shingles; #15 felt building paper; aluminum gutters, downspouts, drip edge and flashings.	1.17	1.42	2.59	2.2%
6 Interiors	Walls & ceilings, 1/2″ taped & finished gypsum wallboard, primed & painted with 2 coats of finish paint; birch faced hollow core interior doors, frames & hardware, painted finish; medium weight carpeting with pad, 40%; sheet vinyl, 15%; oak hardwood, 40%; ceramic tile, 5%; hardwood tread stairway.	15.75	15.36	31.11	26.9%
7 Specialties	Average grade kitchen cabinets and countertop; stainless steel kitchen sink; 40 gallon electric water heater.	3.93	1.22	5.15	4.4%
8 Mechanical	Three fixture bathroom: bathtub, water closet, vanity and sink; gas fired hot air heating system.	3.49	3.40	6.89	5.9%
9 Electrical	200 amp electric service; wiring, duplex & GFI receptacles, wall switches, door bell, appliance circuits, fans and communications cabling; average grade lighting fixtures	1.55	2.44	3.99	3.4%
10 Overhead	Contractor's overhead and profit and plans.	8.99	7.87	16.86	14.6%
	Total	61.75	54.10	**115.85**	

RESIDENTIAL | Average | 2-1/2 Story

- **Simple design from standard plans**
- **Single family — 1 full bath, 1 kitchen**
- **No basement**
- **Asphalt shingles on roof**
- **Hot air heat**
- **Gypsum wallboard interior finishes**
- **Materials and workmanship are average**

Note: The illustration shown may contain some optional components (for example: garages and/or fireplaces) whose costs are shown in the modifications, adjustments, & alternatives below or at the end of the square foot section.

Base cost per square foot of living area

Exterior Wall	Living Area 1200	1400	1600	1800	2000	2400	2800	3200	3600	4000	4400
Wood Siding - Wood Frame	152.75	142.90	130.35	127.85	122.95	115.20	109.25	103.20	100.10	94.60	92.80
Brick Veneer - Wood Frame	159.70	149.20	136.20	133.60	128.40	120.10	113.95	107.45	104.20	98.35	96.45
Stucco on Wood Frame	146.60	137.35	125.25	122.70	118.25	110.90	105.15	99.50	96.55	91.30	89.65
Solid Masonry	174.80	162.85	148.75	146.20	140.25	130.80	124.15	116.70	112.95	106.50	104.40
Finished Basement, Add	20.00	19.60	18.85	18.75	18.25	17.45	17.15	16.55	16.20	15.85	15.60
Unfinished Basement, Add	7.80	7.20	6.75	6.65	6.35	5.90	5.65	5.30	5.10	4.85	4.75

Modifications

Add to the total cost

Upgrade Kitchen Cabinets	$ + 5907
Solid Surface Countertops (Included)	
Full Bath - including plumbing, wall and floor finishes	+ 8111
Half Bath - including plumbing, wall and floor finishes	+ 4789
One Car Attached Garage	+ 15,383
One Car Detached Garage	+ 20,232
Fireplace & Chimney	+ 8605

Adjustments

For multi family - add to total cost

Additional Kitchen	$ + 10,268
Additional Bath	+ 8111
Additional Entry & Exit	+ 1804
Separate Heating	+ 1748
Separate Electric	+ 1868

For Townhouse/Rowhouse - Multiply cost per square foot by

Inner Unit	.90
End Unit	.95

Alternatives

Add to or deduct from the cost per square foot of living area

Cedar Shake Roof	+ 1.55
Clay Tile Roof	+ 3
Slate Roof	+ 3.40
Upgrade Walls to Skim Coat Plaster	+ .60
Upgrade Ceilings to Textured Finish	+ .61
Air Conditioning, in Heating Ductwork	+ 2.82
In Separate Ductwork	+ 5.77
Heating Systems, Hot Water	+ 1.26
Heat Pump	+ 1.48
Electric Heat	– 1.73
Not Heated	– 3.39

Additional upgrades or components

Kitchen Cabinets & Countertops	Page 93
Bathroom Vanities	94
Fireplaces & Chimneys	94
Windows, Skylights & Dormers	94
Appliances	95
Breezeways & Porches	95
Finished Attic	95
Garages	96
Site Improvements	96
Wings & Ells	56

Important: See the Reference Section for Location Factors (to adjust for your city) and Estimating Forms.

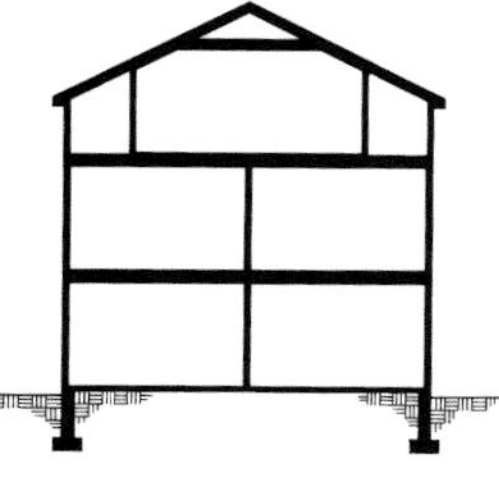

Average 2-1/2 Story

Living Area - 3200 S.F.
Perimeter - 150 L.F.

		Cost Per Square Foot Of Living Area			% of Total
		Mat.	Inst.	Total	(rounded)
1 Site Work	Site preparation for slab; 4′ deep trench excavation for foundation wall.		.82	.82	0.8%
2 Foundation	Continuous reinforced concrete footing, 10″ deep x 20″ wide; dampproofed and insulated 8″ thick reinforced concrete block foundation wall, 4′ deep; trowel finished 4″ thick concrete slab on 4″ crushed stone base and polyethylene vapor barrier.	2.63	3.39	6.02	5.8%
3 Framing	Exterior walls - 2″ x 6″ wood studs, 16″ O.C.; 1/2″ sheathing; gable end roof framing, steep pitch 2″ x 10″ rafters, 16″ O.C. with 1/2″ plywood sheathing; 2″ x 10″ floor joists, 16″ O.C. with bridging and 5/8″ subflooring; 2″ x 4″ interior partitions.	7.73	10.03	17.76	17.2%
4 Exterior Walls	Beveled wood siding and housewrap on insulated wood frame walls; R38 attic insulation; double hung wood windows; flush solid core doors, frame and hardware, painted finish; aluminum storm and screen doors.	13.28	5.20	18.48	17.9%
5 Roofing	25 year asphalt roof shingles; #15 felt building paper; aluminum gutters, downspouts, drip edge and flashings.	1.04	1.26	2.30	2.2%
6 Interiors	Walls & ceilings, 1/2″ taped & finished gypsum wallboard, primed & painted with 2 coats of finish paint; birch faced hollow core interior doors, frames & hardware, painted finish; medium weight carpeting with pad, 40%; sheet vinyl, 15%; oak hardwood, 40%; ceramic tile, 5%; hardwood tread stairway.	15.55	14.86	30.41	29.5%
7 Specialties	Average grade kitchen cabinets and countertop; stainless steel kitchen sink; 40 gallon electric water heater.	2.45	.76	3.21	3.1%
8 Mechanical	Three fixture bathroom: bathtub, water closet, vanity and sink; gas fired hot air heating system.	2.56	3.03	5.59	5.4%
9 Electrical	200 amp electric service; wiring, duplex & GFI receptacles, wall switches, door bell, appliance circuits, fans and communications cabling; average grade lighting fixtures.	1.44	2.17	3.61	3.5%
10 Overhead	Contractor's overhead and profit and plans.	7.92	7.08	15.00	14.5%
	Total	54.60	48.60	**103.20**	

- **Simple design from standard plans**
- **Single family — 1 full bath, 1 kitchen**
- **No basement**
- **Asphalt shingles on roof**
- **Hot air heat**
- **Gypsum wallboard interior finishes**
- **Materials and workmanship are average**

Note: The illustration shown may contain some optional components (for example: garages and/or fireplaces) whose costs are shown in the modifications, adjustments, & alternatives below or at the end of the square foot section.

Base cost per square foot of living area

Exterior Wall	Living Area 1500	1800	2100	2500	3000	3500	4000	4500	5000	5500	6000
Wood Siding - Wood Frame	138.70	125.55	119.55	114.80	106.15	102.35	97.00	91.40	89.55	87.40	85.20
Brick Veneer - Wood Frame	145.00	131.35	124.95	120.05	110.85	106.80	101.05	95.10	93.20	90.85	88.50
Stucco on Wood Frame	133.40	120.60	114.90	110.40	102.15	98.50	93.55	88.15	86.40	84.35	82.40
Solid Masonry	159.60	144.90	137.55	132.10	121.80	117.10	110.45	103.80	101.55	98.95	96.10
Finished Basement, Add	17.15	16.95	16.40	16.05	15.45	15.15	14.60	14.25	14.15	13.95	13.65
Unfinished Basement, Add	6.35	5.95	5.65	5.35	5.00	4.80	4.50	4.25	4.15	4.05	3.90

Modifications

Add to the total cost

Upgrade Kitchen Cabinets	$ + 5907
Solid Surface Countertops (Included)	
Full Bath - including plumbing, wall and floor finishes	+ 8111
Half Bath - including plumbing, wall and floor finishes	+ 4789
One Car Attached Garage	+ 15,383
One Car Detached Garage	+ 20,232
Fireplace & Chimney	+ 8605

Adjustments

For multi family - add to total cost

Additional Kitchen	$ + 10,268
Additional Bath	+ 8111
Additional Entry & Exit	+ 1804
Separate Heating	+ 1748
Separate Electric	+ 1868

For Townhouse/Rowhouse - Multiply cost per square foot by

Inner Unit	.88
End Unit	.94

Alternatives

Add to or deduct from the cost per square foot of living area

Cedar Shake Roof	+ 1.20
Clay Tile Roof	+ 2.30
Slate Roof	+ 2.60
Upgrade Walls to Skim Coat Plaster	+ .63
Upgrade Ceilings to Textured Finish	+ .61
Air Conditioning, in Heating Ductwork	+ 2.82
In Separate Ductwork	+ 5.77
Heating Systems, Hot Water	+ 1.26
Heat Pump	+ 1.48
Electric Heat	– 1.34
Not Heated	– 3.16

Additional upgrades or components

Kitchen Cabinets & Countertops	Page 93
Bathroom Vanities	94
Fireplaces & Chimneys	94
Windows, Skylights & Dormers	94
Appliances	95
Breezeways & Porches	95
Finished Attic	95
Garages	96
Site Improvements	96
Wings & Ells	56

Important: See the Reference Section for Location Factors (to adjust for your city) and Estimating Forms.

Average 3 Story

Living Area - 3000 S.F.
Perimeter - 135 L.F.

		Cost Per Square Foot Of Living Area			% of Total
		Mat.	Inst.	Total	(rounded)
1 Site Work	Site preparation for slab; 4′ deep trench excavation for foundation wall.		.88	.88	0.8%
2 Foundation	Continuous reinforced concrete footing, 10″ deep x 20″ wide; dampproofed and insulated 8″ thick reinforced concrete block foundation wall, 4′ deep; trowel finished 4″ thick concrete slab on 4″ crushed stone base and polyethylene vapor barrier.	2.43	3.20	5.63	5.3%
3 Framing	Exterior walls - 2″ x 6″ wood studs, 16″ O.C.; 1/2″ sheathing; gable end roof framing, 2″ x 10″ rafters, 16″ O.C. with 1/2″ plywood sheathing; 2″ x 10″ floor joists, 16″ O.C. with bridging and 5/8″ subflooring; 2″ x 4″ interior partitions.	7.64	9.91	17.55	16.5%
4 Exterior Walls	Beveled wood siding and housewrap on insulated wood frame walls; R38 attic insulation; double hung wood windows; flush solid core doors, frame and hardware, painted finish; aluminum storm and screen doors.	14.71	5.81	20.52	19.3%
5 Roofing	25 year asphalt roof shingles; #15 felt building paper; aluminum gutters, downspouts, drip edge and flashings.	.78	.95	1.73	1.6%
6 Interiors	Walls & ceilings, 1/2″ taped & finished gypsum wallboard, primed & painted with 2 coats of finish paint; birch faced hollow core interior doors, frames & hardware, painted finish; medium weight carpeting with pad, 40%; sheet vinyl, 15%; oak hardwood, 40%; ceramic tile, 5%; hardwood tread stairway.	16.13	15.45	31.58	29.8%
7 Specialties	Average grade kitchen cabinets and countertop; stainless steel kitchen sink; 40 gallon electric water heater.	2.61	.82	3.43	3.2%
8 Mechanical	Three fixture bathroom: bathtub, water closet, vanity and sink; gas fired hot air heating system.	2.67	3.07	5.74	5.4%
9 Electrical	200 amp electric service; wiring, duplex & GFI receptacles, wall switches, door bell, appliance circuits, fans and communications cabling; average grade lighting fixtures.	1.45	2.20	3.65	3.4%
10 Overhead	Contractor's overhead and profit and plans.	8.23	7.21	15.44	14.5%
	Total	56.65	49.50	**106.15**	

- **Simple design from standard plans**
- **Single family — 1 full bath, 1 kitchen**
- **No basement**
- **Asphalt shingles on roof**
- **Hot air heat**
- **Gypsum wallboard interior finishes**
- **Materials and workmanship are average**

Note: The illustration shown may contain some optional components (for example: garages and/or fireplaces) whose costs are shown in the modifications, adjustments, & alternatives below or at the end of the square foot section.

Base cost per square foot of living area

	Living Area										
Exterior Wall	1000	1200	1400	1600	1800	2000	2200	2600	3000	3400	3800
Wood Siding - Wood Frame	142.55	128.60	122.05	117.45	112.95	108.00	104.75	98.75	92.80	90.15	87.70
Brick Veneer - Wood Frame	147.40	133.15	126.30	121.50	116.70	111.65	108.15	101.85	95.70	92.90	90.35
Stucco on Wood Frame	138.35	124.75	118.40	114.00	109.60	104.80	101.70	96.00	90.25	87.75	85.45
Solid Masonry	158.10	143.10	135.55	130.40	125.05	119.65	115.80	108.80	102.10	98.95	96.10
Finished Basement, Add	23.65	23.30	22.55	22.00	21.45	21.05	20.60	19.90	19.30	18.90	18.60
Unfinished Basement, Add	9.40	8.75	8.25	7.95	7.60	7.30	7.10	6.60	6.30	6.05	5.85

Modifications

Add to the total cost

Upgrade Kitchen Cabinets	$ + 5907
Solid Surface Countertops (Included)	
Full Bath - including plumbing, wall and floor finishes	+ 8111
Half Bath - including plumbing, wall and floor finishes	+ 4789
One Car Attached Garage	+ 15,383
One Car Detached Garage	+ 20,232
Fireplace & Chimney	+ 7100

Adjustments

For multi family - add to total cost

Additional Kitchen	$ + 10,268
Additional Bath	+ 8111
Additional Entry & Exit	+ 1804
Separate Heating	+ 1748
Separate Electric	+ 1868

For Townhouse/Rowhouse - Multiply cost per square foot by

Inner Unit	.91
End Unit	.96

Alternatives

Add to or deduct from the cost per square foot of living area

Cedar Shake Roof	+ 1.75
Clay Tile Roof	+ 3.45
Slate Roof	+ 3.90
Upgrade Walls to Skim Coat Plaster	+ .58
Upgrade Ceilings to Textured Finish	+ .61
Air Conditioning, in Heating Ductwork	+ 3.10
In Separate Ductwork	+ 5.96
Heating Systems, Hot Water	+ 1.38
Heat Pump	+ 1.45
Electric Heat	– .97
Not Heated	– 2.90

Additional upgrades or components

Kitchen Cabinets & Countertops	Page 93
Bathroom Vanities	94
Fireplaces & Chimneys	94
Windows, Skylights & Dormers	94
Appliances	95
Breezeways & Porches	95
Finished Attic	95
Garages	96
Site Improvements	96
Wings & Ells	56

Important: See the Reference Section for Location Factors (to adjust for your city) and Estimating Forms.

Average Bi-Level

Living Area - 2000 S.F.
Perimeter - 135 L.F.

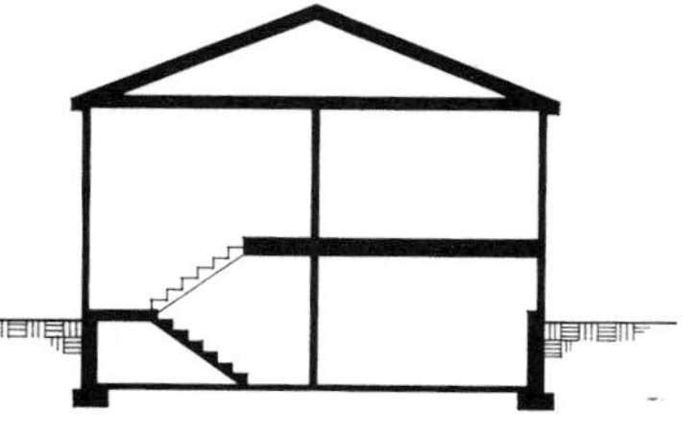

		Cost Per Square Foot Of Living Area			% of Total
		Mat.	Inst.	Total	(rounded)
1 Site Work	Site preparation for slab; 4′ deep trench excavation for foundation wall.		1.33	1.33	1.2%
2 Foundation	Continuous reinforced concrete footing, 10″ deep x 20″ wide; dampproofed and insulated 8″ thick reinforced concrete block foundation wall, 4′ deep; trowel finished 4″ thick concrete slab on 4″ crushed stone base and polyethylene vapor barrier.	3.67	4.79	8.46	7.8%
3 Framing	Exterior walls - 2″ x 6″ wood studs, 16″ O.C.; 1/2″ sheathing; gable end roof framing, 2″ x 10″ rafters, 16″ O.C. with 1/2″ plywood sheathing; 2″ x 10″ floor joists, 16″ O.C. with bridging and 5/8″ subflooring; 2″ x 4″ interior partitions.	7.16	9.38	16.54	15.3%
4 Exterior Walls	Beveled wood siding and housewrap on insulated wood frame walls; R38 attic insulation; double hung wood windows; flush solid core doors, frame and hardware, painted finish; aluminum storm and screen doors.	12.15	4.82	16.97	15.7%
5 Roofing	25 year roof asphalt shingles; #15 felt building paper; aluminum gutters, downspouts, drip edge and flashings.	1.17	1.42	2.59	2.4%
6 Interiors	Walls & ceilings, 1/2″ taped & finished gypsum wallboard, primed & painted with 2 coats of finish paint; birch faced hollow core interior doors, frames & hardware, painted finish; medium weight carpeting with pad, 40%; sheet vinyl, 15%; oak hardwood, 40%; ceramic tile, 5%; hardwood tread stairway.	15.45	14.90	30.35	28.1%
7 Specialties	Average grade kitchen cabinets and countertop; stainless steel kitchen sink; 40 gallon electric water heater.	3.93	1.22	5.15	4.8%
8 Mechanical	Three fixture bathroom: bathtub, water closet, vanity and sink; gas fired hot air heating system.	3.49	3.40	6.89	6.4%
9 Electrical	200 amp electric service; wiring, duplex & GFI receptacles, wall switches, door bell, appliance circuits, fans and communications cabling; average grade lighting fixtures.	1.55	2.44	3.99	3.7%
10 Overhead	Contractor's overhead and profit and plans.	8.28	7.45	15.73	14.6%
	Total	56.85	51.15	**108.00**	

RESIDENTIAL | Average | Tri-Level

- **Simple design from standard plans**
- **Single family — 1 full bath, 1 kitchen**
- **No basement**
- **Asphalt shingles on roof**
- **Hot air heat**
- **Gypsum wallboard interior finishes**
- **Materials and workmanship are average**

Note: The illustration shown may contain some optional components (for example: garages and/or fireplaces) whose costs are shown in the modifications, adjustments, & alternatives below or at the end of the square foot section.

Base cost per square foot of living area

Exterior Wall	Living Area 1200	1500	1800	2100	2400	2700	3000	3400	3800	4200	4600
Wood Siding - Wood Frame	132.60	121.45	113.10	105.90	101.40	98.75	95.75	93.05	88.60	85.00	83.20
Brick Veneer - Wood Frame	137.10	125.50	116.85	109.25	104.50	101.80	98.60	95.80	91.15	87.40	85.50
Stucco on Wood Frame	128.65	117.85	109.85	102.95	98.55	96.10	93.25	90.60	86.35	82.90	81.15
Solid Masonry	146.90	134.40	124.80	116.55	111.30	108.35	104.80	101.75	96.70	92.60	90.55
Finished Basement, Add*	27.15	26.65	25.55	24.65	24.00	23.65	23.20	22.90	22.35	21.95	21.70
Unfinished Basement, Add*	10.45	9.70	8.95	8.40	8.05	7.85	7.55	7.40	7.05	6.80	6.70

*Basement under middle level only.

Modifications

Add to the total cost

Upgrade Kitchen Cabinets	$ + 5907
Solid Surface Countertops (Included)	
Full Bath - including plumbing, wall and floor finishes	+ 8111
Half Bath - including plumbing, wall and floor finishes	+ 4789
One Car Attached Garage	+ 15,383
One Car Detached Garage	+ 20,232
Fireplace & Chimney	+ 7100

Adjustments

For multi family - add to total cost

Additional Kitchen	$ + 10,268
Additional Bath	+ 8111
Additional Entry & Exit	+ 1804
Separate Heating	+ 1748
Separate Electric	+ 1868

For Townhouse/Rowhouse - Multiply cost per square foot by

Inner Unit	.90
End Unit	.95

Alternatives

Add to or deduct from the cost per square foot of living area

Cedar Shake Roof	+ 2.55
Clay Tile Roof	+ 4.95
Slate Roof	+ 5.65
Upgrade Walls to Skim Coat Plaster	+ .50
Upgrade Ceilings to Textured Finish	+ .61
Air Conditioning, in Heating Ductwork	+ 2.61
In Separate Ductwork	+ 5.60
Heating Systems, Hot Water	+ 1.34
Heat Pump	+ 1.51
Electric Heat	– .82
Not Heated	– 2.81

Additional upgrades or components

Kitchen Cabinets & Countertops	Page 93
Bathroom Vanities	94
Fireplaces & Chimneys	94
Windows, Skylights & Dormers	94
Appliances	95
Breezeways & Porches	95
Finished Attic	95
Garages	96
Site Improvements	96
Wings & Ells	56

Important: See the Reference Section for Location Factors (to adjust for your city) and Estimating Forms.

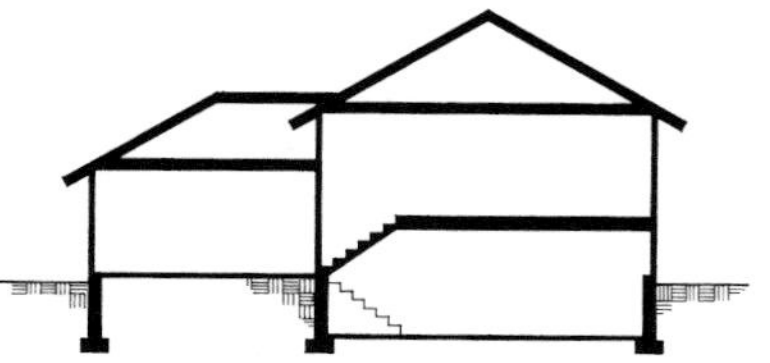

Average Tri-Level

Living Area - 2400 S.F.
Perimeter - 163 L.F.

			Cost Per Square Foot Of Living Area			% of Total
			Mat.	Inst.	Total	(rounded)
1	Site Work	Site preparation for slab; 4′ deep trench excavation for foundation wall.		1.10	1.10	1.1%
2	Foundation	Continuous reinforced concrete footing, 10″ deep x 20″ wide; dampproofed and insulated 8″ thick reinforced concrete block foundation wall, 4′ deep; trowel finished 4″ thick concrete slab on 4″ crushed stone base and polyethylene vapor barrier.	4.11	5.19	9.30	9.2%
3	Framing	Exterior walls - 2″ x 6″ wood studs, 16″ O.C.; 1/2″ sheathing; gable end roof framing, 2″ x 10″ rafters, 16″ O.C. with 1/2″ plywood sheathing; 2″ x 10″ floor joists, 16″ O.C. with bridging and 5/8″ subflooring; 2″ x 4″ interior partitions.	6.88	9.01	15.89	15.7%
4	Exterior Walls	Beveled wood siding and housewrap on insulated wood frame walls; R38 attic insulation; double hung wood windows; flush solid core doors, frame and hardware, painted finish; aluminum storm and screen doors.	10.70	4.21	14.91	14.7%
5	Roofing	25 year asphalt roof shingles; #15 felt building paper; aluminum gutters, downspouts, drip edge and flashings.	1.55	1.89	3.44	3.4%
6	Interiors	Walls & ceilings, 1/2″ taped & finished gypsum wallboard, primed & painted with 2 coats of finish paint; birch faced hollow core interior doors, frames & hardware, painted finish; medium weight carpeting with pad, 40%; sheet vinyl, 15%; oak hardwood, 40%; ceramic tile, 5%; hardwood tread stairway.	14.19	13.50	27.69	27.3%
7	Specialties	Average grade kitchen cabinets and countertop; stainless steel kitchen sink; 40 gallon electric water heater.	3.25	1.02	4.27	4.2%
8	Mechanical	Three fixture bathroom: bathtub, water closet, vanity and sink; gas fired hot air heating system.	2.97	3.24	6.21	6.1%
9	Electrical	200 amp electric service; wiring, duplex & GFI receptacles, wall switches, door bell, appliance circuits, fans and communications cabling; average grade lighting fixtures.	1.50	2.32	3.82	3.8%
10	Overhead	Contractor's overhead and profit and plans.	7.70	7.07	14.77	14.6%
		Total	52.85	48.55	**101.40**	

- **Post and beam frame**
- **Log exterior walls**
- **Simple design from standard plans**
- **Single family — 1 full bath, 1 kitchen**
- **No basement**
- **Asphalt shingles on roof**
- **Hot air heat**
- **Gypsum wallboard interior finishes**
- **Materials and workmanship are average**

Note: The illustration shown may contain some optional components (for example: garages and/or fireplaces) whose costs are shown in the modifications, adjustments, & alternatives below or at the end of the square foot section.

Base cost per square foot of living area

	Living Area										
Exterior Wall	**600**	**800**	**1000**	**1200**	**1400**	**1600**	**1800**	**2000**	**2400**	**2800**	**3200**
6" Log - Solid Wall	194.10	175.20	161.20	150.20	140.75	134.70	131.35	127.30	119.40	113.60	109.55
8" Log - Solid Wall	181.85	164.25	151.25	141.20	132.60	127.05	124.00	120.35	113.05	107.75	104.10
Finished Basement, Add	41.15	39.75	37.90	36.20	34.80	33.90	33.40	32.60	31.55	30.80	30.10
Unfinished Basement, Add	17.15	15.55	14.35	13.30	12.45	11.90	11.55	11.10	10.50	10.05	9.60

Modifications

Add to the total cost

Upgrade Kitchen Cabinets	$ + 5907
Solid Surface Countertops (Included)	
Full Bath - including plumbing, wall and floor finishes	+ 8111
Half Bath - including plumbing, wall and floor finishes	+ 4789
One Car Attached Garage	+ 15,383
One Car Detached Garage	+ 20,232
Fireplace & Chimney	+ 7100

Adjustments

For multi family - add to total cost

Additional Kitchen	$ + 10,268
Additional Bath	+ 8111
Additional Entry & Exit	+ 1804
Separate Heating	+ 1748
Separate Electric	+ 1868

For Townhouse/Rowhouse - Multiply cost per square foot by

Inner Unit	.92
End Unit	.96

Alternatives

Add to or deduct from the cost per square foot of living area

Cedar Shake Roof	+ 3.55
Air Conditioning, in Heating Ductwork	+ 5.11
In Separate Ductwork	+ 7.49
Heating Systems, Hot Water	+ 1.48
Heat Pump	+ 1.24
Electric Heat	– 1.29
Not Heated	– 3.12

Additional upgrades or components

Kitchen Cabinets & Countertops	Page 93
Bathroom Vanities	94
Fireplaces & Chimneys	94
Windows, Skylights & Dormers	94
Appliances	95
Breezeways & Porches	95
Finished Attic	95
Garages	96
Site Improvements	96
Wings & Ells	56

Important: See the Reference Section for Location Factors (to adjust for your city) and Estimating Forms.

Solid Wall 1 Story

Living Area - 1600 S.F.
Perimeter - 163 L.F.

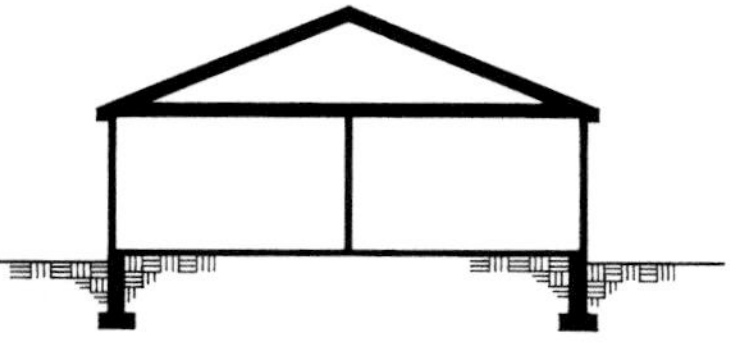

		Cost Per Square Foot Of Living Area			% of Total
		Mat.	Inst.	Total	(rounded)
1 Site Work	Site preparation for slab; 4′ deep trench excavation for foundation wall.		1.66	1.66	1.2%
2 Foundation	Continuous reinforced concrete footing, 10″ deep x 20″ wide; dampproofed and insulated 8″ thick reinforced concrete block foundation wall, 4′ deep; trowel finished 4″ thick concrete slab on 4″ crushed stone base and polyethylene vapor barrier.	6.16	7.80	13.96	10.4%
3 Framing	Exterior walls - precut traditional log home, handcrafted white cedar or pine logs, delivery included; heavy timber roof framing with 2″ thick tongue and groove decking, rigid insulation with 5/8″ sheathing; 2″ x 4″ interior partitions.	26.55	16.51	43.06	32.0%
4 Exterior Walls	R38 attic insulation; double hung wood windows; flush solid core doors, frame and hardware, painted finish; aluminum storm and screen doors.	5.65	2.81	8.46	6.3%
5 Roofing	25 year asphalt roof shingles; #15 felt building paper; aluminum gutters, downspouts, drip edge and flashings.	2.33	2.84	5.17	3.8%
6 Interiors	Walls & ceilings, 1/2″ taped & finished gypsum wallboard, primed & painted with 2 coats of finish paint; birch faced hollow core interior doors, frames & hardware, painted finish; medium weight carpeting with pad, 40%; sheet vinyl, 15%; oak hardwood, 40%; ceramic tile, 5%.	12.41	12.02	24.43	18.1%
7 Specialties	Average grade kitchen cabinets and countertop; stainless steel kitchen sink; 40 gallon electric water heater.	4.88	1.53	6.41	4.8%
8 Mechanical	Three fixture bathroom: bathtub, water closet, vanity and sink; gas fired hot air heating system.	4.14	3.66	7.80	5.8%
9 Electrical	200 amp electric service; wiring, duplex & GFI receptacles, wall switches, door bell, appliance circuits, fans and communications cabling; average grade lighting fixtures.	1.61	2.60	4.21	3.1%
10 Overhead	Contractor's overhead and profit and plans.	10.82	8.72	19.54	14.5%
	Total	74.55	60.15	**134.70**	

- Post and beam frame
- Log exterior walls
- Simple design from standard plans
- Single family — 1 full bath, 1 kitchen
- No basement
- Asphalt shingles on roof
- Hot air heat
- Gypsum wallboard interior finishes
- Materials and workmanship are average

Note: The illustration shown may contain some optional components (for example: garages and/or fireplaces) whose costs are shown in the modifications, adjustments, & alternatives below or at the end of the square foot section.

Base cost per square foot of living area

Exterior Wall	Living Area 1000	1200	1400	1600	1800	2000	2200	2600	3000	3400	3800
6" Log - Solid Wall	165.30	150.15	142.55	137.35	131.95	126.45	122.60	115.45	108.70	105.50	102.55
8" Log - Solid Wall	152.65	138.40	131.55	126.85	122.05	116.95	113.55	107.25	101.10	98.30	95.75
Finished Basement, Add	23.65	23.30	22.55	22.00	21.45	21.05	20.60	19.90	19.30	18.90	18.60
Unfinished Basement, Add	9.40	8.75	8.25	7.95	7.60	7.30	7.10	6.60	6.30	6.05	5.85

Modifications

Add to the total cost

Upgrade Kitchen Cabinets	$ + 5907
Solid Surface Countertops (Included)	
Full Bath - including plumbing, wall and floor finishes	+ 8111
Half Bath - including plumbing, wall and floor finishes	+ 4789
One Car Attached Garage	+ 15,383
One Car Detached Garage	+ 20,232
Fireplace & Chimney	+ 7845

Adjustments

For multi family - add to total cost

Additional Kitchen	$ + 10,268
Additional Bath	+ 8111
Additional Entry & Exit	+ 1804
Separate Heating	+ 1748
Separate Electric	+ 1868

For Townhouse/Rowhouse - Multiply cost per square foot by

Inner Unit	.92
End Unit	.96

Alternatives

Add to or deduct from the cost per square foot of living area

Cedar Shake Roof	+ 1.75
Air Conditioning, in Heating Ductwork	+ 3.10
In Separate Ductwork	+ 5.96
Heating Systems, Hot Water	+ 1.38
Heat Pump	+ 1.45
Electric Heat	– .97
Not Heated	– 2.90

Additional upgrades or components

Kitchen Cabinets & Countertops	Page 93
Bathroom Vanities	94
Fireplaces & Chimneys	94
Windows, Skylights & Dormers	94
Appliances	95
Breezeways & Porches	95
Finished Attic	95
Garages	96
Site Improvements	96
Wings & Ells	56

Important: See the Reference Section for Location Factors (to adjust for your city) and Estimating Forms.

Solid Wall 2 Story

Living Area - 2000 S.F.
Perimeter - 135 L.F.

		Cost Per Square Foot Of Living Area			% of Total
		Mat.	Inst.	Total	(rounded)
1 Site Work	Site preparation for slab; 4′ deep trench excavation for foundation wall.		1.33	1.33	1.1%
2 Foundation	Continuous reinforced concrete footing, 10″ deep x 20″ wide; dampproofed and insulated 8″ thick reinforced concrete block foundation wall, 4′ deep; trowel finished 4″ thick concrete slab on 4″ crushed stone base and polyethylene vapor barrier.	3.67	4.79	8.46	6.7%
3 Framing	Exterior walls - precut traditional log home, handcrafted white cedar or pine logs, delivery included; heavy timber roof framing with 2″ thick T. & G. decking, rigid insulation with 5/8″ sheathing; heavy timber columns, beams & joists with 2″ thick T. & G. decking; 2″ x 4″ interior partitions.	27.13	15.87	43.00	34.0%
4 Exterior Walls	R38 attic insulation; double hung wood windows; flush solid core doors, frame and hardware, painted finish; aluminum storm and screen doors.	5.68	2.97	8.65	6.8%
5 Roofing	25 year asphalt roof shingles; #15 felt building paper; aluminum gutters, downspouts, drip edge and flashings.	1.17	1.42	2.59	2.0%
6 Interiors	Walls & ceilings, 1/2″ taped & finished gypsum wallboard, primed & painted with 2 coats of finish paint; birch faced hollow core interior doors, frames & hardware, painted finish; medium weight carpeting with pad, 40%; sheet vinyl, 15%; oak hardwood, 40%; ceramic tile, 5%; hardwood tread stairway.	14.51	13.53	28.04	22.2%
7 Specialties	Average grade kitchen cabinets and countertop; stainless steel kitchen sink; 40 gallon electric water heater.	3.93	1.22	5.15	4.1%
8 Mechanical	Three fixture bathroom: bathtub, water closet, vanity and sink; gas fired hot air heating system.	3.49	3.40	6.89	5.4%
9 Electrical	200 amp electric service; wiring, duplex & GFI receptacles, wall switches, door bell, appliance circuits, fans and communications cabling; average grade lighting fixtures.	1.55	2.44	3.99	3.2%
10 Overhead	Contractor's overhead and profit and plans.	10.37	7.98	18.35	14.5%
	Total	71.50	54.95	**126.45**	

1 Story — Base cost per square foot of living area

Exterior Wall	Living Area 50	100	200	300	400	500	600	700
Wood Siding - Wood Frame	228.50	176.20	153.45	128.80	121.40	116.85	113.90	114.70
Brick Veneer - Wood Frame	228.90	172.60	148.20	120.65	112.70	107.80	104.70	105.30
Stucco on Wood Frame	216.30	167.40	146.05	123.75	116.85	112.60	109.85	110.75
Solid Masonry	280.35	213.15	184.30	149.30	139.90	134.15	132.15	132.10
Finished Basement, Add	66.45	54.90	49.35	40.00	38.15	37.05	36.30	35.80
Unfinished Basement, Add	31.45	23.75	20.55	15.20	14.15	13.50	13.10	12.75

1-1/2 Story — Base cost per square foot of living area

Exterior Wall	Living Area 100	200	300	400	500	600	700	800
Wood Siding - Wood Frame	185.00	149.05	127.50	114.20	107.65	104.45	100.50	99.30
Brick Veneer - Wood Frame	240.25	180.50	150.10	131.45	122.25	117.30	112.00	110.05
Stucco on Wood Frame	216.90	161.80	134.55	119.30	111.05	106.70	102.00	100.15
Solid Masonry	273.85	207.35	172.50	148.90	138.40	132.50	126.45	124.35
Finished Basement, Add	44.65	40.30	36.60	32.50	31.40	30.65	30.00	29.90
Unfinished Basement, Add	19.70	16.50	14.35	12.00	11.40	10.90	10.55	10.50

2 Story — Base cost per square foot of living area

Exterior Wall	Living Area 100	200	400	600	800	1000	1200	1400
Wood Siding - Wood Frame	184.55	138.45	118.40	98.05	91.50	87.50	84.85	85.90
Brick Veneer - Wood Frame	244.75	171.60	138.30	111.30	102.35	96.80	93.20	93.60
Stucco on Wood Frame	218.60	152.90	122.75	100.90	93.00	88.10	84.95	85.55
Solid Masonry	282.35	198.45	160.70	126.25	115.80	109.35	105.15	105.10
Finished Basement, Add	35.30	29.55	26.85	22.15	21.25	20.70	20.30	20.05
Unfinished Basement, Add	15.90	12.05	10.45	7.80	7.25	6.90	6.70	6.60

Base costs do not include bathroom or kitchen facilities. Use Modifications/Adjustments/Alternatives on pages 93–96 where appropriate.

1 Story

1-1/2 Story

2 Story

2-1/2 Story

Bi-Level

Tri-Level

RESIDENTIAL	Custom	1 Story

- A distinct residence from designer's plans
- Single family — 1 full bath, 1 half bath, 1 kitchen
- No basement
- Asphalt shingles on roof
- Forced hot air heat/air conditioning
- Gypsum wallboard interior finishes
- Materials and workmanship are above average

Note: The illustration shown may contain some optional components (for example: garages and/or fireplaces) whose costs are shown in the modifications, adjustments, & alternatives below or at the end of the square foot section.

Base cost per square foot of living area

Exterior Wall	Living Area 800	1000	1200	1400	1600	1800	2000	2400	2800	3200	3600
Wood Siding - Wood Frame	220.35	198.65	181.90	168.35	159.15	153.85	147.75	136.85	128.80	123.10	117.35
Brick Veneer - Wood Frame	231.40	208.80	191.00	176.65	166.85	161.35	154.75	143.25	134.70	128.60	122.50
Stone Veneer - Wood Frame	241.25	217.65	199.00	183.90	173.70	167.85	160.95	148.85	139.90	133.45	126.95
Solid Masonry	241.85	218.25	199.55	184.40	174.10	168.40	161.45	149.25	140.25	133.70	127.25
Finished Basement, Add	61.85	61.60	58.95	56.80	55.35	54.60	53.35	51.85	50.55	49.50	48.55
Unfinished Basement, Add	27.30	25.80	24.40	23.30	22.55	22.15	21.55	20.70	20.05	19.45	19.00

Modifications

Add to the total cost

Upgrade Kitchen Cabinets	$ + 1876
Solid Surface Countertops (Included)	
Full Bath - including plumbing, wall and floor finishes	+ 9733
Half Bath - including plumbing, wall and floor finishes	+ 5747
Two Car Attached Garage	+ 30,276
Two Car Detached Garage	+ 34,573
Fireplace & Chimney	+ 7426

Adjustments

For multi family - add to total cost

Additional Kitchen	$ + 22,521
Additional Full Bath & Half Bath	+ 15,480
Additional Entry & Exit	+ 1804
Separate Heating & Air Conditioning	+ 7787
Separate Electric	+ 1868

For Townhouse/Rowhouse - Multiply cost per square foot by

Inner Unit	.90
End Unit	.95

Alternatives

Add to or deduct from the cost per square foot of living area

Cedar Shake Roof	+ 2.70
Clay Tile Roof	+ 6.10
Slate Roof	+ 7
Upgrade Ceilings to Textured Finish	+ .61
Air Conditioning, in Heating Ductwork	Base System
Heating Systems, Hot Water	+ 1.52
Heat Pump	+ 1.24
Electric Heat	– 3.63
Not Heated	– 3.91

Additional upgrades or components

Kitchen Cabinets & Countertops	Page 93
Bathroom Vanities	94
Fireplaces & Chimneys	94
Windows, Skylights & Dormers	94
Appliances	95
Breezeways & Porches	95
Finished Attic	95
Garages	96
Site Improvements	96
Wings & Ells	74

Important: See the Reference Section for Location Factors (to adjust for your city) and Estimating Forms.

Custom 1 Story

Living Area - 2400 S.F.
Perimeter - 207 L.F.

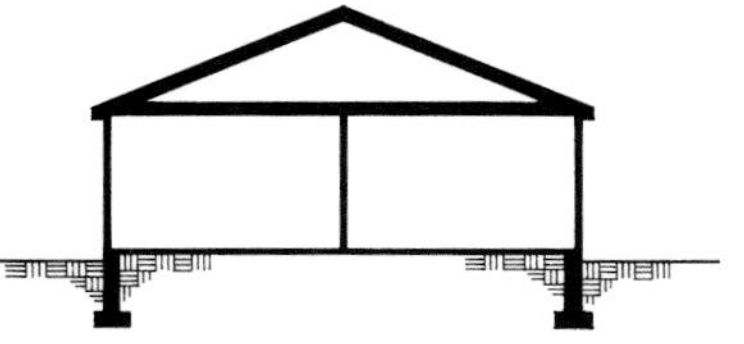

		Description	Cost Per Square Foot Of Living Area			% of Total
			Mat.	Inst.	Total	(rounded)
1	**Site Work**	Site preparation for slab; 4′ deep trench excavation for foundation wall.		1.24	1.24	0.9%
2	**Foundation**	Continuous reinforced concrete footing, 12″ deep x 24″ wide; dampproofed and insulated 12″ thick reinforced concrete block foundation wall, 4′ deep; trowel finished 4″ thick concrete slab on 4″ crushed stone base and polyethylene vapor barrier.	6.82	8.27	15.09	11.0%
3	**Framing**	Exterior walls - 2″ x 6″ wood studs, 16″ O.C.; 1/2″ sheathing; gable end roof framing, 2″ x 10″ rafters, 16″ O.C. with 1/2″ plywood sheathing; 2″ x 4″ interior partitions.	7.10	9.38	16.48	12.0%
4	**Exterior Walls**	1″ x 6″ tongue and groove vertical wood siding and housewrap on insulated wood frame walls; R38 attic insulation; plastic clad double hung wood windows; raised panel exterior doors, frames and hardware, painted finish; wood storm and screen door.	12.03	4.20	16.23	11.9%
5	**Roofing**	Red cedar roof shingles, perfections; #15 felt building paper; aluminum gutters, downspouts and drip edge; copper flashings.	4.90	4.02	8.92	6.5%
6	**Interiors**	Skim coated 1/2″ thick gypsum wallboard walls and ceilings, primed and painted with 2 coats; interior raised panel solid core doors, frames and hardware, painted finish; oak hardwood flooring, 70%; ceramic tile flooring, 30%.	15.54	13.74	29.28	21.4%
7	**Specialties**	Custom grade kitchen cabinets and countertops; double bowl kitchen sink; 75 gallon gas water heater.	8.76	1.65	10.41	7.6%
8	**Mechanical**	Three fixture bathroom: bathtub, water closet, vanity and sink; gas fired heating and air conditioning system.	7.34	3.74	11.08	8.1%
9	**Electrical**	200 amp electric service; wiring, duplex & GFI receptacles, wall switches, door bell, appliance circuits, air conditioning circuit, fans and communications cabling; custom grade lighting fixtures.	2.51	2.79	5.30	3.9%
10	**Overhead**	Contractor's overhead and profit and design.	13.00	9.82	22.82	16.7%
		Total	78.00	58.85	**136.85**	

- A distinct residence from designer's plans
- Single family — 1 full bath, 1 half bath, 1 kitchen
- No basement
- Asphalt shingles on roof
- Forced hot air heat/air conditioning
- Gypsum wallboard interior finishes
- Materials and workmanship are above average

Note: The illustration shown may contain some optional components (for example: garages and/or fireplaces) whose costs are shown in the modifications, adjustments, & alternatives below or at the end of the square foot section.

Base cost per square foot of living area

Exterior Wall	Living Area 1000	1200	1400	1600	1800	2000	2400	2800	3200	3600	4000
Wood Siding - Wood Frame	199.10	184.80	174.50	162.55	155.55	148.80	136.10	130.20	124.75	120.70	114.95
Brick Veneer - Wood Frame	211.10	196.00	185.20	172.30	164.75	157.55	143.85	137.55	131.50	127.20	121.05
Stone Veneer - Wood Frame	221.70	205.90	194.65	180.90	172.95	165.25	150.65	144.00	137.50	132.95	126.40
Solid Masonry	222.20	206.40	195.15	181.25	173.30	165.65	151.00	144.35	137.80	133.25	126.65
Finished Basement, Add	41.05	41.30	40.15	38.50	37.55	36.75	35.10	34.30	33.35	32.90	32.20
Unfinished Basement, Add	18.30	17.55	16.95	16.15	15.60	15.25	14.35	13.90	13.45	13.25	12.85

Modifications

Add to the total cost

Upgrade Kitchen Cabinets	$ + 1876
Solid Surface Countertops (Included)	
Full Bath - including plumbing, wall and floor finishes	+ 9733
Half Bath - including plumbing, wall and floor finishes	+ 5747
Two Car Attached Garage	+ 30,276
Two Car Detached Garage	+ 34,631
Fireplace & Chimney	+ 7426

Adjustments

For multi family - add to total cost

Additional Kitchen	$ + 22,521
Additional Full Bath & Half Bath	+ 15,480
Additional Entry & Exit	+ 1804
Separate Heating & Air Conditioning	+ 7787
Separate Electric	+ 1868

For Townhouse/Rowhouse - Multiply cost per square foot by

Inner Unit	.90
End Unit	.95

Alternatives

Add to or deduct from the cost per square foot of living area

Cedar Shake Roof	+ 1.95
Clay Tile Roof	+ 4.40
Slate Roof	+ 5.05
Upgrade Ceilings to Textured Finish	+ .61
Air Conditioning, in Heating Ductwork	Base System
Heating Systems, Hot Water	+ 1.46
Heat Pump	+ 1.29
Electric Heat	– 3.19
Not Heated	– 3.61

Additional upgrades or components

Important: See the Reference Section for Location Factors (to adjust for your city) and Estimating Forms.

Custom 1-1/2 Story

Living Area - 2800 S.F.
Perimeter - 175 L.F.

		Cost Per Square Foot Of Living Area			% of Total
		Mat.	Inst.	Total	(rounded)
1 Site Work	Site preparation for slab; 4′ deep trench excavation for foundation wall.		1.07	1.07	0.8%
2 Foundation	Continuous reinforced concrete footing, 12″ deep x 24″ wide; dampproofed and insulated 12″ thick reinforced concrete block foundation wall, 4′ deep; trowel finished 4″ thick concrete slab on 4″ crushed stone base and polyethylene vapor barrier.	4.78	5.90	10.68	8.2%
3 Framing	Exterior walls - 2″ x 6″ wood studs, 16″ O.C.; 1/2″ sheathing; gable end roof framing, steep pitch 2″ x 10″ rafters, 16″ O.C. with 1/2″ plywood sheathing; 2″ x 10″ floor joists, 16″ O.C. with bridging and 5/8″ subflooring; 2″ x 4″ interior partitions.	7.72	10.04	17.76	13.6%
4 Exterior Walls	1″ x 6″ tongue and groove vertical wood siding and housewrap on insulated wood frame walls; R38 attic insulation; plastic clad double hung wood windows; raised panel exterior doors, frames and hardware, painted finish; wood storm and screen door.	11.73	4.15	15.88	12.2%
5 Roofing	Red cedar roof shingles, perfections; #15 felt building paper; aluminum gutters, downspouts and drip edge; copper flashings.	3.55	2.92	6.47	5.0%
6 Interiors	Skim coated 1/2″ thick gypsum wallboard walls and ceilings, primed and painted with 2 coats; interior raised panel solid core doors, frames and hardware, painted finish; oak hardwood flooring, 70%; ceramic tile flooring, 30%; hardwood tread stairway.	17.57	15.18	32.75	25.2%
7 Specialties	Custom grade kitchen cabinets and countertops; double bowl kitchen sink; 75 gallon gas water heater.	7.49	1.41	8.90	6.8%
8 Mechanical	Three fixture bathroom: bathtub, water closet, vanity and sink; gas fired heating and air conditioning system.	6.32	3.49	9.81	7.5%
9 Electrical	200 amp electric service; wiring, duplex & GFI receptacles, wall switches, door bell, appliance circuits, air conditioning circuit, fans and communications cabling; custom grade lighting fixtures.	2.48	2.70	5.18	4.0%
10 Overhead	Contractor's overhead and profit and design.	12.31	9.39	21.70	16.7%
	Total	73.95	56.25	**130.20**	

- **A distinct residence from designer's plans**
- **Single family — 1 full bath, 1 half bath, 1 kitchen**
- **No basement**
- **Asphalt shingles on roof**
- **Forced hot air heat/air conditioning**
- **Gypsum wallboard interior finishes**
- **Materials and workmanship are above average**

Note: The illustration shown may contain some optional components (for example: garages and/or fireplaces) whose costs are shown in the modifications, adjustments, & alternatives below or at the end of the square foot section.

Base cost per square foot of living area

Exterior Wall	Living Area 1200	1400	1600	1800	2000	2400	2800	3200	3600	4000	4400
Wood Siding - Wood Frame	181.85	170.45	162.40	155.25	147.70	136.95	127.85	121.80	118.25	114.40	111.25
Brick Veneer - Wood Frame	193.75	181.60	173.05	165.20	157.45	145.70	135.75	129.20	125.45	121.15	117.80
Stone Veneer - Wood Frame	204.25	191.40	182.35	174.05	165.90	153.30	142.75	135.75	131.70	127.05	123.55
Solid Masonry	205.20	192.30	183.25	174.85	166.70	154.05	143.35	136.30	132.35	127.65	124.00
Finished Basement, Add	33.00	33.20	32.35	31.45	30.85	29.45	28.45	27.75	27.35	26.80	26.40
Unfinished Basement, Add	14.75	14.10	13.70	13.20	12.95	12.20	11.60	11.25	11.05	10.80	10.60

Modifications

Add to the total cost

Upgrade Kitchen Cabinets	$ + 1876
Solid Surface Countertops (Included)	
Full Bath - including plumbing, wall and floor finishes	+ 9733
Half Bath - including plumbing, wall and floor finishes	+ 5747
Two Car Attached Garage	+ 30,276
Two Car Detached Garage	+ 34,631
Fireplace & Chimney	+ 8382

Adjustments

For multi family - add to total cost

Additional Kitchen	$ + 22,521
Additional Full Bath & Half Bath	+ 15,480
Additional Entry & Exit	+ 1804
Separate Heating & Air Conditioning	+ 7787
Separate Electric	+ 1868

For Townhouse/Rowhouse - Multiply cost per square foot by

Inner Unit	.87
End Unit	.93

Alternatives

Add to or deduct from the cost per square foot of living area

Cedar Shake Roof	+ 1.35
Clay Tile Roof	+ 3.05
Slate Roof	+ 3.50
Upgrade Ceilings to Textured Finish	+ .61
Air Conditioning, in Heating Ductwork	Base System
Heating Systems, Hot Water	+ 1.40
Heat Pump	+ 1.44
Electric Heat	– 3.19
Not Heated	– 3.40

Additional upgrades or components

Kitchen Cabinets & Countertops	Page 93
Bathroom Vanities	94
Fireplaces & Chimneys	94
Windows, Skylights & Dormers	94
Appliances	95
Breezeways & Porches	95
Finished Attic	95
Garages	96
Site Improvements	96
Wings & Ells	74

Important: See the Reference Section for Location Factors (to adjust for your city) and Estimating Forms.

Custom 2 Story

Living Area - 2800 S.F.
Perimeter - 156 L.F.

		Cost Per Square Foot Of Living Area			% of Total
		Mat.	Inst.	Total	(rounded)
1 Site Work	Site preparation for slab; 4′ deep trench excavation for foundation wall.		1.07	1.07	0.8%
2 Foundation	Continuous reinforced concrete footing, 12″ deep x 24″ wide; dampproofed and insulated 12″ thick reinforced concrete block foundation wall, 4′ deep; trowel finished 4″ thick concrete slab on 4″ crushed stone base and polyethylene vapor barrier.	4.05	5.06	9.11	7.1%
3 Framing	Exterior walls - 2″ x 6″ wood studs, 16″ O.C.; 1/2″ sheathing; gable end roof framing, 2″ x 10″ rafters, 16″ O.C. with 1/2″ plywood sheathing; 2″ x 10″ floor joists, 16″ O.C. with bridging and 5/8″ subflooring; 2″ x 4″ interior partitions.	7.33	9.52	16.85	13.2%
4 Exterior Walls	1″ x 6″ tongue and groove vertical wood siding and housewrap on insulated wood frame walls; R38 attic insulation; plastic clad double hung wood windows; raised panel exterior doors, frames and hardware, painted finish; wood storm and screen door.	12.73	4.51	17.24	13.5%
5 Roofing	Red cedar roof shingles, perfections; #15 felt building paper; aluminum gutters, downspouts and drip edge; copper flashings.	2.46	2.02	4.48	3.5%
6 Interiors	Skim coated 1/2″ thick gypsum wallboard walls and ceilings, primed and painted with 2 coats; interior raised panel solid core doors, frames and hardware, painted finish; oak hardwood flooring, 70%; ceramic tile flooring, 30%; hardwood tread stairway.	18.04	15.62	33.66	26.3%
7 Specialties	Custom grade kitchen cabinets and countertops; double bowl kitchen sink; 75 gallon gas water heater.	7.49	1.41	8.90	7.0%
8 Mechanical	Three fixture bathroom: bathtub, water closet, vanity and sink; gas fired heating and air conditioning system.	6.49	3.54	10.03	7.8%
9 Electrical	200 amp electric service; wiring, duplex & GFI receptacles, wall switches, door bell, appliance circuits, air conditioning circuit, fans and communications cabling; custom grade lighting fixtures.	2.48	2.70	5.18	4.1%
10 Overhead	Contractor's overhead and profit and design.	12.23	9.10	21.33	16.7%
	Total	73.30	54.55	**127.85**	

- **A distinct residence from designer's plans**
- **Single family — 1 full bath, 1 half bath, 1 kitchen**
- **No basement**
- **Asphalt shingles on roof**
- **Forced hot air heat/air conditioning**
- **Gypsum wallboard interior finishes**
- **Materials and workmanship are above average**

Note: The illustration shown may contain some optional components (for example: garages and/or fireplaces) whose costs are shown in the modifications, adjustments, & alternatives below or at the end of the square foot section.

Base cost per square foot of living area

Exterior Wall	Living Area 1500	1800	2100	2400	2800	3200	3600	4000	4500	5000	5500
Wood Siding - Wood Frame	178.90	161.00	150.20	143.70	135.45	127.50	123.10	116.25	112.65	109.40	106.10
Brick Veneer - Wood Frame	191.65	172.65	160.75	153.65	144.90	136.05	131.30	123.90	119.85	116.20	112.60
Stone Veneer - Wood Frame	202.85	183.00	169.95	162.40	153.30	143.65	138.50	130.50	126.15	122.25	118.35
Solid Masonry	203.75	183.80	170.70	163.10	153.95	144.25	139.05	131.05	126.65	122.70	118.75
Finished Basement, Add	26.20	26.15	24.80	24.20	23.60	22.65	22.15	21.55	21.10	20.75	20.40
Unfinished Basement, Add	11.80	11.20	10.50	10.15	9.85	9.35	9.10	8.85	8.60	8.40	8.20

Modifications

Add to the total cost

Upgrade Kitchen Cabinets	$ + 1876
Solid Surface Countertops (Included)	
Full Bath - including plumbing, wall and floor finishes	+ 9733
Half Bath - including plumbing, wall and floor finishes	+ 5747
Two Car Attached Garage	+ 30,276
Two Car Detached Garage	+ 34,631
Fireplace & Chimney	+ 8382

Adjustments

For multi family - add to total cost

Additional Kitchen	$ + 22,521
Additional Full Bath & Half Bath	+ 15,480
Additional Entry & Exit	+ 1804
Separate Heating & Air Conditioning	+ 7787
Separate Electric	+ 1868

For Townhouse/Rowhouse - Multiply cost per square foot by

Inner Unit	.87
End Unit	.94

Alternatives

Add to or deduct from the cost per square foot of living area

Cedar Shake Roof	+ 1.20
Clay Tile Roof	+ 2.65
Slate Roof	+ 3.05
Upgrade Ceilings to Textured Finish	+ .61
Air Conditioning, in Heating Ductwork	Base System
Heating Systems, Hot Water	+ 1.28
Heat Pump	+ 1.48
Electric Heat	– 5.63
Not Heated	– 3.40

Additional upgrades or components

Kitchen Cabinets & Countertops	Page 93
Bathroom Vanities	94
Fireplaces & Chimneys	94
Windows, Skylights & Dormers	94
Appliances	95
Breezeways & Porches	95
Finished Attic	95
Garages	96
Site Improvements	96
Wings & Ells	74

Important: See the Reference Section for Location Factors (to adjust for your city) and Estimating Forms.

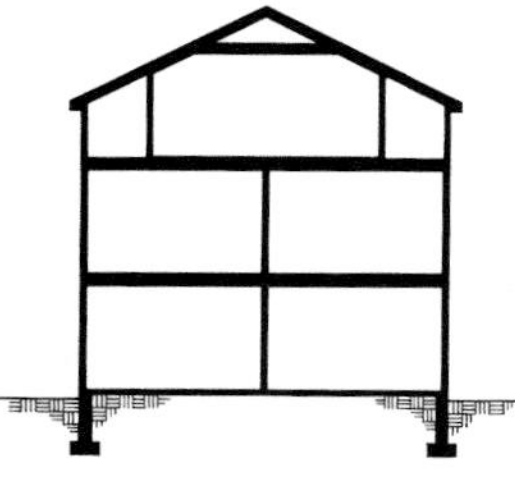

Custom 2-1/2 Story

Living Area - 3200 S.F.
Perimeter - 150 L.F.

		Cost Per Square Foot Of Living Area			% of Total
		Mat.	Inst.	Total	(rounded)
1 Site Work	Site preparation for slab; 4′ deep trench excavation for foundation wall.		.93	.93	0.7%
2 Foundation	Continuous reinforced concrete footing, 12″ deep x 24″ wide; dampproofed and insulated 12″ thick reinforced concrete block foundation wall, 4′ deep; trowel finished 4″ thick concrete slab on 4″ crushed stone base and polyethylene vapor barrier.	3.35	4.20	7.55	5.9%
3 Framing	Exterior walls - 2″ x 6″ wood studs, 16″ O.C.; 1/2″ sheathing; gable end roof framing, steep pitch 2″ x 10″ rafters, 16″ O.C. with 1/2″ plywood sheathing; 2″ x 10″ floor joists, 16″ O.C. with bridging and 5/8″ subflooring; 2″ x 4″ interior partitions.	7.87	10.19	18.06	14.2%
4 Exterior Walls	1″ x 6″ tongue and groove vertical wood siding and housewrap on insulated wood frame walls; R38 attic insulation; plastic clad double hung wood windows; raised panel exterior doors, frames and hardware, painted finish; wood storm and screen door.	13.14	4.68	17.82	14.0%
5 Roofing	Red cedar roof shingles, perfections; #15 felt building paper; aluminum gutters, downspouts and drip edge; copper flashings.	2.18	1.80	3.98	3.1%
6 Interiors	Skim coated 1/2″ thick gypsum wallboard walls and ceilings, primed and painted with 2 coats; interior raised panel solid core doors, frames and hardware, painted finish; oak hardwood flooring, 70%; ceramic tile flooring, 30%; hardwood tread stairway.	19.21	16.57	35.78	28.1%
7 Specialties	Custom grade kitchen cabinets and countertops; double bowl kitchen sink; 75 gallon gas water heater.	6.56	1.23	7.79	6.1%
8 Mechanical	Three fixture bathroom: bathtub, water closet, vanity and sink; gas fired heating and air conditioning system.	5.82	3.41	9.23	7.2%
9 Electrical	200 amp electric service; wiring, duplex & GFI receptacles, wall switches, door bell, appliance circuits, air conditioning circuit, fans and communications cabling; custom grade lighting fixtures.	2.46	2.64	5.10	4.0%
10 Overhead	Contractor's overhead and profit and design.	12.11	9.15	21.26	16.7%
	Total	72.70	54.80	**127.50**	

- **A distinct residence from designer's plans**
- **Single family — 1 full bath, 1 half bath, 1 kitchen**
- **No basement**
- **Asphalt shingles on roof**
- **Forced hot air heat/air conditioning**
- **Gypsum wallboard interior finishes**
- **Materials and workmanship are above average**

Note: The illustration shown may contain some optional components (for example: garages and/or fireplaces) whose costs are shown in the modifications, adjustments, & alternatives below or at the end of the square foot section.

Base cost per square foot of living area

Exterior Wall	Living Area 1500	1800	2100	2500	3000	3500	4000	4500	5000	5500	6000
Wood Siding - Wood Frame	175.65	157.90	149.15	141.80	**130.70**	125.15	118.35	111.40	108.75	106.00	103.25
Brick Veneer - Wood Frame	188.30	169.70	160.15	152.25	140.15	134.15	126.50	119.00	116.05	113.05	109.85
Stone Veneer - Wood Frame	199.45	180.05	169.80	161.50	148.55	142.05	133.70	125.55	122.50	119.20	115.70
Solid Masonry	200.60	181.10	170.80	162.40	149.40	142.80	134.50	126.30	123.10	119.85	116.35
Finished Basement, Add	22.95	23.00	22.15	21.50	20.60	20.00	19.30	18.70	18.45	18.20	17.80
Unfinished Basement, Add	10.35	9.80	9.40	9.10	8.65	8.35	7.95	7.65	7.50	7.40	7.20

Modifications

Add to the total cost

Upgrade Kitchen Cabinets	$ + 1876
Solid Surface Countertops (Included)	
Full Bath - including plumbing, wall and floor finishes	+ 9733
Half Bath - including plumbing, wall and floor finishes	+ 5747
Two Car Attached Garage	+ 30,276
Two Car Detached Garage	+ 34,631
Fireplace & Chimney	+ 9465

Adjustments

For multi family - add to total cost

Additional Kitchen	$ + 22,521
Additional Full Bath & Half Bath	+ 15,480
Additional Entry & Exit	+ 1804
Separate Heating & Air Conditioning	+ 7787
Separate Electric	+ 1868

For Townhouse/Rowhouse - Multiply cost per square foot by

Inner Unit	.85
End Unit	.93

Alternatives

Add to or deduct from the cost per square foot of living area

Cedar Shake Roof	+ .90
Clay Tile Roof	+ 2.05
Slate Roof	+ 2.35
Upgrade Ceilings to Textured Finish	+ .61
Air Conditioning, in Heating Ductwork	Base System
Heating Systems, Hot Water	+ 1.28
Heat Pump	+ 1.48
Electric Heat	– 5.63
Not Heated	– 3.29

Additional upgrades or components

Kitchen Cabinets & Countertops	Page 93
Bathroom Vanities	94
Fireplaces & Chimneys	94
Windows, Skylights & Dormers	94
Appliances	95
Breezeways & Porches	95
Finished Attic	95
Garages	96
Site Improvements	96
Wings & Ells	74

Important: See the Reference Section for Location Factors (to adjust for your city) and Estimating Forms.

Custom 3 Story

Living Area - 3000 S.F.
Perimeter - 135 L.F.

			Cost Per Square Foot Of Living Area			% of Total
			Mat.	Inst.	Total	(rounded)
1	Site Work	Site preparation for slab; 4′ deep trench excavation for foundation wall.		.99	.99	0.8%
2	Foundation	Continuous reinforced concrete footing, 12″ deep x 24″ wide; dampproofed and insulated 12″ thick reinforced concrete block foundation wall, 4′ deep; trowel finished 4″ thick concrete slab on 4″ crushed stone base and polyethylene vapor barrier.	3.13	3.97	7.10	5.4%
3	Framing	Exterior walls - 2″ x 6″ wood studs, 16″ O.C.; 1/2″ sheathing; gable end roof framing, 2″ x 10″ rafters, 16″ O.C. with 1/2″ plywood sheathing; 2″ x 10″ floor joists, 16″ O.C. with bridging and 5/8″ subflooring; 2″ x 4″ interior partitions.	7.77	10.07	17.84	13.6%
4	Exterior Walls	1″ x 6″ tongue and groove vertical wood siding and housewrap on insulated wood frame walls; R38 attic insulation; plastic clad double hung wood windows; raised panel exterior doors, frames and hardware, painted finish; wood storm and screen door.	14.61	5.23	19.84	15.2%
5	Roofing	Red cedar roof shingles, perfections; #15 felt building paper; aluminum gutters, downspouts and drip edge; copper flashings.	1.63	1.34	2.97	2.3%
6	Interiors	Skim coated 1/2″ thick gypsum wallboard walls and ceilings, primed and painted with 2 coats; interior raised panel solid core doors, frames and hardware, painted finish; oak hardwood flooring, 70%; ceramic tile flooring, 30%; hardwood tread stairway.	19.93	17.17	37.10	28.4%
7	Specialties	Custom grade kitchen cabinets and countertops; double bowl kitchen sink; 75 gallon gas water heater.	7.00	1.33	8.33	6.4%
8	Mechanical	Three fixture bathroom: bathtub, water closet, vanity and sink; gas fired heating and air conditioning system.	6.12	3.47	9.59	7.3%
9	Electrical	200 amp electric service; wiring, duplex & GFI receptacles, wall switches, door bell, appliance circuits, air conditioning circuit, fans and communications cabling; custom grade lighting fixtures.	2.47	2.67	5.14	3.9%
10	Overhead	Contractor's overhead and profit and design.	12.54	9.26	21.80	16.7%
		Total	75.20	55.50	**130.70**	

- **A distinct residence from designer's plans**
- **Single family — 1 full bath, 1 half bath, 1 kitchen**
- **No basement**
- **Asphalt shingles on roof**
- **Forced hot air heat/air conditioning**
- **Gypsum wallboard interior finishes**
- **Materials and workmanship are above average**

Note: The illustration shown may contain some optional components (for example: garages and/or fireplaces) whose costs are shown in the modifications, adjustments, & alternatives below or at the end of the square foot section.

Base cost per square foot of living area

Exterior Wall	Living Area 1200	1400	1600	1800	2000	2400	2800	3200	3600	4000	4400
Wood Siding - Wood Frame	172.25	161.55	153.80	147.15	139.95	129.95	121.40	115.80	112.50	109.00	106.10
Brick Veneer - Wood Frame	181.30	170.00	161.90	154.80	147.35	136.55	127.40	121.50	118.00	114.15	111.05
Stone Veneer - Wood Frame	189.25	177.45	169.05	161.55	153.80	142.40	132.75	126.45	122.75	118.65	115.40
Solid Masonry	189.95	178.10	169.65	162.05	154.35	142.95	133.25	126.85	123.20	119.00	115.75
Finished Basement, Add	33.00	33.20	32.35	31.45	30.85	29.45	28.45	27.75	27.35	26.80	26.40
Unfinished Basement, Add	14.75	14.10	13.70	13.20	12.95	12.20	11.60	11.25	11.05	10.80	10.60

Modifications

Add to the total cost

Upgrade Kitchen Cabinets	$ + 1876
Solid Surface Countertops (Included)	
Full Bath - including plumbing, wall and floor finishes	+ 9733
Half Bath - including plumbing, wall and floor finishes	+ 5747
Two Car Attached Garage	+ 30,276
Two Car Detached Garage	+ 34,631
Fireplace & Chimney	+ 7426

Adjustments

For multi family - add to total cost

Additional Kitchen	$ + 22,521
Additional Full Bath & Half Bath	+ 15,480
Additional Entry & Exit	+ 1804
Separate Heating & Air Conditioning	+ 7787
Separate Electric	+ 1868

For Townhouse/Rowhouse - Multiply cost per square foot by

Inner Unit	.89
End Unit	.95

Alternatives

Add to or deduct from the cost per square foot of living area

Cedar Shake Roof	+ 1.35
Clay Tile Roof	+ 3.05
Slate Roof	+ 3.50
Upgrade Ceilings to Textured Finish	+ .61
Air Conditioning, in Heating Ductwork	Base System
Heating Systems, Hot Water	+ 1.40
Heat Pump	+ 1.44
Electric Heat	– 3.19
Not Heated	– 3.29

Additional upgrades or components

Kitchen Cabinets & Countertops	Page 93
Bathroom Vanities	94
Fireplaces & Chimneys	94
Windows, Skylights & Dormers	94
Appliances	95
Breezeways & Porches	95
Finished Attic	95
Garages	96
Site Improvements	96
Wings & Ells	74

Important: See the Reference Section for Location Factors (to adjust for your city) and Estimating Forms.

Custom Bi-Level

Living Area - 2800 S.F.
Perimeter - 156 L.F.

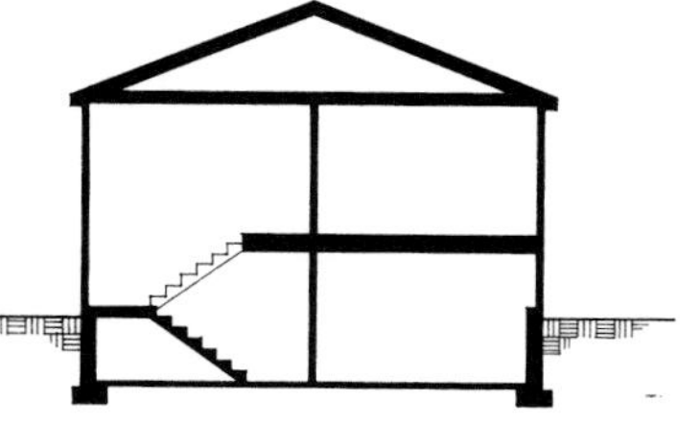

		Cost Per Square Foot Of Living Area			% of Total
		Mat.	Inst.	Total	(rounded)
1 Site Work	Site preparation for slab; 4′ deep trench excavation for foundation wall.		1.07	1.07	0.9%
2 Foundation	Continuous reinforced concrete footing, 12″ deep x 24″ wide; dampproofed and insulated 12″ thick reinforced concrete block foundation wall, 4′ deep; trowel finished 4″ thick concrete slab on 4″ crushed stone base and polyethylene vapor barrier.	4.05	5.06	9.11	7.5%
3 Framing	Exterior walls - 2″ x 6″ wood studs, 16″ O.C.; 1/2″ sheathing; gable end roof framing, 2″ x 10″ rafters, 16″ O.C. with 1/2″ plywood sheathing; 2″ x 10″ floor joists, 16″ O.C. with bridging and 5/8″ subflooring; 2″ x 4″ interior partitions.	6.82	8.88	15.70	12.9%
4 Exterior Walls	1″ x 6″ tongue and groove vertical wood siding and housewrap on insulated wood frame walls; R38 attic insulation; plastic clad double hung wood windows; raised panel exterior doors, frames and hardware, painted finish; wood storm and screen door.	10.14	3.58	13.72	11.3%
5 Roofing	Red cedar roof shingles, perfections; #15 felt building paper; aluminum gutters, downspouts and drip edge; copper flashings.	2.46	2.02	4.48	3.7%
6 Interiors	Skim coated 1/2″ thick gypsum wallboard walls and ceilings, primed and painted with 2 coats; interior raised panel solid core doors, frames and hardware, painted finish; oak hardwood flooring, 70%; ceramic tile flooring, 30%; hardwood tread stairway.	17.77	15.20	32.97	27.2%
7 Specialties	Custom grade kitchen cabinets and countertops; double bowl kitchen sink; 75 gallon gas water heater.	7.49	1.41	8.90	7.3%
8 Mechanical	Three fixture bathroom: bathtub, water closet, vanity and sink; gas fired heating and air conditioning system.	6.49	3.54	10.03	8.3%
9 Electrical	200 amp electric service; wiring, duplex & GFI receptacles, wall switches, door bell, appliance circuits, air conditioning circuit, fans and communications cabling; custom grade lighting fixtures.	2.48	2.70	5.18	4.3%
10 Overhead	Contractor's overhead and profit and design.	11.55	8.69	20.24	16.7%
	Total	69.25	52.15	**121.40**	

- **A distinct residence from designer's plans**
- **Single family — 1 full bath, 1 half bath, 1 kitchen**
- **No basement**
- **Asphalt shingles on roof**
- **Forced hot air heat/air conditioning**
- **Gypsum wallboard interior finishes**
- **Materials and workmanship are above average**

Note: The illustration shown may contain some optional components (for example: garages and/or fireplaces) whose costs are shown in the modifications, adjustments, & alternatives below or at the end of the square foot section.

Base cost per square foot of living area

Exterior Wall	Living Area 1200	1500	1800	2100	2400	2800	3200	3600	4000	4500	5000
Wood Siding - Wood Frame	178.00	161.10	148.55	138.20	131.25	126.05	120.15	114.00	111.20	105.45	102.20
Brick Veneer - Wood Frame	187.10	169.35	156.00	145.00	137.60	132.10	125.75	119.25	116.35	110.15	106.65
Stone Veneer - Wood Frame	195.15	176.65	162.55	150.95	143.20	137.50	130.75	123.90	120.80	114.25	110.50
Solid Masonry	195.75	177.15	163.00	151.40	143.65	138.00	131.10	124.20	121.20	114.60	110.85
Finished Basement, Add*	41.20	41.05	39.30	37.85	36.90	36.25	35.35	34.55	34.20	33.35	32.80
Unfinished Basement, Add*	18.20	17.15	16.25	15.55	15.00	14.70	14.25	13.80	13.65	13.20	12.90

*Basement under middle level only.

Modifications

Add to the total cost

Upgrade Kitchen Cabinets	$ + 1876
Solid Surface Countertops (Included)	
Full Bath - including plumbing, wall and floor finishes	+ 9733
Half Bath - including plumbing, wall and floor finishes	+ 5747
Two Car Attached Garage	+ 30,276
Two Car Detached Garage	+ 34,631
Fireplace & Chimney	+ 7426

Adjustments

For multi family - add to total cost

Additional Kitchen	$ + 22,521
Additional Full Bath & Half Bath	+ 15,480
Additional Entry & Exit	+ 1804
Separate Heating & Air Conditioning	+ 7787
Separate Electric	+ 1868

For Townhouse/Rowhouse - Multiply cost per square foot by

Inner Unit	.87
End Unit	.94

Alternatives

Add to or deduct from the cost per square foot of living area

Cedar Shake Roof	+ 1.95
Clay Tile Roof	+ 4.40
Slate Roof	+ 5.05
Upgrade Ceilings to Textured Finish	+ .61
Air Conditioning, in Heating Ductwork	Base System
Heating Systems, Hot Water	+ 1.36
Heat Pump	+ 1.50
Electric Heat	– 2.86
Not Heated	– 3.29

Additional upgrades or components

Kitchen Cabinets & Countertops	Page 93
Bathroom Vanities	94
Fireplaces & Chimneys	94
Windows, Skylights & Dormers	94
Appliances	95
Breezeways & Porches	95
Finished Attic	95
Garages	96
Site Improvements	96
Wings & Ells	74

Custom Tri-Level

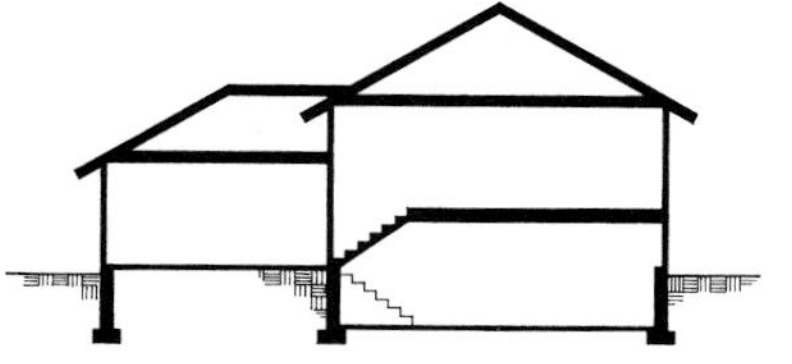

Living Area - 3200 S.F.
Perimeter - 198 L.F.

			Cost Per Square Foot Of Living Area			% of Total
			Mat.	Inst.	Total	(rounded)
1	Site Work	Site preparation for slab; 4′ deep trench excavation for foundation wall.		.93	.93	0.8%
2	Foundation	Continuous reinforced concrete footing, 12″ deep x 24″ wide; damproofed and insulated 12″ thick reinforced concrete block foundation wall, 4′ deep; trowel finished 4″ thick concrete slab on 4″ crushed stone base and polyethylene vapor barrier.	4.76	5.83	10.59	8.8%
3	Framing	Exterior walls - 2″ x 6″ wood studs, 16″ O.C.; 1/2″ sheathing; gable end roof framing, 2″ x 10″ rafters, 16″ O.C. with 1/2″ plywood sheathing; 2″ x 10″ floor joists, 16″ O.C. with bridging and 5/8″ subflooring; 2″ x 4″ interior partitions.	6.81	8.94	15.75	13.1%
4	Exterior Walls	1″ x 6″ tongue and groove vertical wood siding and housewrap on insulated wood frame walls; R38 attic insulation; plastic clad double hung wood windows; raised panel exterior doors, frames and hardware, painted finish; wood storm and screen door.	10.06	3.53	13.59	11.3%
5	Roofing	Red cedar roof shingles, perfections; #15 felt building paper; aluminum gutters, downspouts and drip edge; copper flashings.	3.27	2.69	5.96	5.0%
6	Interiors	Skim coated 1/2″ thick gypsum wallboard walls and ceilings, primed and painted with 2 coats; interior raised panel solid core doors, frames and hardware, painted finish; oak hardwood flooring, 70%; ceramic tile flooring, 30%; hardwood tread stairway.	16.70	14.45	31.15	25.9%
7	Specialties	Custom grade kitchen cabinets and countertops; double bowl kitchen sink; 75 gallon gas water heater.	6.56	1.23	7.79	6.5%
8	Mechanical	Three fixture bathroom: bathtub, water closet, vanity and sink; gas fired heating and air conditioning system.	5.82	3.41	9.23	7.7%
9	Electrical	200 amp electric service; wiring, duplex & GFI receptacles, wall switches, door bell, appliance circuits, air conditioning circuit, fans and communications cabling; custom grade lighting fixtures.	2.46	2.64	5.10	4.2%
10	Overhead	Contractor's overhead and profit and design.	11.31	8.75	20.06	16.7%
		Total	67.75	52.40	**120.15**	

1 Story — Base cost per square foot of living area

Exterior Wall	Living Area 50	100	200	300	400	500	600	700
Wood Siding - Wood Frame	262.10	203.60	178.25	150.95	142.70	137.65	134.30	135.30
Brick Veneer - Wood Frame	290.75	224.10	195.25	162.35	152.90	147.15	143.40	144.10
Stone Veneer - Wood Frame	316.00	242.10	210.30	172.35	161.95	155.55	151.40	151.80
Solid Masonry	321.95	246.35	213.85	174.70	164.00	157.55	153.25	153.65
Finished Basement, Add	99.90	85.10	77.00	63.55	60.90	59.25	58.20	57.45
Unfinished Basement, Add	73.90	51.25	40.80	31.40	28.95	27.40	26.45	25.75

1-1/2 Story — Base cost per square foot of living area

Exterior Wall	Living Area 100	200	300	400	500	600	700	800
Wood Siding - Wood Frame	208.45	170.85	147.95	134.30	127.35	124.05	119.80	118.55
Brick Veneer - Wood Frame	234.00	191.30	165.05	147.60	139.65	135.65	130.75	129.40
Stone Veneer - Wood Frame	256.55	209.35	180.05	159.30	150.40	145.85	140.40	139.00
Solid Masonry	261.80	213.60	183.60	162.05	153.00	148.30	142.70	141.25
Finished Basement, Add	66.65	61.60	56.25	50.25	48.70	47.60	46.55	46.50
Unfinished Basement, Add	44.25	33.85	28.85	24.60	23.15	22.15	21.30	21.00

2 Story — Base cost per square foot of living area

Exterior Wall	Living Area 100	200	400	600	800	1000	1200	1400
Wood Siding - Wood Frame	207.75	158.40	136.75	115.80	108.80	104.50	101.70	103.05
Brick Veneer - Wood Frame	236.45	178.85	153.85	127.25	119.05	114.10	110.80	111.85
Stone Veneer - Wood Frame	261.65	196.90	168.85	137.20	128.10	122.50	118.80	119.55
Solid Masonry	267.60	201.15	172.35	139.60	130.20	124.45	120.65	121.40
Finished Basement, Add	50.00	42.60	38.60	31.85	30.45	29.65	29.15	28.75
Unfinished Basement, Add	36.95	25.65	20.40	15.75	14.45	13.75	13.20	12.90

Base costs do not include bathroom or kitchen facilities. Use Modifications/Adjustments/Alternatives on pages 93–96 where appropriate.

1 Story

1-1/2 Story

2 Story

2-1/2 Story

Bi-Level

Tri-Level

- **Unique residence built from an architect's plan**
- **Single family — 1 full bath, 1 half bath, 1 kitchen**
- **No basement**
- **Cedar shakes on roof**
- **Forced hot air heat/air conditioning**
- **Gypsum wallboard interior finishes**
- **Many special features**
- **Extraordinary materials and workmanship**

Note: The illustration shown may contain some optional components (for example: garages and/or fireplaces) whose costs are shown in the modifications, adjustments, & alternatives below or at the end of the square foot section.

Base cost per square foot of living area

Exterior Wall	Living Area 1000	1200	1400	1600	1800	2000	2400	2800	3200	3600	4000
Wood Siding - Wood Frame	233.90	213.50	197.25	186.05	179.65	172.25	159.20	149.60	142.85	136.05	130.35
Brick Veneer - Wood Frame	245.20	223.70	206.45	194.75	187.95	180.15	166.35	156.20	148.95	141.80	135.75
Solid Brick	258.15	235.40	217.10	204.75	197.60	189.15	174.60	163.80	156.00	148.35	141.90
Solid Stone	265.70	242.15	223.30	210.50	203.20	194.45	179.35	168.30	160.15	152.20	145.50
Finished Basement, Add	60.70	65.30	62.65	60.85	59.85	58.45	56.45	54.95	53.50	52.40	51.40
Unfinished Basement, Add	27.50	25.85	24.50	23.60	23.15	22.35	21.40	20.60	19.85	19.25	18.75

Modifications

Add to the total cost

Upgrade Kitchen Cabinets	$ + 2641
Solid Surface Countertops (Included)	
Full Bath - including plumbing, wall and floor finishes	+ 11,680
Half Bath - including plumbing, wall and floor finishes	+ 6896
Two Car Attached Garage	+ 34,538
Two Car Detached Garage	+ 39,236
Fireplace & Chimney	+ 10,588

Adjustments

For multi family - add to total cost

Additional Kitchen	$ + 30,712
Additional Full Bath & Half Bath	+ 18,576
Additional Entry & Exit	+ 2794
Separate Heating & Air Conditioning	+ 7787
Separate Electric	+ 1868

For Townhouse/Rowhouse - Multiply cost per square foot by

Inner Unit	.90
End Unit	.95

Alternatives

Add to or deduct from the cost per square foot of living area

Heavyweight Asphalt Shingles	– 2.70
Clay Tile Roof	+ 3.35
Slate Roof	+ 4.30
Upgrade Ceilings to Textured Finish	+ .61
Air Conditioning, in Heating Ductwork	Base System
Heating Systems, Hot Water	+ 1.64
Heat Pump	+ 1.33
Electric Heat	– 3.19
Not Heated	– 4.25

Additional upgrades or components

Kitchen Cabinets & Countertops	Page 93
Bathroom Vanities	94
Fireplaces & Chimneys	94
Windows, Skylights & Dormers	94
Appliances	95
Breezeways & Porches	95
Finished Attic	95
Garages	96
Site Improvements	96
Wings & Ells	92

Important: See the Reference Section for Location Factors (to adjust for your city) and Estimating Forms.

Luxury 1 Story

Living Area - 2800 S.F.
Perimeter - 219 L.F.

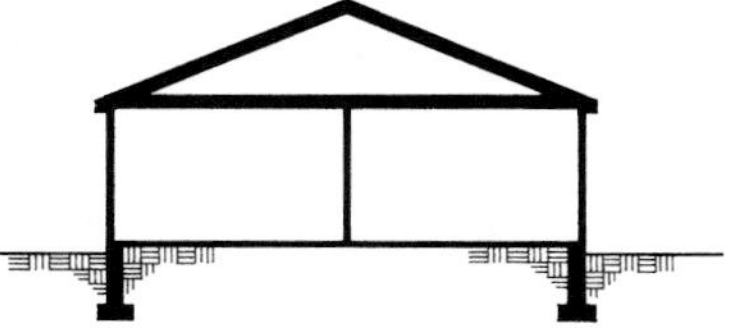

		Cost Per Square Foot Of Living Area			% of Total
		Mat.	Inst.	Total	(rounded)
1 Site Work	Site preparation for slab; 4' deep trench excavation for foundation wall.		1.15	1.15	0.8%
2 Foundation	Continuous reinforced concrete footing, 12" deep x 24" wide; damproofed and insulated 12" thick reinforced concrete block foundation wall, 4' deep; trowel finished 6" thick concrete slab on 4" crushed stone base and polyethylene vapor barrier.	8.07	8.60	16.67	11.1%
3 Framing	Exterior walls - 2" x 6" wood studs, 16" O.C.; 1/2" sheathing; gable end roof framing, 2" x 10" rafters, 16" O.C. with 1/2" plywood sheathing; 2" x 4" interior partitions.	7.46	9.82	17.28	11.6%
4 Exterior Walls	1" x 6" tongue and groove vertical wood siding and housewrap on insulated wood frame walls; R38 attic insulation; metal clad double hung wood windows; raised panel exterior doors, frame and hardware, painted finish; wood storm and screen door.	12.62	4.26	16.88	11.3%
5 Roofing	Red cedar roof shingles, perfections; #15 felt building paper; aluminum gutters, downspouts and drip edge; copper flashings.	5.30	4.35	9.65	6.5%
6 Interiors	Skim coated 1/2" thick gypsum wallboard walls and ceilings, primed and painted with 2 coats; interior raised panel solid core doors, frames and hardware, painted finish; oak hardwood flooring, 70%; ceramic tile flooring, 30%.	14.33	15.14	29.47	19.7%
7 Specialties	Luxury grade kitchen cabinets and countertops; double bowl kitchen sink; 75 gallon gas water heater.	9.90	1.84	11.74	7.8%
8 Mechanical	Three fixture bathroom: bathtub, water closet, vanity and sink; gas fired heating and air conditioning system.	7.22	3.92	11.14	7.4%
9 Electrical	200 amp electric service; wiring, duplex & GFI receptacles, wall switches, dimmer switches, door bell, appliance circuits, air conditioning circuit, fans and communications cabling; luxury grade lighting fixtures.	3.75	2.91	6.66	4.5%
10 Overhead	Contractor's overhead and profit and architect's fees.	16.50	12.46	28.96	19.4%
	Total	85.15	64.45	**149.60**	

- **Unique residence built from an architect's plan**
- **Single family — 1 full bath, 1 half bath, 1 kitchen**
- **No basement**
- **Cedar shakes on roof**
- **Forced hot air heat/air conditioning**
- **Gypsum wallboard interior finishes**
- **Many special features**
- **Extraordinary materials and workmanship**

Note: The illustration shown may contain some optional components (for example: garages and/or fireplaces) whose costs are shown in the modifications, adjustments, & alternatives below or at the end of the square foot section.

Base cost per square foot of living area

Exterior Wall	Living Area 1000	1200	1400	1600	1800	2000	2400	2800	3200	3600	4000
Wood Siding - Wood Frame	235.15	217.40	204.75	190.45	181.95	173.75	158.70	**151.35**	144.85	139.95	133.20
Brick Veneer - Wood Frame	248.55	230.00	216.70	201.30	192.20	183.55	167.40	159.60	152.40	147.30	140.05
Solid Brick	263.30	243.85	229.85	213.25	203.55	194.30	176.85	168.55	160.75	155.25	147.45
Solid Stone	272.75	252.70	238.35	220.95	210.80	201.25	182.95	174.35	166.10	160.45	152.30
Finished Basement, Add	42.70	46.35	44.95	42.90	41.75	40.70	38.55	37.70	36.50	35.95	35.05
Unfinished Basement, Add	19.85	18.90	18.25	17.15	16.65	16.10	15.05	14.55	13.95	13.70	13.25

Modifications

Add to the total cost

Upgrade Kitchen Cabinets	$ + 2641
Solid Surface Countertops (Included)	
Full Bath - including plumbing, wall and floor finishes	+ 11,680
Half Bath - including plumbing, wall and floor finishes	+ 6896
Two Car Attached Garage	+ 34,538
Two Car Detached Garage	+ 39,236
Fireplace & Chimney	+ 10,588

Adjustments

For multi family - add to total cost

Additional Kitchen	$ + 30,712
Additional Full Bath & Half Bath	+ 18,576
Additional Entry & Exit	+ 2794
Separate Heating & Air Conditioning	+ 7787
Separate Electric	+ 1868

For Townhouse/Rowhouse - Multiply cost per square foot by

Inner Unit	.90
End Unit	.95

Alternatives

Add to or deduct from the cost per square foot of living area

Heavyweight Asphalt Shingles	– 1.95
Clay Tile Roof	+ 2.45
Slate Roof	+ 3.10
Upgrade Ceilings to Textured Finish	+ .61
Air Conditioning, in Heating Ductwork	Base System
Heating Systems, Hot Water	+ 1.56
Heat Pump	+ 1.47
Electric Heat	– 3.19
Not Heated	– 3.92

Additional upgrades or components

Kitchen Cabinets & Countertops	Page 93
Bathroom Vanities	94
Fireplaces & Chimneys	94
Windows, Skylights & Dormers	94
Appliances	95
Breezeways & Porches	95
Finished Attic	95
Garages	96
Site Improvements	96
Wings & Ells	92

Important: See the Reference Section for Location Factors (to adjust for your city) and Estimating Forms.

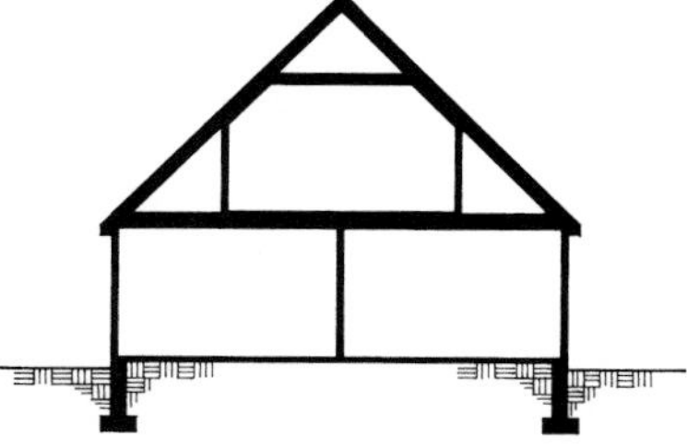

Luxury 1-1/2 Story

Living Area - 2800 S.F.
Perimeter - 175 L.F.

		Cost Per Square Foot Of Living Area			% of Total
		Mat.	Inst.	Total	(rounded)
1 Site Work	Site preparation for slab; 4′ deep trench excavation for foundation wall.		1.15	1.15	0.8%
2 Foundation	Continuous reinforced concrete footing, 12″ deep x 24″ wide; dampproofed and insulated 12″ thick reinforced concrete block foundation wall, 4′ deep; trowel finished 6″ thick concrete slab on 4″ crushed stone base and polyethylene vapor barrier.	5.84	6.48	12.32	8.1%
3 Framing	Exterior walls - 2″ x 6″ wood studs, 16″ O.C.; 1/2″ sheathing; gable end roof framing, steep pitch 2″ x 10″ rafters, 16″ O.C. with 1/2″ plywood sheathing; 2″ x 12″ floor joists, 16″ O.C. with bridging and 5/8″ subflooring; 2″ x 4″ interior partitions.	8.61	10.89	19.50	12.9%
4 Exterior Walls	1″ x 6″ tongue and groove vertical wood siding and housewrap on insulated wood frame walls; R38 attic insulation; metal clad double hung wood windows; raised panel exterior doors, frame and hardware, painted finish; wood storm and screen door.	13.48	4.59	18.07	11.9%
5 Roofing	Red cedar roof shingles, perfections; #15 felt building paper; aluminum gutters, downspouts and drip edge; copper flashings.	3.84	3.15	6.99	4.6%
6 Interiors	Skim coated 1/2″ thick gypsum wallboard walls and ceilings, primed and painted with 2 coats; interior raised panel solid core doors, frames and hardware, painted finish; oak hardwood flooring, 70%; ceramic tile flooring, 30%; hardwood tread stairway.	17.13	17.38	34.51	22.8%
7 Specialties	Luxury grade kitchen cabinets and countertops; double bowl kitchen sink; 75 gallon gas water heater.	9.90	1.84	11.74	7.8%
8 Mechanical	Three fixture bathroom: bathtub, water closet, vanity and sink; gas fired heating and air conditioning system.	7.22	3.92	11.14	7.4%
9 Electrical	200 amp electric service; wiring, duplex & GFI receptacles, wall switches, dimmer switches, door bell, appliance circuits, air conditioning circuit, fans and communications cabling; luxury grade lighting fixtures.	3.75	2.91	6.66	4.4%
10 Overhead	Contractor's overhead and profit and architect's fees.	16.73	12.54	29.27	19.3%
	Total	86.50	64.85	**151.35**	

RESIDENTIAL | Luxury | 2 Story

- Unique residence built from an architect's plan
- Single family — 1 full bath, 1 half bath, 1 kitchen
- No basement
- Cedar shakes on roof
- Forced hot air heat/air conditioning
- Gypsum wallboard interior finishes
- Many special features
- Extraordinary materials and workmanship

Note: The illustration shown may contain some optional components (for example: garages and/or fireplaces) whose costs are shown in the modifications, adjustments, & alternatives below or at the end of the square foot section.

Base cost per square foot of living area

Exterior Wall	Living Area 1200	1400	1600	1800	2000	2400	2800	3200	3600	4000	4400
Wood Siding - Wood Frame	213.55	199.75	189.85	181.30	172.35	159.45	148.55	**141.35**	137.15	132.50	128.70
Brick Veneer - Wood Frame	226.85	212.20	201.70	192.45	183.10	169.15	157.35	149.60	145.10	140.00	135.95
Solid Brick	243.30	227.55	216.40	206.25	196.45	181.15	168.30	159.85	155.00	149.25	144.95
Solid Stone	251.40	235.10	223.55	213.05	203.00	187.10	173.65	164.90	159.90	153.85	149.35
Finished Basement, Add	34.30	37.25	36.20	35.00	34.30	32.70	31.35	30.45	30.00	29.30	28.85
Unfinished Basement, Add	15.95	15.20	14.70	14.10	13.80	12.90	12.25	11.80	11.55	11.20	11.00

Modifications

Add to the total cost

Upgrade Kitchen Cabinets	$ + 2641
Solid Surface Countertops (Included)	
Full Bath - including plumbing, wall and floor finishes	+ 11,680
Half Bath - including plumbing, wall and floor finishes	+ 6896
Two Car Attached Garage	+ 34,538
Two Car Detached Garage	+ 39,236
Fireplace & Chimney	+ 11,606

Adjustments

For multi family - add to total cost

Additional Kitchen	$ + 30,712
Additional Full Bath & Half Bath	+ 18,576
Additional Entry & Exit	+ 2794
Separate Heating & Air Conditioning	+ 7787
Separate Electric	+ 1868

For Townhouse/Rowhouse - Multiply cost per square foot by

Inner Unit	.86
End Unit	.93

Alternatives

Add to or deduct from the cost per square foot of living area

Heavyweight Asphalt Shingles	– 1.35
Clay Tile Roof	+ 1.70
Slate Roof	+ 2.15
Upgrade Ceilings to Textured Finish	+ .61
Air Conditioning, in Heating Ductwork	Base System
Heating Systems, Hot Water	+ 1.52
Heat Pump	+ 1.55
Electric Heat	– 2.90
Not Heated	– 3.71

Additional upgrades or components

Kitchen Cabinets & Countertops	Page 93
Bathroom Vanities	94
Fireplaces & Chimneys	94
Windows, Skylights & Dormers	94
Appliances	95
Breezeways & Porches	95
Finished Attic	95
Garages	96
Site Improvements	96
Wings & Ells	92

Important: See the Reference Section for Location Factors (to adjust for your city) and Estimating Forms.

Luxury 2 Story

Living Area - 3200 S.F.
Perimeter - 163 L.F.

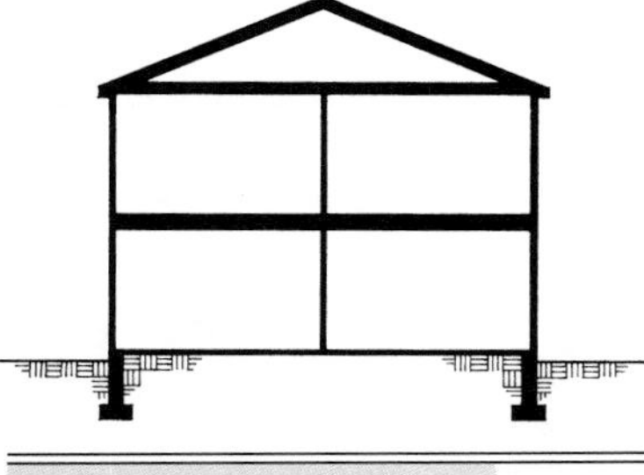

		Cost Per Square Foot Of Living Area			% of Total
		Mat.	Inst.	Total	(rounded)
1 Site Work	Site preparation for slab; 4′ deep trench excavation for foundation wall.		1.01	1.01	0.7%
2 Foundation	Continuous reinforced concrete footing, 12″ deep x 24″ wide; dampproofed and insulated 12″ thick reinforced concrete block foundation wall, 4′ deep; trowel finished 6″ thick concrete slab on 4″ crushed stone base and polyethylene vapor barrier.	4.72	5.28	10.00	7.1%
3 Framing	Exterior walls - 2″ x 6″ wood studs, 16″ O.C.; 1/2″ sheathing; gable end roof framing, 2″ x 10″ rafters, 16″ O.C. with 1/2″ plywood sheathing; 2″ x 12″ floor joists, 16″ O.C. with bridging and 5/8″ subflooring; 2″ x 4″ interior partitions.	8.15	10.16	18.31	13.0%
4 Exterior Walls	1″ x 6″ tongue and groove vertical wood siding and housewrap on insulated wood frame walls; R38 attic insulation; metal clad double hung wood windows; raised panel exterior doors, frame and hardware, painted finish; wood storm and screen door.	13.65	4.66	18.31	13.0%
5 Roofing	Red cedar roof shingles, perfections; #15 felt building paper; aluminum gutters, downspouts and drip edge; copper flashings.	2.66	2.18	4.84	3.4%
6 Interiors	Skim coated 1/2″ thick gypsum wallboard walls and ceilings, primed and painted with 2 coats; interior raised panel solid core doors, frames and hardware, painted finish; oak hardwood flooring, 70%; ceramic tile flooring, 30%; hardwood tread stairway.	17.02	17.39	34.41	24.3%
7 Specialties	Luxury grade kitchen cabinets and countertops; double bowl kitchen sink; 75 gallon gas water heater.	8.67	1.60	10.27	7.3%
8 Mechanical	Three fixture bathroom: bathtub, water closet, vanity and sink; gas fired heating and air conditioning system.	6.50	3.78	10.28	7.3%
9 Electrical	200 amp electric service; wiring, duplex & GFI receptacles, wall switches, dimmer switches, door bell, appliance circuits, air conditioning circuit, fans and communications cabling; luxury grade lighting fixtures.	3.73	2.84	6.57	4.6%
10 Overhead	Contractor's overhead and profit and architect's fees.	15.60	11.75	27.35	19.3%
	Total	80.70	60.65	**141.35**	

- **Unique residence built from an architect's plan**
- **Single family — 1 full bath, 1 half bath, 1 kitchen**
- **No basement**
- **Cedar shakes on roof**
- **Forced hot air heat/air conditioning**
- **Gypsum wallboard interior finishes**
- **Many special features**
- **Extraordinary materials and workmanship**

Note: The illustration shown may contain some optional components (for example: garages and/or fireplaces) whose costs are shown in the modifications, adjustments, & alternatives below or at the end of the square foot section.

Base cost per square foot of living area

Exterior Wall	Living Area 1500	1800	2100	2500	3000	3500	4000	4500	5000	5500	6000
Wood Siding - Wood Frame	208.75	187.70	174.80	165.30	**152.55**	143.30	134.55	130.20	126.40	122.50	118.55
Brick Veneer - Wood Frame	223.00	200.85	186.55	176.30	162.55	152.45	143.05	138.20	134.00	129.80	125.50
Solid Brick	239.85	216.35	200.40	189.45	174.40	163.25	153.05	147.70	143.00	138.40	133.75
Solid Stone	249.00	224.80	208.00	196.60	181.00	169.25	158.50	152.90	147.95	143.05	138.30
Finished Basement, Add	27.40	29.45	27.75	26.90	25.60	24.55	23.75	23.15	22.75	22.30	21.95
Unfinished Basement, Add	12.90	12.15	11.30	10.80	10.20	9.65	9.25	8.95	8.75	8.50	8.35

Modifications

Add to the total cost

Upgrade Kitchen Cabinets	$ + 2641
Solid Surface Countertops (Included)	
Full Bath - including plumbing, wall and floor finishes	+ 11,680
Half Bath - including plumbing, wall and floor finishes	+ 6896
Two Car Attached Garage	+ 34,538
Two Car Detached Garage	+ 39,236
Fireplace & Chimney	+ 12,701

Adjustments

For multi family - add to total cost

Additional Kitchen	$ + 30,712
Additional Full Bath & Half Bath	+ 18,576
Additional Entry & Exit	+ 2794
Separate Heating & Air Conditioning	+ 7787
Separate Electric	+ 1868

For Townhouse/Rowhouse - Multiply cost per square foot by

Inner Unit	.86
End Unit	.93

Alternatives

Add to or deduct from the cost per square foot of living area

Heavyweight Asphalt Shingles	– 1.20
Clay Tile Roof	+ 1.45
Slate Roof	+ 1.85
Upgrade Ceilings to Textured Finish	+ .61
Air Conditioning, in Heating Ductwork	Base System
Heating Systems, Hot Water	+ 1.37
Heat Pump	+ 1.60
Electric Heat	– 5.69
Not Heated	– 3.71

Additional upgrades or components

Kitchen Cabinets & Countertops	Page 93
Bathroom Vanities	94
Fireplaces & Chimneys	94
Windows, Skylights & Dormers	94
Appliances	95
Breezeways & Porches	95
Finished Attic	95
Garages	96
Site Improvements	96
Wings & Ells	92

Important: See the Reference Section for Location Factors (to adjust for your city) and Estimating Forms.

Luxury 2-1/2 Story

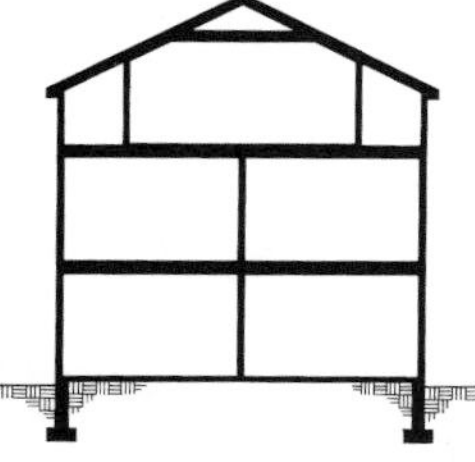

Living Area - 3000 S.F.
Perimeter - 148 L.F.

		Cost Per Square Foot Of Living Area			% of Total
		Mat.	Inst.	Total	(rounded)
1 Site Work	Site preparation for slab; 4′ deep trench excavation for foundation wall.		1.07	1.07	0.7%
2 Foundation	Continuous reinforced concrete footing, 12″ deep x 24″ wide; dampproofed and insulated 12″ thick reinforced concrete block foundation wall, 4′ deep; trowel finished 6″ thick concrete slab on 4″ crushed stone base and polyethylene vapor barrier.	4.15	4.79	8.94	5.9%
3 Framing	Exterior walls - 2″ x 6″ wood studs, 16″ O.C.; 1/2″ sheathing; gable end roof framing, steep pitch 2″ x 10″ rafters, 16″ O.C. with 1/2″ plywood sheathing; 2″ x 12″ floor joists, 16″ O.C. with bridging and 5/8″ subflooring; 2″ x 4″ interior partitions.	9.03	11.22	20.25	13.3%
4 Exterior Walls	1″ x 6″ tongue and groove vertical wood siding and housewrap on insulated wood frame walls; R38 attic insulation; metal clad double hung wood windows; raised panel exterior doors, frame and hardware, painted finish; wood storm and screen door.	15.84	5.43	21.27	13.9%
5 Roofing	Red cedar roof shingles, perfections; #15 felt building paper; aluminum gutters, downspouts and drip edge; copper flashings.	2.36	1.94	4.30	2.8%
6 Interiors	Skim coated 1/2″ thick gypsum wallboard walls and ceilings, primed and painted with 2 coats; interior raised panel solid core doors, frames and hardware, painted finish; oak hardwood flooring, 70%; ceramic tile flooring, 30%; hardwood tread stairway.	19.55	19.36	38.91	25.5%
7 Specialties	Luxury grade kitchen cabinets and countertops; double bowl kitchen sink; 75 gallon gas water heater.	9.26	1.72	10.98	7.2%
8 Mechanical	Three fixture bathroom: bathtub, water closet, vanity and sink; gas fired heating and air conditioning system.	6.85	3.83	10.68	7.0%
9 Electrical	200 amp electric service; wiring, duplex & GFI receptacles, wall switches, dimmer switches, door bell, appliance circuits, air conditioning circuit, fans and communications cabling; luxury grade lighting fixtures.	3.74	2.88	6.62	4.3%
10 Overhead	Contractor's overhead and profit and architect's fees.	16.97	12.56	29.53	19.4%
	Total	87.75	64.80	**152.55**	

- **Unique residence built from an architect's plan**
- **Single family — 1 full bath, 1 half bath, 1 kitchen**
- **No basement**
- **Cedar shakes on roof**
- **Forced hot air heat/air conditioning**
- **Gypsum wallboard interior finishes**
- **Many special features**
- **Extraordinary materials and workmanship**

Note: The illustration shown may contain some optional components (for example: garages and/or fireplaces) whose costs are shown in the modifications, adjustments, & alternatives below or at the end of the square foot section.

Base cost per square foot of living area

Exterior Wall	Living Area 1500	1800	2100	2500	3000	3500	4000	4500	5000	5500	6000
Wood Siding - Wood Frame	205.20	184.35	173.80	164.80	151.70	144.95	136.95	128.90	125.75	122.50	119.30
Brick Veneer - Wood Frame	219.40	197.50	186.00	176.45	162.35	155.05	146.10	137.35	133.90	130.35	126.65
Solid Brick	237.30	214.10	201.50	191.30	175.80	167.75	157.65	148.05	144.25	140.35	135.95
Solid Stone	245.60	221.80	208.70	198.10	182.00	173.60	162.95	152.90	149.00	144.95	140.35
Finished Basement, Add	24.00	25.80	24.75	24.05	22.80	22.15	21.25	20.55	20.20	19.90	19.45
Unfinished Basement, Add	11.25	10.60	10.10	9.80	9.15	8.85	8.40	8.05	7.85	7.65	7.45

Modifications

Add to the total cost

Upgrade Kitchen Cabinets	$ + 2641
Solid Surface Countertops (Included)	
Full Bath - including plumbing, wall and floor finishes	+ 11,680
Half Bath - including plumbing, wall and floor finishes	+ 6896
Two Car Attached Garage	+ 34,538
Two Car Detached Garage	+ 39,236
Fireplace & Chimney	+ 12,701

Adjustments

For multi family - add to total cost

Additional Kitchen	$ + 30,712
Additional Full Bath & Half Bath	+ 18,576
Additional Entry & Exit	+ 2794
Separate Heating & Air Conditioning	+ 7787
Separate Electric	+ 1868

For Townhouse/Rowhouse - Multiply cost per square foot by

Inner Unit	.84
End Unit	.92

Alternatives

Add to or deduct from the cost per square foot of living area

Heavyweight Asphalt Shingles	– .90
Clay Tile Roof	+ 1.10
Slate Roof	+ 1.45
Upgrade Ceilings to Textured Finish	+ .61
Air Conditioning, in Heating Ductwork	Base System
Heating Systems, Hot Water	+ 1.37
Heat Pump	+ 1.60
Electric Heat	– 5.69
Not Heated	– 3.61

Additional upgrades or components

Kitchen Cabinets & Countertops	Page 93
Bathroom Vanities	94
Fireplaces & Chimneys	94
Windows, Skylights & Dormers	94
Appliances	95
Breezeways & Porches	95
Finished Attic	95
Garages	96
Site Improvements	96
Wings & Ells	92

Important: See the Reference Section for Location Factors (to adjust for your city) and Estimating Forms.

Luxury 3 Story

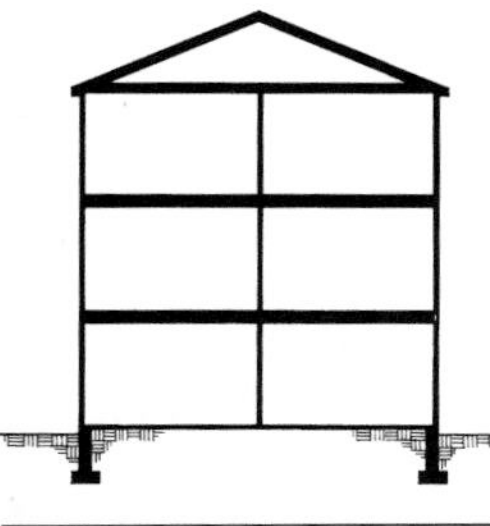

Living Area - 3000 S.F.
Perimeter - 135 L.F.

		Cost Per Square Foot Of Living Area			% of Total
		Mat.	Inst.	Total	(rounded)
1 Site Work	Site preparation for slab; 4′ deep trench excavation for foundation wall.		1.07	1.07	0.7%
2 Foundation	Continuous reinforced concrete footing, 12″ deep x 24″ wide; dampproofed and insulated 12″ thick reinforced concrete block foundation wall, 4′ deep; trowel finished 6″ thick concrete slab on 4″ crushed stone base and polyethylene vapor barrier.	3.74	4.36	8.10	5.3%
3 Framing	Exterior walls - 2″ x 6″ wood studs, 16″ O.C.; 1/2″ sheathing; gable end roof framing, 2″ x 10″ rafters, 16″ O.C. with 1/2″ plywood sheathing; 2″ x 12″ floor joists, 16″ O.C. with bridging and 5/8″ subflooring; 2″ x 4″ interior partitions.	8.88	10.95	19.83	13.1%
4 Exterior Walls	1″ x 6″ tongue and groove vertical wood siding and housewrap on insulated wood frame walls; R38 attic insulation; metal clad double hung wood windows; raised panel exterior doors, frame and hardware, painted finish; wood storm and screen door.	16.86	5.79	22.65	14.9%
5 Roofing	Red cedar roof shingles, perfections; #15 felt building paper; aluminum gutters, downspouts and drip edge; copper flashings.	1.76	1.45	3.21	2.1%
6 Interiors	Skim coated 1/2″ thick gypsum wallboard walls and ceilings, primed and painted with 2 coats; interior raised panel solid core doors, frames and hardware, painted finish; oak hardwood flooring, 70%; ceramic tile flooring, 30%; hardwood tread stairway.	19.66	19.54	39.20	25.8%
7 Specialties	Luxury grade kitchen cabinets and countertops; double bowl kitchen sink; 75 gallon gas water heater.	9.26	1.72	10.98	7.2%
8 Mechanical	Three fixture bathroom: bathtub, water closet, vanity and sink; gas fired heating and air conditioning system.	6.85	3.83	10.68	7.0%
9 Electrical	200 amp electric service; wiring, duplex & GFI receptacles, wall switches, dimmer switches, door bell, appliance circuits, air conditioning circuit, fans and communications cabling; luxury grade lighting fixtures.	3.74	2.88	6.62	4.4%
10 Overhead	Contractor's overhead and profit and architect's fees.	17.00	12.36	29.36	19.4%
	Total	87.75	63.95	**151.70**	

- Unique residence built from an architect's plan
- Single family — 1 full bath, 1 half bath, 1 kitchen
- No basement
- Cedar shakes on roof
- Forced hot air heat/air conditioning
- Gypsum wallboard interior finishes
- Many special features
- Extraordinary materials and workmanship

Note: The illustration shown may contain some optional components (for example: garages and/or fireplaces) whose costs are shown in the modifications, adjustments, & alternatives below or at the end of the square foot section.

Base cost per square foot of living area

Exterior Wall	Living Area 1200	1400	1600	1800	2000	2400	2800	3200	3600	4000	4400
Wood Siding - Wood Frame	202.40	189.35	179.90	171.90	163.30	151.30	141.05	134.40	130.50	126.25	122.65
Brick Veneer - Wood Frame	212.55	198.85	188.95	180.45	171.50	158.65	147.80	140.75	136.50	131.95	128.20
Solid Brick	224.80	210.20	199.85	190.75	181.45	167.60	155.90	148.40	143.85	138.85	134.85
Solid Stone	231.20	216.20	205.55	196.05	186.60	172.25	160.20	152.30	147.70	142.40	138.35
Finished Basement, Add	34.30	37.25	36.20	35.00	34.30	32.70	31.35	30.45	30.00	29.30	28.85
Unfinished Basement, Add	15.95	15.20	14.70	14.10	13.80	12.90	12.25	11.80	11.55	11.20	11.00

Modifications

Add to the total cost

Upgrade Kitchen Cabinets	$ + 2641
Solid Surface Countertops (Included)	
Full Bath - including plumbing, wall and floor finishes	+ 11,680
Half Bath - including plumbing, wall and floor finishes	+ 6896
Two Car Attached Garage	+ 34,538
Two Car Detached Garage	+ 39,236
Fireplace & Chimney	+ 10,588

Adjustments

For multi family - add to total cost

Additional Kitchen	$ + 30,712
Additional Full Bath & Half Bath	+ 18,576
Additional Entry & Exit	+ 2794
Separate Heating & Air Conditioning	+ 7787
Separate Electric	+ 1868

For Townhouse/Rowhouse - Multiply cost per square foot by

Inner Unit	.89
End Unit	.94

Alternatives

Add to or deduct from the cost per square foot of living area

Heavyweight Asphalt Shingles	– 1.35
Clay Tile Roof	+ 1.70
Slate Roof	+ 2.15
Upgrade Ceilings to Textured Finish	+ .61
Air Conditioning, in Heating Ductwork	Base System
Heating Systems, Hot Water	+ 1.52
Heat Pump	+ 1.55
Electric Heat	– 2.90
Not Heated	– 3.71

Additional upgrades or components

Kitchen Cabinets & Countertops	Page 93
Bathroom Vanities	94
Fireplaces & Chimneys	94
Windows, Skylights & Dormers	94
Appliances	95
Breezeways & Porches	95
Finished Attic	95
Garages	96
Site Improvements	96
Wings & Ells	92

Luxury Bi-Level

Living Area - 3200 S.F.
Perimeter - 163 L.F.

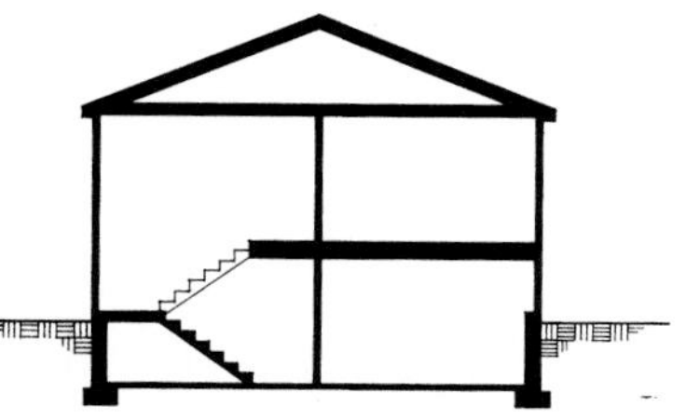

		Cost Per Square Foot Of Living Area			% of Total
		Mat.	Inst.	Total	(rounded)
1 Site Work	Site preparation for slab; 4′ deep trench excavation for foundation wall.		1.01	1.01	0.8%
2 Foundation	Continuous reinforced concrete footing, 12″ deep x 24″ wide; dampproofed and insulated 12″ thick reinforced concrete block foundation wall, 4′ deep; trowel finished 6″ thick concrete slab on 4″ crushed stone base and polyethylene vapor barrier.	4.72	5.28	10.00	7.4%
3 Framing	Exterior walls - 2″ x 6″ wood studs, 16″ O.C.; 1/2″ sheathing; gable end roof framing, 2″ x 10″ rafters, 16″ O.C. with 1/2″ plywood sheathing; 2″ x 12″ floor joists, 16″ O.C. with bridging and 5/8″ subflooring; 2″ x 4″ interior partitions.	7.65	9.53	17.18	12.8%
4 Exterior Walls	1″ x 6″ tongue and groove vertical wood siding and housewrap on insulated wood frame walls; R38 attic insulation; metal clad double hung wood windows; raised panel exterior doors, frame and hardware, painted finish; wood storm and screen door.	10.86	3.67	14.53	10.8%
5 Roofing	Red cedar roof shingles, perfections: #15 felt building paper; aluminum gutters, downspouts and drip edge; copper flashings.	2.66	2.18	4.84	3.6%
6 Interiors	Skim coated 1/2″ thick gypsum wallboard walls and ceilings, primed and painted with 2 coats; interior raised panel solid core doors, frames and hardware, painted finish; oak hardwood flooring, 70%; ceramic tile flooring, 30%; hardwood tread stairway.	16.74	16.98	33.72	25.1%
7 Specialties	Luxury grade kitchen cabinets and countertops; double bowl kitchen sink; 75 gallon gas water heater.	8.67	1.60	10.27	7.6%
8 Mechanical	Three fixture bathroom: bathtub, water closet, vanity and sink; gas fired heating and air conditioning system.	6.50	3.78	10.28	7.6%
9 Electrical	200 amp electric service; wiring, duplex & GFI receptacles, wall switches, dimmer switches, door bell, appliance circuits, air conditioning circuit, fans and communications cabling; luxury grade lighting fixtures	3.73	2.84	6.57	4.9%
10 Overhead	Contractor's overhead and profit and architect's fees.	14.77	11.23	26.00	19.3%
	Total	76.30	58.10	**134.40**	

- **Unique residence built from an architect's plan**
- **Single family — 1 full bath, 1 half bath, 1 kitchen**
- **No basement**
- **Cedar shakes on roof**
- **Forced hot air heat/air conditioning**
- **Gypsum wallboard interior finishes**
- **Many special features**
- **Extraordinary materials and workmanship**

Note: The illustration shown may contain some optional components (for example: garages and/or fireplaces) whose costs are shown in the modifications, adjustments, & alternatives below or at the end of the square foot section.

Base cost per square foot of living area

	Living Area										
Exterior Wall	**1500**	**1800**	**2100**	**2400**	**2800**	**3200**	**3600**	**4000**	**4500**	**5000**	**5500**
Wood Siding - Wood Frame	188.50	173.45	161.05	152.75	146.40	139.30	**132.15**	128.75	122.00	118.15	114.05
Brick Veneer - Wood Frame	197.75	181.80	168.65	159.85	153.20	145.65	138.00	134.45	127.20	123.15	118.75
Solid Brick	208.60	191.65	177.60	168.25	161.30	153.15	144.95	141.20	133.45	129.00	124.30
Solid Stone	214.65	197.00	182.50	172.80	165.70	157.20	148.75	144.90	136.85	132.10	127.35
Finished Basement, Add*	40.45	43.50	41.75	40.60	39.80	38.60	37.60	37.25	36.15	35.50	34.90
Unfinished Basement, Add*	18.35	17.25	16.30	15.75	15.40	14.75	14.25	14.05	13.55	13.15	12.85

*Basement under middle level only.

Modifications

Add to the total cost

Upgrade Kitchen Cabinets	$ + 2641
Solid Surface Countertops (Included)	
Full Bath - including plumbing, wall and floor finishes	+ 11,680
Half Bath - including plumbing, wall and floor finishes	+ 6896
Two Car Attached Garage	+ 34,538
Two Car Detached Garage	+ 39,236
Fireplace & Chimney	+ 10,588

Adjustments

For multi family - add to total cost

Additional Kitchen	$ + 30,712
Additional Full Bath & Half Bath	+ 18,576
Additional Entry & Exit	+ 2794
Separate Heating & Air Conditioning	+ 7787
Separate Electric	+ 1868

For Townhouse/Rowhouse - Multiply cost per square foot by

Inner Unit	.86
End Unit	.93

Alternatives

Add to or deduct from the cost per square foot of living area

Heavyweight Asphalt Shingles	– 1.95
Clay Tile Roof	+ 2.45
Slate Roof	+ 3.10
Upgrade Ceilings to Textured Finish	+ .61
Air Conditioning, in Heating Ductwork	Base System
Heating Systems, Hot Water	+ 1.47
Heat Pump	+ 1.62
Electric Heat	– 2.54
Not Heated	– 3.61

Additional upgrades or components

Important: See the Reference Section for Location Factors (to adjust for your city) and Estimating Forms.

Luxury Tri-Level

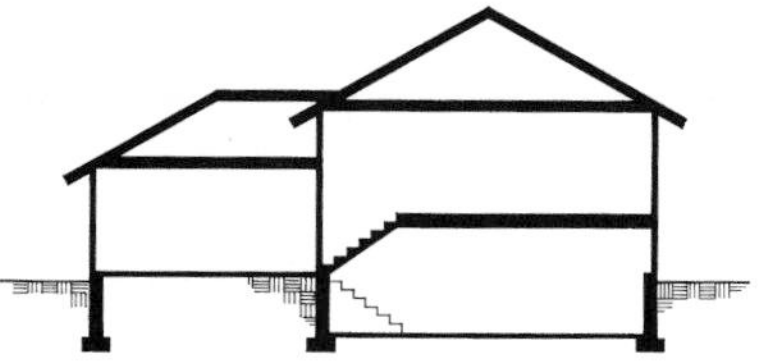

Living Area - 3600 S.F.
Perimeter - 207 L.F.

			Cost Per Square Foot Of Living Area			% of Total
			Mat.	Inst.	Total	(rounded)
1	Site Work	Site preparation for slab; 4′ deep trench excavation for foundation wall.		.89	.89	0.7%
2	Foundation	Continuous reinforced concrete footing, 12″ deep x 24″ wide; dampproofed and insulated 12″ thick reinforced concrete block foundation wall, 4′ deep; trowel finished 6″ thick concrete slab on 4″ crushed stone base and polyethylene vapor barrier.	5.63	6.06	11.69	8.8%
3	Framing	Exterior walls - 2″ x 6″ wood studs, 16″ O.C.; 1/2″ sheathing; gable end roof framing, 2″ x 10″ rafters, 16″ O.C. with 1/2″ plywood sheathing; 2″ x 12″ floor joists, 16″ O.C. with bridging and 5/8″ subflooring; 2″ x 4″ interior partitions.	7.41	9.37	16.78	12.7%
4	Exterior Walls	1″ x 6″ tongue and groove vertical wood siding and housewrap on insulated wood frame walls; R38 attic insulation; metal clad double hung wood windows; raised panel exterior doors, frame and hardware, painted finish; wood storm and screen door.	10.71	3.62	14.33	10.8%
5	Roofing	Red cedar roof shingles, perfections; #15 felt building paper; aluminum gutters, downspouts and drip edge; copper flashings.	3.54	2.91	6.45	4.9%
6	Interiors	Skim coated 1/2″ thick gypsum wallboard walls and ceilings, primed and painted with 2 coats; interior raised panel solid core doors, frames and hardware, painted finish; oak hardwood flooring, 70%; ceramic tile flooring, 30%; hardwood tread stairway.	15.41	15.76	31.17	23.6%
7	Specialties	Luxury grade kitchen cabinets and countertops; double bowl kitchen sink; 75 gallon gas water heater.	7.71	1.43	9.14	6.9%
8	Mechanical	Three fixture bathroom: bathtub, water closet, vanity and sink; gas fired heating and air conditioning system.	5.96	3.65	9.61	7.3%
9	Electrical	200 amp electric service; wiring, duplex & GFI receptacles, wall switches, dimmer switches, door bell, appliance circuits, air conditioning circuit, fans and communications cabling; luxury grade lighting fixtures.	3.71	2.80	6.51	4.9%
10	Overhead	Contractor's overhead and profit and architect's fees.	14.42	11.16	25.58	19.4%
		Total	74.50	57.65	**132.15**	

1 Story — Base cost per square foot of living area

Exterior Wall	Living Area 50	100	200	300	400	500	600	700
Wood Siding - Wood Frame	286.10	220.95	192.60	162.30	153.10	147.55	143.75	144.90
Brick Veneer - Wood Frame	318.85	244.40	212.10	175.30	164.75	158.40	154.20	154.90
Solid Brick	365.55	277.75	239.90	193.90	181.50	174.00	169.00	169.20
Solid Stone	382.85	290.10	250.25	200.70	187.65	179.75	174.50	174.55
Finished Basement, Add	109.30	99.00	88.70	71.55	68.20	66.10	64.75	63.75
Unfinished Basement, Add	57.55	44.35	38.90	29.70	27.90	26.75	26.05	25.50

1-1/2 Story — Base cost per square foot of living area

Exterior Wall	Living Area 100	200	300	400	500	600	700	800
Wood Siding - Wood Frame	226.65	184.80	159.30	144.15	136.45	132.75	128.05	126.65
Brick Veneer - Wood Frame	255.90	208.25	178.85	159.35	150.50	146.10	140.55	139.05
Solid Brick	297.60	241.60	206.65	181.05	170.50	165.00	158.45	156.80
Solid Stone	313.05	254.00	216.95	189.05	177.95	171.95	165.05	163.40
Finished Basement, Add	71.85	71.00	64.10	56.60	54.55	53.15	51.95	51.75
Unfinished Basement, Add	36.55	31.10	27.40	23.40	22.30	21.55	20.90	20.80

2 Story — Base cost per square foot of living area

Exterior Wall	Living Area 100	200	400	600	800	1000	1200	1400
Wood Siding - Wood Frame	222.65	167.65	143.55	120.30	112.55	107.75	104.60	106.15
Brick Veneer - Wood Frame	255.40	191.10	163.10	133.35	124.25	118.65	115.05	116.15
Solid Brick	302.10	224.45	190.90	151.90	140.90	134.25	129.85	130.45
Solid Stone	319.40	236.85	201.15	158.70	147.05	140.00	135.30	135.80
Finished Basement, Add	54.75	49.55	44.40	35.85	34.15	33.05	32.40	31.90
Unfinished Basement, Add	28.80	22.20	19.45	14.90	13.95	13.35	13.05	12.80

Base costs do not include bathroom or kitchen facilities. Use Modifications/Adjustments/Alternatives on pages 93–96 where appropriate.

Important: See the Reference Section for Location Factors (to adjust for your city) and Estimating Forms.

Kitchen cabinets - Base units, hardwood *(Cost per Unit)*

	Economy	Average	Custom	Luxury
24" deep, 35" high,				
One top drawer,				
One door below				
12" wide	$ 293	$ 390	$ 519	$ 683
15" wide	300	400	532	700
18" wide	319	425	565	744
21" wide	326	435	579	761
24" wide	394	525	698	919
Four drawers				
12" wide	296	395	525	691
15" wide	300	400	532	700
18" wide	326	435	579	761
24" wide	368	490	652	858
Two top drawers,				
Two doors below				
27" wide	401	535	712	936
30" wide	450	600	798	1050
33" wide	469	625	831	1094
36" wide	484	645	858	1129
42" wide	514	685	911	1199
48" wide	548	730	971	1278
Range or sink base				
(Cost per unit)				
Two doors below				
30" wide	383	510	678	893
33" wide	405	540	718	945
36" wide	424	565	751	989
42" wide	450	600	798	1050
48" wide	469	625	831	1094
Corner Base Cabinet				
(Cost per unit)				
36" wide	656	875	1164	1531
Lazy Susan *(Cost per unit)*				
With revolving door	863	1150	1530	2013

Kitchen cabinets - Wall cabinets, hardwood *(Cost per Unit)*

	Economy	Average	Custom	Luxury
12" deep, 2 doors				
12" high				
30" wide	$ 263	$ 350	$ 466	$ 613
36" wide	95	126	168	221
15" high				
30" wide	263	350	466	613
33" wide	323	430	572	753
36" wide	315	420	559	735
24" high				
30" wide	341	455	605	796
36" wide	368	490	652	858
42" wide	274	365	485	639
30" high, 1 door				
12" wide	255	340	452	595
15" wide	270	360	479	630
18" wide	293	390	519	683
24" wide	334	445	592	779
30" high, 2 doors				
27" wide	371	495	658	866
30" wide	386	515	685	901
36" wide	431	575	765	1006
42" wide	469	625	831	1094
48" wide	525	700	931	1225
Corner wall, 30" high				
24" wide	386	515	685	901
30" wide	398	530	705	928
36" wide	450	600	798	1050
Broom closet				
84" high, 24" deep				
18" wide	713	950	1264	1663
Oven Cabinet				
84" high, 24" deep				
27" wide	1069	1425	1895	2494

Kitchen countertops *(Cost per L.F.)*

	Economy	Average	Custom	Luxury
Solid Surface				
24" wide, no backsplash	$ 111	$ 148	$ 197	$ 259
with backsplash	126	168	223	294
Stock plastic laminate, 24" wide				
with backsplash	26	35	47	61
Custom plastic laminate, no splash				
7/8" thick, alum. molding	41	55	73	96
1-1/4" thick, no splash	45	59	79	104
Marble				
1/2" - 3/4" thick w/splash	63	84	112	148
Maple, laminated				
1-1/2" thick w/splash	134	179	238	313
Stainless steel				
(per S.F.)	166	221	294	387
Cutting blocks, recessed				
16" x 20" x 1" (each)	136	181	241	317

Vanity bases *(Cost per Unit)*

	Economy	Average	Custom	Luxury
2 door, 30″ high, 21″ deep				
24″ wide	$ 341	$ 455	$ 605	$ 796
30″ wide	409	545	725	954
36″ wide	383	510	678	893
48″ wide	555	740	984	1295

Solid surface vanity tops *(Cost Each)*

	Economy	Average	Custom	Luxury
Center bowl				
22″ x 25″	$ 415	$ 448	$ 484	$ 523
22″ x 31″	475	513	554	598
22″ x 37″	545	589	636	687
22″ x 49″	670	724	782	845

Fireplaces & Chimneys *(Cost per Unit)*

	1-1/2 Story	2 Story	3 Story
Economy (prefab metal)			
Exterior chimney & 1 fireplace	$ 6742	$ 7450	$ 8172
Interior chimney & 1 fireplace	6461	7183	7515
Average (masonry)			
Exterior chimney & 1 fireplace	7100	7845	8605
Interior chimney & 1 fireplace	6803	7563	7913
For more than 1 flue, add	483	822	1378
For more than 1 fireplace, add	4763	4763	4763
Custom (masonry)			
Exterior chimney & 1 fireplace	7426	8382	9465
Interior chimney & 1 fireplace	6962	7881	8494
For more than 1 flue, add	582	1008	1687
For more than 1 fireplace, add	5334	5334	5334
Luxury (masonry)			
Exterior chimney & 1 fireplace	10588	11606	12701
Interior chimney & 1 fireplace	10101	11044	11690
For more than 1 flue, add	874	1459	2037
For more than 1 fireplace, add	8345	8345	8345

Windows and Skylights *(Cost Each)*

	Economy	Average	Custom	Luxury
Fixed Picture Windows				
3′-6″ x 4′-0″	$ 375	$ 500	$ 665	$ 875
4′-0″ x 6′-0″	648	864	1150	1513
5′-0″ x 6′-0″	733	977	1300	1711
6′-0″ x 6′-0″	747	996	1325	1743
Bay/Bow Windows				
8′-0″ x 5′-0″	888	1184	1575	2072
10′-0″ x 5′-0″	930	1240	1650	2171
10′-0″ x 6′-0″	1537	2048	2725	3586
12′-0″ x 6′-0″	2002	2669	3550	4671
Palladian Windows				
3′-2″ x 6′-4″		1541	2050	2697
4′-0″ x 6′-0″		1691	2250	2961
5′-5″ x 6′-10″		2312	3075	4046
8′-0″ x 6′-0″		2781	3700	4868
Skylights				
46″ x 21-1/2″	618	950	1160	1276
46″ x 28″	640	985	1215	1337
57″ x 44″	702	1080	1230	1353

Dormers *(Cost/S.F. of plan area)*

	Economy	Average	Custom	Luxury
Framing and Roofing Only				
Gable dormer, 2″ x 6″ roof frame	$ 34	$ 38	$ 42	$ 68
2″ x 8″ roof frame	35	39	43	71
Shed dormer, 2″ x 6″ roof frame	21	26	29	44
2″ x 8″ roof frame	23	27	30	45
2″ x 10″ roof frame	25	29	31	46

Appliances *(Cost per Unit)*

	Economy	Average	Custom	Luxury
Range				
30″ free standing, 1 oven	$ 595	$ 1673	$ 2212	$ 2750
2 oven	1200	2500	3150	3800
30″ built-in, 1 oven	1025	1713	2057	2400
2 oven	1575	1812	1930	2048
21″ free standing				
1 oven	595	758	839	920
Counter Top Ranges				
4 burner standard	445	1335	1780	2225
As above with griddle	1675	3088	3794	4500
Microwave Oven	241	506	638	770
Compactor				
4 to 1 compaction	875	1150	1288	1425
Deep Freeze				
15 to 23 C.F.	780	903	964	1025
30 C.F.	965	1095	1160	1225
Dehumidifier, portable, auto.				
15 pint	365	420	448	475
30 pint	410	472	503	533
Washing Machine, automatic	825	1425	1725	2025
Water Heater				
Electric, glass lined				
30 gal.	1125	1363	1482	1600
80 gal.	2150	2688	2957	3225
Water Heater, Gas, glass lined				
30 gal.	1950	2375	2588	2800
50 gal.	2325	2825	3075	3325
Dishwasher, built-in				
2 cycles	500	668	752	835
4 or more cycles	630	780	1603	2425
Dryer, automatic	790	1433	1754	2075
Garage Door Opener	550	658	712	765
Garbage Disposal	186	241	269	296
Heater, Electric, built-in				
1250 watt ceiling type	266	326	356	385
1250 watt wall type	355	375	385	395
Wall type w/blower	345			720
1500 watt	345	397	423	449
3000 watt	720	828	882	936
Hood For Range, 2 speed				
30″ wide	226	738	994	1250
42″ wide	291	1333	1854	2375
Humidifier, portable				
7 gal. per day	180	207	221	234
15 gal. per day	218	251	267	283
Ice Maker, automatic				
13 lb. per day	1600	1840	1960	2080
51 lb. per day	1825	2099	2236	2373
Refrigerator, no frost				
10-12 C.F.	570	645	683	720
14-16 C.F.	700	950	1075	1200
18-20 C.F.	880	1453	1739	2025
21-29 C.F.	1300	2125	2538	2950
Sump Pump, 1/3 H.P.	330	430	480	530

Breezeway *(Cost per S.F.)*

Class	Type	Area (S.F.)			
		50	100	150	200
Economy	Open	$ 48.99	$ 40.60	$ 35.56	$ 33.05
	Enclosed	179.86	127.84	105.08	90.31
Average	Open	57.78	46.95	41.11	38.18
	Enclosed	183.86	132.86	109.99	98.31
Custom	Open	69.82	57.16	50.20	46.73
	Enclosed	249.29	178.95	147.40	131.35
Luxury	Open	77.98	64.09	56.73	53.05
	Enclosed	293.49	208.20	170.11	150.79

Porches *(Cost per S.F.)*

Class	Type	Area (S.F.)				
		25	50	100	200	300
Economy	Open	$ 86.67	$ 64.12	$ 48.21	$ 36.73	$ 32.90
	Enclosed	189.71	150.21	99.29	75.20	65.40
Average	Open	90.29	69.25	50.78	38.02	33.75
	Enclosed	209.50	165.24	106.19	78.03	66.89
Custom	Open	154.10	111.61	85.66	63.90	62.19
	Enclosed	255.28	203.70	134.47	101.48	108.37
Luxury	Open	167.69	121.97	93.07	67.81	61.59
	Enclosed	293.21	265.38	171.05	124.30	102.43

Finished attic *(Cost per S.F.)*

Class	Area (S.F.)				
	400	500	600	800	1000
Economy	$ 23.61	$ 22.83	$ 21.87	$ 21.51	$ 20.71
Average	36.57	35.77	34.89	34.45	33.49
Custom	46.76	45.71	44.67	43.99	43.09
Luxury	59.38	57.94	56.58	55.24	54.33

Alarm system *(Cost per System)*

	Burglar Alarm	Smoke Detector
Economy	$ 445	$ 75
Average	505	87
Custom	895	222
Luxury	1525	256

Sauna, prefabricated
(Cost per unit, including heater and controls—7′ high)

Size	Cost
6′ x 4′	$ 5450
6′ x 5′	6325
6′ x 6′	6750
6′ x 9′	7925
8′ x 10′	10,300
8′ x 12′	11,000
10′ x 12′	16,300

Garages *

(Costs include exterior wall systems comparable with the quality of the residence. Included in the cost is an allowance for one personnel door, manual overhead door(s) and electrical fixture.)

Class	Type									
	Detached			Attached			Built-in		Basement	
	One Car	Two Car	Three Car	One Car	Two Car	Three Car	One Car	Two Car	One Car	Two Car
Economy										
Wood	$18,467	$28,175	$37,884	$14,279	$24,532	$34,241	$-1984	$-3968	$1851	$2550
Masonry	24,997	36,348	47,699	18,366	30,261	41,611	-2786	-5572		
Average										
Wood	20,232	30,384	40,536	15,383	26,080	36,232	-2201	-4401	2109	3067
Masonry	25,136	36,521	47,907	18,452	30,382	41,768	-2803	-4465		
Custom										
Wood	22,565	34,631	46,697	17,507	30,276	42,341	-3144	-2944	3096	5039
Masonry	27,465	40,762	54,060	20,573	34,573	47,871	-3746	-4148		
Luxury										
Wood	25,210	39,236	53,262	19,809	34,538	48,565	-3256	-3169	4184	6693
Masonry	32,653	48,551	64,449	24,467	41,068	56,966	-4170	-4997		

*See the Introduction to this section for definitions of garage types.

Swimming pools *(Cost per S.F.)*

Residential	
In-ground	$ 40.50 - 97.50
Deck equipment	1.30
Paint pool, preparation & 3 coats (epoxy)	5.38
Rubber base paint	4.84
Pool Cover	1.97
Swimming Pool Heaters	
(not including wiring, external piping, base or pad)	
Gas	
155 MBH	$ 3125.00
190 MBH	3775.00
500 MBH	13,100.
Electric	
15 KW 7200 gallon pool	2900.00
24 KW 9600 gallon pool	3375.00
54 KW 24,000 gallon pool	5325.00

Wood and coal stoves

Wood Only	
Free Standing (minimum)	$ 2295
Fireplace Insert (minimum)	1886
Coal Only	
Free Standing	$ 2132
Fireplace Insert	2334
Wood and Coal	
Free Standing	$ 4385
Fireplace Insert	4493

Sidewalks *(Cost per S.F.)*

Concrete, 3000 psi with wire mesh	4" thick	$ 4.60
	5" thick	5.40
	6" thick	6.06
Precast concrete patio blocks (natural)	2" thick	7.10
Precast concrete patio blocks (colors)	2" thick	7.35
Flagstone, bluestone	1" thick	22.20
Flagstone, bluestone	1-1/2" thick	31.15
Slate (natural, irregular)	3/4" thick	20.00
Slate (random rectangular)	1/2" thick	31.25
Seeding		
Fine grading & seeding includes lime, fertilizer & seed	per S.Y.	3.17
Lawn Sprinkler System	per S.F.	1.09

Fencing *(Cost per L.F.)*

Chain Link, 4' high, galvanized	$ 18.20
Gate, 4' high (each)	245.00
Cedar Picket, 3' high, 2 rail	16.25
Gate (each)	240.00
3 Rail, 4' high	19.60
Gate (each)	259.00
Cedar Stockade, 3 Rail, 6' high	18.95
Gate (each)	262.00
Board & Battens, 2 sides 6' high, pine	36.50
6' high, cedar	40.00
No. 1 Cedar, basketweave, 6' high	37.00
Gate, 6' high (each)	320.00

Carport *(Cost per S.F.)*

Economy	$ 10.80
Average	16.15
Custom	23.67
Luxury	27.38

Assemblies Section

Table of Contents

RSMeans data: Assemblies—How They Work

The following is a detailed explanation of a sample Assemblies Cost Table. Included are an illustration and accompanying system descriptions. Additionally, related systems and price sheets may be included. Next to each bold number below is the item being described with the appropriate component of the sample entry in parentheses. General contractors should add an additional markup to the figures shown in the Assemblies section. Note: Throughout this section, the words assembly and system will be used interchangeably.

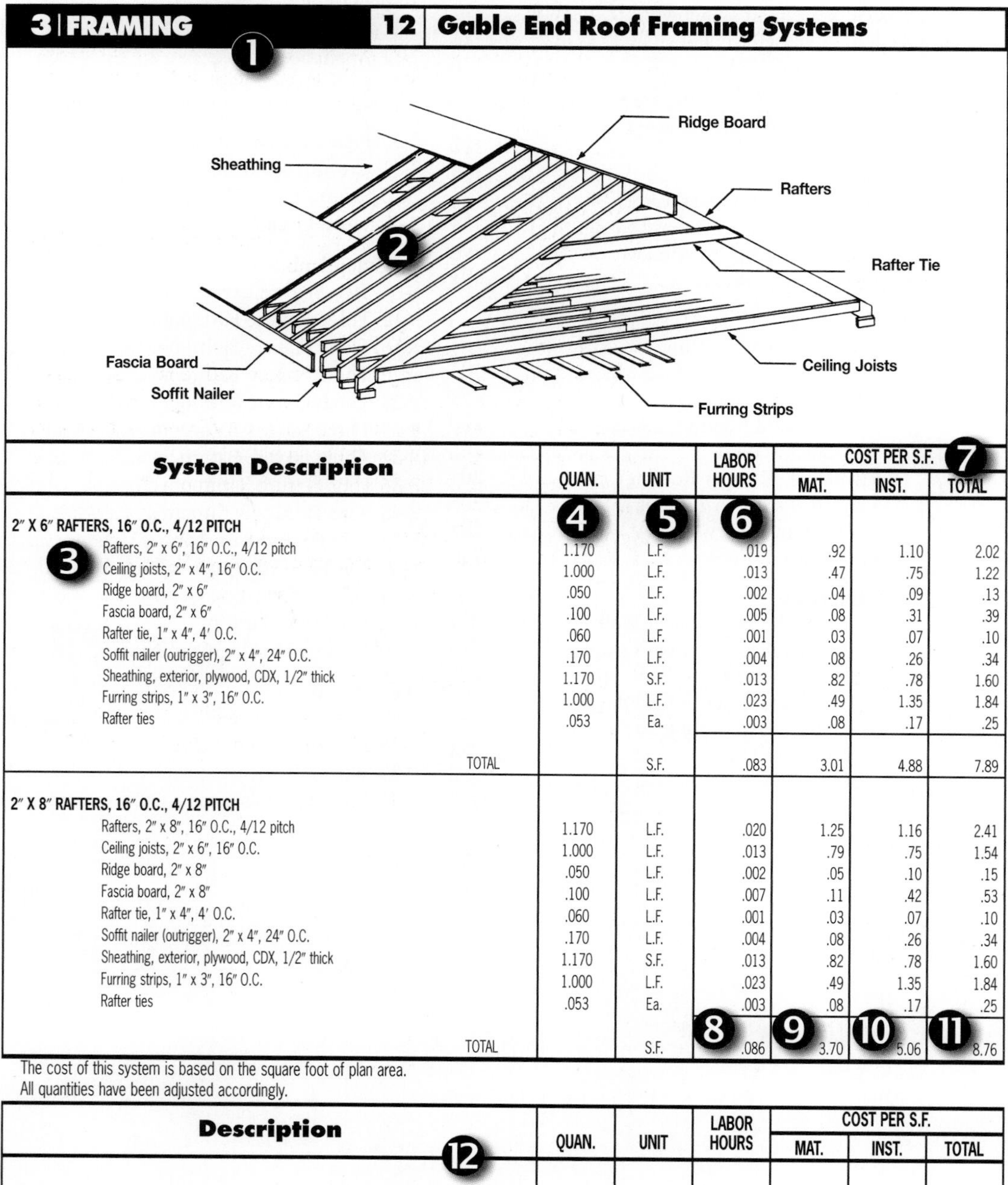

System Description	QUAN.	UNIT	LABOR HOURS	COST PER S.F. 7 MAT.	INST.	TOTAL
	4	5	6			
2″ X 6″ RAFTERS, 16″ O.C., 4/12 PITCH						
3 Rafters, 2″ x 6″, 16″ O.C., 4/12 pitch	1.170	L.F.	.019	.92	1.10	2.02
Ceiling joists, 2″ x 4″, 16″ O.C.	1.000	L.F.	.013	.47	.75	1.22
Ridge board, 2″ x 6″	.050	L.F.	.002	.04	.09	.13
Fascia board, 2″ x 6″	.100	L.F.	.005	.08	.31	.39
Rafter tie, 1″ x 4″, 4′ O.C.	.060	L.F.	.001	.03	.07	.10
Soffit nailer (outrigger), 2″ x 4″, 24″ O.C.	.170	L.F.	.004	.08	.26	.34
Sheathing, exterior, plywood, CDX, 1/2″ thick	1.170	S.F.	.013	.82	.78	1.60
Furring strips, 1″ x 3″, 16″ O.C.	1.000	L.F.	.023	.49	1.35	1.84
Rafter ties	.053	Ea.	.003	.08	.17	.25
TOTAL		S.F.	.083	3.01	4.88	7.89
2″ X 8″ RAFTERS, 16″ O.C., 4/12 PITCH						
Rafters, 2″ x 8″, 16″ O.C., 4/12 pitch	1.170	L.F.	.020	1.25	1.16	2.41
Ceiling joists, 2″ x 6″, 16″ O.C.	1.000	L.F.	.013	.79	.75	1.54
Ridge board, 2″ x 8″	.050	L.F.	.002	.05	.10	.15
Fascia board, 2″ x 8″	.100	L.F.	.007	.11	.42	.53
Rafter tie, 1″ x 4″, 4′ O.C.	.060	L.F.	.001	.03	.07	.10
Soffit nailer (outrigger), 2″ x 4″, 24″ O.C.	.170	L.F.	.004	.08	.26	.34
Sheathing, exterior, plywood, CDX, 1/2″ thick	1.170	S.F.	.013	.82	.78	1.60
Furring strips, 1″ x 3″, 16″ O.C.	1.000	L.F.	.023	.49	1.35	1.84
Rafter ties	.053	Ea.	.003	.08	.17	.25
TOTAL		S.F.	8 .086	9 3.70	10 5.06	11 8.76

The cost of this system is based on the square foot of plan area.
All quantities have been adjusted accordingly.

Description 12	QUAN.	UNIT	LABOR HOURS	COST PER S.F. MAT.	INST.	TOTAL

❶ System Identification (3 Framing 12)

Each Assemblies section has been assigned a unique identification number, component category, system number, and system description.

❷ Illustration

Included with most assemblies is an illustration with individual components labeled. Elements involved in the total system function are shown.

❸ System Description (2" x 6" Rafters, 16" O.C., 4/12 Pitch)

The components of a typical system are listed separately to show what has been included in the development of the total system price. Each assembly includes a brief outline of any special conditions to be used when pricing a system. Alternative components can also be found. Simply insert any chosen new element into the chart to develop a custom system.

❹ Quantities for Each Component

Each material in a system is shown with the quantity required for the system unit. For example, there are 1.170 L.F. of rafter per S.F. of plan area.

❺ Unit of Measure for Each Component

The abbreviated designation indicates the unit of measure, as defined by industry standards, upon which the individual component has been priced. In this example, items are priced by the linear foot (L.F.) or the square foot (S.F.).

❻ Labor-Hours

This is the amount of time it takes to install the quantity of the individual component.

❼ Unit of Measure (Cost per S.F.)

In the three right-hand columns, each cost figure is adjusted to agree with the unit of measure for the entire system. In this case, cost per S.F. is the common unit of measure.

❽ Labor-Hours (0.086)

The labor-hours column shows the amount of time necessary to install the system per the unit of measure. For example, it takes 0.086 labor-hours to install one square foot (plan area) of this roof framing system.

❾ Materials (3.70)

This column contains the material cost of each element. These cost figures include 10% for profit.

❿ Installation (5.06)

This column contains labor and equipment costs. Labor rates include bare cost and the installing contractor's overhead and profit. On the average, the labor cost will be 69.5% over the bare labor cost. Equipment costs include 10% for profit.

⓫ Totals (8.76)

The figure in this column is the sum of the material and installation costs.

⓬ Work Sheet

Using the selective price sheet, it is possible to create estimates with alternative items for any number of systems.

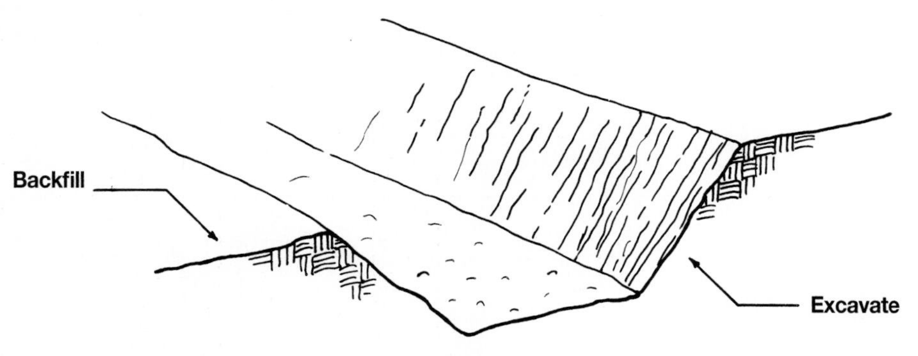

System Description	QUAN.	UNIT	LABOR HOURS	COST EACH		
				MAT.	INST.	TOTAL
BUILDING, 24′ X 38′, 4′ DEEP						
Cut & chip light trees to 6″ diam.	.190	Acre	9.120		793.25	793.25
Excavator, hydraulic, crawler mtd., 1 C.Y. cap. = 100 C.Y./hr.	174.000	C.Y.	3.480		367.14	367.14
Backfill, dozer, 4″ lifts, no compaction	87.000	C.Y.	.580		138.33	138.33
Rough grade, dozer, 30′ from building	87.000	C.Y.	.580		138.33	138.33
Mobilize and demobilize equipment	4.000	Ea.	12.000		948	948
TOTAL		Ea.	25.760		2,385.05	2,385.05
BUILDING, 26′ X 46′, 4′ DEEP						
Cut & chip light trees to 6″ diam.	.210	Acre	10.080		876.75	876.75
Excavator, hydraulic, crawler mtd., 1 C.Y. cap. = 100 C.Y./hr.	201.000	C.Y.	4.020		424.11	424.11
Backfill, dozer, 4″ lifts, no compaction	100.000	C.Y.	.667		159	159
Rough grade, dozer, 30′ from building	100.000	C.Y.	.667		159	159
Mobilize and demobilize equipment	4.000	Ea.	12.000		948	948
TOTAL		Ea.	27.434		2,566.86	2,566.86
BUILDING, 26′ X 60′, 4′ DEEP						
Cut & chip light trees to 6″ diam.	.240	Acre	11.520		1,002	1,002
Excavator, hydraulic, crawler mtd., 1 C.Y. cap. = 100 C.Y./hr.	240.000	C.Y.	4.800		506.40	506.40
Backfill, dozer, 4″ lifts, no compaction	120.000	C.Y.	.800		190.80	190.80
Rough grade, dozer, 30′ from building	120.000	C.Y.	.800		190.80	190.80
Mobilize and demobilize equipment	4.000	Ea.	12.000		948	948
TOTAL		Ea.	29.920		2,838	2,838
BUILDING, 30′ X 66′, 4′ DEEP						
Cut & chip light trees to 6″ diam.	.260	Acre	12.480		1,085.50	1,085.50
Excavator, hydraulic, crawler mtd., 1 C.Y. cap. = 100 C.Y./hr.	268.000	C.Y.	5.360		565.48	565.48
Backfill, dozer, 4″ lifts, no compaction	134.000	C.Y.	.894		213.06	213.06
Rough grade, dozer, 30′ from building	134.000	C.Y.	.894		213.06	213.06
Mobilize and demobilize equipment	4.000	Ea.	12.000		948	948
TOTAL		Ea.	31.628		3,025.10	3,025.10

The costs in this system are on a cost each basis.
Quantities are based on 1′-0″ clearance on each side of footing.

Description	QUAN.	UNIT	LABOR HOURS	COST EACH		
				MAT.	INST.	TOTAL

Footing Excavation Price Sheet	QUAN.	UNIT	LABOR HOURS	COST EACH		
				MAT.	INST.	TOTAL
Clear and grub, medium brush, 30′ from building, 24′ x 38′	.190	Acre	9.120		790	790
26′ x 46′	.210	Acre	10.080		880	880
26′ x 60′	.240	Acre	11.520		1,000	1,000
30′ x 66′	.260	Acre	12.480		1,075	1,075
Light trees, to 6″ dia. cut & chip, 24′ x 38′	.190	Acre	9.120		790	790
26′ x 46′	.210	Acre	10.080		880	880
26′ x 60′	.240	Acre	11.520		1,000	1,000
30′ x 66′	.260	Acre	12.480		1,075	1,075
Medium trees, to 10″ dia. cut & chip, 24′ x 38′	.190	Acre	13.029		1,125	1,125
26′ x 46′	.210	Acre	14.400		1,250	1,250
26′ x 60′	.240	Acre	16.457		1,425	1,425
30′ x 66′	.260	Acre	17.829		1,550	1,550
Excavation, footing, 24′ x 38′, 2′ deep	68.000	C.Y.	.906		144	144
4′ deep	174.000	C.Y.	2.319		365	365
8′ deep	384.000	C.Y.	5.119		810	810
26′ x 46′, 2′ deep	79.000	C.Y.	1.053		167	167
4′ deep	201.000	C.Y.	2.679		425	425
8′ deep	404.000	C.Y.	5.385		850	850
26′ x 60′, 2′ deep	94.000	C.Y.	1.253		199	199
4′ deep	240.000	C.Y.	3.199		505	505
8′ deep	483.000	C.Y.	6.438		1,025	1,025
30′ x 66′, 2′ deep	105.000	C.Y.	1.400		221	221
4′ deep	268.000	C.Y.	3.572		565	565
8′ deep	539.000	C.Y.	7.185		1,125	1,125
Backfill, 24′ x 38′, 2″ lifts, no compaction	34.000	C.Y.	.227		54	54
Compaction, air tamped, add	34.000	C.Y.	2.267		490	490
4″ lifts, no compaction	87.000	C.Y.	.580		139	139
Compaction, air tamped, add	87.000	C.Y.	5.800		1,250	1,250
8″ lifts, no compaction	192.000	C.Y.	1.281		305	305
Compaction, air tamped, add	192.000	C.Y.	12.801		2,750	2,750
26′ x 46′, 2″ lifts, no compaction	40.000	C.Y.	.267		63.50	63.50
Compaction, air tamped, add	40.000	C.Y.	2.667		575	575
4″ lifts, no compaction	100.000	C.Y.	.667		159	159
Compaction, air tamped, add	100.000	C.Y.	6.667		1,450	1,450
8″ lifts, no compaction	202.000	C.Y.	1.347		320	320
Compaction, air tamped, add	202.000	C.Y.	13.467		2,900	2,900
26′ x 60′, 2″ lifts, no compaction	47.000	C.Y.	.313		75	75
Compaction, air tamped, add	47.000	C.Y.	3.133		675	675
4″ lifts, no compaction	120.000	C.Y.	.800		191	191
Compaction, air tamped, add	120.000	C.Y.	8.000		1,725	1,725
8″ lifts, no compaction	242.000	C.Y.	1.614		385	385
Compaction, air tamped, add	242.000	C.Y.	16.134		3,475	3,475
30′ x 66′, 2″ lifts, no compaction	53.000	C.Y.	.354		84	84
Compaction, air tamped, add	53.000	C.Y.	3.534		760	760
4″ lifts, no compaction	134.000	C.Y.	.894		213	213
Compaction, air tamped, add	134.000	C.Y.	8.934		1,925	1,925
8″ lifts, no compaction	269.000	C.Y.	1.794		425	425
Compaction, air tamped, add	269.000	C.Y.	17.934		3,875	3,875
Rough grade, 30′ from building, 24′ x 38′	87.000	C.Y.	.580		139	139
26′ x 46′	100.000	C.Y.	.667		159	159
26′ x 60′	120.000	C.Y.	.800		191	191
30′ x 66′	134.000	C.Y.	.894		213	213
Mobilize and demobilize equipment	4.000	Ea.	12.000		950	950

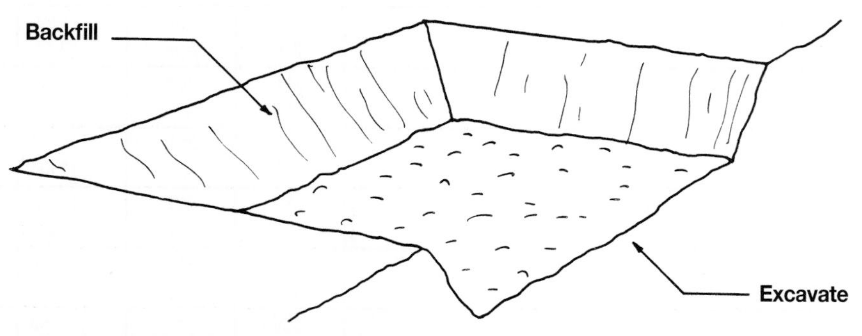

System Description	QUAN.	UNIT	LABOR HOURS	COST EACH MAT.	COST EACH INST.	COST EACH TOTAL
BUILDING, 24′ X 38′, 8′ DEEP						
Medium clearing	.190	Acre	2.027		287.85	287.85
Excavate, track loader, 1-1/2 C.Y. bucket	550.000	C.Y.	7.860		1,122	1,122
Backfill, dozer, 8″ lifts, no compaction	180.000	C.Y.	1.201		286.20	286.20
Rough grade, dozer, 30′ from building	280.000	C.Y.	1.868		445.20	445.20
Mobilize and demobilize equipment	4.000	Ea.	12.000		948	948
TOTAL		Ea.	24.956		3,089.25	3,089.25
BUILDING, 26′ X 46′, 8′ DEEP						
Medium clearing	.210	Acre	2.240		318.15	318.15
Excavate, track loader, 1-1/2 C.Y. bucket	672.000	C.Y.	9.603		1,370.88	1,370.88
Backfill, dozer, 8″ lifts, no compaction	220.000	C.Y.	1.467		349.80	349.80
Rough grade, dozer, 30′ from building	340.000	C.Y.	2.268		540.60	540.60
Mobilize and demobilize equipment	4.000	Ea.	12.000		948	948
TOTAL		Ea.	27.578		3,527.43	3,527.43
BUILDING, 26′ X 60′, 8′ DEEP						
Medium clearing	.240	Acre	2.560		363.60	363.60
Excavate, track loader, 1-1/2 C.Y. bucket	829.000	C.Y.	11.846		1,691.16	1,691.16
Backfill, dozer, 8″ lifts, no compaction	270.000	C.Y.	1.801		429.30	429.30
Rough grade, dozer, 30′ from building	420.000	C.Y.	2.801		667.80	667.80
Mobilize and demobilzie equipment	4.000	Ea.	12.000		948	948
TOTAL		Ea.	31.008		4,099.86	4,099.86
BUILDING, 30′ X 66′, 8′ DEEP						
Medium clearing	.260	Acre	2.773		393.90	393.90
Excavate, track loader, 1-1/2 C.Y. bucket	990.000	C.Y.	14.147		2,019.60	2,019.60
Backfill dozer, 8″ lifts, no compaction	320.000	C.Y.	2.134		508.80	508.80
Rough grade, dozer, 30′ from building	500.000	C.Y.	3.335		795	795
Mobilize and demobilize equipment	4.000	Ea.	12.000		948	948
TOTAL		Ea.	34.389		4,665.30	4,665.30

The costs in this system are on a cost each basis.
Quantities are based on 1′-0″ clearance beyond footing projection.

Description	QUAN.	UNIT	LABOR HOURS	COST EACH MAT.	COST EACH INST.	COST EACH TOTAL

Foundation Excavation Price Sheet	QUAN.	UNIT	LABOR HOURS	COST EACH		
				MAT.	INST.	TOTAL
Clear & grub, medium brush, 30′ from building, 24′ x 38′	.190	Acre	2.027		288	288
26′ x 46′	.210	Acre	2.240		320	320
26′ x 60′	.240	Acre	2.560		365	365
30′ x 66′	.260	Acre	2.773		395	395
Light trees, to 6″ dia. cut & chip, 24′ x 38′	.190	Acre	9.120		790	790
26′ x 46′	.210	Acre	10.080		880	880
26′ x 60′	.240	Acre	11.520		1,000	1,000
30′ x 66′	.260	Acre	12.480		1,075	1,075
Medium trees, to 10″ dia. cut & chip, 24′ x 38′	.190	Acre	13.029		1,125	1,125
26′ x 46′	.210	Acre	14.400		1,250	1,250
26′ x 60′	.240	Acre	16.457		1,425	1,425
30′ x 66′	.260	Acre	17.829		1,550	1,550
Excavation, basement, 24′ x 38′, 2′ deep	98.000	C.Y.	1.400		200	200
4′ deep	220.000	C.Y.	3.144		450	450
8′ deep	550.000	C.Y.	7.860		1,125	1,125
26′ x 46′, 2′ deep	123.000	C.Y.	1.758		251	251
4′ deep	274.000	C.Y.	3.915		560	560
8′ deep	672.000	C.Y.	9.603		1,375	1,375
26′ x 60′, 2′ deep	157.000	C.Y.	2.244		320	320
4′ deep	345.000	C.Y.	4.930		705	705
8′ deep	829.000	C.Y.	11.846		1,700	1,700
30′ x 66′, 2′ deep	192.000	C.Y.	2.744		390	390
4′ deep	419.000	C.Y.	5.988		855	855
8′ deep	990.000	C.Y.	14.147		2,025	2,025
Backfill, 24′ x 38′, 2″ lifts, no compaction	32.000	C.Y.	.213		51	51
Compaction, air tamped, add	32.000	C.Y.	2.133		460	460
4″ lifts, no compaction	72.000	C.Y.	.480		115	115
Compaction, air tamped, add	72.000	C.Y.	4.800		1,025	1,025
8″ lifts, no compaction	180.000	C.Y.	1.201		286	286
Compaction, air tamped, add	180.000	C.Y.	12.001		2,575	2,575
26′ x 46′, 2″ lifts, no compaction	40.000	C.Y.	.267		63.50	63.50
Compaction, air tamped, add	40.000	C.Y.	2.667		575	575
4″ lifts, no compaction	90.000	C.Y.	.600		143	143
Compaction, air tamped, add	90.000	C.Y.	6.000		1,300	1,300
8″ lifts, no compaction	220.000	C.Y.	1.467		350	350
Compacton, air tamped, add	220.000	C.Y.	14.667		3,175	3,175
26′ x 60′, 2″ lifts, no compaction	50.000	C.Y.	.334		79.50	79.50
Compaction, air tamped, add	50.000	C.Y.	3.334		720	720
4″ lifts, no compaction	110.000	C.Y.	.734		175	175
Compaction, air tamped, add	110.000	C.Y.	7.334		1,575	1,575
8″ lifts, no compaction	270.000	C.Y.	1.801		430	430
Compaction, air tamped, add	270.000	C.Y.	18.001		3,875	3,875
30′ x 66′, 2″ lifts, no compaction	60.000	C.Y.	.400		95.50	95.50
Compaction, air tamped, add	60.000	C.Y.	4.000		860	860
4″ lifts, no compaction	130.000	C.Y.	.867		207	207
Compaction, air tamped, add	130.000	C.Y.	8.667		1,875	1,875
8″ lifts, no compaction	320.000	C.Y.	2.134		510	510
Compaction, air tamped, add	320.000	C.Y.	21.334		4,600	4,600
Rough grade, 30′ from building, 24′ x 38′	280.000	C.Y.	1.868		445	445
26′ x 46′	340.000	C.Y.	2.268		540	540
26′ x 60′	420.000	C.Y.	2.801		665	665
30′ x 66′	500.000	C.Y.	3.335		795	795
Mobilize and demobilize equipment	4.000	Ea.	12.000		950	950

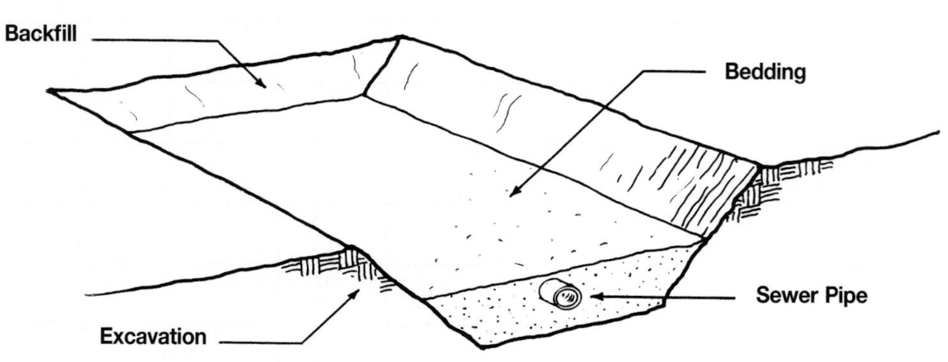

System Description	QUAN.	UNIT	LABOR HOURS	COST PER L.F.		
				MAT.	INST.	TOTAL
2′ DEEP						
Excavation, backhoe	.296	C.Y.	.032		2.39	2.39
Alternate pricing method, 4″ deep	.111	C.Y.	.044	3.72	2.30	6.02
Utility, sewer, 6″ cast iron	1.000	L.F.	.283	39.56	16.25	55.81
Compaction in 12″ layers, hand tamp, add to above	.185	C.Y.	.044		1.98	1.98
TOTAL		L.F.	.403	43.28	22.92	66.20
4′ DEEP						
Excavation, backhoe	.889	C.Y.	.095		7.16	7.16
Alternate pricing method, 4″ deep	.111	C.Y.	.044	3.72	2.30	6.02
Utility, sewer, 6″ cast iron	1.000	L.F.	.283	39.56	16.25	55.81
Compaction in 12″ layers, hand tamp, add to above	.778	C.Y.	.183		8.32	8.32
TOTAL		L.F.	.605	43.28	34.03	77.31
6′ DEEP						
Excavation, backhoe	1.770	C.Y.	.189		14.25	14.25
Alternate pricing method, 4″ deep	.111	C.Y.	.044	3.72	2.30	6.02
Utility, sewer, 6″ cast iron	1.000	L.F.	.283	39.56	16.25	55.81
Compaction in 12″ layers, hand tamp, add to above	1.660	C.Y.	.391		17.76	17.76
TOTAL		L.F.	.907	43.28	50.56	93.84
8′ DEEP						
Excavation, backhoe	2.960	C.Y.	.316		23.83	23.83
Alternate pricing method, 4″ deep	.111	C.Y.	.044	3.72	2.30	6.02
Utility, sewer, 6″ cast iron	1.000	L.F.	.283	39.56	16.25	55.81
Compaction in 12″ layers, hand tamp, add to above	2.850	C.Y.	.671		30.50	30.50
TOTAL		L.F.	1.314	43.28	72.88	116.16

The costs in this system are based on a cost per linear foot of trench, and based on 2′ wide at bottom of trench up to 6′ deep.

Description	QUAN.	UNIT	LABOR HOURS	COST PER L.F.		
				MAT.	INST.	TOTAL

Utility Trenching Price Sheet	QUAN.	UNIT	LABOR HOURS	COST PER UNIT		
				MAT.	INST.	TOTAL
Excavation, bottom of trench 2′ wide, 2′ deep	.296	C.Y.	.032		2.39	2.39
4′ deep	.889	C.Y.	.095		7.15	7.15
6′ deep	1.770	C.Y.	.142		11.35	11.35
8′ deep	2.960	C.Y.	.105		21	21
Bedding, sand, bottom of trench 2′ wide, no compaction, pipe, 2″ diameter	.070	C.Y.	.028	2.35	1.45	3.80
4″ diameter	.084	C.Y.	.034	2.81	1.74	4.55
6″ diameter	.105	C.Y.	.042	3.52	2.17	5.69
8″ diameter	.122	C.Y.	.049	4.09	2.52	6.61
Compacted, pipe, 2″ diameter	.074	C.Y.	.030	2.48	1.52	4
4″ diameter	.092	C.Y.	.037	3.08	1.90	4.98
6″ diameter	.111	C.Y.	.044	3.72	2.30	6.02
8″ diameter	.129	C.Y.	.052	4.32	2.66	6.98
3/4″ stone, bottom of trench 2′ wide, pipe, 4″ diameter	.082	C.Y.	.033	2.75	1.69	4.44
6″ diameter	.099	C.Y.	.040	3.32	2.05	5.37
3/8″ stone, bottom of trench 2′ wide, pipe, 4″ diameter	.084	C.Y.	.034	2.81	1.74	4.55
6″ diameter	.102	C.Y.	.041	3.42	2.10	5.52
Utilities, drainage & sewerage, corrugated plastic, 6″ diameter	1.000	L.F.	.069	4.04	3.19	7.23
8″ diameter	1.000	L.F.	.072	7.25	3.33	10.58
Concrete, non-reinforced, 6″ diameter	1.000	L.F.	.181	8.45	10.10	18.55
8″ diameter	1.000	L.F.	.214	9.25	11.85	21.10
PVC, SDR 35, 4″ diameter	1.000	L.F.	.064	1.84	2.98	4.82
6″ diameter	1.000	L.F.	.069	4.04	3.19	7.23
8″ diameter	1.000	L.F.	.072	7.25	3.33	10.58
Gas & service, polyethylene, 1-1/4″ diameter	1.000	L.F.	.059	1.80	3.14	4.94
Steel sched.40, 1″ diameter	1.000	L.F.	.107	5.70	6.90	12.60
2″ diameter	1.000	L.F.	.114	8.95	7.40	16.35
Sub-drainage, PVC, perforated, 3″ diameter	1.000	L.F.	.064	1.84	2.98	4.82
4″ diameter	1.000	L.F.	.064	1.84	2.98	4.82
5″ diameter	1.000	L.F.	.069	4.04	3.19	7.23
6″ diameter	1.000	L.F.	.069	4.04	3.19	7.23
Water service, copper, type K, 3/4″	1.000	L.F.	.083	7.25	5.50	12.75
1″ diameter	1.000	L.F.	.093	10.75	6.20	16.95
PVC, 3/4″	1.000	L.F.	.121	4.47	8	12.47
1″ diameter	1.000	L.F.	.134	7.65	8.90	16.55
Backfill, bottom of trench 2′ wide no compact, 2′ deep, pipe, 2″ diameter	.226	L.F.	.053		2.42	2.42
4″ diameter	.212	L.F.	.050		2.27	2.27
6″ diameter	.185	L.F.	.044		1.98	1.98
4′ deep, pipe, 2″ diameter	.819	C.Y.	.193		8.75	8.75
4″ diameter	.805	C.Y.	.189		8.60	8.60
6″ diameter	.778	C.Y.	.183		8.30	8.30
6′ deep, pipe, 2″ diameter	1.700	C.Y.	.400		18.20	18.20
4″ diameter	1.690	C.Y.	.398		18.10	18.10
6″ diameter	1.660	C.Y.	.391		17.75	17.75
8′ deep, pipe, 2″ diameter	2.890	C.Y.	.680		31	31
4″ diameter	2.870	C.Y.	.675		30.50	30.50
6″ diameter	2.850	C.Y.	.671		30.50	30.50

16 Sidewalk Systems

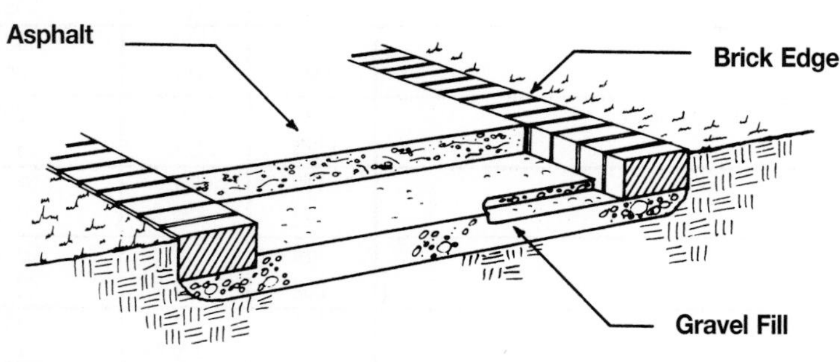

System Description	QUAN.	UNIT	LABOR HOURS	COST PER S.F.		
				MAT.	INST.	TOTAL
ASPHALT SIDEWALK SYSTEM, 3′ WIDE WALK						
Gravel fill, 4″ deep	1.000	S.F.	.001	.29	.05	.34
Compact fill	.012	C.Y.			.02	.02
Handgrade	1.000	S.F.	.004		.19	.19
Walking surface, bituminous paving, 2″ thick	1.000	S.F.	.007	.83	.39	1.22
Edging, brick, laid on edge	.670	L.F.	.079	2.09	4.29	6.38
TOTAL		S.F.	.091	3.21	4.94	8.15
CONCRETE SIDEWALK SYSTEM, 3′ WIDE WALK						
Gravel fill, 4″ deep	1.000	S.F.	.001	.29	.05	.34
Compact fill	.012	C.Y.			.02	.02
Handgrade	1.000	S.F.	.004		.19	.19
Walking surface, concrete, 4″ thick	1.000	S.F.	.040	2.44	2.16	4.60
Edging, brick, laid on edge	.670	L.F.	.079	2.09	4.29	6.38
TOTAL		S.F.	.124	4.82	6.71	11.53
PAVERS, BRICK SIDEWALK SYSTEM, 3′ WIDE WALK						
Sand base fill, 4″ deep	1.000	S.F.	.001	.48	.08	.56
Compact fill	.012	C.Y.			.02	.02
Handgrade	1.000	S.F.	.004		.19	.19
Walking surface, brick pavers	1.000	S.F.	.160	3.20	8.65	11.85
Edging, redwood, untreated, 1″ x 4″	.670	L.F.	.032	1.68	1.90	3.58
TOTAL		S.F.	.197	5.36	10.84	16.20

The costs in this system are based on a cost per square foot of sidewalk area. Concrete used is 3000 p.s.i.

Description	QUAN.	UNIT	LABOR HOURS	COST PER S.F.		
				MAT.	INST.	TOTAL

Sidewalk Price Sheet	QUAN.	UNIT	LABOR HOURS	COST PER S.F.		
				MAT.	INST.	TOTAL
Base, crushed stone, 3″ deep	1.000	S.F.	.001	.33	.11	.44
6″ deep	1.000	S.F.	.001	.65	.11	.76
9″ deep	1.000	S.F.	.002	.95	.15	1.10
12″ deep	1.000	S.F.	.002	2.64	.18	2.82
Bank run gravel, 6″ deep	1.000	S.F.	.001	.44	.07	.51
9″ deep	1.000	S.F.	.001	.65	.11	.76
12″ deep	1.000	S.F.	.001	.89	.13	1.02
Compact base, 3″ deep	.009	C.Y.	.001		.02	.02
6″ deep	.019	C.Y.	.001		.03	.03
9″ deep	.028	C.Y.	.001		.05	.05
Handgrade	1.000	S.F.	.004		.19	.19
Surface, brick, pavers dry joints, laid flat, running bond	1.000	S.F.	.160	3.20	8.65	11.85
Basket weave	1.000	S.F.	.168	4.33	9.10	13.43
Herringbone	1.000	S.F.	.174	4.33	9.40	13.73
Laid on edge, running bond	1.000	S.F.	.229	4.88	12.35	17.23
Mortar jts. laid flat, running bond	1.000	S.F.	.192	3.84	10.40	14.24
Basket weave	1.000	S.F.	.202	5.20	10.90	16.10
Herringbone	1.000	S.F.	.209	5.20	11.30	16.50
Laid on edge, running bond	1.000	S.F.	.274	5.85	14.80	20.65
Bituminous paving, 1-1/2″ thick	1.000	S.F.	.006	.62	.29	.91
2″ thick	1.000	S.F.	.007	.83	.39	1.22
2-1/2″ thick	1.000	S.F.	.008	1.05	.42	1.47
Sand finish, 3/4″ thick	1.000	S.F.	.001	.34	.12	.46
1″ thick	1.000	S.F.	.001	.43	.14	.57
Concrete, reinforced, broom finish, 4″ thick	1.000	S.F.	.040	2.44	2.16	4.60
5″ thick	1.000	S.F.	.044	3.02	2.38	5.40
6″ thick	1.000	S.F.	.047	3.52	2.54	6.06
Crushed stone, white marble, 3″ thick	1.000	S.F.	.009	.54	.43	.97
Bluestone, 3″ thick	1.000	S.F.	.009	.22	.43	.65
Flagging, bluestone, 1″	1.000	S.F.	.198	11.55	10.65	22.20
1-1/2″	1.000	S.F.	.188	21	10.15	31.15
Slate, natural cleft, 3/4″	1.000	S.F.	.174	10.60	9.40	20
Random rect., 1/2″	1.000	S.F.	.152	23	8.25	31.25
Granite blocks	1.000	S.F.	.174	22.50	9.40	31.90
Edging, corrugated aluminum, 4″, 3′ wide walk	.666	L.F.	.008	1.62	.49	2.11
4′ wide walk	.500	L.F.	.006	1.22	.37	1.59
6″, 3′ wide walk	.666	L.F.	.010	2.02	.57	2.59
4′ wide walk	.500	L.F.	.007	1.52	.43	1.95
Redwood-cedar-cypress, 1″ x 4″, 3′ wide walk	.666	L.F.	.021	.87	1.25	2.12
4′ wide walk	.500	L.F.	.016	.65	.94	1.59
2″ x 4″, 3′ wide walk	.666	L.F.	.032	1.68	1.90	3.58
4′ wide walk	.500	L.F.	.024	1.27	1.43	2.70
Brick, dry joints, 3′ wide walk	.666	L.F.	.079	2.09	4.29	6.38
4′ wide walk	.500	L.F.	.059	1.56	3.20	4.76
Mortar joints, 3′ wide walk	.666	L.F.	.095	2.51	5.15	7.66
4′ wide walk	.500	L.F.	.071	1.87	3.84	5.71

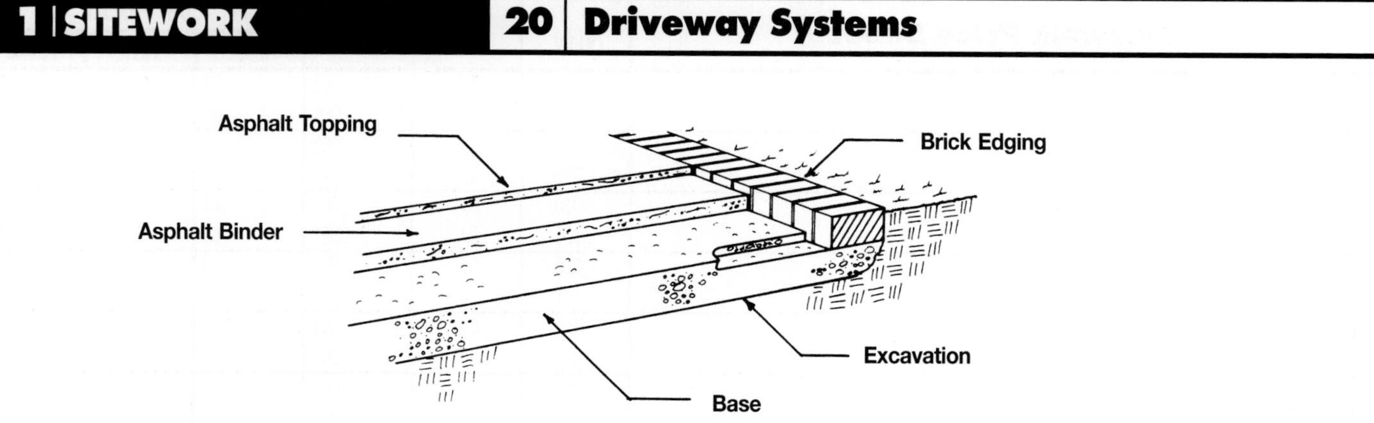

System Description	QUAN.	UNIT	LABOR HOURS	COST PER S.F. MAT.	COST PER S.F. INST.	COST PER S.F. TOTAL
ASPHALT DRIVEWAY TO 10′ WIDE						
Excavation, driveway to 10′ wide, 6″ deep	.019	C.Y.			.05	.05
Base, 6″ crushed stone	1.000	S.F.	.001	.65	.11	.76
Handgrade base	1.000	S.F.	.004		.19	.19
2″ thick base	1.000	S.F.	.002	.83	.20	1.03
1″ topping	1.000	S.F.	.001	.43	.14	.57
Edging, brick pavers	.200	L.F.	.024	.62	1.28	1.90
TOTAL		S.F.	.032	2.53	1.97	4.50
CONCRETE DRIVEWAY TO 10′ WIDE						
Excavation, driveway to 10′ wide, 6″ deep	.019	C.Y.			.05	.05
Base, 6″ crushed stone	1.000	S.F.	.001	.65	.11	.76
Handgrade base	1.000	S.F.	.004		.19	.19
Surface, concrete, 4″ thick	1.000	S.F.	.040	2.44	2.16	4.60
Edging, brick pavers	.200	L.F.	.024	.62	1.28	1.90
TOTAL		S.F.	.069	3.71	3.79	7.50
PAVERS, BRICK DRIVEWAY TO 10′ WIDE						
Excavation, driveway to 10′ wide, 6″ deep	.019	C.Y.			.05	.05
Base, 6″ sand	1.000	S.F.	.001	.76	.13	.89
Handgrade base	1.000	S.F.	.004		.19	.19
Surface, pavers, brick laid flat, running bond	1.000	S.F.	.160	3.20	8.65	11.85
Edging, redwood, untreated, 2″ x 4″	.200	L.F.	.010	.51	.57	1.08
TOTAL		S.F.	.175	4.47	9.59	14.06

Description	QUAN.	UNIT	LABOR HOURS	COST PER S.F. MAT.	COST PER S.F. INST.	COST PER S.F. TOTAL

Driveway Price Sheet	QUAN.	UNIT	LABOR HOURS	COST PER S.F.		
				MAT.	INST.	TOTAL
Excavation, by machine, 10′ wide, 6″ deep	.019	C.Y.	.001		.05	.05
12″ deep	.037	C.Y.	.001		.08	.08
18″ deep	.055	C.Y.	.001		.13	.13
20′ wide, 6″ deep	.019	C.Y.	.001		.05	.05
12″ deep	.037	C.Y.	.001		.08	.08
18″ deep	.055	C.Y.	.001		.13	.13
Base, crushed stone, 10′ wide, 3″ deep	1.000	S.F.	.001	.33	.06	.39
6″ deep	1.000	S.F.	.001	.65	.11	.76
9″ deep	1.000	S.F.	.002	.95	.15	1.10
20′ wide, 3″ deep	1.000	S.F.	.001	.33	.06	.39
6″ deep	1.000	S.F.	.001	.65	.11	.76
9″ deep	1.000	S.F.	.002	.95	.15	1.10
Bank run gravel, 10′ wide, 3″ deep	1.000	S.F.	.001	.22	.04	.26
6″ deep	1.000	S.F.	.001	.44	.07	.51
9″ deep	1.000	S.F.	.001	.65	.11	.76
20′ wide, 3″ deep	1.000	S.F.	.001	.22	.04	.26
6″ deep	1.000	S.F.	.001	.44	.07	.51
9″ deep	1.000	S.F.	.001	.65	.11	.76
Handgrade, 10′ wide	1.000	S.F.	.004		.19	.19
20′ wide	1.000	S.F.	.004		.19	.19
Surface, asphalt, 10′ wide, 3/4″ topping, 1″ base	1.000	S.F.	.002	.98	.26	1.24
2″ base	1.000	S.F.	.003	1.17	.32	1.49
1″ topping, 1″ base	1.000	S.F.	.002	1.07	.28	1.35
2″ base	1.000	S.F.	.003	1.26	.34	1.60
20′ wide, 3/4″ topping, 1″ base	1.000	S.F.	.002	.98	.26	1.24
2″ base	1.000	S.F.	.003	1.17	.32	1.49
1″ topping, 1″ base	1.000	S.F.	.002	1.07	.28	1.35
2″ base	1.000	S.F.	.003	1.26	.34	1.60
Concrete, 10′ wide, 4″ thick	1.000	S.F.	.040	2.44	2.16	4.60
6″ thick	1.000	S.F.	.047	3.52	2.54	6.06
20′ wide, 4″ thick	1.000	S.F.	.040	2.44	2.16	4.60
6″ thick	1.000	S.F.	.047	3.52	2.54	6.06
Paver, brick 10′ wide dry joints, running bond, laid flat	1.000	S.F.	.160	3.20	8.65	11.85
Laid on edge	1.000	S.F.	.229	4.88	12.35	17.23
Mortar joints, laid flat	1.000	S.F.	.192	3.84	10.40	14.24
Laid on edge	1.000	S.F.	.274	5.85	14.80	20.65
20′ wide, running bond, dry jts., laid flat	1.000	S.F.	.160	3.20	8.65	11.85
Laid on edge	1.000	S.F.	.229	4.88	12.35	17.23
Mortar joints, laid flat	1.000	S.F.	.192	3.84	10.40	14.24
Laid on edge	1.000	S.F.	.274	5.85	14.80	20.65
Crushed stone, 10′ wide, white marble, 3″	1.000	S.F.	.009	.54	.43	.97
Bluestone, 3″	1.000	S.F.	.009	.22	.43	.65
20′ wide, white marble, 3″	1.000	S.F.	.009	.54	.43	.97
Bluestone, 3″	1.000	S.F.	.009	.22	.43	.65
Soil cement, 10′ wide	1.000	S.F.	.007	.41	1.01	1.42
20′ wide	1.000	S.F.	.007	.41	1.01	1.42
Granite blocks, 10′ wide	1.000	S.F.	.174	22.50	9.40	31.90
20′ wide	1.000	S.F.	.174	22.50	9.40	31.90
Asphalt block, solid 1-1/4″ thick	1.000	S.F.	.119	10.35	6.40	16.75
Solid 3″ thick	1.000	S.F.	.123	14.45	6.65	21.10
Edging, brick, 10′ wide	.200	L.F.	.024	.62	1.28	1.90
20′ wide	.100	L.F.	.012	.31	.64	.95
Redwood, untreated 2″ x 4″, 10′ wide	.200	L.F.	.010	.51	.57	1.08
20′ wide	.100	L.F.	.005	.25	.29	.54
Granite, 4 1/2″ x 12″ straight, 10′ wide	.200	L.F.	.032	1.65	2.08	3.73
20′ wide	.100	L.F.	.016	.83	1.05	1.88
Finishes, asphalt sealer, 10′ wide	1.000	S.F.	.023	.94	1.06	2
20′ wide	1.000	S.F.	.023	.94	1.06	2
Concrete, exposed aggregate 10′ wide	1.000	S.F.	.013	.22	.74	.96
20′ wide	1.000	S.F.	.013	.22	.74	.96

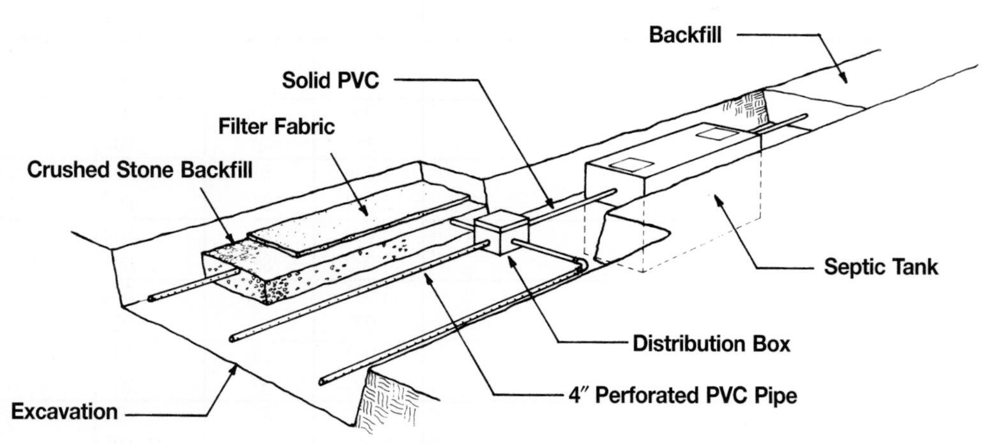

System Description	QUAN.	UNIT	LABOR HOURS	COST EACH		
				MAT.	INST.	TOTAL
SEPTIC SYSTEM WITH 600 S.F. LEACHING FIELD, 1000 GALLON TANK						
Mobilization	2.000	Ea.	16.000		1,380	1,380
Tank, precast, 1000 gallon	1.000	Ea.	3.500	1,150	188.85	1,338.85
Effluent filter	1.000	Ea.	1.000	46	60.50	106.50
Distribution box, precast	1.000	Ea.	1.000	104	45.50	149.50
Flow leveler	3.000	Ea.	.480	8.16	21.75	29.91
PVC pipe, 4″ diameter	25.000	L.F.	1.600	46	74.50	120.50
Tee	1.000	Ea.	1.000	21.50	60.50	82
Elbow	2.000	Ea.	1.333	24.50	81	105.50
Viewport cap	1.000	Ea.	.333	9.15	20	29.15
Filter fabric	67.000	S.Y.	.447	115.91	20.10	136.01
Detectable marking tape	1.600	C.L.F.	.085	15.84	3.87	19.71
PVC perforated pipe, 4″ diameter	135.000	L.F.	8.640	248.40	402.30	650.70
Excavation	160.000	C.Y.	17.654		1,776	1,776
Backfill	133.000	L.C.Y.	2.660		272.65	272.65
Spoil	55.000	L.C.Y.	3.056		308	308
Compaction	113.000	E.C.Y.	15.066		748.06	748.06
Stone fill	39.000	C.Y.	6.240	1,638	401.70	2,039.70
TOTAL		Ea.	80.094	3,427.46	5,865.28	9,292.74
SEPTIC SYSTEM WITH 750 S.F. LEACHING FIELD, 1500 GALLON TANK						
Mobilization	2.000	Ea.	16.000		1,380	1,380
Tank, precast, 1500 gallon	1.000	Ea.	4.000	1,725	215.50	1,940.50
Effluent filter	1.000	Ea.	1.000	46	60.50	106.50
Distribution box, precast	1.000	Ea.	1.000	104	45.50	149.50
Flow leveler	3.000	Ea.	.480	8.16	21.75	29.91
PVC pipe, 4″ diameter	25.000	L.F.	1.600	46	74.50	120.50
Tee	125.000	Ea.	1.000	21.50	60.50	82
Elbow	2.000	Ea.	1.333	24.50	81	105.50
Viewport cap	1.000	Ea.	.333	9.15	20	29.15
Filter fabric	84.000	S.Y.	.560	145.32	25.20	170.52
Detectable marking tape	1.900	C.L.F.	.101	18.81	4.60	23.41
PVC perforated pipe, 4″ diameter	165.000	L.F.	10.560	303.60	491.70	795.30
Excavation	199.000	C.Y.	21.958		2,208.90	2,208.90
Backfill	162.000	L.C.Y.	3.240		332.10	332.10
Spoil	73.500	L.C.Y.	4.084		411.61	411.61
Compaction	137.000	E.C.Y.	18.266		906.94	906.94
Stone fill	48.500	C.Y.	7.760	2,037	499.56	2,536.56
TOTAL		Ea.	93.275	4,489.04	6,839.86	11,328.90

The costs in this system include all necessary piping and excavation.

System Description	QUAN.	UNIT	LABOR HOURS	COST EACH		
				MAT.	INST.	TOTAL
SEPTIC SYSTEM WITH 1000 S.F. LEACHING FIELD, 1500 GALLON TANK						
Mobilization	2.000	Ea.	16.000		1,380	1,380
Tank, precast, 1500 gallon	1.000	Ea.	4.000	1,725	215.50	1,940.50
Effluent filter	1.000	Ea.	1.000	46	60.50	106.50
Distribution box, precast	1.000	Ea.	1.000	104	45.50	149.50
Flow leveler	4.000	Ea.	.640	10.88	29	39.88
PVC pipe, 4″ diameter	25.000	L.F.	1.600	46	74.50	120.50
Tee	1.000	Ea.	1.000	21.50	60.50	82
Elbow	4.000	Ea.	2.667	49	162	211
Viewport cap	1.000	Ea.	.333	9.15	20	29.15
Filter fabric	111.000	S.Y.	.740	192.03	33.30	225.33
Detectable marking tape	2.400	C.L.F.	.128	23.76	5.81	29.57
PVC perforated pipe, 4″ diameter	215.000	L.F.	13.760	395.60	640.70	1,036.30
Excavation	229.000	C.Y.	25.268		2,541.90	2,541.90
Backfill	178.000	L.C.Y.	3.560		364.90	364.90
Spoil	91.500	L.C.Y.	5.084		512.41	512.41
Compaction	151.000	E.C.Y.	20.133		999.62	999.62
Stone fill	64.000	C.Y.	10.240	2,688	659.20	3,347.20
TOTAL		Ea.	107.153	5,310.92	7,805.34	13,116.26

The costs in this system include all necessary piping and excavation.

Septic Systems Price Sheet	QUAN.	UNIT	LABOR HOURS	COST EACH		
				MAT.	INST.	TOTAL
Tank, precast concrete, 1000 gallon	1.000	Ea.	3.500	1,150	189	1,339
1500 gallon	1.000	Ea.	4.000	1,725	216	1,941
Distribution box, concrete, 5 outlets	1.000	Ea.	1.000	104	45.50	149.50
12 outlets	1.000	Ea.	2.000	610	91	701
4″ pipe, PVC, solid	25.000	L.F.	1.600	46	74.50	120.50
Tank and field excavation, 600 S.F. field	160.000	C.Y.	17.654		1,775	1,775
750 S.F. field	199.000	C.Y.	21.958		2,200	2,200
1000 S.F. field	229.000	C.Y.	25.268		2,550	2,550
Tank excavation only, 1000 gallon tank	20.000	C.Y.	2.206		222	222
1500 gallon tank	26.000	C.Y.	2.869		289	289
Backfill, crushed stone, 600 S.F. field	39.000	C.Y.	12.160	1,650	400	2,050
750 S.F. field	48.500	C.Y.	22.400	2,025	500	2,525
1000 S.F. field	64.000	C.Y.	.240	2,700	660	3,360
Backfill with excavated material, 600 S.F. field	133.000	L.C.Y.	.400		273	273
750 S.F. field	162.000	L.C.Y.	.367		330	330
1000 S.F. field	178.000	L.C.Y.	.280		365	365
Filter fabric, 600 S.F. field	67.000	S.Y.	2.376	116	20	136
750 S.F. field	84.000	S.Y.	4.860	145	25	170
1000 S.F. field	111.000	S.Y.	.740	192	33.50	225.50
4″ pipe, PVC, perforated, 600 S.F. field	135.000	L.F.	9.280	248	400	648
750 S.F. field	165.000	L.F.	16.960	305	490	795
1000 S.F. field	215.000	L.F.	1.939	395	640	1,035
Pipe fittings, PVC, 600 S.F. field	2.000	Ea.	3.879	24.50	81	105.50
750 S.F. field	2.000	Ea.		24.50	81	105.50
1000 S.F. field	4.000	Ea.		49	162	211
Mobilization	2.000	Ea.	16.000		1,375	1,375
Effluent filter	1.000	Ea.	1.000	46	60.50	106.50
Flow leveler, 600 S.F. field	3.000	Ea.	.480	8.15	22	30.15
750 S.F. field	3.000	Ea.	.480	8.15	22	30.15
1000 S.F. field	4.000	Ea.	.640	10.90	29	39.90
Viewport cap	1.000	Ea.	.333	9.15	20	29.15
Detectable marking tape, 600 S.F. field	1.600	C.L.F.	.085	15.85	3.87	19.72
750 S.F. field	1.900	C.L.F.	.101	18.80	4.60	23.40
1000 S.F. field	2.400	C.L.F.	.128	24	5.80	29.80

System Description	QUAN.	UNIT	LABOR HOURS	COST PER UNIT		
				MAT.	INST.	TOTAL
Chain link fence						
Galv.9ga. wire, 1-5/8"post 10'O.C., 1-3/8"top rail, 2"corner post, 3'hi	1.000	L.F.	.130	10.95	6.05	17
4' high	1.000	L.F.	.141	11.65	6.55	18.20
6' high	1.000	L.F.	.209	13.50	9.70	23.20
Add for gate 3' wide 1-3/8" frame 3' high	1.000	Ea.	2.000	107	93	200
4' high	1.000	Ea.	2.400	133	112	245
6' high	1.000	Ea.	2.400	158	112	270
Add for gate 4' wide 1-3/8" frame 3' high	1.000	Ea.	2.667	129	124	253
4' high	1.000	Ea.	2.667	140	124	264
6' high	1.000	Ea.	3.000	165	140	305
Alum.9ga. wire, 1-5/8"post, 10'O.C., 1-3/8"top rail, 2"corner post,3'hi	1.000	L.F.	.130	9.95	6.05	16
4' high	1.000	L.F.	.141	10.50	6.55	17.05
6' high	1.000	L.F.	.209	13.90	9.70	23.60
Add for gate 3' wide 1-3/8" frame 3' high	1.000	Ea.	2.000	170	93	263
4' high	1.000	Ea.	2.400	185	112	297
6' high	1.000	Ea.	2.400	205	112	317
Add for gate 4' wide 1-3/8" frame 3' high	1.000	Ea.	2.400	179	112	291
4' high	1.000	Ea.	2.667	166	124	290
6' high	1.000	Ea.	3.000	221	140	361
Vinyl 9ga. wire, 1-5/8"post 10'O.C., 1-3/8"top rail, 2"corner post,3'hi	1.000	L.F.	.130	9.55	6.05	15.60
4' high	1.000	L.F.	.141	8.10	6.55	14.65
6' high	1.000	L.F.	.209	10.90	9.70	20.60
Add for gate 3' wide 1-3/8" frame 3' high	1.000	Ea.	2.000	121	93	214
4' high	1.000	Ea.	2.400	142	112	254
6' high	1.000	Ea.	2.400	170	112	282
Add for gate 4' wide 1-3/8" frame 3' high	1.000	Ea.	2.400	132	112	244
4' high	1.000	Ea.	2.667	137	124	261
6' high	1.000	Ea.	3.000	189	140	329
Tennis court, chain link fence, 10' high						
Galv.11ga.wire, 2"post 10'O.C., 1-3/8"top rail, 2-1/2"corner post	1.000	L.F.	.253	9.75	11.75	21.50
Add for gate 3' wide 1-3/8" frame	1.000	Ea.	2.400	254	112	366
Alum.11ga.wire, 2"post 10'O.C., 1-3/8"top rail, 2-1/2"corner post	1.000	L.F.	.253	12.85	11.75	24.60
Add for gate 3' wide 1-3/8" frame	1.000	Ea.	2.400	195	112	307
Vinyl 11ga.wire,2"post 10' O.C.,1-3/8"top rail,2-1/2"corner post	1.000	L.F.	.253	11.20	11.75	22.95
Add for gate 3' wide 1-3/8" frame	1.000	Ea.	2.400	370	112	482
Railings, commercial						
Aluminum balcony rail, 1-1/2" posts with pickets	1.000	L.F.	.164	87	12	99
With expanded metal panels	1.000	L.F.	.164	113	12	125
With porcelain enamel panel inserts	1.000	L.F.	.164	179	12	191
Mild steel, ornamental rounded top rail	1.000	L.F.	.164	163	12	175
As above, but pitch down stairs	1.000	L.F.	.183	212	13.35	225.35
Steel pipe, welded, 1-1/2" round, painted	1.000	L.F.	.160	35	11.70	46.70
Galvanized	1.000	L.F.	.160	50	11.70	61.70
Residential, stock units, mild steel, deluxe	1.000	L.F.	.102	19.10	7.45	26.55
Economy	1.000	L.F.	.102	13.95	7.45	21.40

64 Wood Fence Systems

System Description	QUAN.	UNIT	LABOR HOURS	COST PER UNIT		
				MAT.	INST.	TOTAL
Basketweave, 3/8"x4" boards, 2"x4" stringers on spreaders, 4"x4" posts						
No. 1 cedar, 6' high	1.000	L.F.	.150	28.50	8.60	37.10
Treated pine, 6' high	1.000	L.F.	.160	41.50	9.20	50.70
Board fence, 1"x4" boards, 2"x4" rails, 4"x4" posts						
Preservative treated, 2 rail, 3' high	1.000	L.F.	.166	11.45	9.50	20.95
4' high	1.000	L.F.	.178	13.40	10.20	23.60
3 rail, 5' high	1.000	L.F.	.185	13.95	10.55	24.50
6' high	1.000	L.F.	.192	17.35	11	28.35
Western cedar, No. 1, 2 rail, 3' high	1.000	L.F.	.166	13.80	9.50	23.30
3 rail, 4' high	1.000	L.F.	.178	13.30	10.20	23.50
5' high	1.000	L.F.	.185	15.50	10.55	26.05
6' high	1.000	L.F.	.192	16.45	11	27.45
No. 1 cedar, 2 rail, 3' high	1.000	L.F.	.166	15	9.50	24.50
4' high	1.000	L.F.	.178	16.50	10.20	26.70
3 rail, 5' high	1.000	L.F.	.185	19.70	10.55	30.25
6' high	1.000	L.F.	.192	23.50	11	34.50
Shadow box, 1"x6" boards, 2"x4" rails, 4"x4" posts						
Fir, pine or spruce, treated, 3 rail, 6' high	1.000	L.F.	.160	27.50	9.20	36.70
No. 1 cedar, 3 rail, 4' high	1.000	L.F.	.185	21	10.55	31.55
6' high	1.000	L.F.	.192	29	11	40
Open rail, split rails, No. 1 cedar, 2 rail, 3' high	1.000	L.F.	.150	10.65	8.60	19.25
3 rail, 4' high	1.000	L.F.	.160	13.45	9.20	22.65
No. 2 cedar, 2 rail, 3' high	1.000	L.F.	.150	9.25	8.60	17.85
3 rail, 4' high	1.000	L.F.	.160	8.90	9.20	18.10
Open rail, rustic rails, No. 1 cedar, 2 rail, 3' high	1.000	L.F.	.150	15.40	8.60	24
3 rail, 4' high	1.000	L.F.	.160	13.75	9.20	22.95
No. 2 cedar, 2 rail, 3' high	1.000	L.F.	.150	12.75	8.60	21.35
3 rail, 4' high	1.000	L.F.	.160	9.30	9.20	18.50
Rustic picket, molded pine pickets, 2 rail, 3' high	1.000	L.F.	.171	9.55	9.85	19.40
3 rail, 4' high	1.000	L.F.	.197	11	11.35	22.35
No. 1 cedar, 2 rail, 3' high	1.000	L.F.	.171	11.65	9.85	21.50
3 rail, 4' high	1.000	L.F.	.197	13.40	11.35	24.75
Picket fence, fir, pine or spruce, preserved, treated						
2 rail, 3' high	1.000	L.F.	.171	8.70	9.85	18.55
3 rail, 4' high	1.000	L.F.	.185	10.90	10.55	21.45
Western cedar, 2 rail, 3' high	1.000	L.F.	.171	10.10	9.85	19.95
3 rail, 4' high	1.000	L.F.	.185	10.25	10.55	20.80
No. 1 cedar, 2 rail, 3' high	1.000	L.F.	.171	15.30	9.85	25.15
3 rail, 4' high	1.000	L.F.	.185	20.50	10.55	31.05
Stockade, No. 1 cedar, 3-1/4" rails, 6' high	1.000	L.F.	.150	14.85	8.60	23.45
8' high	1.000	L.F.	.155	20.50	8.90	29.40
No. 2 cedar, treated rails, 6' high	1.000	L.F.	.150	15.25	8.60	23.85
Treated pine, treated rails, 6' high	1.000	L.F.	.150	15.35	8.60	23.95
Gates, No. 2 cedar, picket, 3'-6" wide 4' high	1.000	Ea.	2.667	89.50	153	242.50
No. 2 cedar, rustic round, 3' wide, 3' high	1.000	Ea.	2.667	115	153	268
No. 2 cedar, stockade screen, 3'-6" wide, 6' high	1.000	Ea.	3.000	98.50	172	270.50
General, wood, 3'-6" wide, 4' high	1.000	Ea.	2.400	119	138	257
6' high	1.000	Ea.	3.000	149	172	321

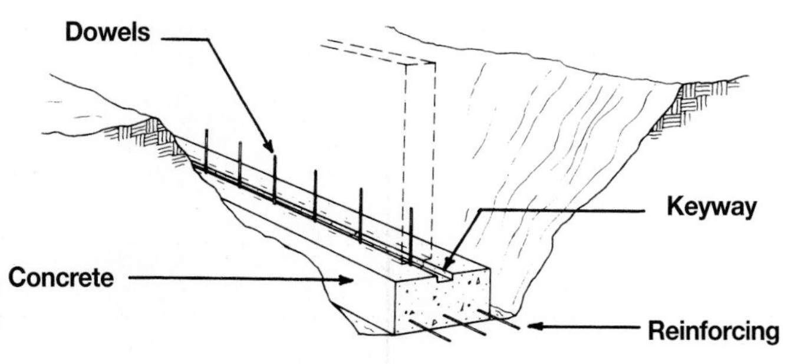

System Description	QUAN.	UNIT	LABOR HOURS	COST PER L.F. MAT.	INST.	TOTAL
8″ THICK BY 18″ WIDE FOOTING						
Concrete, 3000 psi	.040	C.Y.		5.44		5.44
Place concrete, direct chute	.040	C.Y.	.016		.79	.79
Forms, footing, 4 uses	1.330	SFCA	.103	1.10	5.37	6.47
Reinforcing, 1/2″ diameter bars, 2 each	1.380	Lb.	.011	.79	.68	1.47
Keyway, 2″ x 4″, beveled, 4 uses	1.000	L.F.	.015	.25	.89	1.14
Dowels, 1/2″ diameter bars, 2′ long, 6′ O.C.	.166	Ea.	.006	.14	.36	.50
TOTAL		L.F.	.151	7.72	8.09	15.81
12″ THICK BY 24″ WIDE FOOTING						
Concrete, 3000 psi	.070	C.Y.		9.52		9.52
Place concrete, direct chute	.070	C.Y.	.028		1.37	1.37
Forms, footing, 4 uses	2.000	SFCA	.155	1.66	8.08	9.74
Reinforcing, 1/2″ diameter bars, 2 each	1.380	Lb.	.011	.79	.68	1.47
Keyway, 2″ x 4″, beveled, 4 uses	1.000	L.F.	.015	.25	.89	1.14
Dowels, 1/2″ diameter bars, 2′ long, 6′ O.C.	.166	Ea.	.006	.14	.36	.50
TOTAL		L.F.	.215	12.36	11.38	23.74
12″ THICK BY 36″ WIDE FOOTING						
Concrete, 3000 psi	.110	C.Y.		14.96		14.96
Place concrete, direct chute	.110	C.Y.	.044		2.16	2.16
Forms, footing, 4 uses	2.000	SFCA	.155	1.66	8.08	9.74
Reinforcing, 1/2″ diameter bars, 2 each	1.380	Lb.	.011	.79	.68	1.47
Keyway, 2″ x 4″, beveled, 4 uses	1.000	L.F.	.015	.25	.89	1.14
Dowels, 1/2″ diameter bars, 2′ long, 6′ O.C.	.166	Ea.	.006	.14	.36	.50
TOTAL		L.F.	.231	17.80	12.17	29.97

The footing costs in this system are on a cost per linear foot basis.

Description	QUAN.	UNIT	LABOR HOURS	COST PER L.F. MAT.	INST.	TOTAL

Footing Price Sheet	QUAN.	UNIT	LABOR HOURS	COST PER L.F.		
				MAT.	INST.	TOTAL
Concrete, 8″ thick by 18″ wide footing						
2000 psi concrete	.040	C.Y.		4.84		4.84
2500 psi concrete	.040	C.Y.		5		5
3000 psi concrete	.040	C.Y.		5.45		5.45
3500 psi concrete	.040	C.Y.		5.50		5.50
4000 psi concrete	.040	C.Y.		5.65		5.65
12″ thick by 24″ wide footing						
2000 psi concrete	.070	C.Y.		8.45		8.45
2500 psi concrete	.070	C.Y.		8.75		8.75
3000 psi concrete	.070	C.Y.		9.50		9.50
3500 psi concrete	.070	C.Y.		9.60		9.60
4000 psi concrete	.070	C.Y.		9.85		9.85
12″ thick by 36″ wide footing						
2000 psi concrete	.110	C.Y.		13.30		13.30
2500 psi concrete	.110	C.Y.		13.75		13.75
3000 psi concrete	.110	C.Y.		14.95		14.95
3500 psi concrete	.110	C.Y.		15.05		15.05
4000 psi concrete	.110	C.Y.		15.50		15.50
Place concrete, 8″ thick by 18″ wide footing, direct chute	.040	C.Y.	.016		.79	.79
Pumped concrete	.040	C.Y.	.017		1.10	1.10
Crane & bucket	.040	C.Y.	.032		2.12	2.12
12″ thick by 24″ wide footing, direct chute	.070	C.Y.	.028		1.37	1.37
Pumped concrete	.070	C.Y.	.030		1.93	1.93
Crane & bucket	.070	C.Y.	.056		3.71	3.71
12″ thick by 36″ wide footing, direct chute	.110	C.Y.	.044		2.16	2.16
Pumped concrete	.110	C.Y.	.047		3.04	3.04
Crane & bucket	.110	C.Y.	.088		5.80	5.80
Forms, 8″ thick footing, 1 use	1.330	SFCA	.140	3.39	7.30	10.69
4 uses	1.330	SFCA	.103	1.10	5.35	6.45
12″ thick footing, 1 use	2.000	SFCA	.211	5.10	11	16.10
4 uses	2.000	SFCA	.155	1.66	8.10	9.76
Reinforcing, 3/8″ diameter bar, 1 each	.400	Lb.	.003	.23	.20	.43
2 each	.800	Lb.	.006	.46	.39	.85
3 each	1.200	Lb.	.009	.68	.59	1.27
1/2″ diameter bar, 1 each	.700	Lb.	.005	.40	.34	.74
2 each	1.380	Lb.	.011	.79	.68	1.47
3 each	2.100	Lb.	.016	1.20	1.03	2.23
5/8″ diameter bar, 1 each	1.040	Lb.	.008	.59	.51	1.10
2 each	2.080	Lb.	.016	1.19	1.02	2.21
Keyway, beveled, 2″ x 4″, 1 use	1.000	L.F.	.030	.50	1.78	2.28
2 uses	1.000	L.F.	.023	.38	1.34	1.72
2″ x 6″, 1 use	1.000	L.F.	.032	.78	1.88	2.66
2 uses	1.000	L.F.	.024	.59	1.41	2
Dowels, 2 feet long, 6′ O.C., 3/8″ bar	.166	Ea.	.005	.08	.33	.41
1/2″ bar	.166	Ea.	.006	.14	.36	.50
5/8″ bar	.166	Ea.	.006	.22	.39	.61
3/4″ bar	.166	Ea.	.006	.22	.39	.61

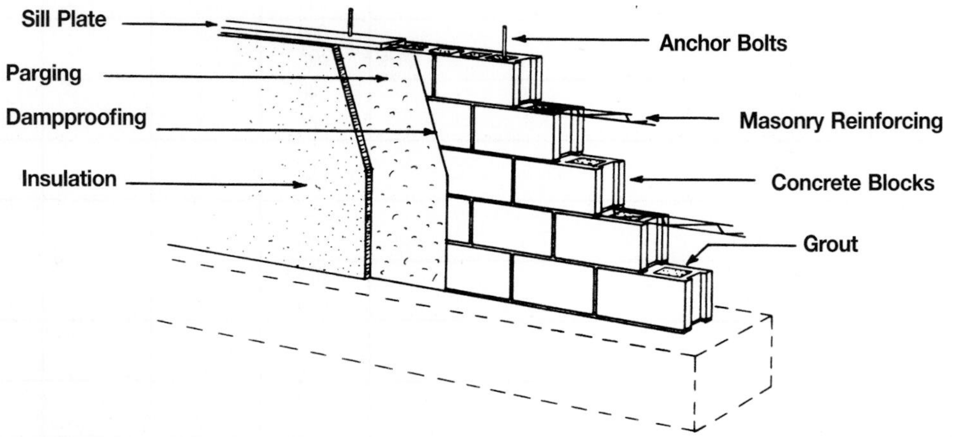

System Description	QUAN.	UNIT	LABOR HOURS	COST PER S.F.		
				MAT.	INST.	TOTAL
8″ WALL, GROUTED, FULL HEIGHT						
Concrete block, 8″ x 16″ x 8″	1.000	S.F.	.094	3.47	5.15	8.62
Masonry reinforcing, every second course	.750	L.F.	.002	.18	.12	.30
Parging, plastering with portland cement plaster, 1 coat	1.000	S.F.	.014	.37	.83	1.20
Dampproofing, bituminous coating, 1 coat	1.000	S.F.	.012	.25	.67	.92
Insulation, 1″ rigid polystyrene	1.000	S.F.	.010	.62	.59	1.21
Grout, solid, pumped	1.000	S.F.	.059	1.48	3.16	4.64
Anchor bolts, 1/2″ diameter, 8″ long, 4′ O.C.	.060	Ea.	.004	.10	.21	.31
Sill plate, 2″ x 4″, treated	.250	L.F.	.007	.16	.43	.59
TOTAL		S.F.	.202	6.63	11.16	17.79
12″ WALL, GROUTED, FULL HEIGHT						
Concrete block, 8″ x 16″ x 12″	1.000	S.F.	.160	5.25	8.65	13.90
Masonry reinforcing, every second course	.750	L.F.	.003	.22	.18	.40
Parging, plastering with portland cement plaster, 1 coat	1.000	S.F.	.014	.37	.83	1.20
Dampproofing, bituminous coating, 1 coat	1.000	S.F.	.012	.25	.67	.92
Insulation, 1″ rigid polystyrene	1.000	S.F.	.010	.62	.59	1.21
Grout, solid, pumped	1.000	S.F.	.063	2.42	3.37	5.79
Anchor bolts, 1/2″ diameter, 8″ long, 4′ O.C.	.060	Ea.	.004	.10	.21	.31
Sill plate, 2″ x 4″, treated	.250	L.F.	.007	.16	.43	.59
TOTAL		S.F.	.273	9.39	14.93	24.32

The costs in this system are based on a square foot of wall. Do not subtract for window or door openings.

Description	QUAN.	UNIT	LABOR HOURS	COST PER S.F.		
				MAT.	INST.	TOTAL

Block Wall Price Sheet	QUAN.	UNIT	LABOR HOURS	COST PER S.F.		
				MAT.	INST.	TOTAL
Concrete, block, 8″ x 16″ x, 6″ thick	1.000	S.F.	.089	3.33	4.83	8.16
8″ thick	1.000	S.F.	.093	3.47	5.15	8.62
10″ thick	1.000	S.F.	.095	4.02	6.30	10.32
12″ thick	1.000	S.F.	.122	5.25	8.65	13.90
Solid block, 8″ x 16″ x, 6″ thick	1.000	S.F.	.091	3.69	4.99	8.68
8″ thick	1.000	S.F.	.096	4.73	5.30	10.03
10″ thick	1.000	S.F.	.096	4.73	5.30	10.03
12″ thick	1.000	S.F.	.126	6.60	7.40	14
Masonry reinforcing, wire strips, to 8″ wide, every course	1.500	L.F.	.004	.36	.24	.60
Every 2nd course	.750	L.F.	.002	.18	.12	.30
Every 3rd course	.500	L.F.	.001	.12	.08	.20
Every 4th course	.400	L.F.	.001	.10	.06	.16
Wire strips to 12″ wide, every course	1.500	L.F.	.006	.44	.36	.80
Every 2nd course	.750	L.F.	.003	.22	.18	.40
Every 3rd course	.500	L.F.	.002	.15	.12	.27
Every 4th course	.400	L.F.	.002	.12	.10	.22
Parging, plastering with portland cement plaster, 1 coat	1.000	S.F.	.014	.37	.83	1.20
2 coats	1.000	S.F.	.022	.57	1.28	1.85
Dampproofing, bituminous, brushed on, 1 coat	1.000	S.F.	.012	.25	.67	.92
2 coats	1.000	S.F.	.016	.50	.89	1.39
Sprayed on, 1 coat	1.000	S.F.	.010	.25	.53	.78
2 coats	1.000	S.F.	.016	.49	.89	1.38
Troweled on, 1/16″ thick	1.000	S.F.	.016	.44	.89	1.33
1/8″ thick	1.000	S.F.	.020	.77	1.11	1.88
1/2″ thick	1.000	S.F.	.023	2.51	1.27	3.78
Insulation, rigid, fiberglass, 1.5#/C.F., unfaced						
1-1/2″ thick R 6.2	1.000	S.F.	.008	.44	.47	.91
2″ thick R 8.5	1.000	S.F.	.008	.55	.47	1.02
3″ thick R 13	1.000	S.F.	.010	.66	.59	1.25
Perlite, 1″ thick R 2.77	1.000	S.F.	.010	.53	.59	1.12
2″ thick R 5.55	1.000	S.F.	.011	.86	.65	1.51
Polystyrene, extruded, 1″ thick R 5.4	1.000	S.F.	.010	.62	.59	1.21
2″ thick R 10.8	1.000	S.F.	.011	1.77	.65	2.42
Molded 1″ thick R 3.85	1.000	S.F.	.010	.30	.59	.89
2″ thick R 7.7	1.000	S.F.	.011	.89	.65	1.54
Grout, concrete block cores, 6″ thick	1.000	S.F.	.044	1.11	2.37	3.48
8″ thick	1.000	S.F.	.059	1.48	3.16	4.64
10″ thick	1.000	S.F.	.061	1.95	3.26	5.21
12″ thick	1.000	S.F.	.063	2.42	3.37	5.79
Anchor bolts, 2′ on center, 1/2″ diameter, 8″ long	.120	Ea.	.005	.19	.43	.62
12″ long	.120	Ea.	.005	.22	.43	.65
3/4″ diameter, 8″ long	.120	Ea.	.006	.67	.44	1.11
12″ long	.120	Ea.	.006	.83	.45	1.28
4′ on center, 1/2″ diameter, 8″ long	.060	Ea.	.002	.10	.21	.31
12″ long	.060	Ea.	.003	.11	.22	.33
3/4″ diameter, 8″ long	.060	Ea.	.003	.33	.22	.55
12″ long	.060	Ea.	.003	.42	.23	.65
Sill plates, treated, 2″ x 4″	.250	L.F.	.007	.16	.43	.59
4″ x 4″	.250	L.F.	.007	.34	.40	.74

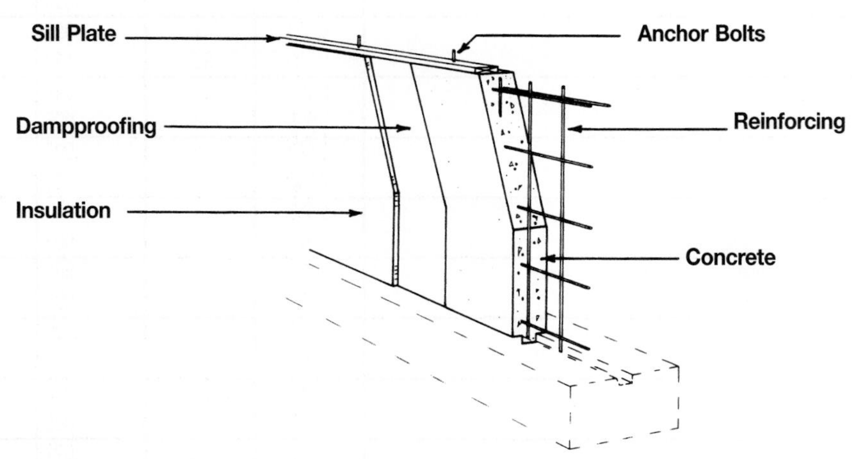

System Description	QUAN.	UNIT	LABOR HOURS	COST PER S.F. MAT.	INST.	TOTAL
8″ THICK, POURED CONCRETE WALL						
Concrete, 8″ thick , 3000 psi	.025	C.Y.		3.40		3.40
Forms, prefabricated plywood, up to 8′ high	2.000	SFCA	.120	2.50	6.34	8.84
Reinforcing, light	.670	Lb.	.004	.38	.23	.61
Placing concrete, direct chute	.025	C.Y.	.013		.66	.66
Dampproofing, brushed on, 2 coats	1.000	S.F.	.016	.50	.89	1.39
Rigid insulation, 1″ polystyrene	1.000	S.F.	.010	.62	.59	1.21
Anchor bolts, 1/2″ diameter, 12″ long, 4′ O.C.	.060	Ea.	.004	.11	.22	.33
Sill plates, 2″ x 4″, treated	.250	L.F.	.007	.16	.43	.59
TOTAL		S.F.	.174	7.67	9.36	17.03
12″ THICK, POURED CONCRETE WALL						
Concrete, 12″ thick, 3000 psi	.040	C.Y.		5.44		5.44
Forms, prefabricated plywood, up to 8′ high	2.000	SFCA	.120	2.50	6.34	8.84
Reinforcing, light	1.000	Lb.	.005	.57	.34	.91
Placing concrete, direct chute	.040	C.Y.	.019		.94	.94
Dampproofing, brushed on, 2 coats	1.000	S.F.	.016	.50	.89	1.39
Rigid insulation, 1″ polystyrene	1.000	S.F.	.010	.62	.59	1.21
Anchor bolts, 1/2″ diameter, 12″ long, 4′ O.C.	.060	Ea.	.004	.11	.22	.33
Sill plates, 2″ x 4″ treated	.250	L.F.	.007	.16	.43	.59
TOTAL		S.F.	.181	9.90	9.75	19.65

The costs in this system are based on sq. ft. of wall. Do not subtract for window and door openings. The costs assume a 4′ high wall.

Description	QUAN.	UNIT	LABOR HOURS	COST PER S.F. MAT.	INST.	TOTAL

Concrete Wall Price Sheet	QUAN.	UNIT	LABOR HOURS	COST PER S.F.		
				MAT.	INST.	TOTAL
Formwork, prefabricated plywood, up to 8′ high	2.000	SFCA	.081	2.50	6.35	8.85
Over 8′ to 16′ high	2.000	SFCA	.076	2.62	8.45	11.07
Job built forms, 1 use per month	2.000	SFCA	.320	6.10	13.70	19.80
4 uses per month	2.000	SFCA	.221	2.30	10	12.30
Reinforcing, 8″ wall, light reinforcing	.670	Lb.	.004	.38	.23	.61
Heavy reinforcing	1.500	Lb.	.008	.86	.51	1.37
10″ wall, light reinforcing	.850	Lb.	.005	.48	.29	.77
Heavy reinforcing	2.000	Lb.	.011	1.14	.68	1.82
12″ wall light reinforcing	1.000	Lb.	.005	.57	.34	.91
Heavy reinforcing	2.250	Lb.	.012	1.28	.77	2.05
Placing concrete, 8″ wall, direct chute	.025	C.Y.	.013		.66	.66
Pumped concrete	.025	C.Y.	.016		1.04	1.04
Crane & bucket	.025	C.Y.	.023		1.49	1.49
10″ wall, direct chute	.030	C.Y.	.016		.79	.79
Pumped concrete	.030	C.Y.	.019		1.25	1.25
Crane & bucket	.030	C.Y.	.027		1.79	1.79
12″ wall, direct chute	.040	C.Y.	.019		.94	.94
Pumped concrete	.040	C.Y.	.023		1.50	1.50
Crane & bucket	.040	C.Y.	.032		2.12	2.12
Dampproofing, bituminous, brushed on, 1 coat	1.000	S.F.	.012	.25	.67	.92
2 coats	1.000	S.F.	.016	.50	.89	1.39
Sprayed on, 1 coat	1.000	S.F.	.010	.25	.53	.78
2 coats	1.000	S.F.	.016	.49	.89	1.38
Troweled on, 1/16″ thick	1.000	S.F.	.016	.44	.89	1.33
1/8″ thick	1.000	S.F.	.020	.77	1.11	1.88
1/2″ thick	1.000	S.F.	.023	2.51	1.27	3.78
Insulation rigid, fiberglass, 1.5#/C.F., unfaced						
1-1/2″ thick, R 6.2	1.000	S.F.	.008	.44	.47	.91
2″ thick, R 8.3	1.000	S.F.	.008	.55	.47	1.02
3″ thick, R 12.4	1.000	S.F.	.010	.66	.59	1.25
Perlite, 1″ thick R 2.77	1.000	S.F.	.010	.53	.59	1.12
2″ thick R 5.55	1.000	S.F.	.011	.86	.65	1.51
Polystyrene, extruded, 1″ thick R 5.40	1.000	S.F.	.010	.62	.59	1.21
2″ thick R 10.8	1.000	S.F.	.011	1.77	.65	2.42
Molded, 1″ thick R 3.85	1.000	S.F.	.010	.30	.59	.89
2″ thick R 7.70	1.000	S.F.	.011	.89	.65	1.54
Anchor bolts, 2′ on center, 1/2″ diameter, 8″ long	.120	Ea.	.005	.19	.43	.62
12″ long	.120	Ea.	.005	.22	.43	.65
3/4″ diameter, 8″ long	.120	Ea.	.006	.67	.44	1.11
12″ long	.120	Ea.	.006	.83	.45	1.28
Sill plates, treated lumber, 2″ x 4″	.250	L.F.	.007	.16	.43	.59
4″ x 4″	.250	L.F.	.007	.34	.40	.74

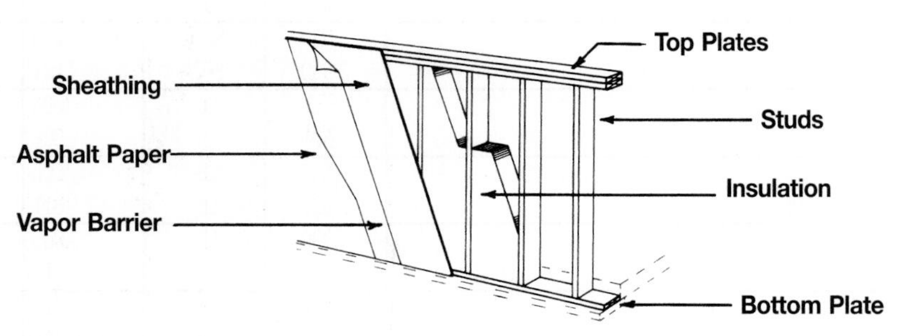

System Description	QUAN.	UNIT	LABOR HOURS	COST PER S.F.		
				MAT.	INST.	TOTAL
2″ X 4″ STUDS, 16″ O.C., WALL						
Studs, 2″ x 4″, 16″ O.C., treated	1.000	L.F.	.015	.66	.86	1.52
Plates, double top plate, single bottom plate, treated, 2″ x 4″	.750	L.F.	.011	.50	.65	1.15
Sheathing, 1/2″, exterior grade, CDX, treated	1.000	S.F.	.014	.94	.84	1.78
Asphalt paper, 15# roll	1.100	S.F.	.002	.07	.14	.21
Vapor barrier, 4 mil polyethylene	1.000	S.F.	.002	.04	.13	.17
Fiberglass insulation, 3-1/2″ thick	1.000	S.F.	.007	.37	.41	.78
TOTAL		S.F.	.051	2.58	3.03	5.61
2″ X 6″ STUDS, 16″ O.C., WALL						
Studs, 2″ x 6″, 16″ O.C., treated	1.000	L.F.	.016	.83	.94	1.77
Plates, double top plate, single bottom plate, treated, 2″ x 6″	.750	L.F.	.012	.62	.71	1.33
Sheathing, 5/8″ exterior grade, CDX, treated	1.000	S.F.	.015	1.42	.90	2.32
Asphalt paper, 15# roll	1.100	S.F.	.002	.07	.14	.21
Vapor barrier, 4 mil polyethylene	1.000	S.F.	.002	.04	.13	.17
Fiberglass insulation, 6″ thick	1.000	S.F.	.007	.47	.41	.88
TOTAL		S.F.	.054	3.45	3.23	6.68
2″ X 8″ STUDS, 16″ O.C., WALL						
Studs, 2″ x 8″, 16″ O.C. treated	1.000	L.F.	.018	1.24	1.05	2.29
Plates, double top plate, single bottom plate, treated, 2″ x 8″	.750	L.F.	.013	.93	.79	1.72
Sheathing, 3/4″ exterior grade, CDX, treated	1.000	S.F.	.016	1.51	.97	2.48
Asphalt paper, 15# roll	1.100	S.F.	.002	.07	.14	.21
Vapor barrier, 4 mil polyethylene	1.000	S.F.	.002	.04	.13	.17
Fiberglass insulation, 9″ thick	1.000	S.F.	.006	.80	.35	1.15
TOTAL		S.F.	.057	4.59	3.43	8.02

The costs in this system are based on a sq. ft. of wall area. Do not subtract for window or door openings. The costs assume a 4′ high wall.

Description	QUAN.	UNIT	LABOR HOURS	COST PER S.F.		
				MAT.	INST.	TOTAL

Wood Wall Foundation Price Sheet	QUAN.	UNIT	LABOR HOURS	COST PER S.F.		
				MAT.	INST.	TOTAL
Studs, treated, 2" x 4", 12" O.C.	1.250	L.F.	.018	.83	1.08	1.91
16" O.C.	1.000	L.F.	.015	.66	.86	1.52
2" x 6", 12" O.C.	1.250	L.F.	.020	1.04	1.18	2.22
16" O.C.	1.000	L.F.	.016	.83	.94	1.77
2" x 8", 12" O.C.	1.250	L.F.	.022	1.55	1.31	2.86
16" O.C.	1.000	L.F.	.018	1.24	1.05	2.29
Plates, treated double top single bottom, 2" x 4"	.750	L.F.	.011	.50	.65	1.15
2" x 6"	.750	L.F.	.012	.62	.71	1.33
2" x 8"	.750	L.F.	.013	.93	.79	1.72
Sheathing, treated exterior grade CDX, 1/2" thick	1.000	S.F.	.014	.94	.84	1.78
5/8" thick	1.000	S.F.	.015	1.42	.90	2.32
3/4" thick	1.000	S.F.	.016	1.51	.97	2.48
Asphalt paper, 15# roll	1.100	S.F.	.002	.07	.14	.21
Vapor barrier, polyethylene, 4 mil	1.000	S.F.	.002	.03	.13	.16
10 mil	1.000	S.F.	.002	.10	.13	.23
Insulation, rigid, fiberglass, 1.5#/C.F., unfaced	1.000	S.F.	.008	.37	.47	.84
1-1/2" thick, R 6.2	1.000	S.F.	.008	.44	.47	.91
2" thick, R 8.3	1.000	S.F.	.008	.55	.47	1.02
3" thick, R 12.4	1.000	S.F.	.010	.67	.60	1.27
Perlite 1" thick, R 2.77	1.000	S.F.	.010	.53	.59	1.12
2" thick, R 5.55	1.000	S.F.	.011	.86	.65	1.51
Polystyrene, extruded, 1" thick, R 5.40	1.000	S.F.	.010	.62	.59	1.21
2" thick, R 10.8	1.000	S.F.	.011	1.77	.65	2.42
Molded 1" thick, R 3.85	1.000	S.F.	.010	.30	.59	.89
2" thick, R 7.7	1.000	S.F.	.011	.89	.65	1.54
Non rigid, batts, fiberglass, paper backed, 3-1/2" thick roll, R 11	1.000	S.F.	.005	.37	.41	.78
6", R 19	1.000	S.F.	.006	.47	.41	.88
9", R 30	1.000	S.F.	.006	.80	.35	1.15
12", R 38	1.000	S.F.	.006	1.17	.35	1.52
Mineral fiber, paper backed, 3-1/2", R 13	1.000	S.F.	.005	.87	.29	1.16
6", R 19	1.000	S.F.	.005	1.37	.29	1.66
10", R 30	1.000	S.F.	.006	1.80	.35	2.15

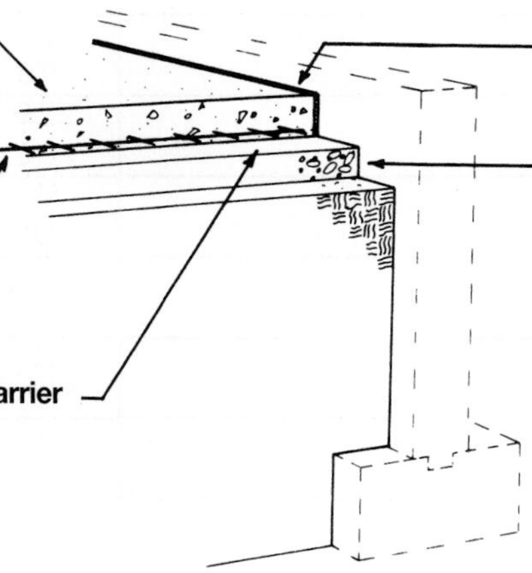

Concrete Slab
Expansion Material
Bank Run Gravel
Welded Wire Fabric
Vapor Barrier

System Description	QUAN.	UNIT	LABOR HOURS	COST PER S.F. MAT.	COST PER S.F. INST.	COST PER S.F. TOTAL
4″ THICK SLAB						
Concrete, 4″ thick, 3000 psi concrete	.012	C.Y.		1.63		1.63
Place concrete, direct chute	.012	C.Y.	.005		.26	.26
Bank run gravel, 4″ deep	1.000	S.F.	.001	.33	.05	.38
Polyethylene vapor barrier, .006″ thick	1.000	S.F.	.002	.04	.13	.17
Edge forms, expansion material	.100	L.F.	.005	.03	.28	.31
Welded wire fabric, 6 x 6, 10/10 (W1.4/W1.4)	1.100	S.F.	.005	.20	.32	.52
Steel trowel finish	1.000	S.F.	.014		.78	.78
TOTAL		S.F.	.032	2.23	1.82	4.05
6″ THICK SLAB						
Concrete, 6″ thick, 3000 psi concrete	.019	C.Y.		2.58		2.58
Place concrete, direct chute	.019	C.Y.	.008		.41	.41
Bank run gravel, 4″ deep	1.000	S.F.	.001	.33	.05	.38
Polyethylene vapor barrier, .006″ thick	1.000	S.F.	.002	.04	.13	.17
Edge forms, expansion material	.100	L.F.	.005	.03	.28	.31
Welded wire fabric, 6 x 6, 10/10 (W1.4/W1.4)	1.100	S.F.	.005	.20	.32	.52
Steel trowel finish	1.000	S.F.	.014		.78	.78
TOTAL		S.F.	.035	3.18	1.97	5.15

The slab costs in this section are based on a cost per square foot of floor area.

Description	QUAN.	UNIT	LABOR HOURS	COST PER S.F. MAT.	COST PER S.F. INST.	COST PER S.F. TOTAL

Floor Slab Price Sheet	QUAN.	UNIT	LABOR HOURS	COST PER S.F.		
				MAT.	INST.	TOTAL
Concrete, 4" thick slab, 2000 psi concrete	.012	C.Y.		1.45		1.45
2500 psi concrete	.012	C.Y.		1.50		1.50
3000 psi concrete	.012	C.Y.		1.63		1.63
3500 psi concrete	.012	C.Y.		1.64		1.64
4000 psi concrete	.012	C.Y.		1.69		1.69
4500 psi concrete	.012	C.Y.		1.74		1.74
5" thick slab, 2000 psi concrete	.015	C.Y.		1.82		1.82
2500 psi concrete	.015	C.Y.		1.88		1.88
3000 psi concrete	.015	C.Y.		2.04		2.04
3500 psi concrete	.015	C.Y.		2.06		2.06
4000 psi concrete	.015	C.Y.		2.12		2.12
4500 psi concrete	.015	C.Y.		2.18		2.18
6" thick slab, 2000 psi concrete	.019	C.Y.		2.30		2.30
2500 psi concrete	.019	C.Y.		2.38		2.38
3000 psi concrete	.019	C.Y.		2.58		2.58
3500 psi concrete	.019	C.Y.		2.60		2.60
4000 psi concrete	.019	C.Y.		2.68		2.68
4500 psi concrete	.019	C.Y.		2.76		2.76
Place concrete, 4" slab, direct chute	.012	C.Y.	.005		.26	.26
Pumped concrete	.012	C.Y.	.006		.38	.38
Crane & bucket	.012	C.Y.	.008		.52	.52
5" slab, direct chute	.015	C.Y.	.007		.33	.33
Pumped concrete	.015	C.Y.	.007		.48	.48
Crane & bucket	.015	C.Y.	.010		.65	.65
6" slab, direct chute	.019	C.Y.	.008		.41	.41
Pumped concrete	.019	C.Y.	.009		.61	.61
Crane & bucket	.019	C.Y.	.012		.82	.82
Gravel, bank run, 4" deep	1.000	S.F.	.001	.33	.05	.38
6" deep	1.000	S.F.	.001	.44	.07	.51
9" deep	1.000	S.F.	.001	.65	.11	.76
12" deep	1.000	S.F.	.001	.89	.13	1.02
3/4" crushed stone, 3" deep	1.000	S.F.	.001	.33	.06	.39
6" deep	1.000	S.F.	.001	.65	.11	.76
9" deep	1.000	S.F.	.002	.95	.15	1.10
12" deep	1.000	S.F.	.002	2.64	.18	2.82
Vapor barrier polyethylene, .004" thick	1.000	S.F.	.002	.03	.13	.16
.006" thick	1.000	S.F.	.002	.04	.13	.17
Edge forms, expansion material, 4" thick slab	.100	L.F.	.004	.02	.18	.20
6" thick slab	.100	L.F.	.005	.03	.28	.31
Welded wire fabric 6 x 6, 10/10 (W1.4/W1.4)	1.100	S.F.	.005	.20	.32	.52
6 x 6, 6/6 (W2.9/W2.9)	1.100	S.F.	.006	.31	.40	.71
4 x 4, 10/10 (W1.4/W1.4)	1.100	S.F.	.006	.29	.36	.65
Finish concrete, screed finish	1.000	S.F.	.009		.32	.32
Float finish	1.000	S.F.	.011		.32	.32
Steel trowel, for resilient floor	1.000	S.F.	.013		1.02	1.02
For finished floor	1.000	S.F.	.015		.78	.78

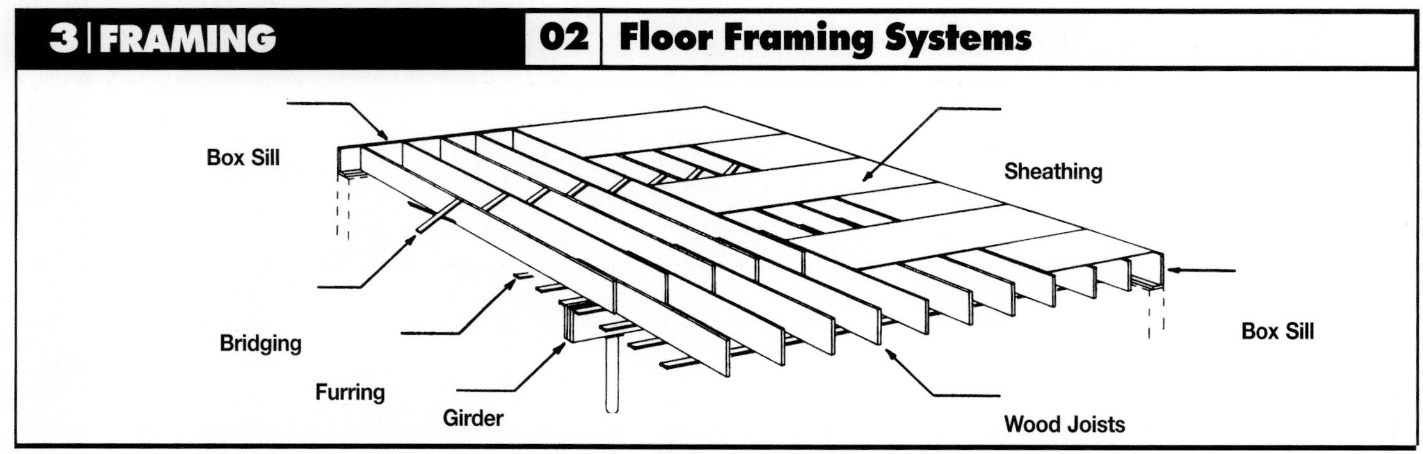

System Description	QUAN.	UNIT	LABOR HOURS	COST PER S.F.		
				MAT.	INST.	TOTAL
2″ X 8″, 16″ O.C.						
Wood joists, 2″ x 8″, 16″ O.C.	1.000	L.F.	.015	1.07	.86	1.93
Bridging, 1″ x 3″, 6′ O.C.	.080	Pr.	.005	.06	.29	.35
Box sills, 2″ x 8″	.150	L.F.	.002	.16	.13	.29
Concrete filled steel column, 4″ diameter	.125	L.F.	.002	.21	.14	.35
Girder, built up from three 2″ x 8″	.125	L.F.	.013	.40	.78	1.18
Sheathing, plywood, subfloor, 5/8″ CDX	1.000	S.F.	.012	.86	.70	1.56
Furring, 1″ x 3″, 16″ O.C.	1.000	L.F.	.023	.49	1.35	1.84
Joist hangers	.036	Ea.	.002	.06	.10	.16
TOTAL		S.F.	.074	3.31	4.35	7.66
2″ X 10″, 16″ O.C.						
Wood joists, 2″ x 10″, 16″ OC	1.000	L.F.	.018	1.60	1.05	2.65
Bridging, 1″ x 3″, 6′ OC	.080	Pr.	.005	.06	.29	.35
Box sills, 2″ x 10″	.150	L.F.	.003	.24	.16	.40
Concrete filled steel column, 4″ diameter	.125	L.F.	.002	.21	.14	.35
5/8″ thick	.125	L.F.	.014	.60	.84	1.44
Sheathing, plywood, subfloor, 5/8″ CDX	1.000	S.F.	.012	.86	.70	1.56
Furring, 1″ x 3″,16″ OC	1.000	L.F.	.023	.49	1.35	1.84
Joist hangers	.036	Ea.	.002	.06	.10	.16
TOTAL		S.F.	.079	4.12	4.63	8.75
2″ X 12″, 16″ O.C.						
Wood joists, 2″ x 12″, 16″ O.C.	1.000	L.F.	.018	2.01	1.08	3.09
Bridging, 1″ x 3″, 6′ O.C.	.080	Pr.	.005	.06	.29	.35
Box sills, 2″ x 12″	.150	L.F.	.003	.30	.16	.46
Concrete filled steel column, 4″ diameter	.125	L.F.	.002	.21	.14	.35
Girder, built up from three 2″ x 12″	.125	L.F.	.015	.76	.89	1.65
Sheathing, plywood, subfloor, 5/8″ CDX	1.000	S.F.	.012	.86	.70	1.56
Furring, 1″ x 3″, 16″ O.C.	1.000	L.F.	.023	.49	1.35	1.84
Joist hangers	.036	Ea.	.002	.06	.10	.16
TOTAL		S.F.	.080	4.75	4.71	9.46

Floor costs on this page are given on a cost per square foot basis.

Description	QUAN.	UNIT	LABOR HOURS	COST PER S.F.		
				MAT.	INST.	TOTAL

Floor Framing Price Sheet (Wood)	QUAN.	UNIT	LABOR HOURS	COST PER S.F.		
				MAT.	INST.	TOTAL
Joists, #2 or better, pine, 2″ x 4″, 12″ O.C.	1.250	L.F.	.016	.59	.94	1.53
16″ O.C.	1.000	L.F.	.013	.47	.75	1.22
2″ x 6″, 12″ O.C.	1.250	L.F.	.016	.99	.94	1.93
16″ O.C.	1.000	L.F.	.013	.79	.75	1.54
2″ x 8″, 12″ O.C.	1.250	L.F.	.018	1.34	1.08	2.42
16″ O.C.	1.000	L.F.	.015	1.07	.86	1.93
2″ x 10″, 12″ O.C.	1.250	L.F.	.022	2	1.31	3.31
16″ O.C.	1.000	L.F.	.018	1.60	1.05	2.65
2″x 12″, 12″ O.C.	1.250	L.F.	.023	2.51	1.35	3.86
16″ O.C.	1.000	L.F.	.018	2.01	1.08	3.09
Bridging, wood 1″ x 3″, joists 12″ O.C.	.100	Pr.	.006	.08	.36	.44
16″ O.C.	.080	Pr.	.005	.06	.29	.35
Metal, galvanized, joists 12″ O.C.	.100	Pr.	.006	.21	.36	.57
16″ O.C.	.080	Pr.	.005	.16	.29	.45
Compression type, joists 12″ O.C.	.100	Pr.	.004	.14	.24	.38
16″ O.C.	.080	Pr.	.003	.11	.19	.30
Box sills, #2 or better, 2″ x 4″	.150	L.F.	.002	.07	.11	.18
2″ x 6″	.150	L.F.	.002	.12	.11	.23
2″ x 8″	.150	L.F.	.002	.16	.13	.29
2″ x 10″	.150	L.F.	.003	.24	.16	.40
2″ x 12″	.150	L.F.	.003	.30	.16	.46
Girders, including lally columns, 3 pieces spiked together, 2″ x 8″	.125	L.F.	.015	.61	.92	1.53
2″ x 10″	.125	L.F.	.016	.81	.98	1.79
2″ x 12″	.125	L.F.	.017	.97	1.03	2
Solid girders, 3″ x 8″	.040	L.F.	.004	.34	.24	.58
3″ x 10″	.040	L.F.	.004	.38	.25	.63
3″ x 12″	.040	L.F.	.004	.41	.26	.67
4″ x 8″	.040	L.F.	.004	.41	.26	.67
4″ x 10″	.040	L.F.	.004	.44	.27	.71
4″ x 12″	.040	L.F.	.004	.46	.28	.74
Steel girders, bolted & including fabrication, wide flange shapes						
12″ deep, 14#/l.f.	.040	L.F.	.003	1.04	.28	1.32
10″ deep, 15#/l.f.	.040	L.F.	.003	1.04	.28	1.32
8″ deep, 10#/l.f.	.040	L.F.	.003	.70	.28	.98
6″ deep, 9#/l.f.	.040	L.F.	.003	.63	.28	.91
5″ deep, 16#/l.f.	.040	L.F.	.003	1.04	.28	1.32
Sheathing, plywood exterior grade CDX, 1/2″ thick	1.000	S.F.	.011	.70	.67	1.37
5/8″ thick	1.000	S.F.	.012	.86	.70	1.56
3/4″ thick	1.000	S.F.	.013	1.03	.75	1.78
Boards, 1″ x 8″ laid regular	1.000	S.F.	.016	2.36	.94	3.30
Laid diagonal	1.000	S.F.	.019	2.36	1.11	3.47
1″ x 10″ laid regular	1.000	S.F.	.015	2.36	.86	3.22
Laid diagonal	1.000	S.F.	.018	2.36	1.05	3.41
Furring, 1″ x 3″, 12″ O.C.	1.250	L.F.	.029	.61	1.69	2.30
16″ O.C.	1.000	L.F.	.023	.49	1.35	1.84
24″ O.C.	.750	L.F.	.017	.37	1.01	1.38

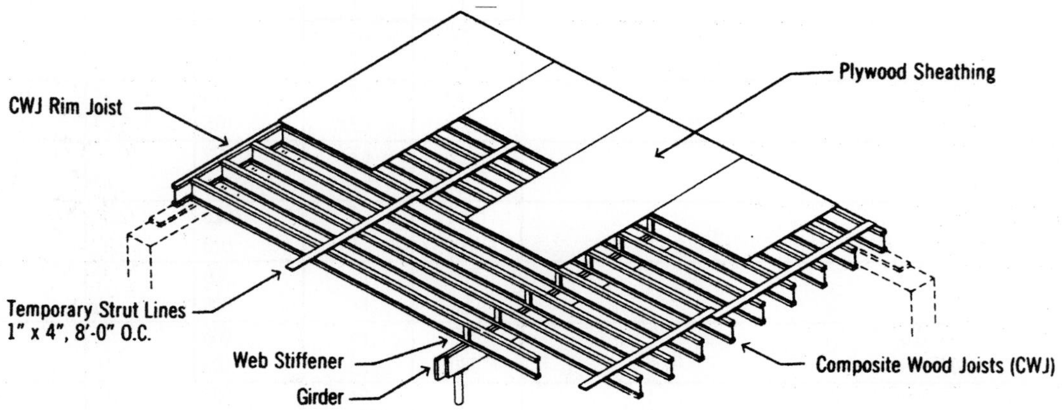

System Description	QUAN.	UNIT	LABOR HOURS	COST PER S.F. MAT.	COST PER S.F. INST.	COST PER S.F. TOTAL
9-1/2" COMPOSITE WOOD JOISTS, 16" O.C.						
CWJ, 9-1/2", 16" O.C., 15' span	1.000	L.F.	.018	1.95	1.05	3
Temp. strut line, 1" x 4", 8' O.C.	.160	L.F.	.003	.09	.19	.28
CWJ rim joist, 9-1/2"	.150	L.F.	.003	.29	.16	.45
Concrete filled steel column, 4" diameter	.125	L.F.	.002	.21	.14	.35
Girder, built up from three 2" x 8"	.125	L.F.	.013	.40	.78	1.18
Sheathing, plywood, subfloor, 5/8" CDX	1.000	S.F.	.012	.86	.70	1.56
TOTAL		S.F.	.051	3.80	3.02	6.82
11-1/2" COMPOSITE WOOD JOISTS, 16" O.C.						
CWJ, 11-1/2", 16" O.C., 18' span	1.000	L.F.	.018	2.23	1.08	3.31
Temp. strut line, 1" x 4", 8' O.C.	.160	L.F.	.003	.09	.19	.28
CWJ rim joist, 11-1/2"	.150	L.F.	.003	.33	.16	.49
Concrete filled steel column, 4" diameter	.125	L.F.	.002	.21	.14	.35
Girder, built up from three 2" x 10"	.125	L.F.	.014	.60	.84	1.44
Sheathing, plywood, subfloor, 5/8" CDX	1.000	S.F.	.012	.86	.70	1.56
TOTAL		S.F.	.052	4.32	3.11	7.43
14" COMPOSITE WOOD JOISTS, 16" O.C.						
CWJ, 14", 16" O.C., 22' span	1.000	L.F.	.020	2.85	1.15	4
Temp. strut line, 1" x 4", 8' O.C.	.160	L.F.	.003	.09	.19	.28
CWJ rim joist, 14"	.150	L.F.	.003	.43	.17	.60
Concrete filled steel column, 4" diameter	.600	L.F.	.002	.21	.14	.35
Girder, built up from three 2" x 12"	.600	L.F.	.015	.76	.89	1.65
Sheathing, plywood, subfloor, 5/8" CDX	1.000	S.F.	.012	.86	.70	1.56
TOTAL		S.F.	.055	5.20	3.24	8.44

Floor costs on this page are given on a cost per square foot basis.

Description	QUAN.	UNIT	LABOR HOURS	COST PER S.F. MAT.	COST PER S.F. INST.	COST PER S.F. TOTAL

Floor Framing Price Sheet (Wood)	QUAN.	UNIT	LABOR HOURS	COST PER S.F.		
				MAT.	INST.	TOTAL
Composite wood joist 9-1/2" deep, 12" O.C.	1.250	L.F.	.022	2.44	1.31	3.75
16" O.C.	1.000	L.F.	.018	1.95	1.05	3
11-1/2" deep, 12" O.C.	1.250	L.F.	.023	2.78	1.34	4.12
16" O.C.	1.000	L.F.	.018	2.23	1.08	3.31
14" deep, 12" O.C.	1.250	L.F.	.024	3.56	1.44	5
16" O.C.	1.000	L.F.	.020	2.85	1.15	4
16 " deep, 12" O.C.	1.250	L.F.	.026	5.70	1.50	7.20
16" O.C.	1.000	L.F.	.021	4.55	1.20	5.75
CWJ rim joist, 9-1/2"	.150	L.F.	.003	.29	.16	.45
11-1/2"	.150	L.F.	.003	.33	.16	.49
14"	.150	L.F.	.003	.43	.17	.60
16"	.150	L.F.	.003	.68	.18	.86
Girders, including lally columns, 3 pieces spiked together, 2" x 8"	.125	L.F.	.015	.61	.92	1.53
2" x 10"	.125	L.F.	.016	.81	.98	1.79
2" x 12"	.125	L.F.	.017	.97	1.03	2
Solid girders, 3" x 8"	.040	L.F.	.004	.34	.24	.58
3" x 10"	.040	L.F.	.004	.38	.25	.63
3" x 12"	.040	L.F.	.004	.41	.26	.67
4" x 8"	.040	L.F.	.004	.41	.26	.67
4" x 10"	.040	L.F.	.004	.44	.27	.71
4" x 12"	.040	L.F.	.004	.46	.28	.74
Steel girders, bolted & including fabrication, wide flange shapes						
12" deep, 14#/l.f.	.040	L.F.	.061	24	6.55	30.55
10" deep, 15#/l.f.	.040	L.F.	.067	26	7.15	33.15
8" deep, 10#/l.f.	.040	L.F.	.067	17.45	7.15	24.60
6" deep, 9#/l.f.	.040	L.F.	.067	15.70	7.15	22.85
5" deep, 16#/l.f.	.040	L.F.	.064	25	6.90	31.90
Sheathing, plywood exterior grade CDX, 1/2" thick	1.000	S.F.	.011	.70	.67	1.37
5/8" thick	1.000	S.F.	.012	.86	.70	1.56
3/4" thick	1.000	S.F.	.013	1.03	.75	1.78
Boards, 1" x 8" laid regular	1.000	S.F.	.016	2.36	.94	3.30
Laid diagonal	1.000	S.F.	.019	2.36	1.11	3.47
1" x 10" laid regular	1.000	S.F.	.015	2.36	.86	3.22
Laid diagonal	1.000	S.F.	.018	2.36	1.05	3.41
Furring, 1" x 3", 12" O.C.	1.250	L.F.	.029	.61	1.69	2.30
16" O.C.	1.000	L.F.	.023	.49	1.35	1.84
24" O.C.	.750	L.F.	.017	.37	1.01	1.38

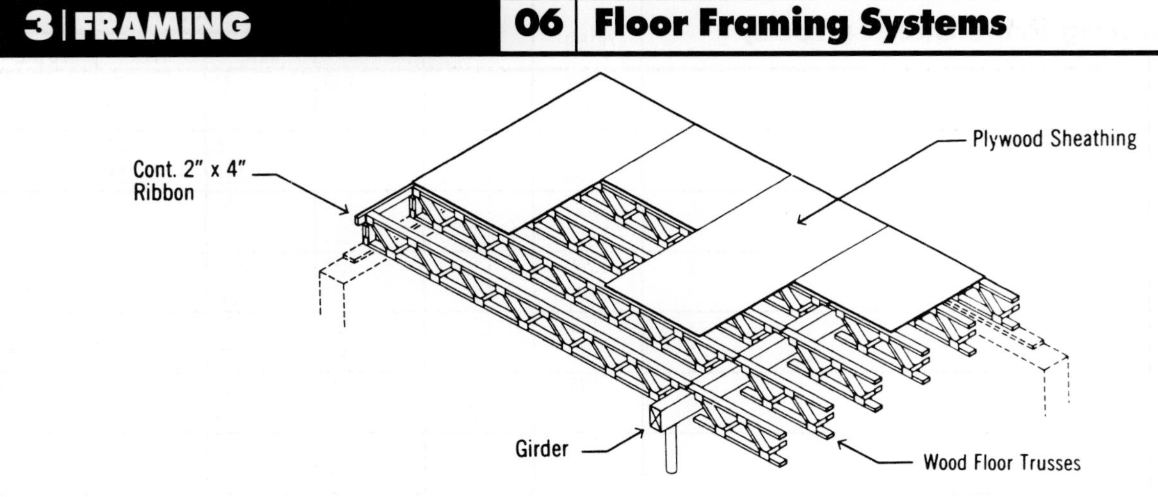

System Description	QUAN.	UNIT	LABOR HOURS	COST PER S.F. MAT.	INST.	TOTAL
12″ OPEN WEB JOISTS, 16″ O.C.						
OWJ 12″, 16″ O.C., 21′ span	1.000	L.F.	.018	4	1.08	5.08
Continuous ribbing, 2″ x 4″	.150	L.F.	.002	.07	.11	.18
Concrete filled steel column, 4″ diameter	.125	L.F.	.002	.21	.14	.35
Girder, built up from three 2″ x 8″	.125	L.F.	.013	.40	.78	1.18
Sheathing, plywood, subfloor, 5/8″ CDX	1.000	S.F.	.012	.86	.70	1.56
Furring, 1″ x 3″, 16″ O.C.	1.000	L.F.	.023	.49	1.35	1.84
TOTAL		S.F.	.070	6.03	4.16	10.19
14″ OPEN WEB WOOD JOISTS, 16″ O.C.						
OWJ 14″, 16″ O.C., 22′ span	1.000	L.F.	.020	4.28	1.15	5.43
Continuous ribbing, 2″ x 4″	.150	L.F.	.002	.07	.11	.18
Concrete filled steel column, 4″ diameter	.125	L.F.	.002	.21	.14	.35
Girder, built up from three 2″ x 10″	.125	L.F.	.014	.60	.84	1.44
Sheathing, plywood, subfloor, 5/8″ CDX	1.000	S.F.	.012	.86	.70	1.56
Furring, 1″ x 3″,16″ O.C.	1.000	L.F.	.023	.49	1.35	1.84
TOTAL		S.F.	.073	6.51	4.29	10.80
16″ OPEN WEB WOOD JOISTS, 16″ O.C.						
OWJ 16″, 16″ O.C., 24′ span	1.000	L.F.	.021	4.25	1.20	5.45
Continuous ribbing, 2″ x 4″	.150	L.F.	.002	.07	.11	.18
Concrete filled steel column, 4″ diameter	.125	L.F.	.002	.21	.14	.35
Girder, built up from three 2″ x 12″	.125	L.F.	.015	.76	.89	1.65
Sheathing, plywood, subfloor, 5/8″ CDX	1.000	S.F.	.012	.86	.70	1.56
Furring, 1″ x 3″, 16″ O.C.	1.000	L.F.	.023	.49	1.35	1.84
TOTAL		S.F.	.075	6.64	4.39	11.03

Floor costs on this page are given on a cost per square foot basis.

Description	QUAN.	UNIT	LABOR HOURS	COST PER S.F. MAT.	INST.	TOTAL

Floor Framing Price Sheet (Wood)	QUAN.	UNIT	LABOR HOURS	COST PER S.F.		
				MAT.	INST.	TOTAL
Open web joists, 12" deep, 12" O.C.	1.250	L.F.	.023	5	1.34	6.34
16" O.C.	1.000	L.F.	.018	4	1.08	5.08
14" deep, 12" O.C.	1.250	L.F.	.024	5.35	1.44	6.79
16" O.C.	1.000	L.F.	.020	4.28	1.15	5.43
16 " deep, 12" O.C.	1.250	L.F.	.026	5.30	1.50	6.80
16" O.C.	1.000	L.F.	.021	4.25	1.20	5.45
18" deep, 12" O.C.	1.250	L.F.	.027	5.75	1.59	7.34
16" O.C.	1.000	L.F.	.022	4.60	1.28	5.88
Continuous ribbing, 2" x 4"	.150	L.F.	.002	.07	.11	.18
2" x 6"	.150	L.F.	.002	.12	.11	.23
2" x 8"	.150	L.F.	.002	.16	.13	.29
2" x 10"	.150	L.F.	.003	.24	.16	.40
2" x 12"	.150	L.F.	.003	.30	.16	.46
Girders, including lally columns, 3 pieces spiked together, 2" x 8"	.125	L.F.	.015	.61	.92	1.53
2" x 10"	.125	L.F.	.016	.81	.98	1.79
2" x 12"	.125	L.F.	.017	.97	1.03	2
Solid girders, 3" x 8"	.040	L.F.	.004	.34	.24	.58
3" x 10"	.040	L.F.	.004	.38	.25	.63
3" x 12"	.040	L.F.	.004	.41	.26	.67
4" x 8"	.040	L.F.	.004	.41	.26	.67
4" x 10"	.040	L.F.	.004	.44	.27	.71
4" x 12"	.040	L.F.	.004	.46	.28	.74
Steel girders, bolted & including fabrication, wide flange shapes						
12" deep, 14#/l.f.	.040	L.F.	.061	24	6.55	30.55
10" deep, 15#/l.f.	.040	L.F.	.067	26	7.15	33.15
8" deep, 10#/l.f.	.040	L.F.	.067	17.45	7.15	24.60
6" deep, 9#/l.f.	.040	L.F.	.067	15.70	7.15	22.85
5" deep, 16#/l.f.	.040	L.F.	.064	25	6.90	31.90
Sheathing, plywood exterior grade CDX, 1/2" thick	1.000	S.F.	.011	.70	.67	1.37
5/8" thick	1.000	S.F.	.012	.86	.70	1.56
3/4" thick	1.000	S.F.	.013	1.03	.75	1.78
Boards, 1" x 8" laid regular	1.000	S.F.	.016	2.36	.94	3.30
Laid diagonal	1.000	S.F.	.019	2.36	1.11	3.47
1" x 10" laid regular	1.000	S.F.	.015	2.36	.86	3.22
Laid diagonal	1.000	S.F.	.018	2.36	1.05	3.41
Furring, 1" x 3", 12" O.C.	1.250	L.F.	.029	.61	1.69	2.30
16" O.C.	1.000	L.F.	.023	.49	1.35	1.84
24" O.C.	.750	L.F.	.017	.37	1.01	1.38

08 Exterior Wall Framing Systems

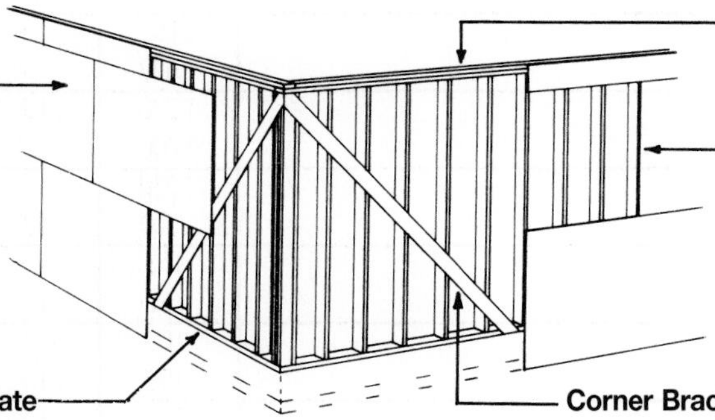

Sheathing
Top Plates
Studs
Bottom Plate
Corner Bracing

System Description	QUAN.	UNIT	LABOR HOURS	COST PER S.F. MAT.	COST PER S.F. INST.	COST PER S.F. TOTAL
2″ X 4″, 16″ O.C.						
2″ x 4″ studs, 16″ O.C.	1.000	L.F.	.015	.46	.86	1.32
Plates, 2″ x 4″, double top, single bottom	.375	L.F.	.005	.17	.32	.49
Corner bracing, let-in, 1″ x 6″	.063	L.F.	.003	.05	.20	.25
Sheathing, 1/2″ plywood, CDX	1.000	S.F.	.011	.70	.67	1.37
Framing connectors, holddowns	.013	Ea.	.013	.58	.77	1.35
TOTAL		S.F.	.047	1.96	2.82	4.78
2″ X 4″, 24″ O.C.						
2″ x 4″ studs, 24″ O.C.	.750	L.F.	.011	.35	.65	1
Plates, 2″ x 4″, double top, single bottom	.375	L.F.	.005	.17	.32	.49
Corner bracing, let-in, 1″ x 6″	.063	L.F.	.002	.05	.13	.18
Sheathing, 1/2″ plywood, CDX	1.000	S.F.	.011	.70	.67	1.37
Framing connectors, holddowns	.013	Ea.	.013	.58	.77	1.35
TOTAL		S.F.	.042	1.85	2.54	4.39
2″ X 6″, 16″ O.C.						
2″ x 6″ studs, 16″ O.C.	1.000	L.F.	.016	.79	.94	1.73
Plates, 2″ x 6″, double top, single bottom	.375	L.F.	.006	.30	.35	.65
Corner bracing, let-in, 1″ x 6″	.063	L.F.	.003	.05	.20	.25
Sheathing, 1/2″ plywood, CDX	1.000	S.F.	.014	.70	.84	1.54
Framing connectors, holddowns	.013	Ea.	.013	.58	.77	1.35
TOTAL		S.F.	.052	2.42	3.10	5.52
2″ X 6″, 24″ O.C.						
2″ x 6″ studs, 24″ O.C.	.750	L.F.	.012	.59	.71	1.30
Plates, 2″ x 6″, double top, single bottom	.375	L.F.	.006	.30	.35	.65
Corner bracing, let-in, 1″ x 6″	.063	L.F.	.002	.05	.13	.18
Sheathing, 1/2″ plywood, CDX	1.000	S.F.	.011	.70	.67	1.37
Framing connectors, holddowns	.013	Ea.	.013	.58	.77	1.35
TOTAL		S.F.	.044	2.22	2.63	4.85

The wall costs on this page are given in cost per square foot of wall.
For window and door openings see below.

Description	QUAN.	UNIT	LABOR HOURS	COST PER S.F. MAT.	COST PER S.F. INST.	COST PER S.F. TOTAL

Exterior Wall Framing Price Sheet	QUAN.	UNIT	LABOR HOURS	COST PER S.F.		
				MAT.	INST.	TOTAL
Studs, #2 or better, 2″ x 4″, 12″ O.C.	1.250	L.F.	.018	.58	1.08	1.66
16″ O.C.	1.000	L.F.	.015	.46	.86	1.32
24″ O.C.	.750	L.F.	.011	.35	.65	1
32″ O.C.	.600	L.F.	.009	.28	.52	.80
2″ x 6″, 12″ O.C.	1.250	L.F.	.020	.99	1.18	2.17
16″ O.C.	1.000	L.F.	.016	.79	.94	1.73
24″ O.C.	.750	L.F.	.012	.59	.71	1.30
32″ O.C.	.600	L.F.	.010	.47	.56	1.03
2″ x 8″, 12″ O.C.	1.250	L.F.	.025	1.69	1.48	3.17
16″ O.C.	1.000	L.F.	.020	1.35	1.18	2.53
24″ O.C.	.750	L.F.	.015	1.01	.89	1.90
32″ O.C.	.600	L.F.	.012	.81	.71	1.52
Plates, #2 or better, double top, single bottom, 2″ x 4″	.375	L.F.	.005	.17	.32	.49
2″ x 6″	.375	L.F.	.006	.30	.35	.65
2″ x 8″	.375	L.F.	.008	.51	.44	.95
Corner bracing, let-in 1″ x 6″ boards, studs, 12″ O.C.	.070	L.F.	.004	.06	.22	.28
16″ O.C.	.063	L.F.	.003	.05	.20	.25
24″ O.C.	.063	L.F.	.002	.05	.13	.18
32″ O.C.	.057	L.F.	.002	.05	.12	.17
Let-in steel ("T" shape), studs, 12″ O.C.	.070	L.F.	.001	.06	.06	.12
16″ O.C.	.063	L.F.	.001	.06	.05	.11
24″ O.C.	.063	L.F.	.001	.06	.05	.11
32″ O.C.	.057	L.F.	.001	.05	.05	.10
Sheathing, plywood CDX, 3/8″ thick	1.000	S.F.	.010	.68	.62	1.30
1/2″ thick	1.000	S.F.	.011	.70	.67	1.37
5/8″ thick	1.000	S.F.	.012	.86	.73	1.59
3/4″ thick	1.000	S.F.	.013	1.03	.79	1.82
Boards, 1″ x 6″, laid regular	1.000	S.F.	.025	1.88	1.45	3.33
Laid diagonal	1.000	S.F.	.027	1.88	1.61	3.49
1″ x 8″, laid regular	1.000	S.F.	.021	2.36	1.23	3.59
Laid diagonal	1.000	S.F.	.025	2.36	1.45	3.81
Wood fiber, regular, no vapor barrier, 1/2″ thick	1.000	S.F.	.013	.70	.79	1.49
5/8″ thick	1.000	S.F.	.013	.78	.79	1.57
Asphalt impregnated 25/32″ thick	1.000	S.F.	.013	.35	.79	1.14
1/2″ thick	1.000	S.F.	.013	.42	.79	1.21
Polystyrene, regular, 3/4″ thick	1.000	S.F.	.010	.62	.59	1.21
2″ thick	1.000	S.F.	.011	1.77	.65	2.42
Fiberglass, foil faced, 1″ thick	1.000	S.F.	.008	.92	.47	1.39
2″ thick	1.000	S.F.	.009	1.74	.53	2.27

Window & Door Openings	QUAN.	UNIT	LABOR HOURS	COST EACH		
				MAT.	INST.	TOTAL
The following costs are to be added to the total costs of the wall for each opening. Do not subtract the area of the openings.						
Headers, 2″ x 6″ double, 2′ long	4.000	L.F.	.178	3.16	10.50	13.66
3′ long	6.000	L.F.	.267	4.74	15.70	20.44
4′ long	8.000	L.F.	.356	6.30	21	27.30
5′ long	10.000	L.F.	.444	7.90	26	33.90
2″ x 8″ double, 4′ long	8.000	L.F.	.376	8.55	22	30.55
5′ long	10.000	L.F.	.471	10.70	27.50	38.20
6′ long	12.000	L.F.	.565	12.85	33	45.85
8′ long	16.000	L.F.	.753	17.10	44.50	61.60
2″ x 10″ double, 4′ long	8.000	L.F.	.400	12.80	23.50	36.30
6′ long	12.000	L.F.	.600	19.20	35.50	54.70
8′ long	16.000	L.F.	.800	25.50	47	72.50
10′ long	20.000	L.F.	1.000	32	59	91
2″ x 12″ double, 8′ long	16.000	L.F.	.853	32	50	82
12′ long	24.000	L.F.	1.280	48	75.50	123.50

12 Gable End Roof Framing Systems

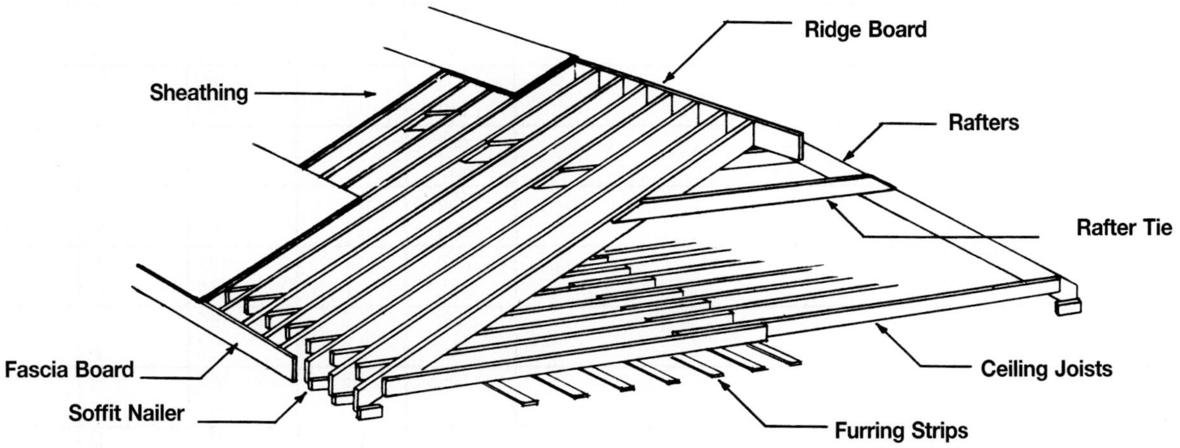

Ridge Board
Sheathing
Rafters
Rafter Tie
Fascia Board
Soffit Nailer
Ceiling Joists
Furring Strips

System Description	QUAN.	UNIT	LABOR HOURS	COST PER S.F. MAT.	COST PER S.F. INST.	COST PER S.F. TOTAL
2″ X 6″ RAFTERS, 16″ O.C., 4/12 PITCH						
Rafters, 2″ x 6″, 16″ O.C., 4/12 pitch	1.170	L.F.	.019	.92	1.10	2.02
Ceiling joists, 2″ x 4″, 16″ O.C.	1.000	L.F.	.013	.47	.75	1.22
Ridge board, 2″ x 6″	.050	L.F.	.002	.04	.09	.13
Fascia board, 2″ x 6″	.100	L.F.	.005	.08	.31	.39
Rafter tie, 1″ x 4″, 4′ O.C.	.060	L.F.	.001	.03	.07	.10
Soffit nailer (outrigger), 2″ x 4″, 24″ O.C.	.170	L.F.	.004	.08	.26	.34
Sheathing, exterior, plywood, CDX, 1/2″ thick	1.170	S.F.	.013	.82	.78	1.60
Furring strips, 1″ x 3″, 16″ O.C.	1.000	L.F.	.023	.49	1.35	1.84
Rafter ties	.053	Ea.	.003	.08	.17	.25
TOTAL		S.F.	.083	3.01	4.88	7.89
2″ X 8″ RAFTERS, 16″ O.C., 4/12 PITCH						
Rafters, 2″ x 8″, 16″ O.C., 4/12 pitch	1.170	L.F.	.020	1.25	1.16	2.41
Ceiling joists, 2″ x 6″, 16″ O.C.	1.000	L.F.	.013	.79	.75	1.54
Ridge board, 2″ x 8″	.050	L.F.	.002	.05	.10	.15
Fascia board, 2″ x 8″	.100	L.F.	.007	.11	.42	.53
Rafter tie, 1″ x 4″, 4′ O.C.	.060	L.F.	.001	.03	.07	.10
Soffit nailer (outrigger), 2″ x 4″, 24″ O.C.	.170	L.F.	.004	.08	.26	.34
Sheathing, exterior, plywood, CDX, 1/2″ thick	1.170	S.F.	.013	.82	.78	1.60
Furring strips, 1″ x 3″, 16″ O.C.	1.000	L.F.	.023	.49	1.35	1.84
Rafter ties	.053	Ea.	.003	.08	.17	.25
TOTAL		S.F.	.086	3.70	5.06	8.76

The cost of this system is based on the square foot of plan area.
All quantities have been adjusted accordingly.

Description	QUAN.	UNIT	LABOR HOURS	COST PER S.F. MAT.	COST PER S.F. INST.	COST PER S.F. TOTAL

Gable End Roof Framing Price Sheet	QUAN.	UNIT	LABOR HOURS	COST PER S.F.		
				MAT.	INST.	TOTAL
Rafters, #2 or better, 16″ O.C., 2″ x 6″, 4/12 pitch	1.170	L.F.	.019	.92	1.10	2.02
8/12 pitch	1.330	L.F.	.027	1.05	1.57	2.62
2″ x 8″, 4/12 pitch	1.170	L.F.	.020	1.25	1.16	2.41
8/12 pitch	1.330	L.F.	.028	1.42	1.68	3.10
2″ x 10″, 4/12 pitch	1.170	L.F.	.030	1.87	1.76	3.63
8/12 pitch	1.330	L.F.	.043	2.13	2.53	4.66
24″ O.C., 2″ x 6″, 4/12 pitch	.940	L.F.	.015	.74	.88	1.62
8/12 pitch	1.060	L.F.	.021	.84	1.25	2.09
2″ x 8″, 4/12 pitch	.940	L.F.	.016	1.01	.93	1.94
8/12 pitch	1.060	L.F.	.023	1.13	1.34	2.47
2″ x 10″, 4/12 pitch	.940	L.F.	.024	1.50	1.41	2.91
8/12 pitch	1.060	L.F.	.034	1.70	2.01	3.71
Ceiling joist, #2 or better, 2″ x 4″, 16″ O.C.	1.000	L.F.	.013	.47	.75	1.22
24″ O.C.	.750	L.F.	.010	.35	.56	.91
2″ x 6″, 16″ O.C.	1.000	L.F.	.013	.79	.75	1.54
24″ O.C.	.750	L.F.	.010	.59	.56	1.15
2″ x 8″, 16″ O.C.	1.000	L.F.	.015	1.07	.86	1.93
24″ O.C.	.750	L.F.	.011	.80	.65	1.45
2″ x 10″, 16″ O.C.	1.000	L.F.	.018	1.60	1.05	2.65
24″ O.C.	.750	L.F.	.013	1.20	.79	1.99
Ridge board, #2 or better, 1″ x 6″	.050	L.F.	.001	.04	.08	.12
1″ x 8″	.050	L.F.	.001	.07	.09	.16
1″ x 10″	.050	L.F.	.002	.09	.09	.18
2″ x 6″	.050	L.F.	.002	.04	.09	.13
2″ x 8″	.050	L.F.	.002	.05	.10	.15
2″ x 10″	.050	L.F.	.002	.08	.12	.20
Fascia board, #2 or better, 1″ x 6″	.100	L.F.	.004	.06	.23	.29
1″ x 8″	.100	L.F.	.005	.07	.27	.34
1″ x 10″	.100	L.F.	.005	.08	.30	.38
2″ x 6″	.100	L.F.	.006	.09	.34	.43
2″ x 8″	.100	L.F.	.007	.11	.42	.53
2″ x 10″	.100	L.F.	.004	.32	.21	.53
Rafter tie, #2 or better, 4′ O.C., 1″ x 4″	.060	L.F.	.001	.03	.07	.10
1″ x 6″	.060	L.F.	.001	.04	.08	.12
2″ x 4″	.060	L.F.	.002	.05	.09	.14
2″ x 6″	.060	L.F.	.002	.06	.12	.18
Soffit nailer (outrigger), 2″ x 4″, 16″ O.C.	.220	L.F.	.006	.10	.33	.43
24″ O.C.	.170	L.F.	.004	.08	.26	.34
2″ x 6″, 16″ O.C.	.220	L.F.	.006	.12	.38	.50
24″ O.C.	.170	L.F.	.005	.09	.30	.39
Sheathing, plywood CDX, 4/12 pitch, 3/8″ thick.	1.170	S.F.	.012	.80	.73	1.53
1/2″ thick	1.170	S.F.	.013	.82	.78	1.60
5/8″ thick	1.170	S.F.	.014	1.01	.85	1.86
8/12 pitch, 3/8″	1.330	S.F.	.014	.90	.82	1.72
1/2″ thick	1.330	S.F.	.015	.93	.89	1.82
5/8″ thick	1.330	S.F.	.016	1.14	.97	2.11
Boards, 4/12 pitch roof, 1″ x 6″	1.170	S.F.	.026	2.20	1.52	3.72
1″ x 8″	1.170	S.F.	.021	2.76	1.26	4.02
8/12 pitch roof, 1″ x 6″	1.330	S.F.	.029	2.50	1.73	4.23
1″ x 8″	1.330	S.F.	.024	3.14	1.44	4.58
Furring, 1″ x 3″, 12″ O.C.	1.200	L.F.	.027	.59	1.62	2.21
16″ O.C.	1.000	L.F.	.023	.49	1.35	1.84
24″ O.C.	.800	L.F.	.018	.39	1.08	1.47

16 Truss Roof Framing Systems

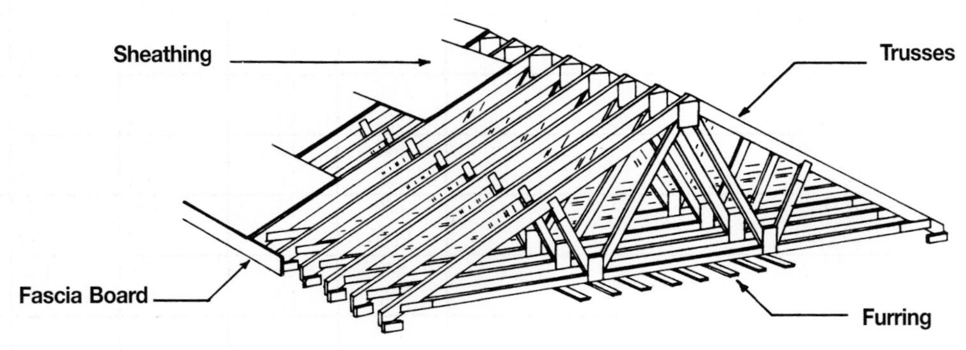

System Description	QUAN.	UNIT	LABOR HOURS	COST PER S.F. MAT.	COST PER S.F. INST.	COST PER S.F. TOTAL
TRUSS, 16″ O.C., 4/12 PITCH, 1′ OVERHANG, 26′ SPAN						
Truss, 40# loading, 16″ O.C., 4/12 pitch, 26′ span	.030	Ea.	.021	2.57	1.41	3.98
Fascia board, 2″ x 6″	.100	L.F.	.005	.08	.31	.39
Sheathing, exterior, plywood, CDX, 1/2″ thick	1.170	S.F.	.013	.82	.78	1.60
Furring, 1″ x 3″, 16″ O.C.	1.000	L.F.	.023	.49	1.35	1.84
Rafter ties	.053	Ea.	.003	.08	.17	.25
TOTAL		S.F.	.065	4.04	4.02	8.06
TRUSS, 16″ O.C., 8/12 PITCH, 1′ OVERHANG, 26′ SPAN						
Truss, 40# loading, 16″ O.C., 8/12 pitch, 26′ span	.030	Ea.	.023	3.63	1.55	5.18
Fascia board, 2″ x 6″	.100	L.F.	.005	.08	.31	.39
Sheathing, exterior, plywood, CDX, 1/2″ thick	1.330	S.F.	.015	.93	.89	1.82
Furring, 1″ x 3″, 16″ O.C.	1.000	L.F.	.023	.49	1.35	1.84
Rafter ties	.053	Ea.	.003	.08	.17	.25
TOTAL		S.F.	.069	5.21	4.27	9.48
TRUSS, 24″ O.C., 4/12 PITCH, 1′ OVERHANG, 26′ SPAN						
Truss, 40# loading, 24″ O.C., 4/12 pitch, 26′ span	.020	Ea.	.014	1.71	.94	2.65
Fascia board, 2″ x 6″	.100	L.F.	.005	.08	.31	.39
Sheathing, exterior, plywood, CDX, 1/2″ thick	1.170	S.F.	.013	.82	.78	1.60
Furring, 1″ x 3″, 16″ O.C.	1.000	L.F.	.023	.49	1.35	1.84
Rafter ties	.035	Ea.	.002	.05	.11	.16
TOTAL		S.F.	.057	3.15	3.49	6.64
TRUSS, 24″ O.C., 8/12 PITCH, 1′ OVERHANG, 26′ SPAN						
Truss, 40# loading, 24″ O.C., 8/12 pitch, 26′ span	.020	Ea.	.015	2.42	1.03	3.45
Fascia board, 2″ x 6″	.100	L.F.	.005	.08	.31	.39
Sheathing, exterior, plywood, CDX, 1/2″ thick	1.330	S.F.	.015	.93	.89	1.82
Furring, 1″ x 3″, 16″ O.C.	1.000	L.F.	.023	.49	1.35	1.84
Rafter ties	.035	Ea.	.002	.05	.11	.16
TOTAL		S.F.	.060	3.97	3.69	7.66

The cost of this system is based on the square foot of plan area.
A one foot overhang is included.

Description	QUAN.	UNIT	LABOR HOURS	COST PER S.F. MAT.	COST PER S.F. INST.	COST PER S.F. TOTAL

Truss Roof Framing Price Sheet	QUAN.	UNIT	LABOR HOURS	COST PER S.F.		
				MAT.	INST.	TOTAL
Truss, 40# loading, including 1′ overhang, 4/12 pitch, 24′ span, 16″ O.C.	.033	Ea.	.022	2.84	1.48	4.32
24″ O.C.	.022	Ea.	.015	1.89	.98	2.87
26′ span, 16″ O.C.	.030	Ea.	.021	2.57	1.41	3.98
24″ O.C.	.020	Ea.	.014	1.71	.94	2.65
28′ span, 16″ O.C.	.027	Ea.	.020	2.92	1.37	4.29
24″ O.C.	.019	Ea.	.014	2.05	.97	3.02
32′ span, 16″ O.C.	.024	Ea.	.019	3	1.29	4.29
24″ O.C.	.016	Ea.	.013	2	.87	2.87
36′ span, 16″ O.C.	.022	Ea.	.019	3.34	1.28	4.62
24″ O.C.	.015	Ea.	.013	2.28	.88	3.16
8/12 pitch, 24′ span, 16″ O.C.	.033	Ea.	.024	3.80	1.61	5.41
24″ O.C.	.022	Ea.	.016	2.53	1.08	3.61
26′ span, 16″ O.C.	.030	Ea.	.023	3.63	1.55	5.18
24″ O.C.	.020	Ea.	.015	2.42	1.03	3.45
28′ span, 16″ O.C.	.027	Ea.	.022	3.67	1.48	5.15
24″ O.C.	.019	Ea.	.016	2.58	1.04	3.62
32′ span, 16″ O.C.	.024	Ea.	.021	3.82	1.43	5.25
24″ O.C.	.016	Ea.	.014	2.54	.95	3.49
36′ span, 16″ O.C.	.022	Ea.	.021	4.11	1.45	5.56
24″ O.C.	.015	Ea.	.015	2.81	.99	3.80
Fascia board, #2 or better, 1″ x 6″	.100	L.F.	.004	.06	.23	.29
1″ x 8″	.100	L.F.	.005	.07	.27	.34
1″ x 10″	.100	L.F.	.005	.08	.30	.38
2″ x 6″	.100	L.F.	.006	.09	.34	.43
2″ x 8″	.100	L.F.	.007	.11	.42	.53
2″ x 10″	.100	L.F.	.009	.16	.53	.69
Sheathing, plywood CDX, 4/12 pitch, 3/8″ thick	1.170	S.F.	.012	.80	.73	1.53
1/2″ thick	1.170	S.F.	.013	.82	.78	1.60
5/8″ thick	1.170	S.F.	.014	1.01	.85	1.86
8/12 pitch, 3/8″ thick	1.330	S.F.	.014	.90	.82	1.72
1/2″ thick	1.330	S.F.	.015	.93	.89	1.82
5/8″ thick	1.330	S.F.	.016	1.14	.97	2.11
Boards, 4/12 pitch, 1″ x 6″	1.170	S.F.	.026	2.20	1.52	3.72
1″ x 8″	1.170	S.F.	.021	2.76	1.26	4.02
8/12 pitch, 1″ x 6″	1.330	S.F.	.029	2.50	1.73	4.23
1″ x 8″	1.330	S.F.	.024	3.14	1.44	4.58
Furring, 1″ x 3″, 12″ O.C.	1.200	L.F.	.027	.59	1.62	2.21
16″ O.C.	1.000	L.F.	.023	.49	1.35	1.84
24″ O.C.	.800	L.F.	.018	.39	1.08	1.47

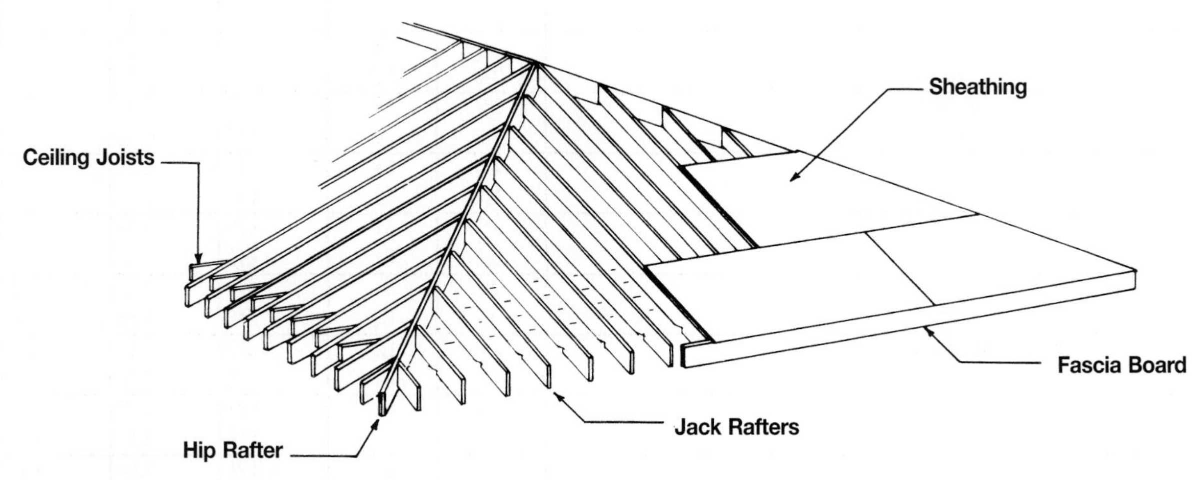

System Description	QUAN.	UNIT	LABOR HOURS	COST PER S.F. MAT.	INST.	TOTAL
2″ X 6″, 16″ O.C., 4/12 PITCH						
Hip rafters, 2″ x 8″, 4/12 pitch	.160	L.F.	.004	.17	.21	.38
Jack rafters, 2″ x 6″, 16″ O.C., 4/12 pitch	1.430	L.F.	.038	1.13	2.25	3.38
Ceiling joists, 2″ x 6″, 16″ O.C.	1.000	L.F.	.013	.79	.75	1.54
Fascia board, 2″ x 8″	.220	L.F.	.016	.24	.92	1.16
Soffit nailer (outrigger), 2″ x 4″, 24″ O.C.	.220	L.F.	.006	.10	.33	.43
Sheathing, 1/2″ exterior plywood, CDX	1.570	S.F.	.018	1.10	1.05	2.15
Furring strips, 1″ x 3″, 16″ O.C.	1.000	L.F.	.023	.49	1.35	1.84
Rafter ties	.070	Ea.	.004	.11	.23	.34
TOTAL		S.F.	.122	4.13	7.09	11.22
2″ X 8″, 16″ O.C., 4/12 PITCH						
Hip rafters, 2″ x 10″, 4/12 pitch	.160	L.F.	.004	.26	.26	.52
Jack rafters, 2″ x 8″, 16″ O.C., 4/12 pitch	1.430	L.F.	.047	1.53	2.75	4.28
Ceiling joists, 2″ x 6″, 16″ O.C.	1.000	L.F.	.013	.79	.75	1.54
Fascia board, 2″ x 8″	.220	L.F.	.012	.18	.71	.89
Soffit nailer (outrigger), 2″ x 4″, 24″ O.C.	.220	L.F.	.006	.10	.33	.43
Sheathing, 1/2″ exterior plywood, CDX	1.570	S.F.	.018	1.10	1.05	2.15
Furring strips, 1″ x 3″, 16″ O.C.	1.000	L.F.	.023	.49	1.35	1.84
Rafter ties	.070	Ea.	.004	.11	.23	.34
TOTAL		S.F.	.127	4.56	7.43	11.99

The cost of this system is based on S.F. of plan area. Measurement is area under the hip roof only. See gable roof system for added costs.

Description	QUAN.	UNIT	LABOR HOURS	COST PER S.F. MAT.	INST.	TOTAL

Hip Roof Framing Price Sheet	QUAN.	UNIT	LABOR HOURS	COST PER S.F.		
				MAT.	INST.	TOTAL
Hip rafters, #2 or better, 2″ x 6″, 4/12 pitch	.160	L.F.	.003	.13	.20	.33
8/12 pitch	.210	L.F.	.006	.17	.34	.51
2″ x 8″, 4/12 pitch	.160	L.F.	.004	.17	.21	.38
8/12 pitch	.210	L.F.	.006	.22	.36	.58
2″ x 10″, 4/12 pitch	.160	L.F.	.004	.26	.26	.52
8/12 pitch roof	.210	L.F.	.008	.34	.45	.79
Jack rafters, #2 or better, 16″ O.C., 2″ x 6″, 4/12 pitch	1.430	L.F.	.038	1.13	2.25	3.38
8/12 pitch	1.800	L.F.	.061	1.42	3.56	4.98
2″ x 8″, 4/12 pitch	1.430	L.F.	.047	1.53	2.75	4.28
8/12 pitch	1.800	L.F.	.075	1.93	4.41	6.34
2″ x 10″, 4/12 pitch	1.430	L.F.	.051	2.29	2.99	5.28
8/12 pitch	1.800	L.F.	.082	2.88	4.84	7.72
24″ O.C., 2″ x 6″, 4/12 pitch	1.150	L.F.	.031	.91	1.81	2.72
8/12 pitch	1.440	L.F.	.048	1.14	2.85	3.99
2″ x 8″, 4/12 pitch	1.150	L.F.	.038	1.23	2.21	3.44
8/12 pitch	1.440	L.F.	.060	1.54	3.53	5.07
2″ x 10″, 4/12 pitch	1.150	L.F.	.041	1.84	2.40	4.24
8/12 pitch	1.440	L.F.	.066	2.30	3.87	6.17
Ceiling joists, #2 or better, 2″ x 4″, 16″ O.C.	1.000	L.F.	.013	.47	.75	1.22
24″ O.C.	.750	L.F.	.010	.35	.56	.91
2″ x 6″, 16″ O.C.	1.000	L.F.	.013	.79	.75	1.54
24″ O.C.	.750	L.F.	.010	.59	.56	1.15
2″ x 8″, 16″ O.C.	1.000	L.F.	.015	1.07	.86	1.93
24″ O.C.	.750	L.F.	.011	.80	.65	1.45
2″ x 10″, 16″ O.C.	1.000	L.F.	.018	1.60	1.05	2.65
24″ O.C.	.750	L.F.	.013	1.20	.79	1.99
Fascia board, #2 or better, 1″ x 6″	.220	L.F.	.009	.13	.51	.64
1″ x 8″	.220	L.F.	.010	.15	.59	.74
1″ x 10″	.220	L.F.	.011	.17	.66	.83
2″ x 6″	.220	L.F.	.013	.19	.74	.93
2″ x 8″	.220	L.F.	.016	.24	.92	1.16
2″ x 10″	.220	L.F.	.020	.35	1.16	1.51
Soffit nailer (outrigger), 2″ x 4″, 16″ O.C.	.280	L.F.	.007	.13	.43	.56
24″ O.C.	.220	L.F.	.006	.10	.33	.43
2″ x 8″, 16″ O.C.	.280	L.F.	.007	.22	.39	.61
24″ O.C.	.220	L.F.	.005	.18	.32	.50
Sheathing, plywood CDX, 4/12 pitch, 3/8″ thick	1.570	S.F.	.016	1.07	.97	2.04
1/2″ thick	1.570	S.F.	.018	1.10	1.05	2.15
5/8″ thick	1.570	S.F.	.019	1.35	1.15	2.50
8/12 pitch, 3/8″ thick	1.900	S.F.	.020	1.29	1.18	2.47
1/2″ thick	1.900	S.F.	.022	1.33	1.27	2.60
5/8″ thick	1.900	S.F.	.023	1.63	1.39	3.02
Boards, 4/12 pitch, 1″ x 6″ boards	1.450	S.F.	.032	2.73	1.89	4.62
1″ x 8″ boards	1.450	S.F.	.027	3.42	1.57	4.99
8/12 pitch, 1″ x 6″ boards	1.750	S.F.	.039	3.29	2.28	5.57
1″ x 8″ boards	1.750	S.F.	.032	4.13	1.89	6.02
Furring, 1″ x 3″, 12″ O.C.	1.200	L.F.	.027	.59	1.62	2.21
16″ O.C.	1.000	L.F.	.023	.49	1.35	1.84
24″ O.C.	.800	L.F.	.018	.39	1.08	1.47

24 Gambrel Roof Framing Systems

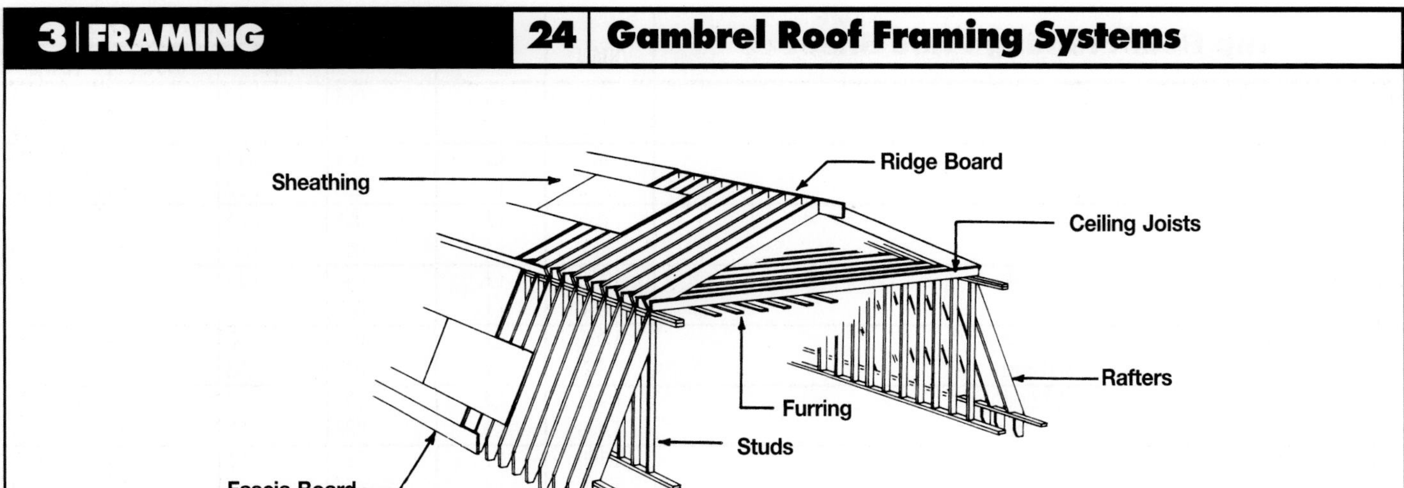

System Description	QUAN.	UNIT	LABOR HOURS	COST PER S.F. MAT.	COST PER S.F. INST.	COST PER S.F. TOTAL
2″ X 6″ RAFTERS, 16″ O.C.						
Roof rafters, 2″ x 6″, 16″ O.C.	1.430	L.F.	.029	1.13	1.69	2.82
Ceiling joists, 2″ x 6″, 16″ O.C.	.710	L.F.	.009	.56	.53	1.09
Stud wall, 2″ x 4″, 16″ O.C., including plates	.790	L.F.	.012	.37	.73	1.10
Furring strips, 1″ x 3″, 16″ O.C.	.710	L.F.	.016	.35	.96	1.31
Ridge board, 2″ x 8″	.050	L.F.	.002	.05	.10	.15
Fascia board, 2″ x 6″	.100	L.F.	.006	.09	.34	.43
Sheathing, exterior grade plywood, 1/2″ thick	1.450	S.F.	.017	1.02	.97	1.99
Rafter ties	.106	Ea.	.006	.16	.34	.50
TOTAL		S.F.	.097	3.73	5.66	9.39
2″ X 8″ RAFTERS, 16″ O.C.						
Roof rafters, 2″ x 8″, 16″ O.C.	1.430	L.F.	.031	1.53	1.80	3.33
Ceiling joists, 2″ x 6″, 16″ O.C.	.710	L.F.	.009	.56	.53	1.09
Stud wall, 2″ x 4″, 16″ O.C., including plates	.790	L.F.	.012	.37	.73	1.10
Furring strips, 1″ x 3″, 16″ O.C.	.710	L.F.	.016	.35	.96	1.31
Ridge board, 2″ x 8″	.050	L.F.	.002	.05	.10	.15
Fascia board, 2″ x 8″	.100	L.F.	.007	.11	.42	.53
Sheathing, exterior grade plywood, 1/2″ thick	1.450	S.F.	.017	1.02	.97	1.99
Rafter ties	.106	Ea.	.006	.16	.34	.50
TOTAL		S.F.	.100	4.15	5.85	10

The cost of this system is based on the square foot of plan area on the first floor.

Description	QUAN.	UNIT	LABOR HOURS	COST PER S.F. MAT.	COST PER S.F. INST.	COST PER S.F. TOTAL

Gambrel Roof Framing Price Sheet	QUAN.	UNIT	LABOR HOURS	COST PER S.F.		
				MAT.	INST.	TOTAL
Roof rafters, #2 or better, 2" x 6", 16" O.C.	1.430	L.F.	.029	1.13	1.69	2.82
24" O.C.	1.140	L.F.	.023	.90	1.35	2.25
2" x 8", 16" O.C.	1.430	L.F.	.031	1.53	1.80	3.33
24" O.C.	1.140	L.F.	.024	1.22	1.44	2.66
2" x 10", 16" O.C.	1.430	L.F.	.046	2.29	2.72	5.01
24" O.C.	1.140	L.F.	.037	1.82	2.17	3.99
Ceiling joist, #2 or better, 2" x 4", 16" O.C.	.710	L.F.	.009	.33	.53	.86
24" O.C.	.570	L.F.	.007	.27	.43	.70
2" x 6", 16" O.C.	.710	L.F.	.009	.56	.53	1.09
24" O.C.	.570	L.F.	.007	.45	.43	.88
2" x 8", 16" O.C.	.710	L.F.	.010	.76	.61	1.37
24" O.C.	.570	L.F.	.008	.61	.49	1.10
Stud wall, #2 or better, 2" x 4", 16" O.C.	.790	L.F.	.012	.37	.73	1.10
24" O.C.	.630	L.F.	.010	.30	.59	.89
2" x 6", 16" O.C.	.790	L.F.	.014	.62	.84	1.46
24" O.C.	.630	L.F.	.011	.50	.67	1.17
Furring, 1" x 3", 16" O.C.	.710	L.F.	.016	.35	.96	1.31
24" O.C.	.590	L.F.	.013	.29	.80	1.09
Ridge board, #2 or better, 1" x 6"	.050	L.F.	.001	.04	.08	.12
1" x 8"	.050	L.F.	.001	.07	.09	.16
1" x 10"	.050	L.F.	.002	.09	.09	.18
2" x 6"	.050	L.F.	.002	.04	.09	.13
2" x 8"	.050	L.F.	.002	.05	.10	.15
2" x 10"	.050	L.F.	.002	.08	.12	.20
Fascia board, #2 or better, 1" x 6"	.100	L.F.	.004	.06	.23	.29
1" x 8"	.100	L.F.	.005	.07	.27	.34
1" x 10"	.100	L.F.	.005	.08	.30	.38
2" x 6"	.100	L.F.	.006	.09	.34	.43
2" x 8"	.100	L.F.	.007	.11	.42	.53
2" x 10"	.100	L.F.	.009	.16	.53	.69
Sheathing, plywood, exterior grade CDX, 3/8" thick	1.450	S.F.	.015	.99	.90	1.89
1/2" thick	1.450	S.F.	.017	1.02	.97	1.99
5/8" thick	1.450	S.F.	.018	1.25	1.06	2.31
3/4" thick	1.450	S.F.	.019	1.49	1.15	2.64
Boards, 1" x 6", laid regular	1.450	S.F.	.032	2.73	1.89	4.62
Laid diagonal	1.450	S.F.	.036	2.73	2.10	4.83
1" x 8", laid regular	1.450	S.F.	.027	3.42	1.57	4.99
Laid diagonal	1.450	S.F.	.032	3.42	1.89	5.31

28 Mansard Roof Framing Systems

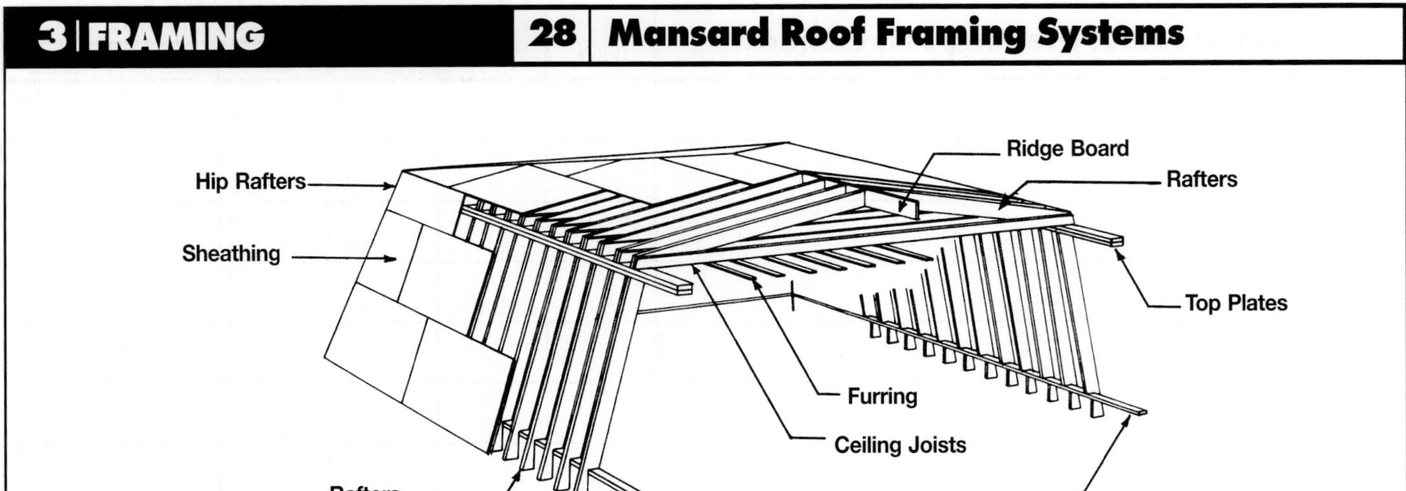

System Description	QUAN.	UNIT	LABOR HOURS	COST PER S.F. MAT.	INST.	TOTAL
2″ X 6″ RAFTERS, 16″ O.C.						
Roof rafters, 2″ x 6″, 16″ O.C.	1.210	L.F.	.033	.96	1.94	2.90
Rafter plates, 2″ x 6″, double top, single bottom	.364	L.F.	.010	.29	.58	.87
Ceiling joists, 2″ x 4″, 16″ O.C.	.920	L.F.	.012	.43	.69	1.12
Hip rafter, 2″ x 6″	.070	L.F.	.002	.06	.13	.19
Jack rafter, 2″ x 6″, 16″ O.C.	1.000	L.F.	.039	.79	2.30	3.09
Ridge board, 2″ x 6″	.018	L.F.	.001	.01	.03	.04
Sheathing, exterior grade plywood, 1/2″ thick	2.210	S.F.	.025	1.55	1.48	3.03
Furring strips, 1″ x 3″, 16″ O.C.	.920	L.F.	.021	.45	1.24	1.69
Rafter ties	.140	Ea.	.008	.21	.46	.67
TOTAL		S.F.	.151	4.75	8.85	13.60
2″ X 8″ RAFTERS, 16″ O.C.						
Roof rafters, 2″ x 8″, 16″ O.C.	1.210	L.F.	.036	1.29	2.12	3.41
Rafter plates, 2″ x 8″, double top, single bottom	.364	L.F.	.011	.39	.64	1.03
Ceiling joists, 2″ x 6″, 16″ O.C.	.920	L.F.	.012	.73	.69	1.42
Hip rafter, 2″ x 8″	.070	L.F.	.002	.07	.14	.21
Jack rafter, 2″ x 8″, 16″ O.C.	1.000	L.F.	.048	1.07	2.81	3.88
Ridge board, 2″ x 8″	.018	L.F.	.001	.02	.04	.06
Sheathing, exterior grade plywood, 1/2″ thick	2.210	S.F.	.025	1.55	1.48	3.03
Furring strips, 1″ x 3″, 16″ O.C.	.920	L.F.	.021	.45	1.24	1.69
Rafter ties	.140	Ea.	.008	.21	.46	.67
TOTAL		S.F.	.164	5.78	9.62	15.40

The cost of this system is based on the square foot of plan area.

Description	QUAN.	UNIT	LABOR HOURS	COST PER S.F. MAT.	INST.	TOTAL

Mansard Roof Framing Price Sheet	QUAN.	UNIT	LABOR HOURS	COST PER S.F.		
				MAT.	INST.	TOTAL
Roof rafters, #2 or better, 2″ x 6″, 16″ O.C.	1.210	L.F.	.033	.96	1.94	2.90
24″ O.C.	.970	L.F.	.026	.77	1.55	2.32
2″ x 8″, 16″ O.C.	1.210	L.F.	.036	1.29	2.12	3.41
24″ O.C.	.970	L.F.	.029	1.04	1.70	2.74
2″ x 10″, 16″ O.C.	1.210	L.F.	.046	1.94	2.69	4.63
24″ O.C.	.970	L.F.	.037	1.55	2.15	3.70
Rafter plates, #2 or better double top single bottom, 2″ x 6″	.364	L.F.	.010	.29	.58	.87
2″ x 8″	.364	L.F.	.011	.39	.64	1.03
2″ x 10″	.364	L.F.	.014	.58	.81	1.39
Ceiling joist, #2 or better, 2″ x 4″, 16″ O.C.	.920	L.F.	.012	.43	.69	1.12
24″ O.C.	.740	L.F.	.009	.35	.56	.91
2″ x 6″, 16″ O.C.	.920	L.F.	.012	.73	.69	1.42
24″ O.C.	.740	L.F.	.009	.58	.56	1.14
2″ x 8″, 16″ O.C.	.920	L.F.	.013	.98	.79	1.77
24″ O.C.	.740	L.F.	.011	.79	.64	1.43
Hip rafter, #2 or better, 2″ x 6″	.070	L.F.	.002	.06	.13	.19
2″ x 8″	.070	L.F.	.002	.07	.14	.21
2″ x 10″	.070	L.F.	.003	.11	.17	.28
Jack rafter, #2 or better, 2″ x 6″, 16″ O.C.	1.000	L.F.	.039	.79	2.30	3.09
24″ O.C.	.800	L.F.	.031	.63	1.84	2.47
2″ x 8″, 16″ O.C.	1.000	L.F.	.048	1.07	2.81	3.88
24″ O.C.	.800	L.F.	.038	.86	2.25	3.11
Ridge board, #2 or better, 1″ x 6″	.018	L.F.	.001	.02	.03	.05
1″ x 8″	.018	L.F.	.001	.03	.03	.06
1″ x 10″	.018	L.F.	.001	.03	.03	.06
2″ x 6″	.018	L.F.	.001	.01	.03	.04
2″ x 8″	.018	L.F.	.001	.02	.04	.06
2″ x 10″	.018	L.F.	.001	.03	.04	.07
Sheathing, plywood exterior grade CDX, 3/8″ thick	2.210	S.F.	.023	1.50	1.37	2.87
1/2″ thick	2.210	S.F.	.025	1.55	1.48	3.03
5/8″ thick	2.210	S.F.	.027	1.90	1.61	3.51
3/4″ thick	2.210	S.F.	.029	2.28	1.75	4.03
Boards, 1″ x 6″, laid regular	2.210	S.F.	.049	4.15	2.87	7.02
Laid diagonal	2.210	S.F.	.054	4.15	3.20	7.35
1″ x 8″, laid regular	2.210	S.F.	.040	5.20	2.39	7.59
Laid diagonal	2.210	S.F.	.049	5.20	2.87	8.07
Furring, 1″ x 3″, 12″ O.C.	1.150	L.F.	.026	.56	1.55	2.11
24″ O.C.	.740	L.F.	.017	.36	1	1.36

32 Shed/Flat Roof Framing Systems

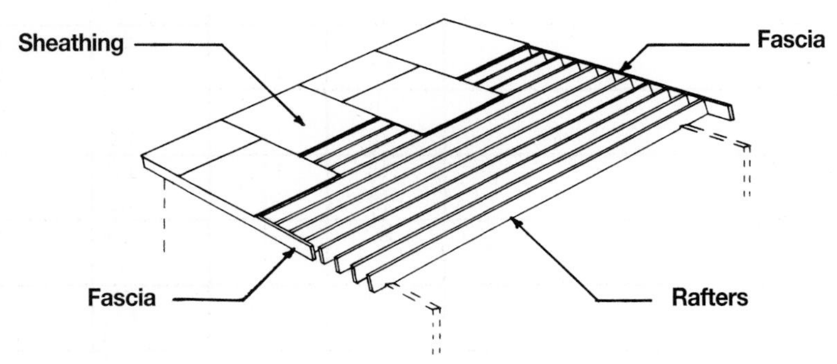

System Description	QUAN.	UNIT	LABOR HOURS	COST PER S.F. MAT.	COST PER S.F. INST.	COST PER S.F. TOTAL
2″ X 6″,16″ O.C., 4/12 PITCH						
Rafters, 2″ x 6″, 16″ O.C., 4/12 pitch	1.170	L.F.	.019	.92	1.10	2.02
Fascia, 2″ x 6″	.100	L.F.	.006	.09	.34	.43
Bridging, 1″ x 3″, 6′ O.C.	.080	Pr.	.005	.06	.29	.35
Sheathing, exterior grade plywood, 1/2″ thick	1.230	S.F.	.014	.86	.82	1.68
Rafter ties	.053	Ea.	.003	.08	.17	.25
TOTAL		S.F.	.047	2.01	2.72	4.73
2″ X 6″, 24″ O.C., 4/12 PITCH						
Rafters, 2″ x 6″, 24″ O.C., 4/12 pitch	.940	L.F.	.015	.74	.88	1.62
Fascia, 2″ x 6″	.100	L.F.	.006	.09	.34	.43
Bridging, 1″ x 3″, 6′ O.C.	.060	Pr.	.004	.05	.22	.27
Sheathing, exterior grade plywood, 1/2″ thick	1.230	S.F.	.014	.86	.82	1.68
Rafter ties	.035	Ea.	.002	.05	.11	.16
TOTAL		S.F.	.041	1.79	2.37	4.16
2″ X 8″, 16″ O.C., 4/12 PITCH						
Rafters, 2″ x 8″, 16″ O.C., 4/12 pitch	1.170	L.F.	.020	1.25	1.16	2.41
Fascia, 2″ x 8″	.100	L.F.	.007	.11	.42	.53
Bridging, 1″ x 3″, 6′ O.C.	.080	Pr.	.005	.06	.29	.35
Sheathing, exterior grade plywood, 1/2″ thick	1.230	S.F.	.014	.86	.82	1.68
Rafter ties	.053	Ea.	.003	.08	.17	.25
TOTAL		S.F.	.049	2.36	2.86	5.22
2″ X 8″, 24″ O.C., 4/12 PITCH						
Rafters, 2″ x 8″, 24″ O.C., 4/12 pitch	.940	L.F.	.016	1.01	.93	1.94
Fascia, 2″ x 8″	.100	L.F.	.007	.11	.42	.53
Bridging, 1″ x 3″, 6′ O.C.	.060	Pr.	.004	.05	.22	.27
Sheathing, exterior grade plywood, 1/2″ thick	1.230	S.F.	.014	.86	.82	1.68
Rafter ties	.035	Ea.	.002	.05	.11	.16
TOTAL		S.F.	.043	2.08	2.50	4.58

The cost of this system is based on the square foot of plan area.
A 1′ overhang is assumed. No ceiling joists or furring are included.

Description	QUAN.	UNIT	LABOR HOURS	COST PER S.F. MAT.	COST PER S.F. INST.	COST PER S.F. TOTAL

Shed/Flat Roof Framing Price Sheet	QUAN.	UNIT	LABOR HOURS	COST PER S.F.		
				MAT.	INST.	TOTAL
Rafters, #2 or better, 16″ O.C., 2″ x 4″, 0 - 4/12 pitch	1.170	L.F.	.014	.69	.82	1.51
5/12 - 8/12 pitch	1.330	L.F.	.020	.79	1.18	1.97
2″ x 6″, 0 - 4/12 pitch	1.170	L.F.	.019	.92	1.10	2.02
5/12 - 8/12 pitch	1.330	L.F.	.027	1.05	1.57	2.62
2″ x 8″, 0 - 4/12 pitch	1.170	L.F.	.020	1.25	1.16	2.41
5/12 - 8/12 pitch	1.330	L.F.	.028	1.42	1.68	3.10
2″ x 10″, 0 - 4/12 pitch	1.170	L.F.	.030	1.87	1.76	3.63
5/12 - 8/12 pitch	1.330	L.F.	.043	2.13	2.53	4.66
24″ O.C., 2″ x 4″, 0 - 4/12 pitch	.940	L.F.	.011	.56	.67	1.23
5/12 - 8/12 pitch	1.060	L.F.	.021	.84	1.25	2.09
2″ x 6″, 0 - 4/12 pitch	.940	L.F.	.015	.74	.88	1.62
5/12 - 8/12 pitch	1.060	L.F.	.021	.84	1.25	2.09
2″ x 8″, 0 - 4/12 pitch	.940	L.F.	.016	1.01	.93	1.94
5/12 - 8/12 pitch	1.060	L.F.	.023	1.13	1.34	2.47
2″ x 10″, 0 - 4/12 pitch	.940	L.F.	.024	1.50	1.41	2.91
5/12 - 8/12 pitch	1.060	L.F.	.034	1.70	2.01	3.71
Fascia, #2 or better,, 1″ x 4″	.100	L.F.	.003	.04	.17	.21
1″ x 6″	.100	L.F.	.004	.06	.23	.29
1″ x 8″	.100	L.F.	.005	.07	.27	.34
1″ x 10″	.100	L.F.	.005	.08	.30	.38
2″ x 4″	.100	L.F.	.005	.07	.28	.35
2″ x 6″	.100	L.F.	.006	.09	.34	.43
2″ x 8″	.100	L.F.	.007	.11	.42	.53
2″ x 10″	.100	L.F.	.009	.16	.53	.69
Bridging, wood 6′ O.C., 1″ x 3″, rafters, 16″ O.C.	.080	Pr.	.005	.06	.29	.35
24″ O.C.	.060	Pr.	.004	.05	.22	.27
Metal, galvanized, rafters, 16″ O.C.	.080	Pr.	.005	.16	.29	.45
24″ O.C.	.060	Pr.	.003	.17	.20	.37
Compression type, rafters, 16″ O.C.	.080	Pr.	.003	.11	.19	.30
24″ O.C.	.060	Pr.	.002	.09	.14	.23
Sheathing, plywood, exterior grade, 3/8″ thick, flat 0 - 4/12 pitch	1.230	S.F.	.013	.84	.76	1.60
5/12 - 8/12 pitch	1.330	S.F.	.014	.90	.82	1.72
1/2″ thick, flat 0 - 4/12 pitch	1.230	S.F.	.014	.86	.82	1.68
5/12 - 8/12 pitch	1.330	S.F.	.015	.93	.89	1.82
5/8″ thick, flat 0 - 4/12 pitch	1.230	S.F.	.015	1.06	.90	1.96
5/12 - 8/12 pitch	1.330	S.F.	.016	1.14	.97	2.11
3/4″ thick, flat 0 - 4/12 pitch	1.230	S.F.	.016	1.27	.97	2.24
5/12 - 8/12 pitch	1.330	S.F.	.018	1.37	1.05	2.42
Boards, 1″ x 6″, laid regular, flat 0 - 4/12 pitch	1.230	S.F.	.027	2.31	1.60	3.91
5/12 - 8/12 pitch	1.330	S.F.	.041	2.50	2.41	4.91
Laid diagonal, flat 0 - 4/12 pitch	1.230	S.F.	.030	2.31	1.78	4.09
5/12 - 8/12 pitch	1.330	S.F.	.044	2.50	2.61	5.11
1″ x 8″, laid regular, flat 0 - 4/12 pitch	1.230	S.F.	.022	2.90	1.33	4.23
5/12 - 8/12 pitch	1.330	S.F.	.034	3.14	1.97	5.11
Laid diagonal, flat 0 - 4/12 pitch	1.230	S.F.	.027	2.90	1.60	4.50
5/12 - 8/12 pitch	1.330	S.F.	.044	2.50	2.61	5.11

40 Gable Dormer Framing Systems

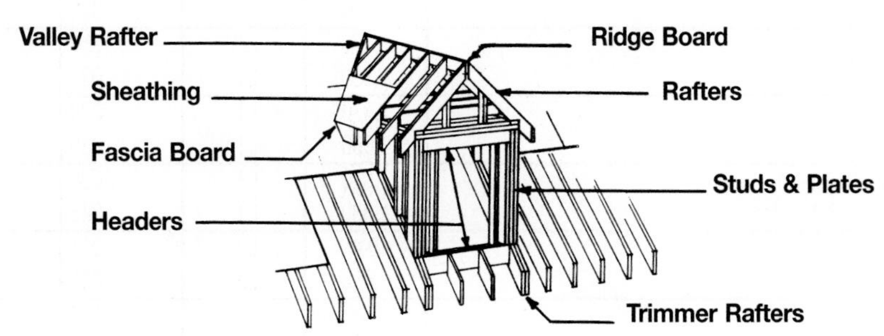

System Description	QUAN.	UNIT	LABOR HOURS	COST PER S.F.		
				MAT.	INST.	TOTAL
2″ X 6″, 16″ O.C.						
Dormer rafter, 2″ x 6″, 16″ O.C.	1.330	L.F.	.036	1.05	2.13	3.18
Ridge board, 2″ x 6″	.280	L.F.	.009	.22	.53	.75
Trimmer rafters, 2″ x 6″	.880	L.F.	.014	.70	.83	1.53
Wall studs & plates, 2″ x 4″, 16″ O.C.	3.160	L.F.	.056	1.49	3.29	4.78
Fascia, 2″ x 6″	.220	L.F.	.012	.18	.71	.89
Valley rafter, 2″ x 6″, 16″ O.C.	.280	L.F.	.009	.22	.52	.74
Cripple rafter, 2″ x 6″, 16″ O.C.	.560	L.F.	.022	.44	1.29	1.73
Headers, 2″ x 6″, doubled	.670	L.F.	.030	.53	1.76	2.29
Ceiling joist, 2″ x 4″, 16″ O.C.	1.000	L.F.	.013	.47	.75	1.22
Sheathing, exterior grade plywood, 1/2″ thick	3.610	S.F.	.041	2.53	2.42	4.95
TOTAL		S.F.	.242	7.83	14.23	22.06
2″ X 8″, 16″ O.C.						
Dormer rafter, 2″ x 8″, 16″ O.C.	1.330	L.F.	.039	1.42	2.33	3.75
Ridge board, 2″ x 8″	.280	L.F.	.010	.30	.59	.89
Trimmer rafter, 2″ x 8″	.880	L.F.	.015	.94	.87	1.81
Wall studs & plates, 2″ x 4″, 16″ O.C.	3.160	L.F.	.056	1.49	3.29	4.78
Fascia, 2″ x 8″	.220	L.F.	.016	.24	.92	1.16
Valley rafter, 2″ x 8″, 16″ O.C.	.280	L.F.	.010	.30	.56	.86
Cripple rafter, 2″ x 8″, 16″ O.C.	.560	L.F.	.027	.60	1.57	2.17
Headers, 2″ x 8″, doubled	.670	L.F.	.032	.72	1.86	2.58
Ceiling joist, 2″ x 4″, 16″ O.C.	1.000	L.F.	.013	.47	.75	1.22
Sheathing,, exterior grade plywood, 1/2″ thick	3.610	S.F.	.041	2.53	2.42	4.95
TOTAL		S.F.	.259	9.01	15.16	24.17

The cost in this system is based on the square foot of plan area.
The measurement being the plan area of the dormer only.

Description	QUAN.	UNIT	LABOR HOURS	COST PER S.F.		
				MAT.	INST.	TOTAL

Gable Dormer Framing Price Sheet	QUAN.	UNIT	LABOR HOURS	COST PER S.F.		
				MAT.	INST.	TOTAL
Dormer rafters, #2 or better, 2″ x 4″, 16″ O.C.	1.330	L.F.	.029	.84	1.70	2.54
24″ O.C.	1.060	L.F.	.023	.67	1.36	2.03
2″ x 6″, 16″ O.C.	1.330	L.F.	.036	1.05	2.13	3.18
24″ O.C.	1.060	L.F.	.029	.84	1.70	2.54
2″ x 8″, 16″ O.C.	1.330	L.F.	.039	1.42	2.33	3.75
24″ O.C.	1.060	L.F.	.031	1.13	1.86	2.99
Ridge board, #2 or better, 1″ x 4″	.280	L.F.	.006	.19	.35	.54
1″ x 6″	.280	L.F.	.007	.24	.44	.68
1″ x 8″	.280	L.F.	.008	.41	.48	.89
2″ x 4″	.280	L.F.	.007	.18	.42	.60
2″ x 6″	.280	L.F.	.009	.22	.53	.75
2″ x 8″	.280	L.F.	.010	.30	.59	.89
Trimmer rafters, #2 or better, 2″ x 4″	.880	L.F.	.011	.56	.66	1.22
2″ x 6″	.880	L.F.	.014	.70	.83	1.53
2″ x 8″	.880	L.F.	.015	.94	.87	1.81
2″ x 10″	.880	L.F.	.022	1.41	1.32	2.73
Wall studs & plates, #2 or better, 2″ x 4″ studs, 16″ O.C.	3.160	L.F.	.056	1.49	3.29	4.78
24″ O.C.	2.800	L.F.	.050	1.32	2.91	4.23
2″ x 6″ studs, 16″ O.C.	3.160	L.F.	.063	2.50	3.73	6.23
24″ O.C.	2.800	L.F.	.056	2.21	3.30	5.51
Fascia, #2 or better, 1″ x 4″	.220	L.F.	.006	.10	.38	.48
1″ x 6″	.220	L.F.	.008	.12	.46	.58
1″ x 8″	.220	L.F.	.009	.14	.54	.68
2″ x 4″	.220	L.F.	.011	.16	.63	.79
2″ x 6″	.220	L.F.	.014	.20	.80	1
2″ x 8″	.220	L.F.	.016	.24	.92	1.16
Valley rafter, #2 or better, 2″ x 4″	.280	L.F.	.007	.18	.41	.59
2″ x 6″	.280	L.F.	.009	.22	.52	.74
2″ x 8″	.280	L.F.	.010	.30	.56	.86
2″ x 10″	.280	L.F.	.012	.45	.69	1.14
Cripple rafter, #2 or better, 2″ x 4″, 16″ O.C.	.560	L.F.	.018	.36	1.04	1.40
24″ O.C.	.450	L.F.	.014	.28	.83	1.11
2″ x 6″, 16″ O.C.	.560	L.F.	.022	.44	1.29	1.73
24″ O.C.	.450	L.F.	.018	.36	1.04	1.40
2″ x 8″, 16″ O.C.	.560	L.F.	.027	.60	1.57	2.17
24″ O.C.	.450	L.F.	.021	.48	1.26	1.74
Headers, #2 or better double header, 2″ x 4″	.670	L.F.	.024	.43	1.41	1.84
2″ x 6″	.670	L.F.	.030	.53	1.76	2.29
2″ x 8″	.670	L.F.	.032	.72	1.86	2.58
2″ x 10″	.670	L.F.	.034	1.07	1.98	3.05
Ceiling joist, #2 or better, 2″ x 4″, 16″ O.C.	1.000	L.F.	.013	.47	.75	1.22
24″ O.C.	.800	L.F.	.010	.38	.60	.98
2″ x 6″, 16″ O.C.	1.000	L.F.	.013	.79	.75	1.54
24″ O.C.	.800	L.F.	.010	.63	.60	1.23
Sheathing, plywood exterior grade, 3/8″ thick	3.610	S.F.	.038	2.45	2.24	4.69
1/2″ thick	3.610	S.F.	.041	2.53	2.42	4.95
5/8″ thick	3.610	S.F.	.044	3.10	2.64	5.74
3/4″ thick	3.610	S.F.	.048	3.72	2.85	6.57
Boards, 1″ x 6″, laid regular	3.610	S.F.	.089	6.80	5.25	12.05
Laid diagonal	3.610	S.F.	.099	6.80	5.80	12.60
1″ x 8″, laid regular	3.610	S.F.	.076	8.50	4.44	12.94
Laid diagonal	3.610	S.F.	.089	8.50	5.25	13.75

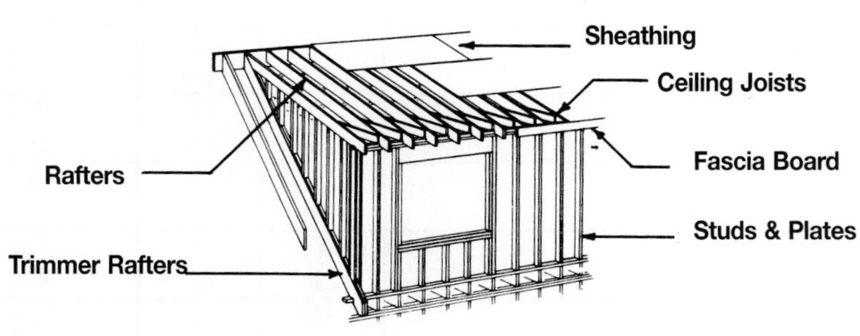

System Description	QUAN.	UNIT	LABOR HOURS	COST PER S.F. MAT.	COST PER S.F. INST.	COST PER S.F. TOTAL
2″ X 6″ RAFTERS, 16″ O.C.						
Dormer rafter, 2″ x 6″, 16″ O.C.	1.080	L.F.	.029	.85	1.73	2.58
Trimmer rafter, 2″ x 6″	.400	L.F.	.006	.32	.38	.70
Studs & plates, 2″ x 4″, 16″ O.C.	2.750	L.F.	.049	1.29	2.86	4.15
Fascia, 2″ x 6″	.250	L.F.	.014	.20	.80	1
Ceiling joist, 2″ x 4″, 16″ O.C.	1.000	L.F.	.013	.47	.75	1.22
Sheathing, exterior grade plywood, CDX, 1/2″ thick	2.940	S.F.	.034	2.06	1.97	4.03
TOTAL		S.F.	.145	5.19	8.49	13.68
2″ X 8″ RAFTERS, 16″ O.C.						
Dormer rafter, 2″ x 8″, 16″ O.C.	1.080	L.F.	.032	1.16	1.89	3.05
Trimmer rafter, 2″ x 8″	.400	L.F.	.007	.43	.40	.83
Studs & plates, 2″ x 4″, 16″ O.C.	2.750	L.F.	.049	1.29	2.86	4.15
Fascia, 2″ x 8″	.250	L.F.	.018	.27	1.05	1.32
Ceiling joist, 2″ x 6″, 16″ O.C.	1.000	L.F.	.013	.79	.75	1.54
Sheathing, exterior grade plywood, CDX, 1/2″ thick	2.940	S.F.	.034	2.06	1.97	4.03
TOTAL		S.F.	.153	6	8.92	14.92
2″ X 10″ RAFTERS, 16″ O.C.						
Dormer rafter, 2″ x 10″, 16″ O.C.	1.080	L.F.	.041	1.73	2.40	4.13
Trimmer rafter, 2″ x 10″	.400	L.F.	.010	.64	.60	1.24
Studs & plates, 2″ x 4″, 16″ O.C.	2.750	L.F.	.049	1.29	2.86	4.15
Fascia, 2″ x 10″	.250	L.F.	.022	.40	1.31	1.71
Ceiling joist, 2″ x 6″, 16″ O.C.	1.000	L.F.	.013	.79	.75	1.54
Sheathing, exterior grade plywood, CDX, 1/2″ thick	2.940	S.F.	.034	2.06	1.97	4.03
TOTAL		S.F.	.169	6.91	9.89	16.80

The cost in this system is based on the square foot of plan area.
The measurement is the plan area of the dormer only.

Description	QUAN.	UNIT	LABOR HOURS	COST PER S.F. MAT.	COST PER S.F. INST.	COST PER S.F. TOTAL

Shed Dormer Framing Price Sheet	QUAN.	UNIT	LABOR HOURS	COST PER S.F.		
				MAT.	INST.	TOTAL
Dormer rafters, #2 or better, 2" x 4", 16" O.C.	1.080	L.F.	.023	.68	1.38	2.06
24" O.C.	.860	L.F.	.019	.54	1.10	1.64
2" x 6", 16" O.C.	1.080	L.F.	.029	.85	1.73	2.58
24" O.C.	.860	L.F.	.023	.68	1.38	2.06
2" x 8", 16" O.C.	1.080	L.F.	.032	1.16	1.89	3.05
24" O.C.	.860	L.F.	.025	.92	1.51	2.43
2" x 10", 16" O.C.	1.080	L.F.	.041	1.73	2.40	4.13
24" O.C.	.860	L.F.	.032	1.38	1.91	3.29
Trimmer rafter, #2 or better, 2" x 4"	.400	L.F.	.005	.25	.30	.55
2" x 6"	.400	L.F.	.006	.32	.38	.70
2" x 8"	.400	L.F.	.007	.43	.40	.83
2" x 10"	.400	L.F.	.010	.64	.60	1.24
Studs & plates, #2 or better, 2" x 4", 16" O.C.	2.750	L.F.	.049	1.29	2.86	4.15
24" O.C.	2.200	L.F.	.039	1.03	2.29	3.32
2" x 6", 16" O.C.	2.750	L.F.	.055	2.17	3.25	5.42
24" O.C.	2.200	L.F.	.044	1.74	2.60	4.34
Fascia, #2 or better, 1" x 4"	.250	L.F.	.006	.10	.38	.48
1" x 6"	.250	L.F.	.008	.12	.46	.58
1" x 8"	.250	L.F.	.009	.14	.54	.68
2" x 4"	.250	L.F.	.011	.16	.63	.79
2" x 6"	.250	L.F.	.014	.20	.80	1
2" x 8"	.250	L.F.	.018	.27	1.05	1.32
Ceiling joist, #2 or better, 2" x 4", 16" O.C.	1.000	L.F.	.013	.47	.75	1.22
24" O.C.	.800	L.F.	.010	.38	.60	.98
2" x 6", 16" O.C.	1.000	L.F.	.013	.79	.75	1.54
24" O.C.	.800	L.F.	.010	.63	.60	1.23
2" x 8", 16" O.C.	1.000	L.F.	.015	1.07	.86	1.93
24" O.C.	.800	L.F.	.012	.86	.69	1.55
Sheathing, plywood exterior grade, 3/8" thick	2.940	S.F.	.031	2	1.82	3.82
1/2" thick	2.940	S.F.	.034	2.06	1.97	4.03
5/8" thick	2.940	S.F.	.036	2.53	2.15	4.68
3/4" thick	2.940	S.F.	.039	3.03	2.32	5.35
Boards, 1" x 6", laid regular	2.940	S.F.	.072	5.55	4.26	9.81
Laid diagonal	2.940	S.F.	.080	5.55	4.73	10.28
1" x 8", laid regular	2.940	S.F.	.062	6.95	3.62	10.57
Laid diagonal	2.940	S.F.	.072	6.95	4.26	11.21

Window Openings	QUAN.	UNIT	LABOR HOURS	COST EACH		
				MAT.	INST.	TOTAL
The following are to be added to the total cost of the dormers for window openings. Do not subtract window area from the stud wall quantities.						
Headers, 2" x 6" doubled, 2' long	4.000	L.F.	.178	3.16	10.50	13.66
3' long	6.000	L.F.	.267	4.74	15.70	20.44
4' long	8.000	L.F.	.356	6.30	21	27.30
5' long	10.000	L.F.	.444	7.90	26	33.90
2" x 8" doubled, 4' long	8.000	L.F.	.376	8.55	22	30.55
5' long	10.000	L.F.	.471	10.70	27.50	38.20
6' long	12.000	L.F.	.565	12.85	33	45.85
8' long	16.000	L.F.	.753	17.10	44.50	61.60
2" x 10" doubled, 4' long	8.000	L.F.	.400	12.80	23.50	36.30
6' long	12.000	L.F.	.600	19.20	35.50	54.70
8' long	16.000	L.F.	.800	25.50	47	72.50
10' long	20.000	L.F.	1.000	32	59	91

48 Partition Framing Systems

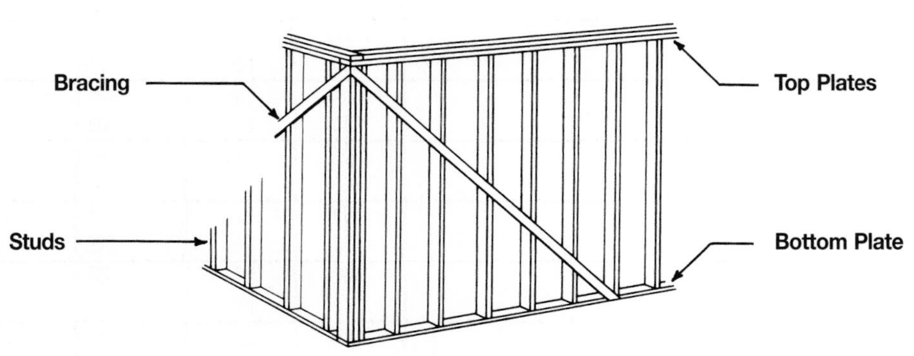

System Description	QUAN.	UNIT	LABOR HOURS	COST PER S.F.		
				MAT.	INST.	TOTAL
2" X 4", 16" O.C.						
2" x 4" studs, #2 or better, 16" O.C.	1.000	L.F.	.015	.46	.86	1.32
Plates, double top, single bottom	.375	L.F.	.005	.17	.32	.49
Cross bracing, let-in, 1" x 6"	.080	L.F.	.004	.07	.25	.32
TOTAL		S.F.	.024	.70	1.43	2.13
2" X 4", 24" O.C.						
2" x 4" studs, #2 or better, 24" O.C.	.800	L.F.	.012	.37	.69	1.06
Plates, double top, single bottom	.375	L.F.	.005	.17	.32	.49
Cross bracing, let-in, 1" x 6"	.080	L.F.	.003	.07	.16	.23
TOTAL		S.F.	.020	.61	1.17	1.78
2" X 6", 16" O.C.						
2" x 6" studs, #2 or better, 16" O.C.	1.000	L.F.	.016	.79	.94	1.73
Plates, double top, single bottom	.375	L.F.	.006	.30	.35	.65
Cross bracing, let-in, 1" x 6"	.080	L.F.	.004	.07	.25	.32
TOTAL		S.F.	.026	1.16	1.54	2.70
2" X 6", 24" O.C.						
2" x 6" studs, #2 or better, 24" O.C.	.800	L.F.	.013	.63	.75	1.38
Plates, double top, single bottom	.375	L.F.	.006	.30	.35	.65
Cross bracing, let-in, 1" x 6"	.080	L.F.	.003	.07	.16	.23
TOTAL		S.F.	.022	1	1.26	2.26

The costs in this system are based on a square foot of wall area. Do not subtract for door or window openings.

Description	QUAN.	UNIT	LABOR HOURS	COST PER S.F.		
				MAT.	INST.	TOTAL

Partition Framing Price Sheet	QUAN.	UNIT	LABOR HOURS	COST PER S.F.		
				MAT.	INST.	TOTAL
Wood studs, #2 or better, 2″ x 4″, 12″ O.C.	1.250	L.F.	.018	.58	1.08	1.66
16″ O.C.	1.000	L.F.	.015	.46	.86	1.32
24″ O.C.	.800	L.F.	.012	.37	.69	1.06
32″ O.C.	.650	L.F.	.009	.30	.56	.86
2″ x 6″, 12″ O.C.	1.250	L.F.	.020	.99	1.18	2.17
16″ O.C.	1.000	L.F.	.016	.79	.94	1.73
24″ O.C.	.800	L.F.	.013	.63	.75	1.38
32″ O.C.	.650	L.F.	.010	.51	.61	1.12
Plates, #2 or better double top single bottom, 2″ x 4″	.375	L.F.	.005	.17	.32	.49
2″ x 6″	.375	L.F.	.006	.30	.35	.65
2″ x 8″	.375	L.F.	.005	.40	.32	.72
Cross bracing, let-in, 1″ x 6″ boards studs, 12″ O.C.	.080	L.F.	.005	.09	.31	.40
16″ O.C.	.080	L.F.	.004	.07	.25	.32
24″ O.C.	.080	L.F.	.003	.07	.16	.23
32″ O.C.	.080	L.F.	.002	.06	.13	.19
Let-in steel (T shaped) studs, 12″ O.C.	.080	L.F.	.001	.09	.08	.17
16″ O.C.	.080	L.F.	.001	.07	.06	.13
24″ O.C.	.080	L.F.	.001	.07	.06	.13
32″ O.C.	.080	L.F.	.001	.06	.05	.11
Steel straps studs, 12″ O.C.	.080	L.F.	.001	.10	.07	.17
16″ O.C.	.080	L.F.	.001	.09	.06	.15
24″ O.C.	.080	L.F.	.001	.09	.06	.15
32″ O.C.	.080	L.F.	.001	.09	.06	.15
Metal studs, load bearing 24″ O.C., 20 ga. galv., 2-1/2″ wide	1.000	S.F.	.015	.70	.88	1.58
3-5/8″ wide	1.000	S.F.	.015	.84	.90	1.74
4″ wide	1.000	S.F.	.016	.87	.92	1.79
6″ wide	1.000	S.F.	.016	1.12	.94	2.06
16 ga., 2-1/2″ wide	1.000	S.F.	.017	.81	1.01	1.82
3-5/8″ wide	1.000	S.F.	.017	.94	1.03	1.97
4″ wide	1.000	S.F.	.018	.98	1.05	2.03
6″ wide	1.000	S.F.	.018	1.24	1.07	2.31
Non-load bearing 24″ O.C., 25 ga. galv., 1-5/8″ wide	1.000	S.F.	.011	.20	.62	.82
2-1/2″ wide	1.000	S.F.	.011	.27	.63	.90
3-5/8″ wide	1.000	S.F.	.011	.30	.64	.94
4″ wide	1.000	S.F.	.011	.34	.64	.98
6″ wide	1.000	S.F.	.011	.42	.65	1.07
20 ga., 2-1/2″ wide	1.000	S.F.	.013	.33	.79	1.12
3-5/8″ wide	1.000	S.F.	.014	.38	.80	1.18
4″ wide	1.000	S.F.	.014	.46	.80	1.26
6″ wide	1.000	S.F.	.014	.55	.81	1.36

Window & Door Openings	QUAN.	UNIT	LABOR HOURS	COST EACH		
				MAT.	INST.	TOTAL
The following costs are to be added to the total costs of the walls. Do not subtract openings from total wall area.						
Headers, 2″ x 6″ double, 2′ long	4.000	L.F.	.178	3.16	10.50	13.66
3′ long	6.000	L.F.	.267	4.74	15.70	20.44
4′ long	8.000	L.F.	.356	6.30	21	27.30
5′ long	10.000	L.F.	.444	7.90	26	33.90
2″ x 8″ double, 4′ long	8.000	L.F.	.376	8.55	22	30.55
5′ long	10.000	L.F.	.471	10.70	27.50	38.20
6′ long	12.000	L.F.	.565	12.85	33	45.85
8′ long	16.000	L.F.	.753	17.10	44.50	61.60
2″ x 10″ double, 4′ long	8.000	L.F.	.400	12.80	23.50	36.30
6′ long	12.000	L.F.	.600	19.20	35.50	54.70
8′ long	16.000	L.F.	.800	25.50	47	72.50
10′ long	20.000	L.F.	1.000	32	59	91
2″ x 12″ double, 8′ long	16.000	L.F.	.853	32	50	82
12′ long	24.000	L.F.	1.280	48	75.50	123.50

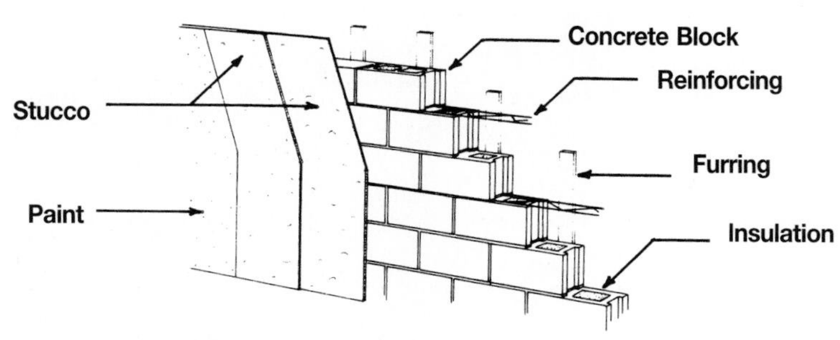

System Description	QUAN.	UNIT	LABOR HOURS	COST PER S.F. MAT.	COST PER S.F. INST.	COST PER S.F. TOTAL
6″ THICK CONCRETE BLOCK WALL						
6″ thick concrete block, 6″ x 8″ x 16″	1.000	S.F.	.100	2.72	5.50	8.22
Masonry reinforcing, truss strips every other course	.625	L.F.	.002	.15	.10	.25
Furring, 1″ x 3″, 16″ O.C.	1.000	L.F.	.016	.52	.95	1.47
Masonry insulation, poured perlite	1.000	S.F.	.013	1.95	.79	2.74
Stucco, 2 coats	1.000	S.F.	.069	.74	3.96	4.70
Masonry paint, 2 coats	1.000	S.F.	.016	.23	.78	1.01
TOTAL		S.F.	.216	6.31	12.08	18.39
8″ THICK CONCRETE BLOCK WALL						
8″ thick concrete block, 8″ x 8″ x 16″	1.000	S.F.	.107	2.85	5.85	8.70
Masonry reinforcing, truss strips every other course	.625	L.F.	.002	.15	.10	.25
Furring, 1″ x 3″, 16″ O.C.	1.000	L.F.	.016	.52	.95	1.47
Masonry insulation, poured perlite	1.000	S.F.	.018	2.57	1.04	3.61
Stucco, 2 coats	1.000	S.F.	.069	.74	3.96	4.70
Masonry paint, 2 coats	1.000	S.F.	.016	.23	.78	1.01
TOTAL		S.F.	.228	7.06	12.68	19.74
12″ THICK CONCRETE BLOCK WALL						
12″ thick concrete block, 12″ x 8″ x 16″	1.000	S.F.	.141	4.63	7.60	12.23
Masonry reinforcing, truss strips every other course	.625	L.F.	.003	.18	.15	.33
Furring, 1″ x 3″, 16″ O.C.	1.000	L.F.	.016	.52	.95	1.47
Masonry insulation, poured perlite	1.000	S.F.	.026	3.80	1.53	5.33
Stucco, 2 coats	1.000	S.F.	.069	.74	3.96	4.70
Masonry paint, 2 coats	1.000	S.F.	.016	.23	.78	1.01
TOTAL		S.F.	.271	10.10	14.97	25.07

Costs for this system are based on a square foot of wall area. Do not subtract for window openings.

Description	QUAN.	UNIT	LABOR HOURS	COST PER S.F. MAT.	COST PER S.F. INST.	COST PER S.F. TOTAL

Masonry Block Price Sheet	QUAN.	UNIT	LABOR HOURS	COST PER S.F. MAT.	INST.	TOTAL
Block concrete, 8″ x 16″ regular, 4″ thick	1.000	S.F.	.093	2	5.10	7.10
6″ thick	1.000	S.F.	.100	2.72	5.50	8.22
8″ thick	1.000	S.F.	.107	2.85	5.85	8.70
10″ thick	1.000	S.F.	.111	3.42	6.10	9.52
12″ thick	1.000	S.F.	.141	4.63	7.60	12.23
Solid block, 4″ thick	1.000	S.F.	.096	2.29	5.30	7.59
6″ thick	1.000	S.F.	.104	3.08	5.70	8.78
8″ thick	1.000	S.F.	.111	4.11	6.10	10.21
10″ thick	1.000	S.F.	.133	5.40	7.15	12.55
12″ thick	1.000	S.F.	.148	6	7.95	13.95
Lightweight, 4″ thick	1.000	S.F.	.093	2	5.10	7.10
6″ thick	1.000	S.F.	.100	2.72	5.50	8.22
8″ thick	1.000	S.F.	.107	2.85	5.85	8.70
10″ thick	1.000	S.F.	.111	3.42	6.10	9.52
12″ thick	1.000	S.F.	.141	4.63	7.60	12.23
Split rib profile, 4″ thick	1.000	S.F.	.116	4.58	6.35	10.93
6″ thick	1.000	S.F.	.123	5.20	6.75	11.95
8″ thick	1.000	S.F.	.131	6.40	7.30	13.70
10″ thick	1.000	S.F.	.157	6.70	8.45	15.15
12″ thick	1.000	S.F.	.175	7.45	9.40	16.85
Masonry reinforcing, wire truss strips, every course, 8″ block	1.375	L.F.	.004	.33	.22	.55
12″ block	1.375	L.F.	.006	.40	.33	.73
Every other course, 8″ block	.625	L.F.	.002	.15	.10	.25
12″ block	.625	L.F.	.003	.18	.15	.33
Furring, wood, 1″ x 3″, 12″ O.C.	1.250	L.F.	.020	.65	1.19	1.84
16″ O.C.	1.000	L.F.	.016	.52	.95	1.47
24″ O.C.	.800	L.F.	.013	.42	.76	1.18
32″ O.C.	.640	L.F.	.010	.33	.61	.94
Steel, 3/4″ channels, 12″ O.C.	1.250	L.F.	.034	.43	1.93	2.36
16″ O.C.	1.000	L.F.	.030	.38	1.71	2.09
24″ O.C.	.800	L.F.	.023	.26	1.29	1.55
32″ O.C.	.640	L.F.	.018	.21	1.03	1.24
Masonry insulation, vermiculite or perlite poured 4″ thick	1.000	S.F.	.009	1.26	.51	1.77
6″ thick	1.000	S.F.	.013	1.93	.78	2.71
8″ thick	1.000	S.F.	.018	2.57	1.04	3.61
10″ thick	1.000	S.F.	.021	3.12	1.26	4.38
12″ thick	1.000	S.F.	.026	3.80	1.53	5.33
Block inserts polystyrene, 6″ thick	1.000	S.F.		1.34		1.34
8″ thick	1.000	S.F.		1.51		1.51
10″ thick	1.000	S.F.		1.56		1.56
12″ thick	1.000	S.F.		1.74		1.74
Stucco, 1 coat	1.000	S.F.	.057	.61	3.26	3.87
2 coats	1.000	S.F.	.069	.74	3.96	4.70
3 coats	1.000	S.F.	.081	.87	4.65	5.52
Painting, 1 coat	1.000	S.F.	.011	.15	.54	.69
2 coats	1.000	S.F.	.016	.23	.78	1.01
Primer & 1 coat	1.000	S.F.	.013	.22	.64	.86
2 coats	1.000	S.F.	.018	.30	.89	1.19
Lath, metal lath expanded 2.5 lb/S.Y., painted	1.000	S.F.	.010	.45	.59	1.04
Galvanized	1.000	S.F.	.012	.50	.65	1.15

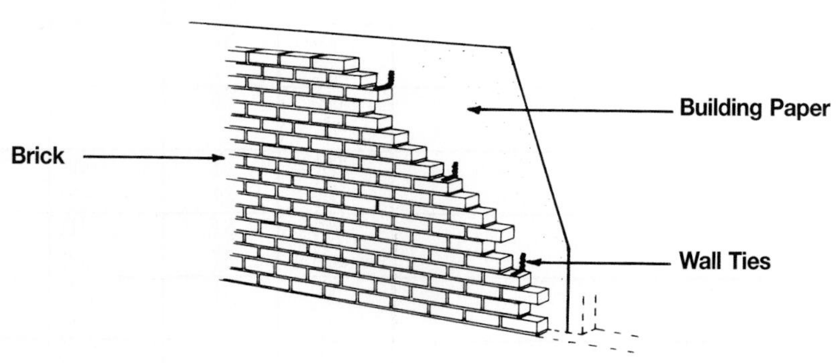

System Description	QUAN.	UNIT	LABOR HOURS	COST PER S.F.		
				MAT.	INST.	TOTAL
SELECT COMMON BRICK						
Brick, select common, running bond	1.000	S.F.	.174	4.75	9.55	14.30
Wall ties, 7/8" x 7", 22 gauge	1.000	Ea.	.008	.17	.45	.62
Building paper, spunbonded polypropylene	1.100	S.F.	.002	.17	.13	.30
Trim, pine, painted	.125	L.F.	.004	.08	.24	.32
TOTAL		S.F.	.188	5.17	10.37	15.54
RED FACED COMMON BRICK						
Brick, common, red faced, running bond	1.000	S.F.	.182	4.57	10	14.57
Wall ties, 7/8" x 7", 22 gauge	1.000	Ea.	.008	.17	.45	.62
Building paper, spundbonded polypropylene	1.100	S.F.	.002	.17	.13	.30
Trim, pine, painted	.125	L.F.	.004	.08	.24	.32
TOTAL		S.F.	.196	4.99	10.82	15.81
BUFF OR GREY FACE BRICK						
Brick, buff or grey	1.000	S.F.	.182	4.83	10	14.83
Wall ties, 7/8" x 7", 22 gauge	1.000	Ea.	.008	.17	.45	.62
Building paper, spundbonded polypropylene	1.100	S.F.	.002	.17	.13	.30
Trim, pine, painted	.125	L.F.	.004	.08	.24	.32
TOTAL		S.F.	.196	5.25	10.82	16.07
STONE WORK, ROUGH STONE, AVERAGE						
Field stone veneer	1.000	S.F.	.223	9.31	12.26	21.57
Wall ties, 7/8" x 7", 22 gauge	1.000	Ea.	.008	.17	.45	.62
Building paper, spundbonded polypropylene	1.000	S.F.	.002	.17	.13	.30
Trim, pine, painted	.125	L.F.	.004	.08	.24	.32
TOTAL		S.F.	.237	9.73	13.08	22.81

The costs in this system are based on a square foot of wall area. Do not subtract area for window & door openings.

Description	QUAN.	UNIT	LABOR HOURS	COST PER S.F.		
				MAT.	INST.	TOTAL

Brick/Stone Veneer Price Sheet	QUAN.	UNIT	LABOR HOURS	COST PER S.F. MAT.	INST.	TOTAL
Brick						
Select common, running bond	1.000	S.F.	.174	4.75	9.55	14.30
Red faced, running bond	1.000	S.F.	.182	4.57	10	14.57
Buff or grey faced, running bond	1.000	S.F.	.182	4.83	10	14.83
Header every 6th course	1.000	S.F.	.216	5.30	11.90	17.20
English bond	1.000	S.F.	.286	6.80	15.70	22.50
Flemish bond	1.000	S.F.	.195	4.82	10.70	15.52
Common bond	1.000	S.F.	.267	6.05	14.65	20.70
Stack bond	1.000	S.F.	.182	4.83	10	14.83
Jumbo, running bond	1.000	S.F.	.092	5.90	5.05	10.95
Norman, running bond	1.000	S.F.	.125	7.25	6.85	14.10
Norwegian, running bond	1.000	S.F.	.107	6.35	5.85	12.20
Economy, running bond	1.000	S.F.	.129	5	7.10	12.10
Engineer, running bond	1.000	S.F.	.154	4.30	8.45	12.75
Roman, running bond	1.000	S.F.	.160	7.95	8.80	16.75
Utility, running bond	1.000	S.F.	.089	5.65	6.10	11.75
Glazed, running bond	1.000	S.F.	.190	14.15	10.45	24.60
Stone work, rough stone, average	1.000	S.F.	.179	9.30	12.25	21.55
Maximum	1.000	S.F.	.267	13.90	18.30	32.20
Wall ties, galvanized, corrugated 7/8" x 7", 22 gauge	1.000	Ea.	.008	.17	.45	.62
16 gauge	1.000	Ea.	.008	.34	.45	.79
Cavity wall, every 3rd course 6" long Z type, 1/4" diameter	1.330	L.F.	.010	.57	.59	1.16
3/16" diameter	1.330	L.F.	.010	.47	.59	1.06
8" long, Z type, 1/4" diameter	1.330	L.F.	.010	.69	.59	1.28
3/16" diameter	1.330	L.F.	.010	.38	.59	.97
Building paper, aluminum and kraft laminated foil, 1 side	1.000	S.F.	.002	.16	.13	.29
2 sides	1.000	S.F.	.002	.17	.13	.30
#15 asphalt paper	1.100	S.F.	.002	.07	.14	.21
Polyethylene, .002" thick	1.000	S.F.	.002	.02	.13	.15
.004" thick	1.000	S.F.	.002	.03	.13	.16
.006" thick	1.000	S.F.	.002	.04	.13	.17
.010" thick	1.000	S.F.	.002	.10	.13	.23
Trim, 1" x 4", cedar	.125	L.F.	.005	.17	.30	.47
Fir	.125	L.F.	.005	.11	.30	.41
Redwood	.125	L.F.	.005	.17	.30	.47
White pine	.125	L.F.	.005	.11	.30	.41

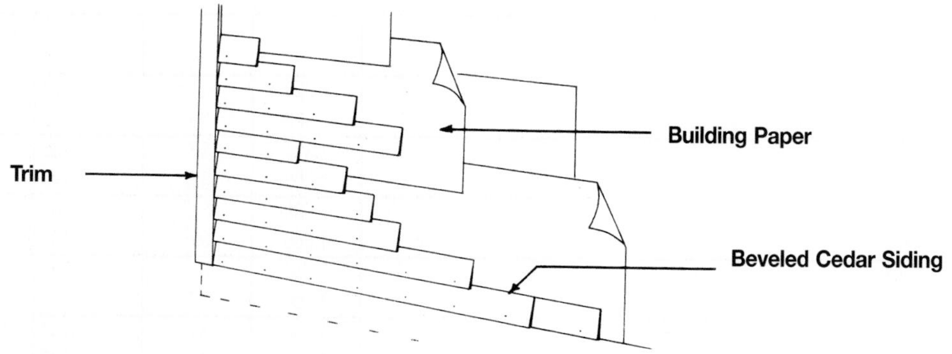

Trim
Building Paper
Beveled Cedar Siding

System Description	QUAN.	UNIT	LABOR HOURS	COST PER S.F.		
				MAT.	INST.	TOTAL
1/2" X 6" BEVELED CEDAR SIDING, "A" GRADE						
1/2" x 6" beveled cedar siding	1.000	S.F.	.027	4.87	1.60	6.47
Building wrap, spundbonded polypropylene	1.100	S.F.	.002	.17	.13	.30
Trim, cedar	.125	L.F.	.005	.17	.30	.47
Paint, primer & 2 coats	1.000	S.F.	.017	.22	.84	1.06
TOTAL		S.F.	.051	5.43	2.87	8.30
1/2" X 8" BEVELED CEDAR SIDING, "A" GRADE						
1/2" x 8" beveled cedar siding	1.000	S.F.	.024	8.55	1.43	9.98
Building wrap, spundbonded polypropylene	1.100	S.F.	.002	.17	.13	.30
Trim, cedar	.125	L.F.	.005	.17	.30	.47
Paint, primer & 2 coats	1.000	S.F.	.017	.22	.84	1.06
TOTAL		S.F.	.048	9.11	2.70	11.81
1" X 4" TONGUE & GROOVE, REDWOOD, VERTICAL GRAIN						
Redwood, clear, vertical grain, 1" x 10"	1.000	S.F.	.020	5.41	1.16	6.57
Building wrap, spunbonded polypropylene	1.100	S.F.	.002	.17	.13	.30
Trim, redwood	.125	L.F.	.005	.17	.30	.47
Sealer, 1 coat, stain, 1 coat	1.000	S.F.	.013	.14	.65	.79
TOTAL		S.F.	.040	5.89	2.24	8.13
1" X 6" TONGUE & GROOVE, REDWOOD, VERTICAL GRAIN						
Redwood, clear, vertical grain, 1" x 10"	1.000	S.F.	.020	5.57	1.19	6.76
Building wrap, spunbonded polypropylene	1.100	S.F.	.002	.17	.13	.30
Trim, redwood	.125	L.F.	.005	.17	.30	.47
Sealer, 1 coat, stain, 1 coat	1.000	S.F.	.013	.14	.65	.79
TOTAL		S.F.	.040	6.05	2.27	8.32

The costs in this system are based on a square foot of wall area.
Do not subtract area for door or window openings.

Description	QUAN.	UNIT	LABOR HOURS	COST PER S.F.		
				MAT.	INST.	TOTAL

Wood Siding Price Sheet	QUAN.	UNIT	LABOR HOURS	COST PER S.F.		
				MAT.	INST.	TOTAL
Siding, beveled cedar, "A" grade, 1/2" x 6"	1.000	S.F.	.028	4.87	1.60	6.47
1/2" x 8"	1.000	S.F.	.023	8.55	1.43	9.98
"B" grade, 1/2" x 6"	1.000	S.F.	.032	5.40	1.78	7.18
1/2" x 8"	1.000	S.F.	.029	9.50	1.59	11.09
Clear grade, 1/2" x 6"	1.000	S.F.	.028	6.10	2	8.10
1/2" x 8"	1.000	S.F.	.023	10.70	1.79	12.49
Redwood, clear vertical grain, 1/2" x 6"	1.000	S.F.	.036	5.50	1.60	7.10
1/2" x 8"	1.000	S.F.	.032	6	1.43	7.43
Clear all heart vertical grain, 1/2" x 6"	1.000	S.F.	.028	6.10	1.78	7.88
1/2" x 8"	1.000	S.F.	.023	6.65	1.59	8.24
Siding board & batten, cedar, "B" grade, 1" x 10"	1.000	S.F.	.031	5.25	1.12	6.37
1" x 12"	1.000	S.F.	.031	5.25	1.12	6.37
Redwood, clear vertical grain, 1" x 6"	1.000	S.F.	.043	5.55	1.60	7.15
1" x 8"	1.000	S.F.	.018	8.80	1.43	10.23
White pine, #2 & better, 1" x 10"	1.000	S.F.	.029	2.76	1.43	4.19
1" x 12"	1.000	S.F.	.029	2.76	1.43	4.19
Siding vertical, tongue & groove, cedar "B" grade, 1" x 4"	1.000	S.F.	.033	6.70	1.16	7.86
1" x 6"	1.000	S.F.	.024	6.90	1.19	8.09
1" x 8"	1.000	S.F.	.024	7.10	1.22	8.32
1" x 10"	1.000	S.F.	.021	7.30	1.26	8.56
"A" grade, 1" x 4"	1.000	S.F.	.033	6.15	1.06	7.21
1" x 6"	1.000	S.F.	.024	6.30	1.09	7.39
1" x 8"	1.000	S.F.	.024	6.45	1.12	7.57
1" x 10"	1.000	S.F.	.021	6.65	1.15	7.80
Clear vertical grain, 1" x 4"	1.000	S.F.	.033	5.65	.98	6.63
1" x 6"	1.000	S.F.	.024	5.80	1	6.80
1" x 8"	1.000	S.F.	.024	5.95	1.02	6.97
1" x 10"	1.000	S.F.	.021	6.10	1.05	7.15
Redwood, clear vertical grain, 1" x 4"	1.000	S.F.	.033	5.40	1.16	6.56
1" x 6"	1.000	S.F.	.024	5.55	1.19	6.74
1" x 8"	1.000	S.F.	.024	5.75	1.22	6.97
1" x 10"	1.000	S.F.	.021	5.90	1.26	7.16
Clear all heart vertical grain, 1" x 4"	1.000	S.F.	.033	4.96	1.06	6.02
1" x 6"	1.000	S.F.	.024	5.10	1.09	6.19
1" x 8"	1.000	S.F.	.024	5.20	1.12	6.32
1" x 10"	1.000	S.F.	.021	5.35	1.15	6.50
White pine, 1" x 10"	1.000	S.F.	.024	4.04	1.26	5.30
Siding plywood, texture 1-11 cedar, 3/8" thick	1.000	S.F.	.024	1.41	1.40	2.81
5/8" thick	1.000	S.F.	.024	2.88	1.40	4.28
Redwood, 3/8" thick	1.000	S.F.	.024	1.41	1.40	2.81
5/8" thick	1.000	S.F.	.024	2.23	1.40	3.63
Fir, 3/8" thick	1.000	S.F.	.024	1.02	1.40	2.42
5/8" thick	1.000	S.F.	.024	1.50	1.40	2.90
Southern yellow pine, 3/8" thick	1.000	S.F.	.024	1.02	1.40	2.42
5/8" thick	1.000	S.F.	.024	1.60	1.40	3
Paper, #15 asphalt felt	1.100	S.F.	.002	.07	.14	.21
Trim, cedar	.125	L.F.	.005	.17	.30	.47
Fir	.125	L.F.	.005	.11	.30	.41
Redwood	.125	L.F.	.005	.17	.30	.47
White pine	.125	L.F.	.005	.11	.30	.41
Painting, primer, & 1 coat	1.000	S.F.	.013	.14	.65	.79
2 coats	1.000	S.F.	.017	.22	.84	1.06
Stain, sealer, & 1 coat	1.000	S.F.	.017	.18	.85	1.03
2 coats	1.000	S.F.	.019	.29	.92	1.21

12 Shingle Siding Systems

Trim

Building Paper

White Cedar Shingles

System Description	QUAN.	UNIT	LABOR HOURS	COST PER S.F. MAT.	INST.	TOTAL
WHITE CEDAR SHINGLES, 5″ EXPOSURE						
White cedar shingles, 16″ long, grade "A", 5″ exposure	1.000	S.F.	.033	2.19	1.96	4.15
Building wrap, spunbonded polypropylene	1.100	S.F.	.002	.17	.13	.30
Trim, cedar	.125	S.F.	.005	.17	.30	.47
Paint, primer & 1 coat	1.000	S.F.	.017	.18	.85	1.03
TOTAL		S.F.	.057	2.71	3.24	5.95
RESQUARED & REBUTTED PERFECTIONS, 5-1/2″ EXPOSURE						
Resquared & rebutted perfections, 5-1/2″ exposure	1.000	S.F.	.027	3.15	1.57	4.72
Building wrap, spunbonded polypropylene	1.100	S.F.	.002	.17	.13	.30
Trim, cedar	.125	S.F.	.005	.17	.30	.47
Stain, sealer & 1 coat	1.000	S.F.	.017	.18	.85	1.03
TOTAL		S.F.	.051	3.67	2.85	6.52
HAND-SPLIT SHAKES, 8-1/2″ EXPOSURE						
Hand-split red cedar shakes, 18″ long, 8-1/2″ exposure	1.000	S.F.	.040	3.05	2.36	5.41
Building wrap, spunbonded polypropylene	1.100	S.F.	.002	.17	.13	.30
Trim, cedar	.125	S.F.	.005	.17	.30	.47
Stain, sealer & 1 coat	1.000	S.F.	.017	.18	.85	1.03
TOTAL		S.F.	.064	3.57	3.64	7.21

The costs in this system are based on a square foot of wall area.
Do not subtract area for door or window openings.

Description	QUAN.	UNIT	LABOR HOURS	COST PER S.F. MAT.	INST.	TOTAL

Shingle Siding Price Sheet	QUAN.	UNIT	LABOR HOURS	COST PER S.F. MAT.	INST.	TOTAL
Shingles wood, white cedar 16" long, "A" grade, 5" exposure	1.000	S.F.	.033	2.19	1.96	4.15
7" exposure	1.000	S.F.	.030	1.97	1.76	3.73
8-1/2" exposure	1.000	S.F.	.032	1.25	1.89	3.14
10" exposure	1.000	S.F.	.028	1.09	1.65	2.74
"B" grade, 5" exposure	1.000	S.F.	.040	1.89	2.36	4.25
7" exposure	1.000	S.F.	.028	1.32	1.65	2.97
8-1/2" exposure	1.000	S.F.	.024	1.13	1.42	2.55
10" exposure	1.000	S.F.	.020	.95	1.18	2.13
Fire retardant, "A" grade, 5" exposure	1.000	S.F.	.033	2.80	1.96	4.76
7" exposure	1.000	S.F.	.028	1.75	1.65	3.40
8-1/2" exposure	1.000	S.F.	.032	1.85	1.88	3.73
10" exposure	1.000	S.F.	.025	1.44	1.46	2.90
Fire retardant, 5" exposure	1.000	S.F.	.029	3.40	1.71	5.11
7" exposure	1.000	S.F.	.036	2.66	2.09	4.75
8-1/2" exposure	1.000	S.F.	.032	2.40	1.88	4.28
10" exposure	1.000	S.F.	.025	1.87	1.46	3.33
Resquared & rebutted, 5-1/2" exposure	1.000	S.F.	.027	3.15	1.57	4.72
7" exposure	1.000	S.F.	.024	2.84	1.41	4.25
8-1/2" exposure	1.000	S.F.	.021	2.52	1.26	3.78
10" exposure	1.000	S.F.	.019	2.21	1.10	3.31
Fire retardant, 5" exposure	1.000	S.F.	.027	3.76	1.57	5.33
7" exposure	1.000	S.F.	.024	3.39	1.41	4.80
8-1/2" exposure	1.000	S.F.	.021	3.01	1.26	4.27
10" exposure	1.000	S.F.	.023	2.05	1.34	3.39
Hand-split, red cedar, 24" long, 7" exposure	1.000	S.F.	.045	4.62	2.63	7.25
8-1/2" exposure	1.000	S.F.	.038	3.96	2.26	6.22
10" exposure	1.000	S.F.	.032	3.30	1.88	5.18
12" exposure	1.000	S.F.	.026	2.64	1.50	4.14
Fire retardant, 7" exposure	1.000	S.F.	.045	5.45	2.63	8.08
8-1/2" exposure	1.000	S.F.	.038	4.69	2.26	6.95
10" exposure	1.000	S.F.	.032	3.91	1.88	5.79
12" exposure	1.000	S.F.	.026	3.13	1.50	4.63
18" long, 5" exposure	1.000	S.F.	.068	5.20	4.01	9.21
7" exposure	1.000	S.F.	.048	3.66	2.83	6.49
8-1/2" exposure	1.000	S.F.	.040	3.05	2.36	5.41
10" exposure	1.000	S.F.	.036	2.75	2.12	4.87
Fire retardant, 5" exposure	1.000	S.F.	.068	6.25	4.01	10.26
7" exposure	1.000	S.F.	.048	4.39	2.83	7.22
8-1/2" exposure	1.000	S.F.	.040	3.66	2.36	6.02
10" exposure	1.000	S.F.	.036	3.30	2.12	5.42
Paper, #15 asphalt felt	1.100	S.F.	.002	.06	.13	.19
Trim, cedar	.125	S.F.	.005	.17	.30	.47
Fir	.125	S.F.	.005	.11	.30	.41
Redwood	.125	S.F.	.005	.17	.30	.47
White pine	.125	S.F.	.005	.11	.30	.41
Painting, primer, & 1 coat	1.000	S.F.	.013	.14	.65	.79
2 coats	1.000	S.F.	.017	.22	.84	1.06
Staining, sealer, & 1 coat	1.000	S.F.	.017	.18	.85	1.03
2 coats	1.000	S.F.	.019	.29	.92	1.21

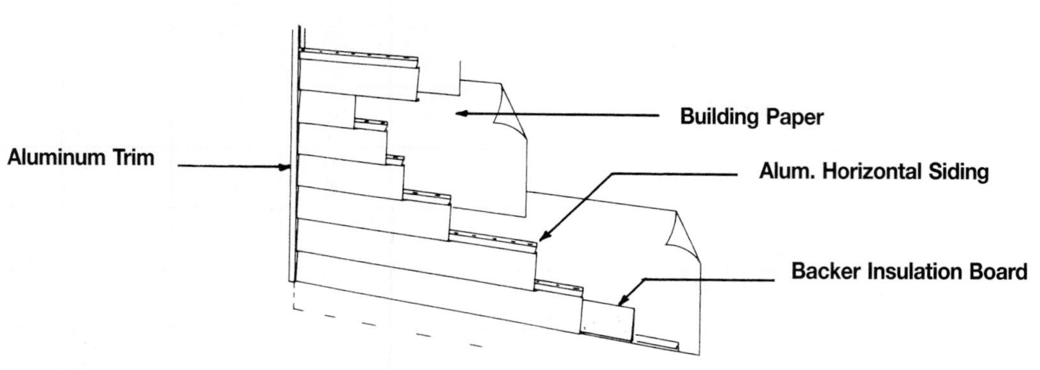

System Description	QUAN.	UNIT	LABOR HOURS	COST PER S.F.		
				MAT.	INST.	TOTAL
ALUMINUM CLAPBOARD SIDING, 8″ WIDE, WHITE						
Aluminum horizontal siding, 8″ clapboard	1.000	S.F.	.031	3.29	1.83	5.12
Backer, insulation board	1.000	S.F.	.008	.44	.47	.91
Trim, aluminum	.600	L.F.	.016	1.14	.92	2.06
Building wrap, spunbonded polypropylene	1.100	S.F.	.002	.17	.13	.30
TOTAL		S.F.	.057	5.04	3.35	8.39
ALUMINUM VERTICAL BOARD & BATTEN, WHITE						
Aluminum vertical board & batten	1.000	S.F.	.027	2.66	1.60	4.26
Backer insulation board	1.000	S.F.	.008	.44	.47	.91
Trim, aluminum	.600	L.F.	.016	1.14	.92	2.06
Building wrap, spunbonded polypropylene	1.100	S.F.	.002	.17	.13	.30
TOTAL		S.F.	.053	4.41	3.12	7.53
VINYL CLAPBOARD SIDING, 8″ WIDE, WHITE						
Vinyl siding, clabboard profile, smooth texture, .042 thick, single 8	1.000	S.F.	.032	.90	1.90	2.80
Backer, insulation board	1.000	S.F.	.008	.44	.47	.91
Vinyl siding, access., outside corner, woodgrain, 4″ face, 3/4″ pocket	.600	L.F.	.014	1.43	.81	2.24
Building wrap, spunbonded polypropylene	1.100	S.F.	.002	.17	.13	.30
TOTAL		S.F.	.056	2.94	3.31	6.25
VINYL VERTICAL BOARD & BATTEN, WHITE						
Vinyl siding, vertical pattern, .046 thick, double 5	1.000	S.F.	.029	1.74	1.71	3.45
Backer, insulation board	1.000	S.F.	.008	.44	.47	.91
Vinyl siding, access., outside corner, woodgrain, 4″ face, 3/4″ pocket	.600	L.F.	.014	1.43	.81	2.24
Building wrap, spunbonded polypropylene	1.100	S.F.	.002	.17	.13	.30
TOTAL		S.F.	.053	3.78	3.12	6.90

The costs in this system are on a square foot of wall basis.
Subtract openings from wall area.

Description	QUAN.	UNIT	LABOR HOURS	COST PER S.F.		
				MAT.	INST.	TOTAL

Metal & Plastic Siding Price Sheet	QUAN.	UNIT	LABOR HOURS	COST PER S.F.		
				MAT.	INST.	TOTAL
Siding, aluminum, .024" thick, smooth, 8" wide, white	1.000	S.F.	.031	3.29	1.83	5.12
Color	1.000	S.F.	.031	3.46	1.83	5.29
Double 4" pattern, 8" wide, white	1.000	S.F.	.031	3.28	1.83	5.11
Color	1.000	S.F.	.031	3.45	1.83	5.28
Double 5" pattern, 10" wide, white	1.000	S.F.	.029	3.05	1.71	4.76
Color	1.000	S.F.	.029	3.22	1.71	4.93
Embossed, single, 8" wide, white	1.000	S.F.	.031	3.08	1.83	4.91
Color	1.000	S.F.	.031	3.25	1.83	5.08
Double 4" pattern, 8" wide, white	1.000	S.F.	.031	3.17	1.83	5
Color	1.000	S.F.	.031	3.34	1.83	5.17
Double 5" pattern, 10" wide, white	1.000	S.F.	.029	3.19	1.71	4.90
Color	1.000	S.F.	.029	3.36	1.71	5.07
Alum siding with insulation board, smooth, 8" wide, white	1.000	S.F.	.031	2.79	1.83	4.62
Color	1.000	S.F.	.031	2.96	1.83	4.79
Double 4" pattern, 8" wide, white	1.000	S.F.	.031	2.77	1.83	4.60
Color	1.000	S.F.	.031	2.94	1.83	4.77
Double 5" pattern, 10" wide, white	1.000	S.F.	.029	2.78	1.71	4.49
Color	1.000	S.F.	.029	2.95	1.71	4.66
Embossed, single, 8" wide, white	1.000	S.F.	.031	3.22	1.83	5.05
Color	1.000	S.F.	.031	3.39	1.83	5.22
Double 4" pattern, 8" wide, white	1.000	S.F.	.031	3.25	1.83	5.08
Color	1.000	S.F.	.031	3.42	1.83	5.25
Double 5" pattern, 10" wide, white	1.000	S.F.	.029	3.25	1.71	4.96
Color	1.000	S.F.	.029	3.42	1.71	5.13
Aluminum, shake finish, 10" wide, white	1.000	S.F.	.029	3.51	1.71	5.22
Color	1.000	S.F.	.029	3.68	1.71	5.39
Aluminum, vertical, 12" wide, white	1.000	S.F.	.027	2.66	1.60	4.26
Color	1.000	S.F.	.027	2.83	1.60	4.43
Vinyl siding, 8" wide, smooth, white	1.000	S.F.	.032	.90	1.90	2.80
Color	1.000	S.F.	.032	1.07	1.90	2.97
10" wide, Dutch lap, smooth, white	1.000	S.F.	.029	1.22	1.71	2.93
Color	1.000	S.F.	.029	1.39	1.71	3.10
Double 4" pattern, 8" wide, white	1.000	S.F.	.032	.90	1.90	2.80
Color	1.000	S.F.	.032	1.07	1.90	2.97
Double 5" pattern, 10" wide, white	1.000	S.F.	.029	.89	1.71	2.60
Color	1.000	S.F.	.029	1.06	1.71	2.77
Embossed, single, 8" wide, white	1.000	S.F.	.032	1.53	1.90	3.43
Color	1.000	S.F.	.032	1.70	1.90	3.60
10" wide, white	1.000	S.F.	.029	1.84	1.71	3.55
Color	1.000	S.F.	.029	2.01	1.71	3.72
Double 4" pattern, 8" wide, white	1.000	S.F.	.032	1.19	1.90	3.09
Color	1.000	S.F.	.032	1.36	1.90	3.26
Double 5" pattern, 10" wide, white	1.000	S.F.	.029	1.19	1.71	2.90
Color	1.000	S.F.	.029	1.36	1.71	3.07
Vinyl, shake finish, 10" wide, white	1.000	S.F.	.029	4.20	2.36	6.56
Color	1.000	S.F.	.029	4.37	2.36	6.73
Vinyl, vertical, double 5" pattern, 10" wide, white	1.000	S.F.	.029	1.74	1.71	3.45
Color	1.000	S.F.	.029	1.91	1.71	3.62
Backer board, installed in siding panels 8" or 10" wide	1.000	S.F.	.008	.44	.47	.91
4' x 8' sheets, polystyrene, 3/4" thick	1.000	S.F.	.010	.62	.59	1.21
4' x 8' fiberboard, plain	1.000	S.F.	.008	.44	.47	.91
Trim, aluminum, white	.600	L.F.	.016	1.14	.92	2.06
Color	.600	L.F.	.016	1.24	.92	2.16
Vinyl, white	.600	L.F.	.014	1.43	.81	2.24
Color	.600	L.F.	.014	1.63	.81	2.44
Paper, #15 asphalt felt	1.100	S.F.	.002	.07	.14	.21
Kraft paper, plain	1.100	S.F.	.002	.18	.14	.32
Foil backed	1.100	S.F.	.002	.19	.14	.33

Description	QUAN.	UNIT	LABOR HOURS	COST PER S.F. MAT.	COST PER S.F. INST.	COST PER S.F. TOTAL
Poured insulation, cellulose fiber, R3.8 per inch (1" thick)	1.000	S.F.	.003	.06	.20	.26
Fiberglass , R4.0 per inch (1" thick)	1.000	S.F.	.003	.06	.20	.26
Mineral wool, R3.0 per inch (1" thick)	1.000	S.F.	.003	.05	.20	.25
Polystyrene, R4.0 per inch (1" thick)	1.000	S.F.	.003	.14	.20	.34
Vermiculite, R2.7 per inch (1" thick)	1.000	S.F.	.003	.49	.20	.69
Perlite, R2.7 per inch (1" thick)	1.000	S.F.	.003	.49	.20	.69
Reflective insulation, aluminum foil reinforced with scrim	1.000	S.F.	.004	.17	.25	.42
Reinforced with woven polyolefin	1.000	S.F.	.004	.25	.25	.50
With single bubble air space, R8.8	1.000	S.F.	.005	.29	.32	.61
With double bubble air space, R9.8	1.000	S.F.	.005	.35	.32	.67
Rigid insulation, fiberglass, unfaced, 1-1/2" thick, R6.2	1.000	S.F.	.008	.44	.47	.91
2" thick, R8.3	1.000	S.F.	.008	.55	.47	1.02
2-1/2" thick, R10.3	1.000	S.F.	.010	.66	.59	1.25
3" thick, R12.4	1.000	S.F.	.010	.66	.59	1.25
Foil faced, 1" thick, R4.3	1.000	S.F.	.008	.92	.47	1.39
1-1/2" thick, R6.2	1.000	S.F.	.008	1.38	.47	1.85
2" thick, R8.7	1.000	S.F.	.009	1.74	.53	2.27
2-1/2" thick, R10.9	1.000	S.F.	.010	2.05	.59	2.64
3" thick, R13.0	1.000	S.F.	.010	2.30	.59	2.89
Perlite, 1" thick R2.77	1.000	S.F.	.010	.53	.59	1.12
2" thick R5.55	1.000	S.F.	.011	.86	.65	1.51
Polystyrene, extruded, blue, 2.2#/C.F., 3/4" thick R4	1.000	S.F.	.010	.62	.59	1.21
1-1/2" thick R8.1	1.000	S.F.	.011	1.23	.65	1.88
2" thick R10.8	1.000	S.F.	.011	1.77	.65	2.42
Molded bead board, white, 1" thick R3.85	1.000	S.F.	.010	.30	.59	.89
1-1/2" thick, R5.6	1.000	S.F.	.011	.59	.65	1.24
2" thick, R7.7	1.000	S.F.	.011	.89	.65	1.54
Non-rigid insulation, batts Fiberglass, kraft faced, 3-1/2" thick, R13, 11" wide	1.000	S.F.	.005	.37	.41	.78
15" wide	1.000	S.F.	.005	.37	.41	.78
23" wide	1.000	S.F.	.005	.37	.41	.78
6" thick, R19, 11" wide	1.000	S.F.	.006	.47	.41	.88
15" wide	1.000	S.F.	.006	.47	.41	.88
23" wide	1.000	S.F.	.006	.47	.41	.88
9" thick, R30, 15" wide	1.000	S.F.	.006	.80	.35	1.15
23" wide	1.000	S.F.	.006	.80	.35	1.15
12" thick, R38, 15" wide	1.000	S.F.	.006	1.17	.35	1.52
23" wide	1.000	S.F.	.006	1.17	.35	1.52
Fiberglass, foil faced, 3-1/2" thick, R13, 15" wide	1.000	S.F.	.005	.52	.35	.87
23" wide	1.000	S.F.	.005	.52	.35	.87
6" thick, R19, 15" thick	1.000	S.F.	.005	.67	.29	.96
23" wide	1.000	S.F.	.005	.67	.29	.96
9" thick, R30, 15" wide	1.000	S.F.	.006	1.02	.35	1.37
23" wide	1.000	S.F.	.006	1.02	.35	1.37

Insulation Systems	QUAN.	UNIT	LABOR HOURS	COST PER S.F.		
				MAT.	INST.	TOTAL
Non-rigid insulation batts						
Fiberglass unfaced, 3-1/2" thick, R13, 15" wide	1.000	S.F.	.005	.36	.29	.65
23" wide	1.000	S.F.	.005	.36	.29	.65
6" thick, R19, 15" wide	1.000	S.F.	.006	.43	.35	.78
23" wide	1.000	S.F.	.006	.43	.35	.78
9" thick, R19, 15" wide	1.000	S.F.	.007	.67	.41	1.08
23" wide	1.000	S.F.	.007	.67	.41	1.08
12" thick, R38, 15" wide	1.000	S.F.	.007	.81	.41	1.22
23" wide	1.000	S.F.	.007	.81	.41	1.22
Mineral fiber batts, 3" thick, R11	1.000	S.F.	.005	.87	.29	1.16
3-1/2" thick, R13	1.000	S.F.	.005	.87	.29	1.16
6" thick, R19	1.000	S.F.	.005	1.37	.29	1.66
6-1/2" thick, R22	1.000	S.F.	.005	1.37	.29	1.66
10" thick, R30	1.000	S.F.	.006	1.80	.35	2.15

28 Double Hung Window Systems

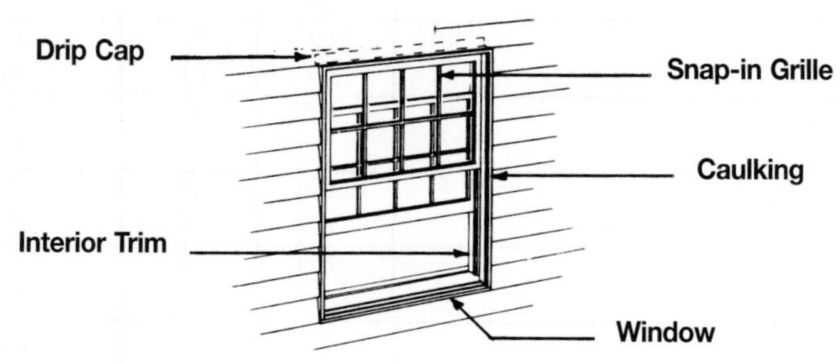

System Description	QUAN.	UNIT	LABOR HOURS	COST EACH		
				MAT.	INST.	TOTAL
BUILDER'S QUALITY WOOD WINDOW 2′ X 3′, DOUBLE HUNG						
Window, primed, builder's quality, 2′ x 3′, insulating glass	1.000	Ea.	.800	265	47	312
Trim, interior casing	11.000	L.F.	.352	17.16	20.68	37.84
Paint, interior & exterior, primer & 2 coats	2.000	Face	1.778	2.56	87	89.56
Caulking	10.000	L.F.	.278	2.50	16.30	18.80
Snap-in grille	1.000	Set	.333	63	19.65	82.65
Drip cap, metal	2.000	L.F.	.040	1.30	2.36	3.66
TOTAL		Ea.	3.581	351.52	192.99	544.51
PLASTIC CLAD WOOD WINDOW 3′ X 4′, DOUBLE HUNG						
Window, plastic clad, premium, 3′ x 4′, insulating glass	1.000	Ea.	.889	445	52.50	497.50
Trim, interior casing	15.000	L.F.	.480	23.40	28.20	51.60
Paint, interior, primer & 2 coats	1.000	Face	.889	1.28	43.50	44.78
Caulking	14.000	L.F.	.389	3.50	22.82	26.32
Snap-in grille	1.000	Set	.333	63	19.65	82.65
TOTAL		Ea.	2.980	536.18	166.67	702.85
METAL CLAD WOOD WINDOW, 3′ X 5′, DOUBLE HUNG						
Window, metal clad, deluxe, 3′ x 5′, insulating glass	1.000	Ea.	1.000	425	59	484
Trim, interior casing	17.000	L.F.	.544	26.52	31.96	58.48
Paint, interior, primer & 2 coats	1.000	Face	.889	1.28	43.50	44.78
Caulking	16.000	L.F.	.444	4	26.08	30.08
Snap-in grille	1.000	Set	.235	155	13.85	168.85
Drip cap, metal	3.000	L.F.	.060	1.95	3.54	5.49
TOTAL		Ea.	3.172	613.75	177.93	791.68

The cost of this system is on a cost per each window basis.

Description	QUAN.	UNIT	LABOR HOURS	COST EACH		
				MAT.	INST.	TOTAL

Double Hung Window Price Sheet	QUAN.	UNIT	LABOR HOURS	COST EACH		
				MAT.	INST.	TOTAL
Windows, double-hung, builder's quality, 2′ x 3′, single glass	1.000	Ea.	.800	212	47	259
Insulating glass	1.000	Ea.	.800	265	47	312
3′ x 4′, single glass	1.000	Ea.	.889	320	52.50	372.50
Insulating glass	1.000	Ea.	.889	325	52.50	377.50
4′ x 4′-6″, single glass	1.000	Ea.	1.000	370	59	429
Insulating glass	1.000	Ea.	1.000	400	59	459
Plastic clad premium insulating glass, 2′-6″ x 3′	1.000	Ea.	.800	375	47	422
3′ x 3′-6″	1.000	Ea.	.800	380	47	427
3′ x 4′	1.000	Ea.	.889	445	52.50	497.50
3′ x 4′-6″	1.000	Ea.	.889	470	52.50	522.50
3′ x 5′	1.000	Ea.	1.000	505	59	564
3′-6″ x 6′	1.000	Ea.	1.000	570	59	629
Metal clad deluxe insulating glass, 2′-6″ x 3′	1.000	Ea.	.800	315	47	362
3′ x 3′-6″	1.000	Ea.	.800	360	47	407
3′ x 4′	1.000	Ea.	.889	375	52.50	427.50
3′ x 4′-6″	1.000	Ea.	.889	395	52.50	447.50
3′ x 5′	1.000	Ea.	1.000	425	59	484
3′-6″ x 6′	1.000	Ea.	1.000	515	59	574
Trim, interior casing, window 2′ x 3′	11.000	L.F.	.367	17.15	20.50	37.65
2′-6″ x 3′	12.000	L.F.	.400	18.70	22.50	41.20
3′ x 3′-6″	14.000	L.F.	.467	22	26.50	48.50
3′ x 4′	15.000	L.F.	.500	23.50	28	51.50
3′ x 4′-6″	16.000	L.F.	.533	25	30	55
3′ x 5′	17.000	L.F.	.567	26.50	32	58.50
3′-6″ x 6′	20.000	L.F.	.667	31	37.50	68.50
4′ x 4′-6″	18.000	L.F.	.600	28	34	62
Paint or stain, interior or exterior, 2′ x 3′ window, 1 coat	1.000	Face	.444	.48	21.50	21.98
2 coats	1.000	Face	.727	.95	35.50	36.45
Primer & 1 coat	1.000	Face	.727	.87	35.50	36.37
Primer & 2 coats	1.000	Face	.889	1.28	43.50	44.78
3′ x 4′ window, 1 coat	1.000	Face	.667	1.07	32.50	33.57
2 coats	1.000	Face	.667	1.84	32.50	34.34
Primer & 1 coat	1.000	Face	.727	1.31	35.50	36.81
Primer & 2 coats	1.000	Face	.889	1.28	43.50	44.78
4′ x 4′-6″ window, 1 coat	1.000	Face	.667	1.07	32.50	33.57
2 coats	1.000	Face	.667	1.84	32.50	34.34
Primer & 1 coat	1.000	Face	.727	1.31	35.50	36.81
Primer & 2 coats	1.000	Face	.889	1.28	43.50	44.78
Caulking, window, 2′ x 3′	10.000	L.F.	.323	2.50	16.30	18.80
2′-6″ x 3′	11.000	L.F.	.355	2.75	17.95	20.70
3′ x 3′-6″	13.000	L.F.	.419	3.25	21	24.25
3′ x 4′	14.000	L.F.	.452	3.50	23	26.50
3′ x 4′-6″	15.000	L.F.	.484	3.75	24.50	28.25
3′ x 5′	16.000	L.F.	.516	4	26	30
3′-6″ x 6′	19.000	L.F.	.613	4.75	31	35.75
4′ x 4′-6″	17.000	L.F.	.548	4.25	27.50	31.75
Grilles, glass size to, 16″ x 24″ per sash	1.000	Set	.333	63	19.65	82.65
32″ x 32″ per sash	1.000	Set	.235	155	13.85	168.85
Drip cap, aluminum, 2′ long	2.000	L.F.	.040	1.30	2.36	3.66
3′ long	3.000	L.F.	.060	1.95	3.54	5.49
4′ long	4.000	L.F.	.080	2.60	4.72	7.32
Wood, 2′ long	2.000	L.F.	.067	3.12	3.76	6.88
3′ long	3.000	L.F.	.100	4.68	5.65	10.33
4′ long	4.000	L.F.	.133	6.25	7.50	13.75

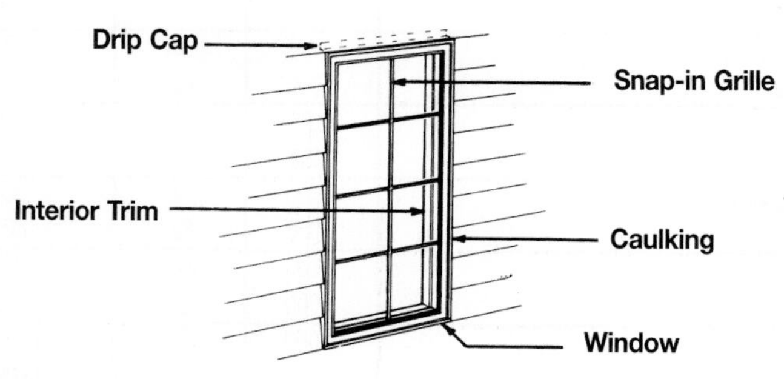

System Description	QUAN.	UNIT	LABOR HOURS	COST EACH		
				MAT.	INST.	TOTAL
BUILDER'S QUALITY WINDOW, WOOD, 2′ BY 3′, CASEMENT						
Window, primed, builder's quality, 2′ x 3′, insulating glass	1.000	Ea.	.800	305	47	352
Trim, interior casing	11.000	L.F.	.352	17.16	20.68	37.84
Paint, interior & exterior, primer & 2 coats	2.000	Face	1.778	2.56	87	89.56
Caulking	10.000	L.F.	.278	2.50	16.30	18.80
Snap-in grille	1.000	Ea.	.267	37.50	15.70	53.20
Drip cap, metal	2.000	L.F.	.040	1.30	2.36	3.66
TOTAL		Ea.	3.515	366.02	189.04	555.06
PLASTIC CLAD WOOD WINDOW, 2′ X 4′, CASEMENT						
Window, plastic clad, premium, 2′ x 4′, insulating glass	1.000	Ea.	.889	365	52.50	417.50
Trim, interior casing	13.000	L.F.	.416	20.28	24.44	44.72
Paint, interior, primer & 2 coats	1.000	Ea.	.889	1.28	43.50	44.78
Caulking	12.000	L.F.	.333	3	19.56	22.56
Snap-in grille	1.000	Ea.	.267	37.50	15.70	53.20
TOTAL		Ea.	2.794	427.06	155.70	582.76
METAL CLAD WOOD WINDOW, 2′ X 5′, CASEMENT						
Window, metal clad, deluxe, 2′ x 5′, insulating glass	1.000	Ea.	1.000	375	59	434
Trim, interior casing	15.000	L.F.	.480	23.40	28.20	51.60
Paint, interior, primer & 2 coats	1.000	Ea.	.889	1.28	43.50	44.78
Caulking	14.000	L.F.	.389	3.50	22.82	26.32
Snap-in grille	1.000	Ea.	.250	51	14.75	65.75
Drip cap, metal	12.000	L.F.	.040	1.30	2.36	3.66
TOTAL		Ea.	3.048	455.48	170.63	626.11

The cost of this system is on a cost per each window basis.

Description	QUAN.	UNIT	LABOR HOURS	COST EACH		
				MAT.	INST.	TOTAL

Casement Window Price Sheet	QUAN.	UNIT	LABOR HOURS	COST EACH		
				MAT.	INST.	TOTAL
Window, casement, builders quality, 2′ x 3′, double insulated glass	1.000	Ea.	.800	310	47	357
Insulating glass	1.000	Ea.	.800	305	47	352
2′ x 4′-6″, double insulated glass	1.000	Ea.	.727	1,025	43	1,068
Insulating glass	1.000	Ea.	.727	610	43	653
2′ x 6′, double insulated glass	1.000	Ea.	.889	495	59	554
Insulating glass	1.000	Ea.	.889	485	59	544
Plastic clad premium insulating glass, 2′ x 3′	1.000	Ea.	.800	299	47	346
2′ x 4′	1.000	Ea.	.889	460	52.50	512.50
2′ x 5′	1.000	Ea.	1.000	615	59	674
2′ x 6′	1.000	Ea.	1.000	450	59	509
Metal clad deluxe insulating glass, 2′ x 3′	1.000	Ea.	.800	325	47	372
2′ x 4′	1.000	Ea.	.889	310	52.50	362.50
2′ x 5′	1.000	Ea.	1.000	355	59	414
2′ x 6′	1.000	Ea.	1.000	410	59	469
Trim, interior casing, window 2′ x 3′	11.000	L.F.	.367	17.15	20.50	37.65
2′ x 4′	13.000	L.F.	.433	20.50	24.50	45
2′ x 4′-6″	14.000	L.F.	.467	22	26.50	48.50
2′ x 5′	15.000	L.F.	.500	23.50	28	51.50
2′ x 6′	17.000	L.F.	.567	26.50	32	58.50
Paint or stain, interior or exterior, 2′ x 3′ window, 1 coat	1.000	Face	.444	.48	21.50	21.98
2 coats	1.000	Face	.727	.95	35.50	36.45
Primer & 1 coat	1.000	Face	.727	.87	35.50	36.37
Primer & 2 coats	1.000	Face	.889	1.28	43.50	44.78
2′ x 4′ window, 1 coat	1.000	Face	.444	.48	21.50	21.98
2 coats	1.000	Face	.727	.95	35.50	36.45
Primer & 1 coat	1.000	Face	.727	.87	35.50	36.37
Primer & 2 coats	1.000	Face	.889	1.28	43.50	44.78
2′ x 6′ window, 1 coat	1.000	Face	.667	1.07	32.50	33.57
2 coats	1.000	Face	.667	1.84	32.50	34.34
Primer & 1 coat	1.000	Face	.727	1.31	35.50	36.81
Primer & 2 coats	1.000	Face	.889	1.28	43.50	44.78
Caulking, window, 2′ x 3′	10.000	L.F.	.323	2.50	16.30	18.80
2′ x 4′	12.000	L.F.	.387	3	19.55	22.55
2′ x 4′-6″	13.000	L.F.	.419	3.25	21	24.25
2′ x 5′	14.000	L.F.	.452	3.50	23	26.50
2′ x 6′	16.000	L.F.	.516	4	26	30
Grilles, glass size, to 20″ x 36″	1.000	Ea.	.267	37.50	15.70	53.20
To 20″ x 56″	1.000	Ea.	.250	51	14.75	65.75
Drip cap, metal, 2′ long	2.000	L.F.	.040	1.30	2.36	3.66
Wood, 2′ long	2.000	L.F.	.067	3.12	3.76	6.88

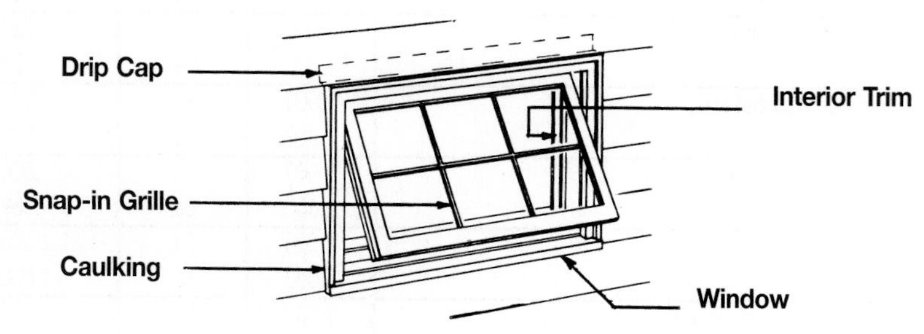

System Description	QUAN.	UNIT	LABOR HOURS	COST EACH		
				MAT.	INST.	TOTAL
BUILDER'S QUALITY WINDOW, WOOD, 34" X 22", AWNING						
Window, 34" x 22", insulating glass	1.000	Ea.	.800	345	47	392
Trim, interior casing	10.500	L.F.	.336	16.38	19.74	36.12
Paint, interior & exterior, primer & 2 coats	2.000	Face	1.778	2.56	87	89.56
Caulking	9.500	L.F.	.264	2.38	15.49	17.87
Snap-in grille	1.000	Ea.	.267	34	15.70	49.70
Drip cap, metal	3.000	L.F.	.060	1.95	3.54	5.49
TOTAL		Ea.	3.505	402.27	188.47	590.74
PLASTIC CLAD WOOD WINDOW, 40" X 28", AWNING						
Window, plastic clad, premium, 40" x 28", insulating glass	1.000	Ea.	.889	380	52.50	432.50
Trim interior casing	13.500	L.F.	.432	21.06	25.38	46.44
Paint, interior, primer & 2 coats	1.000	Face	.889	1.28	43.50	44.78
Caulking	12.500	L.F.	.347	3.13	20.38	23.51
Snap-in grille	1.000	Ea.	.267	34	15.70	49.70
TOTAL		Ea.	2.824	439.47	157.46	596.93
METAL CLAD WOOD WINDOW, 48" X 36", AWNING						
Window, metal clad, deluxe, 48" x 36", insulating glass	1.000	Ea.	1.000	415	59	474
Trim, interior casing	15.000	L.F.	.480	23.40	28.20	51.60
Paint, interior, primer & 2 coats	1.000	Face	.889	1.28	43.50	44.78
Caulking	14.000	L.F.	.389	3.50	22.82	26.32
Snap-in grille	1.000	Ea.	.250	49.50	14.75	64.25
Drip cap, metal	4.000	L.F.	.080	2.60	4.72	7.32
TOTAL		Ea.	3.088	495.28	172.99	668.27

The cost of this system is on a cost per each window basis.

Description	QUAN.	UNIT	LABOR HOURS	COST EACH		
				MAT.	INST.	TOTAL

Awning Window Price Sheet	QUAN.	UNIT	LABOR HOURS	COST EACH		
				MAT.	INST.	TOTAL
Windows, awning, builder's quality, 34″ x 22″, insulated glass	1.000	Ea.	.800	305	47	352
Low E glass	1.000	Ea.	.800	345	47	392
40″ x 28″, insulated glass	1.000	Ea.	.889	345	52.50	397.50
Low E glass	1.000	Ea.	.889	380	52.50	432.50
48″ x 36″, insulated glass	1.000	Ea.	1.000	520	59	579
Low E glass	1.000	Ea.	1.000	550	59	609
Plastic clad premium insulating glass, 34″ x 22″	1.000	Ea.	.800	296	47	343
40″ x 22″	1.000	Ea.	.800	325	47	372
36″ x 28″	1.000	Ea.	.889	340	52.50	392.50
36″ x 36″	1.000	Ea.	.889	380	52.50	432.50
48″ x 28″	1.000	Ea.	1.000	410	59	469
60″ x 36″	1.000	Ea.	1.000	570	59	629
Metal clad deluxe insulating glass, 34″ x 22″	1.000	Ea.	.800	278	52.50	330.50
40″ x 22″	1.000	Ea.	.800	325	52.50	377.50
36″ x 25″	1.000	Ea.	.889	305	52.50	357.50
40″ x 30″	1.000	Ea.	.889	380	52.50	432.50
48″ x 28″	1.000	Ea.	1.000	390	59	449
60″ x 36″	1.000	Ea.	1.000	415	59	474
Trim, interior casing window, 34″ x 22″	10.500	L.F.	.350	16.40	19.75	36.15
40″ x 22″	11.500	L.F.	.383	17.95	21.50	39.45
36″ x 28″	12.500	L.F.	.417	19.50	23.50	43
40″ x 28″	13.500	L.F.	.450	21	25.50	46.50
48″ x 28″	14.500	L.F.	.483	22.50	27.50	50
48″ x 36″	15.000	L.F.	.500	23.50	28	51.50
Paint or stain, interior or exterior, 34″ x 22″, 1 coat	1.000	Face	.444	.48	21.50	21.98
2 coats	1.000	Face	.727	.95	35.50	36.45
Primer & 1 coat	1.000	Face	.727	.87	35.50	36.37
Primer & 2 coats	1.000	Face	.889	1.28	43.50	44.78
36″ x 28″, 1 coat	1.000	Face	.444	.48	21.50	21.98
2 coats	1.000	Face	.727	.95	35.50	36.45
Primer & 1 coat	1.000	Face	.727	.87	35.50	36.37
Primer & 2 coats	1.000	Face	.889	1.28	43.50	44.78
48″ x 36″, 1 coat	1.000	Face	.667	1.07	32.50	33.57
2 coats	1.000	Face	.667	1.84	32.50	34.34
Primer & 1 coat	1.000	Face	.727	1.31	35.50	36.81
Primer & 2 coats	1.000	Face	.889	1.28	43.50	44.78
Caulking, window, 34″ x 22″	9.500	L.F.	.306	2.38	15.50	17.88
40″ x 22″	10.500	L.F.	.339	2.63	17.10	19.73
36″ x 28″	11.500	L.F.	.371	2.88	18.75	21.63
40″ x 28″	12.500	L.F.	.403	3.13	20.50	23.63
48″ x 28″	13.500	L.F.	.436	3.38	22	25.38
48″ x 36″	14.000	L.F.	.452	3.50	23	26.50
Grilles, glass size, to 28″ by 16″	1.000	Ea.	.267	34	15.70	49.70
To 44″ by 24″	1.000	Ea.	.250	49.50	14.75	64.25
Drip cap, aluminum, 3′ long	3.000	L.F.	.060	1.95	3.54	5.49
3′-6″ long	3.500	L.F.	.070	2.28	4.13	6.41
4′ long	4.000	L.F.	.080	2.60	4.72	7.32
Wood, 3′ long	3.000	L.F.	.100	4.68	5.65	10.33
3′-6″ long	3.500	L.F.	.117	5.45	6.60	12.05
4′ long	4.000	L.F.	.133	6.25	7.50	13.75

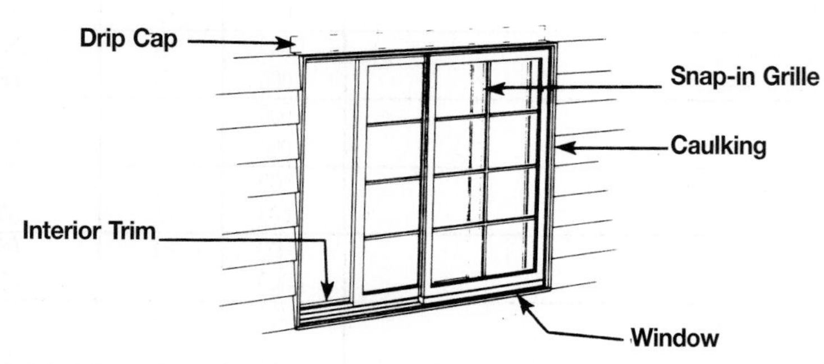

System Description	QUAN.	UNIT	LABOR HOURS	COST EACH		
				MAT.	INST.	TOTAL
BUILDER'S QUALITY WOOD WINDOW, 3′ X 2′, SLIDING						
Window, primed, builder's quality, 3′ x 3′, insul. glass	1.000	Ea.	.800	355	47	402
Trim, interior casing	11.000	L.F.	.352	17.16	20.68	37.84
Paint, interior & exterior, primer & 2 coats	2.000	Face	1.778	2.56	87	89.56
Caulking	10.000	L.F.	.278	2.50	16.30	18.80
Snap-in grille	1.000	Set	.333	42	19.65	61.65
Drip cap, metal	3.000	L.F.	.060	1.95	3.54	5.49
TOTAL		Ea.	3.601	421.17	194.17	615.34
PLASTIC CLAD WOOD WINDOW, 4′ X 3′-6″, SLIDING						
Window, plastic clad, premium, 4′ x 3′-6″, insulating glass	1.000	Ea.	.889	765	52.50	817.50
Trim, interior casing	16.000	L.F.	.512	24.96	30.08	55.04
Paint, interior, primer & 2 coats	1.000	Face	.889	1.28	43.50	44.78
Caulking	17.000	L.F.	.472	4.25	27.71	31.96
Snap-in grille	1.000	Set	.333	42	19.65	61.65
TOTAL		Ea.	3.095	837.49	173.44	1,010.93
METAL CLAD WOOD WINDOW, 6′ X 5′, SLIDING						
Window, metal clad, deluxe, 6′ x 5′, insulating glass	1.000	Ea.	1.000	830	59	889
Trim, interior casing	23.000	L.F.	.736	35.88	43.24	79.12
Paint, interior, primer & 2 coats	1.000	Face	.889	1.28	43.50	44.78
Caulking	22.000	L.F.	.611	5.50	35.86	41.36
Snap-in grille	1.000	Set	.364	50	21.50	71.50
Drip cap, metal	6.000	L.F.	.120	3.90	7.08	10.98
TOTAL		Ea.	3.720	926.56	210.18	1,136.74

The cost of this system is on a cost per each window basis.

Description	QUAN.	UNIT	LABOR HOURS	COST EACH		
				MAT.	INST.	TOTAL

Sliding Window Price Sheet	QUAN.	UNIT	LABOR HOURS	COST EACH		
				MAT.	INST.	TOTAL
Windows, sliding, builder's quality, 3′ x 3′, single glass	1.000	Ea.	.800	325	47	372
Insulating glass	1.000	Ea.	.800	355	47	402
4′ x 3′-6″, single glass	1.000	Ea.	.889	435	52.50	487.50
Insulating glass	1.000	Ea.	.889	430	52.50	482.50
6′ x 5′, single glass	1.000	Ea.	1.000	565	59	624
Insulating glass	1.000	Ea.	1.000	635	59	694
Plastic clad premium insulating glass, 3′ x 3′	1.000	Ea.	.800	690	47	737
4′ x 3′-6″	1.000	Ea.	.889	765	52.50	817.50
5′ x 4′	1.000	Ea.	.889	1,025	52.50	1,077.50
6′ x 5′	1.000	Ea.	1.000	1,275	59	1,334
Metal clad deluxe insulating glass, 3′ x 3′	1.000	Ea.	.800	365	47	412
4′ x 3′-6″	1.000	Ea.	.889	440	52.50	492.50
5′ x 4′	1.000	Ea.	.889	530	52.50	582.50
6′ x 5′	1.000	Ea.	1.000	830	59	889
Trim, interior casing, window 3′ x 2′	11.000	L.F.	.367	17.15	20.50	37.65
3′ x 3′	13.000	L.F.	.433	20.50	24.50	45
4′ x 3′-6″	16.000	L.F.	.533	25	30	55
5′ x 4′	19.000	L.F.	.633	29.50	35.50	65
6′ x 5′	23.000	L.F.	.767	36	43	79
Paint or stain, interior or exterior, 3′ x 2′ window, 1 coat	1.000	Face	.444	.48	21.50	21.98
2 coats	1.000	Face	.727	.95	35.50	36.45
Primer & 1 coat	1.000	Face	.727	.87	35.50	36.37
Primer & 2 coats	1.000	Face	.889	1.28	43.50	44.78
4′ x 3′-6″ window, 1 coat	1.000	Face	.667	1.07	32.50	33.57
2 coats	1.000	Face	.667	1.84	32.50	34.34
Primer & 1 coat	1.000	Face	.727	1.31	35.50	36.81
Primer & 2 coats	1.000	Face	.889	1.28	43.50	44.78
6′ x 5′ window, 1 coat	1.000	Face	.889	2.70	43.50	46.20
2 coats	1.000	Face	1.333	4.93	65	69.93
Primer & 1 coat	1.000	Face	1.333	4.64	65	69.64
Primer & 2 coats	1.000	Face	1.600	6.85	78	84.85
Caulking, window, 3′ x 2′	10.000	L.F.	.323	2.50	16.30	18.80
3′ x 3′	12.000	L.F.	.387	3	19.55	22.55
4′ x 3′-6″	15.000	L.F.	.484	3.75	24.50	28.25
5′ x 4′	18.000	L.F.	.581	4.50	29.50	34
6′ x 5′	22.000	L.F.	.710	5.50	36	41.50
Grilles, glass size, to 14″ x 36″	1.000	Set	.333	42	19.65	61.65
To 36″ x 36″	1.000	Set	.364	50	21.50	71.50
Drip cap, aluminum, 3′ long	3.000	L.F.	.060	1.95	3.54	5.49
4′ long	4.000	L.F.	.080	2.60	4.72	7.32
5′ long	5.000	L.F.	.100	3.25	5.90	9.15
6′ long	6.000	L.F.	.120	3.90	7.10	11
Wood, 3′ long	3.000	L.F.	.100	4.68	5.65	10.33
4′ long	4.000	L.F.	.133	6.25	7.50	13.75
5′ long	5.000	L.F.	.167	7.80	9.40	17.20
6′ long	6.000	L.F.	.200	9.35	11.30	20.65

44 Bow/Bay Window Systems

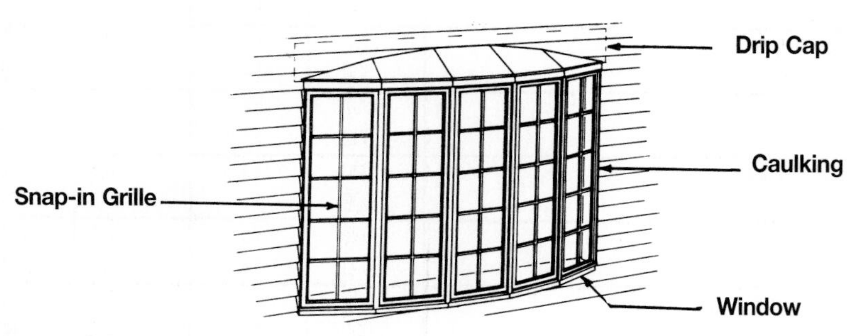

System Description	QUAN.	UNIT	LABOR HOURS	COST EACH MAT.	COST EACH INST.	COST EACH TOTAL
AWNING TYPE BOW WINDOW, BUILDER'S QUALITY, 8′ X 5′						
Window, primed, builder's quality, 8′ x 5′, insulating glass	1.000	Ea.	1.600	1,475	94	1,569
Trim, interior casing	27.000	L.F.	.864	42.12	50.76	92.88
Paint, interior & exterior, primer & 1 coat	2.000	Face	3.200	13.70	156	169.70
Caulking	26.000	L.F.	.722	6.50	42.38	48.88
Snap-in grilles	1.000	Set	1.067	150	62.80	212.80
TOTAL		Ea.	7.453	1,687.32	405.94	2,093.26
CASEMENT TYPE BOW WINDOW, PLASTIC CLAD, 10′ X 6′						
Window, plastic clad, premium, 10′ x 6′, insulating glass	1.000	Ea.	2.286	3,125	135	3,260
Trim, interior casing	33.000	L.F.	1.056	51.48	62.04	113.52
Paint, interior, primer & 1 coat	1.000	Face	1.778	2.56	87	89.56
Caulking	32.000	L.F.	.889	8	52.16	60.16
Snap-in grilles	1.000	Set	1.333	187.50	78.50	266
TOTAL		Ea.	7.342	3,374.54	414.70	3,789.24
DOUBLE HUNG TYPE, METAL CLAD, 9′ X 5′						
Window, metal clad, deluxe, 9′ x 5′, insulating glass	1.000	Ea.	2.667	1,625	157	1,782
Trim, interior casing	29.000	L.F.	.928	45.24	54.52	99.76
Paint, interior, primer & 1 coat	1.000	Face	1.778	2.56	87	89.56
Caulking	28.000	L.F.	.778	7	45.64	52.64
Snap-in grilles	1.000	Set	1.067	150	62.80	212.80
TOTAL		Ea.	7.218	1,829.80	406.96	2,236.76

The cost of this system is on a cost per each window basis.

Description	QUAN.	UNIT	LABOR HOURS	COST EACH MAT.	COST EACH INST.	COST EACH TOTAL

Bow/Bay Window Price Sheet	QUAN.	UNIT	LABOR HOURS	COST EACH		
				MAT.	INST.	TOTAL
Windows, bow awning type, builder's quality, 8′ x 5′, insulating glass	1.000	Ea.	1.600	1,750	94	1,844
Low E glass	1.000	Ea.	1.600	1,475	94	1,569
12′ x 6′, insulating glass	1.000	Ea.	2.667	1,500	157	1,657
Low E glass	1.000	Ea.	2.667	1,625	157	1,782
Plastic clad premium insulating glass, 6′ x 4′	1.000	Ea.	1.600	1,150	94	1,244
9′ x 4′	1.000	Ea.	2.000	1,575	118	1,693
10′ x 5′	1.000	Ea.	2.286	2,600	135	2,735
12′ x 6′	1.000	Ea.	2.667	3,400	157	3,557
Metal clad deluxe insulating glass, 6′ x 4′	1.000	Ea.	1.600	1,350	94	1,444
9′ x 4′	1.000	Ea.	2.000	1,700	118	1,818
10′ x 5′	1.000	Ea.	2.286	2,325	135	2,460
12′ x 6′	1.000	Ea.	2.667	2,975	157	3,132
Bow casement type, builder's quality, 8′ x 5′, single glass	1.000	Ea.	1.600	2,125	94	2,219
Insulating glass	1.000	Ea.	1.600	2,575	94	2,669
12′ x 6′, single glass	1.000	Ea.	2.667	2,650	157	2,807
Insulating glass	1.000	Ea.	2.667	3,600	157	3,757
Plastic clad premium insulating glass, 8′ x 5′	1.000	Ea.	1.600	2,000	94	2,094
10′ x 5′	1.000	Ea.	2.000	2,675	118	2,793
10′ x 6′	1.000	Ea.	2.286	3,125	135	3,260
12′ x 6′	1.000	Ea.	2.667	3,725	157	3,882
Metal clad deluxe insulating glass, 8′ x 5′	1.000	Ea.	1.600	1,875	94	1,969
10′ x 5′	1.000	Ea.	2.000	2,025	118	2,143
10′ x 6′	1.000	Ea.	2.286	2,375	135	2,510
12′ x 6′	1.000	Ea.	2.667	3,250	157	3,407
Bow, double hung type, builder's quality, 8′ x 4′, single glass	1.000	Ea.	1.600	1,500	94	1,594
Insulating glass	1.000	Ea.	1.600	1,600	94	1,694
9′ x 5′, single glass	1.000	Ea.	2.667	1,600	157	1,757
Insulating glass	1.000	Ea.	2.667	1,700	157	1,857
Plastic clad premium insulating glass, 7′ x 4′	1.000	Ea.	1.600	1,550	94	1,644
8′ x 4′	1.000	Ea.	2.000	1,575	118	1,693
8′ x 5′	1.000	Ea.	2.286	1,650	135	1,785
9′ x 5′	1.000	Ea.	2.667	1,700	157	1,857
Metal clad deluxe insulating glass, 7′ x 4′	1.000	Ea.	1.600	1,425	94	1,519
8′ x 4′	1.000	Ea.	2.000	1,475	118	1,593
8′ x 5′	1.000	Ea.	2.286	1,525	135	1,660
9′ x 5′	1.000	Ea.	2.667	1,625	157	1,782
Trim, interior casing, window 7′ x 4′	1.000	Ea.	.767	36	43	79
8′ x 5′	1.000	Ea.	.900	42	51	93
10′ x 6′	1.000	Ea.	1.100	51.50	62	113.50
12′ x 6′	1.000	Ea.	1.233	57.50	69.50	127
Paint or stain, interior, or exterior, 7′ x 4′ window, 1 coat	1.000	Face	.889	2.70	43.50	46.20
Primer & 1 coat	1.000	Face	1.333	4.64	65	69.64
8′ x 5′ window, 1 coat	1.000	Face	.889	2.70	43.50	46.20
Primer & 1 coat	1.000	Face	1.333	4.64	65	69.64
10′ x 6′ window, 1 coat	1.000	Face	1.333	2.14	65	67.14
Primer & 1 coat	1.000	Face	1.778	2.56	87	89.56
12′ x 6′ window, 1 coat	1.000	Face	1.778	5.40	87	92.40
Primer & 1 coat	1.000	Face	2.667	9.30	130	139.30
Drip cap, vinyl moulded window, 7′ long	1.000	Ea.	.533	125	24.50	149.50
8′ long	1.000	Ea.	.533	143	28	171
Caulking, window, 7′ x 4′	1.000	Ea.	.710	5.50	36	41.50
8′ x 5′	1.000	Ea.	.839	6.50	42.50	49
10′ x 6′	1.000	Ea.	1.032	8	52	60
12′ x 6′	1.000	Ea.	1.161	9	58.50	67.50
Grilles, window, 7′ x 4′	1.000	Set	.800	113	47	160
8′ x 5′	1.000	Set	1.067	150	63	213
10′ x 6′	1.000	Set	1.333	188	78.50	266.50
12′ x 6′	1.000	Set	1.600	225	94	319

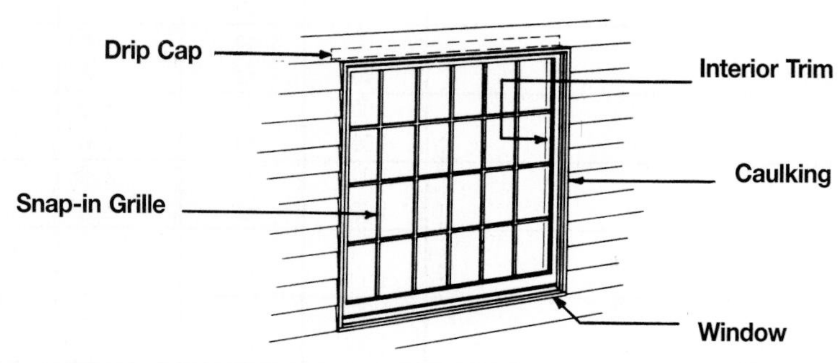

System Description	QUAN.	UNIT	LABOR HOURS	COST EACH		
				MAT.	INST.	TOTAL
BUILDER'S QUALITY PICTURE WINDOW, 4′ X 4′						
Window, primed, builder's quality, 3′-0″ x 4′, insulating glass	1.000	Ea.	1.333	490	78.50	568.50
Trim, interior casing	17.000	L.F.	.544	26.52	31.96	58.48
Paint, interior & exterior, primer & 2 coats	2.000	Face	1.778	2.56	87	89.56
Caulking	16.000	L.F.	.444	4	26.08	30.08
Snap-in grille	1.000	Ea.	.267	142	15.70	157.70
Drip cap, metal	4.000	L.F.	.080	2.60	4.72	7.32
TOTAL		Ea.	4.446	667.68	243.96	911.64
PLASTIC CLAD WOOD WINDOW, 4′-6″ X 6′-6″						
Window, plastic clad, prem., 4′-6″ x 6′-6″, insul. glass	1.000	Ea.	1.455	1,075	85.50	1,160.50
Trim, interior casing	23.000	L.F.	.736	35.88	43.24	79.12
Paint, interior, primer & 2 coats	1.000	Face	.889	1.28	43.50	44.78
Caulking	22.000	L.F.	.611	5.50	35.86	41.36
Snap-in grille	1.000	Ea.	.267	142	15.70	157.70
TOTAL		Ea.	3.958	1,259.66	223.80	1,483.46
METAL CLAD WOOD WINDOW, 6′-6″ X 6′-6″						
Window, metal clad, deluxe, 6′-0″ x 6′-0″, insulating glass	1.000	Ea.	1.600	785	94	879
Trim interior casing	27.000	L.F.	.864	42.12	50.76	92.88
Paint, interior, primer & 2 coats	1.000	Face	1.600	6.85	78	84.85
Caulking	26.000	L.F.	.722	6.50	42.38	48.88
Snap-in grille	1.000	Ea.	.267	142	15.70	157.70
Drip cap, metal	6.500	L.F.	.130	4.23	7.67	11.90
TOTAL		Ea.	5.183	986.70	288.51	1,275.21

The cost of this system is on a cost per each window basis.

Description	QUAN.	UNIT	LABOR HOURS	COST EACH		
				MAT.	INST.	TOTAL

Fixed Window Price Sheet	QUAN.	UNIT	LABOR HOURS	COST EACH		
				MAT.	INST.	TOTAL
Window-picture, builder's quality, 4′ x 4′, single glass	1.000	Ea.	1.333	480	78.50	558.50
Insulating glass	1.000	Ea.	1.333	490	78.50	568.50
4′ x 4′-6″, single glass	1.000	Ea.	1.455	605	85.50	690.50
Insulating glass	1.000	Ea.	1.455	585	85.50	670.50
5′ x 4′, single glass	1.000	Ea.	1.455	640	85.50	725.50
Insulating glass	1.000	Ea.	1.455	670	85.50	755.50
6′ x 4′-6″, single glass	1.000	Ea.	1.600	705	94	799
Insulating glass	1.000	Ea.	1.600	710	94	804
Plastic clad premium insulating glass, 4′ x 4′	1.000	Ea.	1.333	585	78.50	663.50
4′-6″ x 6′-6″	1.000	Ea.	1.455	1,075	85.50	1,160.50
5′-6″ x 6′-6″	1.000	Ea.	1.600	1,200	94	1,294
6′-6″ x 6′-6″	1.000	Ea.	1.600	1,225	94	1,319
Metal clad deluxe insulating glass, 4′ x 4′	1.000	Ea.	1.333	420	78.50	498.50
4′-6″ x 6′-6″	1.000	Ea.	1.455	620	85.50	705.50
5′-6″ x 6′-6″	1.000	Ea.	1.600	685	94	779
6′-6″ x 6′-6″	1.000	Ea.	1.600	785	94	879
Trim, interior casing, window 4′ x 4′	17.000	L.F.	.567	26.50	32	58.50
4′-6″ x 4′-6″	19.000	L.F.	.633	29.50	35.50	65
5′-0″ x 4′-0″	19.000	L.F.	.633	29.50	35.50	65
4′-6″ x 6′-6″	23.000	L.F.	.767	36	43	79
5′-6″ x 6′-6″	25.000	L.F.	.833	39	47	86
6′-6″ x 6′-6″	27.000	L.F.	.900	42	51	93
Paint or stain, interior or exterior, 4′ x 4′ window, 1 coat	1.000	Face	.667	1.07	32.50	33.57
2 coats	1.000	Face	.667	1.84	32.50	34.34
Primer & 1 coat	1.000	Face	.727	1.31	35.50	36.81
Primer & 2 coats	1.000	Face	.889	1.28	43.50	44.78
4′-6″ x 6′-6″ window, 1 coat	1.000	Face	.667	1.07	32.50	33.57
2 coats	1.000	Face	.667	1.84	32.50	34.34
Primer & 1 coat	1.000	Face	.727	1.31	35.50	36.81
Primer & 2 coats	1.000	Face	.889	1.28	43.50	44.78
6′-6″ x 6′-6″ window, 1 coat	1.000	Face	.889	2.70	43.50	46.20
2 coats	1.000	Face	1.333	4.93	65	69.93
Primer & 1 coat	1.000	Face	1.333	4.64	65	69.64
Primer & 2 coats	1.000	Face	1.600	6.85	78	84.85
Caulking, window, 4′ x 4′	1.000	Ea.	.516	4	26	30
4′-6″ x 4′-6″	1.000	Ea.	.581	4.50	29.50	34
5′-0″ x 4′-0″	1.000	Ea.	.581	4.50	29.50	34
4′-6″ x 6′-6″	1.000	Ea.	.710	5.50	36	41.50
5′-6″ x 6′-6″	1.000	Ea.	.774	6	39	45
6′-6″ x 6′-6″	1.000	Ea.	.839	6.50	42.50	49
Grilles, glass size, to 48″ x 48″	1.000	Ea.	.267	142	15.70	157.70
To 60″ x 68″	1.000	Ea.	.286	221	16.85	237.85
Drip cap, aluminum, 4′ long	4.000	L.F.	.080	2.60	4.72	7.32
4′-6″ long	4.500	L.F.	.090	2.93	5.30	8.23
5′ long	5.000	L.F.	.100	3.25	5.90	9.15
6′ long	6.000	L.F.	.120	3.90	7.10	11
Wood, 4′ long	4.000	L.F.	.133	6.25	7.50	13.75
4′-6″ long	4.500	L.F.	.150	7	8.45	15.45
5′ long	5.000	L.F.	.167	7.80	9.40	17.20
6′ long	6.000	L.F.	.200	9.35	11.30	20.65

52 Entrance Door Systems

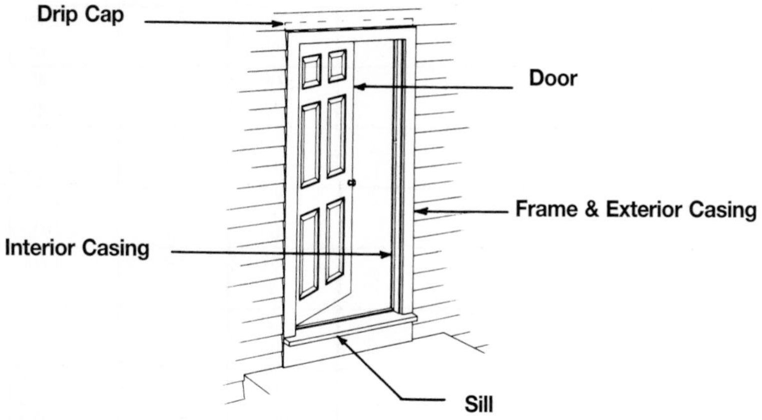

Drip Cap
Door
Frame & Exterior Casing
Interior Casing
Sill

System Description	QUAN.	UNIT	LABOR HOURS	COST EACH MAT.	COST EACH INST.	COST EACH TOTAL
COLONIAL, 6 PANEL, 3′ X 6′-8″, WOOD						
Door, 3′ x 6′-8″ x 1-3/4″ thick, pine, 6 panel colonial	1.000	Ea.	1.067	660	63	723
Frame, 5-13/16″ deep, incl. exterior casing & drip cap	17.000	L.F.	.725	154.70	42.67	197.37
Interior casing, 2-1/2″ wide	18.000	L.F.	.576	28.08	33.84	61.92
Sill, 8/4 x 8″ deep	3.000	L.F.	.480	70.50	28.20	98.70
Butt hinges, brass, 4-1/2″ x 4-1/2″	1.500	Pr.		33		33
Average quality	1.000	Ea.	.571	59	33.50	92.50
Weatherstripping, metal, spring type, bronze	1.000	Set	1.053	28	62	90
Paint, interior & exterior, primer & 2 coats	2.000	Face	1.778	13.50	87	100.50
TOTAL		Ea.	6.250	1,046.78	350.21	1,396.99
SOLID CORE BIRCH, FLUSH, 3′ X 6′-8″						
Door, 3′ x 6′-8″, 1-3/4″ thick, birch, flush solid core	1.000	Ea.	1.067	166	63	229
Frame, 5-13/16″ deep, incl. exterior casing & drip cap	17.000	L.F.	.725	154.70	42.67	197.37
Interior casing, 2-1/2″ wide	18.000	L.F.	.576	28.08	33.84	61.92
Sill, 8/4 x 8″ deep	3.000	L.F.	.480	70.50	28.20	98.70
Butt hinges, brass, 4-1/2″ x 4-1/2″	1.500	Pr.		33		33
Average quality	1.000	Ea.	.571	59	33.50	92.50
Weatherstripping, metal, spring type, bronze	1.000	Set	1.053	28	62	90
Paint, Interior & exterior, primer & 2 coats	2.000	Face	1.778	12.70	87	99.70
TOTAL		Ea.	6.250	551.98	350.21	902.19

These systems are on a cost per each door basis.

Description	QUAN.	UNIT	LABOR HOURS	COST EACH MAT.	COST EACH INST.	COST EACH TOTAL

Entrance Door Price Sheet	QUAN.	UNIT	LABOR HOURS	COST EACH		
				MAT.	INST.	TOTAL
Door exterior wood 1-3/4″ thick, pine, dutch door, 2′-8″ x 6′-8″ minimum	1.000	Ea.	1.333	825	78.50	903.50
Maximum	1.000	Ea.	1.600	1,075	94	1,169
3′-0″ x 6′-8″, minimum	1.000	Ea.	1.333	700	78.50	778.50
Maximum	1.000	Ea.	1.600	1,125	94	1,219
Colonial, 6 panel, 2′-8″ x 6′-8″	1.000	Ea.	1.000	650	59	709
3′-0″ x 6′-8″	1.000	Ea.	1.067	660	63	723
8 panel, 2′-6″ x 6′-8″	1.000	Ea.	1.000	795	59	854
3′-0″ x 6′-8″	1.000	Ea.	1.067	730	63	793
Flush, birch, solid core, 2′-8″ x 6′-8″	1.000	Ea.	1.000	171	59	230
3′-0″ x 6′-8″	1.000	Ea.	1.067	166	63	229
Porch door, 2′-8″ x 6′-8″	1.000	Ea.	1.000	835	59	894
3′-0″ x 6′-8″	1.000	Ea.	1.067	770	63	833
Hand carved mahogany, 2′-8″ x 6′-8″	1.000	Ea.	1.067	2,300	63	2,363
3′-0″ x 6′-8″	1.000	Ea.	1.067	1,600	63	1,663
Rosewood, 2′-8″ x 6′-8″	1.000	Ea.	1.067	825	57.50	882.50
3′-0″ x 6-8″	1.000	Ea.	1.067	775	63	838
Door, metal clad wood 1-3/8″ thick raised panel, 2′-8″ x 6′-8″	1.000	Ea.	1.067	455	55.50	510.50
3′-0″ x 6′-8″	1.000	Ea.	1.067	320	63	383
Deluxe metal door, 3′-0″ x 6′-8″	1.000	Ea.	1.231	320	63	383
3′-0″ x 6′-8″	1.000	Ea.	1.231	320	63	383
Frame, pine, including exterior trim & drip cap, 5/4, x 4-9/16″ deep	17.000	L.F.	.725	126	42.50	168.50
5-13/16″ deep	17.000	L.F.	.725	155	42.50	197.50
6-9/16″ deep	17.000	L.F.	.725	188	42.50	230.50
Safety glass lites, add	1.000	Ea.		94.50		94.50
Interior casing, 2′-8″ x 6′-8″ door	18.000	L.F.	.600	28	34	62
3′-0″ x 6′-8″ door	19.000	L.F.	.633	29.50	35.50	65
Sill, oak, 8/4 x 8″ deep	3.000	L.F.	.480	70.50	28	98.50
8/4 x 10″ deep	3.000	L.F.	.533	88.50	31.50	120
Butt hinges, steel plated, 4-1/2″ x 4-1/2″, plain	1.500	Pr.		33		33
Ball bearing	1.500	Pr.		60		60
Bronze, 4-1/2″ x 4-1/2″, plain	1.500	Pr.		46		46
Ball bearing	1.500	Pr.		61.50		61.50
Lockset, minimum	1.000	Ea.	.571	59	33.50	92.50
Maximum	1.000	Ea.	1.000	262	59	321
Weatherstripping, metal, interlocking, zinc	1.000	Set	2.667	54	157	211
Bronze	1.000	Set	2.667	68	157	225
Spring type, bronze	1.000	Set	1.053	28	62	90
Rubber, minimum	1.000	Set	1.053	12.70	62	74.70
Maximum	1.000	Set	1.143	14.70	67.50	82.20
Felt minimum	1.000	Set	.571	4.95	33.50	38.45
Maximum	1.000	Set	.615	5.70	36.50	42.20
Paint or stain, flush door, interior or exterior, 1 coat	2.000	Face	.941	4.52	46	50.52
2 coats	2.000	Face	1.455	9.05	71	80.05
Primer & 1 coat	2.000	Face	1.455	8.40	71	79.40
Primer & 2 coats	2.000	Face	1.778	12.70	87	99.70
Paneled door, interior & exterior, 1 coat	2.000	Face	1.143	4.84	56	60.84
2 coats	2.000	Face	2.000	9.65	98	107.65
Primer & 1 coat	2.000	Face	1.455	9	71	80
Primer & 2 coats	2.000	Face	1.778	13.50	87	100.50

53 Sliding Door Systems

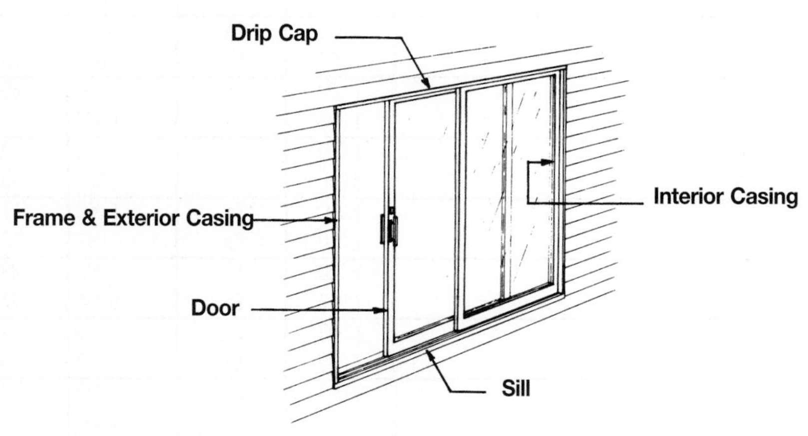

System Description	QUAN.	UNIT	LABOR HOURS	COST EACH		
				MAT.	INST.	TOTAL
WOOD SLIDING DOOR, 8′ WIDE, PREMIUM						
Wood, 5/8″ thick tempered insul. glass, 8′ wide, premium	1.000	Ea.	5.333	2,100	315	2,415
Interior casing	22.000	L.F.	.704	34.32	41.36	75.68
Exterior casing	22.000	L.F.	.704	34.32	41.36	75.68
Sill, oak, 8/4 x 8″ deep	8.000	L.F.	1.280	188	75.20	263.20
Drip cap	8.000	L.F.	.160	5.20	9.44	14.64
Paint, interior & exterior, primer & 2 coats	2.000	Face	2.816	19.36	137.28	156.64
TOTAL		Ea.	10.997	2,381.20	619.64	3,000.84
ALUMINUM SLIDING DOOR, 8′ WIDE, PREMIUM						
Aluminum, 5/8″ tempered insul. glass, 8′ wide, premium	1.000	Ea.	5.333	1,950	315	2,265
Interior casing	22.000	L.F.	.704	34.32	41.36	75.68
Exterior casing	22.000	L.F.	.704	34.32	41.36	75.68
Sill, oak, 8/4 x 8″ deep	8.000	L.F.	1.280	188	75.20	263.20
Drip cap	8.000	L.F.	.160	5.20	9.44	14.64
Paint, interior & exterior, primer & 2 coats	2.000	Face	2.816	19.36	137.28	156.64
TOTAL		Ea.	10.997	2,231.20	619.64	2,850.84

The cost of this system is on a cost per each door basis.

Description	QUAN.	UNIT	LABOR HOURS	COST EACH		
				MAT.	INST.	TOTAL

Sliding Door Price Sheet	QUAN.	UNIT	LABOR HOURS	COST PER UNIT		
				MAT.	INST.	TOTAL
Sliding door, wood, 5/8" thick, tempered insul. glass, 6' wide, premium	1.000	Ea.	4.000	1,650	236	1,886
Economy	1.000	Ea.	4.000	1,375	236	1,611
8'wide, wood premium	1.000	Ea.	5.333	2,100	315	2,415
Economy	1.000	Ea.	5.333	1,725	315	2,040
12' wide, wood premium	1.000	Ea.	6.400	3,525	375	3,900
Economy	1.000	Ea.	6.400	2,825	375	3,200
Aluminum, 5/8" thick, tempered insul. glass, 6'wide, premium	1.000	Ea.	4.000	1,775	236	2,011
Economy	1.000	Ea.	4.000	940	236	1,176
8'wide, premium	1.000	Ea.	5.333	1,950	315	2,265
Economy	1.000	Ea.	5.333	1,675	315	1,990
12' wide, premium	1.000	Ea.	6.400	3,425	375	3,800
Economy	1.000	Ea.	6.400	1,700	375	2,075
Interior casing, 6' wide door	20.000	L.F.	.667	31	37.50	68.50
8' wide door	22.000	L.F.	.733	34.50	41.50	76
12' wide door	26.000	L.F.	.867	40.50	49	89.50
Exterior casing, 6' wide door	20.000	L.F.	.667	31	37.50	68.50
8' wide door	22.000	L.F.	.733	34.50	41.50	76
12' wide door	26.000	L.F.	.867	40.50	49	89.50
Sill, oak, 8/4 x 8" deep, 6' wide door	6.000	L.F.	.960	141	56.50	197.50
8' wide door	8.000	L.F.	1.280	188	75	263
12' wide door	12.000	L.F.	1.920	282	113	395
8/4 x 10" deep, 6' wide door	6.000	L.F.	1.067	177	62.50	239.50
8' wide door	8.000	L.F.	1.422	236	83.50	319.50
12' wide door	12.000	L.F.	2.133	355	125	480
Drip cap, 6' wide door	6.000	L.F.	.120	3.90	7.10	11
8' wide door	8.000	L.F.	.160	5.20	9.45	14.65
12' wide door	12.000	L.F.	.240	7.80	14.15	21.95
Paint or stain, interior & exterior, 6' wide door, 1 coat	2.000	Face	1.600	6.40	78.50	84.90
2 coats	2.000	Face	1.600	6.40	78.50	84.90
Primer & 1 coat	2.000	Face	1.778	12	87	99
Primer & 2 coats	2.000	Face	2.560	17.60	125	142.60
8' wide door, 1 coat	2.000	Face	1.760	7.05	86	93.05
2 coats	2.000	Face	1.760	7.05	86	93.05
Primer & 1 coat	2.000	Face	1.955	13.20	96	109.20
Primer & 2 coats	2.000	Face	2.816	19.35	137	156.35
12' wide door, 1 coat	2.000	Face	2.080	8.30	102	110.30
2 coats	2.000	Face	2.080	8.30	102	110.30
Primer & 1 coat	2.000	Face	2.311	15.60	113	128.60
Primer & 2 coats	2.000	Face	3.328	23	162	185
Aluminum door, trim only, interior & exterior, 6' door, 1 coat	2.000	Face	.800	3.20	39	42.20
2 coats	2.000	Face	.800	3.20	39	42.20
Primer & 1 coat	2.000	Face	.889	6	43.50	49.50
Primer & 2 coats	2.000	Face	1.280	8.80	62.50	71.30
8' wide door, 1 coat	2.000	Face	.880	3.52	43	46.52
2 coats	2.000	Face	.880	3.52	43	46.52
Primer & 1 coat	2.000	Face	.978	6.60	48	54.60
Primer & 2 coats	2.000	Face	1.408	9.70	68.50	78.20
12' wide door, 1 coat	2.000	Face	1.040	4.16	51	55.16
2 coats	2.000	Face	1.040	4.16	51	55.16
Primer & 1 coat	2.000	Face	1.155	7.80	56.50	64.30
Primer & 2 coats	2.000	Face	1.664	11.45	81	92.45

56 Residential Garage Door Systems

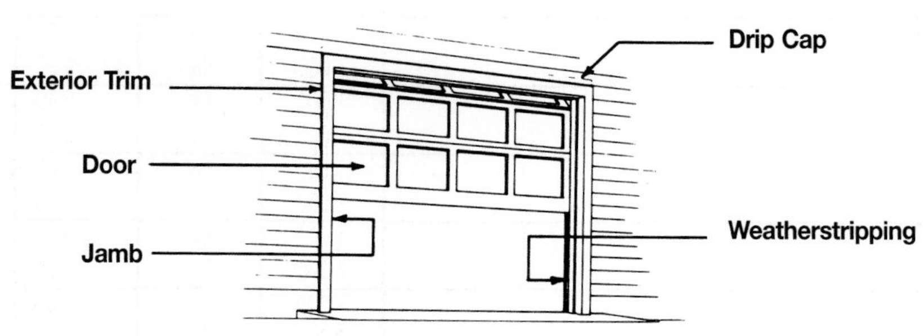

System Description	QUAN.	UNIT	LABOR HOURS	COST EACH MAT.	COST EACH INST.	COST EACH TOTAL
OVERHEAD, SECTIONAL GARAGE DOOR, 9′ X 7′						
Wood, overhead sectional door, std., incl. hardware, 9′ x 7′	1.000	Ea.	2.000	1,150	118	1,268
Jamb & header blocking, 2″ x 6″	25.000	L.F.	.901	19.75	53	72.75
Exterior trim	25.000	L.F.	.800	39	47	86
Paint, interior & exterior, primer & 2 coats	2.000	Face	3.556	27	174	201
Weatherstripping, molding type	1.000	Set	.736	35.88	43.24	79.12
Drip cap	9.000	L.F.	.180	5.85	10.62	16.47
TOTAL		Ea.	8.173	1,277.48	445.86	1,723.34
OVERHEAD, SECTIONAL GARAGE DOOR, 16′ X 7′						
Wood, overhead sectional, std., incl. hardware, 16′ x 7′	1.000	Ea.	2.667	1,875	157	2,032
Jamb & header blocking, 2″ x 6″	30.000	L.F.	1.081	23.70	63.60	87.30
Exterior trim	30.000	L.F.	.960	46.80	56.40	103.20
Paint, interior & exterior, primer & 2 coats	2.000	Face	5.333	40.50	261	301.50
Weatherstripping, molding type	1.000	Set	.960	46.80	56.40	103.20
Drip cap	16.000	L.F.	.320	10.40	18.88	29.28
TOTAL		Ea.	11.321	2,043.20	613.28	2,656.48
OVERHEAD, SWING-UP TYPE, GARAGE DOOR, 16′ X 7′						
Wood, overhead, swing-up, std., incl. hardware, 16′ x 7′	1.000	Ea.	2.667	1,050	157	1,207
Jamb & header blocking, 2″ x 6″	30.000	L.F.	1.081	23.70	63.60	87.30
Exterior trim	30.000	L.F.	.960	46.80	56.40	103.20
Paint, interior & exterior, primer & 2 coats	2.000	Face	5.333	40.50	261	301.50
Weatherstripping, molding type	1.000	Set	.960	46.80	56.40	103.20
Drip cap	16.000	L.F.	.320	10.40	18.88	29.28
TOTAL		Ea.	11.321	1,218.20	613.28	1,831.48

This system is on a cost per each door basis.

Description	QUAN.	UNIT	LABOR HOURS	COST EACH MAT.	COST EACH INST.	COST EACH TOTAL

Resi Garage Door Price Sheet	QUAN.	UNIT	LABOR HOURS	COST EACH		
				MAT.	INST.	TOTAL
Overhead, sectional, including hardware, fiberglass, 9′ x 7′, standard	1.000	Ea.	3.030	1,125	188	1,313
Deluxe	1.000	Ea.	3.030	1,350	188	1,538
16′ x 7′, standard	1.000	Ea.	2.667	1,825	157	1,982
Deluxe	1.000	Ea.	2.667	2,525	157	2,682
Hardboard, 9′ x 7′, standard	1.000	Ea.	2.000	810	118	928
Deluxe	1.000	Ea.	2.000	960	118	1,078
16′ x 7′, standard	1.000	Ea.	2.667	1,425	157	1,582
Deluxe	1.000	Ea.	2.667	1,700	157	1,857
Metal, 9′ x 7′, standard	1.000	Ea.	3.030	1,025	118	1,143
Deluxe	1.000	Ea.	2.000	1,150	157	1,307
16′ x 7′, standard	1.000	Ea.	5.333	1,200	157	1,357
Deluxe	1.000	Ea.	2.667	1,625	188	1,813
Wood, 9′ x 7′, standard	1.000	Ea.	2.000	1,150	118	1,268
Deluxe	1.000	Ea.	2.000	2,525	118	2,643
16′ x 7′, standard	1.000	Ea.	2.667	1,875	157	2,032
Deluxe	1.000	Ea.	2.667	3,500	157	3,657
Overhead swing-up type including hardware, fiberglass, 9′ x 7′, standard	1.000	Ea.	2.000	1,150	118	1,268
Deluxe	1.000	Ea.	2.000	1,275	118	1,393
16′ x 7′, standard	1.000	Ea.	2.667	1,450	157	1,607
Deluxe	1.000	Ea.	2.667	1,850	157	2,007
Hardboard, 9′ x 7′, standard	1.000	Ea.	2.000	640	118	758
Deluxe	1.000	Ea.	2.000	755	118	873
16′ x 7′, standard	1.000	Ea.	2.667	775	157	932
Deluxe	1.000	Ea.	2.667	985	157	1,142
Metal, 9′ x 7′, standard	1.000	Ea.	2.000	700	118	818
Deluxe	1.000	Ea.	2.000	1,225	118	1,343
16′ x 7′, standard	1.000	Ea.	2.667	925	157	1,082
Deluxe	1.000	Ea.	2.667	1,275	157	1,432
Wood, 9′ x 7′, standard	1.000	Ea.	2.000	820	118	938
Deluxe	1.000	Ea.	2.000	1,375	118	1,493
16′ x 7′, standard	1.000	Ea.	2.667	1,050	157	1,207
Deluxe	1.000	Ea.	2.667	2,525	157	2,682
Jamb & header blocking, 2″ x 6″, 9′ x 7′ door	25.000	L.F.	.901	19.75	53	72.75
16′ x 7′ door	30.000	L.F.	1.081	23.50	63.50	87
2″ x 8″, 9′ x 7′ door	25.000	L.F.	1.000	27	59	86
16′ x 7′ door	30.000	L.F.	1.200	32	71	103
Exterior trim, 9′ x 7′ door	25.000	L.F.	.833	39	47	86
16′ x 7′ door	30.000	L.F.	1.000	47	56.50	103.50
Paint or stain, interior & exterior, 9′ x 7′ door, 1 coat	1.000	Face	2.286	9.70	112	121.70
2 coats	1.000	Face	4.000	19.30	196	215.30
Primer & 1 coat	1.000	Face	2.909	17.95	142	159.95
Primer & 2 coats	1.000	Face	3.556	27	174	201
16′ x 7′ door, 1 coat	1.000	Face	3.429	14.50	168	182.50
2 coats	1.000	Face	6.000	29	294	323
Primer & 1 coat	1.000	Face	4.364	27	213	240
Primer & 2 coats	1.000	Face	5.333	40.50	261	301.50
Weatherstripping, molding type, 9′ x 7′ door	1.000	Set	.767	36	43	79
16′ x 7′ door	1.000	Set	1.000	47	56.50	103.50
Drip cap, 9′ door	9.000	L.F.	.180	5.85	10.60	16.45
16′ door	16.000	L.F.	.320	10.40	18.90	29.30
Garage door opener, economy	1.000	Ea.	1.000	490	59	549
Deluxe, including remote control	1.000	Ea.	1.000	705	59	764

58 Aluminum Window Systems

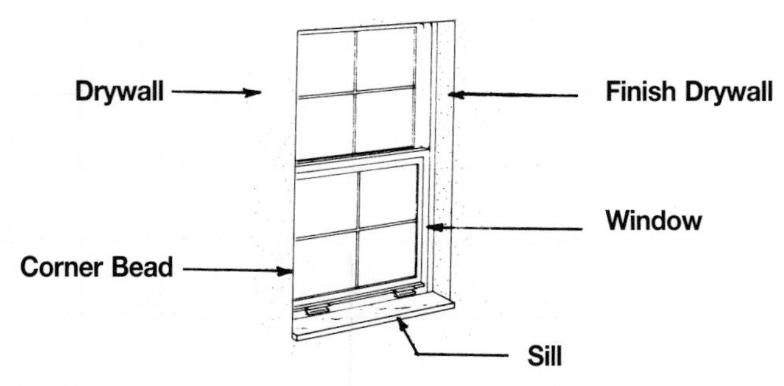

System Description	QUAN.	UNIT	LABOR HOURS	COST EACH		
				MAT.	INST.	TOTAL
SINGLE HUNG, 2′ X 3′ OPENING						
Window, 2′ x 3′ opening, enameled, insulating glass	1.000	Ea.	1.600	286	110	396
Blocking, 1″ x 3″ furring strip nailers	10.000	L.F.	.146	4.90	8.60	13.50
Drywall, 1/2″ thick, standard	5.000	S.F.	.040	1.85	2.35	4.20
Corner bead, 1″ x 1″, galvanized steel	8.000	L.F.	.160	1.36	9.44	10.80
Finish drywall, tape and finish corners inside and outside	16.000	L.F.	.269	1.76	15.84	17.60
Sill, slate	2.000	L.F.	.400	26.40	21.60	48
TOTAL		Ea.	2.615	322.27	167.83	490.10
SLIDING, 3′ X 2′ OPENING						
Window, 3′ x 2′ opening, enameled, insulating glass	1.000	Ea.	1.600	260	110	370
Blocking, 1″ x 3″ furring strip nailers	10.000	L.F.	.146	4.90	8.60	13.50
Drywall, 1/2″ thick, standard	5.000	S.F.	.040	1.85	2.35	4.20
Corner bead, 1″ x 1″, galvanized steel	7.000	L.F.	.140	1.19	8.26	9.45
Finish drywall, tape and finish corners inside and outside	14.000	L.F.	.236	1.54	13.86	15.40
Sill, slate	3.000	L.F.	.600	39.60	32.40	72
TOTAL		Ea.	2.762	309.08	175.47	484.55
AWNING, 3′-1″ X 3′-2″						
Window, 3′-1″ x 3′-2″ opening, enameled, insul. glass	1.000	Ea.	1.600	430	110	540
Blocking, 1″ x 3″ furring strip, nailers	12.500	L.F.	.182	6.13	10.75	16.88
Drywall, 1/2″ thick, standard	4.500	S.F.	.036	1.67	2.12	3.79
Corner bead, 1″ x 1″, galvanized steel	9.250	L.F.	.185	1.57	10.92	12.49
Finish drywall, tape and finish corners, inside and outside	18.500	L.F.	.312	2.04	18.32	20.36
Sill, slate	3.250	L.F.	.650	42.90	35.10	78
TOTAL		Ea.	2.965	484.31	187.21	671.52

Description	QUAN.	UNIT	LABOR HOURS	COST EACH		
				MAT.	INST.	TOTAL

Aluminum Window Price Sheet	QUAN.	UNIT	LABOR HOURS	COST EACH		
				MAT.	INST.	TOTAL
Window, aluminum, awning, 3'-1" x 3'-2", standard glass	1.000	Ea.	1.600	400	110	510
Insulating glass	1.000	Ea.	1.600	430	110	540
4'-5" x 5'-3", standard glass	1.000	Ea.	2.000	440	138	578
Insulating glass	1.000	Ea.	2.000	530	138	668
Casement, 3'-1" x 3'-2", standard glass	1.000	Ea.	1.600	425	110	535
Insulating glass	1.000	Ea.	1.600	575	110	685
Single hung, 2' x 3', standard glass	1.000	Ea.	1.600	236	110	346
Insulating glass	1.000	Ea.	1.600	286	110	396
2'-8" x 6'-8", standard glass	1.000	Ea.	2.000	410	138	548
Insulating glass	1.000	Ea.	2.000	520	138	658
3'-4" x 5'-0", standard glass	1.000	Ea.	1.778	340	123	463
Insulating glass	1.000	Ea.	1.778	370	123	493
Sliding, 3' x 2', standard glass	1.000	Ea.	1.600	242	110	352
Insulating glass	1.000	Ea.	1.600	260	110	370
5' x 3', standard glass	1.000	Ea.	1.778	370	123	493
Insulating glass	1.000	Ea.	1.778	430	123	553
8' x 4', standard glass	1.000	Ea.	2.667	395	184	579
Insulating glass	1.000	Ea.	2.667	635	184	819
Blocking, 1" x 3" furring, opening 3' x 2'	10.000	L.F.	.146	4.90	8.60	13.50
3' x 3'	12.500	L.F.	.182	6.15	10.75	16.90
3' x 5'	16.000	L.F.	.233	7.85	13.75	21.60
4' x 4'	16.000	L.F.	.233	7.85	13.75	21.60
4' x 5'	18.000	L.F.	.262	8.80	15.50	24.30
4' x 6'	20.000	L.F.	.291	9.80	17.20	27
4' x 8'	24.000	L.F.	.349	11.75	20.50	32.25
6'-8" x 2'-8"	19.000	L.F.	.276	9.30	16.35	25.65
Drywall, 1/2" thick, standard, opening 3' x 2'	5.000	S.F.	.040	1.85	2.35	4.20
3' x 3'	6.000	S.F.	.048	2.22	2.82	5.04
3' x 5'	8.000	S.F.	.064	2.96	3.76	6.72
4' x 4'	8.000	S.F.	.064	2.96	3.76	6.72
4' x 5'	9.000	S.F.	.072	3.33	4.23	7.56
4' x 6'	10.000	S.F.	.080	3.70	4.70	8.40
4' x 8'	12.000	S.F.	.096	4.44	5.65	10.09
6'-8" x 2'	9.500	S.F.	.076	3.52	4.47	7.99
Corner bead, 1" x 1", galvanized steel, opening 3' x 2'	7.000	L.F.	.140	1.19	8.25	9.44
3' x 3'	9.000	L.F.	.180	1.53	10.60	12.13
3' x 5'	11.000	L.F.	.220	1.87	13	14.87
4' x 4'	12.000	L.F.	.240	2.04	14.15	16.19
4' x 5'	13.000	L.F.	.260	2.21	15.35	17.56
4' x 6'	14.000	L.F.	.280	2.38	16.50	18.88
4' x 8'	16.000	L.F.	.320	2.72	18.90	21.62
6'-8" x 2'	15.000	L.F.	.300	2.55	17.70	20.25
Tape and finish corners, inside and outside, opening 3' x 2'	14.000	L.F.	.204	1.54	13.85	15.39
3' x 3'	18.000	L.F.	.262	1.98	17.80	19.78
3' x 5'	22.000	L.F.	.320	2.42	22	24.42
4' x 4'	24.000	L.F.	.349	2.64	24	26.64
4' x 5'	26.000	L.F.	.378	2.86	25.50	28.36
4' x 6'	28.000	L.F.	.407	3.08	27.50	30.58
4' x 8'	32.000	L.F.	.466	3.52	31.50	35.02
6'-8" x 2'	30.000	L.F.	.437	3.30	29.50	32.80
Sill, slate, 2' long	2.000	L.F.	.400	26.50	21.50	48
3' long	3.000	L.F.	.600	39.50	32.50	72
4' long	4.000	L.F.	.800	53	43	96
Wood, 1-5/8" x 6-1/4", 2' long	2.000	L.F.	.128	12	7.55	19.55
3' long	3.000	L.F.	.192	18	11.30	29.30
4' long	4.000	L.F.	.256	24	15.10	39.10

System Description	QUAN.	UNIT	LABOR HOURS	COST EACH		
				MAT.	INST.	TOTAL
Storm door, aluminum, combination, storm & screen, anodized, 2′-6″ x 6′-8″	1.000	Ea.	1.067	231	63	294
2′-8″ x 6′-8″	1.000	Ea.	1.143	242	67.50	309.50
3′-0″ x 6′-8″	1.000	Ea.	1.143	210	67.50	277.50
Mill finish, 2′-6″ x 6′-8″	1.000	Ea.	1.067	275	63	338
2′-8″ x 6′-8″	1.000	Ea.	1.143	275	67.50	342.50
3′-0″ x 6′-8″	1.000	Ea.	1.143	296	67.50	363.50
Painted, 2′-6″ x 6′-8″	1.000	Ea.	1.067	257	63	320
2′-8″ x 6′-8″	1.000	Ea.	1.143	281	67.50	348.50
3′-0″ x 6′-8″	1.000	Ea.	1.143	340	67.50	407.50
Wood, combination, storm & screen, crossbuck, 2′-6″ x 6′-9″	1.000	Ea.	1.455	385	85.50	470.50
2′-8″ x 6′-9″	1.000	Ea.	1.600	355	94	449
3′-0″ x 6′-9″	1.000	Ea.	1.778	370	105	475
Full lite, 2′-6″ x 6′-9″	1.000	Ea.	1.455	375	85.50	460.50
2′-8″ x 6′-9″	1.000	Ea.	1.600	375	94	469
3′-0″ x 6′-9″	1.000	Ea.	1.778	380	105	485
Windows, aluminum, combination storm & screen, basement, 1′-10″ x 1′-0″	1.000	Ea.	.533	40	31.50	71.50
2′-9″ x 1′-6″	1.000	Ea.	.533	43	31.50	74.50
3′-4″ x 2′-0″	1.000	Ea.	.533	51	31.50	82.50
Double hung, anodized, 2′-0″ x 3′-5″	1.000	Ea.	.533	110	31.50	141.50
2′-6″ x 5′-0″	1.000	Ea.	.571	138	33.50	171.50
4′-0″ x 6′-0″	1.000	Ea.	.640	261	37.50	298.50
Painted, 2′-0″ x 3′-5″	1.000	Ea.	.533	128	31.50	159.50
2′-6″ x 5′-0″	1.000	Ea.	.571	194	33.50	227.50
4′-0″ x 6′-0″	1.000	Ea.	.640	315	37.50	352.50
Fixed window, anodized, 4′-6″ x 4′-6″	1.000	Ea.	.640	157	37.50	194.50
5′-8″ x 4′-6″	1.000	Ea.	.800	173	47	220
Painted, 4′-6″ x 4′-6″	1.000	Ea.	.640	157	37.50	194.50
5′-8″ x 4′-6″	1.000	Ea.	.800	178	47	225

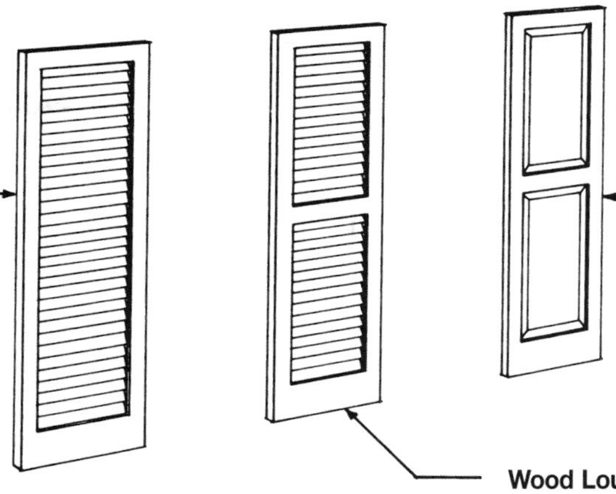

64 Shutters/Blinds Systems

Aluminum Louvered

Raised Panel

Wood Louvered

System Description	QUAN.	UNIT	LABOR HOURS	COST PER PAIR		
				MAT.	INST.	TOTAL
Shutters, exterior blinds, aluminum, louvered, 1′-4″ wide, 3″-0″ long	1.000	Set	.800	220	47	267
4′-0″ long	1.000	Set	.800	265	47	312
5′-4″ long	1.000	Set	.800	320	47	367
6′-8″ long	1.000	Set	.889	395	52.50	447.50
Wood, louvered, 1′-2″ wide, 3′-3″ long	1.000	Set	.800	293	47	340
4′-7″ long	1.000	Set	.800	325	47	372
5′-3″ long	1.000	Set	.800	415	47	462
1′-6″ wide, 3′-3″ long	1.000	Set	.800	305	47	352
4′-7″ long	1.000	Set	.800	410	47	457
Polystyrene, louvered, 1′-2″ wide, 3′-3″ long	1.000	Set	.800	43	47	90
4′-7″ long	1.000	Set	.800	50	47	97
5′-3″ long	1.000	Set	.800	59	47	106
6′-8″ long	1.000	Set	.889	76.50	52.50	129
Vinyl, louvered, 1′-2″ wide, 4′-7″ long	1.000	Set	.720	61	42.50	103.50
1′-4″ x 6′-8″ long	1.000	Set	.889	90.50	52.50	143

04 Gable End Roofing Systems

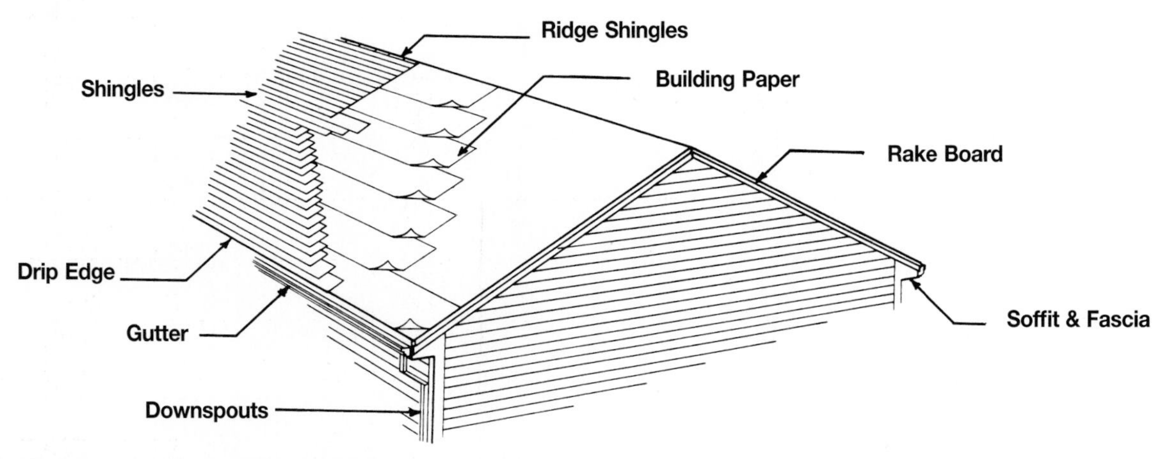

System Description	QUAN.	UNIT	LABOR HOURS	COST PER S.F.		
				MAT.	INST.	TOTAL
ASPHALT, ROOF SHINGLES, CLASS A						
Shingles, inorganic class A, 210-235 lb./sq., 4/12 pitch	1.160	S.F.	.017	1.04	.97	2.01
Drip edge, metal, 5″ wide	.150	L.F.	.003	.10	.18	.28
Building paper, #15 felt	1.300	S.F.	.002	.08	.10	.18
Ridge shingles, asphalt	.042	L.F.	.001	.11	.06	.17
Soffit & fascia, white painted aluminum, 1′ overhang	.083	L.F.	.012	.42	.71	1.13
Rake trim, 1″ x 6″	.040	L.F.	.002	.05	.09	.14
Rake trim, prime and paint	.040	L.F.	.002	.01	.09	.10
Gutter, seamless, aluminum painted	.083	L.F.	.005	.27	.34	.61
Downspouts, aluminum painted	.035	L.F.	.002	.08	.10	.18
Ridge vent	.042	L.F.	.002	.11	.12	.23
TOTAL		S.F.	.048	2.27	2.76	5.03
WOOD, CEDAR SHINGLES NO. 1 PERFECTIONS, 18″ LONG						
Shingles, wood, cedar, No. 1 perfections, 4/12 pitch	1.160	S.F.	.035	3.35	2.05	5.40
Drip edge, metal, 5″ wide	.150	L.F.	.003	.10	.18	.28
Building paper, #15 felt	1.300	S.F.	.002	.08	.10	.18
Ridge shingles, cedar	.042	L.F.	.001	.22	.07	.29
Soffit & fascia, white painted aluminum, 1′ overhang	.083	L.F.	.012	.42	.71	1.13
Rake trim, 1″ x 6″	.040	L.F.	.002	.05	.09	.14
Rake trim, prime and paint	.040	L.F.	.002	.01	.09	.10
Gutter, seamless, aluminum, painted	.083	L.F.	.005	.27	.34	.61
Downspouts, aluminum, painted	.035	L.F.	.002	.08	.10	.18
Ridge vent	.042	L.F.	.002	.11	.12	.23
TOTAL		S.F.	.066	4.69	3.85	8.54

The prices in these systems are based on a square foot of plan area.
All quantities have been adjusted accordingly.

Description	QUAN.	UNIT	LABOR HOURS	COST PER S.F.		
				MAT.	INST.	TOTAL

Gable End Roofing Price Sheet	QUAN.	UNIT	LABOR HOURS	COST PER S.F.		
				MAT.	INST.	TOTAL
Shingles, asphalt, inorganic, class A, 210-235 lb./sq., 4/12 pitch	1.160	S.F.	.017	1.04	.97	2.01
8/12 pitch	1.330	S.F.	.019	1.12	1.05	2.17
Laminated, multi-layered, 240-260 lb./sq., 4/12 pitch	1.160	S.F.	.021	1.45	1.18	2.63
8/12 pitch	1.330	S.F.	.023	1.57	1.28	2.85
Premium laminated, multi-layered, 260-300 lb./sq., 4/12 pitch	1.160	S.F.	.027	1.92	1.52	3.44
8/12 pitch	1.330	S.F.	.030	2.08	1.65	3.73
Clay tile, Spanish tile, red, 4/12 pitch	1.160	S.F.	.053	6.60	2.92	9.52
8/12 pitch	1.330	S.F.	.058	7.15	3.16	10.31
Mission tile, red, 4/12 pitch	1.160	S.F.	.083	5.45	2.92	8.37
8/12 pitch	1.330	S.F.	.090	5.90	3.16	9.06
French tile, red, 4/12 pitch	1.160	S.F.	.071	15.30	2.68	17.98
8/12 pitch	1.330	S.F.	.077	16.60	2.90	19.50
Slate, Buckingham, Virginia, black, 4/12 pitch	1.160	S.F.	.055	7.40	3.05	10.45
8/12 pitch	1.330	S.F.	.059	8	3.30	11.30
Vermont, black or grey, 4/12 pitch	1.160	S.F.	.055	6.40	3.05	9.45
8/12 pitch	1.330	S.F.	.059	6.95	3.30	10.25
Wood, No. 1 red cedar, 5X, 16″ long, 5″ exposure, 4/12 pitch	1.160	S.F.	.038	3.90	2.26	6.16
8/12 pitch	1.330	S.F.	.042	4.23	2.44	6.67
Fire retardant, 4/12 pitch	1.160	S.F.	.038	4.63	2.26	6.89
8/12 pitch	1.330	S.F.	.042	5	2.44	7.44
18″ long, No.1 perfections, 5″ exposure, 4/12 pitch	1.160	S.F.	.035	3.35	2.05	5.40
8/12 pitch	1.330	S.F.	.038	3.63	2.22	5.85
Fire retardant, 4/12 pitch	1.160	S.F.	.035	4.08	2.05	6.13
8/12 pitch	1.330	S.F.	.038	4.42	2.22	6.64
Resquared & rebutted, 18″ long, 6″ exposure, 4/12 pitch	1.160	S.F.	.032	3.78	1.88	5.66
8/12 pitch	1.330	S.F.	.035	4.10	2.04	6.14
Fire retardant, 4/12 pitch	1.160	S.F.	.032	4.51	1.88	6.39
8/12 pitch	1.330	S.F.	.035	4.89	2.04	6.93
Wood shakes hand split, 24″ long, 10″ exposure, 4/12 pitch	1.160	S.F.	.038	3.96	2.26	6.22
8/12 pitch	1.330	S.F.	.042	4.29	2.44	6.73
Fire retardant, 4/12 pitch	1.160	S.F.	.038	4.69	2.26	6.95
8/12 pitch	1.330	S.F.	.042	5.10	2.44	7.54
18″ long, 8″ exposure, 4/12 pitch	1.160	S.F.	.048	3.66	2.83	6.49
8/12 pitch	1.330	S.F.	.052	3.97	3.07	7.04
Fire retardant, 4/12 pitch	1.160	S.F.	.048	4.39	2.83	7.22
8/12 pitch	1.330	S.F.	.052	4.76	3.07	7.83
Drip edge, metal, 5″ wide	.150	L.F.	.003	.10	.18	.28
8″ wide	.150	L.F.	.003	.14	.18	.32
Building paper, #15 asphalt felt	1.300	S.F.	.002	.08	.10	.18
Ridge shingles, asphalt	.042	L.F.	.001	.11	.06	.17
Clay	.042	L.F.	.002	.23	.28	.51
Slate	.042	L.F.	.002	.47	.09	.56
Wood, shingles	.042	L.F.	.001	.22	.07	.29
Shakes	.042	L.F.	.001	.22	.07	.29
Soffit & fascia, aluminum, vented, 1′ overhang	.083	L.F.	.012	.42	.71	1.13
2′ overhang	.083	L.F.	.013	.60	.78	1.38
Vinyl, vented, 1′ overhang	.083	L.F.	.011	.44	.65	1.09
2′ overhang	.083	L.F.	.012	.59	.78	1.37
Wood, board fascia, plywood soffit, 1′ overhang	.083	L.F.	.004	.02	.18	.20
2′ overhang	.083	L.F.	.006	.04	.27	.31
Rake trim, painted, 1″ x 6″	.040	L.F.	.004	.06	.18	.24
1″ x 8″	.040	L.F.	.004	.23	.17	.40
Gutter, 5″ box, aluminum, seamless, painted	.083	L.F.	.006	.27	.34	.61
Vinyl	.083	L.F.	.006	.15	.34	.49
Downspout, 2″ x 3″, aluminum, one story house	.035	L.F.	.001	.05	.10	.15
Two story house	.060	L.F.	.003	.09	.16	.25
Vinyl, one story house	.035	L.F.	.002	.08	.10	.18
Two story house	.060	L.F.	.003	.09	.16	.25

08 Hip Roof Roofing Systems

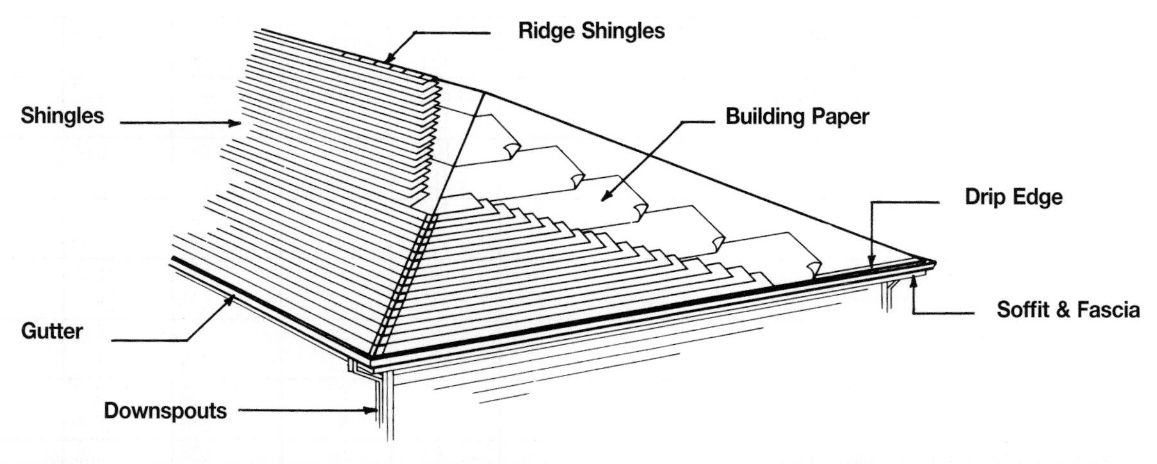

System Description	QUAN.	UNIT	LABOR HOURS	COST PER S.F.		
				MAT.	INST.	TOTAL
ASPHALT, ROOF SHINGLES, CLASS A						
Shingles, inorganic, class A, 210-235 lb./sq. 4/12 pitch	1.570	S.F.	.023	1.38	1.29	2.67
Drip edge, metal, 5″ wide	.122	L.F.	.002	.08	.14	.22
Building paper, #15 asphalt felt	1.800	S.F.	.002	.11	.14	.25
Ridge shingles, asphalt	.075	L.F.	.002	.19	.10	.29
Soffit & fascia, white painted aluminum, 1′ overhang	.120	L.F.	.017	.60	1.03	1.63
Gutter, seamless, aluminum, painted	.120	L.F.	.008	.39	.50	.89
Downspouts, aluminum, painted	.035	L.F.	.002	.08	.10	.18
Ridge vent	.028	L.F.	.001	.07	.08	.15
TOTAL		S.F.	.057	2.90	3.38	6.28
WOOD, CEDAR SHINGLES, NO. 1 PERFECTIONS, 18″ LONG						
Shingles, red cedar, No. 1 perfections, 5″ exp., 4/12 pitch	1.570	S.F.	.047	4.46	2.74	7.20
Drip edge, metal, 5″ wide	.122	L.F.	.002	.08	.14	.22
Building paper, #15 asphalt felt	1.800	S.F.	.002	.11	.14	.25
Ridge shingles, wood, cedar	.075	L.F.	.002	.39	.13	.52
Soffit & fascia, white painted aluminum, 1′ overhang	.120	L.F.	.017	.60	1.03	1.63
Gutter, seamless, aluminum, painted	.120	L.F.	.008	.39	.50	.89
Downspouts, aluminum, painted	.035	L.F.	.002	.08	.10	.18
Ridge vent	.028	L.F.	.001	.07	.08	.15
TOTAL		S.F.	.081	6.18	4.86	11.04

The prices in these systems are based on a square foot of plan area.
All quantities have been adjusted accordingly.

Description	QUAN.	UNIT	LABOR HOURS	COST PER S.F.		
				MAT.	INST.	TOTAL

Hip Roof - Roofing Price Sheet	QUAN.	UNIT	LABOR HOURS	COST PER S.F.		
				MAT.	INST.	TOTAL
Shingles, asphalt, inorganic, class A, 210-235 lb./sq., 4/12 pitch	1.570	S.F.	.023	1.38	1.29	2.67
8/12 pitch	1.850	S.F.	.028	1.64	1.53	3.17
Laminated, multi-layered, 240-260 lb./sq., 4/12 pitch	1.570	S.F.	.028	1.94	1.58	3.52
8/12 pitch	1.850	S.F.	.034	2.30	1.87	4.17
Prem. laminated, multi-layered, 260-300 lb./sq., 4/12 pitch	1.570	S.F.	.037	2.56	2.03	4.59
8/12 pitch	1.850	S.F.	.043	3.04	2.41	5.45
Clay tile, Spanish tile, red, 4/12 pitch	1.570	S.F.	.071	8.80	3.89	12.69
8/12 pitch	1.850	S.F.	.084	10.45	4.62	15.07
Mission tile, red, 4/12 pitch	1.570	S.F.	.111	7.30	3.89	11.19
8/12 pitch	1.850	S.F.	.132	8.65	4.62	13.27
French tile, red, 4/12 pitch	1.570	S.F.	.095	20.50	3.57	24.07
8/12 pitch	1.850	S.F.	.113	24	4.24	28.24
Slate, Buckingham, Virginia, black, 4/12 pitch	1.570	S.F.	.073	9.85	4.06	13.91
8/12 pitch	1.850	S.F.	.087	11.70	4.83	16.53
Vermont, black or grey, 4/12 pitch	1.570	S.F.	.073	8.55	4.06	12.61
8/12 pitch	1.850	S.F.	.087	10.15	4.83	14.98
Wood, red cedar, No.1 5X, 16" long, 5" exposure, 4/12 pitch	1.570	S.F.	.051	5.20	3.01	8.21
8/12 pitch	1.850	S.F.	.061	6.20	3.57	9.77
Fire retardant, 4/12 pitch	1.570	S.F.	.051	6.20	3.01	9.21
8/12 pitch	1.850	S.F.	.061	7.35	3.57	10.92
18" long, No.1 perfections, 5" exposure, 4/12 pitch	1.570	S.F.	.047	4.46	2.74	7.20
8/12 pitch	1.850	S.F.	.055	5.30	3.25	8.55
Fire retardant, 4/12 pitch	1.570	S.F.	.047	5.45	2.74	8.19
8/12 pitch	1.850	S.F.	.055	6.45	3.25	9.70
Resquared & rebutted, 18" long, 6" exposure, 4/12 pitch	1.570	S.F.	.043	5.05	2.51	7.56
8/12 pitch	1.850	S.F.	.051	6	2.98	8.98
Fire retardant, 4/12 pitch	1.570	S.F.	.043	6	2.51	8.51
8/12 pitch	1.850	S.F.	.051	7.15	2.98	10.13
Wood shakes hand split, 24" long, 10" exposure, 4/12 pitch	1.570	S.F.	.051	5.30	3.01	8.31
8/12 pitch	1.850	S.F.	.061	6.25	3.57	9.82
Fire retardant, 4/12 pitch	1.570	S.F.	.051	6.25	3.01	9.26
8/12 pitch	1.850	S.F.	.061	7.45	3.57	11.02
18" long, 8" exposure, 4/12 pitch	1.570	S.F.	.064	4.88	3.78	8.66
8/12 pitch	1.850	S.F.	.076	5.80	4.48	10.28
Fire retardant, 4/12 pitch	1.570	S.F.	.064	5.85	3.78	9.63
8/12 pitch	1.850	S.F.	.076	6.95	4.48	11.43
Drip edge, metal, 5" wide	.122	L.F.	.002	.08	.14	.22
8" wide	.122	L.F.	.002	.11	.14	.25
Building paper, #15 asphalt felt	1.800	S.F.	.002	.11	.14	.25
Ridge shingles, asphalt	.075	L.F.	.002	.19	.10	.29
Clay	.075	L.F.	.003	.41	.50	.91
Slate	.075	L.F.	.003	.84	.17	1.01
Wood, shingles	.075	L.F.	.002	.39	.13	.52
Shakes	.075	L.F.	.002	.39	.13	.52
Soffit & fascia, aluminum, vented, 1′ overhang	.120	L.F.	.017	.60	1.03	1.63
2′ overhang	.120	L.F.	.019	.87	1.13	2
Vinyl, vented, 1′ overhang	.120	L.F.	.016	.64	.94	1.58
2′ overhang	.120	L.F.	.017	.85	1.13	1.98
Wood, board fascia, plywood soffit, 1′ overhang	.120	L.F.	.004	.02	.18	.20
2′ overhang	.120	L.F.	.006	.04	.27	.31
Gutter, 5" box, aluminum, seamless, painted	.120	L.F.	.008	.39	.50	.89
Vinyl	.120	L.F.	.009	.22	.49	.71
Downspout, 2" x 3", aluminum, one story house	.035	L.F.	.002	.08	.10	.18
Two story house	.060	L.F.	.003	.09	.16	.25
Vinyl, one story house	.035	L.F.	.001	.05	.10	.15
Two story house	.060	L.F.	.003	.09	.16	.25

12 Gambrel Roofing Systems

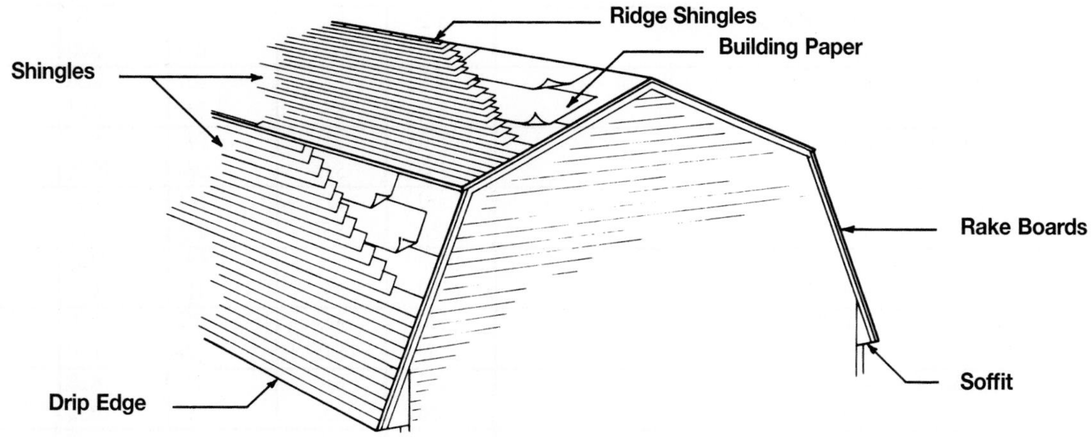

Ridge Shingles
Building Paper
Shingles
Rake Boards
Soffit
Drip Edge

System Description	QUAN.	UNIT	LABOR HOURS	COST PER S.F.		
				MAT.	INST.	TOTAL
ASPHALT, ROOF SHINGLES, CLASS A						
Shingles, asphalt, inorganic, class A, 210-235 lb./sq.	1.450	S.F.	.022	1.30	1.21	2.51
Drip edge, metal, 5″ wide	.146	L.F.	.003	.10	.17	.27
Building paper, #15 asphalt felt	1.500	S.F.	.002	.09	.11	.20
Ridge shingles, asphalt	.042	L.F.	.001	.11	.06	.17
Soffit & fascia, painted aluminum, 1′ overhang	.083	L.F.	.012	.42	.71	1.13
Rake trim, 1″ x 6″	.063	L.F.	.003	.08	.15	.23
Rake trim, prime and paint	.063	L.F.	.003	.02	.14	.16
Gutter, seamless, alumunum, painted	.083	L.F.	.005	.27	.34	.61
Downspouts, aluminum, painted	.042	L.F.	.002	.10	.12	.22
Ridge vent	.042	L.F.	.002	.11	.12	.23
TOTAL		S.F.	.055	2.60	3.13	5.73
WOOD, CEDAR SHINGLES, NO. 1 PERFECTIONS, 18″ LONG						
Shingles, wood, red cedar, No. 1 perfections, 5″ exposure	1.450	S.F.	.044	4.19	2.57	6.76
Drip edge, metal, 5″ wide	.146	L.F.	.003	.10	.17	.27
Building paper, #15 asphalt felt	1.500	S.F.	.002	.09	.11	.20
Ridge shingles, wood	.042	L.F.	.001	.22	.07	.29
Soffit & fascia, white painted aluminum, 1′ overhang	.083	L.F.	.012	.42	.71	1.13
Rake trim, 1″ x 6″	.063	L.F.	.003	.08	.15	.23
Rake trim, prime and paint	.063	L.F.	.001	.02	.06	.08
Gutter, seamless, aluminum, painted	.083	L.F.	.005	.27	.34	.61
Downspouts, aluminum, painted	.042	L.F.	.002	.10	.12	.22
Ridge vent	.042	L.F.	.002	.11	.12	.23
TOTAL		S.F.	.075	5.60	4.42	10.02

The prices in this system are based on a square foot of plan area.
All quantities have been adjusted accordingly.

Description	QUAN.	UNIT	LABOR HOURS	COST PER S.F.		
				MAT.	INST.	TOTAL

Gambrel Roofing Price Sheet	QUAN.	UNIT	LABOR HOURS	COST PER S.F.		
				MAT.	INST.	TOTAL
Shingles, asphalt, standard, inorganic, class A, 210-235 lb./sq.	1.450	S.F.	.022	1.30	1.21	2.51
Laminated, multi-layered, 240-260 lb./sq.	1.450	S.F.	.027	1.82	1.48	3.30
Premium laminated, multi-layered, 260-300 lb./sq.	1.450	S.F.	.034	2.40	1.91	4.31
Slate, Buckingham, Virginia, black	1.450	S.F.	.069	9.25	3.81	13.06
Vermont, black or grey	1.450	S.F.	.069	8.05	3.81	11.86
Wood, red cedar, No.1 5X, 16″ long, 5″ exposure, plain	1.450	S.F.	.048	4.88	2.82	7.70
Fire retardant	1.450	S.F.	.048	5.80	2.82	8.62
18″ long, No.1 perfections, 6″ exposure, plain	1.450	S.F.	.044	4.19	2.57	6.76
Fire retardant	1.450	S.F.	.044	5.10	2.57	7.67
Resquared & rebutted, 18″ long, 6″ exposure, plain	1.450	S.F.	.040	4.73	2.36	7.09
Fire retardant	1.450	S.F.	.040	5.65	2.36	8.01
Shakes, hand split, 24″ long, 10″ exposure, plain	1.450	S.F.	.048	4.95	2.82	7.77
Fire retardant	1.450	S.F.	.048	5.85	2.82	8.67
18″ long, 8″ exposure, plain	1.450	S.F.	.060	4.58	3.54	8.12
Fire retardant	1.450	S.F.	.060	5.50	3.54	9.04
Drip edge, metal, 5″ wide	.146	L.F.	.003	.10	.17	.27
8″ wide	.146	L.F.	.003	.13	.17	.30
Building paper, #15 asphalt felt	1.500	S.F.	.002	.09	.11	.20
Ridge shingles, asphalt	.042	L.F.	.001	.11	.06	.17
Slate	.042	L.F.	.002	.47	.09	.56
Wood, shingles	.042	L.F.	.001	.22	.07	.29
Soffit & fascia, aluminum, vented, 1′ overhang	.083	L.F.	.012	.42	.71	1.13
2′ overhang	.083	L.F.	.013	.60	.78	1.38
Vinyl vented, 1′ overhang	.083	L.F.	.011	.44	.65	1.09
2′ overhang	.083	L.F.	.012	.59	.78	1.37
Wood board fascia, plywood soffit, 1′ overhang	.083	L.F.	.004	.02	.18	.20
2′ overhang	.083	L.F.	.006	.04	.27	.31
Rake trim, painted, 1″ x 6″	.063	L.F.	.006	.10	.29	.39
1″ x 8″	.063	L.F.	.007	.12	.38	.50
Gutter, 5″ box, aluminum, seamless, painted	.083	L.F.	.006	.27	.34	.61
Vinyl	.083	L.F.	.006	.15	.34	.49
Downspout 2″ x 3″, aluminum, one story house	.042	L.F.	.002	.06	.11	.17
Two story house	.070	L.F.	.003	.11	.19	.30
Vinyl, one story house	.042	L.F.	.002	.06	.11	.17
Two story house	.070	L.F.	.003	.11	.19	.30

16 Mansard Roofing Systems

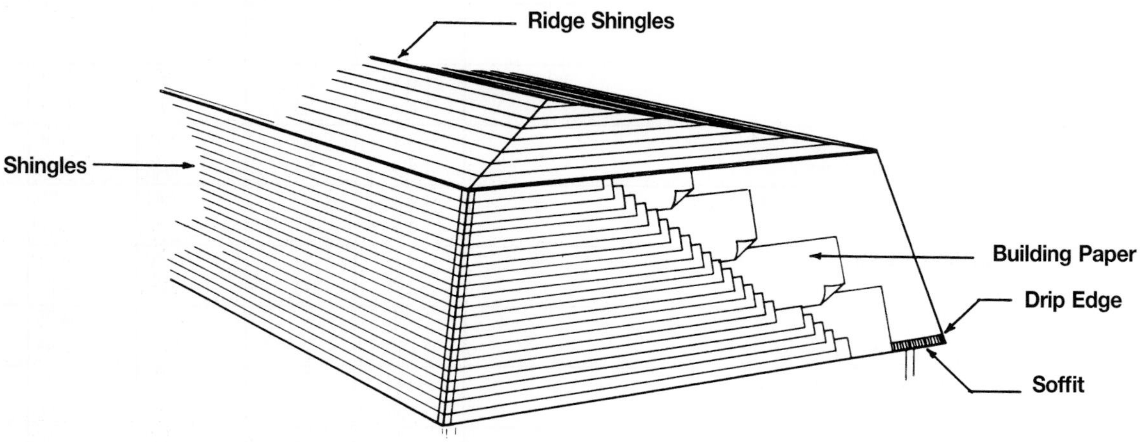

Ridge Shingles
Shingles
Building Paper
Drip Edge
Soffit

System Description	QUAN.	UNIT	LABOR HOURS	COST PER S.F. MAT.	COST PER S.F. INST.	COST PER S.F. TOTAL
ASPHALT, ROOF SHINGLES, CLASS A						
Shingles, standard inorganic class A 210-235 lb./sq.	2.210	S.F.	.032	1.90	1.77	3.67
Drip edge, metal, 5″ wide	.122	L.F.	.002	.08	.14	.22
Building paper, #15 asphalt felt	2.300	S.F.	.003	.14	.18	.32
Ridge shingles, asphalt	.090	L.F.	.002	.23	.12	.35
Soffit & fascia, white painted aluminum, 1′ overhang	.122	L.F.	.018	.61	1.04	1.65
Gutter, seamless, aluminum, painted	.122	L.F.	.008	.39	.51	.90
Downspouts, aluminum, painted	.042	L.F.	.002	.10	.12	.22
Ridge vent	.028	L.F.	.001	.07	.08	.15
TOTAL		S.F.	.068	3.52	3.96	7.48
WOOD, CEDAR SHINGLES, NO. 1 PERFECTIONS, 18″ LONG						
Shingles, wood, red cedar, No. 1 perfections, 5″ exposure	2.210	S.F.	.064	6.14	3.76	9.90
Drip edge, metal, 5″ wide	.122	L.F.	.002	.08	.14	.22
Building paper, #15 asphalt felt	2.300	S.F.	.003	.14	.18	.32
Ridge shingles, wood	.090	L.F.	.003	.47	.15	.62
Soffit & fascia, white painted aluminum, 1′ overhang	.122	L.F.	.018	.61	1.04	1.65
Gutter, seamless, aluminum, painted	.122	L.F.	.008	.39	.51	.90
Downspouts, aluminum, painted	.042	L.F.	.002	.10	.12	.22
Ridge vent	.028	L.F.	.001	.07	.08	.15
TOTAL		S.F.	.101	8	5.98	13.98

The prices in these systems are based on a square foot of plan area.
All quantities have been adjusted accordingly.

Description	QUAN.	UNIT	LABOR HOURS	COST PER S.F. MAT.	COST PER S.F. INST.	COST PER S.F. TOTAL

Mansard Roofing Price Sheet	QUAN.	UNIT	LABOR HOURS	COST PER S.F. MAT.	INST.	TOTAL
Shingles, asphalt, standard, inorganic, class A, 210-235 lb./sq.	2.210	S.F.	.032	1.90	1.77	3.67
Laminated, multi-layered, 240-260 lb./sq.	2.210	S.F.	.039	2.66	2.17	4.83
Premium laminated, multi-layered, 260-300 lb./sq.	2.210	S.F.	.050	3.52	2.79	6.31
Slate Buckingham, Virginia, black	2.210	S.F.	.101	13.55	5.60	19.15
Vermont, black or grey	2.210	S.F.	.101	11.75	5.60	17.35
Wood, red cedar, No.1 5X, 16″ long, 5″ exposure, plain	2.210	S.F.	.070	7.15	4.14	11.29
Fire retardant	2.210	S.F.	.070	8.50	4.14	12.64
18″ long, No.1 perfections 6″ exposure, plain	2.210	S.F.	.064	6.15	3.76	9.91
Fire retardant	2.210	S.F.	.064	7.50	3.76	11.26
Resquared & rebutted, 18″ long, 6″ exposure, plain	2.210	S.F.	.059	6.95	3.45	10.40
Fire retardant	2.210	S.F.	.059	8.25	3.45	11.70
Shakes, hand split, 24″ long 10″ exposure, plain	2.210	S.F.	.070	7.25	4.14	11.39
Fire retardant	2.210	S.F.	.070	8.60	4.14	12.74
18″ long, 8″ exposure, plain	2.210	S.F.	.088	6.70	5.20	11.90
Fire retardant	2.210	S.F.	.088	8.05	5.20	13.25
Drip edge, metal, 5″ wide	.122	S.F.	.002	.08	.14	.22
8″ wide	.122	S.F.	.002	.11	.14	.25
Building paper, #15 asphalt felt	2.300	S.F.	.003	.14	.18	.32
Ridge shingles, asphalt	.090	L.F.	.002	.23	.12	.35
Slate	.090	L.F.	.004	1.01	.20	1.21
Wood, shingles	.090	L.F.	.003	.47	.15	.62
Soffit & fascia, aluminum vented, 1′ overhang	.122	L.F.	.018	.61	1.04	1.65
2′ overhang	.122	L.F.	.020	.88	1.15	2.03
Vinyl vented, 1′ overhang	.122	L.F.	.016	.65	.96	1.61
2′ overhang	.122	L.F.	.018	.86	1.15	2.01
Wood board fascia, plywood soffit, 1′ overhang	.122	L.F.	.013	.37	.72	1.09
2′ overhang	.122	L.F.	.019	.53	1.10	1.63
Gutter, 5″ box, aluminum, seamless, painted	.122	L.F.	.008	.39	.51	.90
Vinyl	.122	L.F.	.009	.22	.50	.72
Downspout 2″ x 3″, aluminum, one story house	.042	L.F.	.002	.06	.11	.17
Two story house	.070	L.F.	.003	.10	.19	.29
Vinyl, one story house	.042	L.F.	.002	.06	.11	.17
Two story house	.070	L.F.	.003	.10	.19	.29

20 Shed Roofing Systems

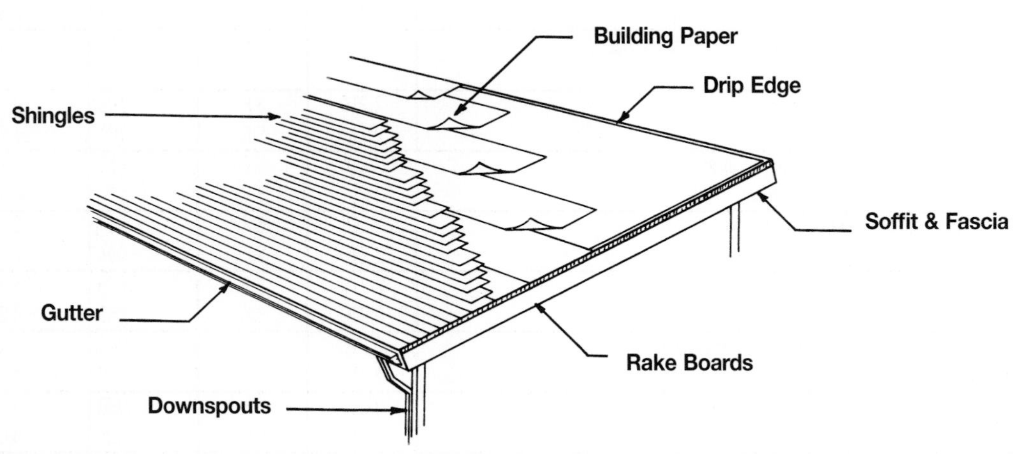

System Description	QUAN.	UNIT	LABOR HOURS	COST PER S.F. MAT.	INST.	TOTAL
ASPHALT, ROOF SHINGLES, CLASS A						
Shingles, inorganic class A 210-235 lb./sq. 4/12 pitch	1.230	S.F.	.019	1.12	1.05	2.17
Drip edge, metal, 5″ wide	.100	L.F.	.002	.07	.12	.19
Building paper, #15 asphalt felt	1.300	S.F.	.002	.08	.10	.18
Soffit & fascia, white painted aluminum, 1′ overhang	.080	L.F.	.012	.40	.68	1.08
Rake trim, 1″ x 6″	.043	L.F.	.002	.05	.10	.15
Rake trim, prime and paint	.043	L.F.	.002	.01	.09	.10
Gutter, seamless, aluminum, painted	.040	L.F.	.003	.13	.17	.30
Downspouts, painted aluminum	.020	L.F.	.001	.05	.06	.11
TOTAL		S.F.	.043	1.91	2.37	4.28
WOOD, CEDAR SHINGLES, NO. 1 PERFECTIONS, 18″ LONG						
Shingles, red cedar, No. 1 perfections, 5″ exp., 4/12 pitch	1.230	S.F.	.035	3.35	2.05	5.40
Drip edge, metal, 5″ wide	.100	L.F.	.002	.07	.12	.19
Building paper, #15 asphalt felt	1.300	S.F.	.002	.08	.10	.18
Soffit & fascia, white painted aluminum, 1′ overhang	.080	L.F.	.012	.40	.68	1.08
Rake trim, 1″ x 6″	.043	L.F.	.002	.05	.10	.15
Rake trim, prime and paint	.043	L.F.	.001	.01	.04	.05
Gutter, seamless, aluminum, painted	.040	L.F.	.003	.13	.17	.30
Downspouts, painted aluminum	.020	L.F.	.001	.05	.06	.11
TOTAL		S.F.	.058	4.14	3.32	7.46

The prices in these systems are based on a square foot of plan area.
All quantities have been adjusted accordingly.

Description	QUAN.	UNIT	LABOR HOURS	COST PER S.F. MAT.	INST.	TOTAL

Shed Roofing Price Sheet	QUAN.	UNIT	LABOR HOURS	COST PER S.F.		
				MAT.	INST.	TOTAL
Shingles, asphalt, inorganic, class A, 210-235 lb./sq., 4/12 pitch	1.230	S.F.	.017	1.04	.97	2.01
8/12 pitch	1.330	S.F.	.019	1.12	1.05	2.17
Laminated, multi-layered, 240-260 lb./sq. 4/12 pitch	1.230	S.F.	.021	1.45	1.18	2.63
8/12 pitch	1.330	S.F.	.023	1.57	1.28	2.85
Premium laminated, multi-layered, 260-300 lb./sq. 4/12 pitch	1.230	S.F.	.027	1.92	1.52	3.44
8/12 pitch	1.330	S.F.	.030	2.08	1.65	3.73
Clay tile, Spanish tile, red, 4/12 pitch	1.230	S.F.	.053	6.60	2.92	9.52
8/12 pitch	1.330	S.F.	.058	7.15	3.16	10.31
Mission tile, red, 4/12 pitch	1.230	S.F.	.083	5.45	2.92	8.37
8/12 pitch	1.330	S.F.	.090	5.90	3.16	9.06
French tile, red, 4/12 pitch	1.230	S.F.	.071	15.30	2.68	17.98
8/12 pitch	1.330	S.F.	.077	16.60	2.90	19.50
Slate, Buckingham, Virginia, black, 4/12 pitch	1.230	S.F.	.055	7.40	3.05	10.45
8/12 pitch	1.330	S.F.	.059	8	3.30	11.30
Vermont, black or grey, 4/12 pitch	1.230	S.F.	.055	6.40	3.05	9.45
8/12 pitch	1.330	S.F.	.059	6.95	3.30	10.25
Wood, red cedar, No.1 5X, 16″ long, 5″ exposure, 4/12 pitch	1.230	S.F.	.038	3.90	2.26	6.16
8/12 pitch	1.330	S.F.	.042	4.23	2.44	6.67
Fire retardant, 4/12 pitch	1.230	S.F.	.038	4.63	2.26	6.89
8/12 pitch	1.330	S.F.	.042	5	2.44	7.44
18″ long, 6″ exposure, 4/12 pitch	1.230	S.F.	.035	3.35	2.05	5.40
8/12 pitch	1.330	S.F.	.038	3.63	2.22	5.85
Fire retardant, 4/12 pitch	1.230	S.F.	.035	4.08	2.05	6.13
8/12 pitch	1.330	S.F.	.038	4.42	2.22	6.64
Resquared & rebutted, 18″ long, 6″ exposure, 4/12 pitch	1.230	S.F.	.032	3.78	1.88	5.66
8/12 pitch	1.330	S.F.	.035	4.10	2.04	6.14
Fire retardant, 4/12 pitch	1.230	S.F.	.032	4.51	1.88	6.39
8/12 pitch	1.330	S.F.	.035	4.89	2.04	6.93
Wood shakes, hand split, 24″ long, 10″ exposure, 4/12 pitch	1.230	S.F.	.038	3.96	2.26	6.22
8/12 pitch	1.330	S.F.	.042	4.29	2.44	6.73
Fire retardant, 4/12 pitch	1.230	S.F.	.038	4.69	2.26	6.95
8/12 pitch	1.330	S.F.	.042	5.10	2.44	7.54
18″ long, 8″ exposure, 4/12 pitch	1.230	S.F.	.048	3.66	2.83	6.49
8/12 pitch	1.330	S.F.	.052	3.97	3.07	7.04
Fire retardant, 4/12 pitch	1.230	S.F.	.048	4.39	2.83	7.22
8/12 pitch	1.330	S.F.	.052	4.76	3.07	7.83
Drip edge, metal, 5″ wide	.100	L.F.	.002	.07	.12	.19
8″ wide	.100	L.F.	.002	.09	.12	.21
Building paper, #15 asphalt felt	1.300	S.F.	.002	.08	.10	.18
Soffit & fascia, aluminum vented, 1′ overhang	.080	L.F.	.012	.40	.68	1.08
2′ overhang	.080	L.F.	.013	.58	.75	1.33
Vinyl vented, 1′ overhang	.080	L.F.	.011	.43	.63	1.06
2′ overhang	.080	L.F.	.012	.56	.75	1.31
Wood board fascia, plywood soffit, 1′ overhang	.080	L.F.	.010	.25	.52	.77
2′ overhang	.080	L.F.	.014	.36	.80	1.16
Rake, trim, painted, 1″ x 6″	.043	L.F.	.004	.06	.19	.25
1″ x 8″	.043	L.F.	.004	.06	.19	.25
Gutter, 5″ box, aluminum, seamless, painted	.040	L.F.	.003	.13	.17	.30
Vinyl	.040	L.F.	.003	.07	.16	.23
Downspout 2″ x 3″, aluminum, one story house	.020	L.F.	.001	.03	.05	.08
Two story house	.020	L.F.	.001	.05	.09	.14
Vinyl, one story house	.020	L.F.	.001	.03	.05	.08
Two story house	.020	L.F.	.001	.05	.09	.14

24 Gable Dormer Roofing Systems

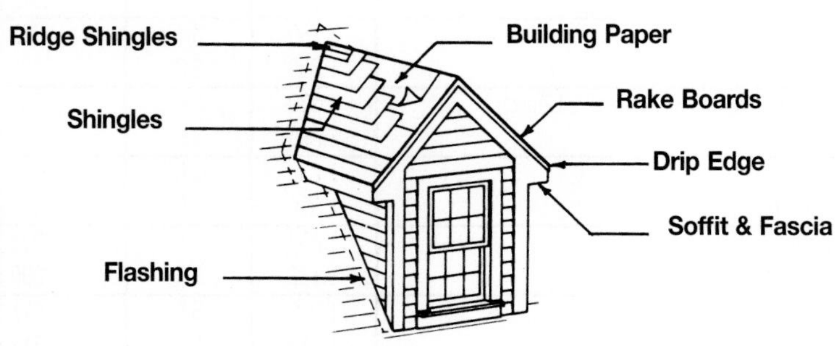

System Description	QUAN.	UNIT	LABOR HOURS	COST PER S.F. MAT.	INST.	TOTAL
ASPHALT, ROOF SHINGLES, CLASS A						
Shingles, standard inorganic class A 210-235 lb./sq	1.400	S.F.	.020	1.21	1.13	2.34
Drip edge, metal, 5" wide	.220	L.F.	.004	.15	.26	.41
Building paper, #15 asphalt felt	1.500	S.F.	.002	.09	.11	.20
Ridge shingles, asphalt	.280	L.F.	.007	.70	.38	1.08
Soffit & fascia, aluminum, vented	.220	L.F.	.032	1.10	1.88	2.98
Flashing, aluminum, mill finish, .013" thick	1.500	S.F.	.083	1.43	4.58	6.01
TOTAL		S.F.	.148	4.68	8.34	13.02
WOOD, CEDAR, NO. 1 PERFECTIONS						
Shingles, red cedar, No.1 perfections, 18" long, 5" exp.	1.400	S.F.	.041	3.91	2.39	6.30
Drip edge, metal, 5" wide	.220	L.F.	.004	.15	.26	.41
Building paper, #15 asphalt felt	1.500	S.F.	.002	.09	.11	.20
Ridge shingles, wood	.280	L.F.	.008	1.47	.47	1.94
Soffit & fascia, aluminum, vented	.220	L.F.	.032	1.10	1.88	2.98
Flashing, aluminum, mill finish, .013" thick	1.500	S.F.	.083	1.43	4.58	6.01
TOTAL		S.F.	.170	8.15	9.69	17.84
SLATE, BUCKINGHAM, BLACK						
Shingles, Buckingham, Virginia, black	1.400	S.F.	.064	8.61	3.56	12.17
Drip edge, metal, 5" wide	.220	L.F.	.004	.15	.26	.41
Building paper, #15 asphalt felt	1.500	S.F.	.002	.09	.11	.20
Ridge shingles, slate	.280	L.F.	.011	3.14	.62	3.76
Soffit & fascia, aluminum, vented	.220	L.F.	.032	1.10	1.88	2.98
Flashing, copper, 16 oz.	1.500	S.F.	.104	13.35	5.78	19.13
TOTAL		S.F.	.217	26.44	12.21	38.65

The prices in these systems are based on a square foot of plan area under the dormer roof.

Description	QUAN.	UNIT	LABOR HOURS	COST PER S.F. MAT.	INST.	TOTAL

Gable Dormer Roofing Price Sheet	QUAN.	UNIT	LABOR HOURS	COST PER S.F.		
				MAT.	INST.	TOTAL
Shingles, asphalt, standard, inorganic, class A, 210-235 lb./sq.	1.400	S.F.	.020	1.21	1.13	2.34
Laminated, multi-layered, 240-260 lb./sq.	1.400	S.F.	.025	1.69	1.38	3.07
Premium laminated, multi-layered, 260-300 lb./sq.	1.400	S.F.	.032	2.24	1.78	4.02
Clay tile, Spanish tile, red	1.400	S.F.	.062	7.70	3.40	11.10
Mission tile, red	1.400	S.F.	.097	6.35	3.40	9.75
French tile, red	1.400	S.F.	.083	17.85	3.12	20.97
Slate Buckingham, Virginia, black	1.400	S.F.	.064	8.60	3.56	12.16
Vermont, black or grey	1.400	S.F.	.064	7.50	3.56	11.06
Wood, red cedar, No.1 5X, 16″ long, 5″ exposure	1.400	S.F.	.045	4.55	2.63	7.18
Fire retardant	1.400	S.F.	.045	5.40	2.63	8.03
18″ long, No.1 perfections, 5″ exposure	1.400	S.F.	.041	3.91	2.39	6.30
Fire retardant	1.400	S.F.	.041	4.76	2.39	7.15
Resquared & rebutted, 18″ long, 5″ exposure	1.400	S.F.	.037	4.41	2.20	6.61
Fire retardant	1.400	S.F.	.037	5.25	2.20	7.45
Shakes hand split, 24″ long, 10″ exposure	1.400	S.F.	.045	4.62	2.63	7.25
Fire retardant	1.400	S.F.	.045	5.45	2.63	8.08
18″ long, 8″ exposure	1.400	S.F.	.056	4.27	3.30	7.57
Fire retardant	1.400	S.F.	.056	5.10	3.30	8.40
Drip edge, metal, 5″ wide	.220	L.F.	.004	.15	.26	.41
8″ wide	.220	L.F.	.004	.20	.26	.46
Building paper, #15 asphalt felt	1.500	S.F.	.002	.09	.11	.20
Ridge shingles, asphalt	.280	L.F.	.007	.70	.38	1.08
Clay	.280	L.F.	.011	1.51	1.88	3.39
Slate	.280	L.F.	.011	3.14	.62	3.76
Wood	.280	L.F.	.008	1.47	.47	1.94
Soffit & fascia, aluminum, vented	.220	L.F.	.032	1.10	1.88	2.98
Vinyl, vented	.220	L.F.	.029	1.18	1.73	2.91
Wood, board fascia, plywood soffit	.220	L.F.	.026	.70	1.45	2.15
Flashing, aluminum, .013″ thick	1.500	S.F.	.083	1.43	4.58	6.01
.032″ thick	1.500	S.F.	.083	2.25	4.58	6.83
.040″ thick	1.500	S.F.	.083	3.80	4.58	8.38
.050″ thick	1.500	S.F.	.083	4.55	4.58	9.13
Copper, 16 oz.	1.500	S.F.	.104	13.35	5.80	19.15
20 oz.	1.500	S.F.	.109	17.80	6.05	23.85
24 oz.	1.500	S.F.	.114	24.50	6.35	30.85
32 oz.	1.500	S.F.	.120	32.50	6.65	39.15

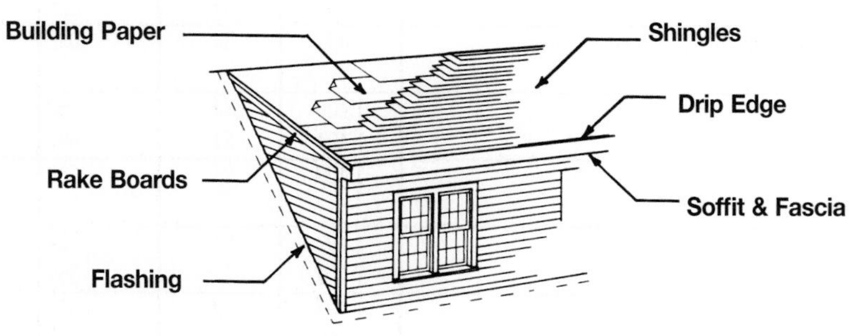

System Description		QUAN.	UNIT	LABOR HOURS	COST PER S.F.		
					MAT.	INST.	TOTAL
ASPHALT, ROOF SHINGLES, CLASS A							
Shingles, standard inorganic class A 210-235 lb./sq.		1.100	S.F.	.016	.95	.89	1.84
Drip edge, aluminum, 5" wide		.250	L.F.	.005	.16	.30	.46
Building paper, #15 asphalt felt		1.200	S.F.	.002	.07	.09	.16
Soffit & fascia, aluminum, vented, 1' overhang		.250	L.F.	.036	1.25	2.14	3.39
Flashing, aluminum, mill finish, 0.013" thick		.800	L.F.	.044	.76	2.44	3.20
	TOTAL		S.F.	.103	3.19	5.86	9.05
WOOD, CEDAR, NO. 1 PERFECTIONS, 18" LONG							
Shingles, wood, red cedar, #1 perfections, 5" exposure		1.100	S.F.	.032	3.07	1.88	4.95
Drip edge, aluminum, 5" wide		.250	L.F.	.005	.16	.30	.46
Building paper, #15 asphalt felt		1.200	S.F.	.002	.07	.09	.16
Soffit & fascia, aluminum, vented, 1' overhang		.250	L.F.	.036	1.25	2.14	3.39
Flashing, aluminum, mill finish, 0.013" thick		.800	L.F.	.044	.76	2.44	3.20
	TOTAL		S.F.	.119	5.31	6.85	12.16
SLATE, BUCKINGHAM, BLACK							
Shingles, slate, Buckingham, black		1.100	S.F.	.050	6.77	2.79	9.56
Drip edge, aluminum, 5" wide		.250	L.F.	.005	.16	.30	.46
Building paper, #15 asphalt felt		1.200	S.F.	.002	.07	.09	.16
Soffit & fascia, aluminum, vented, 1' overhang		.250	L.F.	.036	1.25	2.14	3.39
Flashing, copper, 16 oz.		.800	L.F.	.056	7.12	3.08	10.20
	TOTAL		S.F.	.149	15.37	8.40	23.77

The prices in this system are based on a square foot of plan area under the dormer roof.

Description	QUAN.	UNIT	LABOR HOURS	COST PER S.F.		
				MAT.	INST.	TOTAL

Shed Dormer Roofing Price Sheet	QUAN.	UNIT	LABOR HOURS	COST PER S.F.		
				MAT.	INST.	TOTAL
Shingles, asphalt, standard, inorganic, class A, 210-235 lb./sq.	1.100	S.F.	.016	.95	.89	1.84
Laminated, multi-layered, 240-260 lb./sq.	1.100	S.F.	.020	1.33	1.08	2.41
Premium laminated, multi-layered, 260-300 lb./sq.	1.100	S.F.	.025	1.76	1.40	3.16
Clay tile, Spanish tile, red	1.100	S.F.	.049	6.05	2.67	8.72
Mission tile, red	1.100	S.F.	.077	5	2.67	7.67
French tile, red	1.100	S.F.	.065	14.05	2.45	16.50
Slate Buckingham, Virginia, black	1.100	S.F.	.050	6.75	2.79	9.54
Vermont, black or grey	1.100	S.F.	.050	5.90	2.79	8.69
Wood, red cedar, No. 1 5X, 16″ long, 5″ exposure	1.100	S.F.	.035	3.58	2.07	5.65
Fire retardant	1.100	S.F.	.035	4.25	2.07	6.32
18″ long, No.1 perfections, 5″ exposure	1.100	S.F.	.032	3.07	1.88	4.95
Fire retardant	1.100	S.F.	.032	3.74	1.88	5.62
Resquared & rebutted, 18″ long, 5″ exposure	1.100	S.F.	.029	3.47	1.73	5.20
Fire retardant	1.100	S.F.	.029	4.14	1.73	5.87
Shakes hand split, 24″ long, 10″ exposure	1.100	S.F.	.035	3.63	2.07	5.70
Fire retardant	1.100	S.F.	.035	4.30	2.07	6.37
18″ long, 8″ exposure	1.100	S.F.	.044	3.36	2.60	5.96
Fire retardant	1.100	S.F.	.044	4.03	2.60	6.63
Drip edge, metal, 5″ wide	.250	L.F.	.005	.16	.30	.46
8″ wide	.250	L.F.	.005	.23	.30	.53
Building paper, #15 asphalt felt	1.200	S.F.	.002	.07	.09	.16
Soffit & fascia, aluminum, vented	.250	L.F.	.036	1.25	2.14	3.39
Vinyl, vented	.250	L.F.	.033	1.34	1.96	3.30
Wood, board fascia, plywood soffit	.250	L.F.	.030	.81	1.65	2.46
Flashing, aluminum, .013″ thick	.800	L.F.	.044	.76	2.44	3.20
.032″ thick	.800	L.F.	.044	1.20	2.44	3.64
.040″ thick	.800	L.F.	.044	2.02	2.44	4.46
.050″ thick	.800	L.F.	.044	2.42	2.44	4.86
Copper, 16 oz.	.800	L.F.	.056	7.10	3.08	10.18
20 oz.	.800	L.F.	.058	9.50	3.22	12.72
24 oz.	.800	L.F.	.061	13	3.38	16.38
32 oz.	.800	L.F.	.064	17.20	3.54	20.74

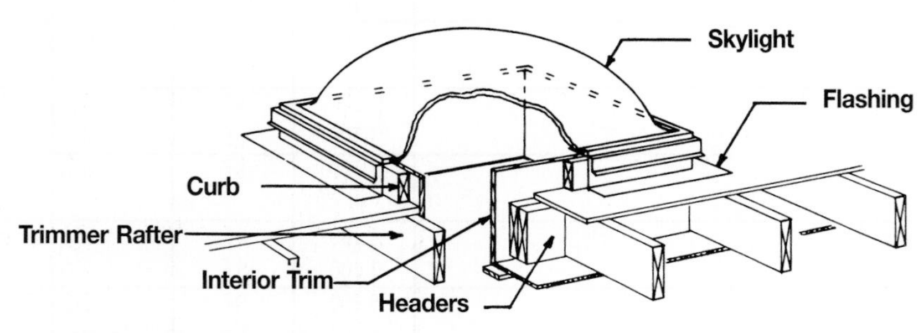

System Description	QUAN.	UNIT	LABOR HOURS	COST EACH		
				MAT.	INST.	TOTAL
SKYLIGHT, FIXED, 32" X 32"						
Skylight, fixed bubble, insulating, 32" x 32"	1.000	Ea.	1.422	234.66	78.22	312.88
Trimmer rafters, 2" x 6"	28.000	L.F.	.448	22.12	26.32	48.44
Headers, 2" x 6"	6.000	L.F.	.267	4.74	15.72	20.46
Curb, 2" x 4"	12.000	L.F.	.154	5.64	9	14.64
Flashing, aluminum, .013" thick	13.500	S.F.	.745	12.83	41.18	54.01
Moldings, casing, ogee, 11/16" x 2-1/2", pine	12.000	L.F.	.384	18.72	22.56	41.28
Trim primer coat, oil base, brushwork	12.000	L.F.	.148	.48	7.20	7.68
Trim paint, 1 coat, brushwork	12.000	L.F.	.148	.72	7.20	7.92
TOTAL		Ea.	3.716	299.91	207.40	507.31
SKYLIGHT, FIXED, 48" X 48"						
Skylight, fixed bubble, insulating, 48" x 48"	1.000	Ea.	1.296	552	71.36	623.36
Trimmer rafters, 2" x 6"	28.000	L.F.	.448	22.12	26.32	48.44
Headers, 2" x 6"	8.000	L.F.	.356	6.32	20.96	27.28
Curb, 2" x 4"	16.000	L.F.	.205	7.52	12	19.52
Flashing, aluminum, .013" thick	16.000	S.F.	.883	15.20	48.80	64
Moldings, casing, ogee, 11/16" x 2-1/2", pine	16.000	L.F.	.512	24.96	30.08	55.04
Trim primer coat, oil base, brushwork	16.000	L.F.	.197	.64	9.60	10.24
Trim paint, 1 coat, brushwork	16.000	L.F.	.197	.96	9.60	10.56
TOTAL		Ea.	4.094	629.72	228.72	858.44
SKYWINDOW, OPERATING, 24" X 48"						
Skywindow, operating, thermopane glass, 24" x 48"	1.000	Ea.	3.200	655	176	831
Trimmer rafters, 2" x 6"	28.000	L.F.	.448	22.12	26.32	48.44
Headers, 2" x 6"	8.000	L.F.	.267	4.74	15.72	20.46
Curb, 2" x 4"	14.000	L.F.	.179	6.58	10.50	17.08
Flashing, aluminum, .013" thick	14.000	S.F.	.772	13.30	42.70	56
Moldings, casing, ogee, 11/16" x 2-1/2", pine	14.000	L.F.	.448	21.84	26.32	48.16
Trim primer coat, oil base, brushwork	14.000	L.F.	.172	.56	8.40	8.96
Trim paint, 1 coat, brushwork	14.000	L.F.	.172	.84	8.40	9.24
TOTAL		Ea.	5.658	724.98	314.36	1,039.34

The prices in these systems are on a cost each basis.

Description	QUAN.	UNIT	LABOR HOURS	COST EACH		
				MAT.	INST.	TOTAL

Skylight/Skywindow Price Sheet	QUAN.	UNIT	LABOR HOURS	COST EACH		
				MAT.	INST.	TOTAL
Skylight, fixed bubble insulating, 24" x 24"	1.000	Ea.	.800	132	44	176
32" x 32"	1.000	Ea.	1.422	235	78	313
32" x 48"	1.000	Ea.	.864	370	47.50	417.50
48" x 48"	1.000	Ea.	1.296	550	71.50	621.50
Ventilating bubble insulating, 36" x 36"	1.000	Ea.	2.667	525	147	672
52" x 52"	1.000	Ea.	2.667	735	147	882
28" x 52"	1.000	Ea.	3.200	545	176	721
36" x 52"	1.000	Ea.	3.200	595	176	771
Skywindow, operating, thermopane glass, 24" x 48"	1.000	Ea.	3.200	655	176	831
32" x 48"	1.000	Ea.	3.556	685	196	881
Trimmer rafters, 2" x 6"	28.000	L.F.	.448	22	26.50	48.50
2" x 8"	28.000	L.F.	.472	30	27.50	57.50
2" x 10"	28.000	L.F.	.711	45	42	87
Headers, 24" window, 2" x 6"	4.000	L.F.	.178	3.16	10.50	13.66
2" x 8"	4.000	L.F.	.188	4.28	11.10	15.38
2" x 10"	4.000	L.F.	.200	6.40	11.80	18.20
32" window, 2" x 6"	6.000	L.F.	.267	4.74	15.70	20.44
2" x 8"	6.000	L.F.	.282	6.40	16.60	23
2" x 10"	6.000	L.F.	.300	9.60	17.70	27.30
48" window, 2" x 6"	8.000	L.F.	.356	6.30	21	27.30
2" x 8"	8.000	L.F.	.376	8.55	22	30.55
2" x 10"	8.000	L.F.	.400	12.80	23.50	36.30
Curb, 2" x 4", skylight, 24" x 24"	8.000	L.F.	.102	3.76	6	9.76
32" x 32"	12.000	L.F.	.154	5.65	9	14.65
32" x 48"	14.000	L.F.	.179	6.60	10.50	17.10
48" x 48"	16.000	L.F.	.205	7.50	12	19.50
Flashing, aluminum .013" thick, skylight, 24" x 24"	9.000	S.F.	.497	8.55	27.50	36.05
32" x 32"	13.500	S.F.	.745	12.85	41	53.85
32" x 48"	14.000	S.F.	.772	13.30	42.50	55.80
48" x 48"	16.000	S.F.	.883	15.20	49	64.20
Copper 16 oz., skylight, 24" x 24"	9.000	S.F.	.626	80	34.50	114.50
32" x 32"	13.500	S.F.	.939	120	52	172
32" x 48"	14.000	S.F.	.974	125	54	179
48" x 48"	16.000	S.F.	1.113	142	61.50	203.50
Trim, interior casing painted, 24" x 24"	8.000	L.F.	.347	13.70	18.95	32.65
32" x 32"	12.000	L.F.	.520	20.50	28.50	49
32" x 48"	14.000	L.F.	.607	24	33	57
48" x 48"	16.000	L.F.	.693	27.50	38	65.50

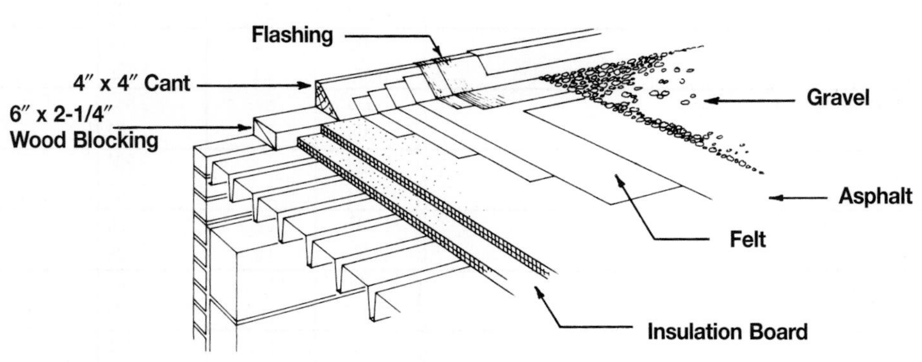

System Description	QUAN.	UNIT	LABOR HOURS	COST PER S.F. MAT.	COST PER S.F. INST.	COST PER S.F. TOTAL
ASPHALT, ORGANIC, 4-PLY, INSULATED DECK						
Membrane, asphalt, 4-plies #15 felt, gravel surfacing	1.000	S.F.	.025	1.52	1.58	3.10
Insulation board, 2-layers of 1-1/16″ glass fiber	2.000	S.F.	.012	2.38	.68	3.06
Roof deck insulation, fastening alternatives, coated screws, 4″ long	1.000	S.F	.003	.12	.14	.26
Wood blocking, 2″ x 6″	.040	L.F.	.004	.09	.25	.34
Treated 4″ x 4″ cant strip	.040	L.F.	.001	.08	.06	.14
Flashing, aluminum, 0.040″ thick	.050	S.F.	.003	.13	.15	.28
TOTAL		S.F.	.048	4.32	2.86	7.18
ASPHALT, INORGANIC, 3-PLY, INSULATED DECK						
Membrane, asphalt, 3-plies type IV glass felt, gravel surfacing	1.000	S.F.	.028	1.53	1.74	3.27
Insulation board, 2-layers of 1-1/16″ glass fiber	2.000	S.F.	.012	2.38	.68	3.06
Roof deck insulation, fastening alternatives, coated screws, 4″ long	1.000	S.F	.003	.12	.14	.26
Wood blocking, 2″ x 6″	.040	L.F.	.004	.09	.25	.34
Treated 4″ x 4″ cant strip	.040	L.F.	.001	.08	.06	.14
Flashing, aluminum, 0.040″ thick	.050	S.F.	.003	.13	.15	.28
TOTAL		S.F.	.051	4.33	3.02	7.35
COAL TAR, ORGANIC, 4-PLY, INSULATED DECK						
Membrane, coal tar, 4-plies #15 felt, gravel surfacing	1.000	S.F.	.027	2.28	1.65	3.93
Insulation board, 2-layers of 1-1/16″ glass fiber	2.000	S.F.	.012	2.38	.68	3.06
Roof deck insulation, fastening alternatives, coated screws, 4″ long	1.000	S.F	.003	.12	.14	.26
Wood blocking, 2″ x 6″	.040	L.F.	.004	.09	.25	.34
Treated 4″ x 4″ cant strip	.040	L.F.	.001	.08	.06	.14
Flashing, aluminum, 0.040″ thick	.050	S.F.	.003	.13	.15	.28
TOTAL		S.F.	.050	5.08	2.93	8.01
COAL TAR, INORGANIC, 3-PLY, INSULATED DECK						
Membrane, coal tar, 3-plies type IV glass felt, gravel surfacing	1.000	S.F.	.029	1.89	1.83	3.72
Insulation board, 2-layers of 1-1/16″ glass fiber	2.000	S.F.	.012	2.38	.68	3.06
Roof deck insulation, fastening alternatives, coated screws, 4″ long	1.000	S.F	.003	.12	.14	.26
Wood blocking, 2″ x 6″	.040	L.F.	.004	.09	.25	.34
Treated 4″ x 4″ cant strip	.040	L.F.	.001	.08	.06	.14
Flashing, aluminum, 0.040″ thick	.050	S.F.	.003	.13	.15	.28
TOTAL		S.F.	.052	4.69	3.11	7.80

Built-Up Roofing Price Sheet	QUAN.	UNIT	LABOR HOURS	COST PER S.F.		
				MAT.	INST.	TOTAL
Membrane, asphalt, 4-plies #15 organic felt, gravel surfacing	1.000	S.F.	.025	1.52	1.58	3.10
Asphalt base sheet & 3-plies #15 asphalt felt	1.000	S.F.	.025	1.18	1.58	2.76
3-plies type IV glass fiber felt	1.000	S.F.	.028	1.53	1.74	3.27
4-plies type IV glass fiber felt	1.000	S.F.	.028	1.88	1.74	3.62
Coal tar, 4-plies #15 organic felt, gravel surfacing	1.000	S.F.	.027			
4-plies tarred felt	1.000	S.F.	.027	2.28	1.65	3.93
3-plies type IV glass fiber felt	1.000	S.F.	.029	1.89	1.83	3.72
4-plies type IV glass fiber felt	1.000	S.F.	.027	2.61	1.65	4.26
Roll, asphalt, 1-ply #15 organic felt, 2-plies mineral surfaced	1.000	S.F.	.021	.79	1.29	2.08
3-plies type IV glass fiber, 1-ply mineral surfaced	1.000	S.F.	.022	1.37	1.39	2.76
Insulation boards, glass fiber, 1-1/16" thick	1.000	S.F.	.008	1.31	.48	1.79
2-1/16" thick	1.000	S.F.	.010	1.88	.57	2.45
2-7/16" thick	1.000	S.F.	.010	2.02	.57	2.59
Expanded perlite, 1" thick	1.000	S.F.	.010	.73	.57	1.30
1-1/2" thick	1.000	S.F.	.010	1.01	.57	1.58
2" thick	1.000	S.F.	.011	1.33	.63	1.96
Fiberboard, 1" thick	1.000	S.F.	.010	.77	.57	1.34
1-1/2" thick	1.000	S.F.	.010	1.11	.57	1.68
2" thick	1.000	S.F.	.010	1.40	.57	1.97
Extruded polystyrene, 15 PSI compressive strength, 2" thick R10	1.000	S.F.	.006	.92	.41	1.33
3" thick R15	1.000	S.F.	.008	1.72	.48	2.20
4" thick R20	1.000	S.F.	.008	2.28	.48	2.76
Tapered for drainage	1.000	S.F.	.005	.68	.37	1.05
40 PSI compressive strength, 1" thick R5	1.000	S.F.	.005	1.08	.37	1.45
2" thick R10	1.000	S.F.	.006	1.94	.41	2.35
3" thick R15	1.000	S.F.	.008	2.75	.48	3.23
4" thick R20	1.000	S.F.	.008	3.57	.48	4.05
Fiberboard high density, 1/2" thick R1.3	1.000	S.F.	.008	.42	.48	.90
1" thick R2.5	1.000	S.F.	.010	.75	.57	1.32
1 1/2" thick R3.8	1.000	S.F.	.010	1.04	.57	1.61
Polyisocyanurate, 1 1/2" thick	1.000	S.F.	.006	.78	.41	1.19
2" thick	1.000	S.F.	.007	.99	.45	1.44
3 1/2" thick	1.000	S.F.	.008	2.17	.48	2.65
Tapered for drainage	1.000	S.F.	.006	.69	.38	1.07
Expanded polystyrene, 1" thick	1.000	S.F.	.005	.42	.37	.79
2" thick R10	1.000	S.F.	.006	.71	.41	1.12
3" thick R11	1.000	S.F.	.006	1.01	.41	1.42
Wood blocking, treated, 6" x 2" & 4" x 4" cant	.040	L.F.	.002	.14	.14	.28
6" x 4-1/2" & 4" x 4" cant	.040	L.F.	.005	.21	.31	.52
6" x 5" & 4" x 4" cant	.040	L.F.	.007	.27	.40	.67
Flashing, aluminum, 0.019" thick	.050	S.F.	.003	.08	.15	.23
0.032" thick	.050	S.F.	.003	.08	.15	.23
0.040" thick	.050	S.F.	.003	.13	.15	.28
Copper sheets, 16 oz., under 500 lbs.	.050	S.F.	.003	.45	.19	.64
Over 500 lbs.	.050	S.F.	.003	.45	.14	.59
20 oz., under 500 lbs.	.050	S.F.	.004	.59	.20	.79
Over 500 lbs.	.050	S.F.	.003	.56	.15	.71
Stainless steel, 32 gauge	.050	S.F.	.003	.19	.14	.33
28 gauge	.050	S.F.	.003	.26	.14	.40
26 gauge	.050	S.F.	.003	.26	.14	.40
24 gauge	.050	S.F.	.003	.29	.14	.43

Did you know?

RSMeans data is available through our online application:

- Search for costs by keyword
- Leverage the most up-to-date data
- Build and export estimates

Try it free
rsmeans.com/2019freetrial

04 Drywall & Thincoat Wall Systems

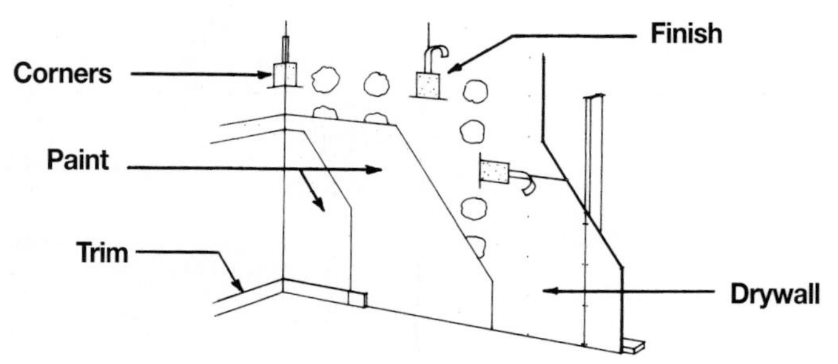

System Description	QUAN.	UNIT	LABOR HOURS	COST PER S.F.		
				MAT.	INST.	TOTAL
1/2″ DRYWALL, TAPED & FINISHED						
Gypsum wallboard, 1/2″ thick, standard	1.000	S.F.	.008	.37	.47	.84
Finish, taped & finished joints	1.000	S.F.	.008	.05	.47	.52
Corners, taped & finished, 32 L.F. per 12′ x 12′ room	.083	L.F.	.002	.01	.09	.10
Painting, primer & 2 coats	1.000	S.F.	.011	.22	.51	.73
Paint trim, to 6″ wide, primer + 1 coat enamel	.125	L.F.	.001	.02	.06	.08
Moldings, base, ogee profile, 9/16″ x 4-1/2, red oak	.125	L.F.	.005	.60	.27	.87
TOTAL		S.F.	.035	1.27	1.87	3.14
THINCOAT, SKIM-COAT, ON 1/2″ BACKER DRYWALL						
Gypsum wallboard, 1/2″ thick, thincoat backer	1.000	S.F.	.008	.37	.47	.84
Thincoat plaster	1.000	S.F.	.011	.13	.64	.77
Corners, taped & finished, 32 L.F. per 12′ x 12′ room	.083	L.F.	.002	.01	.09	.10
Painting, primer & 2 coats	1.000	S.F.	.011	.22	.51	.73
Paint trim, to 6″ wide, primer + 1 coat enamel	.125	L.F.	.001	.02	.06	.08
Moldings, base, ogee profile, 9/16″ x 4-1/2, red oak	.125	L.F.	.005	.60	.27	.87
TOTAL		S.F.	.038	1.35	2.04	3.39
5/8″ DRYWALL, TAPED & FINISHED						
Gypsum wallboard, 5/8″ thick, standard	1.000	S.F.	.008	.39	.47	.86
Finish, taped & finished joints	1.000	S.F.	.008	.05	.47	.52
Corners, taped & finished, 32 L.F. per 12′ x 12′ room	.083	L.F.	.002	.01	.09	.10
Painting, primer & 2 coats	1.000	S.F.	.011	.22	.51	.73
Moldings, base, ogee profile, 9/16″ x 4-1/2, red oak	.125	L.F.	.005	.60	.27	.87
Paint trim, to 6″ wide, primer + 1 coat enamel	.125	L.F.	.001	.02	.06	.08
TOTAL		S.F.	.035	1.29	1.87	3.16

The costs in this system are based on a square foot of wall.
Do not deduct for openings.

Description	QUAN.	UNIT	LABOR HOURS	COST PER S.F.		
				MAT.	INST.	TOTAL

Drywall & Thincoat Wall Price Sheet	QUAN.	UNIT	LABOR HOURS	COST PER S.F.		
				MAT.	INST.	TOTAL
Gypsum wallboard, 1/2″ thick, standard	1.000	S.F.	.008	.37	.47	.84
Fire resistant	1.000	S.F.	.008	.41	.47	.88
Water resistant	1.000	S.F.	.008	.46	.47	.93
5/8″ thick, standard	1.000	S.F.	.008	.39	.47	.86
Fire resistant	1.000	S.F.	.008	.40	.47	.87
Water resistant	1.000	S.F.	.008	.55	.47	1.02
Gypsum wallboard backer for thincoat system, 1/2″ thick	1.000	S.F.	.008	.37	.47	.84
5/8″ thick	1.000	S.F.	.008	.39	.47	.86
Gypsum wallboard, taped & finished	1.000	S.F.	.008	.05	.47	.52
Texture spray	1.000	S.F.	.010	.04	.57	.61
Thincoat plaster, including tape	1.000	S.F.	.011	.13	.64	.77
Gypsum wallboard corners, taped & finished, 32 L.F. per 4′ x 4′ room	.250	L.F.	.004	.03	.25	.28
6′ x 6′ room	.110	L.F.	.002	.01	.11	.12
10′ x 10′ room	.100	L.F.	.001	.01	.10	.11
12′ x 12′ room	.083	L.F.	.001	.01	.08	.09
16′ x 16′ room	.063	L.F.	.001	.01	.06	.07
Thincoat system, 32 L.F. per 4′ x 4′ room	.250	L.F.	.003	.03	.16	.19
6′ x 6′ room	.110	L.F.	.001	.01	.07	.08
10′ x 10′ room	.100	L.F.	.001	.01	.06	.07
12′ x 12′ room	.083	L.F.	.001	.01	.05	.06
16′ x 16′ room	.063	L.F.	.001	.01	.04	.05
Painting, primer, & 1 coat	1.000	S.F.	.008	.14	.40	.54
& 2 coats	1.000	S.F.	.011	.22	.51	.73
Wallpaper, $7/double roll	1.000	S.F.	.013	.65	.61	1.26
$17/double roll	1.000	S.F.	.015	1.34	.73	2.07
$40/double roll	1.000	S.F.	.018	2.35	.90	3.25
Wallcovering, medium weight vinyl	1.000	S.F.	.017	1.09	.82	1.91
Tile, ceramic adhesive thin set, 4 1/4″ x 4 1/4″ tiles	1.000	S.F.	.084	2.85	4.14	6.99
6″ x 6″ tiles	1.000	S.F.	.080	4	4.49	8.49
Pregrouted sheets	1.000	S.F.	.067	6.25	3.28	9.53
Trim, painted or stained, baseboard	.125	L.F.	.006	.62	.33	.95
Base shoe	.125	L.F.	.005	.07	.31	.38
Chair rail	.125	L.F.	.005	.23	.28	.51
Cornice molding	.125	L.F.	.004	.16	.28	.44
Cove base, vinyl	.125	L.F.	.003	.18	.18	.36
Paneling, not including furring or trim						
Plywood, prefinished, 1/4″ thick, 4′ x 8′ sheets, vert. grooves						
Birch faced, minimum	1.000	S.F.	.032	1.64	1.88	3.52
Average	1.000	S.F.	.038	1.33	2.24	3.57
Maximum	1.000	S.F.	.046	1.22	2.69	3.91
Mahogany, African	1.000	S.F.	.040	2.73	2.36	5.09
Philippine (lauan)	1.000	S.F.	.032	.72	1.88	2.60
Oak or cherry, minimum	1.000	S.F.	.032	1.49	1.88	3.37
Maximum	1.000	S.F.	.040	2.18	2.36	4.54
Rosewood	1.000	S.F.	.050	3.48	2.95	6.43
Teak	1.000	S.F.	.040	3.55	2.36	5.91
Chestnut	1.000	S.F.	.043	5.85	2.51	8.36
Pecan	1.000	S.F.	.040	2.76	2.36	5.12
Walnut, minimum	1.000	S.F.	.032	2.83	1.88	4.71
Maximum	1.000	S.F.	.040	2.74	2.36	5.10

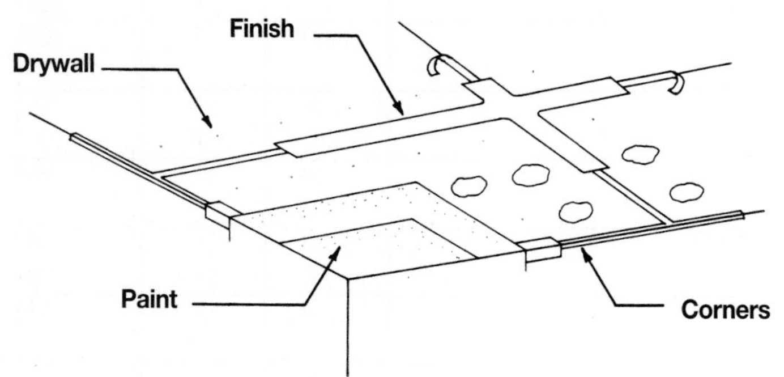

System Description	QUAN.	UNIT	LABOR HOURS	COST PER S.F. MAT.	COST PER S.F. INST.	COST PER S.F. TOTAL
1/2″ GYPSUM WALLBOARD, TAPED & FINISHED						
Gypsum wallboard, 1/2″ thick, standard	1.000	S.F.	.008	.37	.47	.84
Finish, taped & finished	1.000	S.F.	.008	.05	.47	.52
Corners, taped & finished, 12′ x 12′ room	.333	L.F.	.006	.04	.33	.37
Paint, primer & 2 coats	1.000	S.F.	.011	.22	.51	.73
TOTAL		S.F.	.033	.68	1.78	2.46
THINCOAT, SKIM COAT ON 1/2″ GYPSUM WALLBOARD						
Gypsum wallboard, 1/2″ thick, thincoat backer	1.000	S.F.	.008	.37	.47	.84
Thincoat plaster	1.000	S.F.	.011	.13	.64	.77
Corners, taped & finished, 12′ x 12′ room	.333	L.F.	.006	.04	.33	.37
Paint, primer & 2 coats	1.000	S.F.	.011	.22	.51	.73
TOTAL		S.F.	.036	.76	1.95	2.71
WATER-RESISTANT GYPSUM WALLBOARD, 1/2″ THICK, TAPED & FINISHED						
Gypsum wallboard, 1/2″ thick, water-resistant	1.000	S.F.	.008	.46	.47	.93
Finish, taped & finished	1.000	S.F.	.008	.05	.47	.52
Corners, taped & finished, 12′ x 12′ room	.333	L.F.	.006	.04	.33	.37
Paint, primer & 2 coats	1.000	S.F.	.011	.22	.51	.73
TOTAL		S.F.	.033	.77	1.78	2.55
5/8″ GYPSUM WALLBOARD, TAPED & FINISHED						
Gypsum wallboard, 5/8″ thick, standard	1.000	S.F.	.008	.39	.47	.86
Finish, taped & finished	1.000	S.F.	.008	.05	.47	.52
Corners, taped & finished, 12′ x 12′ room	.333	L.F.	.006	.04	.33	.37
Paint, primer & 2 coats	1.000	S.F.	.011	.22	.51	.73
TOTAL		S.F.	.033	.70	1.78	2.48

The costs in this system are based on a square foot of ceiling.

Description	QUAN.	UNIT	LABOR HOURS	COST PER S.F. MAT.	COST PER S.F. INST.	COST PER S.F. TOTAL

Drywall & Thincoat Ceilings Price Sheet	QUAN.	UNIT	LABOR HOURS	COST PER S.F.		
				MAT.	INST.	TOTAL
Gypsum wallboard ceilings, 1/2" thick, standard	1.000	S.F.	.008	.37	.47	.84
Fire resistant	1.000	S.F.	.008	.41	.47	.88
Water resistant	1.000	S.F.	.008	.46	.47	.93
5/8" thick, standard	1.000	S.F.	.008	.39	.47	.86
Fire resistant	1.000	S.F.	.008	.40	.47	.87
Water resistant	1.000	S.F.	.008	.55	.47	1.02
Gypsum wallboard backer for thincoat ceiling system, 1/2" thick	1.000	S.F.	.016	.78	.94	1.72
5/8" thick	1.000	S.F.	.016	.80	.94	1.74
Gypsum wallboard ceilings, taped & finished	1.000	S.F.	.008	.05	.47	.52
Texture spray	1.000	S.F.	.010	.04	.57	.61
Thincoat plaster	1.000	S.F.	.011	.13	.64	.77
Corners taped & finished, 4' x 4' room	1.000	L.F.	.015	.11	.99	1.10
6' x 6' room	.667	L.F.	.010	.07	.66	.73
10' x 10' room	.400	L.F.	.006	.04	.40	.44
12' x 12' room	.333	L.F.	.005	.04	.33	.37
16' x 16' room	.250	L.F.	.003	.02	.19	.21
Thincoat system, 4' x 4' room	1.000	L.F.	.011	.13	.64	.77
6' x 6' room	.667	L.F.	.007	.09	.43	.52
10' x 10' room	.400	L.F.	.004	.05	.26	.31
12' x 12' room	.333	L.F.	.004	.04	.21	.25
16' x 16' room	.250	L.F.	.002	.02	.12	.14
Painting, primer & 1 coat	1.000	S.F.	.008	.14	.40	.54
& 2 coats	1.000	S.F.	.011	.22	.51	.73
Wallpaper, double roll, solid pattern, avg. workmanship	1.000	S.F.	.013	.65	.61	1.26
Basic pattern, avg. workmanship	1.000	S.F.	.015	1.34	.73	2.07
Basic pattern, quality workmanship	1.000	S.F.	.018	2.35	.90	3.25
Tile, ceramic adhesive thin set, 4 1/4" x 4 1/4" tiles	1.000	S.F.	.084	2.85	4.14	6.99
6" x 6" tiles	1.000	S.F.	.080	4	4.49	8.49
Pregrouted sheets	1.000	S.F.	.067	6.25	3.28	9.53

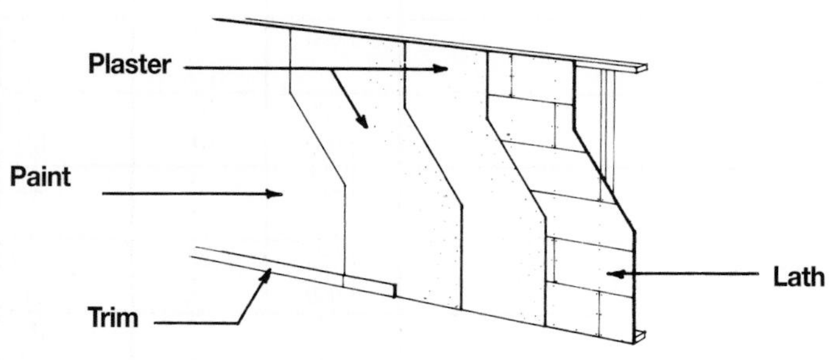

System Description	QUAN.	UNIT	LABOR HOURS	COST PER S.F. MAT.	COST PER S.F. INST.	COST PER S.F. TOTAL
PLASTER ON GYPSUM LATH						
Plaster, gypsum or perlite, 2 coats	1.000	S.F.	.053	.46	3.07	3.53
Lath, 3/8" gypsum	1.000	S.F.	.010	.39	.59	.98
Corners, expanded metal, 32 L.F. per 12' x 12' room	.083	L.F.	.002	.01	.10	.11
Painting, primer & 2 coats	1.000	S.F.	.011	.22	.51	.73
Paint trim, to 6" wide, primer + 1 coat enamel	.125	L.F.	.001	.02	.06	.08
Moldings, base, ogee profile, 9/16" x 4-1/2, red oak	.125	L.F.	.005	.60	.27	.87
TOTAL		S.F.	.082	1.70	4.60	6.30
PLASTER ON METAL LATH						
Plaster, gypsum or perlite, 2 coats	1.000	S.F.	.053	.46	3.07	3.53
Lath, 2.5 Lb. diamond, metal	1.000	S.F.	.010	.45	.59	1.04
Corners, expanded metal, 32 L.F. per 12' x 12' room	.083	L.F.	.002	.01	.10	.11
Painting, primer & 2 coats	1.000	S.F.	.011	.22	.51	.73
Paint trim, to 6" wide, primer + 1 coat enamel	.125	L.F.	.001	.02	.06	.08
Moldings, base, ogee profile, 9/16" x 4-1/2, red oak	.125	L.F.	.005	.60	.27	.87
TOTAL		S.F.	.082	1.76	4.60	6.36
STUCCO ON METAL LATH						
Stucco, 2 coats	1.000	S.F.	.041	.42	2.35	2.77
Lath, 2.5 Lb. diamond, metal	1.000	S.F.	.010	.45	.59	1.04
Corners, expanded metal, 32 L.F. per 12' x 12' room	.083	L.F.	.002	.01	.10	.11
Painting, primer & 2 coats	1.000	S.F.	.011	.22	.51	.73
Paint trim, to 6" wide, primer + 1 coat enamel	.125	L.F.	.001	.02	.06	.08
Moldings, base, ogee profile, 9/16" x 4-1/2, red oak	.125	L.F.	.005	.60	.27	.87
TOTAL		S.F.	.070	1.72	3.88	5.60

The costs in these systems are based on a per square foot of wall area.
Do not deduct for openings.

Description	QUAN.	UNIT	LABOR HOURS	COST PER S.F. MAT.	COST PER S.F. INST.	COST PER S.F. TOTAL

Plaster & Stucco Wall Price Sheet	QUAN.	UNIT	LABOR HOURS	COST PER S.F. MAT.	INST.	TOTAL
Plaster, gypsum or perlite, 2 coats	1.000	S.F.	.053	.46	3.07	3.53
3 coats	1.000	S.F.	.065	.66	3.72	4.38
Lath, gypsum, standard, 3/8″ thick	1.000	S.F.	.010	.39	.59	.98
Fire resistant, 3/8″ thick	1.000	S.F.	.013	.31	.72	1.03
1/2″ thick	1.000	S.F.	.014	.35	.77	1.12
Metal, diamond, 2.5 Lb.	1.000	S.F.	.010	.45	.59	1.04
3.4 Lb.	1.000	S.F.	.012	.49	.67	1.16
Rib, 2.75 Lb.	1.000	S.F.	.012	.40	.67	1.07
3.4 Lb.	1.000	S.F.	.013	.52	.72	1.24
Corners, expanded metal, 32 L.F. per 4′ x 4′ room	.250	L.F.	.005	.04	.30	.34
6′ x 6′ room	.110	L.F.	.002	.02	.13	.15
10′ x 10′ room	.100	L.F.	.002	.02	.12	.14
12′ x 12′ room	.083	L.F.	.002	.01	.10	.11
16′ x 16′ room	.063	L.F.	.001	.01	.07	.08
Painting, primer & 1 coats	1.000	S.F.	.008	.14	.40	.54
Primer & 2 coats	1.000	S.F.	.011	.22	.51	.73
Wallpaper, low price double roll	1.000	S.F.	.013	.65	.61	1.26
Medium price double roll	1.000	S.F.	.015	1.34	.73	2.07
High price double roll	1.000	S.F.	.018	2.35	.90	3.25
Tile, ceramic thin set, 4-1/4″ x 4-1/4″ tiles	1.000	S.F.	.084	2.85	4.14	6.99
6″ x 6″ tiles	1.000	S.F.	.080	4	4.49	8.49
Pregrouted sheets	1.000	S.F.	.067	6.25	3.28	9.53
Trim, painted or stained, baseboard	.125	L.F.	.006	.62	.33	.95
Base shoe	.125	L.F.	.005	.07	.31	.38
Chair rail	.125	L.F.	.005	.23	.28	.51
Cornice molding	.125	L.F.	.004	.16	.28	.44
Cove base, vinyl	.125	L.F.	.003	.18	.18	.36
Paneling not including furring or trim						
Plywood, prefinished, 1/4″ thick, 4′ x 8′ sheets, vert. grooves						
Birch faced, minimum	1.000	S.F.	.032	1.64	1.88	3.52
Average	1.000	S.F.	.038	1.33	2.24	3.57
Maximum	1.000	S.F.	.046	1.22	2.69	3.91
Mahogany, African	1.000	S.F.	.040	2.73	2.36	5.09
Philippine (lauan)	1.000	S.F.	.032	.72	1.88	2.60
Oak or cherry, minimum	1.000	S.F.	.032	1.49	1.88	3.37
Maximum	1.000	S.F.	.040	2.18	2.36	4.54
Rosewood	1.000	S.F.	.050	3.48	2.95	6.43
Teak	1.000	S.F.	.040	3.55	2.36	5.91
Chestnut	1.000	S.F.	.043	5.85	2.51	8.36
Pecan	1.000	S.F.	.040	2.76	2.36	5.12
Walnut, minimum	1.000	S.F.	.032	2.83	1.88	4.71
Maximum	1.000	S.F.	.040	2.74	2.36	5.10

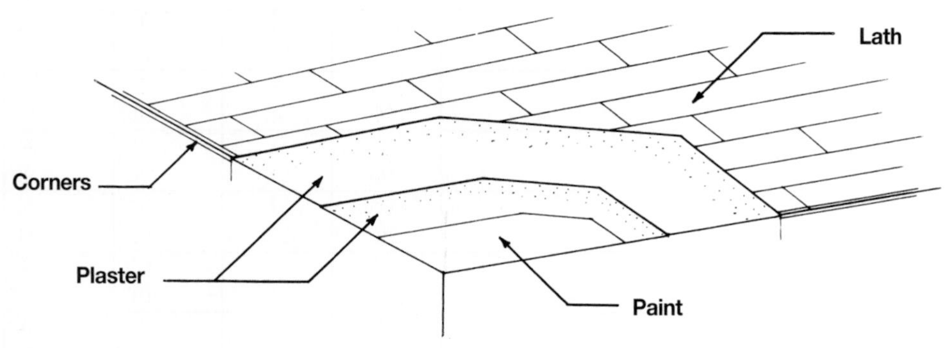

System Description	QUAN.	UNIT	LABOR HOURS	COST PER S.F.		
				MAT.	INST.	TOTAL
PLASTER ON GYPSUM LATH						
Plaster, gypsum or perlite, 2 coats	1.000	S.F.	.061	.46	3.49	3.95
Gypsum lath, plain or perforated, nailed, 3/8" thick	1.000	S.F.	.010	.39	.59	.98
Gypsum lath, ceiling installation adder	1.000	S.F.	.004		.23	.23
Corners, expanded metal, 12' x 12' room	.330	L.F.	.007	.06	.39	.45
Painting, primer & 2 coats	1.000	S.F.	.011	.22	.51	.73
TOTAL		S.F.	.093	1.13	5.21	6.34
PLASTER ON METAL LATH						
Plaster, gypsum or perlite, 2 coats	1.000	S.F.	.061	.46	3.49	3.95
Lath, 2.5 Lb. diamond, metal	1.000	S.F.	.012	.45	.67	1.12
Corners, expanded metal, 12' x 12' room	.330	L.F.	.007	.06	.39	.45
Painting, primer & 2 coats	1.000	S.F.	.011	.22	.51	.73
TOTAL		S.F.	.091	1.19	5.06	6.25
STUCCO ON GYPSUM LATH						
Stucco, 2 coats	1.000	S.F.	.041	.42	2.35	2.77
Gypsum lath, plain or perforated, nailed, 3/8" thick	1.000	S.F.	.010	.39	.59	.98
Gypsum lath, ceiling installation adder	1.000	S.F.	.004		.23	.23
Corners, expanded metal, 12' x 12' room	.330	L.F.	.007	.06	.39	.45
Painting, primer & 2 coats	1.000	S.F.	.011	.22	.51	.73
TOTAL		S.F.	.073	1.09	4.07	5.16
STUCCO ON METAL LATH						
Stucco, 2 coats	1.000	S.F.	.041	.42	2.35	2.77
Lath, 2.5 Lb. diamond, metal	1.000	S.F.	.012	.45	.67	1.12
Corners, expanded metal, 12' x 12' room	.330	L.F.	.007	.06	.39	.45
Painting, primer & 2 coats	1.000	S.F.	.011	.22	.51	.73
TOTAL		S.F.	.071	1.15	3.92	5.07

The costs in these systems are based on a square foot of ceiling area.

Description	QUAN.	UNIT	LABOR HOURS	COST PER S.F.		
				MAT.	INST.	TOTAL

Plaster & Stucco Ceiling Price Sheet	QUAN.	UNIT	LABOR HOURS	COST PER S.F.		
				MAT.	INST.	TOTAL
Plaster, gypsum or perlite, 2 coats	1.000	S.F.	.061	.46	3.49	3.95
3 coats	1.000	S.F.	.065	.66	3.72	4.38
Lath, gypsum, standard, 3/8" thick	1.000	S.F.	.014	.39	.82	1.21
Fire resistant, 3/8" thick	1.000	S.F.	.017	.31	.95	1.26
1/2" thick	1.000	S.F.	.018	.35	1	1.35
Metal, diamond, 2.5 Lb.	1.000	S.F.	.012	.45	.67	1.12
3.4 Lb.	1.000	S.F.	.015	.49	.84	1.33
Rib, 2.75 Lb.	1.000	S.F.	.012	.40	.67	1.07
3.4 Lb.	1.000	S.F.	.013	.52	.72	1.24
Corners expanded metal, 4′ x 4′ room	1.000	L.F.	.020	.17	1.18	1.35
6′ x 6′ room	.667	L.F.	.013	.11	.79	.90
10′ x 10′ room	.400	L.F.	.008	.07	.47	.54
12′ x 12′ room	.333	L.F.	.007	.06	.39	.45
16′ x 16′ room	.250	L.F.	.004	.03	.22	.25
Painting, primer & 1 coat	1.000	S.F.	.008	.14	.40	.54
Primer & 2 coats	1.000	S.F.	.011	.22	.51	.73

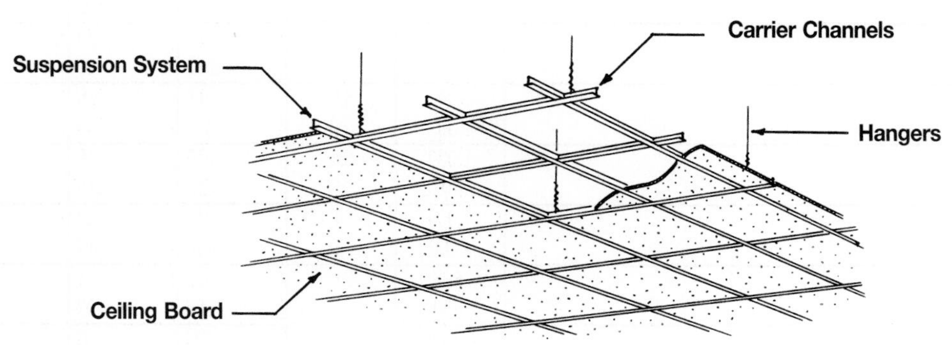

System Description	QUAN.	UNIT	LABOR HOURS	COST PER S.F.		
				MAT.	INST.	TOTAL
2′ X 2′ GRID, FILM FACED FIBERGLASS, 5/8″ THICK						
Suspension system, 2′ x 2′ grid, T bar	1.000	S.F.	.012	1.14	.73	1.87
Ceiling board, film faced fiberglass, 5/8″ thick	1.000	S.F.	.013	1.44	.75	2.19
Carrier channels, 1-1/2″ x 3/4″	1.000	S.F.	.017	.13	1	1.13
Hangers, #12 wire	1.000	S.F.	.002	.01	.10	.11
TOTAL		S.F.	.044	2.72	2.58	5.30
2′ X 4′ GRID, FILM FACED FIBERGLASS, 5/8″ THICK						
Suspension system, 2′ x 4′ grid, T bar	1.000	S.F.	.010	.88	.59	1.47
Ceiling board, film faced fiberglass, 5/8″ thick	1.000	S.F.	.013	1.44	.75	2.19
Carrier channels, 1-1/2″ x 3/4″	1.000	S.F.	.017	.13	1	1.13
Hangers, #12 wire	1.000	S.F.	.002	.01	.10	.11
TOTAL		S.F.	.042	2.46	2.44	4.90
2′ X 2′ GRID, MINERAL FIBER, REVEAL EDGE, 1″ THICK						
Suspension system, 2′ x 2′ grid, T bar	1.000	S.F.	.012	1.14	.73	1.87
Ceiling board, mineral fiber, reveal edge, 1″ thick	1.000	S.F.	.013	2.33	.79	3.12
Carrier channels, 1-1/2″ x 3/4″	1.000	S.F.	.017	.13	1	1.13
Hangers, #12 wire	1.000	S.F.	.002	.01	.10	.11
TOTAL		S.F.	.044	3.61	2.62	6.23
2′ X 4′ GRID, MINERAL FIBER, REVEAL EDGE, 1″ THICK						
Suspension system, 2′ x 4′ grid, T bar	1.000	S.F.	.010	.88	.59	1.47
Ceiling board, mineral fiber, reveal edge, 1″ thick	1.000	S.F.	.013	2.33	.79	3.12
Carrier channels, 1-1/2″ x 3/4″	1.000	S.F.	.017	.13	1	1.13
Hangers, #12 wire	1.000	S.F.	.002	.01	.10	.11
TOTAL		S.F.	.042	3.35	2.48	5.83

Description	QUAN.	UNIT	LABOR HOURS	COST PER S.F.		
				MAT.	INST.	TOTAL

Suspended Ceiling Price Sheet	QUAN.	UNIT	LABOR HOURS	COST PER S.F.		
				MAT.	INST.	TOTAL
Suspension systems, T bar, 2′ x 2′ grid	1.000	S.F.	.012	1.14	.73	1.87
2′ x 4′ grid	1.000	S.F.	.010	.88	.59	1.47
Concealed Z bar, 12″ module	1.000	S.F.	.015	1.05	.91	1.96
Ceiling boards, fiberglass, film faced, 2′ x 2′ or 2′ x 4′, 5/8″ thick	1.000	S.F.	.013	1.44	.75	2.19
3/4″ thick	1.000	S.F.	.013	3.41	.79	4.20
3″ thick thermal R11	1.000	S.F.	.018	4.27	1.05	5.32
Glass cloth faced, 3/4″ thick	1.000	S.F.	.016	3.36	.94	4.30
1″ thick	1.000	S.F.	.016	4.06	.97	5.03
1-1/2″ thick, nubby face	1.000	S.F.	.017	3	.99	3.99
Mineral fiber boards, 5/8″ thick, aluminum face 2′ x 2′	1.000	S.F.	.013	4.37	.79	5.16
2′ x 4′	1.000	S.F.	.012	4.38	.73	5.11
Standard faced, 2′ x 2′ or 2′ x 4′	1.000	S.F.	.012	1.10	.70	1.80
Plastic coated face, 2′ x 2′ or 2′ x 4′	1.000	S.F.	.020	2.89	1.18	4.07
Fire rated, 2 hour rating, 5/8″ thick	1.000	S.F.	.012	1.44	.70	2.14
Tegular edge, 2′ x 2′ or 2′ x 4′, 5/8″ thick, fine textured	1.000	S.F.	.013	1.27	1	2.27
Rough textured	1.000	S.F.	.015	1.51	1	2.51
3/4″ thick, fine textured	1.000	S.F.	.016	2.62	1.05	3.67
Rough textured	1.000	S.F.	.018	1.73	1.05	2.78
Luminous panels, prismatic, acrylic	1.000	S.F.	.020	3.33	1.18	4.51
Polystyrene	1.000	S.F.	.020	1.88	1.18	3.06
Flat or ribbed, acrylic	1.000	S.F.	.020	4.91	1.18	6.09
Polystyrene	1.000	S.F.	.020	2.62	1.18	3.80
Drop pan, white, acrylic	1.000	S.F.	.020	6.25	1.18	7.43
Polystyrene	1.000	S.F.	.020	4.99	1.18	6.17
Carrier channels, 4′-0″ on center, 3/4″ x 1-1/2″	1.000	S.F.	.017	.13	1	1.13
1-1/2″ x 3-1/2″	1.000	S.F.	.017	.33	1	1.33
Hangers, #12 wire	1.000	S.F.	.002	.01	.10	.11

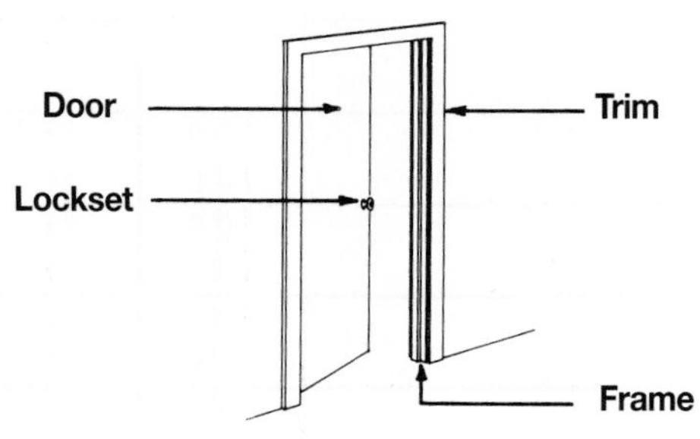

System Description	QUAN.	UNIT	LABOR HOURS	COST EACH MAT.	COST EACH INST.	COST EACH TOTAL
LAUAN, FLUSH DOOR, HOLLOW CORE						
Door, flush, lauan, hollow core, 2′-8″ wide x 6′-8″ high	1.000	Ea.	.889	83	52.50	135.50
Frame, pine, 4-5/8″ jamb	17.000	L.F.	.725	89.25	42.67	131.92
Moldings, casing, ogee, 11/16″ x 2-1/2″, pine	34.000	L.F.	1.088	53.04	63.92	116.96
Paint trim, to 6″ wide, primer + 1 coat enamel	34.000	L.F.	.340	5.10	16.66	21.76
Butt hinges, chrome, 3-1/2″ x 3-1/2″	1.500	Pr.		46.50		46.50
Lockset, passage	1.000	Ea.	.500	43	29.50	72.50
Prime door & frame, oil, primer, brushwork	2.000	Face	1.600	8.88	78	86.88
Paint door and frame, oil, 2 coats	2.000	Face	2.667	10.20	130	140.20
TOTAL		Ea.	7.809	338.97	413.25	752.22
BIRCH, FLUSH DOOR, HOLLOW CORE						
Door, flush, birch, hollow core, 2′-8″ wide x 6′-8″ high	1.000	Ea.	.889	100	52.50	152.50
Frame, pine, 4-5/8″ jamb	17.000	L.F.	.725	89.25	42.67	131.92
Moldings, casing, ogee, 11/16″ x 2-1/2″, pine	34.000	L.F.	1.088	53.04	63.92	116.96
Butt hinges, chrome, 3-1/2″ x 3-1/2″	1.500	Pr.		46.50		46.50
Lockset, passage	1.000	Ea.	.500	43	29.50	72.50
Prime door & frame, oil, primer, brushwork	2.000	Face	1.600	8.88	78	86.88
Paint door and frame, oil, 2 coats	2.000	Face	2.667	10.20	130	140.20
TOTAL		Ea.	7.469	350.87	396.59	747.46
RAISED PANEL, SOLID, PINE DOOR						
Door, pine, raised panel, 2′-8″ wide x 6′-8″ high	1.000	Ea.	.889	297	52.50	349.50
Frame, pine, 4-5/8″ jamb	17.000	L.F.	.725	89.25	42.67	131.92
Moldings, casing, ogee, 11/16″ x 2-1/2″, pine	34.000	L.F.	1.088	53.04	63.92	116.96
Butt hinges, bronze, 3-1/2″ x 3-1/2″	1.500	Pr.		54		54
Lockset, passage	1.000	Ea.	.500	43	29.50	72.50
Prime door & frame, oil, primer, brushwork	2.000	Face	1.600	8.88	78	86.88
Paint door and frame, oil, 2 coats	2.000	Face	2.667	10.20	130	140.20
TOTAL		Ea.	7.469	555.37	396.59	951.96

The costs in these systems are based on a cost per each door.

Description	QUAN.	UNIT	LABOR HOURS	COST EACH MAT.	COST EACH INST.	COST EACH TOTAL

Interior Door Price Sheet	QUAN.	UNIT	LABOR HOURS	COST EACH		
				MAT.	INST.	TOTAL
Door, hollow core, lauan 1-3/8″ thick, 6′-8″ high x 1′-6″ wide	1.000	Ea.	.889	59.50	52.50	112
2′-0″ wide	1.000	Ea.	.889	71	52.50	123.50
2′-6″ wide	1.000	Ea.	.889	80.50	52.50	133
2′-8″ wide	1.000	Ea.	.889	83	52.50	135.50
3′-0″ wide	1.000	Ea.	.941	89	55.50	144.50
Birch 1-3/8″ thick, 6′-8″ high x 1′-6″ wide	1.000	Ea.	.889	73.50	52.50	126
2′-0″ wide	1.000	Ea.	.889	81	52.50	133.50
2′-6″ wide	1.000	Ea.	.889	94.50	52.50	147
2′-8″ wide	1.000	Ea.	.889	100	52.50	152.50
3′-0″ wide	1.000	Ea.	.941	104	55.50	159.50
Louvered pine 1-3/8″ thick, 6′-8″ high x 1′-6″ wide	1.000	Ea.	.842	170	49.50	219.50
2′-0″ wide	1.000	Ea.	.889	192	52.50	244.50
2′-6″ wide	1.000	Ea.	.889	218	52.50	270.50
2′-8″ wide	1.000	Ea.	.889	240	52.50	292.50
3′-0″ wide	1.000	Ea.	.941	259	55.50	314.50
Paneled pine 1-3/8″ thick, 6′-8″ high x 1′-6″ wide	1.000	Ea.	.842	211	49.50	260.50
2′-0″ wide	1.000	Ea.	.889	257	52.50	309.50
2′-6″ wide	1.000	Ea.	.889	288	52.50	340.50
2′-8″ wide	1.000	Ea.	.889	297	52.50	349.50
3′-0″ wide	1.000	Ea.	.941	310	55.50	365.50
Frame, pine, 1′-6″ thru 2′-0″ wide door, 3-5/8″ deep	16.000	L.F.	.683	73.50	40	113.50
4-5/8″ deep	16.000	L.F.	.683	84	40	124
5-5/8″ deep	16.000	L.F.	.683	105	40	145
2′-6″ thru 3′0″ wide door, 3-5/8″ deep	17.000	L.F.	.725	78	42.50	120.50
4-5/8″ deep	17.000	L.F.	.725	89.50	42.50	132
5-5/8″ deep	17.000	L.F.	.725	111	42.50	153.50
Trim, casing, painted, both sides, 1′-6″ thru 2′-6″ wide door	32.000	L.F.	1.855	53	98.50	151.50
2′-6″ thru 3′-0″ wide door	34.000	L.F.	1.971	56.50	105	161.50
Butt hinges 3-1/2″ x 3-1/2″, steel plated, chrome	1.500	Pr.		46.50		46.50
Bronze	1.500	Pr.		54		54
Locksets, passage, minimum	1.000	Ea.	.500	43	29.50	72.50
Maximum	1.000	Ea.	.575	49.50	34	83.50
Privacy, miniumum	1.000	Ea.	.625	54	37	91
Maximum	1.000	Ea.	.675	58	40	98
Paint 2 sides, primer & 2 cts., flush door, 1′-6″ to 2′-0″ wide	2.000	Face	5.547	26.50	270	296.50
2′-6″ thru 3′-0″ wide	2.000	Face	6.933	33	340	373
Louvered door, 1′-6″ thru 2′-0″ wide	2.000	Face	6.400	23.50	310	333.50
2′-6″ thru 3′-0″ wide	2.000	Face	8.000	29.50	390	419.50
Paneled door, 1′-6″ thru 2′-0″ wide	2.000	Face	6.400	23.50	310	333.50
2′-6″ thru 3′-0″ wide	2.000	Face	8.000	29.50	390	419.50

24 Closet Door Systems

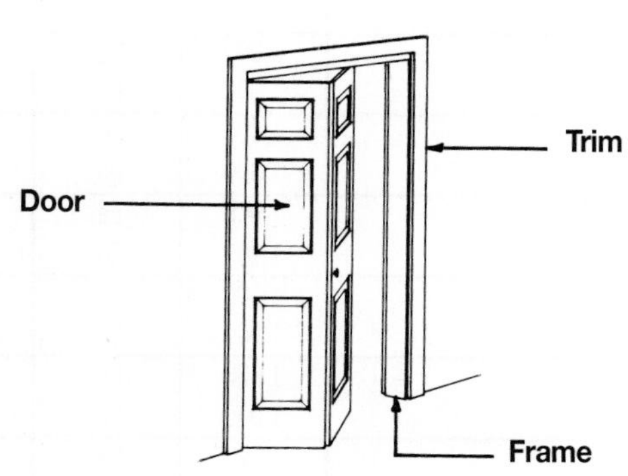

System Description	QUAN.	UNIT	LABOR HOURS	COST EACH		
				MAT.	INST.	TOTAL
BI-PASSING, FLUSH, LAUAN, HOLLOW CORE, 4′-0″ X 6′-8″						
Door, flush, lauan, hollow core, 4′-0″ x 6′-8″ opening	1.000	Ea.	1.333	191	78.50	269.50
Frame, pine, 4-5/8″ jamb	18.000	L.F.	.768	94.50	45.18	139.68
Moldings, casing, ogee, 11/16″ x 2-1/2″, pine	36.000	L.F.	1.152	56.16	67.68	123.84
Prime door & frame, oil, primer, brushwork	2.000	Face	1.600	8.88	78	86.88
Paint door and frame, oil, 2 coats	2.000	Face	2.667	10.20	130	140.20
TOTAL		Ea.	7.520	360.74	399.36	760.10
BI-PASSING, FLUSH, BIRCH, HOLLOW CORE, 6′-0″ X 6′-8″						
Door, flush, birch, hollow core, 6′-0″ x 6′-8″ opening	1.000	Ea.	1.600	375	94	469
Frame, pine, 4-5/8″ jamb	19.000	L.F.	.811	99.75	47.69	147.44
Moldings, casing, ogee, 11/16″ x 2-1/2″, pine	38.000	L.F.	1.216	59.28	71.44	130.72
Prime door & frame, oil, primer, brushwork	2.000	Face	2.000	11.10	97.50	108.60
Paint door and frame, oil, 2 coats	2.000	Face	3.333	12.75	162.50	175.25
TOTAL		Ea.	8.960	557.88	473.13	1,031.01
BI-FOLD, PINE, PANELED, 3′-0″ X 6′-8″						
Door, pine, paneled, 3′-0″ x 6′-8″ opening	1.000	Ea.	1.231	279	72.50	351.50
Frame, pine, 4-5/8″ jamb	17.000	L.F.	.725	89.25	42.67	131.92
Moldings, casing, ogee, 11/16″ x 2-1/2″, pine	34.000	L.F.	1.088	53.04	63.92	116.96
Prime door & frame, oil, primer, brushwork	2.000	Face	1.600	8.88	78	86.88
Paint door and frame, oil, 2 coats	2.000	Face	2.667	10.20	130	140.20
TOTAL		Ea.	7.311	440.37	387.09	827.46
BI-FOLD, PINE, LOUVERED, 6′-0″ X 6′-8″						
Door, pine, louvered, 6′-0″ x 6′-8″ opening	1.000	Ea.	1.600	355	94	449
Frame, pine, 4-5/8″ jamb	19.000	L.F.	.811	99.75	47.69	147.44
Moldings, casing, ogee, 11/16″ x 2-1/2″, pine	38.000	L.F.	1.216	59.28	71.44	130.72
Prime door & frame, oil, primer, brushwork	2.500	Face	2.000	11.10	97.50	108.60
Paint door and frame, oil, 2 coats	2.500	Face	3.333	12.75	162.50	175.25
TOTAL		Ea.	8.960	537.88	473.13	1,011.01

The costs in this system are based on a cost per each door.

Description	QUAN.	UNIT	LABOR HOURS	COST EACH		
				MAT.	INST.	TOTAL

Closet Door Price Sheet	QUAN.	UNIT	LABOR HOURS	COST EACH		
				MAT.	INST.	TOTAL
Doors, bi-passing, pine, louvered, 4′-0″ x 6′-8″ opening	1.000	Ea.	1.333	555	78.50	633.50
6′-0″ x 6′-8″ opening	1.000	Ea.	1.600	825	94	919
Paneled, 4′-0″ x 6′-8″ opening	1.000	Ea.	1.333	535	78.50	613.50
6′-0″ x 6′-8″ opening	1.000	Ea.	1.600	960	94	1,054
Flush, birch, hollow core, 4′-0″ x 6′-8″ opening	1.000	Ea.	1.333	293	78.50	371.50
6′-0″ x 6′-8″ opening	1.000	Ea.	1.600	375	94	469
Flush, lauan, hollow core, 4′-0″ x 6′-8″ opening	1.000	Ea.	1.333	191	78.50	269.50
6′-0″ x 6′-8″ opening	1.000	Ea.	1.600	193	94	287
Bi-fold, pine, louvered, 3′-0″ x 6′-8″ opening	1.000	Ea.	1.231	279	72.50	351.50
6′-0″ x 6′-8″ opening	1.000	Ea.	1.600	355	94	449
Paneled, 3′-0″ x 6′-8″ opening	1.000	Ea.	1.231	279	72.50	351.50
6′-0″ x 6′-8″ opening	1.000	Ea.	1.600	355	94	449
Flush, birch, hollow core, 3′-0″ x 6′-8″ opening	1.000	Ea.	1.231	85	72.50	157.50
6′-0″ x 6′-8″ opening	1.000	Ea.	1.600	147	94	241
Flush, lauan, hollow core, 3′-0″ x 6′8″ opening	1.000	Ea.	1.231	335	72.50	407.50
6′-0″ x 6′-8″ opening	1.000	Ea.	1.600	510	94	604
Frame pine, 3′-0″ door, 3-5/8″ deep	17.000	L.F.	.725	78	42.50	120.50
4-5/8″ deep	17.000	L.F.	.725	89.50	42.50	132
5-5/8″ deep	17.000	L.F.	.725	111	42.50	153.50
4′-0″ door, 3-5/8″ deep	18.000	L.F.	.768	82.50	45	127.50
4-5/8″ deep	18.000	L.F.	.768	94.50	45	139.50
5-5/8″ deep	18.000	L.F.	.768	118	45	163
6′-0″ door, 3-5/8″ deep	19.000	L.F.	.811	87	47.50	134.50
4-5/8″ deep	19.000	L.F.	.811	100	47.50	147.50
5-5/8″ deep	19.000	L.F.	.811	124	47.50	171.50
Trim both sides, painted 3′-0″ x 6′-8″ door	34.000	L.F.	1.971	56.50	105	161.50
4′-0″ x 6′-8″ door	36.000	L.F.	2.086	60	111	171
6′-0″ x 6′-8″ door	38.000	L.F.	2.203	63	117	180
Paint 2 sides, primer & 2 cts., flush door & frame, 3′ x 6′-8″ opng	2.000	Face	2.914	14.30	156	170.30
4′-0″ x 6′-8″ opening	2.000	Face	3.886	19.10	208	227.10
6′-0″ x 6′-8″ opening	2.000	Face	4.857	24	260	284
Paneled door & frame, 3′-0″ x 6′-8″ opening	2.000	Face	6.000	22	293	315
4′-0″ x 6′-8″ opening	2.000	Face	8.000	29.50	390	419.50
6′-0″ x 6′-8″ opening	2.000	Face	10.000	37	490	527
Louvered door & frame, 3′-0″ x 6′-8″ opening	2.000	Face	6.000	22	293	315
4′-0″ x 6′-8″ opening	2.000	Face	8.000	29.50	390	419.50
6′-0″ x 6′-8″ opening	2.000	Face	10.000	37	490	527

60 Carpet Systems

System Description	QUAN.	UNIT	LABOR HOURS	COST PER S.F. MAT.	COST PER S.F. INST.	COST PER S.F. TOTAL
Carpet, direct glue-down, nylon, level loop, 26 oz.	1.000	S.F.	.018	2.85	.66	3.51
32 oz.	1.000	S.F.	.018	4.83	.66	5.49
40 oz.	1.000	S.F.	.018	6	.66	6.66
Nylon, plush, 20 oz.	1.000	S.F.	.018	2.47	.66	3.13
24 oz.	1.000	S.F.	.018	2.39	.66	3.05
30 oz.	1.000	S.F.	.018	3.42	.66	4.08
42 oz.	1.000	S.F.	.022	5.60	.71	6.31
48 oz.	1.000	S.F.	.022	6.45	.71	7.16
54 oz.	1.000	S.F.	.022	7.30	.71	8.01
Olefin, 15 oz.	1.000	S.F.	.018	1.83	.66	2.49
22 oz.	1.000	S.F.	.018	1.86	.66	2.52
Tile, foam backed, needle punch	1.000	S.F.	.014	4.51	.78	5.29
Tufted loop or shag	1.000	S.F.	.014	3.66	.78	4.44
Wool, 36 oz., level loop	1.000	S.F.	.018	13.60	.71	14.31
32 oz., patterned	1.000	S.F.	.020	12.40	.71	13.11
48 oz., patterned	1.000	S.F.	.020	13.60	.71	14.31
Padding, sponge rubber cushion, minimum	1.000	S.F.	.006	.58	.33	.91
Maximum	1.000	S.F.	.006	1.07	.33	1.40
Felt, 32 oz. to 56 oz., minimum	1.000	S.F.	.006	.76	.33	1.09
Maximum	1.000	S.F.	.006	1.42	.33	1.75
Bonded urethane, 3/8" thick, minimum	1.000	S.F.	.006	.73	.33	1.06
Maximum	1.000	S.F.	.006	.99	.33	1.32
Prime urethane, 1/4" thick, minimum	1.000	S.F.	.006	.44	.33	.77
Maximum	1.000	S.F.	.006	.82	.33	1.15
Stairs, for stairs, add to above carpet prices	1.000	Riser	.267		14.85	14.85
Underlayment plywood, 3/8" thick	1.000	S.F.	.011	1.16	.63	1.79
1/2" thick	1.000	S.F.	.011	1.36	.65	2.01
5/8" thick	1.000	S.F.	.011	1.50	.67	2.17
3/4" thick	1.000	S.F.	.012	1.62	.73	2.35
Particle board, 3/8" thick	1.000	S.F.	.011	.44	.63	1.07
1/2" thick	1.000	S.F.	.011	.46	.65	1.11
5/8" thick	1.000	S.F.	.011	.59	.67	1.26
3/4" thick	1.000	S.F.	.012	.74	.73	1.47
Hardboard, 4' x 4', 0.215" thick	1.000	S.F.	.011	.76	.63	1.39

System Description	QUAN.	UNIT	LABOR HOURS	COST PER S.F.		
				MAT.	INST.	TOTAL
Resilient flooring, asphalt tile on concrete, 1/8" thick						
Color group B	1.000	S.F.	.020	1.56	1.11	2.67
Color group C & D	1.000	S.F.	.020	1.72	1.11	2.83
Asphalt tile on wood subfloor, 1/8" thick						
Color group B	1.000	S.F.	.020	1.85	1.11	2.96
Color group C & D	1.000	S.F.	.020	2.01	1.11	3.12
Vinyl composition tile, 12" x 12", 1/16" thick	1.000	S.F.	.016	1.34	.89	2.23
Embossed	1.000	S.F.	.016	2.94	.89	3.83
Marbleized	1.000	S.F.	.016	2.94	.89	3.83
Plain	1.000	S.F.	.016	3.80	.89	4.69
.080" thick, embossed	1.000	S.F.	.016	1.69	.89	2.58
Marbleized	1.000	S.F.	.016	3.38	.89	4.27
Plain	1.000	S.F.	.016	3.15	.89	4.04
1/8" thick, marbleized	1.000	S.F.	.016	2.70	.89	3.59
Plain	1.000	S.F.	.016	1.91	.89	2.80
Vinyl tile, 12" x 12", .050" thick, minimum	1.000	S.F.	.016	4.09	.89	4.98
Maximum	1.000	S.F.	.016	5.70	.89	6.59
1/8" thick, minimum	1.000	S.F.	.016	7.70	.89	8.59
Maximum	1.000	S.F.	.016	3.54	.89	4.43
1/8" thick, solid colors	1.000	S.F.	.016	6.85	.89	7.74
Florentine pattern	1.000	S.F.	.016	7.35	.89	8.24
Marbleized or travertine pattern	1.000	S.F.	.016	7.20	.89	8.09
Vinyl sheet goods, backed, .070" thick, minimum	1.000	S.F.	.032	4.92	1.78	6.70
Maximum	1.000	S.F.	.040	4.26	2.22	6.48
.093" thick, minimum	1.000	S.F.	.035	4.62	1.93	6.55
Maximum	1.000	S.F.	.040	7.30	2.22	9.52
.125" thick, minimum	1.000	S.F.	.035	4.32	1.93	6.25
Maximum	1.000	S.F.	.040	8.05	2.22	10.27
Wood, oak, finished in place, 25/32" x 2-1/2" clear	1.000	S.F.	.074	4.09	4	8.09
Select	1.000	S.F.	.074	5.05	4	9.05
No. 1 common	1.000	S.F.	.074	4.98	4	8.98
Prefinished, oak, 2-1/2" wide	1.000	S.F.	.047	6.45	2.77	9.22
3-1/4" wide	1.000	S.F.	.043	6.30	2.55	8.85
Ranch plank, oak, random width	1.000	S.F.	.055	7.40	3.25	10.65
Parquet, 5/16" thick, finished in place, oak, minimum	1.000	S.F.	.077	6.25	4.18	10.43
Maximum	1.000	S.F.	.107	11.55	5.95	17.50
Teak, minimum	1.000	S.F.	.077	7.40	4.18	11.58
Maximum	1.000	S.F.	.107	12.35	5.95	18.30
Sleepers, treated, 16" O.C., 1" x 2"	1.000	S.F.	.007	.34	.40	.74
1" x 3"	1.000	S.F.	.008	.53	.47	1
2" x 4"	1.000	S.F.	.011	.71	.63	1.34
2" x 6"	1.000	S.F.	.012	.93	.73	1.66
Subfloor, plywood, 1/2" thick	1.000	S.F.	.011	.70	.63	1.33
5/8" thick	1.000	S.F.	.012	.86	.70	1.56
3/4" thick	1.000	S.F.	.013	1.03	.75	1.78
Ceramic tile, color group 2, 1" x 1"	1.000	S.F.	.087	7.15	4.30	11.45
2" x 2" or 2" x 1"	1.000	S.F.	.084	7	4.14	11.14
Color group 1, 8" x 8"	1.000	S.F.	.064	5.80	2.62	8.42
12" x 12"	1.000	S.F.	.049	7.40	2.71	10.11
16" x 16"	1.000	S.F.	.029	8.15	2.81	10.96

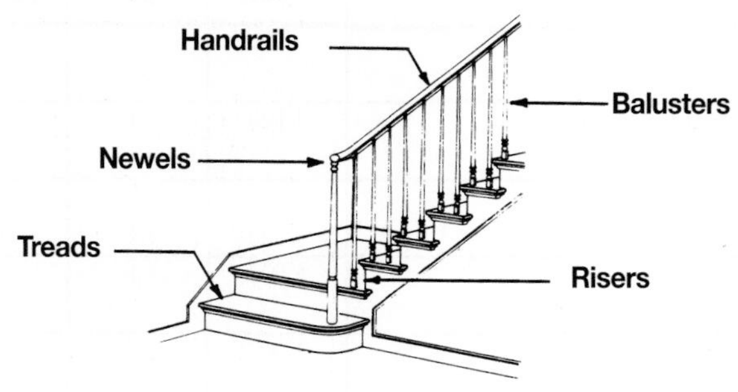

System Description	QUAN.	UNIT	LABOR HOURS	COST EACH		
				MAT.	INST.	TOTAL
7 RISERS, OAK TREADS, BOX STAIRS						
Treads, oak, 1-1/4" x 10" wide, 3' long	6.000	Ea.	2.667	642	156	798
Risers, 3/4" thick, beech	7.000	Ea.	.672	226.80	39.48	266.28
30" primed pine balusters	12.000	Ea.	1.000	48.60	58.92	107.52
Newels, 3" wide, plain, paint grade, square	2.000	Ea.	2.286	131	135	266
Handrails, oak laminated	7.000	L.F.	.933	294	54.95	348.95
Stringers, 2" x 10", 3 each	21.000	L.F.	.306	9.66	18.06	27.72
TOTAL		Ea.	7.864	1,352.06	462.41	1,814.47
14 RISERS, OAK TREADS, BOX STAIRS						
Treads, oak, 1-1/4" x 10" wide, 3' long	13.000	Ea.	5.778	1,391	338	1,729
Risers, 3/4" thick, beech	14.000	Ea.	1.344	453.60	78.96	532.56
30" primed pine balusters	26.000	Ea.	2.167	105.30	127.66	232.96
Newels, 3" wide, plain, paint grade, square	2.000	Ea.	2.286	131	135	266
Handrails, oak, laminated	14.000	L.F.	1.867	588	109.90	697.90
Stair stringers, 2" x 10"	42.000	L.F.	5.169	67.20	304.50	371.70
TOTAL		Ea.	18.611	2,736.10	1,094.02	3,830.12
14 RISERS, PINE TREADS, BOX STAIRS						
Treads, pine, 9-1/2" x 3/4" thick	13.000	Ea.	5.778	244.40	338	582.40
Risers, 3/4" thick, pine	14.000	Ea.	1.344	291.90	78.96	370.86
30" primed pine balusters	26.000	Ea.	2.167	105.30	127.66	232.96
Newels, 3" wide, plain, paint grade, square	2.000	Ea.	2.286	131	135	266
Handrails, oak, laminated	14.000	L.F.	1.867	588	109.90	697.90
Stair stringers, 2" x 10"	42.000	L.F.	5.169	67.20	304.50	371.70
TOTAL		Ea.	18.611	1,427.80	1,094.02	2,521.82

Description	QUAN.	UNIT	LABOR HOURS	COST EACH		
				MAT.	INST.	TOTAL

Stairway Price Sheet	QUAN.	UNIT	LABOR HOURS	COST EACH		
				MAT.	INST.	TOTAL
Treads, oak, 1-1/16" x 9-1/2", 3' long, 7 riser stair	6.000	Ea.	2.667	640	156	796
14 riser stair	13.000	Ea.	5.778	1,400	340	1,740
1-1/16" x 11-1/2", 3' long, 7 riser stair	6.000	Ea.	2.667	710	156	866
14 riser stair	13.000	Ea.	5.778	1,525	340	1,865
Pine, 3/4" x 9-1/2", 3' long, 7 riser stair	6.000	Ea.	2.667	113	156	269
14 riser stair	13.000	Ea.	5.778	244	340	584
3/4" x 11-1/4", 3' long, 7 riser stair	6.000	Ea.	2.667	126	156	282
14 riser stair	13.000	Ea.	5.778	273	340	613
Risers, oak, 3/4" x 7-1/2" high, 7 riser stair	7.000	Ea.	2.625	149	39.50	188.50
14 riser stair	14.000	Ea.	5.250	298	79	377
Beech, 3/4" x 7-1/2" high, 7 riser stair	7.000	Ea.	2.625	227	39.50	266.50
14 riser stair	14.000	Ea.	5.250	455	79	534
Baluster, turned, 30" high, primed pine, 7 riser stair	12.000	Ea.	3.429	48.50	59	107.50
14 riser stair	26.000	Ea.	7.428	105	128	233
30" birch, 7 riser stair	12.000	Ea.	3.429	44	53.50	97.50
14 riser stair	26.000	Ea.	7.428	95.50	116	211.50
42" pine, 7 riser stair	12.000	Ea.	3.556	64	59	123
14 riser stair	26.000	Ea.	7.704	139	128	267
42" birch, 7 riser stair	12.000	Ea.	3.556	64	59	123
14 riser stair	26.000	Ea.	7.704	139	128	267
Newels, 3-1/4" wide, starting, 7 riser stair	2.000	Ea.	2.286	131	135	266
14 riser stair	2.000	Ea.	2.286	131	135	266
Landing, 7 riser stair	2.000	Ea.	3.200	218	188	406
14 riser stair	2.000	Ea.	3.200	218	188	406
Handrails, oak, laminated, 7 riser stair	7.000	L.F.	.933	294	55	349
14 riser stair	14.000	L.F.	1.867	590	110	700
Stringers, fir, 2" x 10" 7 riser stair	21.000	L.F.	2.585	33.50	152	185.50
14 riser stair	42.000	L.F.	5.169	67	305	372
2" x 12", 7 riser stair	21.000	L.F.	2.585	42	152	194
14 riser stair	42.000	L.F.	5.169	84.50	305	389.50

Special Stairways	QUAN.	UNIT	LABOR HOURS	COST EACH		
				MAT.	INST.	TOTAL
Basement stairs, open risers	1.000	Flight	4.000	910	235	1,145
Spiral stairs, oak, 4'-6" diameter, prefabricated, 9' high	1.000	Flight	10.667	2,925	630	3,555
Aluminum, 5'-0" diameter stock unit	1.000	Flight	9.956	11,300	730	12,030
Custom unit	1.000	Flight	9.956	20,300	730	21,030
Cast iron, 4'-0" diameter, minimum	1.000	Flight	9.956	11,300	730	12,030
Maximum	1.000	Flight	17.920	19,600	1,300	20,900
Steel, industrial, pre-erected, 3'-6" wide, bar rail	1.000	Flight	7.724	8,200	825	9,025
Picket rail	1.000	Flight	7.724	9,175	825	10,000

Division 7 - Specialties

System Description	QUAN.	UNIT	LABOR HOURS	COST PER L.F.		
				MAT.	INST.	TOTAL
KITCHEN, ECONOMY GRADE						
Top cabinets, economy grade	1.000	L.F.	.171	74.24	10.08	84.32
Bottom cabinets, economy grade	1.000	L.F.	.256	111.36	15.12	126.48
Square edge, plastic face countertop	1.000	L.F.	.267	38	15.70	53.70
Blocking, wood, 2″ x 4″	1.000	L.F.	.032	.47	1.88	2.35
Soffit, framing, wood, 2″ x 4″	4.000	L.F.	.071	1.88	4.20	6.08
Soffit drywall	2.000	S.F.	.047	.88	2.80	3.68
Drywall painting	2.000	S.F.	.013	.12	.78	.90
TOTAL		L.F.	.857	226.95	50.56	277.51
AVERAGE GRADE						
Top cabinets, average grade	1.000	L.F.	.213	92.80	12.60	105.40
Bottom cabinets, average grade	1.000	L.F.	.320	139.20	18.90	158.10
Solid surface countertop, solid color	1.000	L.F.	.800	79.50	47	126.50
Blocking, wood, 2″ x 4″	1.000	L.F.	.032	.47	1.88	2.35
Soffit framing, wood, 2″ x 4″	4.000	L.F.	.071	1.88	4.20	6.08
Soffit drywall	2.000	S.F.	.047	.88	2.80	3.68
Drywall painting	2.000	S.F.	.013	.12	.78	.90
TOTAL		L.F.	1.496	314.85	88.16	403.01
CUSTOM GRADE						
Top cabinets, custom grade	1.000	L.F.	.256	196	15	211
Bottom cabinets, custom grade	1.000	L.F.	.384	294	22.50	316.50
Solid surface countertop, premium patterned color	1.000	L.F.	1.067	147	63	210
Blocking, wood, 2″ x 4″	1.000	L.F.	.032	.47	1.88	2.35
Soffit framing, wood, 2″ x 4″	4.000	L.F.	.071	1.88	4.20	6.08
Soffit drywall	2.000	S.F.	.047	.88	2.80	3.68
Drywall painting	2.000	S.F.	.013	.12	.78	.90
TOTAL		L.F.	1.870	640.35	110.16	750.51

Description	QUAN.	UNIT	LABOR HOURS	COST PER L.F.		
				MAT.	INST.	TOTAL

Kitchen Price Sheet	QUAN.	UNIT	LABOR HOURS	COST PER L.F.		
				MAT.	INST.	TOTAL
Top cabinets, economy grade	1.000	L.F.	.171	74	10.10	84.10
Average grade	1.000	L.F.	.213	93	12.60	105.60
Custom grade	1.000	L.F.	.256	196	15	211
Bottom cabinets, economy grade	1.000	L.F.	.256	111	15.10	126.10
Average grade	1.000	L.F.	.320	139	18.90	157.90
Custom grade	1.000	L.F.	.384	294	22.50	316.50
Counter top, laminated plastic, 7/8″ thick, no splash	1.000	L.F.	.267	39.50	15.70	55.20
With backsplash	1.000	L.F.	.267	50.50	15.70	66.20
1-1/4″ thick, no splash	1.000	L.F.	.286	42.50	16.85	59.35
With backsplash	1.000	L.F.	.286	48.50	16.85	65.35
Post formed, laminated plastic	1.000	L.F.	.267	13.90	15.70	29.60
Ceramic tile, with backsplash	1.000	L.F.	.427	23	9.40	32.40
Marble, with backsplash, minimum	1.000	L.F.	.471	57	27.50	84.50
Maximum	1.000	L.F.	.615	143	36	179
Maple, solid laminated, no backsplash	1.000	L.F.	.286	101	16.85	117.85
With backsplash	1.000	L.F.	.286	120	16.85	136.85
Solid Surface, with backsplash, minimum	1.000	L.F.	.842	93	49.50	142.50
Maximum	1.000	L.F.	1.067	147	63	210
Blocking, wood, 2″ x 4″	1.000	L.F.	.032	.47	1.88	2.35
2″ x 6″	1.000	L.F.	.036	.79	2.12	2.91
2″ x 8″	1.000	L.F.	.040	1.07	2.36	3.43
Soffit framing, wood, 2″ x 3″	4.000	L.F.	.064	1.84	3.76	5.60
2″ x 4″	4.000	L.F.	.071	1.88	4.20	6.08
Soffit, drywall, painted	2.000	S.F.	.060	1	3.58	4.58
Paneling, standard	2.000	S.F.	.064	3.28	3.76	7.04
Deluxe	2.000	S.F.	.091	2.44	5.40	7.84
Sinks, porcelain on cast iron, single bowl, 21″ x 24″	1.000	Ea.	10.334	655	615	1,270
21″ x 30″	1.000	Ea.	10.334	1,200	615	1,815
Double bowl, 20″ x 32″	1.000	Ea.	10.810	735	645	1,380
Stainless steel, single bowl, 16″ x 20″	1.000	Ea.	10.334	990	615	1,605
22″ x 25″	1.000	Ea.	10.334	1,075	615	1,690
Double bowl, 20″ x 32″	1.000	Ea.	10.810	915	645	1,560

Appliance Price Sheet	QUAN.	UNIT	LABOR HOURS	COST PER L.F.		
				MAT.	INST.	TOTAL
Range, free standing, minimum	1.000	Ea.	3.600	600	208	808
Maximum	1.000	Ea.	6.000	2,650	315	2,965
Built-in, minimum	1.000	Ea.	3.333	1,025	225	1,250
Maximum	1.000	Ea.	10.000	2,000	605	2,605
Counter top range, 4-burner, minimum	1.000	Ea.	3.333	435	225	660
Maximum	1.000	Ea.	4.667	2,125	315	2,440
Compactor, built-in, minimum	1.000	Ea.	2.215	800	136	936
Maximum	1.000	Ea.	3.282	1,300	199	1,499
Dishwasher, built-in, minimum	1.000	Ea.	6.735	590	450	1,040
Maximum	1.000	Ea.	9.235	755	615	1,370
Garbage disposer, minimum	1.000	Ea.	2.810	237	187	424
Maximum	1.000	Ea.	2.810	345	187	532
Microwave oven, minimum	1.000	Ea.	2.615	126	177	303
Maximum	1.000	Ea.	4.615	520	310	830
Range hood, ducted, minimum	1.000	Ea.	4.658	137	286	423
Maximum	1.000	Ea.	5.991	1,075	365	1,440
Ductless, minimum	1.000	Ea.	2.615	144	163	307
Maximum	1.000	Ea.	3.948	1,100	244	1,344
Refrigerator, 16 cu.ft., minimum	1.000	Ea.	2.000	620	90.50	710.50
Maximum	1.000	Ea.	3.200	990	145	1,135
16 cu.ft. with icemaker, minimum	1.000	Ea.	4.210	860	229	1,089
Maximum	1.000	Ea.	5.410	1,225	283	1,508
19 cu.ft., minimum	1.000	Ea.	2.667	600	121	721
Maximum	1.000	Ea.	4.667	1,050	212	1,262
19 cu.ft. with icemaker, minimum	1.000	Ea.	5.143	870	271	1,141
Maximum	1.000	Ea.	7.143	1,325	360	1,685
Sinks, porcelain on cast iron single bowl, 21″ x 24″	1.000	Ea.	10.334	655	615	1,270
21″ x 30″	1.000	Ea.	10.334	1,200	615	1,815
Double bowl, 20″ x 32″	1.000	Ea.	10.810	735	645	1,380
Stainless steel, single bowl 16″ x 20″	1.000	Ea.	10.334	990	615	1,605
22″ x 25″	1.000	Ea.	10.334	1,075	615	1,690
Double bowl, 20″ x 32″	1.000	Ea.	10.810	915	645	1,560
Water heater, electric, 30 gallon	1.000	Ea.	3.636	1,100	241	1,341
40 gallon	1.000	Ea.	4.000	1,175	265	1,440
Gas, 30 gallon	1.000	Ea.	4.000	1,950	265	2,215
75 gallon	1.000	Ea.	5.333	3,050	355	3,405
Wall, packaged terminal heater/air conditioner cabinet, wall sleeve, louver, electric heat, thermostat, manual changeover, 208V						
6000 BTUH cooling, 8800 BTU heating	1.000	Ea.	2.667	775	164	939
9000 BTUH cooling, 13,900 BTU heating	1.000	Ea.	3.200	1,075	196	1,271
12,000 BTUH cooling, 13,900 BTU heating	1.000	Ea.	4.000	1,625	245	1,870
15,000 BTUH cooling, 13,900 BTU heating	1.000	Ea.	5.333	1,600	325	1,925

System Description	QUAN.	UNIT	LABOR HOURS	COST EACH MAT.	COST EACH INST.	COST EACH TOTAL
Curtain rods, stainless, 1″ diameter, 3′ long	1.000	Ea.	.615	30	36.50	66.50
5′ long	1.000	Ea.	.615	30	36.50	66.50
Grab bar, 1″ diameter, 12″ long	1.000	Ea.	.283	28	16.70	44.70
36″ long	1.000	Ea.	.340	32	20	52
1-1/4″ diameter, 12″ long	1.000	Ea.	.333	33	19.65	52.65
36″ long	1.000	Ea.	.400	37.50	23.50	61
1-1/2″ diameter, 12″ long	1.000	Ea.	.383	38	22.50	60.50
36″ long	1.000	Ea.	.460	43	27	70
Mirror, 18″ x 24″	1.000	Ea.	.400	51.50	23.50	75
72″ x 24″	1.000	Ea.	1.333	320	78.50	398.50
Medicine chest with mirror, 18″ x 24″	1.000	Ea.	.400	204	23.50	227.50
36″ x 24″	1.000	Ea.	.600	305	35.50	340.50
Toilet tissue dispenser, surface mounted, minimum	1.000	Ea.	.267	19.45	15.70	35.15
Maximum	1.000	Ea.	.400	29	23.50	52.50
Flush mounted, minimum	1.000	Ea.	.293	21.50	17.25	38.75
Maximum	1.000	Ea.	.427	31	25	56
Towel bar, 18″ long, minimum	1.000	Ea.	.278	37	16.40	53.40
Maximum	1.000	Ea.	.348	46.50	20.50	67
24″ long, minimum	1.000	Ea.	.313	42	18.45	60.45
Maximum	1.000	Ea.	.383	51	22.50	73.50
36″ long, minimum	1.000	Ea.	.381	56	22.50	78.50
Maximum	1.000	Ea.	.419	61.50	25	86.50

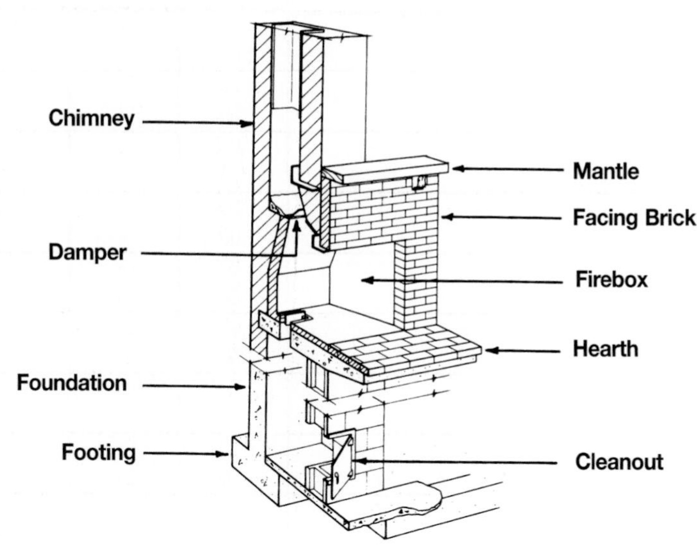

System Description	QUAN.	UNIT	LABOR HOURS	COST EACH		
				MAT.	INST.	TOTAL
MASONRY FIREPLACE						
Footing, 8″ thick, concrete, 4′ x 7′	.700	C.Y.	2.800	148.40	157.53	305.93
Foundation, concrete block, 32″ x 60″ x 4′ deep	1.000	Ea.	5.275	199.80	289.80	489.60
Fireplace, brick firebox, 30″ x 29″ opening	1.000	Ea.	40.000	645	2,150	2,795
Damper, cast iron, 30″ opening	1.000	Ea.	1.333	137	78.50	215.50
Facing brick, standard size brick, 6′ x 5′	30.000	S.F.	5.217	142.50	286.50	429
Hearth, standard size brick, 3′ x 6′	1.000	Ea.	8.000	232	430	662
Chimney, standard size brick, 8″ x 12″ flue, one story house	12.000	V.L.F.	12.000	528	648	1,176
Mantle, 4″ x 8″, wood	6.000	L.F.	1.333	57.30	78.60	135.90
Cleanout, cast iron, 8″ x 8″	1.000	Ea.	.667	60.50	39	99.50
TOTAL		Ea.	76.625	2,150.50	4,157.93	6,308.43

The costs in this system are on a cost each basis.

Description	QUAN.	UNIT	LABOR HOURS	COST EACH		
				MAT.	INST.	TOTAL

Masonry Fireplace Price Sheet

Masonry Fireplace Price Sheet	QUAN.	UNIT	LABOR HOURS	COST EACH MAT.	COST EACH INST.	COST EACH TOTAL
Footing 8" thick, 3' x 6'	.440	C.Y.	1.326	93.50	99	192.50
4' x 7'	.700	C.Y.	2.110	148	158	306
5' x 8'	1.000	C.Y.	3.014	212	225	437
1' thick, 3' x 6'	.670	C.Y.	2.020	142	151	293
4' x 7'	1.030	C.Y.	3.105	218	232	450
5' x 8'	1.480	C.Y.	4.461	315	330	645
Foundation-concrete block, 24" x 48", 4' deep	1.000	Ea.	4.267	160	232	392
8' deep	1.000	Ea.	8.533	320	465	785
24" x 60", 4' deep	1.000	Ea.	4.978	186	270	456
8' deep	1.000	Ea.	9.956	375	540	915
32" x 48", 4' deep	1.000	Ea.	4.711	176	256	432
8' deep	1.000	Ea.	9.422	355	510	865
32" x 60", 4' deep	1.000	Ea.	5.333	200	290	490
8' deep	1.000	Ea.	10.845	405	590	995
32" x 72", 4' deep	1.000	Ea.	6.133	230	335	565
8' deep	1.000	Ea.	12.267	460	665	1,125
Fireplace, brick firebox 30" x 29" opening	1.000	Ea.	40.000	645	2,150	2,795
48" x 30" opening	1.000	Ea.	60.000	970	3,225	4,195
Steel fire box with registers, 25" opening	1.000	Ea.	26.667	1,200	1,450	2,650
48" opening	1.000	Ea.	44.000	2,050	2,375	4,425
Damper, cast iron, 30" opening	1.000	Ea.	1.333	137	78.50	215.50
36" opening	1.000	Ea.	1.556	160	91.50	251.50
Steel, 30" opening	1.000	Ea.	1.333	131	78.50	209.50
36" opening	1.000	Ea.	1.556	153	91.50	244.50
Facing for fireplace, standard size brick, 6' x 5'	30.000	S.F.	5.217	143	287	430
7' x 5'	35.000	S.F.	6.087	166	335	501
8' x 6'	48.000	S.F.	8.348	228	460	688
Fieldstone, 6' x 5'	30.000	S.F.	5.217	445	287	732
7' x 5'	35.000	S.F.	6.087	515	335	850
8' x 6'	48.000	S.F.	8.348	710	460	1,170
Sheetrock on metal, studs, 6' x 5'	30.000	S.F.	.980	26	57.50	83.50
7' x 5'	35.000	S.F.	1.143	30.50	67	97.50
8' x 6'	48.000	S.F.	1.568	42	92	134
Hearth, standard size brick, 3' x 6'	1.000	Ea.	8.000	232	430	662
3' x 7'	1.000	Ea.	9.280	269	500	769
3' x 8'	1.000	Ea.	10.640	310	570	880
Stone, 3' x 6'	1.000	Ea.	8.000	242	430	672
3' x 7'	1.000	Ea.	9.280	280	500	780
3' x 8'	1.000	Ea.	10.640	320	570	890
Chimney, standard size brick , 8" x 12" flue, one story house	12.000	V.L.F.	12.000	530	650	1,180
Two story house	20.000	V.L.F.	20.000	880	1,075	1,955
Mantle wood, beams, 4" x 8"	6.000	L.F.	1.333	57.50	78.50	136
4" x 10"	6.000	L.F.	1.371	72.50	80.50	153
Ornate, prefabricated, 6' x 3'-6" opening, minimum	1.000	Ea.	1.600	510	94	604
Maximum	1.000	Ea.	1.600	700	94	794
Cleanout, door and frame, cast iron, 8" x 8"	1.000	Ea.	.667	60.50	39	99.50
12" x 12"	1.000	Ea.	.800	98	47	145

30 Prefabricated Fireplace Systems

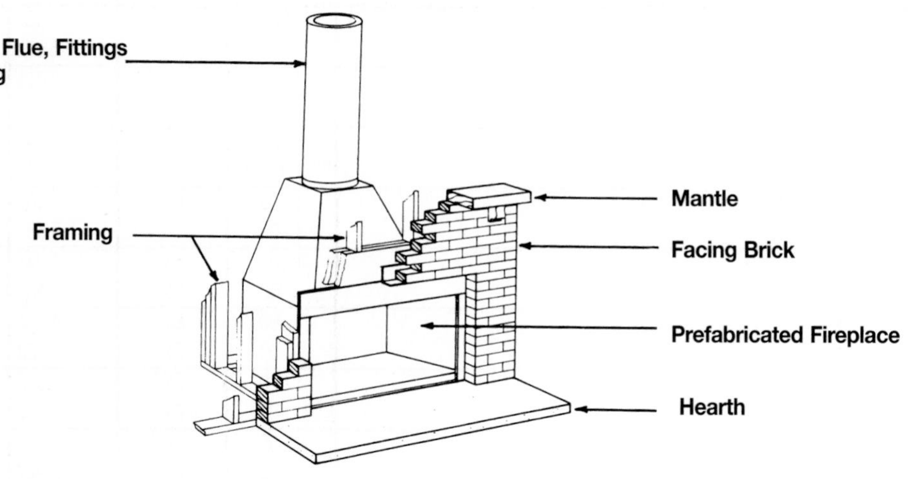

System Description	QUAN.	UNIT	LABOR HOURS	COST EACH		
				MAT.	INST.	TOTAL
PREFABRICATED FIREPLACE						
Prefabricated fireplace, metal, painted	1.000	Ea.	6.154	1,775	360	2,135
Framing, 2″ x 4″ studs, 6′ x 5′	35.000	L.F.	.509	16.10	30.10	46.20
Fire resistant gypsum drywall, unfinished	40.000	S.F.	.320	16.40	18.80	35.20
Drywall finishing adder	40.000	S.F.	.320	2	18.80	20.80
Facing, brick, standard size brick, 6′ x 5′	30.000	S.F.	5.217	142.50	286.50	429
Hearth, standard size brick, 3′ x 6′	1.000	Ea.	8.000	232	430	662
Chimney, one story house, framing, 2″ x 4″ studs	80.000	L.F.	1.164	36.80	68.80	105.60
Sheathing, plywood, 5/8″ thick	32.000	S.F.	.758	92.16	44.80	136.96
Flue, 10″ metal, insulated pipe	12.000	V.L.F.	4.000	588	232.80	820.80
Fittings, ceiling support	1.000	Ea.	.667	160	39	199
Fittings, joist shield	1.000	Ea.	.727	790	42.50	832.50
Fittings, roof flashing	1.000	Ea.	.667	450	39	489
Mantle beam, wood, 4″ x 8″	6.000	L.F.	1.333	57.30	78.60	135.90
TOTAL		Ea.	29.836	4,358.26	1,689.70	6,047.96

The costs in this system are on a cost each basis.

Description	QUAN.	UNIT	LABOR HOURS	COST EACH		
				MAT.	INST.	TOTAL

Prefabricated Fireplace Price Sheet

Prefabricated Fireplace Price Sheet	QUAN.	UNIT	LABOR HOURS	COST EACH		
				MAT.	INST.	TOTAL
Prefabricated fireplace, minimum	1.000	Ea.	6.154	1,775	360	2,135
Average	1.000	Ea.	8.000	2,000	470	2,470
Maximum	1.000	Ea.	8.889	3,550	525	4,075
Framing, 2″ x 4″ studs, fireplace, 6′ x 5′	35.000	L.F.	.509	16.10	30	46.10
7′ x 5′	40.000	L.F.	.582	18.40	34.50	52.90
8′ x 6′	45.000	L.F.	.655	20.50	38.50	59
Sheetrock, 1/2″ thick, fireplace, 6′ x 5′	40.000	S.F.	.640	18.40	37.50	55.90
7′ x 5′	45.000	S.F.	.720	20.50	42.50	63
8′ x 6′	50.000	S.F.	.800	23	47	70
Facing for fireplace, brick, 6′ x 5′	30.000	S.F.	5.217	143	287	430
7′ x 5′	35.000	S.F.	6.087	166	335	501
8′ x 6′	48.000	S.F.	8.348	228	460	688
Fieldstone, 6′ x 5′	30.000	S.F.	5.217	430	287	717
7′ x 5′	35.000	S.F.	6.087	505	335	840
8′ x 6′	48.000	S.F.	8.348	690	460	1,150
Hearth, standard size brick, 3′ x 6′	1.000	Ea.	8.000	232	430	662
3′ x 7′	1.000	Ea.	9.280	269	500	769
3′ x 8′	1.000	Ea.	10.640	310	570	880
Stone, 3′ x 6′	1.000	Ea.	8.000	242	430	672
3′ x 7′	1.000	Ea.	9.280	280	500	780
3′ x 8′	1.000	Ea.	10.640	320	570	890
Chimney, framing, 2″ x 4″, one story house	80.000	L.F.	1.164	37	69	106
Two story house	120.000	L.F.	1.746	55	103	158
Sheathing, plywood, 5/8″ thick	32.000	S.F.	.758	92	45	137
Stucco on plywood	32.000	S.F.	1.125	60	65.50	125.50
Flue, 10″ metal pipe, insulated, one story house	12.000	V.L.F.	4.000	590	233	823
Two story house	20.000	V.L.F.	6.667	980	390	1,370
Fittings, ceiling support	1.000	Ea.	.667	160	39	199
Fittings joist sheild, one story house	1.000	Ea.	.667	790	42.50	832.50
Two story house	2.000	Ea.	1.333	1,575	85	1,660
Fittings roof flashing	1.000	Ea.	.667	450	39	489
Mantle, wood beam, 4″ x 8″	6.000	L.F.	1.333	57.50	78.50	136
4″ x 10″	6.000	L.F.	1.371	72.50	80.50	153
Ornate prefabricated, 6′ x 3′-6″ opening, minimum	1.000	Ea.	1.600	510	94	604
Maximum	1.000	Ea.	1.600	700	94	794

System Description	QUAN.	UNIT	LABOR HOURS	COST EACH MAT.	COST EACH INST.	COST EACH TOTAL
Economy, lean to, shell only, not including 2′ stub wall, fndtn, flrs, heat						
4′ x 16′	1.000	Ea.	26.212	2,700	1,525	4,225
4′ x 24′	1.000	Ea.	30.259	3,125	1,775	4,900
6′ x 10′	1.000	Ea.	16.552	1,900	975	2,875
6′ x 16′	1.000	Ea.	23.034	2,625	1,350	3,975
6′ x 24′	1.000	Ea.	29.793	3,400	1,750	5,150
8′ x 10′	1.000	Ea.	22.069	2,525	1,300	3,825
8′ x 16′	1.000	Ea.	38.400	4,375	2,250	6,625
8′ x 24′	1.000	Ea.	49.655	5,675	2,925	8,600
Free standing, 8′ x 8′	1.000	Ea.	17.356	1,675	1,025	2,700
8′ x 16′	1.000	Ea.	30.211	2,900	1,775	4,675
8′ x 24′	1.000	Ea.	39.051	3,750	2,300	6,050
10′ x 10′	1.000	Ea.	18.824	4,850	1,100	5,950
10′ x 16′	1.000	Ea.	24.095	6,200	1,425	7,625
10′ x 24′	1.000	Ea.	31.624	8,150	1,875	10,025
14′ x 10′	1.000	Ea.	20.741	7,000	1,225	8,225
14′ x 16′	1.000	Ea.	24.889	8,400	1,475	9,875
14′ x 24′	1.000	Ea.	33.349	11,300	1,975	13,275
Standard,lean to,shell only,not incl.2′stub wall, fndtn,flrs, heat 4′x10′	1.000	Ea.	28.235	2,900	1,650	4,550
4′ x 16′	1.000	Ea.	39.341	4,050	2,300	6,350
4′ x 24′	1.000	Ea.	45.412	4,675	2,650	7,325
6′ x 10′	1.000	Ea.	24.827	2,825	1,475	4,300
6′ x 16′	1.000	Ea.	34.538	3,950	2,025	5,975
6′ x 24′	1.000	Ea.	44.689	5,100	2,625	7,725
8′ x 10′	1.000	Ea.	33.103	3,775	1,950	5,725
8′ x 16′	1.000	Ea.	57.600	6,575	3,400	9,975
8′ x 24′	1.000	Ea.	74.482	8,500	4,400	12,900
Free standing, 8′ x 8′	1.000	Ea.	26.034	2,500	1,525	4,025
8′ x 16′	1.000	Ea.	45.316	4,350	2,675	7,025
8′ x 24′	1.000	Ea.	58.577	5,625	3,450	9,075
10′ x 10′	1.000	Ea.	28.236	7,275	1,675	8,950
10′ x 16′	1.000	Ea.	36.142	9,300	2,125	11,425
10′ x 24′	1.000	Ea.	47.436	12,200	2,800	15,000
14′ x 10′	1.000	Ea.	31.112	10,500	1,850	12,350
14′ x 16′	1.000	Ea.	37.334	12,600	2,200	14,800
14′ x 24′	1.000	Ea.	50.030	16,900	2,950	19,850
Deluxe,lean to,shell only,not incl.2′stub wall, fndtn, flrs or heat, 4′x10′	1.000	Ea.	20.645	4,600	1,225	5,825
4′ x 16′	1.000	Ea.	33.032	7,350	1,950	9,300
4′ x 24′	1.000	Ea.	49.548	11,000	2,925	13,925
6′ x 10′	1.000	Ea.	30.968	6,900	1,825	8,725
6′ x 16′	1.000	Ea.	49.548	11,000	2,925	13,925
6′ x 24′	1.000	Ea.	74.323	16,600	4,400	21,000
8′ x 10′	1.000	Ea.	41.290	9,200	2,450	11,650
8′ x 16′	1.000	Ea.	66.065	14,700	3,900	18,600
8′ x 24′	1.000	Ea.	99.097	22,100	5,850	27,950
Freestanding, 8′ x 8′	1.000	Ea.	18.618	5,850	1,100	6,950
8′ x 16′	1.000	Ea.	37.236	11,700	2,200	13,900
8′ x 24′	1.000	Ea.	55.855	17,600	3,300	20,900
10′ x 10′	1.000	Ea.	29.091	9,150	1,725	10,875
10′ x 16′	1.000	Ea.	46.546	14,600	2,750	17,350
10′ x 24′	1.000	Ea.	69.818	22,000	4,125	26,125
14′ x 10′	1.000	Ea.	40.727	12,800	2,400	15,200
14′ x 16′	1.000	Ea.	65.164	20,500	3,850	24,350
14′ x 24′	1.000	Ea.	97.746	30,700	5,750	36,450

System Description	QUAN.	UNIT	LABOR HOURS	COST EACH MAT.	COST EACH INST.	COST EACH TOTAL
Swimming pools, vinyl lined, metal sides, sand bottom, 12′ x 28′	1.000	Ea.	50.177	8,075	3,125	11,200
12′ x 32′	1.000	Ea.	55.366	8,900	3,450	12,350
12′ x 36′	1.000	Ea.	60.061	9,650	3,725	13,375
16′ x 32′	1.000	Ea.	66.798	10,700	4,150	14,850
16′ x 36′	1.000	Ea.	71.190	11,400	4,425	15,825
16′ x 40′	1.000	Ea.	74.703	12,000	4,625	16,625
20′ x 36′	1.000	Ea.	77.860	12,500	4,850	17,350
20′ x 40′	1.000	Ea.	82.135	13,200	5,100	18,300
20′ x 44′	1.000	Ea.	90.348	14,500	5,625	20,125
24′ x 40′	1.000	Ea.	98.562	15,800	6,125	21,925
24′ x 44′	1.000	Ea.	108.418	17,400	6,725	24,125
24′ x 48′	1.000	Ea.	118.274	19,000	7,350	26,350
Vinyl lined, concrete sides, 12′ x 28′	1.000	Ea.	79.447	12,800	4,925	17,725
12′ x 32′	1.000	Ea.	88.818	14,300	5,525	19,825
12′ x 36′	1.000	Ea.	97.656	15,700	6,075	21,775
16′ x 32′	1.000	Ea.	111.393	17,900	6,925	24,825
16′ x 36′	1.000	Ea.	121.354	19,500	7,525	27,025
16′ x 40′	1.000	Ea.	130.445	21,000	8,100	29,100
28′ x 36′	1.000	Ea.	140.585	22,600	8,725	31,325
20′ x 40′	1.000	Ea.	149.336	24,000	9,275	33,275
20′ x 44′	1.000	Ea.	164.270	26,400	10,200	36,600
24′ x 40′	1.000	Ea.	179.203	28,800	11,100	39,900
24′ x 44′	1.000	Ea.	197.124	31,700	12,300	44,000
24′ x 48′	1.000	Ea.	215.044	34,600	13,300	47,900
Gunite, bottom and sides, 12′ x 28′	1.000	Ea.	129.767	18,600	8,050	26,650
12′ x 32′	1.000	Ea.	142.164	20,400	8,825	29,225
12′ x 36′	1.000	Ea.	153.028	22,000	9,500	31,500
16′ x 32′	1.000	Ea.	167.743	24,100	10,400	34,500
16′ x 36′	1.000	Ea.	176.421	25,400	11,000	36,400
16′ x 40′	1.000	Ea.	182.368	26,200	11,300	37,500
20′ x 36′	1.000	Ea.	187.949	27,000	11,700	38,700
20′ x 40′	1.000	Ea.	179.200	35,600	11,200	46,800
20′ x 44′	1.000	Ea.	197.120	39,200	12,300	51,500
24′ x 40′	1.000	Ea.	215.040	42,700	13,400	56,100
24′ x 44′	1.000	Ea.	273.244	39,300	17,000	56,300
24′ x 48′	1.000	Ea.	298.077	42,800	18,500	61,300

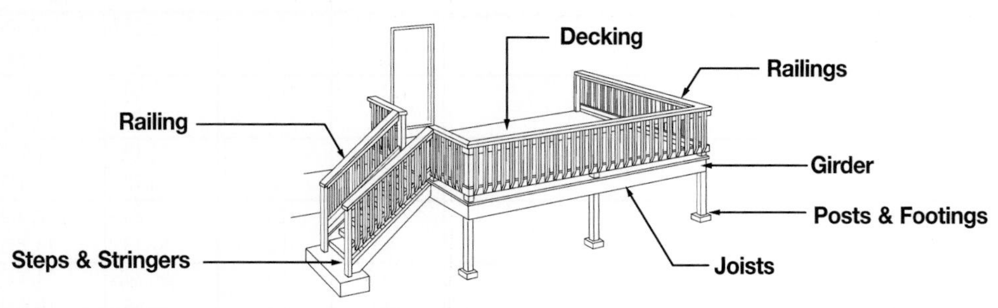

System Description	QUAN.	UNIT	LABOR HOURS	COST PER S.F.		
				MAT.	INST.	TOTAL
8′ X 12′ DECK, PRESSURE TREATED LUMBER, JOISTS 16″ O.C.						
4″ x 4″ Posts	.500	L.F.	.021	.69	1.21	1.90
Ledger, bolted 4′ O.C.	.125	L.F.	.005	.21	.31	.52
Joists, 2″ x 10″, 16″ O.C.	1.080	L.F.	.019	1.72	1.13	2.85
5/4″ x 6″ decking	1.000	S.F.	.025	2.38	1.47	3.85
2″ x 12″ stair stringers	.375	L.F.	.046	.72	2.72	3.44
5/4″ x 6″ stair treads	.500	L.F.	.050	.56	2.95	3.51
4′ x 4″ handrail posts	.210	L.F.	.009	.29	.51	.80
2″ x 6″ handrail cap	.520	L.F.	.014	.44	.82	1.26
2″ x 4″ baluster support	1.040	L.F.	.028	.70	1.63	2.33
2″ x 2″ balusters	3.850	L.F.	.093	1.77	5.51	7.28
Post footings	.042	Ea.	.134	1.45	7.01	8.46
Concrete step	.010	Ea.	.016	.80	.94	1.74
Joist hangers	.210	Ea.	.010	.34	.60	.94
TOTAL		S.F.	.470	12.07	26.81	38.88
12′ X 16′ DECK, PRESSURE TREATED LUMBER, JOISTS 24″ O.C.						
4″ x 4″ Posts	.312	L.F.	.013	.43	.76	1.19
Ledger, bolted 4′ O.C.	.083	L.F.	.003	.12	.20	.32
Joists, 2″ x 8″, 16″ O.C.	.810	L.F.	.014	1.29	.85	2.14
5/4″ x 6″ decking	1.000	S.F.	.025	2.38	1.47	3.85
2″ x 12″ stair stringers	.188	L.F.	.023	.36	1.36	1.72
5/4″ x 6″ stair treads	.250	L.F.	.025	1.84	1.48	3.32
4″ x 4″ handrail posts	.125	L.F.	.005	.17	.30	.47
2″ x 6″ handrail cap	.323	L.F.	.009	.27	.51	.78
2″ x 4″ baluster support	.650	L.F.	.017	.44	1.02	1.46
2″ x 2″ balusters	2.440	L.F.	.059	1.12	3.49	4.61
Post footings	.026	Ea.	.083	.90	4.34	5.24
Concrete step	.005	Ea.	.008	.40	.46	.86
Joist hangers	.135	Ea.	.007	.22	.39	.61
Deck support girder	.083	L.F.	.002	.26	.14	.40
TOTAL		S.F.	.293	10.20	16.77	26.97

The costs in this system are on a square foot basis.

System Description	QUAN.	UNIT	LABOR HOURS	COST PER S.F.		
				MAT.	INST.	TOTAL
12′ X 24′ DECK, REDWOOD OR CEDAR, JOISTS 16″ O.C.						
4″ x 4″ redwood posts	.291	L.F.	.012	2.10	.70	2.80
Redwood joists, 2″ x 8″, 16″ O.C.	.001	M.B.F.	.010	5.67	.59	6.26
5/4″ x 6″ redwood decking	1.000	S.F.	.025	5.35	1.47	6.82
Stair stringers	.125	L.F.	.015	.24	.91	1.15
Stringer trim	.083	L.F.	.003	.45	.20	.65
Stair treads	.135	S.F.	.003	.72	.20	.92
Handrail supports	.083	L.F.	.003	.60	.20	.80
Handrail cap	.152	L.F.	.004	1.28	.24	1.52
Baluster supports	.416	L.F.	.011	3.49	.65	4.14
Balusters	1.875	L.F.	.050	2.46	2.94	5.40
Post footings	.024	Ea.	.077	.83	4.01	4.84
Concrete step	.003	Ea.	.005	.24	.28	.52
Joist hangers	.132	Ea.	.006	.21	.38	.59
Deck support girder		M.B.F.	.003	1.55	.16	1.71
TOTAL		S.F.	.227	25.19	12.93	38.12

The costs in this system are on a square foot basis.

Wood Deck Price Sheet	QUAN.	UNIT	LABOR HOURS	COST PER S.F.		
				MAT.	INST.	TOTAL
Decking, treated lumber, 1″ x 4″	3.430	L.F.	.031	11	5.85	16.85
2″ x 4″	3.430	L.F.	.033	7.70	5.40	13.10
2″ x 6″	2.200	L.F.	.041	4	3.23	7.23
5/4″ x 6″	2.200	L.F.	.027	5.25	3.23	8.48
Redwood or cedar, 1″ x 4″	3.430	L.F.	.035	10.70	5.85	16.55
2″ x 4″	3.430	L.F.	.036	22	5.40	27.40
2″ x 6″	2.200	L.F.	.028	25.50	3.23	28.73
5/4″ x 6″	2.200	L.F.	.027	16.15	3.23	19.38
Joists for deck, treated lumber, 2″ x 8″, 16″ O.C.	1.000	L.F.	.015	1.28	.86	2.14
24″ O.C.	.750	L.F.	.012	.96	.65	1.61
2″ x 10″, 16″ O.C.	1.000	L.F.	.018	1.59	1.05	2.64
24″ O.C.	.750	L.F.	.014	1.19	.79	1.98
Redwood or cedar, 2″ x 8″, 16″ O.C.	1.000	L.F.	.015	5.10	.86	5.96
24″ O.C.	.750	L.F.	.012	3.83	.65	4.48
2″ x 10″, 16″ O.C.	1.000	L.F.	.018	8.60	1.05	9.65
24″ O.C.	.750	L.F.	.014	6.45	.79	7.24
Girder for joists, treated lumber, 2″ x 10″, 8′ x 12′ deck	.250	L.F.	.002	.40	.21	.61
12′ x 16′ deck	.167	L.F.	.001	.26	.14	.40
12′ x 24′ deck	.167	L.F.	.001	.26	.14	.40
Redwood or cedar, 2″ x 10″, 8′ x 12′ deck	.250	L.F.	.002	2.15	.26	2.41
12′ x 16′ deck	.167	L.F.	.001	1.44	.18	1.62
12′ x 24′ deck	.167	L.F.	.001	1.44	.18	1.62
Posts, 4″ x 4″, including concrete footing, 8′ x 12′ deck	.500	L.F.	.022	2.10	8.05	10.15
12′ x 16′ deck	.312	L.F.	.017	1.33	5.10	6.43
12′ x 24′ deck	.291	L.F.	.017	1.23	4.71	5.94
Stairs 2″ x 10″ stringers, treated lumber, 8′ x 12′ deck	1.000	Set	.020	1.60	4.70	6.30
12′ x 16′ deck	1.000	Set	.012	.80	2.35	3.15
12′ x 24′ deck	1.000	Set	.008	.56	1.65	2.21
Redwood or cedar, 8′ x 12′ deck	1.000	Set	.040	4.05	4.70	8.75
12′ x 16′ deck	1.000	Set	.020	2.03	2.35	4.38
12′ x 24′ deck	1.000	Set	.012	1.42	1.65	3.07
Railings 2″ x 4″, treated lumber, 8′ x 12′ deck	.520	L.F.	.026	2.91	7.95	10.86
12′ x 16′ deck	.323	L.F.	.017	1.83	5	6.83
12′ x 24′ deck	.210	L.F.	.014	1.32	3.67	4.99
Redwood or cedar, 8′ x 12′ deck	1.000	L.F.	.009	10.75	8.50	19.25

Wood Deck Price Sheet	QUAN.	UNIT	LABOR HOURS	COST PER S.F. MAT.	INST.	TOTAL
12′ x 16′ deck	.670	L.F.	.006	6.70	5.35	12.05
12′ x 24′ deck	.540	L.F.	.005	4.77	3.94	8.71
Alternative decking, wood/plastic composite, 5/4″ x 6″	2.200	L.F.	.055	8.30	3.23	11.53
1″ x 4″ square edge fir	3.430	L.F.	.100	11.05	5.85	16.90
1″ x 4″ tongue and groove fir	3.430	L.F.	.122	5.80	7.15	12.95
1″ x 4″ mahogany	3.430	L.F.	.100	7.90	5.85	13.75
5/4″ x 6″ PVC	2.200	L.F.	.064	8.10	3.76	11.86

Division 8 - Mechanical

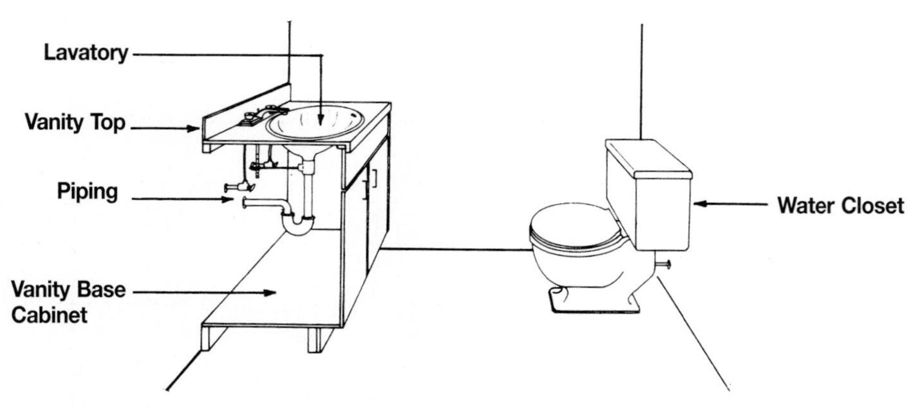

System Description	QUAN.	UNIT	LABOR HOURS	COST EACH		
				MAT.	INST.	TOTAL
LAVATORY INSTALLED WITH VANITY, PLUMBING IN 2 WALLS						
Water closet, floor mounted, 2 piece, close coupled, white	1.000	Ea.	3.019	223	180	403
Rough-in, vent, 2" diameter DWV piping	1.000	Ea.	.955	76	57	133
Waste, 4" diameter DWV piping	1.000	Ea.	.828	111	49.35	160.35
Supply, 1/2" diameter type "L" copper supply piping	1.000	Ea.	.593	22.32	39.30	61.62
Lavatory, 20" x 18", P.E. cast iron white	1.000	Ea.	2.500	325	149	474
Rough-in, vent, 1-1/2" diameter DWV piping	1.000	Ea.	.901	71.40	53.80	125.20
Waste, 2" diameter DWV piping	1.000	Ea.	.955	76	57	133
Supply, 1/2" diameter type "L" copper supply piping	1.000	Ea.	.988	37.20	65.50	102.70
Piping, supply, 1/2" diameter type "L" copper supply piping	10.000	L.F.	.988	37.20	65.50	102.70
Waste, 4" diameter DWV piping	7.000	L.F.	1.931	259	115.15	374.15
Vent, 2" diameter DWV piping	12.000	L.F.	2.866	228	171	399
Vanity base cabinet, 2 door, 30" wide	1.000	Ea.	1.000	485	59	544
Vanity top, plastic & laminated, square edge	2.670	L.F.	.712	122.82	41.92	164.74
TOTAL		Ea.	18.236	2,073.94	1,103.52	3,177.46
LAVATORY WITH WALL-HUNG LAVATORY, PLUMBING IN 2 WALLS						
Water closet, floor mounted, 2 piece close coupled, white	1.000	Ea.	3.019	223	180	403
Rough-in, vent, 2" diameter DWV piping	1.000	Ea.	.955	76	57	133
Waste, 4" diameter DWV piping	1.000	Ea.	.828	111	49.35	160.35
Supply, 1/2" diameter type "L" copper supply piping	1.000	Ea.	.593	22.32	39.30	61.62
Lavatory, 20" x 18", P.E. cast iron, wall hung, white	1.000	Ea.	2.000	264	119	383
Rough-in, vent, 1-1/2" diameter DWV piping	1.000	Ea.	.901	71.40	53.80	125.20
Waste, 2" diameter DWV piping	1.000	Ea.	.955	76	57	133
Supply, 1/2" diameter type "L" copper supply piping	1.000	Ea.	.988	37.20	65.50	102.70
Piping, supply, 1/2" diameter type "L" copper supply piping	10.000	L.F.	.988	37.20	65.50	102.70
Waste, 4" diameter DWV piping	7.000	L.F.	1.931	259	115.15	374.15
Vent, 2" diameter DWV piping	12.000	L.F.	2.866	228	171	399
Carrier, steel for studs, no arms	1.000	Ea.	1.143	66	76	142
TOTAL		Ea.	17.167	1,471.12	1,048.60	2,519.72

Description	QUAN.	UNIT	LABOR HOURS	COST EACH		
				MAT.	INST.	TOTAL

Two Fixture Lavatory Price Sheet	QUAN.	UNIT	LABOR HOURS	COST EACH		
				MAT.	INST.	TOTAL
Water closet, close coupled standard 2 piece, white	1.000	Ea.	3.019	223	180	403
Color	1.000	Ea.	3.019	465	180	645
One piece elongated bowl, white	1.000	Ea.	3.019	850	180	1,030
Color	1.000	Ea.	3.019	1,075	180	1,255
Low profile, one piece elongated bowl, white	1.000	Ea.	3.019	900	180	1,080
Color	1.000	Ea.	3.019	1,075	180	1,255
Rough-in for water closet						
1/2" copper supply, 4" cast iron waste, 2" cast iron vent	1.000	Ea.	2.376	209	146	355
4" PVC waste, 2" PVC vent	1.000	Ea.	2.678	96	164	260
4" copper waste , 2" copper vent	1.000	Ea.	2.520	287	160	447
3" cast iron waste, 1-1/2" cast iron vent	1.000	Ea.	2.244	158	138	296
3" PVC waste, 1-1/2" PVC vent	1.000	Ea.	2.388	86	152	238
3" copper waste, 1-1/2" copper vent	1.000	Ea.	2.524	208	154	362
1/2" PVC supply, 4" PVC waste, 2" PVC vent	1.000	Ea.	2.974	116	183	299
3" PVC waste, 1-1/2" PVC vent	1.000	Ea.	2.684	106	172	278
1/2" steel supply, 4" cast iron waste, 2" cast iron vent	1.000	Ea.	2.545	213	157	370
4" cast iron waste, 2" steel vent	1.000	Ea.	2.590	192	159	351
4" PVC waste, 2" PVC vent	1.000	Ea.	2.847	100	175	275
Lavatory, vanity top mounted, P.E. on cast iron 20" x 18" white	1.000	Ea.	2.500	325	149	474
Color	1.000	Ea.	2.500	530	149	679
Steel, enameled 10" x 17" white	1.000	Ea.	2.759	133	165	298
Color	1.000	Ea.	2.500	130	149	279
Vitreous china 20" x 16", white	1.000	Ea.	2.963	226	177	403
Color	1.000	Ea.	2.963	226	177	403
Wall hung, P.E. on cast iron, 20" x 18", white	1.000	Ea.	2.000	264	119	383
Color	1.000	Ea.	2.000	277	119	396
Vitreous china 19" x 17", white	1.000	Ea.	2.286	128	136	264
Color	1.000	Ea.	2.286	150	136	286
Rough-in supply waste and vent for lavatory						
1/2" copper supply, 2" cast iron waste, 1-1/2" cast iron vent	1.000	Ea.	2.844	185	176	361
2" PVC waste, 1-1/2" PVC vent	1.000	Ea.	2.962	111	189	300
2" copper waste, 1-1/2" copper vent	1.000	Ea.	2.308	172	153	325
1-1/2" PVC waste, 1-1/4" PVC vent	1.000	Ea.	2.639	111	175	286
1-1/2" copper waste, 1-1/4" copper vent	1.000	Ea.	2.114	145	140	285
1/2" PVC supply, 2" PVC waste, 1-1/2" PVC vent	1.000	Ea.	3.456	145	222	367
1-1/2" PVC waste, 1-1/4" PVC vent	1.000	Ea.	3.133	145	208	353
1/2" steel supply, 2" cast iron waste, 1-1/2" cast iron vent	1.000	Ea.	3.126	191	195	386
2" cast iron waste, 2" steel vent	1.000	Ea.	3.225	175	201	376
2" PVC waste, 1-1/2" PVC vent	1.000	Ea.	3.244	118	208	326
1-1/2" PVC waste, 1-1/4" PVC vent	1.000	Ea.	2.921	118	194	312
Piping, supply, 1/2" copper, type "L"	10.000	L.F.	.988	37	65.50	102.50
1/2" steel	10.000	L.F.	1.270	44	84	128
1/2" PVC	10.000	L.F.	1.482	71	98	169
Waste, 4" cast iron	7.000	L.F.	1.931	259	115	374
4" copper	7.000	L.F.	2.800	425	168	593
4" PVC	7.000	L.F.	2.333	80.50	139	219.50
Vent, 2" cast iron	12.000	L.F.	2.866	228	171	399
2" copper	12.000	L.F.	2.182	246	145	391
2" PVC	12.000	L.F.	3.254	117	194	311
2" steel	12.000	Ea.	3.000	166	179	345
Vanity base cabinet, 2 door, 24" x 30"	1.000	Ea.	1.000	485	59	544
24" x 36"	1.000	Ea.	1.200	440	70.50	510.50
Vanity top, laminated plastic, square edge 25" x 32"	2.670	L.F.	.712	123	42	165
25" x 38"	3.170	L.F.	.845	146	50	196
Post formed, laminated plastic, 25" x 32"	2.670	L.F.	.712	37	42	79
25" x 38"	3.170	L.F.	.845	44	50	94
Cultured marble, 25" x 32" with bowl	1.000	Ea.	2.500	177	149	326
25" x 38" with bowl	1.000	Ea.	2.500	206	149	355
Carrier for lavatory, steel for studs	1.000	Ea.	1.143	66	76	142
Wood 2" x 8" blocking	1.330	L.F.	.053	1.42	3.14	4.56

12 Three Fixture Bathroom Systems

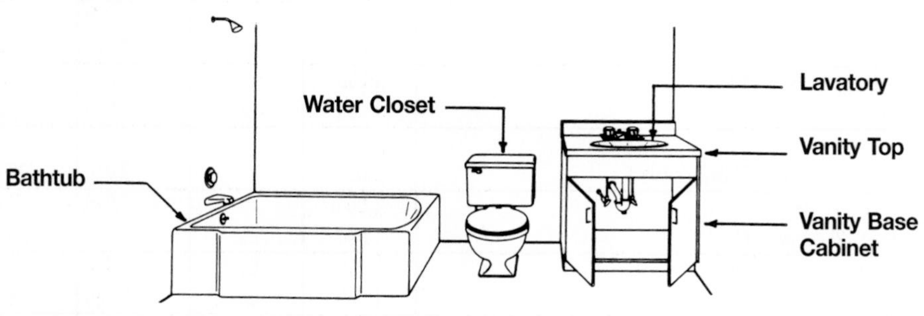

System Description	QUAN.	UNIT	LABOR HOURS	COST EACH MAT.	COST EACH INST.	COST EACH TOTAL
BATHROOM INSTALLED WITH VANITY						
Water closet, floor mounted, 2 piece, close coupled, white	1.000	Ea.	3.019	223	180	403
Rough-in, waste, 4" diameter DWV piping	1.000	Ea.	.828	111	49.35	160.35
Vent, 2" diameter DWV piping	1.000	Ea.	.955	76	57	133
Supply, 1/2" diameter type "L" copper supply piping	1.000	Ea.	.593	22.32	39.30	61.62
Lavatory, 20" x 18", P.E. cast iron with accessories, white	1.000	Ea.	2.500	325	149	474
Rough-in, supply, 1/2" diameter type "L" copper supply piping	1.000	Ea.	.988	37.20	65.50	102.70
Waste, 1-1/2" diameter DWV piping	1.000	Ea.	1.803	142.80	107.60	250.40
Bathtub, P.E. cast iron, 5' long with accessories, white	1.000	Ea.	3.636	1,325	217	1,542
Rough-in, waste, 4" diameter DWV piping	1.000	Ea.	.828	111	49.35	160.35
Vent, 1-1/2" diameter DWV piping	1.000	Ea.	.593	52.80	39.20	92
Supply, 1/2" diameter type "L" copper supply piping	1.000	Ea.	.988	37.20	65.50	102.70
Piping, supply, 1/2" diameter type "L" copper supply piping	20.000	L.F.	1.975	74.40	131	205.40
Waste, 4" diameter DWV piping	9.000	L.F.	2.483	333	148.05	481.05
Vent, 2" diameter DWV piping	6.000	L.F.	1.500	82.80	89.40	172.20
Vanity base cabinet, 2 door, 30" wide	1.000	Ea.	1.000	485	59	544
Vanity top, plastic laminated square edge	2.670	L.F.	.712	101.46	41.92	143.38
TOTAL		Ea.	24.401	3,539.98	1,488.17	5,028.15
BATHROOM WITH WALL HUNG LAVATORY						
Water closet, floor mounted, 2 piece, close coupled, white	1.000	Ea.	3.019	223	180	403
Rough-in, vent, 2" diameter DWV piping	1.000	Ea.	.955	76	57	133
Waste, 4" diameter DWV piping	1.000	Ea.	.828	111	49.35	160.35
Supply, 1/2" diameter type "L" copper supply piping	1.000	Ea.	.593	22.32	39.30	61.62
Lavatory, 20" x 18" P.E. cast iron, wall hung, white	1.000	Ea.	2.000	264	119	383
Rough-in, waste, 1-1/2" diameter DWV piping	1.000	Ea.	1.803	142.80	107.60	250.40
Supply, 1/2" diameter type "L" copper supply piping	1.000	Ea.	.988	37.20	65.50	102.70
Bathtub, P.E. cast iron, 5' long with accessories, white	1.000	Ea.	3.636	1,325	217	1,542
Rough-in, waste, 4" diameter DWV piping	1.000	Ea.	.828	111	49.35	160.35
Supply, 1/2" diameter type "L" copper supply piping	1.000	Ea.	.988	37.20	65.50	102.70
Vent, 1-1/2" diameter DWV piping	1.000	Ea.	1.482	132	98	230
Piping, supply, 1/2" diameter type "L" copper supply piping	20.000	L.F.	1.975	74.40	131	205.40
Waste, 4" diameter DWV piping	9.000	L.F.	2.483	333	148.05	481.05
Vent, 2" diameter DWV piping	6.000	L.F.	1.500	82.80	89.40	172.20
Carrier, steel, for studs, no arms	1.000	Ea.	1.143	66	76	142
TOTAL		Ea.	24.221	3,037.72	1,492.05	4,529.77

The costs in this system are a cost each basis, all necessary piping is included.

Three Fixture Bathroom Price Sheet	QUAN.	UNIT	LABOR HOURS	COST EACH		
				MAT.	INST.	TOTAL
Water closet, close coupled standard 2 piece, white	1.000	Ea.	3.019	223	180	403
Color	1.000	Ea.	3.019	465	180	645
One piece, elongated bowl, white	1.000	Ea.	3.019	850	180	1,030
Color	1.000	Ea.	3.019	1,075	180	1,255
Low profile, one piece elongated bowl, white	1.000	Ea.	3.019	900	180	1,080
Color	1.000	Ea.	3.019	1,075	180	1,255
Rough-in, for water closet						
1/2" copper supply, 4" cast iron waste, 2" cast iron vent	1.000	Ea.	2.376	209	146	355
4" PVC/DWV waste, 2" PVC vent	1.000	Ea.	2.678	96	164	260
4" copper waste, 2" copper vent	1.000	Ea.	2.520	287	160	447
3" cast iron waste, 1-1/2" cast iron vent	1.000	Ea.	2.244	158	138	296
3" PVC waste, 1-1/2" PVC vent	1.000	Ea.	2.388	86	152	238
3" copper waste, 1-1/2" copper vent	1.000	Ea.	2.014	155	128	283
1/2" PVC supply, 4" PVC waste, 2" PVC vent	1.000	Ea.	2.974	116	183	299
3" PVC waste, 1-1/2" PVC supply	1.000	Ea.	2.684	106	172	278
1/2" steel supply, 4" cast iron waste, 2" cast iron vent	1.000	Ea.	2.545	213	157	370
4" cast iron waste, 2" steel vent	1.000	Ea.	2.590	192	159	351
4" PVC waste, 2" PVC vent	1.000	Ea.	2.847	100	175	275
Lavatory, wall hung, P.E. cast iron 20" x 18", white	1.000	Ea.	2.000	264	119	383
Color	1.000	Ea.	2.000	277	119	396
Vitreous china 19" x 17", white	1.000	Ea.	2.286	128	136	264
Color	1.000	Ea.	2.286	150	136	286
Lavatory, for vanity top, P.E. cast iron 20" x 18"", white	1.000	Ea.	2.500	325	149	474
Color	1.000	Ea.	2.500	530	149	679
Steel, enameled 20" x 17", white	1.000	Ea.	2.759	133	165	298
Color	1.000	Ea.	2.500	130	149	279
Vitreous china 20" x 16", white	1.000	Ea.	2.963	226	177	403
Color	1.000	Ea.	2.963	226	177	403
Rough-in, for lavatory						
1/2" copper supply, 1-1/2" C.I. waste, 1-1/2" C.I. vent	1.000	Ea.	2.791	180	173	353
1-1/2" PVC waste, 1-1/4" PVC vent	1.000	Ea.	2.639	111	175	286
1/2" steel supply, 1-1/4" cast iron waste, 1-1/4" steel vent	1.000	Ea.	2.890	141	181	322
1-1/4" PVC waste, 1-1/4" PVC vent	1.000	Ea.	2.794	121	185	306
1/2" PVC supply, 1-1/2" PVC waste, 1-1/2" PVC vent	1.000	Ea.	3.260	141	216	357
Bathtub, P.E. cast iron, 5′ long corner with fittings, white	1.000	Ea.	3.636	1,325	217	1,542
Color	1.000	Ea.	3.636	1,650	217	1,867
Rough-in, for bathtub						
1/2" copper supply, 4" cast iron waste, 1-1/2" copper vent	1.000	Ea.	2.409	201	154	355
4" PVC waste, 1-1/2" PVC vent	1.000	Ea.	2.877	107	184	291
1/2" steel supply, 4" cast iron waste, 1-1/2" steel vent	1.000	Ea.	2.898	183	181	364
4" PVC waste, 1-1/2" PVC vent	1.000	Ea.	3.159	114	203	317
1/2" PVC supply, 4" PVC waste, 1-1/2" PVC vent	1.000	Ea.	3.371	141	217	358
Piping, supply 1/2" copper	20.000	L.F.	1.975	74.50	131	205.50
1/2" steel	20.000	L.F.	2.540	87.50	168	255.50
1/2" PVC	20.000	L.F.	2.963	142	196	338
Piping, waste, 4" cast iron no hub	9.000	L.F.	2.483	335	148	483
4" PVC/DWV	9.000	L.F.	3.000	104	179	283
4" copper/DWV	9.000	L.F.	3.600	550	216	766
Piping, vent 2" cast iron no hub	6.000	L.F.	1.433	114	85.50	199.50
2" copper/DWV	6.000	L.F.	1.091	123	72.50	195.50
2" PVC/DWV	6.000	L.F.	1.627	58.50	97	155.50
2" steel, galvanized	6.000	L.F.	1.500	83	89.50	172.50
Vanity base cabinet, 2 door, 24" x 30"	1.000	Ea.	1.000	485	59	544
24" x 36"	1.000	Ea.	1.200	440	70.50	510.50
Vanity top, laminated plastic square edge 25" x 32"	2.670	L.F.	.712	101	42	143
25" x 38"	3.160	L.F.	.843	120	49.50	169.50
Cultured marble, 25" x 32", with bowl	1.000	Ea.	2.500	177	149	326
25" x 38", with bowl	1.000	Ea.	2.500	206	149	355
Carrier, for lavatory, steel for studs, no arms	1.000	Ea.	1.143	66	76	142
Wood, 2" x 8" blocking	1.300	L.F.	.052	1.39	3.07	4.46

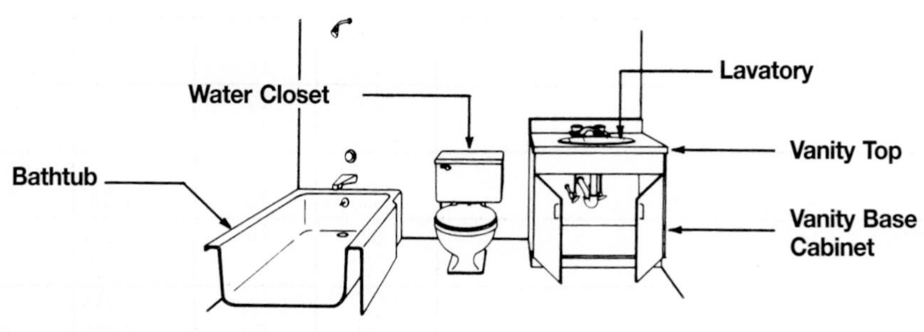

System Description	QUAN.	UNIT	LABOR HOURS	COST EACH		
				MAT.	INST.	TOTAL
BATHROOM WITH LAVATORY INSTALLED IN VANITY						
Water closet, floor mounted, 2 piece, close coupled, white	1.000	Ea.	3.019	223	180	403
Rough-in, waste, 4" diameter DWV piping	1.000	Ea.	.828	111	49.35	160.35
Vent, 2" diameter DWV piping	1.000	Ea.	.955	76	57	133
Supply, 1/2" diameter type "L" copper supply piping	1.000	Ea.	.593	22.32	39.30	61.62
Lavatory, 20" x 18", P.E. cast iron with accessories, white	1.000	Ea.	2.500	325	149	474
Rough-in, waste, 1-1/2" diameter DWV piping	1.000	Ea.	1.803	142.80	107.60	250.40
Supply, 1/2" diameter type "L" copper supply piping	1.000	Ea.	.988	37.20	65.50	102.70
Bathtub, P.E. cast iron 5′ long with accessories, white	1.000	Ea.	3.636	1,325	217	1,542
Rough-in, waste, 4" diameter DWV piping	1.000	Ea.	.828	111	49.35	160.35
Vent, 1-1/2" diameter DWV piping	1.000	Ea.	.593	52.80	39.20	92
Supply, 1/2" diameter type "L" copper supply piping	1.000	Ea.	.988	37.20	65.50	102.70
Piping, supply, 1/2" diameter type "L" copper supply piping	10.000	L.F.	.988	37.20	65.50	102.70
Waste, 4" diameter DWV piping	6.000	L.F.	1.655	222	98.70	320.70
Vent, 2" diameter DWV piping	6.000	L.F.	1.500	82.80	89.40	172.20
Vanity base cabinet, 2 door, 30" wide	1.000	Ea.	1.000	485	59	544
Vanity top, plastic laminated square edge	2.670	L.F.	.712	101.46	41.92	143.38
TOTAL		Ea.	22.586	3,391.78	1,373.32	4,765.10
BATHROOM WITH WALL HUNG LAVATORY						
Water closet, floor mounted, 2 piece, close coupled, white	1.000	Ea.	3.019	223	180	403
Rough-in, vent, 2" diameter DWV piping	1.000	Ea.	.955	76	57	133
Waste, 4" diameter DWV piping	1.000	Ea.	.828	111	49.35	160.35
Supply, 1/2" diameter type "L" copper supply piping	1.000	Ea.	.593	22.32	39.30	61.62
Lavatory, 20" x 18" P.E. cast iron, wall hung, white	1.000	Ea.	2.000	264	119	383
Rough-in, waste, 1-1/2" diameter DWV piping	1.000	Ea.	1.803	142.80	107.60	250.40
Supply, 1/2" diameter type "L" copper supply piping	1.000	Ea.	.988	37.20	65.50	102.70
Bathtub, P.E. cast iron, 5′ long with accessories, white	1.000	Ea.	3.636	1,325	217	1,542
Rough-in, waste, 4" diameter DWV piping	1.000	Ea.	.828	111	49.35	160.35
Supply, 1/2" diameter type "L" copper supply piping	1.000	Ea.	.988	37.20	65.50	102.70
Vent, 1-1/2" diameter DWV piping	1.000	Ea.	.593	52.80	39.20	92
Piping, supply, 1/2" diameter type "L" copper supply piping	10.000	L.F.	.988	37.20	65.50	102.70
Waste, 4" diameter DWV piping	6.000	L.F.	1.655	222	98.70	320.70
Vent, 2" diameter DWV piping	6.000	L.F.	1.500	82.80	89.40	172.20
Carrier, steel, for studs, no arms	1.000	Ea.	1.143	66	76	142
TOTAL		Ea.	21.517	2,810.32	1,318.40	4,128.72

The costs in this system are on a cost each basis. All necessary piping is included.

Three Fixture Bathroom Price Sheet	QUAN.	UNIT	LABOR HOURS	COST EACH		
				MAT.	INST.	TOTAL
Water closet, close coupled standard 2 piece, white	1.000	Ea.	3.019	223	180	403
Color	1.000	Ea.	3.019	465	180	645
One piece elongated bowl, white	1.000	Ea.	3.019	850	180	1,030
Color	1.000	Ea.	3.019	1,075	180	1,255
Low profile, one piece elongated bowl, white	1.000	Ea.	3.019	900	180	1,080
Color	1.000	Ea.	3.019	1,075	180	1,255
Rough-in for water closet						
1/2" copper supply, 4" cast iron waste, 2" cast iron vent	1.000	Ea.	2.376	209	146	355
4" PVC/DWV waste, 2" PVC vent	1.000	Ea.	2.678	96	164	260
4" carrier waste, 2" copper vent	1.000	Ea.	2.520	287	160	447
3" cast iron waste, 1-1/2" cast iron vent	1.000	Ea.	2.244	158	138	296
3" PVC waste, 1-1/2" PVC vent	1.000	Ea.	2.388	86	152	238
3" copper waste, 1-1/2" copper vent	1.000	Ea.	2.014	155	128	283
1/2" PVC supply, 4" PVC waste, 2" PVC vent	1.000	Ea.	2.974	116	183	299
3" PVC waste, 1-1/2" PVC supply	1.000	Ea.	2.684	106	172	278
1/2" steel supply, 4" cast iron waste, 2" cast iron vent	1.000	Ea.	2.545	213	157	370
4" cast iron waste, 2" steel vent	1.000	Ea.	2.590	192	159	351
4" PVC waste, 2" PVC vent	1.000	Ea.	2.847	100	175	275
Lavatory, wall hung, PE cast iron 20" x 18", white	1.000	Ea.	2.000	264	119	383
Color	1.000	Ea.	2.000	277	119	396
Vitreous china 19" x 17", white	1.000	Ea.	2.286	128	136	264
Color	1.000	Ea.	2.286	150	136	286
Lavatory, for vanity top, PE cast iron 20" x 18", white	1.000	Ea.	2.500	325	149	474
Color	1.000	Ea.	2.500	530	149	679
Steel enameled 20" x 17", white	1.000	Ea.	2.759	133	165	298
Color	1.000	Ea.	2.500	130	149	279
Vitreous china 20" x 16", white	1.000	Ea.	2.963	226	177	403
Color	1.000	Ea.	2.963	226	177	403
Rough-in for lavatory						
1/2" copper supply, 1-1/2" cast iron waste, 1-1/2" cast iron vent	1.000	Ea.	2.791	180	173	353
1-1/2" PVC waste, 1-1/4" PVC vent	1.000	Ea.	2.639	111	175	286
1/2" steel supply, 1-1/4" cast iron waste, 1-1/4" steel vent	1.000	Ea.	2.890	141	181	322
1-1/4" PVC waste, 1-1/4" PVC vent	1.000	Ea.	2.794	121	185	306
1/2" PVC supply, 1-1/2" PVC waste, 1-1/2" PVC vent	1.000	Ea.	3.260	141	216	357
Bathtub, PE cast iron, 5′ long corner with fittings, white	1.000	Ea.	3.636	1,325	217	1,542
Color	1.000	Ea.	3.636	1,650	217	1,867
Rough-in for bathtub						
1/2" copper supply, 4" cast iron waste, 1-1/2" copper vent	1.000	Ea.	2.409	201	154	355
4" PVC waste, 1/2" PVC vent	1.000	Ea.	2.877	107	184	291
1/2" steel supply, 4" cast iron waste, 1-1/2" steel vent	1.000	Ea.	2.898	183	181	364
4" PVC waste, 1-1/2" PVC vent	1.000	Ea.	3.159	114	203	317
1/2" PVC supply, 4" PVC waste, 1-1/2" PVC vent	1.000	Ea.	3.371	141	217	358
Piping supply, 1/2" copper	10.000	L.F.	.988	37	65.50	102.50
1/2" steel	10.000	L.F.	1.270	44	84	128
1/2" PVC	10.000	L.F.	1.482	71	98	169
Piping waste, 4" cast iron no hub	6.000	L.F.	1.655	222	98.50	320.50
4" PVC/DWV	6.000	L.F.	2.000	69	119	188
4" copper/DWV	6.000	L.F.	2.400	365	144	509
Piping vent 2" cast iron no hub	6.000	L.F.	1.433	114	85.50	199.50
2" copper/DWV	6.000	L.F.	1.091	123	72.50	195.50
2" PVC/DWV	6.000	L.F.	1.627	58.50	97	155.50
2" steel, galvanized	6.000	L.F.	1.500	83	89.50	172.50
Vanity base cabinet, 2 door, 24" x 30"	1.000	Ea.	1.000	485	59	544
24" x 36"	1.000	Ea.	1.200	440	70.50	510.50
Vanity top, laminated plastic square edge 25" x 32"	2.670	L.F.	.712	101	42	143
25" x 38"	3.160	L.F.	.843	120	49.50	169.50
Cultured marble, 25" x 32", with bowl	1.000	Ea.	2.500	177	149	326
25" x 38", with bowl	1.000	Ea.	2.500	206	149	355
Carrier, for lavatory, steel for studs, no arms	1.000	Ea.	1.143	66	76	142
Wood, 2" x 8" blocking	1.300	L.F.	.052	1.39	3.07	4.46

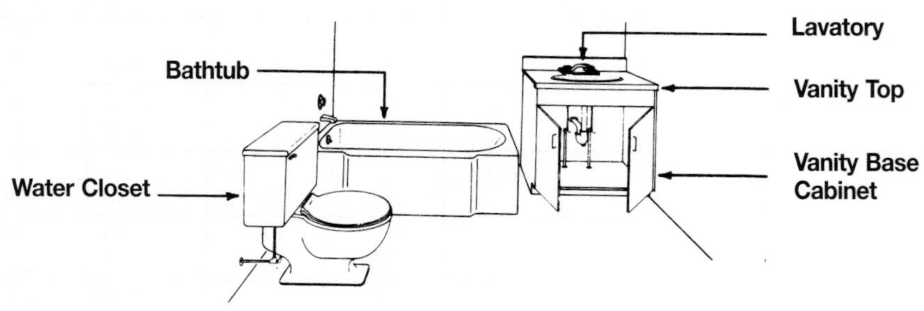

System Description	QUAN.	UNIT	LABOR HOURS	COST EACH MAT.	COST EACH INST.	COST EACH TOTAL
BATHROOM WITH LAVATORY INSTALLED IN VANITY						
Water closet, floor mounted, 2 piece, close coupled, white	1.000	Ea.	3.019	223	180	403
Rough-in, vent, 2" diameter DWV piping	1.000	Ea.	.955	76	57	133
Waste, 4" diameter DWV piping	1.000	Ea.	.828	111	49.35	160.35
Supply, 1/2" diameter type "L" copper supply piping	1.000	Ea.	.593	22.32	39.30	61.62
Lavatory, 20" x 18", PE cast iron with accessories, white	1.000	Ea.	2.500	325	149	474
Rough-in, vent, 1-1/2" diameter DWV piping	1.000	Ea.	1.803	142.80	107.60	250.40
Supply, 1/2" diameter type "L" copper supply piping	1.000	Ea.	.988	37.20	65.50	102.70
Bathtub, P.E. cast iron, 5' long with accessories, white	1.000	Ea.	3.636	1,325	217	1,542
Rough-in, waste, 4" diameter DWV piping	1.000	Ea.	.828	111	49.35	160.35
Supply, 1/2" diameter type "L" copper supply piping	1.000	Ea.	.988	37.20	65.50	102.70
Vent, 1-1/2" diameter DWV piping	1.000	Ea.	.593	52.80	39.20	92
Piping, supply, 1/2" diameter type "L" copper supply piping	32.000	L.F.	3.161	119.04	209.60	328.64
Waste, 4" diameter DWV piping	12.000	L.F.	3.310	444	197.40	641.40
Vent, 2" diameter DWV piping	6.000	L.F.	1.500	82.80	89.40	172.20
Vanity base cabinet, 2 door, 30" wide	1.000	Ea.	1.000	485	59	544
Vanity top, plastic laminated square edge	2.670	L.F.	.712	101.46	41.92	143.38
TOTAL		Ea.	26.414	3,695.62	1,616.12	5,311.74
BATHROOM WITH WALL HUNG LAVATORY						
Water closet, floor mounted, 2 piece, close coupled, white	1.000	Ea.	3.019	223	180	403
Rough-in, vent, 2" diameter DWV piping	1.000	Ea.	.955	76	57	133
Waste, 4" diameter DWV piping	1.000	Ea.	.828	111	49.35	160.35
Supply, 1/2" diameter type "L" copper supply piping	1.000	Ea.	.593	22.32	39.30	61.62
Lavatory, 20" x 18" P.E. cast iron, wall hung, white	1.000	Ea.	2.000	264	119	383
Rough-in, waste, 1-1/2" diameter DWV piping	1.000	Ea.	1.803	142.80	107.60	250.40
Supply, 1/2" diameter type "L" copper supply piping	1.000	Ea.	.988	37.20	65.50	102.70
Bathtub, P.E. cast iron, 5' long with accessories, white	1.000	Ea.	3.636	1,325	217	1,542
Rough-in, waste, 4" diameter DWV piping	1.000	Ea.	.828	111	49.35	160.35
Supply, 1/2" diameter type "L" copper supply piping	1.000	Ea.	.988	37.20	65.50	102.70
Vent, 1-1/2" diameter DWV piping	1.000	Ea.	.593	52.80	39.20	92
Piping, supply, 1/2" diameter type "L" copper supply piping	32.000	L.F.	3.161	119.04	209.60	328.64
Waste, 4" diameter DWV piping	12.000	L.F.	3.310	444	197.40	641.40
Vent, 2" diameter DWV piping	6.000	L.F.	1.500	82.80	89.40	172.20
Carrier steel, for studs, no arms	1.000	Ea.	1.143	66	76	142
TOTAL		Ea.	25.345	3,114.16	1,561.20	4,675.36

The costs in this system are on a cost each basis. All necessary piping is included.

Three Fixture Bathroom Price Sheet	QUAN.	UNIT	LABOR HOURS	COST EACH		
				MAT.	INST.	TOTAL
Water closet, close coupled, standard 2 piece, white	1.000	Ea.	3.019	223	180	403
Color	1.000	Ea.	3.019	465	180	645
One piece, elongated bowl, white	1.000	Ea.	3.019	850	180	1,030
Color	1.000	Ea.	3.019	1,075	180	1,255
Low profile, one piece, elongated bowl, white	1.000	Ea.	3.019	900	180	1,080
Color	1.000	Ea.	3.019	1,075	180	1,255
Rough-in, for water closet						
1/2" copper supply, 4" cast iron waste, 2" cast iron vent	1.000	Ea.	2.376	209	146	355
4" PVC/DWV waste, 2" PVC vent	1.000	Ea.	2.678	96	164	260
4" copper waste, 2" copper vent	1.000	Ea.	2.520	287	160	447
3" cast iron waste, 1-1/2" cast iron vent	1.000	Ea.	2.244	158	138	296
3" PVC waste, 1-1/2" PVC vent	1.000	Ea.	2.388	86	152	238
3" copper waste, 1-1/2" copper vent	1.000	Ea.	2.014	155	128	283
1/2" PVC supply, 4" PVC waste, 2" PVC vent	1.000	Ea.	2.974	116	183	299
3" PVC waste, 1-1/2" PVC supply	1.000	Ea.	2.684	106	172	278
1/2" steel supply, 4" cast iron waste, 2" cast iron vent	1.000	Ea.	2.545	213	157	370
4" cast iron waste, 2" steel vent	1.000	Ea.	2.590	192	159	351
4" PVC waste, 2" PVC vent	1.000	Ea.	2.847	100	175	275
Lavatory wall hung, P.E. cast iron, 20" x 18", white	1.000	Ea.	2.000	264	119	383
Color	1.000	Ea.	2.000	277	119	396
Vitreous china, 19" x 17", white	1.000	Ea.	2.286	128	136	264
Color	1.000	Ea.	2.286	150	136	286
Lavatory, for vanity top, P.E., cast iron, 20" x 18", white	1.000	Ea.	2.500	325	149	474
Color	1.000	Ea.	2.500	530	149	679
Steel, enameled, 20" x 17", white	1.000	Ea.	2.759	133	165	298
Color	1.000	Ea.	2.500	130	149	279
Vitreous china, 20" x 16", white	1.000	Ea.	2.963	226	177	403
Color	1.000	Ea.	2.963	226	177	403
Rough-in, for lavatory						
1/2" copper supply, 1-1/2" C.I. waste, 1-1/2" C.I. vent	1.000	Ea.	2.791	180	173	353
1-1/2" PVC waste, 1-1/4" PVC vent	1.000	Ea.	2.639	111	175	286
1/2" steel supply, 1-1/4" cast iron waste, 1-1/4" steel vent	1.000	Ea.	2.890	141	181	322
1-1/4" PVC waste, 1-1/4" PVC vent	1.000	Ea.	2.794	121	185	306
1/2" PVC supply, 1-1/2" PVC waste, 1-1/2" PVC vent	1.000	Ea.	3.260	141	216	357
Bathtub, P.E. cast iron, 5' long corner with fittings, white	1.000	Ea.	3.636	1,325	217	1,542
Color	1.000	Ea.	3.636	1,650	217	1,867
Rough-in, for bathtub						
1/2" copper supply, 4" cast iron waste, 1-1/2" copper vent	1.000	Ea.	2.409	201	154	355
4" PVC waste, 1/2" PVC vent	1.000	Ea.	2.877	107	184	291
1/2" steel supply, 4" cast iron waste, 1-1/2" steel vent	1.000	Ea.	2.898	183	181	364
4" PVC waste, 1-1/2" PVC vent	1.000	Ea.	3.159	114	203	317
1/2" PVC supply, 4" PVC waste, 1-1/2" PVC vent	1.000	Ea.	3.371	141	217	358
Piping, supply, 1/2" copper	32.000	L.F.	3.161	119	210	329
1/2" steel	32.000	L.F.	4.063	140	269	409
1/2" PVC	32.000	L.F.	4.741	227	315	542
Piping, waste, 4" cast iron no hub	12.000	L.F.	3.310	445	197	642
4" PVC/DWV	12.000	L.F.	4.000	138	239	377
4" copper/DWV	12.000	L.F.	4.800	730	288	1,018
Piping, vent, 2" cast iron no hub	6.000	L.F.	1.433	114	85.50	199.50
2" copper/DWV	6.000	L.F.	1.091	123	72.50	195.50
2" PVC/DWV	6.000	L.F.	1.627	58.50	97	155.50
2" steel, galvanized	6.000	L.F.	1.500	83	89.50	172.50
Vanity base cabinet, 2 door, 24" x 30"	1.000	Ea.	1.000	485	59	544
24" x 36"	1.000	Ea.	1.200	440	70.50	510.50
Vanity top, laminated plastic square edge, 25" x 32"	2.670	L.F.	.712	101	42	143
25" x 38"	3.160	L.F.	.843	120	49.50	169.50
Cultured marble, 25" x 32", with bowl	1.000	Ea.	2.500	177	149	326
25" x 38", with bowl	1.000	Ea.	2.500	206	149	355
Carrier, for lavatory, steel for studs, no arms	1.000	Ea.	1.143	66	76	142
Wood, 2" x 8" blocking	1.300	L.F.	.052	1.39	3.07	4.46

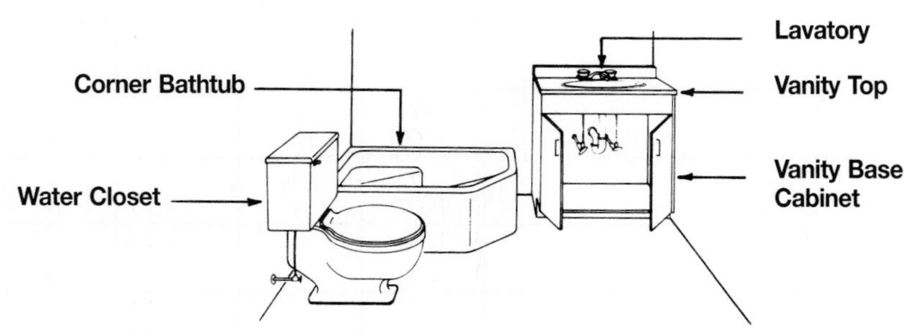

System Description	QUAN.	UNIT	LABOR HOURS	COST EACH		
				MAT.	INST.	TOTAL
BATHROOM WITH LAVATORY INSTALLED IN VANITY						
Water closet, floor mounted, 2 piece, close coupled, white	1.000	Ea.	3.019	223	180	403
Rough-in, vent, 2" diameter DWV piping	1.000	Ea.	.955	76	57	133
Waste, 4" diameter DWV piping	1.000	Ea.	.828	111	49.35	160.35
Supply, 1/2" diameter type "L" copper supply piping	1.000	Ea.	.593	22.32	39.30	61.62
Lavatory, 20" x 18", P.E. cast iron with fittings, white	1.000	Ea.	2.500	325	149	474
Rough-in, waste, 1-1/2" diameter DWV piping	1.000	Ea.	1.803	142.80	107.60	250.40
Supply, 1/2" diameter type "L" copper supply piping	1.000	Ea.	.988	37.20	65.50	102.70
Bathtub, P.E. cast iron, corner with fittings, white	1.000	Ea.	3.636	3,100	217	3,317
Rough-in, waste, 4" diameter DWV piping	1.000	Ea.	.828	111	49.35	160.35
Supply, 1/2" diameter type "L" copper supply piping	1.000	Ea.	.988	37.20	65.50	102.70
Vent, 1-1/2" diameter DWV piping	1.000	Ea.	.593	52.80	39.20	92
Piping, supply, 1/2" diameter type "L" copper supply piping	32.000	L.F.	3.161	119.04	209.60	328.64
Waste, 4" diameter DWV piping	12.000	L.F.	3.310	444	197.40	641.40
Vent, 2" diameter DWV piping	6.000	L.F.	1.500	82.80	89.40	172.20
Vanity base cabinet, 2 door, 30" wide	1.000	Ea.	1.000	485	59	544
Vanity top, plastic laminated, square edge	2.670	L.F.	.712	122.82	41.92	164.74
TOTAL		Ea.	26.414	5,491.98	1,616.12	7,108.10
BATHROOM WITH WALL HUNG LAVATORY						
Water closet, floor mounted, 2 piece, close coupled, white	1.000	Ea.	3.019	223	180	403
Rough-in, vent, 2" diameter DWV piping	1.000	Ea.	.955	76	57	133
Waste, 4" diameter DWV piping	1.000	Ea.	.828	111	49.35	160.35
Supply, 1/2" diameter type "L" copper supply piping	1.000	Ea.	.593	22.32	39.30	61.62
Lavatory, 20" x 18", P.E. cast iron, with fittings, white	1.000	Ea.	2.000	264	119	383
Rough-in, waste, 1-1/2" diameter DWV piping	1.000	Ea.	1.803	142.80	107.60	250.40
Supply, 1/2" diameter type "L" copper supply piping	1.000	Ea.	.988	37.20	65.50	102.70
Bathtub, P.E. cast iron, corner, with fittings, white	1.000	Ea.	3.636	3,100	217	3,317
Rough-in, waste, 4" diameter DWV piping	1.000	Ea.	.828	111	49.35	160.35
Supply, 1/2" diameter type "L" copper supply piping	1.000	Ea.	.988	37.20	65.50	102.70
Vent, 1-1/2" diameter DWV piping	1.000	Ea.	.593	52.80	39.20	92
Piping, supply, 1/2" diameter type "L" copper supply piping	32.000	L.F.	3.161	119.04	209.60	328.64
Waste, 4" diameter DWV piping	12.000	L.F.	3.310	444	197.40	641.40
Vent, 2" diameter DWV piping	6.000	L.F.	1.500	82.80	89.40	172.20
Carrier, steel, for studs, no arms	1.000	Ea.	1.143	66	76	142
TOTAL		Ea.	25.345	4,889.16	1,561.20	6,450.36

The costs in this system are on a cost each basis. All necessary piping is included.

Three Fixture Bathroom Price Sheet	QUAN.	UNIT	LABOR HOURS	COST EACH		
				MAT.	INST.	TOTAL
Water closet, close coupled, standard 2 piece, white	1.000	Ea.	3.019	223	180	403
Color	1.000	Ea.	3.019	465	180	645
One piece elongated bowl, white	1.000	Ea.	3.019	850	180	1,030
Color	1.000	Ea.	3.019	1,075	180	1,255
Low profile, one piece elongated bowl, white	1.000	Ea.	3.019	900	180	1,080
Color	1.000	Ea.	3.019	1,075	180	1,255
Rough-in, for water closet						
1/2" copper supply, 4" cast iron waste, 2" cast iron vent	1.000	Ea.	2.376	209	146	355
4" PVC/DWV waste, 2" PVC vent	1.000	Ea.	2.678	96	164	260
4" copper waste, 2" copper vent	1.000	Ea.	2.520	287	160	447
3" cast iron waste, 1-1/2" cast iron vent	1.000	Ea.	2.244	158	138	296
3" PVC waste, 1-1/2" PVC vent	1.000	Ea.	2.388	86	152	238
3" copper waste, 1-1/2" copper vent	1.000	Ea.	2.014	155	128	283
1/2" PVC supply, 4" PVC waste, 2" PVC vent	1.000	Ea.	2.974	116	183	299
3" PVC waste, 1-1/2" PVC supply	1.000	Ea.	2.684	106	172	278
1/2" steel supply, 4" cast iron waste, 2" cast iron vent	1.000	Ea.	2.545	213	157	370
4" cast iron waste, 2" steel vent	1.000	Ea.	2.590	192	159	351
4" PVC waste, 2" PVC vent	1.000	Ea.	2.847	100	175	275
Lavatory, wall hung P.E. cast iron 20" x 18", white	1.000	Ea.	2.000	264	119	383
Color	1.000	Ea.	2.000	277	119	396
Vitreous china 19" x 17", white	1.000	Ea.	2.286	128	136	264
Color	1.000	Ea.	2.286	150	136	286
Lavatory, for vanity top, P.E., cast iron, 20" x 18", white	1.000	Ea.	2.500	325	149	474
Color	1.000	Ea.	2.500	530	149	679
Steel enameled 20" x 17", white	1.000	Ea.	2.759	133	165	298
Color	1.000	Ea.	2.500	130	149	279
Vitreous china 20" x 16", white	1.000	Ea.	2.963	226	177	403
Color	1.000	Ea.	2.963	226	177	403
Rough-in, for lavatory						
1/2" copper supply, 1-1/2" cast iron waste, 1-1/2" cast iron vent	1.000	Ea.	2.791	180	173	353
1-1/2" PVC waste, 1-1/4" PVC vent	1.000	Ea.	2.639	111	175	286
1/2" steel supply, 1-1/4" cast iron waste, 1-1/4" steel vent	1.000	Ea.	2.890	141	181	322
1-1/4" PVC waste, 1-1/4" PVC vent	1.000	Ea.	2.794	121	185	306
1/2" PVC supply, 1-1/2" PVC waste, 1-1/2" PVC vent	1.000	Ea.	3.260	141	216	357
Bathtub, P.E. cast iron, corner with fittings, white	1.000	Ea.	3.636	3,100	217	3,317
Color	1.000	Ea.	4.000	2,975	239	3,214
Rough-in, for bathtub						
1/2" copper supply, 4" cast iron waste, 1-1/2" copper vent	1.000	Ea.	2.409	201	154	355
4" PVC waste, 1-1/2" PVC vent	1.000	Ea.	2.877	107	184	291
1/2" steel supply, 4" cast iron waste, 1-1/2" steel vent	1.000	Ea.	2.898	183	181	364
4" PVC waste, 1-1/2" PVC vent	1.000	Ea.	3.159	114	203	317
1/2" PVC supply, 4" PVC waste, 1-1/2" PVC vent	1.000	Ea.	3.371	141	217	358
Piping, supply, 1/2" copper	32.000	L.F.	3.161	119	210	329
1/2" steel	32.000	L.F.	4.063	140	269	409
1/2" PVC	32.000	L.F.	4.741	227	315	542
Piping, waste, 4" cast iron, no hub	12.000	L.F.	3.310	445	197	642
4" PVC/DWV	12.000	L.F.	4.000	138	239	377
4" copper/DWV	12.000	L.F.	4.800	730	288	1,018
Piping, vent 2" cast iron, no hub	6.000	L.F.	1.433	114	85.50	199.50
2" copper/DWV	6.000	L.F.	1.091	123	72.50	195.50
2" PVC/DWV	6.000	L.F.	1.627	58.50	97	155.50
2" steel, galvanized	6.000	L.F.	1.500	83	89.50	172.50
Vanity base cabinet, 2 door, 24" x 30"	1.000	Ea.	1.000	485	59	544
24" x 36"	1.000	Ea.	1.200	440	70.50	510.50
Vanity top, laminated plastic square edge 25" x 32"	2.670	L.F.	.712	123	42	165
25" x 38"	3.160	L.F.	.843	145	49.50	194.50
Cultured marble, 25" x 32", with bowl	1.000	Ea.	2.500	177	149	326
25" x 38", with bowl	1.000	Ea.	2.500	206	149	355
Carrier, for lavatory, steel for studs, no arms	1.000	Ea.	1.143	66	76	142
Wood, 2" x 8" blocking	1.300	L.F.	.053	1.42	3.14	4.56

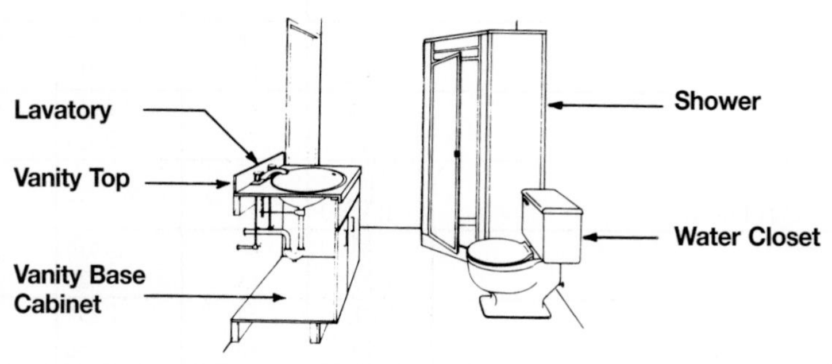

System Description	QUAN.	UNIT	LABOR HOURS	COST EACH MAT.	COST EACH INST.	COST EACH TOTAL
BATHROOM WITH SHOWER, LAVATORY INSTALLED IN VANITY						
Water closet, floor mounted, 2 piece, close coupled, white	1.000	Ea.	3.019	223	180	403
Rough-in, vent, 2″ diameter DWV piping	1.000	Ea.	.955	76	57	133
Waste, 4″ diameter DWV piping	1.000	Ea.	.828	111	49.35	160.35
Supply, 1/2″ diameter type "L" copper supply piping	1.000	Ea.	.593	22.32	39.30	61.62
Lavatory, 20″ x 18″ P.E. cast iron with fittings, white	1.000	Ea.	2.500	325	149	474
Rough-in, waste, 1-1/2″ diameter DWV piping	1.000	Ea.	1.803	142.80	107.60	250.40
Supply, 1/2″ diameter type "L" copper supply piping	1.000	Ea.	.988	37.20	65.50	102.70
Shower, steel enameled, stone base, corner, white	1.000	Ea.	3.200	1,300	191	1,491
Shower mixing valve	1.000	Ea.	1.333	155	88.50	243.50
Shower door	1.000	Ea.	1.000	480	64.50	544.50
Rough-in, vent, 1-1/2″ diameter DWV piping	1.000	Ea.	.225	17.85	13.45	31.30
Waste, 2″ diameter DWV piping	1.000	Ea.	1.433	114	85.50	199.50
Supply, 1/2″ diameter type "L" copper supply piping	1.000	Ea.	1.580	59.52	104.80	164.32
Piping, supply, 1/2″ diameter type "L" copper supply piping	36.000	L.F.	4.148	156.24	275.10	431.34
Waste, 4″ diameter DWV piping	7.000	L.F.	2.759	370	164.50	534.50
Vent, 2″ diameter DWV piping	6.000	L.F.	2.250	124.20	134.10	258.30
Vanity base 2 door, 30″ wide	1.000	Ea.	1.000	485	59	544
Vanity top, plastic laminated, square edge	2.170	L.F.	.712	134.84	41.92	176.76
TOTAL		Ea.	30.326	4,333.97	1,870.12	6,204.09
BATHROOM WITH SHOWER, WALL HUNG LAVATORY						
Water closet, floor mounted, close coupled	1.000	Ea.	3.019	223	180	403
Rough-in, vent, 2″ diameter DWV piping	1.000	Ea.	.955	76	57	133
Waste, 4″ diameter DWV piping	1.000	Ea.	.828	111	49.35	160.35
Supply, 1/2″ diameter type "L" copper supply piping	1.000	Ea.	.593	22.32	39.30	61.62
Lavatory, 20″ x 18″ P.E. cast iron with fittings, white	1.000	Ea.	2.000	264	119	383
Rough-in, waste, 1-1/2″ diameter DWV piping	1.000	Ea.	1.803	142.80	107.60	250.40
Supply, 1/2″ diameter type "L" copper supply piping	1.000	Ea.	.988	37.20	65.50	102.70
Shower, steel enameled, stone base, white	1.000	Ea.	3.200	1,300	191	1,491
Mixing valve	1.000	Ea.	1.333	155	88.50	243.50
Shower door	1.000	Ea.	1.000	480	64.50	544.50
Rough-in, vent, 1-1/2″ diameter DWV piping	1.000	Ea.	.225	17.85	13.45	31.30
Waste, 2″ diameter DWV piping	1.000	Ea.	1.433	114	85.50	199.50
Supply, 1/2″ diameter type "L" copper supply piping	1.000	Ea.	1.580	59.52	104.80	164.32
Piping, supply, 1/2″ diameter type "L" copper supply piping	36.000	L.F.	4.148	156.24	275.10	431.34
Waste, 4″ diameter DWV piping	7.000	L.F.	2.759	370	164.50	534.50
Vent, 2″ diameter DWV piping	6.000	L.F.	2.250	124.20	134.10	258.30
Carrier, steel, for studs, no arms	1.000	Ea.	1.143	66	76	142
TOTAL		Ea.	29.257	3,719.13	1,815.20	5,534.33

The costs in this system are on a cost each basis. All necessary piping is included.

Three Fixture Bathroom Price Sheet	QUAN.	UNIT	LABOR HOURS	COST EACH		
				MAT.	INST.	TOTAL
Water closet, close coupled, standard 2 piece, white	1.000	Ea.	3.019	223	180	403
Color	1.000	Ea.	3.019	465	180	645
One piece elongated bowl, white	1.000	Ea.	3.019	850	180	1,030
Color	1.000	Ea.	3.019	1,075	180	1,255
Low profile, one piece elongated bowl, white	1.000	Ea.	3.019	900	180	1,080
Color	1.000	Ea.	3.019	1,075	180	1,255
Rough-in, for water closet						
1/2" copper supply, 4" cast iron waste, 2" cast iron vent	1.000	Ea.	2.376	209	146	355
4" PVC/DWV waste, 2" PVC vent	1.000	Ea.	2.678	96	164	260
4" copper waste, 2" copper vent	1.000	Ea.	2.520	287	160	447
3" cast iron waste, 1-1/2" cast iron vent	1.000	Ea.	2.244	158	138	296
3" PVC waste, 1-1/2" PVC vent	1.000	Ea.	2.388	86	152	238
3" copper waste, 1-1/2" copper vent	1.000	Ea.	2.014	155	128	283
1/2" PVC supply, 4" PVC waste, 2" PVC vent	1.000	Ea.	2.974	116	183	299
3" PVC waste, 1-1/2" PVC supply	1.000	Ea.	2.684	106	172	278
1/2" steel supply, 4" cast iron waste, 2" cast iron vent	1.000	Ea.	2.545	213	157	370
4" cast iron waste, 2" steel vent	1.000	Ea.	2.590	192	159	351
4" PVC waste, 2" PVC vent	1.000	Ea.	2.847	100	175	275
Lavatory, wall hung, P.E. cast iron 20" x 18", white	1.000	Ea.	2.000	264	119	383
Color	1.000	Ea.	2.000	277	119	396
Vitreous china 19" x 17", white	1.000	Ea.	2.286	128	136	264
Color	1.000	Ea.	2.286	150	136	286
Lavatory, for vanity top, P.E. cast iron 20" x 18", white	1.000	Ea.	2.500	325	149	474
Color	1.000	Ea.	2.500	530	149	679
Steel enameled 20" x 17", white	1.000	Ea.	2.759	133	165	298
Color	1.000	Ea.	2.500	130	149	279
Vitreous china 20" x 16", white	1.000	Ea.	2.963	226	177	403
Color	1.000	Ea.	2.963	226	177	403
Rough-in, for lavatory						
1/2" copper supply, 1-1/2" cast iron waste, 1-1/2" cast iron vent	1.000	Ea.	2.791	180	173	353
1-1/2" PVC waste, 1-1/2" PVC vent	1.000	Ea.	2.639	111	175	286
1/2" steel supply, 1-1/4" cast iron waste, 1-1/4" steel vent	1.000	Ea.	2.890	141	181	322
1-1/4" PVC waste, 1-1/4" PVC vent	1.000	Ea.	2.921	118	194	312
1/2" PVC supply, 1-1/2" PVC waste, 1-1/2" PVC vent	1.000	Ea.	3.260	141	216	357
Shower, steel enameled stone base, 32" x 32", white	1.000	Ea.	8.000	1,300	191	1,491
Color	1.000	Ea.	7.822	1,950	175	2,125
36" x 36" white	1.000	Ea.	8.889	1,625	199	1,824
Color	1.000	Ea.	8.889	2,225	199	2,424
Rough-in, for shower						
1/2" copper supply, 4" cast iron waste, 1-1/2" copper vent	1.000	Ea.	3.238	191	204	395
4" PVC waste, 1-1/2" PVC vent	1.000	Ea.	3.429	127	217	344
1/2" steel supply, 4" cast iron waste, 1-1/2" steel vent	1.000	Ea.	3.665	191	232	423
4" PVC waste, 1-1/2" PVC vent	1.000	Ea.	3.881	137	246	383
1/2" PVC supply, 4" PVC waste, 1-1/2" PVC vent	1.000	Ea.	4.219	181	269	450
Piping, supply, 1/2" copper	36.000	L.F.	4.148	156	275	431
1/2" steel	36.000	L.F.	5.333	184	355	539
1/2" PVC	36.000	L.F.	6.222	298	410	708
Piping, waste, 4" cast iron no hub	7.000	L.F.	2.759	370	165	535
4" PVC/DWV	7.000	L.F.	3.333	115	199	314
4" copper/DWV	7.000	L.F.	4.000	610	240	850
Piping, vent, 2" cast iron no hub	6.000	L.F.	2.149	171	128	299
2" copper/DWV	6.000	L.F.	1.636	185	108	293
2" PVC/DWV	6.000	L.F.	2.441	88	146	234
2" steel, galvanized	6.000	L.F.	2.250	124	134	258
Vanity base cabinet, 2 door, 24" x 30"	1.000	Ea.	1.000	485	59	544
24" x 36"	1.000	Ea.	1.200	440	70.50	510.50
Vanity top, laminated plastic square edge, 25" x 32"	2.170	L.F.	.712	135	42	177
25" x 38"	2.670	L.F.	.845	160	50	210
Carrier, for lavatory, steel for studs, no arms	1.000	Ea.	1.143	66	76	142
Wood, 2" x 8" blocking	1.300	L.F.	.052	1.39	3.07	4.46

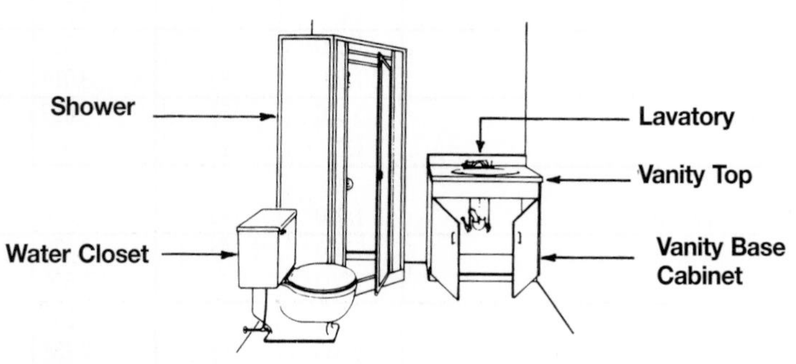

System Description	QUAN.	UNIT	LABOR HOURS	COST EACH		
				MAT.	INST.	TOTAL
BATHROOM WITH LAVATORY INSTALLED IN VANITY						
Water closet, floor mounted, 2 piece, close coupled, white	1.000	Ea.	3.019	223	180	403
Rough-in, vent, 2" diameter DWV piping	1.000	Ea.	.955	76	57	133
Waste, 4" diameter DWV piping	1.000	Ea.	.828	111	49.35	160.35
Supply, 1/2" diameter type "L" copper supply piping	1.000	Ea.	.593	22.32	39.30	61.62
Lavatory, 20" x 18", P.E. cast iron with fittings, white	1.000	Ea.	2.500	325	149	474
Rough-in, waste, 1-1/2" diameter DWV piping	1.000	Ea.	1.803	142.80	107.60	250.40
Supply, 1/2" diameter type "L" copper supply piping	1.000	Ea.	.988	37.20	65.50	102.70
Shower, steel enameled, stone base, corner, white	1.000	Ea.	3.200	1,300	191	1,491
Mixing valve	1.000	Ea.	1.333	155	88.50	243.50
Shower door	1.000	Ea.	1.000	480	64.50	544.50
Rough-in, vent, 1-1/2" diameter DWV piping	1.000	Ea.	.225	17.85	13.45	31.30
Waste, 2" diameter DWV piping	1.000	Ea.	1.433	114	85.50	199.50
Supply, 1/2" diameter type "L" copper supply piping	1.000	Ea.	1.580	59.52	104.80	164.32
Piping, supply, 1/2" diameter type "L" copper supply piping	36.000	L.F.	3.556	133.92	235.80	369.72
Waste, 4" diameter DWV piping	7.000	L.F.	1.931	259	115.15	374.15
Vent, 2" diameter DWV piping	6.000	L.F.	1.500	82.80	89.40	172.20
Vanity base, 2 door, 30" wide	1.000	Ea.	1.000	485	59	544
Vanity top, plastic laminated, square edge	2.670	L.F.	.712	101.46	41.92	143.38
TOTAL		Ea.	28.156	4,125.87	1,736.77	5,862.64
BATHROOM, WITH WALL HUNG LAVATORY						
Water closet, floor mounted, 2 piece, close coupled, white	1.000	Ea.	3.019	223	180	403
Rough-in, vent, 2" diameter DWV piping	1.000	Ea.	.955	76	57	133
Waste, 4" diameter DWV piping	1.000	Ea.	.828	111	49.35	160.35
Supply, 1/2" diameter type "L" copper supply piping	1.000	Ea.	.593	22.32	39.30	61.62
Lavatory, wall hung, 20" x 18" P.E. cast iron with fittings, white	1.000	Ea.	2.000	264	119	383
Rough-in, waste, 1-1/2" diameter DWV piping	1.000	Ea.	1.803	142.80	107.60	250.40
Supply, 1/2" diameter type "L" copper supply piping	1.000	Ea.	.988	37.20	65.50	102.70
Shower, steel enameled, stone base, corner, white	1.000	Ea.	3.200	1,300	191	1,491
Mixing valve	1.000	Ea.	1.333	155	88.50	243.50
Shower door	1.000	Ea.	1.000	480	64.50	544.50
Rough-in, waste, 1-1/2" diameter DWV piping	1.000	Ea.	.225	17.85	13.45	31.30
Waste, 2" diameter DWV piping	1.000	Ea.	1.433	114	85.50	199.50
Supply, 1/2" diameter type "L" copper supply piping	1.000	Ea.	1.580	59.52	104.80	164.32
Piping, supply, 1/2" diameter type "L" copper supply piping	36.000	L.F.	3.556	133.92	235.80	369.72
Waste, 4" diameter DWV piping	7.000	L.F.	1.931	259	115.15	374.15
Vent, 2" diameter DWV piping	6.000	L.F.	1.500	82.80	89.40	172.20
Carrier, steel, for studs, no arms	1.000	Ea.	1.143	66	76	142
TOTAL		Ea.	27.087	3,544.41	1,681.85	5,226.26

The costs in this system are on a cost each basis. All necessary piping is included.

Three Fixture Bathroom Price Sheet	QUAN.	UNIT	LABOR HOURS	COST EACH		
				MAT.	INST.	TOTAL
Water closet, close coupled, standard 2 piece, white	1.000	Ea.	3.019	223	180	403
Color	1.000	Ea.	3.019	465	180	645
One piece elongated bowl, white	1.000	Ea.	3.019	850	180	1,030
Color	1.000	Ea.	3.019	1,075	180	1,255
Low profile one piece elongated bowl, white	1.000	Ea.	3.019	900	180	1,080
Color	1.000	Ea.	3.623	1,075	180	1,255
Rough-in, for water closet						
1/2" copper supply, 4" cast iron waste, 2" cast iron vent	1.000	Ea.	2.376	209	146	355
4" P.V.C./DWV waste, 2" PVC vent	1.000	Ea.	2.678	96	164	260
4" copper waste, 2" copper vent	1.000	Ea.	2.520	287	160	447
3" cast iron waste, 1-1/2" cast iron vent	1.000	Ea.	2.244	158	138	296
3" PVC waste, 1-1/2" PVC vent	1.000	Ea.	2.388	86	152	238
3" copper waste, 1-1/2" copper vent	1.000	Ea.	2.014	155	128	283
1/2" P.V.C. supply, 4" P.V.C. waste, 2" P.V.C. vent	1.000	Ea.	2.974	116	183	299
3" P.V.C. waste, 1-1/2" P.V.C. vent	1.000	Ea.	2.684	106	172	278
1/2" steel supply, 4" cast iron waste, 2" cast iron vent	1.000	Ea.	2.545	213	157	370
4" cast iron waste, 2" steel vent	1.000	Ea.	2.590	192	159	351
4" P.V.C. waste, 2" P.V.C. vent	1.000	Ea.	2.847	100	175	275
Lavatory, wall hung P.E. cast iron 20" x 18", white	1.000	Ea.	2.000	264	119	383
Color	1.000	Ea.	2.000	277	119	396
Vitreous china 19" x 17", white	1.000	Ea.	2.286	128	136	264
Color	1.000	Ea.	2.286	150	136	286
Lavatory, for vanity top P.E. cast iron 20" x 18", white	1.000	Ea.	2.500	325	149	474
Color	1.000	Ea.	2.500	530	149	679
Steel enameled 20" x 17", white	1.000	Ea.	2.759	133	165	298
Color	1.000	Ea.	2.500	130	149	279
Vitreous china 20" x 16", white	1.000	Ea.	2.963	226	177	403
Color	1.000	Ea.	2.963	226	177	403
Rough-in, for lavatory						
1/2" copper supply, 1-1/2" cast iron waste, 1-1/2" cast iron vent	1.000	Ea.	2.791	180	173	353
1-1/2" P.V.C. waste, 1-1/2" P.V.C. vent	1.000	Ea.	2.639	111	175	286
1/2" steel supply, 1-1/2" cast iron waste, 1-1/4" steel vent	1.000	Ea.	2.890	141	181	322
1-1/2" P.V.C. waste, 1-1/4" P.V.C. vent	1.000	Ea.	2.921	118	194	312
1/2" P.V.C. supply, 1-1/2" P.V.C. waste, 1-1/2" P.V.C. vent	1.000	Ea.	3.260	141	216	357
Shower, steel enameled stone base, 32" x 32", white	1.000	Ea.	8.000	1,300	191	1,491
Color	1.000	Ea.	7.822	1,950	175	2,125
36" x 36", white	1.000	Ea.	8.889	1,625	199	1,824
Color	1.000	Ea.	8.889	2,225	199	2,424
Rough-in, for shower						
1/2" copper supply, 2" cast iron waste, 1-1/2" copper vent	1.000	Ea.	3.161	187	200	387
2" P.V.C. waste, 1-1/2" P.V.C. vent	1.000	Ea.	3.429	127	217	344
1/2" steel supply, 2" cast iron waste, 1-1/2" steel vent	1.000	Ea.	3.887	299	245	544
2" P.V.C. waste, 1-1/2" P.V.C. vent	1.000	Ea.	3.881	137	246	383
1/2" P.V.C. supply, 2" P.V.C. waste, 1-1/2" P.V.C. vent	1.000	Ea.	4.219	181	269	450
Piping, supply, 1/2" copper	36.000	L.F.	3.556	134	236	370
1/2" steel	36.000	L.F.	4.571	158	300	458
1/2" P.V.C.	36.000	L.F.	5.333	256	355	611
Waste, 4" cast iron, no hub	7.000	L.F.	1.931	259	115	374
4" P.V.C./DWV	7.000	L.F.	2.333	80.50	139	219.50
4" copper/DWV	7.000	L.F.	2.800	425	168	593
Vent, 2" cast iron, no hub	6.000	L.F.	1.091	123	72.50	195.50
2" copper/DWV	6.000	L.F.	1.091	123	72.50	195.50
2" P.V.C./DWV	6.000	L.F.	1.627	58.50	97	155.50
2" steel, galvanized	6.000	L.F.	1.500	83	89.50	172.50
Vanity base cabinet, 2 door, 24" x 30"	1.000	Ea.	1.000	485	59	544
24" x 36"	1.000	Ea.	1.200	440	70.50	510.50
Vanity top, laminated plastic square edge, 25" x 32"	2.670	L.F.	.712	101	42	143
25" x 38"	3.170	L.F.	.845	120	50	170
Carrier , for lavatory, steel, for studs, no arms	1.000	Ea.	1.143	66	76	142
Wood, 2" x 8" blocking	1.300	L.F.	.052	1.39	3.07	4.46

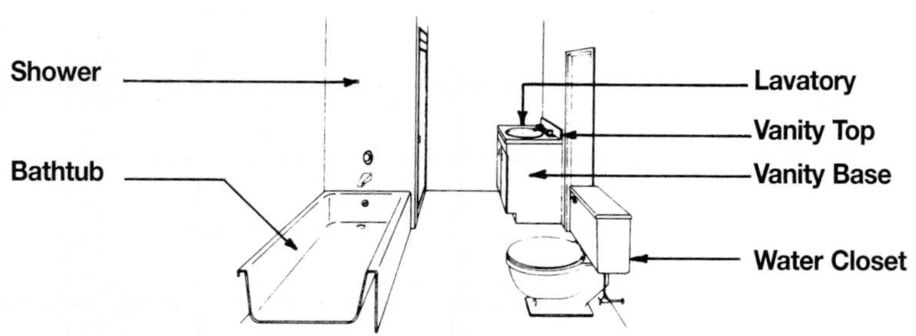

System Description	QUAN.	UNIT	LABOR HOURS	COST EACH MAT.	COST EACH INST.	COST EACH TOTAL
BATHROOM WITH LAVATORY INSTALLED IN VANITY						
Water closet, floor mounted, 2 piece, close coupled, white	1.000	Ea.	3.019	223	180	403
Rough-in, vent, 2″ diameter DWV piping	1.000	Ea.	.955	76	57	133
Waste, 4″ diameter DWV piping	1.000	Ea.	.828	111	49.35	160.35
Supply, 1/2″ diameter type "L" copper supply piping	1.000	Ea.	.593	22.32	39.30	61.62
Lavatory, 20″ x 18″ P.E. cast iron with fittings, white	1.000	Ea.	2.500	325	149	474
Shower, steel, enameled, stone base, corner, white	1.000	Ea.	3.333	1,925	199	2,124
Mixing valve	1.000	Ea.	1.333	155	88.50	243.50
Shower door	1.000	Ea.	1.000	480	64.50	544.50
Rough-in, waste, 1-1/2″ diameter DWV piping	2.000	Ea.	4.507	357	269	626
Supply, 1/2″ diameter type "L" copper supply piping	2.000	Ea.	3.161	119.04	209.60	328.64
Bathtub, P.E. cast iron, 5′ long with fittings, white	1.000	Ea.	3.636	1,325	217	1,542
Rough-in, waste, 4″ diameter DWV piping	1.000	Ea.	.828	111	49.35	160.35
Supply, 1/2″ diameter type "L" copper supply piping	1.000	Ea.	.988	37.20	65.50	102.70
Vent, 1-1/2″ diameter DWV piping	1.000	Ea.	.593	52.80	39.20	92
Piping, supply, 1/2″ diameter type "L" copper supply piping	42.000	L.F.	4.148	156.24	275.10	431.34
Waste, 4″ diameter DWV piping	10.000	L.F.	2.759	370	164.50	534.50
Vent, 2″ diameter DWV piping	13.000	L.F.	3.250	179.40	193.70	373.10
Vanity base, 2 doors, 30″ wide	1.000	Ea.	1.000	485	59	544
Vanity top, plastic laminated, square edge	2.670	L.F.	.712	101.46	41.92	143.38
TOTAL		Ea.	39.143	6,611.46	2,410.52	9,021.98
BATHROOM WITH WALL HUNG LAVATORY						
Water closet, floor mounted, 2 piece, close coupled, white	1.000	Ea.	3.019	223	180	403
Rough-in, vent, 2″ diameter DWV piping	1.000	Ea.	.955	76	57	133
Waste, 4″ diameter DWV piping	1.000	Ea.	.828	111	49.35	160.35
Supply, 1/2″ diameter type "L" copper supply piping	1.000	Ea.	.593	22.32	39.30	61.62
Lavatory, 20″ x 18″ P.E. cast iron with fittings, white	1.000	Ea.	2.000	264	119	383
Shower, steel enameled, stone base, corner , white	1.000	Ea.	3.333	1,925	199	2,124
Mixing valve	1.000	Ea.	1.333	155	88.50	243.50
Shower door	1.000	Ea.	1.000	480	64.50	544.50
Rough-in, waste, 1-1/2″ diameter DWV piping	2.000	Ea.	4.507	357	269	626
Supply, 1/2″ diameter type "L" copper supply piping	2.000	Ea.	3.161	119.04	209.60	328.64
Bathtub, P.E. cast iron, 5′ long with fittings, white	1.000	Ea.	3.636	1,325	217	1,542
Rough-in, waste, 4″ diameter DWV piping	1.000	Ea.	.828	111	49.35	160.35
Supply, 1/2″ diameter type "L" copper supply piping	1.000	Ea.	.988	37.20	65.50	102.70
Vent, 1-1/2″ diameter copper DWV piping	1.000	Ea.	.593	52.80	39.20	92
Piping, supply, 1/2″ diameter type "L" copper supply piping	42.000	L.F.	4.148	156.24	275.10	431.34
Waste, 4″ diameter DWV piping	10.000	L.F.	2.759	370	164.50	534.50
Vent, 2″ diameter DWV piping	13.000	L.F.	3.250	179.40	193.70	373.10
Carrier, steel, for studs, no arms	1.000	Ea.	1.143	66	76	142
TOTAL		Ea.	38.074	6,030	2,355.60	8,385.60

The costs in this system are on a cost each basis. All necessary piping is included.

Four Fixture Bathroom Price Sheet	QUAN.	UNIT	LABOR HOURS	COST EACH		
				MAT.	INST.	TOTAL
Water closet, close coupled, standard 2 piece, white	1.000	Ea.	3.019	223	180	403
Color	1.000	Ea.	3.019	465	180	645
One piece elongated bowl, white	1.000	Ea.	3.019	850	180	1,030
Color	1.000	Ea.	3.019	1,075	180	1,255
Low profile, one piece elongated bowl, white	1.000	Ea.	3.019	900	180	1,080
Color	1.000	Ea.	3.019	1,075	180	1,255
1/2" copper supply, 4" cast iron waste, 2" cast iron vent	1.000	Ea.	2.376	209	146	355
4" PVC/DWV waste, 2" PVC vent	1.000	Ea.	2.678	96	164	260
4" copper waste, 2" copper vent	1.000	Ea.	2.520	287	160	447
3" cast iron waste, 1-1/2" cast iron vent	1.000	Ea.	2.244	158	138	296
3" P.V.C. waste, 1-1/2" P.V.C. vent	1.000	Ea.	2.388	86	152	238
3" copper waste, 1-1/2" copper vent	1.000	Ea.	2.014	155	128	283
1/2" P.V.C. supply, 4" P.V.C. waste, 2" P.V.C. vent	1.000	Ea.	2.974	116	183	299
3" P.V.C. waste, 1-1/2" P.V.C. vent	1.000	Ea.	2.684	106	172	278
1/2" steel supply, 4" cast iron waste, 2" cast iron vent	1.000	Ea.	2.545	213	157	370
4" cast iron waste, 2" steel vent	1.000	Ea.	2.590	192	159	351
4" P.V.C. waste, 2" P.V.C. vent	1.000	Ea.	2.847	100	175	275
Lavatory, wall hung P.E. cast iron 20" x 18", white	1.000	Ea.	2.000	264	119	383
Color	1.000	Ea.	2.000	277	119	396
Vitreous china 19" x 17", white	1.000	Ea.	2.286	128	136	264
Color	1.000	Ea.	2.286	150	136	286
Lavatory for vanity top, P.E. cast iron 20" x 18", white	1.000	Ea.	2.500	325	149	474
Color	1.000	Ea.	2.500	530	149	679
Steel enameled, 20" x 17", white	1.000	Ea.	2.759	133	165	298
Color	1.000	Ea.	2.500	130	149	279
Vitreous china 20" x 16", white	1.000	Ea.	2.963	226	177	403
Color	1.000	Ea.	2.963	226	177	403
Shower, steel enameled stone base, 36" square, white	1.000	Ea.	8.889	1,925	199	2,124
Color	1.000	Ea.	8.889	1,950	199	2,149
Rough-in, for lavatory or shower						
1/2" copper supply, 1-1/2" cast iron waste, 1-1/2" cast iron vent	1.000	Ea.	3.834	238	239	477
1-1/2" P.V.C. waste, 1-1/4" P.V.C. vent	1.000	Ea.	3.675	151	244	395
1/2" steel supply, 1-1/4" cast iron waste, 1-1/4" steel vent	1.000	Ea.	4.103	203	258	461
1-1/4" P.V.C. waste, 1-1/4" P.V.C. vent	1.000	Ea.	3.937	167	261	428
1/2" P.V.C. supply, 1-1/2" P.V.C. waste, 1-1/2" P.V.C. vent	1.000	Ea.	4.592	202	305	507
Bathtub, P.E. cast iron, 5' long with fittings, white	1.000	Ea.	3.636	1,325	217	1,542
Color	1.000	Ea.	3.636	1,650	217	1,867
Steel, enameled 5' long with fittings, white	1.000	Ea.	2.909	555	174	729
Color	1.000	Ea.	2.909	555	174	729
Rough-in, for bathtub						
1/2" copper supply, 4" cast iron waste, 1-1/2" copper vent	1.000	Ea.	2.409	201	154	355
4" P.V.C. waste, 1-1/2" P.V.C. vent	1.000	Ea.	2.877	107	184	291
1/2" steel supply, 4" cast iron waste, 1-1/2" steel vent	1.000	Ea.	2.898	183	181	364
4" P.V.C. waste, 1-1/2" P.V.C. vent	1.000	Ea.	3.159	114	203	317
1/2" P.V.C. supply, 4" P.V.C. waste, 1-1/2" P.V.C. vent	1.000	Ea.	3.371	141	217	358
Piping, supply, 1/2" copper	42.000	L.F.	4.148	156	275	431
1/2" steel	42.000	L.F.	5.333	184	355	539
1/2" P.V.C.	42.000	L.F.	6.222	298	410	708
Waste, 4" cast iron, no hub	10.000	L.F.	2.759	370	165	535
4" P.V.C./DWV	10.000	L.F.	3.333	115	199	314
4" copper/DWV	10.000	Ea.	4.000	610	240	850
Vent 2" cast iron, no hub	13.000	L.F.	3.105	247	185	432
2" copper/DWV	13.000	L.F.	2.364	267	157	424
2" P.V.C./DWV	13.000	L.F.	3.525	127	211	338
2" steel, galvanized	13.000	L.F.	3.250	179	194	373
Vanity base cabinet, 2 doors, 30" wide	1.000	Ea.	1.000	485	59	544
Vanity top, plastic laminated, square edge	2.670	L.F.	.712	101	42	143
Carrier, steel for studs, no arms	1.000	Ea.	1.143	66	76	142
Wood, 2" x 8" blocking	1.300	L.F.	.052	1.39	3.07	4.46

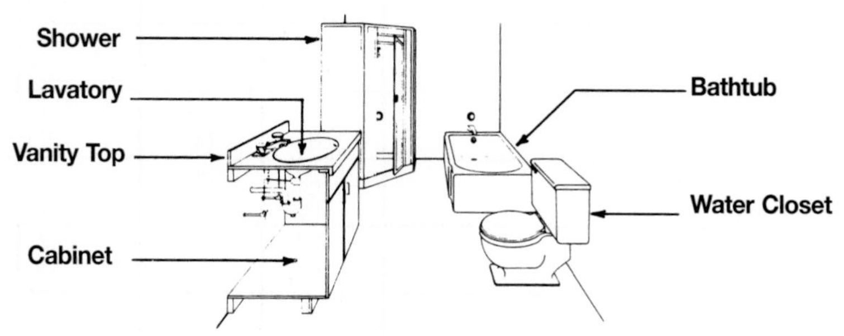

System Description	QUAN.	UNIT	LABOR HOURS	COST EACH MAT.	COST EACH INST.	COST EACH TOTAL
BATHROOM WITH LAVATORY INSTALLED IN VANITY						
Water closet, floor mounted, 2 piece, close coupled, white	1.000	Ea.	3.019	223	180	403
Rough-in, vent, 2" diameter DWV piping	1.000	Ea.	.955	76	57	133
Waste, 4" diameter DWV piping	1.000	Ea.	.828	111	49.35	160.35
Supply, 1/2" diameter type "L" copper supply piping	1.000	Ea.	.593	22.32	39.30	61.62
Lavatory, 20" x 18" P.E. cast iron with fittings, white	1.000	Ea.	2.500	325	149	474
Shower, steel, enameled, stone base, corner, white	1.000	Ea.	3.333	1,925	199	2,124
Mixing valve	1.000	Ea.	1.333	155	88.50	243.50
Shower door	1.000	Ea.	1.000	480	64.50	544.50
Rough-in, waste, 1-1/2" diameter DWV piping	2.000	Ea.	4.507	357	269	626
Supply, 1/2" diameter type "L" copper supply piping	2.000	Ea.	3.161	119.04	209.60	328.64
Bathtub, P.E. cast iron, 5' long with fittings, white	1.000	Ea.	3.636	1,325	217	1,542
Rough-in, waste, 4" diameter DWV piping	1.000	Ea.	.828	111	49.35	160.35
Supply, 1/2" diameter type "L" copper supply piping	1.000	Ea.	.988	37.20	65.50	102.70
Vent, 1-1/2" diameter DWV piping	1.000	Ea.	.593	52.80	39.20	92
Piping, supply, 1/2" diameter type "L" copper supply piping	42.000	L.F.	4.939	186	327.50	513.50
Waste, 4" diameter DWV piping	10.000	L.F.	4.138	555	246.75	801.75
Vent, 2" diameter DWV piping	13.000	L.F.	4.500	248.40	268.20	516.60
Vanity base, 2 doors, 30" wide	1.000	Ea.	1.000	485	59	544
Vanity top, plastic laminated, square edge	2.670	L.F.	.712	134.84	41.92	176.76
TOTAL		Ea.	42.563	6,928.60	2,619.67	9,548.27
BATHROOM WITH WALL HUNG LAVATORY						
Water closet, floor mounted, 2 piece, close coupled, white	1.000	Ea.	3.019	223	180	403
Rough-in, vent, 2" diameter DWV piping	1.000	Ea.	.955	76	57	133
Waste, 4" diameter DWV piping	1.000	Ea.	.828	111	49.35	160.35
Supply, 1/2" diameter type "L" copper supply piping	1.000	Ea.	.593	22.32	39.30	61.62
Lavatory, 20" x 18" P.E. cast iron with fittings, white	1.000	Ea.	2.000	264	119	383
Shower, steel enameled, stone base, corner, white	1.000	Ea.	3.333	1,925	199	2,124
Mixing valve	1.000	Ea.	1.333	155	88.50	243.50
Shower door	1.000	Ea.	1.000	480	64.50	544.50
Rough-in, waste, 1-1/2" diameter DWV piping	2.000	Ea.	4.507	357	269	626
Supply, 1/2" diameter type "L" copper supply piping	2.000	Ea.	3.161	119.04	209.60	328.64
Bathtub, P.E. cast iron, 5" long with fittings, white	1.000	Ea.	3.636	1,325	217	1,542
Rough-in, waste, 4" diameter DWV piping	1.000	Ea.	.828	111	49.35	160.35
Supply, 1/2" diameter type "L" copper supply piping	1.000	Ea.	.988	37.20	65.50	102.70
Vent, 1-1/2" diameter DWV piping	1.000	Ea.	.593	52.80	39.20	92
Piping, supply, 1/2" diameter type "L" copper supply piping	42.000	L.F.	4.939	186	327.50	513.50
Waste, 4" diameter DWV piping	10.000	L.F.	4.138	555	246.75	801.75
Vent, 2" diameter DWV piping	13.000	L.F.	4.500	248.40	268.20	516.60
Carrier, steel for studs, no arms	1.000	Ea.	1.143	66	76	142
TOTAL		Ea.	41.494	6,313.76	2,564.75	8,878.51

The costs in this system are on a cost each basis. All necessary piping is included.

Four Fixture Bathroom Price Sheet	QUAN.	UNIT	LABOR HOURS	COST EACH		
				MAT.	INST.	TOTAL
Water closet, close coupled, standard 2 piece, white	1.000	Ea.	3.019	223	180	403
Color	1.000	Ea.	3.019	465	180	645
One piece, elongated bowl, white	1.000	Ea.	3.019	850	180	1,030
Color	1.000	Ea.	3.019	1,075	180	1,255
Low profile, one piece elongated bowl, white	1.000	Ea.	3.019	900	180	1,080
Color	1.000	Ea.	3.019	1,075	180	1,255
Rough-in, for water closet						
1/2" copper supply, 4" cast iron waste, 2" cast iron vent	1.000	Ea.	2.376	209	146	355
4" PVC/DWV waste, 2" PVC vent	1.000	Ea.	2.678	96	164	260
4" copper waste, 2" copper vent	1.000	Ea.	2.520	287	160	447
3" cast iron waste, 1-1/2" cast iron vent	1.000	Ea.	2.244	158	138	296
3" PVC waste, 1-1/2" PVC vent	1.000	Ea.	2.388	86	152	238
3" PVC waste, 1-1/2" PVC vent	1.000	Ea.	2.014	155	128	283
1/2" PVC supply, 4" PVC waste, 2" PVC vent	1.000	Ea.	2.974	116	183	299
3" PVC waste, 1-1/2" PVC vent	1.000	Ea.	2.684	106	172	278
1/2" steel supply, 4" cast iron waste, 2" cast iron vent	1.000	Ea.	2.545	213	157	370
4" cast iron waste, 2" steel vent	1.000	Ea.	2.590	192	159	351
4" PVC waste, 2" PVC vent	1.000	Ea.	2.847	100	175	275
Lavatory wall hung, P.E. cast iron 20" x 18", white	1.000	Ea.	2.000	264	119	383
Color	1.000	Ea.	2.000	277	119	396
Vitreous china 19" x 17", white	1.000	Ea.	2.286	128	136	264
Color	1.000	Ea.	2.286	150	136	286
Lavatory for vanity top, P.E. cast iron, 20" x 18", white	1.000	Ea.	2.500	325	149	474
Color	1.000	Ea.	2.500	530	149	679
Steel, enameled 20" x 17", white	1.000	Ea.	2.759	133	165	298
Color	1.000	Ea.	2.500	130	149	279
Vitreous china 20" x 16", white	1.000	Ea.	2.963	226	177	403
Color	1.000	Ea.	2.963	226	177	403
Shower, steel enameled, stone base 36" square, white	1.000	Ea.	8.889	1,925	199	2,124
Color	1.000	Ea.	8.889	1,950	199	2,149
Rough-in, for lavatory and shower						
1/2" copper supply, 1-1/2" cast iron waste, 1-1/2" cast iron vent	1.000	Ea.	7.668	475	480	955
1-1/2" PVC waste, 1-1/4" PVC vent	1.000	Ea.	7.352	300	490	790
1/2" steel supply, 1-1/4" cast iron waste, 1-1/4" steel vent	1.000	Ea.	8.205	405	515	920
1-1/4" PVC waste, 1-1/4" PVC vent	1.000	Ea.	7.873	335	520	855
1/2" PVC supply, 1-1/2" PVC waste, 1-1/2" PVC vent	1.000	Ea.	9.185	405	610	1,015
Bathtub, P.E. cast iron, 5' long with fittings, white	1.000	Ea.	3.636	1,325	217	1,542
Color	1.000	Ea.	3.636	1,650	217	1,867
Steel enameled, 5' long with fittings, white	1.000	Ea.	2.909	555	174	729
Color	1.000	Ea.	2.909	555	174	729
Rough-in, for bathtub						
1/2" copper supply, 4" cast iron waste, 1-1/2" copper vent	1.000	Ea.	2.409	201	154	355
4" PVC waste, 1-1/2" PVC vent	1.000	Ea.	2.877	107	184	291
1/2" steel supply, 4" cast iron waste, 1-1/2" steel vent	1.000	Ea.	2.898	183	181	364
4" PVC waste, 1-1/2" PVC vent	1.000	Ea.	3.159	114	203	317
1/2" PVC supply, 4" PVC waste, 1-1/2" PVC vent	1.000	Ea.	3.371	141	217	358
Piping supply, 1/2" copper	42.000	L.F.	4.148	156	275	431
1/2" steel	42.000	L.F.	5.333	184	355	539
1/2" PVC	42.000	L.F.	6.222	298	410	708
Piping, waste, 4" cast iron, no hub	10.000	L.F.	3.586	480	214	694
4" PVC/DWV	10.000	L.F.	4.333	150	259	409
4" copper/DWV	10.000	L.F.	5.200	795	310	1,105
Piping, vent, 2" cast iron, no hub	13.000	L.F.	3.105	247	185	432
2" copper/DWV	13.000	L.F.	2.364	267	157	424
2" PVC/DWV	13.000	L.F.	3.525	127	211	338
2" steel, galvanized	13.000	L.F.	3.250	179	194	373
Vanity base cabinet, 2 doors, 30" wide	1.000	Ea.	1.000	485	59	544
Vanity top, plastic laminated, square edge	3.160	L.F.	.843	120	49.50	169.50
Carrier, steel, for studs, no arms	1.000	Ea.	1.143	66	76	142
Wood, 2" x 8" blocking	1.300	L.F.	.052	1.39	3.07	4.46

44 Five Fixture Bathroom Systems

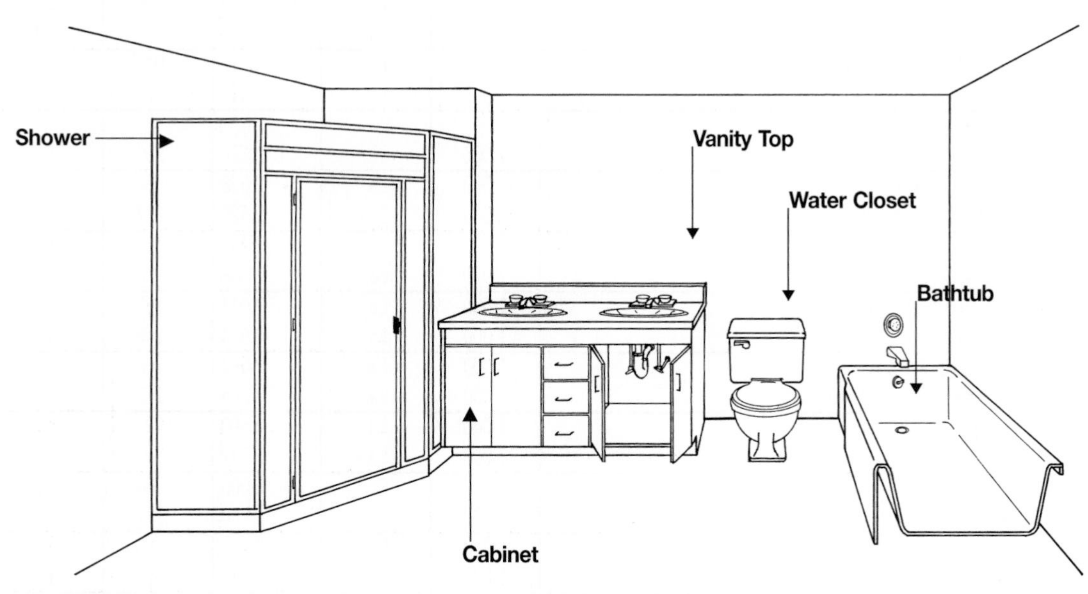

System Description	QUAN.	UNIT	LABOR HOURS	COST EACH MAT.	COST EACH INST.	COST EACH TOTAL
BATHROOM WITH SHOWER, BATHTUB, LAVATORIES IN VANITY						
Water closet, floor mounted, 1 piece, white	1.000	Ea.	3.019	900	180	1,080
Rough-in, vent, 2″ diameter DWV piping	1.000	Ea.	.955	76	57	133
Waste, 4″ diameter DWV piping	1.000	Ea.	.828	111	49.35	160.35
Supply, 1/2″ diameter type "L" copper supply piping	1.000	Ea.	.593	22.32	39.30	61.62
Lavatory, 20″ x 16″, vitreous china oval, with fittings, white	2.000	Ea.	5.926	452	354	806
Shower, steel enameled, stone base, corner, white	1.000	Ea.	3.333	1,925	199	2,124
Mixing valve	1.000	Ea.	1.333	155	88.50	243.50
Shower door	1.000	Ea.	1.000	480	64.50	544.50
Rough-in, waste, 1-1/2″ diameter DWV piping	3.000	Ea.	5.408	428.40	322.80	751.20
Supply, 1/2″ diameter type "L" copper supply piping	3.000	Ea.	2.963	111.60	196.50	308.10
Bathtub, P.E. cast iron, 5′ long with fittings, white	1.000	Ea.	3.636	1,325	217	1,542
Rough-in, waste, 4″ diameter DWV piping	1.000	Ea.	1.103	148	65.80	213.80
Supply, 1/2″ diameter type "L" copper supply piping	1.000	Ea.	.988	37.20	65.50	102.70
Vent, 1-1/2″ diameter copper DWV piping	1.000	Ea.	.593	52.80	39.20	92
Piping, supply, 1/2″ diameter type "L" copper supply piping	42.000	L.F.	4.148	156.24	275.10	431.34
Waste, 4″ diameter DWV piping	10.000	L.F.	2.759	370	164.50	534.50
Vent, 2″ diameter DWV piping	13.000	L.F.	3.250	179.40	193.70	373.10
Vanity base, 2 door, 24″ x 48″	1.000	Ea.	1.400	655	82.50	737.50
Vanity top, plastic laminated, square edge	4.170	L.F.	1.112	158.46	65.47	223.93
TOTAL		Ea.	44.347	7,743.42	2,719.72	10,463.14

The costs in this system are on a cost each basis. All necessary piping is included.

Description	QUAN.	UNIT	LABOR HOURS	COST EACH MAT.	COST EACH INST.	COST EACH TOTAL

Five Fixture Bathroom Price Sheet	QUAN.	UNIT	LABOR HOURS	COST EACH		
				MAT.	INST.	TOTAL
Water closet, close coupled, standard 2 piece, white	1.000	Ea.	3.019	223	180	403
Color	1.000	Ea.	3.019	465	180	645
One piece elongated bowl, white	1.000	Ea.	3.019	850	180	1,030
Color	1.000	Ea.	3.019	1,075	180	1,255
Low profile, one piece elongated bowl, white	1.000	Ea.	3.019	900	180	1,080
Color	1.000	Ea.	3.019	1,075	180	1,255
Rough-in, supply, waste and vent for water closet						
1/2" copper supply, 4" cast iron waste, 2" cast iron vent	1.000	Ea.	2.376	209	146	355
4" P.V.C./DWV waste, 2" P.V.C. vent	1.000	Ea.	2.678	96	164	260
4" copper waste, 2" copper vent	1.000	Ea.	2.520	287	160	447
3" cast iron waste, 1-1/2" cast iron vent	1.000	Ea.	2.244	158	138	296
3" P.V.C. waste, 1-1/2" P.V.C. vent	1.000	Ea.	2.388	86	152	238
3" copper waste, 1-1/2" copper vent	1.000	Ea.	2.014	155	128	283
1/2" P.V.C. supply, 4" P.V.C. waste, 2" P.V.C. vent	1.000	Ea.	2.974	116	183	299
3" P.V.C. waste, 1-1/2" P.V.C. supply	1.000	Ea.	2.684	106	172	278
1/2" steel supply, 4" cast iron waste, 2" cast iron vent	1.000	Ea.	2.545	213	157	370
4" cast iron waste, 2" steel vent	1.000	Ea.	2.590	192	159	351
4" P.V.C. waste, 2" P.V.C. vent	1.000	Ea.	2.847	100	175	275
Lavatory, wall hung, P.E. cast iron 20" x 18", white	2.000	Ea.	4.000	530	238	768
Color	2.000	Ea.	4.000	555	238	793
Vitreous china, 19" x 17", white	2.000	Ea.	4.571	256	272	528
Color	2.000	Ea.	4.571	300	272	572
Lavatory, for vanity top, P.E. cast iron, 20" x 18", white	2.000	Ea.	5.000	650	298	948
Color	2.000	Ea.	5.000	1,050	298	1,348
Steel enameled 20" x 17", white	2.000	Ea.	5.517	266	330	596
Color	2.000	Ea.	5.000	260	298	558
Vitreous china 20" x 16", white	2.000	Ea.	5.926	450	355	805
Color	2.000	Ea.	5.926	450	355	805
Shower, steel enameled, stone base 36" square, white	1.000	Ea.	8.889	1,925	199	2,124
Color	1.000	Ea.	8.889	1,950	199	2,149
Rough-in, for lavatory or shower						
1/2" copper supply, 1-1/2" cast iron waste, 1-1/2" cast iron vent	3.000	Ea.	8.371	540	520	1,060
1-1/2" P.V.C. waste, 1-1/4" P.V.C. vent	3.000	Ea.	7.916	335	525	860
1/2" steel supply, 1-1/4" cast iron waste, 1-1/4" steel vent	3.000	Ea.	8.670	425	540	965
1-1/4" P.V.C. waste, 1-1/4" P.V.C. vent	3.000	Ea.	8.381	365	555	920
1/2" P.V.C. supply, 1-1/2" P.V.C. waste, 1-1/2" P.V.C. vent	3.000	Ea.	9.778	425	650	1,075
Bathtub, P.E. cast iron 5' long with fittings, white	1.000	Ea.	3.636	1,325	217	1,542
Color	1.000	Ea.	3.636	1,650	217	1,867
Steel, enameled 5' long with fittings, white	1.000	Ea.	2.909	555	174	729
Color	1.000	Ea.	2.909	555	174	729
Rough-in, for bathtub						
1/2" copper supply, 4" cast iron waste, 1-1/2" copper vent	1.000	Ea.	2.684	238	171	409
4" P.V.C. waste, 1-1/2" P.V.C. vent	1.000	Ea.	3.210	118	204	322
1/2" steel supply, 4" cast iron waste, 1-1/2" steel vent	1.000	Ea.	3.173	220	198	418
4" P.V.C. waste, 1-1/2" P.V.C. vent	1.000	Ea.	3.492	125	223	348
1/2" P.V.C. supply, 4" P.V.C. waste, 1-1/2" P.V.C. vent	1.000	Ea.	3.704	152	237	389
Piping, supply, 1/2" copper	42.000	L.F.	4.148	156	275	431
1/2" steel	42.000	L.F.	5.333	184	355	539
1/2" P.V.C.	42.000	L.F.	6.222	298	410	708
Piping, waste, 4" cast iron, no hub	10.000	L.F.	2.759	370	165	535
4" P.V.C./DWV	10.000	L.F.	3.333	115	199	314
4" copper/DWV	10.000	L.F.	4.000	610	240	850
Piping, vent, 2" cast iron, no hub	13.000	L.F.	3.105	247	185	432
2" copper/DWV	13.000	L.F.	2.364	267	157	424
2" P.V.C./DWV	13.000	L.F.	3.525	127	211	338
2" steel, galvanized	13.000	L.F.	3.250	179	194	373
Vanity base cabinet, 2 doors, 24" x 48"	1.000	Ea.	1.400	655	82.50	737.50
Vanity top, plastic laminated, square edge	4.170	L.F.	1.112	158	65.50	223.50
Carrier, steel, for studs, no arms	1.000	Ea.	1.143	66	76	142
Wood, 2" x 8" blocking	1.300	L.F.	.052	1.39	3.07	4.46

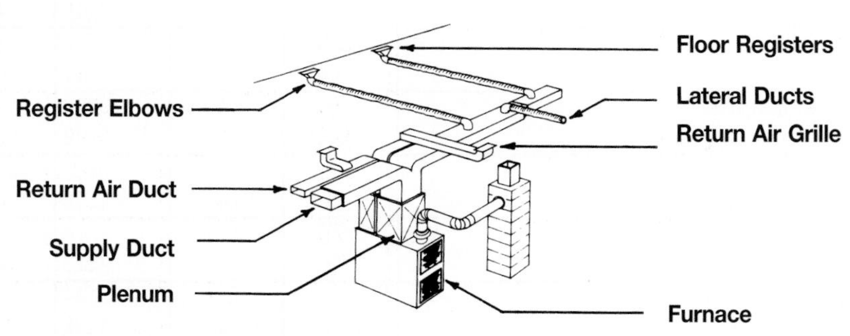

System Description	QUAN.	UNIT	LABOR HOURS	COST PER SYSTEM		
				MAT.	INST.	TOTAL
HEATING ONLY, GAS FIRED HOT AIR, ONE ZONE, 1200 S.F. BUILDING						
Furnace, gas, up flow	1.000	Ea.	5.000	860	291	1,151
Intermittent pilot	1.000	Ea.		283		283
Supply duct, rigid fiberglass	176.000	S.F.	12.068	174.24	728.64	902.88
Return duct, sheet metal, galvanized	158.000	Lb.	16.137	99.54	971.70	1,071.24
Lateral ducts, 6″ flexible fiberglass	144.000	L.F.	8.862	496.80	515.52	1,012.32
Register, elbows	12.000	Ea.	6.400	195.60	372	567.60
Floor registers, enameled steel	12.000	Ea.	3.000	179.40	194.40	373.80
Floor grille, return air	2.000	Ea.	.727	62	47	109
Thermostat	1.000	Ea.	1.000	46.50	68	114.50
Plenum	1.000	Ea.	1.000	108	58	166
TOTAL		System	54.194	2,505.08	3,246.26	5,751.34
HEATING/COOLING, GAS FIRED FORCED AIR, ONE ZONE, 1200 S.F. BUILDING						
Furnace, including plenum, compressor, coil	1.000	Ea.	14.720	6,003	855.60	6,858.60
Intermittent pilot	1.000	Ea.		283		283
Supply duct, rigid fiberglass	176.000	S.F.	12.068	174.24	728.64	902.88
Return duct, sheet metal, galvanized	158.000	Lb.	16.137	99.54	971.70	1,071.24
Lateral duct, 6″ flexible fiberglass	144.000	L.F.	8.862	496.80	515.52	1,012.32
Register elbows	12.000	Ea.	6.400	195.60	372	567.60
Floor registers, enameled steel	12.000	Ea.	3.000	179.40	194.40	373.80
Floor grille return air	2.000	Ea.	.727	62	47	109
Thermostat	1.000	Ea.	1.000	46.50	68	114.50
Refrigeration piping, 25 ft. (pre-charged)	1.000	Ea.		310		310
TOTAL		System	62.914	7,850.08	3,752.86	11,602.94

The costs in these systems are based on complete system basis. For larger buildings use the price sheet on the opposite page.

Description	QUAN.	UNIT	LABOR HOURS	COST PER SYSTEM		
				MAT.	INST.	TOTAL

Gas Heating/Cooling Price Sheet	QUAN.	UNIT	LABOR HOURS	COST EACH		
				MAT.	INST.	TOTAL
Furnace, heating only, 100 MBH, area to 1200 S.F.	1.000	Ea.	5.000	860	291	1,151
120 MBH, area to 1500 S.F.	1.000	Ea.	5.000	860	291	1,151
160 MBH, area to 2000 S.F.	1.000	Ea.	5.714	2,000	335	2,335
200 MBH, area to 2400 S.F.	1.000	Ea.	6.154	3,750	360	4,110
Heating/cooling, 100 MBH heat, 36 MBH cool, to 1200 S.F.	1.000	Ea.	16.000	6,525	930	7,455
120 MBH heat, 42 MBH cool, to 1500 S.F.	1.000	Ea.	18.462	6,975	1,125	8,100
144 MBH heat, 47 MBH cool, to 2000 S.F.	1.000	Ea.	20.000	8,050	1,200	9,250
200 MBH heat, 60 MBH cool, to 2400 S.F.	1.000	Ea.	34.286	8,475	2,075	10,550
Intermittent pilot, 100 MBH furnace	1.000	Ea.		283		283
200 MBH furnace	1.000	Ea.		283		283
Supply duct, rectangular, area to 1200 S.F., rigid fiberglass	176.000	S.F.	12.068	174	730	904
Sheet metal insulated	228.000	Lb.	31.331	1,050	3,275	4,325
Area to 1500 S.F., rigid fiberglass	176.000	S.F.	12.068	174	730	904
Sheet metal insulated	228.000	Lb.	31.331	1,050	3,275	4,325
Area to 2400 S.F., rigid fiberglass	205.000	S.F.	14.057	203	850	1,053
Sheet metal insulated	271.000	Lb.	37.048	1,225	3,850	5,075
Round flexible, insulated 6″ diameter, to 1200 S.F.	156.000	L.F.	9.600	540	560	1,100
To 1500 S.F.	184.000	L.F.	11.323	635	660	1,295
8″ diameter, to 2000 S.F.	269.000	L.F.	23.911	1,175	1,400	2,575
To 2400 S.F.	248.000	L.F.	22.045	1,075	1,300	2,375
Return duct, sheet metal galvanized, to 1500 S.F.	158.000	Lb.	16.137	99.50	970	1,069.50
To 2400 S.F.	191.000	Lb.	19.507	120	1,175	1,295
Lateral ducts, flexible round 6″ insulated, to 1200 S.F.	144.000	L.F.	8.862	495	515	1,010
To 1500 S.F.	172.000	L.F.	10.585	595	615	1,210
To 2000 S.F.	261.000	L.F.	16.062	900	935	1,835
To 2400 S.F.	300.000	L.F.	18.462	1,025	1,075	2,100
Spiral steel insulated, to 1200 S.F.	144.000	L.F.	20.067	1,650	3,225	4,875
To 1500 S.F.	172.000	L.F.	23.952	1,950	3,850	5,800
To 2000 S.F.	261.000	L.F.	36.352	2,975	5,850	8,825
To 2400 S.F.	300.000	L.F.	41.825	3,425	6,725	10,150
Rectangular sheet metal galvanized insulated, to 1200 S.F.	228.000	Lb.	39.056	1,900	5,050	6,950
To 1500 S.F.	344.000	Lb.	53.966	2,325	6,475	8,800
To 2000 S.F.	522.000	Lb.	81.926	3,525	9,850	13,375
To 2400 S.F.	600.000	Lb.	94.189	4,050	11,300	15,350
Register elbows, to 1500 S.F.	12.000	Ea.	6.400	196	370	566
To 2400 S.F.	14.000	Ea.	7.460	228	435	663
Floor registers, enameled steel w/damper, to 1500 S.F.	12.000	Ea.	3.000	179	194	373
To 2400 S.F.	14.000	Ea.	4.308	246	279	525
Return air grille, area to 1500 S.F. 12″ x 12″	2.000	Ea.	.727	62	47	109
Area to 2400 S.F. 8″ x 16″	2.000	Ea.	.444	41.50	29	70.50
Area to 2400 S.F. 8″ x 16″	2.000	Ea.	.727	63	47	110
16″ x 16″	1.000	Ea.	.364	40	23.50	63.50
Thermostat, manual, 1 set back	1.000	Ea.	1.000	46.50	68	114.50
Electric, timed, 1 set back	1.000	Ea.	1.000	83	68	151
2 set back	1.000	Ea.	1.000	247	68	315
Plenum, heating only, 100 M.B.H.	1.000	Ea.	1.000	108	58	166
120 MBH	1.000	Ea.	1.000	108	58	166
160 MBH	1.000	Ea.	1.000	108	58	166
200 MBH	1.000	Ea.	1.000	108	58	166
Refrigeration piping, 3/8″	25.000	L.F.		35.50		35.50
3/4″	25.000	L.F.		78		78
7/8″	25.000	L.F.		91		91
Refrigerant piping, 25 ft. (precharged)	1.000	Ea.		310		310
Diffusers, ceiling, 6″ diameter, to 1500 S.F.	10.000	Ea.	4.444	138	290	428
To 2400 S.F.	12.000	Ea.	6.000	175	390	565
Floor, aluminum, adjustable, 2-1/4″ x 12″ to 1500 S.F.	12.000	Ea.	3.000	137	194	331
To 2400 S.F.	14.000	Ea.	3.500	160	227	387
Side wall, aluminum, adjustable, 8″ x 4″, to 1500 S.F.	12.000	Ea.	3.000	246	194	440
5″ x 10″ to 2400 S.F.	12.000	Ea.	3.692	335	239	574

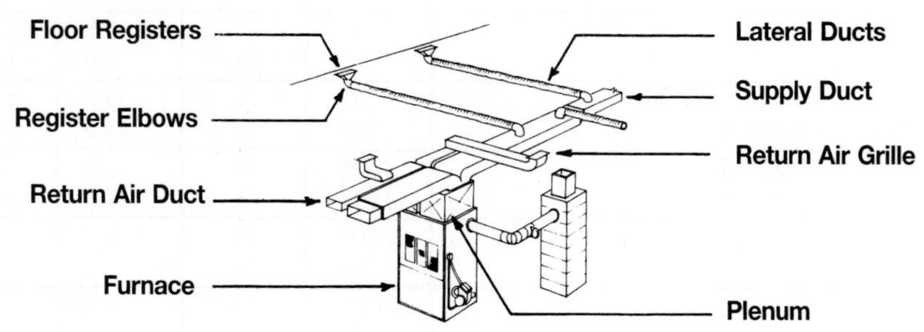

System Description	QUAN.	UNIT	LABOR HOURS	COST PER SYSTEM		
				MAT.	INST.	TOTAL
HEATING ONLY, OIL FIRED HOT AIR, ONE ZONE, 1200 S.F. BUILDING						
Furnace, oil fired, atomizing gun type burner	1.000	Ea.	4.571	3,250	266	3,516
3/8" diameter copper supply pipe	1.000	Ea.	2.759	124.80	183	307.80
Shut off valve	1.000	Ea.	.333	19.10	22	41.10
Oil tank, 275 gallon, on legs	1.000	Ea.	3.200	570	196	766
Supply duct, rigid fiberglass	176.000	S.F.	12.068	174.24	728.64	902.88
Return duct, sheet metal, galvanized	158.000	Lb.	16.137	99.54	971.70	1,071.24
Lateral ducts, 6" flexible fiberglass	144.000	L.F.	8.862	496.80	515.52	1,012.32
Register elbows	12.000	Ea.	6.400	195.60	372	567.60
Floor register, enameled steel	12.000	Ea.	3.000	179.40	194.40	373.80
Floor grille, return air	2.000	Ea.	.727	62	47	109
Thermostat	1.000	Ea.	1.000	46.50	68	114.50
TOTAL		System	59.057	5,217.98	3,564.26	8,782.24
HEATING/COOLING, OIL FIRED, FORCED AIR, ONE ZONE, 1200 S.F. BUILDING						
Furnace, including plenum, compressor, coil	1.000	Ea.	16.000	6,975	930	7,905
3/8" diameter copper supply pipe	1.000	Ea.	2.759	124.80	183	307.80
Shut off valve	1.000	Ea.	.333	19.10	22	41.10
Oil tank, 275 gallon on legs	1.000	Ea.	3.200	570	196	766
Supply duct, rigid fiberglass	176.000	S.F.	12.068	174.24	728.64	902.88
Return duct, sheet metal, galvanized	158.000	Lb.	16.137	99.54	971.70	1,071.24
Lateral ducts, 6" flexible fiberglass	144.000	L.F.	8.862	496.80	515.52	1,012.32
Register elbows	12.000	Ea.	6.400	195.60	372	567.60
Floor registers, enameled steel	12.000	Ea.	3.000	179.40	194.40	373.80
Floor grille, return air	2.000	Ea.	.727	62	47	109
Refrigeration piping (precharged)	25.000	L.F.		310		310
TOTAL		System	69.486	9,206.48	4,160.26	13,366.74

Description	QUAN.	UNIT	LABOR HOURS	COST PER SYSTEM		
				MAT.	INST.	TOTAL

Oil Fired Heating/Cooling Price Sheet	QUAN.	UNIT	LABOR HOURS	COST EACH		
				MAT.	INST.	TOTAL
Furnace, heating, 95.2 MBH,area to 1200 S.F.	1.000	Ea.	4.706	3,175	274	3,449
123.2 MBH, area to 1500 S.F.	1.000	Ea.	5.000	3,200	291	3,491
151.2 MBH, area to 2000 S.F.	1.000	Ea.	5.333	3,550	310	3,860
200 MBH, area to 2400 S.F.	1.000	Ea.	6.154	4,150	360	4,510
Heating/cooling, 95.2 MBH heat, 36 MBH cool, to 1200 S.F.	1.000	Ea.	16.000	6,975	930	7,905
112 MBH heat, 42 MBH cool, to 1500 S.F.	1.000	Ea.	24.000	10,500	1,400	11,900
151 MBH heat, 47 MBH cool, to 2000 S.F.	1.000	Ea.	20.800	9,075	1,200	10,275
184.8 MBH heat, 60 MBH cool, to 2400 S.F.	1.000	Ea.	24.000	5,975	1,450	7,425
Oil piping to furnace, 3/8" dia., copper	1.000	Ea.	3.412	283	224	507
Oil tank, on legs above ground, 275 gallons	1.000	Ea.	3.200	570	196	766
550 gallons	1.000	Ea.	5.926	4,875	365	5,240
Below ground, 275 gallons	1.000	Ea.	3.200	570	196	766
550 gallons	1.000	Ea.	5.926	4,875	365	5,240
1000 gallons	1.000	Ea.	6.400	8,125	395	8,520
Supply duct, rectangular, area to 1200 S.F., rigid fiberglass	176.000	S.F.	12.068	174	730	904
Sheet metal, insulated	228.000	Lb.	31.331	1,050	3,275	4,325
Area to 1500 S.F., rigid fiberglass	176.000	S.F.	12.068	174	730	904
Sheet metal, insulated	228.000	Lb.	31.331	1,050	3,275	4,325
Area to 2400 S.F., rigid fiberglass	205.000	S.F.	14.057	203	850	1,053
Sheet metal, insulated	271.000	Lb.	37.048	1,225	3,850	5,075
Round flexible, insulated, 6" diameter to 1200 S.F.	156.000	L.F.	9.600	540	560	1,100
To 1500 S.F.	184.000	L.F.	11.323	635	660	1,295
8" diameter to 2000 S.F.	269.000	L.F.	23.911	1,175	1,400	2,575
To 2400 S.F.	269.000	L.F.	22.045	1,075	1,300	2,375
Return duct, sheet metal galvanized, to 1500 S.F.	158.000	Lb.	16.137	99.50	970	1,069.50
To 2400 S.F.	191.000	Lb.	19.507	120	1,175	1,295
Lateral ducts, flexible round, 6", insulated to 1200 S.F.	144.000	L.F.	8.862	495	515	1,010
To 1500 S.F.	172.000	L.F.	10.585	595	615	1,210
To 2000 S.F.	261.000	L.F.	16.062	900	935	1,835
To 2400 S.F.	300.000	L.F.	18.462	1,025	1,075	2,100
Spiral steel, insulated to 1200 S.F.	144.000	L.F.	20.067	1,650	3,225	4,875
To 1500 S.F.	172.000	L.F.	23.952	1,950	3,850	5,800
To 2000 S.F.	261.000	L.F.	36.352	2,975	5,850	8,825
To 2400 S.F.	300.000	L.F.	41.825	3,425	6,725	10,150
Rectangular sheet metal galvanized insulated, to 1200 S.F.	288.000	Lb.	45.183	1,950	5,425	7,375
To 1500 S.F.	344.000	Lb.	53.966	2,325	6,475	8,800
To 2000 S.F.	522.000	Lb.	81.926	3,525	9,850	13,375
To 2400 S.F.	600.000	Lb.	94.189	4,050	11,300	15,350
Register elbows, to 1500 S.F.	12.000	Ea.	6.400	196	370	566
To 2400 S.F.	14.000	Ea.	7.470	228	435	663
Floor registers, enameled steel w/damper, to 1500 S.F.	12.000	Ea.	3.000	179	194	373
To 2400 S.F.	14.000	Ea.	4.308	246	279	525
Return air grille, area to 1500 S.F., 12" x 12"	2.000	Ea.	.727	62	47	109
12" x 24"	1.000	Ea.	.444	41.50	29	70.50
Area to 2400 S.F., 8" x 16"	2.000	Ea.	.727	63	47	110
16" x 16"	1.000	Ea.	.364	40	23.50	63.50
Thermostat, manual, 1 set back	1.000	Ea.	1.000	46.50	68	114.50
Electric, timed, 1 set back	1.000	Ea.	1.000	83	68	151
2 set back	1.000	Ea.	1.000	247	68	315
Refrigeration piping, 3/8"	25.000	L.F.		35.50		35.50
3/4"	25.000	L.F.		78		78
Diffusers, ceiling, 6" diameter, to 1500 S.F.	10.000	Ea.	4.444	138	290	428
To 2400 S.F.	12.000	Ea.	6.000	175	390	565
Floor, aluminum, adjustable, 2-1/4" x 12" to 1500 S.F.	12.000	Ea.	3.000	137	194	331
To 2400 S.F.	14.000	Ea.	3.500	160	227	387
Side wall, aluminum, adjustable, 8" x 4", to 1500 S.F.	12.000	Ea.	3.000	246	194	440
5" x 10" to 2400 S.F.	12.000	Ea.	3.692	335	239	574

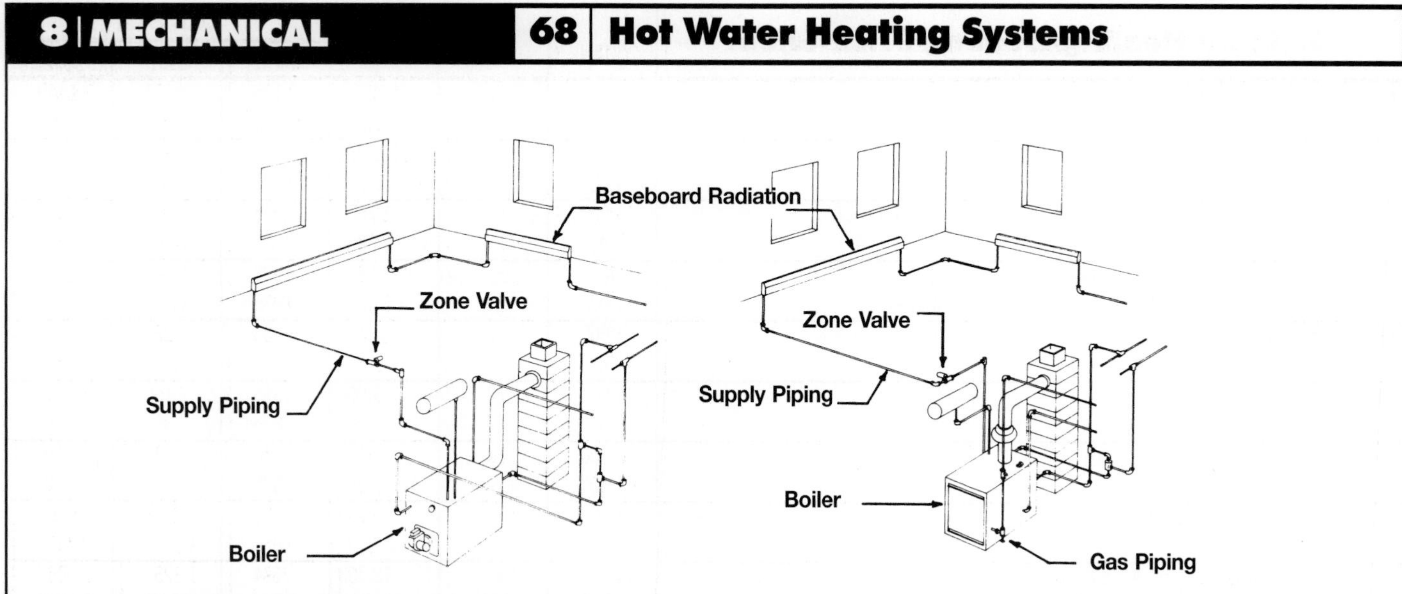

System Description	QUAN.	UNIT	LABOR HOURS	COST EACH		
				MAT.	INST.	TOTAL
OIL FIRED HOT WATER HEATING SYSTEM, AREA TO 1200 S.F.						
Boiler package, oil fired, 97 MBH, area to 1200 S.F. building	1.000	Ea.	15.000	2,050	885	2,935
3/8" diameter copper supply pipe	1.000	Ea.	2.759	124.80	183	307.80
Shut off valve	1.000	Ea.	.333	19.10	22	41.10
Oil tank, 275 gallon, with black iron filler pipe	1.000	Ea.	3.200	570	196	766
Supply piping, 3/4" copper tubing	176.000	L.F.	18.526	848.32	1,232	2,080.32
Supply fittings, copper 3/4"	36.000	Ea.	15.158	95.04	1,008	1,103.04
Supply valves, 3/4"	2.000	Ea.	.800	308	53	361
Baseboard radiation, 3/4"	106.000	L.F.	35.333	911.60	2,173	3,084.60
Zone valve	1.000	Ea.	.400	158	27.50	185.50
TOTAL		Ea.	91.509	5,084.86	5,779.50	10,864.36
OIL FIRED HOT WATER HEATING SYSTEM, AREA TO 2400 S.F.						
Boiler package, oil fired, 225 MBH, area to 2400 S.F. building	1.000	Ea.	25.105	4,925	1,475	6,400
3/8" diameter copper supply pipe	1.000	Ea.	2.759	124.80	183	307.80
Shut off valve	1.000	Ea.	.333	19.10	22	41.10
Oil tank, 550 gallon, with black iron pipe filler pipe	1.000	Ea.	5.926	4,875	365	5,240
Supply piping, 3/4" copper tubing	228.000	L.F.	23.999	1,098.96	1,596	2,694.96
Supply fittings, copper	46.000	Ea.	19.368	121.44	1,288	1,409.44
Supply valves	2.000	Ea.	.800	308	53	361
Baseboard radiation	212.000	L.F.	70.666	1,823.20	4,346	6,169.20
Zone valve	1.000	Ea.	.400	158	27.50	185.50
TOTAL		Ea.	149.356	13,453.50	9,355.50	22,809

The costs in this system are on a cost each basis. The costs represent total cost for the system based on a gross square foot of plan area.

Description	QUAN.	UNIT	LABOR HOURS	COST EACH		
				MAT.	INST.	TOTAL

Hot Water Heating Price Sheet	QUAN.	UNIT	LABOR HOURS	COST EACH		
				MAT.	INST.	TOTAL
Boiler, oil fired, 97 MBH, area to 1200 S.F.	1.000	Ea.	15.000	2,050	885	2,935
118 MBH, area to 1500 S.F.	1.000	Ea.	16.506	2,225	975	3,200
161 MBH, area to 2000 S.F.	1.000	Ea.	18.405	2,500	1,075	3,575
215 MBH, area to 2400 S.F.	1.000	Ea.	19.704	3,975	2,300	6,275
Oil piping, (valve & filter), 3/8" copper	1.000	Ea.	3.289	144	205	349
1/4" copper	1.000	Ea.	3.242	164	214	378
Oil tank, filler pipe and cap on legs, 275 gallon	1.000	Ea.	3.200	570	196	766
550 gallon	1.000	Ea.	5.926	4,875	365	5,240
Buried underground, 275 gallon	1.000	Ea.	3.200	570	196	766
550 gallon	1.000	Ea.	5.926	4,875	365	5,240
1000 gallon	1.000	Ea.	6.400	8,125	395	8,520
Supply piping copper, area to 1200 S.F., 1/2" tubing	176.000	L.F.	17.384	655	1,150	1,805
3/4" tubing	176.000	L.F.	18.526	850	1,225	2,075
Area to 1500 S.F., 1/2" tubing	186.000	L.F.	18.371	690	1,225	1,915
3/4" tubing	186.000	L.F.	19.578	895	1,300	2,195
Area to 2000 S.F., 1/2" tubing	204.000	L.F.	20.149	760	1,325	2,085
3/4" tubing	204.000	L.F.	21.473	985	1,425	2,410
Area to 2400 S.F., 1/2" tubing	228.000	L.F.	22.520	850	1,500	2,350
3/4" tubing	228.000	L.F.	23.999	1,100	1,600	2,700
Supply pipe fittings copper, area to 1200 S.F., 1/2"	36.000	Ea.	14.400	47.50	955	1,002.50
3/4"	36.000	Ea.	15.158	95	1,000	1,095
Area to 1500 S.F., 1/2"	40.000	Ea.	16.000	53	1,050	1,103
3/4"	40.000	Ea.	16.842	106	1,125	1,231
Area to 2000 S.F., 1/2"	44.000	Ea.	17.600	58	1,175	1,233
3/4"	44.000	Ea.	18.526	116	1,225	1,341
Area to 2400, S.F., 1/2"	46.000	Ea.	18.400	60.50	1,225	1,285.50
3/4"	46.000	Ea.	19.368	121	1,300	1,421
Supply valves, 1/2" pipe size	2.000	Ea.	.667	226	44	270
3/4"	2.000	Ea.	.800	310	53	363
Baseboard radiation, area to 1200 S.F., 1/2" tubing	106.000	L.F.	28.267	1,375	1,725	3,100
3/4" tubing	106.000	L.F.	35.333	910	2,175	3,085
Area to 1500 S.F., 1/2" tubing	134.000	L.F.	35.734	1,725	2,200	3,925
3/4" tubing	134.000	L.F.	44.666	1,150	2,750	3,900
Area to 2000 S.F., 1/2" tubing	178.000	L.F.	47.467	2,300	2,900	5,200
3/4" tubing	178.000	L.F.	59.333	1,525	3,650	5,175
Area to 2400 S.F., 1/2" tubing	212.000	L.F.	56.534	2,725	3,475	6,200
3/4" tubing	212.000	L.F.	70.666	1,825	4,350	6,175
Zone valves, 1/2" tubing	1.000	Ea.	.400	158	27.50	185.50
3/4" tubing	1.000	Ea.	.400	153	27.50	180.50

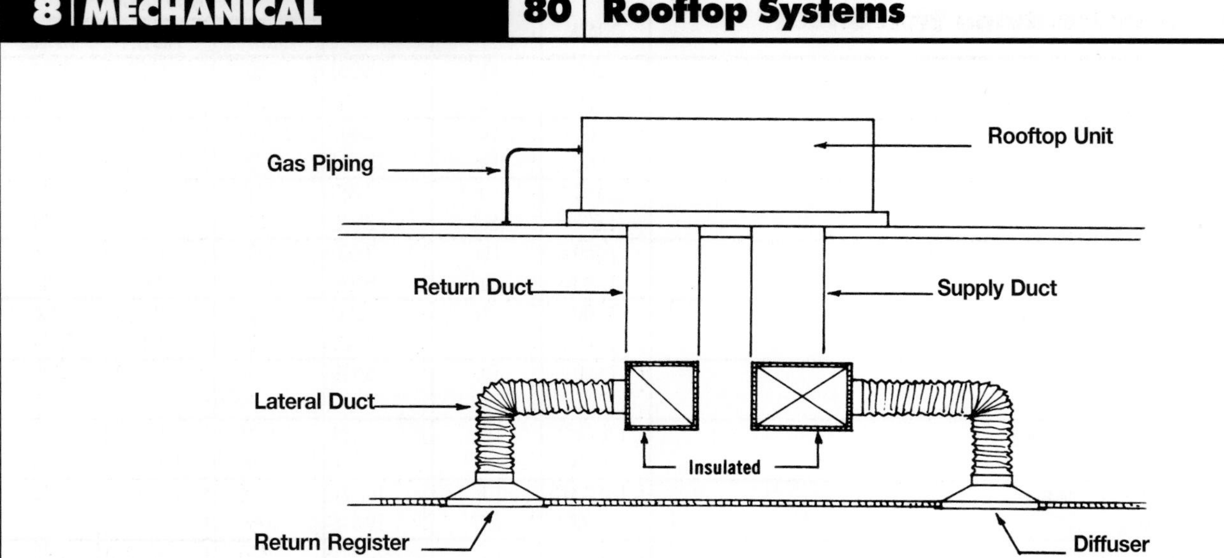

System Description	QUAN.	UNIT	LABOR HOURS	COST EACH		
				MAT.	INST.	TOTAL
ROOFTOP HEATING/COOLING UNIT, AREA TO 2000 S.F.						
Rooftop unit, single zone, electric cool, gas heat, to 2000 S.F.	1.000	Ea.	28.521	4,850	1,750	6,600
Gas piping	34.500	L.F.	5.207	168.02	345	513.02
Duct, supply and return, galvanized steel	38.000	Lb.	3.881	23.94	233.70	257.64
Insulation, ductwork	33.000	S.F.	6.286	168.30	349.80	518.10
Lateral duct, flexible duct 12″ diameter, insulated	72.000	L.F.	11.520	439.20	669.60	1,108.80
Diffusers	4.000	Ea.	4.571	1,240	296	1,536
Return registers	1.000	Ea.	.727	128	47	175
TOTAL		Ea.	60.713	7,017.46	3,691.10	10,708.56
ROOFTOP HEATING/COOLING UNIT, AREA TO 5000 S.F.						
Rooftop unit, single zone, electric cool, gas heat, to 5000 S.F.	1.000	Ea.	42.032	14,400	2,475	16,875
Gas piping	86.250	L.F.	13.019	420.04	862.50	1,282.54
Duct supply and return, galvanized steel	95.000	Lb.	9.702	59.85	584.25	644.10
Insulation, ductwork	82.000	S.F.	15.619	418.20	869.20	1,287.40
Lateral duct, flexible duct, 12″ diameter, insulated	180.000	L.F.	28.800	1,098	1,674	2,772
Diffusers	10.000	Ea.	11.429	3,100	740	3,840
Return registers	3.000	Ea.	2.182	384	141	525
TOTAL		Ea.	122.783	19,880.09	7,345.95	27,226.04

Description	QUAN.	UNIT	LABOR HOURS	COST EACH		
				MAT.	INST.	TOTAL

Rooftop Price Sheet	QUAN.	UNIT	LABOR HOURS	COST EACH		
				MAT.	INST.	TOTAL
Rooftop unit, single zone, electric cool, gas heat to 2000 S.F.	1.000	Ea.	28.521	4,850	1,750	6,600
Area to 3000 S.F.	1.000	Ea.	35.982	10,100	2,125	12,225
Area to 5000 S.F.	1.000	Ea.	42.032	14,400	2,475	16,875
Area to 10000 S.F.	1.000	Ea.	68.376	39,300	4,200	43,500
Gas piping, area 2000 through 4000 S.F.	34.500	L.F.	5.207	168	345	513
Area 5000 to 10000 S.F.	86.250	L.F.	13.019	420	865	1,285
Duct, supply and return, galvanized steel, to 2000 S.F.	38.000	Lb.	3.881	24	234	258
Area to 3000 S.F.	57.000	Lb.	5.821	36	350	386
Area to 5000 S.F.	95.000	Lb.	9.702	60	585	645
Area to 10000 S.F.	190.000	Lb.	19.405	120	1,175	1,295
Rigid fiberglass, area to 2000 S.F.	33.000	S.F.	2.263	32.50	137	169.50
Area to 3000 S.F.	49.000	S.F.	3.360	48.50	203	251.50
Area to 5000 S.F.	82.000	S.F.	5.623	81	340	421
Area to 10000 S.F.	164.000	S.F.	11.245	162	680	842
Insulation, supply and return, blanket type, area to 2000 S.F.	33.000	S.F.	1.508	168	350	518
Area to 3000 S.F.	49.000	S.F.	2.240	250	520	770
Area to 5000 S.F.	82.000	S.F.	3.748	420	870	1,290
Area to 10000 S.F.	164.000	S.F.	7.496	835	1,750	2,585
Lateral ducts, flexible round, 12″ insulated, to 2000 S.F.	72.000	L.F.	11.520	440	670	1,110
Area to 3000 S.F.	108.000	L.F.	17.280	660	1,000	1,660
Area to 5000 S.F.	180.000	L.F.	28.800	1,100	1,675	2,775
Area to 10000 S.F.	360.000	L.F.	57.600	2,200	3,350	5,550
Rectangular, galvanized steel, to 2000 S.F.	239.000	Lb.	24.409	151	1,475	1,626
Area to 3000 S.F.	360.000	Lb.	36.767	227	2,225	2,452
Area to 5000 S.F.	599.000	Lb.	61.176	375	3,675	4,050
Area to 10000 S.F.	998.000	Lb.	101.926	630	6,150	6,780
Diffusers, ceiling, 1 to 4 way blow, 24″ x 24″, to 2000 S.F.	4.000	Ea.	4.571	1,250	296	1,546
Area to 3000 S.F.	6.000	Ea.	6.857	1,850	445	2,295
Area to 5000 S.F.	10.000	Ea.	11.429	3,100	740	3,840
Area to 10000 S.F.	20.000	Ea.	22.857	6,200	1,475	7,675
Return grilles, 24″ x 24″, to 2000 S.F.	1.000	Ea.	.727	128	47	175
Area to 3000 S.F.	2.000	Ea.	1.455	256	94	350
Area to 5000 S.F.	3.000	Ea.	2.182	385	141	526
Area to 10000 S.F.	5.000	Ea.	3.636	640	235	875

Division 9 - Electrical

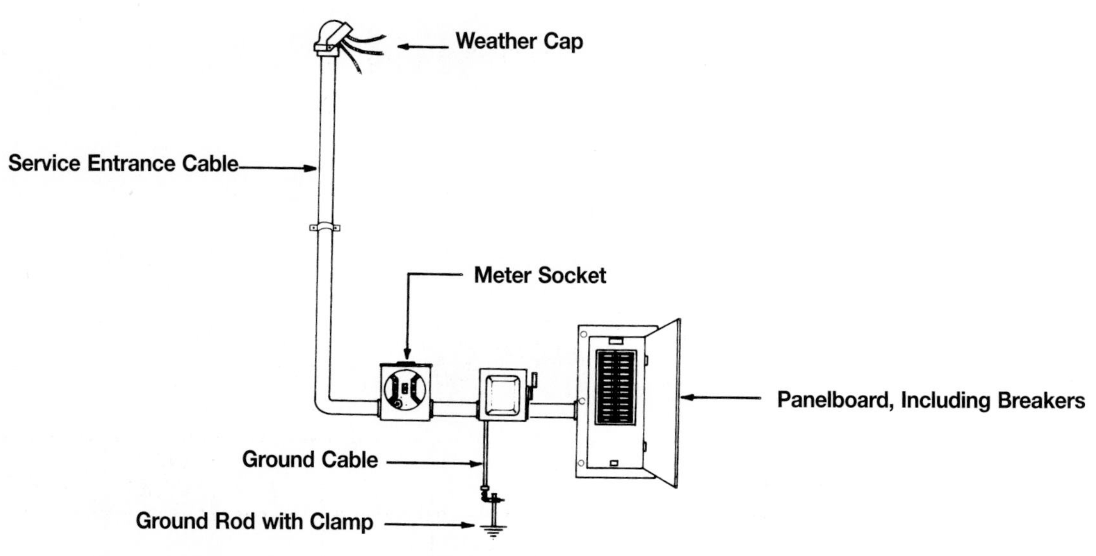

System Description	QUAN.	UNIT	LABOR HOURS	COST EACH		
				MAT.	INST.	TOTAL
100 AMP SERVICE						
Weather cap	1.000	Ea.	.667	8.75	45	53.75
Service entrance cable	10.000	L.F.	.762	36.80	51.50	88.30
Meter socket	1.000	Ea.	2.500	49.50	169	218.50
Ground rod with clamp	1.000	Ea.	1.455	21.50	98.50	120
Ground cable	5.000	L.F.	.250	7.50	16.90	24.40
Panel board, 12 circuit	1.000	Ea.	6.667	135	450	585
TOTAL		Ea.	12.301	259.05	830.90	1,089.95
200 AMP SERVICE						
Weather cap	1.000	Ea.	1.000	18.95	67.50	86.45
Service entrance cable	10.000	L.F.	1.143	31.90	77	108.90
Meter socket	1.000	Ea.	4.211	108	284	392
Ground rod with clamp	1.000	Ea.	1.818	54	123	177
Ground cable	10.000	L.F.	.500	15	33.80	48.80
3/4" EMT	5.000	L.F.	.308	6.50	20.80	27.30
Panel board, 24 circuit	1.000	Ea.	12.308	278	750	1,028
TOTAL		Ea.	21.288	512.35	1,356.10	1,868.45
400 AMP SERVICE						
Weather cap	1.000	Ea.	2.963	183	200	383
Service entrance cable	180.000	L.F.	5.760	455.40	388.80	844.20
Meter socket	1.000	Ea.	4.211	108	284	392
Ground rod with clamp	1.000	Ea.	2.000	54.50	135	189.50
Ground cable	20.000	L.F.	.485	46.60	32.80	79.40
3/4" Greenfield	20.000	L.F.	1.000	12.60	67.60	80.20
Current transformer cabinet	1.000	Ea.	6.154	195	415	610
Panel board, 42 circuit	1.000	Ea.	33.333	5,500	2,250	7,750
TOTAL		Ea.	55.906	6,555.10	3,773.20	10,328.30

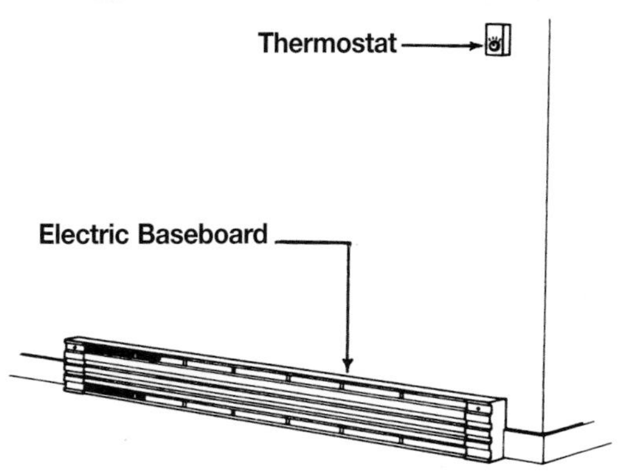

System Description		QUAN.	UNIT	LABOR HOURS	COST EACH		
					MAT.	INST.	TOTAL
4′ BASEBOARD HEATER							
Electric baseboard heater, 4′ long		1.000	Ea.	1.194	43	80.50	123.50
Thermostat, integral		1.000	Ea.	.500	30.50	34	64.50
Romex, 12-3 with ground		40.000	L.F.	1.600	17.60	108	125.60
Panel board breaker, 15-50 amp		1.000	Ea.	.300	16.65	20.25	36.90
	TOTAL		Ea.	3.594	107.75	242.75	350.50
6′ BASEBOARD HEATER							
Electric baseboard heater, 6′ long		1.000	Ea.	1.600	57	108	165
Thermostat, integral		1.000	Ea.	.500	30.50	34	64.50
Romex, 12-3 with ground		40.000	L.F.	1.600	17.60	108	125.60
Panel board breaker, 15-50 amp		1.000	Ea.	.400	22.20	27	49.20
	TOTAL		Ea.	4.100	127.30	277	404.30
8′ BASEBOARD HEATER							
Electric baseboard heater, 8′ long		1.000	Ea.	2.000	72	135	207
Thermostat, integral		1.000	Ea.	.500	30.50	34	64.50
Romex, 12-3 with ground		40.000	L.F.	1.600	17.60	108	125.60
Panel board breaker, 15-50 amp		1.000	Ea.	.500	27.75	33.75	61.50
	TOTAL		Ea.	4.600	147.85	310.75	458.60
10′ BASEBOARD HEATER							
Electric baseboard heater, 10′ long		1.000	Ea.	2.424	170	164	334
Thermostat, integral		1.000	Ea.	.500	30.50	34	64.50
Romex, 12-3 with ground		40.000	L.F.	1.600	17.60	108	125.60
Panel board breaker, 15-50 amp		1.000	Ea.	.750	41.63	50.63	92.26
	TOTAL		Ea.	5.274	259.73	356.63	616.36

The costs in this system are on a cost each basis and include all necessary conduit fittings.

Description	QUAN.	UNIT	LABOR HOURS	COST EACH		
				MAT.	INST.	TOTAL

System Description	QUAN.	UNIT	LABOR HOURS	COST EACH MAT.	COST EACH INST.	COST EACH TOTAL
Air conditioning receptacles						
Using non-metallic sheathed cable	1.000	Ea.	.800	19.65	54	73.65
Using BX cable	1.000	Ea.	.964	30	65	95
Using EMT conduit	1.000	Ea.	1.194	46.50	80.50	127
Disposal wiring						
Using non-metallic sheathed cable	1.000	Ea.	.889	21.50	60	81.50
Using BX cable	1.000	Ea.	1.067	31.50	72	103.50
Using EMT conduit	1.000	Ea.	1.333	49	90	139
Dryer circuit						
Using non-metallic sheathed cable	1.000	Ea.	1.455	31	98.50	129.50
Using BX cable	1.000	Ea.	1.739	41.50	117	158.50
Using EMT conduit	1.000	Ea.	2.162	55.50	146	201.50
Duplex receptacles						
Using non-metallic sheathed cable	1.000	Ea.	.615	19.65	41.50	61.15
Using BX cable	1.000	Ea.	.741	30	50	80
Using EMT conduit	1.000	Ea.	.920	46.50	62	108.50
Exhaust fan wiring						
Using non-metallic sheathed cable	1.000	Ea.	.800	12.40	54	66.40
Using BX cable	1.000	Ea.	.964	23	65	88
Using EMT conduit	1.000	Ea.	1.194	39	80.50	119.50
Furnace circuit & switch						
Using non-metallic sheathed cable	1.000	Ea.	1.333	28.50	90	118.50
Using BX cable	1.000	Ea.	1.600	40.50	108	148.50
Using EMT conduit	1.000	Ea.	2.000	53	135	188
Ground fault						
Using non-metallic sheathed cable	1.000	Ea.	1.000	57	67.50	124.50
Using BX cable	1.000	Ea.	1.212	66.50	82	148.50
Using EMT conduit	1.000	Ea.	1.481	84	100	184
Heater circuits						
Using non-metallic sheathed cable	1.000	Ea.	1.000	23	67.50	90.50
Using BX cable	1.000	Ea.	1.212	29.50	82	111.50
Using EMT conduit	1.000	Ea.	1.481	43.50	100	143.50
Lighting wiring						
Using non-metallic sheathed cable	1.000	Ea.	.500	26.50	34	60.50
Using BX cable	1.000	Ea.	.602	32.50	40.50	73
Using EMT conduit	1.000	Ea.	.748	41.50	50.50	92
Range circuits						
Using non-metallic sheathed cable	1.000	Ea.	2.000	78	135	213
Using BX cable	1.000	Ea.	2.424	114	164	278
Using EMT conduit	1.000	Ea.	2.963	96	200	296
Switches, single pole						
Using non-metallic sheathed cable	1.000	Ea.	.500	12.40	34	46.40
Using BX cable	1.000	Ea.	.602	23	40.50	63.50
Using EMT conduit	1.000	Ea.	.748	39	50.50	89.50
Switches, 3-way						
Using non-metallic sheathed cable	1.000	Ea.	.667	16.60	45	61.60
Using BX cable	1.000	Ea.	.800	23.50	54	77.50
Using EMT conduit	1.000	Ea.	1.333	40.50	90	130.50
Water heater						
Using non-metallic sheathed cable	1.000	Ea.	1.600	21	108	129
Using BX cable	1.000	Ea.	1.905	35.50	129	164.50
Using EMT conduit	1.000	Ea.	2.353	44	159	203
Weatherproof receptacle						
Using non-metallic sheathed cable	1.000	Ea.	1.333	147	90	237
Using BX cable	1.000	Ea.	1.600	153	108	261
Using EMT conduit	1.000	Ea.	2.000	169	135	304

System Description	QUAN.	UNIT	LABOR HOURS	COST EACH MAT.	INST.	TOTAL
Fluorescent strip, 4′ long, 1 light, average	1.000	Ea.	.941	34	63.50	97.50
Deluxe	1.000	Ea.	1.129	41	76	117
2 lights, average	1.000	Ea.	1.000	48	67.50	115.50
Deluxe	1.000	Ea.	1.200	57.50	81	138.50
8′ long, 1 light, average	1.000	Ea.	1.194	61	80.50	141.50
Deluxe	1.000	Ea.	1.433	73	96.50	169.50
2 lights, average	1.000	Ea.	1.290	73	87	160
Deluxe	1.000	Ea.	1.548	87.50	104	191.50
Surface mounted, 4′ x 1′, economy	1.000	Ea.	.914	58	61.50	119.50
Average	1.000	Ea.	1.143	72.50	77	149.50
Deluxe	1.000	Ea.	1.371	87	92.50	179.50
4′ x 2′, economy	1.000	Ea.	1.208	75	81.50	156.50
Average	1.000	Ea.	1.509	94	102	196
Deluxe	1.000	Ea.	1.811	113	122	235
Recessed, 4′x 1′, 2 lamps, economy	1.000	Ea.	1.123	45	76	121
Average	1.000	Ea.	1.404	56.50	95	151.50
Deluxe	1.000	Ea.	1.684	68	114	182
4′ x 2′, 4′ lamps, economy	1.000	Ea.	1.362	55	92	147
Average	1.000	Ea.	1.702	68.50	115	183.50
Deluxe	1.000	Ea.	2.043	82	138	220
Incandescent, exterior, 150W, single spot	1.000	Ea.	.500	38.50	34	72.50
Double spot	1.000	Ea.	1.167	115	79	194
Recessed, 100W, economy	1.000	Ea.	.800	59.50	54	113.50
Average	1.000	Ea.	1.000	74.50	67.50	142
Deluxe	1.000	Ea.	1.200	89.50	81	170.50
150W, economy	1.000	Ea.	.800	92	54	146
Average	1.000	Ea.	1.000	115	67.50	182.50
Deluxe	1.000	Ea.	1.200	138	81	219
Surface mounted, 60W, economy	1.000	Ea.	.800	73	54	127
Average	1.000	Ea.	1.000	81	67.50	148.50
Deluxe	1.000	Ea.	1.194	115	80.50	195.50
Metal halide, recessed 2′ x 2′ 250W	1.000	Ea.	2.500	289	169	458
2′ x 2′, 400W	1.000	Ea.	2.759	405	186	591
Surface mounted, 2′ x 2′, 250W	1.000	Ea.	2.963	375	200	575
2′ x 2′, 400W	1.000	Ea.	3.333	445	225	670
High bay, single, unit, 400W	1.000	Ea.	3.478	455	235	690
Twin unit, 400W	1.000	Ea.	5.000	905	340	1,245
Low bay, 250W	1.000	Ea.	2.500	400	169	569

Unit Price Section

Table of Contents

Table of Contents (cont.)

RSMeans data: Unit Prices— How They Work

All RSMeans data: Unit Prices are organized in the same way.

03 30 Cast-In-Place Concrete

03 30 53 – Miscellaneous Cast-In-Place Concrete

03 30 53.40 Concrete In Place		Crew	Daily Output	Labor-Hours	Unit	Material	Labor	Equipment	Total	Total Incl O&P
						2019 Bare Costs				
0010	CONCRETE IN PLACE									
0020	Including forms (4 uses), Grade 60 rebar, concrete (Portland cement									
0050	Type I), placement and finishing unless otherwise indicated									
0500	Chimney foundations (5000 psi), over 5 C.Y.	C-14C	32.22	3.476	C.Y.	178	118	.83	296.83	390
0510	(3500 psi), under 5 C.Y.	"	23.71	4.724	"	206	160	1.13	367.13	490
3540	Equipment pad (3000 psi), 3′ x 3′ x 6″ thick	C-14H	45	1.067	Ea.	46.50	37.50	.59	84.59	114
3550	4′ x 4′ x 6″ thick		30	1.600		72.50	56.50	.88	129.88	173
3560	5′ x 5′ x 8″ thick		18	2.667		132	94	1.47	227.47	300
3570	6′ x 6′ x 8″ thick		14	3.429		181	121	1.89	303.89	400
3580	8′ x 8′ x 10″ thick		8	6		385	211	3.30	599.30	775
3590	10′ x 10′ x 12″ thick		5	9.600		665	340	5.30	1,010.30	1,300
3800	Footings (3000 psi), spread under 1 C.Y.	C-14C	28	4	C.Y.	193	136	.96	329.96	435
3825	1 C.Y. to 5 C.Y.		43	2.605		227	88.50	.63	316.13	395
3850	Over 5 C.Y.		75	1.493		212	50.50	.36	262.86	320

It is important to understand the structure of RSMeans data: Unit Prices so that you can find information easily and use it correctly.

1 Line Numbers

Line Numbers consist of 12 characters, which identify a unique location in the database for each task. The first 6 or 8 digits conform to the Construction Specifications Institute MasterFormat® 2016. The remainder of the digits are a further breakdown in order to arrange items in understandable groups of similar tasks. Line numbers are consistent across all of our publications, so a line number in any of our products will always refer to the same item of work.

2 Descriptions

Descriptions are shown in a hierarchical structure to make them readable. In order to read a complete description, read up through the indents to the top of the section. Include everything that is above and to the left that is not contradicted by information below. For instance, the complete description for line 03 30 53.40 3550 is "Concrete in place, including forms (4 uses), Grade 60 rebar, concrete (Portland cement Type 1), placement and finishing unless otherwise indicated; Equipment pad (3000 psi), 4' × 4' × 6" thick."

3 RSMeans data

When using **RSMeans data**, it is important to read through an entire section to ensure that you use the data that most closely matches your work. Note that sometimes there is additional information shown in the section that may improve your price. There are frequently lines that further describe, add to, or adjust data for specific situations.

4 Reference Information

Gordian's RSMeans engineers have created **reference** information to assist you in your estimate. **If** there is information that applies to a section, it will be indicated at the start of the section. The Reference Section is located in the back of the data set.

5 Crews

Crews include labor and/or equipment necessary to accomplish each task. In this case, Crew C-14H is used. Gordian's RSMeans staff selects a crew to represent the workers and equipment that are typically used for that task. In this case, Crew C-14H consists of one carpenter foreman (outside), two carpenters, one rodman, one laborer, one cement finisher, and one gas engine vibrator. Details of all crews can be found in the Reference Section.

Crews - Residential

Crew No.	Bare Costs		Incl. Subs O&P		Cost Per Labor-Hour	
Crew C-14H	**Hr.**	**Daily**	**Hr.**	**Daily**	**Bare Costs**	**Incl. O&P**
1 Carpenter Foreman (outside)	$37.65	$301.20	$62.20	$497.60	$35.18	$57.92
2 Carpenters	35.65	570.40	58.90	942.40		
1 Rodman (reinf.)	38.95	311.60	64.40	515.20		
1 Laborer	27.50	220.00	45.45	363.60		
1 Cement Finisher	35.70	285.60	57.70	461.60		
1 Gas Engine Vibrator		26.55		29.20	0.55	0.61
48 L.H., Daily Totals		$1715.35		$2809.61	$35.74	$58.53

6 Daily Output

The **Daily Output** is the amount of work that the crew can do in a normal 8-hour workday, including mobilization, layout, movement of materials, and cleanup. In this case, crew C-14H can install thirty 4' × 4' × 6" thick concrete pads in a day. Daily output is variable and based on many factors, including the size of the job, location, and environmental conditions. RSMeans data represents work done in daylight (or adequate lighting) and temperate conditions.

7 Labor-Hours

The figure in the **Labor-Hours** column is the amount of labor required to perform one unit of work–in this case the amount of labor required to construct one 4' × 4' equipment pad. This figure is calculated by dividing the number of hours of labor in the crew by the daily output (48 labor-hours divided by 30 pads = 1.6 hours of labor per pad). Multiply 1.6 times 60 to see the value in minutes: 60 × 1.6 = 96 minutes. Note: the labor-hour figure is not dependent on the crew size. A change in crew size will result in a corresponding change in daily output, but the labor-hours per unit of work will not change.

8 Unit of Measure

All RSMeans data: Unit Prices include the typical **Unit of Measure** used for estimating that item. For concrete-in-place the typical unit is cubic yards (C.Y.) or each (Ea.). For installing broadloom carpet it is square yard and for gypsum board it is square foot. The estimator needs to take special care that the unit in the data matches the unit in the take-off. Unit conversions may be found in the Reference Section.

9 Bare Costs

Bare Costs are the costs of materials, labor, and equipment that the installing contractor pays. They represent the cost, in U.S. dollars, for one unit of work. They do not include any markups for profit or labor burden.

10 Bare Total

The **Total column** represents the total bare cost for the installing contractor in U.S. dollars. In this case, the sum of $72.50 for material + $56.50 for labor + $.88 for equipment is $129.88.

11 Total Incl O&P

The **Total Incl O&P column** is the total cost, including overhead and profit, that the installing contractor will charge the customer. This represents the cost of materials plus 10% profit, the cost of labor plus labor burden and 10% profit, and the cost of equipment plus 10% profit. It does not include the general contractor's overhead and profit. Note: See the inside back cover of the printed product or the Reference Section of the electronic product for details on how the labor burden is calculated.

National Average

*The RSMeans data in our print publications represent a "national average" cost. This data should be modified to the project location using the **City Cost Indexes** or **Location Factors** tables found in the Reference Section. Use the Location Factors to adjust estimate totals if the project covers multiple trades. Use the City Cost Indexes (CCI) for single trade projects or projects where a more detailed analysis is required. All figures in the two tables are derived from the same research. The last row of data in the CCI—the weighted average—is the same as the numbers reported for each location in the location factor table.*

RSMeans data: Unit Prices—How They Work (Continued)

Project Name: Pre-Engineered Steel Building		Architect: As Shown						
Location:	Anywhere, USA						01/01/19	RESI
(1) Line Number	Description	Qty	Unit	Material	Labor	Equipment	SubContract	Estimate Total
03 30 53.40 3940	Strip footing, 12" x 24", reinforced	15	C.Y.	$2,460.00	$1,185.00	$8.40	$0.00	
03 30 53.40 3950	Strip footing, 12" x 36", reinforced	34	C.Y.	$5,372.00	$2,159.00	$15.30	$0.00	
03 11 13.65 3000	Concrete slab edge forms	500	L.F.	$155.00	$840.00	$0.00	$0.00	
03 22 11.10 0200	Welded wire fabric reinforcing	150	C.S.F.	$3,000.00	$3,000.00	$0.00	$0.00	
03 31 13.35 0300	Ready mix concrete, 4000 psi for slab on grade	278	C.Y.	$35,584.00	$0.00	$0.00	$0.00	
03 31 13.70 4300	Place, strike off & consolidate concrete slab	278	C.Y.	$0.00	$3,544.50	$133.44	$0.00	
03 35 13.30 0250	Machine float & trowel concrete slab	15,000	S.F.	$0.00	$6,900.00	$300.00	$0.00	
03 15 16.20 0140	Cut control joints in concrete slab	950	L.F.	$47.50	$304.00	$57.00	$0.00	
03 39 23.13 0300	Sprayed concrete curing membrane	150	C.S.F.	$1,890.00	$694.50	$0.00	$0.00	
Division 03	**Subtotal**			**$48,508.50**	**$18,627.00**	**$514.14**	**$0.00**	**$67,649.64**
08 36 13.10 2650	Manual 10' x 10' steel sectional overhead door	8	Ea.	$10,600.00	$2,520.00	$0.00	$0.00	
08 36 13.10 2860	Insulation and steel back panel for OH door	800	S.F.	$4,120.00	$0.00	$0.00	$0.00	
Division 08	**Subtotal**			**$14,720.00**	**$2,520.00**	**$0.00**	**$0.00**	**$17,240.00**
13 34 19.50 1100	Pre-Engineered Steel Building, 100' x 150' x 24'	15,000	SF Flr.	$0.00	$0.00	$0.00	$315,000.00	
13 34 19.50 6050	Framing for PESB door opening, 3' x 7'	4	Opng.	$0.00	$0.00	$0.00	$1,920.00	
13 34 19.50 6100	Framing for PESB door opening, 10' x 10'	8	Opng.	$0.00	$0.00	$0.00	$8,200.00	
13 34 19.50 6200	Framing for PESB window opening, 4' x 3'	6	Opng.	$0.00	$0.00	$0.00	$2,910.00	
13 34 19.50 5750	PESB door, 3' x 7', single leaf	4	Opng.	$2,700.00	$508.00	$0.00	$0.00	
13 34 19.50 7750	PESB sliding window, 4' x 3' with screen	6	Opng.	$2,910.00	$294.00	$44.10	$0.00	
13 34 19.50 6550	PESB gutter, eave type, 26 ga., painted	300	L.F.	$2,160.00	$597.00	$0.00	$0.00	
13 34 19.50 8650	PESB roof vent, 12" wide x 10' long	15	Ea.	$562.50	$2,385.00	$0.00	$0.00	
13 34 19.50 6900	PESB insulation, vinyl faced, 4" thick	27,400	S.F.	$12,330.00	$6,850.00	$0.00	$0.00	
Division 13	**Subtotal**			**$20,662.50**	**$10,634.00**	**$44.10**	**$328,030.00**	**$359,370.60**
		Subtotal		**$83,891.00**	**$31,781.00**	**$558.24**	**$328,030.00**	**$444,260.24**
Division 01	(2) **General Requirements @ 7%**			5,872.37	2,224.67	39.08	22,962.10	
		Estimate Subtotal		**$89,763.37**	**$34,005.67**	**$597.32**	**$350,992.10**	**$444,260.24**
	(3)	**Sales Tax @ 5%**		**4,488.17**		**29.87**	**8,774.80**	
		Subtotal A		**94,251.54**	**34,005.67**	**627.18**	**359,766.90**	
	(4)	**GC O & P**		**9,425.15**	**22,477.75**	**62.72**	**35,976.69**	
		Subtotal B		**103,676.69**	**56,483.42**	**689.90**	**395,743.59**	**$556,593.60**
	(5)	**Contingency @ 5%**						**27,829.68**
		Subtotal C						**$584,423.28**
	(6)	**Bond @ $12/1000 +10% O&P**						**7,714.39**
		Subtotal D						**$592,137.67**
	(7)	**Location Adjustment Factor**			**113.90**			**82,307.14**
		Grand Total						**$674,444.81**

This estimate is based on an interactive spreadsheet. You are free to download it and adjust it to your methodology.
A copy of this spreadsheet is available at **RSMeans.com/2019books.**

Sample Estimate

This sample demonstrates the elements of an estimate, including a tally of the RSMeans data lines and a summary of the markups on a contractor's work to arrive at a total cost to the owner. The Location Factor with RSMeans data is added at the bottom of the estimate to adjust the cost of the work to a specific location.

1 Work Performed

The body of the estimate shows the RSMeans data selected, including the line number, a brief description of each item, its take-off unit and quantity, and the bare costs of materials, labor, and equipment. This estimate also includes a column titled "SubContract." This data is taken from the column "Total Incl O&P" and represents the total that a subcontractor would charge a general contractor for the work, including the sub's markup for overhead and profit.

2 Division 1, General Requirements

This is the first division numerically but the last division estimated. Division 1 includes project-wide needs provided by the general contractor. These requirements vary by project but may include temporary facilities and utilities, security, testing, project cleanup, etc. For small projects a percentage can be used—typically between 5% and 15% of project cost. For large projects the costs may be itemized and priced individually.

3 Sales Tax

If the work is subject to state or local sales taxes, the amount must be added to the estimate. Sales tax may be added to material costs, equipment costs, and subcontracted work. In this case, sales tax was added in all three categories. It was assumed that approximately half the subcontracted work would be material cost, so the tax was applied to 50% of the subcontract total.

4 GC O&P

This entry represents the general contractor's markup on material, labor, equipment, and subcontractor costs. Our standard markup on materials, equipment, and subcontracted work is 10%. In this estimate, the markup on the labor performed by the GC's workers uses "Skilled Workers Average" shown in Column F on the table "Installing Contractor's Overhead & Profit," which can be found on the inside back cover of the printed product or in the Reference Section of the electronic product.

5 Contingency

A factor for contingency may be added to any estimate to represent the cost of unknowns that may occur between the time that the estimate is performed and the time the project is constructed. The amount of the allowance will depend on the stage of design at which the estimate is done and the contractor's assessment of the risk involved. Refer to section 01 21 16.50 for contingency allowances.

6 Bonds

Bond costs should be added to the estimate. The figures here represent a typical performance bond, ensuring the owner that if the general contractor does not complete the obligations in the construction contract the bonding company will pay the cost for completion of the work.

7 Location Adjustment

Published prices are based on national average costs. If necessary, adjust the total cost of the project using a location factor from the "Location Factor" table or the "City Cost Index" table. Use location factors if the work is general, covering multiple trades. If the work is by a single trade (e.g., masonry) use the more specific data found in the "City Cost Indexes."

Estimating Tips

01 20 00 Price and Payment Procedures

- Allowances that should be added to estimates to cover contingencies and job conditions that are not included in the national average material and labor costs are shown in Section 01 21.
- When estimating historic preservation projects (depending on the condition of the existing structure and the owner's requirements), a 15–20% contingency or allowance is recommended, regardless of the stage of the drawings.

01 30 00 Administrative Requirements

- Before determining a final cost estimate, it is good practice to review all the items listed in Subdivisions 01 31 and 01 32 to make final adjustments for items that may need customizing to specific job conditions.
- Requirements for initial and periodic submittals can represent a significant cost to the General Requirements of a job. Thoroughly check the submittal specifications when estimating a project to determine any costs that should be included.

01 40 00 Quality Requirements

- All projects will require some degree of quality control. This cost is not included in the unit cost of construction listed in each division. Depending upon the terms of the contract, the various costs of inspection and testing can be the responsibility of either the owner or the contractor. Be sure to include the required costs in your estimate.

01 50 00 Temporary Facilities and Controls

- Barricades, access roads, safety nets, scaffolding, security, and many more requirements for the execution of a safe project are elements of direct cost. These costs can easily be overlooked when preparing an estimate. When looking through the major classifications of this subdivision, determine which items apply to each division in your estimate.
- Construction equipment rental costs can be found in the Reference Section in Section 01 54 33. Operators' wages are not included in equipment rental costs.
- Equipment mobilization and demobilization costs are not included in equipment rental costs and must be considered separately.
- The cost of small tools provided by the installing contractor for his workers is covered in the "Overhead" column on the "Installing Contractor's Overhead and Profit" table that lists labor trades, base rates, and markups. Therefore, it is included in the "Total Incl. O&P" cost of any unit price line item.

01 70 00 Execution and Closeout Requirements

- When preparing an estimate, thoroughly read the specifications to determine the requirements for Contract Closeout. Final cleaning, record documentation, operation and maintenance data, warranties and bonds, and spare parts and maintenance materials can all be elements of cost for the completion of a contract. Do not overlook these in your estimate.

Reference Numbers

Reference numbers are shown at the beginning of some major classifications. These numbers refer to related items in the Reference Section. The reference information may be an estimating procedure, an alternate pricing method, or technical information.

Note: Not all subdivisions listed here necessarily appear. ■

01 11 Summary of Work

01 11 31 – Professional Consultants

01 11 31.10 Architectural Fees		Crew	Daily Output	Labor-Hours	Unit	2019 Bare Costs Material	Labor	Equipment	Total	Total Incl O&P
0010	**ARCHITECTURAL FEES** R011110-10									
0020	For new construction									
0060	Minimum				Project				4.90%	4.90%
0090	Maximum								16%	16%
0100	For alteration work, to $500,000, add to new construction fee								50%	50%
0150	Over $500,000, add to new construction fee				↓				25%	25%
01 11 31.20 Construction Management Fees										
0010	**CONSTRUCTION MANAGEMENT FEES**									
0060	For work to $100,000				Project				10%	10%
0070	To $250,000								9%	9%
0090	To $1,000,000				↓				6%	6%
01 11 31.75 Renderings										
0010	**RENDERINGS** Color, matted, 20" x 30", eye level,									
0050	Average				Ea.	3,175			3,175	3,500

01 21 Allowances

01 21 16 – Contingency Allowances

01 21 16.50 Contingencies		Crew	Daily Output	Labor-Hours	Unit	Material	Labor	Equipment	Total	Total Incl O&P
0010	**CONTINGENCIES**, Add to estimate									
0020	Conceptual stage				Project				20%	20%
0150	Final working drawing stage				"				3%	3%

01 21 55 – Job Conditions Allowance

01 21 55.50 Job Conditions		Crew	Daily Output	Labor-Hours	Unit	Material	Labor	Equipment	Total	Total Incl O&P
0010	**JOB CONDITIONS** Modifications to applicable									
8000	Remove and reset contents of small room	1 Clab	6.50	1.231	Room		34		34	56
8010	Average room		4.70	1.702			47		47	77.50
8020	Large room		3.50	2.286			63		63	104
8030	Extra large room	↓	2.40	3.333	↓		91.50		91.50	152

01 21 63 – Taxes

01 21 63.10 Taxes		Crew	Daily Output	Labor-Hours	Unit	Material	Labor	Equipment	Total	Total Incl O&P
0010	**TAXES** R012909-80									
0020	Sales tax, State, average				%	5.08%				
0050	Maximum R012909-85					7.50%				
0200	Social Security, on first $118,500 of wages						7.65%			
0300	Unemployment, combined Federal and State, minimum R012909-86						.60%			
0350	Average						9.60%			
0400	Maximum				↓		12%			

01 31 Project Management and Coordination

01 31 13 – Project Coordination

01 31 13.30 Insurance		Crew	Daily Output	Labor-Hours	Unit	Material	2019 Bare Costs Labor	Equipment	Total	Total Incl O&P
0010	**INSURANCE** R013113-40									
0020	Builders risk, standard, minimum				Job				.24%	.24%
0050	Maximum R013113-50								.64%	.64%
0200	All-risk type, minimum								.25%	.25%
0250	Maximum R013113-60				▼				.62%	.62%
0400	Contractor's equipment floater, minimum				Value				.50%	.50%
0450	Maximum				"				1.50%	1.50%
0600	Public liability, average				Job				2.02%	2.02%
0800	Workers' compensation & employer's liability, average									
0850	by trade, carpentry, general				Payroll		13.05%			
0900	Clerical						.46%			
0950	Concrete						12.44%			
1000	Electrical						5.52%			
1050	Excavation						9.03%			
1100	Glazing						12.91%			
1150	Insulation						11.13%			
1200	Lathing						8.67%			
1250	Masonry						13.73%			
1300	Painting & decorating						11.68%			
1350	Pile driving						14.53%			
1400	Plastering						10.73%			
1450	Plumbing						6.94%			
1500	Roofing						30.73%			
1550	Sheet metal work (HVAC)						8.82%			
1600	Steel erection, structural						27.45%			
1650	Tile work, interior ceramic						8.85%			
1700	Waterproofing, brush or hand caulking						7.01%			
1800	Wrecking						20.85%			
2000	Range of 35 trades in 50 states, excl. wrecking & clerical, min.						1.41%			
2100	Average						12.25%			
2200	Maximum				▼		108.79%			

01 41 Regulatory Requirements

01 41 26 – Permit Requirements

01 41 26.50 Permits		Crew	Daily Output	Labor-Hours	Unit	Material	Labor	Equipment	Total	Total Incl O&P
0010	**PERMITS**									
0020	Rule of thumb, most cities, minimum				Job				.50%	.50%
0100	Maximum				"				2%	2%

01 54 Construction Aids

01 54 16 – Temporary Hoists

01 54 16.50 Weekly Forklift Crew		Crew	Daily Output	Labor-Hours	Unit	Material	Labor	Equipment	Total	Total Incl O&P
0010	**WEEKLY FORKLIFT CREW**									
0100	All-terrain forklift, 45' lift, 35' reach, 9000 lb. capacity	A-3P	.20	40	Week		1,425	2,325	3,750	4,900

01 54 19 – Temporary Cranes

01 54 19.50 Daily Crane Crews		Crew	Daily Output	Labor-Hours	Unit	Material	2019 Bare Costs Labor	Equipment	Total	Total Incl O&P
0010	**DAILY CRANE CREWS** for small jobs, portal to portal									
0100	12-ton truck-mounted hydraulic crane	A-3H	1	8	Day		305	710	1,015	1,275
0900	If crane is needed on a Saturday, Sunday or Holiday									
0910	At time-and-a-half, add				Day		50%			
0920	At double time, add				"		100%			

01 54 23 – Temporary Scaffolding and Platforms

01 54 23.60 Pump Staging		Crew	Daily Output	Labor-Hours	Unit	Material	Labor	Equipment	Total	Total Incl O&P
0010	**PUMP STAGING**, Aluminum									
1300	System in place, 50' working height, per use based on 50 uses	2 Carp	84.80	.189	C.S.F.	6.55	6.75		13.30	18.30
1400	100 uses R015423-20		84.80	.189		3.27	6.75		10.02	14.70
1500	150 uses		84.80	.189		2.19	6.75		8.94	13.50

01 54 23.70 Scaffolding		Crew	Daily Output	Labor-Hours	Unit	Material	Labor	Equipment	Total	Total Incl O&P
0010	**SCAFFOLDING** R015423-10									
0015	Steel tube, regular, no plank, labor only to erect & dismantle									
0091	Building exterior, wall face, 1 to 5 stories, 6'-4" x 5' frames	3 Clab	8	3	C.S.F.		82.50		82.50	136
0201	6 to 12 stories	4 Clab	8	4			110		110	182
0310	13 to 20 stories	5 Carp	8	5			178		178	295
0461	Building interior, wall face area, up to 16' high	3 Clab	12	2			55		55	91
0561	16' to 40' high		10	2.400			66		66	109
0801	Building interior floor area, up to 30' high		150	.160	C.C.F.		4.40		4.40	7.25
0901	Over 30' high	4 Clab	160	.200	"		5.50		5.50	9.10
0906	Complete system for face of walls, no plank, material only rent/mo				C.S.F.	33			33	36.50
0908	Interior spaces, no plank, material only rent/mo				C.C.F.	3.80			3.80	4.18
0910	Steel tubular, heavy duty shoring, buy									
0920	Frames 5' high 2' wide				Ea.	92			92	101
0925	5' high 4' wide					97.50			97.50	107
0930	6' high 2' wide					99			99	109
0935	6' high 4' wide					116			116	128
0940	Accessories									
0945	Cross braces				Ea.	17.10			17.10	18.80
0950	U-head, 8" x 8"					18.90			18.90	21
0955	J-head, 4" x 8"					13.75			13.75	15.15
0960	Base plate, 8" x 8"					15.15			15.15	16.65
0965	Leveling jack					36			36	39.50
1000	Steel tubular, regular, buy									
1100	Frames 3' high 5' wide				Ea.	89.50			89.50	98.50
1150	5' high 5' wide					108			108	119
1200	6'-4" high 5' wide					106			106	117
1350	7'-6" high 6' wide					157			157	173
1500	Accessories, cross braces					18.05			18.05	19.85
1550	Guardrail post					19.90			19.90	22
1600	Guardrail 7' section					8.15			8.15	8.95
1650	Screw jacks & plates					25.50			25.50	28.50
1700	Sidearm brackets					22.50			22.50	25
1750	8" casters					37.50			37.50	41
1800	Plank 2" x 10" x 16'-0"					67			67	74
1900	Stairway section					284			284	310
1910	Stairway starter bar					32.50			32.50	35.50
1920	Stairway inside handrail					53			53	58.50
1930	Stairway outside handrail					84.50			84.50	93
1940	Walk-thru frame guardrail					42			42	46

01 54 Construction Aids

01 54 23 – Temporary Scaffolding and Platforms

01 54 23.70 Scaffolding		Crew	Daily Output	Labor-Hours	Unit	2019 Bare Costs Material	Labor	Equipment	Total	Total Incl O&P
2000	Steel tubular, regular, rent/mo.									
2100	Frames 3′ high 5′ wide				Ea.	4.45			4.45	4.90
2150	5′ high 5′ wide					4.45			4.45	4.90
2200	6′-4″ high 5′ wide					5.40			5.40	5.95
2250	7′-6″ high 6′ wide					9.85			9.85	10.85
2500	Accessories, cross braces					.89			.89	.98
2550	Guardrail post					.89			.89	.98
2600	Guardrail 7′ section					.89			.89	.98
2650	Screw jacks & plates					1.78			1.78	1.96
2700	Sidearm brackets					1.78			1.78	1.96
2750	8″ casters					7.10			7.10	7.80
2800	Outrigger for rolling tower					2.67			2.67	2.94
2850	Plank 2″ x 10″ x 16′-0″					9.85			9.85	10.85
2900	Stairway section					31.50			31.50	34.50
2940	Walk-thru frame guardrail				↓	2.22			2.22	2.44
3000	Steel tubular, heavy duty shoring, rent/mo.									
3250	5′ high 2′ & 4′ wide				Ea.	8.35			8.35	9.20
3300	6′ high 2′ & 4′ wide					8.35			8.35	9.20
3500	Accessories, cross braces					.89			.89	.98
3600	U-head, 8″ x 8″					2.46			2.46	2.71
3650	J-head, 4″ x 8″					2.46			2.46	2.71
3700	Base plate, 8″ x 8″					.89			.89	.98
3750	Leveling jack					2.46			2.46	2.71
5700	Planks, 2″ x 10″ x 16′-0″, labor only to erect & remove to 50′ H	3 Carp	72	.333			11.90		11.90	19.65
5800	Over 50′ high	4 Carp	80	.400	↓		14.25		14.25	23.50

01 54 23.80 Staging Aids		Crew	Daily Output	Labor-Hours	Unit	Material	Labor	Equipment	Total	Total Incl O&P
0010	**STAGING AIDS** and fall protection equipment									
0100	Sidewall staging bracket, tubular, buy				Ea.	58.50			58.50	64
0110	Cost each per day, based on 250 days use				Day	.23			.23	.26
0200	Guard post, buy				Ea.	55			55	60.50
0210	Cost each per day, based on 250 days use				Day	.22			.22	.24
0300	End guard chains, buy per pair				Pair	46.50			46.50	51
0310	Cost per set per day, based on 250 days use				Day	.23			.23	.26
1010	Cost each per day, based on 250 days use				″	.05			.05	.05
1100	Wood bracket, buy				Ea.	25.50			25.50	28
1110	Cost each per day, based on 250 days use				Day	.10			.10	.11
2010	Cost per pair per day, based on 250 days use					.55			.55	.61
3010	Cost each per day, based on 250 days use				↓	.23			.23	.26
3100	Aluminum scaffolding plank, 20″ wide x 24′ long, buy				Ea.	735			735	810
3110	Cost each per day, based on 250 days use				Day	2.94			2.94	3.24
4010	Cost each per day, based on 250 days use				″	.67			.67	.74
4100	Rope for safety line, 5/8″ x 100′ nylon, buy				Ea.	56.50			56.50	62
4110	Cost each per day, based on 250 days use				Day	.23			.23	.25
4200	Permanent U-Bolt roof anchor, buy				Ea.	30.50			30.50	33.50
4300	Temporary (one use) roof ridge anchor, buy				″	6.40			6.40	7.05
5000	Installation (setup and removal) of staging aids									
5010	Sidewall staging bracket	2 Carp	64	.250	Ea.		8.90		8.90	14.75
5020	Guard post with 2 wood rails	″	64	.250			8.90		8.90	14.75
5030	End guard chains, set	1 Carp	64	.125			4.46		4.46	7.35
5100	Roof shingling bracket		96	.083			2.97		2.97	4.91
5200	Ladder jack	↓	64	.125			4.46		4.46	7.35
5300	Wood plank, 2″ x 10″ x 16′	2 Carp	80	.200	↓		7.15		7.15	11.80

01 54 Construction Aids

01 54 23 – Temporary Scaffolding and Platforms

01 54 23.80 Staging Aids		Crew	Daily Output	Labor-Hours	Unit	Material	2019 Bare Costs Labor	Equipment	Total	Total Incl O&P
5310	Aluminum scaffold plank, 20" x 24'	2 Carp	40	.400	Ea.		14.25		14.25	23.50
5410	Safety rope	1 Carp	40	.200			7.15		7.15	11.80
5420	Permanent U-Bolt roof anchor (install only)	2 Carp	40	.400			14.25		14.25	23.50
5430	Temporary roof ridge anchor (install only)	1 Carp	64	.125			4.46		4.46	7.35

01 54 36 – Equipment Mobilization

01 54 36.50 Mobilization		Crew	Daily Output	Labor-Hours	Unit	Material	2019 Bare Costs Labor	Equipment	Total	Total Incl O&P
0010	**MOBILIZATION** (Use line item again for demobilization) R015436-50									
0015	Up to 25 mi. haul dist. (50 mi. RT for mob/demob crew)									
1200	Small equipment, placed in rear of, or towed by pickup truck	A-3A	4	2	Ea.		71.50	30.50	102	151
1300	Equipment hauled on 3-ton capacity towed trailer	A-3Q	2.67	3			107	55.50	162.50	237
1400	20-ton capacity	B-34U	2	8			274	218	492	690
1500	40-ton capacity	B-34N	2	8			281	335	616	830
1700	Crane, truck-mounted, up to 75 ton (driver only)	1 Eqhv	4	2			76		76	124
2500	For each additional 5 miles haul distance, add						10%	10%		
3000	For large pieces of equipment, allow for assembly/knockdown									
3100	For mob/demob of micro-tunneling equip, see Section 33 05 23.19									

01 56 Temporary Barriers and Enclosures

01 56 13 – Temporary Air Barriers

01 56 13.60 Tarpaulins		Crew	Daily Output	Labor-Hours	Unit	Material	2019 Bare Costs Labor	Equipment	Total	Total Incl O&P
0010	**TARPAULINS**									
0020	Cotton duck, 10-13.13 oz./S.Y., 6' x 8'				S.F.	.86			.86	.95
0050	30' x 30'					.60			.60	.66
0200	Reinforced polyethylene 3 mils thick, white					.04			.04	.04
0300	4 mils thick, white, clear or black					.13			.13	.14
0730	Polyester reinforced w/integral fastening system, 11 mils thick					.19			.19	.21

01 56 16 – Temporary Dust Barriers

01 56 16.10 Dust Barriers, Temporary		Crew	Daily Output	Labor-Hours	Unit	Material	2019 Bare Costs Labor	Equipment	Total	Total Incl O&P
0010	**DUST BARRIERS, TEMPORARY**									
0020	Spring loaded telescoping pole & head, to 12', erect and dismantle	1 Clab	240	.033	Ea.		.92		.92	1.51
0025	Cost per day (based upon 250 days)				Day	.22			.22	.24
0030	To 21', erect and dismantle	1 Clab	240	.033	Ea.		.92		.92	1.51
0035	Cost per day (based upon 250 days)				Day	.68			.68	.74
0040	Accessories, caution tape reel, erect and dismantle	1 Clab	480	.017	Ea.		.46		.46	.76
0045	Cost per day (based upon 250 days)				Day	.32			.32	.35
0060	Foam rail and connector, erect and dismantle	1 Clab	240	.033	Ea.		.92		.92	1.51
0065	Cost per day (based upon 250 days)				Day	.10			.10	.11
0070	Caution tape	1 Clab	384	.021	C.L.F.	2.69	.57		3.26	3.91
0080	Zipper, standard duty		60	.133	Ea.	7.40	3.67		11.07	14.20
0090	Heavy duty		48	.167	"	9.95	4.58		14.53	18.50
0100	Polyethylene sheet, 4 mil		37	.216	Sq.	2.66	5.95		8.61	12.80
0110	6 mil		37	.216	"	3.87	5.95		9.82	14.10
1000	Dust partition, 6 mil polyethylene, 1" x 3" frame	2 Carp	2000	.008	S.F.	.33	.29		.62	.83
1080	2" x 4" frame	"	2000	.008	"	.35	.29		.64	.86

01 66 Product Storage and Handling Requirements

01 66 19 – Material Handling

01 66 19.10 Material Handling		Crew	Daily Output	Labor-Hours	Unit	Material	2019 Bare Costs Labor	Equipment	Total	Total Incl O&P
0010	**MATERIAL HANDLING**									
0020	Above 2nd story, via stairs, per C.Y. of material per floor	2 Clab	145	.110	C.Y.		3.03		3.03	5
0030	Via elevator, per C.Y. of material		240	.067			1.83		1.83	3.03
0050	Distances greater than 200′, per C.Y. of material per each addl 200′	↓	300	.053	↓		1.47		1.47	2.42

01 71 Examination and Preparation

01 71 23 – Field Engineering

01 71 23.13 Construction Layout

		Crew	Daily Output	Labor-Hours	Unit	Material	Labor	Equipment	Total	Total Incl O&P
0010	**CONSTRUCTION LAYOUT**									
1100	Crew for layout of building, trenching or pipe laying, 2 person crew	A-6	1	16	Day		570	45.50	615.50	995
1200	3 person crew	A-7	1	24	"		910	45.50	955.50	1,550

01 74 Cleaning and Waste Management

01 74 13 – Progress Cleaning

01 74 13.20 Cleaning Up

		Crew	Daily Output	Labor-Hours	Unit	Material	Labor	Equipment	Total	Total Incl O&P
0010	**CLEANING UP**									
0020	After job completion, allow, minimum				Job				.30%	.30%
0040	Maximum				"				1%	1%

01 76 Protecting Installed Construction

01 76 13 – Temporary Protection of Installed Construction

01 76 13.20 Temporary Protection

		Crew	Daily Output	Labor-Hours	Unit	Material	Labor	Equipment	Total	Total Incl O&P
0010	**TEMPORARY PROTECTION**									
0020	Flooring, 1/8″ tempered hardboard, taped seams	2 Carp	1500	.011	S.F.	.46	.38		.84	1.14
0030	Peel away carpet protection	1 Clab	3200	.003	"	.13	.07		.20	.25

Division Notes

		CREW	DAILY OUTPUT	LABOR-HOURS	UNIT	BARE COSTS MAT.	LABOR	EQUIP.	TOTAL	TOTAL INCL O&P

Estimating Tips

02 30 00 Subsurface Investigation

In preparing estimates on structures involving earthwork or foundations, all information concerning soil characteristics should be obtained. Look particularly for hazardous waste, evidence of prior dumping of debris, and previous stream beds.

02 40 00 Demolition and Structure Moving

The costs shown for selective demolition do not include rubbish handling or disposal. These items should be estimated separately using RSMeans data or other sources.

- Historic preservation often requires that the contractor remove materials from the existing structure, rehab them, and replace them. The estimator must be aware of any related measures and precautions that must be taken when doing selective demolition and cutting and patching. Requirements may include special handling and storage, as well as security.
- In addition to Subdivision 02 41 00, you can find selective demolition items in each division. Example: Roofing demolition is in Division 7.
- Absent of any other specific reference, an approximate demolish-in-place cost can be obtained by halving the new-install labor cost. To remove for reuse, allow the entire new-install labor figure.

02 40 00 Building Deconstruction

This section provides costs for the careful dismantling and recycling of most low-rise building materials.

02 50 00 Containment of Hazardous Waste

This section addresses on-site hazardous waste disposal costs.

02 80 00 Hazardous Material Disposal/Remediation

This subdivision includes information on hazardous waste handling, asbestos remediation, lead remediation, and mold remediation. See reference numbers R028213-20 and R028319-60 for further guidance in using these unit price lines.

02 90 00 Monitoring Chemical Sampling, Testing Analysis

This section provides costs for on-site sampling and testing hazardous waste.

Reference Numbers

Reference numbers are shown at the beginning of some major classifications. These numbers refer to related items in the Reference Section. The reference information may be an estimating procedure, an alternate pricing method, or technical information.

Note: Not all subdivisions listed here necessarily appear. ■

02 21 Surveys

02 21 13 – Site Surveys

02 21 13.09 Topographical Surveys	Crew	Daily Output	Labor-Hours	Unit	Material	2019 Bare Costs Labor	Equipment	Total	Total Incl O&P
0010 TOPOGRAPHICAL SURVEYS									
0020 Topographical surveying, conventional, minimum	A-7	3.30	7.273	Acre	23	275	13.75	311.75	495
0100 Maximum	A-8	.60	53.333	"	61	1,975	75.50	2,111.50	3,425

02 21 13.13 Boundary and Survey Markers	Crew	Daily Output	Labor-Hours	Unit	Material	Labor	Equipment	Total	Total Incl O&P
0010 BOUNDARY AND SURVEY MARKERS									
0300 Lot location and lines, large quantities, minimum	A-7	2	12	Acre	35.50	455	22.50	513	815
0320 Average	"	1.25	19.200		61.50	725	36.50	823	1,300
0400 Small quantities, maximum	A-8	1	32		76	1,175	45.50	1,296.50	2,075
0600 Monuments, 3′ long	A-7	10	2.400	Ea.	32	91	4.54	127.54	190
0800 Property lines, perimeter, cleared land	"	1000	.024	L.F.	.08	.91	.05	1.04	1.64
0900 Wooded land	A-8	875	.037	"	.10	1.36	.05	1.51	2.41

02 32 Geotechnical Investigations

02 32 13 – Subsurface Drilling and Sampling

02 32 13.10 Boring and Exploratory Drilling	Crew	Daily Output	Labor-Hours	Unit	Material	Labor	Equipment	Total	Total Incl O&P
0010 BORING AND EXPLORATORY DRILLING									
0020 Borings, initial field stake out & determination of elevations	A-6	1	16	Day		570	45.50	615.50	995
0100 Drawings showing boring details				Total		335		335	425
0200 Report and recommendations from P.E.						775		775	970
0300 Mobilization and demobilization	B-55	4	4			120	251	371	475
0350 For over 100 miles, per added mile		450	.036	Mile		1.06	2.23	3.29	4.20
0600 Auger holes in earth, no samples, 2-1/2″ diameter		78.60	.204	L.F.		6.10	12.75	18.85	24
0800 Cased borings in earth, with samples, 2-1/2″ diameter		55.50	.288	"	17.05	8.60	18.05	43.70	53
1400 Borings, earth, drill rig and crew with truck mounted auger		1	16	Day		480	1,000	1,480	1,900
1500 For inner city borings, add, minimum								10%	10%
1510 Maximum								20%	20%

02 41 Demolition

02 41 13 – Selective Site Demolition

02 41 13.17 Demolish, Remove Pavement and Curb	Crew	Daily Output	Labor-Hours	Unit	Material	Labor	Equipment	Total	Total Incl O&P
0010 DEMOLISH, REMOVE PAVEMENT AND CURB R024119-10									
5010 Pavement removal, bituminous roads, up to 3″ thick	B-38	690	.035	S.Y.		1.05	1.62	2.67	3.52
5050 4″-6″ thick		420	.057			1.73	2.67	4.40	5.80
5100 Bituminous driveways		640	.038			1.13	1.75	2.88	3.79
5200 Concrete to 6″ thick, hydraulic hammer, mesh reinforced		255	.094			2.84	4.39	7.23	9.50
5300 Rod reinforced		200	.120			3.63	5.60	9.23	12.15
5600 With hand held air equipment, bituminous, to 6″ thick	B-39	1900	.025	S.F.		.70	.12	.82	1.30
5700 Concrete to 6″ thick, no reinforcing		1600	.030			.83	.15	.98	1.54
5800 Mesh reinforced		1400	.034			.95	.17	1.12	1.76
5900 Rod reinforced		765	.063			1.75	.31	2.06	3.23
6000 Curbs, concrete, plain	B-6	360	.067	L.F.		2.01	.89	2.90	4.30
6100 Reinforced		275	.087			2.64	1.16	3.80	5.65
6200 Granite		360	.067			2.01	.89	2.90	4.30
6300 Bituminous		528	.045			1.37	.61	1.98	2.93

02 41 13.23 Utility Line Removal	Crew	Daily Output	Labor-Hours	Unit	Material	Labor	Equipment	Total	Total Incl O&P
0010 UTILITY LINE REMOVAL									
0015 No hauling, abandon catch basin or manhole	B-6	7	3.429	Ea.		104	45.50	149.50	222
0020 Remove existing catch basin or manhole, masonry		4	6			181	80	261	385
0030 Catch basin or manhole frames and covers, stored		13	1.846			56	24.50	80.50	119

02 41 Demolition

02 41 13 – Selective Site Demolition

02 41 13.23 Utility Line Removal		Crew	Daily Output	Labor-Hours	Unit	Material	2019 Bare Costs Labor	Equipment	Total	Total Incl O&P
0040	Remove and reset	B-6	7	3.429	Ea.		104	45.50	149.50	222
2900	Pipe removal, sewer/water, no excavation, 12" diameter		175	.137	L.F.		4.14	1.83	5.97	8.85
2930	15"-18" diameter	B-12Z	150	.160			4.96	9.60	14.56	18.70
2960	21"-24" diameter	"	120	.200			6.20	12	18.20	23.50

02 41 13.30 Minor Site Demolition		Crew	Daily Output	Labor-Hours	Unit	Material	Labor	Equipment	Total	Total Incl O&P
0010	MINOR SITE DEMOLITION R024119-10									
1000	Masonry walls, block, solid	B-5	1800	.022	C.F.		.66	.80	1.46	1.98
1200	Brick, solid		900	.044			1.33	1.60	2.93	3.95
1400	Stone, with mortar		900	.044			1.33	1.60	2.93	3.95
1500	Dry set		1500	.027			.80	.96	1.76	2.37
4000	Sidewalk removal, bituminous, 2" thick	B-6	350	.069	S.Y.		2.07	.91	2.98	4.42
4010	2-1/2" thick		325	.074			2.23	.99	3.22	4.76
4050	Brick, set in mortar		185	.130			3.92	1.73	5.65	8.35
4100	Concrete, plain, 4"		160	.150			4.53	2	6.53	9.65
4110	Plain, 5"		140	.171			5.20	2.29	7.49	11.05
4120	Plain, 6"		120	.200			6.05	2.67	8.72	12.90
4200	Mesh reinforced, concrete, 4"		150	.160			4.84	2.13	6.97	10.30
4210	5" thick		131	.183			5.55	2.44	7.99	11.80
4220	6" thick		112	.214			6.50	2.86	9.36	13.80
4300	Slab on grade removal, plain	B-5	45	.889	C.Y.		26.50	32	58.50	79
4310	Mesh reinforced		33	1.212			36	43.50	79.50	108
4320	Rod reinforced		25	1.600			48	57.50	105.50	142
4400	For congested sites or small quantities, add up to								200%	200%
4450	For disposal on site, add	B-11A	232	.069			2.24	5.55	7.79	9.80
4500	To 5 miles, add	B-34D	76	.105			3.46	8.35	11.81	14.90

02 41 13.60 Selective Demolition Fencing		Crew	Daily Output	Labor-Hours	Unit	Material	Labor	Equipment	Total	Total Incl O&P
0010	SELECTIVE DEMOLITION FENCING R024119-10									
1600	Fencing, barbed wire, 3 strand	2 Clab	430	.037	L.F.		1.02		1.02	1.69
1650	5 strand	"	280	.057			1.57		1.57	2.60
1700	Chain link, posts & fabric, 8'-10' high, remove only	B-6	445	.054			1.63	.72	2.35	3.48

02 41 16 – Structure Demolition

02 41 16.13 Building Demolition		Crew	Daily Output	Labor-Hours	Unit	Material	Labor	Equipment	Total	Total Incl O&P
0010	BUILDING DEMOLITION Large urban projects, incl. 20 mi. haul R024119-10									
0500	Small bldgs, or single bldgs, no salvage included, steel	B-3	14800	.003	C.F.		.10	.16	.26	.34
0600	Concrete	"	11300	.004	"		.13	.21	.34	.45
0605	Concrete, plain	B-5	33	1.212	C.Y.		36	43.50	79.50	108
0610	Reinforced		25	1.600			48	57.50	105.50	142
0615	Concrete walls		34	1.176			35	42.50	77.50	105
0620	Elevated slabs		26	1.538			46	55.50	101.50	137
0650	Masonry	B-3	14800	.003	C.F.		.10	.16	.26	.34
0700	Wood		14800	.003	"		.10	.16	.26	.34
1000	Demolition single family house, one story, wood 1600 S.F.		1	48	Ea.		1,500	2,325	3,825	5,050
1020	3200 S.F.		.50	96			3,000	4,675	7,675	10,100
1200	Demolition two family house, two story, wood 2400 S.F.		.67	71.964			2,250	3,500	5,750	7,550
1220	4200 S.F.		.38	128			4,000	6,225	10,225	13,500
1300	Demolition three family house, three story, wood 3200 S.F.		.50	96			3,000	4,675	7,675	10,100
1320	5400 S.F.		.30	160			5,000	7,775	12,775	16,800

02 41 Demolition

02 41 16 – Structure Demolition

02 41 16.17 Building Demolition Footings and Foundations		Crew	Daily Output	Labor-Hours	Unit	Material	2019 Bare Costs Labor	Equipment	Total	Total Incl O&P
0010	**BUILDING DEMOLITION FOOTINGS AND FOUNDATIONS** R024119-10									
0200	Floors, concrete slab on grade,									
0240	4" thick, plain concrete	B-13L	5000	.003	S.F.		.12	.40	.52	.64
0280	Reinforced, wire mesh		4000	.004			.15	.50	.65	.80
0300	Rods		4500	.004			.13	.44	.57	.71
0400	6" thick, plain concrete		4000	.004			.15	.50	.65	.80
0420	Reinforced, wire mesh		3200	.005			.19	.62	.81	1
0440	Rods		3600	.004			.17	.55	.72	.89
1000	Footings, concrete, 1' thick, 2' wide		300	.053	L.F.		2.02	6.65	8.67	10.60
1080	1'-6" thick, 2' wide		250	.064			2.43	8	10.43	12.80
1120	3' wide		200	.080			3.03	10	13.03	16
1200	Average reinforcing, add								10%	10%
2000	Walls, block, 4" thick	B-13L	8000	.002	S.F.		.08	.25	.33	.39
2040	6" thick		6000	.003			.10	.33	.43	.54
2080	8" thick		4000	.004			.15	.50	.65	.80
2100	12" thick		3000	.005			.20	.67	.87	1.06
2400	Concrete, plain concrete, 6" thick		4000	.004			.15	.50	.65	.80
2420	8" thick		3500	.005			.17	.57	.74	.91
2440	10" thick		3000	.005			.20	.67	.87	1.06
2500	12" thick		2500	.006			.24	.80	1.04	1.28
2600	For average reinforcing, add								10%	10%
4000	For congested sites or small quantities, add up to								200%	200%
4200	Add for disposal, on site	B-11A	232	.069	C.Y.		2.24	5.55	7.79	9.80
4250	To five miles	B-30	220	.109	"		3.75	9.30	13.05	16.35

02 41 19 – Selective Demolition

02 41 19.13 Selective Building Demolition

Line	Description	Crew	Daily Output	Labor-Hours	Unit	Material	Labor	Equipment	Total	Total Incl O&P
0010	**SELECTIVE BUILDING DEMOLITION**									
0020	Costs related to selective demolition of specific building components									
0025	are included under Common Work Results (XX 05)									
0030	in the component's appropriate division.									

02 41 19.16 Selective Demolition, Cutout

Line	Description	Crew	Daily Output	Labor-Hours	Unit	Material	Labor	Equipment	Total	Total Incl O&P
0010	**SELECTIVE DEMOLITION, CUTOUT** R024119-10									
0020	Concrete, elev. slab, light reinforcement, under 6 C.F.	B-9	65	.615	C.F.		17.15	3.61	20.76	32.50
0050	Light reinforcing, over 6 C.F.		75	.533	"		14.90	3.13	18.03	28
0200	Slab on grade to 6" thick, not reinforced, under 8 S.F.		85	.471	S.F.		13.15	2.76	15.91	24.50
0250	8-16 S.F.		175	.229	"		6.40	1.34	7.74	12.05
0255	For over 16 S.F. see Line 02 41 16.17 0400									
0600	Walls, not reinforced, under 6 C.F.	B-9	60	.667	C.F.		18.60	3.91	22.51	35
0650	6-12 C.F.	"	80	.500	"		13.95	2.94	16.89	26
0655	For over 12 C.F. see Line 02 41 16.17 2500									
1000	Concrete, elevated slab, bar reinforced, under 6 C.F.	B-9	45	.889	C.F.		25	5.20	30.20	47
1050	Bar reinforced, over 6 C.F.		50	.800	"		22.50	4.70	27.20	42
1200	Slab on grade to 6" thick, bar reinforced, under 8 S.F.		75	.533	S.F.		14.90	3.13	18.03	28
1250	8-16 S.F.		150	.267	"		7.45	1.57	9.02	14
1255	For over 16 S.F. see Line 02 41 16.17 0440									
1400	Walls, bar reinforced, under 6 C.F.	B-9	50	.800	C.F.		22.50	4.70	27.20	42
1450	6-12 C.F.	"	70	.571	"		15.95	3.35	19.30	30
1455	For over 12 C.F. see Lines 02 41 16.17 2500 and 2600									
2000	Brick, to 4 S.F. opening, not including toothing									
2040	4" thick	B-9	30	1.333	Ea.		37	7.85	44.85	70
2060	8" thick		18	2.222			62	13.05	75.05	116

02 41 19 – Selective Demolition

02 41 19.16 Selective Demolition, Cutout

	02 41 19.16 Selective Demolition, Cutout	Crew	Daily Output	Labor-Hours	Unit	Material	2019 Bare Costs Labor	Equipment	Total	Total Incl O&P
2080	12" thick	B-9	10	4	Ea.		112	23.50	135.50	210
2400	Concrete block, to 4 S.F. opening, 2" thick		35	1.143			32	6.70	38.70	60
2420	4" thick		30	1.333			37	7.85	44.85	70
2440	8" thick		27	1.481			41.50	8.70	50.20	78
2460	12" thick		24	1.667			46.50	9.80	56.30	88
2600	Gypsum block, to 4 S.F. opening, 2" thick		80	.500			13.95	2.94	16.89	26
2620	4" thick		70	.571			15.95	3.35	19.30	30
2640	8" thick		55	.727			20.50	4.27	24.77	38
2800	Terra cotta, to 4 S.F. opening, 4" thick		70	.571			15.95	3.35	19.30	30
2840	8" thick		65	.615			17.15	3.61	20.76	32.50
2880	12" thick		50	.800			22.50	4.70	27.20	42
3000	Toothing masonry cutouts, brick, soft old mortar	1 Brhe	40	.200	V.L.F.		5.90		5.90	9.85
3100	Hard mortar		30	.267			7.85		7.85	13.10
3200	Block, soft old mortar		70	.114			3.37		3.37	5.60
3400	Hard mortar		50	.160			4.72		4.72	7.85
6000	Walls, interior, not including re-framing,									
6010	openings to 5 S.F.									
6100	Drywall to 5/8" thick	1 Clab	24	.333	Ea.		9.15		9.15	15.15
6200	Paneling to 3/4" thick		20	.400			11		11	18.20
6300	Plaster, on gypsum lath		20	.400			11		11	18.20
6340	On wire lath		14	.571			15.70		15.70	26
7000	Wood frame, not including re-framing, openings to 5 S.F.									
7200	Floors, sheathing and flooring to 2" thick	1 Clab	5	1.600	Ea.		44		44	72.50
7310	Roofs, sheathing to 1" thick, not including roofing		6	1.333			36.50		36.50	60.50
7410	Walls, sheathing to 1" thick, not including siding		7	1.143			31.50		31.50	52

02 41 19.19 Selective Demolition

	02 41 19.19 Selective Demolition	Crew	Daily Output	Labor-Hours	Unit	Material	Labor	Equipment	Total	Total Incl O&P
0010	**SELECTIVE DEMOLITION**, Rubbish Handling R024119-10									
0020	The following are to be added to the demolition prices									
0600	Dumpster, weekly rental, 1 dump/week, 6 C.Y. capacity (2 tons)				Week	415			415	455
0700	10 C.Y. capacity (3 tons)					480			480	530
0725	20 C.Y. capacity (5 tons) R024119-20					565			565	625
0800	30 C.Y. capacity (7 tons)					730			730	800
0840	40 C.Y. capacity (10 tons)					775			775	850
2000	Load, haul, dump and return, 0'-50' haul, hand carried	2 Clab	24	.667	C.Y.		18.35		18.35	30.50
2005	Wheeled		37	.432			11.90		11.90	19.65
2040	0'-100' haul, hand carried		16.50	.970			26.50		26.50	44
2045	Wheeled		25	.640			17.60		17.60	29
2050	Forklift	A-3R	25	.320			11.40	5.75	17.15	25
2080	Haul and return, add per each extra 100' haul, hand carried	2 Clab	35.50	.451			12.40		12.40	20.50
2085	Wheeled		54	.296			8.15		8.15	13.45
2120	For travel in elevators, up to 10 floors, add		140	.114			3.14		3.14	5.20
2130	0'-50' haul, incl. up to 5 riser stairs, hand carried		23	.696			19.15		19.15	31.50
2135	Wheeled		35	.457			12.55		12.55	21
2140	6-10 riser stairs, hand carried		22	.727			20		20	33
2145	Wheeled		34	.471			12.95		12.95	21.50
2150	11-20 riser stairs, hand carried		20	.800			22		22	36.50
2155	Wheeled		31	.516			14.20		14.20	23.50
2160	21-40 riser stairs, hand carried		16	1			27.50		27.50	45.50
2165	Wheeled		24	.667			18.35		18.35	30.50
2170	0'-100' haul, incl. 5 riser stairs, hand carried		15	1.067			29.50		29.50	48.50
2175	Wheeled		23	.696			19.15		19.15	31.50
2180	6-10 riser stairs, hand carried		14	1.143			31.50		31.50	52

02 41 Demolition

02 41 19 – Selective Demolition

02 41 19.19 Selective Demolition		Crew	Daily Output	Labor-Hours	Unit	2019 Bare Costs Material	Labor	Equipment	Total	Total Incl O&P
2185	Wheeled	2 Clab	21	.762	C.Y.		21		21	34.50
2190	11-20 riser stairs, hand carried		12	1.333			36.50		36.50	60.50
2195	Wheeled		18	.889			24.50		24.50	40.50
2200	21-40 riser stairs, hand carried		8	2			55		55	91
2205	Wheeled		12	1.333			36.50		36.50	60.50
2210	Haul and return, add per each extra 100′ haul, hand carried		35.50	.451			12.40		12.40	20.50
2215	Wheeled		54	.296			8.15		8.15	13.45
2220	For each additional flight of stairs, up to 5 risers, add		550	.029	Flight		.80		.80	1.32
2225	6-10 risers, add		275	.058			1.60		1.60	2.64
2230	11-20 risers, add		138	.116			3.19		3.19	5.25
2235	21-40 risers, add		69	.232			6.40		6.40	10.55
3000	Loading & trucking, including 2 mile haul, chute loaded	B-16	45	.711	C.Y.		21	12.60	33.60	48.50
3040	Hand loading truck, 50′ haul	″	48	.667			19.55	11.80	31.35	45.50
3080	Machine loading truck	B-17	120	.267			8.25	5.50	13.75	19.60
5000	Haul, per mile, up to 8 C.Y. truck	B-34B	1165	.007			.23	.49	.72	.91
5100	Over 8 C.Y. truck	″	1550	.005			.17	.37	.54	.68

02 41 19.20 Selective Demolition, Dump Charges

		Crew	Daily Output	Labor-Hours	Unit	Material	Labor	Equipment	Total	Total Incl O&P
0010	SELECTIVE DEMOLITION, DUMP CHARGES R024119-10									
0020	Dump charges, typical urban city, tipping fees only									
0100	Building construction materials				Ton	74			74	81
0200	Trees, brush, lumber					63			63	69.50
0300	Rubbish only					63			63	69.50
0500	Reclamation station, usual charge					74			74	81

02 41 19.21 Selective Demolition, Gutting

		Crew	Daily Output	Labor-Hours	Unit	Material	Labor	Equipment	Total	Total Incl O&P
0010	SELECTIVE DEMOLITION, GUTTING R024119-10									
0020	Building interior, including disposal, dumpster fees not included									
0500	Residential building									
0560	Minimum	B-16	400	.080	SF Flr.		2.35	1.42	3.77	5.45
0580	Maximum	″	360	.089	″		2.61	1.58	4.19	6.05
0900	Commercial building									
1000	Minimum	B-16	350	.091	SF Flr.		2.68	1.62	4.30	6.20
1020	Maximum	″	250	.128	″		3.76	2.27	6.03	8.70

02 83 Lead Remediation

02 83 19 – Lead-Based Paint Remediation

02 83 19.22 Preparation of Lead Containment Area

		Crew	Daily Output	Labor-Hours	Unit	Material	Labor	Equipment	Total	Total Incl O&P
0010	PREPARATION OF LEAD CONTAINMENT AREA									
0020	Lead abatement work area, test kit, per swab	1 Skwk	16	.500	Ea.	3.08	18.25		21.33	34
0025	For dust barriers see Section 01 56 16.10									
0050	Caution sign	1 Skwk	48	.167	Ea.	8.45	6.10		14.55	19.40
0100	Pre-cleaning, HEPA vacuum and wet wipe, floor and wall surfaces	3 Skwk	5000	.005	S.F.	.01	.18		.19	.30
0105	Ceiling, 6′-11′ high		4100	.006		.08	.21		.29	.43
0108	12′-15′ high		3550	.007		.07	.25		.32	.49
0115	Over 15′ high		3000	.008		.09	.29		.38	.59
0500	Cover surfaces with polyethylene sheeting									
0550	Floors, each layer, 6 mil	3 Skwk	8000	.003	S.F.	.04	.11		.15	.22
0560	Walls, each layer, 6 mil	″	6000	.004	″	.04	.15		.19	.28
0570	For heights above 14′, add						20%			
2400	Post abatement cleaning of protective sheeting, HEPA vacuum & wet wipe	3 Skwk	5000	.005	S.F.	.01	.18		.19	.30
2450	Doff, bag and seal protective sheeting		12000	.002		.01	.07		.08	.13

02 83 Lead Remediation

02 83 19 – Lead-Based Paint Remediation

02 83 19.22 Preparation of Lead Containment Area		Crew	Daily Output	Labor-Hours	Unit	2019 Bare Costs Material	Labor	Equipment	Total	Total Incl O&P
2500	Post abatement cleaning, HEPA vacuum & wet wipe	3 Skwk	5000	.005	S.F.	.01	.18		.19	.30

02 83 19.23 Encapsulation of Lead-Based Paint

		Crew	Daily Output	Labor-Hours	Unit	Material	Labor	Equipment	Total	Total Incl O&P
0010	**ENCAPSULATION OF LEAD-BASED PAINT**									
0020	Interior, brushwork, trim, under 6″	1 Pord	240	.033	L.F.	2.37	.99		3.36	4.24
0030	6″-12″ wide		180	.044		3.09	1.33		4.42	5.55
0040	Balustrades		300	.027		1.95	.80		2.75	3.45
0050	Pipe to 4″ diameter		500	.016		1.17	.48		1.65	2.07
0060	To 8″ diameter		375	.021		1.54	.64		2.18	2.73
0070	To 12″ diameter		250	.032		2.28	.96		3.24	4.07
0080	To 16″ diameter		170	.047		3.39	1.40		4.79	6.05
0090	Cabinets, ornate design		200	.040	S.F.	2.92	1.19		4.11	5.15
0100	Simple design		250	.032	″	2.34	.96		3.30	4.13
0110	Doors, 3′ x 7′, both sides, incl. frame & trim									
0120	Flush	1 Pord	6	1.333	Ea.	29	40		69	97
0130	French, 10-15 lite		3	2.667		5.95	79.50		85.45	137
0140	Panel		4	2		36	59.50		95.50	138
0150	Louvered		2.75	2.909		33	87		120	178
0160	Windows, per interior side, per 15 S.F.									
0170	1-6 lite	1 Pord	14	.571	Ea.	20.50	17.05		37.55	50.50
0180	7-10 lite		7.50	1.067		22.50	32		54.50	77
0190	12 lite		5.75	1.391		30.50	41.50		72	102
0200	Radiators		8	1		72	30		102	129
0210	Grilles, vents		275	.029	S.F.	2.05	.87		2.92	3.68
0220	Walls, roller, drywall or plaster		1000	.008		.59	.24		.83	1.04
0230	With spunbonded reinforcing fabric		720	.011		.69	.33		1.02	1.30
0240	Wood		800	.010		.72	.30		1.02	1.28
0250	Ceilings, roller, drywall or plaster		900	.009		.69	.27		.96	1.19
0260	Wood		700	.011		.83	.34		1.17	1.47
0270	Exterior, brushwork, gutters and downspouts		300	.027	L.F.	1.95	.80		2.75	3.45
0280	Columns		400	.020	S.F.	1.44	.60		2.04	2.56
0290	Spray, siding		600	.013	″	.97	.40		1.37	1.72
0300	Miscellaneous									
0310	Electrical conduit, brushwork, to 2″ diameter	1 Pord	500	.016	L.F.	1.17	.48		1.65	2.07
0320	Brick, block or concrete, spray		500	.016	S.F.	1.17	.48		1.65	2.07
0330	Steel, flat surfaces and tanks to 12″		500	.016		1.17	.48		1.65	2.07
0340	Beams, brushwork		400	.020		1.44	.60		2.04	2.56
0350	Trusses		400	.020		1.44	.60		2.04	2.56

02 83 19.26 Removal of Lead-Based Paint

		Crew	Daily Output	Labor-Hours	Unit	Material	Labor	Equipment	Total	Total Incl O&P
0010	**REMOVAL OF LEAD-BASED PAINT**									
0011	By chemicals, per application									
0050	Baseboard, to 6″ wide	1 Pord	64	.125	L.F.	.79	3.73		4.52	6.95
0070	To 12″ wide		32	.250	″	1.49	7.45		8.94	13.90
0200	Balustrades, one side		28	.286	S.F.	1.46	8.55		10.01	15.55
1400	Cabinets, simple design		32	.250		1.33	7.45		8.78	13.70
1420	Ornate design		25	.320		1.56	9.55		11.11	17.35
1600	Cornice, simple design		60	.133		1.59	3.98		5.57	8.25
1620	Ornate design		20	.400		5.50	11.95		17.45	25.50
2800	Doors, one side, flush		84	.095		2.08	2.84		4.92	6.95
2820	Two panel		80	.100		1.33	2.99		4.32	6.35
2840	Four panel		45	.178		1.40	5.30		6.70	10.25
2880	For trim, one side, add		64	.125	L.F.	.76	3.73		4.49	6.95
3000	Fence, picket, one side		30	.267	S.F.	1.33	7.95		9.28	14.50

02 83 Lead Remediation

02 83 19 – Lead-Based Paint Remediation

02 83 19.26 Removal of Lead-Based Paint		Crew	Daily Output	Labor-Hours	Unit	Material	2019 Bare Costs Labor	Equipment	Total	Total Incl O&P
3200	Grilles, one side, simple design	1 Pord	30	.267	S.F.	1.34	7.95		9.29	14.50
3220	Ornate design		25	.320		1.44	9.55		10.99	17.25
3240	Handrails		90	.089	L.F.	1.34	2.65		3.99	5.80
4400	Pipes, to 4" diameter		90	.089		2.08	2.65		4.73	6.65
4420	To 8" diameter		50	.160		4.16	4.78		8.94	12.35
4440	To 12" diameter		36	.222		6.25	6.65		12.90	17.70
4460	To 16" diameter		20	.400		8.30	11.95		20.25	28.50
4500	For hangers, add		40	.200	Ea.	2.52	5.95		8.47	12.55
4800	Siding		90	.089	S.F.	1.32	2.65		3.97	5.80
5000	Trusses, open		55	.145	SF Face	1.99	4.34		6.33	9.30
6200	Windows, one side only, double-hung, 1/1 light, 24" x 48" high		4	2	Ea.	25	59.50		84.50	126
6220	30" x 60" high		3	2.667		33	79.50		112.50	167
6240	36" x 72" high		2.50	3.200		39.50	95.50		135	200
6280	40" x 80" high		2	4		49	119		168	250
6400	Colonial window, 6/6 light, 24" x 48" high		2	4		49	119		168	250
6420	30" x 60" high		1.50	5.333		64.50	159		223.50	330
6440	36" x 72" high		1	8		97	239		336	495
6480	40" x 80" high		1	8		97	239		336	495
6600	8/8 light, 24" x 48" high		2	4		49	119		168	250
6620	40" x 80" high		1	8		97	239		336	495
6800	12/12 light, 24" x 48" high		1	8		97	239		336	495
6820	40" x 80" high		.75	10.667		130	320		450	665
6840	Window frame & trim items, included in pricing above									

02 87 Biohazard Remediation

02 87 13 – Mold Remediation

02 87 13.33 Removal and Disposal of Materials With Mold		Crew	Daily Output	Labor-Hours	Unit	Material	Labor	Equipment	Total	Total Incl O&P
0010	**REMOVAL AND DISPOSAL OF MATERIALS WITH MOLD**									
0015	Demolition in mold contaminated area									
0200	Ceiling, including suspension system, plaster and lath	A-9	2100	.030	S.F.	.08	1.12		1.20	1.98
0210	Finished plaster, leaving wire lath		585	.109		.28	4.02		4.30	7.05
0220	Suspended acoustical tile		3500	.018		.05	.67		.72	1.18
0230	Concealed tile grid system		3000	.021		.06	.78		.84	1.38
0240	Metal pan grid system		1500	.043		.11	1.57		1.68	2.76
0250	Gypsum board		2500	.026		.07	.94		1.01	1.65
0255	Plywood		2500	.026		.07	.94		1.01	1.65
0260	Lighting fixtures up to 2' x 4'		72	.889	Ea.	2.32	32.50		34.82	57.50
0400	Partitions, non load bearing									
0410	Plaster, lath, and studs	A-9	690	.093	S.F.	.88	3.41		4.29	6.70
0450	Gypsum board and studs		1390	.046		.12	1.69		1.81	2.98
0465	Carpet & pad		1390	.046		.12	1.69		1.81	2.98
0600	Pipe insulation, air cell type, up to 4" diameter pipe		900	.071	L.F.	.19	2.61		2.80	4.60
0610	4" to 8" diameter pipe		800	.080		.21	2.94		3.15	5.20
0620	10" to 12" diameter pipe		700	.091		.24	3.36		3.60	5.90
0630	14" to 16" diameter pipe		550	.116		.30	4.28		4.58	7.55
0650	Over 16" diameter pipe		650	.098	S.F.	.26	3.62		3.88	6.40
9000	For type B (supplied air) respirator equipment, add				%				10%	10%

Estimating Tips

General

- Carefully check all the plans and specifications. Concrete often appears on drawings other than structural drawings, including mechanical and electrical drawings for equipment pads. The cost of cutting and patching is often difficult to estimate. See Subdivision 03 81 for Concrete Cutting, Subdivision 02 41 19.16 for Cutout Demolition, Subdivision 03 05 05.10 for Concrete Demolition, and Subdivision 02 41 19.19 for Rubbish Handling (handling, loading, and hauling of debris).
- Always obtain concrete prices from suppliers near the job site. A volume discount can often be negotiated, depending upon competition in the area. Remember to add for waste, particularly for slabs and footings on grade.

03 10 00 Concrete Forming and Accessories

- A primary cost for concrete construction is forming. Most jobs today are constructed with prefabricated forms. The selection of the forms best suited for the job and the total square feet of forms required for efficient concrete forming and placing are key elements in estimating concrete construction. Enough forms must be available for erection to make efficient use of the concrete placing equipment and crew.
- Concrete accessories for forming and placing depend upon the systems used. Study the plans and specifications to ensure that all special accessory requirements have been included in the cost estimate, such as anchor bolts, inserts, and hangers.
- Included within costs for forms-in-place are all necessary bracing and shoring.

03 20 00 Concrete Reinforcing

- Ascertain that the reinforcing steel supplier has included all accessories, cutting, bending, and an allowance for lapping, splicing, and waste. A good rule of thumb is 10% for lapping, splicing, and waste. Also, 10% waste should be allowed for welded wire fabric.
- The unit price items in the subdivisions for Reinforcing In Place, Glass Fiber Reinforcing, and Welded Wire Fabric include the labor to install accessories such as beam and slab bolsters, high chairs, and bar ties and tie wire. The material cost for these accessories is not included; they may be obtained from the Accessories Subdivisions.

03 30 00 Cast-In-Place Concrete

- When estimating structural concrete, pay particular attention to requirements for concrete additives, curing methods, and surface treatments. Special consideration for climate, hot or cold, must be included in your estimate. Be sure to include requirements for concrete placing equipment and concrete finishing.
- For accurate concrete estimating, the estimator must consider each of the following major components individually: forms, reinforcing steel, ready-mix concrete, placement of the concrete, and finishing of the top surface. For faster estimating, Subdivision 03 30 53.40 for Concrete-In-Place can be used; here, various items of concrete work are presented that include the costs of all five major components (unless specifically stated otherwise).

03 40 00 Precast Concrete
03 50 00 Cast Decks and Underlayment

- The cost of hauling precast concrete structural members is often an important factor. For this reason, it is important to get a quote from the nearest supplier. It may become economically feasible to set up precasting beds on the site if the hauling costs are prohibitive.

Reference Numbers

Reference numbers are shown at the beginning of some major classifications. These numbers refer to related items in the Reference Section. The reference information may be an estimating procedure, an alternate pricing method, or technical information.

Note: Not all subdivisions listed here necessarily appear. ■

03 01 Maintenance of Concrete

03 01 30 – Maintenance of Cast-In-Place Concrete

03 01 30.64 Floor Patching		Crew	Daily Output	Labor-Hours	Unit	Material	Labor	Equipment	Total	Total Incl O&P
						2019 Bare Costs				
0010	**FLOOR PATCHING**									
0012	Floor patching, 1/4" thick, small areas, regular	1 Cefi	170	.047	S.F.	3.38	1.68		5.06	6.45
0100	Epoxy	"	100	.080	"	8.60	2.86		11.46	14.05

03 05 Common Work Results for Concrete

03 05 13 – Basic Concrete Materials

03 05 13.85 Winter Protection

		Crew	Daily Output	Labor-Hours	Unit	Material	Labor	Equipment	Total	Total Incl O&P
0010	**WINTER PROTECTION**									
0012	For heated ready mix, add				C.Y.	5.35			5.35	5.90
0100	Temporary heat to protect concrete, 24 hours	2 Clab	50	.320	M.S.F.	204	8.80		212.80	239
0200	Temporary shelter for slab on grade, wood frame/polyethylene sheeting									
0201	Build or remove, light framing for short spans	2 Carp	10	1.600	M.S.F.	315	57		372	440
0210	Large framing for long spans	"	3	5.333	"	420	190		610	775
0710	Electrically heated pads, 15 watts/S.F., 20 uses				S.F.	.56			.56	.61

03 11 Concrete Forming

03 11 13 – Structural Cast-In-Place Concrete Forming

03 11 13.25 Forms In Place, Columns

		Crew	Daily Output	Labor-Hours	Unit	Material	Labor	Equipment	Total	Total Incl O&P
0010	**FORMS IN PLACE, COLUMNS**									
1500	Round fiber tube, recycled paper, 1 use, 8" diameter G	C-1	155	.206	L.F.	2.94	6.50		9.44	14.05
1550	10" diameter G		155	.206		3.64	6.50		10.14	14.80
1600	12" diameter G		150	.213		4.14	6.75		10.89	15.70
1650	14" diameter G		145	.221		4.40	6.95		11.35	16.40
1700	16" diameter G		140	.229		5.50	7.20		12.70	18
1720	18" diameter G		140	.229		6.40	7.20		13.60	19
5000	Job-built plywood, 8" x 8" columns, 1 use		165	.194	SFCA	2.73	6.10		8.83	13.15
5500	12" x 12" columns, 1 use		180	.178	"	2.63	5.60		8.23	12.20
7400	Steel framed plywood, based on 50 uses of purchased									
7420	forms, and 4 uses of bracing lumber									
7500	8" x 8" column	C-1	340	.094	SFCA	2.18	2.97		5.15	7.30
7550	10" x 10"		350	.091		1.94	2.88		4.82	6.90
7600	12" x 12"		370	.086		1.65	2.73		4.38	6.35

03 11 13.45 Forms In Place, Footings

		Crew	Daily Output	Labor-Hours	Unit	Material	Labor	Equipment	Total	Total Incl O&P
0010	**FORMS IN PLACE, FOOTINGS**									
0020	Continuous wall, plywood, 1 use	C-1	375	.085	SFCA	7	2.69		9.69	12.15
0150	4 use	"	485	.066	"	2.28	2.08		4.36	5.95
1500	Keyway, 4 use, tapered wood, 2" x 4"	1 Carp	530	.015	L.F.	.23	.54		.77	1.14
1550	2" x 6"	"	500	.016	"	.36	.57		.93	1.33
5000	Spread footings, job-built lumber, 1 use	C-1	305	.105	SFCA	2.31	3.31		5.62	8.05
5150	4 use	"	414	.077	"	.75	2.44		3.19	4.87

03 11 13.50 Forms In Place, Grade Beam

		Crew	Daily Output	Labor-Hours	Unit	Material	Labor	Equipment	Total	Total Incl O&P
0010	**FORMS IN PLACE, GRADE BEAM**									
0020	Job-built plywood, 1 use	C-2	530	.091	SFCA	3.20	2.88		6.08	8.30
0150	4 use	"	605	.079	"	1.04	2.53		3.57	5.35

03 11 13.65 Forms In Place, Slab On Grade

		Crew	Daily Output	Labor-Hours	Unit	Material	Labor	Equipment	Total	Total Incl O&P
0010	**FORMS IN PLACE, SLAB ON GRADE**									
1000	Bulkhead forms w/keyway, wood, 6" high, 1 use	C-1	510	.063	L.F.	1.13	1.98		3.11	4.52
1400	Bulkhead form for slab, 4-1/2" high, exp metal, incl. keyway & stakes G		1200	.027		.94	.84		1.78	2.42

03 11 Concrete Forming

03 11 13 – Structural Cast-In-Place Concrete Forming

03 11 13.65 Forms In Place, Slab On Grade			Crew	Daily Output	Labor-Hours	Unit	Material	2019 Bare Costs Labor	Equipment	Total	Total Incl O&P
1410	5-1/2″ high	G	C-1	1100	.029	L.F.	1.18	.92		2.10	2.82
1420	7-1/2″ high	G		960	.033		1.41	1.05		2.46	3.29
1430	9-1/2″ high	G		840	.038		1.53	1.20		2.73	3.67
2000	Curb forms, wood, 6″ to 12″ high, on grade, 1 use			215	.149	SFCA	1.94	4.69		6.63	9.95
2150	4 use			275	.116	″	.63	3.67		4.30	6.80
3000	Edge forms, wood, 4 use, on grade, to 6″ high			600	.053	L.F.	.31	1.68		1.99	3.13
3050	7″ to 12″ high			435	.074	SFCA	.73	2.32		3.05	4.64
4000	For slab blockouts, to 12″ high, 1 use			200	.160	L.F.	.84	5.05		5.89	9.25
4100	Plastic (extruded), to 6″ high, multiple use, on grade			800	.040	″	6.65	1.26		7.91	9.40
8760	Void form, corrugated fiberboard, 4″ x 12″, 4′ long	G		3000	.011	S.F.	3.43	.34		3.77	4.33
8770	6″ x 12″, 4′ long			3000	.011		4.59	.34		4.93	5.60
8780	1/4″ thick hardboard protective cover for void form		2 Carp	1500	.011		.69	.38		1.07	1.39

03 11 13.85 Forms In Place, Walls			Crew	Daily Output	Labor-Hours	Unit	Material	Labor	Equipment	Total	Total Incl O&P
0010	**FORMS IN PLACE, WALLS**										
0100	Box out for wall openings, to 16″ thick, to 10 S.F.		C-2	24	2	Ea.	27	63.50		90.50	136
0150	Over 10 S.F. (use perimeter)		″	280	.171	L.F.	2.32	5.45		7.77	11.60
0250	Brick shelf, 4″ w, add to wall forms, use wall area above shelf										
0260	1 use		C-2	240	.200	SFCA	2.51	6.35		8.86	13.30
0350	4 use			300	.160	″	1	5.10		6.10	9.55
0500	Bulkhead, wood with keyway, 1 use, 2 piece			265	.181	L.F.	2.23	5.75		7.98	12
0600	Bulkhead forms with keyway, 1 piece expanded metal, 8″ wall	G	C-1	1000	.032		1.41	1.01		2.42	3.22
0610	10″ wall	G	″	800	.040		1.53	1.26		2.79	3.77
2000	Wall, job-built plywood, to 8′ high, 1 use		C-2	370	.130	SFCA	2.78	4.13		6.91	9.90
2050	2 use			435	.110		1.77	3.51		5.28	7.80
2100	3 use			495	.097		1.29	3.09		4.38	6.50
2150	4 use			505	.095		1.05	3.03		4.08	6.15
2400	Over 8′ to 16′ high, 1 use			280	.171		3.08	5.45		8.53	12.45
2450	2 use			345	.139		1.36	4.43		5.79	8.85
2500	3 use			375	.128		.97	4.08		5.05	7.80
2550	4 use			395	.122		.79	3.87		4.66	7.25
7800	Modular prefabricated plywood, based on 20 uses of purchased										
7820	forms, and 4 uses of bracing lumber										
7860	To 8′ high		C-2	800	.060	SFCA	1.13	1.91		3.04	4.42
8060	Over 8′ to 16′ high		″	600	.080	″	1.19	2.55		3.74	5.55

03 11 19 – Insulating Concrete Forming

03 11 19.10 Insulating Forms, Left In Place		Crew	Daily Output	Labor-Hours	Unit	Material	Labor	Equipment	Total	Total Incl O&P
0010	**INSULATING FORMS, LEFT IN PLACE**									
0020	Forms includes layout, excludes rebar, embedments, bucks for openings,									
0030	scaffolding, wall bracing, concrete, and concrete placing.									
0040	S.F. is for exterior face but includes forms for both faces of wall									
0100	Straight blocks or panels, molded, walls up to 4′ high									
0110	4″ core wall	4 Carp	1984	.016	S.F.	3.61	.58		4.19	4.92
0120	6″ core wall		1808	.018		3.63	.63		4.26	5.05
0130	8″ core wall		1536	.021		3.76	.74		4.50	5.35
0140	10″ core wall		1152	.028		4.29	.99		5.28	6.35
0150	12″ core wall		992	.032		4.87	1.15		6.02	7.25
0200	90 degree corner blocks or panels, molded, walls up to 4′ high									
0210	4″ core wall	4 Carp	1880	.017	S.F.	3.72	.61		4.33	5.10
0220	6″ core wall		1708	.019		3.74	.67		4.41	5.20
0230	8″ core wall		1324	.024		3.87	.86		4.73	5.70
0240	10″ core wall		987	.032		4.47	1.16		5.63	6.85
0250	12″ core wall		884	.036		4.79	1.29		6.08	7.40

03 11 Concrete Forming

03 11 19 – Insulating Concrete Forming

03 11 19.10 Insulating Forms, Left In Place		Crew	Daily Output	Labor-Hours	Unit	Material	2019 Bare Costs Labor	Equipment	Total	Total Incl O&P
0300	45 degree corner blocks or panels, molded, walls up to 4′ high									
0310	4″ core wall	4 Carp	1880	.017	S.F.	3.92	.61		4.53	5.30
0320	6″ core wall		1712	.019		4.05	.67		4.72	5.55
0330	8″ core wall		1324	.024		4.16	.86		5.02	6
0400	T blocks or panels, molded, walls up to 4′ high									
0420	6″ core wall	4 Carp	1540	.021	S.F.	4.77	.74		5.51	6.45
0430	8″ core wall		1325	.024		4.82	.86		5.68	6.70
0440	Non-standard corners or Ts requiring trimming & strapping		192	.167		4.87	5.95		10.82	15.15
0500	Radius blocks or panels, molded, walls up to 4′ high, 6″ core wall									
0520	5′ to 10′ diameter, molded blocks or panels	4 Carp	2400	.013	S.F.	5.90	.48		6.38	7.25
0530	10′ to 15′ diameter, requiring trimming and strapping, add		500	.064		4.43	2.28		6.71	8.65
0540	15′-1″ to 30′ diameter, requiring trimming and strapping, add		1200	.027		4.43	.95		5.38	6.45
0550	30′-1″ to 60′ diameter, requiring trimming and strapping, add		1600	.020		4.43	.71		5.14	6.05
0560	60′-1″ to 100′ diameter, requiring trimming and strapping, add		2800	.011		4.43	.41		4.84	5.55
0600	Additional labor for blocks/panels in higher walls (excludes scaffolding)									
0610	4′-1″ to 9′-4″ high, add						10%			
0620	9′-5″ to 12′-0″ high, add						20%			
0630	12′-1″ to 20′-0″ high, add						35%			
0640	Over 20′-0″ high, add						55%			
0700	Taper block or panels, molded, single course									
0720	6″ core wall	4 Carp	1600	.020	S.F.	3.90	.71		4.61	5.45
0730	8″ core wall	"	1392	.023	"	3.95	.82		4.77	5.70
0800	ICF brick ledge (corbel) block or panels, molded, single course									
0820	6″ core wall	4 Carp	1200	.027	S.F.	4.30	.95		5.25	6.30
0830	8″ core wall	"	1152	.028	"	4.39	.99		5.38	6.45
0900	ICF curb (shelf) block or panels, molded, single course									
0930	8″ core wall	4 Carp	688	.047	S.F.	3.71	1.66		5.37	6.80
0940	10″ core wall	"	544	.059	"	4.24	2.10		6.34	8.10
0950	Wood form to hold back concrete to form shelf, 8″ high	2 Carp	400	.040	L.F.	.85	1.43		2.28	3.30
1000	ICF half height block or panels, molded, single course									
1010	4″ core wall	4 Carp	1248	.026	S.F.	4.70	.91		5.61	6.65
1020	6″ core wall		1152	.028		4.72	.99		5.71	6.85
1030	8″ core wall		942	.034		4.77	1.21		5.98	7.25
1040	10″ core wall		752	.043		5.25	1.52		6.77	8.30
1050	12″ core wall		648	.049		5.20	1.76		6.96	8.60
1100	ICF half height block/panels, made by field sawing full height block/panels									
1110	4″ core wall	4 Carp	800	.040	S.F.	1.81	1.43		3.24	4.35
1120	6″ core wall		752	.043		1.82	1.52		3.34	4.51
1130	8″ core wall		600	.053		1.88	1.90		3.78	5.20
1140	10″ core wall		496	.065		2.15	2.30		4.45	6.15
1150	12″ core wall		400	.080		2.44	2.85		5.29	7.40
1200	Additional insulation inserted into forms between ties									
1210	1 layer (2″ thick)	4 Carp	14000	.002	S.F.	1.03	.08		1.11	1.26
1220	2 layers (4″ thick)		7000	.005		2.06	.16		2.22	2.54
1230	3 layers (6″ thick)		4622	.007		3.09	.25		3.34	3.81
1300	EPS window/door bucks, molded, permanent									
1310	4″ core wall (9″ wide)	2 Carp	200	.080	L.F.	3.46	2.85		6.31	8.50
1320	6″ core wall (11″ wide)		200	.080		3.55	2.85		6.40	8.60
1330	8″ core wall (13″ wide)		176	.091		3.72	3.24		6.96	9.45
1340	10″ core wall (15″ wide)		152	.105		4.84	3.75		8.59	11.50
1350	12″ core wall (17″ wide)		152	.105		5.40	3.75		9.15	12.15
1360	2″ x 6″ temporary buck bracing (includes installing and removing)		400	.040		.71	1.43		2.14	3.14
1400	Wood window/door bucks (instead of EPS bucks), permanent									

03 11 Concrete Forming

03 11 19 – Insulating Concrete Forming

03 11 19.10 Insulating Forms, Left In Place		Crew	Daily Output	Labor-Hours	Unit	2019 Bare Costs Material	Labor	Equipment	Total	Total Incl O&P
1410	4" core wall (9" wide)	2 Carp	400	.040	L.F.	1.43	1.43		2.86	3.94
1420	6" core wall (11" wide)		400	.040		1.81	1.43		3.24	4.35
1430	8" core wall (13" wide)		350	.046		7.55	1.63		9.18	11
1440	10" core wall (15" wide)		300	.053		10.40	1.90		12.30	14.60
1450	12" core wall (17" wide)		300	.053		10.40	1.90		12.30	14.60
1460	2" x 6" temporary buck bracing (includes installing and removing)		800	.020		.71	.71		1.42	1.96
1500	ICF alignment brace (incl. stiff-back, diagonal kick-back, work platform									
1501	bracket & guard rail post), fastened to one face of wall forms @ 6' O.C.									
1510	1st tier up to 10' tall									
1520	Rental of ICF alignment brace set, per set				Week	10			10	11
1530	Labor (includes installing & removing)	2 Carp	30	.533	Ea.		19		19	31.50
1560	2nd tier from 10' to 20' tall (excludes mason's scaffolding up to 10' high)									
1570	Rental of ICF alignment brace set, per set				Week	10			10	11
1580	Labor (includes installing & removing)	4 Carp	30	1.067	Ea.		38		38	63
1600	2" x 10" wood plank for work platform, 16' long									
1610	Plank material cost pro-rated over 20 uses				Ea.	1.15			1.15	1.26
1620	Labor (includes installing & removing)	2 Carp	48	.333	"		11.90		11.90	19.65
1700	2" x 4" lumber for top & middle rails for work platform									
1710	Railing material cost pro-rated over 20 uses				Ea.	.02			.02	.02
1720	Labor (includes installing & removing)	2 Carp	2400	.007	L.F.		.24		.24	.39
1800	ICF accessories									
1810	Wire clip to secure forms in place	2 Carp	2100	.008	Ea.	.39	.27		.66	.88
1820	Masonry anchor embedment (excludes ties by mason)		1600	.010		4.23	.36		4.59	5.25
1830	Ledger anchor embedment (excludes timber hanger & screws)		128	.125		7.55	4.46		12.01	15.70
1900	See section 01 54 23.70 for mason's scaffolding components									
1910	See section 03 15 19.05 for anchor bolt sleeves									
1920	See section 03 15 19.10 for anchor bolts									
1930	See section 03 15 19.20 for dovetail anchor components									
1940	See section 03 15 19.30 for embedded inserts									
1950	See section 03 21 05.10 for rebar accessories									
1960	See section 03 21 11.60 for reinforcing bars in place									
1970	See section 03 31 13.35 for ready-mix concrete material									
1980	See section 03 31 13.70 for placement and consolidation of concrete									
1990	See section 06 05 23.60 for timber connectors									

03 11 23 – Permanent Stair Forming

03 11 23.75 Forms In Place, Stairs		Crew	Daily Output	Labor-Hours	Unit	Material	Labor	Equipment	Total	Total Incl O&P
0010	**FORMS IN PLACE, STAIRS**									
0015	(Slant length x width), 1 use	C-2	165	.291	S.F.	6.10	9.25		15.35	22
0150	4 use		190	.253		2.15	8.05		10.20	15.70
2000	Stairs, cast on sloping ground (length x width), 1 use		220	.218		2.43	6.95		9.38	14.25
2025	2 use		232	.207		1.34	6.60		7.94	12.40
2050	3 use		244	.197		.97	6.25		7.22	11.45
2100	4 use		256	.188		.79	5.95		6.74	10.75

03 15 Concrete Accessories

03 15 05 – Concrete Forming Accessories

03 15 05.12 Chamfer Strips

	03 15 05.12 Chamfer Strips	Crew	Daily Output	Labor-Hours	Unit	2019 Bare Costs Material	Labor	Equipment	Total	Total Incl O&P
0010	**CHAMFER STRIPS**									
5000	Wood, 1/2" wide	1 Carp	535	.015	L.F.	.13	.53		.66	1.02
5200	3/4" wide		525	.015		.17	.54		.71	1.09
5400	1" wide		515	.016		.27	.55		.82	1.21

03 15 05.15 Column Form Accessories

		Crew	Daily Output	Labor-Hours	Unit	Material	Labor	Equipment	Total	Total Incl O&P
0010	**COLUMN FORM ACCESSORIES**									
1000	Column clamps, adjustable to 24" x 24", buy G				Set	166			166	183
1100	Rent per month G				"	13.25			13.25	14.60

03 15 05.75 Sleeves and Chases

		Crew	Daily Output	Labor-Hours	Unit	Material	Labor	Equipment	Total	Total Incl O&P
0010	**SLEEVES AND CHASES**									
0100	Plastic, 1 use, 12" long, 2" diameter	1 Carp	100	.080	Ea.	1.89	2.85		4.74	6.80
0150	4" diameter		90	.089		5.30	3.17		8.47	11.10
0200	6" diameter		75	.107		9.30	3.80		13.10	16.55

03 15 05.80 Snap Ties

		Crew	Daily Output	Labor-Hours	Unit	Material	Labor	Equipment	Total	Total Incl O&P
0010	**SNAP TIES**, 8-1/4" L&W (Lumber and wedge)									
0100	2250 lb., w/flat washer, 8" wall G				C	94			94	104
0150	10" wall G					138			138	152
0200	12" wall G					141			141	155
0500	With plastic cone, 8" wall G					83			83	91.50
0550	10" wall G					87			87	96
0600	12" wall G					93			93	103

03 15 05.95 Wall and Foundation Form Accessories

		Crew	Daily Output	Labor-Hours	Unit	Material	Labor	Equipment	Total	Total Incl O&P
0010	**WALL AND FOUNDATION FORM ACCESSORIES**									
4000	Nail stakes, 3/4" diameter, 18" long G				Ea.	2			2	2.20
4050	24" long G					2.68			2.68	2.95
4200	30" long G					3.38			3.38	3.72
4250	36" long G					4.26			4.26	4.69

03 15 13 – Waterstops

03 15 13.50 Waterstops

		Crew	Daily Output	Labor-Hours	Unit	Material	Labor	Equipment	Total	Total Incl O&P
0010	**WATERSTOPS**, PVC and Rubber									
0020	PVC, ribbed 3/16" thick, 4" wide	1 Carp	155	.052	L.F.	1.43	1.84		3.27	4.61
0050	6" wide		145	.055		2.47	1.97		4.44	5.95
0500	With center bulb, 6" wide, 3/16" thick		135	.059		2.56	2.11		4.67	6.30
0550	3/8" thick		130	.062		4.59	2.19		6.78	8.65
0600	9" wide x 3/8" thick		125	.064		7.80	2.28		10.08	12.35

03 15 16 – Concrete Construction Joints

03 15 16.20 Control Joints, Saw Cut

		Crew	Daily Output	Labor-Hours	Unit	Material	Labor	Equipment	Total	Total Incl O&P
0010	**CONTROL JOINTS, SAW CUT**									
0100	Sawcut control joints in green concrete									
0120	1" depth	C-27	2000	.008	L.F.	.04	.29	.06	.39	.56
0140	1-1/2" depth		1800	.009		.05	.32	.06	.43	.64
0160	2" depth		1600	.010		.07	.36	.07	.50	.74
0180	Sawcut joint reservoir in cured concrete									
0182	3/8" wide x 3/4" deep, with single saw blade	C-27	1000	.016	L.F.	.05	.57	.11	.73	1.10
0184	1/2" wide x 1" deep, with double saw blades		900	.018		.10	.63	.12	.85	1.27
0186	3/4" wide x 1-1/2" deep, with double saw blades		800	.020		.21	.71	.14	1.06	1.53
0190	Water blast joint to wash away laitance, 2 passes	C-29	2500	.003			.09	.03	.12	.18
0200	Air blast joint to blow out debris and air dry, 2 passes	C-28	2000	.004			.14	.01	.15	.24
0300	For backer rod, see Section 07 91 23.10									
0340	For joint sealant, see Section 03 15 16.30 or 07 92 13.20									

03 15 Concrete Accessories

03 15 16 – Concrete Construction Joints

03 15 16.30 Expansion Joints			Crew	Daily Output	Labor-Hours	Unit	Material	2019 Bare Costs Labor	Equipment	Total	Total Incl O&P
0010	**EXPANSION JOINTS**										
0020	Keyed, cold, 24 ga., incl. stakes, 3-1/2" high	G	1 Carp	200	.040	L.F.	.88	1.43		2.31	3.33
0050	4-1/2" high	G		200	.040		.94	1.43		2.37	3.39
0100	5-1/2" high	G		195	.041		1.18	1.46		2.64	3.72
2000	Premolded, bituminous fiber, 1/2" x 6"			375	.021		.43	.76		1.19	1.73
2050	1" x 12"			300	.027		2.02	.95		2.97	3.79
2140	Concrete expansion joint, recycled paper and fiber, 1/2" x 6"	G		390	.021		.43	.73		1.16	1.68
2150	1/2" x 12"	G		360	.022		.86	.79		1.65	2.25
2500	Neoprene sponge, closed cell, 1/2" x 6"			375	.021		2.34	.76		3.10	3.83
2550	1" x 12"			300	.027		8.85	.95		9.80	11.30
5000	For installation in walls, add							75%			
5250	For installation in boxouts, add							25%			

03 15 19 – Cast-In Concrete Anchors

03 15 19.05 Anchor Bolt Accessories			Crew	Daily Output	Labor-Hours	Unit	Material	Labor	Equipment	Total	Total Incl O&P
0010	**ANCHOR BOLT ACCESSORIES**										
0015	For anchor bolts set in fresh concrete, see Section 03 15 19.10										
8150	Anchor bolt sleeve, plastic, 1" diameter bolts		1 Carp	60	.133	Ea.	13.60	4.75		18.35	23
8500	1-1/2" diameter			28	.286		20	10.20		30.20	39
8600	2" diameter			24	.333		19.95	11.90		31.85	41.50
8650	3" diameter			20	.400		36.50	14.25		50.75	63.50

03 15 19.10 Anchor Bolts			Crew	Daily Output	Labor-Hours	Unit	Material	Labor	Equipment	Total	Total Incl O&P
0010	**ANCHOR BOLTS**										
0015	Made from recycled materials										
0025	Single bolts installed in fresh concrete, no templates										
0030	Hooked w/nut and washer, 1/2" diameter, 8" long	G	1 Carp	132	.061	Ea.	1.48	2.16		3.64	5.20
0040	12" long	G		131	.061		1.64	2.18		3.82	5.40
0070	3/4" diameter, 8" long	G		127	.063		5.05	2.25		7.30	9.25
0080	12" long	G		125	.064		6.30	2.28		8.58	10.70

03 15 19.30 Inserts			Crew	Daily Output	Labor-Hours	Unit	Material	Labor	Equipment	Total	Total Incl O&P
0010	**INSERTS**										
1000	Inserts, slotted nut type for 3/4" bolts, 4" long	G	1 Carp	84	.095	Ea.	21	3.40		24.40	28.50
2100	6" long	G		84	.095		23.50	3.40		26.90	31
2150	8" long	G		84	.095		31	3.40		34.40	39.50
2200	Slotted, strap type, 4" long	G		84	.095		22.50	3.40		25.90	30
2300	8" long	G		84	.095		33.50	3.40		36.90	42.50
9950	For galvanized inserts, add						30%				

03 21 Reinforcement Bars

03 21 11 – Plain Steel Reinforcement Bars

03 21 11.60 Reinforcing In Place			Crew	Daily Output	Labor-Hours	Unit	Material	Labor	Equipment	Total	Total Incl O&P
0010	**REINFORCING IN PLACE**, 50-60 ton lots, A615 Grade 60										
0020	Includes labor, but not material cost, to install accessories										
0030	Made from recycled materials										
0502	Footings, #4 to #7	G	4 Rodm	4200	.008	Lb.	.51	.30		.81	1.06
0550	#8 to #18	G		3.60	8.889	Ton	1,025	345		1,370	1,700
0602	Slab on grade, #3 to #7	G		4200	.008	Lb.	.51	.30		.81	1.06
0702	Walls, #3 to #7	G		6000	.005	"	.51	.21		.72	.91
0750	#8 to #18	G		4	8	Ton	1,025	310		1,335	1,650
0900	For other than 50-60 ton lots										

03 21 Reinforcement Bars

03 21 11 – Plain Steel Reinforcement Bars

03 21 11.60 Reinforcing In Place		Crew	Daily Output	Labor-Hours	Unit	2019 Bare Costs Material	Labor	Equipment	Total	Total Incl O&P
1000	Under 10 ton job, #3 to #7, add					25%	10%			
1010	#8 to #18, add					20%	10%			
1050	10-50 ton job, #3 to #7, add					10%				
1060	#8 to #18, add					5%				
1100	60-100 ton job, #3 to #7, deduct					5%				
1110	#8 to #18, deduct					10%				
1150	Over 100 ton job, #3 to #7, deduct					10%				
1160	#8 to #18, deduct					15%				
2400	Dowels, 2 feet long, deformed, #3 G	2 Rodm	520	.031	Ea.	.43	1.20		1.63	2.45
2410	#4 G		480	.033		.76	1.30		2.06	2.98
2420	#5 G		435	.037		1.18	1.43		2.61	3.67
2430	#6 G		360	.044		1.70	1.73		3.43	4.73
2600	Dowel sleeves for CIP concrete, 2-part system									
2610	Sleeve base, plastic, for 5/8" smooth dowel sleeve, fasten to edge form	1 Rodm	200	.040	Ea.	.55	1.56		2.11	3.19
2615	Sleeve, plastic, 12" long, for 5/8" smooth dowel, snap onto base		400	.020		1.46	.78		2.24	2.90
2620	Sleeve base, for 3/4" smooth dowel sleeve		175	.046		.54	1.78		2.32	3.53
2625	Sleeve, 12" long, for 3/4" smooth dowel		350	.023		1.27	.89		2.16	2.87
2630	Sleeve base, for 1" smooth dowel sleeve		150	.053		.70	2.08		2.78	4.20
2635	Sleeve, 12" long, for 1" smooth dowel		300	.027		1.44	1.04		2.48	3.30
2700	Dowel caps, visual warning only, plastic, #3 to #8	2 Rodm	800	.020		.32	.78		1.10	1.64
2720	#8 to #18		750	.021		.84	.83		1.67	2.29
2750	Impalement protective, plastic, #4 to #9		800	.020		1.16	.78		1.94	2.57

03 22 Fabric and Grid Reinforcing

03 22 11 – Plain Welded Wire Fabric Reinforcing

03 22 11.10 Welded Wire Fabric		Crew	Daily Output	Labor-Hours	Unit	Material	Labor	Equipment	Total	Total Incl O&P
0011	**WELDED WIRE FABRIC**, 6 x 6 - W1.4 x W1.4 (10 x 10) G	2 Rodm	3500	.005	S.F.	.16	.18		.34	.47
0301	6 x 6 - W2.9 x W2.9 (6 x 6) 42 lb./C.S.F. G		2900	.006		.26	.22		.48	.64
0501	4 x 4 - W1.4 x W1.4 (10 x 10) 31 lb./C.S.F. G		3100	.005		.24	.20		.44	.59
0750	Rolls									
0901	2 x 2 - #12 galv. for gunite reinforcing G	2 Rodm	650	.025	S.F.	.70	.96		1.66	2.36

03 22 13 – Galvanized Welded Wire Fabric Reinforcing

03 22 13.10 Galvanized Welded Wire Fabric		Crew	Daily Output	Labor-Hours	Unit	Material	Labor	Equipment	Total	Total Incl O&P
0010	**GALVANIZED WELDED WIRE FABRIC**									
0100	Add to plain welded wire pricing for galvanized welded wire				Lb.	.24			.24	.27

03 22 16 – Epoxy-Coated Welded Wire Fabric Reinforcing

03 22 16.10 Epoxy-Coated Welded Wire Fabric		Crew	Daily Output	Labor-Hours	Unit	Material	Labor	Equipment	Total	Total Incl O&P
0010	**EPOXY-COATED WELDED WIRE FABRIC**									
0100	Add to plain welded wire pricing for epoxy-coated welded wire				Lb.	.42			.42	.46

03 23 Stressed Tendon Reinforcing

03 23 05 – Prestressing Tendons

03 23 05.50 Prestressing Steel		Crew	Daily Output	Labor-Hours	Unit	Material	2019 Bare Costs Labor	Equipment	Total	Total Incl O&P
0010	**PRESTRESSING STEEL**									
3000	Slabs on grade, 0.5-inch diam. non-bonded strands, HDPE sheathed,									
3050	attached dead-end anchors, loose stressing-end anchors									
3100	25′ x 30′ slab, strands @ 36″ OC, placing	2 Rodm	2940	.005	S.F.	.66	.21		.87	1.07
3105	Stressing	C-4A	3750	.004			.17	.01	.18	.29
3110	42″ OC, placing	2 Rodm	3200	.005		.58	.19		.77	.96
3115	Stressing	C-4A	4040	.004			.15	.01	.16	.27
3120	48″ OC, placing	2 Rodm	3510	.005		.51	.18		.69	.85
3125	Stressing	C-4A	4390	.004			.14	.01	.15	.24
3150	25′ x 40′ slab, strands @ 36″ OC, placing	2 Rodm	3370	.005		.64	.19		.83	1.01
3155	Stressing	C-4A	4360	.004			.14	.01	.15	.25
3160	42″ OC, placing	2 Rodm	3760	.004		.55	.17		.72	.88
3165	Stressing	C-4A	4820	.003			.13	.01	.14	.22
3170	48″ OC, placing	2 Rodm	4090	.004		.49	.15		.64	.79
3175	Stressing	C-4A	5190	.003			.12	.01	.13	.21
3200	30′ x 30′ slab, strands @ 36″ OC, placing	2 Rodm	3260	.005		.63	.19		.82	1.02
3205	Stressing	C-4A	4190	.004			.15	.01	.16	.26
3210	42″ OC, placing	2 Rodm	3530	.005		.57	.18		.75	.92
3215	Stressing	C-4A	4500	.004			.14	.01	.15	.24
3220	48″ OC, placing	2 Rodm	3840	.004		.51	.16		.67	.83
3225	Stressing	C-4A	4850	.003			.13	.01	.14	.22
3230	30′ x 40′ slab, strands @ 36″ OC, placing	2 Rodm	3780	.004		.61	.16		.77	.94
3235	Stressing	C-4A	4920	.003			.13	.01	.14	.22
3240	42″ OC, placing	2 Rodm	4190	.004		.53	.15		.68	.84
3245	Stressing	C-4A	5410	.003			.12	.01	.13	.20
3250	48″ OC, placing	2 Rodm	4520	.004		.48	.14		.62	.76
3255	Stressing	C-4A	5790	.003			.11	.01	.12	.19
3260	30′ x 50′ slab, strands @ 36″ OC, placing	2 Rodm	4300	.004		.57	.14		.71	.87
3265	Stressing	C-4A	5650	.003			.11	.01	.12	.19
3270	42″ OC, placing	2 Rodm	4720	.003		.51	.13		.64	.78
3275	Stressing	C-4A	6150	.003			.10	.01	.11	.18
3280	48″ OC, placing	2 Rodm	5240	.003		.45	.12		.57	.69
3285	Stressing	C-4A	6760	.002	↓		.09	.01	.10	.16

03 24 Fibrous Reinforcing

03 24 05 – Reinforcing Fibers

03 24 05.30 Synthetic Fibers		Crew	Daily Output	Labor-Hours	Unit	Material	Labor	Equipment	Total	Total Incl O&P
0010	**SYNTHETIC FIBERS**									
0100	Synthetic fibers, add to concrete				Lb.	4.71			4.71	5.20
0110	1-1/2 lb./C.Y.				C.Y.	7.30			7.30	8

03 24 05.70 Steel Fibers		Crew	Daily Output	Labor-Hours	Unit	Material	Labor	Equipment	Total	Total Incl O&P
0010	**STEEL FIBERS**									
0140	ASTM A850, Type V, continuously deformed, 1-1/2″ long x 0.045″ diam.									
0150	Add to price of ready mix concrete G				Lb.	1.25			1.25	1.38
0205	Alternate pricing, dosing at 5 lb./C.Y., add to price of RMC G				C.Y.	6.25			6.25	6.90
0210	10 lb./C.Y. G					12.50			12.50	13.75
0215	15 lb./C.Y. G					18.75			18.75	20.50
0220	20 lb./C.Y. G					25			25	27.50
0225	25 lb./C.Y. G					31.50			31.50	34.50
0230	30 lb./C.Y. G					37.50			37.50	41.50
0235	35 lb./C.Y. G				↓	44			44	48

03 24 Fibrous Reinforcing

03 24 05 – Reinforcing Fibers

03 24 05.70 Steel Fibers		Crew	Daily Output	Labor-Hours	Unit	2019 Bare Costs Material	Labor	Equipment	Total	Total Incl O&P
0240	40 lb./C.Y. G				C.Y.	50			50	55
0250	50 lb./C.Y. G					62.50			62.50	69
0275	75 lb./C.Y. G					94			94	103
0300	100 lb./C.Y. G					125			125	138

03 30 Cast-In-Place Concrete

03 30 53 – Miscellaneous Cast-In-Place Concrete

03 30 53.40 Concrete In Place

Line	Description	Crew	Daily Output	Labor-Hours	Unit	Material	Labor	Equipment	Total	Total Incl O&P
0010	**CONCRETE IN PLACE**									
0020	Including forms (4 uses), Grade 60 rebar, concrete (Portland cement									
0050	Type I), placement and finishing unless otherwise indicated									
0500	Chimney foundations (5000 psi), over 5 C.Y.	C-14C	32.22	3.476	C.Y.	178	118	.83	296.83	390
0510	(3500 psi), under 5 C.Y.	"	23.71	4.724	"	206	160	1.13	367.13	490
3540	Equipment pad (3000 psi), 3′ x 3′ x 6″ thick	C-14H	45	1.067	Ea.	46.50	37.50	.59	84.59	114
3550	4′ x 4′ x 6″ thick		30	1.600		72.50	56.50	.88	129.88	173
3560	5′ x 5′ x 8″ thick		18	2.667		132	94	1.47	227.47	300
3570	6′ x 6′ x 8″ thick		14	3.429		181	121	1.89	303.89	400
3580	8′ x 8′ x 10″ thick		8	6		385	211	3.30	599.30	775
3590	10′ x 10′ x 12″ thick		5	9.600		665	340	5.30	1,010.30	1,300
3800	Footings (3000 psi), spread under 1 C.Y.	C-14C	28	4	C.Y.	193	136	.96	329.96	435
3825	1 C.Y. to 5 C.Y.		43	2.605		227	88.50	.63	316.13	395
3850	Over 5 C.Y.		75	1.493		212	50.50	.36	262.86	320
3900	Footings, strip (3000 psi), 18″ x 9″, unreinforced	C-14L	40	2.400		149	79.50	.67	229.17	296
3920	18″ x 9″, reinforced	C-14C	35	3.200		174	109	.77	283.77	370
3925	20″ x 10″, unreinforced	C-14L	45	2.133		146	70.50	.60	217.10	277
3930	20″ x 10″, reinforced	C-14C	40	2.800		166	95	.67	261.67	340
3935	24″ x 12″, unreinforced	C-14L	55	1.745		143	58	.49	201.49	254
3940	24″ x 12″, reinforced	C-14C	48	2.333		164	79	.56	243.56	310
3945	36″ x 12″, unreinforced	C-14L	70	1.371		139	45.50	.38	184.88	228
3950	36″ x 12″, reinforced	C-14C	60	1.867		158	63.50	.45	221.95	278
4000	Foundation mat (3000 psi), under 10 C.Y.		38.67	2.896		231	98.50	.70	330.20	415
4050	Over 20 C.Y.		56.40	1.986		205	67.50	.48	272.98	335
4520	Handicap access ramp (4000 psi), railing both sides, 3′ wide	C-14H	14.58	3.292	L.F.	370	116	1.81	487.81	600
4525	5′ wide		12.22	3.928		380	138	2.16	520.16	650
4530	With 6″ curb and rails both sides, 3′ wide		8.55	5.614		380	198	3.09	581.09	745
4535	5′ wide		7.31	6.566		385	231	3.61	619.61	810
4650	Slab on grade (3500 psi), not including finish, 4″ thick	C-14E	60.75	1.449	C.Y.	147	50.50	.43	197.93	244
4700	6″ thick	"	92	.957	"	141	33.50	.29	174.79	211
4701	Thickened slab edge (3500 psi), for slab on grade poured									
4702	monolithically with slab; depth is in addition to slab thickness;									
4703	formed vertical outside edge, earthen bottom and inside slope									
4705	8″ deep x 8″ wide bottom, unreinforced	C-14L	2190	.044	L.F.	3.99	1.45	.01	5.45	6.80
4710	8″ x 8″, reinforced	C-14C	1670	.067		6.40	2.28	.02	8.70	10.85
4715	12″ deep x 12″ wide bottom, unreinforced	C-14L	1800	.053		8.20	1.77	.01	9.98	11.95
4720	12″ x 12″, reinforced	C-14C	1310	.086		12.60	2.90	.02	15.52	18.65
4725	16″ deep x 16″ wide bottom, unreinforced	C-14L	1440	.067		13.90	2.21	.02	16.13	18.95
4730	16″ x 16″, reinforced	C-14C	1120	.100		19.15	3.39	.02	22.56	26.50
4735	20″ deep x 20″ wide bottom, unreinforced	C-14L	1150	.083		21	2.76	.02	23.78	27.50
4740	20″ x 20″, reinforced	C-14C	920	.122		28	4.13	.03	32.16	37.50
4745	24″ deep x 24″ wide bottom, unreinforced	C-14L	930	.103		30	3.42	.03	33.45	38.50
4750	24″ x 24″, reinforced	C-14C	740	.151		38.50	5.15	.04	43.69	51

03 30 Cast-In-Place Concrete

03 30 53 – Miscellaneous Cast-In-Place Concrete

03 30 53.40 Concrete In Place		Crew	Daily Output	Labor-Hours	Unit	Material	2019 Bare Costs Labor	Equipment	Total	Total Incl O&P
4751	Slab on grade (3500 psi), incl. troweled finish, not incl. forms									
4760	or reinforcing, over 10,000 S.F., 4" thick	C-14F	3425	.021	S.F.	1.62	.70	.01	2.33	2.92
4820	6" thick	"	3350	.021	"	2.37	.71	.01	3.09	3.78
5000	Slab on grade (3000 psi), incl. broom finish, not incl. forms									
5001	or reinforcing, 4" thick	C-14G	2873	.019	S.F.	1.60	.63	.01	2.24	2.81
5010	6" thick		2590	.022		2.51	.70	.01	3.22	3.91
5020	8" thick		2320	.024		3.28	.78	.01	4.07	4.89
6800	Stairs (3500 psi), not including safety treads, free standing, 3'-6" wide	C-14H	83	.578	LF Nose	6.25	20.50	.32	27.07	41
6850	Cast on ground		125	.384	"	5.30	13.50	.21	19.01	28.50
7000	Stair landings, free standing		200	.240	S.F.	5	8.45	.13	13.58	19.60
7050	Cast on ground		475	.101	"	4.03	3.55	.06	7.64	10.35

03 31 Structural Concrete

03 31 13 – Heavyweight Structural Concrete

03 31 13.25 Concrete, Hand Mix

		Crew	Daily Output	Labor-Hours	Unit	Material	Labor	Equipment	Total	Total Incl O&P
0010	**CONCRETE, HAND MIX** for small quantities or remote areas									
0050	Includes bulk local aggregate, bulk sand, bagged Portland									
0060	cement (Type I) and water, using gas powered cement mixer									
0125	2500 psi	C-30	135	.059	C.F.	4.03	1.63	1.20	6.86	8.45
0130	3000 psi		135	.059		4.35	1.63	1.20	7.18	8.80
0135	3500 psi		135	.059		4.54	1.63	1.20	7.37	9
0140	4000 psi		135	.059		4.77	1.63	1.20	7.60	9.25
0145	4500 psi		135	.059		5.05	1.63	1.20	7.88	9.55
0150	5000 psi		135	.059		5.40	1.63	1.20	8.23	9.90
0300	Using pre-bagged dry mix and wheelbarrow (80-lb. bag = 0.6 C.F.)									
0340	4000 psi	1 Clab	48	.167	C.F.	7.95	4.58		12.53	16.35

03 31 13.30 Concrete, Volumetric Site-Mixed

		Crew	Daily Output	Labor-Hours	Unit	Material	Labor	Equipment	Total	Total Incl O&P
0010	**CONCRETE, VOLUMETRIC SITE-MIXED**									
0015	Mixed on-site in volumetric truck									
0020	Includes local aggregate, sand, Portland cement (Type I) and water									
0025	Excludes all additives and treatments									
0100	3000 psi, 1 C.Y. mixed and discharged				C.Y.	215			215	236
0110	2 C.Y.					159			159	174
0120	3 C.Y.					140			140	154
0130	4 C.Y.					124			124	136
0140	5 C.Y.					114			114	125
0200	For truck holding/waiting time past first 2 on-site hours, add				Hr.	97			97	107
0210	For trip charge beyond first 20 miles, each way, add				Mile	3.65			3.65	4.02
0220	For each additional increase of 500 psi, add				Ea.	4.59			4.59	5.05

03 31 13.35 Heavyweight Concrete, Ready Mix

		Crew	Daily Output	Labor-Hours	Unit	Material	Labor	Equipment	Total	Total Incl O&P
0010	**HEAVYWEIGHT CONCRETE, READY MIX**, delivered									
0012	Includes local aggregate, sand, Portland cement (Type I) and water									
0015	Excludes all additives and treatments									
0020	2000 psi				C.Y.	110			110	121
0100	2500 psi					113			113	125
0150	3000 psi					124			124	136
0200	3500 psi					125			125	137
0300	4000 psi					128			128	141
0350	4500 psi					132			132	145
0400	5000 psi					139			139	153

03 31 Structural Concrete

03 31 13 – Heavyweight Structural Concrete

03 31 13.35 Heavyweight Concrete, Ready Mix		Crew	Daily Output	Labor-Hours	Unit	Material	2019 Bare Costs Labor	Equipment	Total	Total Incl O&P
0411	6000 psi				C.Y.	143			143	157
0412	8000 psi					150			150	166
0413	10,000 psi					158			158	174
0414	12,000 psi					166			166	182
1000	For high early strength (Portland cement Type III), add					10%				
1300	For winter concrete (hot water), add					5.35			5.35	5.90
1410	For mid-range water reducer, add					4.01			4.01	4.41
1420	For high-range water reducer/superplasticizer, add					6.25			6.25	6.90
1430	For retarder, add					3.25			3.25	3.58
1440	For non-Chloride accelerator, add					6.40			6.40	7.05
1450	For Chloride accelerator, per 1%, add					3.91			3.91	4.30
1460	For fiber reinforcing, synthetic (1 lb./C.Y.), add					8.05			8.05	8.85
1500	For Saturday delivery, add					8.25			8.25	9.10
1510	For truck holding/waiting time past 1st hour per load, add				Hr.	105			105	115
1520	For short load (less than 4 C.Y.), add per load				Ea.	92			92	101
2000	For all lightweight aggregate, add				C.Y.	45%				

03 31 13.70 Placing Concrete		Crew	Daily Output	Labor-Hours	Unit	Material	Labor	Equipment	Total	Total Incl O&P
0010	**PLACING CONCRETE**									
0020	Includes labor and equipment to place, level (strike off) and consolidate									
1900	Footings, continuous, shallow, direct chute	C-6	120	.400	C.Y.		11.70	.44	12.14	19.70
1950	Pumped	C-20	150	.427			12.80	6	18.80	27.50
2000	With crane and bucket	C-7	90	.800			24.50	11.80	36.30	53
2400	Footings, spread, under 1 C.Y., direct chute	C-6	55	.873			25.50	.97	26.47	43
2600	Over 5 C.Y., direct chute		120	.400			11.70	.44	12.14	19.70
2900	Foundation mats, over 20 C.Y., direct chute		350	.137			4	.15	4.15	6.75
4300	Slab on grade, up to 6" thick, direct chute		110	.436			12.75	.48	13.23	21.50
4350	Pumped	C-20	130	.492			14.80	6.90	21.70	32
4400	With crane and bucket	C-7	110	.655			19.85	9.65	29.50	43
4900	Walls, 8" thick, direct chute	C-6	90	.533			15.55	.59	16.14	26
4950	Pumped	C-20	100	.640			19.20	8.95	28.15	41.50
5000	With crane and bucket	C-7	80	.900			27.50	13.25	40.75	59.50
5050	12" thick, direct chute	C-6	100	.480			14	.53	14.53	23.50
5100	Pumped	C-20	110	.582			17.45	8.15	25.60	37.50
5200	With crane and bucket	C-7	90	.800			24.50	11.80	36.30	53
5600	Wheeled concrete dumping, add to placing costs above									
5610	Walking cart, 50' haul, add	C-18	32	.281	C.Y.		7.80	1.82	9.62	14.90
5620	150' haul, add		24	.375			10.40	2.42	12.82	19.85
5700	250' haul, add		18	.500			13.85	3.23	17.08	26.50
5800	Riding cart, 50' haul, add	C-19	80	.113			3.12	1.22	4.34	6.50
5810	150' haul, add		60	.150			4.16	1.63	5.79	8.65
5900	250' haul, add		45	.200			5.55	2.17	7.72	11.55

03 35 Concrete Finishing

03 35 13 – High-Tolerance Concrete Floor Finishing

03 35 13.30 Finishing Floors, High Tolerance		Crew	Daily Output	Labor-Hours	Unit	Material	2019 Bare Costs Labor	Equipment	Total	Total Incl O&P
0010	**FINISHING FLOORS, HIGH TOLERANCE**									
0012	Finishing of fresh concrete flatwork requires that concrete									
0013	first be placed, struck off & consolidated									
0015	Basic finishing for various unspecified flatwork									
0100	Bull float only	C-10	4000	.006	S.F.		.20		.20	.32
0125	Bull float & manual float		2000	.012			.40		.40	.64
0150	Bull float, manual float & broom finish, w/edging & joints		1850	.013			.43		.43	.70
0200	Bull float, manual float & manual steel trowel		1265	.019			.63		.63	1.02
0210	For specified Random Access Floors in ACI Classes 1, 2, 3 and 4 to achieve									
0215	Composite Overall Floor Flatness and Levelness values up to FF35/FL25									
0250	Bull float, machine float & machine trowel (walk-behind)	C-10C	1715	.014	S.F.		.46	.02	.48	.78
0300	Power screed, bull float, machine float & trowel (walk-behind)	C-10D	2400	.010			.33	.05	.38	.60
0350	Power screed, bull float, machine float & trowel (ride-on)	C-10E	4000	.006			.20	.06	.26	.39

03 35 23 – Exposed Aggregate Concrete Finishing

03 35 23.30 Finishing Floors, Exposed Aggregate		Crew	Daily Output	Labor-Hours	Unit	Material	Labor	Equipment	Total	Total Incl O&P
0010	**FINISHING FLOORS, EXPOSED AGGREGATE**									
1600	Exposed local aggregate finish, seeded on fresh concrete, 3 lb./S.F.	1 Cefi	625	.013	S.F.	.20	.46		.66	.96
1650	4 lb./S.F.	"	465	.017	"	.34	.61		.95	1.36

03 35 29 – Tooled Concrete Finishing

03 35 29.30 Finishing Floors, Tooled		Crew	Daily Output	Labor-Hours	Unit	Material	Labor	Equipment	Total	Total Incl O&P
0010	**FINISHING FLOORS, TOOLED**									
4400	Stair finish, fresh concrete, float finish	1 Cefi	275	.029	S.F.		1.04		1.04	1.68
4500	Steel trowel finish	"	200	.040	"		1.43		1.43	2.31

03 35 29.60 Finishing Walls		Crew	Daily Output	Labor-Hours	Unit	Material	Labor	Equipment	Total	Total Incl O&P
0010	**FINISHING WALLS**									
0020	Break ties and patch voids	1 Cefi	540	.015	S.F.	.04	.53		.57	.90
0050	Burlap rub with grout	"	450	.018		.04	.63		.67	1.08
0300	Bush hammer, green concrete	B-39	1000	.048			1.34	.23	1.57	2.47
0350	Cured concrete	"	650	.074			2.06	.36	2.42	3.80
0500	Acid etch	1 Cefi	575	.014		.13	.50		.63	.94
0850	Grind form fins flush	1 Clab	700	.011	L.F.		.31		.31	.52

03 35 33 – Stamped Concrete Finishing

03 35 33.50 Slab Texture Stamping		Crew	Daily Output	Labor-Hours	Unit	Material	Labor	Equipment	Total	Total Incl O&P
0010	**SLAB TEXTURE STAMPING**									
0050	Stamping requires that concrete first be placed, struck off, consolidated,									
0060	bull floated and free of bleed water. Decorative stamping tasks include:									
0100	Step 1 - first application of dry shake colored hardener	1 Cefi	6400	.001	S.F.	.42	.04		.46	.53
0110	Step 2 - bull float		6400	.001			.04		.04	.07
0130	Step 3 - second application of dry shake colored hardener		6400	.001		.21	.04		.25	.30
0140	Step 4 - bull float, manual float & steel trowel	3 Cefi	1280	.019			.67		.67	1.08
0150	Step 5 - application of dry shake colored release agent	1 Cefi	6400	.001		.10	.04		.14	.18
0160	Step 6 - place, tamp & remove mats	3 Cefi	2400	.010		.82	.36		1.18	1.49
0170	Step 7 - touch up edges, mat joints & simulated grout lines	1 Cefi	1280	.006			.22		.22	.36
0300	Alternate stamping estimating method includes all tasks above	4 Cefi	800	.040		1.55	1.43		2.98	4.01
0400	Step 8 - pressure wash @ 3000 psi after 24 hours	1 Cefi	1600	.005			.18		.18	.29
0500	Step 9 - roll 2 coats cure/seal compound when dry	"	800	.010		.63	.36		.99	1.27

03 39 Concrete Curing

03 39 13 – Water Concrete Curing

03 39 13.50 Water Curing		Crew	Daily Output	Labor-Hours	Unit	Material	Labor	Equipment	Total	Total Incl O&P
						2019 Bare Costs				
0011	**WATER CURING**									
0020	With burlap, 4 uses assumed, 7.5 oz.	2 Clab	5500	.003	S.F.	.15	.08		.23	.30
0101	10 oz.	"	5500	.003	"	.28	.08		.36	.43

03 39 23 – Membrane Concrete Curing

03 39 23.13 Chemical Compound Membrane Concrete Curing

		Crew	Daily Output	Labor-Hours	Unit	Material	Labor	Equipment	Total	Total Incl O&P
0010	**CHEMICAL COMPOUND MEMBRANE CONCRETE CURING**									
0301	Sprayed membrane curing compound	2 Clab	9500	.002	S.F.	.13	.05		.18	.22

03 39 23.23 Sheet Membrane Concrete Curing

		Crew	Daily Output	Labor-Hours	Unit	Material	Labor	Equipment	Total	Total Incl O&P
0010	**SHEET MEMBRANE CONCRETE CURING**									
0201	Curing blanket, burlap/poly, 2-ply	2 Clab	7000	.002	S.F.	.21	.06		.27	.33

03 41 Precast Structural Concrete

03 41 23 – Precast Concrete Stairs

03 41 23.50 Precast Stairs

		Crew	Daily Output	Labor-Hours	Unit	Material	Labor	Equipment	Total	Total Incl O&P
0010	**PRECAST STAIRS**									
0020	Precast concrete treads on steel stringers, 3′ wide	C-12	75	.640	Riser	153	22.50	6.30	181.80	212
0300	Front entrance, 5′ wide with 48″ platform, 2 risers		16	3	Flight	675	105	29.50	809.50	945
0350	5 risers		12	4		1,075	140	39.50	1,254.50	1,450
0500	6′ wide, 2 risers		15	3.200		745	112	31.50	888.50	1,050
0550	5 risers		11	4.364		1,175	153	43	1,371	1,600
0700	7′ wide, 2 risers	↓	14	3.429		935	120	33.50	1,088.50	1,250
1200	Basement entrance stairwell, 6 steps, incl. steel bulkhead door	B-51	22	2.182		1,875	62.50	8.90	1,946.40	2,175
1250	14 steps	"	11	4.364	↓	3,125	125	17.75	3,267.75	3,650

03 48 Precast Concrete Specialties

03 48 43 – Precast Concrete Trim

03 48 43.40 Precast Lintels

		Crew	Daily Output	Labor-Hours	Unit	Material	Labor	Equipment	Total	Total Incl O&P
0010	**PRECAST LINTELS**, smooth gray, prestressed, stock units only									
0800	4″ wide x 8″ high x 4′ long	D-10	28	1.143	Ea.	35.50	40	13.30	88.80	120
0850	8′ long		24	1.333		81.50	46.50	15.50	143.50	184
1000	6″ wide x 8″ high x 4′ long		26	1.231		54	43	14.30	111.30	147
1050	10′ long	↓	22	1.455	↓	132	51	16.90	199.90	248

03 48 43.90 Precast Window Sills

		Crew	Daily Output	Labor-Hours	Unit	Material	Labor	Equipment	Total	Total Incl O&P
0010	**PRECAST WINDOW SILLS**									
0600	Precast concrete, 4″ tapers to 3″, 9″ wide	D-1	70	.229	L.F.	20.50	7.40		27.90	35
0650	11″ wide	"	60	.267	"	33	8.65		41.65	51

03 54 Cast Underlayment

03 54 13 – Gypsum Cement Underlayment

03 54 13.50 Poured Gypsum Underlayment		Crew	Daily Output	Labor-Hours	Unit	Material	2019 Bare Costs Labor	Equipment	Total	Total Incl O&P
0010	POURED GYPSUM UNDERLAYMENT									
0400	Underlayment, gypsum based, self-leveling 2500 psi, pumped, 1/2" thick	C-8	24000	.002	S.F.	.44	.07	.04	.55	.65
0500	3/4" thick		20000	.003		.66	.09	.04	.79	.92
0600	1" thick		16000	.004		.88	.11	.05	1.04	1.21
1400	Hand placed, 1/2" thick	C-18	450	.020		.44	.55	.13	1.12	1.55
1500	3/4" thick	"	300	.030		.66	.83	.19	1.68	2.31

03 63 Epoxy Grouting

03 63 05 – Grouting of Dowels and Fasteners

03 63 05.10 Epoxy Only		Crew	Daily Output	Labor-Hours	Unit	Material	Labor	Equipment	Total	Total Incl O&P
0010	EPOXY ONLY									
1500	Chemical anchoring, epoxy cartridge, excludes layout, drilling, fastener									
1530	For fastener 3/4" diam. x 6" embedment	2 Skwk	72	.222	Ea.	5.20	8.10		13.30	19.20
1535	1" diam. x 8" embedment		66	.242		7.75	8.85		16.60	23.50
1540	1-1/4" diam. x 10" embedment		60	.267		15.55	9.75		25.30	33.50
1545	1-3/4" diam. x 12" embedment		54	.296		26	10.80		36.80	46.50
1550	14" embedment		48	.333		31	12.15		43.15	54
1555	2" diam. x 12" embedment		42	.381		41.50	13.90		55.40	68.50
1560	18" embedment		32	.500		52	18.25		70.25	87.50

03 82 Concrete Boring

03 82 16 – Concrete Drilling

03 82 16.10 Concrete Impact Drilling		Crew	Daily Output	Labor-Hours	Unit	Material	Labor	Equipment	Total	Total Incl O&P
0010	CONCRETE IMPACT DRILLING									
0020	Includes bit cost, layout and set-up time, no anchors									
0050	Up to 4" deep in concrete/brick floors/walls									
0100	Holes, 1/4" diameter	1 Carp	75	.107	Ea.	.07	3.80		3.87	6.40
0150	For each additional inch of depth in same hole, add		430	.019		.02	.66		.68	1.12
0200	3/8" diameter		63	.127		.05	4.53		4.58	7.55
0250	For each additional inch of depth in same hole, add		340	.024		.01	.84		.85	1.41
0300	1/2" diameter		50	.160		.06	5.70		5.76	9.45
0350	For each additional inch of depth in same hole, add		250	.032		.01	1.14		1.15	1.90
0400	5/8" diameter		48	.167		.10	5.95		6.05	9.90
0450	For each additional inch of depth in same hole, add		240	.033		.03	1.19		1.22	1.99
0500	3/4" diameter		45	.178		.13	6.35		6.48	10.60
0550	For each additional inch of depth in same hole, add		220	.036		.03	1.30		1.33	2.18
0600	7/8" diameter		43	.186		.18	6.65		6.83	11.15
0650	For each additional inch of depth in same hole, add		210	.038		.05	1.36		1.41	2.29
0700	1" diameter		40	.200		.17	7.15		7.32	12
0750	For each additional inch of depth in same hole, add		190	.042		.04	1.50		1.54	2.53
0800	1-1/4" diameter		38	.211		.28	7.50		7.78	12.70
0850	For each additional inch of depth in same hole, add		180	.044		.07	1.58		1.65	2.70
0900	1-1/2" diameter		35	.229		.51	8.15		8.66	14
0950	For each additional inch of depth in same hole, add		165	.048		.13	1.73		1.86	3
1000	For ceiling installations, add						40%			

Division Notes

		CREW	DAILY OUTPUT	LABOR-HOURS	UNIT	BARE COSTS MAT.	LABOR	EQUIP.	TOTAL	TOTAL INCL O&P

Estimating Tips

04 05 00 Common Work Results for Masonry

- The terms mortar and grout are often used interchangeably—and incorrectly. Mortar is used to bed masonry units, seal the entry of air and moisture, provide architectural appearance, and allow for size variations in the units. Grout is used primarily in reinforced masonry construction and to bond the masonry to the reinforcing steel. Common mortar types are M (2500 psi), S (1800 psi), N (750 psi), and O (350 psi), and they conform to ASTM C270. Grout is either fine or coarse and conforms to ASTM C476, and in-place strengths generally exceed 2500 psi. Mortar and grout are different components of masonry construction and are placed by entirely different methods. An estimator should be aware of their unique uses and costs.
- Mortar is included in all assembled masonry line items. The mortar cost, part of the assembled masonry material cost, includes all ingredients, all labor, and all equipment required. Please see reference number R040513-10.
- Waste, specifically the loss/droppings of mortar and the breakage of brick and block, is included in all unit cost lines that include mortar and masonry units in this division. A factor of 25% is added for mortar and 3% for brick and concrete masonry units.
- Scaffolding or staging is not included in any of the Division 4 costs. Refer to Subdivision 01 54 23 for scaffolding and staging costs.

04 20 00 Unit Masonry

- The most common types of unit masonry are brick and concrete masonry. The major classifications of brick are building brick (ASTM C62), facing brick (ASTM C216), glazed brick, fire brick, and pavers. Many varieties of texture and appearance can exist within these classifications, and the estimator would be wise to check local custom and availability within the project area. For repair and remodeling jobs, matching the existing brick may be the most important criteria.
- Brick and concrete block are priced by the piece and then converted into a price per square foot of wall. Openings less than two square feet are generally ignored by the estimator because any savings in units used are offset by the cutting and trimming required.
- It is often difficult and expensive to find and purchase small lots of historic brick. Costs can vary widely. Many design issues affect costs, selection of mortar mix, and repairs or replacement of masonry materials. Cleaning techniques must be reflected in the estimate.
- All masonry walls, whether interior or exterior, require bracing. The cost of bracing walls during construction should be included by the estimator, and this bracing must remain in place until permanent bracing is complete. Permanent bracing of masonry walls is accomplished by masonry itself, in the form of pilasters or abutting wall corners, or by anchoring the walls to the structural frame. Accessories in the form of anchors, anchor slots, and ties are used, but their supply and installation can be by different trades. For instance, anchor slots on spandrel beams and columns are supplied and welded in place by the steel fabricator, but the ties from the slots into the masonry are installed by the bricklayer. Regardless of the installation method, the estimator must be certain that these accessories are accounted for in pricing.

Reference Numbers

Reference numbers are shown at the beginning of some major classifications. These numbers refer to related items in the Reference Section. The reference information may be an estimating procedure, an alternate pricing method, or technical information.

Note: Not all subdivisions listed here necessarily appear. ■

04 01 Maintenance of Masonry

04 01 20 – Maintenance of Unit Masonry

04 01 20.20 Pointing Masonry		Crew	Daily Output	Labor-Hours	Unit	Material	2019 Bare Costs Labor	Equipment	Total	Total Incl O&P
0010	**POINTING MASONRY**									
0300	Cut and repoint brick, hard mortar, running bond	1 Bric	80	.100	S.F.	.56	3.53		4.09	6.50
0320	Common bond		77	.104		.56	3.66		4.22	6.70
0360	Flemish bond		70	.114		.59	4.03		4.62	7.35
0400	English bond		65	.123		.59	4.34		4.93	7.90
0600	Soft old mortar, running bond		100	.080		.56	2.82		3.38	5.30
0620	Common bond		96	.083		.56	2.94		3.50	5.50
0640	Flemish bond		90	.089		.59	3.13		3.72	5.85
0680	English bond		82	.098		.59	3.44		4.03	6.40
0700	Stonework, hard mortar		140	.057	L.F.	.74	2.01		2.75	4.18
0720	Soft old mortar		160	.050	"	.74	1.76		2.50	3.76
1000	Repoint, mask and grout method, running bond		95	.084	S.F.	.74	2.97		3.71	5.75
1020	Common bond		90	.089		.74	3.13		3.87	6
1040	Flemish bond		86	.093		.78	3.28		4.06	6.30
1060	English bond		77	.104		.78	3.66		4.44	6.95
2000	Scrub coat, sand grout on walls, thin mix, brushed		120	.067		3.49	2.35		5.84	7.75
2020	Troweled		98	.082		4.85	2.88		7.73	10.15

04 01 20.30 Pointing CMU		Crew	Daily Output	Labor-Hours	Unit	Material	Labor	Equipment	Total	Total Incl O&P
0010	**POINTING CMU**									
0300	Cut and repoint block, hard mortar, running bond	1 Bric	190	.042	S.F.	.23	1.48		1.71	2.72
0310	Stacked bond		200	.040		.23	1.41		1.64	2.60
0600	Soft old mortar, running bond		230	.035		.23	1.23		1.46	2.29
0610	Stacked bond		245	.033		.23	1.15		1.38	2.17

04 01 30 – Unit Masonry Cleaning

04 01 30.60 Brick Washing		Crew	Daily Output	Labor-Hours	Unit	Material	Labor	Equipment	Total	Total Incl O&P
0010	**BRICK WASHING** R040130-10									
0012	Acid cleanser, smooth brick surface	1 Bric	560	.014	S.F.	.05	.50		.55	.90
0050	Rough brick		400	.020		.07	.71		.78	1.26
0060	Stone, acid wash		600	.013		.09	.47		.56	.88
1000	Muriatic acid, price per gallon in 5 gallon lots				Gal.	10.95			10.95	12.05

04 05 Common Work Results for Masonry

04 05 05 – Selective Demolition for Masonry

04 05 05.10 Selective Demolition		Crew	Daily Output	Labor-Hours	Unit	Material	Labor	Equipment	Total	Total Incl O&P
0010	**SELECTIVE DEMOLITION** R024119-10									
0200	Bond beams, 8" block with #4 bar	2 Clab	32	.500	L.F.		13.75		13.75	22.50
0300	Concrete block walls, unreinforced, 2" thick		1200	.013	S.F.		.37		.37	.61
0310	4" thick		1150	.014			.38		.38	.63
0320	6" thick		1100	.015			.40		.40	.66
0330	8" thick		1050	.015			.42		.42	.69
0340	10" thick		1000	.016			.44		.44	.73
0360	12" thick		950	.017			.46		.46	.77
0380	Reinforced alternate courses, 2" thick		1130	.014			.39		.39	.64
0390	4" thick		1080	.015			.41		.41	.67
0400	6" thick		1035	.015			.43		.43	.70
0410	8" thick		990	.016			.44		.44	.73
0420	10" thick		940	.017			.47		.47	.77
0430	12" thick		890	.018			.49		.49	.82
0440	Reinforced alternate courses & vertically 48" OC, 4" thick		900	.018			.49		.49	.81
0450	6" thick		850	.019			.52		.52	.86

04 05 Common Work Results for Masonry

04 05 05 – Selective Demolition for Masonry

04 05 05.10 Selective Demolition		Crew	Daily Output	Labor-Hours	Unit	Material	2019 Bare Costs Labor	Equipment	Total	Total Incl O&P
0460	8" thick	2 Clab	800	.020	S.F.		.55		.55	.91
0480	10" thick		750	.021			.59		.59	.97
0490	12" thick		700	.023			.63		.63	1.04
1000	Chimney, 16" x 16", soft old mortar	1 Clab	55	.145	C.F.		4		4	6.60
1020	Hard mortar		40	.200			5.50		5.50	9.10
1030	16" x 20", soft old mortar		55	.145			4		4	6.60
1040	Hard mortar		40	.200			5.50		5.50	9.10
1050	16" x 24", soft old mortar		55	.145			4		4	6.60
1060	Hard mortar		40	.200			5.50		5.50	9.10
1080	20" x 20", soft old mortar		55	.145			4		4	6.60
1100	Hard mortar		40	.200			5.50		5.50	9.10
1110	20" x 24", soft old mortar		55	.145			4		4	6.60
1120	Hard mortar		40	.200			5.50		5.50	9.10
1140	20" x 32", soft old mortar		55	.145			4		4	6.60
1160	Hard mortar		40	.200			5.50		5.50	9.10
1200	48" x 48", soft old mortar		55	.145			4		4	6.60
1220	Hard mortar		40	.200			5.50		5.50	9.10
1250	Metal, high temp steel jacket, 24" diameter	E-2	130	.369	V.L.F.		14.70	13.05	27.75	40
1260	60" diameter	"	60	.800			32	28.50	60.50	85.50
1280	Flue lining, up to 12" x 12"	1 Clab	200	.040			1.10		1.10	1.82
1282	Up to 24" x 24"		150	.053			1.47		1.47	2.42
2000	Columns, 8" x 8", soft old mortar		48	.167			4.58		4.58	7.60
2020	Hard mortar		40	.200			5.50		5.50	9.10
2060	16" x 16", soft old mortar		16	.500			13.75		13.75	22.50
2100	Hard mortar		14	.571			15.70		15.70	26
2140	24" x 24", soft old mortar		8	1			27.50		27.50	45.50
2160	Hard mortar		6	1.333			36.50		36.50	60.50
2200	36" x 36", soft old mortar		4	2			55		55	91
2220	Hard mortar		3	2.667			73.50		73.50	121
2230	Alternate pricing method, soft old mortar		30	.267	C.F.		7.35		7.35	12.10
2240	Hard mortar		23	.348	"		9.55		9.55	15.80
3000	Copings, precast or masonry, to 8" wide									
3020	Soft old mortar	1 Clab	180	.044	L.F.		1.22		1.22	2.02
3040	Hard mortar	"	160	.050	"		1.38		1.38	2.27
3100	To 12" wide									
3120	Soft old mortar	1 Clab	160	.050	L.F.		1.38		1.38	2.27
3140	Hard mortar	"	140	.057	"		1.57		1.57	2.60
4000	Fireplace, brick, 30" x 24" opening									
4020	Soft old mortar	1 Clab	2	4	Ea.		110		110	182
4040	Hard mortar		1.25	6.400			176		176	291
4100	Stone, soft old mortar		1.50	5.333			147		147	242
4120	Hard mortar		1	8			220		220	365
5000	Veneers, brick, soft old mortar		140	.057	S.F.		1.57		1.57	2.60
5020	Hard mortar		125	.064			1.76		1.76	2.91
5050	Glass block, up to 4" thick		500	.016			.44		.44	.73
5100	Granite and marble, 2" thick		180	.044			1.22		1.22	2.02
5120	4" thick		170	.047			1.29		1.29	2.14
5140	Stone, 4" thick		180	.044			1.22		1.22	2.02
5160	8" thick		175	.046			1.26		1.26	2.08
5400	Alternate pricing method, stone, 4" thick		60	.133	C.F.		3.67		3.67	6.05
5420	8" thick		85	.094	"		2.59		2.59	4.28

04 05 Common Work Results for Masonry

04 05 13 – Masonry Mortaring

04 05 13.10 Cement		Crew	Daily Output	Labor-Hours	Unit	Material	Labor	Equipment	Total	Total Incl O&P
						2019 Bare Costs				
0010	CEMENT R040513-10									
0100	Masonry, 70 lb. bag, T.L. lots				Bag	14.30			14.30	15.75
0150	L.T.L. lots					15.20			15.20	16.70
0200	White, 70 lb. bag, T.L. lots					18.05			18.05	19.85
0250	L.T.L. lots					20			20	22

04 05 16 – Masonry Grouting

04 05 16.30 Grouting		Crew	Daily Output	Labor-Hours	Unit	Material	Labor	Equipment	Total	Total Incl O&P
0010	GROUTING									
0011	Bond beams & lintels, 8" deep, 6" thick, 0.15 C.F./L.F.	D-4	1480	.027	L.F.	.78	.82	.08	1.68	2.31
0020	8" thick, 0.2 C.F./L.F.		1400	.029		1.27	.86	.09	2.22	2.93
0050	10" thick, 0.25 C.F./L.F.		1200	.033		1.30	1.01	.10	2.41	3.23
0060	12" thick, 0.3 C.F./L.F.		1040	.038		1.57	1.16	.12	2.85	3.79
0200	Concrete block cores, solid, 4" thk., by hand, 0.067 C.F./S.F. of wall	D-8	1100	.036	S.F.	.35	1.20		1.55	2.38
0210	6" thick, pumped, 0.175 C.F./S.F.	D-4	720	.056		.91	1.68	.17	2.76	3.99
0250	8" thick, pumped, 0.258 C.F./S.F.		680	.059		1.35	1.78	.18	3.31	4.64
0300	10" thick, pumped, 0.340 C.F./S.F.		660	.061		1.77	1.83	.19	3.79	5.20
0350	12" thick, pumped, 0.422 C.F./S.F.		640	.063		2.20	1.89	.20	4.29	5.80

04 05 19 – Masonry Anchorage and Reinforcing

04 05 19.05 Anchor Bolts		Crew	Daily Output	Labor-Hours	Unit	Material	Labor	Equipment	Total	Total Incl O&P
0010	ANCHOR BOLTS									
0015	Installed in fresh grout in CMU bond beams or filled cores, no templates									
0020	Hooked, with nut and washer, 1/2" diameter, 8" long	1 Bric	132	.061	Ea.	1.48	2.14		3.62	5.20
0030	12" long		131	.061		1.64	2.15		3.79	5.40
0040	5/8" diameter, 8" long		129	.062		4.10	2.19		6.29	8.15
0050	12" long		127	.063		5.05	2.22		7.27	9.25
0060	3/4" diameter, 8" long		127	.063		5.05	2.22		7.27	9.25
0070	12" long		125	.064		6.30	2.26		8.56	10.70

04 05 19.16 Masonry Anchors		Crew	Daily Output	Labor-Hours	Unit	Material	Labor	Equipment	Total	Total Incl O&P
0010	MASONRY ANCHORS									
0020	For brick veneer, galv., corrugated, 7/8" x 7", 22 ga.	1 Bric	10.50	.762	C	15.75	27		42.75	62.50
0100	24 ga.		10.50	.762		10.25	27		37.25	56.50
0150	16 ga.		10.50	.762		31	27		58	79
0200	Buck anchors, galv., corrugated, 16 ga., 2" bend, 8" x 2"		10.50	.762		64.50	27		91.50	116
0250	8" x 3"		10.50	.762		69.50	27		96.50	121
0660	Cavity wall, Z-type, galvanized, 6" long, 1/8" diam.		10.50	.762		24	27		51	71.50
0670	3/16" diameter		10.50	.762		33.50	27		60.50	81.50
0680	1/4" diameter		10.50	.762		40	27		67	89
0850	8" long, 3/16" diameter		10.50	.762		26.50	27		53.50	74
0855	1/4" diameter		10.50	.762		48.50	27		75.50	98
1000	Rectangular type, galvanized, 1/4" diameter, 2" x 6"		10.50	.762		76	27		103	129
1050	4" x 6"		10.50	.762		91.50	27		118.50	145
1100	3/16" diameter, 2" x 6"		10.50	.762		49	27		76	99
1150	4" x 6"		10.50	.762		56	27		83	107
1500	Rigid partition anchors, plain, 8" long, 1" x 1/8"		10.50	.762		242	27		269	310
1550	1" x 1/4"		10.50	.762		289	27		316	365
1580	1-1/2" x 1/8"		10.50	.762		258	27		285	330
1600	1-1/2" x 1/4"		10.50	.762		330	27		357	410
1650	2" x 1/8"		10.50	.762		305	27		332	380
1700	2" x 1/4"		10.50	.762		415	27		442	500

04 05 Common Work Results for Masonry

04 05 19 – Masonry Anchorage and Reinforcing

04 05 19.26 Masonry Reinforcing Bars		Crew	Daily Output	Labor-Hours	Unit	Material	Labor	Equipment	Total	Total Incl O&P
						2019 Bare Costs				
0010	**MASONRY REINFORCING BARS** R040519-50									
0015	Steel bars A615, placed horiz., #3 & #4 bars	1 Bric	450	.018	Lb.	.51	.63		1.14	1.61
0050	Placed vertical, #3 & #4 bars		350	.023		.51	.81		1.32	1.91
0060	#5 & #6 bars		650	.012		.51	.43		.94	1.29
0200	Joint reinforcing, regular truss, to 6" wide, mill std galvanized		30	.267	C.L.F.	26	9.40		35.40	44
0250	12" wide		20	.400		28.50	14.10		42.60	55
0400	Cavity truss with drip section, to 6" wide		30	.267		24	9.40		33.40	41.50
0450	12" wide		20	.400		27.50	14.10		41.60	53.50

04 21 Clay Unit Masonry

04 21 13 – Brick Masonry

04 21 13.13 Brick Veneer Masonry		Crew	Daily Output	Labor-Hours	Unit	Material	Labor	Equipment	Total	Total Incl O&P
0010	**BRICK VENEER MASONRY**, T.L. lots, excl. scaff., grout & reinforcing									
0015	Material costs incl. 3% brick and 25% mortar waste R042110-10									
2000	Standard, sel. common, 4" x 2-2/3" x 8" (6.75/S.F.) R042110-20	D-8	230	.174	S.F.	4.32	5.75		10.07	14.30
2020	Red, 4" x 2-2/3" x 8", running bond		220	.182		4.15	6		10.15	14.55
2050	Full header every 6th course (7.88/S.F.) R042110-50		185	.216		4.84	7.10		11.94	17.20
2100	English, full header every 2nd course (10.13/S.F.)		140	.286		6.20	9.40		15.60	22.50
2150	Flemish, alternate header every course (9.00/S.F.)		150	.267		5.50	8.80		14.30	20.50
2200	Flemish, alt. header every 6th course (7.13/S.F.)		205	.195		4.38	6.45		10.83	15.50
2250	Full headers throughout (13.50/S.F.)		105	.381		8.25	12.55		20.80	30
2300	Rowlock course (13.50/S.F.)		100	.400		8.25	13.20		21.45	31
2350	Rowlock stretcher (4.50/S.F.)		310	.129		2.79	4.25		7.04	10.15
2400	Soldier course (6.75/S.F.)		200	.200		4.15	6.60		10.75	15.55
2450	Sailor course (4.50/S.F.)		290	.138		2.79	4.54		7.33	10.65
2600	Buff or gray face, running bond (6.75/S.F.)		220	.182		4.39	6		10.39	14.85
2700	Glazed face brick, running bond		210	.190		12.85	6.30		19.15	24.50
2750	Full header every 6th course (7.88/S.F.)		170	.235		15	7.75		22.75	29.50
3000	Jumbo, 6" x 4" x 12" running bond (3.00/S.F.)		435	.092		5.35	3.03		8.38	10.95
3050	Norman, 4" x 2-2/3" x 12" running bond (4.5/S.F.)		320	.125		6.60	4.12		10.72	14.10
3100	Norwegian, 4" x 3-1/5" x 12" (3.75/S.F.)		375	.107		5.75	3.51		9.26	12.20
3150	Economy, 4" x 4" x 8" (4.50/S.F.)		310	.129		4.55	4.25		8.80	12.10
3200	Engineer, 4" x 3-1/5" x 8" (5.63/S.F.)		260	.154		3.91	5.05		8.96	12.75
3250	Roman, 4" x 2" x 12" (6.00/S.F.)		250	.160		7.20	5.25		12.45	16.75
3300	S.C.R., 6" x 2-2/3" x 12" (4.50/S.F.)		310	.129		6.30	4.25		10.55	14.05
3350	Utility, 4" x 4" x 12" (3.00/S.F.)		360	.111		5.15	3.66		8.81	11.75
3360	For less than truck load lots, add					.10%				
3400	For cavity wall construction, add						15%			
3450	For stacked bond, add						10%			
3500	For interior veneer construction, add						15%			
3550	For curved walls, add						30%			

04 21 13.14 Thin Brick Veneer		Crew	Daily Output	Labor-Hours	Unit	Material	Labor	Equipment	Total	Total Incl O&P
0010	**THIN BRICK VENEER**									
0015	Material costs incl. 3% brick and 25% mortar waste									
0020	On & incl. metal panel support sys, modular, 2-2/3" x 5/8" x 8", red	D-7	92	.174	S.F.	9.50	5.30		14.80	19
0100	Closure, 4" x 5/8" x 8"		110	.145		9.25	4.43		13.68	17.30
0110	Norman, 2-2/3" x 5/8" x 12"		110	.145		10.05	4.43		14.48	18.25
0120	Utility, 4" x 5/8" x 12"		125	.128		8.95	3.90		12.85	16.15
0130	Emperor, 4" x 3/4" x 16"		175	.091		10.15	2.78		12.93	15.65
0140	Super emperor, 8" x 3/4" x 16"		195	.082		11.50	2.50		14	16.70

04 21 13.14 Thin Brick Veneer		Crew	Daily Output	Labor-Hours	Unit	Material	2019 Bare Costs Labor	Equipment	Total	Total Incl O&P
0150	For L shaped corners with 4" return, add				L.F.	9.25			9.25	10.20
0200	On masonry/plaster back-up, modular, 2-2/3" x 5/8" x 8", red	D-7	137	.117	S.F.	4.55	3.56		8.11	10.75
0210	Closure, 4" x 5/8" x 8"		165	.097		4.30	2.95		7.25	9.50
0220	Norman, 2-2/3" x 5/8" x 12"		165	.097		5.15	2.95		8.10	10.40
0230	Utility, 4" x 5/8" x 12"		185	.086		4.03	2.63		6.66	8.70
0240	Emperor, 4" x 3/4" x 16"		260	.062		5.20	1.87		7.07	8.70
0250	Super emperor, 8" x 3/4" x 16"		285	.056		6.55	1.71		8.26	9.95
0260	For L shaped corners with 4" return, add				L.F.	9.25			9.25	10.20
0270	For embedment into pre-cast concrete panels, add				S.F.	14.40			14.40	15.85
04 21 13.15 Chimney										
0010	**CHIMNEY**, excludes foundation, scaffolding, grout and reinforcing									
0100	Brick, 16" x 16", 8" flue	D-1	18.20	.879	V.L.F.	25.50	28.50		54	75.50
0150	16" x 20" with one 8" x 12" flue		16	1		40	32.50		72.50	98
0200	16" x 24" with two 8" x 8" flues		14	1.143		58.50	37		95.50	126
0250	20" x 20" with one 12" x 12" flue		13.70	1.168		49	38		87	117
0300	20" x 24" with two 8" x 12" flues		12	1.333		66.50	43		109.50	145
0350	20" x 32" with two 12" x 12" flues		10	1.600		86.50	52		138.50	182
04 21 13.18 Columns										
0010	**COLUMNS**, solid, excludes scaffolding, grout and reinforcing									
0050	Brick, 8" x 8", 9 brick/V.L.F.	D-1	56	.286	V.L.F.	5.40	9.25		14.65	21.50
0100	12" x 8", 13.5 brick/V.L.F.		37	.432		8.10	14		22.10	32.50
0200	12" x 12", 20 brick/V.L.F.		25	.640		12	20.50		32.50	47.50
0300	16" x 12", 27 brick/V.L.F.		19	.842		16.25	27.50		43.75	63.50
0400	16" x 16", 36 brick/V.L.F.		14	1.143		21.50	37		58.50	85.50
0500	20" x 16", 45 brick/V.L.F.		11	1.455		27	47		74	109
0600	20" x 20", 56 brick/V.L.F.		9	1.778		33.50	57.50		91	133
0700	24" x 20", 68 brick/V.L.F.		7	2.286		41	74		115	168
0800	24" x 24", 81 brick/V.L.F.		6	2.667		48.50	86.50		135	198
1000	36" x 36", 182 brick/V.L.F.		3	5.333		109	173		282	410
04 21 13.30 Oversized Brick										
0010	**OVERSIZED BRICK**, excludes scaffolding, grout and reinforcing									
0100	Veneer, 4" x 2.25" x 16"	D-8	387	.103	S.F.	5.25	3.41		8.66	11.50
0102	8" x 2.25" x 16", multicell		265	.151		17.25	4.97		22.22	27.50
0105	4" x 2.75" x 16"		412	.097		5.65	3.20		8.85	11.55
0107	8" x 2.75" x 16", multicell		295	.136		17.20	4.47		21.67	26.50
0110	4" x 4" x 16"		460	.087		3.74	2.87		6.61	8.90
0120	4" x 8" x 16"		533	.075		4.49	2.47		6.96	9.05
0122	4" x 8" x 16" multicell		327	.122		15.45	4.03		19.48	23.50
0125	Loadbearing, 6" x 4" x 16", grouted and reinforced		387	.103		11.75	3.41		15.16	18.65
0130	8" x 4" x 16", grouted and reinforced		327	.122		12.55	4.03		16.58	20.50
0132	10" x 4" x 16", grouted and reinforced		327	.122		25.50	4.03		29.53	34.50
0135	6" x 8" x 16", grouted and reinforced		440	.091		14.80	3		17.80	21.50
0140	8" x 8" x 16", grouted and reinforced		400	.100		15.90	3.30		19.20	23
0145	Curtainwall/reinforced veneer, 6" x 4" x 16"		387	.103		16.95	3.41		20.36	24.50
0150	8" x 4" x 16"		327	.122		20	4.03		24.03	29
0152	10" x 4" x 16"		327	.122		28	4.03		32.03	37.50
0155	6" x 8" x 16"		440	.091		20.50	3		23.50	27.50
0160	8" x 8" x 16"		400	.100		28.50	3.30		31.80	37
0200	For 1 to 3 slots in face, add					15%				
0210	For 4 to 7 slots in face, add					25%				
0220	For bond beams, add					20%				
0230	For bullnose shapes, add					20%				

04 21 Clay Unit Masonry

04 21 13 – Brick Masonry

04 21 13.30 Oversized Brick		Crew	Daily Output	Labor-Hours	Unit	2019 Bare Costs Material	Labor	Equipment	Total	Total Incl O&P
0240	For open end knockout, add				S.F.	10%				
0250	For white or gray color group, add					10%				
0260	For 135 degree corner, add					250%				

04 21 13.32 Brick Veneer Masonry		Crew	Daily Output	Labor-Hours	Unit	Material	Labor	Equipment	Total	Total Incl O&P
0010	**BRICK VENEER MASONRY**, for residential installations									
0020	Residential brick veneer, queen size	D-14	337	.095	S.F.	2.97	3.21		6.18	8.60
0030	Modular size	″	270	.119	″	4.11	4.01		8.12	11.20

04 21 13.35 Common Building Brick		Crew	Daily Output	Labor-Hours	Unit	Material	Labor	Equipment	Total	Total Incl O&P
0010	**COMMON BUILDING BRICK**, C62, T.L. lots, material only R042110-20									
0020	Standard				M	565			565	625
0050	Select				″	530			530	585

04 21 13.45 Face Brick		Crew	Daily Output	Labor-Hours	Unit	Material	Labor	Equipment	Total	Total Incl O&P
0010	**FACE BRICK** Material Only, C216, T.L. lots R042110-20									
0300	Standard modular, 4″ x 2-2/3″ x 8″				M	505			505	560
2170	For less than truck load lots, add					15			15	16.50
2180	For buff or gray brick, add					16			16	17.60

04 22 Concrete Unit Masonry

04 22 10 – Concrete Masonry Units

04 22 10.11 Autoclave Aerated Concrete Block		Crew	Daily Output	Labor-Hours	Unit	Material	Labor	Equipment	Total	Total Incl O&P
0010	**AUTOCLAVE AERATED CONCRETE BLOCK**, excl. scaffolding, grout & reinforcing									
0050	Solid, 4″ x 8″ x 24″, incl. mortar G	D-8	600	.067	S.F.	1.61	2.20		3.81	5.45
0060	6″ x 8″ x 24″ G		600	.067		2.38	2.20		4.58	6.30
0070	8″ x 8″ x 24″ G		575	.070		3.19	2.29		5.48	7.35
0080	10″ x 8″ x 24″ G		575	.070		3.88	2.29		6.17	8.10
0090	12″ x 8″ x 24″ G		550	.073		4.79	2.40		7.19	9.25

04 22 10.14 Concrete Block, Back-Up		Crew	Daily Output	Labor-Hours	Unit	Material	Labor	Equipment	Total	Total Incl O&P
0010	**CONCRETE BLOCK, BACK-UP**, C90, 2000 psi R042210-20									
0020	Normal weight, 8″ x 16″ units, tooled joint 1 side									
0050	Not-reinforced, 2000 psi, 2″ thick	D-8	475	.084	S.F.	1.62	2.77		4.39	6.40
0200	4″ thick		460	.087		1.92	2.87		4.79	6.90
0300	6″ thick		440	.091		2.57	3		5.57	7.80
0350	8″ thick		400	.100		2.69	3.30		5.99	8.45
0400	10″ thick		330	.121		3.21	3.99		7.20	10.20
0450	12″ thick	D-9	310	.155		4.30	5		9.30	13.10
1000	Reinforced, alternate courses, 4″ thick	D-8	450	.089		2.10	2.93		5.03	7.20
1100	6″ thick		430	.093		2.76	3.07		5.83	8.15
1150	8″ thick		395	.101		2.89	3.34		6.23	8.75
1200	10″ thick		320	.125		3.39	4.12		7.51	10.55
1250	12″ thick	D-9	300	.160		4.49	5.20		9.69	13.60

04 22 10.16 Concrete Block, Bond Beam		Crew	Daily Output	Labor-Hours	Unit	Material	Labor	Equipment	Total	Total Incl O&P
0010	**CONCRETE BLOCK, BOND BEAM**, C90, 2000 psi									
0020	Not including grout or reinforcing									
0125	Regular block, 6″ thick	D-8	584	.068	L.F.	2.78	2.26		5.04	6.80
0130	8″ high, 8″ thick	″	565	.071		2.90	2.33		5.23	7.10
0150	12″ thick	D-9	510	.094		4.07	3.05		7.12	9.60
0525	Lightweight, 6″ thick	D-8	592	.068		3.03	2.23		5.26	7.05

04 22 Concrete Unit Masonry

04 22 10 – Concrete Masonry Units

04 22 10.19 Concrete Block, Insulation Inserts		Crew	Daily Output	Labor-Hours	Unit	2019 Bare Costs Material	Labor	Equipment	Total	Total Incl O&P
0010	**CONCRETE BLOCK, INSULATION INSERTS**									
0100	Styrofoam, plant installed, add to block prices									
0200	8″ x 16″ units, 6″ thick				S.F.	1.22			1.22	1.34
0250	8″ thick					1.37			1.37	1.51
0300	10″ thick					1.42			1.42	1.56
0350	12″ thick					1.58			1.58	1.74
0500	8″ x 8″ units, 8″ thick					1.21			1.21	1.33
0550	12″ thick					1.44			1.44	1.58

04 22 10.23 Concrete Block, Decorative

		Crew	Daily Output	Labor-Hours	Unit	Material	Labor	Equipment	Total	Total Incl O&P
0010	**CONCRETE BLOCK, DECORATIVE**, C90, 2000 psi									
5000	Split rib profile units, 1″ deep ribs, 8 ribs									
5100	8″ x 16″ x 4″ thick	D-8	345	.116	S.F.	4.16	3.82		7.98	10.95
5150	6″ thick		325	.123		4.73	4.06		8.79	11.95
5200	8″ thick		300	.133		5.85	4.39		10.24	13.70
5250	12″ thick	D-9	275	.175		6.75	5.65		12.40	16.85
5400	For special deeper colors, 4″ thick, add					1.39			1.39	1.53
5450	12″ thick, add					1.39			1.39	1.53
5600	For white, 4″ thick, add					1.39			1.39	1.53
5650	6″ thick, add					1.39			1.39	1.53
5700	8″ thick, add					1.45			1.45	1.59
5750	12″ thick, add					1.45			1.45	1.59

04 22 10.24 Concrete Block, Exterior

		Crew	Daily Output	Labor-Hours	Unit	Material	Labor	Equipment	Total	Total Incl O&P
0010	**CONCRETE BLOCK, EXTERIOR**, C90, 2000 psi									
0020	Reinforced alt courses, tooled joints 2 sides									
0100	Normal weight, 8″ x 16″ x 6″ thick	D-8	395	.101	S.F.	2.56	3.34		5.90	8.35
0200	8″ thick		360	.111		3.96	3.66		7.62	10.45
0250	10″ thick		290	.138		4.42	4.54		8.96	12.45
0300	12″ thick	D-9	250	.192		5.30	6.20		11.50	16.20

04 22 10.26 Concrete Block Foundation Wall

		Crew	Daily Output	Labor-Hours	Unit	Material	Labor	Equipment	Total	Total Incl O&P
0010	**CONCRETE BLOCK FOUNDATION WALL**, C90/C145									
0050	Normal-weight, cut joints, horiz joint reinf, no vert reinf.									
0200	Hollow, 8″ x 16″ x 6″ thick	D-8	455	.088	S.F.	3.03	2.90		5.93	8.15
0250	8″ thick		425	.094		3.16	3.10		6.26	8.60
0300	10″ thick		350	.114		3.66	3.77		7.43	10.30
0350	12″ thick	D-9	300	.160		4.76	5.20		9.96	13.90
0500	Solid, 8″ x 16″ block, 6″ thick	D-8	440	.091		3.35	3		6.35	8.70
0550	8″ thick	″	415	.096		4.30	3.18		7.48	10.05
0600	12″ thick	D-9	350	.137		6	4.44		10.44	14

04 22 10.32 Concrete Block, Lintels

		Crew	Daily Output	Labor-Hours	Unit	Material	Labor	Equipment	Total	Total Incl O&P
0010	**CONCRETE BLOCK, LINTELS**, C90, normal weight									
0100	Including grout and horizontal reinforcing									
0200	8″ x 8″ x 8″, 1 #4 bar	D-4	300	.133	L.F.	4.10	4.03	.42	8.55	11.65
0250	2 #4 bars		295	.136		4.34	4.10	.43	8.87	12.10
0400	8″ x 16″ x 8″, 1 #4 bar		275	.145		4.13	4.40	.46	8.99	12.35
0450	2 #4 bars		270	.148		4.36	4.48	.47	9.31	12.75
1000	12″ x 8″ x 8″, 1 #4 bar		275	.145		5.65	4.40	.46	10.51	14
1100	2 #4 bars		270	.148		5.85	4.48	.47	10.80	14.40
1150	2 #5 bars		270	.148		6.10	4.48	.47	11.05	14.70
1200	2 #6 bars		265	.151		6.45	4.57	.47	11.49	15.15
1500	12″ x 16″ x 8″, 1 #4 bar		250	.160		7.05	4.84	.50	12.39	16.35
1600	2 #3 bars		245	.163		7.05	4.94	.51	12.50	16.55

04 22 Concrete Unit Masonry

04 22 10 – Concrete Masonry Units

04 22 10.32 Concrete Block, Lintels		Crew	Daily Output	Labor-Hours	Unit	Material	2019 Bare Costs Labor	Equipment	Total	Total Incl O&P
1650	2 #4 bars	D-4	245	.163	L.F.	7.30	4.94	.51	12.75	16.75
1700	2 #5 bars	↓	240	.167	↓	7.55	5.05	.52	13.12	17.30
04 22 10.33 Lintel Block										
0010	**LINTEL BLOCK**									
3481	Lintel block 6″ x 8″ x 8″	D-1	300	.053	Ea.	1.38	1.73		3.11	4.40
3501	6″ x 16″ x 8″		275	.058		2.13	1.88		4.01	5.50
3521	8″ x 8″ x 8″		275	.058		1.29	1.88		3.17	4.56
3561	8″ x 16″ x 8″	↓	250	.064	↓	2	2.07		4.07	5.65
04 22 10.34 Concrete Block, Partitions										
0010	**CONCRETE BLOCK, PARTITIONS**, excludes scaffolding									
1000	Lightweight block, tooled joints, 2 sides, hollow									
1100	Not reinforced, 8″ x 16″ x 4″ thick	D-8	440	.091	S.F.	1.95	3		4.95	7.15
1150	6″ thick		410	.098		2.79	3.21		6	8.40
1200	8″ thick		385	.104		3.44	3.42		6.86	9.50
1250	10″ thick	↓	370	.108		4.11	3.56		7.67	10.45
1300	12″ thick	D-9	350	.137	↓	4.29	4.44		8.73	12.10
4000	Regular block, tooled joints, 2 sides, hollow									
4100	Not reinforced, 8″ x 16″ x 4″ thick	D-8	430	.093	S.F.	1.82	3.07		4.89	7.10
4150	6″ thick		400	.100		2.48	3.30		5.78	8.20
4200	8″ thick		375	.107		2.59	3.51		6.10	8.70
4250	10″ thick	↓	360	.111		3.11	3.66		6.77	9.50
4300	12″ thick	D-9	340	.141	↓	4.21	4.57		8.78	12.25

04 23 Glass Unit Masonry

04 23 13 – Vertical Glass Unit Masonry

04 23 13.10 Glass Block		Crew	Daily Output	Labor-Hours	Unit	Material	Labor	Equipment	Total	Total Incl O&P
0010	**GLASS BLOCK**									
0100	Plain, 4″ thick, under 1,000 S.F., 6″ x 6″	D-8	115	.348	S.F.	23.50	11.45		34.95	44.50
0150	8″ x 8″		160	.250		13.30	8.25		21.55	28.50
0160	end block		160	.250		62.50	8.25		70.75	83
0170	90 degree corner		160	.250		66.50	8.25		74.75	87.50
0180	45 degree corner		160	.250		60	8.25		68.25	80
0200	12″ x 12″		175	.229		21.50	7.55		29.05	36.50
0210	4″ x 8″		160	.250		36.50	8.25		44.75	54
0220	6″ x 8″	↓	160	.250	↓	23	8.25		31.25	39
0700	For solar reflective blocks, add					100%				
1000	Thinline, plain, 3-1/8″ thick, under 1,000 S.F., 6″ x 6″	D-8	115	.348	S.F.	24	11.45		35.45	45.50
1050	8″ x 8″		160	.250		13.25	8.25		21.50	28.50
1400	For cleaning block after installation (both sides), add	↓	1000	.040	↓	.16	1.32		1.48	2.38

04 24 Adobe Unit Masonry

04 24 16 – Manufactured Adobe Unit Masonry

04 24 16.06 Adobe Brick		Crew	Daily Output	Labor-Hours	Unit	Material	Labor	Equipment	Total	Total Incl O&P
						2019 Bare Costs				
0010	**ADOBE BRICK**, Semi-stabilized, with cement mortar									
0060	Brick, 10" x 4" x 14", 2.6/S.F. G	D-8	560	.071	S.F.	4.75	2.35		7.10	9.10
0080	12" x 4" x 16", 2.3/S.F. G		580	.069		6.95	2.27		9.22	11.45
0100	10" x 4" x 16", 2.3/S.F. G		590	.068		6.50	2.23		8.73	10.85
0120	8" x 4" x 16", 2.3/S.F. G		560	.071		4.96	2.35		7.31	9.35
0140	4" x 4" x 16", 2.3/S.F. G		540	.074		4.85	2.44		7.29	9.40
0160	6" x 4" x 16", 2.3/S.F. G		540	.074		4.38	2.44		6.82	8.90
0180	4" x 4" x 12", 3.0/S.F. G		520	.077		5.60	2.53		8.13	10.45
0200	8" x 4" x 12", 3.0/S.F. G	↓	520	.077	↓	4.36	2.53		6.89	9

04 27 Multiple-Wythe Unit Masonry

04 27 10 – Multiple-Wythe Masonry

04 27 10.10 Cornices		Crew	Daily Output	Labor-Hours	Unit	Material	Labor	Equipment	Total	Total Incl O&P
0010	**CORNICES**									
0110	Face bricks, 12 brick/S.F.	D-1	30	.533	SF Face	5.75	17.25		23	35.50
0150	15 brick/S.F.	"	23	.696	"	6.90	22.50		29.40	45

04 27 10.30 Brick Walls		Crew	Daily Output	Labor-Hours	Unit	Material	Labor	Equipment	Total	Total Incl O&P
0010	**BRICK WALLS**, including mortar, excludes scaffolding									
0800	Face brick, 4" thick wall, 6.75 brick/S.F.	D-8	215	.186	S.F.	4.07	6.15		10.22	14.70
0850	Common brick, 4" thick wall, 6.75 brick/S.F.		240	.167		4.48	5.50		9.98	14.10
0900	8" thick, 13.50 brick/S.F.		135	.296		9.20	9.75		18.95	26.50
1000	12" thick, 20.25 brick/S.F.		95	.421		13.85	13.85		27.70	38.50
1050	16" thick, 27.00 brick/S.F.		75	.533		18.70	17.55		36.25	50
1200	Reinforced, face brick, 4" thick wall, 6.75 brick/S.F.		210	.190		4.24	6.30		10.54	15.10
1220	Common brick, 4" thick wall, 6.75 brick/S.F.		235	.170		4.65	5.60		10.25	14.45
1250	8" thick, 13.50 brick/S.F.		130	.308		9.55	10.15		19.70	27.50
1260	8" thick, 2.25 brick/S.F.		130	.308		1.71	10.15		11.86	18.80
1300	12" thick, 20.25 brick/S.F.		90	.444		14.35	14.65		29	40.50
1350	16" thick, 27.00 brick/S.F.	↓	70	.571	↓	19.40	18.85		38.25	53

04 27 10.40 Steps		Crew	Daily Output	Labor-Hours	Unit	Material	Labor	Equipment	Total	Total Incl O&P
0010	**STEPS**									
0012	Entry steps, select common brick	D-1	.30	53.333	M	530	1,725		2,255	3,450

04 41 Dry-Placed Stone

04 41 10 – Dry Placed Stone

04 41 10.10 Rough Stone Wall		Crew	Daily Output	Labor-Hours	Unit	Material	Labor	Equipment	Total	Total Incl O&P
0011	**ROUGH STONE WALL**, Dry									
0012	Dry laid (no mortar), under 18" thick G	D-1	60	.267	C.F.	13.90	8.65		22.55	29.50
0100	Random fieldstone, under 18" thick G	D-12	60	.533		13.90	17.25		31.15	44.50
0150	Over 18" thick G	"	63	.508	↓	16.65	16.45		33.10	46
0500	Field stone veneer G	D-8	120	.333	S.F.	12.65	11		23.65	32
0510	Valley stone veneer G		120	.333		12.65	11		23.65	32
0520	River stone veneer G	↓	120	.333	↓	12.65	11		23.65	32
0600	Rubble stone walls, in mortar bed, up to 18" thick G	D-11	75	.320	C.F.	16.75	10.65		27.40	36.50

04 43 Stone Masonry

04 43 10 – Masonry with Natural and Processed Stone

04 43 10.45 Granite

	04 43 10.45 Granite	Crew	Daily Output	Labor-Hours	Unit	2019 Bare Costs Material	Labor	Equipment	Total	Total Incl O&P
0010	**GRANITE**, cut to size									
2500	Steps, copings, etc., finished on more than one surface									
2550	Low price, gray, light gray, etc.	D-10	50	.640	C.F.	92.50	22.50	7.45	122.45	146
2575	Medium price, pink, brown, etc.		50	.640		120	22.50	7.45	149.95	177
2600	High price, red, black, etc.		50	.640		148	22.50	7.45	177.95	207
2800	Pavers, 4″ x 4″ x 4″ blocks, split face and joints									
2850	Low price, gray, light gray, etc.	D-11	80	.300	S.F.	13.15	10		23.15	31
2875	Medium price, pink, brown, etc.		80	.300		21	10		31	40
2900	High price, red, black, etc.		80	.300		29	10		39	48.50
5000	Reclaimed or antique									
5010	Treads, up to 12″ wide	D-10	100	.320	L.F.	42.50	11.20	3.72	57.42	69
5020	Up to 18″ wide		100	.320		40	11.20	3.72	54.92	66.50
5030	Capstone, size varies		50	.640		29	22.50	7.45	58.95	77
5040	Posts		30	1.067	V.L.F.	30	37.50	12.40	79.90	109

04 43 10.55 Limestone

	04 43 10.55 Limestone	Crew	Daily Output	Labor-Hours	Unit	Material	Labor	Equipment	Total	Total Incl O&P
0010	**LIMESTONE**, cut to size									
0020	Veneer facing panels									
0500	Texture finish, light stick, 4-1/2″ thick, 5′ x 12′	D-4	300	.133	S.F.	20.50	4.03	.42	24.95	29.50
0750	5″ thick, 5′ x 14′ panels	D-10	275	.116		21.50	4.07	1.35	26.92	32
1000	Sugarcube finish, 2″ thick, 3′ x 5′ panels		275	.116		28.50	4.07	1.35	33.92	39.50
1050	3″ thick, 4′ x 9′ panels		275	.116		25.50	4.07	1.35	30.92	36
1200	4″ thick, 5′ x 11′ panels		275	.116		29.50	4.07	1.35	34.92	40.50
1400	Sugarcube, textured finish, 4-1/2″ thick, 5′ x 12′		275	.116		32	4.07	1.35	37.42	43
1450	5″ thick, 5′ x 14′ panels		275	.116		33	4.07	1.35	38.42	44.50
2000	Coping, sugarcube finish, top & 2 sides		30	1.067	C.F.	66.50	37.50	12.40	116.40	149
2100	Sills, lintels, jambs, trim, stops, sugarcube finish, simple		20	1.600		66.50	56	18.60	141.10	187
2150	Detailed		20	1.600		66.50	56	18.60	141.10	187
2300	Steps, extra hard, 14″ wide, 6″ rise		50	.640	L.F.	28.50	22.50	7.45	58.45	76
3000	Quoins, plain finish, 6″ x 12″ x 12″	D-12	25	1.280	Ea.	39.50	41.50		81	113
3050	6″ x 16″ x 24″	″	25	1.280	″	53	41.50		94.50	127

04 43 10.60 Marble

	04 43 10.60 Marble	Crew	Daily Output	Labor-Hours	Unit	Material	Labor	Equipment	Total	Total Incl O&P
0011	**MARBLE**, ashlar, split face, +/- 4″ thick, random									
0040	Lengths 1′ to 4′ & heights 2″ to 7-1/2″, average	D-8	175	.229	S.F.	17.80	7.55		25.35	32
0100	Base, polished, 3/4″ or 7/8″ thick, polished, 6″ high	D-10	65	.492	L.F.	11.70	17.20	5.70	34.60	47.50
1000	Facing, polished finish, cut to size, 3/4″ to 7/8″ thick									
1050	Carrara or equal	D-10	130	.246	S.F.	22	8.60	2.86	33.46	41.50
1100	Arabescato or equal	″	130	.246	″	38.50	8.60	2.86	49.96	60
2200	Window sills, 6″ x 3/4″ thick	D-1	85	.188	L.F.	11.95	6.10		18.05	23.50
2500	Flooring, polished tiles, 12″ x 12″ x 3/8″ thick									
2510	Thin set, Giallo Solare or equal	D-11	90	.267	S.F.	16.95	8.90		25.85	33.50
2600	Sky Blue or equal		90	.267		15.20	8.90		24.10	31.50
2700	Mortar bed, Giallo Solare or equal		65	.369		16.95	12.30		29.25	39
2740	Sky Blue or equal		65	.369		15.20	12.30		27.50	37.50
2780	Travertine, 3/8″ thick, Sierra or equal	D-10	130	.246		9.25	8.60	2.86	20.71	27.50
2790	Silver or equal	″	130	.246		25.50	8.60	2.86	36.96	46
3500	Thresholds, 3′ long, 7/8″ thick, 4″ to 5″ wide, plain	D-12	24	1.333	Ea.	34	43		77	110
3550	Beveled		24	1.333	″	76	43		119	156
3700	Window stools, polished, 7/8″ thick, 5″ wide		85	.376	L.F.	21	12.20		33.20	44

04 43 10.75 Sandstone or Brownstone

	04 43 10.75 Sandstone or Brownstone	Crew	Daily Output	Labor-Hours	Unit	Material	Labor	Equipment	Total	Total Incl O&P
0011	**SANDSTONE OR BROWNSTONE**									
0100	Sawed face veneer, 2-1/2″ thick, to 2′ x 4′ panels	D-10	130	.246	S.F.	21	8.60	2.86	32.46	41
0150	4″ thick, to 3′-6″ x 8′ panels		100	.320		21	11.20	3.72	35.92	46

04 43 Stone Masonry

04 43 10 – Masonry with Natural and Processed Stone

04 43 10.75 Sandstone or Brownstone		Crew	Daily Output	Labor-Hours	Unit	Material	Labor	Equipment	Total	Total Incl O&P
						2019 Bare Costs				
0300	Split face, random sizes	D-10	100	.320	S.F.	13.65	11.20	3.72	28.57	37.50
0350	Cut stone trim (limestone)									
0360	Ribbon stone, 4″ thick, 5′ pieces	D-8	120	.333	Ea.	161	11		172	195
0370	Cove stone, 4″ thick, 5′ pieces		105	.381		162	12.55		174.55	199
0380	Cornice stone, 10″ to 12″ wide		90	.444		200	14.65		214.65	245
0390	Band stone, 4″ thick, 5′ pieces		145	.276		106	9.10		115.10	132
0410	Window and door trim, 3″ to 4″ wide		160	.250		90.50	8.25		98.75	114
0420	Key stone, 18″ long		60	.667		93	22		115	139

04 43 10.80 Slate		Crew	Daily Output	Labor-Hours	Unit	Material	Labor	Equipment	Total	Total Incl O&P
0010	**SLATE**									
3100	Stair landings, 1″ thick, black, clear	D-1	65	.246	S.F.	21	7.95		28.95	36.50
3200	Ribbon	″	65	.246	″	23	7.95		30.95	39
3500	Stair treads, sand finish, 1″ thick x 12″ wide									
3600	3 L.F. to 6 L.F.	D-10	120	.267	L.F.	25	9.35	3.10	37.45	46.50
3700	Ribbon, sand finish, 1″ thick x 12″ wide									
3750	To 6 L.F.	D-10	120	.267	L.F.	21	9.35	3.10	33.45	42

04 43 10.85 Window Sill		Crew	Daily Output	Labor-Hours	Unit	Material	Labor	Equipment	Total	Total Incl O&P
0010	**WINDOW SILL**									
0020	Bluestone, thermal top, 10″ wide, 1-1/2″ thick	D-1	85	.188	S.F.	8.90	6.10		15	19.95
0050	2″ thick		75	.213	″	9.25	6.90		16.15	21.50
0100	Cut stone, 5″ x 8″ plain		48	.333	L.F.	12.60	10.80		23.40	32
0200	Face brick on edge, brick, 8″ wide		80	.200		5.55	6.50		12.05	16.90
0400	Marble, 9″ wide, 1″ thick		85	.188		9	6.10		15.10	20
0900	Slate, colored, unfading, honed, 12″ wide, 1″ thick		85	.188		9.30	6.10		15.40	20.50
0950	2″ thick		70	.229		8.95	7.40		16.35	22

04 51 Flue Liner Masonry

04 51 10 – Clay Flue Lining

04 51 10.10 Flue Lining		Crew	Daily Output	Labor-Hours	Unit	Material	Labor	Equipment	Total	Total Incl O&P
0010	**FLUE LINING**, including mortar									
0020	Clay, 8″ x 8″	D-1	125	.128	V.L.F.	6.10	4.14		10.24	13.60
0100	8″ x 12″		103	.155		8.75	5.05		13.80	18
0200	12″ x 12″		93	.172		11.95	5.55		17.50	22.50
0300	12″ x 18″		84	.190		23.50	6.15		29.65	36.50
0400	18″ x 18″		75	.213		30.50	6.90		37.40	45.50
0500	20″ x 20″		66	.242		45.50	7.85		53.35	63
0600	24″ x 24″		56	.286		58.50	9.25		67.75	80
1000	Round, 18″ diameter		66	.242		40	7.85		47.85	57
1100	24″ diameter		47	.340		79	11		90	105

04 57 Masonry Fireplaces

04 57 10 – Brick or Stone Fireplaces

04 57 10.10 Fireplace		Crew	Daily Output	Labor-Hours	Unit	Material	2019 Bare Costs Labor	Equipment	Total	Total Incl O&P
0010	**FIREPLACE**									
0100	Brick fireplace, not incl. foundations or chimneys									
0110	30" x 29" opening, incl. chamber, plain brickwork	D-1	.40	40	Ea.	585	1,300		1,885	2,800
0200	Fireplace box only (110 brick)	"	2	8	"	162	259		421	610
0300	For elaborate brickwork and details, add					35%	35%			
0400	For hearth, brick & stone, add	D-1	2	8	Ea.	211	259		470	660
0410	For steel, damper, cleanouts, add		4	4		15.45	130		145.45	233
0600	Plain brickwork, incl. metal circulator		.50	32		905	1,025		1,930	2,725
0800	Face brick only, standard size, 8" x 2-2/3" x 4"		.30	53.333	M	595	1,725		2,320	3,525
0900	Stone fireplace, fieldstone, add				SF Face	9.05			9.05	10
1000	Cut stone, add				"	8.80			8.80	9.65

04 72 Cast Stone Masonry

04 72 10 – Cast Stone Masonry Features

04 72 10.10 Coping

		Crew	Daily Output	Labor-Hours	Unit	Material	Labor	Equipment	Total	Total Incl O&P
0010	**COPING**, stock units									
0050	Precast concrete, 10" wide, 4" tapers to 3-1/2", 8" wall	D-1	75	.213	L.F.	24	6.90		30.90	38
0100	12" wide, 3-1/2" tapers to 3", 10" wall		70	.229		26	7.40		33.40	41
0110	14" wide, 4" tapers to 3-1/2", 12" wall		65	.246		30	7.95		37.95	46.50
0150	16" wide, 4" tapers to 3-1/2", 14" wall		60	.267		32	8.65		40.65	50
0300	Limestone for 12" wall, 4" thick		90	.178		16.80	5.75		22.55	28
0350	6" thick		80	.200		24	6.50		30.50	37.50
0500	Marble, to 4" thick, no wash, 9" wide		90	.178		12.70	5.75		18.45	23.50
0550	12" wide		80	.200		19.35	6.50		25.85	32.50
0700	Terra cotta, 9" wide		90	.178		7.75	5.75		13.50	18.10
0750	12" wide		80	.200		8.70	6.50		15.20	20.50
0800	Aluminum, for 12" wall		80	.200		9.20	6.50		15.70	21

04 72 20 – Cultured Stone Veneer

04 72 20.10 Cultured Stone Veneer Components

		Crew	Daily Output	Labor-Hours	Unit	Material	Labor	Equipment	Total	Total Incl O&P
0010	**CULTURED STONE VENEER COMPONENTS**									
0110	On wood frame and sheathing substrate, random sized cobbles, corner stones	D-8	70	.571	V.L.F.	10.55	18.85		29.40	43
0120	Field stones		140	.286	S.F.	7.60	9.40		17	24
0130	Random sized flats, corner stones		70	.571	V.L.F.	10.70	18.85		29.55	43.50
0140	Field stones		140	.286	S.F.	8.85	9.40		18.25	25.50
0150	Horizontal lined ledgestones, corner stones		75	.533	V.L.F.	10.55	17.55		28.10	41
0160	Field stones		150	.267	S.F.	7.60	8.80		16.40	23
0170	Random shaped flats, corner stones		65	.615	V.L.F.	10.55	20.50		31.05	45.50
0180	Field stones		150	.267	S.F.	7.55	8.80		16.35	23
0190	Random shaped/textured face, corner stones		65	.615	V.L.F.	10.55	20.50		31.05	45.50
0200	Field stones		130	.308	S.F.	7.50	10.15		17.65	25
0210	Random shaped river rock, corner stones		65	.615	V.L.F.	10.55	20.50		31.05	45.50
0220	Field stones		130	.308	S.F.	7.50	10.15		17.65	25
0240	On concrete or CMU substrate, random sized cobbles, corner stones		70	.571	V.L.F.	9.70	18.85		28.55	42
0250	Field stones		140	.286	S.F.	7.20	9.40		16.60	23.50
0260	Random sized flats, corner stones		70	.571	V.L.F.	9.85	18.85		28.70	42.50
0270	Field stones		140	.286	S.F.	8.45	9.40		17.85	25
0280	Horizontal lined ledgestones, corner stones		75	.533	V.L.F.	9.70	17.55		27.25	40
0290	Field stones		150	.267	S.F.	7.20	8.80		16	22.50
0300	Random shaped flats, corner stones		70	.571	V.L.F.	9.70	18.85		28.55	42
0310	Field stones		140	.286	S.F.	7.15	9.40		16.55	23.50

04 72 Cast Stone Masonry

04 72 20 – Cultured Stone Veneer

04 72 20.10 Cultured Stone Veneer Components		Crew	Daily Output	Labor-Hours	Unit	2019 Bare Costs Material	Labor	Equipment	Total	Total Incl O&P
0320	Random shaped/textured face, corner stones	D-8	65	.615	V.L.F.	9.70	20.50		30.20	44.50
0330	Field stones		130	.308	S.F.	7.10	10.15		17.25	24.50
0340	Random shaped river rock, corner stones		65	.615	V.L.F.	9.70	20.50		30.20	44.50
0350	Field stones	↓	130	.308	S.F.	7.10	10.15		17.25	24.50
0360	Cultured stone veneer, #15 felt weather resistant barrier	1 Clab	3700	.002	Sq.	5.40	.06		5.46	6.05
0390	Water table or window sill, 18" long	1 Bric	80	.100	Ea.	10	3.53		13.53	16.90

Estimating Tips

05 05 00 Common Work Results for Metals

- Nuts, bolts, washers, connection angles, and plates can add a significant amount to both the tonnage of a structural steel job and the estimated cost. As a rule of thumb, add 10% to the total weight to account for these accessories.
- Type 2 steel construction, commonly referred to as "simple construction," consists generally of field-bolted connections with lateral bracing supplied by other elements of the building, such as masonry walls or x-bracing. The estimator should be aware, however, that shop connections may be accomplished by welding or bolting. The method may be particular to the fabrication shop and may have an impact on the estimated cost.

05 10 00 Structural Steel

- Steel items can be obtained from two sources: a fabrication shop or a metals service center. Fabrication shops can fabricate items under more controlled conditions than crews in the field can. They are also more efficient and can produce items more economically. Metal service centers serve as a source of long mill shapes to both fabrication shops and contractors.
- Most line items in this structural steel subdivision, and most items in 05 50 00 Metal Fabrications, are indicated as being shop fabricated. The bare material cost for these shop fabricated items is the "Invoice Cost" from the shop and includes the mill base price of steel plus mill extras, transportation to the shop, shop drawings and detailing where warranted, shop fabrication and handling, sandblasting and a shop coat of primer paint, all necessary structural bolts, and delivery to the job site. The bare labor cost and bare equipment cost for these shop fabricated items are for field installation or erection.
- Line items in Subdivision 05 12 23.40 Lightweight Framing, and other items scattered in Division 5, are indicated as being field fabricated. The bare material cost for these field fabricated items is the "Invoice Cost" from the metals service center and includes the mill base price of steel plus mill extras, transportation to the metals service center, material handling, and delivery of long lengths of mill shapes to the job site. Material costs for structural bolts and welding rods should be added to the estimate. The bare labor cost and bare equipment cost for these items are for both field fabrication and field installation or erection, and include time for cutting, welding, and drilling in the fabricated metal items. Drilling into concrete and fasteners to fasten field fabricated items to other work is not included and should be added to the estimate.

05 20 00 Steel Joist Framing

- In any given project the total weight of open web steel joists is determined by the loads to be supported and the design. However, economies can be realized in minimizing the amount of labor used to place the joists. This is done by maximizing the joist spacing and therefore minimizing the number of joists required to be installed on the job. Certain spacings and locations may be required by the design, but in other cases maximizing the spacing and keeping it as uniform as possible will keep the costs down.

05 30 00 Steel Decking

- The takeoff and estimating of a metal deck involve more than the area of the floor or roof and the type of deck specified or shown on the drawings. Many different sizes and types of openings may exist. Small openings for individual pipes or conduits may be drilled after the floor/roof is installed, but larger openings may require special deck lengths as well as reinforcing or structural support. The estimator should determine who will be supplying this reinforcing. Additionally, some deck terminations are part of the deck package, such as screed angles and pour stops, and others will be part of the steel contract, such as angles attached to structural members and cast-in-place angles and plates. The estimator must ensure that all pieces are accounted for in the complete estimate.

05 50 00 Metal Fabrications

- The most economical steel stairs are those that use common materials, standard details, and most importantly, a uniform and relatively simple method of field assembly. Commonly available A36/A992 channels and plates are very good choices for the main stringers of the stairs, as are angles and tees for the carrier members. Risers and treads are usually made by specialty shops, and it is most economical to use a typical detail in as many places as possible. The stairs should be pre-assembled and shipped directly to the site. The field connections should be simple and straightforward enough to be accomplished efficiently, and with minimum equipment and labor.

Reference Numbers

Reference numbers are shown at the beginning of some major classifications. These numbers refer to related items in the Reference Section. The reference information may be an estimating procedure, an alternate pricing method, or technical information.

Note: Not all subdivisions listed here necessarily appear. ■

05 05 Common Work Results for Metals

05 05 19 – Post-Installed Concrete Anchors

05 05 19.10 Chemical Anchors		Crew	Daily Output	Labor-Hours	Unit	2019 Bare Costs Material	Labor	Equipment	Total	Total Incl O&P
0010	**CHEMICAL ANCHORS**									
0020	Includes layout & drilling									
1430	Chemical anchor, w/rod & epoxy cartridge, 3/4" diameter x 9-1/2" long	B-89A	27	.593	Ea.	10.90	18.95	4.15	34	48
1435	1" diameter x 11-3/4" long		24	.667		18.80	21.50	4.67	44.97	61
1440	1-1/4" diameter x 14" long		21	.762		37.50	24.50	5.35	67.35	88
1445	1-3/4" diameter x 15" long		20	.800		72	25.50	5.60	103.10	128
1450	18" long		17	.941		86	30	6.60	122.60	152
1455	2" diameter x 18" long		16	1		114	32	7	153	186
1460	24" long		15	1.067		148	34	7.50	189.50	228

05 05 19.20 Expansion Anchors

Line	Description		Crew	Daily Output	Labor-Hours	Unit	Material	Labor	Equipment	Total	Total Incl O&P
0010	**EXPANSION ANCHORS**										
0100	Anchors for concrete, brick or stone, no layout and drilling										
0200	Expansion shields, zinc, 1/4" diameter, 1-5/16" long, single	G	1 Carp	90	.089	Ea.	.45	3.17		3.62	5.75
0300	1-3/8" long, double	G		85	.094		.59	3.36		3.95	6.20
0400	3/8" diameter, 1-1/2" long, single	G		85	.094		.70	3.36		4.06	6.30
0500	2" long, double	G		80	.100		1.18	3.57		4.75	7.20
0600	1/2" diameter, 2-1/16" long, single	G		80	.100		1.37	3.57		4.94	7.40
0700	2-1/2" long, double	G		75	.107		2.17	3.80		5.97	8.70
0800	5/8" diameter, 2-5/8" long, single	G		75	.107		2.19	3.80		5.99	8.70
0900	2-3/4" long, double	G		70	.114		2.97	4.07		7.04	10
1000	3/4" diameter, 2-3/4" long, single	G		70	.114		3.78	4.07		7.85	10.90
1100	3-15/16" long, double	G		65	.123		5.55	4.39		9.94	13.35
2100	Hollow wall anchors for gypsum wall board, plaster or tile										
2500	3/16" diameter, short	G	1 Carp	150	.053	Ea.	.52	1.90		2.42	3.71
3000	Toggle bolts, bright steel, 1/8" diameter, 2" long	G		85	.094		.26	3.36		3.62	5.85
3100	4" long	G		80	.100		.29	3.57		3.86	6.20
3200	3/16" diameter, 3" long	G		80	.100		.34	3.57		3.91	6.25
3300	6" long	G		75	.107		.45	3.80		4.25	6.80
3400	1/4" diameter, 3" long	G		75	.107		.46	3.80		4.26	6.80
3500	6" long	G		70	.114		.59	4.07		4.66	7.40
3600	3/8" diameter, 3" long	G		70	.114		.86	4.07		4.93	7.70
3700	6" long	G		60	.133		1.71	4.75		6.46	9.75
3800	1/2" diameter, 4" long	G		60	.133		1.68	4.75		6.43	9.70
3900	6" long	G		50	.160		2.25	5.70		7.95	11.90
4000	Nailing anchors										
4100	Nylon nailing anchor, 1/4" diameter, 1" long		1 Carp	3.20	2.500	C	24	89		113	174
4200	1-1/2" long			2.80	2.857		25.50	102		127.50	197
4300	2" long			2.40	3.333		28	119		147	227
4400	Metal nailing anchor, 1/4" diameter, 1" long	G		3.20	2.500		22.50	89		111.50	172
4500	1-1/2" long	G		2.80	2.857		29.50	102		131.50	201
4600	2" long	G		2.40	3.333		36.50	119		155.50	236
5000	Screw anchors for concrete, masonry,										
5100	stone & tile, no layout or drilling included										
5700	Lag screw shields, 1/4" diameter, short	G	1 Carp	90	.089	Ea.	.48	3.17		3.65	5.80
5800	Long	G		85	.094		.55	3.36		3.91	6.15
5900	3/8" diameter, short	G		85	.094		.73	3.36		4.09	6.35
6000	Long	G		80	.100		.93	3.57		4.50	6.90
6100	1/2" diameter, short	G		80	.100		1.04	3.57		4.61	7.05
6200	Long	G		75	.107		1.38	3.80		5.18	7.80
6300	5/8" diameter, short	G		70	.114		1.56	4.07		5.63	8.45
6400	Long	G		65	.123		2.08	4.39		6.47	9.55
6600	Lead, #6 & #8, 3/4" long	G		260	.031		.19	1.10		1.29	2.02

05 05 Common Work Results for Metals

05 05 19 – Post-Installed Concrete Anchors

05 05 19.20 Expansion Anchors			Crew	Daily Output	Labor-Hours	Unit	Material	Labor	Equipment	Total	Total Incl O&P
							2019 Bare Costs				
6700	#10 - #14, 1-1/2" long	G	1 Carp	200	.040	Ea.	.40	1.43		1.83	2.80
6800	#16 & #18, 1-1/2" long	G		160	.050		.42	1.78		2.20	3.41
6900	Plastic, #6 & #8, 3/4" long			260	.031		.05	1.10		1.15	1.87
7000	#8 & #10, 7/8" long			240	.033		.06	1.19		1.25	2.03
7100	#10 & #12, 1" long			220	.036		.08	1.30		1.38	2.23
7200	#14 & #16, 1-1/2" long			160	.050		.07	1.78		1.85	3.03
8950	Self-drilling concrete screw, hex washer head, 3/16" diam. x 1-3/4" long	G		300	.027		.19	.95		1.14	1.78
8960	2-1/4" long	G		250	.032		.21	1.14		1.35	2.11
8970	Phillips flat head, 3/16" diam. x 1-3/4" long	G		300	.027		.19	.95		1.14	1.78
8980	2-1/4" long	G		250	.032		.21	1.14		1.35	2.11

05 05 21 – Fastening Methods for Metal

05 05 21.15 Drilling Steel		Crew	Daily Output	Labor-Hours	Unit	Material	Labor	Equipment	Total	Total Incl O&P
0010	**DRILLING STEEL**									
1910	Drilling & layout for steel, up to 1/4" deep, no anchor									
1920	Holes, 1/4" diameter	1 Sswk	112	.071	Ea.	.09	2.84		2.93	5.05
1925	For each additional 1/4" depth, add		336	.024		.09	.95		1.04	1.74
1930	3/8" diameter		104	.077		.09	3.06		3.15	5.40
1935	For each additional 1/4" depth, add		312	.026		.09	1.02		1.11	1.87
1940	1/2" diameter		96	.083		.10	3.31		3.41	5.85
1945	For each additional 1/4" depth, add		288	.028		.10	1.10		1.20	2.03
1950	5/8" diameter		88	.091		.17	3.61		3.78	6.45
1955	For each additional 1/4" depth, add		264	.030		.17	1.20		1.37	2.27
1960	3/4" diameter		80	.100		.19	3.98		4.17	7.10
1965	For each additional 1/4" depth, add		240	.033		.19	1.32		1.51	2.51
1970	7/8" diameter		72	.111		.27	4.42		4.69	7.95
1975	For each additional 1/4" depth, add		216	.037		.27	1.47		1.74	2.85
1980	1" diameter		64	.125		.25	4.97		5.22	8.85
1985	For each additional 1/4" depth, add		192	.042		.25	1.66		1.91	3.14
1990	For drilling up, add						40%			

05 05 23 – Metal Fastenings

05 05 23.10 Bolts and Hex Nuts			Crew	Daily Output	Labor-Hours	Unit	Material	Labor	Equipment	Total	Total Incl O&P
0010	**BOLTS & HEX NUTS**, Steel, A307										
0100	1/4" diameter, 1/2" long	G	1 Sswk	140	.057	Ea.	.06	2.27		2.33	4.01
0200	1" long	G		140	.057		.07	2.27		2.34	4.02
0300	2" long	G		130	.062		.10	2.45		2.55	4.35
0400	3" long	G		130	.062		.15	2.45		2.60	4.41
0500	4" long	G		120	.067		.17	2.65		2.82	4.78
0600	3/8" diameter, 1" long	G		130	.062		.14	2.45		2.59	4.40
0700	2" long	G		130	.062		.18	2.45		2.63	4.44
0800	3" long	G		120	.067		.24	2.65		2.89	4.86
0900	4" long	G		120	.067		.30	2.65		2.95	4.93
1000	5" long	G		115	.070		.38	2.77		3.15	5.20
1100	1/2" diameter, 1-1/2" long	G		120	.067		.48	2.65		3.13	5.15
1200	2" long	G		120	.067		.56	2.65		3.21	5.20
1300	4" long	G		115	.070		.92	2.77		3.69	5.80
1400	6" long	G		110	.073		1.30	2.89		4.19	6.45
1500	8" long	G		105	.076		1.73	3.03		4.76	7.15
1600	5/8" diameter, 1-1/2" long	G		120	.067		.85	2.65		3.50	5.55
1700	2" long	G		120	.067		.94	2.65		3.59	5.65
1800	4" long	G		115	.070		1.35	2.77		4.12	6.30
1900	6" long	G		110	.073		1.73	2.89		4.62	6.90
2000	8" long	G		105	.076		2.56	3.03		5.59	8.05

05 05 Common Work Results for Metals

05 05 23 – Metal Fastenings

05 05 23.10 Bolts and Hex Nuts

Line	Description		Crew	Daily Output	Labor-Hours	Unit	2019 Bare Costs Material	Labor	Equipment	Total	Total Incl O&P
2100	10″ long	G	1 Sswk	100	.080	Ea.	3.22	3.18		6.40	9.05
2200	3/4″ diameter, 2″ long	G		120	.067		1.13	2.65		3.78	5.85
2300	4″ long	G		110	.073		1.61	2.89		4.50	6.75
2400	6″ long	G		105	.076		2.07	3.03		5.10	7.55
2500	8″ long	G		95	.084		3.12	3.35		6.47	9.25
2600	10″ long	G		85	.094		4.09	3.74		7.83	11
2700	12″ long	G		80	.100		4.79	3.98		8.77	12.15
2800	1″ diameter, 3″ long	G		105	.076		2.91	3.03		5.94	8.45
2900	6″ long	G		90	.089		4.24	3.53		7.77	10.80
3000	12″ long	G		75	.107		7.60	4.24		11.84	15.70
3100	For galvanized, add						75%				
3200	For stainless, add						350%				

05 05 23.30 Lag Screws

Line	Description		Crew	Daily Output	Labor-Hours	Unit	Material	Labor	Equipment	Total	Total Incl O&P
0010	**LAG SCREWS**										
0020	Steel, 1/4″ diameter, 2″ long	G	1 Carp	200	.040	Ea.	.10	1.43		1.53	2.47
0100	3/8″ diameter, 3″ long	G		150	.053		.28	1.90		2.18	3.45
0200	1/2″ diameter, 3″ long	G		130	.062		.69	2.19		2.88	4.38
0300	5/8″ diameter, 3″ long	G		120	.067		1.31	2.38		3.69	5.35

05 05 23.50 Powder Actuated Tools and Fasteners

Line	Description		Crew	Daily Output	Labor-Hours	Unit	Material	Labor	Equipment	Total	Total Incl O&P
0010	**POWDER ACTUATED TOOLS & FASTENERS**										
0020	Stud driver, .22 caliber, single shot					Ea.	162			162	178
0100	.27 caliber, semi automatic, strip					″	460			460	505
0300	Powder load, single shot, .22 cal, power level 2, brown					C	6.50			6.50	7.15
0400	Strip, .27 cal, power level 4, red						9.30			9.30	10.20
0600	Drive pin, .300 x 3/4″ long	G	1 Carp	4.80	1.667		4.43	59.50		63.93	103
0700	.300 x 3″ long with washer	G	″	4	2		13.50	71.50		85	133

05 05 23.55 Rivets

Line	Description		Crew	Daily Output	Labor-Hours	Unit	Material	Labor	Equipment	Total	Total Incl O&P
0010	**RIVETS**										
0100	Aluminum rivet & mandrel, 1/2″ grip length x 1/8″ diameter	G	1 Carp	4.80	1.667	C	8.05	59.50		67.55	107
0200	3/16″ diameter	G		4	2		11	71.50		82.50	130
0300	Aluminum rivet, steel mandrel, 1/8″ diameter	G		4.80	1.667		10.35	59.50		69.85	109
0400	3/16″ diameter	G		4	2		16.80	71.50		88.30	137
0500	Copper rivet, steel mandrel, 1/8″ diameter	G		4.80	1.667		10.20	59.50		69.70	109
0800	Stainless rivet & mandrel, 1/8″ diameter	G		4.80	1.667		23.50	59.50		83	124
0900	3/16″ diameter	G		4	2		39	71.50		110.50	161
1000	Stainless rivet, steel mandrel, 1/8″ diameter	G		4.80	1.667		15.50	59.50		75	115
1100	3/16″ diameter	G		4	2		26	71.50		97.50	147
1200	Steel rivet and mandrel, 1/8″ diameter	G		4.80	1.667		8.15	59.50		67.65	107
1300	3/16″ diameter	G		4	2		11.50	71.50		83	131
1400	Hand riveting tool, standard					Ea.	75.50			75.50	83
1500	Deluxe						425			425	470
1600	Power riveting tool, standard						540			540	595
1700	Deluxe						1,550			1,550	1,700

05 12 Structural Steel Framing

05 12 23 – Structural Steel for Buildings

05 12 23.10 Ceiling Supports

Line	Description		Crew	Daily Output	Labor-Hours	Unit	2019 Bare Costs Material	Labor	Equipment	Total	Total Incl O&P
0010	**CEILING SUPPORTS**										
1000	Entrance door/folding partition supports, shop fabricated	G	E-4	60	.533	L.F.	26.50	21.50	1.59	49.59	68
1100	Linear accelerator door supports	G		14	2.286		120	92	6.85	218.85	300
1200	Lintels or shelf angles, hung, exterior hot dipped galv.	G		267	.120		18.05	4.82	.36	23.23	28.50
1250	Two coats primer paint instead of galv.	G		267	.120		15.65	4.82	.36	20.83	26
1400	Monitor support, ceiling hung, expansion bolted	G		4	8	Ea.	420	320	24	764	1,050
1450	Hung from pre-set inserts	G		6	5.333		450	215	15.95	680.95	885
1600	Motor supports for overhead doors	G		4	8		213	320	24	557	820
1700	Partition support for heavy folding partitions, without pocket	G		24	1.333	L.F.	60	53.50	3.99	117.49	163
1750	Supports at pocket only	G		12	2.667		120	107	7.95	234.95	325
2000	Rolling grilles & fire door supports	G		34	.941		51.50	38	2.81	92.31	125
2100	Spider-leg light supports, expansion bolted to ceiling slab	G		8	4	Ea.	172	161	11.95	344.95	480
2150	Hung from pre-set inserts	G		12	2.667	"	185	107	7.95	299.95	400
2400	Toilet partition support	G		36	.889	L.F.	60	36	2.66	98.66	131
2500	X-ray travel gantry support	G		12	2.667	"	206	107	7.95	320.95	420

05 12 23.15 Columns, Lightweight

Line	Description		Crew	Daily Output	Labor-Hours	Unit	Material	Labor	Equipment	Total	Total Incl O&P
0010	**COLUMNS, LIGHTWEIGHT**										
8000	Lally columns, to 8′, 3-1/2″ diameter		2 Carp	24	.667	Ea.	45	24		69	89
8080	4″ diameter		"	20	.800	"	64	28.50		92.50	117

05 12 23.17 Columns, Structural

Line	Description		Crew	Daily Output	Labor-Hours	Unit	Material	Labor	Equipment	Total	Total Incl O&P
0010	**COLUMNS, STRUCTURAL**										
0015	Made from recycled materials										
0020	Shop fab'd for 100-ton, 1-2 story project, bolted connections										
0800	Steel, concrete filled, extra strong pipe, 3-1/2″ diameter		E-2	660	.073	L.F.	43.50	2.89	2.57	48.96	56
0830	4″ diameter			780	.062		48.50	2.45	2.18	53.13	60
0890	5″ diameter			1020	.047		58	1.87	1.67	61.54	68.50
0930	6″ diameter			1200	.040		76.50	1.59	1.42	79.51	89
0940	8″ diameter			1100	.044		76.50	1.74	1.54	79.78	89
1100	For galvanizing, add					Lb.	.25			.25	.28
1300	For web ties, angles, etc., add per added lb.		1 Sswk	945	.008		1.32	.34		1.66	2.03
1500	Steel pipe, extra strong, no concrete, 3″ to 5″ diameter	G	E-2	16000	.003		1.32	.12	.11	1.55	1.78
1600	6″ to 12″ diameter	G		14000	.003		1.32	.14	.12	1.58	1.81
5100	Structural tubing, rect., 5″ to 6″ wide, light section	G		8000	.006		1.32	.24	.21	1.77	2.09
5200	Heavy section	G		12000	.004		1.32	.16	.14	1.62	1.88
8090	For projects 75 to 99 tons, add					%	10%				
8092	50 to 74 tons, add						20%				
8094	25 to 49 tons, add						30%	10%			
8096	10 to 24 tons, add						50%	25%			
8098	2 to 9 tons, add						75%	50%			
8099	Less than 2 tons, add						100%	100%			

05 12 23.45 Lintels

Line	Description		Crew	Daily Output	Labor-Hours	Unit	Material	Labor	Equipment	Total	Total Incl O&P
0010	**LINTELS**										
0015	Made from recycled materials										
0020	Plain steel angles, shop fabricated, under 500 lb.	G	1 Bric	550	.015	Lb.	1.02	.51		1.53	1.97
0100	500 to 1,000 lb.	G		640	.013	"	.99	.44		1.43	1.82
2000	Steel angles, 3-1/2″ x 3″, 1/4″ thick, 2′-6″ long	G		47	.170	Ea.	14.25	6		20.25	25.50
2100	4′-6″ long	G		26	.308		25.50	10.85		36.35	46.50
2600	4″ x 3-1/2″, 1/4″ thick, 5′-0″ long	G		21	.381		33	13.45		46.45	58.50
2700	9′-0″ long	G		12	.667		59	23.50		82.50	104

05 12 Structural Steel Framing

05 12 23 – Structural Steel for Buildings

05 12 23.65 Plates			Crew	Daily Output	Labor-Hours	Unit	2019 Bare Costs Material	Labor	Equipment	Total	Total Incl O&P
0010	PLATES										
0015	Made from recycled materials										
0020	For connections & stiffener plates, shop fabricated										
0050	1/8" thick (5.1 lb./S.F.)	G				S.F.	6.75			6.75	7.40
0100	1/4" thick (10.2 lb./S.F.)	G					13.50			13.50	14.85
0300	3/8" thick (15.3 lb./S.F.)	G					20			20	22
0400	1/2" thick (20.4 lb./S.F.)	G					27			27	29.50
0450	3/4" thick (30.6 lb./S.F.)	G					40.50			40.50	44.50
0500	1" thick (40.8 lb./S.F.)	G				↓	54			54	59.50
2000	Steel plate, warehouse prices, no shop fabrication										
2100	1/4" thick (10.2 lb./S.F.)	G				S.F.	7.55			7.55	8.30

05 12 23.79 Structural Steel			Crew	Daily Output	Labor-Hours	Unit	Material	Labor	Equipment	Total	Total Incl O&P
0010	STRUCTURAL STEEL										
0020	Shop fab'd for 100-ton, 1-2 story project, bolted conn's.										
0050	Beams, W 6 x 9	G	E-2	720	.067	L.F.	14.25	2.65	2.36	19.26	23
0100	W 8 x 10	G		720	.067		15.85	2.65	2.36	20.86	24.50
0200	Columns, W 6 x 15	G		540	.089		26	3.54	3.14	32.68	38
0250	W 8 x 31	G	↓	540	.089	↓	53.50	3.54	3.14	60.18	68
7990	For projects 75 to 99 tons, add					All	10%				
7992	50 to 75 tons, add						20%				
7994	25 to 49 tons, add						30%	10%			
7996	10 to 24 tons, add						50%	25%			
7998	2 to 9 tons, add						75%	50%			
7999	Less than 2 tons, add					↓	100%	100%			

05 31 Steel Decking

05 31 23 – Steel Roof Decking

05 31 23.50 Roof Decking			Crew	Daily Output	Labor-Hours	Unit	Material	Labor	Equipment	Total	Total Incl O&P
0010	ROOF DECKING										
0015	Made from recycled materials										
2100	Open type, 1-1/2" deep, Type B, wide rib, galv., 22 ga., under 50 sq.	G	E-4	4500	.007	S.F.	2.20	.29	.02	2.51	2.94
2600	20 ga., under 50 squares	G		3865	.008		2.58	.33	.02	2.93	3.44
2900	18 ga., under 50 squares	G		3800	.008		3.31	.34	.03	3.68	4.26
3050	16 ga., under 50 squares	G	↓	3700	.009	↓	4.47	.35	.03	4.85	5.55

05 31 33 – Steel Form Decking

05 31 33.50 Form Decking			Crew	Daily Output	Labor-Hours	Unit	Material	Labor	Equipment	Total	Total Incl O&P
0010	FORM DECKING										
0015	Made from recycled materials										
6100	Slab form, steel, 28 ga., 9/16" deep, Type UFS, uncoated	G	E-4	4000	.008	S.F.	1.66	.32	.02	2	2.42
6200	Galvanized	G		4000	.008		1.47	.32	.02	1.81	2.21
6220	24 ga., 1" deep, Type UF1X, uncoated	G		3900	.008		1.59	.33	.02	1.94	2.35
6240	Galvanized	G		3900	.008		1.87	.33	.02	2.22	2.66
6300	24 ga., 1-5/16" deep, Type UFX, uncoated	G		3800	.008		1.69	.34	.03	2.06	2.48
6400	Galvanized	G		3800	.008		1.99	.34	.03	2.36	2.81
6500	22 ga., 1-5/16" deep, uncoated	G		3700	.009		2.15	.35	.03	2.53	2.99
6600	Galvanized	G		3700	.009		2.19	.35	.03	2.57	3.04
6700	22 ga., 2" deep, uncoated	G		3600	.009		2.78	.36	.03	3.17	3.71
6800	Galvanized	G	↓	3600	.009	↓	2.73	.36	.03	3.12	3.65

05 41 Structural Metal Stud Framing

05 41 13 – Load-Bearing Metal Stud Framing

05 41 13.05 Bracing			Crew	Daily Output	Labor-Hours	Unit	Material	2019 Bare Costs Labor	Equipment	Total	Total Incl O&P
0010	**BRACING**, shear wall X-bracing, per 10′ x 10′ bay, one face										
0015	Made of recycled materials										
0120	Metal strap, 20 ga. x 4″ wide	G	2 Carp	18	.889	Ea.	18.90	31.50		50.40	73.50
0130	6″ wide	G		18	.889		30.50	31.50		62	86
0160	18 ga. x 4″ wide	G		16	1		31	35.50		66.50	93
0170	6″ wide	G		16	1		46	35.50		81.50	110
0410	Continuous strap bracing, per horizontal row on both faces										
0420	Metal strap, 20 ga. x 2″ wide, studs 12″ OC	G	1 Carp	7	1.143	C.L.F.	55	40.50		95.50	128
0430	16″ OC	G		8	1		55	35.50		90.50	120
0440	24″ OC	G		10	.800		55	28.50		83.50	108
0450	18 ga. x 2″ wide, studs 12″ OC	G		6	1.333		75	47.50		122.50	161
0460	16″ OC	G		7	1.143		75	40.50		115.50	150
0470	24″ OC	G		8	1		75	35.50		110.50	142

05 41 13.10 Bridging			Crew	Daily Output	Labor-Hours	Unit	Material	Labor	Equipment	Total	Total Incl O&P
0010	**BRIDGING**, solid between studs w/1-1/4″ leg track, per stud bay										
0015	Made from recycled materials										
0200	Studs 12″ OC, 18 ga. x 2-1/2″ wide	G	1 Carp	125	.064	Ea.	.90	2.28		3.18	4.76
0210	3-5/8″ wide	G		120	.067		1.16	2.38		3.54	5.20
0220	4″ wide	G		120	.067		1.22	2.38		3.60	5.25
0230	6″ wide	G		115	.070		1.60	2.48		4.08	5.85
0240	8″ wide	G		110	.073		1.85	2.59		4.44	6.30
0300	16 ga. x 2-1/2″ wide	G		115	.070		1.15	2.48		3.63	5.35
0310	3-5/8″ wide	G		110	.073		1.39	2.59		3.98	5.80
0320	4″ wide	G		110	.073		1.48	2.59		4.07	5.90
0330	6″ wide	G		105	.076		1.89	2.72		4.61	6.55
0340	8″ wide	G		100	.080		2.61	2.85		5.46	7.60
1200	Studs 16″ OC, 18 ga. x 2-1/2″ wide	G		125	.064		1.16	2.28		3.44	5.05
1210	3-5/8″ wide	G		120	.067		1.49	2.38		3.87	5.55
1220	4″ wide	G		120	.067		1.57	2.38		3.95	5.65
1230	6″ wide	G		115	.070		2.05	2.48		4.53	6.35
1240	8″ wide	G		110	.073		2.38	2.59		4.97	6.90
1300	16 ga. x 2-1/2″ wide	G		115	.070		1.47	2.48		3.95	5.70
1310	3-5/8″ wide	G		110	.073		1.78	2.59		4.37	6.25
1320	4″ wide	G		110	.073		1.90	2.59		4.49	6.35
1330	6″ wide	G		105	.076		2.43	2.72		5.15	7.15
1340	8″ wide	G		100	.080		3.35	2.85		6.20	8.40
2200	Studs 24″ OC, 18 ga. x 2-1/2″ wide	G		125	.064		1.67	2.28		3.95	5.60
2210	3-5/8″ wide	G		120	.067		2.15	2.38		4.53	6.30
2220	4″ wide	G		120	.067		2.27	2.38		4.65	6.40
2230	6″ wide	G		115	.070		2.96	2.48		5.44	7.35
2240	8″ wide	G		110	.073		3.44	2.59		6.03	8.05
2300	16 ga. x 2-1/2″ wide	G		115	.070		2.12	2.48		4.60	6.45
2310	3-5/8″ wide	G		110	.073		2.58	2.59		5.17	7.10
2320	4″ wide	G		110	.073		2.75	2.59		5.34	7.30
2330	6″ wide	G		105	.076		3.51	2.72		6.23	8.35
2340	8″ wide	G		100	.080		4.85	2.85		7.70	10.05
3000	Continuous bridging, per row										
3100	16 ga. x 1-1/2″ channel thru studs 12″ OC	G	1 Carp	6	1.333	C.L.F.	49	47.50		96.50	133
3110	16″ OC	G		7	1.143		49	40.50		89.50	122
3120	24″ OC	G		8.80	.909		49	32.50		81.50	108
4100	2″ x 2″ angle x 18 ga., studs 12″ OC	G		7	1.143		76	40.50		116.50	151
4110	16″ OC	G		9	.889		76	31.50		107.50	136

05 41 Structural Metal Stud Framing

05 41 13 – Load-Bearing Metal Stud Framing

05 41 13.10 Bridging

Line	Description		Crew	Daily Output	Labor-Hours	Unit	Material	Labor	Equipment	Total	Total Incl O&P
							2019 Bare Costs				
4120	24" OC	G	1 Carp	12	.667	C.L.F.	76	24		100	123
4200	16 ga., studs 12" OC	G		5	1.600		95.50	57		152.50	199
4210	16" OC	G		7	1.143		95.50	40.50		136	173
4220	24" OC	G		10	.800		95.50	28.50		124	152

05 41 13.25 Framing, Boxed Headers/Beams

Line	Description		Crew	Daily Output	Labor-Hours	Unit	Material	Labor	Equipment	Total	Total Incl O&P
0010	**FRAMING, BOXED HEADERS/BEAMS**										
0015	Made from recycled materials										
0200	Double, 18 ga. x 6" deep	G	2 Carp	220	.073	L.F.	5.65	2.59		8.24	10.50
0210	8" deep	G		210	.076		5.85	2.72		8.57	10.90
0220	10" deep	G		200	.080		7	2.85		9.85	12.40
0230	12" deep	G		190	.084		7.60	3		10.60	13.35
0300	16 ga. x 8" deep	G		180	.089		6.70	3.17		9.87	12.65
0310	10" deep	G		170	.094		8.50	3.36		11.86	14.90
0320	12" deep	G		160	.100		9.25	3.57		12.82	16.10
0400	14 ga. x 10" deep	G		140	.114		9.20	4.07		13.27	16.85
0410	12" deep	G		130	.123		10.05	4.39		14.44	18.30
1210	Triple, 18 ga. x 8" deep	G		170	.094		8.45	3.36		11.81	14.85
1220	10" deep	G		165	.097		10	3.46		13.46	16.70
1230	12" deep	G		160	.100		10.95	3.57		14.52	17.95
1300	16 ga. x 8" deep	G		145	.110		9.75	3.93		13.68	17.25
1310	10" deep	G		140	.114		12.30	4.07		16.37	20.50
1320	12" deep	G		135	.119		13.45	4.23		17.68	22
1400	14 ga. x 10" deep	G		115	.139		12.50	4.96		17.46	22
1410	12" deep	G		110	.145		13.85	5.20		19.05	24

05 41 13.30 Framing, Stud Walls

Line	Description		Crew	Daily Output	Labor-Hours	Unit	Material	Labor	Equipment	Total	Total Incl O&P
0010	**FRAMING, STUD WALLS** w/top & bottom track, no openings,										
0020	Headers, beams, bridging or bracing										
0025	Made from recycled materials										
4100	8' high walls, 18 ga. x 2-1/2" wide, studs 12" OC	G	2 Carp	54	.296	L.F.	9.10	10.55		19.65	27.50
4110	16" OC	G		77	.208		7.25	7.40		14.65	20
4120	24" OC	G		107	.150		5.40	5.35		10.75	14.75
4130	3-5/8" wide, studs 12" OC	G		53	.302		10.85	10.75		21.60	30
4140	16" OC	G		76	.211		8.70	7.50		16.20	22
4150	24" OC	G		105	.152		6.55	5.45		12	16.20
4160	4" wide, studs 12" OC	G		52	.308		11.30	10.95		22.25	30.50
4170	16" OC	G		74	.216		9.05	7.70		16.75	22.50
4180	24" OC	G		103	.155		6.80	5.55		12.35	16.65
4190	6" wide, studs 12" OC	G		51	.314		14.40	11.20		25.60	34.50
4200	16" OC	G		73	.219		11.55	7.80		19.35	25.50
4210	24" OC	G		101	.158		8.70	5.65		14.35	18.95
4220	8" wide, studs 12" OC	G		50	.320		16.40	11.40		27.80	37
4230	16" OC	G		72	.222		13.20	7.90		21.10	27.50
4240	24" OC	G		100	.160		9.95	5.70		15.65	20.50
4300	16 ga. x 2-1/2" wide, studs 12" OC	G		47	.340		10.75	12.15		22.90	32
4310	16" OC	G		68	.235		8.50	8.40		16.90	23
4320	24" OC	G		94	.170		6.25	6.05		12.30	16.90
4330	3-5/8" wide, studs 12" OC	G		46	.348		12.35	12.40		24.75	34
4340	16" OC	G		66	.242		9.85	8.65		18.50	25
4350	24" OC	G		92	.174		7.30	6.20		13.50	18.25
4360	4" wide, studs 12" OC	G		45	.356		12.90	12.70		25.60	35
4370	16" OC	G		65	.246		10.25	8.80		19.05	26
4380	24" OC	G		90	.178		7.60	6.35		13.95	18.80

05 41 13 – Load-Bearing Metal Stud Framing

05 41 13.30 Framing, Stud Walls		Crew	Daily Output	Labor-Hours	Unit	Material	Labor	Equipment	Total	Total Incl O&P
						2019 Bare Costs				
4390 6″ wide, studs 12″ OC	G	2 Carp	44	.364	L.F.	16.25	12.95		29.20	39.50
4400 16″ OC	G		64	.250		12.95	8.90		21.85	29
4410 24″ OC	G		88	.182		9.65	6.50		16.15	21.50
4420 8″ wide, studs 12″ OC	G		43	.372		19.70	13.25		32.95	43.50
4430 16″ OC	G		63	.254		15.65	9.05		24.70	32
4440 24″ OC	G		86	.186		11.60	6.65		18.25	23.50
5100 10′ high walls, 18 ga. x 2-1/2″ wide, studs 12″ OC	G		54	.296		10.90	10.55		21.45	29.50
5110 16″ OC	G		77	.208		8.60	7.40		16	22
5120 24″ OC	G		107	.150		6.30	5.35		11.65	15.75
5130 3-5/8″ wide, studs 12″ OC	G		53	.302		13	10.75		23.75	32
5140 16″ OC	G		76	.211		10.30	7.50		17.80	24
5150 24″ OC	G		105	.152		7.60	5.45		13.05	17.35
5160 4″ wide, studs 12″ OC	G		52	.308		13.55	10.95		24.50	33
5170 16″ OC	G		74	.216		10.75	7.70		18.45	24.50
5180 24″ OC	G		103	.155		7.95	5.55		13.50	17.85
5190 6″ wide, studs 12″ OC	G		51	.314		17.25	11.20		28.45	37.50
5200 16″ OC	G		73	.219		13.70	7.80		21.50	28
5210 24″ OC	G		101	.158		10.15	5.65		15.80	20.50
5220 8″ wide, studs 12″ OC	G		50	.320		19.65	11.40		31.05	40.50
5230 16″ OC	G		72	.222		15.60	7.90		23.50	30.50
5240 24″ OC	G		100	.160		11.60	5.70		17.30	22
5300 16 ga. x 2-1/2″ wide, studs 12″ OC	G		47	.340		13	12.15		25.15	34.50
5310 16″ OC	G		68	.235		10.20	8.40		18.60	25
5320 24″ OC	G		94	.170		7.35	6.05		13.40	18.15
5330 3-5/8″ wide, studs 12″ OC	G		46	.348		14.90	12.40		27.30	37
5340 16″ OC	G		66	.242		11.75	8.65		20.40	27
5350 24″ OC	G		92	.174		8.55	6.20		14.75	19.65
5360 4″ wide, studs 12″ OC	G		45	.356		15.55	12.70		28.25	38
5370 16″ OC	G		65	.246		12.25	8.80		21.05	28
5380 24″ OC	G		90	.178		8.95	6.35		15.30	20.50
5390 6″ wide, studs 12″ OC	G		44	.364		19.55	12.95		32.50	43
5400 16″ OC	G		64	.250		15.40	8.90		24.30	31.50
5410 24″ OC	G		88	.182		11.30	6.50		17.80	23
5420 8″ wide, studs 12″ OC	G		43	.372		23.50	13.25		36.75	48
5430 16″ OC	G		63	.254		18.70	9.05		27.75	35.50
5440 24″ OC	G		86	.186		13.65	6.65		20.30	26
6190 12′ high walls, 18 ga. x 6″ wide, studs 12″ OC	G		41	.390		20	13.90		33.90	45
6200 16″ OC	G		58	.276		15.80	9.85		25.65	33.50
6210 24″ OC	G		81	.198		11.55	7.05		18.60	24.50
6220 8″ wide, studs 12″ OC	G		40	.400		23	14.25		37.25	48.50
6230 16″ OC	G		57	.281		18	10		28	36.50
6240 24″ OC	G		80	.200		13.20	7.15		20.35	26.50
6390 16 ga. x 6″ wide, studs 12″ OC	G		35	.457		23	16.30		39.30	52
6400 16″ OC	G		51	.314		17.90	11.20		29.10	38
6410 24″ OC	G		70	.229		12.95	8.15		21.10	27.50
6420 8″ wide, studs 12″ OC	G		34	.471		28	16.80		44.80	58
6430 16″ OC	G		50	.320		21.50	11.40		32.90	43
6440 24″ OC	G		69	.232		15.65	8.25		23.90	31
6530 14 ga. x 3-5/8″ wide, studs 12″ OC	G		34	.471		21.50	16.80		38.30	51
6540 16″ OC	G		48	.333		16.80	11.90		28.70	38
6550 24″ OC	G		65	.246		12.05	8.80		20.85	28
6560 4″ wide, studs 12″ OC	G		33	.485		22.50	17.30		39.80	53.50
6570 16″ OC	G	↓	47	.340	↓	17.65	12.15		29.80	39.50

05 41 Structural Metal Stud Framing

05 41 13 – Load-Bearing Metal Stud Framing

05 41 13.30 Framing, Stud Walls		Crew	Daily Output	Labor-Hours	Unit	Material	2019 Bare Costs Labor	Equipment	Total	Total Incl O&P
6580	24" OC	G	2 Carp	64	.250	L.F.	12.70	8.90		21.60	29
6730	12 ga. x 3-5/8" wide, studs 12" OC	G		31	.516		30	18.40		48.40	63
6740	16" OC	G		43	.372		23	13.25		36.25	47.50
6750	24" OC	G		59	.271		16.20	9.65		25.85	34
6760	4" wide, studs 12" OC	G		30	.533		31.50	19		50.50	66.50
6770	16" OC	G		42	.381		24.50	13.60		38.10	49.50
6780	24" OC	G		58	.276		17.30	9.85		27.15	35.50
7390	16' high walls, 16 ga. x 6" wide, studs 12" OC	G		33	.485		29.50	17.30		46.80	61
7400	16" OC	G		48	.333		23	11.90		34.90	44.50
7410	24" OC	G		67	.239		16.25	8.50		24.75	32
7420	8" wide, studs 12" OC	G		32	.500		36	17.85		53.85	69
7430	16" OC	G		47	.340		28	12.15		40.15	50.50
7440	24" OC	G		66	.242		19.70	8.65		28.35	36
7560	14 ga. x 4" wide, studs 12" OC	G		31	.516		29	18.40		47.40	62.50
7570	16" OC	G		45	.356		22.50	12.70		35.20	46
7580	24" OC	G		61	.262		16	9.35		25.35	33
7590	6" wide, studs 12" OC	G		30	.533		37	19		56	72
7600	16" OC	G		44	.364		28.50	12.95		41.45	53
7610	24" OC	G		60	.267		20	9.50		29.50	38
7760	12 ga. x 4" wide, studs 12" OC	G		29	.552		41.50	19.65		61.15	78
7770	16" OC	G		40	.400		31.50	14.25		45.75	58.50
7780	24" OC	G		55	.291		22	10.35		32.35	41.50
7790	6" wide, studs 12" OC	G		28	.571		52	20.50		72.50	91
7800	16" OC	G		39	.410		40	14.65		54.65	68
7810	24" OC	G		54	.296		28	10.55		38.55	48
8590	20' high walls, 14 ga. x 6" wide, studs 12" OC	G		29	.552		45	19.65		64.65	82
8600	16" OC	G		42	.381		35	13.60		48.60	61
8610	24" OC	G		57	.281		24.50	10		34.50	43.50
8620	8" wide, studs 12" OC	G		28	.571		52	20.50		72.50	90.50
8630	16" OC	G		41	.390		40	13.90		53.90	67.50
8640	24" OC	G		56	.286		28.50	10.20		38.70	48.50
8790	12 ga. x 6" wide, studs 12" OC	G		27	.593		64.50	21		85.50	106
8800	16" OC	G		37	.432		49	15.40		64.40	79.50
8810	24" OC	G		51	.314		34	11.20		45.20	56
8820	8" wide, studs 12" OC	G		26	.615		78.50	22		100.50	123
8830	16" OC	G		36	.444		60	15.85		75.85	92
8840	24" OC	G	↓	50	.320	↓	41.50	11.40		52.90	65

05 42 Cold-Formed Metal Joist Framing

05 42 13 – Cold-Formed Metal Floor Joist Framing

05 42 13.05 Bracing			Crew	Daily Output	Labor-Hours	Unit	Material	Labor	Equipment	Total	Total Incl O&P
0010	**BRACING**, continuous, per row, top & bottom										
0015	Made from recycled materials										
0120	Flat strap, 20 ga. x 2" wide, joists at 12" OC	G	1 Carp	4.67	1.713	C.L.F.	57.50	61		118.50	165
0130	16" OC	G		5.33	1.501		55.50	53.50		109	150
0140	24" OC	G		6.66	1.201		53.50	43		96.50	130
0150	18 ga. x 2" wide, joists at 12" OC	G		4	2		74	71.50		145.50	200
0160	16" OC	G		4.67	1.713		73	61		134	181
0170	24" OC	G	↓	5.33	1.501	↓	71.50	53.50		125	167

05 42 Cold-Formed Metal Joist Framing

05 42 13 – Cold-Formed Metal Floor Joist Framing

05 42 13.10 Bridging

	05 42 13.10 Bridging		Crew	Daily Output	Labor-Hours	Unit	Material	Labor	Equipment	Total	Total Incl O&P
							2019 Bare Costs				
0010	**BRIDGING**, solid between joists w/1-1/4" leg track, per joist bay										
0015	Made from recycled materials										
0230	Joists 12" OC, 18 ga. track x 6" wide	G	1 Carp	80	.100	Ea.	1.60	3.57		5.17	7.65
0240	8" wide	G		75	.107		1.85	3.80		5.65	8.35
0250	10" wide	G		70	.114		2.32	4.07		6.39	9.30
0260	12" wide	G		65	.123		2.63	4.39		7.02	10.15
0330	16 ga. track x 6" wide	G		70	.114		1.89	4.07		5.96	8.85
0340	8" wide	G		65	.123		2.61	4.39		7	10.10
0350	10" wide	G		60	.133		2.93	4.75		7.68	11.10
0360	12" wide	G		55	.145		3.37	5.20		8.57	12.25
0440	14 ga. track x 8" wide	G		60	.133		2.95	4.75		7.70	11.10
0450	10" wide	G		55	.145		4.05	5.20		9.25	13
0460	12" wide	G		50	.160		4.68	5.70		10.38	14.55
0550	12 ga. track x 10" wide	G		45	.178		5.95	6.35		12.30	17
0560	12" wide	G		40	.200		5.65	7.15		12.80	18
1230	16" OC, 18 ga. track x 6" wide	G		80	.100		2.05	3.57		5.62	8.15
1240	8" wide	G		75	.107		2.38	3.80		6.18	8.90
1250	10" wide	G		70	.114		2.97	4.07		7.04	10
1260	12" wide	G		65	.123		3.37	4.39		7.76	10.95
1330	16 ga. track x 6" wide	G		70	.114		2.43	4.07		6.50	9.40
1340	8" wide	G		65	.123		3.35	4.39		7.74	10.95
1350	10" wide	G		60	.133		3.76	4.75		8.51	12
1360	12" wide	G		55	.145		4.32	5.20		9.52	13.30
1440	14 ga. track x 8" wide	G		60	.133		3.78	4.75		8.53	12
1450	10" wide	G		55	.145		5.20	5.20		10.40	14.25
1460	12" wide	G		50	.160		6	5.70		11.70	16
1550	12 ga. track x 10" wide	G		45	.178		7.60	6.35		13.95	18.80
1560	12" wide	G		40	.200		7.25	7.15		14.40	19.75
2230	24" OC, 18 ga. track x 6" wide	G		80	.100		2.96	3.57		6.53	9.15
2240	8" wide	G		75	.107		3.44	3.80		7.24	10.10
2250	10" wide	G		70	.114		4.30	4.07		8.37	11.50
2260	12" wide	G		65	.123		4.87	4.39		9.26	12.60
2330	16 ga. track x 6" wide	G		70	.114		3.51	4.07		7.58	10.60
2340	8" wide	G		65	.123		4.85	4.39		9.24	12.60
2350	10" wide	G		60	.133		5.45	4.75		10.20	13.85
2360	12" wide	G		55	.145		6.25	5.20		11.45	15.45
2440	14 ga. track x 8" wide	G		60	.133		5.45	4.75		10.20	13.85
2450	10" wide	G		55	.145		7.50	5.20		12.70	16.80
2460	12" wide	G		50	.160		8.70	5.70		14.40	18.95
2550	12 ga. track x 10" wide	G		45	.178		11	6.35		17.35	22.50
2560	12" wide	G	↓	40	.200	↓	10.50	7.15		17.65	23.50

05 42 13.25 Framing, Band Joist

	05 42 13.25 Framing, Band Joist		Crew	Daily Output	Labor-Hours	Unit	Material	Labor	Equipment	Total	Total Incl O&P
0010	**FRAMING, BAND JOIST** (track) fastened to bearing wall										
0015	Made from recycled materials										
0220	18 ga. track x 6" deep	G	2 Carp	1000	.016	L.F.	1.30	.57		1.87	2.37
0230	8" deep	G		920	.017		1.51	.62		2.13	2.68
0240	10" deep	G		860	.019		1.89	.66		2.55	3.18
0320	16 ga. track x 6" deep	G		900	.018		1.54	.63		2.17	2.75
0330	8" deep	G		840	.019		2.13	.68		2.81	3.46
0340	10" deep	G		780	.021		2.39	.73		3.12	3.84
0350	12" deep	G		740	.022		2.75	.77		3.52	4.30
0430	14 ga. track x 8" deep	G	↓	750	.021	↓	2.40	.76		3.16	3.90

05 42 13 – Cold-Formed Metal Floor Joist Framing

05 42 13.25 Framing, Band Joist			Crew	Daily Output	Labor-Hours	Unit	2019 Bare Costs Material	Labor	Equipment	Total	Total Incl O&P
0440	10″ deep	G	2 Carp	720	.022	L.F.	3.31	.79		4.10	4.95
0450	12″ deep	G		700	.023		3.82	.82		4.64	5.55
0540	12 ga. track x 10″ deep	G		670	.024		4.84	.85		5.69	6.70
0550	12″ deep	G	↓	650	.025	↓	4.61	.88		5.49	6.50

05 42 13.30 Framing, Boxed Headers/Beams			Crew	Daily Output	Labor-Hours	Unit	Material	Labor	Equipment	Total	Total Incl O&P
0010	**FRAMING, BOXED HEADERS/BEAMS**										
0015	Made from recycled materials										
0200	Double, 18 ga. x 6″ deep	G	2 Carp	220	.073	L.F.	5.65	2.59		8.24	10.50
0210	8″ deep	G		210	.076		5.85	2.72		8.57	10.90
0220	10″ deep	G		200	.080		7	2.85		9.85	12.40
0230	12″ deep	G		190	.084		7.60	3		10.60	13.35
0300	16 ga. x 8″ deep	G		180	.089		6.70	3.17		9.87	12.65
0310	10″ deep	G		170	.094		8.50	3.36		11.86	14.90
0320	12″ deep	G		160	.100		9.25	3.57		12.82	16.10
0400	14 ga. x 10″ deep	G		140	.114		9.20	4.07		13.27	16.85
0410	12″ deep	G		130	.123		10.05	4.39		14.44	18.30
0500	12 ga. x 10″ deep	G		110	.145		12.10	5.20		17.30	22
0510	12″ deep	G		100	.160		13.40	5.70		19.10	24
1210	Triple, 18 ga. x 8″ deep	G		170	.094		8.45	3.36		11.81	14.85
1220	10″ deep	G		165	.097		10	3.46		13.46	16.70
1230	12″ deep	G		160	.100		10.95	3.57		14.52	17.95
1300	16 ga. x 8″ deep	G		145	.110		9.75	3.93		13.68	17.25
1310	10″ deep	G		140	.114		12.30	4.07		16.37	20.50
1320	12″ deep	G		135	.119		13.45	4.23		17.68	22
1400	14 ga. x 10″ deep	G		115	.139		13.30	4.96		18.26	23
1410	12″ deep	G		110	.145		14.65	5.20		19.85	24.50
1500	12 ga. x 10″ deep	G		90	.178		17.70	6.35		24.05	30
1510	12″ deep	G	↓	85	.188	↓	19.60	6.70		26.30	32.50

05 42 13.40 Framing, Joists			Crew	Daily Output	Labor-Hours	Unit	Material	Labor	Equipment	Total	Total Incl O&P
0010	**FRAMING, JOISTS**, no band joists (track), web stiffeners, headers,										
0020	Beams, bridging or bracing										
0025	Made from recycled materials										
0030	Joists (2″ flange) and fasteners, materials only										
0220	18 ga. x 6″ deep	G				L.F.	1.76			1.76	1.94
0230	8″ deep	G					1.87			1.87	2.06
0240	10″ deep	G					2.22			2.22	2.44
0320	16 ga. x 6″ deep	G					2.16			2.16	2.38
0330	8″ deep	G					2.33			2.33	2.56
0340	10″ deep	G					3.01			3.01	3.31
0350	12″ deep	G					3.41			3.41	3.75
0430	14 ga. x 8″ deep	G					2.93			2.93	3.22
0440	10″ deep	G					3.37			3.37	3.71
0450	12″ deep	G					3.83			3.83	4.22
0540	12 ga. x 10″ deep	G					4.90			4.90	5.40
0550	12″ deep	G				↓	5.60			5.60	6.15
1010	Installation of joists to band joists, beams & headers, labor only										
1220	18 ga. x 6″ deep		2 Carp	110	.145	Ea.		5.20		5.20	8.55
1230	8″ deep			90	.178			6.35		6.35	10.45
1240	10″ deep			80	.200			7.15		7.15	11.80
1320	16 ga. x 6″ deep			95	.168			6		6	9.90
1330	8″ deep			70	.229			8.15		8.15	13.45
1340	10″ deep		↓	60	.267	↓		9.50		9.50	15.70

05 42 Cold-Formed Metal Joist Framing

05 42 13 – Cold-Formed Metal Floor Joist Framing

05 42 13.40 Framing, Joists		Crew	Daily Output	Labor-Hours	Unit	Material	2019 Bare Costs Labor	Equipment	Total	Total Incl O&P
1350	12" deep	2 Carp	55	.291	Ea.		10.35		10.35	17.15
1430	14 ga. x 8" deep		65	.246			8.80		8.80	14.50
1440	10" deep		45	.356			12.70		12.70	21
1450	12" deep		35	.457			16.30		16.30	27
1540	12 ga. x 10" deep		40	.400			14.25		14.25	23.50
1550	12" deep	↓	30	.533	↓		19		19	31.50

05 42 13.45 Framing, Web Stiffeners		Crew	Daily Output	Labor-Hours	Unit	Material	Labor	Equipment	Total	Total Incl O&P
0010	**FRAMING, WEB STIFFENERS** at joist bearing, fabricated from									
0020	Stud piece (1-5/8" flange) to stiffen joist (2" flange)									
0025	Made from recycled materials									
2120	For 6" deep joist, with 18 ga. x 2-1/2" stud G	1 Carp	120	.067	Ea.	.92	2.38		3.30	4.94
2130	3-5/8" stud G		110	.073		1.08	2.59		3.67	5.45
2140	4" stud G		105	.076		1.12	2.72		3.84	5.70
2150	6" stud G		100	.080		1.42	2.85		4.27	6.25
2160	8" stud G		95	.084		1.61	3		4.61	6.75
2220	8" deep joist, with 2-1/2" stud G		120	.067		1.23	2.38		3.61	5.30
2230	3-5/8" stud G		110	.073		1.45	2.59		4.04	5.85
2240	4" stud G		105	.076		1.50	2.72		4.22	6.15
2250	6" stud G		100	.080		1.90	2.85		4.75	6.80
2260	8" stud G		95	.084		2.16	3		5.16	7.35
2320	10" deep joist, with 2-1/2" stud G		110	.073		1.53	2.59		4.12	5.95
2330	3-5/8" stud G		100	.080		1.79	2.85		4.64	6.70
2340	4" stud G		95	.084		1.86	3		4.86	7
2350	6" stud G		90	.089		2.36	3.17		5.53	7.85
2360	8" stud G		85	.094		2.67	3.36		6.03	8.50
2420	12" deep joist, with 2-1/2" stud G		110	.073		1.84	2.59		4.43	6.30
2430	3-5/8" stud G		100	.080		2.16	2.85		5.01	7.10
2440	4" stud G		95	.084		2.24	3		5.24	7.40
2450	6" stud G		90	.089		2.84	3.17		6.01	8.35
2460	8" stud G		85	.094		3.22	3.36		6.58	9.10
3130	For 6" deep joist, with 16 ga. x 3-5/8" stud G		100	.080		1.27	2.85		4.12	6.10
3140	4" stud G		95	.084		1.32	3		4.32	6.40
3150	6" stud G		90	.089		1.65	3.17		4.82	7.05
3160	8" stud G		85	.094		2.02	3.36		5.38	7.75
3230	8" deep joist, with 3-5/8" stud G		100	.080		1.70	2.85		4.55	6.60
3240	4" stud G		95	.084		1.77	3		4.77	6.90
3250	6" stud G		90	.089		2.21	3.17		5.38	7.70
3260	8" stud G		85	.094		2.71	3.36		6.07	8.55
3330	10" deep joist, with 3-5/8" stud G		85	.094		2.11	3.36		5.47	7.85
3340	4" stud G		80	.100		2.19	3.57		5.76	8.30
3350	6" stud G		75	.107		2.74	3.80		6.54	9.30
3360	8" stud G		70	.114		3.35	4.07		7.42	10.45
3430	12" deep joist, with 3-5/8" stud G		85	.094		2.54	3.36		5.90	8.35
3440	4" stud G		80	.100		2.64	3.57		6.21	8.80
3450	6" stud G		75	.107		3.30	3.80		7.10	9.95
3460	8" stud G		70	.114		4.04	4.07		8.11	11.20
4230	For 8" deep joist, with 14 ga. x 3-5/8" stud G		90	.089		2.10	3.17		5.27	7.55
4240	4" stud G		85	.094		2.21	3.36		5.57	8
4250	6" stud G		80	.100		2.79	3.57		6.36	8.95
4260	8" stud G		75	.107		3.15	3.80		6.95	9.75
4330	10" deep joist, with 3-5/8" stud G		75	.107		2.61	3.80		6.41	9.15
4340	4" stud G	↓	70	.114	↓	2.74	4.07		6.81	9.75

05 42 Cold-Formed Metal Joist Framing

05 42 13 – Cold-Formed Metal Floor Joist Framing

05 42 13.45 Framing, Web Stiffeners			Crew	Daily Output	Labor-Hours	Unit	Material	Labor (2019 Bare Costs)	Equipment	Total	Total Incl O&P
4350	6" stud	G	1 Carp	65	.123	Ea.	3.45	4.39		7.84	11.05
4360	8" stud	G		60	.133		3.90	4.75		8.65	12.15
4430	12" deep joist, with 3-5/8" stud	G		75	.107		3.14	3.80		6.94	9.75
4440	4" stud	G		70	.114		3.30	4.07		7.37	10.40
4450	6" stud	G		65	.123		4.16	4.39		8.55	11.85
4460	8" stud	G		60	.133		4.70	4.75		9.45	13
5330	For 10" deep joist, with 12 ga. x 3-5/8" stud	G		65	.123		3.75	4.39		8.14	11.40
5340	4" stud	G		60	.133		4	4.75		8.75	12.25
5350	6" stud	G		55	.145		5.05	5.20		10.25	14.10
5360	8" stud	G		50	.160		6.10	5.70		11.80	16.10
5430	12" deep joist, with 3-5/8" stud	G		65	.123		4.52	4.39		8.91	12.20
5440	4" stud	G		60	.133		4.82	4.75		9.57	13.15
5450	6" stud	G		55	.145		6.10	5.20		11.30	15.25
5460	8" stud	G	↓	50	.160	↓	7.35	5.70		13.05	17.45

05 42 23 – Cold-Formed Metal Roof Joist Framing

05 42 23.05 Framing, Bracing			Crew	Daily Output	Labor-Hours	Unit	Material	Labor	Equipment	Total	Total Incl O&P
0010	**FRAMING, BRACING**										
0015	Made from recycled materials										
0020	Continuous bracing, per row										
0100	16 ga. x 1-1/2" channel thru rafters/trusses @ 16" OC	G	1 Carp	4.50	1.778	C.L.F.	49	63.50		112.50	159
0120	24" OC	G		6	1.333		49	47.50		96.50	133
0300	2" x 2" angle x 18 ga., rafters/trusses @ 16" OC	G		6	1.333		76	47.50		123.50	162
0320	24" OC	G		8	1		76	35.50		111.50	143
0400	16 ga., rafters/trusses @ 16" OC	G		4.50	1.778		95.50	63.50		159	210
0420	24" OC	G	↓	6.50	1.231	↓	95.50	44		139.50	178

05 42 23.10 Framing, Bridging			Crew	Daily Output	Labor-Hours	Unit	Material	Labor	Equipment	Total	Total Incl O&P
0010	**FRAMING, BRIDGING**										
0015	Made from recycled materials										
0020	Solid, between rafters w/1-1/4" leg track, per rafter bay										
1200	Rafters 16" OC, 18 ga. x 4" deep	G	1 Carp	60	.133	Ea.	1.57	4.75		6.32	9.55
1210	6" deep	G		57	.140		2.05	5		7.05	10.50
1220	8" deep	G		55	.145		2.38	5.20		7.58	11.15
1230	10" deep	G		52	.154		2.97	5.50		8.47	12.30
1240	12" deep	G		50	.160		3.37	5.70		9.07	13.10
2200	24" OC, 18 ga. x 4" deep	G		60	.133		2.27	4.75		7.02	10.35
2210	6" deep	G		57	.140		2.96	5		7.96	11.50
2220	8" deep	G		55	.145		3.44	5.20		8.64	12.35
2230	10" deep	G		52	.154		4.30	5.50		9.80	13.80
2240	12" deep	G	↓	50	.160	↓	4.87	5.70		10.57	14.75

05 42 23.50 Framing, Parapets			Crew	Daily Output	Labor-Hours	Unit	Material	Labor	Equipment	Total	Total Incl O&P
0010	**FRAMING, PARAPETS**										
0015	Made from recycled materials										
0100	3' high installed on 1st story, 18 ga. x 4" wide studs, 12" OC	G	2 Carp	100	.160	L.F.	5.70	5.70		11.40	15.65
0110	16" OC	G		150	.107		4.85	3.80		8.65	11.65
0120	24" OC	G		200	.080		4.01	2.85		6.86	9.10
0200	6" wide studs, 12" OC	G		100	.160		7.30	5.70		13	17.45
0210	16" OC	G		150	.107		6.25	3.80		10.05	13.15
0220	24" OC	G		200	.080		5.15	2.85		8	10.40
1100	Installed on 2nd story, 18 ga. x 4" wide studs, 12" OC	G		95	.168		5.70	6		11.70	16.15
1110	16" OC	G		145	.110		4.85	3.93		8.78	11.85
1120	24" OC	G		190	.084		4.01	3		7.01	9.35
1200	6" wide studs, 12" OC	G	↓	95	.168	↓	7.30	6		13.30	17.95

05 42 Cold-Formed Metal Joist Framing

05 42 23 – Cold-Formed Metal Roof Joist Framing

05 42 23.50 Framing, Parapets			Crew	Daily Output	Labor-Hours	Unit	Material	2019 Bare Costs Labor	Equipment	Total	Total Incl O&P
1210	16" OC	G	2 Carp	145	.110	L.F.	6.25	3.93		10.18	13.35
1220	24" OC	G		190	.084		5.15	3		8.15	10.65
2100	Installed on gable, 18 ga. x 4" wide studs, 12" OC	G		85	.188		5.70	6.70		12.40	17.35
2110	16" OC	G		130	.123		4.85	4.39		9.24	12.60
2120	24" OC	G		170	.094		4.01	3.36		7.37	9.95
2200	6" wide studs, 12" OC	G		85	.188		7.30	6.70		14	19.15
2210	16" OC	G		130	.123		6.25	4.39		10.64	14.10
2220	24" OC	G		170	.094		5.15	3.36		8.51	11.25

05 42 23.60 Framing, Roof Rafters			Crew	Daily Output	Labor-Hours	Unit	Material	Labor	Equipment	Total	Total Incl O&P
0010	**FRAMING, ROOF RAFTERS**										
0015	Made from recycled materials										
0100	Boxed ridge beam, double, 18 ga. x 6" deep	G	2 Carp	160	.100	L.F.	5.65	3.57		9.22	12.10
0110	8" deep	G		150	.107		5.85	3.80		9.65	12.70
0120	10" deep	G		140	.114		7	4.07		11.07	14.45
0130	12" deep	G		130	.123		7.60	4.39		11.99	15.65
0200	16 ga. x 6" deep	G		150	.107		6.40	3.80		10.20	13.35
0210	8" deep	G		140	.114		6.70	4.07		10.77	14.15
0220	10" deep	G		130	.123		8.50	4.39		12.89	16.60
0230	12" deep	G		120	.133		9.25	4.75		14	18.05
1100	Rafters, 2" flange, material only, 18 ga. x 6" deep	G					1.76			1.76	1.94
1110	8" deep	G					1.87			1.87	2.06
1120	10" deep	G					2.22			2.22	2.44
1130	12" deep	G					2.55			2.55	2.81
1200	16 ga. x 6" deep	G					2.16			2.16	2.38
1210	8" deep	G					2.33			2.33	2.56
1220	10" deep	G					3.01			3.01	3.31
1230	12" deep	G					3.41			3.41	3.75
2100	Installation only, ordinary rafter to 4:12 pitch, 18 ga. x 6" deep		2 Carp	35	.457	Ea.		16.30		16.30	27
2110	8" deep			30	.533			19		19	31.50
2120	10" deep			25	.640			23		23	37.50
2130	12" deep			20	.800			28.50		28.50	47
2200	16 ga. x 6" deep			30	.533			19		19	31.50
2210	8" deep			25	.640			23		23	37.50
2220	10" deep			20	.800			28.50		28.50	47
2230	12" deep			15	1.067			38		38	63
8100	Add to labor, ordinary rafters on steep roofs							25%			
8110	Dormers & complex roofs							50%			
8200	Hip & valley rafters to 4:12 pitch							25%			
8210	Steep roofs							50%			
8220	Dormers & complex roofs							75%			
8300	Hip & valley jack rafters to 4:12 pitch							50%			
8310	Steep roofs							75%			
8320	Dormers & complex roofs							100%			

05 42 23.70 Framing, Soffits and Canopies			Crew	Daily Output	Labor-Hours	Unit	Material	Labor	Equipment	Total	Total Incl O&P
0010	**FRAMING, SOFFITS & CANOPIES**										
0015	Made from recycled materials										
0130	Continuous ledger track @ wall, studs @ 16" OC, 18 ga. x 4" wide	G	2 Carp	535	.030	L.F.	1.05	1.07		2.12	2.91
0140	6" wide	G		500	.032		1.36	1.14		2.50	3.38
0150	8" wide	G		465	.034		1.58	1.23		2.81	3.77
0160	10" wide	G		430	.037		1.98	1.33		3.31	4.37
0230	Studs @ 24" OC, 18 ga. x 4" wide	G		800	.020		1	.71		1.71	2.28
0240	6" wide	G		750	.021		1.30	.76		2.06	2.69

05 42 Cold-Formed Metal Joist Framing

05 42 23 – Cold-Formed Metal Roof Joist Framing

05 42 23.70 Framing, Soffits and Canopies		Crew	Daily Output	Labor-Hours	Unit	Material	Labor	Equipment	Total	Total Incl O&P
						2019 Bare Costs				
0250	8" wide G	2 Carp	700	.023	L.F.	1.51	.82		2.33	3.01
0260	10" wide G		650	.025		1.89	.88		2.77	3.53
1000	Horizontal soffit and canopy members, material only									
1030	1-5/8" flange studs, 18 ga. x 4" deep G				L.F.	1.34			1.34	1.48
1040	6" deep G					1.70			1.70	1.87
1050	8" deep G					1.93			1.93	2.13
1140	2" flange joists, 18 ga. x 6" deep G					2.02			2.02	2.22
1150	8" deep G					2.14			2.14	2.35
1160	10" deep G					2.53			2.53	2.79
4030	Installation only, 18 ga., 1-5/8" flange x 4" deep	2 Carp	130	.123	Ea.		4.39		4.39	7.25
4040	6" deep		110	.145			5.20		5.20	8.55
4050	8" deep		90	.178			6.35		6.35	10.45
4140	2" flange, 18 ga. x 6" deep		110	.145			5.20		5.20	8.55
4150	8" deep		90	.178			6.35		6.35	10.45
4160	10" deep		80	.200			7.15		7.15	11.80
6010	Clips to attach fascia to rafter tails, 2" x 2" x 18 ga. angle G	1 Carp	120	.067		.90	2.38		3.28	4.92
6020	16 ga. angle G	"	100	.080		1.13	2.85		3.98	5.95

05 44 Cold-Formed Metal Trusses

05 44 13 – Cold-Formed Metal Roof Trusses

05 44 13.60 Framing, Roof Trusses		Crew	Daily Output	Labor-Hours	Unit	Material	Labor	Equipment	Total	Total Incl O&P
0010	**FRAMING, ROOF TRUSSES**									
0015	Made from recycled materials									
0020	Fabrication of trusses on ground, Fink (W) or King Post, to 4:12 pitch									
0120	18 ga. x 4" chords, 16' span G	2 Carp	12	1.333	Ea.	62.50	47.50		110	148
0130	20' span G		11	1.455		78.50	52		130.50	172
0140	24' span G		11	1.455		94	52		146	189
0150	28' span G		10	1.600		110	57		167	215
0160	32' span G		10	1.600		125	57		182	232
0250	6" chords, 28' span G		9	1.778		139	63.50		202.50	258
0260	32' span G		9	1.778		159	63.50		222.50	280
0270	36' span G		8	2		179	71.50		250.50	315
0280	40' span G		8	2		199	71.50		270.50	335
1120	5:12 to 8:12 pitch, 18 ga. x 4" chords, 16' span G		10	1.600		71.50	57		128.50	173
1130	20' span G		9	1.778		89.50	63.50		153	204
1140	24' span G		9	1.778		108	63.50		171.50	223
1150	28' span G		8	2		125	71.50		196.50	256
1160	32' span G		8	2		143	71.50		214.50	276
1250	6" chords, 28' span G		7	2.286		159	81.50		240.50	310
1260	32' span G		7	2.286		182	81.50		263.50	335
1270	36' span G		6	2.667		204	95		299	380
1280	40' span G		6	2.667		227	95		322	405
2120	9:12 to 12:12 pitch, 18 ga. x 4" chords, 16' span G		8	2		89.50	71.50		161	217
2130	20' span G		7	2.286		112	81.50		193.50	258
2140	24' span G		7	2.286		134	81.50		215.50	283
2150	28' span G		6	2.667		157	95		252	330
2160	32' span G		6	2.667		179	95		274	355
2250	6" chords, 28' span G		5	3.200		199	114		313	405
2260	32' span G		5	3.200		227	114		341	440
2270	36' span G		4	4		256	143		399	515
2280	40' span G		4	4		284	143		427	545

05 44 Cold-Formed Metal Trusses

05 44 13 – Cold-Formed Metal Roof Trusses

05 44 13.60 Framing, Roof Trusses		Crew	Daily Output	Labor-Hours	Unit	2019 Bare Costs Material	Labor	Equipment	Total	Total Incl O&P
5120	Erection only of roof trusses, to 4:12 pitch, 16' span	F-6	48	.833	Ea.		27.50	9.85	37.35	56
5130	20' span		46	.870			28.50	10.25	38.75	58.50
5140	24' span		44	.909			30	10.70	40.70	61.50
5150	28' span		42	.952			31.50	11.25	42.75	64
5160	32' span		40	1			33	11.80	44.80	67
5170	36' span		38	1.053			34.50	12.40	46.90	70.50
5180	40' span		36	1.111			36.50	13.10	49.60	74.50
5220	5:12 to 8:12 pitch, 16' span		42	.952			31.50	11.25	42.75	64
5230	20' span		40	1			33	11.80	44.80	67
5240	24' span		38	1.053			34.50	12.40	46.90	70.50
5250	28' span		36	1.111			36.50	13.10	49.60	74.50
5260	32' span		34	1.176			38.50	13.85	52.35	79
5270	36' span		32	1.250			41	14.75	55.75	83.50
5280	40' span		30	1.333			44	15.70	59.70	89.50
5320	9:12 to 12:12 pitch, 16' span		36	1.111			36.50	13.10	49.60	74.50
5330	20' span		34	1.176			38.50	13.85	52.35	79
5340	24' span		32	1.250			41	14.75	55.75	83.50
5350	28' span		30	1.333			44	15.70	59.70	89.50
5360	32' span		28	1.429			47	16.85	63.85	96
5370	36' span		26	1.538			50.50	18.15	68.65	103
5380	40' span		24	1.667			54.50	19.65	74.15	112

05 51 Metal Stairs

05 51 13 – Metal Pan Stairs

05 51 13.50 Pan Stairs		Crew	Daily Output	Labor-Hours	Unit	Material	Labor	Equipment	Total	Total Incl O&P
0010	**PAN STAIRS**, shop fabricated, steel stringers									
0015	Made from recycled materials									
1700	Pre-erected, steel pan tread, 3'-6" wide, 2 line pipe rail G	E-2	87	.552	Riser	535	22	19.50	576.50	645
1800	With flat bar picket rail G	"	87	.552	"	595	22	19.50	636.50	715

05 51 23 – Metal Fire Escapes

05 51 23.50 Fire Escape Stairs		Crew	Daily Output	Labor-Hours	Unit	Material	Labor	Equipment	Total	Total Incl O&P
0010	**FIRE ESCAPE STAIRS**, portable									
0100	Portable ladder				Ea.	123			123	135

05 52 Metal Railings

05 52 13 – Pipe and Tube Railings

05 52 13.50 Railings, Pipe		Crew	Daily Output	Labor-Hours	Unit	Material	Labor	Equipment	Total	Total Incl O&P
0010	**RAILINGS, PIPE**, shop fab'd, 3'-6" high, posts @ 5' OC									
0015	Made from recycled materials									
0020	Aluminum, 2 rail, satin finish, 1-1/4" diameter G	E-4	160	.200	L.F.	52	8.05	.60	60.65	71.50
0030	Clear anodized G		160	.200		63.50	8.05	.60	72.15	84.50
0040	Dark anodized G		160	.200		70.50	8.05	.60	79.15	92
0080	1-1/2" diameter, satin finish G		160	.200		61	8.05	.60	69.65	81.50
0090	Clear anodized G		160	.200		68.50	8.05	.60	77.15	90
0100	Dark anodized G		160	.200		75.50	8.05	.60	84.15	97.50
0140	Aluminum, 3 rail, 1-1/4" diam., satin finish G		137	.234		67.50	9.40	.70	77.60	91.50
0150	Clear anodized G		137	.234		84	9.40	.70	94.10	110
0160	Dark anodized G		137	.234		92.50	9.40	.70	102.60	119
0200	1-1/2" diameter, satin finish G		137	.234		80	9.40	.70	90.10	105

05 52 Metal Railings

05 52 13 – Pipe and Tube Railings

05 52 13.50 Railings, Pipe			Crew	Daily Output	Labor-Hours	Unit	2019 Bare Costs Material	Labor	Equipment	Total	Total Incl O&P
0210	Clear anodized	G	E-4	137	.234	L.F.	91	9.40	.70	101.10	117
0220	Dark anodized	G		137	.234		100	9.40	.70	110.10	127
0500	Steel, 2 rail, on stairs, primed, 1-1/4" diameter	G		160	.200		29.50	8.05	.60	38.15	47
0520	1-1/2" diameter	G		160	.200		32	8.05	.60	40.65	49.50
0540	Galvanized, 1-1/4" diameter	G		160	.200		39	8.05	.60	47.65	57.50
0560	1-1/2" diameter	G		160	.200		45.50	8.05	.60	54.15	64.50
0580	Steel, 3 rail, primed, 1-1/4" diameter	G		137	.234		43	9.40	.70	53.10	64
0600	1-1/2" diameter	G		137	.234		46.50	9.40	.70	56.60	68
0620	Galvanized, 1-1/4" diameter	G		137	.234		60	9.40	.70	70.10	83
0640	1-1/2" diameter	G		137	.234		72	9.40	.70	82.10	96
0700	Stainless steel, 2 rail, 1-1/4" diam., #4 finish	G		137	.234		131	9.40	.70	141.10	161
0720	High polish	G		137	.234		213	9.40	.70	223.10	251
0740	Mirror polish	G		137	.234		265	9.40	.70	275.10	310
0760	Stainless steel, 3 rail, 1-1/2" diam., #4 finish	G		120	.267		197	10.75	.80	208.55	236
0770	High polish	G		120	.267		330	10.75	.80	341.55	380
0780	Mirror finish	G		120	.267		400	10.75	.80	411.55	460
0900	Wall rail, alum. pipe, 1-1/4" diam., satin finish	G		213	.150		25	6.05	.45	31.50	38.50
0905	Clear anodized	G		213	.150		31.50	6.05	.45	38	45.50
0910	Dark anodized	G		213	.150		37	6.05	.45	43.50	51.50
0915	1-1/2" diameter, satin finish	G		213	.150		28	6.05	.45	34.50	41.50
0920	Clear anodized	G		213	.150		34.50	6.05	.45	41	49
0925	Dark anodized	G		213	.150		44	6.05	.45	50.50	59
0930	Steel pipe, 1-1/4" diameter, primed	G		213	.150		17.25	6.05	.45	23.75	30
0935	Galvanized	G		213	.150		25	6.05	.45	31.50	38.50
0940	1-1/2" diameter	G		176	.182		18.15	7.30	.54	25.99	33.50
0945	Galvanized	G		213	.150		25	6.05	.45	31.50	38.50
0955	Stainless steel pipe, 1-1/2" diam., #4 finish	G		107	.299		105	12.05	.89	117.94	138
0960	High polish	G		107	.299		214	12.05	.89	226.94	257
0965	Mirror polish	G		107	.299		253	12.05	.89	265.94	300
2000	2-line pipe rail (1-1/2" T&B) with 1/2" pickets @ 4-1/2" OC,										
2005	attached handrail on brackets										
2010	42" high aluminum, satin finish, straight & level	G	E-4	120	.267	L.F.	277	10.75	.80	288.55	325
2050	42" high steel, primed, straight & level	G	"	120	.267		148	10.75	.80	159.55	182
4000	For curved and level rails, add						10%	10%			
4100	For sloped rails for stairs, add						30%	30%			

05 58 Formed Metal Fabrications

05 58 25 – Formed Lamp Posts

05 58 25.40 Lamp Posts			Crew	Daily Output	Labor-Hours	Unit	Material	Labor	Equipment	Total	Total Incl O&P
0010	**LAMP POSTS**										
0020	Aluminum, 7' high, stock units, post only	G	1 Carp	16	.500	Ea.	82.50	17.85		100.35	120
0100	Mild steel, plain	G	"	16	.500	"	75	17.85		92.85	112

05 71 Decorative Metal Stairs

05 71 13 – Fabricated Metal Spiral Stairs

05 71 13.50 Spiral Stairs			Crew	Daily Output	Labor-Hours	Unit	Material	2019 Bare Costs Labor	Equipment	Total	Total Incl O&P
0010	**SPIRAL STAIRS**										
1805	Shop fabricated, custom ordered										
1810	Aluminum, 5'-0" diameter, plain units	G	E-4	45	.711	Riser	730	28.50	2.13	760.63	855
1820	Fancy units	G		45	.711		1,300	28.50	2.13	1,330.63	1,500
1900	Cast iron, 4'-0" diameter, plain units	G		45	.711		735	28.50	2.13	765.63	855
1920	Fancy units	G		25	1.280		1,275	51.50	3.83	1,330.33	1,500
3100	Spiral stair kits, 12 stacking risers to fit exact floor height										
3110	Steel, flat metal treads, primed, 3'-6" diameter	G	2 Carp	1.60	10	Flight	1,325	355		1,680	2,050
3120	4'-0" diameter	G		1.45	11.034		1,525	395		1,920	2,325
3130	4'-6" diameter	G		1.35	11.852		1,675	425		2,100	2,550
3140	5'-0" diameter	G		1.25	12.800		1,850	455		2,305	2,775
3310	Checkered plate tread, primed, 3'-6" diameter	G		1.45	11.034		1,575	395		1,970	2,375
3320	4'-0" diameter	G		1.35	11.852		1,775	425		2,200	2,650
3330	4'-6" diameter	G		1.25	12.800		1,925	455		2,380	2,875
3340	5'-0" diameter	G		1.15	13.913		2,075	495		2,570	3,125
3510	Red oak covers on flat metal treads, 3'-6" diameter			1.35	11.852		2,400	425		2,825	3,350
3520	4'-0" diameter			1.25	12.800		2,875	455		3,330	3,925
3530	4'-6" diameter			1.15	13.913		3,100	495		3,595	4,250
3540	5'-0" diameter			1.05	15.238		3,350	545		3,895	4,575

05 75 Decorative Formed Metal

05 75 13 – Columns

05 75 13.10 Aluminum Columns			Crew	Daily Output	Labor-Hours	Unit	Material	Labor	Equipment	Total	Total Incl O&P
0010	**ALUMINUM COLUMNS**										
0015	Made from recycled materials										
0020	Aluminum, extruded, stock units, no cap or base, 6" diameter	G	E-4	240	.133	L.F.	15.25	5.35	.40	21	26.50
0100	8" diameter	G	"	170	.188	"	19.55	7.60	.56	27.71	35.50
0410	Caps and bases, plain, 6" diameter	G				Set	26			26	28.50
0420	8" diameter	G				"	30.50			30.50	33.50
0500	For square columns, add to column prices above					L.F.	50%				
0700	Residential, flat, 8' high, plain	G	E-4	20	1.600	Ea.	103	64.50	4.78	172.28	230
0720	Fancy	G		20	1.600		200	64.50	4.78	269.28	335
0740	Corner type, plain	G		20	1.600		177	64.50	4.78	246.28	310
0760	Fancy	G		20	1.600		350	64.50	4.78	419.28	500

05 75 13.20 Columns, Ornamental			Crew	Daily Output	Labor-Hours	Unit	Material	Labor	Equipment	Total	Total Incl O&P
0010	**COLUMNS, ORNAMENTAL**, shop fabricated										
6400	Mild steel, flat, 9" wide, stock units, painted, plain	G	E-4	160	.200	V.L.F.	9.65	8.05	.60	18.30	25
6450	Fancy	G		160	.200		18.75	8.05	.60	27.40	35
6500	Corner columns, painted, plain	G		160	.200		16.60	8.05	.60	25.25	33
6550	Fancy	G		160	.200		33	8.05	.60	41.65	50.50

Division Notes

	CREW	DAILY OUTPUT	LABOR-HOURS	UNIT	BARE COSTS MAT.	BARE COSTS LABOR	BARE COSTS EQUIP.	BARE COSTS TOTAL	TOTAL INCL O&P

Estimating Tips

06 05 00 Common Work Results for Wood, Plastics, and Composites

- Common to any wood-framed structure are the accessory connector items such as screws, nails, adhesives, hangers, connector plates, straps, angles, and hold-downs. For typical wood-framed buildings, such as residential projects, the aggregate total for these items can be significant, especially in areas where seismic loading is a concern. For floor and wall framing, the material cost is based on 10 to 25 lbs. of accessory connectors per MBF. Hold-downs, hangers, and other connectors should be taken off by the piece.

  Included with material costs are fasteners for a normal installation. Gordian's RSMeans engineers use manufacturers' recommendations, written specifications, and/or standard construction practice for the sizing and spacing of fasteners. Prices for various fasteners are shown for informational purposes only. Adjustments should be made if unusual fastening conditions exist.

06 10 00 Carpentry

- Lumber is a traded commodity and therefore sensitive to supply and demand in the marketplace. Even with "budgetary" estimating of wood-framed projects, it is advisable to call local suppliers for the latest market pricing.
- The common quantity unit for wood-framed projects is "thousand board feet" (MBF). A board foot is a volume of wood—1" x 1' x 1' or 144 cubic inches. Board-foot quantities are generally calculated using nominal material dimensions—dressed sizes are ignored. Board foot per lineal foot of any stick of lumber can be calculated by dividing the nominal cross-sectional area by 12. As an example, 2,000 lineal feet of 2 x 12 equates to 4 MBF by dividing the nominal area, 2 x 12, by 12, which equals 2, and multiplying that by 2,000 to give 4,000 board feet. This simple rule applies to all nominal dimensioned lumber.
- Waste is an issue of concern at the quantity takeoff for any area of construction. Framing lumber is sold in even foot lengths, i.e., 8', 10', 12', 14', 16', and depending on spans, wall heights, and the grade of lumber, waste is inevitable. A rule of thumb for lumber waste is 5–10% depending on material quality and the complexity of the framing.
- Wood in various forms and shapes is used in many projects, even where the main structural framing is steel, concrete, or masonry. Plywood as a back-up partition material and 2x boards used as blocking and cant strips around roof edges are two common examples. The estimator should ensure that the costs of all wood materials are included in the final estimate.

06 20 00 Finish Carpentry

- It is necessary to consider the grade of workmanship when estimating labor costs for erecting millwork and an interior finish. In practice, there are three grades: premium, custom, and economy. The RSMeans daily output for base and case moldings is in the range of 200 to 250 L.F. per carpenter per day. This is appropriate for most average custom-grade projects. For premium projects, an adjustment to productivity of 25–50% should be made, depending on the complexity of the job.

Reference Numbers

Reference numbers are shown at the beginning of some major classifications. These numbers refer to related items in the Reference Section. The reference information may be an estimating procedure, an alternate pricing method, or technical information.

Note: Not all subdivisions listed here necessarily appear. ■

06 05 Common Work Results for Wood, Plastics, and Composites

06 05 05 – Selective Demolition for Wood, Plastics, and Composites

06 05 05.10 Selective Demolition Wood Framing		Crew	Daily Output	Labor-Hours	Unit	2019 Bare Costs Material	Labor	Equipment	Total	Total Incl O&P
0010	**SELECTIVE DEMOLITION WOOD FRAMING** R024119-10									
0100	Timber connector, nailed, small	1 Clab	96	.083	Ea.		2.29		2.29	3.79
0110	Medium		60	.133			3.67		3.67	6.05
0120	Large		48	.167			4.58		4.58	7.60
0130	Bolted, small		48	.167			4.58		4.58	7.60
0140	Medium		32	.250			6.90		6.90	11.35
0150	Large		24	.333			9.15		9.15	15.15
2958	Beams, 2″ x 6″	2 Clab	1100	.015	L.F.		.40		.40	.66
2960	2″ x 8″		825	.019			.53		.53	.88
2965	2″ x 10″		665	.024			.66		.66	1.09
2970	2″ x 12″		550	.029			.80		.80	1.32
2972	2″ x 14″		470	.034			.94		.94	1.55
2975	4″ x 8″	B-1	413	.058			1.64		1.64	2.71
2980	4″ x 10″		330	.073			2.05		2.05	3.39
2985	4″ x 12″		275	.087			2.46		2.46	4.06
3000	6″ x 8″		275	.087			2.46		2.46	4.06
3040	6″ x 10″		220	.109			3.07		3.07	5.10
3080	6″ x 12″		185	.130			3.65		3.65	6.05
3120	8″ x 12″		140	.171			4.83		4.83	8
3160	10″ x 12″		110	.218			6.15		6.15	10.15
3162	Alternate pricing method		1.10	21.818	M.B.F.		615		615	1,025
3170	Blocking, in 16″ OC wall framing, 2″ x 4″	1 Clab	600	.013	L.F.		.37		.37	.61
3172	2″ x 6″		400	.020			.55		.55	.91
3174	In 24″ OC wall framing, 2″ x 4″		600	.013			.37		.37	.61
3176	2″ x 6″		400	.020			.55		.55	.91
3178	Alt method, wood blocking removal from wood framing		.40	20	M.B.F.		550		550	910
3179	Wood blocking removal from steel framing		.36	22.222	″		610		610	1,000
3180	Bracing, let in, 1″ x 3″, studs 16″ OC		1050	.008	L.F.		.21		.21	.35
3181	Studs 24″ OC		1080	.007			.20		.20	.34
3182	1″ x 4″, studs 16″ OC		1050	.008			.21		.21	.35
3183	Studs 24″ OC		1080	.007			.20		.20	.34
3184	1″ x 6″, studs 16″ OC		1050	.008			.21		.21	.35
3185	Studs 24″ OC		1080	.007			.20		.20	.34
3186	2″ x 3″, studs 16″ OC		800	.010			.28		.28	.45
3187	Studs 24″ OC		830	.010			.27		.27	.44
3188	2″ x 4″, studs 16″ OC		800	.010			.28		.28	.45
3189	Studs 24″ OC		830	.010			.27		.27	.44
3190	2″ x 6″, studs 16″ OC		800	.010			.28		.28	.45
3191	Studs 24″ OC		830	.010			.27		.27	.44
3192	2″ x 8″, studs 16″ OC		800	.010			.28		.28	.45
3193	Studs 24″ OC		830	.010			.27		.27	.44
3194	"T" shaped metal bracing, studs at 16″ OC		1060	.008			.21		.21	.34
3195	Studs at 24″ OC		1200	.007			.18		.18	.30
3196	Metal straps, studs at 16″ OC		1200	.007			.18		.18	.30
3197	Studs at 24″ OC		1240	.006			.18		.18	.29
3200	Columns, round, 8′ to 14′ tall		40	.200	Ea.		5.50		5.50	9.10
3202	Dimensional lumber sizes	2 Clab	1.10	14.545	M.B.F.		400		400	660
3250	Blocking, between joists	1 Clab	320	.025	Ea.		.69		.69	1.14
3252	Bridging, metal strap, between joists		320	.025	Pr.		.69		.69	1.14
3254	Wood, between joists		320	.025	″		.69		.69	1.14
3260	Door buck, studs, header & access., 8′ high 2″ x 4″ wall, 3′ wide		32	.250	Ea.		6.90		6.90	11.35
3261	4′ wide		32	.250			6.90		6.90	11.35
3262	5′ wide		32	.250			6.90		6.90	11.35

06 05 Common Work Results for Wood, Plastics, and Composites

06 05 05 – Selective Demolition for Wood, Plastics, and Composites

06 05 05.10 Selective Demolition Wood Framing		Crew	Daily Output	Labor-Hours	Unit	Material	2019 Bare Costs Labor	Equipment	Total	Total Incl O&P
3263	6′ wide	1 Clab	32	.250	Ea.		6.90		6.90	11.35
3264	8′ wide		30	.267			7.35		7.35	12.10
3265	10′ wide		30	.267			7.35		7.35	12.10
3266	12′ wide		30	.267			7.35		7.35	12.10
3267	2″ x 6″ wall, 3′ wide		32	.250			6.90		6.90	11.35
3268	4′ wide		32	.250			6.90		6.90	11.35
3269	5′ wide		32	.250			6.90		6.90	11.35
3270	6′ wide		32	.250			6.90		6.90	11.35
3271	8′ wide		30	.267			7.35		7.35	12.10
3272	10′ wide		30	.267			7.35		7.35	12.10
3273	12′ wide		30	.267			7.35		7.35	12.10
3274	Window buck, studs, header & access., 8′ high 2″ x 4″ wall, 2′ wide		24	.333			9.15		9.15	15.15
3275	3′ wide		24	.333			9.15		9.15	15.15
3276	4′ wide		24	.333			9.15		9.15	15.15
3277	5′ wide		24	.333			9.15		9.15	15.15
3278	6′ wide		24	.333			9.15		9.15	15.15
3279	7′ wide		24	.333			9.15		9.15	15.15
3280	8′ wide		22	.364			10		10	16.55
3281	10′ wide		22	.364			10		10	16.55
3282	12′ wide		22	.364			10		10	16.55
3283	2″ x 6″ wall, 2′ wide		24	.333			9.15		9.15	15.15
3284	3′ wide		24	.333			9.15		9.15	15.15
3285	4′ wide		24	.333			9.15		9.15	15.15
3286	5′ wide		24	.333			9.15		9.15	15.15
3287	6′ wide		24	.333			9.15		9.15	15.15
3288	7′ wide		24	.333			9.15		9.15	15.15
3289	8′ wide		22	.364			10		10	16.55
3290	10′ wide		22	.364			10		10	16.55
3291	12′ wide		22	.364			10		10	16.55
3360	Deck or porch decking		825	.010	L.F.		.27		.27	.44
3400	Fascia boards, 1″ x 6″		500	.016			.44		.44	.73
3440	1″ x 8″		450	.018			.49		.49	.81
3480	1″ x 10″		400	.020			.55		.55	.91
3490	2″ x 6″		450	.018			.49		.49	.81
3500	2″ x 8″		400	.020			.55		.55	.91
3510	2″ x 10″		350	.023			.63		.63	1.04
3610	Furring, on wood walls or ceiling		4000	.002	S.F.		.06		.06	.09
3620	On masonry or concrete walls or ceiling		1200	.007	″		.18		.18	.30
3800	Headers over openings, 2 @ 2″ x 6″		110	.073	L.F.		2		2	3.31
3840	2 @ 2″ x 8″		100	.080			2.20		2.20	3.64
3880	2 @ 2″ x 10″		90	.089			2.44		2.44	4.04
3885	Alternate pricing method		.26	30.651	M.B.F.		845		845	1,400
3920	Joists, 1″ x 4″		1250	.006	L.F.		.18		.18	.29
3930	1″ x 6″		1135	.007			.19		.19	.32
3940	1″ x 8″		1000	.008			.22		.22	.36
3950	1″ x 10″		895	.009			.25		.25	.41
3960	1″ x 12″		765	.010			.29		.29	.48
4200	2″ x 4″	2 Clab	1000	.016			.44		.44	.73
4230	2″ x 6″		970	.016			.45		.45	.75
4240	2″ x 8″		940	.017			.47		.47	.77
4250	2″ x 10″		910	.018			.48		.48	.80
4280	2″ x 12″		880	.018			.50		.50	.83
4281	2″ x 14″		850	.019			.52		.52	.86

06 05 05 – Selective Demolition for Wood, Plastics, and Composites

06 05 05.10 Selective Demolition Wood Framing		Crew	Daily Output	Labor-Hours	Unit	Material	2019 Bare Costs Labor	Equipment	Total	Total Incl O&P
4282	Composite joists, 9-1/2"	2 Clab	960	.017	L.F.		.46		.46	.76
4283	11-7/8"		930	.017			.47		.47	.78
4284	14"		897	.018			.49		.49	.81
4285	16"		865	.019			.51		.51	.84
4290	Wood joists, alternate pricing method		1.50	10.667	M.B.F.		293		293	485
4500	Open web joist, 12" deep		500	.032	L.F.		.88		.88	1.45
4505	14" deep		475	.034			.93		.93	1.53
4510	16" deep		450	.036			.98		.98	1.62
4520	18" deep		425	.038			1.04		1.04	1.71
4530	24" deep		400	.040			1.10		1.10	1.82
4550	Ledger strips, 1" x 2"	1 Clab	1200	.007			.18		.18	.30
4560	1" x 3"		1200	.007			.18		.18	.30
4570	1" x 4"		1200	.007			.18		.18	.30
4580	2" x 2"		1100	.007			.20		.20	.33
4590	2" x 4"		1000	.008			.22		.22	.36
4600	2" x 6"		1000	.008			.22		.22	.36
4601	2" x 8 or 2" x 10"		800	.010			.28		.28	.45
4602	4" x 6"		600	.013			.37		.37	.61
4604	4" x 8"		450	.018			.49		.49	.81
5400	Posts, 4" x 4"	2 Clab	800	.020			.55		.55	.91
5405	4" x 6"		550	.029			.80		.80	1.32
5410	4" x 8"		440	.036			1		1	1.65
5425	4" x 10"		390	.041			1.13		1.13	1.86
5430	4" x 12"		350	.046			1.26		1.26	2.08
5440	6" x 6"		400	.040			1.10		1.10	1.82
5445	6" x 8"		350	.046			1.26		1.26	2.08
5450	6" x 10"		320	.050			1.38		1.38	2.27
5455	6" x 12"		290	.055			1.52		1.52	2.51
5480	8" x 8"		300	.053			1.47		1.47	2.42
5500	10" x 10"		240	.067			1.83		1.83	3.03
5660	T&G floor planks		2	8	M.B.F.		220		220	365
5682	Rafters, ordinary, 16" OC, 2" x 4"		880	.018	S.F.		.50		.50	.83
5683	2" x 6"		840	.019			.52		.52	.87
5684	2" x 8"		820	.020			.54		.54	.89
5685	2" x 10"		820	.020			.54		.54	.89
5686	2" x 12"		810	.020			.54		.54	.90
5687	24" OC, 2" x 4"		1170	.014			.38		.38	.62
5688	2" x 6"		1117	.014			.39		.39	.65
5689	2" x 8"		1091	.015			.40		.40	.67
5690	2" x 10"		1091	.015			.40		.40	.67
5691	2" x 12"		1077	.015			.41		.41	.68
5795	Rafters, ordinary, 2" x 4" (alternate method)		862	.019	L.F.		.51		.51	.84
5800	2" x 6" (alternate method)		850	.019			.52		.52	.86
5840	2" x 8" (alternate method)		837	.019			.53		.53	.87
5855	2" x 10" (alternate method)		825	.019			.53		.53	.88
5865	2" x 12" (alternate method)		812	.020			.54		.54	.90
5870	Sill plate, 2" x 4"	1 Clab	1170	.007			.19		.19	.31
5871	2" x 6"		780	.010			.28		.28	.47
5872	2" x 8"		586	.014			.38		.38	.62
5873	Alternate pricing method		.78	10.256	M.B.F.		282		282	465
5885	Ridge board, 1" x 4"	2 Clab	900	.018	L.F.		.49		.49	.81
5886	1" x 6"		875	.018			.50		.50	.83
5887	1" x 8"		850	.019			.52		.52	.86

06 05 Common Work Results for Wood, Plastics, and Composites

06 05 05 – Selective Demolition for Wood, Plastics, and Composites

06 05 05.10 Selective Demolition Wood Framing		Crew	Daily Output	Labor-Hours	Unit	2019 Bare Costs Material	Labor	Equipment	Total	Total Incl O&P
5888	1″ x 10″	2 Clab	825	.019	L.F.		.53		.53	.88
5889	1″ x 12″		800	.020			.55		.55	.91
5890	2″ x 4″		900	.018			.49		.49	.81
5892	2″ x 6″		875	.018			.50		.50	.83
5894	2″ x 8″		850	.019			.52		.52	.86
5896	2″ x 10″		825	.019			.53		.53	.88
5898	2″ x 12″		800	.020			.55		.55	.91
6050	Rafter tie, 1″ x 4″		1250	.013			.35		.35	.58
6052	1″ x 6″		1135	.014			.39		.39	.64
6054	2″ x 4″		1000	.016			.44		.44	.73
6056	2″ x 6″		970	.016			.45		.45	.75
6070	Sleepers, on concrete, 1″ x 2″	1 Clab	4700	.002			.05		.05	.08
6075	1″ x 3″		4000	.002			.06		.06	.09
6080	2″ x 4″		3000	.003			.07		.07	.12
6085	2″ x 6″		2600	.003			.08		.08	.14
6086	Sheathing from roof, 5/16″	2 Clab	1600	.010	S.F.		.28		.28	.45
6088	3/8″		1525	.010			.29		.29	.48
6090	1/2″		1400	.011			.31		.31	.52
6092	5/8″		1300	.012			.34		.34	.56
6094	3/4″		1200	.013			.37		.37	.61
6096	Board sheathing from roof		1400	.011			.31		.31	.52
6100	Sheathing, from walls, 1/4″		1200	.013			.37		.37	.61
6110	5/16″		1175	.014			.37		.37	.62
6120	3/8″		1150	.014			.38		.38	.63
6130	1/2″		1125	.014			.39		.39	.65
6140	5/8″		1100	.015			.40		.40	.66
6150	3/4″		1075	.015			.41		.41	.68
6152	Board sheathing from walls		1500	.011			.29		.29	.49
6158	Subfloor/roof deck, with boards		2200	.007			.20		.20	.33
6159	Subfloor/roof deck, with tongue & groove boards		2000	.008			.22		.22	.36
6160	Plywood, 1/2″ thick		768	.021			.57		.57	.95
6162	5/8″ thick		760	.021			.58		.58	.96
6164	3/4″ thick		750	.021			.59		.59	.97
6165	1-1/8″ thick		720	.022			.61		.61	1.01
6166	Underlayment, particle board, 3/8″ thick	1 Clab	780	.010			.28		.28	.47
6168	1/2″ thick		768	.010			.29		.29	.47
6170	5/8″ thick		760	.011			.29		.29	.48
6172	3/4″ thick		750	.011			.29		.29	.49
6200	Stairs and stringers, straight run	2 Clab	40	.400	Riser		11		11	18.20
6240	With platforms, winders or curves	″	26	.615	″		16.90		16.90	28
6300	Components, tread	1 Clab	110	.073	Ea.		2		2	3.31
6320	Riser		80	.100	″		2.75		2.75	4.55
6390	Stringer, 2″ x 10″		260	.031	L.F.		.85		.85	1.40
6400	2″ x 12″		260	.031			.85		.85	1.40
6410	3″ x 10″		250	.032			.88		.88	1.45
6420	3″ x 12″		250	.032			.88		.88	1.45
6590	Wood studs, 2″ x 3″	2 Clab	3076	.005			.14		.14	.24
6600	2″ x 4″		2000	.008			.22		.22	.36
6640	2″ x 6″		1600	.010			.28		.28	.45
6720	Wall framing, including studs, plates and blocking, 2″ x 4″	1 Clab	600	.013	S.F.		.37		.37	.61
6740	2″ x 6″		480	.017	″		.46		.46	.76
6750	Headers, 2″ x 4″		1125	.007	L.F.		.20		.20	.32
6755	2″ x 6″		1125	.007			.20		.20	.32

06 05 05 – Selective Demolition for Wood, Plastics, and Composites

06 05 05.10 Selective Demolition Wood Framing		Crew	Daily Output	Labor-Hours	Unit	Material	2019 Bare Costs Labor	Equipment	Total	Total Incl O&P
6760	2" x 8"	1 Clab	1050	.008	L.F.		.21		.21	.35
6765	2" x 10"		1050	.008			.21		.21	.35
6770	2" x 12"		1000	.008			.22		.22	.36
6780	4" x 10"		525	.015			.42		.42	.69
6785	4" x 12"		500	.016			.44		.44	.73
6790	6" x 8"		560	.014			.39		.39	.65
6795	6" x 10"		525	.015			.42		.42	.69
6797	6" x 12"		500	.016			.44		.44	.73
7000	Trusses									
7050	12' span	2 Clab	74	.216	Ea.		5.95		5.95	9.85
7150	24' span	F-3	66	.606			19.85	7.15	27	41
7200	26' span		64	.625			20.50	7.35	27.85	42
7250	28' span		62	.645			21	7.60	28.60	43.50
7300	30' span		58	.690			22.50	8.15	30.65	46.50
7350	32' span		56	.714			23.50	8.40	31.90	48.50
7400	34' span		54	.741			24.50	8.75	33.25	49.50
7450	36' span		52	.769			25	9.05	34.05	52
8000	Soffit, T&G wood	1 Clab	520	.015	S.F.		.42		.42	.70
8010	Hardboard, vinyl or aluminum	"	640	.013			.34		.34	.57
8030	Plywood	2 Carp	315	.051			1.81		1.81	2.99
9500	See Section 02 41 19.19 for rubbish handling									

06 05 05.20 Selective Demolition Millwork and Trim		Crew	Daily Output	Labor-Hours	Unit	Material	Labor	Equipment	Total	Total Incl O&P
0010	**SELECTIVE DEMOLITION MILLWORK AND TRIM** R024119-10									
1000	Cabinets, wood, base cabinets, per L.F.	2 Clab	80	.200	L.F.		5.50		5.50	9.10
1020	Wall cabinets, per L.F.	"	80	.200	"		5.50		5.50	9.10
1060	Remove and reset, base cabinets	2 Carp	18	.889	Ea.		31.50		31.50	52.50
1070	Wall cabinets		20	.800			28.50		28.50	47
1072	Oven cabinet, 7' high		11	1.455			52		52	85.50
1074	Cabinet door, up to 2' high	1 Clab	66	.121			3.33		3.33	5.50
1076	2' - 4' high	"	46	.174			4.78		4.78	7.90
1100	Steel, painted, base cabinets	2 Clab	60	.267	L.F.		7.35		7.35	12.10
1120	Wall cabinets		60	.267	"		7.35		7.35	12.10
1200	Casework, large area		320	.050	S.F.		1.38		1.38	2.27
1220	Selective		200	.080	"		2.20		2.20	3.64
1500	Counter top, straight runs		200	.080	L.F.		2.20		2.20	3.64
1510	L, U or C shapes		120	.133	"		3.67		3.67	6.05
2000	Paneling, 4' x 8' sheets		2000	.008	S.F.		.22		.22	.36
2100	Boards, 1" x 4"		700	.023			.63		.63	1.04
2120	1" x 6"		750	.021			.59		.59	.97
2140	1" x 8"		800	.020			.55		.55	.91
3000	Trim, baseboard, to 6" wide		1200	.013	L.F.		.37		.37	.61
3040	Greater than 6" and up to 12" wide		1000	.016			.44		.44	.73
3080	Remove and reset, minimum	2 Carp	400	.040			1.43		1.43	2.36
3090	Maximum	"	300	.053			1.90		1.90	3.14
3100	Ceiling trim	2 Clab	1000	.016			.44		.44	.73
3120	Chair rail		1200	.013			.37		.37	.61
3140	Railings with balusters		240	.067			1.83		1.83	3.03
3160	Wainscoting		700	.023	S.F.		.63		.63	1.04
4000	Curtain rod	1 Clab	80	.100	L.F.		2.75		2.75	4.55

06 05 Common Work Results for Wood, Plastics, and Composites

06 05 23 – Wood, Plastic, and Composite Fastenings

06 05 23.10 Nails		Crew	Daily Output	Labor-Hours	Unit	2019 Bare Costs Material	Labor	Equipment	Total	Total Incl O&P
0010	**NAILS**, material only, based upon 50# box purchase									
0020	Copper nails, plain				Lb.	11.45			11.45	12.60
0400	Stainless steel, plain					8.70			8.70	9.55
0500	Box, 3d to 20d, bright					1.46			1.46	1.61
0520	Galvanized					2.39			2.39	2.63
0600	Common, 3d to 60d, plain					1.16			1.16	1.28
0700	Galvanized					2.40			2.40	2.64
0800	Aluminum					11			11	12.10
1000	Annular or spiral thread, 4d to 60d, plain					3.28			3.28	3.61
1200	Galvanized					3.07			3.07	3.38
1400	Drywall nails, plain					1.82			1.82	2
1600	Galvanized					1.88			1.88	2.07
1800	Finish nails, 4d to 10d, plain					1.26			1.26	1.39
2000	Galvanized					1.88			1.88	2.07
2100	Aluminum					8.10			8.10	8.90
2300	Flooring nails, hardened steel, 2d to 10d, plain					3.63			3.63	3.99
2400	Galvanized					4.03			4.03	4.43
2500	Gypsum lath nails, 1-1/8", 13 ga. flathead, blued					3.27			3.27	3.60
2600	Masonry nails, hardened steel, 3/4" to 3" long, plain					2.39			2.39	2.63
2700	Galvanized					3.90			3.90	4.29
2900	Roofing nails, threaded, galvanized					2.78			2.78	3.06
3100	Aluminum					7.55			7.55	8.30
3300	Compressed lead head, threaded, galvanized					3.04			3.04	3.34
3600	Siding nails, plain shank, galvanized					2.54			2.54	2.79
3800	Aluminum					5.85			5.85	6.40
5000	Add to prices above for cement coating					.14			.14	.15
5200	Zinc or tin plating					.26			.26	.29
5500	Vinyl coated sinkers, 8d to 16d				↓	2.63			2.63	2.89
06 05 23.50 Wood Screws										
0010	**WOOD SCREWS**									
0020	#8, 1" long, steel				C	4.71			4.71	5.20
0100	Brass					12.20			12.20	13.40
0200	#8, 2" long, steel					8.15			8.15	9
0300	Brass					25			25	27
0400	#10, 1" long, steel					3.38			3.38	3.72
0500	Brass					15.65			15.65	17.25
0600	#10, 2" long, steel					5.40			5.40	5.95
0700	Brass					27			27	29.50
0800	#10, 3" long, steel					9.05			9.05	9.95
1000	#12, 2" long, steel					7.60			7.60	8.35
1100	Brass					35			35	38.50
1500	#12, 3" long, steel					11.25			11.25	12.40
2000	#12, 4" long, steel				↓	25			25	27.50
06 05 23.60 Timber Connectors										
0010	**TIMBER CONNECTORS**									
0020	Add up cost of each part for total cost of connection									
0100	Connector plates, steel, with bolts, straight	2 Carp	75	.213	Ea.	28	7.60		35.60	43
0110	Tee, 7 ga.		50	.320		40	11.40		51.40	63
0120	T- Strap, 14 ga., 12" x 8" x 2"		50	.320		40	11.40		51.40	63
0150	Anchor plates, 7 ga., 9" x 7"	↓	75	.213		28	7.60		35.60	43
0200	Bolts, machine, sq. hd. with nut & washer, 1/2" diameter, 4" long	1 Carp	140	.057		.92	2.04		2.96	4.38
0300	7-1/2" long	↓	130	.062	↓	1.71	2.19		3.90	5.50

06 05 23 – Wood, Plastic, and Composite Fastenings

06 05 23.60 Timber Connectors		Crew	Daily Output	Labor-Hours	Unit	2019 Bare Costs Material	Labor	Equipment	Total	Total Incl O&P
0500	3/4″ diameter, 7-1/2″ long	1 Carp	130	.062	Ea.	3.12	2.19		5.31	7.05
0610	Machine bolts, w/nut, washer, 3/4″ diameter, 15″ L, HD's & beam hangers		95	.084		5.80	3		8.80	11.35
0720	Machine bolts, sq. hd. w/nut & wash		150	.053	Lb.	3.79	1.90		5.69	7.30
0800	Drilling bolt holes in timber, 1/2″ diameter		450	.018	Inch		.63		.63	1.05
0900	1″ diameter		350	.023	″		.82		.82	1.35
1100	Framing anchor, angle, 3″ x 3″ x 1-1/2″, 12 ga.		175	.046	Ea.	2.44	1.63		4.07	5.35
1150	Framing anchors, 18 ga., 4-1/2″ x 2-3/4″		175	.046		2.44	1.63		4.07	5.35
1160	Framing anchors, 18 ga., 4-1/2″ x 3″		175	.046		2.44	1.63		4.07	5.35
1170	Clip anchors plates, 18 ga., 12″ x 1-1/8″		175	.046		2.44	1.63		4.07	5.35
1250	Holdowns, 3 ga. base, 10 ga. body		8	1		40.50	35.50		76	104
1260	Holdowns, 7 ga. 11-1/16″ x 3-1/4″		8	1		40.50	35.50		76	104
1270	Holdowns, 7 ga. 14-3/8″ x 3-1/8″		8	1		40.50	35.50		76	104
1275	Holdowns, 12 ga. 8″ x 2-1/2″		8	1		40.50	35.50		76	104
1300	Joist and beam hangers, 18 ga. galv., for 2″ x 4″ joist		175	.046		.77	1.63		2.40	3.54
1400	2″ x 6″ to 2″ x 10″ joist		165	.048		1.46	1.73		3.19	4.47
1600	16 ga. galv., 3″ x 6″ to 3″ x 10″ joist		160	.050		2.91	1.78		4.69	6.15
1700	3″ x 10″ to 3″ x 14″ joist		160	.050		4.31	1.78		6.09	7.70
1800	4″ x 6″ to 4″ x 10″ joist		155	.052		3.36	1.84		5.20	6.75
1900	4″ x 10″ to 4″ x 14″ joist		155	.052		4.81	1.84		6.65	8.35
2000	Two-2″ x 6″ to two-2″ x 10″ joists		150	.053		3.89	1.90		5.79	7.40
2100	Two-2″ x 10″ to two-2″ x 14″ joists		150	.053		4.35	1.90		6.25	7.95
2300	3/16″ thick, 6″ x 8″ joist		145	.055		70	1.97		71.97	80.50
2400	6″ x 10″ joist		140	.057		73	2.04		75.04	83.50
2500	6″ x 12″ joist		135	.059		74	2.11		76.11	85
2700	1/4″ thick, 6″ x 14″ joist		130	.062		77	2.19		79.19	88.50
2900	Plywood clips, extruded aluminum H clip, for 3/4″ panels					.23			.23	.25
3000	Galvanized 18 ga. back-up clip					.18			.18	.20
3200	Post framing, 16 ga. galv. for 4″ x 4″ base, 2 piece	1 Carp	130	.062		15.35	2.19		17.54	20.50
3300	Cap		130	.062		24.50	2.19		26.69	30.50
3500	Rafter anchors, 18 ga. galv., 1-1/2″ wide, 5-1/4″ long		145	.055		.45	1.97		2.42	3.75
3600	10-3/4″ long		145	.055		1.38	1.97		3.35	4.77
3800	Shear plates, 2-5/8″ diameter		120	.067		2.87	2.38		5.25	7.10
3900	4″ diameter		115	.070		2.83	2.48		5.31	7.20
4000	Sill anchors, embedded in concrete or block, 25-1/2″ long		115	.070		13.65	2.48		16.13	19.10
4100	Spike grids, 3″ x 6″		120	.067		1.27	2.38		3.65	5.35
4400	Split rings, 2-1/2″ diameter		120	.067		2.83	2.38		5.21	7.05
4500	4″ diameter		110	.073		3.27	2.59		5.86	7.90
4550	Tie plate, 20 ga., 7″ x 3-1/8″		110	.073		3.27	2.59		5.86	7.90
4560	5″ x 4-1/8″		110	.073		3.27	2.59		5.86	7.90
4575	Twist straps, 18 ga., 12″ x 1-1/4″		110	.073		3.27	2.59		5.86	7.90
4580	16″ x 1-1/4″		110	.073		3.27	2.59		5.86	7.90
4600	Strap ties, 20 ga., 2-1/16″ wide, 12-13/16″ long		180	.044		.89	1.58		2.47	3.60
4700	16 ga., 1-3/8″ wide, 12″ long		180	.044		.89	1.58		2.47	3.60
4800	1-1/4″ wide, 21-5/8″ long		160	.050		2.73	1.78		4.51	5.95
5000	Toothed rings, 2-5/8″ or 4″ diameter		90	.089		2.39	3.17		5.56	7.90
5200	Truss plates, nailed, 20 ga., up to 32′ span		17	.471	Truss	15.15	16.80		31.95	44
5400	Washers, 2″ x 2″ x 1/8″				Ea.	.48			.48	.53
5500	3″ x 3″ x 3/16″				″	1.29			1.29	1.42
6000	Angles and gussets, painted									
6012	7 ga., 3-1/4″ x 3-1/4″ x 2-1/2″ long	1 Carp	1.90	4.211	C	1,100	150		1,250	1,450
6014	3-1/4″ x 3-1/4″ x 5″ long		1.90	4.211		2,125	150		2,275	2,600
6016	3-1/4″ x 3-1/4″ x 7-1/2″ long		1.85	4.324		4,025	154		4,179	4,675
6018	5-3/4″ x 5-3/4″ x 2-1/2″ long		1.85	4.324		2,600	154		2,754	3,100

06 05 23 – Wood, Plastic, and Composite Fastenings

06 05 23.60 Timber Connectors		Crew	Daily Output	Labor-Hours	Unit	Material	2019 Bare Costs Labor	Equipment	Total	Total Incl O&P
6020	5-3/4" x 5-3/4" x 5" long	1 Carp	1.85	4.324	C	4,125	154		4,279	4,800
6022	5-3/4" x 5-3/4" x 7-1/2" long		1.80	4.444		6,125	158		6,283	6,975
6024	3 ga., 4-1/4" x 4-1/4" x 3" long		1.85	4.324		2,750	154		2,904	3,275
6026	4-1/4" x 4-1/4" x 6" long		1.85	4.324		5,950	154		6,104	6,800
6028	4-1/4" x 4-1/4" x 9" long		1.80	4.444		6,700	158		6,858	7,600
6030	7-1/4" x 7-1/4" x 3" long		1.80	4.444		4,775	158		4,933	5,500
6032	7-1/4" x 7-1/4" x 6" long		1.80	4.444		6,450	158		6,608	7,350
6034	7-1/4" x 7-1/4" x 9" long	↓	1.75	4.571	↓	14,500	163		14,663	16,200
6036	Gussets									
6038	7 ga., 8-1/8" x 8-1/8" x 2-3/4" long	1 Carp	1.80	4.444	C	4,850	158		5,008	5,575
6040	3 ga., 9-3/4" x 9-3/4" x 3-1/4" long	"	1.80	4.444	"	8,450	158		8,608	9,550
6101	Beam hangers, polymer painted									
6102	Bolted, 3 ga. (W x H x L)									
6104	3-1/4" x 9" x 12" top flange	1 Carp	1	8	C	19,400	285		19,685	21,900
6106	5-1/4" x 9" x 12" top flange		1	8		20,300	285		20,585	22,900
6108	5-1/4" x 11" x 11-3/4" top flange		1	8		21,000	285		21,285	23,600
6110	6-7/8" x 9" x 12" top flange		1	8		21,000	285		21,285	23,600
6112	6-7/8" x 11" x 13-1/2" top flange		1	8		28,700	285		28,985	32,000
6114	8-7/8" x 11" x 15-1/2" top flange	↓	1	8	↓	23,900	285		24,185	26,700
6116	Nailed, 3 ga. (W x H x L)									
6118	3-1/4" x 10-1/2" x 10" top flange	1 Carp	1.80	4.444	C	21,700	158		21,858	24,200
6120	3-1/4" x 10-1/2" x 12" top flange		1.80	4.444		21,400	158		21,558	23,900
6122	5-1/4" x 9-1/2" x 10" top flange		1.80	4.444		17,900	158		18,058	19,900
6124	5-1/4" x 9-1/2" x 12" top flange		1.80	4.444		31,000	158		31,158	34,300
6128	6-7/8" x 8-1/2" x 12" top flange	↓	1.80	4.444	↓	21,000	158		21,158	23,400
6134	Saddle hangers, glu-lam (W x H x L)									
6136	3-1/4" x 10-1/2" x 5-1/4" x 6" saddle	1 Carp	.50	16	C	14,900	570		15,470	17,300
6138	3-1/4" x 10-1/2" x 6-7/8" x 6" saddle		.50	16		15,600	570		16,170	18,100
6140	3-1/4" x 10-1/2" x 8-7/8" x 6" saddle		.50	16		17,600	570		18,170	20,300
6142	3-1/4" x 19-1/2" x 5-1/4" x 10-1/8" saddle		.40	20		17,300	715		18,015	20,200
6144	3-1/4" x 19-1/2" x 6-7/8" x 10-1/8" saddle		.40	20		15,900	715		16,615	18,700
6146	3-1/4" x 19-1/2" x 8-7/8" x 10-1/8" saddle		.40	20		16,800	715		17,515	19,700
6148	5-1/4" x 9-1/2" x 5-1/4" x 12" saddle		.50	16		20,700	570		21,270	23,600
6150	5-1/4" x 9-1/2" x 6-7/8" x 9" saddle		.50	16		19,600	570		20,170	22,500
6152	5-1/4" x 10-1/2" x spec x 12" saddle		.50	16		24,600	570		25,170	28,000
6154	5-1/4" x 18" x 5-1/4" x 12-1/8" saddle		.40	20		18,100	715		18,815	21,100
6156	5-1/4" x 18" x 6-7/8" x 12-1/8" saddle		.40	20		19,800	715		20,515	22,900
6158	5-1/4" x 18" x spec x 12-1/8" saddle		.40	20		21,500	715		22,215	24,800
6160	6-7/8" x 8-1/2" x 6-7/8" x 12" saddle		.50	16		23,200	570		23,770	26,500
6162	6-7/8" x 8-1/2" x 8-7/8" x 12" saddle		.50	16		24,000	570		24,570	27,300
6164	6-7/8" x 10-1/2" x spec x 12" saddle		.50	16		23,200	570		23,770	26,500
6166	6-7/8" x 18" x 6-7/8" x 13-3/4" saddle		.40	20		23,500	715		24,215	27,100
6168	6-7/8" x 18" x 8-7/8" x 13-3/4" saddle		.40	20		22,100	715		22,815	25,600
6170	6-7/8" x 18" x spec x 13-3/4" saddle		.40	20		26,300	715		27,015	30,100
6172	8-7/8" x 18" x spec x 15-3/4" saddle	↓	.40	20	↓	40,500	715		41,215	45,700
6201	Beam and purlin hangers, galvanized, 12 ga.									
6202	Purlin or joist size, 3" x 8"	1 Carp	1.70	4.706	C	2,175	168		2,343	2,675
6204	3" x 10"		1.70	4.706		2,125	168		2,293	2,625
6206	3" x 12"		1.65	4.848		2,675	173		2,848	3,200
6208	3" x 14"		1.65	4.848		2,825	173		2,998	3,400
6210	3" x 16"		1.65	4.848		3,000	173		3,173	3,575
6212	4" x 8"		1.65	4.848		2,200	173		2,373	2,700
6214	4" x 10"	↓	1.65	4.848	↓	2,150	173		2,323	2,625

06 05 23 – Wood, Plastic, and Composite Fastenings

06 05 23.60 Timber Connectors		Crew	Daily Output	Labor-Hours	Unit	Material	Labor	Equipment	Total	Total Incl O&P
						2019 Bare Costs				
6216	4" x 12"	1 Carp	1.60	5	C	2,525	178		2,703	3,100
6218	4" x 14"		1.60	5		2,950	178		3,128	3,525
6220	4" x 16"		1.60	5		2,850	178		3,028	3,450
6222	6" x 8"		1.60	5		2,325	178		2,503	2,875
6224	6" x 10"		1.55	5.161		2,475	184		2,659	3,000
6226	6" x 12"		1.55	5.161		4,500	184		4,684	5,250
6228	6" x 14"		1.50	5.333		4,750	190		4,940	5,575
6230	6" x 16"	↓	1.50	5.333	↓	5,000	190		5,190	5,825
6250	Beam seats									
6252	Beam size, 5-1/4" wide									
6254	5" x 7" x 1/4"	1 Carp	1.80	4.444	C	7,500	158		7,658	8,500
6256	6" x 7" x 3/8"		1.80	4.444		9,800	158		9,958	11,100
6258	7" x 7" x 3/8"		1.80	4.444		9,050	158		9,208	10,200
6260	8" x 7" x 3/8"	↓	1.80	4.444	↓	10,700	158		10,858	12,000
6262	Beam size, 6-7/8" wide									
6264	5" x 9" x 1/4"	1 Carp	1.80	4.444	C	9,000	158		9,158	10,200
6266	6" x 9" x 3/8"		1.80	4.444		11,700	158		11,858	13,200
6268	7" x 9" x 3/8"		1.80	4.444		13,800	158		13,958	15,400
6270	8" x 9" x 3/8"	↓	1.80	4.444	↓	16,300	158		16,458	18,200
6272	Special beams, over 6-7/8" wide									
6274	5" x 10" x 3/8"	1 Carp	1.80	4.444	C	14,100	158		14,258	15,900
6276	6" x 10" x 3/8"		1.80	4.444		16,500	158		16,658	18,400
6278	7" x 10" x 3/8"		1.80	4.444		14,900	158		15,058	16,700
6280	8" x 10" x 3/8"		1.75	4.571		16,100	163		16,263	18,100
6282	5-1/4" x 12" x 5/16"		1.75	4.571		12,500	163		12,663	14,000
6284	6-1/2" x 12" x 3/8"		1.75	4.571		20,700	163		20,863	23,100
6286	5-1/4" x 16" x 5/16"		1.70	4.706		18,300	168		18,468	20,400
6288	6-1/2" x 16" x 3/8"		1.70	4.706		27,800	168		27,968	30,900
6290	5-1/4" x 20" x 5/16"		1.70	4.706		24,700	168		24,868	27,500
6292	6-1/2" x 20" x 3/8"	↓	1.65	4.848	↓	32,700	173		32,873	36,200
6300	Column bases									
6302	4" x 4", 16 ga.	1 Carp	1.80	4.444	C	840	158		998	1,175
6306	7 ga.		1.80	4.444		3,350	158		3,508	3,950
6308	4" x 6", 16 ga.		1.80	4.444		1,675	158		1,833	2,100
6312	7 ga.		1.80	4.444		2,900	158		3,058	3,425
6314	6" x 6", 16 ga.		1.75	4.571		1,925	163		2,088	2,400
6318	7 ga.		1.75	4.571		3,675	163		3,838	4,300
6320	6" x 8", 7 ga.		1.70	4.706		3,425	168		3,593	4,025
6322	6" x 10", 7 ga.		1.70	4.706		3,900	168		4,068	4,550
6324	6" x 12", 7 ga.		1.70	4.706		4,225	168		4,393	4,925
6326	8" x 8', 7 ga.		1.65	4.848		6,700	173		6,873	7,650
6330	8" x 10", 7 ga		1.65	4.848		7,350	173		7,523	8,375
6332	8" x 12", 7 ga.		1.60	5		8,000	178		8,178	9,100
6334	10" x 10", 3 ga.		1.60	5		8,150	178		8,328	9,275
6336	10" x 12", 3 ga.		1.60	5		9,375	178		9,553	10,600
6338	12" x 12", 3 ga.	↓	1.55	5.161	↓	10,200	184		10,384	11,500
6350	Column caps, painted, 3 ga.									
6352	3-1/4" x 3-5/8"	1 Carp	1.80	4.444	C	11,300	158		11,458	12,700
6354	3-1/4" x 5-1/2"		1.80	4.444		11,300	158		11,458	12,700
6356	3-5/8" x 3-5/8"		1.80	4.444		7,800	158		7,958	8,825
6358	3-5/8" x 5-1/2"		1.80	4.444		7,800	158		7,958	8,825
6360	5-1/4" x 5-1/2"		1.75	4.571		10,300	163		10,463	11,700
6362	5-1/4" x 7-1/2"	↓	1.75	4.571	↓	11,100	163		11,263	12,500

06 05 23.60 Timber Connectors		Crew	Daily Output	Labor-Hours	Unit	2019 Bare Costs Material	Labor	Equipment	Total	Total Incl O&P
6364	5-1/2" x 3-5/8"	1 Carp	1.75	4.571	C	10,300	163		10,463	11,700
6366	5-1/2" x 5-1/2"		1.75	4.571		10,300	163		10,463	11,700
6368	5-1/2" x 7-1/2"		1.70	4.706		11,900	168		12,068	13,400
6370	6-7/8" x 5-1/2"		1.70	4.706		12,500	168		12,668	14,000
6372	6-7/8" x 6-7/8"		1.70	4.706		12,500	168		12,668	14,000
6374	6-7/8" x 7-1/2"		1.70	4.706		12,500	168		12,668	14,000
6376	7-1/2" x 5-1/2"		1.65	4.848		13,000	173		13,173	14,600
6378	7-1/2" x 7-1/2"		1.65	4.848		13,000	173		13,173	14,600
6380	8-7/8" x 5-1/2"		1.60	5		13,800	178		13,978	15,400
6382	8-7/8" x 7-1/2"		1.60	5		13,800	178		13,978	15,400
6384	9-1/2" x 5-1/2"	↓	1.60	5	↓	18,600	178		18,778	20,700
6400	Floor tie anchors, polymer paint									
6402	10 ga., 3" x 37-1/2"	1 Carp	1.80	4.444	C	4,950	158		5,108	5,700
6404	3-1/2" x 45-1/2"		1.75	4.571		5,175	163		5,338	5,975
6406	3 ga., 3-1/2" x 56"	↓	1.70	4.706	↓	9,100	168		9,268	10,300
6410	Girder hangers									
6412	6" wall thickness, 4" x 6"	1 Carp	1.80	4.444	C	2,775	158		2,933	3,300
6414	4" x 8"		1.80	4.444		3,100	158		3,258	3,675
6416	8" wall thickness, 4" x 6"		1.80	4.444		3,100	158		3,258	3,675
6418	4" x 8"	↓	1.80	4.444	↓	3,550	158		3,708	4,150
6420	Hinge connections, polymer painted									
6422	3/4" thick top plate									
6424	5-1/4" x 12" w/5" x 5" top	1 Carp	1	8	C	38,700	285		38,985	43,100
6426	5-1/4" x 15" w/6" x 6" top		.80	10		41,100	355		41,455	45,800
6428	5-1/4" x 18" w/7" x 7" top		.70	11.429		43,200	405		43,605	48,300
6430	5-1/4" x 26" w/9" x 9" top	↓	.60	13.333	↓	46,000	475		46,475	51,500
6432	1" thick top plate									
6434	6-7/8" x 14" w/5" x 5" top	1 Carp	.80	10	C	47,100	355		47,455	52,500
6436	6-7/8" x 17" w/6" x 6" top		.80	10		52,500	355		52,855	58,000
6438	6-7/8" x 21" w/7" x 7" top		.70	11.429		57,000	405		57,405	63,500
6440	6-7/8" x 31" w/9" x 9" top	↓	.60	13.333	↓	62,500	475		62,975	69,500
6442	1-1/4" thick top plate									
6444	8-7/8" x 16" w/5" x 5" top	1 Carp	.60	13.333	C	59,000	475		59,475	65,500
6446	8-7/8" x 21" w/6" x 6" top		.50	16		64,500	570		65,070	72,000
6448	8-7/8" x 26" w/7" x 7" top		.40	20		73,500	715		74,215	81,500
6450	8-7/8" x 39" w/9" x 9" top	↓	.30	26.667	↓	91,500	950		92,450	102,000
6460	Holddowns									
6462	Embedded along edge									
6464	26" long, 12 ga.	1 Carp	.90	8.889	C	1,200	315		1,515	1,850
6466	35" long, 12 ga.		.85	9.412		1,650	335		1,985	2,375
6468	35" long, 10 ga.	↓	.85	9.412	↓	1,800	335		2,135	2,550
6470	Embedded away from edge									
6472	Medium duty, 12 ga.									
6474	18-1/2" long	1 Carp	.95	8.421	C	835	300		1,135	1,400
6476	23-3/4" long		.90	8.889		1,050	315		1,365	1,675
6478	28" long		.85	9.412		1,050	335		1,385	1,700
6480	35" long	↓	.85	9.412	↓	1,450	335		1,785	2,150
6482	Heavy duty, 10 ga.									
6484	28" long	1 Carp	.85	9.412	C	2,025	335		2,360	2,775
6486	35" long	"	.85	9.412	"	2,200	335		2,535	2,975
6490	Surface mounted (W x H)									
6492	2-1/2" x 5-3/4", 7 ga.	1 Carp	1	8	C	2,375	285		2,660	3,100
6494	2-1/2" x 8", 12 ga.	↓	1	8	↓	1,450	285		1,735	2,075

06 05 23 – Wood, Plastic, and Composite Fastenings

06 05 23.60 Timber Connectors		Crew	Daily Output	Labor-Hours	Unit	Material	Labor	Equipment	Total	Total Incl O&P
						2019 Bare Costs				
6496	2-7/8" x 6-3/8", 7 ga.	1 Carp	1	8	C	4,700	285		4,985	5,625
6498	2-7/8" x 12-1/2", 3 ga.		1	8		4,075	285		4,360	4,950
6500	3-3/16" x 9-3/8", 10 ga.		1	8		4,525	285		4,810	5,450
6502	3-1/2" x 11-5/8", 3 ga.		1	8		5,750	285		6,035	6,800
6504	3-1/2" x 14-3/4", 3 ga.		1	8		12,600	285		12,885	14,400
6506	3-1/2" x 16-1/2", 3 ga.		1	8		16,600	285		16,885	18,800
6508	3-1/2" x 20-1/2", 3 ga.		.90	8.889		17,400	315		17,715	19,600
6510	3-1/2" x 24-1/2", 3 ga.		.90	8.889		26,100	315		26,415	29,200
6512	4-1/4" x 20-3/4", 3 ga.		.90	8.889		24,300	315		24,615	27,200
6520	Joist hangers									
6522	Sloped, field adjustable, 18 ga.									
6524	2" x 6"	1 Carp	1.65	4.848	C	540	173		713	875
6526	2" x 8"		1.65	4.848		1,125	173		1,298	1,525
6528	2" x 10" and up		1.65	4.848		1,150	173		1,323	1,525
6530	3" x 10" and up		1.60	5		1,200	178		1,378	1,625
6532	4" x 10" and up		1.55	5.161		1,350	184		1,534	1,800
6536	Skewed 45°, 16 ga.									
6538	2" x 4"	1 Carp	1.75	4.571	C	970	163		1,133	1,350
6540	2" x 6" or 2" x 8"		1.65	4.848		930	173		1,103	1,300
6542	2" x 10" or 2" x 12"		1.65	4.848		1,125	173		1,298	1,500
6544	2" x 14" or 2" x 16"		1.60	5		1,950	178		2,128	2,450
6546	(2) 2" x 6" or (2) 2" x 8"		1.60	5		1,775	178		1,953	2,250
6548	(2) 2" x 10" or (2) 2" x 12"		1.55	5.161		1,900	184		2,084	2,400
6550	(2) 2" x 14" or (2) 2" x 16"		1.50	5.333		3,000	190		3,190	3,625
6552	4" x 6" or 4" x 8"		1.60	5		1,650	178		1,828	2,125
6554	4" x 10" or 4" x 12"		1.55	5.161		1,775	184		1,959	2,250
6556	4" x 14" or 4" x 16"		1.55	5.161		2,700	184		2,884	3,275
6560	Skewed 45°, 14 ga.									
6562	(2) 2" x 6" or (2) 2" x 8"	1 Carp	1.60	5	C	2,025	178		2,203	2,550
6564	(2) 2" x 10" or (2) 2" x 12"		1.55	5.161		2,825	184		3,009	3,400
6566	(2) 2" x 14" or (2) 2" x 16"		1.50	5.333		4,150	190		4,340	4,875
6568	4" x 6" or 4" x 8"		1.60	5		2,350	178		2,528	2,900
6570	4" x 10" or 4" x 12"		1.55	5.161		2,550	184		2,734	3,100
6572	4" x 14" or 4" x 16"		1.55	5.161		3,425	184		3,609	4,075
6590	Joist hangers, heavy duty 12 ga., galvanized									
6592	2" x 4"	1 Carp	1.75	4.571	C	1,375	163		1,538	1,800
6594	2" x 6"		1.65	4.848		1,500	173		1,673	1,925
6595	2" x 6", 16 ga.		1.65	4.848		1,425	173		1,598	1,850
6596	2" x 8"		1.65	4.848		2,300	173		2,473	2,800
6597	2" x 8", 16 ga.		1.65	4.848		2,175	173		2,348	2,675
6598	2" x 10"		1.65	4.848		2,350	173		2,523	2,850
6600	2" x 12"		1.65	4.848		2,875	173		3,048	3,450
6602	2" x 14"		1.65	4.848		3,175	173		3,348	3,775
6604	2" x 16"		1.65	4.848		3,375	173		3,548	3,975
6606	3" x 4"		1.65	4.848		2,025	173		2,198	2,500
6608	3" x 6"		1.65	4.848		2,725	173		2,898	3,250
6610	3" x 8"		1.65	4.848		2,750	173		2,923	3,300
6612	3" x 10"		1.60	5		3,200	178		3,378	3,825
6614	3" x 12"		1.60	5		3,825	178		4,003	4,500
6616	3" x 14"		1.60	5		4,475	178		4,653	5,225
6618	3" x 16"		1.60	5		4,950	178		5,128	5,725
6620	(2) 2" x 4"		1.75	4.571		2,250	163		2,413	2,750
6622	(2) 2" x 6"		1.60	5		2,700	178		2,878	3,275

	06 05 23.60 Timber Connectors	Crew	Daily Output	Labor-Hours	Unit	2019 Bare Costs Material	Labor	Equipment	Total	Total Incl O&P
6624	(2) 2" x 8"	1 Carp	1.60	5	C	2,750	178		2,928	3,325
6626	(2) 2" x 10"		1.55	5.161		3,000	184		3,184	3,600
6628	(2) 2" x 12"		1.55	5.161		3,825	184		4,009	4,500
6630	(2) 2" x 14"		1.50	5.333		4,075	190		4,265	4,825
6632	(2) 2" x 16"		1.50	5.333		4,125	190		4,315	4,850
6634	4" x 4"		1.65	4.848		1,725	173		1,898	2,175
6636	4" x 6"		1.60	5		1,900	178		2,078	2,375
6638	4" x 8"		1.60	5		2,200	178		2,378	2,700
6640	4" x 10"		1.55	5.161		2,700	184		2,884	3,250
6642	4" x 12"		1.55	5.161		2,850	184		3,034	3,450
6644	4" x 14"		1.55	5.161		3,425	184		3,609	4,050
6646	4" x 16"		1.55	5.161		3,750	184		3,934	4,425
6648	(3) 2" x 10"		1.50	5.333		3,925	190		4,115	4,625
6650	(3) 2" x 12"		1.50	5.333		4,375	190		4,565	5,125
6652	(3) 2" x 14"		1.45	5.517		4,950	197		5,147	5,775
6654	(3) 2" x 16"		1.45	5.517		5,050	197		5,247	5,875
6656	6" x 6"		1.60	5		2,025	178		2,203	2,525
6658	6" x 8"		1.60	5		2,300	178		2,478	2,825
6660	6" x 10"		1.55	5.161		2,750	184		2,934	3,325
6662	6" x 12"		1.55	5.161		3,125	184		3,309	3,725
6664	6" x 14"		1.50	5.333		3,925	190		4,115	4,625
6666	6" x 16"		1.50	5.333		4,625	190		4,815	5,400
6690	Knee braces, galvanized, 12 ga.									
6692	Beam depth, 10" x 15" x 5' long	1 Carp	1.80	4.444	C	6,000	158		6,158	6,850
6694	15" x 22-1/2" x 7' long		1.70	4.706		6,875	168		7,043	7,850
6696	22-1/2" x 28-1/2" x 8' long		1.60	5		7,400	178		7,578	8,425
6698	28-1/2" x 36" x 10' long		1.55	5.161		7,700	184		7,884	8,775
6700	36" x 42" x 12' long		1.50	5.333		8,500	190		8,690	9,675
6710	Mudsill anchors									
6714	2" x 4" or 3" x 4"	1 Carp	115	.070	C	179	2.48		181.48	201
6716	2" x 6" or 3" x 6"		115	.070		179	2.48		181.48	201
6718	Block wall, 13-1/4" long		115	.070		79	2.48		81.48	91
6720	21-1/4" long		115	.070		128	2.48		130.48	145
6730	Post bases, 12 ga. galvanized									
6732	Adjustable, 3-9/16" x 3-9/16"	1 Carp	1.30	6.154	C	1,150	219		1,369	1,625
6734	3-9/16" x 5-1/2"		1.30	6.154		1,975	219		2,194	2,525
6736	4" x 4"		1.30	6.154		925	219		1,144	1,375
6738	4" x 6"		1.30	6.154		1,350	219		1,569	1,825
6740	5-1/2" x 5-1/2"		1.30	6.154		1,750	219		1,969	2,275
6742	6" x 6"		1.30	6.154		3,450	219		3,669	4,150
6744	Elevated, 3-9/16" x 3-1/4"		1.30	6.154		1,300	219		1,519	1,775
6746	5-1/2" x 3-5/16"		1.30	6.154		1,725	219		1,944	2,250
6748	5-1/2" x 5"		1.30	6.154		2,650	219		2,869	3,275
6750	Regular, 3-9/16" x 3-3/8"		1.30	6.154		955	219		1,174	1,400
6752	4" x 3-3/8"		1.30	6.154		1,425	219		1,644	1,925
6754	18 ga., 5-1/4" x 3-1/8"		1.30	6.154		1,400	219		1,619	1,900
6755	5-1/2" x 3-3/8"		1.30	6.154		1,400	219		1,619	1,900
6756	5-1/2" x 5-3/8"		1.30	6.154		2,025	219		2,244	2,575
6758	6" x 3-3/8"		1.30	6.154		2,275	219		2,494	2,850
6760	6" x 5-3/8"		1.30	6.154		2,875	219		3,094	3,525
6762	Post combination cap/bases									
6764	3-9/16" x 3-9/16"	1 Carp	1.20	6.667	C	440	238		678	880
6766	3-9/16" x 5-1/2"		1.20	6.667		1,100	238		1,338	1,600

06 05 23.60 Timber Connectors		Crew	Daily Output	Labor-Hours	Unit	Material	Labor	Equipment	Total	Total Incl O&P
						2019 Bare Costs				
6768	4" x 4"	1 Carp	1.20	6.667	C	2,200	238		2,438	2,825
6770	5-1/2" x 5-1/2"		1.20	6.667		1,125	238		1,363	1,650
6772	6" x 6"		1.20	6.667		4,775	238		5,013	5,650
6774	7-1/2" x 7-1/2"		1.20	6.667		4,975	238		5,213	5,875
6776	8" x 8"		1.20	6.667		5,225	238		5,463	6,150
6790	Post-beam connection caps									
6792	Beam size 3-9/16"									
6794	12 ga. post, 4" x 4"	1 Carp	1	8	C	2,925	285		3,210	3,700
6796	4" x 6"		1	8		4,425	285		4,710	5,325
6798	4" x 8"		1	8		6,525	285		6,810	7,650
6800	16 ga. post, 4" x 4"		1	8		1,150	285		1,435	1,725
6802	4" x 6"		1	8		1,975	285		2,260	2,650
6804	4" x 8"		1	8		3,550	285		3,835	4,375
6805	18 ga. post, 2-7/8" x 3"		1	8		3,550	285		3,835	4,375
6806	Beam size 5-1/2"									
6808	12 ga. post, 6" x 4"	1 Carp	1	8	C	3,650	285		3,935	4,475
6810	6" x 6"		1	8		4,950	285		5,235	5,925
6812	6" x 8"		1	8		5,025	285		5,310	6,000
6816	16 ga. post, 6" x 4"		1	8		1,975	285		2,260	2,650
6818	6" x 6"		1	8		2,025	285		2,310	2,700
6820	Beam size 7-1/2"									
6822	12 ga. post, 8" x 4"	1 Carp	1	8	C	5,400	285		5,685	6,400
6824	8" x 6"		1	8		5,300	285		5,585	6,300
6826	8" x 8"		1	8		7,950	285		8,235	9,225
6840	Purlin anchors, embedded									
6842	Heavy duty, 10 ga.									
6844	Straight, 28" long	1 Carp	1.60	5	C	1,900	178		2,078	2,400
6846	35" long		1.50	5.333		2,000	190		2,190	2,525
6848	Twisted, 28" long		1.60	5		1,900	178		2,078	2,400
6850	35" long		1.50	5.333		2,000	190		2,190	2,525
6852	Regular duty, 12 ga.									
6854	Straight, 18-1/2" long	1 Carp	1.80	4.444	C	965	158		1,123	1,325
6856	23-3/4" long		1.70	4.706		1,225	168		1,393	1,600
6858	29" long		1.60	5		1,250	178		1,428	1,675
6860	35" long		1.50	5.333		1,700	190		1,890	2,200
6862	Twisted, 18" long		1.80	4.444		965	158		1,123	1,325
6866	28" long		1.60	5		1,150	178		1,328	1,575
6868	35" long		1.50	5.333		1,700	190		1,890	2,200
6870	Straight, plastic coated									
6872	23-1/2" long	1 Carp	1.60	5	C	2,375	178		2,553	2,925
6874	26-7/8" long		1.60	5		2,800	178		2,978	3,375
6876	32-1/2" long		1.50	5.333		2,975	190		3,165	3,600
6878	35-7/8" long		1.50	5.333		3,100	190		3,290	3,725
6890	Purlin hangers, painted									
6892	12 ga., 2" x 6"	1 Carp	1.80	4.444	C	2,000	158		2,158	2,450
6894	2" x 8"		1.80	4.444		2,175	158		2,333	2,650
6896	2" x 10"		1.80	4.444		2,350	158		2,508	2,825
6898	2" x 12"		1.75	4.571		2,525	163		2,688	3,050
6900	2" x 14"		1.75	4.571		2,700	163		2,863	3,225
6902	2" x 16"		1.75	4.571		2,875	163		3,038	3,425
6904	3" x 6"		1.70	4.706		2,025	168		2,193	2,500
6906	3" x 8"		1.70	4.706		2,175	168		2,343	2,675
6908	3" x 10"		1.70	4.706		2,350	168		2,518	2,875

06 05 23 – Wood, Plastic, and Composite Fastenings

06 05 23.60 Timber Connectors		Crew	Daily Output	Labor-Hours	Unit	2019 Bare Costs Material	Labor	Equipment	Total	Total Incl O&P
6910	3" x 12"	1 Carp	1.65	4.848	C	2,700	173		2,873	3,250
6912	3" x 14"		1.65	4.848		2,875	173		3,048	3,425
6914	3" x 16"		1.65	4.848		3,050	173		3,223	3,625
6916	4" x 6"		1.65	4.848		2,025	173		2,198	2,500
6918	4" x 8"		1.65	4.848		2,200	173		2,373	2,700
6920	4" x 10"		1.65	4.848		2,375	173		2,548	2,875
6922	4" x 12"		1.60	5		2,800	178		2,978	3,375
6924	4" x 14"		1.60	5		2,975	178		3,153	3,575
6926	4" x 16"		1.60	5		3,150	178		3,328	3,750
6928	6" x 6"		1.60	5		2,650	178		2,828	3,225
6930	6" x 8"		1.60	5		2,850	178		3,028	3,425
6932	6" x 10"		1.55	5.161		3,050	184		3,234	3,650
6934	double 2" x 6"		1.70	4.706		2,175	168		2,343	2,675
6936	double 2" x 8"		1.70	4.706		2,350	168		2,518	2,875
6938	double 2" x 10"		1.70	4.706		2,525	168		2,693	3,050
6940	double 2" x 12"		1.65	4.848		2,700	173		2,873	3,250
6942	double 2" x 14"		1.65	4.848		2,875	173		3,048	3,425
6944	double 2" x 16"		1.65	4.848		3,050	173		3,223	3,625
6960	11 ga., 4" x 6"		1.65	4.848		4,000	173		4,173	4,675
6962	4" x 8"		1.65	4.848		4,300	173		4,473	5,000
6964	4" x 10"		1.65	4.848		4,600	173		4,773	5,325
6966	6" x 6"		1.60	5		4,050	178		4,228	4,750
6968	6" x 8"		1.60	5		4,350	178		4,528	5,075
6970	6" x 10"		1.55	5.161		4,650	184		4,834	5,400
6972	6" x 12"		1.55	5.161		4,950	184		5,134	5,725
6974	6" x 14"		1.55	5.161		5,225	184		5,409	6,050
6976	6" x 16"		1.50	5.333		5,525	190		5,715	6,400
6978	7 ga., 8" x 6"		1.60	5		4,400	178		4,578	5,125
6980	8" x 8"		1.60	5		4,700	178		4,878	5,450
6982	8" x 10"		1.55	5.161		4,975	184		5,159	5,775
6984	8" x 12"		1.55	5.161		5,275	184		5,459	6,100
6986	8" x 14"		1.50	5.333		5,575	190		5,765	6,450
6988	8" x 16"		1.50	5.333		5,875	190		6,065	6,800
7000	Strap connectors, galvanized									
7002	12 ga., 2-1/16" x 36"	1 Carp	1.55	5.161	C	1,275	184		1,459	1,700
7004	2-1/16" x 47"		1.50	5.333		1,775	190		1,965	2,275
7005	10 ga., 2-1/16" x 72"		1.50	5.333		1,850	190		2,040	2,375
7006	7 ga., 2-1/16" x 34"		1.55	5.161		2,675	184		2,859	3,250
7008	2-1/16" x 45"		1.50	5.333		4,125	190		4,315	4,850
7010	3 ga., 3" x 32"		1.55	5.161		5,325	184		5,509	6,150
7012	3" x 41"		1.55	5.161		5,525	184		5,709	6,375
7014	3" x 50"		1.50	5.333		8,425	190		8,615	9,600
7016	3" x 59"		1.50	5.333		10,300	190		10,490	11,600
7018	3-1/2" x 68"		1.45	5.517		10,400	197		10,597	11,800
7030	Tension ties									
7032	19-1/8" long, 16 ga., 3/4" anchor bolt	1 Carp	1.80	4.444	C	1,475	158		1,633	1,850
7034	20" long, 12 ga., 1/2" anchor bolt		1.80	4.444		1,900	158		2,058	2,325
7036	20" long, 12 ga., 3/4" anchor bolt		1.80	4.444		1,900	158		2,058	2,325
7038	27-3/4" long, 12 ga., 3/4" anchor bolt		1.75	4.571		3,350	163		3,513	3,975
7050	Truss connectors, galvanized									
7052	Adjustable hanger									
7054	18 ga., 2" x 6"	1 Carp	1.65	4.848	C	490	173		663	820
7056	4" x 6"		1.65	4.848		750	173		923	1,100

06 05 Common Work Results for Wood, Plastics, and Composites

06 05 23 – Wood, Plastic, and Composite Fastenings

06 05 23.60 Timber Connectors		Crew	Daily Output	Labor-Hours	Unit	Material	Labor	Equipment	Total	Total Incl O&P
						2019 Bare Costs				
7058	16 ga., 4″ x 10″	1 Carp	1.60	5	C	1,100	178		1,278	1,500
7060	(2) 2″ x 10″	↓	1.60	5	↓	1,100	178		1,278	1,500
7062	Connectors to plate									
7064	16 ga., 2″ x 4″ plate	1 Carp	1.80	4.444	C	480	158		638	790
7066	2″ x 6″ plate	″	1.80	4.444	″	730	158		888	1,050
7068	Hip jack connector									
7070	14 ga.	1 Carp	1.50	5.333	C	3,000	190		3,190	3,625

06 05 23.80 Metal Bracing		Crew	Daily Output	Labor-Hours	Unit	Material	Labor	Equipment	Total	Total Incl O&P
0010	**METAL BRACING**									
0302	Let-in, "T" shaped, 22 ga. galv. steel, studs at 16″ OC	1 Carp	580	.014	L.F.	.82	.49		1.31	1.71
0402	Studs at 24″ OC		600	.013		.82	.48		1.30	1.69
0502	Steel straps, 16 ga. galv. steel, studs at 16″ OC		600	.013		1.06	.48		1.54	1.96
0602	Studs at 24″ OC	↓	620	.013	↓	1.06	.46		1.52	1.93

06 11 Wood Framing

06 11 10 – Framing with Dimensional, Engineered or Composite Lumber

06 11 10.01 Forest Stewardship Council Certification		Crew	Daily Output	Labor-Hours	Unit	Material	Labor	Equipment	Total	Total Incl O&P
0010	**FOREST STEWARDSHIP COUNCIL CERTIFICATION**									
0020	For Forest Stewardship Council (FSC) cert dimension lumber, add G					65%				

06 11 10.02 Blocking		Crew	Daily Output	Labor-Hours	Unit	Material	Labor	Equipment	Total	Total Incl O&P
0010	**BLOCKING**									
1790	Bolted to concrete									
1798	Ledger board, 2″ x 4″	1 Carp	180	.044	L.F.	4.82	1.58		6.40	7.90
1800	2″ x 6″		160	.050		5.10	1.78		6.88	8.55
1810	4″ x 6″		140	.057		8.90	2.04		10.94	13.15
1820	4″ x 8″	↓	120	.067	↓	10.05	2.38		12.43	15
1950	Miscellaneous, to wood construction									
2000	2″ x 4″	1 Carp	250	.032	L.F.	.43	1.14		1.57	2.35
2005	Pneumatic nailed		305	.026		.43	.94		1.37	2.01
2050	2″ x 6″		222	.036		.72	1.28		2	2.91
2055	Pneumatic nailed		271	.030		.73	1.05		1.78	2.54
2100	2″ x 8″		200	.040		.97	1.43		2.40	3.43
2105	Pneumatic nailed		244	.033		.98	1.17		2.15	3.01
2150	2″ x 10″		178	.045		1.45	1.60		3.05	4.25
2155	Pneumatic nailed		217	.037		1.47	1.31		2.78	3.78
2200	2″ x 12″		151	.053		1.83	1.89		3.72	5.15
2205	Pneumatic nailed	↓	185	.043	↓	1.85	1.54		3.39	4.58
2300	To steel construction									
2320	2″ x 4″	1 Carp	208	.038	L.F.	.43	1.37		1.80	2.74
2340	2″ x 6″		180	.044		.72	1.58		2.30	3.42
2360	2″ x 8″		158	.051		.97	1.81		2.78	4.05
2380	2″ x 10″		136	.059		1.45	2.10		3.55	5.05
2400	2″ x 12″	↓	109	.073	↓	1.83	2.62		4.45	6.35

06 11 10.04 Wood Bracing		Crew	Daily Output	Labor-Hours	Unit	Material	Labor	Equipment	Total	Total Incl O&P
0010	**WOOD BRACING**									
0012	Let-in, with 1″ x 6″ boards, studs @ 16″ OC	1 Carp	150	.053	L.F.	.79	1.90		2.69	4.01
0202	Studs @ 24″ OC	″	230	.035	″	.79	1.24		2.03	2.92

06 11 Wood Framing

06 11 10 – Framing with Dimensional, Engineered or Composite Lumber

06 11 10.06 Bridging		Crew	Daily Output	Labor-Hours	Unit	Material	Labor	Equipment	Total	Total Incl O&P
						2019 Bare Costs				
0010	BRIDGING									
0012	Wood, for joists 16″ OC, 1″ x 3″	1 Carp	130	.062	Pr.	.71	2.19		2.90	4.40
0017	Pneumatic nailed		170	.047		.79	1.68		2.47	3.64
0102	2″ x 3″ bridging		130	.062		.72	2.19		2.91	4.41
0107	Pneumatic nailed		170	.047		.75	1.68		2.43	3.60
0302	Steel, galvanized, 18 ga., for 2″ x 10″ joists at 12″ OC		130	.062		1.86	2.19		4.05	5.65
0352	16″ OC		135	.059		1.75	2.11		3.86	5.40
0402	24″ OC		140	.057		2.57	2.04		4.61	6.20
0602	For 2″ x 14″ joists at 16″ OC		130	.062		1.88	2.19		4.07	5.70
0902	Compression type, 16″ OC, 2″ x 8″ joists		200	.040		1.29	1.43		2.72	3.78
1002	2″ x 12″ joists		200	.040		1.28	1.43		2.71	3.77

06 11 10.10 Beam and Girder Framing		Crew	Daily Output	Labor-Hours	Unit	Material	Labor	Equipment	Total	Total Incl O&P
0010	BEAM AND GIRDER FRAMING R061110-30									
1000	Single, 2″ x 6″	2 Carp	700	.023	L.F.	.72	.82		1.54	2.14
1005	Pneumatic nailed		812	.020		.73	.70		1.43	1.96
1020	2″ x 8″		650	.025		.97	.88		1.85	2.52
1025	Pneumatic nailed		754	.021		.98	.76		1.74	2.33
1040	2″ x 10″		600	.027		1.45	.95		2.40	3.17
1045	Pneumatic nailed		696	.023		1.47	.82		2.29	2.96
1060	2″ x 12″		550	.029		1.83	1.04		2.87	3.72
1065	Pneumatic nailed		638	.025		1.85	.89		2.74	3.51
1080	2″ x 14″		500	.032		2.32	1.14		3.46	4.43
1085	Pneumatic nailed		580	.028		2.34	.98		3.32	4.20
1100	3″ x 8″		550	.029		2.95	1.04		3.99	4.96
1120	3″ x 10″		500	.032		3.70	1.14		4.84	5.95
1140	3″ x 12″		450	.036		4.37	1.27		5.64	6.90
1160	3″ x 14″		400	.040		5.05	1.43		6.48	7.90
1170	4″ x 6″	F-3	1100	.036		3.15	1.19	.43	4.77	5.90
1180	4″ x 8″		1000	.040		4.29	1.31	.47	6.07	7.40
1200	4″ x 10″		950	.042		5.05	1.38	.50	6.93	8.40
1220	4″ x 12″		900	.044		5.55	1.46	.52	7.53	9.10
1240	4″ x 14″		850	.047		6.55	1.54	.55	8.64	10.35
1250	6″ x 8″		525	.076		8.05	2.50	.90	11.45	14
1260	6″ x 10″		500	.080		7	2.62	.94	10.56	13.10
1290	8″ x 12″		300	.133		16.85	4.37	1.57	22.79	27.50
2000	Double, 2″ x 6″	2 Carp	625	.026		1.44	.91		2.35	3.10
2005	Pneumatic nailed		725	.022		1.46	.79		2.25	2.91
2020	2″ x 8″		575	.028		1.95	.99		2.94	3.78
2025	Pneumatic nailed		667	.024		1.97	.86		2.83	3.57
2040	2″ x 10″		550	.029		2.90	1.04		3.94	4.90
2045	Pneumatic nailed		638	.025		2.93	.89		3.82	4.71
2060	2″ x 12″		525	.030		3.66	1.09		4.75	5.85
2065	Pneumatic nailed		610	.026		3.69	.94		4.63	5.60
2080	2″ x 14″		475	.034		4.64	1.20		5.84	7.10
2085	Pneumatic nailed		551	.029		4.68	1.04		5.72	6.85
3000	Triple, 2″ x 6″		550	.029		2.16	1.04		3.20	4.09
3005	Pneumatic nailed		638	.025		2.19	.89		3.08	3.89
3020	2″ x 8″		525	.030		2.92	1.09		4.01	5
3025	Pneumatic nailed		609	.026		2.95	.94		3.89	4.80
3040	2″ x 10″		500	.032		4.35	1.14		5.49	6.65
3045	Pneumatic nailed		580	.028		4.40	.98		5.38	6.45
3060	2″ x 12″		475	.034		5.50	1.20		6.70	8.05

06 11 Wood Framing

06 11 10 – Framing with Dimensional, Engineered or Composite Lumber

06 11 10.10 Beam and Girder Framing		Crew	Daily Output	Labor-Hours	Unit	2019 Bare Costs Material	Labor	Equipment	Total	Total Incl O&P
3065	Pneumatic nailed	2 Carp	551	.029	L.F.	5.55	1.04		6.59	7.80
3080	2″ x 14″		450	.036		7.55	1.27		8.82	10.40
3085	Pneumatic nailed		522	.031		7	1.09		8.09	9.50

06 11 10.12 Ceiling Framing

		Crew	Daily Output	Labor-Hours	Unit	Material	Labor	Equipment	Total	Total Incl O&P
0010	**CEILING FRAMING**									
6000	Suspended, 2″ x 3″	2 Carp	1000	.016	L.F.	.42	.57		.99	1.40
6050	2″ x 4″		900	.018		.43	.63		1.06	1.52
6100	2″ x 6″		800	.020		.72	.71		1.43	1.97
6150	2″ x 8″		650	.025		.97	.88		1.85	2.52

06 11 10.14 Posts and Columns

		Crew	Daily Output	Labor-Hours	Unit	Material	Labor	Equipment	Total	Total Incl O&P
0010	**POSTS AND COLUMNS**									
0100	4″ x 4″	2 Carp	390	.041	L.F.	1.86	1.46		3.32	4.47
0150	4″ x 6″		275	.058		3.15	2.07		5.22	6.90
0200	4″ x 8″		220	.073		4.29	2.59		6.88	9
0250	6″ x 6″		215	.074		5.45	2.65		8.10	10.40
0300	6″ x 8″		175	.091		8.05	3.26		11.31	14.25
0350	6″ x 10″		150	.107		7	3.80		10.80	14

06 11 10.18 Joist Framing

		Crew	Daily Output	Labor-Hours	Unit	Material	Labor	Equipment	Total	Total Incl O&P
0010	**JOIST FRAMING**									
2000	Joists, 2″ x 4″	2 Carp	1250	.013	L.F.	.43	.46		.89	1.22
2005	Pneumatic nailed		1438	.011		.43	.40		.83	1.13
2100	2″ x 6″		1250	.013		.72	.46		1.18	1.54
2105	Pneumatic nailed		1438	.011		.73	.40		1.13	1.46
2150	2″ x 8″		1100	.015		.97	.52		1.49	1.93
2155	Pneumatic nailed		1265	.013		.98	.45		1.43	1.83
2200	2″ x 10″		900	.018		1.45	.63		2.08	2.65
2205	Pneumatic nailed		1035	.015		1.47	.55		2.02	2.52
2250	2″ x 12″		875	.018		1.83	.65		2.48	3.09
2255	Pneumatic nailed		1006	.016		1.85	.57		2.42	2.97
2300	2″ x 14″		770	.021		2.32	.74		3.06	3.77
2305	Pneumatic nailed		886	.018		2.34	.64		2.98	3.63
2350	3″ x 6″		925	.017		2.09	.62		2.71	3.32
2400	3″ x 10″		780	.021		3.70	.73		4.43	5.30
2450	3″ x 12″		600	.027		4.37	.95		5.32	6.40
2500	4″ x 6″		800	.020		3.15	.71		3.86	4.64
2550	4″ x 10″		600	.027		5.05	.95		6	7.10
2600	4″ x 12″		450	.036		5.55	1.27		6.82	8.20
2605	Sister joist, 2″ x 6″		800	.020		.72	.71		1.43	1.97
2606	Pneumatic nailed		960	.017		.73	.59		1.32	1.78
3000	Composite wood joist 9-1/2″ deep		.90	17.778	M.L.F.	1,775	635		2,410	3,000
3010	11-1/2″ deep		.88	18.182		2,025	650		2,675	3,300
3020	14″ deep		.82	19.512		2,600	695		3,295	4,000
3030	16″ deep		.78	20.513		4,125	730		4,855	5,750
4000	Open web joist 12″ deep		.88	18.182		3,625	650		4,275	5,075
4002	Per linear foot		880	.018	L.F.	3.64	.65		4.29	5.05
4004	Treated, per linear foot		880	.018	″	4.58	.65		5.23	6.10
4010	14″ deep		.82	19.512	M.L.F.	3,875	695		4,570	5,425
4012	Per linear foot		820	.020	L.F.	3.88	.70		4.58	5.40
4014	Treated, per linear foot		820	.020	″	4.98	.70		5.68	6.65
4020	16″ deep		.78	20.513	M.L.F.	3,850	730		4,580	5,450
4022	Per linear foot		780	.021	L.F.	3.86	.73		4.59	5.45
4024	Treated, per linear foot		780	.021	″	5.10	.73		5.83	6.85

06 11 10 – Framing with Dimensional, Engineered or Composite Lumber

06 11 10.18 Joist Framing		Crew	Daily Output	Labor-Hours	Unit	2019 Bare Costs Material	Labor	Equipment	Total	Total Incl O&P
4030	18″ deep	2 Carp	.74	21.622	M.L.F.	4,200	770		4,970	5,875
4032	Per linear foot		740	.022	L.F.	4.19	.77		4.96	5.90
4034	Treated, per linear foot		740	.022	″	5.60	.77		6.37	7.40
6000	Composite rim joist, 1-1/4″ x 9-1/2″		.90	17.778	M.L.F.	2,050	635		2,685	3,325
6010	1-1/4″ x 11-1/2″		.88	18.182		2,300	650		2,950	3,600
6020	1-1/4″ x 14-1/2″		.82	19.512		3,000	695		3,695	4,450
6030	1-1/4″ x 16-1/2″		.78	20.513		2,875	730		3,605	4,350

06 11 10.24 Miscellaneous Framing		Crew	Daily Output	Labor-Hours	Unit	Material	Labor	Equipment	Total	Total Incl O&P
0010	**MISCELLANEOUS FRAMING**									
2000	Firestops, 2″ x 4″	2 Carp	780	.021	L.F.	.43	.73		1.16	1.68
2005	Pneumatic nailed		952	.017		.43	.60		1.03	1.46
2100	2″ x 6″		600	.027		.72	.95		1.67	2.36
2105	Pneumatic nailed		732	.022		.73	.78		1.51	2.09
5000	Nailers, treated, wood construction, 2″ x 4″		800	.020		.60	.71		1.31	1.84
5005	Pneumatic nailed		960	.017		.61	.59		1.20	1.65
5100	2″ x 6″		750	.021		.76	.76		1.52	2.09
5105	Pneumatic nailed		900	.018		.77	.63		1.40	1.89
5120	2″ x 8″		700	.023		1.12	.82		1.94	2.59
5125	Pneumatic nailed		840	.019		1.14	.68		1.82	2.37
5200	Steel construction, 2″ x 4″		750	.021		.60	.76		1.36	1.92
5220	2″ x 6″		700	.023		.76	.82		1.58	2.19
5240	2″ x 8″		650	.025		1.13	.88		2.01	2.69
7000	Rough bucks, treated, for doors or windows, 2″ x 6″		400	.040		.76	1.43		2.19	3.19
7005	Pneumatic nailed		480	.033		.77	1.19		1.96	2.80
7100	2″ x 8″		380	.042		1.12	1.50		2.62	3.72
7105	Pneumatic nailed		456	.035		1.14	1.25		2.39	3.32
8000	Stair stringers, 2″ x 10″		130	.123		1.45	4.39		5.84	8.85
8100	2″ x 12″		130	.123		1.83	4.39		6.22	9.25
8150	3″ x 10″		125	.128		3.70	4.56		8.26	11.60
8200	3″ x 12″		125	.128		4.37	4.56		8.93	12.35
8870	Laminated structural lumber, 1-1/4″ x 11-1/2″		130	.123		2.30	4.39		6.69	9.80
8880	1-1/4″ x 14-1/2″		130	.123		2.99	4.39		7.38	10.55

06 11 10.26 Partitions		Crew	Daily Output	Labor-Hours	Unit	Material	Labor	Equipment	Total	Total Incl O&P
0010	**PARTITIONS**									
0020	Single bottom and double top plate, no waste, std. & better lumber									
0180	2″ x 4″ studs, 8′ high, studs 12″ OC	2 Carp	80	.200	L.F.	4.69	7.15		11.84	16.95
0185	12″ OC, pneumatic nailed		96	.167		4.75	5.95		10.70	15
0200	16″ OC		100	.160		3.84	5.70		9.54	13.60
0205	16″ OC, pneumatic nailed		120	.133		3.88	4.75		8.63	12.10
0300	24″ OC		125	.128		2.98	4.56		7.54	10.85
0305	24″ OC, pneumatic nailed		150	.107		3.02	3.80		6.82	9.60
0380	10′ high, studs 12″ OC		80	.200		5.55	7.15		12.70	17.90
0385	12″ OC, pneumatic nailed		96	.167		5.60	5.95		11.55	15.95
0400	16″ OC		100	.160		4.48	5.70		10.18	14.30
0405	16″ OC, pneumatic nailed		120	.133		4.53	4.75		9.28	12.85
0500	24″ OC		125	.128		3.41	4.56		7.97	11.30
0505	24″ OC, pneumatic nailed		150	.107		3.45	3.80		7.25	10.10
0580	12′ high, studs 12″ OC		65	.246		6.40	8.80		15.20	21.50
0585	12″ OC, pneumatic nailed		78	.205		6.45	7.30		13.75	19.20
0600	16″ OC		80	.200		5.10	7.15		12.25	17.45
0605	16″ OC, pneumatic nailed		96	.167		5.20	5.95		11.15	15.50
0700	24″ OC		100	.160		3.84	5.70		9.54	13.60

06 11 10 – Framing with Dimensional, Engineered or Composite Lumber

06 11 10.26 Partitions		Crew	Daily Output	Labor-Hours	Unit	2019 Bare Costs Material	Labor	Equipment	Total	Total Incl O&P
0705	24" OC, pneumatic nailed	2 Carp	120	.133	L.F.	3.88	4.75		8.63	12.10
0780	2" x 6" studs, 8' high, studs 12" OC		70	.229		7.95	8.15		16.10	22
0785	12" OC, pneumatic nailed		84	.190		8.05	6.80		14.85	20
0800	16" OC		90	.178		6.50	6.35		12.85	17.60
0805	16" OC, pneumatic nailed		108	.148		6.55	5.30		11.85	15.95
0900	24" OC		115	.139		5.05	4.96		10.01	13.75
0905	24" OC, pneumatic nailed		138	.116		5.10	4.13		9.23	12.45
0980	10' high, studs 12" OC		70	.229		9.40	8.15		17.55	24
0985	12" OC, pneumatic nailed		84	.190		9.50	6.80		16.30	21.50
1000	16" OC		90	.178		7.60	6.35		13.95	18.80
1005	16" OC, pneumatic nailed		108	.148		7.65	5.30		12.95	17.20
1100	24" OC		115	.139		5.75	4.96		10.71	14.55
1105	24" OC, pneumatic nailed		138	.116		5.85	4.13		9.98	13.25
1180	12' high, studs 12" OC		55	.291		10.80	10.35		21.15	29
1185	12" OC, pneumatic nailed		66	.242		10.95	8.65		19.60	26.50
1200	16" OC		70	.229		8.65	8.15		16.80	23
1205	16" OC, pneumatic nailed		84	.190		8.75	6.80		15.55	21
1300	24" OC		90	.178		6.50	6.35		12.85	17.60
1305	24" OC, pneumatic nailed		108	.148		6.55	5.30		11.85	15.95
1400	For horizontal blocking, 2" x 4", add		600	.027		.43	.95		1.38	2.04
1500	2" x 6", add		600	.027		.72	.95		1.67	2.36
1600	For openings, add		250	.064			2.28		2.28	3.77
1702	Headers for above openings, material only, add				B.F.	.81			.81	.89

06 11 10.28 Porch or Deck Framing		Crew	Daily Output	Labor-Hours	Unit	Material	Labor	Equipment	Total	Total Incl O&P
0010	**PORCH OR DECK FRAMING**									
0100	Treated lumber, posts or columns, 4" x 4"	2 Carp	390	.041	L.F.	1.25	1.46		2.71	3.79
0110	4" x 6"		275	.058		1.95	2.07		4.02	5.55
0120	4" x 8"		220	.073		4.06	2.59		6.65	8.75
0130	Girder, single, 4" x 4"		675	.024		1.25	.84		2.09	2.77
0140	4" x 6"		600	.027		1.95	.95		2.90	3.71
0150	4" x 8"		525	.030		4.06	1.09		5.15	6.25
0160	Double, 2" x 4"		625	.026		1.24	.91		2.15	2.87
0170	2" x 6"		600	.027		1.57	.95		2.52	3.30
0180	2" x 8"		575	.028		2.32	.99		3.31	4.19
0190	2" x 10"		550	.029		2.89	1.04		3.93	4.89
0200	2" x 12"		525	.030		4.28	1.09		5.37	6.50
0210	Triple, 2" x 4"		575	.028		1.86	.99		2.85	3.68
0220	2" x 6"		550	.029		2.35	1.04		3.39	4.30
0230	2" x 8"		525	.030		3.48	1.09		4.57	5.65
0240	2" x 10"		500	.032		4.34	1.14		5.48	6.65
0250	2" x 12"		475	.034		6.40	1.20		7.60	9.05
0260	Ledger, bolted 4' OC, 2" x 4"		400	.040		.77	1.43		2.20	3.20
0270	2" x 6"		395	.041		.92	1.44		2.36	3.40
0280	2" x 8"		390	.041		1.29	1.46		2.75	3.83
0290	2" x 10"		385	.042		1.56	1.48		3.04	4.16
0300	2" x 12"		380	.042		2.24	1.50		3.74	4.94
0310	Joists, 2" x 4"		1250	.013		.62	.46		1.08	1.43
0320	2" x 6"		1250	.013		.78	.46		1.24	1.61
0330	2" x 8"		1100	.015		1.16	.52		1.68	2.14
0340	2" x 10"		900	.018		1.45	.63		2.08	2.64
0350	2" x 12"		875	.018		1.73	.65		2.38	2.99
0360	Railings and trim, 1" x 4"	1 Carp	300	.027		.53	.95		1.48	2.15

06 11 10 – Framing with Dimensional, Engineered or Composite Lumber

06 11 10.28 Porch or Deck Framing		Crew	Daily Output	Labor-Hours	Unit	2019 Bare Costs Material	Labor	Equipment	Total	Total Incl O&P
0370	2" x 2"	1 Carp	300	.027	L.F.	.42	.95		1.37	2.03
0380	2" x 4"		300	.027		.61	.95		1.56	2.24
0390	2" x 6"		300	.027		.77	.95		1.72	2.41
0400	Decking, 1" x 4"		275	.029	S.F.	2.92	1.04		3.96	4.92
0410	2" x 4"		300	.027		2.05	.95		3	3.82
0420	2" x 6"		320	.025		1.65	.89		2.54	3.29
0430	5/4" x 6"		320	.025		2.17	.89		3.06	3.85
0440	Balusters, square, 2" x 2"	2 Carp	660	.024	L.F.	.42	.86		1.28	1.89
0450	Turned, 2" x 2"		420	.038		.56	1.36		1.92	2.85
0460	Stair stringer, 2" x 10"		130	.123		1.45	4.39		5.84	8.85
0470	2" x 12"		130	.123		1.73	4.39		6.12	9.15
0480	Stair treads, 1" x 4"		140	.114		2.92	4.07		6.99	9.95
0490	2" x 4"		140	.114		.62	4.07		4.69	7.45
0500	2" x 6"		160	.100		.93	3.57		4.50	6.90
0510	5/4" x 6"		160	.100		1.01	3.57		4.58	7
0520	Turned handrail post, 4" x 4"		64	.250	Ea.	37	8.90		45.90	55.50
0530	Lattice panel, 4' x 8', 1/2"		1600	.010	S.F.	.73	.36		1.09	1.39
0535	3/4"		1600	.010	"	1.08	.36		1.44	1.78
0540	Cedar, posts or columns, 4" x 4"		390	.041	L.F.	3.76	1.46		5.22	6.55
0550	4" x 6"		275	.058		5.85	2.07		7.92	9.90
0560	4" x 8"		220	.073		11.15	2.59		13.74	16.60
0800	Decking, 1" x 4"		550	.029		2.84	1.04		3.88	4.83
0810	2" x 4"		600	.027		5.80	.95		6.75	7.90
0820	2" x 6"		640	.025		10.55	.89		11.44	13.05
0830	5/4" x 6"		640	.025		6.65	.89		7.54	8.80
0840	Railings and trim, 1" x 4"		600	.027		2.84	.95		3.79	4.69
0860	2" x 4"		600	.027		5.80	.95		6.75	7.90
0870	2" x 6"		600	.027		10.55	.95		11.50	13.15
0920	Stair treads, 1" x 4"		140	.114		2.84	4.07		6.91	9.85
0930	2" x 4"		140	.114		5.80	4.07		9.87	13.10
0940	2" x 6"		160	.100		10.55	3.57		14.12	17.50
0950	5/4" x 6"		160	.100		6.65	3.57		10.22	13.25
0980	Redwood, posts or columns, 4" x 4"		390	.041		6.55	1.46		8.01	9.60
0990	4" x 6"		275	.058		12.80	2.07		14.87	17.55
1000	4" x 8"		220	.073		23.50	2.59		26.09	30.50
1240	Decking, 1" x 4"	1 Carp	275	.029	S.F.	4.04	1.04		5.08	6.15
1260	2" x 6"		340	.024		7.60	.84		8.44	9.75
1270	5/4" x 6"		320	.025		4.85	.89		5.74	6.80
1280	Railings and trim, 1" x 4"	2 Carp	600	.027	L.F.	1.19	.95		2.14	2.88
1310	2" x 6"		600	.027		7.60	.95		8.55	9.95
1420	Alternative decking, wood/plastic composite, 5/4" x 6" G		640	.025		3.43	.89		4.32	5.25
1440	1" x 4" square edge fir		550	.029		2.93	1.04		3.97	4.93
1450	1" x 4" tongue and groove fir		450	.036		1.54	1.27		2.81	3.78
1460	1" x 4" mahogany		550	.029		2.09	1.04		3.13	4.01
1462	5/4" x 6" PVC		550	.029		3.35	1.04		4.39	5.40
1465	Framing, porch or deck, alt deck fastening, screws, add	1 Carp	240	.033	S.F.		1.19		1.19	1.96
1470	Accessories, joist hangers, 2" x 4"		160	.050	Ea.	.77	1.78		2.55	3.80
1480	2" x 6" through 2" x 12"		150	.053		1.46	1.90		3.36	4.75
1530	Post footing, incl excav, backfill, tube form & concrete, 4' deep, 8" diam.	F-7	12	2.667		17.95	84		101.95	159
1540	10" diameter		11	2.909		24.50	92		116.50	179
1550	12" diameter		10	3.200		31.50	101		132.50	202

06 11 Wood Framing

06 11 10 – Framing with Dimensional, Engineered or Composite Lumber

06 11 10.30 Roof Framing		Crew	Daily Output	Labor-Hours	Unit	Material	Labor	Equipment	Total	Total Incl O&P
						2019 Bare Costs				
0010	**ROOF FRAMING**									
1900	Rough fascia, 2″ x 6″	2 Carp	250	.064	L.F.	.72	2.28		3	4.56
2000	2″ x 8″		225	.071		.97	2.54		3.51	5.25
2100	2″ x 10″		180	.089		1.45	3.17		4.62	6.85
5002	Rafters, to 4 in 12 pitch, 2″ x 6″, ordinary		1000	.016		.72	.57		1.29	1.73
5021	On steep roofs		800	.020		.72	.71		1.43	1.97
5041	On dormers or complex roofs		590	.027		.72	.97		1.69	2.39
5062	2″ x 8″, ordinary		950	.017		.97	.60		1.57	2.06
5081	On steep roofs		750	.021		.97	.76		1.73	2.33
5101	On dormers or complex roofs		540	.030		.97	1.06		2.03	2.82
5122	2″ x 10″, ordinary		630	.025		1.45	.91		2.36	3.10
5141	On steep roofs		495	.032		1.45	1.15		2.60	3.50
5161	On dormers or complex roofs		425	.038		1.45	1.34		2.79	3.82
5182	2″ x 12″, ordinary		575	.028		1.83	.99		2.82	3.65
5201	On steep roofs		455	.035		1.83	1.25		3.08	4.08
5221	On dormers or complex roofs		395	.041		1.83	1.44		3.27	4.40
5250	Composite rafter, 9-1/2″ deep		575	.028		1.77	.99		2.76	3.58
5260	11-1/2″ deep		575	.028		2.03	.99		3.02	3.87
5301	Hip and valley rafters, 2″ x 6″, ordinary		760	.021		.72	.75		1.47	2.03
5321	On steep roofs		585	.027		.72	.98		1.70	2.40
5341	On dormers or complex roofs		510	.031		.72	1.12		1.84	2.64
5361	2″ x 8″, ordinary		720	.022		.97	.79		1.76	2.38
5381	On steep roofs		545	.029		.97	1.05		2.02	2.80
5401	On dormers or complex roofs		470	.034		.97	1.21		2.18	3.08
5421	2″ x 10″, ordinary		570	.028		1.45	1		2.45	3.25
5441	On steep roofs		440	.036		1.45	1.30		2.75	3.74
5461	On dormers or complex roofs		380	.042		1.45	1.50		2.95	4.08
5481	2″ x 12″, ordinary		525	.030		1.83	1.09		2.92	3.81
5501	On steep roofs		410	.039		1.83	1.39		3.22	4.31
5521	On dormers or complex roofs		355	.045		1.83	1.61		3.44	4.66
5541	Hip and valley jacks, 2″ x 6″, ordinary		600	.027		.72	.95		1.67	2.36
5561	On steep roofs		475	.034		.72	1.20		1.92	2.77
5581	On dormers or complex roofs		410	.039		.72	1.39		2.11	3.09
5601	2″ x 8″, ordinary		490	.033		.97	1.16		2.13	2.99
5621	On steep roofs		385	.042		.97	1.48		2.45	3.52
5641	On dormers or complex roofs		335	.048		.97	1.70		2.67	3.88
5661	2″ x 10″, ordinary		450	.036		1.45	1.27		2.72	3.69
5681	On steep roofs		350	.046		1.45	1.63		3.08	4.29
5701	On dormers or complex roofs		305	.052		1.45	1.87		3.32	4.69
5721	2″ x 12″, ordinary		375	.043		1.83	1.52		3.35	4.52
5741	On steep roofs		295	.054		1.83	1.93		3.76	5.20
5762	On dormers or complex roofs		255	.063		1.83	2.24		4.07	5.70
5781	Rafter tie, 1″ x 4″, #3		800	.020		.53	.71		1.24	1.76
5791	2″ x 4″, #3		800	.020		.43	.71		1.14	1.65
5801	Ridge board, #2 or better, 1″ x 6″		600	.027		.79	.95		1.74	2.44
5821	1″ x 8″		550	.029		1.35	1.04		2.39	3.19
5841	1″ x 10″		500	.032		1.71	1.14		2.85	3.76
5861	2″ x 6″		500	.032		.72	1.14		1.86	2.67
5881	2″ x 8″		450	.036		.97	1.27		2.24	3.16
5901	2″ x 10″		400	.040		1.45	1.43		2.88	3.96
5921	Roof cants, split, 4″ x 4″		650	.025		1.86	.88		2.74	3.50
5941	6″ x 6″		600	.027		5.45	.95		6.40	7.55

06 11 10 – Framing with Dimensional, Engineered or Composite Lumber

06 11 10.30 Roof Framing		Crew	Daily Output	Labor-Hours	Unit	2019 Bare Costs Material	Labor	Equipment	Total	Total Incl O&P
5961	Roof curbs, untreated, 2″ x 6″	2 Carp	520	.031	L.F.	.72	1.10		1.82	2.60
5981	2″ x 12″		400	.040		1.83	1.43		3.26	4.37
6001	Sister rafters, 2″ x 6″		800	.020		.72	.71		1.43	1.97
6021	2″ x 8″		640	.025		.96	.89		1.85	2.53
6041	2″ x 10″		535	.030		1.45	1.07		2.52	3.36
6061	2″ x 12″		455	.035		1.83	1.25		3.08	4.08

06 11 10.32 Sill and Ledger Framing		Crew	Daily Output	Labor-Hours	Unit	Material	Labor	Equipment	Total	Total Incl O&P
0010	**SILL AND LEDGER FRAMING**									
0020	Extruded polystyrene sill sealer, 5-1/2″ wide	1 Carp	1600	.005	L.F.	.14	.18		.32	.44
2002	Ledgers, nailed, 2″ x 4″	2 Carp	755	.021		.43	.76		1.19	1.72
2052	2″ x 6″		600	.027		.72	.95		1.67	2.36
2102	Bolted, not including bolts, 3″ x 6″		325	.049		2.07	1.76		3.83	5.20
2152	3″ x 12″		233	.069		4.34	2.45		6.79	8.80
2602	Mud sills, redwood, construction grade, 2″ x 4″		895	.018		2.30	.64		2.94	3.58
2622	2″ x 6″		780	.021		3.45	.73		4.18	5
4002	Sills, 2″ x 4″		600	.027		.42	.95		1.37	2.03
4052	2″ x 6″		550	.029		.71	1.04		1.75	2.49
4082	2″ x 8″		500	.032		.96	1.14		2.10	2.94
4101	2″ x 10″		450	.036		1.43	1.27		2.70	3.67
4121	2″ x 12″		400	.040		1.81	1.43		3.24	4.35
4202	Treated, 2″ x 4″		550	.029		.60	1.04		1.64	2.36
4222	2″ x 6″		500	.032		.75	1.14		1.89	2.70
4242	2″ x 8″		450	.036		1.11	1.27		2.38	3.31
4261	2″ x 10″		400	.040		1.39	1.43		2.82	3.88
4281	2″ x 12″		350	.046		2.07	1.63		3.70	4.96
4402	4″ x 4″		450	.036		1.20	1.27		2.47	3.41
4422	4″ x 6″		350	.046		1.87	1.63		3.50	4.75
4462	4″ x 8″		300	.053		3.96	1.90		5.86	7.50
4480	4″ x 10″		260	.062		5.75	2.19		7.94	9.90

06 11 10.34 Sleepers		Crew	Daily Output	Labor-Hours	Unit	Material	Labor	Equipment	Total	Total Incl O&P
0010	**SLEEPERS**									
0100	On concrete, treated, 1″ x 2″	2 Carp	2350	.007	L.F.	.31	.24		.55	.74
0150	1″ x 3″		2000	.008		.48	.29		.77	1
0200	2″ x 4″		1500	.011		.65	.38		1.03	1.34
0250	2″ x 6″		1300	.012		.85	.44		1.29	1.66

06 11 10.36 Soffit and Canopy Framing		Crew	Daily Output	Labor-Hours	Unit	Material	Labor	Equipment	Total	Total Incl O&P
0010	**SOFFIT AND CANOPY FRAMING**									
1002	Canopy or soffit framing, 1″ x 4″	2 Carp	900	.018	L.F.	.53	.63		1.16	1.63
1021	1″ x 6″		850	.019		.79	.67		1.46	1.98
1042	1″ x 8″		750	.021		1.35	.76		2.11	2.74
1102	2″ x 4″		620	.026		.43	.92		1.35	1.99
1121	2″ x 6″		560	.029		.72	1.02		1.74	2.47
1142	2″ x 8″		500	.032		.97	1.14		2.11	2.95
1202	3″ x 4″		500	.032		1.22	1.14		2.36	3.23
1221	3″ x 6″		400	.040		2.09	1.43		3.52	4.66
1242	3″ x 10″		300	.053		3.70	1.90		5.60	7.20

06 11 10.38 Treated Lumber Framing Material		Crew	Daily Output	Labor-Hours	Unit	Material	Labor	Equipment	Total	Total Incl O&P
0010	**TREATED LUMBER FRAMING MATERIAL**									
0100	2″ x 4″				M.B.F.	895			895	980
0110	2″ x 6″					750			750	825
0120	2″ x 8″					835			835	920
0130	2″ x 10″					830			830	915

06 11 Wood Framing

06 11 10 – Framing with Dimensional, Engineered or Composite Lumber

06 11 10.38 Treated Lumber Framing Material	Crew	Daily Output	Labor-Hours	Unit	Material	2019 Bare Costs Labor	Equipment	Total	Total Incl O&P
0140 2″ x 12″				M.B.F.	1,025			1,025	1,125
0200 4″ x 4″					900			900	990
0210 4″ x 6″					935			935	1,025
0220 4″ x 8″					1,475			1,475	1,625

06 11 10.40 Wall Framing	Crew	Daily Output	Labor-Hours	Unit	Material	Labor	Equipment	Total	Total Incl O&P
0010 WALL FRAMING R061110-30									
0100 Door buck, studs, header, access, 8′ high, 2″ x 4″ wall, 3′ wide	1 Carp	32	.250	Ea.	18.30	8.90		27.20	35
0110 4′ wide		32	.250		19.75	8.90		28.65	36.50
0120 5′ wide		32	.250		24	8.90		32.90	41
0130 6′ wide		32	.250		25.50	8.90		34.40	43.50
0140 8′ wide		30	.267		37.50	9.50		47	56.50
0150 10′ wide		30	.267		51	9.50		60.50	71.50
0160 12′ wide		30	.267		70	9.50		79.50	92.50
0170 2″ x 6″ wall, 3′ wide		32	.250		28	8.90		36.90	45.50
0180 4′ wide		32	.250		29	8.90		37.90	47
0190 5′ wide		32	.250		33.50	8.90		42.40	51.50
0200 6′ wide		32	.250		35	8.90		43.90	53.50
0210 8′ wide		30	.267		47	9.50		56.50	67
0220 10′ wide		30	.267		60.50	9.50		70	82
0230 12′ wide		30	.267		79.50	9.50		89	103
0240 Window buck, studs, header & access, 8′ high 2″ x 4″ wall, 2′ wide		24	.333		18.95	11.90		30.85	40.50
0250 3′ wide		24	.333		22.50	11.90		34.40	44.50
0260 4′ wide		24	.333		25	11.90		36.90	47
0270 5′ wide		24	.333		29	11.90		40.90	51.50
0280 6′ wide		24	.333		32	11.90		43.90	55
0290 7′ wide		24	.333		42	11.90		53.90	65.50
0300 8′ wide		22	.364		46.50	12.95		59.45	72.50
0310 10′ wide		22	.364		61	12.95		73.95	89
0320 12′ wide		22	.364		83	12.95		95.95	113
0330 2″ x 6″ wall, 2′ wide		24	.333		31	11.90		42.90	53.50
0340 3′ wide		24	.333		34.50	11.90		46.40	57.50
0350 4′ wide		24	.333		37.50	11.90		49.40	61
0360 5′ wide		24	.333		42	11.90		53.90	65.50
0370 6′ wide		24	.333		46	11.90		57.90	70
0380 7′ wide		24	.333		57	11.90		68.90	82
0390 8′ wide		22	.364		62	12.95		74.95	90
0400 10′ wide		22	.364		77.50	12.95		90.45	107
0410 12′ wide		22	.364		101	12.95		113.95	133
2002 Headers over openings, 2″ x 6″	2 Carp	360	.044	L.F.	.72	1.58		2.30	3.41
2007 2″ x 6″, pneumatic nailed		432	.037		.73	1.32		2.05	2.98
2052 2″ x 8″		340	.047		.97	1.68		2.65	3.84
2057 2″ x 8″, pneumatic nailed		408	.039		.98	1.40		2.38	3.39
2101 2″ x 10″		320	.050		1.45	1.78		3.23	4.55
2106 2″ x 10″, pneumatic nailed		384	.042		1.47	1.49		2.96	4.06
2152 2″ x 12″		300	.053		1.83	1.90		3.73	5.15
2157 2″ x 12″, pneumatic nailed		360	.044		1.85	1.58		3.43	4.65
2180 4″ x 8″		260	.062		4.29	2.19		6.48	8.35
2185 4″ x 8″, pneumatic nailed		312	.051		4.31	1.83		6.14	7.75
2191 4″ x 10″		240	.067		5.05	2.38		7.43	9.50
2196 4″ x 10″, pneumatic nailed		288	.056		5.10	1.98		7.08	8.85
2202 4″ x 12″		190	.084		5.55	3		8.55	11.05
2207 4″ x 12″, pneumatic nailed		228	.070		5.60	2.50		8.10	10.30

06 11 10.40 Wall Framing		Crew	Daily Output	Labor-Hours	Unit	Material	Labor	Equipment	Total	Total Incl O&P
						2019 Bare Costs				
2241	6" x 10"	2 Carp	165	.097	L.F.	7	3.46		10.46	13.40
2246	6" x 10", pneumatic nailed		198	.081		7.05	2.88		9.93	12.50
2251	6" x 12"		140	.114		8.85	4.07		12.92	16.50
2256	6" x 12", pneumatic nailed		168	.095		8.90	3.40		12.30	15.40
5002	Plates, untreated, 2" x 3"		850	.019		.42	.67		1.09	1.57
5007	2" x 3", pneumatic nailed		1020	.016		.42	.56		.98	1.39
5022	2" x 4"		800	.020		.43	.71		1.14	1.65
5027	2" x 4", pneumatic nailed		960	.017		.43	.59		1.02	1.45
5041	2" x 6"		750	.021		.72	.76		1.48	2.05
5045	2" x 6", pneumatic nailed		900	.018		.73	.63		1.36	1.85
5061	Treated, 2" x 3"		850	.019		.52	.67		1.19	1.68
5066	2" x 3", treated, pneumatic nailed		1020	.016		.53	.56		1.09	1.50
5081	2" x 4"		800	.020		.60	.71		1.31	1.84
5086	2" x 4", treated, pneumatic nailed		960	.017		.61	.59		1.20	1.65
5101	2" x 6"		750	.021		.76	.76		1.52	2.09
5106	2" x 6", treated, pneumatic nailed		900	.018		.77	.63		1.40	1.89
5122	Studs, 8' high wall, 2" x 3"		1200	.013		.42	.48		.90	1.25
5127	2" x 3", pneumatic nailed		1440	.011		.42	.40		.82	1.12
5142	2" x 4"		1100	.015		.42	.52		.94	1.32
5147	2" x 4", pneumatic nailed		1320	.012		.43	.43		.86	1.18
5162	2" x 6"		1000	.016		.72	.57		1.29	1.73
5167	2" x 6", pneumatic nailed		1200	.013		.73	.48		1.21	1.59
5182	3" x 4"		800	.020		1.22	.71		1.93	2.53
5187	3" x 4", pneumatic nailed		960	.017		1.23	.59		1.82	2.34
5201	Installed on second story, 2" x 3"		1170	.014		.42	.49		.91	1.27
5206	2" x 3", pneumatic nailed		1200	.013		.42	.48		.90	1.26
5221	2" x 4"		1015	.016		.43	.56		.99	1.40
5226	2" x 4", pneumatic nailed		1080	.015		.43	.53		.96	1.34
5241	2" x 6"		890	.018		.72	.64		1.36	1.85
5246	2" x 6", pneumatic nailed		1020	.016		.73	.56		1.29	1.72
5261	3" x 4"		800	.020		1.22	.71		1.93	2.53
5266	3" x 4", pneumatic nailed		960	.017		1.23	.59		1.82	2.34
5281	Installed on dormer or gable, 2" x 3"		1045	.015		.42	.55		.97	1.36
5286	2" x 3", pneumatic nailed		1254	.013		.42	.45		.87	1.22
5301	2" x 4"		905	.018		.43	.63		1.06	1.51
5306	2" x 4", pneumatic nailed		1086	.015		.43	.53		.96	1.34
5321	2" x 6"		800	.020		.72	.71		1.43	1.97
5326	2" x 6", pneumatic nailed		960	.017		.73	.59		1.32	1.78
5341	3" x 4"		700	.023		1.22	.82		2.04	2.70
5346	3" x 4", pneumatic nailed		840	.019		1.23	.68		1.91	2.48
5361	6' high wall, 2" x 3"		970	.016		.42	.59		1.01	1.43
5366	2" x 3", pneumatic nailed		1164	.014		.42	.49		.91	1.28
5381	2" x 4"		850	.019		.43	.67		1.10	1.58
5386	2" x 4", pneumatic nailed		1020	.016		.43	.56		.99	1.39
5401	2" x 6"		740	.022		.72	.77		1.49	2.06
5406	2" x 6", pneumatic nailed		888	.018		.73	.64		1.37	1.86
5421	3" x 4"		600	.027		1.22	.95		2.17	2.92
5426	3" x 4", pneumatic nailed		720	.022		1.23	.79		2.02	2.67
5441	Installed on second story, 2" x 3"		950	.017		.43	.60		1.03	1.46
5446	2" x 3", pneumatic nailed		1140	.014		.42	.50		.92	1.30
5461	2" x 4"		810	.020		.43	.70		1.13	1.63
5466	2" x 4", pneumatic nailed		972	.016		.43	.59		1.02	1.44
5481	2" x 6"		700	.023		.72	.82		1.54	2.14

06 11 10 – Framing with Dimensional, Engineered or Composite Lumber

06 11 10.40 Wall Framing		Crew	Daily Output	Labor-Hours	Unit	Material	Labor	Equipment	Total	Total Incl O&P
						2019 Bare Costs				
5486	2" x 6", pneumatic nailed	2 Carp	840	.019	L.F.	.73	.68		1.41	1.92
5501	3" x 4"		550	.029		1.22	1.04		2.26	3.06
5506	3" x 4", pneumatic nailed		660	.024		1.23	.86		2.09	2.79
5521	Installed on dormer or gable, 2" x 3"		850	.019		.42	.67		1.09	1.57
5526	2" x 3", pneumatic nailed		1020	.016		.42	.56		.98	1.39
5541	2" x 4"		720	.022		.43	.79		1.22	1.78
5546	2" x 4", pneumatic nailed		864	.019		.43	.66		1.09	1.56
5561	2" x 6"		620	.026		.72	.92		1.64	2.31
5566	2" x 6", pneumatic nailed		744	.022		.73	.77		1.50	2.07
5581	3" x 4"		480	.033		1.22	1.19		2.41	3.31
5586	3" x 4", pneumatic nailed		576	.028		1.23	.99		2.22	3
5601	3' high wall, 2" x 3"		740	.022		.42	.77		1.19	1.73
5606	2" x 3", pneumatic nailed		888	.018		.42	.64		1.06	1.53
5621	2" x 4"		640	.025		.43	.89		1.32	1.94
5626	2" x 4", pneumatic nailed		768	.021		.43	.74		1.17	1.70
5641	2" x 6"		550	.029		.72	1.04		1.76	2.50
5646	2" x 6", pneumatic nailed		660	.024		.73	.86		1.59	2.23
5661	3" x 4"		440	.036		1.22	1.30		2.52	3.49
5666	3" x 4", pneumatic nailed		528	.030		1.23	1.08		2.31	3.14
5681	Installed on second story, 2" x 3"		700	.023		.42	.82		1.24	1.81
5686	2" x 3", pneumatic nailed		840	.019		.42	.68		1.10	1.59
5701	2" x 4"		610	.026		.43	.94		1.37	2.01
5706	2" x 4", pneumatic nailed		732	.022		.43	.78		1.21	1.76
5721	2" x 6"		520	.031		.72	1.10		1.82	2.60
5726	2" x 6", pneumatic nailed		624	.026		.73	.91		1.64	2.31
5741	3" x 4"		430	.037		1.22	1.33		2.55	3.54
5746	3" x 4", pneumatic nailed		516	.031		1.23	1.11		2.34	3.19
5761	Installed on dormer or gable, 2" x 3"		625	.026		.42	.91		1.33	1.97
5766	2" x 3", pneumatic nailed		750	.021		.42	.76		1.18	1.73
5781	2" x 4"		545	.029		.43	1.05		1.48	2.20
5786	2" x 4", pneumatic nailed		654	.024		.43	.87		1.30	1.91
5801	2" x 6"		465	.034		.72	1.23		1.95	2.82
5806	2" x 6", pneumatic nailed		558	.029		.73	1.02		1.75	2.49
5821	3" x 4"		380	.042		1.22	1.50		2.72	3.83
5826	3" x 4", pneumatic nailed		456	.035		1.23	1.25		2.48	3.43
8250	For second story & above, add						5%			
8300	For dormer & gable, add						15%			

06 11 10.42 Furring		Crew	Daily Output	Labor-Hours	Unit	Material	Labor	Equipment	Total	Total Incl O&P
0010	**FURRING**									
0012	Wood strips, 1" x 2", on walls, on wood	1 Carp	550	.015	L.F.	.27	.52		.79	1.16
0015	On wood, pneumatic nailed		710	.011		.27	.40		.67	.96
0300	On masonry		495	.016		.29	.58		.87	1.27
0400	On concrete		260	.031		.29	1.10		1.39	2.13
0600	1" x 3", on walls, on wood		550	.015		.44	.52		.96	1.35
0605	On wood, pneumatic nailed		710	.011		.44	.40		.84	1.15
0700	On masonry		495	.016		.48	.58		1.06	1.47
0800	On concrete		260	.031		.48	1.10		1.58	2.33
0850	On ceilings, on wood		350	.023		.44	.82		1.26	1.84
0855	On wood, pneumatic nailed		450	.018		.44	.63		1.07	1.54
0900	On masonry		320	.025		.48	.89		1.37	1.99
0950	On concrete		210	.038		.48	1.36		1.84	2.76

06 11 Wood Framing

06 11 10 – Framing with Dimensional, Engineered or Composite Lumber

06 11 10.44 Grounds		Crew	Daily Output	Labor-Hours	Unit	Material	2019 Bare Costs Labor	Equipment	Total	Total Incl O&P
0010	**GROUNDS**									
0020	For casework, 1" x 2" wood strips, on wood	1 Carp	330	.024	L.F.	.27	.86		1.13	1.73
0100	On masonry		285	.028		.29	1		1.29	1.97
0200	On concrete		250	.032		.29	1.14		1.43	2.20
0400	For plaster, 3/4" deep, on wood		450	.018		.27	.63		.90	1.35
0500	On masonry		225	.036		.29	1.27		1.56	2.41
0600	On concrete		175	.046		.29	1.63		1.92	3.01
0700	On metal lath	↓	200	.040	↓	.29	1.43		1.72	2.68

06 12 Structural Panels

06 12 10 – Structural Insulated Panels

06 12 10.10 OSB Faced Panels

			Crew	Daily Output	Labor-Hours	Unit	Material	Labor	Equipment	Total	Total Incl O&P
0010	**OSB FACED PANELS**										
0100	Structural insul. panels, 7/16" OSB both faces, EPS insul., 3-5/8" T	G	F-3	2075	.019	S.F.	3.74	.63	.23	4.60	5.40
0110	5-5/8" thick	G		1725	.023		4.20	.76	.27	5.23	6.20
0120	7-3/8" thick	G		1425	.028		5.30	.92	.33	6.55	7.75
0130	9-3/8" thick	G		1125	.036		5.65	1.16	.42	7.23	8.60
0140	7/16" OSB one face, EPS insul., 3-5/8" thick	G		2175	.018		3.89	.60	.22	4.71	5.50
0150	5-5/8" thick	G		1825	.022		4.51	.72	.26	5.49	6.45
0160	7-3/8" thick	G		1525	.026		4.77	.86	.31	5.94	7
0170	9-3/8" thick	G		1225	.033		5.55	1.07	.38	7	8.30
0190	7/16" OSB - 1/2" GWB faces, EPS insul., 3-5/8" T	G		2075	.019		3.56	.63	.23	4.42	5.20
0200	5-5/8" thick	G		1725	.023		4.87	.76	.27	5.90	6.90
0210	7-3/8" thick	G		1425	.028		4.85	.92	.33	6.10	7.25
0220	9-3/8" thick	G		1125	.036		5.45	1.16	.42	7.03	8.35
0240	7/16" OSB - 1/2" MRGWB faces, EPS insul., 3-5/8" T	G		2075	.019		4.38	.63	.23	5.24	6.10
0250	5-5/8" thick	G		1725	.023		4.39	.76	.27	5.42	6.40
0260	7-3/8" thick	G		1425	.028		4.86	.92	.33	6.11	7.25
0270	9-3/8" thick	G	↓	1125	.036		5.55	1.16	.42	7.13	8.50
0300	For 1/2" GWB added to OSB skin, add	G					1.44			1.44	1.58
0310	For 1/2" MRGWB added to OSB skin, add	G					1.63			1.63	1.79
0320	For one T1-11 skin, add to OSB-OSB	G					2.06			2.06	2.27
0330	For one 19/32" CDX skin, add to OSB-OSB	G				↓	1.54			1.54	1.69
0500	Structural insulated panel, 7/16" OSB both sides, straw core										
0510	4-3/8" T, walls (w/sill, splines, plates)	G	F-6	2400	.017	S.F.	7.50	.55	.20	8.25	9.35
0520	Floors (w/splines)	G		2400	.017		7.50	.55	.20	8.25	9.35
0530	Roof (w/splines)	G		2400	.017		7.50	.55	.20	8.25	9.35
0550	7-7/8" T, walls (w/sill, splines, plates)	G		2400	.017		11.70	.55	.20	12.45	13.95
0560	Floors (w/splines)	G		2400	.017		11.70	.55	.20	12.45	13.95
0570	Roof (w/splines)	G	↓	2400	.017	↓	11.70	.55	.20	12.45	13.95

06 12 19 – Composite Shearwall Panels

06 12 19.10 Steel and Wood Composite Shearwall Panels

		Crew	Daily Output	Labor-Hours	Unit	Material	Labor	Equipment	Total	Total Incl O&P
0010	**STEEL & WOOD COMPOSITE SHEARWALL PANELS**									
0020	Anchor bolts, 36" long (must be placed in wet concrete)	1 Carp	150	.053	Ea.	40.50	1.90		42.40	47.50
0030	On concrete, 2" x 4" & 2" x 6" walls, 7'-10' high, 360 lb. shear, 12" wide	2 Carp	8	2		475	71.50		546.50	645
0040	715 lb. shear, 15" wide		8	2		570	71.50		641.50	745
0050	1860 lb. shear, 18" wide		8	2		635	71.50		706.50	815
0060	2780 lb. shear, 21" wide		8	2		740	71.50		811.50	935
0070	3790 lb. shear, 24" wide		8	2		795	71.50		866.50	995
0080	2" x 6" walls, 11'-13' high, 1180 lb. shear, 18" wide	↓	6	2.667	↓	725	95		820	955

06 12 Structural Panels

06 12 19 – Composite Shearwall Panels

06 12 19.10 Steel and Wood Composite Shearwall Panels		Crew	Daily Output	Labor-Hours	Unit	Material	Labor	Equipment	Total	Total Incl O&P
						2019 Bare Costs				
0090	1555 lb. shear, 21" wide	2 Carp	6	2.667	Ea.	740	95		835	965
0100	2280 lb. shear, 24" wide	↓	6	2.667	↓	885	95		980	1,125
0110	For installing above on wood floor frame, add									
0120	Coupler nuts, threaded rods, bolts, shear transfer plate kit	1 Carp	16	.500	Ea.	60.50	17.85		78.35	96
0130	Framing anchors, angle (2 required)	"	96	.083	"	2.44	2.97		5.41	7.60
0140	For blocking see Section 06 11 10.02									
0150	For installing above, first floor to second floor, wood floor frame, add									
0160	Add stack option to first floor wall panel				Ea.	66.50			66.50	73
0170	Threaded rods, bolts, shear transfer plate kit	1 Carp	16	.500		76.50	17.85		94.35	114
0180	Framing anchors, angle (2 required)	"	96	.083	↓	2.44	2.97		5.41	7.60
0190	For blocking see section 06 11 10.02									
0200	For installing stacked panels, balloon framing									
0210	Add stack option to first floor wall panel				Ea.	66.50			66.50	73
0220	Threaded rods, bolts kit	1 Carp	16	.500	"	43	17.85		60.85	76.50

06 13 Heavy Timber Construction

06 13 13 – Log Construction

06 13 13.10 Log Structures

		Crew	Daily Output	Labor-Hours	Unit	Material	Labor	Equipment	Total	Total Incl O&P
0010	**LOG STRUCTURES**									
0020	Exterior walls, pine, D logs, with double T&G, 6" x 6"	2 Carp	500	.032	L.F.	4.83	1.14		5.97	7.20
0030	6" x 8"		375	.043		5.50	1.52		7.02	8.55
0040	8" x 6"		375	.043		4.83	1.52		6.35	7.80
0050	8" x 7"		322	.050		4.83	1.77		6.60	8.25
0060	Square/rectangular logs, with double T&G, 6" x 6"		500	.032		4.83	1.14		5.97	7.20
0070	6" x 8"		375	.043		4.83	1.52		6.35	7.80
0080	8" x 6"		375	.043		4.83	1.52		6.35	7.80
0090	8" x 7"		322	.050		4.83	1.77		6.60	8.25
0100	Round logs, with double T&G, 6" x 8"		375	.043		4.83	1.52		6.35	7.80
0110	8" x 7"		322	.050	↓	4.83	1.77		6.60	8.25
0120	Log siding, ship lapped, 2" x 6"		225	.071	S.F.	2.99	2.54		5.53	7.50
0130	2" x 8"		200	.080		2.53	2.85		5.38	7.50
0140	2" x 12"	↓	180	.089	↓	2.27	3.17		5.44	7.75
0150	Foam sealant strip, 3/8" x 3/8"	1 Carp	1920	.004	L.F.	.16	.15		.31	.43
0152	Chinking, 2" - 3" wide joint, 1/4" to 3/8" deep		600	.013		1.36	.48		1.84	2.29
0154	Caulking, 1/4" to 1/2" joint		900	.009		.49	.32		.81	1.06
0156	Backer rod, 1/4"	↓	900	.009	↓	.18	.32		.50	.71
0157	Penetrating wood preservative	1 Pord	2000	.004	S.F.	.16	.12		.28	.37
0158	Insect treatment	"	4000	.002	"	.38	.06		.44	.52
0160	Upper floor framing, pine, posts/columns, 4" x 6"	2 Carp	750	.021	L.F.	3.13	.76		3.89	4.70
0180	4" x 8"		562	.028		4.26	1.02		5.28	6.35
0190	6" x 6"		500	.032		5.40	1.14		6.54	7.85
0200	6" x 8"		375	.043		8	1.52		9.52	11.30
0210	8" x 8"		281	.057		9.70	2.03		11.73	14
0220	8" x 10"		225	.071		12.10	2.54		14.64	17.50
0230	Beams, 4" x 8"		562	.028		4.26	1.02		5.28	6.35
0240	4" x 10"		449	.036		5	1.27		6.27	7.65
0250	4" x 12"		375	.043		5.50	1.52		7.02	8.55
0260	6" x 8"		375	.043		8	1.52		9.52	11.30
0270	6" x 10"		300	.053		7	1.90		8.90	10.80
0280	6" x 12"		250	.064		8.80	2.28		11.08	13.45
0290	8" x 10"	↓	225	.071	↓	12.10	2.54		14.64	17.50

06 13 13 – Log Construction

06 13 13.10 Log Structures		Crew	Daily Output	Labor-Hours	Unit	Material	Labor (2019 Bare Costs)	Equipment	Total	Total Incl O&P
0300	8" x 12"	2 Carp	188	.085	L.F.	16.75	3.03		19.78	23.50
0310	Joists, 4" x 8"		562	.028		4.26	1.02		5.28	6.35
0320	4" x 10"		449	.036		5	1.27		6.27	7.65
0330	4" x 12"		375	.043		5.50	1.52		7.02	8.55
0340	6" x 8"		375	.043		8	1.52		9.52	11.30
0350	6" x 10"		300	.053		7	1.90		8.90	10.80
0360	6" x 12"		250	.064		8.80	2.28		11.08	13.45
0370	8" x 10"		225	.071		12.10	2.54		14.64	17.50
0380	8" x 12"		188	.085		14.50	3.03		17.53	21
0390	Decking, 1" x 6" T&G		964	.017	S.F.	1.62	.59		2.21	2.76
0400	1" x 8" T&G		700	.023		1.48	.82		2.30	2.98
0410	2" x 6" T&G		482	.033		3.52	1.18		4.70	5.85
0420	Gable end roof framing, rafters, 4" x 8"		562	.028	L.F.	4.26	1.02		5.28	6.35
0430	4" x 10"		449	.036		5	1.27		6.27	7.65
0450	4" x 12"		375	.043		5.50	1.52		7.02	8.55
0460	6" x 8"		375	.043		8	1.52		9.52	11.30
0470	6" x 10"		300	.053		7	1.90		8.90	10.80
0480	6" x 12"		250	.064		8.80	2.28		11.08	13.45
0490	8" x 10"		225	.071		12.10	2.54		14.64	17.50
0500	8" x 12"		188	.085		14.50	3.03		17.53	21
0510	Purlins, 4" x 8"		562	.028		4.26	1.02		5.28	6.35
0520	6" x 8"		375	.043		8	1.52		9.52	11.30
0530	Roof decking, 1" x 6" T&G		640	.025	S.F.	1.62	.89		2.51	3.25
0540	1" x 8" T&G		430	.037		1.48	1.33		2.81	3.82
0550	2" x 6" T&G		320	.050		3.52	1.78		5.30	6.80

06 13 23 – Heavy Timber Framing

06 13 23.10 Heavy Framing		Crew	Daily Output	Labor-Hours	Unit	Material	Labor (2019 Bare Costs)	Equipment	Total	Total Incl O&P
0010	**HEAVY FRAMING**									
0020	Beams, single 6" x 10"	2 Carp	1.10	14.545	M.B.F.	1,550	520		2,070	2,575
0100	Single 8" x 16"		1.20	13.333	"	1,900	475		2,375	2,875
0202	Built from 2" lumber, multiple 2" x 14"		900	.018	B.F.	.99	.63		1.62	2.13
0212	Built from 3" lumber, multiple 3" x 6"		700	.023		1.38	.82		2.20	2.87
0222	Multiple 3" x 8"		800	.020		1.47	.71		2.18	2.79
0232	Multiple 3" x 10"		900	.018		1.47	.63		2.10	2.67
0242	Multiple 3" x 12"		1000	.016		1.45	.57		2.02	2.53
0252	Built from 4" lumber, multiple 4" x 6"		800	.020		1.56	.71		2.27	2.90
0262	Multiple 4" x 8"		900	.018		1.60	.63		2.23	2.81
0272	Multiple 4" x 10"		1000	.016		1.51	.57		2.08	2.60
0282	Multiple 4" x 12"		1100	.015		1.38	.52		1.90	2.38
0292	Columns, structural grade, 1500f, 4" x 4"		450	.036	L.F.	2.46	1.27		3.73	4.79
0302	6" x 6"		225	.071		4.10	2.54		6.64	8.70
0402	8" x 8"		240	.067		7.95	2.38		10.33	12.70
0502	10" x 10"		90	.178		14.65	6.35		21	26.50
0602	12" x 12"		70	.229		18.35	8.15		26.50	33.50
0802	Floor planks, 2" thick, T&G, 2" x 6"		1050	.015	B.F.	1.56	.54		2.10	2.62
0902	2" x 10"		1100	.015		1.58	.52		2.10	2.60
1102	3" thick, 3" x 6"		1050	.015		1.61	.54		2.15	2.68
1202	3" x 10"		1100	.015		1.64	.52		2.16	2.67
1402	Girders, structural grade, 12" x 12"		800	.020		1.53	.71		2.24	2.86
1502	10" x 16"		1000	.016		2.52	.57		3.09	3.72
2302	Roof purlins, 4" thick, structural grade		1050	.015		1.60	.54		2.14	2.66

06 15 Wood Decking

06 15 16 – Wood Roof Decking

06 15 16.10 Solid Wood Roof Decking		Crew	Daily Output	Labor-Hours	Unit	Material	2019 Bare Costs Labor	Equipment	Total	Total Incl O&P
0010	SOLID WOOD ROOF DECKING									
0350	Cedar planks, 2" thick	2 Carp	350	.046	S.F.	6.85	1.63		8.48	10.20
0400	3" thick		320	.050		10.25	1.78		12.03	14.20
0500	4" thick		250	.064		13.65	2.28		15.93	18.75
0550	6" thick		200	.080		20.50	2.85		23.35	27
0650	Douglas fir, 2" thick		350	.046		3.07	1.63		4.70	6.05
0700	3" thick		320	.050		4.60	1.78		6.38	8
0800	4" thick		250	.064		6.15	2.28		8.43	10.50
0850	6" thick		200	.080		9.20	2.85		12.05	14.80
0950	Hemlock, 2" thick		350	.046		3.12	1.63		4.75	6.10
1000	3" thick		320	.050		4.68	1.78		6.46	8.10
1100	4" thick		250	.064		6.25	2.28		8.53	10.60
1150	6" thick		200	.080		9.35	2.85		12.20	15
1250	Western white spruce, 2" thick		350	.046		1.99	1.63		3.62	4.88
1300	3" thick		320	.050		2.99	1.78		4.77	6.25
1400	4" thick		250	.064		3.99	2.28		6.27	8.15
1450	6" thick	↓	200	.080	↓	6	2.85		8.85	11.30

06 15 23 – Laminated Wood Decking

06 15 23.10 Laminated Roof Deck		Crew	Daily Output	Labor-Hours	Unit	Material	Labor	Equipment	Total	Total Incl O&P
0010	LAMINATED ROOF DECK									
0020	Pine or hemlock, 3" thick	2 Carp	425	.038	S.F.	5.65	1.34		6.99	8.40
0100	4" thick		325	.049		7.35	1.76		9.11	10.95
0300	Cedar, 3" thick		425	.038		7.70	1.34		9.04	10.70
0400	4" thick		325	.049		10.30	1.76		12.06	14.25
0600	Fir, 3" thick		425	.038		6.60	1.34		7.94	9.45
0700	4" thick	↓	325	.049	↓	8.40	1.76		10.16	12.15

06 16 Sheathing

06 16 13 – Insulating Sheathing

06 16 13.10 Insulating Sheathing		Crew	Daily Output	Labor-Hours	Unit	Material	Labor	Equipment	Total	Total Incl O&P
0010	INSULATING SHEATHING									
0020	Expanded polystyrene, 1#/C.F. density, 3/4" thick, R2.89 G	2 Carp	1400	.011	S.F.	.36	.41		.77	1.07
0030	1" thick, R3.85 G		1300	.012		.43	.44		.87	1.20
0040	2" thick, R7.69 G		1200	.013		.70	.48		1.18	1.56
0050	Extruded polystyrene, 15 psi compressive strength, 1" thick, R5 G		1300	.012		.72	.44		1.16	1.52
0060	2" thick, R10 G		1200	.013		.89	.48		1.37	1.76
0070	Polyisocyanurate, 2#/C.F. density, 3/4" thick G		1400	.011		.66	.41		1.07	1.39
0080	1" thick G		1300	.012		.60	.44		1.04	1.39
0090	1-1/2" thick G		1250	.013		.76	.46		1.22	1.58
0100	2" thick G	↓	1200	.013	↓	.95	.48		1.43	1.83

06 16 23 – Subflooring

06 16 23.10 Subfloor		Crew	Daily Output	Labor-Hours	Unit	Material	Labor	Equipment	Total	Total Incl O&P
0010	SUBFLOOR									
0011	Plywood, CDX, 1/2" thick	2 Carp	1500	.011	SF Flr.	.63	.38		1.01	1.33
0015	Pneumatic nailed		1860	.009		.63	.31		.94	1.21
0102	5/8" thick		1350	.012		.78	.42		1.20	1.56
0107	Pneumatic nailed		1674	.010		.78	.34		1.12	1.42
0202	3/4" thick		1250	.013		.94	.46		1.40	1.78
0207	Pneumatic nailed		1550	.010		.94	.37		1.31	1.64
0302	1-1/8" thick, 2-4-1 including underlayment	↓	1050	.015	↓	2.12	.54		2.66	3.23

06 16 Sheathing

06 16 23 – Subflooring

06 16 23.10 Subfloor		Crew	Daily Output	Labor-Hours	Unit	Material	Labor	Equipment	Total	Total Incl O&P
						2019 Bare Costs				
0440	With boards, 1" x 6", S4S, laid regular	2 Carp	900	.018	SF Flr.	1.71	.63		2.34	2.93
0452	1" x 8", laid regular		1000	.016		2.15	.57		2.72	3.30
0462	Laid diagonal		850	.019		2.15	.67		2.82	3.47
0502	1" x 10", laid regular		1100	.015		2.14	.52		2.66	3.22
0602	Laid diagonal		900	.018		2.14	.63		2.77	3.41
1500	OSB, 5/8" thick		1330	.012	S.F.	.54	.43		.97	1.30
1600	3/4" thick		1230	.013	"	.67	.46		1.13	1.51
8990	Subfloor adhesive, 3/8" bead	1 Carp	2300	.003	L.F.	.11	.12		.23	.33

06 16 26 – Underlayment

06 16 26.10 Wood Product Underlayment		Crew	Daily Output	Labor-Hours	Unit	Material	Labor	Equipment	Total	Total Incl O&P
0010	**WOOD PRODUCT UNDERLAYMENT**									
0015	Plywood, underlayment grade, 1/4" thick	2 Carp	1500	.011	S.F.	.95	.38		1.33	1.67
0018	Pneumatic nailed		1860	.009		.95	.31		1.26	1.55
0030	3/8" thick		1500	.011		1.05	.38		1.43	1.79
0070	Pneumatic nailed		1860	.009		1.05	.31		1.36	1.67
0102	1/2" thick		1450	.011		1.24	.39		1.63	2.01
0107	Pneumatic nailed		1798	.009		1.24	.32		1.56	1.88
0202	5/8" thick		1400	.011		1.36	.41		1.77	2.17
0207	Pneumatic nailed		1736	.009		1.36	.33		1.69	2.04
0302	3/4" thick		1300	.012		1.47	.44		1.91	2.35
0306	Pneumatic nailed		1612	.010		1.47	.35		1.82	2.20
0502	Particle board, 3/8" thick		1500	.011		.40	.38		.78	1.07
0507	Pneumatic nailed		1860	.009		.40	.31		.71	.95
0602	1/2" thick		1450	.011		.42	.39		.81	1.11
0607	Pneumatic nailed		1798	.009		.42	.32		.74	.98
0802	5/8" thick		1400	.011		.54	.41		.95	1.26
0807	Pneumatic nailed		1736	.009		.54	.33		.87	1.13
0902	3/4" thick		1300	.012		.67	.44		1.11	1.47
0907	Pneumatic nailed		1612	.010		.67	.35		1.02	1.32
1100	Hardboard, underlayment grade, 4' x 4', .215" thick G		1500	.011		.69	.38		1.07	1.39

06 16 33 – Wood Board Sheathing

06 16 33.10 Board Sheathing		Crew	Daily Output	Labor-Hours	Unit	Material	Labor	Equipment	Total	Total Incl O&P
0009	**BOARD SHEATHING**									
0010	Roof, 1" x 6" boards, laid horizontal	2 Carp	725	.022	S.F.	1.71	.79		2.50	3.18
0020	On steep roof		520	.031		1.71	1.10		2.81	3.69
0040	On dormers, hips, & valleys		480	.033		1.71	1.19		2.90	3.84
0050	Laid diagonal		650	.025		1.71	.88		2.59	3.33
0070	1" x 8" boards, laid horizontal		875	.018		2.15	.65		2.80	3.44
0080	On steep roof		635	.025		2.15	.90		3.05	3.84
0090	On dormers, hips, & valleys		580	.028		2.20	.98		3.18	4.05
0100	Laid diagonal		725	.022		2.15	.79		2.94	3.66
0110	Skip sheathing, 1" x 4", 7" OC	1 Carp	1200	.007		.66	.24		.90	1.11
0120	1" x 6", 9" OC		1450	.006		.79	.20		.99	1.19
0180	T&G sheathing/decking, 1" x 6"		1000	.008		1.78	.29		2.07	2.43
0190	2" x 6"		1000	.008		3.87	.29		4.16	4.73
0200	Walls, 1" x 6" boards, laid regular	2 Carp	650	.025		1.71	.88		2.59	3.33
0210	Laid diagonal		585	.027		1.71	.98		2.69	3.49
0220	1" x 8" boards, laid regular		765	.021		2.15	.75		2.90	3.59
0230	Laid diagonal		650	.025		2.15	.88		3.03	3.81

06 16 Sheathing

06 16 36 – Wood Panel Product Sheathing

	06 16 36.10 Sheathing	Crew	Daily Output	Labor-Hours	Unit	2019 Bare Costs Material	Labor	Equipment	Total	Total Incl O&P
0010	**SHEATHING** R061636-20									
0012	Plywood on roofs, CDX									
0032	5/16" thick	2 Carp	1600	.010	S.F.	.59	.36		.95	1.24
0037	Pneumatic nailed		1952	.008		.59	.29		.88	1.13
0052	3/8" thick		1525	.010		.61	.37		.98	1.30
0057	Pneumatic nailed		1860	.009		.61	.31		.92	1.19
0102	1/2" thick		1400	.011		.63	.41		1.04	1.37
0103	Pneumatic nailed		1708	.009		.63	.33		.96	1.25
0202	5/8" thick		1300	.012		.78	.44		1.22	1.59
0207	Pneumatic nailed		1586	.010		.78	.36		1.14	1.45
0302	3/4" thick		1200	.013		.94	.48		1.42	1.82
0307	Pneumatic nailed		1464	.011		.94	.39		1.33	1.67
0502	Plywood on walls, with exterior CDX, 3/8" thick		1200	.013		.61	.48		1.09	1.47
0507	Pneumatic nailed		1488	.011		.61	.38		.99	1.31
0603	1/2" thick		1125	.014		.63	.51		1.14	1.54
0608	Pneumatic nailed		1395	.011		.63	.41		1.04	1.38
0702	5/8" thick		1050	.015		.78	.54		1.32	1.76
0707	Pneumatic nailed		1302	.012		.78	.44		1.22	1.58
0803	3/4" thick		975	.016		.94	.59		1.53	2
0808	Pneumatic nailed		1209	.013		.94	.47		1.41	1.81
1000	For shear wall construction, add						20%			
1200	For structural 1 exterior plywood, add				S.F.	10%				
3000	Wood fiber, regular, no vapor barrier, 1/2" thick	2 Carp	1200	.013		.64	.48		1.12	1.49
3100	5/8" thick		1200	.013		.71	.48		1.19	1.57
3300	No vapor barrier, in colors, 1/2" thick		1200	.013		.81	.48		1.29	1.68
3400	5/8" thick		1200	.013		.85	.48		1.33	1.73
3600	With vapor barrier one side, white, 1/2" thick		1200	.013		.62	.48		1.10	1.47
3700	Vapor barrier 2 sides, 1/2" thick		1200	.013		.85	.48		1.33	1.73
3800	Asphalt impregnated, 25/32" thick		1200	.013		.32	.48		.80	1.14
3850	Intermediate, 1/2" thick		1200	.013		.38	.48		.86	1.21
4500	Oriented strand board, on roof, 7/16" thick G		1460	.011		.47	.39		.86	1.17
4505	Pneumatic nailed G		1780	.009		.47	.32		.79	1.05
4550	1/2" thick G		1400	.011		.47	.41		.88	1.19
4555	Pneumatic nailed G		1736	.009		.47	.33		.80	1.06
4600	5/8" thick G		1300	.012		.64	.44		1.08	1.43
4605	Pneumatic nailed G		1586	.010		.64	.36		1	1.29
4610	On walls, 7/16" thick G		1200	.013		.47	.48		.95	1.31
4615	Pneumatic nailed G		1488	.011		.47	.38		.85	1.15
4620	1/2" thick G		1195	.013		.47	.48		.95	1.31
4625	Pneumatic nailed G		1325	.012		.47	.43		.90	1.23
4630	5/8" thick G		1050	.015		.64	.54		1.18	1.60
4635	Pneumatic nailed G		1302	.012		.64	.44		1.08	1.42
4700	Oriented strand board, factory laminated W.R. barrier, on roof, 1/2" thick G		1400	.011		.77	.41		1.18	1.52
4705	Pneumatic nailed G		1736	.009		.77	.33		1.10	1.39
4720	5/8" thick G		1300	.012		.93	.44		1.37	1.75
4725	Pneumatic nailed G		1586	.010		.93	.36		1.29	1.61
4730	5/8" thick, T&G G		1150	.014		1.24	.50		1.74	2.18
4735	Pneumatic nailed, T&G G		1400	.011		1.24	.41		1.65	2.03
4740	On walls, 7/16" thick G		1200	.013		.71	.48		1.19	1.57
4745	Pneumatic nailed G		1488	.011		.71	.38		1.09	1.41
4750	1/2" thick G		1195	.013		.77	.48		1.25	1.64
4755	Pneumatic nailed G		1325	.012		.77	.43		1.20	1.56

06 16 Sheathing

06 16 36 – Wood Panel Product Sheathing

06 16 36.10 Sheathing		Crew	Daily Output	Labor-Hours	Unit	Material	Labor	Equipment	Total	Total Incl O&P
						2019 Bare Costs				
4800	Joint sealant tape, 3-1/2"	2 Carp	7600	.002	L.F.	.30	.08		.38	.45
4810	Joint sealant tape, 6"	↓	7600	.002	"	.41	.08		.49	.57

06 16 43 – Gypsum Sheathing

06 16 43.10 Gypsum Sheathing		Crew	Daily Output	Labor-Hours	Unit	Material	Labor	Equipment	Total	Total Incl O&P
0010	**GYPSUM SHEATHING**									
0020	Gypsum, weatherproof, 1/2" thick	2 Carp	1125	.014	S.F.	.46	.51		.97	1.35
0040	With embedded glass mats	"	1100	.015	"	.71	.52		1.23	1.64

06 17 Shop-Fabricated Structural Wood

06 17 33 – Wood I-Joists

06 17 33.10 Wood and Composite I-Joists		Crew	Daily Output	Labor-Hours	Unit	Material	Labor	Equipment	Total	Total Incl O&P
0010	**WOOD AND COMPOSITE I-JOISTS**									
0100	Plywood webs, incl. bridging & blocking, panels 24" OC									
1200	15' to 24' span, 50 psf live load	F-5	2400	.013	SF Flr.	1.99	.42		2.41	2.89
1300	55 psf live load		2250	.014		2.29	.45		2.74	3.26
1400	24' to 30' span, 45 psf live load		2600	.012		2.93	.39		3.32	3.86
1500	55 psf live load	↓	2400	.013	↓	4.68	.42		5.10	5.85

06 17 53 – Shop-Fabricated Wood Trusses

06 17 53.10 Roof Trusses		Crew	Daily Output	Labor-Hours	Unit	Material	Labor	Equipment	Total	Total Incl O&P
0010	**ROOF TRUSSES**									
5000	Common wood, 2" x 4" metal plate connected, 24" OC, 4/12 slope									
5010	1' overhang, 12' span	F-5	55	.582	Ea.	36.50	18.30		54.80	70.50
5050	20' span	F-6	62	.645		74	21	7.60	102.60	125
5100	24' span		60	.667		78	22	7.85	107.85	131
5150	26' span		57	.702		78	23	8.25	109.25	133
5200	28' span		53	.755		98.50	25	8.90	132.40	159
5240	30' span		51	.784		107	26	9.25	142.25	171
5250	32' span		50	.800		114	26.50	9.45	149.95	179
5280	34' span		48	.833		107	27.50	9.85	144.35	174
5350	8/12 pitch, 1' overhang, 20' span		57	.702		89	23	8.25	120.25	145
5400	24' span		55	.727		104	24	8.55	136.55	164
5450	26' span		52	.769		110	25.50	9.05	144.55	173
5500	28' span		49	.816		123	27	9.60	159.60	191
5550	32' span		45	.889		144	29	10.50	183.50	219
5600	36' span		41	.976		170	32	11.50	213.50	253
5650	38' span		40	1		193	33	11.80	237.80	280
5700	40' span	↓	40	1	↓	201	33	11.80	245.80	288

06 18 13 – Glued-Laminated Beams

06 18 13.10 Laminated Beams		Crew	Daily Output	Labor-Hours	Unit	2019 Bare Costs Material	Labor	Equipment	Total	Total Incl O&P
0010	**LAMINATED BEAMS**									
0050	3-1/2" x 18"	F-3	480	.083	L.F.	28	2.73	.98	31.71	36
0100	5-1/4" x 11-7/8"		450	.089		27.50	2.91	1.05	31.46	36
0150	5-1/4" x 16"		360	.111		38.50	3.64	1.31	43.45	50
0200	5-1/4" x 18"		290	.138		42	4.52	1.63	48.15	55.50
0250	5-1/4" x 24"		220	.182		61.50	5.95	2.14	69.59	79.50
0300	7" x 11-7/8"		320	.125		42	4.10	1.47	47.57	54.50
0350	7" x 16"		260	.154		58	5.05	1.81	64.86	74.50
0400	7" x 18"	↓	210	.190		67.50	6.25	2.25	76	87.50
0500	For premium appearance, add to L.F. prices					5%				
0550	For industrial type, deduct					15%				
0600	For stain and varnish, add					5%				
0650	For 3/4" laminations, add				↓	25%				

06 18 13.20 Laminated Framing		Crew	Daily Output	Labor-Hours	Unit	Material	Labor	Equipment	Total	Total Incl O&P
0010	**LAMINATED FRAMING**									
0020	30 lb., short term live load, 15 lb. dead load									
0200	Straight roof beams, 20' clear span, beams 8' OC	F-3	2560	.016	SF Flr.	2.20	.51	.18	2.89	3.47
0300	Beams 16' OC		3200	.013		1.61	.41	.15	2.17	2.61
0500	40' clear span, beams 8' OC		3200	.013		4.17	.41	.15	4.73	5.45
0600	Beams 16' OC	↓	3840	.010		3.45	.34	.12	3.91	4.51
0800	60' clear span, beams 8' OC	F-4	2880	.014		7.15	.46	.34	7.95	8.95
0900	Beams 16' OC	"	3840	.010		5.35	.34	.26	5.95	6.75
1100	Tudor arches, 30' to 40' clear span, frames 8' OC	F-3	1680	.024		9.30	.78	.28	10.36	11.85
1200	Frames 16' OC	"	2240	.018		7.25	.59	.21	8.05	9.20
1400	50' to 60' clear span, frames 8' OC	F-4	2200	.018		10	.60	.45	11.05	12.50
1500	Frames 16' OC		2640	.015		8.55	.50	.37	9.42	10.65
1700	Radial arches, 60' clear span, frames 8' OC		1920	.021		9.40	.68	.51	10.59	12
1800	Frames 16' OC		2880	.014		7.50	.46	.34	8.30	9.35
2000	100' clear span, frames 8' OC		1600	.025		9.65	.82	.61	11.08	12.70
2100	Frames 16' OC		2400	.017		8.55	.55	.41	9.51	10.75
2300	120' clear span, frames 8' OC		1440	.028		12.85	.91	.68	14.44	16.35
2400	Frames 16' OC	↓	1920	.021		11.75	.68	.51	12.94	14.60
2600	Bowstring trusses, 20' OC, 40' clear span	F-3	2400	.017		5.80	.55	.20	6.55	7.55
2700	60' clear span	F-4	3600	.011		5.25	.36	.27	5.88	6.65
2800	100' clear span		4000	.010		7.40	.33	.25	7.98	8.95
2900	120' clear span	↓	3600	.011		7.85	.36	.27	8.48	9.55
3100	For premium appearance, add to S.F. prices					5%				
3300	For industrial type, deduct					15%				
3500	For stain and varnish, add					5%				
3900	For 3/4" laminations, add to straight					25%				
4100	Add to curved				↓	15%				
4300	Alternate pricing method: (use nominal footage of									
4310	components). Straight beams, camber less than 6"	F-3	3.50	11.429	M.B.F.	3,150	375	135	3,660	4,225
4400	Columns, including hardware		2	20		3,375	655	236	4,266	5,025
4600	Curved members, radius over 32'		2.50	16		3,450	525	189	4,164	4,875
4700	Radius 10' to 32'	↓	3	13.333		3,425	435	157	4,017	4,675
4900	For complicated shapes, add maximum					100%				
5100	For pressure treating, add to straight					35%				
5200	Add to curved				↓	45%				
6000	Laminated veneer members, southern pine or western species									
6050	1-3/4" wide x 5-1/2" deep	2 Carp	480	.033	L.F.	3.49	1.19		4.68	5.80
6100	9-1/2" deep	↓	480	.033	↓	4.36	1.19		5.55	6.75

06 18 13 – Glued-Laminated Beams

06 18 13.20 Laminated Framing		Crew	Daily Output	Labor-Hours	Unit	Material	Labor	Equipment	Total	Total Incl O&P
						2019 Bare Costs				
6150	14" deep	2 Carp	450	.036	L.F.	7.55	1.27		8.82	10.40
6200	18" deep	↓	450	.036	↓	10.40	1.27		11.67	13.55
6300	Parallel strand members, southern pine or western species									
6350	1-3/4" wide x 9-1/4" deep	2 Carp	480	.033	L.F.	4.65	1.19		5.84	7.05
6400	11-1/4" deep		450	.036		5.10	1.27		6.37	7.70
6450	14" deep		400	.040		7.55	1.43		8.98	10.65
6500	3-1/2" wide x 9-1/4" deep		480	.033		16.05	1.19		17.24	19.60
6550	11-1/4" deep		450	.036		19.90	1.27		21.17	24
6600	14" deep		400	.040		23	1.43		24.43	27.50
6650	7" wide x 9-1/4" deep		450	.036		33.50	1.27		34.77	38.50
6700	11-1/4" deep		420	.038		42	1.36		43.36	48.50
6750	14" deep	↓	400	.040	↓	50	1.43		51.43	57.50
8000	Straight beams									
8102	20' span									
8104	3-1/8" x 9"	F-3	30	1.333	Ea.	147	43.50	15.70	206.20	252
8106	x 10-1/2"		30	1.333		172	43.50	15.70	231.20	279
8108	x 12"		30	1.333		196	43.50	15.70	255.20	305
8110	x 13-1/2"		30	1.333		221	43.50	15.70	280.20	335
8112	x 15"		29	1.379		246	45	16.25	307.25	365
8114	5-1/8" x 10-1/2"		30	1.333		282	43.50	15.70	341.20	400
8116	x 12"		30	1.333		320	43.50	15.70	379.20	445
8118	x 13-1/2"		30	1.333		360	43.50	15.70	419.20	490
8120	x 15"		29	1.379		405	45	16.25	466.25	540
8122	x 16-1/2"		29	1.379		445	45	16.25	506.25	580
8124	x 18"		29	1.379		485	45	16.25	546.25	625
8126	x 19-1/2"		29	1.379		525	45	16.25	586.25	670
8128	x 21"		28	1.429		565	47	16.85	628.85	715
8130	x 22-1/2"		28	1.429		605	47	16.85	668.85	760
8132	x 24"		28	1.429		645	47	16.85	708.85	805
8134	6-3/4" x 12"		29	1.379		425	45	16.25	486.25	560
8136	x 13-1/2"		29	1.379		475	45	16.25	536.25	620
8138	x 15"		29	1.379		530	45	16.25	591.25	680
8140	x 16-1/2"		28	1.429		585	47	16.85	648.85	735
8142	x 18"		28	1.429		635	47	16.85	698.85	795
8144	x 19-1/2"		28	1.429		690	47	16.85	753.85	855
8146	x 21"		27	1.481		740	48.50	17.45	805.95	915
8148	x 22-1/2"		27	1.481		795	48.50	17.45	860.95	975
8150	x 24"		27	1.481		850	48.50	17.45	915.95	1,025
8152	x 25-1/2"		27	1.481		900	48.50	17.45	965.95	1,100
8154	x 27"		26	1.538		955	50.50	18.15	1,023.65	1,150
8156	x 28-1/2"		26	1.538		1,000	50.50	18.15	1,068.65	1,200
8158	x 30"	↓	26	1.538	↓	1,050	50.50	18.15	1,118.65	1,275
8200	30' span									
8250	3-1/8" x 9"	F-3	30	1.333	Ea.	221	43.50	15.70	280.20	335
8252	x 10-1/2"		30	1.333		258	43.50	15.70	317.20	375
8254	x 12"		30	1.333		295	43.50	15.70	354.20	415
8256	x 13-1/2"		30	1.333		330	43.50	15.70	389.20	455
8258	x 15"		29	1.379		370	45	16.25	431.25	500
8260	5-1/8" x 10-1/2"		30	1.333		425	43.50	15.70	484.20	555
8262	x 12"		30	1.333		485	43.50	15.70	544.20	620
8264	x 13-1/2"		30	1.333		545	43.50	15.70	604.20	690
8266	x 15"		29	1.379		605	45	16.25	666.25	760
8268	x 16-1/2"	↓	29	1.379	↓	665	45	16.25	726.25	825

06 18 13.20 Laminated Framing		Crew	Daily Output	Labor-Hours	Unit	2019 Bare Costs Material	Labor	Equipment	Total	Total Incl O&P
8270	x 18"	F-3	29	1.379	Ea.	725	45	16.25	786.25	890
8272	x 19-1/2"		29	1.379		785	45	16.25	846.25	960
8274	x 21"		28	1.429		845	47	16.85	908.85	1,025
8276	x 22-1/2"		28	1.429		905	47	16.85	968.85	1,100
8278	x 24"		28	1.429		965	47	16.85	1,028.85	1,175
8280	6-3/4" x 12"		29	1.379		635	45	16.25	696.25	795
8282	x 13-1/2"		29	1.379		715	45	16.25	776.25	885
8284	x 15"		29	1.379		795	45	16.25	856.25	970
8286	x 16-1/2"		28	1.429		875	47	16.85	938.85	1,050
8288	x 18"		28	1.429		955	47	16.85	1,018.85	1,150
8290	x 19-1/2"		28	1.429		1,025	47	16.85	1,088.85	1,250
8292	x 21"		27	1.481		1,125	48.50	17.45	1,190.95	1,325
8294	x 22-1/2"		27	1.481		1,200	48.50	17.45	1,265.95	1,425
8296	x 24"		27	1.481		1,275	48.50	17.45	1,340.95	1,500
8298	x 25-1/2"		27	1.481		1,350	48.50	17.45	1,415.95	1,600
8300	x 27"		26	1.538		1,425	50.50	18.15	1,493.65	1,675
8302	x 28-1/2"		26	1.538		1,500	50.50	18.15	1,568.65	1,775
8304	x 30"	↓	26	1.538	↓	1,600	50.50	18.15	1,668.65	1,850
8400	40' span									
8402	3-1/8" x 9"	F-3	30	1.333	Ea.	295	43.50	15.70	354.20	415
8404	x 10-1/2"		30	1.333		345	43.50	15.70	404.20	470
8406	x 12"		30	1.333		395	43.50	15.70	454.20	520
8408	x 13-1/2"		30	1.333		440	43.50	15.70	499.20	575
8410	x 15"		29	1.379		490	45	16.25	551.25	635
8412	5-1/8" x 10-1/2"		30	1.333		565	43.50	15.70	624.20	710
8414	x 12"		30	1.333		645	43.50	15.70	704.20	800
8416	x 13-1/2"		30	1.333		725	43.50	15.70	784.20	885
8418	x 15"		29	1.379		805	45	16.25	866.25	980
8420	x 16-1/2"		29	1.379		885	45	16.25	946.25	1,075
8422	x 18"		29	1.379		965	45	16.25	1,026.25	1,175
8424	x 19-1/2"		29	1.379		1,050	45	16.25	1,111.25	1,250
8426	x 21"		28	1.429		1,125	47	16.85	1,188.85	1,350
8428	x 22-1/2"		28	1.429		1,200	47	16.85	1,263.85	1,425
8430	x 24"		28	1.429		1,300	47	16.85	1,363.85	1,525
8432	6-3/4" x 12"		29	1.379		850	45	16.25	911.25	1,025
8434	x 13-1/2"		29	1.379		955	45	16.25	1,016.25	1,150
8436	x 15"		29	1.379		1,050	45	16.25	1,111.25	1,275
8438	x 16-1/2"		28	1.429		1,175	47	16.85	1,238.85	1,375
8440	x 18"		28	1.429		1,275	47	16.85	1,338.85	1,500
8442	x 19-1/2"		28	1.429		1,375	47	16.85	1,438.85	1,625
8444	x 21"		27	1.481		1,475	48.50	17.45	1,540.95	1,725
8446	x 22-1/2"		27	1.481		1,600	48.50	17.45	1,665.95	1,850
8448	x 24"		27	1.481		1,700	48.50	17.45	1,765.95	1,975
8450	x 25-1/2"		27	1.481		1,800	48.50	17.45	1,865.95	2,075
8452	x 27"		26	1.538		1,900	50.50	18.15	1,968.65	2,200
8454	x 28-1/2"		26	1.538		2,025	50.50	18.15	2,093.65	2,325
8456	x 30"	↓	26	1.538	↓	2,125	50.50	18.15	2,193.65	2,425

06 22 13 – Standard Pattern Wood Trim

06 22 13.10 Millwork		Crew	Daily Output	Labor-Hours	Unit	2019 Bare Costs Material	Labor	Equipment	Total	Total Incl O&P
0010	**MILLWORK**									
0020	Rule of thumb, milled material equals rough lumber cost x 3									
1020	1" x 12", custom birch				L.F.	4.70			4.70	5.15
1040	Cedar					5.55			5.55	6.10
1060	Oak					5.25			5.25	5.75
1080	Redwood					4.79			4.79	5.25
1100	Southern yellow pine					4.11			4.11	4.52
1120	Sugar pine					6.30			6.30	6.95
1140	Teak					34			34	37.50
1160	Walnut					7.60			7.60	8.40
1180	White pine					5.40			5.40	5.95

06 22 13.15 Moldings, Base		Crew	Daily Output	Labor-Hours	Unit	Material	Labor	Equipment	Total	Total Incl O&P
0010	**MOLDINGS, BASE**									
5100	Classic profile, 5/8" x 5-1/2", finger jointed and primed	1 Carp	250	.032	L.F.	1.63	1.14		2.77	3.67
5105	Poplar		240	.033		1.90	1.19		3.09	4.05
5110	Red oak		220	.036		2.51	1.30		3.81	4.90
5115	Maple		220	.036		3.79	1.30		5.09	6.30
5120	Cherry		220	.036		4.49	1.30		5.79	7.10
5125	3/4" x 7-1/2", finger jointed and primed		250	.032		2.02	1.14		3.16	4.10
5130	Poplar		240	.033		2.61	1.19		3.80	4.83
5135	Red oak		220	.036		3.63	1.30		4.93	6.15
5140	Maple		220	.036		5.05	1.30		6.35	7.70
5145	Cherry		220	.036		6	1.30		7.30	8.75
5150	Modern profile, 5/8" x 3-1/2", finger jointed and primed		250	.032		.97	1.14		2.11	2.95
5155	Poplar		240	.033		1.04	1.19		2.23	3.10
5160	Red oak		220	.036		1.71	1.30		3.01	4.02
5165	Maple		220	.036		2.65	1.30		3.95	5.05
5170	Cherry		220	.036		2.92	1.30		4.22	5.35
5175	Ogee profile, 7/16" x 3", finger jointed and primed		250	.032		.68	1.14		1.82	2.63
5180	Poplar		240	.033		.77	1.19		1.96	2.81
5185	Red oak		220	.036		.97	1.30		2.27	3.21
5200	9/16" x 3-1/2", finger jointed and primed		250	.032		.67	1.14		1.81	2.62
5205	Pine		240	.033		1.21	1.19		2.40	3.29
5210	Red oak		220	.036		2.89	1.30		4.19	5.30
5215	9/16" x 4-1/2", red oak		220	.036		4.33	1.30		5.63	6.90
5220	5/8" x 3-1/2", finger jointed and primed		250	.032		1.01	1.14		2.15	2.99
5225	Poplar		240	.033		1.04	1.19		2.23	3.10
5230	Red oak		220	.036		1.71	1.30		3.01	4.02
5235	Maple		220	.036		2.65	1.30		3.95	5.05
5240	Cherry		220	.036		2.87	1.30		4.17	5.30
5245	5/8" x 4", finger jointed and primed		250	.032		1.26	1.14		2.40	3.26
5250	Poplar		240	.033		1.34	1.19		2.53	3.43
5255	Red oak		220	.036		1.92	1.30		3.22	4.25
5260	Maple		220	.036		2.97	1.30		4.27	5.40
5265	Cherry		220	.036		3.11	1.30		4.41	5.55
5270	Rectangular profile, oak, 3/8" x 1-1/4"		260	.031		1.27	1.10		2.37	3.20
5275	1/2" x 2-1/2"		255	.031		2.25	1.12		3.37	4.32
5280	1/2" x 3-1/2"		250	.032		2.83	1.14		3.97	4.99
5285	1" x 6"		240	.033		4.09	1.19		5.28	6.45
5290	1" x 8"		240	.033		5.05	1.19		6.24	7.55
5295	Pine, 3/8" x 1-3/4"		260	.031		.50	1.10		1.60	2.36
5300	7/16" x 2-1/2"		255	.031		.79	1.12		1.91	2.72

06 22 Millwork

06 22 13 – Standard Pattern Wood Trim

06 22 13.15 Moldings, Base		Crew	Daily Output	Labor-Hours	Unit	Material	Labor	Equipment	Total	Total Incl O&P
						2019 Bare Costs				
5305	1" x 6"	1 Carp	240	.033	L.F.	.73	1.19		1.92	2.76
5310	1" x 8"		240	.033		.94	1.19		2.13	2.99
5315	Shoe, 1/2" x 3/4", primed		260	.031		.53	1.10		1.63	2.39
5320	Pine		240	.033		.35	1.19		1.54	2.34
5325	Poplar		240	.033		.42	1.19		1.61	2.42
5330	Red oak		220	.036		.56	1.30		1.86	2.75
5335	Maple		220	.036		.84	1.30		2.14	3.06
5340	Cherry		220	.036		.85	1.30		2.15	3.07
5345	11/16" x 1-1/2", pine		240	.033		.74	1.19		1.93	2.77
5350	Caps, 11/16" x 1-3/8", pine		240	.033		.61	1.19		1.80	2.63
5355	3/4" x 1-3/4", finger jointed and primed		260	.031		.88	1.10		1.98	2.78
5360	Poplar		240	.033		1.05	1.19		2.24	3.11
5365	Red oak		220	.036		1.34	1.30		2.64	3.61
5370	Maple		220	.036		1.65	1.30		2.95	3.95
5375	Cherry		220	.036		3.34	1.30		4.64	5.80
5380	Combination base & shoe, 9/16" x 3-1/2" & 1/2" x 3/4", pine		125	.064		1.56	2.28		3.84	5.50
5385	Three piece oak, 6" high		80	.100		6	3.57		9.57	12.50
5390	Including 3/4" x 1" base shoe		70	.114		6.50	4.07		10.57	13.90
5395	Flooring cant strip, 3/4" x 3/4", pre-finished pine	↓	260	.031		.45	1.10		1.55	2.30
5400	For pre-finished, stain and clear coat, add					.57			.57	.63
5405	Clear coat only, add				↓	.44			.44	.48

06 22 13.30 Moldings, Casings		Crew	Daily Output	Labor-Hours	Unit	Material	Labor	Equipment	Total	Total Incl O&P
0010	**MOLDINGS, CASINGS**									
0085	Apron, 9/16" x 2-1/2", pine	1 Carp	250	.032	L.F.	1.44	1.14		2.58	3.46
0090	5/8" x 2-1/2", pine		250	.032		1.68	1.14		2.82	3.73
0110	5/8" x 3-1/2", pine		220	.036		1.99	1.30		3.29	4.33
0300	Band, 11/16" x 1-1/8", pine		270	.030		.76	1.06		1.82	2.58
0310	11/16" x 1-1/2", finger jointed and primed		270	.030		.76	1.06		1.82	2.58
0320	Pine		270	.030		1.02	1.06		2.08	2.87
0330	11/16" x 1-3/4", finger jointed and primed		270	.030		.95	1.06		2.01	2.79
0350	Pine		270	.030		1.03	1.06		2.09	2.88
0355	Beaded, 3/4" x 3-1/2", finger jointed and primed		220	.036		.97	1.30		2.27	3.21
0360	Poplar		220	.036		1.16	1.30		2.46	3.41
0365	Red oak		220	.036		1.71	1.30		3.01	4.02
0370	Maple		220	.036		2.65	1.30		3.95	5.05
0375	Cherry		220	.036		3.21	1.30		4.51	5.65
0380	3/4" x 4", finger jointed and primed		220	.036		1.25	1.30		2.55	3.51
0385	Poplar		220	.036		1.65	1.30		2.95	3.95
0390	Red oak		220	.036		2.35	1.30		3.65	4.72
0395	Maple		220	.036		2.68	1.30		3.98	5.10
0400	Cherry		220	.036		3.85	1.30		5.15	6.35
0405	3/4" x 5-1/2", finger jointed and primed		200	.040		1.49	1.43		2.92	4
0410	Poplar		200	.040		2.01	1.43		3.44	4.57
0415	Red oak		200	.040		2.95	1.43		4.38	5.60
0420	Maple		200	.040		3.89	1.43		5.32	6.65
0425	Cherry		200	.040		4.47	1.43		5.90	7.30
0430	Classic profile, 3/4" x 2-3/4", finger jointed and primed		250	.032		.87	1.14		2.01	2.84
0435	Poplar		250	.032		1.05	1.14		2.19	3.03
0440	Red oak		250	.032		1.54	1.14		2.68	3.57
0445	Maple		250	.032		2.18	1.14		3.32	4.28
0450	Cherry		250	.032		2.51	1.14		3.65	4.64
0455	Fluted, 3/4" x 3-1/2", poplar	↓	220	.036	↓	1.18	1.30		2.48	3.44

06 22 13 – Standard Pattern Wood Trim

06 22 13.30 Moldings, Casings		Crew	Daily Output	Labor-Hours	Unit	Material	Labor	Equipment	Total	Total Incl O&P
						2019 Bare Costs				
0460	Red oak	1 Carp	220	.036	L.F.	1.72	1.30		3.02	4.03
0465	Maple		220	.036		3.15	1.30		4.45	5.60
0470	Cherry		220	.036		3.21	1.30		4.51	5.65
0475	3/4" x 4", poplar		220	.036		1.45	1.30		2.75	3.73
0480	Red oak		220	.036		2.05	1.30		3.35	4.39
0485	Maple		220	.036		2.68	1.30		3.98	5.10
0490	Cherry		220	.036		3.85	1.30		5.15	6.35
0495	3/4" x 5-1/2", poplar		200	.040		2.01	1.43		3.44	4.57
0500	Red oak		200	.040		3.07	1.43		4.50	5.75
0505	Maple		200	.040		3.89	1.43		5.32	6.65
0510	Cherry		200	.040		4.47	1.43		5.90	7.30
0515	3/4" x 7-1/2", poplar		190	.042		1.28	1.50		2.78	3.89
0520	Red oak		190	.042		3.96	1.50		5.46	6.85
0525	Maple		190	.042		6.75	1.50		8.25	9.95
0530	Cherry		190	.042		8.10	1.50		9.60	11.40
0535	3/4" x 9-1/2", poplar		180	.044		4.16	1.58		5.74	7.20
0540	Red oak		180	.044		6.60	1.58		8.18	9.85
0545	Maple		180	.044		10.80	1.58		12.38	14.45
0550	Cherry		180	.044		11.75	1.58		13.33	15.55
0555	Modern profile, 9/16" x 2-1/4", poplar		250	.032		.84	1.14		1.98	2.80
0560	Red oak		250	.032		.94	1.14		2.08	2.91
0565	11/16" x 2-1/2", finger jointed & primed		250	.032		.86	1.14		2	2.82
0570	Pine		250	.032		1.38	1.14		2.52	3.40
0575	3/4" x 2-1/2", poplar		250	.032		.93	1.14		2.07	2.90
0580	Red oak		250	.032		1.24	1.14		2.38	3.24
0585	Maple		250	.032		1.91	1.14		3.05	3.98
0590	Cherry		250	.032		2.73	1.14		3.87	4.88
0595	Mullion, 5/16" x 2", pine		270	.030		.91	1.06		1.97	2.75
0600	9/16" x 2-1/2", finger jointed and primed		250	.032		.98	1.14		2.12	2.96
0605	Pine		250	.032		1.35	1.14		2.49	3.36
0610	Red oak		250	.032		3.37	1.14		4.51	5.60
0615	1-1/16" x 3-3/4", red oak		220	.036		7	1.30		8.30	9.85
0620	Ogee, 7/16" x 2-1/2", poplar		250	.032		.74	1.14		1.88	2.69
0625	Red oak		250	.032		.90	1.14		2.04	2.87
0630	9/16" x 2-1/4", finger jointed and primed		250	.032		.60	1.14		1.74	2.54
0635	Poplar		250	.032		.62	1.14		1.76	2.56
0640	Red oak		250	.032		.83	1.14		1.97	2.79
0645	11/16" x 2-1/2", finger jointed and primed		250	.032		.72	1.14		1.86	2.67
0700	Pine		250	.032		1.42	1.14		2.56	3.44
0701	Red oak		250	.032		3.28	1.14		4.42	5.50
0730	11/16" x 3-1/2", finger jointed and primed		220	.036		1.37	1.30		2.67	3.65
0750	Pine		220	.036		1.85	1.30		3.15	4.17
0755	3/4" x 2-1/2", finger jointed and primed		250	.032		.75	1.14		1.89	2.70
0760	Poplar		250	.032		.96	1.14		2.10	2.93
0765	Red oak		250	.032		1.29	1.14		2.43	3.30
0770	Maple		250	.032		1.88	1.14		3.02	3.95
0775	Cherry		250	.032		2.36	1.14		3.50	4.47
0780	3/4" x 3-1/2", finger jointed and primed		220	.036		.98	1.30		2.28	3.22
0785	Poplar		220	.036		1.18	1.30		2.48	3.44
0790	Red oak		220	.036		1.72	1.30		3.02	4.03
0795	Maple		220	.036		2.65	1.30		3.95	5.05
0800	Cherry		220	.036		3.26	1.30		4.56	5.70
4700	Square profile, 1" x 1", teak		215	.037		2.31	1.33		3.64	4.73

06 22 Millwork

06 22 13 – Standard Pattern Wood Trim

06 22 13.30 Moldings, Casings		Crew	Daily Output	Labor-Hours	Unit	Material	Labor	Equipment	Total	Total Incl O&P
						2019 Bare Costs				
4800	Rectangular profile, 1" x 3", teak	1 Carp	200	.040	L.F.	6.75	1.43		8.18	9.80

06 22 13.35 Moldings, Ceilings		Crew	Daily Output	Labor-Hours	Unit	Material	Labor	Equipment	Total	Total Incl O&P
0010	**MOLDINGS, CEILINGS**									
0600	Bed, 9/16" x 1-3/4", pine	1 Carp	270	.030	L.F.	1.21	1.06		2.27	3.08
0650	9/16" x 2", pine		270	.030		1.22	1.06		2.28	3.09
0710	9/16" x 1-3/4", oak		270	.030		2.38	1.06		3.44	4.37
1200	Cornice, 9/16" x 1-3/4", pine		270	.030		.99	1.06		2.05	2.84
1300	9/16" x 2-1/4", pine		265	.030		1.32	1.08		2.40	3.23
1350	Cove, 1/2" x 2-1/4", poplar		265	.030		1.17	1.08		2.25	3.06
1360	Red oak		265	.030		1.56	1.08		2.64	3.49
1370	Hard maple		265	.030		1.67	1.08		2.75	3.61
1380	Cherry		265	.030		2.28	1.08		3.36	4.28
2400	9/16" x 1-3/4", pine		270	.030		1	1.06		2.06	2.85
2401	Oak		270	.030		.99	1.06		2.05	2.84
2500	11/16" x 2-3/4", pine		265	.030		1.90	1.08		2.98	3.87
2510	Crown, 5/8" x 5/8", poplar		300	.027		.47	.95		1.42	2.09
2520	Red oak		300	.027		.55	.95		1.50	2.17
2530	Hard maple		300	.027		.76	.95		1.71	2.40
2540	Cherry		300	.027		.80	.95		1.75	2.45
2600	9/16" x 3-5/8", pine		250	.032		2.13	1.14		3.27	4.22
2700	11/16" x 4-1/4", pine		250	.032		3.02	1.14		4.16	5.20
2705	Oak		250	.032		6.40	1.14		7.54	8.95
2710	3/4" x 1-3/4", poplar		270	.030		.75	1.06		1.81	2.57
2720	Red oak		270	.030		1.11	1.06		2.17	2.97
2730	Hard maple		270	.030		1.46	1.06		2.52	3.35
2740	Cherry		270	.030		1.79	1.06		2.85	3.72
2750	3/4" x 2", poplar		270	.030		.99	1.06		2.05	2.84
2760	Red oak		270	.030		1.32	1.06		2.38	3.20
2770	Hard maple		270	.030		1.96	1.06		3.02	3.90
2780	Cherry		270	.030		1.99	1.06		3.05	3.94
2790	3/4" x 2-3/4", poplar		265	.030		1.07	1.08		2.15	2.95
2800	Red oak		265	.030		1.65	1.08		2.73	3.59
2810	Hard maple		265	.030		2.24	1.08		3.32	4.24
2820	Cherry		265	.030		2.54	1.08		3.62	4.57
2830	3/4" x 3-1/2", poplar		250	.032		1.38	1.14		2.52	3.39
2840	Red oak		250	.032		2.05	1.14		3.19	4.13
2850	Hard maple		250	.032		3	1.14		4.14	5.20
2860	Cherry		250	.032		3.12	1.14		4.26	5.30
2870	FJP poplar		250	.032		1.05	1.14		2.19	3.03
2880	3/4" x 5", poplar		245	.033		2.01	1.16		3.17	4.13
2890	Red oak		245	.033		2.99	1.16		4.15	5.20
2900	Hard maple		245	.033		3.95	1.16		5.11	6.25
2910	Cherry		245	.033		4.77	1.16		5.93	7.15
2920	FJP poplar		245	.033		1.47	1.16		2.63	3.53
2930	3/4" x 6-1/4", poplar		240	.033		2.42	1.19		3.61	4.62
2940	Red oak		240	.033		3.64	1.19		4.83	5.95
2950	Hard maple		240	.033		4.96	1.19		6.15	7.40
2960	Cherry		240	.033		5.90	1.19		7.09	8.45
2970	7/8" x 8-3/4", poplar		220	.036		4.50	1.30		5.80	7.10
2980	Red oak		220	.036		6.15	1.30		7.45	8.90
2990	Hard maple		220	.036		8.50	1.30		9.80	11.50
3000	Cherry		220	.036		10.05	1.30		11.35	13.20

06 22 13 – Standard Pattern Wood Trim

06 22 13.35 Moldings, Ceilings		Crew	Daily Output	Labor-Hours	Unit	Material	Labor	Equipment	Total	Total Incl O&P
						2019 Bare Costs				
3010	1" x 7-1/4", poplar	1 Carp	220	.036	L.F.	4.57	1.30		5.87	7.15
3020	Red oak		220	.036		6.15	1.30		7.45	8.90
3030	Hard maple		220	.036		9.10	1.30		10.40	12.15
3040	Cherry		220	.036		10	1.30		11.30	13.15
3050	1-1/16" x 4-1/4", poplar		250	.032		2.56	1.14		3.70	4.69
3060	Red oak		250	.032		2.82	1.14		3.96	4.98
3070	Hard maple		250	.032		4.30	1.14		5.44	6.60
3080	Cherry		250	.032		5.35	1.14		6.49	7.80
3090	Dentil crown, 3/4" x 5", poplar		250	.032		2.01	1.14		3.15	4.09
3100	Red oak		250	.032		2.99	1.14		4.13	5.15
3110	Hard maple		250	.032		3.93	1.14		5.07	6.20
3120	Cherry		250	.032		4.54	1.14		5.68	6.85
3130	Dentil piece for above, 1/2" x 1/2", poplar		300	.027		3.31	.95		4.26	5.20
3140	Red oak		300	.027		3.29	.95		4.24	5.20
3150	Hard maple		300	.027		4.26	.95		5.21	6.25
3160	Cherry	↓	300	.027	↓	4.10	.95		5.05	6.10

06 22 13.40 Moldings, Exterior		Crew	Daily Output	Labor-Hours	Unit	Material	Labor	Equipment	Total	Total Incl O&P
0010	**MOLDINGS, EXTERIOR**									
0100	Band board, cedar, rough sawn, 1" x 2"	1 Carp	300	.027	L.F.	.69	.95		1.64	2.33
0110	1" x 3"		300	.027		1.03	.95		1.98	2.70
0120	1" x 4"		250	.032		1.37	1.14		2.51	3.38
0130	1" x 6"		250	.032		2.05	1.14		3.19	4.14
0140	1" x 8"		225	.036		2.74	1.27		4.01	5.10
0150	1" x 10"		225	.036		3.40	1.27		4.67	5.85
0160	1" x 12"		200	.040		4.09	1.43		5.52	6.85
0240	STK, 1" x 2"		300	.027		.46	.95		1.41	2.08
0250	1" x 3"		300	.027		.50	.95		1.45	2.12
0260	1" x 4"		250	.032		.82	1.14		1.96	2.78
0270	1" x 6"		250	.032		1.34	1.14		2.48	3.35
0280	1" x 8"		225	.036		2.29	1.27		3.56	4.61
0290	1" x 10"		225	.036		2.81	1.27		4.08	5.20
0300	1" x 12"		200	.040		4.42	1.43		5.85	7.20
0310	Pine, #2, 1" x 2"		300	.027		.29	.95		1.24	1.89
0320	1" x 3"		300	.027		.46	.95		1.41	2.07
0330	1" x 4"		250	.032		.56	1.14		1.70	2.50
0340	1" x 6"		250	.032		.82	1.14		1.96	2.79
0350	1" x 8"		225	.036		1.38	1.27		2.65	3.61
0360	1" x 10"		225	.036		1.76	1.27		3.03	4.03
0370	1" x 12"		200	.040		2.20	1.43		3.63	4.78
0380	D & better, 1" x 2"		300	.027		.50	.95		1.45	2.12
0390	1" x 3"		300	.027		.65	.95		1.60	2.28
0400	1" x 4"		250	.032		.83	1.14		1.97	2.79
0410	1" x 6"		250	.032		1.10	1.14		2.24	3.09
0420	1" x 8"		225	.036		1.63	1.27		2.90	3.88
0430	1" x 10"		225	.036		2.20	1.27		3.47	4.51
0440	1" x 12"		200	.040		2.71	1.43		4.14	5.35
0450	Redwood, clear all heart, 1" x 2"		300	.027		.65	.95		1.60	2.28
0460	1" x 3"		300	.027		.96	.95		1.91	2.63
0470	1" x 4"		250	.032		1.22	1.14		2.36	3.22
0480	1" x 6"		252	.032		1.81	1.13		2.94	3.87
0490	1" x 8"		225	.036		2.41	1.27		3.68	4.74
0500	1" x 10"	↓	225	.036	↓	4.02	1.27		5.29	6.50

06 22 13.40 Moldings, Exterior		Crew	Daily Output	Labor-Hours	Unit	Material	Labor	Equipment	Total	Total Incl O&P
						2019 Bare Costs				
0510	1″ x 12″	1 Carp	200	.040	L.F.	4.85	1.43		6.28	7.70
0530	Corner board, cedar, rough sawn, 1″ x 2″		225	.036		.69	1.27		1.96	2.85
0540	1″ x 3″		225	.036		1.03	1.27		2.30	3.22
0550	1″ x 4″		200	.040		1.37	1.43		2.80	3.86
0560	1″ x 6″		200	.040		2.05	1.43		3.48	4.62
0570	1″ x 8″		200	.040		2.74	1.43		4.17	5.35
0580	1″ x 10″		175	.046		3.40	1.63		5.03	6.45
0590	1″ x 12″		175	.046		4.09	1.63		5.72	7.20
0670	STK, 1″ x 2″		225	.036		.46	1.27		1.73	2.60
0680	1″ x 3″		225	.036		.50	1.27		1.77	2.64
0690	1″ x 4″		200	.040		.80	1.43		2.23	3.24
0700	1″ x 6″		200	.040		1.34	1.43		2.77	3.83
0710	1″ x 8″		200	.040		2.29	1.43		3.72	4.88
0720	1″ x 10″		175	.046		2.81	1.63		4.44	5.80
0730	1″ x 12″		175	.046		4.42	1.63		6.05	7.55
0740	Pine, #2, 1″ x 2″		225	.036		.29	1.27		1.56	2.41
0750	1″ x 3″		225	.036		.46	1.27		1.73	2.59
0760	1″ x 4″		200	.040		.56	1.43		1.99	2.98
0770	1″ x 6″		200	.040		.82	1.43		2.25	3.27
0780	1″ x 8″		200	.040		1.38	1.43		2.81	3.88
0790	1″ x 10″		175	.046		1.76	1.63		3.39	4.63
0800	1″ x 12″		175	.046		2.20	1.63		3.83	5.10
0810	D & better, 1″ x 2″		225	.036		.50	1.27		1.77	2.64
0820	1″ x 3″		225	.036		.65	1.27		1.92	2.80
0830	1″ x 4″		200	.040		.83	1.43		2.26	3.27
0840	1″ x 6″		200	.040		1.10	1.43		2.53	3.57
0850	1″ x 8″		200	.040		1.63	1.43		3.06	4.15
0860	1″ x 10″		175	.046		2.20	1.63		3.83	5.10
0870	1″ x 12″		175	.046		2.71	1.63		4.34	5.65
0880	Redwood, clear all heart, 1″ x 2″		225	.036		.65	1.27		1.92	2.80
0890	1″ x 3″		225	.036		.96	1.27		2.23	3.15
0900	1″ x 4″		200	.040		1.22	1.43		2.65	3.70
0910	1″ x 6″		200	.040		1.81	1.43		3.24	4.36
0920	1″ x 8″		200	.040		2.41	1.43		3.84	5
0930	1″ x 10″		175	.046		4.02	1.63		5.65	7.10
0940	1″ x 12″		175	.046		4.85	1.63		6.48	8.05
0950	Cornice board, cedar, rough sawn, 1″ x 2″		330	.024		.69	.86		1.55	2.19
0960	1″ x 3″		290	.028		1.03	.98		2.01	2.76
0970	1″ x 4″		250	.032		1.37	1.14		2.51	3.38
0980	1″ x 6″		250	.032		2.05	1.14		3.19	4.14
0990	1″ x 8″		200	.040		2.74	1.43		4.17	5.35
1000	1″ x 10″		180	.044		3.40	1.58		4.98	6.35
1010	1″ x 12″		180	.044		4.09	1.58		5.67	7.10
1020	STK, 1″ x 2″		330	.024		.46	.86		1.32	1.94
1030	1″ x 3″		290	.028		.50	.98		1.48	2.18
1040	1″ x 4″		250	.032		.82	1.14		1.96	2.78
1050	1″ x 6″		250	.032		1.34	1.14		2.48	3.35
1060	1″ x 8″		200	.040		2.29	1.43		3.72	4.88
1070	1″ x 10″		180	.044		2.81	1.58		4.39	5.70
1080	1″ x 12″		180	.044		4.42	1.58		6	7.50
1500	Pine, #2, 1″ x 2″		330	.024		.29	.86		1.15	1.75
1510	1″ x 3″		290	.028		.31	.98		1.29	1.97
1600	1″ x 4″	▼	250	.032	▼	.56	1.14		1.70	2.50

06 22 13.40 Moldings, Exterior		Crew	Daily Output	Labor-Hours	Unit	2019 Bare Costs Material	Labor	Equipment	Total	Total Incl O&P
1700	1" x 6"	1 Carp	250	.032	L.F.	.82	1.14		1.96	2.79
1800	1" x 8"		200	.040		1.38	1.43		2.81	3.88
1900	1" x 10"		180	.044		1.76	1.58		3.34	4.56
2000	1" x 12"		180	.044		2.20	1.58		3.78	5.05
2020	D & better, 1" x 2"		330	.024		.50	.86		1.36	1.98
2030	1" x 3"		290	.028		.65	.98		1.63	2.34
2040	1" x 4"		250	.032		.83	1.14		1.97	2.79
2050	1" x 6"		250	.032		1.10	1.14		2.24	3.09
2060	1" x 8"		200	.040		1.63	1.43		3.06	4.15
2070	1" x 10"		180	.044		2.20	1.58		3.78	5.05
2080	1" x 12"		180	.044		2.71	1.58		4.29	5.60
2090	Redwood, clear all heart, 1" x 2"		330	.024		.65	.86		1.51	2.14
2100	1" x 3"		290	.028		.96	.98		1.94	2.69
2110	1" x 4"		250	.032		1.22	1.14		2.36	3.22
2120	1" x 6"		250	.032		1.81	1.14		2.95	3.88
2130	1" x 8"		200	.040		2.41	1.43		3.84	5
2140	1" x 10"		180	.044		4.02	1.58		5.60	7.05
2150	1" x 12"		180	.044		4.85	1.58		6.43	7.95
2160	3 piece, 1" x 2", 1" x 4", 1" x 6", rough sawn cedar		80	.100		4.13	3.57		7.70	10.45
2180	STK cedar		80	.100		2.62	3.57		6.19	8.80
2200	#2 pine		80	.100		1.67	3.57		5.24	7.75
2210	D & better pine		80	.100		2.43	3.57		6	8.55
2220	Clear all heart redwood		80	.100		3.68	3.57		7.25	9.95
2230	1" x 8", 1" x 10", 1" x 12", rough sawn cedar		65	.123		10.20	4.39		14.59	18.45
2240	STK cedar		65	.123		9.45	4.39		13.84	17.65
2300	#2 pine		65	.123		5.30	4.39		9.69	13.10
2320	D & better pine		65	.123		6.50	4.39		10.89	14.40
2330	Clear all heart redwood		65	.123		11.25	4.39		15.64	19.60
2340	Door/window casing, cedar, rough sawn, 1" x 2"		275	.029		.69	1.04		1.73	2.47
2350	1" x 3"		275	.029		1.03	1.04		2.07	2.84
2360	1" x 4"		250	.032		1.37	1.14		2.51	3.38
2370	1" x 6"		250	.032		2.05	1.14		3.19	4.14
2380	1" x 8"		230	.035		2.74	1.24		3.98	5.05
2390	1" x 10"		230	.035		3.40	1.24		4.64	5.80
2395	1" x 12"		210	.038		4.09	1.36		5.45	6.75
2410	STK, 1" x 2"		275	.029		.46	1.04		1.50	2.22
2420	1" x 3"		275	.029		.50	1.04		1.54	2.26
2430	1" x 4"		250	.032		.82	1.14		1.96	2.78
2440	1" x 6"		250	.032		1.34	1.14		2.48	3.35
2450	1" x 8"		230	.035		2.29	1.24		3.53	4.57
2460	1" x 10"		230	.035		2.81	1.24		4.05	5.15
2470	1" x 12"		210	.038		4.42	1.36		5.78	7.10
2550	Pine, #2, 1" x 2"		275	.029		.29	1.04		1.33	2.03
2560	1" x 3"		275	.029		.46	1.04		1.50	2.21
2570	1" x 4"		250	.032		.56	1.14		1.70	2.50
2580	1" x 6"		250	.032		.82	1.14		1.96	2.79
2590	1" x 8"		230	.035		1.38	1.24		2.62	3.57
2600	1" x 10"		230	.035		1.76	1.24		3	3.99
2610	1" x 12"		210	.038		2.20	1.36		3.56	4.66
2620	Pine, D & better, 1" x 2"		275	.029		.50	1.04		1.54	2.26
2630	1" x 3"		275	.029		.65	1.04		1.69	2.42
2640	1" x 4"		250	.032		.83	1.14		1.97	2.79
2650	1" x 6"		250	.032		1.10	1.14		2.24	3.09

06 22 13.40 Moldings, Exterior		Crew	Daily Output	Labor-Hours	Unit	Material	2019 Bare Costs Labor	Equipment	Total	Total Incl O&P
2660	1″ x 8″	1 Carp	230	.035	L.F.	1.63	1.24		2.87	3.84
2670	1″ x 10″		230	.035		2.20	1.24		3.44	4.47
2680	1″ x 12″		210	.038		2.71	1.36		4.07	5.20
2690	Redwood, clear all heart, 1″ x 2″		275	.029		.65	1.04		1.69	2.42
2695	1″ x 3″		275	.029		.96	1.04		2	2.77
2710	1″ x 4″		250	.032		1.22	1.14		2.36	3.22
2715	1″ x 6″		250	.032		1.81	1.14		2.95	3.88
2730	1″ x 8″		230	.035		2.41	1.24		3.65	4.70
2740	1″ x 10″		230	.035		4.02	1.24		5.26	6.45
2750	1″ x 12″		210	.038		4.85	1.36		6.21	7.60
3500	Bellyband, pine, 11/16″ x 4-1/4″		250	.032		2.84	1.14		3.98	5
3610	Brickmold, pine, 1-1/4″ x 2″		200	.040		2.11	1.43		3.54	4.68
3620	FJP, 1-1/4″ x 2″		200	.040		.99	1.43		2.42	3.45
5100	Fascia, cedar, rough sawn, 1″ x 2″		275	.029		.69	1.04		1.73	2.47
5110	1″ x 3″		275	.029		1.03	1.04		2.07	2.84
5120	1″ x 4″		250	.032		1.37	1.14		2.51	3.38
5200	1″ x 6″		250	.032		2.05	1.14		3.19	4.14
5300	1″ x 8″		230	.035		2.74	1.24		3.98	5.05
5310	1″ x 10″		230	.035		3.40	1.24		4.64	5.80
5320	1″ x 12″		210	.038		4.09	1.36		5.45	6.75
5400	2″ x 4″		220	.036		1.11	1.30		2.41	3.36
5500	2″ x 6″		220	.036		1.66	1.30		2.96	3.97
5600	2″ x 8″		200	.040		2.21	1.43		3.64	4.79
5700	2″ x 10″		180	.044		2.75	1.58		4.33	5.65
5800	2″ x 12″		170	.047		6.55	1.68		8.23	9.95
6120	STK, 1″ x 2″		275	.029		.46	1.04		1.50	2.22
6130	1″ x 3″		275	.029		.50	1.04		1.54	2.26
6140	1″ x 4″		250	.032		.82	1.14		1.96	2.78
6150	1″ x 6″		250	.032		1.34	1.14		2.48	3.35
6160	1″ x 8″		230	.035		2.29	1.24		3.53	4.57
6170	1″ x 10″		230	.035		2.81	1.24		4.05	5.15
6180	1″ x 12″		210	.038		4.42	1.36		5.78	7.10
6185	2″ x 2″		260	.031		.75	1.10		1.85	2.63
6190	Pine, #2, 1″ x 2″		275	.029		.29	1.04		1.33	2.03
6200	1″ x 3″		275	.029		.46	1.04		1.50	2.21
6210	1″ x 4″		250	.032		.56	1.14		1.70	2.50
6220	1″ x 6″		250	.032		.82	1.14		1.96	2.79
6230	1″ x 8″		230	.035		1.38	1.24		2.62	3.57
6240	1″ x 10″		230	.035		1.76	1.24		3	3.99
6250	1″ x 12″		210	.038		2.20	1.36		3.56	4.66
6260	D & better, 1″ x 2″		275	.029		.50	1.04		1.54	2.26
6270	1″ x 3″		275	.029		.65	1.04		1.69	2.42
6280	1″ x 4″		250	.032		.83	1.14		1.97	2.79
6290	1″ x 6″		250	.032		1.10	1.14		2.24	3.09
6300	1″ x 8″		230	.035		1.63	1.24		2.87	3.84
6310	1″ x 10″		230	.035		2.20	1.24		3.44	4.47
6312	1″ x 12″		210	.038		2.71	1.36		4.07	5.20
6330	Southern yellow, 1-1/4″ x 5″		240	.033		3.07	1.19		4.26	5.35
6340	1-1/4″ x 6″		240	.033		2.60	1.19		3.79	4.82
6350	1-1/4″ x 8″		215	.037		3.68	1.33		5.01	6.25
6360	1-1/4″ x 12″		190	.042		5.35	1.50		6.85	8.40
6370	Redwood, clear all heart, 1″ x 2″		275	.029		.65	1.04		1.69	2.42
6380	1″ x 3″		275	.029		1.22	1.04		2.26	3.05

06 22 13.40 Moldings, Exterior		Crew	Daily Output	Labor-Hours	Unit	Material	Labor	Equipment	Total	Total Incl O&P
						2019 Bare Costs				
6390	1" x 4"	1 Carp	250	.032	L.F.	1.22	1.14		2.36	3.22
6400	1" x 6"		250	.032		1.81	1.14		2.95	3.88
6410	1" x 8"		230	.035		2.41	1.24		3.65	4.70
6420	1" x 10"		230	.035		4.02	1.24		5.26	6.45
6430	1" x 12"		210	.038		4.85	1.36		6.21	7.60
6440	1-1/4" x 5"		240	.033		1.90	1.19		3.09	4.05
6450	1-1/4" x 6"		240	.033		2.27	1.19		3.46	4.45
6460	1-1/4" x 8"		215	.037		3.60	1.33		4.93	6.15
6470	1-1/4" x 12"		190	.042		7.25	1.50		8.75	10.50
6580	Frieze, cedar, rough sawn, 1" x 2"		275	.029		.69	1.04		1.73	2.47
6590	1" x 3"		275	.029		1.03	1.04		2.07	2.84
6600	1" x 4"		250	.032		1.37	1.14		2.51	3.38
6610	1" x 6"		250	.032		2.05	1.14		3.19	4.14
6620	1" x 8"		250	.032		2.74	1.14		3.88	4.89
6630	1" x 10"		225	.036		3.40	1.27		4.67	5.85
6640	1" x 12"		200	.040		4.05	1.43		5.48	6.80
6650	STK, 1" x 2"		275	.029		.46	1.04		1.50	2.22
6660	1" x 3"		275	.029		.50	1.04		1.54	2.26
6670	1" x 4"		250	.032		.82	1.14		1.96	2.78
6680	1" x 6"		250	.032		1.34	1.14		2.48	3.35
6690	1" x 8"		250	.032		2.29	1.14		3.43	4.40
6700	1" x 10"		225	.036		2.81	1.27		4.08	5.20
6710	1" x 12"		200	.040		4.42	1.43		5.85	7.20
6790	Pine, #2, 1" x 2"		275	.029		.29	1.04		1.33	2.03
6800	1" x 3"		275	.029		.46	1.04		1.50	2.21
6810	1" x 4"		250	.032		.56	1.14		1.70	2.50
6820	1" x 6"		250	.032		.82	1.14		1.96	2.79
6830	1" x 8"		250	.032		1.38	1.14		2.52	3.40
6840	1" x 10"		225	.036		1.76	1.27		3.03	4.03
6850	1" x 12"		200	.040		2.20	1.43		3.63	4.78
6860	D & better, 1" x 2"		275	.029		.50	1.04		1.54	2.26
6870	1" x 3"		275	.029		.65	1.04		1.69	2.42
6880	1" x 4"		250	.032		.83	1.14		1.97	2.79
6890	1" x 6"		250	.032		1.10	1.14		2.24	3.09
6900	1" x 8"		250	.032		1.63	1.14		2.77	3.67
6910	1" x 10"		225	.036		2.20	1.27		3.47	4.51
6920	1" x 12"		200	.040		2.71	1.43		4.14	5.35
6930	Redwood, clear all heart, 1" x 2"		275	.029		.65	1.04		1.69	2.42
6940	1" x 3"		275	.029		.96	1.04		2	2.77
6950	1" x 4"		250	.032		1.22	1.14		2.36	3.22
6960	1" x 6"		250	.032		1.81	1.14		2.95	3.88
6970	1" x 8"		250	.032		2.41	1.14		3.55	4.53
6980	1" x 10"		225	.036		4.02	1.27		5.29	6.50
6990	1" x 12"		200	.040		4.85	1.43		6.28	7.70
7000	Grounds, 1" x 1", cedar, rough sawn		300	.027		.35	.95		1.30	1.96
7010	STK		300	.027		.28	.95		1.23	1.88
7020	Pine, #2		300	.027		.18	.95		1.13	1.77
7030	D & better		300	.027		.31	.95		1.26	1.91
7050	Redwood		300	.027		.40	.95		1.35	2.01
7060	Rake/verge board, cedar, rough sawn, 1" x 2"		225	.036		.69	1.27		1.96	2.85
7070	1" x 3"		225	.036		1.03	1.27		2.30	3.22
7080	1" x 4"		200	.040		1.37	1.43		2.80	3.86
7090	1" x 6"		200	.040		2.05	1.43		3.48	4.62

06 22 13 – Standard Pattern Wood Trim

06 22 13.40 Moldings, Exterior		Crew	Daily Output	Labor-Hours	Unit	Material	2019 Bare Costs Labor	Equipment	Total	Total Incl O&P
7100	1" x 8"	1 Carp	190	.042	L.F.	2.74	1.50		4.24	5.50
7110	1" x 10"		190	.042		3.40	1.50		4.90	6.20
7120	1" x 12"		180	.044		4.09	1.58		5.67	7.10
7130	STK, 1" x 2"		225	.036		.46	1.27		1.73	2.60
7140	1" x 3"		225	.036		.50	1.27		1.77	2.64
7150	1" x 4"		200	.040		.82	1.43		2.25	3.26
7160	1" x 6"		200	.040		1.34	1.43		2.77	3.83
7170	1" x 8"		190	.042		2.29	1.50		3.79	5
7180	1" x 10"		190	.042		2.81	1.50		4.31	5.55
7190	1" x 12"		180	.044		4.42	1.58		6	7.50
7200	Pine, #2, 1" x 2"		225	.036		.29	1.27		1.56	2.41
7210	1" x 3"		225	.036		.46	1.27		1.73	2.59
7220	1" x 4"		200	.040		.56	1.43		1.99	2.98
7230	1" x 6"		200	.040		.82	1.43		2.25	3.27
7240	1" x 8"		190	.042		1.38	1.50		2.88	4
7250	1" x 10"		190	.042		1.76	1.50		3.26	4.42
7260	1" x 12"		180	.044		2.20	1.58		3.78	5.05
7340	D & better, 1" x 2"		225	.036		.50	1.27		1.77	2.64
7350	1" x 3"		225	.036		.65	1.27		1.92	2.80
7360	1" x 4"		200	.040		.83	1.43		2.26	3.27
7370	1" x 6"		200	.040		1.10	1.43		2.53	3.57
7380	1" x 8"		190	.042		1.63	1.50		3.13	4.27
7390	1" x 10"		190	.042		2.20	1.50		3.70	4.90
7400	1" x 12"		180	.044		2.71	1.58		4.29	5.60
7410	Redwood, clear all heart, 1" x 2"		225	.036		.65	1.27		1.92	2.80
7420	1" x 3"		225	.036		.96	1.27		2.23	3.15
7430	1" x 4"		200	.040		1.22	1.43		2.65	3.70
7440	1" x 6"		200	.040		1.81	1.43		3.24	4.36
7450	1" x 8"		190	.042		2.41	1.50		3.91	5.15
7460	1" x 10"		190	.042		4.02	1.50		5.52	6.90
7470	1" x 12"		180	.044		4.85	1.58		6.43	7.95
7480	2" x 4"		200	.040		2.33	1.43		3.76	4.93
7490	2" x 6"		182	.044		3.48	1.57		5.05	6.40
7500	2" x 8"		165	.048		4.64	1.73		6.37	7.95
7630	Soffit, cedar, rough sawn, 1" x 2"	2 Carp	440	.036		.69	1.30		1.99	2.90
7640	1" x 3"		440	.036		1.03	1.30		2.33	3.27
7650	1" x 4"		420	.038		1.37	1.36		2.73	3.74
7660	1" x 6"		420	.038		2.05	1.36		3.41	4.50
7670	1" x 8"		420	.038		2.74	1.36		4.10	5.25
7680	1" x 10"		400	.040		3.40	1.43		4.83	6.10
7690	1" x 12"		400	.040		4.09	1.43		5.52	6.85
7700	STK, 1" x 2"		440	.036		.46	1.30		1.76	2.65
7710	1" x 3"		440	.036		.50	1.30		1.80	2.69
7720	1" x 4"		420	.038		.82	1.36		2.18	3.14
7730	1" x 6"		420	.038		1.34	1.36		2.70	3.71
7740	1" x 8"		420	.038		2.29	1.36		3.65	4.76
7750	1" x 10"		400	.040		2.81	1.43		4.24	5.45
7760	1" x 12"		400	.040		4.42	1.43		5.85	7.20
7770	Pine, #2, 1" x 2"		440	.036		.29	1.30		1.59	2.46
7780	1" x 3"		440	.036		.46	1.30		1.76	2.64
7790	1" x 4"		420	.038		.56	1.36		1.92	2.86
7800	1" x 6"		420	.038		.82	1.36		2.18	3.15
7810	1" x 8"		420	.038		1.38	1.36		2.74	3.76

06 22 Millwork

06 22 13 – Standard Pattern Wood Trim

06 22 13.40 Moldings, Exterior		Crew	Daily Output	Labor-Hours	Unit	Material	Labor	Equipment	Total	Total Incl O&P
						2019 Bare Costs				
7820	1″ x 10″	2 Carp	400	.040	L.F.	1.76	1.43		3.19	4.30
7830	1″ x 12″		400	.040		2.20	1.43		3.63	4.78
7840	D & better, 1″ x 2″		440	.036		.50	1.30		1.80	2.69
7850	1″ x 3″		440	.036		.65	1.30		1.95	2.85
7860	1″ x 4″		420	.038		.83	1.36		2.19	3.15
7870	1″ x 6″		420	.038		1.10	1.36		2.46	3.45
7880	1″ x 8″		420	.038		1.63	1.36		2.99	4.03
7890	1″ x 10″		400	.040		2.20	1.43		3.63	4.78
7900	1″ x 12″		400	.040		2.71	1.43		4.14	5.35
7910	Redwood, clear all heart, 1″ x 2″		440	.036		.65	1.30		1.95	2.85
7920	1″ x 3″		440	.036		.96	1.30		2.26	3.20
7930	1″ x 4″		420	.038		1.22	1.36		2.58	3.58
7940	1″ x 6″		420	.038		1.81	1.36		3.17	4.24
7950	1″ x 8″		420	.038		2.41	1.36		3.77	4.89
7960	1″ x 10″		400	.040		4.02	1.43		5.45	6.80
7970	1″ x 12″		400	.040		4.85	1.43		6.28	7.70
8050	Trim, crown molding, pine, 11/16″ x 4-1/4″	1 Carp	250	.032		3.96	1.14		5.10	6.25
8060	Back band, 11/16″ x 1-1/16″		250	.032		.89	1.14		2.03	2.86
8070	Insect screen frame stock, 1-1/16″ x 1-3/4″		395	.020		2.24	.72		2.96	3.65
8080	Dentils, 2-1/2″ x 2-1/2″ x 4″, 6″ OC		30	.267		1.29	9.50		10.79	17.10
8100	Fluted, 5-1/2″		165	.048		5.10	1.73		6.83	8.45
8110	Stucco bead, 1-3/8″ x 1-5/8″		250	.032		2.35	1.14		3.49	4.46

06 22 13.45 Moldings, Trim		Crew	Daily Output	Labor-Hours	Unit	Material	Labor	Equipment	Total	Total Incl O&P
0010	**MOLDINGS, TRIM**									
0200	Astragal, stock pine, 11/16″ x 1-3/4″	1 Carp	255	.031	L.F.	1.25	1.12		2.37	3.23
0250	1-5/16″ x 2-3/16″		240	.033		2.63	1.19		3.82	4.85
0800	Chair rail, stock pine, 5/8″ x 2-1/2″		270	.030		1.54	1.06		2.60	3.44
0900	5/8″ x 3-1/2″		240	.033		2.31	1.19		3.50	4.50
1000	Closet pole, stock pine, 1-1/8″ diameter		200	.040		1.16	1.43		2.59	3.64
1100	Fir, 1-5/8″ diameter		200	.040		2.05	1.43		3.48	4.62
1150	Corner, inside, 5/16″ x 1″		225	.036		.34	1.27		1.61	2.46
1160	Outside, 1-1/16″ x 1-1/16″		240	.033		1.35	1.19		2.54	3.45
1161	1-5/16″ x 1-5/16″		240	.033		1.61	1.19		2.80	3.73
3300	Half round, stock pine, 1/4″ x 1/2″		270	.030		.25	1.06		1.31	2.03
3350	1/2″ x 1″		255	.031		.64	1.12		1.76	2.55
3400	Handrail, fir, single piece, stock, hardware not included									
3450	1-1/2″ x 1-3/4″	1 Carp	80	.100	L.F.	2.13	3.57		5.70	8.25
3470	Pine, 1-1/2″ x 1-3/4″		80	.100		1.99	3.57		5.56	8.10
3500	1-1/2″ x 2-1/2″		76	.105		2.71	3.75		6.46	9.20
3600	Lattice, stock pine, 1/4″ x 1-1/8″		270	.030		.41	1.06		1.47	2.20
3700	1/4″ x 1-3/4″		250	.032		.62	1.14		1.76	2.56
3800	Miscellaneous, custom, pine, 1″ x 1″		270	.030		.45	1.06		1.51	2.25
3850	1″ x 2″		265	.030		.90	1.08		1.98	2.77
3900	1″ x 3″		240	.033		1.35	1.19		2.54	3.45
4100	Birch or oak, nominal 1″ x 1″		240	.033		.39	1.19		1.58	2.39
4200	Nominal 1″ x 3″		215	.037		1.18	1.33		2.51	3.48
4400	Walnut, nominal 1″ x 1″		215	.037		.64	1.33		1.97	2.89
4500	Nominal 1″ x 3″		200	.040		1.91	1.43		3.34	4.46
4700	Teak, nominal 1″ x 1″		215	.037		2.85	1.33		4.18	5.30
4800	Nominal 1″ x 3″		200	.040		8.55	1.43		9.98	11.75
4900	Quarter round, stock pine, 1/4″ x 1/4″		275	.029		.25	1.04		1.29	1.99
4950	3/4″ x 3/4″		255	.031		.53	1.12		1.65	2.43

06 22 Millwork

06 22 13 – Standard Pattern Wood Trim

06 22 13.45 Moldings, Trim		Crew	Daily Output	Labor-Hours	Unit	2019 Bare Costs Material	Labor	Equipment	Total	Total Incl O&P
5600	Wainscot moldings, 1-1/8" x 9/16", 2' high, minimum	1 Carp	76	.105	S.F.	12.55	3.75		16.30	20
5700	Maximum		65	.123	"	16.30	4.39		20.69	25

06 22 13.50 Moldings, Window and Door

Line	Description	Crew	Daily Output	Labor-Hours	Unit	Material	Labor	Equipment	Total	Total Incl O&P
0010	**MOLDINGS, WINDOW AND DOOR**									
2800	Door moldings, stock, decorative, 1-1/8" wide, plain	1 Carp	17	.471	Set	48	16.80		64.80	80
2900	Detailed		17	.471	"	111	16.80		127.80	150
2960	Clear pine door jamb, no stops, 11/16" x 4-9/16"		240	.033	L.F.	4.93	1.19		6.12	7.35
3150	Door trim set, 1 head and 2 sides, pine, 2-1/2" wide		12	.667	Opng.	24	24		48	66
3170	3-1/2" wide		11	.727	"	31.50	26		57.50	77.50
3250	Glass beads, stock pine, 3/8" x 1/2"		275	.029	L.F.	.36	1.04		1.40	2.11
3270	3/8" x 7/8"		270	.030		.40	1.06		1.46	2.19
4850	Parting bead, stock pine, 3/8" x 3/4"		275	.029		.47	1.04		1.51	2.23
4870	1/2" x 3/4"		255	.031		.42	1.12		1.54	2.31
5000	Stool caps, stock pine, 11/16" x 3-1/2"		200	.040		2.40	1.43		3.83	5
5100	1-1/16" x 3-1/4"		150	.053		3.76	1.90		5.66	7.30
5300	Threshold, oak, 3' long, inside, 5/8" x 3-5/8"		32	.250	Ea.	12.95	8.90		21.85	29
5400	Outside, 1-1/2" x 7-5/8"		16	.500	"	47.50	17.85		65.35	81.50
5900	Window trim sets, including casings, header, stops,									
5910	stool and apron, 2-1/2" wide, FJP	1 Carp	13	.615	Opng.	33	22		55	72.50
5950	Pine		10	.800		38.50	28.50		67	89.50
6000	Oak		6	1.333		78	47.50		125.50	165

06 22 13.60 Moldings, Soffits

Line	Description	Crew	Daily Output	Labor-Hours	Unit	Material	Labor	Equipment	Total	Total Incl O&P
0010	**MOLDINGS, SOFFITS**									
0200	Soffits, pine, 1" x 4"	2 Carp	420	.038	L.F.	.52	1.36		1.88	2.82
0210	1" x 6"		420	.038		.79	1.36		2.15	3.10
0220	1" x 8"		420	.038		1.34	1.36		2.70	3.72
0230	1" x 10"		400	.040		1.70	1.43		3.13	4.23
0240	1" x 12"		400	.040		2.14	1.43		3.57	4.72
0250	STK cedar, 1" x 4"		420	.038		.78	1.36		2.14	3.10
0260	1" x 6"		420	.038		1.30	1.36		2.66	3.67
0270	1" x 8"		420	.038		2.25	1.36		3.61	4.72
0280	1" x 10"		400	.040		2.75	1.43		4.18	5.40
0290	1" x 12"		400	.040		4.36	1.43		5.79	7.15
1000	Exterior AC plywood, 1/4" thick		400	.040	S.F.	1	1.43		2.43	3.46
1050	3/8" thick		400	.040		1.05	1.43		2.48	3.52
1100	1/2" thick		400	.040		1.24	1.43		2.67	3.72
1150	Polyvinyl chloride, white, solid	1 Carp	230	.035		2.19	1.24		3.43	4.46
1160	Perforated	"	230	.035		2.19	1.24		3.43	4.46
1170	Accessories, "J" channel 5/8"	2 Carp	700	.023	L.F.	.51	.82		1.33	1.91

06 25 Prefinished Paneling

06 25 13 – Prefinished Hardboard Paneling

06 25 13.10 Paneling, Hardboard

Line	Description	Crew	Daily Output	Labor-Hours	Unit	Material	Labor	Equipment	Total	Total Incl O&P
0010	**PANELING, HARDBOARD**									
0050	Not incl. furring or trim, hardboard, tempered, 1/8" thick G	2 Carp	500	.032	S.F.	.46	1.14		1.60	2.39
0100	1/4" thick G		500	.032		.64	1.14		1.78	2.58
0300	Tempered pegboard, 1/8" thick G		500	.032		.44	1.14		1.58	2.36
0400	1/4" thick G		500	.032		.70	1.14		1.84	2.65
0600	Untempered hardboard, natural finish, 1/8" thick G		500	.032		.44	1.14		1.58	2.36
0700	1/4" thick G		500	.032		.52	1.14		1.66	2.45

06 25 Prefinished Paneling

06 25 13 – Prefinished Hardboard Paneling

06 25 13.10 Paneling, Hardboard		Crew	Daily Output	Labor-Hours	Unit	Material	Labor	Equipment	Total	Total Incl O&P
						2019 Bare Costs				
0900	Untempered pegboard, 1/8" thick G	2 Carp	500	.032	S.F.	.49	1.14		1.63	2.42
1000	1/4" thick G		500	.032		.48	1.14		1.62	2.41
1200	Plastic faced hardboard, 1/8" thick G		500	.032		.62	1.14		1.76	2.56
1300	1/4" thick G		500	.032		.87	1.14		2.01	2.84
1500	Plastic faced pegboard, 1/8" thick G		500	.032		.64	1.14		1.78	2.58
1600	1/4" thick G		500	.032		.80	1.14		1.94	2.76
1800	Wood grained, plain or grooved, 1/8" thick G		500	.032		.65	1.14		1.79	2.60
1900	1/4" thick G		425	.038	↓	1.42	1.34		2.76	3.78
2100	Moldings, wood grained MDF		500	.032	L.F.	.41	1.14		1.55	2.33
2200	Pine	↓	425	.038	"	1.41	1.34		2.75	3.77

06 25 16 – Prefinished Plywood Paneling

06 25 16.10 Paneling, Plywood		Crew	Daily Output	Labor-Hours	Unit	Material	Labor	Equipment	Total	Total Incl O&P
0010	**PANELING, PLYWOOD**									
2400	Plywood, prefinished, 1/4" thick, 4' x 8' sheets									
2410	with vertical grooves. Birch faced, economy	2 Carp	500	.032	S.F.	1.49	1.14		2.63	3.52
2420	Average		420	.038		1.21	1.36		2.57	3.57
2430	Custom		350	.046		1.11	1.63		2.74	3.91
2600	Mahogany, African		400	.040		2.48	1.43		3.91	5.10
2700	Philippine (Lauan)		500	.032		.65	1.14		1.79	2.60
2900	Oak		500	.032		1.35	1.14		2.49	3.37
3000	Cherry		400	.040		1.98	1.43		3.41	4.54
3200	Rosewood		320	.050		3.16	1.78		4.94	6.45
3400	Teak		400	.040		3.23	1.43		4.66	5.90
3600	Chestnut		375	.043		5.35	1.52		6.87	8.35
3800	Pecan		400	.040		2.51	1.43		3.94	5.10
3900	Walnut, average		500	.032		2.57	1.14		3.71	4.71
3950	Custom		400	.040		2.49	1.43		3.92	5.10
4000	Plywood, prefinished, 3/4" thick, stock grades, economy		320	.050		1.42	1.78		3.20	4.51
4100	Average		224	.071		4.98	2.55		7.53	9.70
4300	Architectural grade, custom		224	.071		5.35	2.55		7.90	10.10
4400	Luxury		160	.100		5.45	3.57		9.02	11.90
4600	Plywood, "A" face, birch, VC, 1/2" thick, natural		450	.036		1.39	1.27		2.66	3.62
4700	Select		450	.036		1.92	1.27		3.19	4.20
4900	Veneer core, 3/4" thick, natural		320	.050		2.27	1.78		4.05	5.45
5000	Select		320	.050		2.63	1.78		4.41	5.85
5200	Lumber core, 3/4" thick, natural		320	.050		3.14	1.78		4.92	6.40
5500	Plywood, knotty pine, 1/4" thick, A2 grade		450	.036		1.59	1.27		2.86	3.84
5600	A3 grade		450	.036		2.10	1.27		3.37	4.40
5800	3/4" thick, veneer core, A2 grade		320	.050		2.33	1.78		4.11	5.50
5900	A3 grade		320	.050		2.40	1.78		4.18	5.60
6100	Aromatic cedar, 1/4" thick, plywood		400	.040		2.26	1.43		3.69	4.85
6200	1/4" thick, particle board	↓	400	.040	↓	1.16	1.43		2.59	3.64

06 25 26 – Panel System

06 25 26.10 Panel Systems		Crew	Daily Output	Labor-Hours	Unit	Material	Labor	Equipment	Total	Total Incl O&P
0010	**PANEL SYSTEMS**									
0100	Raised panel, eng. wood core w/wood veneer, std., paint grade	2 Carp	300	.053	S.F.	12.10	1.90		14	16.45
0110	Oak veneer		300	.053		24.50	1.90		26.40	30
0120	Maple veneer		300	.053		30	1.90		31.90	36
0130	Cherry veneer		300	.053		36.50	1.90		38.40	43.50
0300	Class I fire rated, paint grade		300	.053		13.80	1.90		15.70	18.30
0310	Oak veneer		300	.053		30.50	1.90		32.40	36.50
0320	Maple veneer	↓	300	.053	↓	41.50	1.90		43.40	48.50

06 25 Prefinished Paneling

06 25 26 – Panel System

06 25 26.10 Panel Systems		Crew	Daily Output	Labor-Hours	Unit	Material	2019 Bare Costs Labor	Equipment	Total	Total Incl O&P
0330	Cherry veneer	2 Carp	300	.053	S.F.	49.50	1.90		51.40	57.50
0510	Beadboard, 5/8" MDF, standard, primed		300	.053		9.50	1.90		11.40	13.60
0520	Oak veneer, unfinished		300	.053		14.90	1.90		16.80	19.55
0530	Maple veneer, unfinished		300	.053		16	1.90		17.90	20.50
0610	Rustic paneling, 5/8" MDF, standard, maple veneer, unfinished		300	.053		18.60	1.90		20.50	23.50

06 26 Board Paneling

06 26 13 – Profile Board Paneling

06 26 13.10 Paneling, Boards		Crew	Daily Output	Labor-Hours	Unit	Material	Labor	Equipment	Total	Total Incl O&P
0010	**PANELING, BOARDS**									
6400	Wood board paneling, 3/4" thick, knotty pine	2 Carp	300	.053	S.F.	2.04	1.90		3.94	5.40
6500	Rough sawn cedar		300	.053		3.40	1.90		5.30	6.90
6700	Redwood, clear, 1" x 4" boards		300	.053		5.05	1.90		6.95	8.70
6900	Aromatic cedar, closet lining, boards		275	.058		2.46	2.07		4.53	6.15
8950	On ceiling, wood board, install	1 Carp	225	.036		3.40	1.27		4.67	5.85

06 43 Wood Stairs and Railings

06 43 13 – Wood Stairs

06 43 13.20 Prefabricated Wood Stairs		Crew	Daily Output	Labor-Hours	Unit	Material	Labor	Equipment	Total	Total Incl O&P
0010	**PREFABRICATED WOOD STAIRS**									
0100	Box stairs, prefabricated, 3'-0" wide									
0110	Oak treads, up to 14 risers	2 Carp	39	.410	Riser	98.50	14.65		113.15	133
0600	With pine treads for carpet, up to 14 risers	"	39	.410	"	63.50	14.65		78.15	94
1100	For 4' wide stairs, add				Flight	25%				
1550	Stairs, prefabricated stair handrail with balusters	1 Carp	30	.267	L.F.	82	9.50		91.50	106
1700	Basement stairs, prefabricated, pine treads									
1710	Pine risers, 3' wide, up to 14 risers	2 Carp	52	.308	Riser	63.50	10.95		74.45	88
4000	Residential, wood, oak treads, prefabricated		1.50	10.667	Flight	1,275	380		1,655	2,025
4200	Built in place		.44	36.364	"	2,325	1,300		3,625	4,700
4400	Spiral, oak, 4'-6" diameter, unfinished, prefabricated,									
4500	incl. railing, 9' high	2 Carp	1.50	10.667	Flight	2,675	380		3,055	3,550

06 43 13.40 Wood Stair Parts		Crew	Daily Output	Labor-Hours	Unit	Material	Labor	Equipment	Total	Total Incl O&P
0010	**WOOD STAIR PARTS**									
0020	Pin top balusters, 1-1/4", oak, 34"	1 Carp	96	.083	Ea.	5.20	2.97		8.17	10.60
0030	38"		96	.083		5.80	2.97		8.77	11.25
0040	42"		96	.083		6.30	2.97		9.27	11.85
0050	Poplar, 34"		96	.083		3.35	2.97		6.32	8.60
0060	38"		96	.083		5.90	2.97		8.87	11.40
0070	42"		96	.083		6.55	2.97		9.52	12.10
0080	Maple, 34"		96	.083		4.96	2.97		7.93	10.35
0090	38"		96	.083		5.60	2.97		8.57	11.05
0100	42"		96	.083		7.15	2.97		10.12	12.75
0130	Primed, 34"		96	.083		3.68	2.97		6.65	8.95
0140	38"		96	.083		4.22	2.97		7.19	9.55
0150	42"		96	.083		4.86	2.97		7.83	10.25
0180	Box top balusters, 1-1/4", oak, 34"		60	.133		8.70	4.75		13.45	17.40
0190	38"		60	.133		11.80	4.75		16.55	21
0200	42"		60	.133		12.90	4.75		17.65	22
0210	Poplar, 34"		60	.133		7.70	4.75		12.45	16.30

06 43 Wood Stairs and Railings

06 43 13 – Wood Stairs

06 43 13.40 Wood Stair Parts		Crew	Daily Output	Labor-Hours	Unit	Material	Labor	Equipment	Total	Total Incl O&P
						2019 Bare Costs				
0220	38"	1 Carp	60	.133	Ea.	8.30	4.75		13.05	17
0230	42"		60	.133		9.15	4.75		13.90	17.90
0240	Maple, 34"		60	.133		9.75	4.75		14.50	18.60
0250	38"		60	.133		10.70	4.75		15.45	19.60
0260	42"		60	.133		11.80	4.75		16.55	21
0290	Primed, 34"		60	.133		8.25	4.75		13	16.90
0300	38"		60	.133		12.05	4.75		16.80	21
0310	42"		60	.133		11.35	4.75		16.10	20.50
0340	Square balusters, cut from lineal stock, pine, 1-1/16" x 1-1/16"		180	.044	L.F.	1.30	1.58		2.88	4.05
0350	1-5/16" x 1-5/16"		180	.044		1.99	1.58		3.57	4.81
0360	1-5/8" x 1-5/8"		180	.044		2.55	1.58		4.13	5.45
0370	Turned newel, oak, 3-1/2" square, 48" high		8	1	Ea.	104	35.50		139.50	173
0380	62" high		8	1		100	35.50		135.50	169
0390	Poplar, 3-1/2" square, 48" high		8	1		46.50	35.50		82	110
0400	62" high		8	1		57	35.50		92.50	122
0410	Maple, 3-1/2" square, 48" high		8	1		64.50	35.50		100	130
0420	62" high		8	1		79.50	35.50		115	147
0430	Square newel, oak, 3-1/2" square, 48" high		8	1		54.50	35.50		90	119
0440	58" high		8	1		65.50	35.50		101	131
0450	Poplar, 3-1/2" square, 48" high		8	1		42	35.50		77.50	105
0460	58" high		8	1		51	35.50		86.50	116
0470	Maple, 3" square, 48" high		8	1		53.50	35.50		89	118
0480	58" high		8	1		67.50	35.50		103	133
0490	Railings, oak, economy		96	.083	L.F.	9.30	2.97		12.27	15.15
0500	Average		96	.083		15.05	2.97		18.02	21.50
0510	Custom		96	.083		18	2.97		20.97	24.50
0520	Maple, economy		96	.083		11.75	2.97		14.72	17.85
0530	Average		96	.083		13.40	2.97		16.37	19.65
0540	Custom		96	.083		21	2.97		23.97	28
0550	Oak, for bending rail, economy		48	.167		26.50	5.95		32.45	39
0560	Average		48	.167		29.50	5.95		35.45	42.50
0570	Custom		48	.167		33	5.95		38.95	46.50
0580	Maple, for bending rail, economy		48	.167		28.50	5.95		34.45	41.50
0590	Average		48	.167		31.50	5.95		37.45	44.50
0600	Custom		48	.167		34.50	5.95		40.45	48
0610	Risers, oak, 3/4" x 8", 36" long		80	.100	Ea.	12.90	3.57		16.47	20
0620	42" long		70	.114		15.05	4.07		19.12	23.50
0630	48" long		63	.127		17.20	4.53		21.73	26.50
0640	54" long		56	.143		19.35	5.10		24.45	30
0650	60" long		50	.160		21.50	5.70		27.20	33
0660	72" long		42	.190		26	6.80		32.80	39.50
0670	Poplar, 3/4" x 8", 36" long		80	.100		12.65	3.57		16.22	19.80
0680	42" long		71	.113		14.75	4.02		18.77	23
0690	48" long		63	.127		16.85	4.53		21.38	26
0700	54" long		56	.143		18.95	5.10		24.05	29.50
0710	60" long		50	.160		21	5.70		26.70	32.50
0720	72" long		42	.190		25.50	6.80		32.30	39
0730	Pine, 1" x 8", 36" long		80	.100		4.03	3.57		7.60	10.35
0740	42" long		70	.114		4.70	4.07		8.77	11.90
0750	48" long		63	.127		5.35	4.53		9.88	13.40
0760	54" long		56	.143		6.05	5.10		11.15	15.05
0770	60" long		50	.160		6.70	5.70		12.40	16.80
0780	72" long		42	.190		8.05	6.80		14.85	20

06 43 Wood Stairs and Railings

06 43 13 – Wood Stairs

06 43 13.40 Wood Stair Parts		Crew	Daily Output	Labor-Hours	Unit	Material	2019 Bare Costs Labor	Equipment	Total	Total Incl O&P
0790	Treads, oak, no returns, 1-1/32" x 11-1/2" x 36" long	1 Carp	32	.250	Ea.	29	8.90		37.90	47
0800	42" long		32	.250		34	8.90		42.90	52.50
0810	48" long		32	.250		39	8.90		47.90	57.50
0820	54" long		32	.250		43.50	8.90		52.40	63
0830	60" long		32	.250		48.50	8.90		57.40	68.50
0840	72" long		32	.250		58	8.90		66.90	79
0850	Mitred return one end, 1-1/32" x 11-1/2" x 36" long		24	.333		38.50	11.90		50.40	62
0860	42" long		24	.333		45	11.90		56.90	69
0870	48" long		24	.333		51.50	11.90		63.40	76
0880	54" long		24	.333		57.50	11.90		69.40	83
0890	60" long		24	.333		64	11.90		75.90	90
0900	72" long		24	.333		77	11.90		88.90	104
0910	Mitred return two ends, 1-1/32" x 11-1/2" x 36" long		12	.667		48.50	24		72.50	93
0920	42" long		12	.667		56.50	24		80.50	102
0930	48" long		12	.667		64.50	24		88.50	111
0940	54" long		12	.667		73	24		97	120
0950	60" long		12	.667		81	24		105	129
0960	72" long		12	.667		97	24		121	147
0970	Starting step, oak, 48", bullnose		8	1		171	35.50		206.50	247
0980	Double end bullnose		8	1		259	35.50		294.50	345
1030	Skirt board, pine, 1" x 10"		55	.145	L.F.	1.70	5.20		6.90	10.40
1040	1" x 12"		52	.154	"	2.14	5.50		7.64	11.40
1050	Oak landing tread, 1-1/16" thick		54	.148	S.F.	7.90	5.30		13.20	17.45
1060	Oak cove molding		96	.083	L.F.	.99	2.97		3.96	6
1070	Oak stringer molding		96	.083	"	3.95	2.97		6.92	9.25
1090	Rail bolt, 5/16" x 3-1/2"		48	.167	Ea.	3.50	5.95		9.45	13.65
1100	5/16" x 4-1/2"		48	.167		3.03	5.95		8.98	13.15
1120	Newel post anchor		16	.500		12.90	17.85		30.75	43.50
1130	Tapered plug, 1/2"		240	.033		.99	1.19		2.18	3.05
1140	1"		240	.033		.98	1.19		2.17	3.04

06 43 16 – Wood Railings

06 43 16.10 Wood Handrails and Railings		Crew	Daily Output	Labor-Hours	Unit	Material	Labor	Equipment	Total	Total Incl O&P
0010	**WOOD HANDRAILS AND RAILINGS**									
0020	Custom design, architectural grade, hardwood, plain	1 Carp	38	.211	L.F.	12.65	7.50		20.15	26.50
0100	Shaped		30	.267		52.50	9.50		62	73
0300	Stock interior railing with spindles 4" OC, 4' long		40	.200		48	7.15		55.15	64.50
0400	8' long		48	.167		48	5.95		53.95	62.50

06 44 Ornamental Woodwork

06 44 19 – Wood Grilles

06 44 19.10 Grilles		Crew	Daily Output	Labor-Hours	Unit	Material	Labor	Equipment	Total	Total Incl O&P
0010	**GRILLES** and panels, hardwood, sanded									
0020	2' x 4' to 4' x 8', custom designs, unfinished, economy	1 Carp	38	.211	S.F.	57.50	7.50		65	75.50
0050	Average		30	.267		71.50	9.50		81	94
0100	Custom		19	.421		75	15		90	108

06 44 33 – Wood Mantels

06 44 33.10 Fireplace Mantels		Crew	Daily Output	Labor-Hours	Unit	Material	Labor	Equipment	Total	Total Incl O&P
0010	**FIREPLACE MANTELS**									
0015	6" molding, 6' x 3'-6" opening, plain, paint grade	1 Carp	5	1.600	Opng.	465	57		522	605
0100	Ornate, oak		5	1.600		635	57		692	795

06 44 Ornamental Woodwork

06 44 33 – Wood Mantels

06 44 33.10 Fireplace Mantels		Crew	Daily Output	Labor-Hours	Unit	Material	Labor	Equipment	Total	Total Incl O&P
						2019 Bare Costs				
0300	Prefabricated pine, colonial type, stock, deluxe	1 Carp	2	4	Opng.	1,900	143		2,043	2,300
0400	Economy	↓	3	2.667	↓	820	95		915	1,050

06 44 33.20 Fireplace Mantel Beam

		Crew	Daily Output	Labor-Hours	Unit	Material	Labor	Equipment	Total	Total Incl O&P
0010	**FIREPLACE MANTEL BEAM**									
0020	Rough texture wood, 4" x 8"	1 Carp	36	.222	L.F.	8.70	7.90		16.60	22.50
0100	4" x 10"		35	.229	"	11	8.15		19.15	25.50
0300	Laminated hardwood, 2-1/4" x 10-1/2" wide, 6' long		5	1.600	Ea.	108	57		165	213
0400	8' long		5	1.600	"	151	57		208	260
0600	Brackets for above, rough sawn		12	.667	Pr.	10.45	24		34.45	51
0700	Laminated	↓	12	.667	"	21.50	24		45.50	63

06 44 39 – Wood Posts and Columns

06 44 39.10 Decorative Beams

		Crew	Daily Output	Labor-Hours	Unit	Material	Labor	Equipment	Total	Total Incl O&P
0010	**DECORATIVE BEAMS**									
0020	Rough sawn cedar, non-load bearing, 4" x 4"	2 Carp	180	.089	L.F.	1.75	3.17		4.92	7.20
0100	4" x 6"		170	.094		1.88	3.36		5.24	7.60
0200	4" x 8"		160	.100		2.49	3.57		6.06	8.65
0300	4" x 10"		150	.107		4.35	3.80		8.15	11.10
0400	4" x 12"		140	.114		5	4.07		9.07	12.25
0500	8" x 8"		130	.123		4.89	4.39		9.28	12.65
0600	Plastic beam, "hewn finish", 6" x 2"		240	.067		3.57	2.38		5.95	7.85
0601	6" x 4"	↓	220	.073	↓	3.94	2.59		6.53	8.60

06 44 39.20 Columns

		Crew	Daily Output	Labor-Hours	Unit	Material	Labor	Equipment	Total	Total Incl O&P
0010	**COLUMNS**									
0050	Aluminum, round colonial, 6" diameter	2 Carp	80	.200	V.L.F.	18.80	7.15		25.95	32.50
0100	8" diameter		62.25	.257		18.65	9.15		27.80	35.50
0200	10" diameter		55	.291		23	10.35		33.35	42
0250	Fir, stock units, hollow round, 6" diameter		80	.200		26.50	7.15		33.65	41
0300	8" diameter		80	.200		34	7.15		41.15	49.50
0350	10" diameter		70	.229		42	8.15		50.15	60
0360	12" diameter		65	.246		61.50	8.80		70.30	82
0400	Solid turned, to 8' high, 3-1/2" diameter		80	.200		11.60	7.15		18.75	24.50
0500	4-1/2" diameter		75	.213		12.60	7.60		20.20	26.50
0600	5-1/2" diameter		70	.229		16.95	8.15		25.10	32
0800	Square columns, built-up, 5" x 5"		65	.246		35	8.80		43.80	53
0900	Solid, 3-1/2" x 3-1/2"		130	.123		10.30	4.39		14.69	18.60
1600	Hemlock, tapered, T&G, 12" diam., 10' high		100	.160		51	5.70		56.70	65.50
1700	16' high		65	.246		85.50	8.80		94.30	109
1900	14" diameter, 10' high		100	.160		103	5.70		108.70	123
2000	18' high		65	.246		106	8.80		114.80	131
2200	18" diameter, 12' high		65	.246		179	8.80		187.80	212
2300	20' high		50	.320		134	11.40		145.40	166
2500	20" diameter, 14' high		40	.400		194	14.25		208.25	238
2600	20' high	↓	35	.457		185	16.30		201.30	230
2800	For flat pilasters, deduct				↓	33%				
3000	For splitting into halves, add				Ea.	115			115	126
4000	Rough sawn cedar posts, 4" x 4"	2 Carp	250	.064	V.L.F.	4.37	2.28		6.65	8.60
4100	4" x 6"		235	.068		8.90	2.43		11.33	13.80
4200	6" x 6"		220	.073		15.20	2.59		17.79	21
4300	8" x 8"	↓	200	.080	↓	19.15	2.85		22	25.50

06 48 Wood Frames

06 48 13 – Exterior Wood Door Frames

06 48 13.10	Exterior Wood Door Frames and Accessories	Crew	Daily Output	Labor-Hours	Unit	Material	2019 Bare Costs Labor	Equipment	Total	Total Incl O&P
0010	**EXTERIOR WOOD DOOR FRAMES AND ACCESSORIES**									
0400	Exterior frame, incl. ext. trim, pine, 5/4 x 4-9/16″ deep	2 Carp	375	.043	L.F.	6.75	1.52		8.27	9.90
0420	5-3/16″ deep		375	.043		8.25	1.52		9.77	11.60
0440	6-9/16″ deep		375	.043		10.05	1.52		11.57	13.55
0600	Oak, 5/4 x 4-9/16″ deep		350	.046		12.65	1.63		14.28	16.60
0620	5-3/16″ deep		350	.046		13.90	1.63		15.53	18
0640	6-9/16″ deep		350	.046		18.70	1.63		20.33	23
1000	Sills, 8/4 x 8″ deep, oak, no horns		100	.160		7.15	5.70		12.85	17.25
1020	2″ horns		100	.160		21.50	5.70		27.20	33
1040	3″ horns		100	.160		21.50	5.70		27.20	33
1100	8/4 x 10″ deep, oak, no horns		90	.178		6.55	6.35		12.90	17.65
1120	2″ horns		90	.178		27	6.35		33.35	40
1140	3″ horns		90	.178		27	6.35		33.35	40
2000	Wood frame & trim, ext., colonial, 3′ opng., fluted pilasters, flat head		22	.727	Ea.	505	26		531	600
2010	Dentil head		21	.762		600	27		627	705
2020	Ram's head		20	.800		675	28.50		703.50	790
2100	5′-4″ opening, in-swing, fluted pilasters, flat head		17	.941		450	33.50		483.50	555
2120	Ram's head		15	1.067		1,325	38		1,363	1,550
2140	Out-swing, fluted pilasters, flat head		17	.941		480	33.50		513.50	580
2160	Ram's head		15	1.067		1,525	38		1,563	1,750
2400	6′-0″ opening, in-swing, fluted pilasters, flat head		16	1		470	35.50		505.50	575
2420	Ram's head		10	1.600		1,500	57		1,557	1,750
2460	Out-swing, fluted pilasters, flat head		16	1		480	35.50		515.50	590
2480	Ram's head		10	1.600		1,525	57		1,582	1,775
2600	For two sidelights, flat head, add		30	.533	Opng.	300	19		319	360
2620	Ram's head, add		20	.800	″	910	28.50		938.50	1,050
2700	Custom birch frame, 3′-0″ opening		16	1	Ea.	235	35.50		270.50	315
2750	6′-0″ opng.		16	1		390	35.50		425.50	490
2900	Exterior, modern, plain trim, 3′ opng., in-swing, FJP		26	.615		48	22		70	89.50
2920	Fir		24	.667		56.50	24		80.50	102
2940	Oak		22	.727		66.50	26		92.50	116

06 48 16 – Interior Wood Door Frames

06 48 16.10	Interior Wood Door Jamb and Frames	Crew	Daily Output	Labor-Hours	Unit	Material	2019 Bare Costs Labor	Equipment	Total	Total Incl O&P
0010	**INTERIOR WOOD DOOR JAMB AND FRAMES**									
3000	Interior frame, pine, 11/16″ x 3-5/8″ deep	2 Carp	375	.043	L.F.	4.16	1.52		5.68	7.10
3020	4-9/16″ deep		375	.043		4.75	1.52		6.27	7.75
3040	5-3/16″ deep		375	.043		5.95	1.52		7.47	9.05
3200	Oak, 11/16″ x 3-5/8″ deep		350	.046		2.78	1.63		4.41	5.75
3220	4-9/16″ deep		350	.046		11.15	1.63		12.78	14.95
3240	5-3/16″ deep		350	.046		18.20	1.63		19.83	22.50
3400	Walnut, 11/16″ x 3-5/8″ deep		350	.046		9.55	1.63		11.18	13.20
3420	4-9/16″ deep		350	.046		10.85	1.63		12.48	14.60
3440	5-3/16″ deep		350	.046		9.85	1.63		11.48	13.50
3600	Pocket door frame		16	1	Ea.	101	35.50		136.50	170
3800	Threshold, oak, 5/8″ x 3-5/8″ deep		200	.080	L.F.	3.57	2.85		6.42	8.65
3820	4-5/8″ deep		190	.084		4.16	3		7.16	9.55
3840	5-5/8″ deep		180	.089		6.65	3.17		9.82	12.55

06 49 19 – Exterior Wood Shutters

06 49 19.10 Shutters, Exterior		Crew	Daily Output	Labor-Hours	Unit	Material	Labor	Equipment	Total	Total Incl O&P
						2019 Bare Costs				
0010	**SHUTTERS, EXTERIOR**									
0012	Aluminum, louvered, 1'-4" wide, 3'-0" long	1 Carp	10	.800	Pr.	200	28.50		228.50	267
0200	4'-0" long		10	.800		241	28.50		269.50	310
0300	5'-4" long		10	.800		291	28.50		319.50	365
0400	6'-8" long		9	.889		360	31.50		391.50	450
1000	Pine, louvered, primed, each 1'-2" wide, 3'-3" long		10	.800		267	28.50		295.50	340
1100	4'-7" long		10	.800		294	28.50		322.50	370
1250	Each 1'-4" wide, 3'-0" long		10	.800		288	28.50		316.50	360
1350	5'-3" long		10	.800		375	28.50		403.50	460
1500	Each 1'-6" wide, 3'-3" long		10	.800		279	28.50		307.50	350
1600	4'-7" long		10	.800		370	28.50		398.50	455
1620	Cedar, louvered, 1'-2" wide, 5'-7" long		10	.800		345	28.50		373.50	425
1630	Each 1'-4" wide, 2'-2" long		10	.800		185	28.50		213.50	250
1640	3'-0" long		10	.800		240	28.50		268.50	310
1650	3'-3" long		10	.800		251	28.50		279.50	325
1660	3'-11" long		10	.800		290	28.50		318.50	365
1670	4'-3" long		10	.800		320	28.50		348.50	400
1680	5'-3" long		10	.800		395	28.50		423.50	480
1690	5'-11" long		10	.800		415	28.50		443.50	505
1700	Door blinds, 6'-9" long, each 1'-3" wide		9	.889		465	31.50		496.50	570
1710	1'-6" wide		9	.889		420	31.50		451.50	515
1720	Cedar, solid raised panel, each 1'-4" wide, 3'-3" long		10	.800		350	28.50		378.50	430
1730	3'-11" long		10	.800		350	28.50		378.50	430
1740	4'-3" long		10	.800		360	28.50		388.50	440
1750	4'-7" long		10	.800		395	28.50		423.50	480
1760	4'-11" long		10	.800		425	28.50		453.50	515
1770	5'-11" long		10	.800		560	28.50		588.50	660
1800	Door blinds, 6'-9" long, each 1'-3" wide		9	.889		555	31.50		586.50	665
1900	1'-6" wide		9	.889		655	31.50		686.50	775
2500	Polystyrene, solid raised panel, each 1'-4" wide, 3'-3" long		10	.800		87.50	28.50		116	143
2600	3'-11" long		10	.800		93	28.50		121.50	149
2700	4'-7" long		10	.800		105	28.50		133.50	163
2800	5'-3" long		10	.800		120	28.50		148.50	179
2900	6'-8" long		9	.889		149	31.50		180.50	217
4500	Polystyrene, louvered, each 1'-2" wide, 3'-3" long		10	.800		39	28.50		67.50	90
4600	4'-7" long		10	.800		45.50	28.50		74	97
4750	5'-3" long		10	.800		53.50	28.50		82	106
4850	6'-8" long		9	.889		69.50	31.50		101	129
6000	Vinyl, louvered, each 1'-2" x 4'-7" long		10	.800		61.50	28.50		90	115
6200	Each 1'-4" x 6'-8" long	▼	9	.889	▼	82.50	31.50		114	143
8000	PVC exterior rolling shutters									
8100	including crank control	1 Carp	8	1	Ea.	760	35.50		795.50	895
8500	Insulative - 6' x 6'-8" stock unit	"	8	1	"	1,075	35.50		1,110.50	1,250

06 51 13.10 Recycled Plastic Lumber		Crew	Daily Output	Labor-Hours	Unit	Material	Labor	Equipment	Total	Total Incl O&P
						2019 Bare Costs				
0010	RECYCLED PLASTIC LUMBER									
4000	Sheeting, recycled plastic, black or white, 4′ x 8′ x 1/8″ G	2 Carp	1100	.015	S.F.	1.07	.52		1.59	2.04
4010	4′ x 8′ x 3/16″ G		1100	.015		1.49	.52		2.01	2.50
4020	4′ x 8′ x 1/4″ G		950	.017		1.41	.60		2.01	2.54
4030	4′ x 8′ x 3/8″ G		950	.017		2.85	.60		3.45	4.13
4040	4′ x 8′ x 1/2″ G		900	.018		3.78	.63		4.41	5.20
4050	4′ x 8′ x 5/8″ G		900	.018		6.50	.63		7.13	8.20
4060	4′ x 8′ x 3/4″ G		850	.019		8.05	.67		8.72	9.95
4070	Add for colors G				Ea.	5%				
8500	100% recycled plastic, var colors, NLB, 2″ x 2″ G				L.F.	1.96			1.96	2.16
8510	2″ x 4″ G					3.96			3.96	4.36
8520	2″ x 6″ G					6.20			6.20	6.85
8530	2″ x 8″ G					8.80			8.80	9.70
8540	2″ x 10″ G					12.45			12.45	13.70
8550	5/4″ x 4″ G					4.73			4.73	5.20
8560	5/4″ x 6″ G					6.55			6.55	7.25
8570	1″ x 6″ G					2.83			2.83	3.11
8580	1/2″ x 8″ G					3.41			3.41	3.75
8590	2″ x 10″ T&G G					12.60			12.60	13.90
8600	3″ x 10″ T&G G					19.45			19.45	21.50
8610	Add for premium colors G					20%				

06 51 13.12 Structural Plastic Lumber		Crew	Daily Output	Labor-Hours	Unit	Material	Labor	Equipment	Total	Total Incl O&P
0010	STRUCTURAL PLASTIC LUMBER									
1320	Plastic lumber, posts or columns, 4″ x 4″	2 Carp	390	.041	L.F.	11.05	1.46		12.51	14.55
1325	4″ x 6″		275	.058		13.55	2.07		15.62	18.35
1330	4″ x 8″		220	.073		19.80	2.59		22.39	26.50
1340	Girder, single, 4″ x 4″		675	.024		11.05	.84		11.89	13.55
1345	4″ x 6″		600	.027		13.55	.95		14.50	16.45
1350	4″ x 8″		525	.030		19.80	1.09		20.89	24
1352	Double, 2″ x 4″		625	.026		8.45	.91		9.36	10.80
1354	2″ x 6″		600	.027		9.95	.95		10.90	12.50
1356	2″ x 8″		575	.028		17.05	.99		18.04	20.50
1358	2″ x 10″		550	.029		24.50	1.04		25.54	28.50
1360	2″ x 12″		525	.030		25.50	1.09		26.59	30
1362	Triple, 2″ x 4″		575	.028		12.70	.99		13.69	15.60
1364	2″ x 6″		550	.029		14.95	1.04		15.99	18.15
1366	2″ x 8″		525	.030		25.50	1.09		26.59	30
1368	2″ x 10″		500	.032		36.50	1.14		37.64	42.50
1370	2″ x 12″		475	.034		38.50	1.20		39.70	44.50
1372	Ledger, bolted 4′ OC, 2″ x 4″		400	.040		4.38	1.43		5.81	7.20
1374	2″ x 6″		550	.029		5.05	1.04		6.09	7.25
1376	2″ x 8″		390	.041		8.65	1.46		10.11	11.90
1378	2″ x 10″		385	.042		12.30	1.48		13.78	16
1380	2″ x 12″		380	.042		12.90	1.50		14.40	16.70
1382	Joists, 2″ x 4″		1250	.013		4.23	.46		4.69	5.40
1384	2″ x 6″		1250	.013		4.99	.46		5.45	6.25
1386	2″ x 8″		1100	.015		8.55	.52		9.07	10.25
1388	2″ x 10″		500	.032		12.30	1.14		13.44	15.45
1390	2″ x 12″		875	.018		12.80	.65		13.45	15.20
1392	Railings and trim, 5/4″ x 4″	1 Carp	300	.027		4.95	.95		5.90	7
1394	2″ x 2″		300	.027		2.10	.95		3.05	3.88
1396	2″ x 4″		300	.027		4.21	.95		5.16	6.20

06 51 Structural Plastic Shapes and Plates

06 51 13 – Plastic Lumber

06 51 13.12 Structural Plastic Lumber		Crew	Daily Output	Labor-Hours	Unit	Material	Labor	Equipment	Total	Total Incl O&P
						2019 Bare Costs				
1398	2" x 6"	1 Carp	300	.027	L.F.	4.95	.95		5.90	7

06 63 Plastic Railings

06 63 10 – Plastic (PVC) Railings

06 63 10.10 Plastic Railings		Crew	Daily Output	Labor-Hours	Unit	Material	Labor	Equipment	Total	Total Incl O&P
0010	PLASTIC RAILINGS									
0100	Horizontal PVC handrail with balusters, 3-1/2" wide, 36" high	1 Carp	96	.083	L.F.	29.50	2.97		32.47	37.50
0150	42" high		96	.083		29.50	2.97		32.47	37.50
0200	Angled PVC handrail with balusters, 3-1/2" wide, 36" high		72	.111		20	3.96		23.96	28.50
0250	42" high		72	.111		29.50	3.96		33.46	39
0300	Post sleeve for 4 x 4 post		96	.083		15.70	2.97		18.67	22
0400	Post cap for 4 x 4 post, flat profile		48	.167	Ea.	13.55	5.95		19.50	25
0450	Newel post style profile		48	.167		26.50	5.95		32.45	39
0500	Raised corbeled profile		48	.167		37	5.95		42.95	50.50
0550	Post base trim for 4 x 4 post		96	.083		21.50	2.97		24.47	28.50

06 65 Plastic Trim

06 65 10 – PVC Trim

06 65 10.10 PVC Trim, Exterior		Crew	Daily Output	Labor-Hours	Unit	Material	Labor	Equipment	Total	Total Incl O&P
0010	PVC TRIM, EXTERIOR									
0100	Cornerboards, 5/4" x 6" x 6"	1 Carp	240	.033	L.F.	6.95	1.19		8.14	9.55
0110	Door/window casing, 1" x 4"		200	.040		1.63	1.43		3.06	4.15
0120	1" x 6"		200	.040		2.11	1.43		3.54	4.68
0130	1" x 8"		195	.041		2.98	1.46		4.44	5.70
0140	1" x 10"		195	.041		3.84	1.46		5.30	6.65
0150	1" x 12"		190	.042		4.20	1.50		5.70	7.10
0160	5/4" x 4"		195	.041		1.81	1.46		3.27	4.41
0170	5/4" x 6"		195	.041		2.78	1.46		4.24	5.50
0180	5/4" x 8"		190	.042		3.66	1.50		5.16	6.50
0190	5/4" x 10"		190	.042		4.64	1.50		6.14	7.60
0200	5/4" x 12"		185	.043		5.35	1.54		6.89	8.45
0210	Fascia, 1" x 4"		250	.032		1.63	1.14		2.77	3.67
0220	1" x 6"		250	.032		2.11	1.14		3.25	4.20
0230	1" x 8"		225	.036		2.98	1.27		4.25	5.35
0240	1" x 10"		225	.036		3.84	1.27		5.11	6.30
0250	1" x 12"		200	.040		4.20	1.43		5.63	7
0260	5/4" x 4"		240	.033		1.81	1.19		3	3.95
0270	5/4" x 6"		240	.033		2.78	1.19		3.97	5
0280	5/4" x 8"		215	.037		3.66	1.33		4.99	6.20
0290	5/4" x 10"		215	.037		4.64	1.33		5.97	7.30
0300	5/4" x 12"		190	.042		5.35	1.50		6.85	8.40
0310	Frieze, 1" x 4"		250	.032		1.63	1.14		2.77	3.67
0320	1" x 6"		250	.032		2.11	1.14		3.25	4.20
0330	1" x 8"		225	.036		2.98	1.27		4.25	5.35
0340	1" x 10"		225	.036		3.84	1.27		5.11	6.30
0350	1" x 12"		200	.040		4.20	1.43		5.63	7
0360	5/4" x 4"		240	.033		1.81	1.19		3	3.95
0370	5/4" x 6"		240	.033		2.78	1.19		3.97	5
0380	5/4" x 8"		215	.037		3.66	1.33		4.99	6.20

06 65 Plastic Trim

06 65 10 – PVC Trim

06 65 10.10 PVC Trim, Exterior		Crew	Daily Output	Labor-Hours	Unit	Material	2019 Bare Costs Labor	Equipment	Total	Total Incl O&P
0390	5/4" x 10"	1 Carp	215	.037	L.F.	4.64	1.33		5.97	7.30
0400	5/4" x 12"		190	.042		5.35	1.50		6.85	8.40
0410	Rake, 1" x 4"		200	.040		1.63	1.43		3.06	4.15
0420	1" x 6"		200	.040		2.11	1.43		3.54	4.68
0430	1" x 8"		190	.042		2.98	1.50		4.48	5.75
0440	1" x 10"		190	.042		3.84	1.50		5.34	6.70
0450	1" x 12"		180	.044		4.20	1.58		5.78	7.25
0460	5/4" x 4"		195	.041		1.81	1.46		3.27	4.41
0470	5/4" x 6"		195	.041		2.78	1.46		4.24	5.50
0480	5/4" x 8"		185	.043		3.66	1.54		5.20	6.60
0490	5/4" x 10"		185	.043		4.64	1.54		6.18	7.65
0500	5/4" x 12"		175	.046		5.35	1.63		6.98	8.60
0510	Rake trim, 1" x 4"		225	.036		1.63	1.27		2.90	3.88
0520	1" x 6"		225	.036		2.11	1.27		3.38	4.41
0560	5/4" x 4"		220	.036		1.81	1.30		3.11	4.13
0570	5/4" x 6"		220	.036		2.78	1.30		4.08	5.20
0610	Soffit, 1" x 4"	2 Carp	420	.038		1.63	1.36		2.99	4.03
0620	1" x 6"		420	.038		2.11	1.36		3.47	4.56
0630	1" x 8"		420	.038		2.98	1.36		4.34	5.50
0640	1" x 10"		400	.040		3.84	1.43		5.27	6.60
0650	1" x 12"		400	.040		4.20	1.43		5.63	7
0660	5/4" x 4"		410	.039		1.81	1.39		3.20	4.29
0670	5/4" x 6"		410	.039		2.78	1.39		4.17	5.35
0680	5/4" x 8"		410	.039		3.66	1.39		5.05	6.35
0690	5/4" x 10"		390	.041		4.64	1.46		6.10	7.50
0700	5/4" x 12"		390	.041		5.35	1.46		6.81	8.30

06 80 Composite Fabrications

06 80 10 – Composite Decking

06 80 10.10 Woodgrained Composite Decking		Crew	Daily Output	Labor-Hours	Unit	Material	Labor	Equipment	Total	Total Incl O&P
0010	**WOODGRAINED COMPOSITE DECKING**									
0100	Woodgrained composite decking, 1" x 6"	2 Carp	640	.025	L.F.	4.13	.89		5.02	6
0110	Grooved edge		660	.024		4.30	.86		5.16	6.15
0120	2" x 6"		640	.025		3.95	.89		4.84	5.80
0130	Encased, 1" x 6"		640	.025		4.35	.89		5.24	6.25
0140	Grooved edge		660	.024		4.52	.86		5.38	6.40
0150	2" x 6"		640	.025		5.35	.89		6.24	7.35

06 81 Composite Railings

06 81 10 – Encased Railings

06 81 10.10 Encased Composite Railings		Crew	Daily Output	Labor-Hours	Unit	Material	Labor	Equipment	Total	Total Incl O&P
0010	**ENCASED COMPOSITE RAILINGS**									
0100	Encased composite railing, 6' long, 36" high, incl. balusters	1 Carp	16	.500	Ea.	147	17.85		164.85	191
0110	42" high, incl. balusters		16	.500		219	17.85		236.85	271
0120	8' long, 36" high, incl. balusters		12	.667		165	24		189	221
0130	42" high, incl. balusters		12	.667		163	24		187	219
0140	Accessories, post sleeve, 4" x 4", 39" long		32	.250		28.50	8.90		37.40	46
0150	96" long		24	.333		75.50	11.90		87.40	103
0160	6" x 6", 39" long		32	.250		54	8.90		62.90	74.50

06 81 Composite Railings

06 81 10 – Encased Railings

06 81 10.10 Encased Composite Railings		Crew	Daily Output	Labor-Hours	Unit	2019 Bare Costs Material	Labor	Equipment	Total	Total Incl O&P
0170	96" long	1 Carp	24	.333	Ea.	157	11.90		168.90	192
0180	Accessories, post skirt, 4" x 4"		96	.083		5.30	2.97		8.27	10.70
0190	6" x 6"		96	.083		6.90	2.97		9.87	12.50
0200	Post cap, 4" x 4", flat		48	.167		9.05	5.95		15	19.75
0210	Pyramid		48	.167		7.75	5.95		13.70	18.35
0220	Post cap, 6" x 6", flat		48	.167		12.85	5.95		18.80	24
0230	Pyramid	▼	48	.167	▼	10.65	5.95		16.60	21.50

Division Notes

		CREW	DAILY OUTPUT	LABOR-HOURS	UNIT	BARE COSTS MAT.	LABOR	EQUIP.	TOTAL	TOTAL INCL O&P

Estimating Tips

07 10 00 Dampproofing and Waterproofing

- Be sure of the job specifications before pricing this subdivision. The difference in cost between waterproofing and dampproofing can be great. Waterproofing will hold back standing water. Dampproofing prevents the transmission of water vapor. Also included in this section are vapor retarding membranes.

07 20 00 Thermal Protection

- Insulation and fireproofing products are measured by area, thickness, volume, or R-value. Specifications may give only what the specific R-value should be in a certain situation. The estimator may need to choose the type of insulation to meet that R-value.

07 30 00 Steep Slope Roofing 07 40 00 Roofing and Siding Panels

- Many roofing and siding products are bought and sold by the square. One square is equal to an area that measures 100 square feet.

  This simple change in unit of measure could create a large error if the estimator is not observant. Accessories necessary for a complete installation must be figured into any calculations for both material and labor.

07 50 00 Membrane Roofing 07 60 00 Flashing and Sheet Metal 07 70 00 Roofing and Wall Specialties and Accessories

- The items in these subdivisions compose a roofing system. No one component completes the installation, and all must be estimated. Built-up or single-ply membrane roofing systems are made up of many products and installation trades. Wood blocking at roof perimeters or penetrations, parapet coverings, reglets, roof drains, gutters, downspouts, sheet metal flashing, skylights, smoke vents, and roof hatches all need to be considered along with the roofing material. Several different installation trades will need to work together on the roofing system. Inherent difficulties in the scheduling and coordination of various trades must be accounted for when estimating labor costs.

07 90 00 Joint Protection

- To complete the weather-tight shell, the sealants and caulkings must be estimated. Where different materials meet—at expansion joints, at flashing penetrations, and at hundreds of other locations throughout a construction project—caulking and sealants provide another line of defense against water penetration. Often, an entire system is based on the proper location and placement of caulking or sealants. The detailed drawings that are included as part of a set of architectural plans show typical locations for these materials. When caulking or sealants are shown at typical locations, this means the estimator must include them for all the locations where this detail is applicable. Be careful to keep different types of sealants separate, and remember to consider backer rods and primers if necessary.

Reference Numbers

Reference numbers are shown at the beginning of some major classifications. These numbers refer to related items in the Reference Section. The reference information may be an estimating procedure, an alternate pricing method, or technical information.

Note: Not all subdivisions listed here necessarily appear. ■

07 01 Operation and Maint. of Thermal and Moisture Protection

07 01 50 – Maintenance of Membrane Roofing

07 01 50.10 Roof Coatings		Crew	Daily Output	Labor-Hours	Unit	2019 Bare Costs Material	Labor	Equipment	Total	Total Incl O&P
0010	**ROOF COATINGS**									
0012	Asphalt, brush grade, material only				Gal.	9.05			9.05	9.95
0800	Glass fibered roof & patching cement, 5 gal.					9.15			9.15	10.05
1100	Roof patch & flashing cement, 5 gal.					9.10			9.10	10

07 05 Common Work Results for Thermal and Moisture Protection

07 05 05 – Selective Demolition for Thermal and Moisture Protection

07 05 05.10 Selective Demo., Thermal and Moist. Protection

		Crew	Daily Output	Labor-Hours	Unit	Material	Labor	Equipment	Total	Total Incl O&P
0010	**SELECTIVE DEMO., THERMAL AND MOISTURE PROTECTION**									
0020	Caulking/sealant, to 1" x 1" joint R024119-10	1 Clab	600	.013	L.F.		.37		.37	.61
0120	Downspouts, including hangers		350	.023	"		.63		.63	1.04
0220	Flashing, sheet metal		290	.028	S.F.		.76		.76	1.25
0420	Gutters, aluminum or wood, edge hung		240	.033	L.F.		.92		.92	1.51
0520	Built-in		100	.080	"		2.20		2.20	3.64
0620	Insulation, air/vapor barrier		3500	.002	S.F.		.06		.06	.10
0670	Batts or blankets		1400	.006	C.F.		.16		.16	.26
0720	Foamed or sprayed in place	2 Clab	1000	.016	B.F.		.44		.44	.73
0770	Loose fitting	1 Clab	3000	.003	C.F.		.07		.07	.12
0870	Rigid board		3450	.002	B.F.		.06		.06	.11
1120	Roll roofing, cold adhesive		12	.667	Sq.		18.35		18.35	30.50
1170	Roof accessories, adjustable metal chimney flashing		9	.889	Ea.		24.50		24.50	40.50
1325	Plumbing vent flashing		32	.250	"		6.90		6.90	11.35
1375	Ridge vent strip, aluminum		310	.026	L.F.		.71		.71	1.17
1620	Skylight to 10 S.F.		8	1	Ea.		27.50		27.50	45.50
2120	Roof edge, aluminum soffit and fascia		570	.014	L.F.		.39		.39	.64
2170	Concrete coping, up to 12" wide	2 Clab	160	.100			2.75		2.75	4.55
2220	Drip edge	1 Clab	1000	.008			.22		.22	.36
2270	Gravel stop		950	.008			.23		.23	.38
2370	Sheet metal coping, up to 12" wide		240	.033			.92		.92	1.51
2470	Roof insulation board, over 2" thick	B-2	7800	.005	B.F.		.14		.14	.24
2520	Up to 2" thick	"	3900	.010	S.F.		.29		.29	.47
2620	Roof ventilation, louvered gable vent	1 Clab	16	.500	Ea.		13.75		13.75	22.50
2670	Remove, roof hatch	G-3	15	2.133			71		71	118
2675	Rafter vents	1 Clab	960	.008			.23		.23	.38
2720	Soffit vent and/or fascia vent		575	.014	L.F.		.38		.38	.63
2775	Soffit vent strip, aluminum, 3" to 4" wide		160	.050			1.38		1.38	2.27
2820	Roofing accessories, shingle moulding, to 1" x 4"		1600	.005			.14		.14	.23
2870	Cant strip	B-2	2000	.020			.56		.56	.92
2920	Concrete block walkway	1 Clab	230	.035			.96		.96	1.58
3070	Roofing, felt paper, #15		70	.114	Sq.		3.14		3.14	5.20
3125	#30 felt		30	.267	"		7.35		7.35	12.10
3170	Asphalt shingles, 1 layer	B-2	3500	.011	S.F.		.32		.32	.53
3180	2 layers		1750	.023	"		.64		.64	1.05
3370	Modified bitumen		26	1.538	Sq.		43		43	71
3420	Built-up, no gravel, 3 ply		25	1.600			44.50		44.50	74
3470	4 ply		21	1.905			53		53	88
3620	5 ply		1600	.025	S.F.		.70		.70	1.15
3720	5 ply, with gravel		890	.045			1.25		1.25	2.07
3725	Loose gravel removal		5000	.008			.22		.22	.37
3730	Embedded gravel removal		2000	.020			.56		.56	.92
3870	Fiberglass sheet		1200	.033			.93		.93	1.54

07 05 Common Work Results for Thermal and Moisture Protection

07 05 05 – Selective Demolition for Thermal and Moisture Protection

07 05 05.10 Selective Demo., Thermal and Moist. Protection		Crew	Daily Output	Labor-Hours	Unit	Material	2019 Bare Costs Labor	Equipment	Total	Total Incl O&P
4120	Slate shingles	B-2	1900	.021	S.F.		.59		.59	.97
4170	Ridge shingles, clay or slate		2000	.020	L.F.		.56		.56	.92
4320	Single ply membrane, attached at seams		52	.769	Sq.		21.50		21.50	35.50
4370	Ballasted		75	.533			14.90		14.90	24.50
4420	Fully adhered		39	1.026			28.50		28.50	47.50
4550	Roof hatch, 2′-6″ x 3′-0″	1 Clab	10	.800	Ea.		22		22	36.50
4670	Wood shingles	B-2	2200	.018	S.F.		.51		.51	.84
4820	Sheet metal roofing	"	2150	.019			.52		.52	.86
4970	Siding, horizontal wood clapboards	1 Clab	380	.021			.58		.58	.96
5025	Exterior insulation finish system	"	120	.067			1.83		1.83	3.03
5070	Tempered hardboard, remove and reset	1 Carp	380	.021			.75		.75	1.24
5120	Tempered hardboard sheet siding	"	375	.021			.76		.76	1.26
5170	Metal, corner strips	1 Clab	850	.009	L.F.		.26		.26	.43
5225	Horizontal strips		444	.018	S.F.		.50		.50	.82
5320	Vertical strips		400	.020			.55		.55	.91
5520	Wood shingles		350	.023			.63		.63	1.04
5620	Stucco siding		360	.022			.61		.61	1.01
5670	Textured plywood		725	.011			.30		.30	.50
5720	Vinyl siding		510	.016			.43		.43	.71
5770	Corner strips		900	.009	L.F.		.24		.24	.40
5870	Wood, boards, vertical		400	.020	S.F.		.55		.55	.91
5880	Steel siding, corrugated/ribbed		402.50	.020	"		.55		.55	.90
5920	Waterproofing, protection/drain board	2 Clab	3900	.004	B.F.		.11		.11	.19
5970	Over 1/2″ thick		1750	.009	S.F.		.25		.25	.42
6020	To 1/2″ thick		2000	.008	"		.22		.22	.36

07 11 Dampproofing

07 11 13 – Bituminous Dampproofing

07 11 13.10 Bituminous Asphalt Coating		Crew	Daily Output	Labor-Hours	Unit	Material	Labor	Equipment	Total	Total Incl O&P
0010	**BITUMINOUS ASPHALT COATING**									
0030	Brushed on, below grade, 1 coat	1 Rofc	665	.012	S.F.	.23	.37		.60	.92
0100	2 coat		500	.016		.45	.49		.94	1.39
0300	Sprayed on, below grade, 1 coat		830	.010		.23	.30		.53	.78
0400	2 coat		500	.016		.44	.49		.93	1.38
0500	Asphalt coating, with fibers				Gal.	9.15			9.15	10.05
0600	Troweled on, asphalt with fibers, 1/16″ thick	1 Rofc	500	.016	S.F.	.40	.49		.89	1.33
0700	1/8″ thick		400	.020		.70	.61		1.31	1.88
1000	1/2″ thick		350	.023		2.29	.70		2.99	3.78

07 11 16 – Cementitious Dampproofing

07 11 16.20 Cementitious Parging		Crew	Daily Output	Labor-Hours	Unit	Material	Labor	Equipment	Total	Total Incl O&P
0010	**CEMENTITIOUS PARGING**									
0020	Portland cement, 2 coats, 1/2″ thick	D-1	250	.064	S.F.	.39	2.07		2.46	3.88
0100	Waterproofed Portland cement, 1/2″ thick, 2 coats	"	250	.064	"	6.55	2.07		8.62	10.65

07 19 Water Repellents

07 19 19 – Silicone Water Repellents

07 19 19.10 Silicone Based Water Repellents			Crew	Daily Output	Labor-Hours	Unit	2019 Bare Costs Material	Labor	Equipment	Total	Total Incl O&P
0010	**SILICONE BASED WATER REPELLENTS**										
0020	Water base liquid, roller applied		2 Rofc	7000	.002	S.F.	.42	.07		.49	.60
0200	Silicone or stearate, sprayed on CMU, 1 coat		1 Rofc	4000	.002		.40	.06		.46	.55
0300	2 coats		"	3000	.003		.80	.08		.88	1.03

07 21 Thermal Insulation

07 21 13 – Board Insulation

07 21 13.10 Rigid Insulation

			Crew	Daily Output	Labor-Hours	Unit	Material	Labor	Equipment	Total	Total Incl O&P
0010	**RIGID INSULATION**, for walls										
0040	Fiberglass, 1.5#/C.F., unfaced, 1″ thick, R4.1	G	1 Carp	1000	.008	S.F.	.34	.29		.63	.84
0060	1-1/2″ thick, R6.2	G		1000	.008		.40	.29		.69	.91
0080	2″ thick, R8.3	G		1000	.008		.50	.29		.79	1.02
0120	3″ thick, R12.4	G		800	.010		.60	.36		.96	1.25
0370	3#/C.F., unfaced, 1″ thick, R4.3	G		1000	.008		.56	.29		.85	1.09
0390	1-1/2″ thick, R6.5	G		1000	.008		.80	.29		1.09	1.35
0400	2″ thick, R8.7	G		890	.009		1.07	.32		1.39	1.71
0420	2-1/2″ thick, R10.9	G		800	.010		1.11	.36		1.47	1.81
0440	3″ thick, R13	G		800	.010		1.63	.36		1.99	2.38
0520	Foil faced, 1″ thick, R4.3	G		1000	.008		.84	.29		1.13	1.39
0540	1-1/2″ thick, R6.5	G		1000	.008		1.25	.29		1.54	1.85
0560	2″ thick, R8.7	G		890	.009		1.58	.32		1.90	2.27
0580	2-1/2″ thick, R10.9	G		800	.010		1.86	.36		2.22	2.64
0600	3″ thick, R13	G		800	.010		2.09	.36		2.45	2.89
1600	Isocyanurate, 4′ x 8′ sheet, foil faced, both sides										
1610	1/2″ thick	G	1 Carp	800	.010	S.F.	.31	.36		.67	.93
1620	5/8″ thick	G		800	.010		.55	.36		.91	1.20
1630	3/4″ thick	G		800	.010		.46	.36		.82	1.10
1640	1″ thick	G		800	.010		.61	.36		.97	1.26
1650	1-1/2″ thick	G		730	.011		.67	.39		1.06	1.39
1660	2″ thick	G		730	.011		.90	.39		1.29	1.64
1670	3″ thick	G		730	.011		2.84	.39		3.23	3.77
1680	4″ thick	G		730	.011		2.56	.39		2.95	3.47
1700	Perlite, 1″ thick, R2.77	G		800	.010		.48	.36		.84	1.12
1750	2″ thick, R5.55	G		730	.011		.78	.39		1.17	1.51
1900	Extruded polystyrene, 25 psi compressive strength, 1″ thick, R5	G		800	.010		.56	.36		.92	1.21
1940	2″ thick, R10	G		730	.011		1.12	.39		1.51	1.88
1960	3″ thick, R15	G		730	.011		1.61	.39		2	2.42
2100	Expanded polystyrene, 1″ thick, R3.85	G		800	.010		.27	.36		.63	.89
2120	2″ thick, R7.69	G		730	.011		.54	.39		.93	1.24
2140	3″ thick, R11.49	G		730	.011		.81	.39		1.20	1.54

07 21 13.13 Foam Board Insulation

			Crew	Daily Output	Labor-Hours	Unit	Material	Labor	Equipment	Total	Total Incl O&P
0010	**FOAM BOARD INSULATION**										
0600	Polystyrene, expanded, 1″ thick, R4	G	1 Carp	680	.012	S.F.	.27	.42		.69	.99
0700	2″ thick, R8	G	"	675	.012	"	.54	.42		.96	1.29

07 21 16 – Blanket Insulation

07 21 16.10 Blanket Insulation for Floors/Ceilings

			Crew	Daily Output	Labor-Hours	Unit	Material	Labor	Equipment	Total	Total Incl O&P
0010	**BLANKET INSULATION FOR FLOORS/CEILINGS**										
0020	Including spring type wire fasteners										
2000	Fiberglass, blankets or batts, paper or foil backing										
2100	3-1/2″ thick, R13	G	1 Carp	700	.011	S.F.	.41	.41		.82	1.12

07 21 Thermal Insulation

07 21 16 – Blanket Insulation

07 21 16.10 Blanket Insulation for Floors/Ceilings			Crew	Daily Output	Labor-Hours	Unit	2019 Bare Costs Material	Labor	Equipment	Total	Total Incl O&P
2150	6-1/4" thick, R19	G	1 Carp	600	.013	S.F.	.50	.48		.98	1.34
2210	9-1/2" thick, R30	G		500	.016		.73	.57		1.30	1.74
2220	12" thick, R38	G		475	.017		1.06	.60		1.66	2.16
3000	Unfaced, 3-1/2" thick, R13	G		600	.013		.33	.48		.81	1.15
3010	6-1/4" thick, R19	G		500	.016		.39	.57		.96	1.37
3020	9-1/2" thick, R30	G		450	.018		.61	.63		1.24	1.72
3030	12" thick, R38	G		425	.019		.74	.67		1.41	1.92

07 21 16.20 Blanket Insulation for Walls			Crew	Daily Output	Labor-Hours	Unit	Material	Labor	Equipment	Total	Total Incl O&P
0010	**BLANKET INSULATION FOR WALLS**										
0020	Kraft faced fiberglass, 3-1/2" thick, R11, 15" wide	G	1 Carp	1350	.006	S.F.	.30	.21		.51	.68
0030	23" wide	G		1600	.005		.30	.18		.48	.62
0060	R13, 11" wide	G		1150	.007		.34	.25		.59	.78
0080	15" wide	G		1350	.006		.34	.21		.55	.72
0100	23" wide	G		1600	.005		.34	.18		.52	.66
0110	R15, 11" wide	G		1150	.007		.53	.25		.78	.99
0120	15" wide	G		1350	.006		.53	.21		.74	.93
0140	6" thick, R19, 11" wide	G		1150	.007		.43	.25		.68	.88
0160	15" wide	G		1350	.006		.43	.21		.64	.82
0180	23" wide	G		1600	.005		.43	.18		.61	.76
0201	9" thick, R30, 15" wide	G		1350	.006		.73	.21		.94	1.15
0241	12" thick, R38, 15" wide	G		1350	.006		1.06	.21		1.27	1.52
0410	Foil faced fiberglass, 3-1/2" thick, R13, 11" wide	G		1150	.007		.47	.25		.72	.93
0420	15" wide	G		1350	.006		.47	.21		.68	.87
0442	R15, 11" wide	G		1150	.007		.48	.25		.73	.94
0444	15" wide	G		1350	.006		.48	.21		.69	.88
0461	6" thick, R19, 15" wide	G		1600	.005		.61	.18		.79	.96
0482	R21, 11" wide	G		1150	.007		.64	.25		.89	1.11
0501	9" thick, R30, 15" wide	G		1350	.006		.93	.21		1.14	1.37
0620	Unfaced fiberglass, 3-1/2" thick, R13, 11" wide	G		1150	.007		.33	.25		.58	.77
0821	15" wide	G		1600	.005		.33	.18		.51	.65
0832	R15, 11" wide	G		1150	.007		.46	.25		.71	.92
0834	15" wide	G		1350	.006		.46	.21		.67	.86
0861	6" thick, R19, 15" wide	G		1350	.006		.39	.21		.60	.78
0901	9" thick, R30, 15" wide	G		1150	.007		.61	.25		.86	1.08
0941	12" thick, R38, 15" wide	G		1150	.007		.74	.25		.99	1.22
1300	Wall or ceiling insulation, mineral wool batts										
1320	3-1/2" thick, R15	G	1 Carp	1600	.005	S.F.	.79	.18		.97	1.16
1340	5-1/2" thick, R23	G		1600	.005		1.24	.18		1.42	1.66
1380	7-1/4" thick, R30	G		1350	.006		1.64	.21		1.85	2.15
1700	Non-rigid insul., recycled blue cotton fiber, unfaced batts, R13, 16" wide	G		1600	.005		1.10	.18		1.28	1.50
1710	R19, 16" wide	G		1600	.005		1.35	.18		1.53	1.78
1850	Friction fit wire insulation supports, 16" OC			960	.008	Ea.	.07	.30		.37	.57

07 21 19 – Foamed In Place Insulation

07 21 19.10 Masonry Foamed In Place Insulation			Crew	Daily Output	Labor-Hours	Unit	Material	Labor	Equipment	Total	Total Incl O&P
0010	**MASONRY FOAMED IN PLACE INSULATION**										
0100	Amino-plast foam, injected into block core, 6" block	G	G-2A	6000	.004	Ea.	.17	.11	.10	.38	.49
0110	8" block	G		5000	.005		.20	.13	.12	.45	.59
0120	10" block	G		4000	.006		.25	.16	.16	.57	.73
0130	12" block	G		3000	.008		.34	.22	.21	.77	.99
0140	Injected into cavity wall	G		13000	.002	B.F.	.06	.05	.05	.16	.21
0150	Preparation, drill holes into mortar joint every 4 V.L.F., 5/8" diameter		1 Clab	960	.008	Ea.		.23		.23	.38
0160	7/8" diameter			680	.012			.32		.32	.53

07 21 Thermal Insulation

07 21 19 – Foamed In Place Insulation

07 21 19.10 Masonry Foamed In Place Insulation			Crew	Daily Output	Labor-Hours	Unit	2019 Bare Costs Material	Labor	Equipment	Total	Total Incl O&P
0170	Patch drilled holes, 5/8" diameter		1 Clab	1800	.004	Ea.	.04	.12		.16	.24
0180	7/8" diameter			1200	.007		.05	.18		.23	.35

07 21 23 – Loose-Fill Insulation

07 21 23.10 Poured Loose-Fill Insulation			Crew	Daily Output	Labor-Hours	Unit	Material	Labor	Equipment	Total	Total Incl O&P
0010	**POURED LOOSE-FILL INSULATION**										
0020	Cellulose fiber, R3.8 per inch	G	1 Carp	200	.040	C.F.	.70	1.43		2.13	3.13
0021	4" thick	G		1000	.008	S.F.	.17	.29		.46	.65
0022	6" thick	G		800	.010	"	.28	.36		.64	.90
0080	Fiberglass wool, R4 per inch	G		200	.040	C.F.	.70	1.43		2.13	3.13
0081	4" thick	G		600	.013	S.F.	.24	.48		.72	1.05
0082	6" thick	G		400	.020	"	.34	.71		1.05	1.55
0100	Mineral wool, R3 per inch	G		200	.040	C.F.	.53	1.43		1.96	2.94
0101	4" thick	G		600	.013	S.F.	.17	.48		.65	.98
0102	6" thick	G		400	.020	"	.27	.71		.98	1.47
0300	Polystyrene, R4 per inch	G		200	.040	C.F.	1.51	1.43		2.94	4.02
0301	4" thick	G		600	.013	S.F.	.50	.48		.98	1.34
0302	6" thick	G		400	.020	"	.76	.71		1.47	2.01
0400	Perlite, R2.78 per inch	G		200	.040	C.F.	5.30	1.43		6.73	8.20
0401	4" thick	G		1000	.008	S.F.	1.77	.29		2.06	2.42
0402	6" thick	G		800	.010	"	2.66	.36		3.02	3.52

07 21 23.20 Masonry Loose-Fill Insulation			Crew	Daily Output	Labor-Hours	Unit	Material	Labor	Equipment	Total	Total Incl O&P
0010	**MASONRY LOOSE-FILL INSULATION**, vermiculite or perlite										
0100	In cores of concrete block, 4" thick wall, .115 C.F./S.F.	G	D-1	4800	.003	S.F.	.61	.11		.72	.85
0700	Foamed in place, urethane in 2-5/8" cavity	G	G-2A	1035	.023		1.35	.63	.60	2.58	3.24
0800	For each 1" added thickness, add	G	"	2372	.010		.51	.27	.26	1.04	1.34

07 21 26 – Blown Insulation

07 21 26.10 Blown Insulation			Crew	Daily Output	Labor-Hours	Unit	Material	Labor	Equipment	Total	Total Incl O&P
0010	**BLOWN INSULATION** Ceilings, with open access										
0020	Cellulose, 3-1/2" thick, R13	G	G-4	5000	.005	S.F.	.24	.14	.06	.44	.55
0030	5-3/16" thick, R19	G		3800	.006		.35	.18	.08	.61	.77
0050	6-1/2" thick, R22	G		3000	.008		.45	.23	.10	.78	.98
0100	8-11/16" thick, R30	G		2600	.009		.61	.26	.12	.99	1.23
0120	10-7/8" thick, R38	G		1800	.013		.78	.38	.17	1.33	1.67
1000	Fiberglass, 5.5" thick, R11	G		3800	.006		.24	.18	.08	.50	.64
1050	6" thick, R12	G		3000	.008		.34	.23	.10	.67	.85
1100	8.8" thick, R19	G		2200	.011		.42	.31	.14	.87	1.13
1200	10" thick, R22	G		1800	.013		.49	.38	.17	1.04	1.35
1300	11.5" thick, R26	G		1500	.016		.58	.45	.21	1.24	1.61
1350	13" thick, R30	G		1400	.017		.68	.48	.22	1.38	1.78
1450	16" thick, R38	G		1145	.021		.86	.59	.27	1.72	2.23
1500	20" thick, R49	G		920	.026		1.14	.74	.34	2.22	2.84

07 21 27 – Reflective Insulation

07 21 27.10 Reflective Insulation Options			Crew	Daily Output	Labor-Hours	Unit	Material	Labor	Equipment	Total	Total Incl O&P
0010	**REFLECTIVE INSULATION OPTIONS**										
0020	Aluminum foil on reinforced scrim	G	1 Carp	19	.421	C.S.F.	15	15		30	41.50
0100	Reinforced with woven polyolefin	G		19	.421		22.50	15		37.50	49.50
0500	With single bubble air space, R8.8	G		15	.533		26.50	19		45.50	60.50
0600	With double bubble air space, R9.8	G		15	.533		31.50	19		50.50	66

07 21 Thermal Insulation

07 21 29 – Sprayed Insulation

07 21 29.10 Sprayed-On Insulation		Crew	Daily Output	Labor-Hours	Unit	2019 Bare Costs Material	Labor	Equipment	Total	Total Incl O&P
0010	**SPRAYED-ON INSULATION**									
0300	Closed cell, spray polyurethane foam, 2 lb./C.F. density									
0310	1" thick G	G-2A	6000	.004	S.F.	.51	.11	.10	.72	.87
0320	2" thick G		3000	.008		1.03	.22	.21	1.46	1.74
0330	3" thick G		2000	.012		1.54	.32	.31	2.17	2.61
0335	3-1/2" thick G		1715	.014		1.80	.38	.36	2.54	3.04
0340	4" thick G		1500	.016		2.06	.43	.42	2.91	3.48
0350	5" thick G		1200	.020		2.57	.54	.52	3.63	4.35
0355	5-1/2" thick G		1090	.022		2.83	.60	.57	4	4.78
0360	6" thick G		1000	.024		3.08	.65	.62	4.35	5.20

07 22 Roof and Deck Insulation

07 22 16 – Roof Board Insulation

07 22 16.10 Roof Deck Insulation		Crew	Daily Output	Labor-Hours	Unit	Material	Labor	Equipment	Total	Total Incl O&P
0010	**ROOF DECK INSULATION**, fastening excluded									
0016	Asphaltic cover board, fiberglass lined, 1/8" thick	1 Rofc	1400	.006	S.F.	.48	.18		.66	.85
0018	1/4" thick		1400	.006		.97	.18		1.15	1.39
0020	Fiberboard low density, 1/2" thick, R1.39 G		1300	.006		.35	.19		.54	.73
0030	1" thick, R2.78 G		1040	.008		.59	.24		.83	1.08
0080	1-1/2" thick, R4.17 G		1040	.008		.90	.24		1.14	1.42
0100	2" thick, R5.56 G		1040	.008		1.16	.24		1.40	1.71
0110	Fiberboard high density, 1/2" thick, R1.3 G		1300	.006		.27	.19		.46	.64
0120	1" thick, R2.5 G		1040	.008		.57	.24		.81	1.06
0130	1-1/2" thick, R3.8 G		1040	.008		.84	.24		1.08	1.35
0200	Fiberglass, 3/4" thick, R2.78 G		1300	.006		.62	.19		.81	1.02
0400	15/16" thick, R3.70 G		1300	.006		.82	.19		1.01	1.24
0460	1-1/16" thick, R4.17 G		1300	.006		1.08	.19		1.27	1.53
0600	1-5/16" thick, R5.26 G		1300	.006		1.43	.19		1.62	1.91
0650	2-1/16" thick, R8.33 G		1040	.008		1.60	.24		1.84	2.19
0700	2-7/16" thick, R10 G		1040	.008		1.73	.24		1.97	2.33
0800	Gypsum cover board, fiberglass mat facer, 1/4" thick		1400	.006		.48	.18		.66	.85
0810	1/2" thick		1300	.006		.59	.19		.78	.99
0820	5/8" thick		1200	.007		.62	.20		.82	1.05
0830	Primed fiberglass mat facer, 1/4" thick		1400	.006		.50	.18		.68	.87
0840	1/2" thick		1300	.006		.58	.19		.77	.98
0850	5/8" thick		1200	.007		.61	.20		.81	1.04
1650	Perlite, 1/2" thick, R1.32 G		1365	.006		.27	.18		.45	.62
1655	3/4" thick, R2.08 G		1040	.008		.38	.24		.62	.85
1660	1" thick, R2.78 G		1040	.008		.55	.24		.79	1.04
1670	1-1/2" thick, R4.17 G		1040	.008		.81	.24		1.05	1.32
1680	2" thick, R5.56 G		910	.009		1.10	.27		1.37	1.70
1685	2-1/2" thick, R6.67 G		910	.009		1.45	.27		1.72	2.09
1690	Tapered for drainage G		1040	.008	B.F.	1.05	.24		1.29	1.59
1700	Polyisocyanurate, 2#/C.F. density, 3/4" thick G		1950	.004	S.F.	.50	.13		.63	.78
1705	1" thick G		1820	.004		.44	.13		.57	.72
1715	1-1/2" thick G		1625	.005		.60	.15		.75	.93
1725	2" thick G		1430	.006		.79	.17		.96	1.18
1735	2-1/2" thick G		1365	.006		1.02	.18		1.20	1.44
1745	3" thick G		1300	.006		1.12	.19		1.31	1.57
1755	3-1/2" thick G		1300	.006		1.86	.19		2.05	2.39

07 22 16.10 Roof Deck Insulation			Crew	Daily Output	Labor-Hours	Unit	Material	2019 Bare Costs Labor	Equipment	Total	Total Incl O&P
1765	Tapered for drainage	G	1 Rofc	1820	.004	B.F.	.52	.13		.65	.81
1900	Extruded polystyrene										
1910	15 psi compressive strength, 1" thick, R5	G	1 Rofc	1950	.004	S.F.	.56	.13		.69	.85
1920	2" thick, R10	G		1625	.005		.73	.15		.88	1.07
1930	3" thick, R15	G		1300	.006		1.46	.19		1.65	1.94
1932	4" thick, R20	G		1300	.006		1.96	.19		2.15	2.50
1934	Tapered for drainage	G		1950	.004	B.F.	.51	.13		.64	.79
1940	25 psi compressive strength, 1" thick, R5	G		1950	.004	S.F.	1.12	.13		1.25	1.46
1942	2" thick, R10	G		1625	.005		2.13	.15		2.28	2.61
1944	3" thick, R15	G		1300	.006		3.25	.19		3.44	3.91
1946	4" thick, R20	G		1300	.006		4.48	.19		4.67	5.25
1948	Tapered for drainage	G		1950	.004	B.F.	.56	.13		.69	.85
1950	40 psi compressive strength, 1" thick, R5	G		1950	.004	S.F.	.87	.13		1	1.19
1952	2" thick, R10	G		1625	.005		1.65	.15		1.80	2.09
1954	3" thick, R15	G		1300	.006		2.39	.19		2.58	2.97
1956	4" thick, R20	G		1300	.006		3.13	.19		3.32	3.79
1958	Tapered for drainage	G		1820	.004	B.F.	.87	.13		1	1.20
1960	60 psi compressive strength, 1" thick, R5	G		1885	.004	S.F.	1.05	.13		1.18	1.39
1962	2" thick, R10	G		1560	.005		2	.16		2.16	2.47
1964	3" thick, R15	G		1270	.006		3.26	.19		3.45	3.93
1966	4" thick, R20	G		1235	.006		4.04	.20		4.24	4.81
1968	Tapered for drainage	G		1820	.004	B.F.	1.01	.13		1.14	1.35
2010	Expanded polystyrene, 1#/C.F. density, 3/4" thick, R2.89	G		1950	.004	S.F.	.20	.13		.33	.45
2020	1" thick, R3.85	G		1950	.004		.27	.13		.40	.53
2100	2" thick, R7.69	G		1625	.005		.54	.15		.69	.86
2110	3" thick, R11.49	G		1625	.005		.81	.15		.96	1.16
2120	4" thick, R15.38	G		1625	.005		1.08	.15		1.23	1.46
2130	5" thick, R19.23	G		1495	.005		1.35	.16		1.51	1.79
2140	6" thick, R23.26	G		1495	.005		1.62	.16		1.78	2.08
2150	Tapered for drainage	G		1950	.004	B.F.	.53	.13		.66	.81
2400	Composites with 2" EPS										
2410	1" fiberboard	G	1 Rofc	1325	.006	S.F.	1.47	.19		1.66	1.95
2420	7/16" oriented strand board	G		1040	.008		1.24	.24		1.48	1.79
2430	1/2" plywood	G		1040	.008		1.47	.24		1.71	2.05
2440	1" perlite	G		1040	.008		1.16	.24		1.40	1.71
2450	Composites with 1-1/2" polyisocyanurate										
2460	1" fiberboard	G	1 Rofc	1040	.008	S.F.	1.07	.24		1.31	1.61
2470	1" perlite	G		1105	.007		.94	.22		1.16	1.43
2480	7/16" oriented strand board	G		1040	.008		.85	.24		1.09	1.37
3000	Fastening alternatives, coated screws, 2" long			3744	.002	Ea.	.06	.07		.13	.19
3010	4" long			3120	.003		.11	.08		.19	.26
3020	6" long			2675	.003		.19	.09		.28	.38
3030	8" long			2340	.003		.28	.10		.38	.50
3040	10" long			1872	.004		.47	.13		.60	.76
3050	Pre-drill and drive wedge spike, 2-1/2"			1248	.006		.40	.20		.60	.79
3060	3-1/2"			1101	.007		.58	.22		.80	1.04
3070	4-1/2"			936	.009		.65	.26		.91	1.19
3075	3" galvanized deck plates			7488	.001		.09	.03		.12	.16
3080	Spot mop asphalt		G-1	295	.190	Sq.	5.80	5.45	1.74	12.99	18.15
3090	Full mop asphalt		"	192	.292		11.65	8.40	2.67	22.72	31

07 24 Exterior Insulation and Finish Systems

07 24 13 – Polymer-Based Exterior Insulation and Finish System

07 24 13.10 Exterior Insulation and Finish Systems		Crew	Daily Output	Labor-Hours	Unit	2019 Bare Costs Material	Labor	Equipment	Total	Total Incl O&P
0010	**EXTERIOR INSULATION AND FINISH SYSTEMS**									
0095	Field applied, 1" EPS insulation G	J-1	390	.103	S.F.	1.80	3.38	.35	5.53	7.85
0100	With 1/2" cement board sheathing G		268	.149		2.59	4.91	.50	8	11.45
0105	2" EPS insulation G		390	.103		2.07	3.38	.35	5.80	8.15
0110	With 1/2" cement board sheathing G		268	.149		2.86	4.91	.50	8.27	11.75
0115	3" EPS insulation G		390	.103		2.34	3.38	.35	6.07	8.45
0120	With 1/2" cement board sheathing G		268	.149		3.13	4.91	.50	8.54	12.05
0125	4" EPS insulation G		390	.103		2.61	3.38	.35	6.34	8.75
0130	With 1/2" cement board sheathing G		268	.149		4.19	4.91	.50	9.60	13.20
0140	Premium finish add		1265	.032		.41	1.04	.11	1.56	2.27
0150	Heavy duty reinforcement add		914	.044		.60	1.44	.15	2.19	3.18
0160	2.5#/S.Y. metal lath substrate add	1 Lath	75	.107	S.Y.	3.83	3.75		7.58	10.25
0170	3.4#/S.Y. metal lath substrate add	"	75	.107	"	4.39	3.75		8.14	10.90
0180	Color or texture change	J-1	1265	.032	S.F.	.81	1.04	.11	1.96	2.71
0190	With substrate leveling base coat	1 Plas	530	.015		.82	.53		1.35	1.76
0210	With substrate sealing base coat	1 Pord	1224	.007		.15	.20		.35	.49
0370	V groove shape in panel face				L.F.	.69			.69	.76
0380	U groove shape in panel face				"	.85			.85	.94

07 25 Weather Barriers

07 25 10 – Weather Barriers or Wraps

07 25 10.10 Weather Barriers		Crew	Daily Output	Labor-Hours	Unit	Material	Labor	Equipment	Total	Total Incl O&P
0010	**WEATHER BARRIERS**									
0400	Asphalt felt paper, #15	1 Carp	37	.216	Sq.	5.40	7.70		13.10	18.70
0401	Per square foot	"	3700	.002	S.F.	.05	.08		.13	.19
0450	Housewrap, exterior, spun bonded polypropylene									
0470	Small roll	1 Carp	3800	.002	S.F.	.16	.08		.24	.29
0480	Large roll	"	4000	.002	"	.14	.07		.21	.27
2100	Asphalt felt roof deck vapor barrier, class 1 metal decks	1 Rofc	37	.216	Sq.	22.50	6.65		29.15	36.50
2200	For all other decks	"	37	.216		17	6.65		23.65	30.50
2800	Asphalt felt, 50% recycled content, 15 lb., 4 sq./roll	1 Carp	36	.222		5.90	7.90		13.80	19.60
2810	30 lb., 2 sq./roll	"	36	.222		10	7.90		17.90	24
3000	Building wrap, spun bonded polyethylene	2 Carp	8000	.002	S.F.	.17	.07		.24	.31

07 26 Vapor Retarders

07 26 10 – Above-Grade Vapor Retarders

07 26 10.10 Building Paper		Crew	Daily Output	Labor-Hours	Unit	Material	Labor	Equipment	Total	Total Incl O&P
0011	**BUILDING PAPER**, aluminum and kraft laminated, foil 1 side	1 Carp	3700	.002	S.F.	.14	.08		.22	.29
0101	Foil 2 sides G		3700	.002		.15	.08		.23	.30
0601	Polyethylene vapor barrier, standard, 2 mil G		3700	.002		.02	.08		.10	.15
0701	4 mil G		3700	.002		.03	.08		.11	.16
0901	6 mil G		3700	.002		.04	.08		.12	.17
1201	10 mil G		3700	.002		.09	.08		.17	.23
1801	Reinf. waterproof, 2 mil polyethylene backing, 1 side		3700	.002		.11	.08		.19	.25
1901	2 sides		3700	.002		.14	.08		.22	.29

07 27 Air Barriers

07 27 13 – Modified Bituminous Sheet Air Barriers

07 27 13.10 Modified Bituminous Sheet Air Barrier		Crew	Daily Output	Labor-Hours	Unit	2019 Bare Costs Material	Labor	Equipment	Total	Total Incl O&P
0010	**MODIFIED BITUMINOUS SHEET AIR BARRIER**									
0100	SBS modified sheet laminated to polyethylene sheet, 40 mils, 4" wide	1 Carp	1200	.007	L.F.	.33	.24		.57	.75
0120	6" wide		1100	.007		.45	.26		.71	.92
0140	9" wide		1000	.008		.62	.29		.91	1.16
0160	12" wide	↓	900	.009	↓	.80	.32		1.12	1.40
0180	18" wide	2 Carp	1700	.009	S.F.	.75	.34		1.09	1.37
0200	36" wide	"	1800	.009		.73	.32		1.05	1.32
0220	Adhesive for above	1 Carp	1400	.006	↓	.32	.20		.52	.69

07 27 26 – Fluid-Applied Membrane Air Barriers

07 27 26.10 Fluid Applied Membrane Air Barrier		Crew	Daily Output	Labor-Hours	Unit	Material	Labor	Equipment	Total	Total Incl O&P
0010	**FLUID APPLIED MEMBRANE AIR BARRIER**									
0100	Spray applied vapor barrier, 25 S.F./gallon	1 Pord	1375	.006	S.F.	.02	.17		.19	.30

07 31 Shingles and Shakes

07 31 13 – Asphalt Shingles

07 31 13.10 Asphalt Roof Shingles		Crew	Daily Output	Labor-Hours	Unit	Material	Labor	Equipment	Total	Total Incl O&P
0010	**ASPHALT ROOF SHINGLES**									
0100	Standard strip shingles									
0150	Inorganic, class A, 25 year	1 Rofc	5.50	1.455	Sq.	79	44.50		123.50	167
0155	Pneumatic nailed		7	1.143		79	35		114	150
0200	30 year		5	1.600		96	49		145	195
0205	Pneumatic nailed	↓	6.25	1.280	↓	96	39		135	177
0250	Standard laminated multi-layered shingles									
0300	Class A, 240-260 lb./square	1 Rofc	4.50	1.778	Sq.	110	54.50		164.50	220
0305	Pneumatic nailed		5.63	1.422		110	43.50		153.50	200
0350	Class A, 250-270 lb./square		4	2		110	61.50		171.50	232
0355	Pneumatic nailed	↓	5	1.600	↓	110	49		159	210
0400	Premium, laminated multi-layered shingles									
0450	Class A, 260-300 lb./square	1 Rofc	3.50	2.286	Sq.	145	70		215	287
0455	Pneumatic nailed		4.37	1.831		145	56		201	261
0500	Class A, 300-385 lb./square		3	2.667		274	81.50		355.50	450
0505	Pneumatic nailed		3.75	2.133		274	65.50		339.50	420
0800	#15 felt underlayment		64	.125		5.40	3.83		9.23	12.85
0825	#30 felt underlayment		58	.138		10.55	4.23		14.78	19.25
0850	Self adhering polyethylene and rubberized asphalt underlayment		22	.364	↓	79	11.15		90.15	107
0900	Ridge shingles		330	.024	L.F.	2.28	.74		3.02	3.85
0905	Pneumatic nailed	↓	412.50	.019	"	2.28	.59		2.87	3.58
1000	For steep roofs (7 to 12 pitch or greater), add						50%			

07 31 26 – Slate Shingles

07 31 26.10 Slate Roof Shingles		Crew	Daily Output	Labor-Hours	Unit	Material	Labor	Equipment	Total	Total Incl O&P
0010	**SLATE ROOF SHINGLES** R073126-20									
0100	Buckingham Virginia black, 3/16" - 1/4" thick G	1 Rots	1.75	4.571	Sq.	560	141		701	870
0200	1/4" thick G		1.75	4.571		560	141		701	870
0900	Pennsylvania black, Bangor, #1 clear G		1.75	4.571		495	141		636	800
1200	Vermont, unfading, green, mottled green G		1.75	4.571		510	141		651	815
1300	Semi-weathering green & gray G		1.75	4.571		405	141		546	700
1400	Purple G		1.75	4.571		435	141		576	735
1500	Black or gray G		1.75	4.571	↓	490	141		631	790
2700	Ridge shingles, slate	↓	200	.040	L.F.	10.15	1.23		11.38	13.45

07 31 29.13 Wood Shingles		Crew	Daily Output	Labor-Hours	Unit	2019 Bare Costs Material	Labor	Equipment	Total	Total Incl O&P
0010	**WOOD SHINGLES**									
0012	16" No. 1 red cedar shingles, 5" exposure, on roof	1 Carp	2.50	3.200	Sq.	296	114		410	515
0015	Pneumatic nailed		3.25	2.462		296	88		384	470
0200	7-1/2" exposure, on walls		2.05	3.902		197	139		336	445
0205	Pneumatic nailed		2.67	2.996		197	107		304	395
0300	18" No. 1 red cedar perfections, 5-1/2" exposure, on roof		2.75	2.909		254	104		358	450
0305	Pneumatic nailed		3.57	2.241		254	80		334	410
0500	7-1/2" exposure, on walls		2.25	3.556		187	127		314	415
0505	Pneumatic nailed		2.92	2.740		187	97.50		284.50	365
0600	Resquared and rebutted, 5-1/2" exposure, on roof		3	2.667		286	95		381	470
0605	Pneumatic nailed		3.90	2.051		286	73		359	435
0900	7-1/2" exposure, on walls		2.45	3.265		210	116		326	425
0905	Pneumatic nailed	▼	3.18	2.516		210	89.50		299.50	380
1000	Add to above for fire retardant shingles				▼	55.50			55.50	61
1060	Preformed ridge shingles	1 Carp	400	.020	L.F.	3.95	.71		4.66	5.55
2000	White cedar shingles, 16" long, extras, 5" exposure, on roof		2.40	3.333	Sq.	199	119		318	415
2005	Pneumatic nailed		3.12	2.564		199	91.50		290.50	370
2050	5" exposure on walls		2	4		199	143		342	455
2055	Pneumatic nailed		2.60	3.077		199	110		309	400
2100	7-1/2" exposure, on walls		2	4		142	143		285	390
2105	Pneumatic nailed		2.60	3.077		142	110		252	335
2150	"B" grade, 5" exposure on walls		2	4		172	143		315	425
2155	Pneumatic nailed		2.60	3.077		172	110		282	370
2300	For #15 organic felt underlayment on roof, 1 layer, add		64	.125		5.40	4.46		9.86	13.30
2400	2 layers, add	▼	32	.250		10.80	8.90		19.70	26.50
2600	For steep roofs (7/12 pitch or greater), add to above				▼		50%			
2700	Panelized systems, No.1 cedar shingles on 5/16" CDX plywood									
2800	On walls, 8' strips, 7" or 14" exposure	2 Carp	700	.023	S.F.	6.45	.82		7.27	8.45
3000	Ridge shakes or shingle, wood	1 Carp	280	.029	L.F.	4.79	1.02		5.81	6.95

07 31 29.16 Wood Shakes		Crew	Daily Output	Labor-Hours	Unit	Material	Labor	Equipment	Total	Total Incl O&P
0010	**WOOD SHAKES**									
1100	Hand-split red cedar shakes, 1/2" thick x 24" long, 10" exp. on roof	1 Carp	2.50	3.200	Sq.	300	114		414	520
1105	Pneumatic nailed		3.25	2.462		300	88		388	475
1110	3/4" thick x 24" long, 10" exp. on roof		2.25	3.556		300	127		427	540
1115	Pneumatic nailed		2.92	2.740		300	97.50		397.50	490
1200	1/2" thick, 18" long, 8-1/2" exp. on roof		2	4		279	143		422	540
1205	Pneumatic nailed		2.60	3.077		279	110		389	485
1210	3/4" thick x 18" long, 8-1/2" exp. on roof		1.80	4.444		279	158		437	565
1215	Pneumatic nailed		2.34	3.419		279	122		401	505
1255	10" exposure on walls		2	4		269	143		412	530
1260	10" exposure on walls, pneumatic nailed	▼	2.60	3.077		269	110		379	475
1700	Add to above for fire retardant shakes, 24" long					55.50			55.50	61
1800	18" long				▼	55.50			55.50	61
1810	Ridge shakes	1 Carp	350	.023	L.F.	5.75	.82		6.57	7.70

07 32 Roof Tiles

07 32 13 – Clay Roof Tiles

07 32 13.10 Clay Tiles		Crew	Daily Output	Labor-Hours	Unit	2019 Bare Costs Material	Labor	Equipment	Total	Total Incl O&P
0010	**CLAY TILES**, including accessories									
0300	Flat shingle, interlocking, 15", 166 pcs./sq., fireflashed blend	3 Rots	6	4	Sq.	475	123		598	745
0500	Terra cotta red		6	4		495	123		618	765
0600	Roman pan and top, 18", 102 pcs./sq., fireflashed blend		5.50	4.364		500	134		634	795
0640	Terra cotta red	1 Rots	2.40	3.333		560	103		663	800
1100	Barrel mission tile, 18", 166 pcs./sq., fireflashed blend	3 Rots	5.50	4.364		415	134		549	700
1140	Terra cotta red		5.50	4.364		420	134		554	705
1700	Scalloped edge flat shingle, 14", 145 pcs./sq., fireflashed blend		6	4		1,150	123		1,273	1,500
1800	Terra cotta red		6	4		1,050	123		1,173	1,375
3010	#15 felt underlayment	1 Rofc	64	.125		5.40	3.83		9.23	12.85
3020	#30 felt underlayment		58	.138		10.55	4.23		14.78	19.25
3040	Polyethylene and rubberized asph. underlayment		22	.364		79	11.15		90.15	107

07 32 16 – Concrete Roof Tiles

07 32 16.10 Concrete Tiles		Crew	Daily Output	Labor-Hours	Unit	Material	Labor	Equipment	Total	Total Incl O&P
0010	**CONCRETE TILES**									
0020	Corrugated, 13" x 16-1/2", 90 per sq., 950 lb./sq.									
0050	Earthtone colors, nailed to wood deck	1 Rots	1.35	5.926	Sq.	104	183		287	445
0150	Blues		1.35	5.926		105	183		288	445
0200	Greens		1.35	5.926		105	183		288	445
0250	Premium colors		1.35	5.926		105	183		288	445
0500	Shakes, 13" x 16-1/2", 90 per sq., 950 lb./sq.									
0600	All colors, nailed to wood deck	1 Rots	1.50	5.333	Sq.	131	164		295	440
1500	Accessory pieces, ridge & hip, 10" x 16-1/2", 8 lb. each	"	120	.067	Ea.	3.78	2.05		5.83	7.85
1700	Rake, 6-1/2" x 16-3/4", 9 lb. each					3.78			3.78	4.16
1800	Mansard hip, 10" x 16-1/2", 9.2 lb. each					3.78			3.78	4.16
1900	Hip starter, 10" x 16-1/2", 10.5 lb. each					10.60			10.60	11.65
2000	3 or 4 way apex, 10" each side, 11.5 lb. each					12.10			12.10	13.30

07 33 Natural Roof Coverings

07 33 63 – Vegetated Roofing

07 33 63.10 Green Roof Systems		Crew	Daily Output	Labor-Hours	Unit	Material	Labor	Equipment	Total	Total Incl O&P
0010	**GREEN ROOF SYSTEMS**									
0020	Soil mixture for green roof 30% sand, 55% gravel, 15% soil									
0100	Hoist and spread soil mixture 4" depth up to 5 stories tall roof G	B-13B	4000	.014	S.F.	.23	.42	.25	.90	1.22
0150	6" depth G		2667	.021		.35	.63	.37	1.35	1.83
0200	8" depth G		2000	.028		.47	.84	.49	1.80	2.44
0250	10" depth G		1600	.035		.58	1.05	.61	2.24	3.04
0300	12" depth G		1335	.042		.70	1.26	.74	2.70	3.66
0310	Alt. man-made soil mix, hoist & spread, 4" deep up to 5 stories tall roof G		4000	.014		1.89	.42	.25	2.56	3.04
0350	Mobilization 55 ton crane to site G	1 Eqhv	3.60	2.222	Ea.		84		84	138
0355	Hoisting cost to 5 stories per day (Avg. 28 picks per day) G	B-13B	1	56	Day		1,675	980	2,655	3,850
0360	Mobilization or demobilization, 100 ton crane to site driver & escort G	A-3E	2.50	6.400	Ea.		227	48.50	275.50	425
0365	Hoisting cost 6-10 stories per day (Avg. 21 picks per day) G	B-13C	1	56	Day		1,675	1,875	3,550	4,850
0370	Hoist and spread soil mixture 4" depth 6-10 stories tall roof G		4000	.014	S.F.	.23	.42	.47	1.12	1.47
0375	6" depth G		2667	.021		.35	.63	.70	1.68	2.21
0380	8" depth G		2000	.028		.47	.84	.94	2.25	2.93
0385	10" depth G		1600	.035		.58	1.05	1.17	2.80	3.66
0390	12" depth G		1335	.042		.70	1.26	1.41	3.37	4.40
0400	Green roof edging treated lumber 4" x 4", no hoisting included G	2 Carp	400	.040	L.F.	1.25	1.43		2.68	3.73
0410	4" x 6" G		400	.040		1.95	1.43		3.38	4.50

07 33 Natural Roof Coverings

07 33 63 – Vegetated Roofing

07 33 63.10 Green Roof Systems		Crew	Daily Output	Labor-Hours	Unit	Material	Labor	Equipment	Total	Total Incl O&P
						2019 Bare Costs				
0420	4" x 8" G	2 Carp	360	.044	L.F.	4.06	1.58		5.64	7.10
0430	4" x 6" double stacked G		300	.053		3.89	1.90		5.79	7.40
0500	Green roof edging redwood lumber 4" x 4", no hoisting included G		400	.040		6.55	1.43		7.98	9.55
0510	4" x 6" G		400	.040		12.80	1.43		14.23	16.45
0520	4" x 8" G		360	.044		23.50	1.58		25.08	28.50
0530	4" x 6" double stacked G	↓	300	.053	↓	25.50	1.90		27.40	31
0550	Components, not including membrane or insulation:									
0560	Fluid applied rubber membrane, reinforced, 215 mil thick G	G-5	350	.114	S.F.	.30	3.20	.52	4.02	6.70
0570	Root barrier G	2 Rofc	775	.021		.69	.63		1.32	1.90
0580	Moisture retention barrier and reservoir G	"	900	.018		2.63	.55		3.18	3.87
0600	Planting sedum, light soil, potted, 2-1/4" diameter, 2 per S.F. G	1 Clab	420	.019		6.30	.52		6.82	7.75
0610	1 per S.F. G	"	840	.010		3.14	.26		3.40	3.88
0630	Planting sedum mat per S.F. including shipping (4000 S.F. min) G	4 Clab	4000	.008		7.55	.22		7.77	8.65
0640	Installation sedum mat system (no soil required) per S.F. (4000 S.F. min) G	"	4000	.008	↓	10.50	.22		10.72	11.90
0645	Note: pricing of sedum mats shipped in full truck loads (4000-5000 S.F.)									

07 41 Roof Panels

07 41 13 – Metal Roof Panels

07 41 13.10 Aluminum Roof Panels		Crew	Daily Output	Labor-Hours	Unit	Material	Labor	Equipment	Total	Total Incl O&P
0010	**ALUMINUM ROOF PANELS**									
0020	Corrugated or ribbed, .0155" thick, natural	G-3	1200	.027	S.F.	1.02	.89		1.91	2.59
0300	Painted	"	1200	.027	"	1.48	.89		2.37	3.10

07 41 33 – Plastic Roof Panels

07 41 33.10 Fiberglass Panels		Crew	Daily Output	Labor-Hours	Unit	Material	Labor	Equipment	Total	Total Incl O&P
0010	**FIBERGLASS PANELS**									
0012	Corrugated panels, roofing, 8 oz./S.F.	G-3	1000	.032	S.F.	2.60	1.06		3.66	4.62
0100	12 oz./S.F.		1000	.032		4.62	1.06		5.68	6.85
0300	Corrugated siding, 6 oz./S.F.		880	.036		2.02	1.21		3.23	4.22
0400	8 oz./S.F.		880	.036		2.60	1.21		3.81	4.86
0500	Fire retardant		880	.036		4.05	1.21		5.26	6.45
0600	12 oz. siding, textured		880	.036		3.95	1.21		5.16	6.35
0700	Fire retardant		880	.036		4.57	1.21		5.78	7.05
0900	Flat panels, 6 oz./S.F., clear or colors		880	.036		2.58	1.21		3.79	4.84
1100	Fire retardant, class A		880	.036		3.64	1.21		4.85	6
1300	8 oz./S.F., clear or colors	↓	880	.036	↓	2.48	1.21		3.69	4.73

07 42 Wall Panels

07 42 13 – Metal Wall Panels

07 42 13.20 Aluminum Siding		Crew	Daily Output	Labor-Hours	Unit	Material	Labor	Equipment	Total	Total Incl O&P
0011	**ALUMINUM SIDING**									
6040	0.024" thick smooth white single 8" wide	2 Carp	515	.031	S.F.	2.99	1.11		4.10	5.10
6060	Double 4" pattern		515	.031		2.98	1.11		4.09	5.10
6080	Double 5" pattern		550	.029		2.77	1.04		3.81	4.76
6120	Embossed white, 8" wide		515	.031		2.80	1.11		3.91	4.91
6140	Double 4" pattern		515	.031		2.88	1.11		3.99	5
6160	Double 5" pattern		550	.029		2.90	1.04		3.94	4.90
6170	Vertical, embossed white, 12" wide		590	.027		2.91	.97		3.88	4.80
6320	0.019" thick, insulated, smooth white, 8" wide		515	.031		2.54	1.11		3.65	4.62
6340	Double 4" pattern	↓	515	.031	↓	2.52	1.11		3.63	4.60

07 42 Wall Panels

07 42 13 – Metal Wall Panels

07 42 13.20 Aluminum Siding		Crew	Daily Output	Labor-Hours	Unit	2019 Bare Costs Material	Labor	Equipment	Total	Total Incl O&P
6360	Double 5" pattern	2 Carp	550	.029	S.F.	2.53	1.04		3.57	4.49
6400	Embossed white, 8" wide		515	.031		2.93	1.11		4.04	5.05
6420	Double 4" pattern		515	.031		2.95	1.11		4.06	5.10
6440	Double 5" pattern		550	.029		2.95	1.04		3.99	4.96
6500	Shake finish 10" wide white		550	.029		3.19	1.04		4.23	5.20
6600	Vertical pattern, 12" wide, white		590	.027		2.42	.97		3.39	4.26
6640	For colors add					.15			.15	.17
6700	Accessories, white									
6720	Starter strip 2-1/8"	2 Carp	610	.026	L.F.	.48	.94		1.42	2.07
6740	Sill trim		450	.036		.67	1.27		1.94	2.83
6760	Inside corner		610	.026		1.73	.94		2.67	3.44
6780	Outside corner post		610	.026		3.71	.94		4.65	5.60
6800	Door & window trim		440	.036		.65	1.30		1.95	2.86
6820	For colors add					.15			.15	.17
6900	Soffit & fascia 1' overhang solid	2 Carp	110	.145		4.52	5.20		9.72	13.50
6920	Vented		110	.145		4.55	5.20		9.75	13.55
6940	2' overhang solid		100	.160		6.60	5.70		12.30	16.65
6960	Vented		100	.160		6.60	5.70		12.30	16.65

07 42 13.30 Steel Siding		Crew	Daily Output	Labor-Hours	Unit	Material	Labor	Equipment	Total	Total Incl O&P
0010	**STEEL SIDING**									
0020	Beveled, vinyl coated, 8" wide	1 Carp	265	.030	S.F.	1.79	1.08		2.87	3.75
0050	10" wide	"	275	.029		1.96	1.04		3	3.87
0081	Galv., corrugated or ribbed, on steel frame, 30 ga.	G-3	775	.041		1.24	1.37		2.61	3.63
0101	28 ga.		775	.041		1.35	1.37		2.72	3.76
0301	26 ga.		775	.041		1.75	1.37		3.12	4.20
0401	24 ga.		775	.041		2.33	1.37		3.70	4.83
0601	22 ga.		775	.041		2.41	1.37		3.78	4.92
0701	Colored, corrugated/ribbed, on steel frame, 10 yr. finish, 28 ga.		775	.041		2.02	1.37		3.39	4.49
0901	26 ga.		775	.041		2.08	1.37		3.45	4.56
1001	24 ga.		775	.041		2.32	1.37		3.69	4.82

07 46 Siding

07 46 23 – Wood Siding

07 46 23.10 Wood Board Siding		Crew	Daily Output	Labor-Hours	Unit	Material	Labor	Equipment	Total	Total Incl O&P
0010	**WOOD BOARD SIDING**									
2000	Board & batten, cedar, "B" grade, 1" x 10"	1 Carp	375	.021	S.F.	6.65	.76		7.41	8.55
2200	Redwood, clear, vertical grain, 1" x 10"		375	.021		5.40	.76		6.16	7.15
2400	White pine, #2 & better, 1" x 10"		375	.021		3.67	.76		4.43	5.30
2410	White pine, #2 & better, 1" x 12"		420	.019		3.67	.68		4.35	5.15
3200	Wood, cedar bevel, A grade, 1/2" x 6"		295	.027		4.43	.97		5.40	6.45
3300	1/2" x 8"		330	.024		7.75	.86		8.61	10
3500	3/4" x 10", clear grade		375	.021		7.20	.76		7.96	9.15
3600	"B" grade		375	.021		4.10	.76		4.86	5.75
3800	Cedar, rough sawn, 1" x 4", A grade, natural		220	.036		7.50	1.30		8.80	10.40
3900	Stained		220	.036		7.65	1.30		8.95	10.55
4100	1" x 12", board & batten, #3 & Btr., natural		420	.019		4.79	.68		5.47	6.35
4200	Stained		420	.019		5.15	.68		5.83	6.75
4400	1" x 8" channel siding, #3 & Btr., natural		330	.024		4.92	.86		5.78	6.85
4500	Stained		330	.024		5.05	.86		5.91	7
4700	Redwood, clear, beveled, vertical grain, 1/2" x 4"		220	.036		4.87	1.30		6.17	7.50
4750	1/2" x 6"		295	.027		4.99	.97		5.96	7.10

07 46 23 – Wood Siding

07 46 23.10 Wood Board Siding		Crew	Daily Output	Labor-Hours	Unit	Material	Labor	Equipment	Total	Total Incl O&P
						2019 Bare Costs				
4800	1/2" x 8"	1 Carp	330	.024	S.F.	5.45	.86		6.31	7.45
5000	3/4" x 10"		375	.021		5.10	.76		5.86	6.85
5200	Channel siding, 1" x 10", B grade		375	.021		4.65	.76		5.41	6.35
5250	Redwood, T&G boards, B grade, 1" x 4"		220	.036		6.85	1.30		8.15	9.70
5270	1" x 8"		330	.024		8	.86		8.86	10.25
5400	White pine, rough sawn, 1" x 8", natural		330	.024		2.51	.86		3.37	4.19
5500	Stained		330	.024		2.40	.86		3.26	4.07
5600	T&G, 1" x 8"		330	.024		2.54	.86		3.40	4.22

07 46 29 – Plywood Siding

07 46 29.10 Plywood Siding Options

Code	Description	Crew	Daily Output	Labor-Hours	Unit	Material	Labor	Equipment	Total	Total Incl O&P
0010	**PLYWOOD SIDING OPTIONS**									
0900	Plywood, medium density overlaid, 3/8" thick	2 Carp	750	.021	S.F.	1.38	.76		2.14	2.78
1000	1/2" thick		700	.023		1.57	.82		2.39	3.08
1100	3/4" thick		650	.025		1.93	.88		2.81	3.57
1600	Texture 1-11, cedar, 5/8" thick, natural		675	.024		2.62	.84		3.46	4.28
1700	Factory stained		675	.024		2.92	.84		3.76	4.61
1900	Texture 1-11, fir, 5/8" thick, natural		675	.024		1.36	.84		2.20	2.90
2000	Factory stained		675	.024		1.93	.84		2.77	3.52
2050	Texture 1-11, S.Y.P., 5/8" thick, natural		675	.024		1.45	.84		2.29	3
2100	Factory stained		675	.024		1.53	.84		2.37	3.08
2200	Rough sawn cedar, 3/8" thick, natural		675	.024		1.28	.84		2.12	2.81
2300	Factory stained		675	.024		1.59	.84		2.43	3.15
2500	Rough sawn fir, 3/8" thick, natural		675	.024		.93	.84		1.77	2.42
2600	Factory stained		675	.024		1.10	.84		1.94	2.61
2800	Redwood, textured siding, 5/8" thick		675	.024		2.03	.84		2.87	3.63

07 46 33 – Plastic Siding

07 46 33.10 Vinyl Siding

Code	Description	Crew	Daily Output	Labor-Hours	Unit	Material	Labor	Equipment	Total	Total Incl O&P
0010	**VINYL SIDING**									
3995	Clapboard profile, woodgrain texture, .048 thick, double 4	2 Carp	495	.032	S.F.	1.08	1.15		2.23	3.09
4000	Double 5		550	.029		1.08	1.04		2.12	2.90
4005	Single 8		495	.032		1.39	1.15		2.54	3.43
4010	Single 10		550	.029		1.67	1.04		2.71	3.55
4015	.044 thick, double 4		495	.032		1.07	1.15		2.22	3.08
4020	Double 5		550	.029		1.09	1.04		2.13	2.91
4025	.042 thick, double 4		495	.032		1.09	1.15		2.24	3.10
4030	Double 5		550	.029		1.09	1.04		2.13	2.91
4035	Cross sawn texture, .040 thick, double 4		495	.032		.74	1.15		1.89	2.72
4040	Double 5		550	.029		.67	1.04		1.71	2.45
4045	Smooth texture, .042 thick, double 4		495	.032		.81	1.15		1.96	2.80
4050	Double 5		550	.029		.80	1.04		1.84	2.60
4055	Single 8		495	.032		.81	1.15		1.96	2.80
4060	Cedar texture, .044 thick, double 4		495	.032		1.16	1.15		2.31	3.18
4065	Double 6		600	.027		1.17	.95		2.12	2.86
4070	Dutch lap profile, woodgrain texture, .048 thick, double 5		550	.029		1.10	1.04		2.14	2.93
4075	.044 thick, double 4.5		525	.030		1.08	1.09		2.17	2.99
4080	.042 thick, double 4.5		525	.030		.93	1.09		2.02	2.83
4085	.040 thick, double 4.5		525	.030		.74	1.09		1.83	2.62
4100	Shake profile, 10" wide		400	.040		3.81	1.43		5.24	6.55
4105	Vertical pattern, .046 thick, double 5		550	.029		1.58	1.04		2.62	3.45
4110	.044 thick, triple 3		550	.029		1.79	1.04		2.83	3.68
4115	.040 thick, triple 4		550	.029		1.73	1.04		2.77	3.62
4120	.040 thick, triple 2.66		550	.029		1.80	1.04		2.84	3.70

07 46 33 – Plastic Siding

07 46 33.10 Vinyl Siding		Crew	Daily Output	Labor-Hours	Unit	2019 Bare Costs Material	Labor	Equipment	Total	Total Incl O&P
4125	Insulation, fan folded extruded polystyrene, 1/4"	2 Carp	2000	.008	S.F.	.29	.29		.58	.79
4130	3/8"		2000	.008		.32	.29		.61	.82
4135	Accessories, J channel, 5/8" pocket		700	.023	L.F.	.52	.82		1.34	1.92
4140	3/4" pocket		695	.023		.56	.82		1.38	1.97
4145	1-1/4" pocket		680	.024		1.05	.84		1.89	2.54
4150	Flexible, 3/4" pocket		600	.027		2.44	.95		3.39	4.25
4155	Under sill finish trim		500	.032		.57	1.14		1.71	2.50
4160	Vinyl starter strip		700	.023		.70	.82		1.52	2.12
4165	Aluminum starter strip		700	.023		.30	.82		1.12	1.68
4170	Window casing, 2-1/2" wide, 3/4" pocket		510	.031		1.79	1.12		2.91	3.82
4175	Outside corner, woodgrain finish, 4" face, 3/4" pocket		700	.023		2.16	.82		2.98	3.73
4180	5/8" pocket		700	.023		2.17	.82		2.99	3.74
4185	Smooth finish, 4" face, 3/4" pocket		700	.023		2.15	.82		2.97	3.72
4190	7/8" pocket		690	.023		2.02	.83		2.85	3.60
4195	1-1/4" pocket		700	.023		1.42	.82		2.24	2.92
4200	Soffit and fascia, 1' overhang, solid		120	.133		4.84	4.75		9.59	13.20
4205	Vented		120	.133		4.84	4.75		9.59	13.20
4207	18" overhang, solid		110	.145		5.65	5.20		10.85	14.75
4208	Vented		110	.145		5.65	5.20		10.85	14.75
4210	2' overhang, solid		100	.160		6.40	5.70		12.10	16.45
4215	Vented		100	.160		6.40	5.70		12.10	16.45
4217	3' overhang, solid		100	.160		8	5.70		13.70	18.20
4218	Vented		100	.160		8	5.70		13.70	18.20
4220	Colors for siding and soffits, add				S.F.	.15			.15	.17
4225	Colors for accessories and trim, add				L.F.	.31			.31	.34

07 46 33.20 Polypropylene Siding		Crew	Daily Output	Labor-Hours	Unit	Material	Labor	Equipment	Total	Total Incl O&P
0010	**POLYPROPYLENE SIDING**									
4090	Shingle profile, random grooves, double 7	2 Carp	400	.040	S.F.	3.47	1.43		4.90	6.20
4092	Cornerpost for above	1 Carp	365	.022	L.F.	13	.78		13.78	15.60
4095	Triple 5	2 Carp	400	.040	S.F.	3.48	1.43		4.91	6.20
4097	Cornerpost for above	1 Carp	365	.022	L.F.	12.35	.78		13.13	14.85
5000	Staggered butt, double 7"	2 Carp	400	.040	S.F.	3.68	1.43		5.11	6.40
5002	Cornerpost for above	1 Carp	365	.022	L.F.	13.25	.78		14.03	15.85
5010	Half round, double 6-1/4"	2 Carp	360	.044	S.F.	4.05	1.58		5.63	7.10
5020	Shake profile, staggered butt, double 9"	"	510	.031	"	3.81	1.12		4.93	6.05
5022	Cornerpost for above	1 Carp	365	.022	L.F.	9.90	.78		10.68	12.20
5030	Straight butt, double 7"	2 Carp	400	.040	S.F.	4.05	1.43		5.48	6.80
5032	Cornerpost for above	1 Carp	365	.022	L.F.	13.90	.78		14.68	16.60
6000	Accessories, J channel, 5/8" pocket	2 Carp	700	.023		.52	.82		1.34	1.92
6010	3/4" pocket		695	.023		.56	.82		1.38	1.97
6020	1-1/4" pocket		680	.024		1.05	.84		1.89	2.54
6030	Aluminum starter strip		700	.023		.30	.82		1.12	1.68

07 46 46 – Fiber Cement Siding

07 46 46.10 Fiber Cement Siding		Crew	Daily Output	Labor-Hours	Unit	Material	Labor	Equipment	Total	Total Incl O&P
0010	**FIBER CEMENT SIDING**									
0020	Lap siding, 5/16" thick, 6" wide, 4-3/4" exposure, smooth texture	2 Carp	415	.039	S.F.	1.31	1.37		2.68	3.71
0025	Woodgrain texture		415	.039		1.31	1.37		2.68	3.71
0030	7-1/2" wide, 6-1/4" exposure, smooth texture		425	.038		1.82	1.34		3.16	4.23
0035	Woodgrain texture		425	.038		1.82	1.34		3.16	4.23
0040	8" wide, 6-3/4" exposure, smooth texture		425	.038		1.33	1.34		2.67	3.69
0045	Rough sawn texture		425	.038		1.33	1.34		2.67	3.69
0050	9-1/2" wide, 8-1/4" exposure, smooth texture		440	.036		1.31	1.30		2.61	3.58

07 46 Siding

07 46 46 – Fiber Cement Siding

07 46 46.10 Fiber Cement Siding		Crew	Daily Output	Labor-Hours	Unit	2019 Bare Costs Material	Labor	Equipment	Total	Total Incl O&P
0055	Woodgrain texture	2 Carp	440	.036	S.F.	1.31	1.30		2.61	3.58
0060	12" wide, 10-3/8" exposure, smooth texture		455	.035		2.12	1.25		3.37	4.40
0065	Woodgrain texture		455	.035		2.12	1.25		3.37	4.40
0070	Panel siding, 5/16" thick, smooth texture		750	.021		1.22	.76		1.98	2.60
0075	Stucco texture		750	.021		1.22	.76		1.98	2.60
0080	Grooved woodgrain texture		750	.021		1.22	.76		1.98	2.60
0085	V - grooved woodgrain texture		750	.021		1.22	.76		1.98	2.60
0088	Shingle siding, 48" x 15-1/4" panels, 7" exposure		700	.023		4.30	.82		5.12	6.10
0090	Wood starter strip		400	.040	L.F.	.44	1.43		1.87	2.85
0200	Fiber cement siding, accessories, fascia, 5/4" x 3-1/2"		275	.058		.85	2.07		2.92	4.37
0210	5/4" x 5-1/2"		250	.064		1.99	2.28		4.27	5.95
0220	5/4" x 7-1/2"		225	.071		2.75	2.54		5.29	7.20
0230	5/4" x 9-1/2"		210	.076		3.30	2.72		6.02	8.10

07 46 73 – Soffit

07 46 73.10 Soffit Options		Crew	Daily Output	Labor-Hours	Unit	Material	Labor	Equipment	Total	Total Incl O&P
0010	**SOFFIT OPTIONS**									
0012	Aluminum, residential, .020" thick	1 Carp	210	.038	S.F.	2.08	1.36		3.44	4.53
0100	Baked enamel on steel, 16 or 18 ga.		105	.076		6.15	2.72		8.87	11.30
0300	Polyvinyl chloride, white, solid		230	.035		2.19	1.24		3.43	4.46
0400	Perforated		230	.035		2.19	1.24		3.43	4.46
0500	For colors, add					.15			.15	.17

07 51 Built-Up Bituminous Roofing

07 51 13 – Built-Up Asphalt Roofing

07 51 13.10 Built-Up Roofing Components		Crew	Daily Output	Labor-Hours	Unit	Material	Labor	Equipment	Total	Total Incl O&P
0010	**BUILT-UP ROOFING COMPONENTS**									
0012	Asphalt saturated felt, #30, 2 sq./roll	1 Rofc	58	.138	Sq.	10.55	4.23		14.78	19.25
0200	#15, 4 sq./roll, plain or perforated, not mopped		58	.138		5.40	4.23		9.63	13.60
0250	Perforated		58	.138		5.40	4.23		9.63	13.60
0300	Roll roofing, smooth, #65		15	.533		10.55	16.35		26.90	41
0500	#90		12	.667		38	20.50		58.50	79
0520	Mineralized		12	.667		35.50	20.50		56	76
0540	D.C. (double coverage), 19" selvage edge		10	.800		48	24.50		72.50	97
0580	Adhesive (lap cement)				Gal.	8.65			8.65	9.50

07 51 13.20 Built-Up Roofing Systems		Crew	Daily Output	Labor-Hours	Unit	Material	Labor	Equipment	Total	Total Incl O&P
0010	**BUILT-UP ROOFING SYSTEMS**									
0120	Asphalt flood coat with gravel/slag surfacing, not including									
0140	Insulation, flashing or wood nailers									
0200	Asphalt base sheet, 3 plies #15 asphalt felt, mopped	G-1	22	2.545	Sq.	107	73	23.50	203.50	276
0350	On nailable decks		21	2.667		110	76.50	24.50	211	286
0500	4 plies #15 asphalt felt, mopped		20	2.800		145	80.50	25.50	251	335
0550	On nailable decks		19	2.947		129	84.50	27	240.50	325
2000	Asphalt flood coat, smooth surface									
2200	Asphalt base sheet & 3 plies #15 asphalt felt, mopped	G-1	24	2.333	Sq.	110	67	21.50	198.50	266
2400	On nailable decks		23	2.435		102	70	22.50	194.50	263
2600	4 plies #15 asphalt felt, mopped		24	2.333		129	67	21.50	217.50	287
2700	On nailable decks		23	2.435		121	70	22.50	213.50	284
4500	Coal tar pitch with gravel/slag surfacing									
4600	4 plies #15 tarred felt, mopped	G-1	21	2.667	Sq.	207	76.50	24.50	308	395
4800	3 plies glass fiber felt (type IV), mopped	"	19	2.947	"	172	84.50	27	283.50	370

07 51 Built-Up Bituminous Roofing

07 51 13 – Built-Up Asphalt Roofing

07 51 13.30 Cants	Crew	Daily Output	Labor-Hours	Unit	Material	2019 Bare Costs Labor	Equipment	Total	Total Incl O&P
0010 **CANTS**									
0012 Lumber, treated, 4" x 4" cut diagonally	1 Rofc	325	.025	L.F.	1.96	.75		2.71	3.52
0300 Mineral or fiber, trapezoidal, 1" x 4" x 48"		325	.025		.31	.75		1.06	1.70
0400 1-1/2" x 5-5/8" x 48"		325	.025		.49	.75		1.24	1.90

07 52 Modified Bituminous Membrane Roofing

07 52 13 – Atactic-Polypropylene-Modified Bituminous Membrane Roofing

07 52 13.10 APP Modified Bituminous Membrane	Crew	Daily Output	Labor-Hours	Unit	Material	Labor	Equipment	Total	Total Incl O&P
0010 **APP MODIFIED BITUMINOUS MEMBRANE** R075213-30									
0020 Base sheet, #15 glass fiber felt, nailed to deck	1 Rofc	58	.138	Sq.	11.50	4.23		15.73	20.50
0030 Spot mopped to deck	G-1	295	.190		15.80	5.45	1.74	22.99	29
0040 Fully mopped to deck	"	192	.292		21.50	8.40	2.67	32.57	42
0050 #15 organic felt, nailed to deck	1 Rofc	58	.138		6.95	4.23		11.18	15.30
0060 Spot mopped to deck	G-1	295	.190		11.20	5.45	1.74	18.39	24
0070 Fully mopped to deck	"	192	.292		17	8.40	2.67	28.07	37
2100 APP mod., smooth surf. cap sheet, poly. reinf., torched, 160 mils	G-5	2100	.019	S.F.	.80	.53	.09	1.42	1.93
2150 170 mils		2100	.019		.78	.53	.09	1.40	1.91
2200 Granule surface cap sheet, poly. reinf., torched, 180 mils		2000	.020		.96	.56	.09	1.61	2.17
2250 Smooth surface flashing, torched, 160 mils		1260	.032		.80	.89	.14	1.83	2.65
2300 170 mils		1260	.032		.78	.89	.14	1.81	2.63
2350 Granule surface flashing, torched, 180 mils		1260	.032		.96	.89	.14	1.99	2.83
2400 Fibrated aluminum coating	1 Rofc	3800	.002		.09	.06		.15	.22
2450 Seam heat welding	"	205	.039	L.F.	.09	1.20		1.29	2.26

07 52 16 – Styrene-Butadiene-Styrene Modified Bituminous Membrane Roofing

07 52 16.10 SBS Modified Bituminous Membrane	Crew	Daily Output	Labor-Hours	Unit	Material	Labor	Equipment	Total	Total Incl O&P
0010 **SBS MODIFIED BITUMINOUS MEMBRANE**									
0080 Mod. bit. rfng., SBS mod, gran surf. cap sheet, poly. reinf.									
0650 120 to 149 mils thick	G-1	2000	.028	S.F.	1.35	.81	.26	2.42	3.22
0750 150 to 160 mils	"	2000	.028		1.78	.81	.26	2.85	3.69
1150 For reflective granules, add									
1600 Smooth surface cap sheet, mopped, 145 mils	G-1	2100	.027		.83	.77	.24	1.84	2.57
1620 Lightweight base sheet, fiberglass reinforced, 35 to 47 mil		2100	.027		.29	.77	.24	1.30	1.98
1625 Heavyweight base/ply sheet, reinforced, 87 to 120 mil thick		2100	.027		.92	.77	.24	1.93	2.67
1650 Granulated walkpad, 180 to 220 mils	1 Rofc	400	.020		1.85	.61		2.46	3.14
1700 Smooth surface flashing, 145 mils	G-1	1260	.044		.83	1.28	.41	2.52	3.67
1800 150 mils		1260	.044		.51	1.28	.41	2.20	3.32
1900 Granular surface flashing, 150 mils		1260	.044		.73	1.28	.41	2.42	3.56
2000 160 mils		1260	.044		.76	1.28	.41	2.45	3.60
2010 Elastomeric asphalt primer	1 Rofc	2600	.003		.17	.09		.26	.36
2015 Roofing asphalt, 30 lb./square	G-1	19000	.003		.15	.08	.03	.26	.34
2020 Cold process adhesive, 20 to 30 mils thick	1 Rofc	750	.011		.25	.33		.58	.87
2025 Self adhering vapor retarder, 30 to 45 mils thick	G-5	2150	.019		1.05	.52	.08	1.65	2.19
2050 Seam heat welding	1 Rofc	205	.039	L.F.	.09	1.20		1.29	2.26

07 57 Coated Foamed Roofing

07 57 13 – Sprayed Polyurethane Foam Roofing

07 57 13.10 Sprayed Polyurethane Foam Roofing (S.P.F.)		Crew	Daily Output	Labor-Hours	Unit	Material	2019 Bare Costs Labor	Equipment	Total	Total Incl O&P
0010	**SPRAYED POLYURETHANE FOAM ROOFING (S.P.F.)**									
0100	Primer for metal substrate (when required)	G-2A	3000	.008	S.F.	.50	.22	.21	.93	1.16
0200	Primer for non-metal substrate (when required)		3000	.008		.19	.22	.21	.62	.82
0300	Closed cell spray, polyurethane foam, 3 lb./C.F. density, 1", R6.7		15000	.002		.64	.04	.04	.72	.84
0400	2", R13.4		13125	.002		1.29	.05	.05	1.39	1.56
0500	3", R18.6		11485	.002		1.93	.06	.05	2.04	2.28
0550	4", R24.8		10080	.002		2.58	.06	.06	2.70	3.01
0700	Spray-on silicone coating	↓	2500	.010		1.27	.26	.25	1.78	2.12
0800	Warranty 5-20 year manufacturer's								.15	.15
0900	Warranty 20 year, no dollar limit				↓				.20	.20

07 58 Roll Roofing

07 58 10 – Asphalt Roll Roofing

07 58 10.10 Roll Roofing		Crew	Daily Output	Labor-Hours	Unit	Material	Labor	Equipment	Total	Total Incl O&P
0010	**ROLL ROOFING**									
0100	Asphalt, mineral surface									
0200	1 ply #15 organic felt, 1 ply mineral surfaced									
0300	Selvage roofing, lap 19", nailed & mopped	G-1	27	2.074	Sq.	71	59.50	19	149.50	208
0400	3 plies glass fiber felt (type IV), 1 ply mineral surfaced									
0500	Selvage roofing, lapped 19", mopped	G-1	25	2.240	Sq.	124	64.50	20.50	209	276
0600	Coated glass fiber base sheet, 2 plies of glass fiber									
0700	Felt (type IV), 1 ply mineral surfaced selvage									
0800	Roofing, lapped 19", mopped	G-1	25	2.240	Sq.	133	64.50	20.50	218	285
0900	On nailable decks	"	24	2.333	"	121	67	21.50	209.50	278
1000	3 plies glass fiber felt (type III), 1 ply mineral surfaced									
1100	Selvage roofing, lapped 19", mopped	G-1	25	2.240	Sq.	124	64.50	20.50	209	276

07 61 Sheet Metal Roofing

07 61 13 – Standing Seam Sheet Metal Roofing

07 61 13.10 Standing Seam Sheet Metal Roofing, Field Fab.		Crew	Daily Output	Labor-Hours	Unit	Material	Labor	Equipment	Total	Total Incl O&P
0010	**STANDING SEAM SHEET METAL ROOFING, FIELD FABRICATED**									
0400	Copper standing seam roofing, over 10 squares, 16 oz., 125 lb./sq.	1 Shee	1.30	6.154	Sq.	1,000	240		1,240	1,500
0600	18 oz., 140 lb./sq.	"	1.20	6.667		1,125	260		1,385	1,675
1200	For abnormal conditions or small areas, add					25%	100%			
1300	For lead-coated copper, add				↓	25%				

07 61 16 – Batten Seam Sheet Metal Roofing

07 61 16.10 Batten Seam Sheet Metal Roofing, Field Fab.		Crew	Daily Output	Labor-Hours	Unit	Material	Labor	Equipment	Total	Total Incl O&P
0010	**BATTEN SEAM SHEET METAL ROOFING, FIELD FABRICATED**									
0012	Copper batten seam roofing, over 10 sq., 16 oz., 130 lb./sq.	1 Shee	1.10	7.273	Sq.	1,275	284		1,559	1,875
0100	Zinc/copper alloy batten seam roofing, .020" thick		1.20	6.667		1,475	260		1,735	2,050
0200	Copper roofing, batten seam, over 10 sq., 18 oz., 145 lb./sq.		1	8		1,425	310		1,735	2,100
0800	Zinc, copper alloy roofing, batten seam, .027" thick		1.15	6.957		1,925	271		2,196	2,575
0900	.032" thick		1.10	7.273		1,950	284		2,234	2,625
1000	.040" thick	↓	1.05	7.619	↓	2,500	297		2,797	3,250

07 61 Sheet Metal Roofing

07 61 19 – Flat Seam Sheet Metal Roofing

07 61 19.10 Flat Seam Sheet Metal Roofing, Field Fabricated		Crew	Daily Output	Labor-Hours	Unit	2019 Bare Costs Material	Labor	Equipment	Total	Total Incl O&P
0010	FLAT SEAM SHEET METAL ROOFING, FIELD FABRICATED									
0900	Copper flat seam roofing, over 10 squares, 16 oz., 115 lb./sq.	1 Shee	1.20	6.667	Sq.	940	260		1,200	1,450
0950	18 oz., 130 lb./sq.		1.15	6.957		1,050	271		1,321	1,600
1000	20 oz., 145 lb./sq.		1.10	7.273		1,250	284		1,534	1,850
1008	Zinc flat seam roofing, .020" thick		1.20	6.667		1,275	260		1,535	1,825
1010	.027" thick		1.15	6.957		1,650	271		1,921	2,250
1020	.032" thick		1.12	7.143		1,675	279		1,954	2,275
1030	.040" thick		1.05	7.619		2,125	297		2,422	2,825
1100	Lead flat seam roofing, 5 lb./S.F.	↓	1.30	6.154	↓	1,475	240		1,715	2,025

07 62 Sheet Metal Flashing and Trim

07 62 10 – Sheet Metal Trim

07 62 10.10 Sheet Metal Cladding

		Crew	Daily Output	Labor-Hours	Unit	Material	Labor	Equipment	Total	Total Incl O&P
0010	SHEET METAL CLADDING									
0100	Aluminum, up to 6 bends, .032" thick, window casing	1 Carp	180	.044	S.F.	1.86	1.58		3.44	4.67
0200	Window sill		72	.111	L.F.	1.86	3.96		5.82	8.60
0300	Door casing		180	.044	S.F.	1.86	1.58		3.44	4.67
0400	Fascia		250	.032		1.86	1.14		3	3.93
0500	Rake trim		225	.036		1.86	1.27		3.13	4.14
0700	.024" thick, window casing		180	.044	↓	1.42	1.58		3	4.18
0800	Window sill		72	.111	L.F.	1.42	3.96		5.38	8.10
0900	Door casing		180	.044	S.F.	1.42	1.58		3	4.18
1000	Fascia		250	.032		1.42	1.14		2.56	3.44
1100	Rake trim		225	.036		1.42	1.27		2.69	3.65
1200	Vinyl coated aluminum, up to 6 bends, window casing		180	.044	↓	1.93	1.58		3.51	4.74
1300	Window sill		72	.111	L.F.	1.93	3.96		5.89	8.65
1400	Door casing		180	.044	S.F.	1.93	1.58		3.51	4.74
1500	Fascia		250	.032		1.93	1.14		3.07	4
1600	Rake trim	↓	225	.036	↓	1.93	1.27		3.20	4.21

07 65 Flexible Flashing

07 65 10 – Sheet Metal Flashing

07 65 10.10 Sheet Metal Flashing and Counter Flashing

		Crew	Daily Output	Labor-Hours	Unit	Material	Labor	Equipment	Total	Total Incl O&P
0010	SHEET METAL FLASHING AND COUNTER FLASHING									
0011	Including up to 4 bends									
0020	Aluminum, mill finish, .013" thick	1 Rofc	145	.055	S.F.	.86	1.69		2.55	4
0030	.016" thick		145	.055		1.01	1.69		2.70	4.16
0060	.019" thick		145	.055		1.42	1.69		3.11	4.61
0100	.032" thick		145	.055		1.36	1.69		3.05	4.55
0200	.040" thick		145	.055		2.30	1.69		3.99	5.60
0300	.050" thick		145	.055	↓	2.75	1.69		4.44	6.10
0325	Mill finish 5" x 7" step flashing, .016" thick		1920	.004	Ea.	.15	.13		.28	.40
0350	Mill finish 12" x 12" step flashing, .016" thick	↓	1600	.005	"	.56	.15		.71	.90
0400	Painted finish, add				S.F.	.33			.33	.36
1600	Copper, 16 oz. sheets, under 1000 lb.	1 Rofc	115	.070		8.10	2.13		10.23	12.75
1900	20 oz. sheets, under 1000 lb.		110	.073		10.75	2.23		12.98	15.90
2200	24 oz. sheets, under 1000 lb.		105	.076		14.75	2.34		17.09	20.50
2500	32 oz. sheets, under 1000 lb.		100	.080	↓	19.60	2.45		22.05	26
2700	W shape for valleys, 16 oz., 24" wide	↓	100	.080	L.F.	16.25	2.45		18.70	22.50

07 65 Flexible Flashing

07 65 10 – Sheet Metal Flashing

07 65 10.10 Sheet Metal Flashing and Counter Flashing		Crew	Daily Output	Labor-Hours	Unit	Material	2019 Bare Costs Labor	Equipment	Total	Total Incl O&P
5800	Lead, 2.5 lb./S.F., up to 12" wide	1 Rofc	135	.059	S.F.	6.20	1.82		8.02	10.10
5900	Over 12" wide		135	.059		3.99	1.82		5.81	7.65
8900	Stainless steel sheets, 32 ga.		155	.052		3.41	1.58		4.99	6.60
9000	28 ga.		155	.052		4.75	1.58		6.33	8.10
9100	26 ga.		155	.052		4.70	1.58		6.28	8
9200	24 ga.		155	.052		5.25	1.58		6.83	8.60
9290	For mechanically keyed flashing, add					40%				
9320	Steel sheets, galvanized, 20 ga.	1 Rofc	130	.062	S.F.	1.20	1.89		3.09	4.73
9322	22 ga.		135	.059		1.26	1.82		3.08	4.67
9324	24 ga.		140	.057		.96	1.75		2.71	4.22
9326	26 ga.		148	.054		.84	1.66		2.50	3.91
9328	28 ga.		155	.052		.72	1.58		2.30	3.66
9340	30 ga.		160	.050		.61	1.53		2.14	3.44
9400	Terne coated stainless steel, .015" thick, 28 ga.		155	.052		8.20	1.58		9.78	11.85
9500	.018" thick, 26 ga.		155	.052		9.15	1.58		10.73	12.90
9600	Zinc and copper alloy (brass), .020" thick		155	.052		10.10	1.58		11.68	14
9700	.027" thick		155	.052		12.50	1.58		14.08	16.60
9800	.032" thick		155	.052		15.80	1.58		17.38	20.50
9900	.040" thick		155	.052		20.50	1.58		22.08	25.50

07 65 13 – Laminated Sheet Flashing

07 65 13.10 Laminated Sheet Flashing		Crew	Daily Output	Labor-Hours	Unit	Material	Labor	Equipment	Total	Total Incl O&P
0010	**LAMINATED SHEET FLASHING**, Including up to 4 bends									
8550	Shower pan, 3 ply copper and fabric, 3 oz.	1 Rofc	155	.052	S.F.	4.05	1.58		5.63	7.30
8600	7 oz.	"	155	.052	"	4.87	1.58		6.45	8.20

07 65 19 – Plastic Sheet Flashing

07 65 19.10 Plastic Sheet Flashing and Counter Flashing		Crew	Daily Output	Labor-Hours	Unit	Material	Labor	Equipment	Total	Total Incl O&P
0010	**PLASTIC SHEET FLASHING AND COUNTER FLASHING**									
7300	Polyvinyl chloride, black, 10 mil	1 Rofc	285	.028	S.F.	.27	.86		1.13	1.85
7400	20 mil		285	.028		.26	.86		1.12	1.84
7600	30 mil		285	.028		.33	.86		1.19	1.91
7700	60 mil		285	.028		.86	.86		1.72	2.50
8060	PVC tape, 5" x 45 mils, for joint covers, 100 L.F./roll				Ea.	180			180	198

07 65 23 – Rubber Sheet Flashing

07 65 23.10 Rubber Sheet Flashing and Counter Flashing		Crew	Daily Output	Labor-Hours	Unit	Material	Labor	Equipment	Total	Total Incl O&P
0010	**RUBBER SHEET FLASHING AND COUNTER FLASHING**									
4810	EPDM 90 mils, 1" diameter pipe flashing	1 Rofc	32	.250	Ea.	20.50	7.65		28.15	36.50
4820	2" diameter		30	.267		20	8.15		28.15	37
4830	3" diameter		28	.286		21.50	8.75		30.25	40
4840	4" diameter		24	.333		29	10.20		39.20	50.50
4850	6" diameter		22	.364		29	11.15		40.15	52
8100	Rubber, butyl, 1/32" thick		285	.028	S.F.	2.27	.86		3.13	4.05
8200	1/16" thick		285	.028		3.37	.86		4.23	5.25
8300	Neoprene, cured, 1/16" thick		285	.028		2.66	.86		3.52	4.48
8400	1/8" thick		285	.028		6.15	.86		7.01	8.35

07 65 26 – Self-Adhering Sheet Flashing

07 65 26.10 Self-Adhering Sheet or Roll Flashing		Crew	Daily Output	Labor-Hours	Unit	Material	Labor	Equipment	Total	Total Incl O&P
0010	**SELF-ADHERING SHEET OR ROLL FLASHING**									
0020	Self-adhered flashing, 25 mil cross laminated HDPE, 4" wide	1 Rofc	960	.008	L.F.	.20	.26		.46	.68
0040	6" wide		896	.009		.31	.27		.58	.83
0060	9" wide		832	.010		.46	.29		.75	1.03

07 65 Flexible Flashing

07 65 26 – Self-Adhering Sheet Flashing

07 65 26.10 Self-Adhering Sheet or Roll Flashing		Crew	Daily Output	Labor-Hours	Unit	Material	2019 Bare Costs Labor	Equipment	Total	Total Incl O&P
0080	12″ wide	1 Rofc	768	.010	L.F.	.61	.32		.93	1.25

07 71 Roof Specialties

07 71 19 – Manufactured Gravel Stops and Fasciae

07 71 19.10 Gravel Stop

		Crew	Daily Output	Labor-Hours	Unit	Material	Labor	Equipment	Total	Total Incl O&P
0010	**GRAVEL STOP**									
0020	Aluminum, .050″ thick, 4″ face height, mill finish	1 Shee	145	.055	L.F.	6.55	2.15		8.70	10.80
0080	Duranodic finish		145	.055		7.60	2.15		9.75	11.90
0100	Painted		145	.055		7.60	2.15		9.75	11.90
1200	Copper, 16 oz., 3″ face height		145	.055		25	2.15		27.15	31
1300	6″ face height		135	.059		34	2.31		36.31	41
1350	Galv steel, 24 ga., 4″ leg, plain, with continuous cleat, 4″ face		145	.055		6.60	2.15		8.75	10.85
1360	6″ face height		145	.055		6.70	2.15		8.85	10.95
1500	Polyvinyl chloride, 6″ face height		135	.059		5.80	2.31		8.11	10.25
1800	Stainless steel, 24 ga., 6″ face height		135	.059		15.60	2.31		17.91	21

07 71 19.30 Fascia

		Crew	Daily Output	Labor-Hours	Unit	Material	Labor	Equipment	Total	Total Incl O&P
0010	**FASCIA**									
0100	Aluminum, reverse board and batten, .032″ thick, colored, no furring incl.	1 Shee	145	.055	S.F.	7.15	2.15		9.30	11.40
0200	Residential type, aluminum	1 Carp	200	.040	L.F.	2.13	1.43		3.56	4.70
0220	Vinyl	″	200	.040	″	2.25	1.43		3.68	4.84
0300	Steel, galv and enameled, stock, no furring, long panels	1 Shee	145	.055	S.F.	5.55	2.15		7.70	9.65
0600	Short panels	″	115	.070	″	5.25	2.71		7.96	10.30

07 71 23 – Manufactured Gutters and Downspouts

07 71 23.10 Downspouts

		Crew	Daily Output	Labor-Hours	Unit	Material	Labor	Equipment	Total	Total Incl O&P
0010	**DOWNSPOUTS**									
0020	Aluminum, embossed, .020″ thick, 2″ x 3″	1 Shee	190	.042	L.F.	.93	1.64		2.57	3.74
0100	Enameled		190	.042		1.37	1.64		3.01	4.23
0300	.024″ thick, 2″ x 3″		180	.044		2.14	1.73		3.87	5.25
0400	3″ x 4″		140	.057		2.14	2.23		4.37	6.05
0600	Round, corrugated aluminum, 3″ diameter, .020″ thick		190	.042		2.10	1.64		3.74	5.05
0700	4″ diameter, .025″ thick		140	.057		3.24	2.23		5.47	7.25
0900	Wire strainer, round, 2″ diameter		155	.052	Ea.	1.90	2.01		3.91	5.45
1000	4″ diameter		155	.052		2.42	2.01		4.43	6
1200	Rectangular, perforated, 2″ x 3″		145	.055		2.34	2.15		4.49	6.15
1300	3″ x 4″		145	.055		3.35	2.15		5.50	7.25
1500	Copper, round, 16 oz., stock, 2″ diameter		190	.042	L.F.	8.50	1.64		10.14	12.05
1600	3″ diameter		190	.042		9.05	1.64		10.69	12.65
1800	4″ diameter		145	.055		9.75	2.15		11.90	14.30
1900	5″ diameter		130	.062		14.85	2.40		17.25	20.50
2100	Rectangular, corrugated copper, stock, 2″ x 3″		190	.042		8.20	1.64		9.84	11.75
2200	3″ x 4″		145	.055		9.35	2.15		11.50	13.85
2400	Rectangular, plain copper, stock, 2″ x 3″		190	.042		11.35	1.64		12.99	15.20
2500	3″ x 4″		145	.055		14.15	2.15		16.30	19.15
2700	Wire strainers, rectangular, 2″ x 3″		145	.055	Ea.	17.40	2.15		19.55	22.50
2800	3″ x 4″		145	.055		18.20	2.15		20.35	23.50
3000	Round, 2″ diameter		145	.055		7.20	2.15		9.35	11.45
3100	3″ diameter		145	.055		9.20	2.15		11.35	13.65
3300	4″ diameter		145	.055		12.10	2.15		14.25	16.85
3400	5″ diameter		115	.070		23	2.71		25.71	30
3600	Lead-coated copper, round, stock, 2″ diameter		190	.042	L.F.	23	1.64		24.64	28

07 71 23 – Manufactured Gutters and Downspouts

07 71 23.10 Downspouts		Crew	Daily Output	Labor-Hours	Unit	2019 Bare Costs Material	Labor	Equipment	Total	Total Incl O&P
3700	3" diameter	1 Shee	190	.042	L.F.	23	1.64		24.64	28
3900	4" diameter		145	.055		24	2.15		26.15	30
4000	5" diameter, corrugated		130	.062		24	2.40		26.40	30.50
4200	6" diameter, corrugated		105	.076		31.50	2.97		34.47	40
4300	Rectangular, corrugated, stock, 2" x 3"		190	.042		15.15	1.64		16.79	19.35
4500	Plain, stock, 2" x 3"		190	.042		24.50	1.64		26.14	29.50
4600	3" x 4"		145	.055		34.50	2.15		36.65	41.50
4800	Steel, galvanized, round, corrugated, 2" or 3" diameter, 28 ga.		190	.042		2.11	1.64		3.75	5.05
4900	4" diameter, 28 ga.		145	.055		2.12	2.15		4.27	5.90
5700	Rectangular, corrugated, 28 ga., 2" x 3"		190	.042		2.23	1.64		3.87	5.15
5800	3" x 4"		145	.055		2.24	2.15		4.39	6.05
6000	Rectangular, plain, 28 ga., galvanized, 2" x 3"		190	.042		4.04	1.64		5.68	7.15
6100	3" x 4"		145	.055		4.37	2.15		6.52	8.40
6300	Epoxy painted, 24 ga., corrugated, 2" x 3"		190	.042		2.46	1.64		4.10	5.45
6400	3" x 4"		145	.055		2.98	2.15		5.13	6.85
6600	Wire strainers, rectangular, 2" x 3"		145	.055	Ea.	17.50	2.15		19.65	23
6700	3" x 4"		145	.055		19.40	2.15		21.55	25
6900	Round strainers, 2" or 3" diameter		145	.055		4.38	2.15		6.53	8.40
7000	4" diameter		145	.055		6.05	2.15		8.20	10.25
8200	Vinyl, rectangular, 2" x 3"		210	.038	L.F.	1.94	1.49		3.43	4.60
8300	Round, 2-1/2"		220	.036	"	1.40	1.42		2.82	3.89

07 71 23.20 Downspout Elbows		Crew	Daily Output	Labor-Hours	Unit	Material	Labor	Equipment	Total	Total Incl O&P
0010	**DOWNSPOUT ELBOWS**									
0020	Aluminum, embossed, 2" x 3", .020" thick	1 Shee	100	.080	Ea.	.93	3.12		4.05	6.20
0100	Enameled		100	.080		2.17	3.12		5.29	7.60
0200	Embossed, 3" x 4", .025" thick		100	.080		2.98	3.12		6.10	8.50
0300	Enameled		100	.080		4	3.12		7.12	9.60
0400	Embossed, corrugated, 3" diameter, .020" thick		100	.080		3.13	3.12		6.25	8.65
0500	4" diameter, .025" thick		100	.080		6.70	3.12		9.82	12.55
0600	Copper, 16 oz., 2" diameter		100	.080		10.15	3.12		13.27	16.35
0700	3" diameter		100	.080		9.95	3.12		13.07	16.15
0800	4" diameter		100	.080		13.90	3.12		17.02	20.50
1000	Rectangular, 2" x 3" corrugated		100	.080		9.50	3.12		12.62	15.65
1100	3" x 4" corrugated		100	.080		15.40	3.12		18.52	22
1300	Vinyl, 2-1/2" diameter, 45 or 75 degree bend		100	.080		3.99	3.12		7.11	9.60
1400	Tee Y junction		75	.107		13.10	4.16		17.26	21.50

07 71 23.30 Gutters		Crew	Daily Output	Labor-Hours	Unit	Material	Labor	Equipment	Total	Total Incl O&P
0010	**GUTTERS**									
0012	Aluminum, stock units, 5" K type, .027" thick, plain	1 Shee	125	.064	L.F.	2.84	2.50		5.34	7.25
0100	Enameled		125	.064		2.92	2.50		5.42	7.35
0300	5" K type, .032" thick, plain		125	.064		3.59	2.50		6.09	8.10
0400	Enameled		125	.064		3.58	2.50		6.08	8.05
0700	Copper, half round, 16 oz., stock units, 4" wide		125	.064		8.45	2.50		10.95	13.40
0900	5" wide		125	.064		7.20	2.50		9.70	12.10
1000	6" wide		118	.068		11.40	2.64		14.04	16.95
1200	K type, 16 oz., stock, 5" wide		125	.064		8.40	2.50		10.90	13.40
1300	6" wide		125	.064		8.80	2.50		11.30	13.80
1500	Lead coated copper, 16 oz., half round, stock, 4" wide		125	.064		15.75	2.50		18.25	21.50
1600	6" wide		118	.068		18.65	2.64		21.29	25
1800	K type, stock, 5" wide		125	.064		18.65	2.50		21.15	24.50
1900	6" wide		125	.064		18.85	2.50		21.35	24.50
2100	Copper clad stainless steel, K type, 5" wide		125	.064		7.85	2.50		10.35	12.80

07 71 Roof Specialties

07 71 23 – Manufactured Gutters and Downspouts

07 71 23.30 Gutters		Crew	Daily Output	Labor-Hours	Unit	2019 Bare Costs Material	Labor	Equipment	Total	Total Incl O&P
2200	6″ wide	1 Shee	125	.064	L.F.	10.05	2.50		12.55	15.20
2400	Steel, galv, half round or box, 28 ga., 5″ wide, plain		125	.064		2.16	2.50		4.66	6.50
2500	Enameled		125	.064		2.23	2.50		4.73	6.60
2700	26 ga., stock, 5″ wide		125	.064		2.49	2.50		4.99	6.90
2800	6″ wide		125	.064		2.67	2.50		5.17	7.10
3000	Vinyl, O.G., 4″ wide	1 Carp	115	.070		1.36	2.48		3.84	5.60
3100	5″ wide		115	.070		1.64	2.48		4.12	5.90
3200	4″ half round, stock units		115	.070		1.43	2.48		3.91	5.65
3250	Joint connectors				Ea.	3.06			3.06	3.37
3300	Wood, clear treated cedar, fir or hemlock, 3″ x 4″	1 Carp	100	.080	L.F.	11.20	2.85		14.05	17.05
3400	4″ x 5″	″	100	.080	″	22	2.85		24.85	28.50
5000	Accessories, end cap, K type, aluminum 5″	1 Shee	625	.013	Ea.	.74	.50		1.24	1.64
5010	6″		625	.013		1.52	.50		2.02	2.50
5020	Copper, 5″		625	.013		3.45	.50		3.95	4.63
5030	6″		625	.013		3.67	.50		4.17	4.87
5040	Lead coated copper, 5″		625	.013		13.05	.50		13.55	15.20
5050	6″		625	.013		13.95	.50		14.45	16.20
5060	Copper clad stainless steel, 5″		625	.013		3.55	.50		4.05	4.74
5070	6″		625	.013		3.55	.50		4.05	4.74
5080	Galvanized steel, 5″		625	.013		1.50	.50		2	2.48
5090	6″		625	.013		2.59	.50		3.09	3.68
5100	Vinyl, 4″	1 Carp	625	.013		6.40	.46		6.86	7.80
5110	5″	″	625	.013		6.75	.46		7.21	8.20
5120	Half round, copper, 4″	1 Shee	625	.013		4.62	.50		5.12	5.95
5130	5″		625	.013		4.95	.50		5.45	6.30
5140	6″		625	.013		8.20	.50		8.70	9.90
5150	Lead coated copper, 5″		625	.013		14.70	.50		15.20	17.05
5160	6″		625	.013		22	.50		22.50	25
5170	Copper clad stainless steel, 5″		625	.013		4.85	.50		5.35	6.20
5180	6″		625	.013		4.57	.50		5.07	5.90
5190	Galvanized steel, 5″		625	.013		2.66	.50		3.16	3.76
5200	6″		625	.013		3.31	.50		3.81	4.47
5210	Outlet, aluminum, 2″ x 3″		420	.019		.68	.74		1.42	1.98
5220	3″ x 4″		420	.019		1.07	.74		1.81	2.41
5230	2-3/8″ round		420	.019		.62	.74		1.36	1.91
5240	Copper, 2″ x 3″		420	.019		7.75	.74		8.49	9.75
5250	3″ x 4″		420	.019		8.70	.74		9.44	10.85
5260	2-3/8″ round		420	.019		4.76	.74		5.50	6.50
5270	Lead coated copper, 2″ x 3″		420	.019		26	.74		26.74	29.50
5280	3″ x 4″		420	.019		29.50	.74		30.24	33.50
5290	2-3/8″ round		420	.019		26	.74		26.74	29.50
5300	Copper clad stainless steel, 2″ x 3″		420	.019		7.25	.74		7.99	9.25
5310	3″ x 4″		420	.019		8.25	.74		8.99	10.35
5320	2-3/8″ round		420	.019		4.76	.74		5.50	6.50
5330	Galvanized steel, 2″ x 3″		420	.019		3.77	.74		4.51	5.40
5340	3″ x 4″		420	.019		5.80	.74		6.54	7.60
5350	2-3/8″ round		420	.019		4.71	.74		5.45	6.45
5360	K type mitres, aluminum		65	.123		4.89	4.80		9.69	13.35
5370	Copper		65	.123		16	4.80		20.80	25.50
5380	Lead coated copper		65	.123		55.50	4.80		60.30	69
5390	Copper clad stainless steel		65	.123		27.50	4.80		32.30	38.50
5400	Galvanized steel		65	.123		26.50	4.80		31.30	37.50
5420	Half round mitres, copper		65	.123		68	4.80		72.80	83

07 71 Roof Specialties

07 71 23 – Manufactured Gutters and Downspouts

07 71 23.30 Gutters		Crew	Daily Output	Labor-Hours	Unit	Material	Labor	Equipment	Total	Total Incl O&P
						2019 Bare Costs				
5430	Lead coated copper	1 Shee	65	.123	Ea.	90.50	4.80		95.30	107
5440	Copper clad stainless steel		65	.123		57	4.80		61.80	70.50
5450	Galvanized steel		65	.123		32	4.80		36.80	43
5460	Vinyl mitres and outlets		65	.123		10.80	4.80		15.60	19.85
5470	Sealant		940	.009	L.F.	.01	.33		.34	.56
5480	Soldering		96	.083	″	.27	3.25		3.52	5.70

07 71 23.35 Gutter Guard

		Crew	Daily Output	Labor-Hours	Unit	Material	Labor	Equipment	Total	Total Incl O&P
0010	GUTTER GUARD									
0020	6″ wide strip, aluminum mesh	1 Carp	500	.016	L.F.	2.54	.57		3.11	3.73
0100	Vinyl mesh	″	500	.016	″	2.88	.57		3.45	4.11

07 71 43 – Drip Edge

07 71 43.10 Drip Edge, Rake Edge, Ice Belts

		Crew	Daily Output	Labor-Hours	Unit	Material	Labor	Equipment	Total	Total Incl O&P
0010	DRIP EDGE, RAKE EDGE, ICE BELTS									
0020	Aluminum, .016″ thick, 5″ wide, mill finish	1 Carp	400	.020	L.F.	.59	.71		1.30	1.83
0100	White finish		400	.020		.66	.71		1.37	1.91
0200	8″ wide, mill finish		400	.020		1.53	.71		2.24	2.86
0300	Ice belt, 28″ wide, mill finish		100	.080		7.90	2.85		10.75	13.40
0310	Vented, mill finish		400	.020		2.26	.71		2.97	3.67
0320	Painted finish		400	.020		2.53	.71		3.24	3.96
0400	Galvanized, 5″ wide		400	.020		.61	.71		1.32	1.85
0500	8″ wide, mill finish		400	.020		.84	.71		1.55	2.10
0510	Rake edge, aluminum, 1-1/2″ x 1-1/2″		400	.020		.34	.71		1.05	1.55
0520	3-1/2″ x 1-1/2″		400	.020		.44	.71		1.15	1.66

07 72 Roof Accessories

07 72 23 – Relief Vents

07 72 23.20 Vents

		Crew	Daily Output	Labor-Hours	Unit	Material	Labor	Equipment	Total	Total Incl O&P
0010	VENTS									
0100	Soffit or eave, aluminum, mill finish, strips, 2-1/2″ wide	1 Carp	200	.040	L.F.	.45	1.43		1.88	2.86
0200	3″ wide		200	.040		.46	1.43		1.89	2.87
0300	Enamel finish, 3″ wide		200	.040		.53	1.43		1.96	2.94
0400	Mill finish, rectangular, 4″ x 16″		72	.111	Ea.	1.64	3.96		5.60	8.35
0500	8″ x 16″		72	.111		2.63	3.96		6.59	9.45
2420	Roof ventilator	Q-9	16	1		47.50	35		82.50	110
2500	Vent, roof vent	1 Rofc	24	.333		23.50	10.20		33.70	44.50

07 72 26 – Ridge Vents

07 72 26.10 Ridge Vents and Accessories

		Crew	Daily Output	Labor-Hours	Unit	Material	Labor	Equipment	Total	Total Incl O&P
0010	RIDGE VENTS AND ACCESSORIES									
0100	Aluminum strips, mill finish	1 Rofc	160	.050	L.F.	2.43	1.53		3.96	5.45
0150	Painted finish		160	.050	″	4.18	1.53		5.71	7.35
0200	Connectors		48	.167	Ea.	5.15	5.10		10.25	14.90
0300	End caps		48	.167	″	2.45	5.10		7.55	11.95
0400	Galvanized strips		160	.050	L.F.	3.75	1.53		5.28	6.90
0430	Molded polyethylene, shingles not included		160	.050	″	2.85	1.53		4.38	5.90
0440	End plugs		48	.167	Ea.	2.45	5.10		7.55	11.95
0450	Flexible roll, shingles not included		160	.050	L.F.	2.36	1.53		3.89	5.35
2300	Ridge vent strip, mill finish	1 Shee	155	.052	″	3.94	2.01		5.95	7.65

07 72 Roof Accessories

07 72 53 – Snow Guards

07 72 53.10 Snow Guard Options		Crew	Daily Output	Labor-Hours	Unit	Material	2019 Bare Costs Labor	Equipment	Total	Total Incl O&P
0010	**SNOW GUARD OPTIONS**									
0100	Slate & asphalt shingle roofs, fastened with nails	1 Rofc	160	.050	Ea.	12.40	1.53		13.93	16.40
0200	Standing seam metal roofs, fastened with set screws		48	.167		17.70	5.10		22.80	29
0300	Surface mount for metal roofs, fastened with solder		48	.167		7.50	5.10		12.60	17.50
0400	Double rail pipe type, including pipe		130	.062	L.F.	34	1.89		35.89	41

07 72 80 – Vents

07 72 80.30 Vent Options

		Crew	Daily Output	Labor-Hours	Unit	Material	Labor	Equipment	Total	Total Incl O&P
0010	**VENT OPTIONS**									
0800	Polystyrene baffles, 12" wide for 16" OC rafter spacing	1 Carp	90	.089	Ea.	.49	3.17		3.66	5.80
0900	For 24" OC rafter spacing	"	110	.073	"	.79	2.59		3.38	5.15

07 76 Roof Pavers

07 76 16 – Roof Decking Pavers

07 76 16.10 Roof Pavers and Supports

		Crew	Daily Output	Labor-Hours	Unit	Material	Labor	Equipment	Total	Total Incl O&P
0010	**ROOF PAVERS AND SUPPORTS**									
1000	Roof decking pavers, concrete blocks, 2" thick, natural	1 Clab	115	.070	S.F.	3.56	1.91		5.47	7.10
1100	Colors		115	.070	"	3.82	1.91		5.73	7.35
1200	Support pedestal, bottom cap		960	.008	Ea.	2.64	.23		2.87	3.28
1300	Top cap		960	.008		4.84	.23		5.07	5.70
1400	Leveling shims, 1/16"		1920	.004		1.21	.11		1.32	1.52
1500	1/8"		1920	.004		1.21	.11		1.32	1.52
1600	Buffer pad		960	.008		2.52	.23		2.75	3.15
1700	PVC legs (4" SDR 35)		2880	.003	Inch	.14	.08		.22	.28
2000	Alternate pricing method, system in place		101	.079	S.F.	7.15	2.18		9.33	11.45

07 91 Preformed Joint Seals

07 91 13 – Compression Seals

07 91 13.10 Compression Seals

		Crew	Daily Output	Labor-Hours	Unit	Material	Labor	Equipment	Total	Total Incl O&P
0010	**COMPRESSION SEALS**									
4900	O-ring type cord, 1/4"	1 Bric	472	.017	L.F.	.41	.60		1.01	1.45
4910	1/2"		440	.018		1.06	.64		1.70	2.24
4920	3/4"		424	.019		2.05	.67		2.72	3.37
4930	1"		408	.020		4.15	.69		4.84	5.70
4940	1-1/4"		384	.021		7.90	.73		8.63	9.90
4950	1-1/2"		368	.022		9.70	.77		10.47	11.95
4960	1-3/4"		352	.023		13.60	.80		14.40	16.30
4970	2"		344	.023		22.50	.82		23.32	26

07 91 16 – Joint Gaskets

07 91 16.10 Joint Gaskets

		Crew	Daily Output	Labor-Hours	Unit	Material	Labor	Equipment	Total	Total Incl O&P
0010	**JOINT GASKETS**									
4400	Joint gaskets, neoprene, closed cell w/adh, 1/8" x 3/8"	1 Bric	240	.033	L.F.	.33	1.17		1.50	2.32
4500	1/4" x 3/4"		215	.037		.64	1.31		1.95	2.89
4700	1/2" x 1"		200	.040		1.51	1.41		2.92	4.01
4800	3/4" x 1-1/2"		165	.048		1.65	1.71		3.36	4.67

07 91 Preformed Joint Seals

07 91 23 – Backer Rods

07 91 23.10 Backer Rods		Crew	Daily Output	Labor-Hours	Unit	Material	Labor	Equipment	Total	Total Incl O&P
						2019 Bare Costs				
0010	**BACKER RODS**									
0030	Backer rod, polyethylene, 1/4" diameter	1 Bric	4.60	1.739	C.L.F.	2.42	61.50		63.92	105
0050	1/2" diameter		4.60	1.739		4.09	61.50		65.59	107
0070	3/4" diameter		4.60	1.739		6.95	61.50		68.45	110
0090	1" diameter	↓	4.60	1.739	↓	12.45	61.50		73.95	116

07 91 26 – Joint Fillers

07 91 26.10 Joint Fillers		Crew	Daily Output	Labor-Hours	Unit	Material	Labor	Equipment	Total	Total Incl O&P
0010	**JOINT FILLERS**									
4360	Butyl rubber filler, 1/4" x 1/4"	1 Bric	290	.028	L.F.	.23	.97		1.20	1.87
4365	1/2" x 1/2"		250	.032		.91	1.13		2.04	2.88
4370	1/2" x 3/4"		210	.038		1.37	1.34		2.71	3.74
4375	3/4" x 3/4"		230	.035		2.05	1.23		3.28	4.29
4380	1" x 1"	↓	180	.044	↓	2.73	1.57		4.30	5.60
4390	For coloring, add					12%				
4980	Polyethylene joint backing, 1/4" x 2"	1 Bric	2.08	3.846	C.L.F.	13.85	136		149.85	241
4990	1/4" x 6"		1.28	6.250	"	29.50	220		249.50	400
5600	Silicone, room temp vulcanizing foam seal, 1/4" x 1/2"		1312	.006	L.F.	.46	.22		.68	.87
5610	1/2" x 1/2"		656	.012		.92	.43		1.35	1.74
5620	1/2" x 3/4"		442	.018		1.38	.64		2.02	2.58
5630	3/4" x 3/4"		328	.024		2.08	.86		2.94	3.72
5640	1/8" x 1"		1312	.006		.46	.22		.68	.87
5650	1/8" x 3"		442	.018		1.38	.64		2.02	2.58
5670	1/4" x 3"		295	.027		2.77	.96		3.73	4.64
5680	1/4" x 6"		148	.054		5.55	1.91		7.46	9.30
5690	1/2" x 6"		82	.098		11.10	3.44		14.54	17.95
5700	1/2" x 9"		52.50	.152		16.60	5.35		21.95	27.50
5710	1/2" x 12"	↓	33	.242	↓	22	8.55		30.55	39

07 92 Joint Sealants

07 92 13 – Elastomeric Joint Sealants

07 92 13.20 Caulking and Sealant Options		Crew	Daily Output	Labor-Hours	Unit	Material	Labor	Equipment	Total	Total Incl O&P
0010	**CAULKING AND SEALANT OPTIONS**									
0050	Latex acrylic based, bulk				Gal.	30			30	33
0055	Bulk in place 1/4" x 1/4" bead	1 Bric	300	.027	L.F.	.09	.94		1.03	1.67
0060	1/4" x 3/8"		294	.027		.16	.96		1.12	1.77
0065	1/4" x 1/2"		288	.028		.21	.98		1.19	1.86
0075	3/8" x 3/8"		284	.028		.24	.99		1.23	1.92
0080	3/8" x 1/2"		280	.029		.32	1.01		1.33	2.03
0085	3/8" x 5/8"		276	.029		.39	1.02		1.41	2.13
0095	3/8" x 3/4"		272	.029		.47	1.04		1.51	2.25
0100	1/2" x 1/2"		275	.029		.42	1.03		1.45	2.17
0105	1/2" x 5/8"		269	.030		.53	1.05		1.58	2.33
0110	1/2" x 3/4"		263	.030		.63	1.07		1.70	2.48
0115	1/2" x 7/8"		256	.031		.74	1.10		1.84	2.65
0120	1/2" x 1"		250	.032		.84	1.13		1.97	2.81
0125	3/4" x 3/4"		244	.033		.95	1.16		2.11	2.97
0130	3/4" x 1"		225	.036		1.26	1.25		2.51	3.48
0135	1" x 1"	↓	200	.040	↓	1.68	1.41		3.09	4.20
0190	Cartridges				Gal.	33			33	36
0200	11 fl. oz. cartridge				Ea.	2.82			2.82	3.10

07 92 Joint Sealants

07 92 13 – Elastomeric Joint Sealants

07 92 13.20 Caulking and Sealant Options		Crew	Daily Output	Labor-Hours	Unit	Material	Labor	Equipment	Total	Total Incl O&P
						2019 Bare Costs				
0500	1/4" x 1/2"	1 Bric	288	.028	L.F.	.23	.98		1.21	1.88
0600	1/2" x 1/2"		275	.029		.46	1.03		1.49	2.22
0800	3/4" x 3/4"		244	.033		1.04	1.16		2.20	3.07
0900	3/4" x 1"		225	.036		1.38	1.25		2.63	3.61
1000	1" x 1"	↓	200	.040	↓	1.73	1.41		3.14	4.25
1400	Butyl based, bulk				Gal.	37.50			37.50	41
1500	Cartridges				"	37.50			37.50	41
1700	1/4" x 1/2", 154 L.F./gal.	1 Bric	288	.028	L.F.	.24	.98		1.22	1.90
1800	1/2" x 1/2", 77 L.F./gal.	"	275	.029	"	.49	1.03		1.52	2.24
2300	Polysulfide compounds, 1 component, bulk				Gal.	84.50			84.50	93
2400	Cartridges				"	131			131	144
2600	1 or 2 component, in place, 1/4" x 1/4", 308 L.F./gal.	1 Bric	300	.027	L.F.	.27	.94		1.21	1.87
2700	1/2" x 1/4", 154 L.F./gal.		288	.028		.55	.98		1.53	2.23
2900	3/4" x 3/8", 68 L.F./gal.		272	.029		1.24	1.04		2.28	3.09
3000	1" x 1/2", 38 L.F./gal.	↓	250	.032	↓	2.22	1.13		3.35	4.32
3200	Polyurethane, 1 or 2 component				Gal.	55.50			55.50	61
3300	Cartridges				"	68			68	75
3500	Bulk, in place, 1/4" x 1/4"	1 Bric	300	.027	L.F.	.18	.94		1.12	1.77
3655	1/2" x 1/4"		288	.028		.36	.98		1.34	2.03
3800	3/4" x 3/8"		272	.029		.82	1.04		1.86	2.63
3900	1" x 1/2"	↓	250	.032	↓	1.44	1.13		2.57	3.47
4100	Silicone rubber, bulk				Gal.	58.50			58.50	64
4200	Cartridges				"	52.50			52.50	58

07 92 19 – Acoustical Joint Sealants

07 92 19.10 Acoustical Sealant		Crew	Daily Output	Labor-Hours	Unit	Material	Labor	Equipment	Total	Total Incl O&P
0010	**ACOUSTICAL SEALANT**									
0020	Acoustical sealant, elastomeric, cartridges				Ea.	8.60			8.60	9.45
0025	In place, 1/4" x 1/4"	1 Bric	300	.027	L.F.	.35	.94		1.29	1.96
0030	1/4" x 1/2"		288	.028		.70	.98		1.68	2.40
0035	1/2" x 1/2"		275	.029		1.40	1.03		2.43	3.25
0040	1/2" x 3/4"		263	.030		2.11	1.07		3.18	4.11
0045	3/4" x 3/4"		244	.033		3.16	1.16		4.32	5.40
0050	1" x 1"	↓	200	.040	↓	5.60	1.41		7.01	8.55

Estimating Tips

08 10 00 Doors and Frames

All exterior doors should be addressed for their energy conservation (insulation and seals).

- Most metal doors and frames look alike, but there may be significant differences among them. When estimating these items, be sure to choose the line item that most closely compares to the specification or door schedule requirements regarding:
  - ☐ type of metal
  - ☐ metal gauge
  - ☐ door core material
  - ☐ fire rating
  - ☐ finish
- Wood and plastic doors vary considerably in price. The primary determinant is the veneer material. Lauan, birch, and oak are the most common veneers. Other variables include the following:
  - ☐ hollow or solid core
  - ☐ fire rating
  - ☐ flush or raised panel
  - ☐ finish
- Door pricing includes bore for cylindrical locksets and mortise for hinges.

08 30 00 Specialty Doors and Frames

- There are many varieties of special doors, and they are usually priced per each. Add frames, hardware, or operators required for a complete installation.

08 40 00 Entrances, Storefronts, and Curtain Walls

- Glazed curtain walls consist of the metal tube framing and the glazing material. The cost data in this subdivision is presented for the metal tube framing alone or the composite wall. If your estimate requires a detailed takeoff of the framing, be sure to add the glazing cost and any tints.

08 50 00 Windows

- Steel windows are unglazed and aluminum can be glazed or unglazed. Some metal windows are priced without glass. Refer to 08 80 00 Glazing for glass pricing. The grade C indicates commercial grade windows, usually ASTM C-35.
- All wood windows and vinyl are priced preglazed. The glazing is insulating glass. Add the cost of screens and grills if required and not already included.

08 70 00 Hardware

- Hardware costs add considerably to the cost of a door. The most efficient method to determine the hardware requirements for a project is to review the door and hardware schedule together. One type of door may have different hardware, depending on the door usage.
- Door hinges are priced by the pair, with most doors requiring 1-1/2 pairs per door. The hinge prices do not include installation labor because it is included in door installation. Hinges are classified according to the frequency of use, base material, and finish.

08 80 00 Glazing

- Different openings require different types of glass. The most common types are:
  - ☐ float
  - ☐ tempered
  - ☐ insulating
  - ☐ impact-resistant
  - ☐ ballistic-resistant
- Most exterior windows are glazed with insulating glass. Entrance doors and window walls, where the glass is less than 18" from the floor, are generally glazed with tempered glass. Interior windows and some residential windows are glazed with float glass.
- Coastal communities require the use of impact-resistant glass, dependent on wind speed.
- The insulation or 'u' value is a strong consideration, along with solar heat gain, to determine total energy efficiency.

Reference Numbers

Reference numbers are shown at the beginning of some major classifications. These numbers refer to related items in the Reference Section. The reference information may be an estimating procedure, an alternate pricing method, or technical information.

Note: Not all subdivisions listed here necessarily appear. ■

08 01 53.81 Solid Vinyl Replacement Windows		Crew	Daily Output	Labor-Hours	Unit	Material	2019 Bare Costs Labor	Equipment	Total	Total Incl O&P
0010	SOLID VINYL REPLACEMENT WINDOWS R085313-20									
0020	Double-hung, insulated glass, up to 83 united inches G	2 Carp	8	2	Ea.	181	71.50		252.50	315
0040	84 to 93 G		8	2		203	71.50		274.50	340
0060	94 to 101 G		6	2.667		203	95		298	380
0080	102 to 111 G		6	2.667		213	95		308	390
0100	112 to 120 G		6	2.667		232	95		327	410
0120	For each united inch over 120, add G		800	.020	Inch	2.85	.71		3.56	4.32
0140	Casement windows, one operating sash, 42 to 60 united inches G		8	2	Ea.	244	71.50		315.50	385
0160	61 to 70 G		8	2		270	71.50		341.50	415
0180	71 to 80 G		8	2		305	71.50		376.50	455
0200	81 to 96 G		8	2		330	71.50		401.50	485
0220	Two operating sash, 58 to 78 united inches G		8	2		600	71.50		671.50	785
0240	79 to 88 G		8	2		680	71.50		751.50	870
0260	89 to 98 G		8	2		690	71.50		761.50	880
0280	99 to 108 G		6	2.667		735	95		830	960
0300	109 to 121 G		6	2.667		805	95		900	1,050
0320	Two operating, one fixed sash, 73 to 108 united inches G		8	2		560	71.50		631.50	740
0340	109 to 118 G		8	2		615	71.50		686.50	795
0360	119 to 128 G		6	2.667		720	95		815	945
0380	129 to 138 G		6	2.667		775	95		870	1,000
0400	139 to 156 G		6	2.667		940	95		1,035	1,175
0420	Four operating sash, 98 to 118 united inches G		8	2		1,175	71.50		1,246.50	1,400
0440	119 to 128 G		8	2		1,225	71.50		1,296.50	1,475
0460	129 to 138 G		6	2.667		1,275	95		1,370	1,550
0480	139 to 148 G		6	2.667		1,325	95		1,420	1,600
0500	149 to 168 G		6	2.667		1,425	95		1,520	1,725
0520	169 to 178 G		6	2.667		1,475	95		1,570	1,775
0560	Fixed picture window, up to 63 united inches G		8	2		163	71.50		234.50	298
0580	64 to 83 G		8	2		196	71.50		267.50	335
0600	84 to 101 G		8	2		247	71.50		318.50	390
0620	For each united inch over 101, add G		900	.018	Inch	3	.63		3.63	4.35
0800	Cellulose fiber insulation, poured into sash balance cavity G	1 Carp	36	.222	C.F.	.70	7.90		8.60	13.85
0820	Silicone caulking at perimeter G	"	800	.010	L.F.	.17	.36		.53	.78
2000	Impact resistant replacement windows									
2005	Laminated glass, 120 MPH rating, measure in united inches									
2010	Installation labor does not cover any rework of the window opening									
2020	Double-hung, insulated glass, up to 101 united inches	2 Carp	8	2	Ea.	530	71.50		601.50	700
2025	For each united inch over 101, add		80	.200	Inch	3.31	7.15		10.46	15.45
2100	Casement windows, impact resistant, up to 60 united inches		8	2	Ea.	490	71.50		561.50	660
2120	61 to 70		8	2		515	71.50		586.50	690
2130	71 to 80		8	2		545	71.50		616.50	720
2140	81 to 100		8	2		560	71.50		631.50	735
2150	For each united inch over 100, add		80	.200	Inch	5.05	7.15		12.20	17.35
2200	Awning windows, impact resistant, up to 60 united inches		8	2	Ea.	500	71.50		571.50	670
2220	61 to 70		8	2		525	71.50		596.50	695
2230	71 to 80		8	2		535	71.50		606.50	710
2240	For each united inch over 80, add		80	.200	Inch	4.75	7.15		11.90	17.05
2300	Picture windows, impact resistant, up to 63 united inches		8	2	Ea.	360	71.50		431.50	520
2320	63 to 83		8	2		390	71.50		461.50	550
2330	84 to 101		8	2		425	71.50		496.50	590
2340	For each united inch over 101, add		80	.200	Inch	3.19	7.15		10.34	15.30

08 05 05 – Selective Demolition for Openings

08 05 05.10 Selective Demolition Doors

		Crew	Daily Output	Labor-Hours	Unit	Material	2019 Bare Costs Labor	Equipment	Total	Total Incl O&P
0010	**SELECTIVE DEMOLITION DOORS** R024119-10									
0200	Doors, exterior, 1-3/4" thick, single, 3' x 7' high	1 Clab	16	.500	Ea.		13.75		13.75	22.50
0210	3' x 8' high		10	.800			22		22	36.50
0215	Double, 3' x 8' high		6	1.333			36.50		36.50	60.50
0220	Double, 6' x 7' high		12	.667			18.35		18.35	30.50
0500	Interior, 1-3/8" thick, single, 3' x 7' high		20	.400			11		11	18.20
0520	Double, 6' x 7' high		16	.500			13.75		13.75	22.50
0700	Bi-folding, 3' x 6'-8" high		20	.400			11		11	18.20
0720	6' x 6'-8" high		18	.444			12.20		12.20	20
0900	Bi-passing, 3' x 6'-8" high		16	.500			13.75		13.75	22.50
0940	6' x 6'-8" high		14	.571			15.70		15.70	26
0960	Interior metal door 1-3/4" thick, 3'-0" x 6'-8"		18	.444			12.20		12.20	20
0980	Interior metal door 1-3/4" thick, 3'-0" x 7'-0"		18	.444			12.20		12.20	20
1000	Interior wood door 1-3/4" thick, 3'-0" x 6'-8"		20	.400			11		11	18.20
1020	Interior wood door 1-3/4" thick, 3'-0" x 7'-0"	↓	20	.400			11		11	18.20
1100	Door demo, floor door	2 Sswk	5	3.200			127		127	221
1500	Remove and reset, hollow core	1 Carp	8	1			35.50		35.50	59
1520	Solid		6	1.333			47.50		47.50	78.50
2000	Frames, including trim, metal	↓	8	1			35.50		35.50	59
2200	Wood	2 Carp	32	.500	↓		17.85		17.85	29.50
2201	Alternate pricing method	1 Carp	200	.040	L.F.		1.43		1.43	2.36
3000	Special doors, counter doors	2 Carp	6	2.667	Ea.		95		95	157
3300	Glass, sliding, including frames		12	1.333			47.50		47.50	78.50
3400	Overhead, commercial, 12' x 12' high		4	4			143		143	236
3500	Residential, 9' x 7' high		8	2			71.50		71.50	118
3540	16' x 7' high		7	2.286			81.50		81.50	135
3600	Remove and reset, small		4	4			143		143	236
3620	Large	↓	2.50	6.400			228		228	375
3660	Remove and reset elec. garage door opener	1 Carp	8	1			35.50		35.50	59
4000	Residential lockset, exterior		28	.286			10.20		10.20	16.85
4010	Residential lockset, exterior w/deadbolt		26	.308			10.95		10.95	18.10
4020	Residential lockset, interior		30	.267			9.50		9.50	15.70
4200	Deadbolt lock		32	.250			8.90		8.90	14.75
4224	Pocket door, no frame	↓	8	1			35.50		35.50	59
5590	Remove mail slot	1 Clab	45	.178			4.89		4.89	8.10
5600	Remove door sidelight	1 Carp	6	1.333	↓		47.50		47.50	78.50

08 05 05.20 Selective Demolition of Windows

		Crew	Daily Output	Labor-Hours	Unit	Material	2019 Bare Costs Labor	Equipment	Total	Total Incl O&P
0010	**SELECTIVE DEMOLITION OF WINDOWS** R024119-10									
0200	Aluminum, including trim, to 12 S.F.	1 Clab	16	.500	Ea.		13.75		13.75	22.50
0240	To 25 S.F.		11	.727			20		20	33
0280	To 50 S.F.		5	1.600			44		44	72.50
0320	Storm windows/screens, to 12 S.F.		27	.296			8.15		8.15	13.45
0360	To 25 S.F.		21	.381			10.50		10.50	17.30
0400	To 50 S.F.		16	.500			13.75		13.75	22.50
0500	Screens, incl. aluminum frame, small		20	.400			11		11	18.20
0510	Large		16	.500	↓		13.75		13.75	22.50
0600	Glass, up to 10 S.F./window		200	.040	S.F.		1.10		1.10	1.82
0620	Over 10 S.F./window		150	.053	"		1.47		1.47	2.42
2000	Wood, including trim, to 12 S.F.		22	.364	Ea.		10		10	16.55
2020	To 25 S.F.		18	.444			12.20		12.20	20
2060	To 50 S.F.		13	.615			16.90		16.90	28
2065	To 180 S.F.	↓	8	1	↓		27.50		27.50	45.50

08 05 Common Work Results for Openings

08 05 05 – Selective Demolition for Openings

08 05 05.20 Selective Demolition of Windows		Crew	Daily Output	Labor-Hours	Unit	2019 Bare Costs Material	Labor	Equipment	Total	Total Incl O&P
4300	Remove bay/bow window	2 Carp	6	2.667	Ea.		95		95	157
4410	Remove skylight, plstc domes, flush/curb mtd	″	395	.041	S.F.		1.44		1.44	2.39
4420	Remove skylight, plstc/glass up to 2′ x 3′	1 Carp	15	.533	Ea.		19		19	31.50
4440	Remove skylight, plstc/glass up to 4′ x 6′	2 Carp	10	1.600			57		57	94
4480	Remove roof window up to 3′ x 4′	1 Carp	8	1			35.50		35.50	59
4500	Remove roof window up to 4′ x 6′	2 Carp	6	2.667			95		95	157
5020	Remove and reset window, up to a 2′ x 2′ window	1 Carp	6	1.333			47.50		47.50	78.50
5040	Up to a 3′ x 3′ window		4	2			71.50		71.50	118
5080	Up to a 4′ x 5′ window		2	4			143		143	236
6000	Screening only	1 Clab	4000	.002	S.F.		.06		.06	.09
9100	Window awning, residential	″	80	.100	L.F.		2.75		2.75	4.55

08 11 Metal Doors and Frames

08 11 63 – Metal Screen and Storm Doors and Frames

08 11 63.23 Aluminum Screen and Storm Doors and Frames

Line	Description	Crew	Daily Output	Labor-Hours	Unit	Material	Labor	Equipment	Total	Total Incl O&P
0010	**ALUMINUM SCREEN AND STORM DOORS AND FRAMES**									
0020	Combination storm and screen									
0420	Clear anodic coating, 2′-8″ wide	2 Carp	14	1.143	Ea.	220	40.50		260.50	310
0440	3′-0″ wide	″	14	1.143	″	191	40.50		231.50	278
0500	For 7′-0″ door height, add					8%				
1020	Mill finish, 2′-8″ wide	2 Carp	14	1.143	Ea.	250	40.50		290.50	345
1040	3′-0″ wide	″	14	1.143		269	40.50		309.50	365
1100	For 7′-0″ door, add					8%				
1520	White painted, 2′-8″ wide	2 Carp	14	1.143		256	40.50		296.50	350
1540	3′-0″ wide		14	1.143		310	40.50		350.50	410
1541	Storm door, painted, alum., insul., 6′-8″ x 2′-6″ wide		14	1.143		184	40.50		224.50	270
1545	2′-8″ wide		14	1.143		291	40.50		331.50	390
1600	For 7′-0″ door, add					8%				
1800	Aluminum screen door, 6′-8″ x 2′-8″ wide	2 Carp	14	1.143		209	40.50		249.50	298
1810	3′-0″ wide	″	14	1.143		283	40.50		323.50	380
2000	Wood door & screen, see Section 08 14 33.20									

08 12 Metal Frames

08 12 13 – Hollow Metal Frames

08 12 13.13 Standard Hollow Metal Frames

Line	Description	Crew	Daily Output	Labor-Hours	Unit	Material	Labor	Equipment	Total	Total Incl O&P
0010	**STANDARD HOLLOW METAL FRAMES**									
0020	16 ga., up to 5-3/4″ jamb depth									
0025	3′-0″ x 6′-8″ single G	2 Carp	16	1	Ea.	153	35.50		188.50	227
0028	3′-6″ wide, single G		16	1		263	35.50		298.50	350
0030	4′-0″ wide, single G		16	1		288	35.50		323.50	375
0040	6′-0″ wide, double G		14	1.143		236	40.50		276.50	330
0045	8′-0″ wide, double G		14	1.143		229	40.50		269.50	320
0100	3′-0″ x 7′-0″ single G		16	1		205	35.50		240.50	284
0110	3′-6″ wide, single G		16	1		208	35.50		243.50	288
0112	4′-0″ wide, single G		16	1		256	35.50		291.50	340
0140	6′-0″ wide, double G		14	1.143		237	40.50		277.50	330
0145	8′-0″ wide, double G		14	1.143		247	40.50		287.50	340
1000	16 ga., up to 4-7/8″ deep, 3′-0″ x 7′-0″ single G		16	1		191	35.50		226.50	269
1140	6′-0″ wide, double G		14	1.143		216	40.50		256.50	305

08 12 Metal Frames

08 12 13 – Hollow Metal Frames

08 12 13.13 Standard Hollow Metal Frames			Crew	Daily Output	Labor-Hours	Unit	Material	2019 Bare Costs Labor	Equipment	Total	Total Incl O&P
1200	16 ga., 8-3/4" deep, 3'-0" x 7'-0" single	G	2 Carp	16	1	Ea.	206	35.50		241.50	286
1240	6'-0" wide, double	G		14	1.143		275	40.50		315.50	370
2800	14 ga., up to 3-7/8" deep, 3'-0" x 7'-0" single	G		16	1		265	35.50		300.50	350
2840	6'-0" wide, double	G		14	1.143		310	40.50		350.50	415
3000	14 ga., up to 5-3/4" deep, 3'-0" x 6'-8" single	G		16	1		160	35.50		195.50	235
3002	3'-6" wide, single	G		16	1		257	35.50		292.50	340
3005	4'-0" wide, single	G		16	1		167	35.50		202.50	243
3600	up to 5-3/4" jamb depth, 4'-0" x 7'-0" single	G		15	1.067		255	38		293	345
3620	6'-0" wide, double	G		12	1.333		190	47.50		237.50	288
3640	8'-0" wide, double	G		12	1.333		300	47.50		347.50	410
3700	8'-0" high, 4'-0" wide, single	G		15	1.067		290	38		328	385
3740	8'-0" wide, double	G		12	1.333		340	47.50		387.50	450
4000	6-3/4" deep, 4'-0" x 7'-0" single	G		15	1.067		282	38		320	375
4020	6'-0" wide, double	G		12	1.333		320	47.50		367.50	430
4040	8'-0" wide, double	G		12	1.333		241	47.50		288.50	345
4100	8'-0" high, 4'-0" wide, single	G		15	1.067		189	38		227	271
4140	8'-0" wide, double	G		12	1.333		440	47.50		487.50	565
4400	8-3/4" deep, 4'-0" x 7'-0", single	G		15	1.067		410	38		448	515
4440	8'-0" wide, double	G		12	1.333		440	47.50		487.50	560
4500	4'-0" x 8'-0", single	G		15	1.067		465	38		503	580
4540	8'-0" wide, double	G	↓	12	1.333		475	47.50		522.50	605
4900	For welded frames, add						51.50			51.50	56.50
5400	14 ga., "B" label, up to 5-3/4" deep, 4'-0" x 7'-0" single	G	2 Carp	15	1.067		180	38		218	261
5440	8'-0" wide, double	G		12	1.333		235	47.50		282.50	340
5800	6-3/4" deep, 7'-0" high, 4'-0" wide, single	G		15	1.067		180	38		218	262
5840	8'-0" wide, double	G		12	1.333		330	47.50		377.50	445
6200	8-3/4" deep, 4'-0" x 7'-0" single	G		15	1.067		266	38		304	355
6240	8'-0" wide, double	G	↓	12	1.333	↓	395	47.50		442.50	515
6300	For "A" label use same price as "B" label										
6400	For baked enamel finish, add						30%	15%			
6500	For galvanizing, add						20%				
6600	For hospital stop, add					Ea.	244			244	268
6620	For hospital stop, stainless steel, add					"	113			113	124
7900	Transom lite frames, fixed, add		2 Carp	155	.103	S.F.	45	3.68		48.68	55.50
8000	Movable, add		"	130	.123	"	57.50	4.39		61.89	71

08 13 Metal Doors

08 13 13 – Hollow Metal Doors

08 13 13.15 Metal Fire Doors			Crew	Daily Output	Labor-Hours	Unit	Material	Labor	Equipment	Total	Total Incl O&P
0010	**METAL FIRE DOORS**	R081313-20									
0015	Steel, flush, "B" label, 90 minutes										
0020	Full panel, 20 ga., 2'-0" x 6'-8"		2 Carp	20	.800	Ea.	425	28.50		453.50	515
0040	2'-8" x 6'-8"			18	.889		440	31.50		471.50	540
0060	3'-0" x 6'-8"			17	.941		440	33.50		473.50	540
0080	3'-0" x 7'-0"			17	.941		460	33.50		493.50	560
0140	18 ga., 3'-0" x 6'-8"			16	1		505	35.50		540.50	615
0160	2'-8" x 7'-0"			17	.941		535	33.50		568.50	645
0180	3'-0" x 7'-0"			16	1		520	35.50		555.50	630
0200	4'-0" x 7'-0"		↓	15	1.067	↓	655	38		693	785
0220	For "A" label, 3 hour, 18 ga., use same price as "B" label										
0240	For vision lite, add					Ea.	165			165	181

08 13 13 – Hollow Metal Doors

08 13 13.15 Metal Fire Doors			Crew	Daily Output	Labor-Hours	Unit	Material	2019 Bare Costs Labor	Equipment	Total	Total Incl O&P
0520	Flush, "B" label, 90 minutes, egress core, 20 ga., 2'-0" x 6'-8"		2 Carp	18	.889	Ea.	670	31.50		701.50	795
0540	2'-8" x 6'-8"			17	.941		675	33.50		708.50	795
0560	3'-0" x 6'-8"			16	1		675	35.50		710.50	805
0580	3'-0" x 7'-0"			16	1		705	35.50		740.50	835
0640	Flush, "A" label, 3 hour, egress core, 18 ga., 3'-0" x 6'-8"			15	1.067		735	38		773	870
0660	2'-8" x 7'-0"			16	1		770	35.50		805.50	905
0680	3'-0" x 7'-0"			15	1.067		755	38		793	895
0700	4'-0" x 7'-0"			14	1.143		915	40.50		955.50	1,075

08 13 13.20 Residential Steel Doors			Crew	Daily Output	Labor-Hours	Unit	Material	Labor	Equipment	Total	Total Incl O&P
0010	**RESIDENTIAL STEEL DOORS**										
0020	Prehung, insulated, exterior										
0030	Embossed, full panel, 2'-8" x 6'-8"	G	2 Carp	17	.941	Ea.	415	33.50		448.50	510
0040	3'-0" x 6'-8"	G		15	1.067		291	38		329	385
0060	3'-0" x 7'-0"	G		15	1.067		350	38		388	450
0070	5'-4" x 6'-8", double	G		8	2		585	71.50		656.50	760
0220	Half glass, 2'-8" x 6'-8"	G		17	.941		335	33.50		368.50	420
0240	3'-0" x 6'-8"	G		16	1		335	35.50		370.50	425
0260	3'-0" x 7'-0"	G		16	1		400	35.50		435.50	500
0270	5'-4" x 6'-8", double	G		8	2		990	71.50		1,061.50	1,225
1320	Flush face, full panel, 2'-8" x 6'-8"	G		16	1		297	35.50		332.50	385
1340	3'-0" x 6'-8"	G		15	1.067		297	38		335	390
1360	3'-0" x 7'-0"	G		15	1.067		310	38		348	405
1380	5'-4" x 6'-8", double	G		8	2		725	71.50		796.50	915
1420	Half glass, 2'-8" x 6'-8"	G		17	.941		350	33.50		383.50	440
1440	3'-0" x 6'-8"	G		16	1		350	35.50		385.50	445
1460	3'-0" x 7'-0"	G		16	1		445	35.50		480.50	545
1480	5'-4" x 6'-8", double	G		8	2		675	71.50		746.50	860
1500	Sidelight, full lite, 1'-0" x 6'-8" with grille	G					350			350	385
1510	1'-0" x 6'-8", low E	G					400			400	440
1520	1'-0" x 6'-8", half lite	G					310			310	340
1530	1'-0" x 6'-8", half lite, low E	G					299			299	330
2300	Interior, residential, closet, bi-fold, 2'-0" x 6'-8"	G	2 Carp	16	1		239	35.50		274.50	320
2330	3'-0" wide	G		16	1		283	35.50		318.50	370
2360	4'-0" wide	G		15	1.067		282	38		320	375
2400	5'-0" wide	G		14	1.143		355	40.50		395.50	460
2420	6'-0" wide	G		13	1.231		300	44		344	405
2510	Bi-passing closet, incl. hardware, no frame or trim incl.										
2511	Mirrored, metal frame, 4'-0" x 6'-8"		2 Carp	10	1.600	Opng.	265	57		322	385
2512	5'-0" wide			10	1.600		267	57		324	385
2513	6'-0" wide			10	1.600		295	57		352	420
2514	7'-0" wide			9	1.778		287	63.50		350.50	420
2515	8'-0" wide			9	1.778		480	63.50		543.50	630
2611	Mirrored, metal, 4'-0" x 8'-0"			10	1.600		315	57		372	445
2612	5'-0" wide			10	1.600		325	57		382	450
2613	6'-0" wide			10	1.600		330	57		387	460
2614	7'-0" wide			9	1.778		355	63.50		418.50	495
2615	8'-0" wide			9	1.778		445	63.50		508.50	595

08 14 Wood Doors

08 14 13 – Carved Wood Doors

08 14 13.10 Types of Wood Doors, Carved		Crew	Daily Output	Labor-Hours	Unit	Material	2019 Bare Costs Labor	Equipment	Total	Total Incl O&P
0010	TYPES OF WOOD DOORS, CARVED									
3000	Solid wood, 1-3/4" thick stile and rail									
3020	Mahogany, 3'-0" x 7'-0", six panel	2 Carp	14	1.143	Ea.	1,150	40.50		1,190.50	1,350
3030	With two lites		10	1.600		3,275	57		3,332	3,700
3040	3'-6" x 8'-0", six panel		10	1.600		1,725	57		1,782	2,000
3050	With two lites		8	2		3,000	71.50		3,071.50	3,425
3100	Pine, 3'-0" x 7'-0", six panel		14	1.143		550	40.50		590.50	675
3110	With two lites		10	1.600		870	57		927	1,050
3120	3'-6" x 8'-0", six panel		10	1.600		1,075	57		1,132	1,300
3130	With two lites		8	2		2,050	71.50		2,121.50	2,375
3200	Red oak, 3'-0" x 7'-0", six panel		14	1.143		1,400	40.50		1,440.50	1,600
3210	With two lites		10	1.600		2,600	57		2,657	2,950
3220	3'-6" x 8'-0", six panel		10	1.600		2,925	57		2,982	3,300
3230	With two lites		8	2		3,475	71.50		3,546.50	3,950
4000	Hand carved door, mahogany									
4020	3'-0" x 7'-0", simple design	2 Carp	14	1.143	Ea.	1,825	40.50		1,865.50	2,075
4030	Intricate design		11	1.455		3,800	52		3,852	4,250
4040	3'-6" x 8'-0", simple design		10	1.600		2,900	57		2,957	3,275
4050	Intricate design		8	2		3,800	71.50		3,871.50	4,300
4400	For custom finish, add					595			595	655
4600	Side light, mahogany, 7'-0" x 1'-6" wide, 4 lites	2 Carp	18	.889		1,150	31.50		1,181.50	1,300
4610	6 lites		14	1.143		2,625	40.50		2,665.50	2,950
4620	8'-0" x 1'-6" wide, 4 lites		14	1.143		2,200	40.50		2,240.50	2,500
4630	6 lites		10	1.600		2,150	57		2,207	2,475
4640	Side light, oak, 7'-0" x 1'-6" wide, 4 lites		18	.889		1,250	31.50		1,281.50	1,425
4650	6 lites		14	1.143		2,175	40.50		2,215.50	2,475
4660	8'-0" x 1'-6" wide, 4 lites		14	1.143		1,300	40.50		1,340.50	1,525
4670	6 lites		10	1.600		2,175	57		2,232	2,500

08 14 16 – Flush Wood Doors

08 14 16.09 Smooth Wood Doors		Crew	Daily Output	Labor-Hours	Unit	Material	Labor	Equipment	Total	Total Incl O&P
0010	SMOOTH WOOD DOORS									
0015	Flush, interior, hollow core									
0025	Lauan face, 1-3/8", 3'-0" x 6'-8"	2 Carp	17	.941	Ea.	57	33.50		90.50	118
0030	4'-0" x 6'-8"		16	1		131	35.50		166.50	203
0140	Birch face, 1-3/8", 2'-6" x 6'-8"		17	.941		116	33.50		149.50	183
0180	3'-0" x 6'-8"		17	.941		101	33.50		134.50	167
0200	4'-0" x 6'-8"		16	1		164	35.50		199.50	239
0202	1-3/4", 2'-0" x 6'-8"		17	.941		62	33.50		95.50	124
0204	2'-4" x 7'-0"		16	1		126	35.50		161.50	198
0206	2'-6" x 7'-0"		16	1		130	35.50		165.50	202
0208	2'-8" x 7'-0"		16	1		147	35.50		182.50	220
0210	3'-0" x 7'-0"		16	1		165	35.50		200.50	241
0212	3'-4" x 7'-0"		15	1.067		221	38		259	305
0214	Pair of 3'-0" x 7'-0"		9	1.778	Pr.	272	63.50		335.50	405
0480	For prefinishing, clear, add				Ea.	51			51	56
0500	For prefinishing, stain, add				"	63.50			63.50	70
0620	For dutch door with shelf, add					140%				
1320	M.D. overlay on hardboard, 1-3/8", 2'-0" x 6'-8"	2 Carp	17	.941	Ea.	129	33.50		162.50	198
1340	2'-6" x 6'-8"		17	.941		131	33.50		164.50	200
1380	3'-0" x 6'-8"		17	.941		136	33.50		169.50	206
1400	4'-0" x 6'-8"		16	1		194	35.50		229.50	272
1720	H.P. plastic laminate, 1-3/8", 2'-0" x 6'-8"		16	1		276	35.50		311.50	365

08 14 Wood Doors

08 14 16 – Flush Wood Doors

08 14 16.09 Smooth Wood Doors		Crew	Daily Output	Labor-Hours	Unit	2019 Bare Costs Material	Labor	Equipment	Total	Total Incl O&P
1740	2'-6" x 6'-8"	2 Carp	16	1	Ea.	273	35.50		308.50	360
1780	3'-0" x 6'-8"		15	1.067		305	38		343	405
1785	Door, plastic laminate, 3'-0" x 6'-8"					305			305	340
1800	4'-0" x 6'-8"	2 Carp	14	1.143		400	40.50		440.50	510
2020	Particle core, lauan face, 1-3/8", 2'-6" x 6'-8"		15	1.067		102	38		140	175
2040	3'-0" x 6'-8"		14	1.143		105	40.50		145.50	184
2120	Birch face, 1-3/8", 2'-6" x 6'-8"		15	1.067		115	38		153	190
2140	3'-0" x 6'-8"		14	1.143		127	40.50		167.50	208
3320	M.D. overlay on hardboard, 1-3/8", 2'-6" x 6'-8"		14	1.143		186	40.50		226.50	273
3340	3'-0" x 6'-8"		13	1.231		210	44		254	305
4000	Exterior, flush, solid core, birch, 1-3/4" x 2'-6" x 7'-0"		15	1.067		174	38		212	254
4020	2'-8" wide		15	1.067		247	38		285	335
4040	3'-0" wide		14	1.143		227	40.50		267.50	315
4045	3'-0" x 8'-0"	1 Carp	8	1		495	35.50		530.50	605
4100	Oak faced 1-3/4" x 2'-6" x 7'-0"	2 Carp	15	1.067		229	38		267	315
4120	2'-8" wide		15	1.067		244	38		282	330
4140	3'-0" wide		14	1.143		231	40.50		271.50	325
4160	Walnut faced, 1-3/4" x 3'-0" x 6'-8"	1 Carp	17	.471		545	16.80		561.80	630
4180	3'-6" wide	"	17	.471		665	16.80		681.80	760
4200	Walnut faced, 1-3/4" x 2'-6" x 7'-0"	2 Carp	15	1.067		330	38		368	430
4220	2'-8" wide		15	1.067		325	38		363	420
4240	3'-0" wide		14	1.143		315	40.50		355.50	415
4250	3'-6" wide	1 Carp	14	.571		805	20.50		825.50	920
4260	3'-0" x 8'-0"		8	1		700	35.50		735.50	830
4270	3'-6" wide		8	1		775	35.50		810.50	915
4285	Cherry faced, flush, sc, 1-3/4" x 3'-0" x 8'-0" wide		8	1		495	35.50		530.50	605

08 14 16.10 Wood Doors Decorator		Crew	Daily Output	Labor-Hours	Unit	Material	Labor	Equipment	Total	Total Incl O&P
0010	**WOOD DOORS DECORATOR**									
1800	Exterior, flush, solid wood core, birch 1-3/4" x 2'-6" x 7'-0"	2 Carp	15	1.067	Ea.	325	38		363	425
1820	2'-8" wide		15	1.067		370	38		408	470
1840	3'-0" wide		14	1.143		370	40.50		410.50	480
1900	Oak faced, 1-3/4" x 2'-6" x 7'-0"		15	1.067		215	38		253	299
1920	2'-8" wide		15	1.067		465	38		503	580
1940	3'-0" wide		14	1.143		480	40.50		520.50	600
2100	Walnut faced, 1-3/4" x 2'-6" x 7'-0"		15	1.067		415	38		453	520
2120	2'-8" wide		15	1.067		430	38		468	535
2140	3'-0" wide		14	1.143		455	40.50		495.50	570

08 14 33 – Stile and Rail Wood Doors

08 14 33.10 Wood Doors Paneled		Crew	Daily Output	Labor-Hours	Unit	Material	Labor	Equipment	Total	Total Incl O&P
0010	**WOOD DOORS PANELED**									
0020	Interior, six panel, hollow core, 1-3/8" thick									
0040	Molded hardboard, 2'-0" x 6'-8"	2 Carp	17	.941	Ea.	64	33.50		97.50	126
0060	2'-6" x 6'-8"		17	.941		66.50	33.50		100	129
0070	2'-8" x 6'-8"		17	.941		69.50	33.50		103	132
0080	3'-0" x 6'-8"		17	.941		76.50	33.50		110	140
0140	Embossed print, molded hardboard, 2'-0" x 6'-8"		17	.941		66.50	33.50		100	129
0160	2'-6" x 6'-8"		17	.941		66.50	33.50		100	129
0180	3'-0" x 6'-8"		17	.941		76.50	33.50		110	140
0540	Six panel, solid, 1-3/8" thick, pine, 2'-0" x 6'-8"		15	1.067		161	38		199	240
0560	2'-6" x 6'-8"		14	1.143		159	40.50		199.50	243
0580	3'-0" x 6'-8"		13	1.231		154	44		198	243
1020	Two panel, bored rail, solid, 1-3/8" thick, pine, 1'-6" x 6'-8"		16	1		253	35.50		288.50	335

08 14 33 – Stile and Rail Wood Doors

08 14 33.10 Wood Doors Paneled		Crew	Daily Output	Labor-Hours	Unit	Material	2019 Bare Costs Labor	Equipment	Total	Total Incl O&P
1040	2'-0" x 6'-8"	2 Carp	15	1.067	Ea.	330	38		368	425
1060	2'-6" x 6'-8"		14	1.143		375	40.50		415.50	485
1340	Two panel, solid, 1-3/8" thick, fir, 2'-0" x 6'-8"		15	1.067		166	38		204	246
1360	2'-6" x 6'-8"		14	1.143		225	40.50		265.50	315
1380	3'-0" x 6'-8"		13	1.231		405	44		449	520
1740	Five panel, solid, 1-3/8" thick, fir, 2'-0" x 6'-8"		15	1.067		300	38		338	395
1760	2'-6" x 6'-8"		14	1.143		395	40.50		435.50	505
1780	3'-0" x 6'-8"		13	1.231		395	44		439	510
4190	Exterior, Knotty pine, paneled, 1-3/4", 3'-0" x 6'-8"		16	1		830	35.50		865.50	975
4195	Double 1-3/4", 3'-0" x 6'-8"		16	1		1,650	35.50		1,685.50	1,875
4200	Ash, paneled, 1-3/4", 3'-0" x 6'-8"		16	1		940	35.50		975.50	1,075
4205	Double 1-3/4", 3'-0" x 6'-8"		16	1		1,875	35.50		1,910.50	2,125
4210	Cherry, paneled, 1-3/4", 3'-0" x 6'-8"		16	1		1,075	35.50		1,110.50	1,225
4215	Double 1-3/4", 3'-0" x 6'-8"		16	1		2,150	35.50		2,185.50	2,400
4230	Ash, paneled, 1-3/4", 3'-0" x 8'-0"		16	1		1,175	35.50		1,210.50	1,325
4235	Double 1-3/4", 3'-0" x 8'-0"		16	1		2,325	35.50		2,360.50	2,625
4240	Hard maple, paneled, 1-3/4", 3'-0" x 8'-0"		16	1		1,275	35.50		1,310.50	1,450
4245	Double 1-3/4", 3'-0" x 8'-0"		16	1		2,550	35.50		2,585.50	2,850
4250	Cherry, paneled, 1-3/4", 3'-0" x 8'-0"		16	1		1,275	35.50		1,310.50	1,450
4255	Double 1-3/4", 3'-0" x 8'-0"		16	1		2,550	35.50		2,585.50	2,875

08 14 33.20 Wood Doors Residential		Crew	Daily Output	Labor-Hours	Unit	Material	Labor	Equipment	Total	Total Incl O&P
0010	**WOOD DOORS RESIDENTIAL**									
0200	Exterior, combination storm & screen, pine									
0260	2'-8" wide	2 Carp	10	1.600	Ea.	320	57		377	450
0280	3'-0" wide		9	1.778		335	63.50		398.50	475
0300	7'-1" x 3'-0" wide		9	1.778		355	63.50		418.50	495
0400	Full lite, 6'-9" x 2'-6" wide		11	1.455		340	52		392	460
0420	2'-8" wide		10	1.600		340	57		397	470
0440	3'-0" wide		9	1.778		345	63.50		408.50	485
0500	7'-1" x 3'-0" wide		9	1.778		370	63.50		433.50	515
0604	Door, screen, plain full		12	1.333		310	47.50		357.50	420
0614	Divided		12	1.333		445	47.50		492.50	565
0634	Decor full		12	1.333		545	47.50		592.50	680
0700	Dutch door, pine, 1-3/4" x 2'-8" x 6'-8", 6 panel		12	1.333		750	47.50		797.50	905
0720	Half glass		10	1.600		985	57		1,042	1,175
0800	3'-0" wide, 6 panel		12	1.333		635	47.50		682.50	780
0820	Half glass		10	1.600		1,025	57		1,082	1,225
1000	Entrance door, colonial, 1-3/4" x 6'-8" x 2'-8" wide		16	1		590	35.50		625.50	710
1020	6 panel pine, 3'-0" wide		15	1.067		600	38		638	725
1100	8 panel pine, 2'-8" wide		16	1		725	35.50		760.50	855
1120	3'-0" wide		15	1.067		660	38		698	795
1200	For tempered safety glass lites (min. of 2), add					86			86	94.50
1300	Flush, birch, solid core, 1-3/4" x 6'-8" x 2'-8" wide	2 Carp	16	1		156	35.50		191.50	230
1320	3'-0" wide		15	1.067		151	38		189	229
1350	7'-0" x 2'-8" wide		16	1		140	35.50		175.50	213
1360	3'-0" wide		15	1.067		163	38		201	242
1420	6'-8" x 3'-0" wide, fir		16	1		530	35.50		565.50	640
1720	Carved mahogany 3'-0" x 6'-8"		15	1.067		1,450	38		1,488	1,675
1760	Mahogany, 3'-0" x 6'-8"		15	1.067		705	38		743	840
1930	For dutch door with shelf, add					140%				
2700	Interior, closet, bi-fold, w/hardware, no frame or trim incl.									
2720	Flush, birch, 2'-6" x 6'-8"	2 Carp	13	1.231	Ea.	75.50	44		119.50	156

08 14 33 – Stile and Rail Wood Doors

08 14 33.20 Wood Doors Residential		Crew	Daily Output	Labor-Hours	Unit	Material	Labor	Equipment	Total	Total Incl O&P
						2019 Bare Costs				
2740	3'-0" wide	2 Carp	13	1.231	Ea.	77.50	44		121.50	158
2760	4'-0" wide		12	1.333		120	47.50		167.50	211
2780	5'-0" wide		11	1.455		117	52		169	215
2800	6'-0" wide		10	1.600		133	57		190	241
2804	Flush lauan 2'-0" x 6'-8"		14	1.143		52	40.50		92.50	125
2810	8'-0" wide		9	1.778		211	63.50		274.50	335
2817	6'-0" wide		9	1.778		165	63.50		228.50	286
2820	Flush, hardboard, primed, 6'-8" x 2'-6" wide		13	1.231		77	44		121	158
2840	3'-0" wide		13	1.231		88.50	44		132.50	170
2860	4'-0" wide		12	1.333		155	47.50		202.50	250
2880	5'-0" wide		11	1.455		185	52		237	290
2900	6'-0" wide		10	1.600		186	57		243	299
2920	Hardboard, primed 7'-0" x 4'-0" wide		12	1.333		205	47.50		252.50	305
2930	6'-0" wide		10	1.600		199	57		256	315
3000	Raised panel pine, 6'-6" or 6'-8" x 2'-6" wide		13	1.231		193	44		237	285
3020	3'-0" wide		13	1.231		300	44		344	410
3040	4'-0" wide		12	1.333		325	47.50		372.50	440
3060	5'-0" wide		11	1.455		420	52		472	550
3080	6'-0" wide		10	1.600		465	57		522	605
3180	Louvered, pine, 6'-6" or 6'-8" x 1'-6" wide		13	1.231		201	44		245	294
3190	2'-0" wide		14	1.143		189	40.50		229.50	275
3200	Louvered, pine, 6'-6" or 6'-8" x 2'-6" wide		13	1.231		162	44		206	251
3220	3'-0" wide		13	1.231		254	44		298	350
3225	Door, interior louvered bi-fold, pine, 3'-0" x 6'-8"					254			254	279
3240	4'-0" wide	2 Carp	12	1.333		264	47.50		311.50	370
3260	5'-0" wide		11	1.455		268	52		320	380
3280	6'-0" wide		10	1.600		320	57		377	450
3290	8'-0" wide		10	1.600		490	57		547	635
3300	7'-0" x 3'-0" wide		12	1.333		294	47.50		341.50	405
3320	6'-0" wide		10	1.600		350	57		407	480
4400	Bi-passing closet, incl. hardware and frame, no trim incl.									
4420	Flush, lauan, 6'-8" x 4'-0" wide	2 Carp	12	1.333	Opng.	173	47.50		220.50	270
4440	5'-0" wide		11	1.455		189	52		241	294
4460	6'-0" wide		10	1.600		176	57		233	287
4600	Flush, birch, 6'-8" x 4'-0" wide		12	1.333		266	47.50		313.50	370
4620	5'-0" wide		11	1.455		225	52		277	335
4640	6'-0" wide		10	1.600		340	57		397	470
4800	Louvered, pine, 6'-8" x 4'-0" wide		12	1.333		505	47.50		552.50	635
4820	5'-0" wide		11	1.455		635	52		687	785
4840	6'-0" wide		10	1.600		750	57		807	920
4900	Mirrored, 6'-8" x 4'-0" wide		12	1.333	Ea.	330	47.50		377.50	445
5000	Paneled, pine, 6'-8" x 4'-0" wide		12	1.333	Opng.	485	47.50		532.50	615
5020	5'-0" wide		11	1.455		655	52		707	805
5040	6'-0" wide		10	1.600		875	57		932	1,050
5042	8'-0" wide		12	1.333		975	47.50		1,022.50	1,150
5061	Hardboard, 6'-8" x 4'-0" wide		10	1.600		209	57		266	325
5062	5'-0" wide		10	1.600		215	57		272	330
5063	6'-0" wide		10	1.600		233	57		290	350
6100	Folding accordion, closet, including track and frame									
6200	Rigid PVC	2 Carp	10	1.600	Ea.	112	57		169	217
7310	Passage doors, flush, no frame included									
7320	Hardboard, hollow core, 1-3/8" x 6'-8" x 1'-6" wide	2 Carp	18	.889	Ea.	42.50	31.50		74	99
7330	2'-0" wide		18	.889		47	31.50		78.50	104

08 14 33 – Stile and Rail Wood Doors

08 14 33.20 Wood Doors Residential		Crew	Daily Output	Labor-Hours	Unit	Material	Labor	Equipment	Total	Total Incl O&P
						2019 Bare Costs				
7340	2′-6″ wide	2 Carp	18	.889	Ea.	54.50	31.50		86	113
7350	2′-8″ wide		18	.889		55.50	31.50		87	114
7360	3′-0″ wide		17	.941		57.50	33.50		91	119
7420	Lauan, hollow core, 1-3/8″ x 6′-8″ x 1′-6″ wide		18	.889		54	31.50		85.50	112
7440	2′-0″ wide		18	.889		64.50	31.50		96	124
7450	2′-4″ wide		18	.889		73	31.50		104.50	133
7460	2′-6″ wide		18	.889		73	31.50		104.50	133
7480	2′-8″ wide		18	.889		75.50	31.50		107	136
7500	3′-0″ wide		17	.941		80.50	33.50		114	145
7540	2′-6″ wide		16	1		89	35.50		124.50	157
7560	2′-8″ wide		16	1		91.50	35.50		127	160
7580	3′-0″ wide		16	1		95.50	35.50		131	164
7595	Pair of 3′-0″ wide		9	1.778	Pr.	190	63.50		253.50	315
7700	Birch, hollow core, 1-3/8″ x 6′-8″ x 1′-6″ wide		18	.889	Ea.	66.50	31.50		98	126
7720	2′-0″ wide		18	.889		74	31.50		105.50	134
7740	2′-6″ wide		18	.889		86	31.50		117.50	147
7760	2′-8″ wide		18	.889		91	31.50		122.50	153
7780	3′-0″ wide		17	.941		94.50	33.50		128	160
7790	2′-6″ ash/oak door with hinges		18	.889		97.50	31.50		129	160
7910	2′-8″ wide		16	1		75.50	35.50		111	142
7920	3′-0″ wide		16	1		182	35.50		217.50	259
7940	Pair of 3′-0″ wide		9	1.778	Pr.	400	63.50		463.50	545
8000	Pine louvered, 1-3/8″ x 6′-8″ x 1′-6″ wide		19	.842	Ea.	154	30		184	220
8020	2′-0″ wide		18	.889		175	31.50		206.50	245
8040	2′-6″ wide		18	.889		198	31.50		229.50	271
8060	2′-8″ wide		18	.889		218	31.50		249.50	293
8080	3′-0″ wide		17	.941		236	33.50		269.50	315
8300	Pine paneled, 1-3/8″ x 6′-8″ x 1′-6″ wide		19	.842		192	30		222	261
8320	2′-0″ wide		18	.889		234	31.50		265.50	310
8330	2′-4″ wide		18	.889		252	31.50		283.50	330
8340	2′-6″ wide		18	.889		261	31.50		292.50	340
8360	2′-8″ wide		18	.889		270	31.50		301.50	350
8380	3′-0″ wide		17	.941		283	33.50		316.50	365
8450	French door, pine, 15 lites, 1-3/8″ x 6′-8″ x 2′-6″ wide		18	.889		250	31.50		281.50	330
8470	2′-8″ wide		18	.889		262	31.50		293.50	340
8490	3′-0″ wide		17	.941		270	33.50		303.50	355
8804	Pocket door, 6 panel pine, 2′-6″ x 6′-8″ with frame		10.50	1.524		380	54.50		434.50	510
8814	2′-8″ x 6′-8″		10.50	1.524		390	54.50		444.50	520
8824	3′-0″ x 6′-8″		10.50	1.524		400	54.50		454.50	530
9000	Passage doors, flush, no frame, birch, solid core, 1-3/8″ x 2′-4″ x 7′-0″		16	1		131	35.50		166.50	204
9020	2′-8″ wide		16	1		139	35.50		174.50	212
9040	3′-0″ wide		16	1		154	35.50		189.50	229
9060	3′-4″ wide		15	1.067		292	38		330	385
9080	Pair of 3′-0″ wide		9	1.778	Pr.	305	63.50		368.50	440
9100	Lauan, solid core, 1-3/8″ x 7′-0″ x 2′-4″ wide		16	1	Ea.	172	35.50		207.50	248
9120	2′-8″ wide		16	1		154	35.50		189.50	228
9140	3′-0″ wide		16	1		203	35.50		238.50	282
9160	3′-4″ wide		15	1.067		216	38		254	300
9180	Pair of 3′-0″ wide		9	1.778	Pr.	405	63.50		468.50	550
9200	Hardboard, solid core, 1-3/8″ x 7′-0″ x 2′-4″ wide		16	1	Ea.	175	35.50		210.50	252
9220	2′-8″ wide		16	1		182	35.50		217.50	260
9240	3′-0″ wide		16	1		188	35.50		223.50	266
9260	3′-4″ wide		15	1.067		360	38		398	460

08 14 Wood Doors

08 14 35 – Torrified Doors

08 14 35.10 Torrified Exterior Doors		Crew	Daily Output	Labor-Hours	Unit	Material	2019 Bare Costs Labor	Equipment	Total	Total Incl O&P
0010	**TORRIFIED EXTERIOR DOORS**									
0020	Wood doors made from torrified wood, exterior									
0030	All doors require a finish be applied, all glass is insulated									
0040	All doors require pilot holes for all fasteners									
0100	6 panel, paint grade poplar, 1-3/4" x 3'-0" x 6'-8"	2 Carp	12	1.333	Ea.	1,375	47.50		1,422.50	1,575
0120	Half glass 3'-0" x 6'-8"	"	12	1.333		1,475	47.50		1,522.50	1,700
0200	Side lite, full glass, 1-3/4" x 1'-2" x 6'-8"					970			970	1,075
0220	Side lite, half glass, 1-3/4" x 1'-2" x 6'-8"					950			950	1,050
0300	Raised face, 2 panel, paint grade poplar, 1-3/4" x 3'-0" x 7'-0"	2 Carp	12	1.333		1,375	47.50		1,422.50	1,575
0320	Side lite, raised face, half glass, 1-3/4" x 1'-2" x 7'-0"					1,100			1,100	1,225
0500	6 panel, Fir, 1-3/4" x 3'-0" x 6'-8"	2 Carp	12	1.333		1,950	47.50		1,997.50	2,225
0520	Half glass 3'-0" x 6'-8"	"	12	1.333		1,925	47.50		1,972.50	2,200
0600	Side lite, full glass, 1-3/4" x 1'-2" x 6'-8"					980			980	1,075
0620	Side lite, half glass, 1-3/4" x 1'-2" x 6'-8"					1,000			1,000	1,100
0700	6 panel, Mahogany, 1-3/4" x 3'-0" x 6'-8"	2 Carp	12	1.333		2,175	47.50		2,222.50	2,475
0800	Side lite, full glass, 1-3/4" x 1'-2" x 6'-8"					1,100			1,100	1,200
0820	Side lite, half glass, 1-3/4" x 1'-2" x 6'-8"				↓	1,100			1,100	1,200

08 14 40 – Interior Cafe Doors

08 14 40.10 Cafe Style Doors		Crew	Daily Output	Labor-Hours	Unit	Material	Labor	Equipment	Total	Total Incl O&P
0010	**CAFE STYLE DOORS**									
6520	Interior cafe doors, 2'-6" opening, stock, panel pine	2 Carp	16	1	Ea.	415	35.50		450.50	515
6540	3'-0" opening	"	16	1	"	460	35.50		495.50	565
6550	Louvered pine									
6560	2'-6" opening	2 Carp	16	1	Ea.	335	35.50		370.50	425
8000	3'-0" opening		16	1		360	35.50		395.50	455
8010	2'-6" opening, hardwood		16	1		365	35.50		400.50	460
8020	3'-0" opening	↓	16	1	↓	420	35.50		455.50	520

08 16 Composite Doors

08 16 13 – Fiberglass Doors

08 16 13.10 Entrance Doors, Fibrous Glass		Crew	Daily Output	Labor-Hours	Unit	Material	Labor	Equipment	Total	Total Incl O&P
0010	**ENTRANCE DOORS, FIBROUS GLASS**									
0020	Exterior, fiberglass, door, 2'-8" wide x 6'-8" high G	2 Carp	15	1.067	Ea.	278	38		316	370
0040	3'-0" wide x 6'-8" high G		15	1.067		280	38		318	375
0060	3'-0" wide x 7'-0" high G		15	1.067		500	38		538	615
0080	3'-0" wide x 6'-8" high, with two lites G		15	1.067		355	38		393	455
0100	3'-0" wide x 8'-0" high, with two lites G		15	1.067		545	38		583	665
0110	Half glass, 3'-0" wide x 6'-8" high G		15	1.067		465	38		503	575
0120	3'-0" wide x 6'-8" high, low E G		15	1.067		500	38		538	615
0130	3'-0" wide x 8'-0" high G		15	1.067		600	38		638	725
0140	3'-0" wide x 8'-0" high, low E G	↓	15	1.067		680	38		718	810
0150	Side lights, 1'-0" wide x 6'-8" high G					291			291	320
0160	1'-0" wide x 6'-8" high, low E G					298			298	325
0180	1'-0" wide x 6'-8" high, full glass G					340			340	375
0190	1'-0" wide x 6'-8" high, low E G				↓	380			380	415

08 16 Composite Doors

08 16 14 – French Doors

08 16 14.10 Exterior Doors With Glass Lites		Crew	Daily Output	Labor-Hours	Unit	Material	Labor	Equipment	Total	Total Incl O&P
						2019 Bare Costs				
0010	**EXTERIOR DOORS WITH GLASS LITES**									
0020	French, Fir, 1-3/4", 3'-0" wide x 6'-8" high	2 Carp	12	1.333	Ea.	635	47.50		682.50	780
0025	Double		12	1.333		1,275	47.50		1,322.50	1,475
0030	Maple, 1-3/4", 3'-0" wide x 6'-8" high		12	1.333		715	47.50		762.50	865
0035	Double		12	1.333		1,425	47.50		1,472.50	1,650
0040	Cherry, 1-3/4", 3'-0" wide x 6'-8" high		12	1.333		835	47.50		882.50	995
0045	Double		12	1.333		1,675	47.50		1,722.50	1,900
0100	Mahogany, 1-3/4", 3'-0" wide x 8'-0" high		10	1.600		855	57		912	1,050
0105	Double		10	1.600		1,725	57		1,782	1,975
0110	Fir, 1-3/4", 3'-0" wide x 8'-0" high		10	1.600		1,275	57		1,332	1,500
0115	Double		10	1.600		2,575	57		2,632	2,925
0120	Oak, 1-3/4", 3'-0" wide x 8'-0" high		10	1.600		1,925	57		1,982	2,225
0125	Double	↓	10	1.600	↓	3,850	57		3,907	4,350

08 17 Integrated Door Opening Assemblies

08 17 23 – Integrated Wood Door Opening Assemblies

08 17 23.10 Pre-Hung Doors

08 17 23.10 Pre-Hung Doors		Crew	Daily Output	Labor-Hours	Unit	Material	Labor	Equipment	Total	Total Incl O&P
0010	**PRE-HUNG DOORS**									
0300	Exterior, wood, comb. storm & screen, 6'-9" x 2'-6" wide	2 Carp	15	1.067	Ea.	257	38		295	345
0320	2'-8" wide		15	1.067		345	38		383	445
0340	3'-0" wide	↓	15	1.067		325	38		363	420
0360	For 7'-0" high door, add				↓	41			41	45
1600	Entrance door, flush, birch, solid core									
1620	4-5/8" solid jamb, 1-3/4" x 6'-8" x 2'-8" wide	2 Carp	16	1	Ea.	300	35.50		335.50	390
1640	3'-0" wide		16	1		400	35.50		435.50	500
1642	5-5/8" jamb	↓	16	1		345	35.50		380.50	440
1680	For 7'-0" high door, add				↓	25			25	27.50
2000	Entrance door, colonial, 6 panel pine									
2020	4-5/8" solid jamb, 1-3/4" x 6'-8" x 2'-8" wide	2 Carp	16	1	Ea.	655	35.50		690.50	785
2040	3'-0" wide	"	16	1		690	35.50		725.50	815
2060	For 7'-0" high door, add					56.50			56.50	62.50
2200	For 5-5/8" solid jamb, add					44.50			44.50	49
2230	French style, exterior, 1 lite, 1-3/4" x 3'-0" x 6'-8"	1 Carp	14	.571		675	20.50		695.50	780
2235	9 lites	"	14	.571		710	20.50		730.50	815
2245	15 lites	2 Carp	14	1.143	↓	770	40.50		810.50	915
2250	Double, 15 lites, 2'-0" x 6'-8", 4'-0" opening		7	2.286	Pr.	1,300	81.50		1,381.50	1,550
2260	2'-6" x 6'-8", 5'-0" opening		7	2.286		1,425	81.50		1,506.50	1,675
2280	3'-0" x 6'-8", 6'-0" opening		7	2.286	↓	1,600	81.50		1,681.50	1,900
2430	3'-0" x 7'-0", 15 lites		14	1.143	Ea.	1,000	40.50		1,040.50	1,175
2432	Two 3'-0" x 7'-0"		7	2.286	Pr.	2,100	81.50		2,181.50	2,425
2435	3'-0" x 8'-0"		14	1.143	Ea.	1,075	40.50		1,115.50	1,250
2437	Two, 3'-0" x 8'-0"	↓	7	2.286	Pr.	2,225	81.50		2,306.50	2,550
2500	Exterior, metal face, insulated, incl. jamb, brickmold and									
2520	Threshold, flush, 2'-8" x 6'-8"	2 Carp	16	1	Ea.	299	35.50		334.50	390
2550	3'-0" x 6'-8"		16	1		300	35.50		335.50	390
3500	Embossed, 6 panel, 2'-8" x 6'-8"		16	1		340	35.50		375.50	435
3550	3'-0" x 6'-8"		16	1		345	35.50		380.50	440
3600	2 narrow lites, 2'-8" x 6'-8"		16	1		315	35.50		350.50	405
3650	3'-0" x 6'-8"		16	1		320	35.50		355.50	415
3700	Half glass, 2'-8" x 6'-8"	↓	16	1	↓	340	35.50		375.50	435

08 17 Integrated Door Opening Assemblies

08 17 23 – Integrated Wood Door Opening Assemblies

08 17 23.10 Pre-Hung Doors		Crew	Daily Output	Labor-Hours	Unit	Material	2019 Bare Costs Labor	Equipment	Total	Total Incl O&P
3750	3'-0" x 6'-8"	2 Carp	16	1	Ea.	355	35.50		390.50	450
3800	2 top lites, 2'-8" x 6'-8"		16	1		325	35.50		360.50	420
3850	3'-0" x 6'-8"		16	1		365	35.50		400.50	460
4000	Interior, passage door, 4-5/8" solid jamb									
4370	Pine, louvered, 2'-8" x 6'-8"	2 Carp	17	.941	Ea.	204	33.50		237.50	281
4380	3'-0"		17	.941		212	33.50		245.50	289
4400	Lauan, flush, solid core, 1-3/8" x 6'-8" x 2'-6" wide		17	.941		195	33.50		228.50	270
4420	2'-8" wide		17	.941		195	33.50		228.50	270
4440	3'-0" wide		16	1		212	35.50		247.50	292
4600	Hollow core, 1-3/8" x 6'-8" x 2'-6" wide		17	.941		134	33.50		167.50	204
4620	2'-8" wide		17	.941		137	33.50		170.50	206
4640	3'-0" wide		16	1		140	35.50		175.50	213
4700	For 7'-0" high door, add					41.50			41.50	46
5000	Birch, flush, solid core, 1-3/8" x 6'-8" x 2'-6" wide	2 Carp	17	.941		300	33.50		333.50	385
5020	2'-8" wide		17	.941		209	33.50		242.50	286
5040	3'-0" wide		16	1		325	35.50		360.50	420
5200	Hollow core, 1-3/8" x 6'-8" x 2'-6" wide		17	.941		247	33.50		280.50	330
5220	2'-8" wide		17	.941		280	33.50		313.50	365
5240	3'-0" wide		16	1		266	35.50		301.50	350
5280	For 7'-0" high door, add					35.50			35.50	39
5500	Hardboard paneled, 1-3/8" x 6'-8" x 2'-6" wide	2 Carp	17	.941		152	33.50		185.50	223
5520	2'-8" wide		17	.941		164	33.50		197.50	236
5540	3'-0" wide		16	1		162	35.50		197.50	237
6000	Pine paneled, 1-3/8" x 6'-8" x 2'-6" wide		17	.941		275	33.50		308.50	355
6020	2'-8" wide		17	.941		293	33.50		326.50	375
6040	3'-0" wide		16	1		300	35.50		335.50	390
7200	Prehung, bifold, mirrored, 6'-8" x 5'-0"	1 Carp	9	.889		425	31.50		456.50	525
7220	6'-8" x 6'-0"		9	.889		425	31.50		456.50	525
7240	6'-8" x 8'-0"		6	1.333		740	47.50		787.50	895
7600	Oak, 6 panel, 1-3/4" x 6'-8" x 3'-0"		17	.471		915	16.80		931.80	1,025
8500	Pocket door frame with lauan, flush, hollow core, 1-3/8" x 3'-0" x 6'-8"		17	.471		295	16.80		311.80	355

08 31 Access Doors and Panels

08 31 13 – Access Doors and Frames

08 31 13.20 Bulkhead/Cellar Doors		Crew	Daily Output	Labor-Hours	Unit	Material	Labor	Equipment	Total	Total Incl O&P
0010	**BULKHEAD/CELLAR DOORS**									
0020	Steel, not incl. sides, 44" x 62"	1 Carp	5.50	1.455	Ea.	655	52		707	805
0100	52" x 73"		5.10	1.569		825	56		881	1,000
0500	With sides and foundation plates, 57" x 45" x 24"		4.70	1.702		895	60.50		955.50	1,075
0600	42" x 49" x 51"		4.30	1.860		590	66.50		656.50	760

08 31 13.40 Kennel Doors		Crew	Daily Output	Labor-Hours	Unit	Material	Labor	Equipment	Total	Total Incl O&P
0010	**KENNEL DOORS**									
0020	2 way, swinging type, 13" x 19" opening	2 Carp	11	1.455	Opng.	89.50	52		141.50	184
0100	17" x 29" opening		11	1.455		131	52		183	231
0200	9" x 9" opening, electronic with accessories		11	1.455		154	52		206	255

08 32 Sliding Glass Doors

08 32 13 – Sliding Aluminum-Framed Glass Doors

08 32 13.10 Sliding Aluminum Doors		Crew	Daily Output	Labor-Hours	Unit	Material	2019 Bare Costs Labor	Equipment	Total	Total Incl O&P
0010	**SLIDING ALUMINUM DOORS**									
0350	Aluminum, 5/8" tempered insulated glass, 6' wide									
0400	Premium	2 Carp	4	4	Ea.	1,600	143		1,743	2,000
0450	Economy		4	4		855	143		998	1,175
0500	8' wide, premium		3	5.333		1,775	190		1,965	2,275
0550	Economy		3	5.333		1,525	190		1,715	2,000
0600	12' wide, premium		2.50	6.400		3,100	228		3,328	3,800
0650	Economy		2.50	6.400		1,550	228		1,778	2,075
4000	Aluminum, baked on enamel, temp glass, 6'-8" x 10'-0" wide		4	4		1,125	143		1,268	1,475
4020	Insulating glass, 6'-8" x 6'-0" wide		4	4		975	143		1,118	1,300
4040	8'-0" wide		3	5.333		1,150	190		1,340	1,600
4060	10'-0" wide		2	8		1,450	285		1,735	2,075
4080	Anodized, temp glass, 6'-8" x 6'-0" wide		4	4		470	143		613	755
4100	8'-0" wide		3	5.333		595	190		785	970
4120	10'-0" wide	↓	2	8	↓	695	285		980	1,225

08 32 19 – Sliding Wood-Framed Glass Doors

08 32 19.10 Sliding Wood Doors		Crew	Daily Output	Labor-Hours	Unit	Material	2019 Bare Costs Labor	Equipment	Total	Total Incl O&P
0010	**SLIDING WOOD DOORS**									
0020	Wood, tempered insul. glass, 6' wide, premium	2 Carp	4	4	Ea.	1,500	143		1,643	1,875
0100	Economy		4	4		1,250	143		1,393	1,600
0150	8' wide, wood, premium		3	5.333		1,925	190		2,115	2,425
0200	Economy		3	5.333		1,550	190		1,740	2,050
0235	10' wide, wood, premium		2.50	6.400		2,750	228		2,978	3,400
0240	Economy		2.50	6.400		2,350	228		2,578	2,950
0250	12' wide, wood, premium		2.50	6.400		3,225	228		3,453	3,900
0300	Economy	↓	2.50	6.400	↓	2,575	228		2,803	3,200

08 32 19.15 Sliding Glass Vinyl-Clad Wood Doors		Crew	Daily Output	Labor-Hours	Unit	Material	2019 Bare Costs Labor	Equipment	Total	Total Incl O&P
0010	**SLIDING GLASS VINYL-CLAD WOOD DOORS**									
0020	Glass, sliding vinyl-clad, insul. glass, 6'-0" x 6'-8" G	2 Carp	4	4	Opng.	1,575	143		1,718	1,950
0025	6'-0" x 6'-10" high G		4	4		1,750	143		1,893	2,150
0030	6'-0" x 8'-0" high G		4	4		2,125	143		2,268	2,575
0050	5'-0" x 6'-8" high G		4	4		1,575	143		1,718	1,975
0100	8'-0" x 6'-10" high G		4	4		2,100	143		2,243	2,525
0104	8'-0" x 6'-8" high G		4	4		2,000	143		2,143	2,450
0150	8'-0" x 8'-0" high G		4	4		2,400	143		2,543	2,875
0500	4 leaf, 9'-0" x 6'-10" high G		3	5.333		3,500	190		3,690	4,175
0550	9'-0" x 8'-0" high G		3	5.333		3,925	190		4,115	4,650
0600	12'-0" x 6'-10" high G	↓	3	5.333	↓	4,200	190		4,390	4,950

08 36 Panel Doors

08 36 13 – Sectional Doors

08 36 13.20 Residential Garage Doors		Crew	Daily Output	Labor-Hours	Unit	Material	2019 Bare Costs Labor	Equipment	Total	Total Incl O&P
0010	**RESIDENTIAL GARAGE DOORS**									
0050	Hinged, wood, custom, double door, 9' x 7'	2 Carp	4	4	Ea.	905	143		1,048	1,225
0070	16' x 7'		3	5.333		1,325	190		1,515	1,775
0200	Overhead, sectional, incl. hardware, fiberglass, 9' x 7', standard		5	3.200		1,025	114		1,139	1,325
0220	Deluxe		5	3.200		1,225	114		1,339	1,550
0300	16' x 7', standard		6	2.667		1,650	95		1,745	1,975
0320	Deluxe		6	2.667		2,300	95		2,395	2,675
0500	Hardboard, 9' x 7', standard	↓	8	2	↓	735	71.50		806.50	930

08 36 Panel Doors

08 36 13 – Sectional Doors

08 36 13.20 Residential Garage Doors		Crew	Daily Output	Labor-Hours	Unit	Material	Labor	Equipment	Total	Total Incl O&P
						2019 Bare Costs				
0520	Deluxe	2 Carp	8	2	Ea.	875	71.50		946.50	1,075
0600	16′ x 7′, standard		6	2.667		1,300	95		1,395	1,575
0620	Deluxe		6	2.667		1,525	95		1,620	1,850
0700	Metal, 9′ x 7′, standard		8	2		935	71.50		1,006.50	1,150
0720	Deluxe		6	2.667		1,050	95		1,145	1,300
0800	16′ x 7′, standard		6	2.667		1,100	95		1,195	1,350
0820	Deluxe		5	3.200		1,475	114		1,589	1,825
0900	Wood, 9′ x 7′, standard		8	2		1,050	71.50		1,121.50	1,275
0920	Deluxe		8	2		2,275	71.50		2,346.50	2,650
1000	16′ x 7′, standard		6	2.667		1,700	95		1,795	2,025
1020	Deluxe		6	2.667		3,175	95		3,270	3,650
1800	Door hardware, sectional	1 Carp	4	2		370	71.50		441.50	530
1810	Door tracks only		4	2		172	71.50		243.50	310
1820	One side only		7	1.143		131	40.50		171.50	212
4000	For electric operator, economy, add		8	1		445	35.50		480.50	550
4100	Deluxe, including remote control		8	1		640	35.50		675.50	765
4500	For transmitter/receiver control, add to operator				Total	117			117	128
4600	Transmitters, additional				"	66			66	72.50
6000	Replace section, on sectional door, fiberglass, 9′ x 7′	1 Carp	4	2	Ea.	675	71.50		746.50	865
6020	16′ x 7′		3.50	2.286		780	81.50		861.50	990
6200	Hardboard, 9′ x 7′		4	2		179	71.50		250.50	315
6220	16′ x 7′		3.50	2.286		241	81.50		322.50	400
6300	Metal, 9′ x 7′		4	2		227	71.50		298.50	370
6320	16′ x 7′		3.50	2.286		335	81.50		416.50	505
6500	Wood, 9′ x 7′		4	2		126	71.50		197.50	257
6520	16′ x 7′		3.50	2.286		253	81.50		334.50	415
7010	Garage doors, row of lites					131			131	144

08 51 Metal Windows

08 51 13 – Aluminum Windows

08 51 13.20 Aluminum Windows		Crew	Daily Output	Labor-Hours	Unit	Material	Labor	Equipment	Total	Total Incl O&P
0010	**ALUMINUM WINDOWS**, incl. frame and glazing, commercial grade									
1000	Stock units, casement, 3′-1″ x 3′-2″ opening	2 Sswk	10	1.600	Ea.	385	63.50		448.50	535
1040	Insulating glass	"	10	1.600		525	63.50		588.50	685
1050	Add for storms					123			123	136
1600	Projected, with screen, 3′-1″ x 3′-2″ opening	2 Sswk	10	1.600		365	63.50		428.50	510
1650	Insulating glass	"	10	1.600		390	63.50		453.50	540
1700	Add for storms					120			120	132
2000	4′-5″ x 5′-3″ opening	2 Sswk	8	2		400	79.50		479.50	580
2050	Insulating glass	"	8	2		480	79.50		559.50	670
2100	Add for storms					129			129	142
2500	Enamel finish windows, 3′-1″ x 3′-2″	2 Sswk	10	1.600		370	63.50		433.50	515
2550	Insulating glass		10	1.600		340	63.50		403.50	485
2600	4′-5″ x 5′-3″		8	2		415	79.50		494.50	600
2700	Insulating glass		8	2		445	79.50		524.50	630
3000	Single-hung, 2′ x 3′ opening, enameled, standard glazed		10	1.600		214	63.50		277.50	345
3100	Insulating glass		10	1.600		260	63.50		323.50	395
3300	2′-8″ x 6′-8″ opening, standard glazed		8	2		375	79.50		454.50	550
3400	Insulating glass		8	2		475	79.50		554.50	660
3700	3′-4″ x 5′-0″ opening, standard glazed		9	1.778		310	70.50		380.50	465
3800	Insulating glass		9	1.778		335	70.50		405.50	495

08 51 Metal Windows

08 51 13 – Aluminum Windows

08 51 13.20 Aluminum Windows		Crew	Daily Output	Labor-Hours	Unit	Material	2019 Bare Costs Labor	Equipment	Total	Total Incl O&P
4000	Sliding aluminum, 3′ x 2′ opening, standard glazed	2 Sswk	10	1.600	Ea.	220	63.50		283.50	350
4100	Insulating glass		10	1.600		236	63.50		299.50	370
4300	5′ x 3′ opening, standard glazed		9	1.778		335	70.50		405.50	495
4400	Insulating glass		9	1.778		390	70.50		460.50	555
4600	8′ x 4′ opening, standard glazed		6	2.667		360	106		466	580
4700	Insulating glass		6	2.667		575	106		681	820
5000	9′ x 5′ opening, standard glazed		4	4		540	159		699	870
5100	Insulating glass		4	4		850	159		1,009	1,200
5500	Sliding, with thermal barrier and screen, 6′ x 4′, 2 track		8	2		725	79.50		804.50	940
5700	4 track	↓	8	2		910	79.50		989.50	1,150
6000	For above units with bronze finish, add					15%				
6200	For installation in concrete openings, add				↓	8%				

08 51 13.30 Impact Resistant Aluminum Windows		Crew	Daily Output	Labor-Hours	Unit	Material	Labor	Equipment	Total	Total Incl O&P
0010	**IMPACT RESISTANT ALUMINUM WINDOWS**, incl. frame and glazing									
0100	Single-hung, impact resistant, 2′-8″ x 5′-0″	2 Carp	9	1.778	Ea.	1,250	63.50		1,313.50	1,475
0120	3′-0″ x 5′-0″		9	1.778		1,375	63.50		1,438.50	1,600
0130	4′-0″ x 5′-0″		9	1.778		1,475	63.50		1,538.50	1,700
0250	Horizontal slider, impact resistant, 5′-5″ x 5′-2″	↓	9	1.778	↓	1,625	63.50		1,688.50	1,900

08 51 23 – Steel Windows

08 51 23.40 Basement Utility Windows		Crew	Daily Output	Labor-Hours	Unit	Material	Labor	Equipment	Total	Total Incl O&P
0010	**BASEMENT UTILITY WINDOWS**									
0015	1′-3″ x 2′-8″	1 Carp	16	.500	Ea.	147	17.85		164.85	192
1100	1′-7″ x 2′-8″		16	.500		149	17.85		166.85	194
1200	1′-11″ x 2′-8″	↓	14	.571	↓	154	20.50		174.50	203

08 51 66 – Metal Window Screens

08 51 66.10 Screens		Crew	Daily Output	Labor-Hours	Unit	Material	Labor	Equipment	Total	Total Incl O&P
0010	**SCREENS**									
0020	For metal sash, aluminum or bronze mesh, flat screen	2 Sswk	1200	.013	S.F.	4.54	.53		5.07	5.90
0500	Wicket screen, inside window	″	1000	.016	″	6.90	.64		7.54	8.70
0600	Residential, aluminum mesh and frame, 2′ x 3′	2 Carp	32	.500	Ea.	17.70	17.85		35.55	49
0610	Rescreen		50	.320		14.10	11.40		25.50	34.50
0620	3′ x 5′		32	.500		59.50	17.85		77.35	95
0630	Rescreen		45	.356		36.50	12.70		49.20	61.50
0640	4′ x 8′		25	.640		91	23		114	138
0650	Rescreen		40	.400		54	14.25		68.25	83
0660	Patio door		25	.640	↓	207	23		230	266
0680	Rescreening	↓	1600	.010	S.F.	2.67	.36		3.03	3.53
1000	Screens for solar louvers	2 Sswk	160	.100	″	25.50	3.98		29.48	35

08 52 Wood Windows

08 52 10 – Plain Wood Windows

08 52 10.10 Wood Windows		Crew	Daily Output	Labor-Hours	Unit	Material	Labor	Equipment	Total	Total Incl O&P
0010	**WOOD WINDOWS**, including frame, screens and grilles									
0020	Residential, stock units									
0050	Awning type, double insulated glass, 2′-10″ x 1′-9″ opening	2 Carp	12	1.333	Opng.	231	47.50		278.50	335
0100	2′-10″ x 6′-0″ opening	1 Carp	8	1		560	35.50		595.50	675
0200	4′-0″ x 3′-6″ single pane		10	.800	↓	375	28.50		403.50	460
0300	6′ x 5′ single pane	↓	8	1	Ea.	515	35.50		550.50	625
1000	Casement, 2′-0″ x 3′-4″ high	2 Carp	20	.800		258	28.50		286.50	330
1020	2′-0″ x 4′-0″	↓	18	.889	↓	273	31.50		304.50	355

08 52 Wood Windows

08 52 10 – Plain Wood Windows

08 52 10.10 Wood Windows		Crew	Daily Output	Labor-Hours	Unit	Material	2019 Bare Costs Labor	Equipment	Total	Total Incl O&P
1040	2'-0" x 5'-0"	2 Carp	17	.941	Ea.	320	33.50		353.50	405
1060	2'-0" x 6'-0"		16	1		315	35.50		350.50	410
1080	4'-0" x 3'-4"		15	1.067		595	38		633	720
1100	4'-0" x 4'-0"		15	1.067		670	38		708	800
1120	4'-0" x 5'-0"		14	1.143		755	40.50		795.50	900
1140	4'-0" x 6'-0"		12	1.333		850	47.50		897.50	1,025
1600	Casement units, 8' x 5', with screens, double insulated glass		2.50	6.400	Opng.	1,500	228		1,728	2,025
1700	Low E glass		2.50	6.400		1,650	228		1,878	2,175
2300	Casements, including screens, 2'-0" x 3'-4", double insulated glass		11	1.455		277	52		329	390
2400	Low E glass		11	1.455		277	52		329	390
2600	2 lite, 4'-0" x 4'-0", double insulated glass		9	1.778		505	63.50		568.50	660
2700	Low E glass		9	1.778		520	63.50		583.50	675
2900	3 lite, 5'-2" x 5'-0", double insulated glass		7	2.286		765	81.50		846.50	975
3000	Low E glass		7	2.286		810	81.50		891.50	1,025
3200	4 lite, 7'-0" x 5'-0", double insulated glass		6	2.667		1,100	95		1,195	1,350
3300	Low E glass		6	2.667		1,175	95		1,270	1,425
3500	5 lite, 8'-6" x 5'-0", double insulated glass		5	3.200		1,450	114		1,564	1,800
3600	Low E glass		5	3.200		1,450	114		1,564	1,800
3800	For removable wood grilles, diamond pattern, add				Leaf	39.50			39.50	43.50
3900	Rectangular pattern, add				"	39			39	43
4000	Bow, fixed lites, 8' x 5', double insulated glass	2 Carp	3	5.333	Opng.	1,450	190		1,640	1,925
4100	Low E glass	"	3	5.333	"	1,975	190		2,165	2,500
4150	6'-0" x 5'-0"	1 Carp	8	1	Ea.	1,275	35.50		1,310.50	1,475
4300	Fixed lites, 9'-9" x 5'-0", double insulated glass	2 Carp	2	8	Opng.	1,000	285		1,285	1,575
4400	Low E glass	"	2	8	"	1,100	285		1,385	1,675
5000	Bow, casement, 8'-1" x 4'-8" high	3 Carp	8	3	Ea.	1,600	107		1,707	1,950
5020	9'-6" x 4'-8"		8	3		1,750	107		1,857	2,100
5040	8'-1" x 5'-1"		8	3		1,950	107		2,057	2,325
5060	9'-6" x 5'-1"		6	4		1,950	143		2,093	2,375
5080	8'-1" x 6'-0"		6	4		1,950	143		2,093	2,375
5100	9'-6" x 6'-0"		6	4		2,050	143		2,193	2,475
5800	Skylights, hatches, vents, and sky roofs, see Section 08 62 13.00									

08 52 10.20 Awning Window		Crew	Daily Output	Labor-Hours	Unit	Material	Labor	Equipment	Total	Total Incl O&P
0010	**AWNING WINDOW**, Including frame, screens and grilles									
0100	34" x 22", insulated glass	1 Carp	10	.800	Ea.	278	28.50		306.50	350
0200	Low E glass		10	.800		315	28.50		343.50	390
0300	40" x 28", insulated glass		9	.889		315	31.50		346.50	400
0400	Low E glass		9	.889		345	31.50		376.50	435
0500	48" x 36", insulated glass		8	1		475	35.50		510.50	580
0600	Low E glass		8	1		500	35.50		535.50	610

08 52 10.30 Wood Windows		Crew	Daily Output	Labor-Hours	Unit	Material	Labor	Equipment	Total	Total Incl O&P
0010	**WOOD WINDOWS**, double-hung									
0020	Including frame, double insulated glass, screens and grilles									
0040	Double-hung, 2'-2" x 3'-4" high	2 Carp	15	1.067	Ea.	218	38		256	305
0060	2'-2" x 4'-4"		14	1.143		236	40.50		276.50	325
0080	2'-6" x 3'-4"		13	1.231		227	44		271	325
0100	2'-6" x 4'-0"		12	1.333		236	47.50		283.50	340
0120	2'-6" x 4'-8"		12	1.333		254	47.50		301.50	360
0140	2'-10" x 3'-4"		10	1.600		229	57		286	345
0160	2'-10" x 4'-0"		10	1.600		254	57		311	375
0180	3'-7" x 3'-4"		9	1.778		261	63.50		324.50	390
0200	3'-7" x 5'-4"		9	1.778		295	63.50		358.50	430

08 52 Wood Windows

08 52 10 – Plain Wood Windows

08 52 10.30 Wood Windows		Crew	Daily Output	Labor-Hours	Unit	Material	Labor	Equipment	Total	Total Incl O&P
						2019 Bare Costs				
0220	3'-10" x 5'-4"	2 Carp	8	2	Ea.	520	71.50		591.50	695
3800	Triple glazing for above, add				↓	25%				

08 52 10.40 Casement Window		Crew	Daily Output	Labor-Hours	Unit	Material	Labor	Equipment	Total	Total Incl O&P
0010	**CASEMENT WINDOW**, including frame, screen and grilles R085216-10									
0100	2'-0" x 3'-0" H, double insulated glass G	1 Carp	10	.800	Ea.	281	28.50		309.50	355
0150	Low E glass G		10	.800		276	28.50		304.50	350
0200	2'-0" x 4'-6" high, double insulated glass G		9	.889		395	31.50		426.50	490
0250	Low E glass G		9	.889		410	31.50		441.50	505
0260	Casement 4'-2" x 4'-2" double insulated glass G		11	.727		925	26		951	1,075
0270	4'-0" x 4'-0" Low E glass G		11	.727		555	26		581	655
0290	6'-4" x 5'-7" Low E glass G		9	.889		1,175	31.50		1,206.50	1,350
0300	2'-4" x 6'-0" high, double insulated glass G		8	1		450	35.50		485.50	555
0350	Low E glass G		8	1		445	35.50		480.50	545
0522	Vinyl-clad, premium, double insulated glass, 2'-0" x 3'-0" G		10	.800		287	28.50		315.50	360
0524	2'-0" x 4'-0" G		9	.889		335	31.50		366.50	420
0525	2'-0" x 5'-0" G		8	1		380	35.50		415.50	480
0528	2'-0" x 6'-0" G		8	1		410	35.50		445.50	510
0600	3'-0" x 5'-0" G		8	1		700	35.50		735.50	835
0700	4'-0" x 3'-0" G		8	1		770	35.50		805.50	905
0710	4'-0" x 4'-0" G		8	1		660	35.50		695.50	785
0720	4'-8" x 4'-0" G		8	1		725	35.50		760.50	860
0730	4'-8" x 5'-0" G		6	1.333		830	47.50		877.50	995
0740	4'-8" x 6'-0" G		6	1.333		935	47.50		982.50	1,100
0750	6'-0" x 4'-0" G		6	1.333		850	47.50		897.50	1,025
0800	6'-0" x 5'-0" G	↓	6	1.333		935	47.50		982.50	1,100
0900	5'-6" x 5'-6" G	2 Carp	15	1.067	↓	1,500	38		1,538	1,725
2000	Bay, casement units, 8' x 5', w/screens, double insulated glass		2.50	6.400	Opng.	1,650	228		1,878	2,175
2100	Low E glass	↓	2.50	6.400	"	1,725	228		1,953	2,275
3020	Vinyl-clad, premium, double insulated glass, multiple leaf units									
3080	Single unit, 1'-6" x 5'-0" G	2 Carp	20	.800	Ea.	330	28.50		358.50	410
3100	2'-0" x 2'-0" G		20	.800		226	28.50		254.50	296
3140	2'-0" x 2'-6" G		20	.800		287	28.50		315.50	360
3220	2'-0" x 3'-6" G		20	.800		292	28.50		320.50	365
3260	2'-0" x 4'-0" G		19	.842		335	30		365	415
3300	2'-0" x 4'-6" G		19	.842		340	30		370	425
3340	2'-0" x 5'-0" G		18	.889		380	31.50		411.50	475
3460	2'-4" x 3'-0" G		20	.800		292	28.50		320.50	365
3500	2'-4" x 4'-0" G		19	.842		365	30		395	450
3540	2'-4" x 5'-0" G		18	.889		430	31.50		461.50	525
3700	Double unit, 2'-8" x 5'-0" G		18	.889		605	31.50		636.50	720
3740	2'-8" x 6'-0" G		17	.941		700	33.50		733.50	825
3840	3'-0" x 4'-6" G		18	.889		545	31.50		576.50	655
3860	3'-0" x 5'-0" G		17	.941		700	33.50		733.50	830
3880	3'-0" x 6'-0" G		17	.941		760	33.50		793.50	890
3980	3'-4" x 2'-6" G		19	.842		465	30		495	560
4000	3'-4" x 3'-0" G		12	1.333		470	47.50		517.50	595
4030	3'-4" x 4'-0" G		18	.889		565	31.50		596.50	675
4050	3'-4" x 5'-0" G		12	1.333		705	47.50		752.50	855
4100	3'-4" x 6'-0" G		11	1.455		760	52		812	920
4200	3'-6" x 3'-0" G		18	.889		545	31.50		576.50	655
4340	4'-0" x 3'-0" G		18	.889		525	31.50		556.50	635
4380	4'-0" x 3'-6" G	↓	17	.941	↓	565	33.50		598.50	675

08 52 Wood Windows

08 52 10 – Plain Wood Windows

08 52 10.40 Casement Window			Crew	Daily Output	Labor-Hours	Unit	2019 Bare Costs Material	Labor	Equipment	Total	Total Incl O&P
4420	4'-0" x 4'-0"	G	2 Carp	16	1	Ea.	660	35.50		695.50	785
4460	4'-0" x 4'-4"	G		16	1		660	35.50		695.50	790
4540	4'-0" x 5'-0"	G		16	1		720	35.50		755.50	855
4580	4'-0" x 6'-0"	G		15	1.067		820	38		858	965
4740	4'-8" x 3'-0"	G		18	.889		585	31.50		616.50	700
4780	4'-8" x 3'-6"	G		17	.941		625	33.50		658.50	740
4820	4'-8" x 4'-0"	G		16	1		725	35.50		760.50	860
4860	4'-8" x 5'-0"	G		15	1.067		830	38		868	980
4900	4'-8" x 6'-0"	G		15	1.067		935	38		973	1,100
5060	5'-0" x 5'-0"	G		15	1.067		1,100	38		1,138	1,275
5100	Triple unit, 5'-6" x 3'-0"	G		17	.941		740	33.50		773.50	870
5140	5'-6" x 3'-6"	G		16	1		830	35.50		865.50	975
5180	5'-6" x 4'-6"	G		15	1.067		895	38		933	1,050
5220	5'-6" x 5'-6"	G		15	1.067		1,175	38		1,213	1,350
5300	6'-0" x 4'-6"	G		15	1.067		905	38		943	1,050
5850	5'-0" x 3'-0"	G		12	1.333		625	47.50		672.50	765
5900	5'-0" x 4'-0"	G		11	1.455		965	52		1,017	1,150
6100	5'-0" x 5'-6"	G		10	1.600		1,100	57		1,157	1,325
6150	5'-0" x 6'-0"	G		10	1.600		1,250	57		1,307	1,475
6200	6'-0" x 3'-0"	G		12	1.333		1,200	47.50		1,247.50	1,400
6250	6'-0" x 3'-4"	G		12	1.333		680	47.50		727.50	825
6300	6'-0" x 4'-0"	G		11	1.455		865	52		917	1,025
6350	6'-0" x 5'-0"	G		10	1.600		955	57		1,012	1,150
6400	6'-0" x 6'-0"	G		10	1.600		1,225	57		1,282	1,450
6500	Quadruple unit, 7'-0" x 4'-0"	G		9	1.778		1,150	63.50		1,213.50	1,350
6700	8'-0" x 4'-6"	G		9	1.778		1,425	63.50		1,488.50	1,675
6950	6'-8" x 4'-0"	G		10	1.600		1,125	57		1,182	1,350
7000	6'-8" x 6'-0"	G		10	1.600		1,525	57		1,582	1,775
8190	For installation, add per leaf							15%			
8200	For multiple leaf units, deduct for stationary sash										
8220	2' high					Ea.	24.50			24.50	26.50
8240	4'-6" high						27.50			27.50	30
8260	6' high						36.50			36.50	40

08 52 10.50 Double-Hung			Crew	Daily Output	Labor-Hours	Unit	Material	Labor	Equipment	Total	Total Incl O&P
0010	**DOUBLE-HUNG**, Including frame, screens and grilles										
0100	2'-0" x 3'-0" high, low E insul. glass	G	1 Carp	10	.800	Ea.	193	28.50		221.50	259
0200	3'-0" x 4'-0" high, double insulated glass	G		9	.889		292	31.50		323.50	375
0300	4'-0" x 4'-6" high, low E insulated glass	G		8	1		340	35.50		375.50	430

08 52 10.55 Picture Window		Crew	Daily Output	Labor-Hours	Unit	Material	Labor	Equipment	Total	Total Incl O&P
0010	**PICTURE WINDOW**, Including frame and grilles									
0100	3'-6" x 4'-0" high, double insulated glass	2 Carp	12	1.333	Ea.	435	47.50		482.50	560
0150	Low E glass		12	1.333		445	47.50		492.50	570
0200	4'-0" x 4'-6" high, double insulated glass		11	1.455		550	52		602	690
0250	Low E glass		11	1.455		535	52		587	670
0300	5'-0" x 4'-0" high, double insulated glass		11	1.455		585	52		637	725
0350	Low E glass		11	1.455		610	52		662	755
0400	6'-0" x 4'-6" high, double insulated glass		10	1.600		640	57		697	800
0450	Low E glass		10	1.600		645	57		702	805

08 52 10.65 Wood Sash		Crew	Daily Output	Labor-Hours	Unit	Material	Labor	Equipment	Total	Total Incl O&P
0010	**WOOD SASH**, Including glazing but not trim									
0050	Custom, 5'-0" x 4'-0", 1" double glazed, 3/16" thick lites	2 Carp	3.20	5	Ea.	234	178		412	555
0100	1/4" thick lites		5	3.200		266	114		380	480

08 52 Wood Windows

08 52 10 – Plain Wood Windows

08 52 10.65 Wood Sash

08 52 10.65 Wood Sash			Crew	Daily Output	Labor-Hours	Unit	Material	Labor	Equipment	Total	Total Incl O&P
							2019 Bare Costs				
0200	1″ thick, triple glazed		2 Carp	5	3.200	Ea.	435	114		549	670
0300	7′-0″ x 4′-6″ high, 1″ double glazed, 3/16″ thick lites			4.30	3.721		435	133		568	695
0400	1/4″ thick lites			4.30	3.721		495	133		628	760
0500	1″ thick, triple glazed			4.30	3.721		565	133		698	840
0600	8′-6″ x 5′-0″ high, 1″ double glazed, 3/16″ thick lites			3.50	4.571		590	163		753	915
0700	1/4″ thick lites			3.50	4.571		645	163		808	980
0800	1″ thick, triple glazed			3.50	4.571		685	163		848	1,025
0900	Window frames only, based on perimeter length					L.F.	4.17			4.17	4.59

08 52 10.70 Sliding Windows

08 52 10.70 Sliding Windows			Crew	Daily Output	Labor-Hours	Unit	Material	Labor	Equipment	Total	Total Incl O&P
0010	**SLIDING WINDOWS**										
0100	3′-0″ x 3′-0″ high, double insulated	G	1 Carp	10	.800	Ea.	297	28.50		325.50	370
0120	Low E glass	G		10	.800		325	28.50		353.50	400
0200	4′-0″ x 3′-6″ high, double insulated	G		9	.889		395	31.50		426.50	490
0220	Low E glass	G		9	.889		390	31.50		421.50	485
0300	6′-0″ x 5′-0″ high, double insulated	G		8	1		515	35.50		550.50	625
0320	Low E glass	G		8	1		580	35.50		615.50	695
6000	Sliding, insulating glass, including screens,										
6100	3′-0″ x 3′-0″		2 Carp	6.50	2.462	Ea.	330	88		418	510
6200	4′-0″ x 3′-6″			6.30	2.540		335	90.50		425.50	520
6300	5′-0″ x 4′-0″			6	2.667		420	95		515	615

08 52 13 – Metal-Clad Wood Windows

08 52 13.10 Awning Windows, Metal-Clad

08 52 13.10 Awning Windows, Metal-Clad			Crew	Daily Output	Labor-Hours	Unit	Material	Labor	Equipment	Total	Total Incl O&P
0010	**AWNING WINDOWS, METAL-CLAD**										
2000	Metal-clad, awning deluxe, double insulated glass, 34″ x 22″		1 Carp	9	.889	Ea.	253	31.50		284.50	330
2050	36″ x 25″			9	.889		278	31.50		309.50	360
2100	40″ x 22″			9	.889		295	31.50		326.50	380
2150	40″ x 30″			9	.889		345	31.50		376.50	435
2200	48″ x 28″			8	1		355	35.50		390.50	450
2250	60″ x 36″			8	1		375	35.50		410.50	475

08 52 13.20 Casement Windows, Metal-Clad

08 52 13.20 Casement Windows, Metal-Clad			Crew	Daily Output	Labor-Hours	Unit	Material	Labor	Equipment	Total	Total Incl O&P
0010	**CASEMENT WINDOWS, METAL-CLAD**										
0100	Metal-clad, deluxe, dbl. insul. glass, 2′-0″ x 3′-0″ high	G	1 Carp	10	.800	Ea.	296	28.50		324.50	370
0120	2′-0″ x 4′-0″ high	G		9	.889		320	31.50		351.50	410
0130	2′-0″ x 5′-0″ high	G		8	1		345	35.50		380.50	435
0140	2′-0″ x 6′-0″ high	G		8	1		385	35.50		420.50	485
0150	Casement window, metal-clad, double insul. glass, 3′-6″ x 3′-6″	G		8.90	.899		510	32		542	615
0300	Metal-clad, casement, bldrs mdl, 6′-0″ x 4′-0″, dbl. insul. glass, 3 panels		2 Carp	10	1.600		1,225	57		1,282	1,450
0310	9′-0″ x 4′-0″, 4 panels			8	2		1,550	71.50		1,621.50	1,825
0320	10′-0″ x 5′-0″, 5 panels			7	2.286		2,125	81.50		2,206.50	2,450
0330	12′-0″ x 6′-0″, 6 panels			6	2.667		2,700	95		2,795	3,125

08 52 13.30 Double-Hung Windows, Metal-Clad

08 52 13.30 Double-Hung Windows, Metal-Clad			Crew	Daily Output	Labor-Hours	Unit	Material	Labor	Equipment	Total	Total Incl O&P
0010	**DOUBLE-HUNG WINDOWS, METAL-CLAD**										
0100	Metal-clad, deluxe, dbl. insul. glass, 2′-6″ x 3′-0″ high	G	1 Carp	10	.800	Ea.	286	28.50		314.50	360
0120	3′-0″ x 3′-6″ high	G		10	.800		325	28.50		353.50	405
0140	3′-0″ x 4′-0″ high	G		9	.889		345	31.50		376.50	430
0160	3′-0″ x 4′-6″ high	G		9	.889		360	31.50		391.50	450
0180	3′-0″ x 5′-0″ high	G		8	1		385	35.50		420.50	485
0200	3′-6″ x 6′-0″ high	G		8	1		470	35.50		505.50	575

08 52 Wood Windows

08 52 13 – Metal-Clad Wood Windows

08 52 13.35 Picture and Sliding Windows Metal-Clad

	08 52 13.35 Picture and Sliding Windows Metal-Clad		Crew	Daily Output	Labor-Hours	Unit	Material	2019 Bare Costs Labor	Equipment	Total	Total Incl O&P
0010	**PICTURE AND SLIDING WINDOWS METAL-CLAD**										
2000	Metal-clad, dlx picture, dbl. insul. glass, 4'-0" x 4'-0" high		2 Carp	12	1.333	Ea.	380	47.50		427.50	500
2100	4'-0" x 6'-0" high			11	1.455		565	52		617	705
2200	5'-0" x 6'-0" high			10	1.600		625	57		682	780
2300	6'-0" x 6'-0" high		↓	10	1.600		715	57		772	880
2400	Metal-clad, dlx sliding, dbl. insul. glass, 3'-0" x 3'-0" high	G	1 Carp	10	.800		330	28.50		358.50	410
2420	4'-0" x 3'-6" high	G		9	.889		400	31.50		431.50	495
2440	5'-0" x 4'-0" high	G		9	.889		480	31.50		511.50	585
2460	6'-0" x 5'-0" high	G	↓	8	1	↓	755	35.50		790.50	890

08 52 13.40 Bow and Bay Windows, Metal-Clad

	08 52 13.40 Bow and Bay Windows, Metal-Clad	Crew	Daily Output	Labor-Hours	Unit	Material	Labor	Equipment	Total	Total Incl O&P
0010	**BOW AND BAY WINDOWS, METAL-CLAD**									
0100	Metal-clad, deluxe, dbl. insul. glass, 8'-0" x 5'-0" high, 4 panels	2 Carp	10	1.600	Ea.	1,700	57		1,757	1,975
0120	10'-0" x 5'-0" high, 5 panels		8	2		1,825	71.50		1,896.50	2,150
0140	10'-0" x 6'-0" high, 5 panels		7	2.286		2,175	81.50		2,256.50	2,500
0160	12'-0" x 6'-0" high, 6 panels		6	2.667		2,950	95		3,045	3,400
0400	Double-hung, bldrs. model, bay, 8' x 4' high, dbl. insul. glass		10	1.600		1,350	57		1,407	1,600
0440	Low E glass		10	1.600		1,450	57		1,507	1,700
0460	9'-0" x 5'-0" high, dbl. insul. glass		6	2.667		1,450	95		1,545	1,750
0480	Low E glass		6	2.667		1,550	95		1,645	1,850
0500	Metal-clad, deluxe, dbl. insul. glass, 7'-0" x 4'-0" high		10	1.600		1,300	57		1,357	1,525
0520	8'-0" x 4'-0" high		8	2		1,350	71.50		1,421.50	1,600
0540	8'-0" x 5'-0" high		7	2.286		1,400	81.50		1,481.50	1,650
0560	9'-0" x 5'-0" high	↓	6	2.667	↓	1,475	95		1,570	1,775

08 52 16 – Plastic-Clad Wood Windows

08 52 16.10 Bow Window

	08 52 16.10 Bow Window	Crew	Daily Output	Labor-Hours	Unit	Material	Labor	Equipment	Total	Total Incl O&P
0010	**BOW WINDOW** including frames, screens, and grilles									
0020	End panels operable									
1000	Bow type, casement, wood, bldrs. mdl., 8' x 5' dbl. insul. glass, 4 panel	2 Carp	10	1.600	Ea.	1,575	57		1,632	1,850
1050	Low E glass		10	1.600		1,325	57		1,382	1,575
1100	10'-0" x 5'-0", dbl. insul. glass, 6 panels		6	2.667		1,375	95		1,470	1,650
1200	Low E glass, 6 panels		6	2.667		1,475	95		1,570	1,775
1300	Vinyl-clad, bldrs. model, dbl. insul. glass, 6'-0" x 4'-0", 3 panel		10	1.600		1,050	57		1,107	1,250
1340	9'-0" x 4'-0", 4 panel		8	2		1,425	71.50		1,496.50	1,700
1380	10'-0" x 6'-0", 5 panels		7	2.286		2,350	81.50		2,431.50	2,725
1420	12'-0" x 6'-0", 6 panels		6	2.667		3,100	95		3,195	3,550
2000	Bay window, 8' x 5', dbl. insul. glass		10	1.600		1,925	57		1,982	2,225
2050	Low E glass		10	1.600		2,325	57		2,382	2,675
2100	12'-0" x 6'-0", dbl. insul. glass, 6 panels		6	2.667		2,400	95		2,495	2,800
2200	Low E glass		6	2.667		3,275	95		3,370	3,750
2280	6'-0" x 4'-0"		11	1.455		1,275	52		1,327	1,475
2300	Vinyl-clad, premium, dbl. insul. glass, 8'-0" x 5'-0"		10	1.600		1,800	57		1,857	2,100
2340	10'-0" x 5'-0"		8	2		2,425	71.50		2,496.50	2,800
2380	10'-0" x 6'-0"		7	2.286		2,825	81.50		2,906.50	3,250
2420	12'-0" x 6'-0"		6	2.667		3,375	95		3,470	3,875
2430	14'-0" x 3'-0"		7	2.286		3,050	81.50		3,131.50	3,500
2440	14'-0" x 6'-0"		5	3.200		5,125	114		5,239	5,850
3300	Vinyl-clad, premium, dbl. insul. glass, 7'-0" x 4'-6"		10	1.600		1,400	57		1,457	1,650
3340	8'-0" x 4'-6"		8	2		1,425	71.50		1,496.50	1,700
3380	8'-0" x 5'-0"		7	2.286		1,500	81.50		1,581.50	1,775
3420	9'-0" x 5'-0"	↓	6	2.667	↓	1,550	95		1,645	1,850

08 52 Wood Windows

08 52 16 – Plastic-Clad Wood Windows

08 52 16.15 Awning Window Vinyl-Clad		Crew	Daily Output	Labor-Hours	Unit	Material	Labor	Equipment	Total	Total Incl O&P
						2019 Bare Costs				
0010	**AWNING WINDOW VINYL-CLAD** including frames, screens, and grilles									
0200	Vinyl-clad, premium, double insulated glass, 24" x 17"	1 Carp	12	.667	Ea.	214	24		238	276
0210	24" x 28"		11	.727		261	26		287	330
0250	36" x 17"		9	.889		269	31.50		300.50	350
0280	36" x 28"		9	.889		310	31.50		341.50	395
0300	36" x 36"		9	.889		345	31.50		376.50	435
0320	36" x 40"		9	.889		400	31.50		431.50	495
0340	40" x 22"		10	.800		296	28.50		324.50	370
0360	48" x 28"		8	1		375	35.50		410.50	470
0380	60" x 36"		8	1		515	35.50		550.50	630
08 52 16.20 Half Round, Vinyl-Clad										
0010	**HALF ROUND, VINYL-CLAD**, double insulated glass, incl. grille									
0800	14" height x 24" base	2 Carp	9	1.778	Ea.	425	63.50		488.50	570
1040	15" height x 25" base		8	2		410	71.50		481.50	570
1060	16" height x 28" base		7	2.286		430	81.50		511.50	610
1080	17" height x 29" base		7	2.286		445	81.50		526.50	625
2000	19" height x 33" base	1 Carp	6	1.333		490	47.50		537.50	620
2100	20" height x 35" base		6	1.333		485	47.50		532.50	615
2200	21" height x 37" base		6	1.333		500	47.50		547.50	630
2250	23" height x 41" base	2 Carp	6	2.667		535	95		630	745
2300	26" height x 48" base		6	2.667		560	95		655	770
2350	30" height x 56" base		6	2.667		650	95		745	870
3000	36" height x 67" base	1 Carp	4	2		1,150	71.50		1,221.50	1,375
3040	38" height x 71" base	2 Carp	5	3.200		1,000	114		1,114	1,300
3050	40" height x 75" base	"	5	3.200		1,325	114		1,439	1,650
5000	Elliptical, 71" x 16"	1 Carp	11	.727		1,000	26		1,026	1,150
5100	95" x 21"	"	10	.800		1,425	28.50		1,453.50	1,625
08 52 16.30 Palladian Windows										
0010	**PALLADIAN WINDOWS**									
0020	Vinyl-clad, double insulated glass, including frame and grilles									
0040	3'-2" x 2'-6" high	2 Carp	11	1.455	Ea.	1,275	52		1,327	1,475
0060	3'-2" x 4'-10"		11	1.455		1,775	52		1,827	2,025
0080	3'-2" x 6'-4"		10	1.600		1,775	57		1,832	2,050
0100	4'-0" x 4'-0"		10	1.600		1,575	57		1,632	1,850
0120	4'-0" x 5'-4"	3 Carp	10	2.400		1,900	85.50		1,985.50	2,250
0140	4'-0" x 6'-0"		9	2.667		1,900	95		1,995	2,250
0160	4'-0" x 7'-4"		9	2.667		2,150	95		2,245	2,500
0180	5'-5" x 4'-10"		9	2.667		2,300	95		2,395	2,700
0200	5'-5" x 6'-10"		9	2.667		2,650	95		2,745	3,075
0220	5'-5" x 7'-9"		9	2.667		2,825	95		2,920	3,250
0240	6'-0" x 7'-11"		8	3		3,600	107		3,707	4,150
0260	8'-0" x 6'-0"		8	3		3,200	107		3,307	3,700
08 52 16.35 Double-Hung Window										
0010	**DOUBLE-HUNG WINDOW** including frames, screens, and grilles									
0300	Vinyl-clad, premium, double insulated glass, 2'-6" x 3'-0" G	1 Carp	10	.800	Ea.	345	28.50		373.50	420
0305	2'-6" x 4'-0" G		10	.800		380	28.50		408.50	460
0400	3'-0" x 3'-6" G		10	.800		345	28.50		373.50	425
0500	3'-0" x 4'-0" G		9	.889		405	31.50		436.50	500
0600	3'-0" x 4'-6" G		9	.889		430	31.50		461.50	525
0700	3'-0" x 5'-0" G		8	1		460	35.50		495.50	565
0790	3'-4" x 5'-0" G		8	1		450	35.50		485.50	555

08 52 Wood Windows

08 52 16 – Plastic-Clad Wood Windows

08 52 16.35 Double-Hung Window			Crew	Daily Output	Labor-Hours	Unit	2019 Bare Costs Material	Labor	Equipment	Total	Total Incl O&P
0800	3'-6" x 6'-0"	G	1 Carp	8	1	Ea.	520	35.50		555.50	630
0820	4'-0" x 5'-0"	G		7	1.143		575	40.50		615.50	705
0830	4'-0" x 6'-0"	G		7	1.143		730	40.50		770.50	870

08 52 16.40 Transom Windows		Crew	Daily Output	Labor-Hours	Unit	Material	Labor	Equipment	Total	Total Incl O&P
0010	**TRANSOM WINDOWS**									
0050	Vinyl-clad, premium, dbl. insul. glass, 32" x 8"	1 Carp	16	.500	Ea.	193	17.85		210.85	242
0100	36" x 8"		16	.500		209	17.85		226.85	259
0110	36" x 12"		16	.500		224	17.85		241.85	276
1000	Vinyl-clad, premium, dbl. insul. glass, 4'-0" x 4'-0"	2 Carp	12	1.333		530	47.50		577.50	665
1100	4'-0" x 6'-0"		11	1.455		980	52		1,032	1,150
1200	5'-0" x 6'-0"		10	1.600		1,075	57		1,132	1,300
1300	6'-0" x 6'-0"		10	1.600		1,125	57		1,182	1,325

08 52 16.45 Trapezoid Windows		Crew	Daily Output	Labor-Hours	Unit	Material	Labor	Equipment	Total	Total Incl O&P
0010	**TRAPEZOID WINDOWS**									
0100	Vinyl-clad, including frame and exterior trim									
0900	20" base x 44" leg x 53" leg	2 Carp	13	1.231	Ea.	400	44		444	520
1000	24" base x 90" leg x 102" leg		8	2		705	71.50		776.50	895
3000	36" base x 40" leg x 22" leg		12	1.333		445	47.50		492.50	565
3010	36" base x 44" leg x 25" leg		13	1.231		465	44		509	590
3050	36" base x 26" leg x 48" leg		9	1.778		470	63.50		533.50	625
3100	36" base x 42" legs, 50" peak		9	1.778		560	63.50		623.50	720
3200	36" base x 60" leg x 81" leg		11	1.455		735	52		787	890
4320	44" base x 23" leg x 56" leg		11	1.455		570	52		622	715
4350	44" base x 59" leg x 92" leg		10	1.600		885	57		942	1,075
4500	46" base x 15" leg x 46" leg		8	2		445	71.50		516.50	605
4550	46" base x 16" leg x 48" leg		8	2		465	71.50		536.50	635
4600	46" base x 50" leg x 80" leg		7	2.286		705	81.50		786.50	910
6600	66" base x 12" leg x 42" leg		8	2		605	71.50		676.50	785
6650	66" base x 12" legs, 28" peak		9	1.778		485	63.50		548.50	635
6700	68" base x 23" legs, 31" peak		8	2		605	71.50		676.50	785

08 52 16.70 Vinyl-Clad, Premium, DBL. Insul. Glass			Crew	Daily Output	Labor-Hours	Unit	Material	Labor	Equipment	Total	Total Incl O&P
0010	**VINYL-CLAD, PREMIUM, DBL. INSUL. GLASS**										
1000	Sliding, 3'-0" x 3'-0"	G	1 Carp	10	.800	Ea.	625	28.50		653.50	735
1020	4'-0" x 1'-11"	G		11	.727		605	26		631	710
1040	4'-0" x 3'-0"	G		10	.800		755	28.50		783.50	875
1050	4'-0" x 3'-6"	G		9	.889		695	31.50		726.50	820
1090	4'-0" x 5'-0"	G		9	.889		945	31.50		976.50	1,100
1100	5'-0" x 4'-0"	G		9	.889		925	31.50		956.50	1,075
1120	5'-0" x 5'-0"	G		8	1		1,075	35.50		1,110.50	1,250
1140	6'-0" x 4'-0"	G		8	1		1,125	35.50		1,160.50	1,275
1150	6'-0" x 5'-0"	G		8	1		1,175	35.50		1,210.50	1,325

08 52 50 – Window Accessories

08 52 50.10 Window Grille or Muntin		Crew	Daily Output	Labor-Hours	Unit	Material	Labor	Equipment	Total	Total Incl O&P
0010	**WINDOW GRILLE OR MUNTIN**, snap in type									
0020	Standard pattern interior grilles									
2000	Wood, awning window, glass size, 28" x 16" high	1 Carp	30	.267	Ea.	31	9.50		40.50	49.50
2060	44" x 24" high		32	.250		45	8.90		53.90	64.50
2100	Casement, glass size, 20" x 36" high		30	.267		34	9.50		43.50	53
2180	20" x 56" high		32	.250		46.50	8.90		55.40	66
2200	Double-hung, glass size, 16" x 24" high		24	.333	Set	57.50	11.90		69.40	82.50
2280	32" x 32" high		34	.235	"	141	8.40		149.40	169

08 52 Wood Windows

08 52 50 – Window Accessories

08 52 50.10 Window Grille or Muntin		Crew	Daily Output	Labor-Hours	Unit	Material	2019 Bare Costs Labor	Equipment	Total	Total Incl O&P
2500	Picture, glass size, 48″ x 48″ high	1 Carp	30	.267	Ea.	129	9.50		138.50	158
2580	60″ x 68″ high		28	.286	″	201	10.20		211.20	238
2600	Sliding, glass size, 14″ x 36″ high		24	.333	Set	38	11.90		49.90	61.50
2680	36″ x 36″ high		22	.364	″	45.50	12.95		58.45	71.50

08 52 66 – Wood Window Screens

08 52 66.10 Wood Screens		Crew	Daily Output	Labor-Hours	Unit	Material	Labor	Equipment	Total	Total Incl O&P
0010	**WOOD SCREENS**									
0020	Over 3 S.F., 3/4″ frames	2 Carp	375	.043	S.F.	4.77	1.52		6.29	7.75
0100	1-1/8″ frames	″	375	.043	″	8.55	1.52		10.07	11.90

08 52 69 – Wood Storm Windows

08 52 69.10 Storm Windows			Crew	Daily Output	Labor-Hours	Unit	Material	Labor	Equipment	Total	Total Incl O&P
0010	**STORM WINDOWS**, aluminum residential										
0300	Basement, mill finish, incl. fiberglass screen										
0320	1′-10″ x 1′-0″ high	G	2 Carp	30	.533	Ea.	36	19		55	71.50
0340	2′-9″ x 1′-6″ high	G		30	.533		39	19		58	74.50
0360	3′-4″ x 2′-0″ high	G		30	.533		46.50	19		65.50	82.50
1600	Double-hung, combination, storm & screen										
1700	Custom, clear anodic coating, 2′-0″ x 3′-5″ high		2 Carp	30	.533	Ea.	99.50	19		118.50	142
1720	2′-6″ x 5′-0″ high			28	.571		125	20.50		145.50	172
1740	4′-0″ x 6′-0″ high			25	.640		237	23		260	299
1800	White painted, 2′-0″ x 3′-5″ high			30	.533		116	19		135	160
1820	2′-6″ x 5′-0″ high			28	.571		176	20.50		196.50	228
1840	4′-0″ x 6′-0″ high			25	.640		288	23		311	355
2000	Clear anodic coating, 2′-0″ x 3′-5″ high	G		30	.533		97	19		116	139
2020	2′-6″ x 5′-0″ high	G		28	.571		125	20.50		145.50	172
2040	4′-0″ x 6′-0″ high	G		25	.640		136	23		159	187
2400	White painted, 2′-0″ x 3′-5″ high	G		30	.533		96	19		115	138
2420	2′-6″ x 5′-0″ high	G		28	.571		100	20.50		120.50	144
2440	4′-0″ x 6′-0″ high	G		25	.640		116	23		139	166
2600	Mill finish, 2′-0″ x 3′-5″ high	G		30	.533		91	19		110	132
2620	2′-6″ x 5′-0″ high	G		28	.571		96	20.50		116.50	140
2640	4′-0″ x 6′-8″ high	G		25	.640		116	23		139	166
4000	Picture window, storm, 1 lite, white or bronze finish										
4020	4′-6″ x 4′-6″ high		2 Carp	25	.640	Ea.	142	23		165	195
4040	5′-8″ x 4′-6″ high			20	.800		157	28.50		185.50	220
4400	Mill finish, 4′-6″ x 4′-6″ high			25	.640		142	23		165	195
4420	5′-8″ x 4′-6″ high			20	.800		162	28.50		190.50	225
4600	3 lite, white or bronze finish										
4620	4′-6″ x 4′-6″ high		2 Carp	25	.640	Ea.	166	23		189	221
4640	5′-8″ x 4′-6″ high			20	.800		176	28.50		204.50	240
4800	Mill finish, 4′-6″ x 4′-6″ high			25	.640		162	23		185	216
4820	5′-8″ x 4′-6″ high			20	.800		166	28.50		194.50	230
6000	Sliding window, storm, 2 lite, white or bronze finish										
6020	3′-4″ x 2′-7″ high		2 Carp	28	.571	Ea.	145	20.50		165.50	194
6040	4′-4″ x 3′-3″ high			25	.640		161	23		184	215
6060	5′-4″ x 6′-0″ high			20	.800		217	28.50		245.50	286
9000	Interior storm window										
9100	Storm window interior glass		1 Glaz	107	.075	S.F.	6.30	2.60		8.90	11.25

08 53 Plastic Windows

08 53 13 – Vinyl Windows

08 53 13.10 Solid Vinyl Windows			Crew	Daily Output	Labor-Hours	Unit	Material	2019 Bare Costs Labor	Equipment	Total	Total Incl O&P
0010	**SOLID VINYL WINDOWS**										
0020	Double-hung, including frame and screen, 2′-0″ x 2′-6″		2 Carp	15	1.067	Ea.	206	38		244	290
0040	2′-0″ x 3′-6″			14	1.143		216	40.50		256.50	305
0060	2′-6″ x 4′-6″			13	1.231		257	44		301	355
0080	3′-0″ x 4′-0″			10	1.600		229	57		286	345
0100	3′-0″ x 4′-6″			9	1.778		294	63.50		357.50	430
0120	3′-6″ x 4′-6″			8	2		310	71.50		381.50	460
0140	3′-6″ x 6′-0″		↓	7	2.286	↓	385	81.50		466.50	560

08 53 13.20 Vinyl Single-Hung Windows			Crew	Daily Output	Labor-Hours	Unit	Material	Labor	Equipment	Total	Total Incl O&P
0010	**VINYL SINGLE-HUNG WINDOWS**, insulated glass										
0020	Grids, low E, J fin, extension jambs										
0130	25″ x 41″	G	2 Carp	20	.800	Ea.	204	28.50		232.50	272
0140	25″ x 49″	G		18	.889		214	31.50		245.50	289
0150	25″ x 57″	G		17	.941		228	33.50		261.50	305
0160	25″ x 65″	G		16	1		247	35.50		282.50	330
0170	29″ x 41″	G		18	.889		206	31.50		237.50	279
0180	29″ x 53″	G		18	.889		225	31.50		256.50	300
0190	29″ x 57″	G		17	.941		238	33.50		271.50	320
0200	29″ x 65″	G		16	1		248	35.50		283.50	330
0210	33″ x 41″	G		20	.800		220	28.50		248.50	289
0220	33″ x 53″	G		18	.889		239	31.50		270.50	315
0230	33″ x 57″	G		17	.941		248	33.50		281.50	330
0240	33″ x 65″	G		16	1		258	35.50		293.50	345
0250	37″ x 41″	G		20	.800		251	28.50		279.50	325
0260	37″ x 53″	G		18	.889		275	31.50		306.50	355
0270	37″ x 57″	G		17	.941		285	33.50		318.50	370
0280	37″ x 65″	G	↓	16	1		300	35.50		335.50	390
0500	Vinyl-clad, premium, double insulated glass, circle, 24″ diameter		1 Carp	6	1.333		485	47.50		532.50	615
1000	2′-4″ diameter			6	1.333		555	47.50		602.50	690
1500	2′-11″ diameter		↓	6	1.333	↓	670	47.50		717.50	820

08 53 13.30 Vinyl Double-Hung Windows			Crew	Daily Output	Labor-Hours	Unit	Material	Labor	Equipment	Total	Total Incl O&P
0010	**VINYL DOUBLE-HUNG WINDOWS**, insulated glass										
0100	Grids, low E, J fin, ext. jambs, 21″ x 53″	G	2 Carp	18	.889	Ea.	216	31.50		247.50	290
0102	21″ x 37″	G		18	.889		229	31.50		260.50	305
0104	21″ x 41″	G		18	.889		239	31.50		270.50	315
0106	21″ x 49″	G		18	.889		249	31.50		280.50	325
0110	21″ x 57″	G		17	.941		268	33.50		301.50	350
0120	21″ x 65″	G		16	1		283	35.50		318.50	370
0128	25″ x 37″	G		20	.800		238	28.50		266.50	310
0130	25″ x 41″	G		20	.800		248	28.50		276.50	320
0140	25″ x 49″	G		18	.889		255	31.50		286.50	335
0145	25″ x 53″	G		18	.889		274	31.50		305.50	355
0150	25″ x 57″	G		17	.941		278	33.50		311.50	360
0160	25″ x 65″	G		16	1		293	35.50		328.50	380
0162	25″ x 69″	G		16	1		297	35.50		332.50	385
0164	25″ x 77″	G		16	1		320	35.50		355.50	410
0168	29″ x 37″	G		18	.889		249	31.50		280.50	325
0170	29″ x 41″	G		18	.889		263	31.50		294.50	340
0172	29″ x 49″	G		18	.889		268	31.50		299.50	350
0180	29″ x 53″	G		18	.889		278	31.50		309.50	360
0190	29″ x 57″	G		17	.941		293	33.50		326.50	375
0200	29″ x 65″	G	↓	16	1	↓	305	35.50		340.50	395

08 53 13 – Vinyl Windows

08 53 13.30 Vinyl Double-Hung Windows			Crew	Daily Output	Labor-Hours	Unit	Material	2019 Bare Costs Labor	Equipment	Total	Total Incl O&P
0202	29" x 69"	G	2 Carp	16	1	Ea.	345	35.50		380.50	440
0205	29" x 77"	G		16	1		335	35.50		370.50	425
0208	33" x 37"	G		20	.800		268	28.50		296.50	340
0210	33" x 41"	G		20	.800		269	28.50		297.50	345
0215	33" x 49"	G		20	.800		283	28.50		311.50	355
0220	33" x 53"	G		18	.889		288	31.50		319.50	370
0230	33" x 57"	G		17	.941		297	33.50		330.50	380
0240	33" x 65"	G		16	1		315	35.50		350.50	410
0242	33" x 69"	G		16	1		325	35.50		360.50	415
0246	33" x 77"	G		16	1		345	35.50		380.50	440
0250	37" x 41"	G		20	.800		289	28.50		317.50	365
0255	37" x 49"	G		20	.800		297	28.50		325.50	370
0260	37" x 53"	G		18	.889		315	31.50		346.50	400
0270	37" x 57"	G		17	.941		340	33.50		373.50	430
0280	37" x 65"	G		16	1		355	35.50		390.50	450
0282	37" x 69"	G		16	1		360	35.50		395.50	455
0286	37" x 77"	G		16	1		375	35.50		410.50	470
0300	Solid vinyl, average quality, double insulated glass, 2'-0" x 3'-0"	G	1 Carp	10	.800		293	28.50		321.50	365
0310	3'-0" x 4'-0"	G		9	.889		217	31.50		248.50	291
0330	Premium, double insulated glass, 2'-6" x 3'-0"	G		10	.800		278	28.50		306.50	350
0340	3'-0" x 3'-6"	G		9	.889		310	31.50		341.50	395
0350	3'-0" x 4'-0"	G		9	.889		335	31.50		366.50	425
0360	3'-0" x 4'-6"	G		9	.889		340	31.50		371.50	430
0370	3'-0" x 5'-0"	G		8	1		365	35.50		400.50	460
0380	3'-6" x 6'-0"	G		8	1		410	35.50		445.50	510

08 53 13.40 Vinyl Casement Windows			Crew	Daily Output	Labor-Hours	Unit	Material	Labor	Equipment	Total	Total Incl O&P
0010	**VINYL CASEMENT WINDOWS**, insulated glass										
0015	Grids, low E, J fin, extension jambs, screens										
0100	One lite, 21" x 41"	G	2 Carp	20	.800	Ea.	320	28.50		348.50	395
0110	21" x 47"	G		20	.800		335	28.50		363.50	415
0120	21" x 53"	G		20	.800		355	28.50		383.50	435
0128	24" x 35"	G		19	.842		305	30		335	385
0130	24" x 41"	G		19	.842		325	30		355	410
0140	24" x 47"	G		19	.842		340	30		370	425
0150	24" x 53"	G		19	.842		370	30		400	455
0158	28" x 35"	G		19	.842		315	30		345	400
0160	28" x 41"	G		19	.842		330	30		360	410
0170	28" x 47"	G		19	.842		365	30		395	450
0180	28" x 53"	G		19	.842		390	30		420	480
0184	28" x 59"	G		19	.842		400	30		430	490
0188	Two lites, 33" x 35"	G		18	.889		530	31.50		561.50	640
0190	33" x 41"	G		18	.889		530	31.50		561.50	635
0200	33" x 47"	G		18	.889		555	31.50		586.50	665
0210	33" x 53"	G		18	.889		570	31.50		601.50	685
0212	33" x 59"	G		18	.889		625	31.50		656.50	745
0215	33" x 72"	G		18	.889		645	31.50		676.50	760
0220	41" x 41"	G		18	.889		565	31.50		596.50	675
0230	41" x 47"	G		18	.889		600	31.50		631.50	715
0240	41" x 53"	G		17	.941		645	33.50		678.50	765
0242	41" x 59"	G		17	.941		680	33.50		713.50	805
0246	41" x 72"	G		17	.941		710	33.50		743.50	835
0250	47" x 41"	G		17	.941		575	33.50		608.50	690

08 53 Plastic Windows

08 53 13 – Vinyl Windows

08 53 13.40 Vinyl Casement Windows

	08 53 13.40 Vinyl Casement Windows		Crew	Daily Output	Labor-Hours	Unit	Material	2019 Bare Costs Labor	Equipment	Total	Total Incl O&P
0260	47″ x 47″	G	2 Carp	17	.941	Ea.	605	33.50		638.50	720
0270	47″ x 53″	G		17	.941		645	33.50		678.50	760
0272	47″ x 59″	G		17	.941		715	33.50		748.50	845
0280	56″ x 41″	G		15	1.067		620	38		658	745
0290	56″ x 47″	G		15	1.067		645	38		683	770
0300	56″ x 53″	G		15	1.067		715	38		753	855
0302	56″ x 59″	G		15	1.067		740	38		778	880
0310	56″ x 72″	G		15	1.067		795	38		833	940
0340	Solid vinyl, premium, double insulated glass, 2′-0″ x 3′-0″ high	G	1 Carp	10	.800		272	28.50		300.50	345
0360	2′-0″ x 4′-0″ high	G		9	.889		300	31.50		331.50	385
0380	2′-0″ x 5′-0″ high	G		8	1		350	35.50		385.50	445

08 53 13.50 Vinyl Picture Windows

	08 53 13.50 Vinyl Picture Windows	Crew	Daily Output	Labor-Hours	Unit	Material	Labor	Equipment	Total	Total Incl O&P
0010	**VINYL PICTURE WINDOWS**, insulated glass									
0120	Grids, low E, J fin, ext. jambs, 47″ x 35″	2 Carp	12	1.333	Ea.	300	47.50		347.50	410
0130	47″ x 41″		12	1.333		455	47.50		502.50	585
0140	47″ x 47″		12	1.333		345	47.50		392.50	460
0150	47″ x 53″		11	1.455		370	52		422	490
0160	71″ x 35″		11	1.455		445	52		497	575
0170	71″ x 41″		11	1.455		465	52		517	600
0180	71″ x 47″		11	1.455		500	52		552	635

08 53 13.60 Vinyl Half Round Windows

	08 53 13.60 Vinyl Half Round Windows	Crew	Daily Output	Labor-Hours	Unit	Material	Labor	Equipment	Total	Total Incl O&P
0010	**VINYL HALF ROUND WINDOWS**, Including grille, J fin, low E, ext. jambs									
0100	10″ height x 20″ base	2 Carp	8	2	Ea.	355	71.50		426.50	510
0110	15″ height x 30″ base		8	2		355	71.50		426.50	510
0120	17″ height x 34″ base		7	2.286		355	81.50		436.50	525
0130	19″ height x 38″ base		7	2.286		565	81.50		646.50	755
0140	19″ height x 33″ base		7	2.286		565	81.50		646.50	755
0150	24″ height x 48″ base	1 Carp	6	1.333		565	47.50		612.50	700
0160	25″ height x 50″ base	″	6	1.333		565	47.50		612.50	700
0170	30″ height x 60″ base	2 Carp	6	2.667		565	95		660	775

08 54 Composite Windows

08 54 13 – Fiberglass Windows

08 54 13.10 Fiberglass Single-Hung Windows

	08 54 13.10 Fiberglass Single-Hung Windows		Crew	Daily Output	Labor-Hours	Unit	Material	Labor	Equipment	Total	Total Incl O&P
0010	**FIBERGLASS SINGLE-HUNG WINDOWS**										
0100	Grids, low E, 18″ x 24″	G	2 Carp	18	.889	Ea.	380	31.50		411.50	470
0110	18″ x 40″	G		17	.941		365	33.50		398.50	460
0130	24″ x 40″	G		20	.800		390	28.50		418.50	470
0230	36″ x 36″	G		17	.941		410	33.50		443.50	510
0250	36″ x 48″	G		20	.800		445	28.50		473.50	535
0260	36″ x 60″	G		18	.889		495	31.50		526.50	600
0280	36″ x 72″	G		16	1		530	35.50		565.50	640
0290	48″ x 40″	G		16	1		530	35.50		565.50	640

08 54 13.30 Fiberglass Slider Windows

	08 54 13.30 Fiberglass Slider Windows		Crew	Daily Output	Labor-Hours	Unit	Material	Labor	Equipment	Total	Total Incl O&P
0010	**FIBERGLASS SLIDER WINDOWS**										
0100	Grids, low E, 36″ x 24″	G	2 Carp	20	.800	Ea.	385	28.50		413.50	465
0110	36″ x 36″	G	″	20	.800	″	385	28.50		413.50	465

08 54 13.50 Fiberglass Bay Windows

	08 54 13.50 Fiberglass Bay Windows		Crew	Daily Output	Labor-Hours	Unit	Material	Labor	Equipment	Total	Total Incl O&P
0010	**FIBERGLASS BAY WINDOWS**										
0150	48″ x 36″	G	2 Carp	11	1.455	Ea.	1,125	52		1,177	1,325

08 61 Roof Windows

08 61 13 – Metal Roof Windows

08 61 13.10 Roof Windows		Crew	Daily Output	Labor-Hours	Unit	Material	Labor	Equipment	Total	Total Incl O&P
						2019 Bare Costs				
0010	**ROOF WINDOWS**, fixed high perf tmpd glazing, metallic framed									
0020	46" x 21-1/2", Flashed for shingled roof	1 Carp	8	1	Ea.	277	35.50		312.50	365
0100	46" x 28"		8	1		310	35.50		345.50	400
0125	57" x 44"		6	1.333		375	47.50		422.50	495
0130	72" x 28"		7	1.143		375	40.50		415.50	485
0150	Fixed, laminated tempered glazing, 46" x 21-1/2"		8	1		465	35.50		500.50	575
0175	46" x 28"		8	1		515	35.50		550.50	630
0200	57" x 44"		6	1.333		485	47.50		532.50	615
0500	Vented flashing set for shingled roof, 46" x 21-1/2"		7	1.143		465	40.50		505.50	585
0525	46" x 28"		6	1.333		515	47.50		562.50	650
0550	57" x 44"		5	1.600		650	57		707	810
0560	72" x 28"		5	1.600		650	57		707	810
0575	Flashing set for low pitched roof, 46" x 21-1/2"		7	1.143		540	40.50		580.50	665
0600	46" x 28"		7	1.143		590	40.50		630.50	720
0625	57" x 44"		5	1.600		735	57		792	900
0650	Flashing set for curb, 46" x 21-1/2"		7	1.143		605	40.50		645.50	740
0675	46" x 28"		7	1.143		660	40.50		700.50	795
0700	57" x 44"	↓	5	1.600	↓	805	57		862	985

08 61 16 – Wood Roof Windows

08 61 16.16 Roof Windows, Wood Framed		Crew	Daily Output	Labor-Hours	Unit	Material	Labor	Equipment	Total	Total Incl O&P
0010	**ROOF WINDOWS, WOOD FRAMED**									
5600	Roof window incl. frame, flashing, double insulated glass & screens,									
5610	complete unit, 22" x 38"	2 Carp	3	5.333	Ea.	780	190		970	1,175
5650	2'-5" x 3'-8"		3.20	5		985	178		1,163	1,375
5700	3'-5" x 4'-9"	↓	3.40	4.706	↓	1,100	168		1,268	1,475

08 62 Unit Skylights

08 62 13 – Domed Unit Skylights

08 62 13.10 Domed Skylights			Crew	Daily Output	Labor-Hours	Unit	Material	Labor	Equipment	Total	Total Incl O&P
0010	**DOMED SKYLIGHTS**										
0020	Skylight, fixed dome type, 22" x 22"	G	G-3	12	2.667	Ea.	216	88.50		304.50	385
0030	22" x 46"	G		10	3.200		271	106		377	475
0040	30" x 30"	G		12	2.667		285	88.50		373.50	460
0050	30" x 46"	G		10	3.200		375	106		481	585
0110	Fixed, double glazed, 22" x 27"	G		12	2.667		275	88.50		363.50	450
0120	22" x 46"	G		10	3.200		291	106		397	495
0130	44" x 46"	G		10	3.200		430	106		536	645
0210	Operable, double glazed, 22" x 27"	G		12	2.667		405	88.50		493.50	590
0220	22" x 46"	G		10	3.200		485	106		591	710
0230	44" x 46"	G	↓	10	3.200	↓	925	106		1,031	1,200

08 62 13.20 Skylights			Crew	Daily Output	Labor-Hours	Unit	Material	Labor	Equipment	Total	Total Incl O&P
0010	**SKYLIGHTS**, flush or curb mounted										
2120	Ventilating insulated plexiglass dome with										
2130	curb mounting, 36" x 36"	G	G-3	12	2.667	Ea.	480	88.50		568.50	670
2150	52" x 52"	G		12	2.667		665	88.50		753.50	880
2160	28" x 52"	G		10	3.200		500	106		606	720
2170	36" x 52"	G	↓	10	3.200		540	106		646	770
2180	For electric opening system, add	G					320			320	350
2210	Operating skylight, with thermopane glass, 24" x 48"	G	G-3	10	3.200		595	106		701	830
2220	32" x 48"	G	"	9	3.556	↓	620	118		738	880

08 62 Unit Skylights

08 62 13 – Domed Unit Skylights

08 62 13.20 Skylights		Crew	Daily Output	Labor-Hours	Unit	Material	2019 Bare Costs Labor	Equipment	Total	Total Incl O&P
2310	Non venting insulated plexiglass dome skylight with									
2320	Flush mount 22" x 46" G	G-3	15.23	2.101	Ea.	335	70		405	485
2330	30" x 30" G		16	2		310	66.50		376.50	450
2340	46" x 46" G		13.91	2.301		570	76.50		646.50	750
2350	Curb mount 22" x 46" G		15.23	2.101		365	70		435	515
2360	30" x 30" G		16	2		405	66.50		471.50	555
2370	46" x 46" G		13.91	2.301		625	76.50		701.50	815
2381	Non-insulated flush mount 22" x 46"		15.23	2.101		228	70		298	365
2382	30" x 30"		16	2		205	66.50		271.50	335
2383	46" x 46"		13.91	2.301		385	76.50		461.50	545
2384	Curb mount 22" x 46"		15.23	2.101		191	70		261	325
2385	30" x 30"		16	2		188	66.50		254.50	315
4000	Skylight, solar tube kit, incl. dome, flashing, diffuser, 1 pipe, 10" diam. G	1 Carp	2	4		305	143		448	570
4010	14" diam. G		2	4		405	143		548	680
4020	21" diam. G		2	4		485	143		628	770
4030	Accessories for, 1' long x 9" diam. pipe G		24	.333		50	11.90		61.90	74.50
4040	2' long x 9" diam. pipe G		24	.333		42.50	11.90		54.40	66.50
4050	4' long x 9" diam. pipe G		20	.400		74	14.25		88.25	105
4060	1' long x 13" diam. pipe G		24	.333		70.50	11.90		82.40	97
4070	2' long x 13" diam. pipe G		24	.333		55	11.90		66.90	80
4080	4' long x 13" diam. pipe G		20	.400		102	14.25		116.25	136
4090	6.5" turret ext. for 21" diam. pipe G		16	.500		109	17.85		126.85	150
4100	12' long x 21" diam. flexible pipe G		12	.667		90	24		114	138
4110	45 degree elbow, 10" G		16	.500		151	17.85		168.85	196
4120	14" G		16	.500		84.50	17.85		102.35	123
4130	Interior decorative ring, 9" G		20	.400		47.50	14.25		61.75	76
4140	13" G		20	.400		70	14.25		84.25	100

08 71 Door Hardware

08 71 20 – Hardware

08 71 20.15 Hardware		Crew	Daily Output	Labor-Hours	Unit	Material	Labor	Equipment	Total	Total Incl O&P
0010	**HARDWARE**									
0020	Average hardware cost									
1000	Door hardware, apartment, interior	1 Carp	4	2	Door	545	71.50		616.50	720
1300	Average, door hardware, motel/hotel interior, with access card		4	2	"	690	71.50		761.50	875
2100	Pocket door		6	1.333	Ea.	113	47.50		160.50	203
4000	Door knocker, bright brass		32	.250		82	8.90		90.90	105
4100	Mail slot, bright brass, 2" x 11"		25	.320		61.50	11.40		72.90	86.50
4200	Peep hole, add to price of door					14.95			14.95	16.45

08 71 20.40 Lockset		Crew	Daily Output	Labor-Hours	Unit	Material	Labor	Equipment	Total	Total Incl O&P
0010	**LOCKSET**, Standard duty									
0020	Non-keyed, passage, w/sect. trim	1 Carp	12	.667	Ea.	79.50	24		103.50	127
0100	Privacy		12	.667		81.50	24		105.50	129
0400	Keyed, single cylinder function		10	.800		151	28.50		179.50	213
0500	Lever handled, keyed, single cylinder function		10	.800		140	28.50		168.50	201

08 71 20.41 Dead Locks		Crew	Daily Output	Labor-Hours	Unit	Material	Labor	Equipment	Total	Total Incl O&P
0010	**DEAD LOCKS**									
1203	Deadlock night latch	1 Carp	7.70	1.039	Ea.	51.50	37		88.50	118
1420	Deadbolt lock, single cylinder		10	.800		47	28.50		75.50	98.50
1440	Double cylinder		10	.800		65	28.50		93.50	119

08 71 20.42 Mortise Locksets

08 71 20.42 Mortise Locksets		Crew	Daily Output	Labor-Hours	Unit	Material	Labor	Equipment	Total	Total Incl O&P
						2019 Bare Costs				
0010	**MORTISE LOCKSETS**, Comm., wrought knobs & full escutcheon trim									
0015	Assumes mortise is cut									
0020	Non-keyed, passage, Grade 3	1 Carp	9	.889	Ea.	167	31.50		198.50	236
0030	Grade 1		8	1		435	35.50		470.50	535
0040	Privacy set, Grade 3		9	.889		177	31.50		208.50	248
0050	Grade 1		8	1		475	35.50		510.50	585
0100	Keyed, office/entrance/apartment, Grade 2		8	1		206	35.50		241.50	286
0110	Grade 1		7	1.143		540	40.50		580.50	665
0120	Single cylinder, typical, Grade 3		8	1		197	35.50		232.50	276
0130	Grade 1		7	1.143		540	40.50		580.50	660
0200	Hotel, room, Grade 3		7	1.143		198	40.50		238.50	286
0210	Grade 1 (see also Section 08 71 20.15)		6	1.333		540	47.50		587.50	675
0300	Double cylinder, Grade 3		8	1		236	35.50		271.50	320
0310	Grade 1		7	1.143		555	40.50		595.50	680
1000	Wrought knobs and sectional trim, non-keyed, passage, Grade 3		10	.800		136	28.50		164.50	196
1010	Grade 1		9	.889		435	31.50		466.50	535
1040	Privacy, Grade 3		10	.800		157	28.50		185.50	220
1050	Grade 1		9	.889		490	31.50		521.50	595
1100	Keyed, entrance, office/apartment, Grade 3		9	.889		234	31.50		265.50	310
1103	Install lockset		6.92	1.156		234	41		275	325
1110	Grade 1		8	1		560	35.50		595.50	675
1120	Single cylinder, Grade 3		9	.889		236	31.50		267.50	315
1130	Grade 1		8	1		520	35.50		555.50	635
2000	Cast knobs and full escutcheon trim									
2010	Non-keyed, passage, Grade 3	1 Carp	9	.889	Ea.	287	31.50		318.50	370
2020	Grade 1		8	1		395	35.50		430.50	495
2040	Privacy, Grade 3		9	.889		335	31.50		366.50	420
2050	Grade 1		8	1		455	35.50		490.50	560
2120	Keyed, single cylinder, Grade 3		8	1		345	35.50		380.50	440
2123	Mortise lock		6.15	1.301		345	46.50		391.50	455
2130	Grade 1		7	1.143		545	40.50		585.50	670

08 71 20.50 Door Stops

08 71 20.50 Door Stops		Crew	Daily Output	Labor-Hours	Unit	Material	Labor	Equipment	Total	Total Incl O&P
0010	**DOOR STOPS**									
0020	Holder & bumper, floor or wall	1 Carp	32	.250	Ea.	36.50	8.90		45.40	55
1300	Wall bumper, 4" diameter, with rubber pad, aluminum		32	.250		13.85	8.90		22.75	30
1600	Door bumper, floor type, aluminum		32	.250		3.46	8.90		12.36	18.55
1620	Brass		32	.250		11.70	8.90		20.60	27.50
1630	Bronze		32	.250		19.80	8.90		28.70	37
1900	Plunger type, door mounted		32	.250		30	8.90		38.90	48
2520	Wall type, aluminum		32	.250		34.50	8.90		43.40	52.50
2540	Plunger type, aluminum		32	.250		30	8.90		38.90	48
2560	Brass		32	.250		45	8.90		53.90	64.50
3020	Floor mounted, US3		3	2.667		345	95		440	535

08 71 20.60 Entrance Locks

08 71 20.60 Entrance Locks		Crew	Daily Output	Labor-Hours	Unit	Material	Labor	Equipment	Total	Total Incl O&P
0010	**ENTRANCE LOCKS**									
0015	Cylinder, grip handle deadlocking latch	1 Carp	9	.889	Ea.	184	31.50		215.50	255
0020	Deadbolt		8	1		195	35.50		230.50	274
0100	Push and pull plate, dead bolt		8	1		239	35.50		274.50	320
0900	For handicapped lever, add					156			156	171

08 71 Door Hardware

08 71 20 – Hardware

08 71 20.65 Thresholds		Crew	Daily Output	Labor-Hours	Unit	Material	Labor	Equipment	Total	Total Incl O&P
						2019 Bare Costs				
0010	**THRESHOLDS**									
0011	Threshold 3' long saddles aluminum	1 Carp	48	.167	L.F.	9.95	5.95		15.90	21
0100	Aluminum, 8" wide, 1/2" thick		12	.667	Ea.	51.50	24		75.50	96.50
0500	Bronze		60	.133	L.F.	47	4.75		51.75	60
0600	Bronze, panic threshold, 5" wide, 1/2" thick		12	.667	Ea.	165	24		189	222
0700	Rubber, 1/2" thick, 5-1/2" wide		20	.400		42.50	14.25		56.75	70.50
0800	2-3/4" wide		20	.400		51.50	14.25		65.75	80
1950	ADA compliant thresholds									
2000	Threshold, wood oak 3-1/2" wide x 24" long	1 Carp	12	.667	Ea.	12.20	24		36.20	53
2010	3-1/2" wide x 36" long		12	.667		16.40	24		40.40	57.50
2020	3-1/2" wide x 48" long		12	.667		22	24		46	63.50
2030	4-1/2" wide x 24" long		12	.667		14.10	24		38.10	55
2040	4-1/2" wide x 36" long		12	.667		20.50	24		44.50	62.50
2050	4-1/2" wide x 48" long		12	.667		27	24		51	69.50
2060	6-1/2" wide x 24" long		12	.667		19.60	24		43.60	61
2070	6-1/2" wide x 36" long		12	.667		29.50	24		53.50	72
2080	6-1/2" wide x 48" long		12	.667		39	24		63	82.50
2090	Threshold, wood cherry 3-1/2" wide x 24" long		12	.667		14.60	24		38.60	55.50
2100	3-1/2" wide x 36" long		12	.667		33.50	24		57.50	76.50
2110	3-1/2" wide x 48" long		12	.667		38.50	24		62.50	82
2120	4-1/2" wide x 24" long		12	.667		18.60	24		42.60	60
2130	4-1/2" wide x 36" long		12	.667		37.50	24		61.50	80.50
2140	4-1/2" wide x 48" long		12	.667		45.50	24		69.50	89.50
2150	6-1/2" wide x 24" long		12	.667		26.50	24		50.50	68.50
2160	6-1/2" wide x 36" long		12	.667		50.50	24		74.50	95
2170	6-1/2" wide x 48" long		12	.667		61	24		85	107
2180	Threshold, wood walnut 3-1/2" wide x 24" long		12	.667		17.55	24		41.55	59
2190	3-1/2" wide x 36" long		12	.667		35	24		59	78
2200	3-1/2" wide x 48" long		12	.667		40.50	24		64.50	84
2210	4-1/2" wide x 24" long		12	.667		28	24		52	70
2220	4-1/2" wide x 36" long		12	.667		47	24		71	91
2230	4-1/2" wide x 48" long		12	.667		57.50	24		81.50	103
2240	6-1/2" wide x 24" long		12	.667		44.50	24		68.50	88.50
2250	6-1/2" wide x 36" long		12	.667		70.50	24		94.50	117
2260	6-1/2" wide x 48" long		12	.667		92	24		116	141
2300	Threshold, aluminum 4" wide x 36" long		12	.667		25.50	24		49.50	67.50
2310	4" wide x 48" long		12	.667		30.50	24		54.50	73
2320	4" wide x 72" long		12	.667		46	24		70	90
2330	5" wide x 36" long		12	.667		33.50	24		57.50	76.50
2340	5" wide x 48" long		12	.667		39.50	24		63.50	83
2350	5" wide x 72" long		12	.667		74	24		98	121
2360	6" wide x 36" long		12	.667		46	24		70	90
2370	6" wide x 48" long		12	.667		49	24		73	93.50
2380	6" wide x 72" long		12	.667		71.50	24		95.50	118
2390	7" wide x 36" long		12	.667		50	24		74	94.50
2400	7" wide x 48" long		12	.667		67	24		91	114
2410	7" wide x 72" long		12	.667		95.50	24		119.50	145
2500	Threshold, ramp, aluminum or rubber 24" x 24"		12	.667		190	24		214	249

08 71 20.75 Door Hardware Accessories

		Crew	Daily Output	Labor-Hours	Unit	Material	Labor	Equipment	Total	Total Incl O&P
0010	**DOOR HARDWARE ACCESSORIES**									
1000	Knockers, brass, standard	1 Carp	16	.500	Ea.	66	17.85		83.85	102
1100	Deluxe	"	10	.800	"	94.50	28.50		123	151

08 71 Door Hardware

08 71 20 – Hardware

08 71 20.75 Door Hardware Accessories		Crew	Daily Output	Labor-Hours	Unit	Material	Labor	Equipment	Total	Total Incl O&P
						2019 Bare Costs				
2000	Torsion springs for overhead doors									
2050	1-3/4" diam., 32" long, 0.243" spring	1 Carp	8	1	Ea.	55.50	35.50		91	120
2060	1-3/4" diam., 32" long, 0.250" spring		8	1		50	35.50		85.50	114
2070	2" diam., 32" long, 0.243" spring		8	1		43	35.50		78.50	107
2080	2" diam., 32" long, 0.253" spring		8	1		51.50	35.50		87	116
4500	Rubber door silencers		540	.015		.44	.53		.97	1.35

08 71 20.90 Hinges		Crew	Daily Output	Labor-Hours	Unit	Material	Labor	Equipment	Total	Total Incl O&P
0010	**HINGES**									
0012	Full mortise, avg. freq., steel base, USP, 4-1/2" x 4-1/2"				Pr.	46			46	50.50
0100	5" x 5", USP					66.50			66.50	73
0200	6" x 6", USP					139			139	153
0400	Brass base, 4-1/2" x 4-1/2", US10					66			66	72.50
0500	5" x 5", US10					76			76	83.50
0600	6" x 6", US10					169			169	186
0800	Stainless steel base, 4-1/2" x 4-1/2", US32					87.50			87.50	96.50
0900	For non removable pin, add (security item)				Ea.	5.65			5.65	6.20
0910	For floating pin, driven tips, add					3.28			3.28	3.61
0930	For hospital type tip on pin, add					14.45			14.45	15.85
0940	For steeple type tip on pin, add					20			20	22
0950	Full mortise, high frequency, steel base, 3-1/2" x 3-1/2", US26D				Pr.	31.50			31.50	34.50
1000	4-1/2" x 4-1/2", USP					66.50			66.50	73.50
1100	5" x 5", USP					54			54	59
1200	6" x 6", USP					140			140	155
1400	Brass base, 3-1/2" x 3-1/2", US4					53.50			53.50	59
1430	4-1/2" x 4-1/2", US10					78			78	85.50
1500	5" x 5", US10					128			128	141
1600	6" x 6", US10					174			174	191
1800	Stainless steel base, 4-1/2" x 4-1/2", US32					107			107	118
1810	5" x 4-1/2", US32					143			143	157
1930	For hospital type tip on pin, add				Ea.	14.60			14.60	16.05
1950	Full mortise, low frequency, steel base, 3-1/2" x 3-1/2", US26D				Pr.	29.50			29.50	32.50
2000	4-1/2" x 4-1/2", USP					24.50			24.50	27
2100	5" x 5", USP					50.50			50.50	55.50
2200	6" x 6", USP					97.50			97.50	107
2300	4-1/2" x 4-1/2", US3					17.95			17.95	19.75
2310	5" x 5", US3					43			43	47.50
2400	Brass bass, 4-1/2" x 4-1/2", US10					57			57	62.50
2500	5" x 5", US10					81.50			81.50	89.50
2800	Stainless steel base, 4-1/2" x 4-1/2", US32					76.50			76.50	84.50
8000	Install hinge	1 Carp	34	.235			8.40		8.40	13.85

08 71 20.91 Special Hinges		Crew	Daily Output	Labor-Hours	Unit	Material	Labor	Equipment	Total	Total Incl O&P
0010	**SPECIAL HINGES**									
8000	Continuous hinges									
8010	Steel, piano, 2" x 72"	1 Carp	20	.400	Ea.	23.50	14.25		37.75	49.50
8020	Brass, piano, 1-1/16" x 30"		30	.267		9.20	9.50		18.70	26
8030	Acrylic, piano, 1-3/4" x 12"		40	.200		15.25	7.15		22.40	28.50

08 71 20.92 Mortised Hinges		Crew	Daily Output	Labor-Hours	Unit	Material	Labor	Equipment	Total	Total Incl O&P
0010	**MORTISED HINGES**									
0200	Average frequency, steel plated, ball bearing, 3-1/2" x 3-1/2"				Pr.	28			28	31
0300	Bronze, ball bearing					33			33	36
0900	High frequency, steel plated, ball bearing					106			106	117
1100	Bronze, ball bearing					106			106	117

08 71 Door Hardware

08 71 20 – Hardware

08 71 20.92 Mortised Hinges		Crew	Daily Output	Labor-Hours	Unit	Material	Labor	Equipment	Total	Total Incl O&P
						2019 Bare Costs				
1300	Average frequency, steel plated, ball bearing, 4-1/2" x 4-1/2"				Pr.	36.50			36.50	40
1500	Bronze, ball bearing, to 36" wide					37.50			37.50	41
1700	Low frequency, steel, plated, plain bearing					20			20	22
1900	Bronze, plain bearing				↓	28			28	30.50
08 71 20.95 Kick Plates										
0010	**KICK PLATES**									
0020	Stainless steel, .050", 16 ga., 8" x 28", US32	1 Carp	15	.533	Ea.	41.50	19		60.50	77.50
0080	Mop/Kick, 4" x 28"		15	.533		37.50	19		56.50	73
0090	4" x 30"		15	.533		39.50	19		58.50	75
0100	4" x 34"		15	.533		45	19		64	81
0110	6" x 28"		15	.533		43	19		62	78.50
0120	6" x 30"		15	.533		52	19		71	88.50
0130	6" x 34"	↓	15	.533	↓	58.50	19		77.50	96
08 71 21 – Astragals										
08 71 21.10 Exterior Mouldings, Astragals										
0010	**EXTERIOR MOULDINGS, ASTRAGALS**									
4170	Astragal for double doors, aluminum	1 Carp	4	2	Opng.	39	71.50		110.50	161
4174	Bronze	"	4	2	"	53	71.50		124.50	177
08 71 25 – Weatherstripping										
08 71 25.10 Mechanical Seals, Weatherstripping										
0010	**MECHANICAL SEALS, WEATHERSTRIPPING**									
1000	Doors, wood frame, interlocking, for 3' x 7' door, zinc	1 Carp	3	2.667	Opng.	49	95		144	211
1100	Bronze		3	2.667		62	95		157	225
1300	6' x 7' opening, zinc		2	4		59.50	143		202.50	300
1400	Bronze		2	4	↓	71	143		214	315
1500	Vinyl V strip	↓	6.40	1.250	Ea.	13.15	44.50		57.65	88
1700	Wood frame, spring type, bronze									
1800	3' x 7' door	1 Carp	7.60	1.053	Opng.	25.50	37.50		63	90
1900	6' x 7' door		7	1.143		32	40.50		72.50	103
1920	Felt, 3' x 7' door		14	.571		4.50	20.50		25	38.50
1930	6' x 7' door		13	.615		5.20	22		27.20	42
1950	Rubber, 3' x 7' door		7.60	1.053		11.55	37.50		49.05	74.50
1960	6' x 7' door	↓	7	1.143	↓	13.40	40.50		53.90	82
2200	Metal frame, spring type, bronze									
2300	3' x 7' door	1 Carp	3	2.667	Opng.	51	95		146	213
2400	6' x 7' door	"	2.50	3.200	"	55	114		169	249
2500	For stainless steel, spring type, add					133%				
2700	Metal frame, extruded sections, 3' x 7' door, aluminum	1 Carp	3	2.667	Opng.	28.50	95		123.50	189
2800	Bronze		3	2.667		84	95		179	250
3100	6' x 7' door, aluminum		1.50	5.333		35	190		225	355
3200	Bronze	↓	1.50	5.333	↓	145	190		335	475
3500	Threshold weatherstripping									
3650	Door sweep, flush mounted, aluminum	1 Carp	25	.320	Ea.	21	11.40		32.40	42
3700	Vinyl		25	.320		18.75	11.40		30.15	39.50
5000	Garage door bottom weatherstrip, 12' aluminum, clear		14	.571		26.50	20.50		47	62.50
5010	Bronze		14	.571		94	20.50		114.50	137
5050	Bottom protection, rubber		14	.571		48.50	20.50		69	87
5100	Threshold	↓	14	.571	↓	70	20.50		90.50	111

08 75 Window Hardware

08 75 10 – Window Handles and Latches

08 75 10.10 Handles and Latches		Crew	Daily Output	Labor-Hours	Unit	Material	Labor	Equipment	Total	Total Incl O&P
						2019 Bare Costs				
0010	**HANDLES AND LATCHES**									
1000	Handles, surface mounted, aluminum	1 Carp	24	.333	Ea.	5.90	11.90		17.80	26
1020	Brass		24	.333		5.80	11.90		17.70	26
1040	Chrome		24	.333		8.65	11.90		20.55	29
1200	Window handles window crank ADA		24	.333		16.40	11.90		28.30	37.50
1500	Recessed, aluminum		12	.667		3.52	24		27.52	43.50
1520	Brass		12	.667		5.35	24		29.35	45.50
1540	Chrome		12	.667		4.43	24		28.43	44.50
2000	Latches, aluminum		20	.400		4.29	14.25		18.54	28
2020	Brass		20	.400		5.60	14.25		19.85	29.50
2040	Chrome	↓	20	.400	↓	3.65	14.25		17.90	27.50

08 75 10.15 Window Opening Control		Crew	Daily Output	Labor-Hours	Unit	Material	Labor	Equipment	Total	Total Incl O&P
0010	**WINDOW OPENING CONTROL**									
0015	Window stops									
0020	For double-hung window	1 Carp	35	.229	Ea.	39.50	8.15		47.65	57
0030	Cam action for sliding window		35	.229		4.42	8.15		12.57	18.30
0040	Thumb screw for sliding window	↓	35	.229	↓	4.77	8.15		12.92	18.70
0100	Window guards, child safety bars for single or double-hung									
0110	14″ to 17″ wide max vert opening 26″	1 Carp	16	.500	Ea.	66.50	17.85		84.35	103
0120	17″ to 23″ wide max vert opening 26″		16	.500		76	17.85		93.85	113
0130	23″ to 36″ wide max vert opening 26″		16	.500		86.50	17.85		104.35	125
0140	35″ to 58″ wide max vert opening 26″		16	.500		100	17.85		117.85	140
0150	58″ to 90″ wide max vert opening 26″		16	.500		135	17.85		152.85	178
0160	73″ to 120″ wide max vert opening 26″	↓	16	.500	↓	236	17.85		253.85	290

08 75 30 – Weatherstripping

08 75 30.10 Mechanical Weather Seals		Crew	Daily Output	Labor-Hours	Unit	Material	Labor	Equipment	Total	Total Incl O&P
0010	**MECHANICAL WEATHER SEALS**, Window, double-hung, 3′ x 5′									
0020	Zinc	1 Carp	7.20	1.111	Opng.	22.50	39.50		62	90.50
0100	Bronze		7.20	1.111		43	39.50		82.50	113
0200	Vinyl V strip		7	1.143		11.10	40.50		51.60	79.50
0500	As above but heavy duty, zinc		4.60	1.739		21	62		83	125
0600	Bronze	↓	4.60	1.739	↓	77.50	62		139.50	187

08 79 Hardware Accessories

08 79 20 – Door Accessories

08 79 20.10 Door Hardware Accessories		Crew	Daily Output	Labor-Hours	Unit	Material	Labor	Equipment	Total	Total Incl O&P
0010	**DOOR HARDWARE ACCESSORIES**									
0140	Door bolt, surface, 4″	1 Carp	32	.250	Ea.	16.20	8.90		25.10	32.50
0160	Door latch	″	12	.667	″	8.75	24		32.75	49
0200	Sliding closet door									
0220	Track and hanger, single	1 Carp	10	.800	Ea.	69.50	28.50		98	124
0240	Double		8	1		84.50	35.50		120	152
0260	Door guide, single		48	.167		32	5.95		37.95	45
0280	Double		48	.167		42.50	5.95		48.45	57
0600	Deadbolt and lock cover plate, brass or stainless steel		30	.267		31	9.50		40.50	49.50
0620	Hole cover plate, brass or chrome		35	.229		8.50	8.15		16.65	23
2240	Mortise lockset, passage, lever handle		9	.889		166	31.50		197.50	236
4000	Security chain, standard		18	.444		13.15	15.85		29	40.50
4100	Deluxe	↓	18	.444	↓	46.50	15.85		62.35	77

08 81 Glass Glazing

08 81 10 – Float Glass

08 81 10.10 Various Types and Thickness of Float Glass		Crew	Daily Output	Labor-Hours	Unit	Material	2019 Bare Costs Labor	Equipment	Total	Total Incl O&P
0010	VARIOUS TYPES AND THICKNESS OF FLOAT GLASS									
0020	3/16" plain	2 Glaz	130	.123	S.F.	7.10	4.28		11.38	14.85
0200	Tempered, clear		130	.123		7.35	4.28		11.63	15.15
0300	Tinted		130	.123		7	4.28		11.28	14.75
0600	1/4" thick, clear, plain		120	.133		10.15	4.63		14.78	18.80
0700	Tinted		120	.133		9.60	4.63		14.23	18.20
0800	Tempered, clear		120	.133		7.10	4.63		11.73	15.50
0900	Tinted		120	.133		11.95	4.63		16.58	21
1600	3/8" thick, clear, plain		75	.213		11.45	7.40		18.85	25
1700	Tinted		75	.213		16.80	7.40		24.20	30.50
1800	Tempered, clear		75	.213		18	7.40		25.40	32
1900	Tinted		75	.213		20.50	7.40		27.90	35
2200	1/2" thick, clear, plain		55	.291		18.90	10.10		29	37.50
2300	Tinted		55	.291		29.50	10.10		39.60	49
2400	Tempered, clear		55	.291		27	10.10		37.10	46
2500	Tinted		55	.291		28	10.10		38.10	47
2800	5/8" thick, clear, plain		45	.356		29.50	12.35		41.85	53
2900	Tempered, clear	↓	45	.356	↓	33.50	12.35		45.85	57.50

08 81 25 – Glazing Variables

08 81 25.10 Applications of Glazing		Crew	Daily Output	Labor-Hours	Unit	Material	Labor	Equipment	Total	Total Incl O&P
0010	APPLICATIONS OF GLAZING									
0600	For glass replacement, add				S.F.		100%			
0700	For gasket settings, add				L.F.	6.30			6.30	6.95
0900	For sloped glazing, add				S.F.		26%			
2000	Fabrication, polished edges, 1/4" thick				Inch	.60			.60	.66
2100	1/2" thick					1.42			1.42	1.56
2500	Mitered edges, 1/4" thick					1.42			1.42	1.56
2600	1/2" thick				↓	2.34			2.34	2.57

08 81 30 – Insulating Glass

08 81 30.10 Reduce Heat Transfer Glass		Crew	Daily Output	Labor-Hours	Unit	Material	Labor	Equipment	Total	Total Incl O&P
0010	REDUCE HEAT TRANSFER GLASS									
0015	2 lites 1/8" float, 1/2" thk under 15 S.F.									
0100	Tinted G	2 Glaz	95	.168	S.F.	14.50	5.85		20.35	25.50
0280	Double glazed, 5/8" thk unit, 3/16" float, 15 to 30 S.F., clear		90	.178		14.10	6.20		20.30	25.50
0400	1" thk, dbl. glazed, 1/4" float, 30 to 70 S.F., clear G		75	.213		17.70	7.40		25.10	31.50
0500	Tinted G		75	.213		24	7.40		31.40	38.50
2000	Both lites, light & heat reflective G		85	.188		32.50	6.55		39.05	46.50
2500	Heat reflective, film inside, 1" thick unit, clear G		85	.188		28.50	6.55		35.05	42
2600	Tinted G		85	.188		30.50	6.55		37.05	44.50
3000	Film on weatherside, clear, 1/2" thick unit G		95	.168		20	5.85		25.85	31.50
3100	5/8" thick unit G		90	.178		21	6.20		27.20	33
3200	1" thick unit G		85	.188		28	6.55		34.55	42
5000	Spectrally selective film, on ext., blocks solar gain/allows 70% of light G	↓	95	.168	↓	15.45	5.85		21.30	26.50

08 81 40 – Plate Glass

08 81 40.10 Plate Glass		Crew	Daily Output	Labor-Hours	Unit	Material	Labor	Equipment	Total	Total Incl O&P
0010	PLATE GLASS Twin ground, polished,									
0015	3/16" thick, material				S.F.	5.85			5.85	6.40
0020	3/16" thick	2 Glaz	100	.160		5.85	5.55		11.40	15.55
0100	1/4" thick		94	.170		8	5.90		13.90	18.55
0200	3/8" thick		60	.267		13.60	9.25		22.85	30
0300	1/2" thick	↓	40	.400	↓	26.50	13.90		40.40	52

08 81 Glass Glazing

08 81 55 – Window Glass

08 81 55.10 Sheet Glass		Crew	Daily Output	Labor-Hours	Unit	Material	2019 Bare Costs Labor	Equipment	Total	Total Incl O&P
0010	**SHEET GLASS** (window), clear float, stops, putty bed									
0015	1/8" thick, clear float	2 Glaz	480	.033	S.F.	3.92	1.16		5.08	6.20
0500	3/16" thick, clear		480	.033		6.45	1.16		7.61	9
0600	Tinted		480	.033		8.05	1.16		9.21	10.75
0700	Tempered	↓	480	.033	↓	9.85	1.16		11.01	12.75

08 83 Mirrors

08 83 13 – Mirrored Glass Glazing

08 83 13.10 Mirrors

		Crew	Daily Output	Labor-Hours	Unit	Material	Labor	Equipment	Total	Total Incl O&P
0010	**MIRRORS**, No frames, wall type, 1/4" plate glass, polished edge									
0100	Up to 5 S.F.	2 Glaz	125	.128	S.F.	9.85	4.45		14.30	18.10
0200	Over 5 S.F.		160	.100		9.45	3.48		12.93	16.10
0500	Door type, 1/4" plate glass, up to 12 S.F.		160	.100		9.35	3.48		12.83	16
1000	Float glass, up to 10 S.F., 1/8" thick		160	.100		6.20	3.48		9.68	12.55
1100	3/16" thick		150	.107		7.70	3.71		11.41	14.55
1500	12" x 12" wall tiles, square edge, clear		195	.082		2.68	2.85		5.53	7.65
1600	Veined		195	.082		6.55	2.85		9.40	11.90
2010	Bathroom, unframed, laminated	↓	160	.100	↓	15.05	3.48		18.53	22.50

08 87 Glazing Surface Films

08 87 26 – Bird Control Film

08 87 26.10 Bird Control Film

		Crew	Daily Output	Labor-Hours	Unit	Material	Labor	Equipment	Total	Total Incl O&P
0010	**BIRD CONTROL FILM**									
0050	Patterned, adhered to glass	2 Glaz	180	.089	S.F.	6.85	3.09		9.94	12.65
0200	Decals small, adhered to glass	1 Glaz	50	.160	Ea.	2.52	5.55		8.07	11.90
0250	Decals large, adhered to glass		50	.160	"	8	5.55		13.55	17.95
0300	Decals set of 4, adhered to glass		25	.320	Set	7.20	11.10		18.30	26.50
0400	Bird control film tape	↓	10	.800	Roll	20.50	28		48.50	68.50

08 87 33 – Electrically Tinted Window Film

08 87 33.20 Window Film

		Crew	Daily Output	Labor-Hours	Unit	Material	Labor	Equipment	Total	Total Incl O&P
0010	**WINDOW FILM** adhered on glass (glass not included)									
0015	Film is pre-wired and can be trimmed									
0100	Window film 4 S.F. adhered on glass	1 Glaz	200	.040	S.F.	224	1.39		225.39	249
0120	6 S.F.		180	.044		335	1.54		336.54	375
0160	9 S.F.		170	.047		505	1.64		506.64	560
0180	10 S.F.		150	.053		570	1.85		571.85	635
0200	12 S.F.		144	.056		670	1.93		671.93	740
0220	14 S.F.		140	.057		780	1.99		781.99	865
0240	16 S.F.		128	.063		895	2.17		897.17	985
0260	18 S.F.		126	.063		1,000	2.21		1,002.21	1,100
0280	20 S.F.	↓	120	.067	↓	1,125	2.32		1,127.32	1,225

08 87 53 – Security Films On Glass

08 87 53.10 Security Film Adhered On Glass

		Crew	Daily Output	Labor-Hours	Unit	Material	Labor	Equipment	Total	Total Incl O&P
0010	**SECURITY FILM ADHERED ON GLASS** (glass not included)									
0020	Security film, clear, 7 mil	1 Glaz	200	.040	S.F.	1.11	1.39		2.50	3.51
0030	8 mil		200	.040		2	1.39		3.39	4.49
0040	9 mil		200	.040		1.10	1.39		2.49	3.50
0050	10 mil	↓	200	.040	↓	1.89	1.39		3.28	4.37

08 87 Glazing Surface Films

08 87 53 – Security Films On Glass

08 87 53.10 Security Film Adhered On Glass		Crew	Daily Output	Labor-Hours	Unit	2019 Bare Costs Material	Labor	Equipment	Total	Total Incl O&P
0060	12 mil	1 Glaz	200	.040	S.F.	1.87	1.39		3.26	4.35
0075	15 mil		200	.040		2.36	1.39		3.75	4.89
0100	Security film, sealed with structural adhesive, 7 mil		180	.044		5.65	1.54		7.19	8.75
0110	8 mil		180	.044		6.55	1.54		8.09	9.75
0140	14 mil	↓	180	.044	↓	7.20	1.54		8.74	10.45

08 91 Louvers

08 91 19 – Fixed Louvers

08 91 19.10 Aluminum Louvers		Crew	Daily Output	Labor-Hours	Unit	Material	Labor	Equipment	Total	Total Incl O&P
0010	**ALUMINUM LOUVERS**									
0020	Aluminum with screen, residential, 8″ x 8″	1 Carp	38	.211	Ea.	21	7.50		28.50	35.50
0100	12″ x 12″		38	.211		16.50	7.50		24	30.50
0200	12″ x 18″		35	.229		22	8.15		30.15	37.50
0250	14″ x 24″		30	.267		30	9.50		39.50	48.50
0300	18″ x 24″		27	.296		33	10.55		43.55	54
0500	24″ x 30″		24	.333		63.50	11.90		75.40	89.50
0700	Triangle, adjustable, small		20	.400		60.50	14.25		74.75	90
0800	Large		15	.533		85	19		104	125
2100	Midget, aluminum, 3/4″ deep, 1″ diameter		85	.094		.94	3.36		4.30	6.60
2150	3″ diameter		60	.133		3.07	4.75		7.82	11.25
2200	4″ diameter		50	.160		5.75	5.70		11.45	15.75
2250	6″ diameter	↓	30	.267	↓	3.80	9.50		13.30	19.90

08 95 Vents

08 95 13 – Soffit Vents

08 95 13.10 Wall Louvers		Crew	Daily Output	Labor-Hours	Unit	Material	Labor	Equipment	Total	Total Incl O&P
0010	**WALL LOUVERS**									
2400	Under eaves vent, aluminum, mill finish, 16″ x 4″	1 Carp	48	.167	Ea.	1.95	5.95		7.90	11.95
2500	16″ x 8″	″	48	.167	″	2.92	5.95		8.87	13

08 95 16 – Wall Vents

08 95 16.10 Louvers		Crew	Daily Output	Labor-Hours	Unit	Material	Labor	Equipment	Total	Total Incl O&P
0010	**LOUVERS**									
0020	Redwood, 2′-0″ diameter, full circle	1 Carp	16	.500	Ea.	294	17.85		311.85	355
0100	Half circle		16	.500		214	17.85		231.85	265
0200	Octagonal		16	.500		206	17.85		223.85	257
0300	Triangular, 5/12 pitch, 5′-0″ at base		16	.500		630	17.85		647.85	720
1000	Rectangular, 1′-4″ x 1′-3″		16	.500		27.50	17.85		45.35	59.50
1100	Rectangular, 1′-4″ x 1′-8″		16	.500		42	17.85		59.85	75.50
1200	1′-4″ x 2′-2″		15	.533		49.50	19		68.50	86
1300	1′-9″ x 2′-2″		15	.533		59.50	19		78.50	96.50
1400	2′-3″ x 2′-2″		14	.571		79.50	20.50		100	121
1700	2′-4″ x 2′-11″		13	.615		79.50	22		101.50	124
2000	Aluminum, 12″ x 16″		25	.320		24	11.40		35.40	45.50
2010	16″ x 20″		25	.320		32	11.40		43.40	54
2020	24″ x 30″		25	.320		65.50	11.40		76.90	91
2100	6′ triangle		12	.667		181	24		205	239
3100	Round, 2′-2″ diameter		16	.500		195	17.85		212.85	245
7000	Vinyl gable vent, 8″ x 8″		38	.211		17.65	7.50		25.15	32
7020	12″ x 12″	↓	38	.211	↓	31.50	7.50		39	47

08 95 Vents

08 95 16 – Wall Vents

08 95 16.10 Louvers		Crew	Daily Output	Labor-Hours	Unit	2019 Bare Costs Material	Labor	Equipment	Total	Total Incl O&P
7080	12" x 18"	1 Carp	35	.229	Ea.	40.50	8.15		48.65	58
7200	18" x 24"	↓	30	.267	↓	60	9.50		69.50	81.50

Division Notes

		CREW	DAILY OUTPUT	LABOR-HOURS	UNIT	BARE COSTS MAT.	BARE COSTS LABOR	BARE COSTS EQUIP.	BARE COSTS TOTAL	TOTAL INCL O&P

Estimating Tips

General

- Room Finish Schedule: A complete set of plans should contain a room finish schedule. If one is not available, it would be well worth the time and effort to obtain one.

09 20 00 Plaster and Gypsum Board

- Lath is estimated by the square yard plus a 5% allowance for waste. Furring, channels, and accessories are measured by the linear foot. An extra foot should be allowed for each accessory miter or stop.
- Plaster is also estimated by the square yard. Deductions for openings vary by preference, from zero deduction to 50% of all openings over 2 feet in width. The estimator should allow one extra square foot for each linear foot of horizontal interior or exterior angle located below the ceiling level. Also, double the areas of small radius work.
- Drywall accessories, studs, track, and acoustical caulking are all measured by the linear foot. Drywall taping is figured by the square foot. Gypsum wallboard is estimated by the square foot. No material deductions should be made for door or window openings under 32 S.F.

09 60 00 Flooring

- Tile and terrazzo areas are taken off on a square foot basis. Trim and base materials are measured by the linear foot. Accent tiles are listed per each. Two basic methods of installation are used. Mud set is approximately 30% more expensive than thin set. The cost of grout is included with tile unit price lines unless otherwise noted. In terrazzo work, be sure to include the linear footage of embedded decorative strips, grounds, machine rubbing, and power cleanup.
- Wood flooring is available in strip, parquet, or block configuration. The latter two types are set in adhesives with quantities estimated by the square foot. The laying pattern will influence labor costs and material waste. In addition to the material and labor for laying wood floors, the estimator must make allowances for sanding and finishing these areas, unless the flooring is prefinished.
- Sheet flooring is measured by the square yard. Roll widths vary, so consideration should be given to use the most economical width, as waste must be figured into the total quantity. Consider also the installation methods available—direct glue down or stretched. Direct glue-down installation is assumed with sheet carpet unit price lines unless otherwise noted.

09 70 00 Wall Finishes

- Wall coverings are estimated by the square foot. The area to be covered is measured—length by height of the wall above the baseboards—to calculate the square footage of each wall. This figure is divided by the number of square feet in the single roll which is being used. Deduct, in full, the areas of openings such as doors and windows. Where a pattern match is required allow 25–30% waste.

09 80 00 Acoustic Treatment

- Acoustical systems fall into several categories. The takeoff of these materials should be by the square foot of area with a 5% allowance for waste. Do not forget about scaffolding, if applicable, when estimating these systems.

09 90 00 Painting and Coating

- A major portion of the work in painting involves surface preparation. Be sure to include cleaning, sanding, filling, and masking costs in the estimate.
- Protection of adjacent surfaces is not included in painting costs. When considering the method of paint application, an important factor is the amount of protection and masking required. These must be estimated separately and may be the determining factor in choosing the method of application.

Reference Numbers

Reference numbers are shown at the beginning of some major classifications. These numbers refer to related items in the Reference Section. The reference information may be an estimating procedure, an alternate pricing method, or technical information.

Note: Not all subdivisions listed here necessarily appear. ■

09 01 Maintenance of Finishes

09 01 70 – Maintenance of Wall Finishes

09 01 70.10 Gypsum Wallboard Repairs		Crew	Daily Output	Labor-Hours	Unit	Material	2019 Bare Costs Labor	Equipment	Total	Total Incl O&P
0010	**GYPSUM WALLBOARD REPAIRS**									
0100	Fill and sand, pin/nail holes	1 Carp	960	.008	Ea.		.30		.30	.49
0110	Screw head pops		480	.017			.59		.59	.98
0120	Dents, up to 2″ square		48	.167		.01	5.95		5.96	9.80
0130	2″ to 4″ square		24	.333		.03	11.90		11.93	19.70
0140	Cut square, patch, sand and finish, holes, up to 2″ square		12	.667		.03	24		24.03	39.50
0150	2″ to 4″ square		11	.727		.09	26		26.09	43
0160	4″ to 8″ square		10	.800		.24	28.50		28.74	47.50
0170	8″ to 12″ square		8	1		.49	35.50		35.99	59.50
0180	12″ to 32″ square		6	1.333		1.61	47.50		49.11	80.50
0210	16″ by 48″		5	1.600		2.76	57		59.76	97
0220	32″ by 48″		4	2		4.53	71.50		76.03	123
0230	48″ square		3.50	2.286		6.40	81.50		87.90	142
0240	60″ square		3.20	2.500		10	89		99	158
0500	Skim coat surface with joint compound		1600	.005	S.F.	.03	.18		.21	.33
0510	Prepare, retape and refinish joints		60	.133	L.F.	.64	4.75		5.39	8.55

09 05 Common Work Results for Finishes

09 05 05 – Selective Demolition for Finishes

09 05 05.10 Selective Demolition, Ceilings

Line	Description	Crew	Daily Output	Labor-Hours	Unit	Material	Labor	Equipment	Total	Total Incl O&P
0010	**SELECTIVE DEMOLITION, CEILINGS** R024119-10									
0200	Ceiling, gypsum wall board, furred and nailed or screwed	2 Clab	800	.020	S.F.		.55		.55	.91
1000	Plaster, lime and horse hair, on wood lath, incl. lath		700	.023			.63		.63	1.04
1200	Suspended ceiling, mineral fiber, 2′ x 2′ or 2′ x 4′		1500	.011			.29		.29	.49
1250	On suspension system, incl. system		1200	.013			.37		.37	.61
1500	Tile, wood fiber, 12″ x 12″, glued		900	.018			.49		.49	.81
1540	Stapled		1500	.011			.29		.29	.49
2000	Wood, tongue and groove, 1″ x 4″		1000	.016			.44		.44	.73
2040	1″ x 8″		1100	.015			.40		.40	.66
2400	Plywood or wood fiberboard, 4′ x 8′ sheets		1200	.013			.37		.37	.61
2500	Remove & refinish textured ceiling	1 Plas	222	.036		.03	1.26		1.29	2.09

09 05 05.20 Selective Demolition, Flooring

Line	Description	Crew	Daily Output	Labor-Hours	Unit	Material	Labor	Equipment	Total	Total Incl O&P
0010	**SELECTIVE DEMOLITION, FLOORING** R024119-10									
0200	Brick with mortar	2 Clab	475	.034	S.F.		.93		.93	1.53
0400	Carpet, bonded, including surface scraping		2000	.008			.22		.22	.36
0480	Tackless		9000	.002			.05		.05	.08
0550	Carpet tile, releasable adhesive		5000	.003			.09		.09	.15
0560	Permanent adhesive		1850	.009			.24		.24	.39
0800	Resilient, sheet goods		1400	.011			.31		.31	.52
0850	Vinyl or rubber cove base	1 Clab	1000	.008	L.F.		.22		.22	.36
0860	Vinyl or rubber cove base, molded corner	"	1000	.008	Ea.		.22		.22	.36
0870	For glued and caulked installation, add to labor						50%			
0900	Vinyl composition tile, 12″ x 12″	2 Clab	1000	.016	S.F.		.44		.44	.73
2000	Tile, ceramic, thin set		675	.024			.65		.65	1.08
2020	Mud set		625	.026			.70		.70	1.16
3000	Wood, block, on end	1 Carp	400	.020			.71		.71	1.18
3200	Parquet		450	.018			.63		.63	1.05
3400	Strip flooring, interior, 2-1/4″ x 25/32″ thick		325	.025			.88		.88	1.45
3500	Exterior, porch flooring, 1″ x 4″		220	.036			1.30		1.30	2.14
3800	Subfloor, tongue and groove, 1″ x 6″		325	.025			.88		.88	1.45
3820	1″ x 8″		430	.019			.66		.66	1.10

09 05 Common Work Results for Finishes

09 05 05 – Selective Demolition for Finishes

09 05 05.20 Selective Demolition, Flooring		Crew	Daily Output	Labor-Hours	Unit	Material	2019 Bare Costs Labor	Equipment	Total	Total Incl O&P
3840	1" x 10"	1 Carp	520	.015	S.F.		.55		.55	.91
4000	Plywood, nailed		600	.013			.48		.48	.79
4100	Glued and nailed		400	.020			.71		.71	1.18
4200	Hardboard, 1/4" thick	↓	760	.011	↓		.38		.38	.62
9050	For grinding concrete floors, see Section 03 35 43.10									

09 05 05.30 Selective Demolition, Walls and Partitions		Crew	Daily Output	Labor-Hours	Unit	Material	Labor	Equipment	Total	Total Incl O&P
0010	**SELECTIVE DEMOLITION, WALLS AND PARTITIONS** R024119-10									
0020	Walls, concrete, reinforced	B-39	120	.400	C.F.		11.15	1.96	13.11	20.50
0025	Plain	"	160	.300	"		8.35	1.47	9.82	15.40
1000	Gypsum wallboard, nailed or screwed	1 Clab	1000	.008	S.F.		.22		.22	.36
1010	2 layers		400	.020			.55		.55	.91
1500	Fiberboard, nailed		900	.009			.24		.24	.40
1568	Plenum barrier, sheet lead	↓	300	.027			.73		.73	1.21
2200	Metal or wood studs, finish 2 sides, fiberboard	B-1	520	.046			1.30		1.30	2.15
2250	Lath and plaster		260	.092			2.60		2.60	4.30
2300	Gypsum wallboard		520	.046			1.30		1.30	2.15
2350	Plywood	↓	450	.053			1.50		1.50	2.48
2800	Paneling, 4' x 8' sheets	1 Clab	475	.017			.46		.46	.77
3000	Plaster, lime and horsehair, on wood lath		400	.020			.55		.55	.91
3020	On metal lath		335	.024	↓		.66		.66	1.09
3450	Plaster, interior gypsum, acoustic, or cement		60	.133	S.Y.		3.67		3.67	6.05
3500	Stucco, on masonry		145	.055			1.52		1.52	2.51
3510	Commercial 3-coat		80	.100			2.75		2.75	4.55
3520	Interior stucco		25	.320	↓		8.80		8.80	14.55
3760	Tile, ceramic, on walls, thin set		300	.027	S.F.		.73		.73	1.21
3765	Mud set	↓	250	.032	"		.88		.88	1.45

09 22 Supports for Plaster and Gypsum Board

09 22 03 – Fastening Methods for Finishes

09 22 03.20 Drilling Plaster/Drywall		Crew	Daily Output	Labor-Hours	Unit	Material	Labor	Equipment	Total	Total Incl O&P
0010	**DRILLING PLASTER/DRYWALL**									
1100	Drilling & layout for drywall/plaster walls, up to 1" deep, no anchor									
1200	Holes, 1/4" diameter	1 Carp	150	.053	Ea.	.01	1.90		1.91	3.15
1300	3/8" diameter		140	.057		.01	2.04		2.05	3.38
1400	1/2" diameter		130	.062		.01	2.19		2.20	3.63
1500	3/4" diameter		120	.067		.02	2.38		2.40	3.95
1600	1" diameter		110	.073		.02	2.59		2.61	4.30
1700	1-1/4" diameter		100	.080		.04	2.85		2.89	4.75
1800	1-1/2" diameter	↓	90	.089		.06	3.17		3.23	5.30
1900	For ceiling installations, add				↓		40%			

09 22 13 – Metal Furring

09 22 13.13 Metal Channel Furring		Crew	Daily Output	Labor-Hours	Unit	Material	Labor	Equipment	Total	Total Incl O&P
0010	**METAL CHANNEL FURRING**									
0030	Beams and columns, 7/8" hat channels, galvanized, 12" OC	1 Lath	155	.052	S.F.	.43	1.81		2.24	3.39
0050	16" OC		170	.047		.35	1.65		2	3.04
0070	24" OC		185	.043		.23	1.52		1.75	2.71
0100	Ceilings, on steel, 7/8" hat channels, galvanized, 12" OC		210	.038		.39	1.34		1.73	2.58
0300	16" OC		290	.028		.35	.97		1.32	1.94
0400	24" OC		420	.019		.23	.67		.90	1.34
0600	1-5/8" hat channels, galvanized, 12" OC	↓	190	.042	↓	.53	1.48		2.01	2.96

09 22 Supports for Plaster and Gypsum Board

09 22 13 – Metal Furring

09 22 13.13 Metal Channel Furring		Crew	Daily Output	Labor-Hours	Unit	Material	Labor	Equipment	Total	Total Incl O&P
						2019 Bare Costs				
0700	16″ OC	1 Lath	260	.031	S.F.	.47	1.08		1.55	2.26
0900	24″ OC		390	.021		.32	.72		1.04	1.51
0930	7/8″ hat channels with sound isolation clips, 12″ OC		120	.067		1.70	2.34		4.04	5.65
0940	16″ OC		100	.080		1.28	2.81		4.09	5.90
0950	24″ OC		165	.048		.85	1.70		2.55	3.68
0960	1-5/8″ hat channels, galvanized, 12″ OC		110	.073		1.84	2.56		4.40	6.15
0970	16″ OC		100	.080		1.38	2.81		4.19	6.05
0980	24″ OC		155	.052		.92	1.81		2.73	3.93
1000	Walls, 7/8″ hat channels, galvanized, 12″ OC		235	.034		.39	1.20		1.59	2.36
1200	16″ OC		265	.030		.35	1.06		1.41	2.09
1300	24″ OC		350	.023		.23	.80		1.03	1.55
1500	1-5/8″ hat channels, galvanized, 12″ OC		210	.038		.53	1.34		1.87	2.73
1600	16″ OC		240	.033		.47	1.17		1.64	2.40
1800	24″ OC		305	.026		.32	.92		1.24	1.83
1920	7/8″ hat channels with sound isolation clips, 12″ OC		125	.064		1.70	2.25		3.95	5.50
1940	16″ OC		100	.080		1.28	2.81		4.09	5.90
1950	24″ OC		150	.053		.85	1.87		2.72	3.96
1960	1-5/8″ hat channels, galvanized, 12″ OC		115	.070		1.84	2.45		4.29	5.95
1970	16″ OC		95	.084		1.38	2.96		4.34	6.30
1980	24″ OC		140	.057		.92	2.01		2.93	4.24
3000	Z Furring, walls, 1″ deep, 25 ga., 24″ OC		350	.023		1.36	.80		2.16	2.79
3010	48″ OC		700	.011		.68	.40		1.08	1.40
3020	1-1/2″ deep, 24″ OC		345	.023		1.59	.82		2.41	3.05
3030	48″ OC		695	.012		.79	.40		1.19	1.52
3040	2″ deep, 24″ OC		340	.024		1.91	.83		2.74	3.43
3050	48″ OC		690	.012		.95	.41		1.36	1.71
3060	1″ deep, 20 ga., 24″ OC		350	.023		2.25	.80		3.05	3.76
3070	48″ OC		700	.011		1.12	.40		1.52	1.88
3080	1-1/2″ deep, 24″ OC		345	.023		2.59	.82		3.41	4.15
3090	48″ OC		695	.012		1.29	.40		1.69	2.07
4000	2″ deep, 24″ OC		340	.024		3.18	.83		4.01	4.82
4010	48″ OC		690	.012		1.59	.41		2	2.41

09 22 16 – Non-Structural Metal Framing

09 22 16.13 Non-Structural Metal Stud Framing		Crew	Daily Output	Labor-Hours	Unit	Material	Labor	Equipment	Total	Total Incl O&P
0010	**NON-STRUCTURAL METAL STUD FRAMING**									
1600	Non-load bearing, galv., 8′ high, 25 ga. 1-5/8″ wide, 16″ OC	1 Carp	619	.013	S.F.	.27	.46		.73	1.05
1610	24″ OC		950	.008		.20	.30		.50	.72
1620	2-1/2″ wide, 16″ OC		613	.013		.35	.47		.82	1.16
1630	24″ OC		938	.009		.26	.30		.56	.79
1640	3-5/8″ wide, 16″ OC		600	.013		.40	.48		.88	1.23
1650	24″ OC		925	.009		.30	.31		.61	.84
1660	4″ wide, 16″ OC		594	.013		.44	.48		.92	1.28
1670	24″ OC		925	.009		.33	.31		.64	.87
1680	6″ wide, 16″ OC		588	.014		.55	.49		1.04	1.40
1690	24″ OC		906	.009		.41	.31		.72	.97
1700	20 ga. studs, 1-5/8″ wide, 16″ OC		494	.016		.34	.58		.92	1.33
1710	24″ OC		763	.010		.26	.37		.63	.90
1720	2-1/2″ wide, 16″ OC		488	.016		.43	.58		1.01	1.44
1730	24″ OC		750	.011		.33	.38		.71	.99
1740	3-5/8″ wide, 16″ OC		481	.017		.49	.59		1.08	1.52
1750	24″ OC		738	.011		.37	.39		.76	1.05
1760	4″ wide, 16″ OC		475	.017		.60	.60		1.20	1.65

09 22 16 – Non-Structural Metal Framing

09 22 16.13 Non-Structural Metal Stud Framing		Crew	Daily Output	Labor-Hours	Unit	2019 Bare Costs Material	Labor	Equipment	Total	Total Incl O&P
1770	24" OC	1 Carp	738	.011	S.F.	.45	.39		.84	1.13
1780	6" wide, 16" OC		469	.017		.72	.61		1.33	1.79
1790	24" OC		725	.011		.54	.39		.93	1.24
2000	Non-load bearing, galv., 10' high, 25 ga. 1-5/8" wide, 16" OC		495	.016		.25	.58		.83	1.23
2100	24" OC		760	.011		.19	.38		.57	.82
2200	2-1/2" wide, 16" OC		490	.016		.34	.58		.92	1.33
2250	24" OC		750	.011		.25	.38		.63	.90
2300	3-5/8" wide, 16" OC		480	.017		.37	.59		.96	1.39
2350	24" OC		740	.011		.28	.39		.67	.94
2400	4" wide, 16" OC		475	.017		.42	.60		1.02	1.45
2450	24" OC		740	.011		.31	.39		.70	.98
2500	6" wide, 16" OC		470	.017		.52	.61		1.13	1.57
2550	24" OC		725	.011		.38	.39		.77	1.07
2600	20 ga. studs, 1-5/8" wide, 16" OC		395	.020		.32	.72		1.04	1.55
2650	24" OC		610	.013		.24	.47		.71	1.03
2700	2-1/2" wide, 16" OC		390	.021		.41	.73		1.14	1.66
2750	24" OC		600	.013		.30	.48		.78	1.12
2800	3-5/8" wide, 16" OC		385	.021		.47	.74		1.21	1.74
2850	24" OC		590	.014		.35	.48		.83	1.18
2900	4" wide, 16" OC		380	.021		.57	.75		1.32	1.86
2950	24" OC		590	.014		.42	.48		.90	1.26
3000	6" wide, 16" OC		375	.021		.68	.76		1.44	2.01
3050	24" OC		580	.014		.50	.49		.99	1.36
3060	Non-load bearing, galv., 12' high, 25 ga. 1-5/8" wide, 16" OC		413	.019		.24	.69		.93	1.41
3070	24" OC		633	.013		.18	.45		.63	.93
3080	2-1/2" wide, 16" OC		408	.020		.32	.70		1.02	1.51
3090	24" OC		625	.013		.23	.46		.69	1.01
3100	3-5/8" wide, 16" OC		400	.020		.36	.71		1.07	1.57
3110	24" OC		617	.013		.26	.46		.72	1.05
3120	4" wide, 16" OC		396	.020		.40	.72		1.12	1.63
3130	24" OC		617	.013		.29	.46		.75	1.08
3140	6" wide, 16" OC		392	.020		.50	.73		1.23	1.75
3150	24" OC		604	.013		.36	.47		.83	1.18
3160	20 ga. studs, 1-5/8" wide, 16" OC		329	.024		.31	.87		1.18	1.77
3170	24" OC		508	.016		.23	.56		.79	1.18
3180	2-1/2" wide, 16" OC		325	.025		.39	.88		1.27	1.88
3190	24" OC		500	.016		.28	.57		.85	1.25
3200	3-5/8" wide, 16" OC		321	.025		.45	.89		1.34	1.96
3210	24" OC		492	.016		.33	.58		.91	1.32
3220	4" wide, 16" OC		317	.025		.54	.90		1.44	2.09
3230	24" OC		492	.016		.40	.58		.98	1.39
3240	6" wide, 16" OC		313	.026		.65	.91		1.56	2.23
3250	24" OC		483	.017		.47	.59		1.06	1.50
3260	Non-load bearing, galv., 16' high, 25 ga. 4" wide, 12" OC		195	.041		.51	1.46		1.97	2.98
3270	16" OC		275	.029		.40	1.04		1.44	2.15
3280	24" OC		400	.020		.29	.71		1	1.50
3290	6" wide, 12" OC		190	.042		.64	1.50		2.14	3.18
3300	16" OC		280	.029		.50	1.02		1.52	2.23
3310	24" OC		400	.020		.37	.71		1.08	1.58
3320	20 ga. studs, 2-1/2" wide, 12" OC		180	.044		.50	1.58		2.08	3.17
3330	16" OC		254	.032		.39	1.12		1.51	2.29
3340	24" OC		390	.021		.29	.73		1.02	1.52
3350	3-5/8" wide, 12" OC		170	.047		.57	1.68		2.25	3.40

09 22 Supports for Plaster and Gypsum Board

09 22 16 – Non-Structural Metal Framing

09 22 16.13 Non-Structural Metal Stud Framing		Crew	Daily Output	Labor-Hours	Unit	Material	2019 Bare Costs Labor	Equipment	Total	Total Incl O&P
3360	16" OC	1 Carp	251	.032	S.F.	.45	1.14		1.59	2.38
3370	24" OC		384	.021		.33	.74		1.07	1.59
3380	4" wide, 12" OC		170	.047		.69	1.68		2.37	3.53
3390	16" OC		247	.032		.55	1.15		1.70	2.51
3400	24" OC		384	.021		.40	.74		1.14	1.67
3410	6" wide, 12" OC		175	.046		.83	1.63		2.46	3.61
3420	16" OC		245	.033		.66	1.16		1.82	2.64
3430	24" OC		400	.020		.48	.71		1.19	1.71
3440	Non-load bearing, galv., 20' high, 25 ga. 6" wide, 12" OC		125	.064		.63	2.28		2.91	4.46
3450	16" OC		220	.036		.49	1.30		1.79	2.68
3460	24" OC		360	.022		.35	.79		1.14	1.70
3470	20 ga. studs, 4" wide, 12" OC		120	.067		.68	2.38		3.06	4.68
3480	16" OC		215	.037		.54	1.33		1.87	2.79
3490	6" wide, 12" OC		115	.070		.82	2.48		3.30	5
3500	16" OC		215	.037		.64	1.33		1.97	2.89
3510	24" OC		331	.024		.46	.86		1.32	1.93
5000	For load bearing studs, see Section 05 41 13.30									

09 22 26 – Suspension Systems

09 22 26.13 Ceiling Suspension Systems		Crew	Daily Output	Labor-Hours	Unit	Material	Labor	Equipment	Total	Total Incl O&P
0010	**CEILING SUSPENSION SYSTEMS** for gypsum board or plaster									
8000	Suspended ceilings, including carriers									
8200	1-1/2" carriers, 24" OC with:									
8300	7/8" channels, 16" OC	1 Lath	275	.029	S.F.	.56	1.02		1.58	2.26
8320	24" OC		310	.026		.44	.91		1.35	1.95
8400	1-5/8" channels, 16" OC		205	.039		.68	1.37		2.05	2.96
8420	24" OC		250	.032		.52	1.12		1.64	2.39
8600	2" carriers, 24" OC with:									
8700	7/8" channels, 16" OC	1 Lath	250	.032	S.F.	.66	1.12		1.78	2.54
8720	24" OC		285	.028		.54	.99		1.53	2.19
8800	1-5/8" channels, 16" OC		190	.042		.78	1.48		2.26	3.24
8820	24" OC		225	.036		.63	1.25		1.88	2.70

09 22 36 – Lath

09 22 36.13 Gypsum Lath		Crew	Daily Output	Labor-Hours	Unit	Material	Labor	Equipment	Total	Total Incl O&P
0011	**GYPSUM LATH** Plain or perforated, nailed, 3/8" thick	1 Lath	765	.010	S.F.	.35	.37		.72	.98
0101	1/2" thick, nailed		720	.011		.30	.39		.69	.96
0301	Clipped to steel studs, 3/8" thick		675	.012		.35	.42		.77	1.06
0401	1/2" thick		630	.013		.30	.45		.75	1.05
0601	Firestop gypsum base, to steel studs, 3/8" thick		630	.013		.28	.45		.73	1.03
0701	1/2" thick		585	.014		.32	.48		.80	1.12
0901	Foil back, to steel studs, 3/8" thick		675	.012		.36	.42		.78	1.07
1001	1/2" thick		630	.013		.37	.45		.82	1.13
1501	For ceiling installations, add		1950	.004			.14		.14	.23
1601	For columns and beams, add		1550	.005			.18		.18	.29

09 22 36.23 Metal Lath		Crew	Daily Output	Labor-Hours	Unit	Material	Labor	Equipment	Total	Total Incl O&P
0010	**METAL LATH** R092000-50									
3601	2.5 lb. diamond painted, on wood framing, on walls	1 Lath	765	.010	S.F.	.41	.37		.78	1.04
3701	On ceilings		675	.012		.41	.42		.83	1.12
4201	3.4 lb. diamond painted, wired to steel framing, on walls		675	.012		.45	.42		.87	1.16
4301	On ceilings		540	.015		.45	.52		.97	1.33
5101	Rib lath, painted, wired to steel, on walls, 2.75 lb.		675	.012		.36	.42		.78	1.07
5201	3.4 lb.		630	.013		.48	.45		.93	1.24

09 22 Supports for Plaster and Gypsum Board

09 22 36 – Lath

09 22 36.23 Metal Lath		Crew	Daily Output	Labor-Hours	Unit	Material	Labor	Equipment	Total	Total Incl O&P
						2019 Bare Costs				
5701	Suspended ceiling system, incl. 3.4 lb. diamond lath, painted	1 Lath	135	.059	S.F.	.46	2.08		2.54	3.86
5801	Galvanized	↓	135	.059	↓	.51	2.08		2.59	3.91

09 22 36.83 Accessories, Plaster		Crew	Daily Output	Labor-Hours	Unit	Material	Labor	Equipment	Total	Total Incl O&P
0010	**ACCESSORIES, PLASTER**									
0020	Casing bead, expanded flange, galvanized	1 Lath	2.70	2.963	C.L.F.	53	104		157	226
0200	Foundation weep screed, galvanized	″	2.70	2.963		53	104		157	226
0900	Channels, cold rolled, 16 ga., 3/4″ deep, galvanized					39			39	42.50
1620	Corner bead, expanded bullnose, 3/4″ radius, #10, galvanized	1 Lath	2.60	3.077		26	108		134	203
1650	#1, galvanized		2.55	3.137		48.50	110		158.50	231
1670	Expanded wing, 2-3/4″ wide, #1, galvanized		2.65	3.019		40.50	106		146.50	216
1700	Inside corner (corner rite), 3″ x 3″, painted		2.60	3.077		21	108		129	197
1750	Strip-ex, 4″ wide, painted		2.55	3.137		25	110		135	205
1800	Expansion joint, 3/4″ grounds, limited expansion, galv., 1 piece		2.70	2.963		70.50	104		174.50	246
2100	Extreme expansion, galvanized, 2 piece	↓	2.60	3.077	↓	140	108		248	330

09 23 Gypsum Plastering

09 23 13 – Acoustical Gypsum Plastering

09 23 13.10 Perlite or Vermiculite Plaster		Crew	Daily Output	Labor-Hours	Unit	Material	Labor	Equipment	Total	Total Incl O&P
0010	**PERLITE OR VERMICULITE PLASTER** R092000-50									
0020	In 100 lb. bags, under 200 bags				Bag	18.45			18.45	20.50
0301	2 coats, no lath included, on walls	J-1	830	.048	S.F.	.67	1.59	.16	2.42	3.52
0401	On ceilings		710	.056		.67	1.86	.19	2.72	3.98
0901	3 coats, no lath included, on walls		665	.060		.72	1.98	.20	2.90	4.25
1001	On ceilings	↓	565	.071	↓	.72	2.33	.24	3.29	4.86
1700	For irregular or curved surfaces, add to above				S.Y.		30%			
1800	For columns and beams, add to above						50%			
1900	For soffits, add to ceiling prices				↓		40%			

09 23 20 – Gypsum Plaster

09 23 20.10 Gypsum Plaster On Walls and Ceilings		Crew	Daily Output	Labor-Hours	Unit	Material	Labor	Equipment	Total	Total Incl O&P
0010	**GYPSUM PLASTER ON WALLS AND CEILINGS** R092000-50									
0020	80# bag, less than 1 ton				Bag	15.70			15.70	17.30
0302	2 coats, no lath included, on walls	J-1	750	.053	S.F.	.42	1.76	.18	2.36	3.53
0402	On ceilings		660	.061		.42	2	.20	2.62	3.95
0903	3 coats, no lath included, on walls		620	.065		.60	2.12	.22	2.94	4.38
1002	On ceilings	↓	560	.071	↓	1.05	2.35	.24	3.64	5.30
1600	For irregular or curved surfaces, add						30%			
1800	For columns & beams, add						50%			

09 24 Cement Plastering

09 24 23 – Cement Stucco

09 24 23.40 Stucco		Crew	Daily Output	Labor-Hours	Unit	Material	Labor	Equipment	Total	Total Incl O&P
0010	**STUCCO** R092000-50									
0011	3 coats 1″ thick, float finish, with mesh, on wood frame	J-2	470	.102	S.F.	1.10	3.40	.29	4.79	7.10
0101	On masonry construction	J-1	495	.081		.79	2.66	.27	3.72	5.50
0151	2 coats, 5/8″ thick, float finish, no lath incl.	″	980	.041		.38	1.34	.14	1.86	2.77
0301	For trowel finish, add	1 Plas	1530	.005	↓		.18		.18	.30
0600	For coloring, add	J-1	685	.058	S.Y.	.42	1.92	.20	2.54	3.82
0700	For special texture, add	↓	200	.200	″	1.47	6.60	.68	8.75	13.10

09 24 Cement Plastering

09 24 23 – Cement Stucco

09 24 23.40 Stucco		Crew	Daily Output	Labor-Hours	Unit	Material	Labor	Equipment	Total	Total Incl O&P
						2019 Bare Costs				
1001	Stucco, with bonding agent, 3 coats, on walls	J-1	1800	.022	S.F.	.52	.73	.08	1.33	1.85
1201	Ceilings		1620	.025		.38	.81	.08	1.27	1.84
1301	Beams		720	.056		.38	1.83	.19	2.40	3.62
1501	Columns		900	.044		.38	1.46	.15	1.99	2.98
1601	Mesh, galvanized, nailed to wood, 1.8 lb.	1 Lath	540	.015		.82	.52		1.34	1.74
1801	3.6 lb.		495	.016		.47	.57		1.04	1.43
1901	Wired to steel, galvanized, 1.8 lb.		477	.017		.82	.59		1.41	1.85
2101	3.6 lb.		450	.018		.47	.63		1.10	1.53

09 25 Other Plastering

09 25 23 – Lime Based Plastering

09 25 23.10 Venetian Plaster		Crew	Daily Output	Labor-Hours	Unit	Material	Labor	Equipment	Total	Total Incl O&P
0010	**VENETIAN PLASTER**									
0100	Walls, 1 coat primer, roller applied	1 Plas	950	.008	S.F.	.17	.29		.46	.67
0210	For pigment, light colors add per S.F. plaster					.02			.02	.02
0220	For pigment, dark colors add per S.F. plaster					.04			.04	.04
0300	For sealer/wax coat incl. burnishing, add	1 Plas	300	.027		.38	.93		1.31	1.93

09 26 Veneer Plastering

09 26 13 – Gypsum Veneer Plastering

09 26 13.20 Blueboard		Crew	Daily Output	Labor-Hours	Unit	Material	Labor	Equipment	Total	Total Incl O&P
0010	**BLUEBOARD** For use with thin coat									
0100	plaster application see Section 09 26 13.80									
1000	3/8" thick, on walls or ceilings, standard, no finish included	2 Carp	1900	.008	S.F.	.35	.30		.65	.89
1100	With thin coat plaster finish		875	.018		.46	.65		1.11	1.59
1400	On beams, columns, or soffits, standard, no finish included		675	.024		.40	.84		1.24	1.84
1450	With thin coat plaster finish		475	.034		.52	1.20		1.72	2.55
3000	1/2" thick, on walls or ceilings, standard, no finish included		1900	.008		.35	.30		.65	.89
3100	With thin coat plaster finish		875	.018		.46	.65		1.11	1.59
3300	Fire resistant, no finish included		1900	.008		.35	.30		.65	.89
3400	With thin coat plaster finish		875	.018		.46	.65		1.11	1.59
3450	On beams, columns, or soffits, standard, no finish included		675	.024		.40	.84		1.24	1.84
3500	With thin coat plaster finish		475	.034		.52	1.20		1.72	2.55
3700	Fire resistant, no finish included		675	.024		.40	.84		1.24	1.84
3800	With thin coat plaster finish		475	.034		.52	1.20		1.72	2.55
5000	5/8" thick, on walls or ceilings, fire resistant, no finish included		1900	.008		.36	.30		.66	.90
5100	With thin coat plaster finish		875	.018		.47	.65		1.12	1.60
5500	On beams, columns, or soffits, no finish included		675	.024		.41	.84		1.25	1.86
5600	With thin coat plaster finish		475	.034		.53	1.20		1.73	2.56
6000	For high ceilings, over 8' high, add		3060	.005			.19		.19	.31
6500	For distribution costs 3 stories and above, add per story		6100	.003			.09		.09	.15

09 26 13.80 Thin Coat Plaster		Crew	Daily Output	Labor-Hours	Unit	Material	Labor	Equipment	Total	Total Incl O&P
0010	**THIN COAT PLASTER** R092000-50									
0012	1 coat veneer, not incl. lath	J-1	3600	.011	S.F.	.11	.37	.04	.52	.77
1000	In 50 lb. bags				Bag	15.35			15.35	16.90

09 28 Backing Boards and Underlayments

09 28 13 – Cementitious Backing Boards

09 28 13.10 Cementitious Backerboard		Crew	Daily Output	Labor-Hours	Unit	Material	2019 Bare Costs Labor	Equipment	Total	Total Incl O&P
0010	**CEMENTITIOUS BACKERBOARD**									
0070	Cementitious backerboard, on floor, 3′ x 4′ x 1/2″ sheets	2 Carp	525	.030	S.F.	.93	1.09		2.02	2.83
0080	3′ x 5′ x 1/2″ sheets		525	.030		.80	1.09		1.89	2.68
0090	3′ x 6′ x 1/2″ sheets		525	.030		.79	1.09		1.88	2.67
0100	3′ x 4′ x 5/8″ sheets		525	.030		.98	1.09		2.07	2.88
0110	3′ x 5′ x 5/8″ sheets		525	.030		1	1.09		2.09	2.90
0120	3′ x 6′ x 5/8″ sheets		525	.030		.98	1.09		2.07	2.88
0150	On wall, 3′ x 4′ x 1/2″ sheets		350	.046		.93	1.63		2.56	3.72
0160	3′ x 5′ x 1/2″ sheets		350	.046		.80	1.63		2.43	3.57
0170	3′ x 6′ x 1/2″ sheets		350	.046		.79	1.63		2.42	3.56
0180	3′ x 4′ x 5/8″ sheets		350	.046		.98	1.63		2.61	3.77
0190	3′ x 5′ x 5/8″ sheets		350	.046		1	1.63		2.63	3.79
0200	3′ x 6′ x 5/8″ sheets		350	.046		.98	1.63		2.61	3.77
0250	On counter, 3′ x 4′ x 1/2″ sheets		180	.089		.93	3.17		4.10	6.30
0260	3′ x 5′ x 1/2″ sheets		180	.089		.80	3.17		3.97	6.15
0270	3′ x 6′ x 1/2″ sheets		180	.089		.79	3.17		3.96	6.10
0300	3′ x 4′ x 5/8″ sheets		180	.089		.98	3.17		4.15	6.35
0310	3′ x 5′ x 5/8″ sheets		180	.089		1	3.17		4.17	6.35
0320	3′ x 6′ x 5/8″ sheets	↓	180	.089	↓	.98	3.17		4.15	6.35

09 29 Gypsum Board

09 29 10 – Gypsum Board Panels

09 29 10.20 Taping and Finishing		Crew	Daily Output	Labor-Hours	Unit	Material	Labor	Equipment	Total	Total Incl O&P
0010	**TAPING AND FINISHING**									
3600	For taping and finishing joints, add	2 Carp	2000	.008	S.F.	.05	.29		.34	.52
4500	For thin coat plaster instead of taping, add	J-1	3600	.011	″	.11	.37	.04	.52	.77

09 29 10.30 Gypsum Board		Crew	Daily Output	Labor-Hours	Unit	Material	Labor	Equipment	Total	Total Incl O&P
0010	**GYPSUM BOARD** on walls & ceilings R092910-10									
0100	Nailed or screwed to studs unless otherwise noted									
0110	1/4″ thick, on walls or ceilings, standard, no finish included	2 Carp	1330	.012	S.F.	.40	.43		.83	1.15
0115	1/4″ thick, on walls or ceilings, flexible, no finish included		1050	.015		.54	.54		1.08	1.49
0117	1/4″ thick, on columns or soffits, flexible, no finish included		1050	.015		.54	.54		1.08	1.49
0130	1/4″ thick, standard, no finish included, less than 800 S.F.		510	.031		.40	1.12		1.52	2.29
0150	3/8″ thick, on walls, standard, no finish included		2000	.008		.36	.29		.65	.87
0200	On ceilings, standard, no finish included		1800	.009		.36	.32		.68	.92
0250	On beams, columns, or soffits, no finish included		675	.024		.36	.84		1.20	1.80
0300	1/2″ thick, on walls, standard, no finish included		2000	.008		.34	.29		.63	.84
0350	Taped and finished (level 4 finish)		965	.017		.39	.59		.98	1.41
0390	With compound skim coat (level 5 finish)		775	.021		.44	.74		1.18	1.70
0400	Fire resistant, no finish included		2000	.008		.37	.29		.66	.88
0450	Taped and finished (level 4 finish)		965	.017		.42	.59		1.01	1.44
0490	With compound skim coat (level 5 finish)		775	.021		.47	.74		1.21	1.74
0500	Water resistant, no finish included		2000	.008		.42	.29		.71	.93
0550	Taped and finished (level 4 finish)		965	.017		.47	.59		1.06	1.49
0590	With compound skim coat (level 5 finish)		775	.021		.52	.74		1.26	1.79
0600	Prefinished, vinyl, clipped to studs		900	.018		.51	.63		1.14	1.61
0700	Mold resistant, no finish included		2000	.008		.43	.29		.72	.94
0710	Taped and finished (level 4 finish)		965	.017		.48	.59		1.07	1.50
0720	With compound skim coat (level 5 finish)		775	.021		.53	.74		1.27	1.80
1000	On ceilings, standard, no finish included		1800	.009		.34	.32		.66	.89
1050	Taped and finished (level 4 finish)	↓	765	.021	↓	.39	.75		1.14	1.66

09 29 Gypsum Board

09 29 10 – Gypsum Board Panels

09 29 10.30 Gypsum Board		Crew	Daily Output	Labor-Hours	Unit	2019 Bare Costs Material	Labor	Equipment	Total	Total Incl O&P
1090	With compound skim coat (level 5 finish)	2 Carp	610	.026	S.F.	.44	.94		1.38	2.02
1100	Fire resistant, no finish included		1800	.009		.37	.32		.69	.93
1150	Taped and finished (level 4 finish)		765	.021		.42	.75		1.17	1.69
1195	With compound skim coat (level 5 finish)		610	.026		.47	.94		1.41	2.06
1200	Water resistant, no finish included		1800	.009		.42	.32		.74	.98
1250	Taped and finished (level 4 finish)		765	.021		.47	.75		1.22	1.74
1290	With compound skim coat (level 5 finish)		610	.026		.52	.94		1.46	2.11
1310	Mold resistant, no finish included		1800	.009		.43	.32		.75	.99
1320	Taped and finished (level 4 finish)		765	.021		.48	.75		1.23	1.75
1330	With compound skim coat (level 5 finish)		610	.026		.53	.94		1.47	2.12
1350	Sag resistant, no finish included		1600	.010		.37	.36		.73	1
1360	Taped and finished (level 4 finish)		765	.021		.42	.75		1.17	1.69
1370	With compound skim coat (level 5 finish)		610	.026		.47	.94		1.41	2.06
1500	On beams, columns, or soffits, standard, no finish included		675	.024		.39	.84		1.23	1.83
1550	Taped and finished (level 4 finish)		540	.030		.39	1.06		1.45	2.18
1590	With compound skim coat (level 5 finish)		475	.034		.44	1.20		1.64	2.46
1600	Fire resistant, no finish included		675	.024		.37	.84		1.21	1.81
1650	Taped and finished (level 4 finish)		540	.030		.42	1.06		1.48	2.21
1690	With compound skim coat (level 5 finish)		475	.034		.47	1.20		1.67	2.50
1700	Water resistant, no finish included		675	.024		.48	.84		1.32	1.93
1750	Taped and finished (level 4 finish)		540	.030		.47	1.06		1.53	2.26
1790	With compound skim coat (level 5 finish)		475	.034		.52	1.20		1.72	2.55
1800	Mold resistant, no finish included		675	.024		.49	.84		1.33	1.94
1810	Taped and finished (level 4 finish)		540	.030		.48	1.06		1.54	2.27
1820	With compound skim coat (level 5 finish)		475	.034		.53	1.20		1.73	2.56
1850	Sag resistant, no finish included		675	.024		.43	.84		1.27	1.87
1860	Taped and finished (level 4 finish)		540	.030		.42	1.06		1.48	2.21
1870	With compound skim coat (level 5 finish)		475	.034		.47	1.20		1.67	2.50
2000	5/8" thick, on walls, standard, no finish included		2000	.008		.35	.29		.64	.86
2050	Taped and finished (level 4 finish)		965	.017		.40	.59		.99	1.42
2090	With compound skim coat (level 5 finish)		775	.021		.45	.74		1.19	1.71
2100	Fire resistant, no finish included		2000	.008		.36	.29		.65	.87
2150	Taped and finished (level 4 finish)		965	.017		.41	.59		1	1.43
2195	With compound skim coat (level 5 finish)		775	.021		.46	.74		1.20	1.73
2200	Water resistant, no finish included		2000	.008		.50	.29		.79	1.02
2250	Taped and finished (level 4 finish)		965	.017		.55	.59		1.14	1.58
2290	With compound skim coat (level 5 finish)		775	.021		.60	.74		1.34	1.88
2300	Prefinished, vinyl, clipped to studs		900	.018		.83	.63		1.46	1.96
2510	Mold resistant, no finish included		2000	.008		.49	.29		.78	1.01
2520	Taped and finished (level 4 finish)		965	.017		.54	.59		1.13	1.57
2530	With compound skim coat (level 5 finish)		775	.021		.59	.74		1.33	1.87
3000	On ceilings, standard, no finish included		1800	.009		.35	.32		.67	.91
3050	Taped and finished (level 4 finish)		765	.021		.40	.75		1.15	1.67
3090	With compound skim coat (level 5 finish)		615	.026		.45	.93		1.38	2.02
3100	Fire resistant, no finish included		1800	.009		.36	.32		.68	.92
3150	Taped and finished (level 4 finish)		765	.021		.41	.75		1.16	1.68
3190	With compound skim coat (level 5 finish)		615	.026		.46	.93		1.39	2.04
3200	Water resistant, no finish included		1800	.009		.50	.32		.82	1.07
3250	Taped and finished (level 4 finish)		765	.021		.55	.75		1.30	1.83
3290	With compound skim coat (level 5 finish)		615	.026		.60	.93		1.53	2.19
3300	Mold resistant, no finish included		1800	.009		.49	.32		.81	1.06
3310	Taped and finished (level 4 finish)		765	.021		.54	.75		1.29	1.82
3320	With compound skim coat (level 5 finish)		615	.026		.59	.93		1.52	2.18

09 29 10 – Gypsum Board Panels

09 29 10.30 Gypsum Board		Crew	Daily Output	Labor-Hours	Unit	Material	Labor	Equipment	Total	Total Incl O&P
						2019 Bare Costs				
3500	On beams, columns, or soffits, no finish included	2 Carp	675	.024	S.F.	.40	.84		1.24	1.84
3550	Taped and finished (level 4 finish)		475	.034		.46	1.20		1.66	2.48
3590	With compound skim coat (level 5 finish)		380	.042		.52	1.50		2.02	3.05
3600	Fire resistant, no finish included		675	.024		.41	.84		1.25	1.86
3650	Taped and finished (level 4 finish)		475	.034		.47	1.20		1.67	2.50
3690	With compound skim coat (level 5 finish)		380	.042		.46	1.50		1.96	2.99
3700	Water resistant, no finish included		675	.024		.58	.84		1.42	2.03
3750	Taped and finished (level 4 finish)		475	.034		.60	1.20		1.80	2.64
3790	With compound skim coat (level 5 finish)		380	.042		.63	1.50		2.13	3.17
3800	Mold resistant, no finish included		675	.024		.56	.84		1.40	2.02
3810	Taped and finished (level 4 finish)		475	.034		.59	1.20		1.79	2.63
3820	With compound skim coat (level 5 finish)		380	.042		.62	1.50		2.12	3.16
4000	Fireproofing, beams or columns, 2 layers, 1/2" thick, incl finish		330	.048		.83	1.73		2.56	3.78
4010	Mold resistant		330	.048		.95	1.73		2.68	3.91
4050	5/8" thick		300	.053		.81	1.90		2.71	4.04
4060	Mold resistant		300	.053		1.07	1.90		2.97	4.32
4100	3 layers, 1/2" thick		225	.071		1.25	2.54		3.79	5.55
4110	Mold resistant		225	.071		1.43	2.54		3.97	5.75
4150	5/8" thick		210	.076		1.22	2.72		3.94	5.85
4160	Mold resistant		210	.076		1.61	2.72		4.33	6.25
5200	For work over 8' high, add		3060	.005			.19		.19	.31
5270	For textured spray, add	2 Lath	1600	.010		.04	.35		.39	.61
5350	For finishing inner corners, add	2 Carp	950	.017	L.F.	.10	.60		.70	1.10
5355	For finishing outer corners, add	"	1250	.013		.23	.46		.69	1.01
5500	For acoustical sealant, add per bead	1 Carp	500	.016		.04	.57		.61	.99
5550	Sealant, 1 quart tube				Ea.	7.10			7.10	7.80
6000	Gypsum sound dampening panels									
6010	1/2" thick on walls, multi-layer, lightweight, no finish included	2 Carp	1500	.011	S.F.	2.22	.38		2.60	3.07
6015	Taped and finished (level 4 finish)		725	.022		2.27	.79		3.06	3.79
6020	With compound skim coat (level 5 finish)		580	.028		2.32	.98		3.30	4.18
6025	5/8" thick on walls, for wood studs, no finish included		1500	.011		2.24	.38		2.62	3.09
6030	Taped and finished (level 4 finish)		725	.022		2.29	.79		3.08	3.82
6035	With compound skim coat (level 5 finish)		580	.028		2.34	.98		3.32	4.20
6040	For metal stud, no finish included		1500	.011		2.31	.38		2.69	3.17
6045	Taped and finished (level 4 finish)		725	.022		2.36	.79		3.15	3.89
6050	With compound skim coat (level 5 finish)		580	.028		2.41	.98		3.39	4.28
6055	Abuse resist, no finish included		1500	.011		4.06	.38		4.44	5.10
6060	Taped and finished (level 4 finish)		725	.022		4.11	.79		4.90	5.80
6065	With compound skim coat (level 5 finish)		580	.028		4.16	.98		5.14	6.20
6070	Shear rated, no finish included		1500	.011		5.35	.38		5.73	6.55
6075	Taped and finished (level 4 finish)		725	.022		5.40	.79		6.19	7.25
6080	With compound skim coat (level 5 finish)		580	.028		5.45	.98		6.43	7.65
6085	For SCIF applications, no finish included		1500	.011		5.15	.38		5.53	6.30
6090	Taped and finished (level 4 finish)		725	.022		5.20	.79		5.99	7
6095	With compound skim coat (level 5 finish)		580	.028		5.25	.98		6.23	7.40
6100	1-3/8" thick on walls, THX certified, no finish included		1500	.011		9.30	.38		9.68	10.90
6105	Taped and finished (level 4 finish)		725	.022		9.35	.79		10.14	11.60
6110	With compound skim coat (level 5 finish)		580	.028		9.40	.98		10.38	12
6115	5/8" thick on walls, score & snap installation, no finish included		2000	.008		1.96	.29		2.25	2.63
6120	Taped and finished (level 4 finish)		965	.017		2.01	.59		2.60	3.19
6125	With compound skim coat (level 5 finish)		775	.021		2.06	.74		2.80	3.49
7020	5/8" thick on ceilings, for wood joists, no finish included		1200	.013		2.24	.48		2.72	3.25
7025	Taped and finished (level 4 finish)		510	.031		2.29	1.12		3.41	4.37

09 29 Gypsum Board

09 29 10 – Gypsum Board Panels

09 29 10.30 Gypsum Board		Crew	Daily Output	Labor-Hours	Unit	2019 Bare Costs Material	Labor	Equipment	Total	Total Incl O&P
7030	With compound skim coat (level 5 finish)	2 Carp	410	.039	S.F.	2.34	1.39		3.73	4.87
7035	For metal joists, no finish included		1200	.013		2.31	.48		2.79	3.33
7040	Taped and finished (level 4 finish)		510	.031		2.36	1.12		3.48	4.44
7045	With compound skim coat (level 5 finish)		410	.039		2.41	1.39		3.80	4.95
7050	Abuse resist, no finish included		1200	.013		4.06	.48		4.54	5.25
7055	Taped and finished (level 4 finish)		510	.031		4.11	1.12		5.23	6.35
7060	With compound skim coat (level 5 finish)		410	.039		4.16	1.39		5.55	6.90
7065	Shear rated, no finish included		1200	.013		5.35	.48		5.83	6.70
7070	Taped and finished (level 4 finish)		510	.031		5.40	1.12		6.52	7.80
7075	With compound skim coat (level 5 finish)		410	.039		5.45	1.39		6.84	8.30
7080	For SCIF applications, no finish included		1200	.013		5.15	.48		5.63	6.45
7085	Taped and finished (level 4 finish)		510	.031		5.20	1.12		6.32	7.55
7090	With compound skim coat (level 5 finish)		410	.039		5.25	1.39		6.64	8.05
8010	5/8" thick on ceilings, score & snap installation, no finish included		1600	.010		1.96	.36		2.32	2.75
8015	Taped and finished (level 4 finish)		680	.024		2.01	.84		2.85	3.60
8020	With compound skim coat (level 5 finish)		545	.029		2.06	1.05		3.11	4

09 29 15 – Gypsum Board Accessories

09 29 15.10 Accessories, Drywall		Crew	Daily Output	Labor-Hours	Unit	Material	Labor	Equipment	Total	Total Incl O&P
0011	**ACCESSORIES, DRYWALL** Casing bead, galvanized steel	1 Carp	290	.028	L.F.	.24	.98		1.22	1.90
0101	Vinyl		290	.028		.23	.98		1.21	1.88
0401	Corner bead, galvanized steel, 1-1/4" x 1-1/4"		350	.023		.17	.82		.99	1.54
0411	1-1/4" x 1-1/4", 10' long		35	.229	Ea.	1.69	8.15		9.84	15.30
0601	Vinyl corner bead		400	.020	L.F.	.20	.71		.91	1.40
0901	Furring channel, galv. steel, 7/8" deep, standard		260	.031		.35	1.10		1.45	2.19
1001	Resilient		260	.031		.27	1.10		1.37	2.10
1101	J trim, galvanized steel, 1/2" wide		300	.027			.95		.95	1.57
1121	5/8" wide		300	.027		.31	.95		1.26	1.91
1160	Screws #6 x 1" A				M	11			11	12.10
1170	#6 x 1-5/8" A				"	14.60			14.60	16.05
1501	Z stud, galvanized steel, 1-1/2" wide	1 Carp	260	.031	L.F.	.43	1.10		1.53	2.28

09 30 Tiling

09 30 13 – Ceramic Tiling

09 30 13.45 Ceramic Tile Accessories		Crew	Daily Output	Labor-Hours	Unit	Material	Labor	Equipment	Total	Total Incl O&P
0010	**CERAMIC TILE ACCESSORIES**									
0100	Spacers, 1/8"				C	1.99			1.99	2.19
1310	Sealer for natural stone tile, installed	1 Tilf	650	.012	S.F.	.05	.42		.47	.74

09 30 29 – Metal Tiling

09 30 29.10 Metal Tile		Crew	Daily Output	Labor-Hours	Unit	Material	Labor	Equipment	Total	Total Incl O&P
0010	**METAL TILE** 4' x 4' sheet, 24 ga., tile pattern, nailed									
0200	Stainless steel	2 Carp	512	.031	S.F.	28	1.11		29.11	32.50
0400	Aluminized steel	"	512	.031	"	20.50	1.11		21.61	24.50

09 30 95 – Tile & Stone Setting Materials and Specialties

09 30 95.10 Moisture Resistant, Anti-Fracture Membrane		Crew	Daily Output	Labor-Hours	Unit	Material	Labor	Equipment	Total	Total Incl O&P
0010	**MOISTURE RESISTANT, ANTI-FRACTURE MEMBRANE**									
0200	Elastomeric membrane, 1/16" thick	D-7	275	.058	S.F.	1.13	1.77		2.90	4.10

09 31 Thin-Set Tiling

09 31 13 – Thin-Set Ceramic Tiling

09 31 13.10 Thin-Set Ceramic Tile		Crew	Daily Output	Labor-Hours	Unit	Material	2019 Bare Costs Labor	Equipment	Total	Total Incl O&P
0010	**THIN-SET CERAMIC TILE**									
0020	Backsplash, average grade tiles	1 Tilf	50	.160	S.F.	2.77	5.50		8.27	11.95
0022	Custom grade tiles		50	.160		5.55	5.50		11.05	15
0024	Luxury grade tiles		50	.160		11.10	5.50		16.60	21
0026	Economy grade tiles		50	.160		2.54	5.50		8.04	11.70
0100	Base, using 1′ x 4″ high piece with 1″ x 1″ tiles	D-7	128	.125	L.F.	5	3.81		8.81	11.65
0300	For 6″ high base, 1″ x 1″ tile face, add					1.23			1.23	1.36
0400	For 2″ x 2″ tile face, add to above					.72			.72	.79
0700	Cove base, 4-1/4″ x 4-1/4″	D-7	128	.125		4.21	3.81		8.02	10.80
1000	6″ x 4-1/4″ high		137	.117		4.26	3.56		7.82	10.45
1300	Sanitary cove base, 6″ x 4-1/4″ high		124	.129		4.68	3.93		8.61	11.50
1600	6″ x 6″ high		117	.137		5.35	4.16		9.51	12.60
1800	Bathroom accessories, average (soap dish, toothbrush holder)		82	.195	Ea.	9.70	5.95		15.65	20.50
1900	Bathtub, 5′, rec. 4-1/4″ x 4-1/4″ tile wainscot, adhesive set 6′ high		2.90	5.517		173	168		341	460
2100	7′ high wainscot		2.50	6.400		202	195		397	535
2200	8′ high wainscot		2.20	7.273		231	221		452	610
2500	Bullnose trim, 4-1/4″ x 4-1/4″		128	.125	L.F.	4.24	3.81		8.05	10.80
2800	2″ x 6″		124	.129	″	4.25	3.93		8.18	11.05
3300	Ceramic tile, porcelain type, 1 color, color group 2, 1″ x 1″		183	.087	S.F.	6.50	2.66		9.16	11.45
3310	2″ x 2″ or 2″ x 1″		190	.084		6.35	2.56		8.91	11.15
3350	For random blend, 2 colors, add					1.01			1.01	1.11
3360	4 colors, add					1.51			1.51	1.66
4300	Specialty tile, 4-1/4″ x 4-1/4″ x 1/2″, decorator finish	D-7	183	.087		12.80	2.66		15.46	18.40
4500	Add for epoxy grout, 1/16″ joint, 1″ x 1″ tile		800	.020		.69	.61		1.30	1.74
4600	2″ x 2″ tile		820	.020		.63	.59		1.22	1.65
4610	Add for epoxy grout, 1/8″ joint, 8″ x 8″ x 3/8″ tile, add		900	.018		1.42	.54		1.96	2.43
4800	Pregrouted sheets, walls, 4-1/4″ x 4-1/4″, 6″ x 4-1/4″									
4810	and 8-1/2″ x 4-1/4″, 4 S.F. sheets, silicone grout	D-7	240	.067	S.F.	5.70	2.03		7.73	9.55
5100	Floors, unglazed, 2 S.F. sheets,									
5110	urethane adhesive	D-7	180	.089	S.F.	2.22	2.71		4.93	6.80
5400	Walls, interior, 4-1/4″ x 4-1/4″ tile		190	.084		2.59	2.56		5.15	7
5500	6″ x 4-1/4″ tile		190	.084		3.17	2.56		5.73	7.65
5700	8-1/2″ x 4-1/4″ tile		190	.084		5.70	2.56		8.26	10.45
5800	6″ x 6″ tile		175	.091		3.63	2.78		6.41	8.50
5810	8″ x 8″ tile		170	.094		5.20	2.87		8.07	10.35
5820	12″ x 12″ tile		160	.100		4.69	3.05		7.74	10.05
5830	16″ x 16″ tile		150	.107		5.15	3.25		8.40	10.90
6000	Decorated wall tile, 4-1/4″ x 4-1/4″, color group 1		270	.059		3.63	1.80		5.43	6.90
6100	Color group 4		180	.089		53.50	2.71		56.21	63.50
9300	Ceramic tiles, recycled glass, standard colors, 2″ x 2″ thru 6″ x 6″ G		190	.084		22.50	2.56		25.06	28.50
9310	6″ x 6″ G		175	.091		22.50	2.78		25.28	29
9320	8″ x 8″ G		170	.094		23.50	2.87		26.37	30.50
9330	12″ x 12″ G		160	.100		23.50	3.05		26.55	31
9340	Earthtones, 2″ x 2″ to 4″ x 8″ G		190	.084		26	2.56		28.56	33
9350	6″ x 6″ G		175	.091		26	2.78		28.78	33.50
9360	8″ x 8″ G		170	.094		27	2.87		29.87	34.50
9370	12″ x 12″ G		160	.100		27	3.05		30.05	35
9380	Deep colors, 2″ x 2″ to 4″ x 8″ G		190	.084		31	2.56		33.56	38
9390	6″ x 6″ G		175	.091		31	2.78		33.78	38.50
9400	8″ x 8″ G		170	.094		32.50	2.87		35.37	40
9410	12″ x 12″ G		160	.100		32.50	3.05		35.55	40.50

09 31 Thin-Set Tiling

09 31 33 – Thin-Set Stone Tiling

09 31 33.10 Tiling, Thin-Set Stone		Crew	Daily Output	Labor-Hours	Unit	Material	2019 Bare Costs Labor	Equipment	Total	Total Incl O&P
0010	**TILING, THIN-SET STONE**									
3000	Floors, natural clay, random or uniform, color group 1	D-7	183	.087	S.F.	4.63	2.66		7.29	9.40
3100	Color group 2		183	.087		5.90	2.66		8.56	10.80
3255	Floors, glazed, 6″ x 6″, color group 1		300	.053		5.30	1.62		6.92	8.45
3260	8″ x 8″ tile		300	.053		5.30	1.62		6.92	8.40
3270	12″ x 12″ tile		290	.055		6.75	1.68		8.43	10.10
3280	16″ x 16″ tile		280	.057		7.40	1.74		9.14	10.95
3281	18″ x 18″ tile		270	.059		7.25	1.80		9.05	10.90
3282	20″ x 20″ tile		260	.062		7.15	1.87		9.02	10.85
3283	24″ x 24″ tile		250	.064		9.10	1.95		11.05	13.20
3285	Border, 6″ x 12″ tile		200	.080		10.95	2.44		13.39	16
3290	3″ x 12″ tile		200	.080		12.60	2.44		15.04	17.85

09 32 Mortar-Bed Tiling

09 32 13 – Mortar-Bed Ceramic Tiling

09 32 13.10 Ceramic Tile

Line	Description	Crew	Daily Output	Labor-Hours	Unit	Material	Labor	Equipment	Total	Total Incl O&P
0010	**CERAMIC TILE**									
0050	Base, using 1′ x 4″ high pc. with 1″ x 1″ tiles	D-7	82	.195	L.F.	5.30	5.95		11.25	15.40
0600	Cove base, 4-1/4″ x 4-1/4″ high		91	.176		4.35	5.35		9.70	13.45
0900	6″ x 4-1/4″ high		100	.160		4.40	4.87		9.27	12.70
1200	Sanitary cove base, 6″ x 4-1/4″ high		93	.172		4.82	5.25		10.07	13.75
1500	6″ x 6″ high		84	.190		5.50	5.80		11.30	15.40
2400	Bullnose trim, 4-1/4″ x 4-1/4″		82	.195		4.34	5.95		10.29	14.35
2700	2″ x 6″ bullnose trim		84	.190		4.33	5.80		10.13	14.10
6210	Wall tile, 4-1/4″ x 4-1/4″, better grade	1 Tilf	50	.160	S.F.	9.55	5.50		15.05	19.40
6240	2″ x 2″		50	.160		7.55	5.50		13.05	17.20
6250	6″ x 6″		55	.145		10.20	5		15.20	19.30
6260	8″ x 8″		60	.133		9.45	4.59		14.04	17.80
6600	Crystalline glazed, 4-1/4″ x 4-1/4″, plain	D-7	100	.160		4.48	4.87		9.35	12.80
6700	4-1/4″ x 4-1/4″, scored tile		100	.160		6.55	4.87		11.42	15.10
6900	6″ x 6″ plain		93	.172		5.80	5.25		11.05	14.80
7000	For epoxy grout, 1/16″ joints, 4-1/4″ tile, add		800	.020		.41	.61		1.02	1.43
7200	For tile set in dry mortar, add		1735	.009			.28		.28	.45
7300	For tile set in Portland cement mortar, add		290	.055		.18	1.68		1.86	2.91

09 32 16 – Mortar-Bed Quarry Tiling

09 32 16.10 Quarry Tile

Line	Description	Crew	Daily Output	Labor-Hours	Unit	Material	Labor	Equipment	Total	Total Incl O&P
0010	**QUARRY TILE**									
0100	Base, cove or sanitary, to 5″ high, 1/2″ thick	D-7	110	.145	L.F.	6.40	4.43		10.83	14.20
0300	Bullnose trim, red, 6″ x 6″ x 1/2″ thick		120	.133		5.40	4.06		9.46	12.50
0400	4″ x 4″ x 1/2″ thick		110	.145		4.97	4.43		9.40	12.60
0600	4″ x 8″ x 1/2″ thick, using 8″ as edge		130	.123		5.40	3.75		9.15	12
0700	Floors, 1,000 S.F. lots, red, 4″ x 4″ x 1/2″ thick		120	.133	S.F.	8.80	4.06		12.86	16.25
0900	6″ x 6″ x 1/2″ thick		140	.114		8.25	3.48		11.73	14.70
1000	4″ x 8″ x 1/2″ thick		130	.123		6.80	3.75		10.55	13.50
1300	For waxed coating, add					.76			.76	.84
1500	For non-standard colors, add					.46			.46	.51
1600	For abrasive surface, add					.53			.53	.58
1800	Brown tile, imported, 6″ x 6″ x 3/4″	D-7	120	.133		7.30	4.06		11.36	14.55
1900	8″ x 8″ x 1″		110	.145		9.65	4.43		14.08	17.75
2100	For thin set mortar application, deduct		700	.023			.70		.70	1.12

09 32 Mortar-Bed Tiling

09 32 16 – Mortar-Bed Quarry Tiling

09 32 16.10 Quarry Tile		Crew	Daily Output	Labor-Hours	Unit	Material	2019 Bare Costs Labor	Equipment	Total	Total Incl O&P
2700	Stair tread, 6" x 6" x 3/4", plain	D-7	50	.320	S.F.	7.30	9.75		17.05	24
2800	Abrasive		47	.340		8.65	10.35		19	26.50
3000	Wainscot, 6" x 6" x 1/2", thin set, red		105	.152		6.40	4.64		11.04	14.55
3100	Non-standard colors		105	.152		6.40	4.64		11.04	14.55
3300	Window sill, 6" wide, 3/4" thick		90	.178	L.F.	8.75	5.40		14.15	18.35
3400	Corners		80	.200	Ea.	6.05	6.10		12.15	16.50

09 32 23 – Mortar-Bed Glass Mosaic Tiling

09 32 23.10 Glass Mosaics		Crew	Daily Output	Labor-Hours	Unit	Material	Labor	Equipment	Total	Total Incl O&P
0010	**GLASS MOSAICS** 3/4" tile on 12" sheets, standard grout									
1020	1" tile on 12" sheets, opalescent finish	D-7	73	.219	S.F.	18.55	6.65		25.20	31.50
1040	1" x 2" tile on 12" sheet, blend		73	.219		21	6.65		27.65	34
1060	2" tile on 12" sheet, blend		73	.219		17.35	6.65		24	30
1080	5/8" x random tile, linear, on 12" sheet, blend		73	.219		25	6.65		31.65	38
1600	Dots on 12" sheet		73	.219		25	6.65		31.65	38.50
1700	For glass mosaic tiles set in dry mortar, add		290	.055		.45	1.68		2.13	3.21
1720	For glass mosaic tiles set in Portland cement mortar, add		290	.055		.01	1.68		1.69	2.72
1730	For polyblend sanded tile grout		96.15	.166	Lb.	2.19	5.05		7.24	10.60

09 34 Waterproofing-Membrane Tiling

09 34 13 – Waterproofing-Membrane Ceramic Tiling

09 34 13.10 Ceramic Tile Waterproofing Membrane		Crew	Daily Output	Labor-Hours	Unit	Material	Labor	Equipment	Total	Total Incl O&P
0010	**CERAMIC TILE WATERPROOFING MEMBRANE**									
0020	On floors, including thinset									
0030	Fleece laminated polyethylene grid, 1/8" thick	D-7	250	.064	S.F.	2.29	1.95		4.24	5.65
0040	5/16" thick	"	250	.064	"	2.60	1.95		4.55	6
0050	On walls, including thinset									
0060	Fleece laminated polyethylene sheet, 8 mil thick	D-7	480	.033	S.F.	2.29	1.01		3.30	4.16
0070	Accessories, including thinset									
0080	Joint and corner sheet, 4 mils thick, 5" wide	1 Tilf	240	.033	L.F.	1.35	1.15		2.50	3.33
0090	7-1/4" wide		180	.044		1.71	1.53		3.24	4.35
0100	10" wide		120	.067		2.08	2.30		4.38	6
0110	Pre-formed corners, inside		32	.250	Ea.	7.85	8.60		16.45	22.50
0120	Outside		32	.250		7.65	8.60		16.25	22.50
0130	2" flanged floor drain with 6" stainless steel grate		16	.500		370	17.25		387.25	440
0140	EPS, sloped shower floor		480	.017	S.F.	5.55	.57		6.12	7.05
0150	Curb		32	.250	L.F.	14.05	8.60		22.65	29.50

09 35 Chemical-Resistant Tiling

09 35 13 – Chemical-Resistant Ceramic Tiling

09 35 13.10 Chemical-Resistant Ceramic Tiling		Crew	Daily Output	Labor-Hours	Unit	Material	Labor	Equipment	Total	Total Incl O&P
0010	**CHEMICAL-RESISTANT CERAMIC TILING**									
0100	4-1/4" x 4-1/4" x 1/4", 1/8" joint	D-7	130	.123	S.F.	12.15	3.75		15.90	19.40
0200	6" x 6" x 1/2" thick		120	.133		9.85	4.06		13.91	17.40
0300	8" x 8" x 1/2" thick		110	.145		11	4.43		15.43	19.25
0400	4-1/4" x 4-1/4" x 1/4", 1/4" joint		130	.123		12.95	3.75		16.70	20.50
0500	6" x 6" x 1/2" thick		120	.133		11	4.06		15.06	18.65
0600	8" x 8" x 1/2" thick		110	.145		11.70	4.43		16.13	20
0700	4-1/4" x 4-1/4" x 1/4", 3/8" joint		130	.123		13.70	3.75		17.45	21
0800	6" x 6" x 1/2" thick		120	.133		12	4.06		16.06	19.80

09 35 Chemical-Resistant Tiling

09 35 13 – Chemical-Resistant Ceramic Tiling

09 35 13.10 Chemical-Resistant Ceramic Tiling		Crew	Daily Output	Labor-Hours	Unit	2019 Bare Costs Material	Labor	Equipment	Total	Total Incl O&P
0900	8″ x 8″ x 1/2″ thick	D-7	110	.145	S.F.	12.90	4.43		17.33	21.50

09 35 16 – Chemical-Resistant Quarry Tiling

09 35 16.10 Chemical-Resistant Quarry Tiling		Crew	Daily Output	Labor-Hours	Unit	Material	Labor	Equipment	Total	Total Incl O&P
0010	**CHEMICAL-RESISTANT QUARRY TILING**									
0100	4″ x 8″ x 1/2″ thick, 1/8″ joint	D-7	130	.123	S.F.	11.35	3.75		15.10	18.55
0200	6″ x 6″ x 1/2″ thick		120	.133		11.40	4.06		15.46	19.10
0300	8″ x 8″ x 1/2″ thick		110	.145		10.45	4.43		14.88	18.65
0400	4″ x 8″ x 1/2″ thick, 1/4″ joint		130	.123		12.65	3.75		16.40	19.95
0500	6″ x 6″ x 1/2″ thick		120	.133		12.55	4.06		16.61	20.50
0600	8″ x 8″ x 1/2″ thick		110	.145		11.10	4.43		15.53	19.40
0700	4″ x 8″ x 1/2″ thick, 3/8″ joint		130	.123		13.80	3.75		17.55	21
0800	6″ x 6″ x 1/2″ thick		120	.133		13.55	4.06		17.61	21.50
0900	8″ x 8″ x 1/2″ thick	↓	110	.145	↓	12.35	4.43		16.78	21

09 51 Acoustical Ceilings

09 51 13 – Acoustical Panel Ceilings

09 51 13.10 Ceiling, Acoustical Panel		Crew	Daily Output	Labor-Hours	Unit	Material	Labor	Equipment	Total	Total Incl O&P
0010	**CEILING, ACOUSTICAL PANEL**									
0100	Fiberglass boards, film faced, 2′ x 2′ or 2′ x 4′, 5/8″ thick	1 Carp	625	.013	S.F.	1.31	.46		1.77	2.19
0120	3/4″ thick		600	.013		3.10	.48		3.58	4.20
0130	3″ thick, thermal, R11	↓	450	.018	↓	3.88	.63		4.51	5.30

09 51 14 – Acoustical Fabric-Faced Panel Ceilings

09 51 14.10 Ceiling, Acoustical Fabric-Faced Panel		Crew	Daily Output	Labor-Hours	Unit	Material	Labor	Equipment	Total	Total Incl O&P
0010	**CEILING, ACOUSTICAL FABRIC-FACED PANEL**									
0100	Glass cloth faced fiberglass, 3/4″ thick	1 Carp	500	.016	S.F.	3.05	.57		3.62	4.30
0120	1″ thick		485	.016		3.69	.59		4.28	5.05
0130	1-1/2″ thick, nubby face	↓	475	.017	↓	2.73	.60		3.33	3.99

09 51 23 – Acoustical Tile Ceilings

09 51 23.10 Suspended Acoustic Ceiling Tiles		Crew	Daily Output	Labor-Hours	Unit	Material	Labor	Equipment	Total	Total Incl O&P
0010	**SUSPENDED ACOUSTIC CEILING TILES**, not including									
0100	suspension system									
1110	Mineral fiber tile, lay-in, 2′ x 2′ or 2′ x 4′, 5/8″ thick, fine texture	1 Carp	625	.013	S.F.	.83	.46		1.29	1.66
1115	Rough textured		625	.013		.75	.46		1.21	1.58
1125	3/4″ thick, fine textured		600	.013		2.17	.48		2.65	3.18
1130	Rough textured		600	.013		1.91	.48		2.39	2.89
1135	Fissured		600	.013		2.15	.48		2.63	3.16
1150	Tegular, 5/8″ thick, fine textured		470	.017		1.15	.61		1.76	2.27
1155	Rough textured		470	.017		1.37	.61		1.98	2.51
1165	3/4″ thick, fine textured		450	.018		2.38	.63		3.01	3.67
1170	Rough textured		450	.018		1.57	.63		2.20	2.78
1175	Fissured	↓	450	.018		2.40	.63		3.03	3.69
1185	For plastic film face, add					.80			.80	.88
1190	For fire rating, add					.53			.53	.58
3750	Wood fiber in cementitious binder, 2′ x 2′ or 4′, painted, 1″ thick	1 Carp	600	.013		2.12	.48		2.60	3.12
3760	2″ thick		550	.015		3.71	.52		4.23	4.94
3770	2-1/2″ thick		500	.016		4.63	.57		5.20	6.05
3780	3″ thick	↓	450	.018	↓	5.55	.63		6.18	7.15

09 51 Acoustical Ceilings

09 51 23 – Acoustical Tile Ceilings

09 51 23.30 Suspended Ceilings, Complete		Crew	Daily Output	Labor-Hours	Unit	Material	2019 Bare Costs Labor	Equipment	Total	Total Incl O&P
0010	**SUSPENDED CEILINGS, COMPLETE**, incl. standard									
0100	suspension system but not incl. 1-1/2" carrier channels									
0600	Fiberglass ceiling board, 2' x 4' x 5/8", plain faced	1 Carp	500	.016	S.F.	2.11	.57		2.68	3.26
0700	Offices, 2' x 4' x 3/4"		380	.021		3.90	.75		4.65	5.55
1800	Tile, Z bar suspension, 5/8" mineral fiber tile		150	.053		2.18	1.90		4.08	5.55
1900	3/4" mineral fiber tile	↓	150	.053	↓	2.45	1.90		4.35	5.85

09 51 53 – Direct-Applied Acoustical Ceilings

09 51 53.10 Ceiling Tile		Crew	Daily Output	Labor-Hours	Unit	Material	Labor	Equipment	Total	Total Incl O&P
0010	**CEILING TILE**, stapled or cemented									
0100	12" x 12" or 12" x 24", not including furring									
0600	Mineral fiber, vinyl coated, 5/8" thick	1 Carp	300	.027	S.F.	2.26	.95		3.21	4.06
0700	3/4" thick		300	.027		3.09	.95		4.04	4.97
0900	Fire rated, 3/4" thick, plain faced		300	.027		1.43	.95		2.38	3.14
1000	Plastic coated face		300	.027		2.15	.95		3.10	3.94
1200	Aluminum faced, 5/8" thick, plain	↓	300	.027		1.88	.95		2.83	3.64
3300	For flameproofing, add					.07			.07	.08
3400	For sculptured 3 dimensional, add					.33			.33	.36
3900	For ceiling primer, add					.12			.12	.13
4000	For ceiling cement, add				↓	.41			.41	.45

09 53 Acoustical Ceiling Suspension Assemblies

09 53 23 – Metal Acoustical Ceiling Suspension Assemblies

09 53 23.30 Ceiling Suspension Systems		Crew	Daily Output	Labor-Hours	Unit	Material	Labor	Equipment	Total	Total Incl O&P
0010	**CEILING SUSPENSION SYSTEMS** for boards and tile									
0050	Class A suspension system, 15/16" T bar, 2' x 4' grid	1 Carp	800	.010	S.F.	.80	.36		1.16	1.47
0300	2' x 2' grid	"	650	.012		1.04	.44		1.48	1.87
0350	For 9/16" grid, add					.16			.16	.18
0360	For fire rated grid, add					.09			.09	.10
0370	For colored grid, add					.22			.22	.24
0400	Concealed Z bar suspension system, 12" module	1 Carp	520	.015		.95	.55		1.50	1.96
0600	1-1/2" carrier channels, 4' OC, add		470	.017		.11	.61		.72	1.13
0650	1-1/2" x 3-1/2" channels	↓	470	.017	↓	.30	.61		.91	1.33
0700	Carrier channels for ceilings with									
0900	recessed lighting fixtures, add	1 Carp	460	.017	S.F.	.21	.62		.83	1.25
5000	Wire hangers, #12 wire	"	300	.027	Ea.	.07	.95		1.02	1.65

09 54 Specialty Ceilings

09 54 16 – Luminous Ceilings

09 54 16.10 Ceiling, Luminous		Crew	Daily Output	Labor-Hours	Unit	Material	Labor	Equipment	Total	Total Incl O&P
0010	**CEILING, LUMINOUS**									
0020	Translucent lay-in panels, 2' x 2'	1 Carp	500	.016	S.F.	23	.57		23.57	26.50
0030	2' x 6'	"	500	.016	"	17.95	.57		18.52	20.50

09 54 23 – Linear Metal Ceilings

09 54 23.10 Metal Ceilings		Crew	Daily Output	Labor-Hours	Unit	Material	Labor	Equipment	Total	Total Incl O&P
0010	**METAL CEILINGS**									
0015	Solid alum. planks, 3-1/4" x 12', open reveal	1 Carp	500	.016	S.F.	2.37	.57		2.94	3.55
0020	Closed reveal		500	.016		3.03	.57		3.60	4.27
0030	7-1/4" x 12', open reveal	↓	500	.016	↓	4.04	.57		4.61	5.40

09 54 Specialty Ceilings

09 54 23 – Linear Metal Ceilings

09 54 23.10 Metal Ceilings		Crew	Daily Output	Labor-Hours	Unit	2019 Bare Costs Material	Labor	Equipment	Total	Total Incl O&P
0040	Closed reveal	1 Carp	500	.016	S.F.	5.15	.57		5.72	6.65
0050	Metal, open cell, 2′ x 2′, 6″ cell		500	.016		8.65	.57		9.22	10.50
0060	8″ cell		500	.016		9.60	.57		10.17	11.50
0070	2′ x 4′, 6″ cell		500	.016		5.75	.57		6.32	7.30
0080	8″ cell	↓	500	.016	↓	5.75	.57		6.32	7.30

09 61 Flooring Treatment

09 61 19 – Concrete Floor Staining

09 61 19.40 Floors, Interior		Crew	Daily Output	Labor-Hours	Unit	Material	Labor	Equipment	Total	Total Incl O&P
0010	**FLOORS, INTERIOR**									
0300	Acid stain and sealer									
0310	Stain, one coat	1 Pord	650	.012	S.F.	.15	.37		.52	.76
0320	Two coats		570	.014		.29	.42		.71	1.01
0330	Acrylic sealer, one coat		2600	.003		.23	.09		.32	.40
0340	Two coats	↓	1400	.006	↓	.45	.17		.62	.78

09 62 Specialty Flooring

09 62 19 – Laminate Flooring

09 62 19.10 Floating Floor		Crew	Daily Output	Labor-Hours	Unit	Material	Labor	Equipment	Total	Total Incl O&P
0010	**FLOATING FLOOR**									
8300	Floating floor, laminate, wood pattern strip, complete	1 Clab	133	.060	S.F.	4.67	1.65		6.32	7.90
8310	Components, T&G wood composite strips					4.18			4.18	4.60
8320	Film					.18			.18	.20
8330	Foam					.27			.27	.30
8340	Adhesive					.47			.47	.52
8350	Installation kit				↓	.20			.20	.22
8360	Trim, 2″ wide x 3′ long				L.F.	4.39			4.39	4.83
8370	Reducer moulding				″	5.80			5.80	6.35

09 62 23 – Bamboo Flooring

09 62 23.10 Flooring, Bamboo		Crew	Daily Output	Labor-Hours	Unit	Material	Labor	Equipment	Total	Total Incl O&P
0010	**FLOORING, BAMBOO**									
8600	Flooring, wood, bamboo strips, unfinished, 5/8″ x 4″ x 3′ G	1 Carp	255	.031	S.F.	5.85	1.12		6.97	8.25
8610	5/8″ x 4″ x 4′ G		275	.029		6.05	1.04		7.09	8.35
8620	5/8″ x 4″ x 6′ G		295	.027		6.65	.97		7.62	8.90
8630	Finished, 5/8″ x 4″ x 3′ G		255	.031		6.40	1.12		7.52	8.90
8640	5/8″ x 4″ x 4′ G		275	.029		6.75	1.04		7.79	9.10
8650	5/8″ x 4″ x 6′ G		295	.027	↓	5.05	.97		6.02	7.15
8660	Stair treads, unfinished, 1-1/16″ x 11-1/2″ x 4′ G		18	.444	Ea.	56	15.85		71.85	87.50
8670	Finished, 1-1/16″ x 11-1/2″ x 4′ G		18	.444		86	15.85		101.85	121
8680	Stair risers, unfinished, 5/8″ x 7-1/2″ x 4′ G		18	.444		20.50	15.85		36.35	49
8690	Finished, 5/8″ x 7-1/2″ x 4′ G		18	.444		39	15.85		54.85	69
8700	Stair nosing, unfinished, 6′ long G		16	.500		46	17.85		63.85	80
8710	Finished, 6′ long G	↓	16	.500	↓	43.50	17.85		61.35	77.50

09 62 29 – Cork Flooring

09 62 29.10 Cork Tile Flooring		Crew	Daily Output	Labor-Hours	Unit	Material	Labor	Equipment	Total	Total Incl O&P
0010	**CORK TILE FLOORING**									
2200	Cork tile, standard finish, 1/8″ thick G	1 Tilf	315	.025	S.F.	5.30	.88		6.18	7.20
2250	3/16″ thick G	↓	315	.025	↓	5.95	.88		6.83	7.95

09 62 Specialty Flooring

09 62 29 – Cork Flooring

09 62 29.10 Cork Tile Flooring		Crew	Daily Output	Labor-Hours	Unit	Material	2019 Bare Costs Labor	Equipment	Total	Total Incl O&P
2300	5/16″ thick G	1 Tilf	315	.025	S.F.	6.50	.88		7.38	8.55
2350	1/2″ thick G		315	.025		7.05	.88		7.93	9.15
2500	Urethane finish, 1/8″ thick G		315	.025		5.20	.88		6.08	7.10
2550	3/16″ thick G		315	.025		7.20	.88		8.08	9.30
2600	5/16″ thick G		315	.025		7.50	.88		8.38	9.65
2650	1/2″ thick G		315	.025		7.60	.88		8.48	9.75

09 63 Masonry Flooring

09 63 13 – Brick Flooring

09 63 13.10 Miscellaneous Brick Flooring

	Description	Crew	Daily Output	Labor-Hours	Unit	Material	Labor	Equipment	Total	Total Incl O&P
0010	**MISCELLANEOUS BRICK FLOORING**									
0020	Acid-proof shales, red, 8″ x 3-3/4″ x 1-1/4″ thick	D-7	.43	37.209	M	735	1,125		1,860	2,625
0050	2-1/4″ thick	D-1	.40	40	″	1,025	1,300		2,325	3,300
0260	Cast ceramic, pressed, 4″ x 8″ x 1/2″, unglazed	D-7	100	.160	S.F.	7	4.87		11.87	15.55
0270	Glazed		100	.160		9.30	4.87		14.17	18.10
0280	Hand molded flooring, 4″ x 8″ x 3/4″, unglazed		95	.168		9.20	5.15		14.35	18.45
0290	Glazed		95	.168		11.60	5.15		16.75	21
0300	8″ hexagonal, 3/4″ thick, unglazed		85	.188		10.10	5.75		15.85	20.50
0310	Glazed		85	.188		18.25	5.75		24	29.50
0450	Acid-proof joints, 1/4″ wide	D-1	65	.246		1.61	7.95		9.56	15.05
0500	Pavers, 8″ x 4″, 1″ to 1-1/4″ thick, red	D-7	95	.168		4.06	5.15		9.21	12.75
0510	Ironspot	″	95	.168		5.75	5.15		10.90	14.60
0540	1-3/8″ to 1-3/4″ thick, red	D-1	95	.168		3.92	5.45		9.37	13.40
0560	Ironspot		95	.168		5.70	5.45		11.15	15.35
0580	2-1/4″ thick, red		90	.178		3.99	5.75		9.74	14
0590	Ironspot		90	.178		6.20	5.75		11.95	16.40
0700	Paver, adobe brick, 6″ x 12″, 1/2″ joint G		42	.381		1.46	12.35		13.81	22
0710	Mexican red, 12″ x 12″ G	1 Tilf	48	.167		1.86	5.75		7.61	11.30
0720	Saltillo, 12″ x 12″ G	″	48	.167		1.50	5.75		7.25	10.90
0800	For sidewalks and patios with pavers, see Section 32 14 16.10									
0870	For epoxy joints, add	D-1	600	.027	S.F.	3.08	.86		3.94	4.83
0880	For Furan underlayment, add	″	600	.027		2.55	.86		3.41	4.25
0890	For waxed surface, steam cleaned, add	A-1H	1000	.008		.21	.22	.07	.50	.67

09 63 40 – Stone Flooring

09 63 40.10 Marble

	Description	Crew	Daily Output	Labor-Hours	Unit	Material	Labor	Equipment	Total	Total Incl O&P
0010	**MARBLE**									
0020	Thin gauge tile, 12″ x 6″, 3/8″, white Carara	D-7	60	.267	S.F.	17.25	8.10		25.35	32
0100	Travertine		60	.267		9.05	8.10		17.15	23
0200	12″ x 12″ x 3/8″, thin set, floors		60	.267		11.45	8.10		19.55	25.50
0300	On walls		52	.308		10	9.35		19.35	26
1000	Marble threshold, 4″ wide x 36″ long x 5/8″ thick, white		60	.267	Ea.	11.45	8.10		19.55	25.50

09 63 40.20 Slate Tile

	Description	Crew	Daily Output	Labor-Hours	Unit	Material	Labor	Equipment	Total	Total Incl O&P
0010	**SLATE TILE**									
0020	Vermont, 6″ x 6″ x 1/4″ thick, thin set	D-7	180	.089	S.F.	7.80	2.71		10.51	12.95

09 64 Wood Flooring

09 64 23 – Wood Parquet Flooring

09 64 23.10 Wood Parquet		Crew	Daily Output	Labor-Hours	Unit	Material	Labor	Equipment	Total	Total Incl O&P
						2019 Bare Costs				
0010	**WOOD PARQUET** flooring									
5200	Parquetry, 5/16" thk, no finish, oak, plain pattern	1 Carp	160	.050	S.F.	5.45	1.78		7.23	8.95
5300	Intricate pattern		100	.080		10.25	2.85		13.10	16
5500	Teak, plain pattern		160	.050		6.50	1.78		8.28	10.10
5600	Intricate pattern		100	.080		11	2.85		13.85	16.80
5650	13/16" thick, select grade oak, plain pattern		160	.050		12.45	1.78		14.23	16.65
5700	Intricate pattern		100	.080		17.75	2.85		20.60	24.50
5800	Custom parquetry, including finish, plain pattern		100	.080		17.50	2.85		20.35	24
5900	Intricate pattern		50	.160		25.50	5.70		31.20	37.50
6700	Parquetry, prefinished white oak, 5/16" thick, plain pattern		160	.050		8.55	1.78		10.33	12.35
6800	Intricate pattern		100	.080		8.95	2.85		11.80	14.55
7000	Walnut or teak, parquetry, plain pattern		160	.050		8.65	1.78		10.43	12.50
7100	Intricate pattern	↓	100	.080	↓	16.80	2.85		19.65	23
7200	Acrylic wood parquet blocks, 12" x 12" x 5/16",									
7210	Irradiated, set in epoxy	1 Carp	160	.050	S.F.	10.65	1.78		12.43	14.65

09 64 29 – Wood Strip and Plank Flooring

09 64 29.10 Wood		Crew	Daily Output	Labor-Hours	Unit	Material	Labor	Equipment	Total	Total Incl O&P
0010	**WOOD**									
0020	Fir, vertical grain, 1" x 4", not incl. finish, grade B & better	1 Carp	255	.031	S.F.	3.55	1.12		4.67	5.75
0100	Grade C & better		255	.031		3.34	1.12		4.46	5.50
0300	Flat grain, 1" x 4", not incl. finish, grade B & better		255	.031		4.04	1.12		5.16	6.30
0400	Grade C & better		255	.031		3.89	1.12		5.01	6.15
4000	Maple, strip, 25/32" x 2-1/4", not incl. finish, select		170	.047		5.05	1.68		6.73	8.30
4100	#2 & better		170	.047		5.05	1.68		6.73	8.30
4300	33/32" x 3-1/4", not incl. finish, #1 grade		170	.047		5.80	1.68		7.48	9.15
4400	#2 & better	↓	170	.047	↓	5.20	1.68		6.88	8.45
4600	Oak, white or red, 25/32" x 2-1/4", not incl. finish									
4700	#1 common	1 Carp	170	.047	S.F.	3.50	1.68		5.18	6.60
4900	Select quartered, 2-1/4" wide		170	.047		4.38	1.68		6.06	7.60
5000	Clear		170	.047		4.31	1.68		5.99	7.50
6100	Prefinished, white oak, prime grade, 2-1/4" wide		170	.047		5.90	1.68		7.58	9.20
6200	3-1/4" wide		185	.043		5.70	1.54		7.24	8.85
6400	Ranch plank		145	.055		6.70	1.97		8.67	10.65
6500	Hardwood blocks, 9" x 9", 25/32" thick		160	.050		7.65	1.78		9.43	11.35
7400	Yellow pine, 3/4" x 3-1/8", T&G, C & better, not incl. finish	↓	200	.040		1.73	1.43		3.16	4.26
7500	Refinish wood floor, sand, 2 coats poly, wax, soft wood	1 Clab	400	.020		.22	.55		.77	1.15
7600	Hardwood		130	.062		.22	1.69		1.91	3.04
7800	Sanding and finishing, 2 coats polyurethane	↓	295	.027	↓	.22	.75		.97	1.47
7900	Subfloor and underlayment, see Section 06 16									
8015	Transition molding, 2-1/4" wide, 5' long	1 Carp	19.20	.417	Ea.	21.50	14.85		36.35	48

09 65 Resilient Flooring

09 65 10 – Resilient Tile Underlayment

09 65 10.10 Latex Underlayment		Crew	Daily Output	Labor-Hours	Unit	Material	2019 Bare Costs Labor	Equipment	Total	Total Incl O&P
0010	**LATEX UNDERLAYMENT**									
3600	Latex underlayment, 1/8" thk., cementitious for resilient flooring	1 Tilf	160	.050	S.F.	1.25	1.72		2.97	4.15
4000	Liquid, fortified				Gal.	32			32	35.50

09 65 13 – Resilient Base and Accessories

09 65 13.13 Resilient Base		Crew	Daily Output	Labor-Hours	Unit	Material	2019 Bare Costs Labor	Equipment	Total	Total Incl O&P
0010	**RESILIENT BASE**									
0690	1/8" vinyl base, 2-1/2" H, straight or cove, standard colors	1 Tilf	315	.025	L.F.	.70	.88		1.58	2.18
0700	4" high		315	.025		1.18	.88		2.06	2.71
0710	6" high		315	.025		1.54	.88		2.42	3.10
0720	Corners, 2-1/2" high		315	.025	Ea.	2.23	.88		3.11	3.86
0730	4" high		315	.025		3.03	.88		3.91	4.74
0740	6" high		315	.025		2.86	.88		3.74	4.56
0800	1/8" rubber base, 2-1/2" H, straight or cove, standard colors		315	.025	L.F.	1.11	.88		1.99	2.63
1100	4" high		315	.025		1.30	.88		2.18	2.84
1110	6" high		315	.025		1.91	.88		2.79	3.52
1150	Corners, 2-1/2" high		315	.025	Ea.	2.48	.88		3.36	4.14
1153	4" high		315	.025		2.56	.88		3.44	4.23
1155	6" high		315	.025		3.18	.88		4.06	4.91
1450	For premium color/finish add					50%				
1500	Millwork profile	1 Tilf	315	.025	L.F.	6.30	.88		7.18	8.35

09 65 13.37 Vinyl Transition Strips		Crew	Daily Output	Labor-Hours	Unit	Material	2019 Bare Costs Labor	Equipment	Total	Total Incl O&P
0010	**VINYL TRANSITION STRIPS**									
0100	Various mats. to various mats., adhesive applied, 1/4" to 1/8"	1 Tilf	315	.025	L.F.	1.48	.88		2.36	3.04
0105	0.08" to 1/8"		315	.025		1.34	.88		2.22	2.88
0110	0.08" to 1/4"		315	.025		1.46	.88		2.34	3.02
0115	1/4" to 3/8"		315	.025		1.36	.88		2.24	2.91
0120	1/4" to 1/2"		315	.025		1.36	.88		2.24	2.91
0125	1/4" to 0.08"		315	.025		1.46	.88		2.34	3.02
0200	Vinyl wheeled trans. strips, carpet to var. mats., 1/4" to 1/8" x 2-1/2"		315	.025		5.05	.88		5.93	6.95
0205	1/4" to 1/8" x 4"		315	.025		6.15	.88		7.03	8.15
0210	Various mats. to various mats. 1/4" to 0.08" x 2-1/2"		315	.025		5.05	.88		5.93	7
0215	Carpet to various materials, 1/4" to flush x 2-1/2"		315	.025		4.22	.88		5.10	6.05
0220	1/4" to flush x 4"		315	.025		6.50	.88		7.38	8.55
0225	Various materials to resilient, 3/8" to 1/8" x 2-1/2"		315	.025		4.22	.88		5.10	6.05
0230	Carpet to various materials, 3/8" to 1/4" x 2-1/2"		315	.025		5.80	.88		6.68	7.80
0235	1/4" to 1/4" x 2-1/2"		315	.025		6.50	.88		7.38	8.55
0240	Various materials to resilient, 1/8" to 1/8" x 2-1/2"		315	.025		4.91	.88		5.79	6.80
0245	Various materials to var. mats., 1/8" to flush x 2-1/2"		315	.025		3.29	.88		4.17	5.05
0250	3/8" to flush x 4"		315	.025		6.60	.88		7.48	8.70
0255	1/2" to flush x 4"		315	.025		8.90	.88		9.78	11.20
0260	Various materials to resilient, 1/8" to 0.08" x 2-1/2"		315	.025		3.81	.88		4.69	5.60
0265	0.08" to 0.08" x 2-1/2"		315	.025		3.62	.88		4.50	5.40
0270	3/8" to 0.08" x 2-1/2"		315	.025		3.62	.88		4.50	5.40

09 65 16 – Resilient Sheet Flooring

09 65 16.10 Rubber and Vinyl Sheet Flooring		Crew	Daily Output	Labor-Hours	Unit	Material	2019 Bare Costs Labor	Equipment	Total	Total Incl O&P
0010	**RUBBER AND VINYL SHEET FLOORING**									
5500	Linoleum, sheet goods G	1 Tilf	360	.022	S.F.	3.60	.77		4.37	5.20
5900	Rubber, sheet goods, 36" wide, 1/8" thick		120	.067		9.05	2.30		11.35	13.65
5950	3/16" thick		100	.080		10.15	2.76		12.91	15.60
6000	1/4" thick		90	.089		11.95	3.06		15.01	18.10
8000	Vinyl sheet goods, backed, .065" thick, plain pattern/colors		250	.032		4.47	1.10		5.57	6.70

09 65 Resilient Flooring

09 65 16 – Resilient Sheet Flooring

09 65 16.10 Rubber and Vinyl Sheet Flooring		Crew	Daily Output	Labor-Hours	Unit	2019 Bare Costs Material	Labor	Equipment	Total	Total Incl O&P
8050	Intricate pattern/colors	1 Tilf	200	.040	S.F.	3.87	1.38		5.25	6.50
8100	.080″ thick, plain pattern/colors		230	.035		4.20	1.20		5.40	6.55
8150	Intricate pattern/colors		200	.040		6.60	1.38		7.98	9.50
8200	.125″ thick, plain pattern/colors		230	.035		3.93	1.20		5.13	6.25
8250	Intricate pattern/colors		200	.040		7.30	1.38		8.68	10.25
8700	Adhesive cement, 1 gallon per 200 to 300 S.F.				Gal.	31.50			31.50	34.50
8800	Asphalt primer, 1 gallon per 300 S.F.					14.85			14.85	16.35
8900	Emulsion, 1 gallon per 140 S.F.					19.10			19.10	21

09 65 19 – Resilient Tile Flooring

09 65 19.19 Vinyl Composition Tile Flooring

Code	Description	Crew	Daily Output	Labor-Hours	Unit	Material	Labor	Equipment	Total	Total Incl O&P
0010	**VINYL COMPOSITION TILE FLOORING**									
7000	Vinyl composition tile, 12″ x 12″, 1/16″ thick	1 Tilf	500	.016	S.F.	1.22	.55		1.77	2.23
7050	Embossed		500	.016		2.67	.55		3.22	3.83
7100	Marbleized		500	.016		2.67	.55		3.22	3.83
7150	Solid		500	.016		3.45	.55		4	4.69
7200	3/32″ thick, embossed		500	.016		1.54	.55		2.09	2.58
7250	Marbleized		500	.016		3.07	.55		3.62	4.27
7300	Solid		500	.016		2.86	.55		3.41	4.04
7350	1/8″ thick, marbleized		500	.016		2.45	.55		3	3.59
7400	Solid		500	.016		1.74	.55		2.29	2.80
7450	Conductive		500	.016		5.90	.55		6.45	7.40

09 65 19.23 Vinyl Tile Flooring

Code	Description	Crew	Daily Output	Labor-Hours	Unit	Material	Labor	Equipment	Total	Total Incl O&P
0010	**VINYL TILE FLOORING**									
7500	Vinyl tile, 12″ x 12″, 3/32″ thick, standard colors/patterns	1 Tilf	500	.016	S.F.	3.72	.55		4.27	4.98
7550	1/8″ thick, standard colors/patterns		500	.016		5.20	.55		5.75	6.60
7600	1/8″ thick, premium colors/patterns		500	.016		7	.55		7.55	8.60
7650	Solid colors		500	.016		3.22	.55		3.77	4.43
7700	Marbleized or Travertine pattern		500	.016		6.20	.55		6.75	7.75
7750	Florentine pattern		500	.016		6.65	.55		7.20	8.25
7800	Premium colors/patterns		500	.016		6.55	.55		7.10	8.10

09 65 19.33 Rubber Tile Flooring

Code	Description	Crew	Daily Output	Labor-Hours	Unit	Material	Labor	Equipment	Total	Total Incl O&P
0010	**RUBBER TILE FLOORING**									
6050	Rubber tile, marbleized colors, 12″ x 12″, 1/8″ thick	1 Tilf	400	.020	S.F.	5.90	.69		6.59	7.60
6100	3/16″ thick		400	.020		8.30	.69		8.99	10.20
6300	Special tile, plain colors, 1/8″ thick		400	.020		8.35	.69		9.04	10.30
6350	3/16″ thick		400	.020		10.30	.69		10.99	12.40

09 65 33 – Conductive Resilient Flooring

09 65 33.10 Conductive Rubber and Vinyl Flooring

Code	Description	Crew	Daily Output	Labor-Hours	Unit	Material	Labor	Equipment	Total	Total Incl O&P
0010	**CONDUCTIVE RUBBER AND VINYL FLOORING**									
1700	Conductive flooring, rubber tile, 1/8″ thick	1 Tilf	315	.025	S.F.	7.15	.88		8.03	9.25
1800	Homogeneous vinyl tile, 1/8″ thick	″	315	.025	″	6.95	.88		7.83	9.05

09 66 Terrazzo Flooring

09 66 13 – Portland Cement Terrazzo Flooring

09 66 13.10 Portland Cement Terrazzo		Crew	Daily Output	Labor-Hours	Unit	2019 Bare Costs Material	Labor	Equipment	Total	Total Incl O&P
0010	**PORTLAND CEMENT TERRAZZO**, cast-in-place									
4300	Stone chips, onyx gemstone, per 50 lb. bag				Bag	19			19	21

09 66 16 – Terrazzo Floor Tile

09 66 16.13 Portland Cement Terrazzo Floor Tile		Crew	Daily Output	Labor-Hours	Unit	Material	Labor	Equipment	Total	Total Incl O&P
0010	**PORTLAND CEMENT TERRAZZO FLOOR TILE**									
1200	Floor tiles, non-slip, 1" thick, 12" x 12"	D-1	60	.267	S.F.	25	8.65		33.65	42
1300	1-1/4" thick, 12" x 12"		60	.267		26	8.65		34.65	43
1500	16" x 16"		50	.320		28	10.35		38.35	48.50
1600	1-1/2" thick, 16" x 16"	↓	45	.356	↓	25.50	11.50		37	47

09 66 16.30 Terrazzo, Precast		Crew	Daily Output	Labor-Hours	Unit	Material	Labor	Equipment	Total	Total Incl O&P
0010	**TERRAZZO, PRECAST**									
0020	Base, 6" high, straight	1 Mstz	70	.114	L.F.	12.60	3.88		16.48	20
0100	Cove		60	.133		17.25	4.53		21.78	26.50
0300	8" high, straight		60	.133		16.55	4.53		21.08	25.50
0400	Cove	↓	50	.160		24.50	5.45		29.95	36
0600	For white cement, add					.59			.59	.65
0700	For 16 ga. zinc toe strip, add					2.20			2.20	2.42
0900	Curbs, 4" x 4" high	1 Mstz	40	.200		42.50	6.80		49.30	58
1000	8" x 8" high		30	.267	↓	46.50	9.05		55.55	65.50
4800	Wainscot, 12" x 12" x 1" tiles		12	.667	S.F.	9.70	22.50		32.20	47
4900	16" x 16" x 1-1/2" tiles	↓	8	1	"	18.65	34		52.65	75.50

09 68 Carpeting

09 68 05 – Carpet Accessories

09 68 05.11 Flooring Transition Strip		Crew	Daily Output	Labor-Hours	Unit	Material	Labor	Equipment	Total	Total Incl O&P
0010	**FLOORING TRANSITION STRIP**									
0107	Clamp down brass divider, 12' strip, vinyl to carpet	1 Tilf	31.25	.256	Ea.	14.65	8.80		23.45	30.50
0117	Vinyl to hard surface	"	31.25	.256	"	14.65	8.80		23.45	30.50

09 68 10 – Carpet Pad

09 68 10.10 Commercial Grade Carpet Pad		Crew	Daily Output	Labor-Hours	Unit	Material	Labor	Equipment	Total	Total Incl O&P
0010	**COMMERCIAL GRADE CARPET PAD**									
9001	Sponge rubber pad, 20 oz./sq. yd.	1 Tilf	1350	.006	S.F.	.53	.20		.73	.91
9101	40 to 62 oz./sq. yd.		1350	.006		.98	.20		1.18	1.40
9201	Felt pad, 20 to 32 oz./sq. yd.		1350	.006		.69	.20		.89	1.09
9301	Maximum		1350	.006		1.29	.20		1.49	1.75
9401	Bonded urethane pad, 2.7 density		1350	.006		.66	.20		.86	1.06
9501	13.0 density		1350	.006		.90	.20		1.10	1.32
9601	Prime urethane pad, 2.7 density		1350	.006		.40	.20		.60	.77
9701	13.0 density	↓	1350	.006	↓	.75	.20		.95	1.15

09 68 13 – Tile Carpeting

09 68 13.10 Carpet Tile		Crew	Daily Output	Labor-Hours	Unit	Material	Labor	Equipment	Total	Total Incl O&P
0010	**CARPET TILE**									
0100	Tufted nylon, 18" x 18", hard back, 20 oz.	1 Tilf	80	.100	S.Y.	28	3.45		31.45	36
0110	26 oz.		80	.100		25.50	3.45		28.95	33.50
0200	Cushion back, 20 oz.		80	.100		25	3.45		28.45	33
0210	26 oz.		80	.100		30.50	3.45		33.95	39
6000	Electrostatic dissapative carpet tile, 24" x 24", 24 oz.		80	.100		38	3.45		41.45	47
6100	Electrostatic dissapative carpet tile for access floors, 24" x 24", 24 oz.	↓	80	.100	↓	47.50	3.45		50.95	58

09 68 Carpeting

09 68 16 – Sheet Carpeting

09 68 16.10 Sheet Carpet		Crew	Daily Output	Labor-Hours	Unit	Material	2019 Bare Costs Labor	Equipment	Total	Total Incl O&P
0010	**SHEET CARPET**									
0701	Nylon, level loop, 26 oz., light to medium traffic	1 Tilf	675	.012	S.F.	2.59	.41		3	3.51
0901	32 oz., medium traffic		675	.012		4.39	.41		4.80	5.50
1101	40 oz., medium to heavy traffic		675	.012		5.45	.41		5.86	6.65
2101	Nylon, plush, 20 oz., light traffic		675	.012		2.24	.41		2.65	3.13
2801	24 oz., light to medium traffic		675	.012		2.17	.41		2.58	3.05
2901	30 oz., medium traffic		675	.012		3.11	.41		3.52	4.08
3001	36 oz., medium traffic		675	.012		4.04	.41		4.45	5.10
3101	42 oz., medium to heavy traffic		630	.013		5.10	.44		5.54	6.30
3201	46 oz., medium to heavy traffic		630	.013		5.85	.44		6.29	7.15
3301	54 oz., heavy traffic		630	.013		6.65	.44		7.09	8
3501	Olefin, 15 oz., light traffic		675	.012		1.66	.41		2.07	2.49
3651	22 oz., light traffic		675	.012		1.69	.41		2.10	2.52
4501	50 oz., medium to heavy traffic, level loop		630	.013		12.35	.44		12.79	14.30
4701	32 oz., medium to heavy traffic, patterned		630	.013		11.30	.44		11.74	13.10
4901	48 oz., heavy traffic, patterned	↓	630	.013	↓	12.40	.44		12.84	14.30
5000	For less than full roll (approx. 1500 S.F.), add					25%				
5100	For small rooms, less than 12′ wide, add						25%			
5200	For large open areas (no cuts), deduct						25%			
5600	For bound carpet baseboard, add	1 Tilf	300	.027	L.F.	1.83	.92		2.75	3.49
5610	For stairs, not incl. price of carpet, add	"	30	.267	Riser		9.20		9.20	14.85
8950	For tackless, stretched installation, add padding from 09 68 10.10 to above									
9850	For brand-named specific fiber, add				S.Y.	25%				

09 68 20 – Athletic Carpet

09 68 20.10 Indoor Athletic Carpet		Crew	Daily Output	Labor-Hours	Unit	Material	Labor	Equipment	Total	Total Incl O&P
0010	**INDOOR ATHLETIC CARPET**									
3700	Polyethylene, in rolls, no base incl., landscape surfaces	1 Tilf	275	.029	S.F.	4.19	1		5.19	6.25
3800	Nylon action surface, 1/8″ thick		275	.029		4	1		5	6
3900	1/4″ thick		275	.029		5.75	1		6.75	7.95
4000	3/8″ thick	↓	275	.029	↓	7.25	1		8.25	9.55

09 72 Wall Coverings

09 72 13 – Cork Wall Coverings

09 72 13.10 Covering, Cork Wall		Crew	Daily Output	Labor-Hours	Unit	Material	Labor	Equipment	Total	Total Incl O&P
0010	**COVERING, CORK WALL**									
0600	Cork tiles, light or dark, 12″ x 12″ x 3/16″	1 Pape	240	.033	S.F.	4.25	1		5.25	6.30
0700	5/16″ thick		235	.034		3.33	1.02		4.35	5.35
0900	1/4″ basket weave		240	.033		3.39	1		4.39	5.35
1000	1/2″ natural, non-directional pattern		240	.033		6.65	1		7.65	8.95
1100	3/4″ natural, non-directional pattern		240	.033		11.70	1		12.70	14.50
1200	Granular surface, 12″ x 36″, 1/2″ thick		385	.021		1.33	.62		1.95	2.48
1300	1″ thick		370	.022		1.69	.65		2.34	2.92
1500	Polyurethane coated, 12″ x 12″ x 3/16″ thick		240	.033		4.11	1		5.11	6.15
1600	5/16″ thick		235	.034		5.90	1.02		6.92	8.15
1800	Cork wallpaper, paperbacked, natural		480	.017		1.60	.50		2.10	2.58
1900	Colors	↓	480	.017	↓	2.89	.50		3.39	4

09 72 Wall Coverings

09 72 16 – Vinyl-Coated Fabric Wall Coverings

	09 72 16.13 Flexible Vinyl Wall Coverings	Crew	Daily Output	Labor-Hours	Unit	Material	2019 Bare Costs Labor	Equipment	Total	Total Incl O&P
0010	**FLEXIBLE VINYL WALL COVERINGS**									
3000	Vinyl wall covering, fabric-backed, lightweight, type 1 (12-15 oz./S.Y.)	1 Pape	640	.013	S.F.	1.36	.38		1.74	2.11
3300	Medium weight, type 2 (20-24 oz./S.Y.)		480	.017		.99	.50		1.49	1.91
3400	Heavy weight, type 3 (28 oz./S.Y.)		435	.018		1.40	.55		1.95	2.44
3600	Adhesive, 5 gal. lots (18 S.Y./gal.)				Gal.	12.90			12.90	14.20

	09 72 16.16 Rigid-Sheet Vinyl Wall Coverings	Crew	Daily Output	Labor-Hours	Unit	Material	Labor	Equipment	Total	Total Incl O&P
0010	**RIGID-SHEET VINYL WALL COVERINGS**									
0100	Acrylic, modified, semi-rigid PVC, .028" thick	2 Carp	330	.048	S.F.	1.30	1.73		3.03	4.29
0110	.040" thick	"	320	.050	"	1.96	1.78		3.74	5.10

09 72 19 – Textile Wall Coverings

	09 72 19.10 Textile Wall Covering	Crew	Daily Output	Labor-Hours	Unit	Material	Labor	Equipment	Total	Total Incl O&P
0010	**TEXTILE WALL COVERING**, including sizing; add 10-30% waste @ takeoff									
0020	Silk	1 Pape	640	.013	S.F.	4.75	.38		5.13	5.85
0030	Cotton		640	.013		7.05	.38		7.43	8.35
0040	Linen		640	.013		1.91	.38		2.29	2.71
0050	Blend		640	.013		3.27	.38		3.65	4.21
0060	Linen wall covering, paper backed									
0070	Flame treatment				S.F.	1.09			1.09	1.20
0080	Stain resistance treatment					1.92			1.92	2.11
0100	Grass cloths with lining paper G	1 Pape	400	.020		1.39	.60		1.99	2.51
0110	Premium texture/color G	"	350	.023		3.27	.69		3.96	4.72

09 72 20 – Natural Fiber Wall Covering

	09 72 20.10 Natural Fiber Wall Covering	Crew	Daily Output	Labor-Hours	Unit	Material	Labor	Equipment	Total	Total Incl O&P
0010	**NATURAL FIBER WALL COVERING**, including sizing; add 10-30% waste @ takeoff									
0015	Bamboo	1 Pape	640	.013	S.F.	2.19	.38		2.57	3.02
0030	Burlap		640	.013		1.84	.38		2.22	2.63
0045	Jute		640	.013		1.32	.38		1.70	2.06
0060	Sisal		640	.013		1.73	.38		2.11	2.51

09 72 23 – Wallpapering

	09 72 23.10 Wallpaper	Crew	Daily Output	Labor-Hours	Unit	Material	Labor	Equipment	Total	Total Incl O&P
0010	**WALLPAPER** including sizing; add 10-30% waste @ takeoff R097223-10									
0050	Aluminum foil	1 Pape	275	.029	S.F.	1.03	.87		1.90	2.56
0100	Copper sheets, .025" thick, vinyl backing		240	.033		5.50	1		6.50	7.70
0300	Phenolic backing		240	.033		7.10	1		8.10	9.50
2400	Gypsum-based, fabric-backed, fire resistant									
2500	for masonry walls, 21 oz./S.Y.	1 Pape	800	.010	S.F.	.78	.30		1.08	1.35
2600	Average		720	.011		1.29	.33		1.62	1.97
2700	Small quantities		640	.013		.78	.38		1.16	1.47
3700	Wallpaper, average workmanship, solid pattern, low cost paper		640	.013		.59	.38		.97	1.26
3900	Basic patterns (matching required), avg. cost paper		535	.015		1.22	.45		1.67	2.07
4000	Paper at $85 per double roll, quality workmanship		435	.018		2.14	.55		2.69	3.25

09 74 Flexible Wood Sheets

09 74 16 – Flexible Wood Veneers

09 74 16.10 Veneer, Flexible Wood		Crew	Daily Output	Labor-Hours	Unit	2019 Bare Costs Material	Labor	Equipment	Total	Total Incl O&P
0010	**VENEER, FLEXIBLE WOOD**									
0100	Flexible wood veneer, 1/32" thick, plain woods	1 Pape	100	.080	S.F.	2.46	2.40		4.86	6.65
0110	Exotic woods	"	95	.084	"	3.71	2.53		6.24	8.20

09 91 Painting

09 91 03 – Paint Restoration

09 91 03.20 Sanding

		Crew	Daily Output	Labor-Hours	Unit	Material	Labor	Equipment	Total	Total Incl O&P
0010	**SANDING** and puttying interior trim, compared to									
0100	Painting 1 coat, on quality work				L.F.		100%			
0300	Medium work						50%			
0400	Industrial grade				↓		25%			
0500	Surface protection, placement and removal									
0510	Surface protection, placement and removal, basic drop cloths	1 Pord	6400	.001	S.F.		.04		.04	.06
0520	Masking with paper		800	.010		.07	.30		.37	.57
0530	Volume cover up (using plastic sheathing or building paper)	↓	16000	.001	↓		.01		.01	.02

09 91 03.30 Exterior Surface Preparation

		Crew	Daily Output	Labor-Hours	Unit	Material	Labor	Equipment	Total	Total Incl O&P
0010	**EXTERIOR SURFACE PREPARATION**									
0015	Doors, per side, not incl. frames or trim									
0020	Scrape & sand									
0030	Wood, flush	1 Pord	616	.013	S.F.		.39		.39	.64
0040	Wood, detail		496	.016			.48		.48	.79
0050	Wood, louvered		280	.029			.85		.85	1.40
0060	Wood, overhead	↓	616	.013	↓		.39		.39	.64
0070	Wire brush									
0080	Metal, flush	1 Pord	640	.013	S.F.		.37		.37	.61
0090	Metal, detail		520	.015			.46		.46	.75
0100	Metal, louvered		360	.022			.66		.66	1.09
0110	Metal or fibr., overhead		640	.013			.37		.37	.61
0120	Metal, roll up		560	.014			.43		.43	.70
0130	Metal, bulkhead	↓	640	.013	↓		.37		.37	.61
0140	Power wash, based on 2500 lb. operating pressure									
0150	Metal, flush	A-1H	2240	.004	S.F.		.10	.03	.13	.20
0160	Metal, detail		2120	.004			.10	.04	.14	.21
0170	Metal, louvered		2000	.004			.11	.04	.15	.22
0180	Metal or fibr., overhead		2400	.003			.09	.03	.12	.18
0190	Metal, roll up		2400	.003			.09	.03	.12	.18
0200	Metal, bulkhead	↓	2200	.004	↓		.10	.03	.13	.21
0400	Windows, per side, not incl. trim									
0410	Scrape & sand									
0420	Wood, 1-2 lite	1 Pord	320	.025	S.F.		.75		.75	1.22
0430	Wood, 3-6 lite		280	.029			.85		.85	1.40
0440	Wood, 7-10 lite		240	.033			.99		.99	1.63
0450	Wood, 12 lite		200	.040			1.19		1.19	1.96
0460	Wood, Bay/Bow	↓	320	.025	↓		.75		.75	1.22
0470	Wire brush									
0480	Metal, 1-2 lite	1 Pord	480	.017	S.F.		.50		.50	.82
0490	Metal, 3-6 lite		400	.020			.60		.60	.98
0500	Metal, Bay/Bow	↓	480	.017	↓		.50		.50	.82
0510	Power wash, based on 2500 lb. operating pressure									
0520	1-2 lite	A-1H	4400	.002	S.F.		.05	.02	.07	.10
0530	3-6 lite	↓	4320	.002	↓		.05	.02	.07	.10

09 91 03 – Paint Restoration

09 91 03.30 Exterior Surface Preparation		Crew	Daily Output	Labor-Hours	Unit	Material	2019 Bare Costs Labor	Equipment	Total	Total Incl O&P
0540	7-10 lite	A-1H	4240	.002	S.F.		.05	.02	.07	.11
0550	12 lite		4160	.002			.05	.02	.07	.11
0560	Bay/Bow	↓	4400	.002	↓		.05	.02	.07	.10
0600	Siding, scrape and sand, light=10-30%, med.=30-70%									
0610	Heavy=70-100% of surface to sand									
0650	Texture 1-11, light	1 Pord	480	.017	S.F.		.50		.50	.82
0660	Med.		440	.018			.54		.54	.89
0670	Heavy		360	.022			.66		.66	1.09
0680	Wood shingles, shakes, light		440	.018			.54		.54	.89
0690	Med.		360	.022			.66		.66	1.09
0700	Heavy		280	.029			.85		.85	1.40
0710	Clapboard, light		520	.015			.46		.46	.75
0720	Med.		480	.017			.50		.50	.82
0730	Heavy	↓	400	.020	↓		.60		.60	.98
0740	Wire brush									
0750	Aluminum, light	1 Pord	600	.013	S.F.		.40		.40	.65
0760	Med.		520	.015			.46		.46	.75
0770	Heavy	↓	440	.018	↓		.54		.54	.89
0780	Pressure wash, based on 2500 lb. operating pressure									
0790	Stucco	A-1H	3080	.003	S.F.		.07	.02	.09	.15
0800	Aluminum or vinyl		3200	.003			.07	.02	.09	.14
0810	Siding, masonry, brick & block	↓	2400	.003	↓		.09	.03	.12	.18
1300	Miscellaneous, wire brush									
1310	Metal, pedestrian gate	1 Pord	100	.080	S.F.		2.39		2.39	3.91

09 91 03.40 Interior Surface Preparation		Crew	Daily Output	Labor-Hours	Unit	Material	Labor	Equipment	Total	Total Incl O&P
0010	**INTERIOR SURFACE PREPARATION**									
0020	Doors, per side, not incl. frames or trim									
0030	Scrape & sand									
0040	Wood, flush	1 Pord	616	.013	S.F.		.39		.39	.64
0050	Wood, detail		496	.016			.48		.48	.79
0060	Wood, louvered	↓	280	.029	↓		.85		.85	1.40
0070	Wire brush									
0080	Metal, flush	1 Pord	640	.013	S.F.		.37		.37	.61
0090	Metal, detail		520	.015			.46		.46	.75
0100	Metal, louvered	↓	360	.022	↓		.66		.66	1.09
0110	Hand wash									
0120	Wood, flush	1 Pord	2160	.004	S.F.		.11		.11	.18
0130	Wood, detail		2000	.004			.12		.12	.20
0140	Wood, louvered		1360	.006			.18		.18	.29
0150	Metal, flush		2160	.004			.11		.11	.18
0160	Metal, detail		2000	.004			.12		.12	.20
0170	Metal, louvered	↓	1360	.006	↓		.18		.18	.29
0400	Windows, per side, not incl. trim									
0410	Scrape & sand									
0420	Wood, 1-2 lite	1 Pord	360	.022	S.F.		.66		.66	1.09
0430	Wood, 3-6 lite		320	.025			.75		.75	1.22
0440	Wood, 7-10 lite		280	.029			.85		.85	1.40
0450	Wood, 12 lite		240	.033			.99		.99	1.63
0460	Wood, Bay/Bow	↓	360	.022	↓		.66		.66	1.09
0470	Wire brush									
0480	Metal, 1-2 lite	1 Pord	520	.015	S.F.		.46		.46	.75
0490	Metal, 3-6 lite	↓	440	.018	↓		.54		.54	.89

09 91 03 – Paint Restoration

09 91 03.40 Interior Surface Preparation		Crew	Daily Output	Labor-Hours	Unit	2019 Bare Costs Material	Labor	Equipment	Total	Total Incl O&P
0500	Metal, Bay/Bow	1 Pord	520	.015	S.F.		.46		.46	.75
0600	Walls, sanding, light=10-30%, medium=30-70%,									
0610	heavy=70-100% of surface to sand									
0650	Walls, sand									
0660	Gypsum board or plaster, light	1 Pord	3077	.003	S.F.		.08		.08	.13
0670	Gypsum board or plaster, medium		2160	.004			.11		.11	.18
0680	Gypsum board or plaster, heavy		923	.009			.26		.26	.42
0690	Wood, T&G, light		2400	.003			.10		.10	.16
0700	Wood, T&G, medium		1600	.005			.15		.15	.24
0710	Wood, T&G, heavy		800	.010			.30		.30	.49
0720	Walls, wash									
0730	Gypsum board or plaster	1 Pord	3200	.003	S.F.		.07		.07	.12
0740	Wood, T&G		3200	.003			.07		.07	.12
0750	Masonry, brick & block, smooth		2800	.003			.09		.09	.14
0760	Masonry, brick & block, coarse		2000	.004			.12		.12	.20
8000	For chemical washing, see Section 04 01 30									

09 91 03.41 Scrape After Fire Damage		Crew	Daily Output	Labor-Hours	Unit	Material	Labor	Equipment	Total	Total Incl O&P
0010	**SCRAPE AFTER FIRE DAMAGE**									
0050	Boards, 1" x 4"	1 Pord	336	.024	L.F.		.71		.71	1.16
0060	1" x 6"		260	.031			.92		.92	1.50
0070	1" x 8"		207	.039			1.15		1.15	1.89
0080	1" x 10"		174	.046			1.37		1.37	2.25
0500	Framing, 2" x 4"		265	.030			.90		.90	1.48
0510	2" x 6"		221	.036			1.08		1.08	1.77
0520	2" x 8"		190	.042			1.26		1.26	2.06
0530	2" x 10"		165	.048			1.45		1.45	2.37
0540	2" x 12"		144	.056			1.66		1.66	2.72
1000	Heavy framing, 3" x 4"		226	.035			1.06		1.06	1.73
1010	4" x 4"		210	.038			1.14		1.14	1.86
1020	4" x 6"		191	.042			1.25		1.25	2.05
1030	4" x 8"		165	.048			1.45		1.45	2.37
1040	4" x 10"		144	.056			1.66		1.66	2.72
1060	4" x 12"		131	.061			1.82		1.82	2.99
2900	For sealing, light damage		825	.010	S.F.	.15	.29		.44	.64
2920	Heavy damage		460	.017	"	.33	.52		.85	1.21

09 91 13 – Exterior Painting

09 91 13.30 Fences		Crew	Daily Output	Labor-Hours	Unit	Material	Labor	Equipment	Total	Total Incl O&P
0010	**FENCES** R099100-20									
0100	Chain link or wire metal, one side, water base									
0110	Roll & brush, first coat	1 Pord	960	.008	S.F.	.07	.25		.32	.49
0120	Second coat		1280	.006		.07	.19		.26	.38
0130	Spray, first coat		2275	.004		.07	.11		.18	.25
0140	Second coat		2600	.003		.07	.09		.16	.23
0150	Picket, water base									
0160	Roll & brush, first coat	1 Pord	865	.009	S.F.	.08	.28		.36	.53
0170	Second coat		1050	.008		.08	.23		.31	.45
0180	Spray, first coat		2275	.004		.08	.11		.19	.25
0190	Second coat		2600	.003		.08	.09		.17	.23
0200	Stockade, water base									
0210	Roll & brush, first coat	1 Pord	1040	.008	S.F.	.08	.23		.31	.46
0220	Second coat		1200	.007		.08	.20		.28	.41
0230	Spray, first coat		2275	.004		.08	.11		.19	.25

09 91 Painting

09 91 13 – Exterior Painting

	09 91 13.30 Fences	Crew	Daily Output	Labor-Hours	Unit	Material	2019 Bare Costs Labor	Equipment	Total	Total Incl O&P
0240	Second coat	1 Pord	2600	.003	S.F.	.08	.09		.17	.23

09 91 13.42 Miscellaneous, Exterior

		Crew	Daily Output	Labor-Hours	Unit	Material	Labor	Equipment	Total	Total Incl O&P
0010	**MISCELLANEOUS, EXTERIOR** R099100-20									
0100	Railing, ext., decorative wood, incl. cap & baluster									
0110	Newels & spindles @ 12″ OC									
0120	Brushwork, stain, sand, seal & varnish									
0130	First coat	1 Pord	90	.089	L.F.	.90	2.65		3.55	5.35
0140	Second coat	″	120	.067	″	.90	1.99		2.89	4.25
0150	Rough sawn wood, 42″ high, 2″ x 2″ verticals, 6″ OC									
0160	Brushwork, stain, each coat	1 Pord	90	.089	L.F.	.31	2.65		2.96	4.69
0170	Wrought iron, 1″ rail, 1/2″ sq. verticals									
0180	Brushwork, zinc chromate, 60″ high, bars 6″ OC									
0190	Primer	1 Pord	130	.062	L.F.	.86	1.84		2.70	3.95
0200	Finish coat		130	.062		1.14	1.84		2.98	4.27
0210	Additional coat		190	.042		1.33	1.26		2.59	3.52
0220	Shutters or blinds, single panel, 2′ x 4′, paint all sides									
0230	Brushwork, primer	1 Pord	20	.400	Ea.	.72	11.95		12.67	20.50
0240	Finish coat, exterior latex		20	.400		.58	11.95		12.53	20
0250	Primer & 1 coat, exterior latex		13	.615		1.13	18.35		19.48	31.50
0260	Spray, primer		35	.229		1.04	6.80		7.84	12.35
0270	Finish coat, exterior latex		35	.229		1.23	6.80		8.03	12.55
0280	Primer & 1 coat, exterior latex		20	.400		1.12	11.95		13.07	21
0290	For louvered shutters, add				S.F.	10%				
0300	Stair stringers, exterior, metal									
0310	Roll & brush, zinc chromate, to 14″, each coat	1 Pord	320	.025	L.F.	.38	.75		1.13	1.64
0320	Rough sawn wood, 4″ x 12″									
0330	Roll & brush, exterior latex, each coat	1 Pord	215	.037	L.F.	.09	1.11		1.20	1.91
0340	Trellis/lattice, 2″ x 2″ @ 3″ OC with 2″ x 8″ supports									
0350	Spray, latex, per side, each coat	1 Pord	475	.017	S.F.	.09	.50		.59	.91
0450	Decking, ext., sealer, alkyd, brushwork, sealer coat		1140	.007		.10	.21		.31	.45
0460	1st coat		1140	.007		.12	.21		.33	.47
0470	2nd coat		1300	.006		.09	.18		.27	.40
0500	Paint, alkyd, brushwork, primer coat		1140	.007		.11	.21		.32	.46
0510	1st coat		1140	.007		.14	.21		.35	.50
0520	2nd coat		1300	.006		.10	.18		.28	.41
0600	Sand paint, alkyd, brushwork, 1 coat		150	.053		.14	1.59		1.73	2.76

09 91 13.60 Siding Exterior

		Crew	Daily Output	Labor-Hours	Unit	Material	Labor	Equipment	Total	Total Incl O&P
0010	**SIDING EXTERIOR**, Alkyd (oil base)									
0450	Steel siding, oil base, paint 1 coat, brushwork	2 Pord	2015	.008	S.F.	.11	.24		.35	.51
0500	Spray		4550	.004		.17	.11		.28	.36
0800	Paint 2 coats, brushwork		1300	.012		.23	.37		.60	.85
1000	Spray		2750	.006		.15	.17		.32	.45
1200	Stucco, rough, oil base, paint 2 coats, brushwork		1300	.012		.23	.37		.60	.85
1400	Roller		1625	.010		.24	.29		.53	.74
1600	Spray		2925	.005		.25	.16		.41	.55
1800	Texture 1-11 or clapboard, oil base, primer coat, brushwork		1300	.012		.14	.37		.51	.76
2000	Spray		4550	.004		.14	.11		.25	.33
2100	Paint 1 coat, brushwork		1300	.012		.17	.37		.54	.78
2200	Spray		4550	.004		.17	.11		.28	.35
2400	Paint 2 coats, brushwork		810	.020		.33	.59		.92	1.33
2600	Spray		2600	.006		.37	.18		.55	.70
3000	Stain 1 coat, brushwork		1520	.011		.10	.31		.41	.62

09 91 13 – Exterior Painting

09 91 13.60 Siding Exterior		Crew	Daily Output	Labor-Hours	Unit	Material	Labor	Equipment	Total	Total Incl O&P
						2019 Bare Costs				
3200	Spray	2 Pord	5320	.003	S.F.	.12	.09		.21	.28
3400	Stain 2 coats, brushwork		950	.017		.21	.50		.71	1.05
4000	Spray		3050	.005		.23	.16		.39	.51
4200	Wood shingles, oil base primer coat, brushwork		1300	.012		.13	.37		.50	.75
4400	Spray		3900	.004		.12	.12		.24	.34
4600	Paint 1 coat, brushwork		1300	.012		.14	.37		.51	.75
4800	Spray		3900	.004		.17	.12		.29	.39
5000	Paint 2 coats, brushwork		810	.020		.28	.59		.87	1.27
5200	Spray		2275	.007		.26	.21		.47	.63
5800	Stain 1 coat, brushwork		1500	.011		.10	.32		.42	.63
6000	Spray		3900	.004		.10	.12		.22	.31
6500	Stain 2 coats, brushwork		950	.017		.21	.50		.71	1.05
7000	Spray	↓	2660	.006		.29	.18		.47	.61
8000	For latex paint, deduct					10%				
8100	For work over 12′ H, from pipe scaffolding, add						15%			
8200	For work over 12′ H, from extension ladder, add						25%			
8300	For work over 12′ H, from swing staging, add				↓		35%			

09 91 13.62 Siding, Misc.		Crew	Daily Output	Labor-Hours	Unit	Material	Labor	Equipment	Total	Total Incl O&P
0010	**SIDING, MISC.**, latex paint R099100-10									
0100	Aluminum siding									
0110	Brushwork, primer	2 Pord	2275	.007	S.F.	.07	.21		.28	.41
0120	Finish coat, exterior latex		2275	.007		.06	.21		.27	.40
0130	Primer & 1 coat exterior latex		1300	.012		.13	.37		.50	.75
0140	Primer & 2 coats exterior latex	↓	975	.016	↓	.19	.49		.68	1.01
0150	Mineral fiber shingles									
0160	Brushwork, primer	2 Pord	1495	.011	S.F.	.14	.32		.46	.68
0170	Finish coat, industrial enamel		1495	.011		.19	.32		.51	.73
0180	Primer & 1 coat enamel		810	.020		.33	.59		.92	1.34
0190	Primer & 2 coats enamel		540	.030		.52	.88		1.40	2.03
0200	Roll, primer		1625	.010		.16	.29		.45	.66
0210	Finish coat, industrial enamel		1625	.010		.21	.29		.50	.71
0220	Primer & 1 coat enamel		975	.016		.37	.49		.86	1.20
0230	Primer & 2 coats enamel		650	.025		.57	.73		1.30	1.83
0240	Spray, primer		3900	.004		.12	.12		.24	.34
0250	Finish coat, industrial enamel		3900	.004		.17	.12		.29	.39
0260	Primer & 1 coat enamel		2275	.007		.30	.21		.51	.66
0270	Primer & 2 coats enamel		1625	.010		.47	.29		.76	.99
0280	Waterproof sealer, first coat		4485	.004		.13	.11		.24	.31
0290	Second coat	↓	5235	.003	↓	.12	.09		.21	.28
0300	Rough wood incl. shingles, shakes or rough sawn siding									
0310	Brushwork, primer	2 Pord	1280	.013	S.F.	.14	.37		.51	.77
0320	Finish coat, exterior latex		1280	.013		.10	.37		.47	.72
0330	Primer & 1 coat exterior latex		960	.017		.25	.50		.75	1.09
0340	Primer & 2 coats exterior latex		700	.023		.35	.68		1.03	1.50
0350	Roll, primer		2925	.005		.19	.16		.35	.48
0360	Finish coat, exterior latex		2925	.005		.12	.16		.28	.41
0370	Primer & 1 coat exterior latex		1790	.009		.32	.27		.59	.79
0380	Primer & 2 coats exterior latex		1300	.012		.44	.37		.81	1.08
0390	Spray, primer		3900	.004		.16	.12		.28	.38
0400	Finish coat, exterior latex		3900	.004		.10	.12		.22	.30
0410	Primer & 1 coat exterior latex		2600	.006		.26	.18		.44	.58
0420	Primer & 2 coats exterior latex	↓	2080	.008	↓	.35	.23		.58	.77

09 91 13 – Exterior Painting

09 91 13.62 Siding, Misc.		Crew	Daily Output	Labor-Hours	Unit	2019 Bare Costs Material	Labor	Equipment	Total	Total Incl O&P
0430	Waterproof sealer, first coat	2 Pord	4485	.004	S.F.	.23	.11		.34	.42
0440	Second coat		4485	.004		.13	.11		.24	.31
0450	Smooth wood incl. butt, T&G, beveled, drop or B&B siding									
0460	Brushwork, primer	2 Pord	2325	.007	S.F.	.10	.21		.31	.45
0470	Finish coat, exterior latex		1280	.013		.10	.37		.47	.72
0480	Primer & 1 coat exterior latex		800	.020		.21	.60		.81	1.21
0490	Primer & 2 coats exterior latex		630	.025		.31	.76		1.07	1.58
0500	Roll, primer		2275	.007		.12	.21		.33	.47
0510	Finish coat, exterior latex		2275	.007		.11	.21		.32	.46
0520	Primer & 1 coat exterior latex		1300	.012		.23	.37		.60	.85
0530	Primer & 2 coats exterior latex		975	.016		.34	.49		.83	1.17
0540	Spray, primer		4550	.004		.09	.11		.20	.27
0550	Finish coat, exterior latex		4550	.004		.10	.11		.21	.27
0560	Primer & 1 coat exterior latex		2600	.006		.18	.18		.36	.50
0570	Primer & 2 coats exterior latex		1950	.008		.28	.25		.53	.71
0580	Waterproof sealer, first coat		5230	.003		.13	.09		.22	.29
0590	Second coat		5980	.003		.13	.08		.21	.27
0600	For oil base paint, add					10%				

09 91 13.70 Doors and Windows, Exterior		Crew	Daily Output	Labor-Hours	Unit	Material	Labor	Equipment	Total	Total Incl O&P
0010	**DOORS AND WINDOWS, EXTERIOR** R099100-10									
0100	Door frames & trim, only									
0110	Brushwork, primer R099100-20	1 Pord	512	.016	L.F.	.07	.47		.54	.83
0120	Finish coat, exterior latex		512	.016		.07	.47		.54	.84
0130	Primer & 1 coat, exterior latex		300	.027		.14	.80		.94	1.45
0140	Primer & 2 coats, exterior latex		265	.030		.21	.90		1.11	1.71
0150	Doors, flush, both sides, incl. frame & trim									
0160	Roll & brush, primer	1 Pord	10	.800	Ea.	4.94	24		28.94	44.50
0170	Finish coat, exterior latex		10	.800		5.50	24		29.50	45
0180	Primer & 1 coat, exterior latex		7	1.143		10.45	34		44.45	67.50
0190	Primer & 2 coats, exterior latex		5	1.600		15.90	48		63.90	95.50
0200	Brushwork, stain, sealer & 2 coats polyurethane		4	2		31	59.50		90.50	133
0210	Doors, French, both sides, 10-15 lite, incl. frame & trim									
0220	Brushwork, primer	1 Pord	6	1.333	Ea.	2.47	40		42.47	67.50
0230	Finish coat, exterior latex		6	1.333		2.74	40		42.74	68
0240	Primer & 1 coat, exterior latex		3	2.667		5.20	79.50		84.70	136
0250	Primer & 2 coats, exterior latex		2	4		7.80	119		126.80	205
0260	Brushwork, stain, sealer & 2 coats polyurethane		2.50	3.200		11.40	95.50		106.90	169
0270	Doors, louvered, both sides, incl. frame & trim									
0280	Brushwork, primer	1 Pord	7	1.143	Ea.	4.94	34		38.94	61.50
0290	Finish coat, exterior latex		7	1.143		5.50	34		39.50	62
0300	Primer & 1 coat, exterior latex		4	2		10.45	59.50		69.95	109
0310	Primer & 2 coats, exterior latex		3	2.667		15.60	79.50		95.10	147
0320	Brushwork, stain, sealer & 2 coats polyurethane		4.50	1.778		31	53		84	122
0330	Doors, panel, both sides, incl. frame & trim									
0340	Roll & brush, primer	1 Pord	6	1.333	Ea.	4.94	40		44.94	70.50
0350	Finish coat, exterior latex		6	1.333		5.50	40		45.50	71
0360	Primer & 1 coat, exterior latex		3	2.667		10.45	79.50		89.95	141
0370	Primer & 2 coats, exterior latex		2.50	3.200		15.60	95.50		111.10	173
0380	Brushwork, stain, sealer & 2 coats polyurethane		3	2.667		31	79.50		110.50	165
0400	Windows, per ext. side, based on 15 S.F.									
0410	1 to 6 lite									
0420	Brushwork, primer	1 Pord	13	.615	Ea.	.98	18.35		19.33	31

09 91 Painting

09 91 13 – Exterior Painting

09 91 13.70 Doors and Windows, Exterior		Crew	Daily Output	Labor-Hours	Unit	Material	2019 Bare Costs Labor	Equipment	Total	Total Incl O&P
0430	Finish coat, exterior latex	1 Pord	13	.615	Ea.	1.08	18.35		19.43	31
0440	Primer & 1 coat, exterior latex		8	1		2.06	30		32.06	51.50
0450	Primer & 2 coats, exterior latex		6	1.333		3.08	40		43.08	68.50
0460	Stain, sealer & 1 coat varnish		7	1.143		4.49	34		38.49	61
0470	7 to 10 lite									
0480	Brushwork, primer	1 Pord	11	.727	Ea.	.98	21.50		22.48	36.50
0490	Finish coat, exterior latex		11	.727		1.08	21.50		22.58	36.50
0500	Primer & 1 coat, exterior latex		7	1.143		2.06	34		36.06	58.50
0510	Primer & 2 coats, exterior latex		5	1.600		3.08	48		51.08	81.50
0520	Stain, sealer & 1 coat varnish		6	1.333		4.49	40		44.49	70
0530	12 lite									
0540	Brushwork, primer	1 Pord	10	.800	Ea.	.98	24		24.98	40
0550	Finish coat, exterior latex		10	.800		1.08	24		25.08	40
0560	Primer & 1 coat, exterior latex		6	1.333		2.06	40		42.06	67.50
0570	Primer & 2 coats, exterior latex		5	1.600		3.08	48		51.08	81.50
0580	Stain, sealer & 1 coat varnish		6	1.333		4.38	40		44.38	70
0590	For oil base paint, add					10%				

09 91 13.80 Trim, Exterior		Crew	Daily Output	Labor-Hours	Unit	Material	Labor	Equipment	Total	Total Incl O&P
0010	**TRIM, EXTERIOR** R099100-10									
0100	Door frames & trim (see Doors, interior or exterior)									
0110	Fascia, latex paint, one coat coverage									
0120	1" x 4", brushwork	1 Pord	640	.013	L.F.	.02	.37		.39	.63
0130	Roll		1280	.006		.02	.19		.21	.34
0140	Spray		2080	.004		.02	.11		.13	.21
0150	1" x 6" to 1" x 10", brushwork		640	.013		.08	.37		.45	.69
0160	Roll		1230	.007		.08	.19		.27	.41
0170	Spray		2100	.004		.06	.11		.17	.26
0180	1" x 12", brushwork		640	.013		.08	.37		.45	.69
0190	Roll		1050	.008		.08	.23		.31	.46
0200	Spray		2200	.004		.06	.11		.17	.25
0210	Gutters & downspouts, metal, zinc chromate paint									
0220	Brushwork, gutters, 5", first coat	1 Pord	640	.013	L.F.	.40	.37		.77	1.05
0230	Second coat		960	.008		.38	.25		.63	.83
0240	Third coat		1280	.006		.31	.19		.50	.65
0250	Downspouts, 4", first coat		640	.013		.40	.37		.77	1.05
0260	Second coat		960	.008		.38	.25		.63	.83
0270	Third coat		1280	.006		.31	.19		.50	.65
0280	Gutters & downspouts, wood									
0290	Brushwork, gutters, 5", primer	1 Pord	640	.013	L.F.	.07	.37		.44	.68
0300	Finish coat, exterior latex		640	.013		.06	.37		.43	.68
0310	Primer & 1 coat exterior latex		400	.020		.14	.60		.74	1.13
0320	Primer & 2 coats exterior latex		325	.025		.21	.73		.94	1.43
0330	Downspouts, 4", primer		640	.013		.07	.37		.44	.68
0340	Finish coat, exterior latex		640	.013		.06	.37		.43	.68
0350	Primer & 1 coat exterior latex		400	.020		.14	.60		.74	1.13
0360	Primer & 2 coats exterior latex		325	.025		.10	.73		.83	1.32
0370	Molding, exterior, up to 14" wide									
0380	Brushwork, primer	1 Pord	640	.013	L.F.	.08	.37		.45	.70
0390	Finish coat, exterior latex		640	.013		.08	.37		.45	.70
0400	Primer & 1 coat exterior latex		400	.020		.16	.60		.76	1.16
0410	Primer & 2 coats exterior latex		315	.025		.16	.76		.92	1.42
0420	Stain & fill		1050	.008		.13	.23		.36	.51

09 91 13 – Exterior Painting

09 91 13.80 Trim, Exterior		Crew	Daily Output	Labor-Hours	Unit	2019 Bare Costs Material	Labor	Equipment	Total	Total Incl O&P
0430	Shellac	1 Pord	1850	.004	L.F.	.15	.13		.28	.38
0440	Varnish		1275	.006		.11	.19		.30	.43

09 91 13.90 Walls, Masonry (CMU), Exterior

		Crew	Daily Output	Labor-Hours	Unit	Material	Labor	Equipment	Total	Total Incl O&P
0010	**WALLS, MASONRY (CMU), EXTERIOR**									
0360	Concrete masonry units (CMU), smooth surface									
0370	Brushwork, latex, first coat	1 Pord	640	.013	S.F.	.06	.37		.43	.68
0380	Second coat		960	.008		.05	.25		.30	.46
0390	Waterproof sealer, first coat		736	.011		.28	.32		.60	.84
0400	Second coat		1104	.007		.28	.22		.50	.66
0410	Roll, latex, paint, first coat		1465	.005		.07	.16		.23	.35
0420	Second coat		1790	.004		.06	.13		.19	.28
0430	Waterproof sealer, first coat		1680	.005		.28	.14		.42	.54
0440	Second coat		2060	.004		.28	.12		.40	.50
0450	Spray, latex, paint, first coat		1950	.004		.06	.12		.18	.26
0460	Second coat		2600	.003		.05	.09		.14	.20
0470	Waterproof sealer, first coat		2245	.004		.28	.11		.39	.48
0480	Second coat		2990	.003		.28	.08		.36	.44
0490	Concrete masonry unit (CMU), porous									
0500	Brushwork, latex, first coat	1 Pord	640	.013	S.F.	.12	.37		.49	.74
0510	Second coat		960	.008		.06	.25		.31	.48
0520	Waterproof sealer, first coat		736	.011		.28	.32		.60	.84
0530	Second coat		1104	.007		.28	.22		.50	.66
0540	Roll latex, first coat		1465	.005		.09	.16		.25	.37
0550	Second coat		1790	.004		.06	.13		.19	.28
0560	Waterproof sealer, first coat		1680	.005		.28	.14		.42	.54
0570	Second coat		2060	.004		.28	.12		.40	.50
0580	Spray latex, first coat		1950	.004		.07	.12		.19	.27
0590	Second coat		2600	.003		.05	.09		.14	.20
0600	Waterproof sealer, first coat		2245	.004		.28	.11		.39	.48
0610	Second coat		2990	.003		.28	.08		.36	.44

09 91 23 – Interior Painting

09 91 23.20 Cabinets and Casework

		Crew	Daily Output	Labor-Hours	Unit	Material	Labor	Equipment	Total	Total Incl O&P
0010	**CABINETS AND CASEWORK**									
1000	Primer coat, oil base, brushwork	1 Pord	650	.012	S.F.	.07	.37		.44	.67
2000	Paint, oil base, brushwork, 1 coat		650	.012		.12	.37		.49	.73
2500	2 coats		400	.020		.23	.60		.83	1.23
3000	Stain, brushwork, wipe off		650	.012		.10	.37		.47	.71
4000	Shellac, 1 coat, brushwork		650	.012		.13	.37		.50	.74
4500	Varnish, 3 coats, brushwork, sand after 1st coat		325	.025		.28	.73		1.01	1.50
5000	For latex paint, deduct					10%				

09 91 23.33 Doors and Windows, Interior Alkyd (Oil Base)

		Crew	Daily Output	Labor-Hours	Unit	Material	Labor	Equipment	Total	Total Incl O&P
0010	**DOORS AND WINDOWS, INTERIOR ALKYD (OIL BASE)**									
0500	Flush door & frame, 3′ x 7′, oil, primer, brushwork	1 Pord	10	.800	Ea.	4.04	24		28.04	43.50
1000	Paint, 1 coat		10	.800		4.24	24		28.24	43.50
1200	2 coats		6	1.333		4.66	40		44.66	70
1400	Stain, brushwork, wipe off		18	.444		2.19	13.25		15.44	24
1600	Shellac, 1 coat, brushwork		25	.320		2.67	9.55		12.22	18.60
1800	Varnish, 3 coats, brushwork, sand after 1st coat		9	.889		5.80	26.50		32.30	50
2000	Panel door & frame, 3′ x 7′, oil, primer, brushwork		6	1.333		2.47	40		42.47	67.50
2200	Paint, 1 coat		6	1.333		4.24	40		44.24	69.50
2400	2 coats		3	2.667		10.95	79.50		90.45	142
2600	Stain, brushwork, panel door, 3′ x 7′, not incl. frame		16	.500		2.19	14.95		17.14	27

09 91 23 – Interior Painting

09 91 23.33 Doors and Windows, Interior Alkyd (Oil Base)		Crew	Daily Output	Labor-Hours	Unit	2019 Bare Costs Material	Labor	Equipment	Total	Total Incl O&P
2800	Shellac, 1 coat, brushwork	1 Pord	22	.364	Ea.	2.67	10.85		13.52	20.50
3000	Varnish, 3 coats, brushwork, sand after 1st coat		7.50	1.067		5.80	32		37.80	58.50
3020	French door, incl. 3′ x 7′, 6 lites, frame & trim									
3022	Paint, 1 coat, over existing paint	1 Pord	5	1.600	Ea.	8.50	48		56.50	87.50
3024	2 coats, over existing paint		5	1.600		16.45	48		64.45	96
3026	Primer & 1 coat		3.50	2.286		13.45	68		81.45	127
3028	Primer & 2 coats		3	2.667		22	79.50		101.50	154
3032	Varnish or polyurethane, 1 coat		5	1.600		8.40	48		56.40	87.50
3034	2 coats, sanding between		3	2.667		16.80	79.50		96.30	149
4400	Windows, including frame and trim, per side									
4600	Colonial type, 6/6 lites, 2′ x 3′, oil, primer, brushwork	1 Pord	14	.571	Ea.	.39	17.05		17.44	28.50
5800	Paint, 1 coat		14	.571		.67	17.05		17.72	28.50
6000	2 coats		9	.889		1.30	26.50		27.80	45
6200	3′ x 5′ opening, 6/6 lites, primer coat, brushwork		12	.667		.98	19.90		20.88	33.50
6400	Paint, 1 coat		12	.667		1.67	19.90		21.57	34.50
6600	2 coats		7	1.143		3.25	34		37.25	59.50
6800	4′ x 8′ opening, 6/6 lites, primer coat, brushwork		8	1		2.08	30		32.08	51.50
7000	Paint, 1 coat		8	1		3.57	30		33.57	53
7200	2 coats		5	1.600		6.95	48		54.95	85.50
8000	Single lite type, 2′ x 3′, oil base, primer coat, brushwork		33	.242		.39	7.25		7.64	12.30
8200	Paint, 1 coat		33	.242		.67	7.25		7.92	12.60
8400	2 coats		20	.400		1.30	11.95		13.25	21
8600	3′ x 5′ opening, primer coat, brushwork		20	.400		.98	11.95		12.93	20.50
8800	Paint, 1 coat		20	.400		1.67	11.95		13.62	21.50
8900	2 coats		13	.615		3.25	18.35		21.60	33.50
9200	4′ x 8′ opening, primer coat, brushwork		14	.571		2.08	17.05		19.13	30.50
9400	Paint, 1 coat		14	.571		3.57	17.05		20.62	32
9600	2 coats		8	1		6.95	30		36.95	56.50

09 91 23.35 Doors and Windows, Interior Latex		Crew	Daily Output	Labor-Hours	Unit	Material	Labor	Equipment	Total	Total Incl O&P
0010	**DOORS & WINDOWS, INTERIOR LATEX** R099100-10									
0100	Doors, flush, both sides, incl. frame & trim									
0110	Roll & brush, primer	1 Pord	10	.800	Ea.	4.22	24		28.22	43.50
0120	Finish coat, latex		10	.800		5.60	24		29.60	45
0130	Primer & 1 coat latex		7	1.143		9.85	34		43.85	67
0140	Primer & 2 coats latex		5	1.600		15.10	48		63.10	94.50
0160	Spray, both sides, primer		20	.400		4.44	11.95		16.39	24.50
0170	Finish coat, latex		20	.400		5.90	11.95		17.85	26
0180	Primer & 1 coat latex		11	.727		10.40	21.50		31.90	47
0190	Primer & 2 coats latex		8	1		16	30		46	66.50
0200	Doors, French, both sides, 10-15 lite, incl. frame & trim									
0210	Roll & brush, primer	1 Pord	6	1.333	Ea.	2.11	40		42.11	67.50
0220	Finish coat, latex		6	1.333		2.81	40		42.81	68
0230	Primer & 1 coat latex		3	2.667		4.92	79.50		84.42	135
0240	Primer & 2 coats latex		2	4		7.55	119		126.55	204
0260	Doors, louvered, both sides, incl. frame & trim									
0270	Roll & brush, primer	1 Pord	7	1.143	Ea.	4.22	34		38.22	60.50
0280	Finish coat, latex		7	1.143		5.60	34		39.60	62
0290	Primer & 1 coat, latex		4	2		9.60	59.50		69.10	109
0300	Primer & 2 coats, latex		3	2.667		15.45	79.50		94.95	147
0320	Spray, both sides, primer		20	.400		4.44	11.95		16.39	24.50
0330	Finish coat, latex		20	.400		5.90	11.95		17.85	26
0340	Primer & 1 coat, latex		11	.727		10.40	21.50		31.90	47

09 91 23.35 Doors and Windows, Interior Latex			Crew	Daily Output	Labor-Hours	Unit	Material	2019 Bare Costs Labor	Equipment	Total	Total Incl O&P
0350	Primer & 2 coats, latex		1 Pord	8	1	Ea.	16.35	30		46.35	67
0360	Doors, panel, both sides, incl. frame & trim										
0370	Roll & brush, primer		1 Pord	6	1.333	Ea.	4.44	40		44.44	70
0380	Finish coat, latex			6	1.333		5.60	40		45.60	71
0390	Primer & 1 coat, latex			3	2.667		9.85	79.50		89.35	141
0400	Primer & 2 coats, latex			2.50	3.200		15.45	95.50		110.95	173
0420	Spray, both sides, primer			10	.800		4.44	24		28.44	44
0430	Finish coat, latex			10	.800		5.90	24		29.90	45.50
0440	Primer & 1 coat, latex			5	1.600		10.40	48		58.40	89.50
0450	Primer & 2 coats, latex			4	2		16.35	59.50		75.85	116
0460	Windows, per interior side, based on 15 S.F.										
0470	1 to 6 lite										
0480	Brushwork, primer		1 Pord	13	.615	Ea.	.83	18.35		19.18	31
0490	Finish coat, enamel			13	.615		1.11	18.35		19.46	31
0500	Primer & 1 coat enamel			8	1		1.94	30		31.94	51
0510	Primer & 2 coats enamel			6	1.333		3.05	40		43.05	68.50
0530	7 to 10 lite										
0540	Brushwork, primer		1 Pord	11	.727	Ea.	.83	21.50		22.33	36.50
0550	Finish coat, enamel			11	.727		1.11	21.50		22.61	36.50
0560	Primer & 1 coat enamel			7	1.143		1.94	34		35.94	58
0570	Primer & 2 coats enamel			5	1.600		3.05	48		51.05	81.50
0590	12 lite										
0600	Brushwork, primer		1 Pord	10	.800	Ea.	.83	24		24.83	40
0610	Finish coat, enamel			10	.800		1.11	24		25.11	40
0620	Primer & 1 coat enamel			6	1.333		1.94	40		41.94	67
0630	Primer & 2 coats enamel			5	1.600		3.05	48		51.05	81.50
0650	For oil base paint, add						10%				

09 91 23.39 Doors and Windows, Interior Latex, Zero Voc			Crew	Daily Output	Labor-Hours	Unit	Material	Labor	Equipment	Total	Total Incl O&P
0010	**DOORS & WINDOWS, INTERIOR LATEX, ZERO VOC**										
0100	Doors flush, both sides, incl. frame & trim										
0110	Roll & brush, primer	G	1 Pord	10	.800	Ea.	6.40	24		30.40	46
0120	Finish coat, latex	G		10	.800		6.95	24		30.95	46.50
0130	Primer & 1 coat latex	G		7	1.143		13.35	34		47.35	70.50
0140	Primer & 2 coats latex	G		5	1.600		19.90	48		67.90	100
0160	Spray, both sides, primer	G		20	.400		6.75	11.95		18.70	27
0170	Finish coat, latex	G		20	.400		7.30	11.95		19.25	27.50
0180	Primer & 1 coat latex	G		11	.727		14.15	21.50		35.65	51
0190	Primer & 2 coats latex	G		8	1		21	30		51	72
0200	Doors, French, both sides, 10-15 lite, incl. frame & trim										
0210	Roll & brush, primer	G	1 Pord	6	1.333	Ea.	3.21	40		43.21	68.50
0220	Finish coat, latex	G		6	1.333		3.47	40		43.47	69
0230	Primer & 1 coat latex	G		3	2.667		6.70	79.50		86.20	137
0240	Primer & 2 coats latex	G		2	4		9.95	119		128.95	207
0360	Doors, panel, both sides, incl. frame & trim										
0370	Roll & brush, primer	G	1 Pord	6	1.333	Ea.	6.75	40		46.75	72.50
0380	Finish coat, latex	G		6	1.333		6.95	40		46.95	72.50
0390	Primer & 1 coat, latex	G		3	2.667		13.35	79.50		92.85	145
0400	Primer & 2 coats, latex	G		2.50	3.200		20.50	95.50		116	179
0420	Spray, both sides, primer	G		10	.800		6.75	24		30.75	46.50
0430	Finish coat, latex	G		10	.800		7.30	24		31.30	47
0440	Primer & 1 coat, latex	G		5	1.600		14.15	48		62.15	93.50
0450	Primer & 2 coats, latex	G		4	2		21.50	59.50		81	122

09 91 23.39 Doors and Windows, Interior Latex, Zero Voc		Crew	Daily Output	Labor-Hours	Unit	Material	2019 Bare Costs Labor	Equipment	Total	Total Incl O&P
0460	Windows, per interior side, based on 15 S.F.									
0470	1 to 6 lite									
0480	Brushwork, primer G	1 Pord	13	.615	Ea.	1.27	18.35		19.62	31.50
0490	Finish coat, enamel G		13	.615		1.37	18.35		19.72	31.50
0500	Primer & 1 coat enamel G		8	1		2.64	30		32.64	52
0510	Primer & 2 coats enamel G		6	1.333		4.01	40		44.01	69.50

09 91 23.40 Floors, Interior		Crew	Daily Output	Labor-Hours	Unit	Material	Labor	Equipment	Total	Total Incl O&P
0010	**FLOORS, INTERIOR**									
0100	Concrete paint, latex									
0110	Brushwork									
0120	1st coat	1 Pord	975	.008	S.F.	.15	.25		.40	.57
0130	2nd coat		1150	.007		.10	.21		.31	.45
0140	3rd coat		1300	.006		.08	.18		.26	.39
0150	Roll									
0160	1st coat	1 Pord	2600	.003	S.F.	.20	.09		.29	.37
0170	2nd coat		3250	.002		.12	.07		.19	.25
0180	3rd coat		3900	.002		.09	.06		.15	.20
0190	Spray									
0200	1st coat	1 Pord	2600	.003	S.F.	.17	.09		.26	.34
0210	2nd coat		3250	.002		.09	.07		.16	.22
0220	3rd coat		3900	.002		.08	.06		.14	.18

09 91 23.52 Miscellaneous, Interior		Crew	Daily Output	Labor-Hours	Unit	Material	Labor	Equipment	Total	Total Incl O&P
0010	**MISCELLANEOUS, INTERIOR**									
2400	Floors, conc./wood, oil base, primer/sealer coat, brushwork	2 Pord	1950	.008	S.F.	.09	.25		.34	.50
2450	Roller		5200	.003		.09	.09		.18	.25
2600	Spray		6000	.003		.09	.08		.17	.23
2650	Paint 1 coat, brushwork		1950	.008		.11	.25		.36	.52
2800	Roller		5200	.003		.11	.09		.20	.27
2850	Spray		6000	.003		.12	.08		.20	.26
3000	Stain, wood floor, brushwork, 1 coat		4550	.004		.10	.11		.21	.28
3200	Roller		5200	.003		.11	.09		.20	.27
3250	Spray		6000	.003		.11	.08		.19	.25
3400	Varnish, wood floor, brushwork		4550	.004		.09	.11		.20	.27
3450	Roller		5200	.003		.10	.09		.19	.26
3600	Spray		6000	.003		.10	.08		.18	.24
3800	Grilles, per side, oil base, primer coat, brushwork	1 Pord	520	.015		.13	.46		.59	.89
3850	Spray		1140	.007		.14	.21		.35	.49
3880	Paint 1 coat, brushwork		520	.015		.22	.46		.68	1
3900	Spray		1140	.007		.25	.21		.46	.61
3920	Paint 2 coats, brushwork		325	.025		.43	.73		1.16	1.68
3940	Spray		650	.012		.50	.37		.87	1.15
4500	Louvers, 1 side, primer, brushwork		524	.015		.09	.46		.55	.85
4520	Paint 1 coat, brushwork		520	.015		.10	.46		.56	.86
4530	Spray		1140	.007		.11	.21		.32	.46
4540	Paint 2 coats, brushwork		325	.025		.20	.73		.93	1.42
4550	Spray		650	.012		.22	.37		.59	.84
4560	Paint 3 coats, brushwork		270	.030		.29	.88		1.17	1.77
4570	Spray		500	.016		.33	.48		.81	1.14
5000	Pipe, 1"-4" diameter, primer or sealer coat, oil base, brushwork	2 Pord	1250	.013	L.F.	.09	.38		.47	.73
5100	Spray		2165	.007		.09	.22		.31	.46
5200	Paint 1 coat, brushwork		1250	.013		.12	.38		.50	.76
5300	Spray		2165	.007		.10	.22		.32	.47

09 91 23 – Interior Painting

09 91 23.52 Miscellaneous, Interior		Crew	Daily Output	Labor-Hours	Unit	Material	Labor	Equipment	Total	Total Incl O&P
						2019 Bare Costs				
5350	Paint 2 coats, brushwork	2 Pord	775	.021	L.F.	.21	.62		.83	1.24
5400	Spray		1240	.013		.23	.39		.62	.88
5450	5"-8" diameter, primer or sealer coat, brushwork		620	.026		.19	.77		.96	1.47
5500	Spray		1085	.015		.31	.44		.75	1.06
5550	Paint 1 coat, brushwork		620	.026		.32	.77		1.09	1.61
5600	Spray		1085	.015		.35	.44		.79	1.11
5650	Paint 2 coats, brushwork		385	.042		.42	1.24		1.66	2.49
5700	Spray		620	.026		.46	.77		1.23	1.77
6600	Radiators, per side, primer, brushwork	1 Pord	520	.015	S.F.	.09	.46		.55	.85
6620	Paint, 1 coat		520	.015		.08	.46		.54	.84
6640	2 coats		340	.024		.20	.70		.90	1.37
6660	3 coats		283	.028		.29	.84		1.13	1.70
7000	Trim, wood, incl. puttying, under 6" wide									
7200	Primer coat, oil base, brushwork	1 Pord	650	.012	L.F.	.03	.37		.40	.64
7250	Paint, 1 coat, brushwork		650	.012		.06	.37		.43	.66
7400	2 coats		400	.020		.11	.60		.71	1.10
7450	3 coats		325	.025		.16	.73		.89	1.38
7500	Over 6" wide, primer coat, brushwork		650	.012		.07	.37		.44	.67
7550	Paint, 1 coat, brushwork		650	.012		.11	.37		.48	.72
7600	2 coats		400	.020		.22	.60		.82	1.22
7650	3 coats		325	.025		.32	.73		1.05	1.55
8000	Cornice, simple design, primer coat, oil base, brushwork		650	.012	S.F.	.07	.37		.44	.67
8250	Paint, 1 coat		650	.012		.11	.37		.48	.72
8300	2 coats		400	.020		.22	.60		.82	1.22
8350	Ornate design, primer coat		350	.023		.07	.68		.75	1.19
8400	Paint, 1 coat		350	.023		.11	.68		.79	1.24
8450	2 coats		400	.020		.22	.60		.82	1.22
8600	Balustrades, primer coat, oil base, brushwork		520	.015		.07	.46		.53	.82
8650	Paint, 1 coat		520	.015		.11	.46		.57	.87
8700	2 coats		325	.025		.22	.73		.95	1.44
8900	Trusses and wood frames, primer coat, oil base, brushwork		800	.010		.07	.30		.37	.56
8950	Spray		1200	.007		.07	.20		.27	.40
9000	Paint 1 coat, brushwork		750	.011		.11	.32		.43	.64
9200	Spray		1200	.007		.12	.20		.32	.47
9220	Paint 2 coats, brushwork		500	.016		.22	.48		.70	1.02
9240	Spray		600	.013		.24	.40		.64	.91
9260	Stain, brushwork, wipe off		600	.013		.10	.40		.50	.76
9280	Varnish, 3 coats, brushwork		275	.029		.28	.87		1.15	1.72
9350	For latex paint, deduct					10%				

09 91 23.72 Walls and Ceilings, Interior		Crew	Daily Output	Labor-Hours	Unit	Material	Labor	Equipment	Total	Total Incl O&P
0010	**WALLS AND CEILINGS, INTERIOR**									
0100	Concrete, drywall or plaster, latex, primer or sealer coat									
0200	Smooth finish, brushwork	1 Pord	1150	.007	S.F.	.06	.21		.27	.41
0240	Roller		1350	.006		.06	.18		.24	.36
0280	Spray		2750	.003		.06	.09		.15	.20
0300	Sand finish, brushwork		975	.008		.06	.25		.31	.47
0340	Roller		1150	.007		.06	.21		.27	.41
0380	Spray		2275	.004		.06	.11		.17	.23
0400	Paint 1 coat, smooth finish, brushwork		1200	.007		.07	.20		.27	.41
0440	Roller		1300	.006		.07	.18		.25	.38
0480	Spray		2275	.004		.06	.11		.17	.24
0500	Sand finish, brushwork		1050	.008		.07	.23		.30	.44

09 91 23.72 Walls and Ceilings, Interior		Crew	Daily Output	Labor-Hours	Unit	2019 Bare Costs Material	Labor	Equipment	Total	Total Incl O&P
0540	Roller	1 Pord	1600	.005	S.F.	.07	.15		.22	.32
0580	Spray		2100	.004		.02	.11		.13	.22
0800	Paint 2 coats, smooth finish, brushwork		680	.012		.15	.35		.50	.74
0840	Roller		800	.010		.15	.30		.45	.65
0880	Spray		1625	.005		.13	.15		.28	.39
0900	Sand finish, brushwork		605	.013		.15	.39		.54	.81
0940	Roller		1020	.008		.15	.23		.38	.54
0980	Spray		1700	.005		.13	.14		.27	.38
1200	Paint 3 coats, smooth finish, brushwork		510	.016		.22	.47		.69	1.01
1240	Roller		650	.012		.22	.37		.59	.84
1280	Spray		850	.009		.20	.28		.48	.68
1300	Sand finish, brushwork		454	.018		.33	.53		.86	1.22
1340	Roller		680	.012		.35	.35		.70	.97
1380	Spray		1133	.007		.30	.21		.51	.68
1600	Glaze coating, 2 coats, spray, clear		1200	.007		.56	.20		.76	.95
1640	Multicolor		1200	.007		.87	.20		1.07	1.28
1660	Painting walls, complete, including surface prep, primer &									
1670	2 coats finish, on drywall or plaster, with roller	1 Pord	325	.025	S.F.	.22	.73		.95	1.44
1700	For oil base paint, add					10%				
1800	For ceiling installations, add						25%			
2000	Masonry or concrete block, primer/sealer, latex paint									
2100	Primer, smooth finish, brushwork	1 Pord	1000	.008	S.F.	.15	.24		.39	.56
2110	Roller		1150	.007		.11	.21		.32	.46
2180	Spray		2400	.003		.10	.10		.20	.27
2200	Sand finish, brushwork		850	.009		.11	.28		.39	.58
2210	Roller		975	.008		.11	.25		.36	.52
2280	Spray		2050	.004		.10	.12		.22	.30
2400	Finish coat, smooth finish, brush		1100	.007		.09	.22		.31	.45
2410	Roller		1300	.006		.09	.18		.27	.39
2480	Spray		2400	.003		.07	.10		.17	.24
2500	Sand finish, brushwork		950	.008		.09	.25		.34	.50
2510	Roller		1090	.007		.09	.22		.31	.45
2580	Spray		2040	.004		.07	.12		.19	.27
2800	Primer plus one finish coat, smooth brush		525	.015		.31	.45		.76	1.09
2810	Roller		615	.013		.20	.39		.59	.86
2880	Spray		1200	.007		.17	.20		.37	.52
2900	Sand finish, brushwork		450	.018		.20	.53		.73	1.09
2910	Roller		515	.016		.20	.46		.66	.98
2980	Spray		1025	.008		.17	.23		.40	.57
3200	Primer plus 2 finish coats, smooth, brush		355	.023		.28	.67		.95	1.41
3210	Roller		415	.019		.28	.58		.86	1.25
3280	Spray		800	.010		.25	.30		.55	.76
3300	Sand finish, brushwork		305	.026		.28	.78		1.06	1.59
3310	Roller		350	.023		.28	.68		.96	1.43
3380	Spray		675	.012		.25	.35		.60	.85
3600	Glaze coating, 3 coats, spray, clear		900	.009		.80	.27		1.07	1.31
3620	Multicolor		900	.009		1.06	.27		1.33	1.60
4000	Block filler, 1 coat, brushwork		425	.019		.13	.56		.69	1.06
4100	Silicone, water repellent, 2 coats, spray		2000	.004		.48	.12		.60	.73
4120	For oil base paint, add					10%				
8200	For work 8'-15' H, add						10%			
8300	For work over 15' H, add						20%			
8400	For light textured surfaces, add						10%			

09 91 23 – Interior Painting

09 91 23.72 Walls and Ceilings, Interior		Crew	Daily Output	Labor-Hours	Unit	Material	2019 Bare Costs Labor	Equipment	Total	Total Incl O&P
8410	Heavy textured, add				S.F.		25%			

09 91 23.74 Walls and Ceilings, Interior, Zero VOC Latex			Crew	Daily Output	Labor-Hours	Unit	Material	Labor	Equipment	Total	Total Incl O&P
0010	**WALLS AND CEILINGS, INTERIOR, ZERO VOC LATEX**										
0100	Concrete, dry wall or plaster, latex, primer or sealer coat										
0200	Smooth finish, brushwork	G	1 Pord	1150	.007	S.F.	.08	.21		.29	.43
0240	Roller	G		1350	.006		.08	.18		.26	.38
0280	Spray	G		2750	.003		.06	.09		.15	.21
0300	Sand finish, brushwork	G		975	.008		.08	.25		.33	.49
0340	Roller	G		1150	.007		.09	.21		.30	.44
0380	Spray	G		2275	.004		.07	.11		.18	.25
0400	Paint 1 coat, smooth finish, brushwork	G		1200	.007		.09	.20		.29	.43
0440	Roller	G		1300	.006		.09	.18		.27	.40
0480	Spray	G		2275	.004		.08	.11		.19	.26
0500	Sand finish, brushwork	G		1050	.008		.09	.23		.32	.47
0540	Roller	G		1600	.005		.09	.15		.24	.34
0580	Spray	G		2100	.004		.08	.11		.19	.28
0800	Paint 2 coats, smooth finish, brushwork	G		680	.012		.18	.35		.53	.78
0840	Roller	G		800	.010		.19	.30		.49	.70
0880	Spray	G		1625	.005		.16	.15		.31	.42
0900	Sand finish, brushwork	G		605	.013		.18	.39		.57	.85
0940	Roller	G		1020	.008		.19	.23		.42	.59
0980	Spray	G		1700	.005		.16	.14		.30	.41
1200	Paint 3 coats, smooth finish, brushwork	G		510	.016		.27	.47		.74	1.07
1240	Roller	G		650	.012		.28	.37		.65	.91
1280	Spray	G	↓	850	.009		.24	.28		.52	.73
1800	For ceiling installations, add	G						25%			
8200	For work 8' - 15' H, add							10%			
8300	For work over 15' H, add					↓		20%			

09 91 23.75 Dry Fall Painting		Crew	Daily Output	Labor-Hours	Unit	Material	Labor	Equipment	Total	Total Incl O&P
0010	**DRY FALL PAINTING**									
0100	Sprayed on walls, gypsum board or plaster									
0220	One coat	1 Pord	2600	.003	S.F.	.08	.09		.17	.24
0250	Two coats		1560	.005		.16	.15		.31	.43
0280	Concrete or textured plaster, one coat		1560	.005		.08	.15		.23	.34
0310	Two coats		1300	.006		.16	.18		.34	.48
0340	Concrete block, one coat		1560	.005		.08	.15		.23	.34
0370	Two coats		1300	.006		.16	.18		.34	.48
0400	Wood, one coat		877	.009		.08	.27		.35	.54
0430	Two coats	↓	650	.012	↓	.16	.37		.53	.78
0440	On ceilings, gypsum board or plaster									
0470	One coat	1 Pord	1560	.005	S.F.	.08	.15		.23	.34
0500	Two coats		1300	.006		.16	.18		.34	.48
0530	Concrete or textured plaster, one coat		1560	.005		.08	.15		.23	.34
0560	Two coats		1300	.006		.16	.18		.34	.48
0570	Structural steel, bar joists or metal deck, one coat		1560	.005		.08	.15		.23	.34
0580	Two coats	↓	1040	.008	↓	.16	.23		.39	.56

09 93 Staining and Transparent Finishing

09 93 23 – Interior Staining and Finishing

09 93 23.10 Varnish		Crew	Daily Output	Labor-Hours	Unit	2019 Bare Costs Material	Labor	Equipment	Total	Total Incl O&P
0010	**VARNISH**									
0012	1 coat + sealer, on wood trim, brush, no sanding included	1 Pord	400	.020	S.F.	.07	.60		.67	1.06
0020	1 coat + sealer, on wood trim, brush, no sanding included, no VOC		400	.020		.22	.60		.82	1.22
0100	Hardwood floors, 2 coats, no sanding included, roller	↓	1890	.004	↓	.15	.13		.28	.38

09 96 High-Performance Coatings

09 96 56 – Epoxy Coatings

09 96 56.20 Wall Coatings

		Crew	Daily Output	Labor-Hours	Unit	Material	Labor	Equipment	Total	Total Incl O&P
0010	**WALL COATINGS**									
0100	Acrylic glazed coatings, matte	1 Pord	525	.015	S.F.	.37	.45		.82	1.16
0200	Gloss		305	.026		.78	.78		1.56	2.14
0300	Epoxy coatings, solvent based		525	.015		.47	.45		.92	1.27
0400	Water based		170	.047		.33	1.40		1.73	2.66
0600	Exposed aggregate, troweled on, 1/16" to 1/4", solvent based		235	.034		.74	1.02		1.76	2.47
0700	Water based (epoxy or polyacrylate)		130	.062		1.58	1.84		3.42	4.75
0900	1/2" to 5/8" aggregate, solvent based		130	.062		1.42	1.84		3.26	4.57
1000	Water based		80	.100		2.47	2.99		5.46	7.60
1200	1" aggregate size, solvent based		90	.089		2.52	2.65		5.17	7.10
1300	Water based		55	.145		3.83	4.34		8.17	11.30
1500	Exposed aggregate, sprayed on, 1/8" aggregate, solvent based		295	.027		.58	.81		1.39	1.97
1600	Water based	↓	145	.055	↓	1.24	1.65		2.89	4.06

09 97 Special Coatings

09 97 35 – Dry Erase Coatings

09 97 35.10 Dry Erase Coatings

		Crew	Daily Output	Labor-Hours	Unit	Material	Labor	Equipment	Total	Total Incl O&P
0010	**DRY ERASE COATINGS**									
0020	Dry erase coatings, clear, roller applied	1 Pord	1325	.006	S.F.	2.09	.18		2.27	2.60

Estimating Tips

General

- The items in this division are usually priced per square foot or each.
- Many items in Division 10 require some type of support system or special anchors that are not usually furnished with the item. The required anchors must be added to the estimate in the appropriate division.
- Some items in Division 10, such as lockers, may require assembly before installation. Verify the amount of assembly required. Assembly can often exceed installation time.

10 20 00 Interior Specialties

- Support angles and blocking are not included in the installation of toilet compartments, shower/dressing compartments, or cubicles. Appropriate line items from Division 5 or 6 may need to be added to support the installations.
- Toilet partitions are priced by the stall. A stall consists of a side wall, pilaster, and door with hardware. Toilet tissue holders and grab bars are extra.
- The required acoustical rating of a folding partition can have a significant impact on costs. Verify the sound transmission coefficient rating of the panel priced against the specification requirements.
- Grab bar installation does not include supplemental blocking or backing to support the required load. When grab bars are installed at an existing facility, provisions must be made to attach the grab bars to a solid structure.

Reference Numbers

Reference numbers are shown at the beginning of some major classifications. These numbers refer to related items in the Reference Section. The reference information may be an estimating procedure, an alternate pricing method, or technical information.

Note: Not all subdivisions listed here necessarily appear. ■

10 28 Toilet, Bath, and Laundry Accessories

10 28 13 – Toilet Accessories

10 28 13.13 Commercial Toilet Accessories		Crew	Daily Output	Labor-Hours	Unit	Material	Labor	Equipment	Total	Total Incl O&P
						2019 Bare Costs				
0010	COMMERCIAL TOILET ACCESSORIES									
0200	Curtain rod, stainless steel, 5′ long, 1″ diameter	1 Carp	13	.615	Ea.	27	22		49	66.50
0300	1-1/4″ diameter		13	.615		29.50	22		51.50	68.50
0800	Grab bar, straight, 1-1/4″ diameter, stainless steel, 18″ long		24	.333		30	11.90		41.90	52.50
1100	36″ long		20	.400		34	14.25		48.25	61
1105	42″ long		20	.400		38.50	14.25		52.75	65.50
1120	Corner, 36″ long		20	.400		86.50	14.25		100.75	119
3000	Mirror, with stainless steel 3/4″ square frame, 18″ x 24″		20	.400		47	14.25		61.25	75
3100	36″ x 24″		15	.533		117	19		136	161
3300	72″ x 24″		6	1.333		291	47.50		338.50	400
4300	Robe hook, single, regular		96	.083		19.95	2.97		22.92	27
4400	Heavy duty, concealed mounting		56	.143		22.50	5.10		27.60	33
6400	Towel bar, stainless steel, 18″ long		23	.348		42	12.40		54.40	67
6500	30″ long		21	.381		51	13.60		64.60	78.50
7400	Tumbler holder, for tumbler only		30	.267		27	9.50		36.50	45
7410	Tumbler holder, recessed		20	.400		7.85	14.25		22.10	32
7500	Soap, tumbler & toothbrush		30	.267		19.80	9.50		29.30	37.50
7510	Tumbler & toothbrush holder		20	.400		13.30	14.25		27.55	38

10 28 16 – Bath Accessories

10 28 16.20 Medicine Cabinets		Crew	Daily Output	Labor-Hours	Unit	Material	Labor	Equipment	Total	Total Incl O&P
0010	MEDICINE CABINETS									
0020	With mirror, sst frame, 16″ x 22″, unlighted	1 Carp	14	.571	Ea.	96.50	20.50		117	140
0100	Wood frame		14	.571		136	20.50		156.50	184
0300	Sliding mirror doors, 20″ x 16″ x 4-3/4″, unlighted		7	1.143		126	40.50		166.50	207
0400	24″ x 19″ x 8-1/2″, lighted		5	1.600		216	57		273	330
0600	Triple door, 30″ x 32″, unlighted, plywood body		7	1.143		355	40.50		395.50	460
0700	Steel body		7	1.143		355	40.50		395.50	460
0900	Oak door, wood body, beveled mirror, single door		7	1.143		176	40.50		216.50	261
1000	Double door		6	1.333		370	47.50		417.50	485

10 28 19 – Tub and Shower Enclosures

10 28 19.10 Partitions, Shower		Crew	Daily Output	Labor-Hours	Unit	Material	Labor	Equipment	Total	Total Incl O&P
0010	**PARTITIONS, SHOWER** floor mounted, no plumbing									
0400	Cabinet, one piece, fiberglass, 32″ x 32″	2 Carp	5	3.200	Ea.	620	114		734	875
0420	36″ x 36″		5	3.200		600	114		714	850
0440	36″ x 48″		5	3.200		1,375	114		1,489	1,725
0460	Acrylic, 32″ x 32″		5	3.200		355	114		469	580
0480	36″ x 36″		5	3.200		1,075	114		1,189	1,375
0500	36″ x 48″		5	3.200		1,300	114		1,414	1,625
0520	Shower door for above, clear plastic, 24″ wide	1 Carp	8	1		186	35.50		221.50	263
0540	28″ wide		8	1		246	35.50		281.50	330
0560	Tempered glass, 24″ wide		8	1		241	35.50		276.50	325
0580	28″ wide		8	1		281	35.50		316.50	370
2400	Glass stalls, with doors, no receptors, chrome on brass	2 Shee	3	5.333		1,650	208		1,858	2,150
2700	Anodized aluminum	″	4	4		1,325	156		1,481	1,700
3200	Receptors, precast terrazzo, 32″ x 32″	2 Marb	14	1.143		360	40		400	460
3300	48″ x 34″		9.50	1.684		470	59		529	615
3500	Plastic, simulated terrazzo receptor, 32″ x 32″		14	1.143		170	40		210	254
3600	32″ x 48″		12	1.333		300	46.50		346.50	410
3800	Precast concrete, colors, 32″ x 32″		14	1.143		250	40		290	340
3900	48″ x 48″		8	2		277	70		347	420
4100	Shower doors, economy plastic, 24″ wide	1 Shee	9	.889		135	34.50		169.50	207
4200	Tempered glass door, economy		8	1		295	39		334	390

10 28 Toilet, Bath, and Laundry Accessories

10 28 19 – Tub and Shower Enclosures

10 28 19.10 Partitions, Shower		Crew	Daily Output	Labor-Hours	Unit	Material	Labor	Equipment	Total	Total Incl O&P
						2019 Bare Costs				
4400	Folding, tempered glass, aluminum frame	1 Shee	6	1.333	Ea.	435	52		487	560
4700	Deluxe, tempered glass, chrome on brass frame, 42" to 44"		8	1		435	39		474	545
4800	39" to 48" wide		1	8		665	310		975	1,250
4850	On anodized aluminum frame, obscure glass		2	4		580	156		736	895
4900	Clear glass		1	8		655	310		965	1,250
5100	Shower enclosure, tempered glass, anodized alum. frame									
5120	2 panel & door, corner unit, 32" x 32"	1 Shee	2	4	Ea.	970	156		1,126	1,325
5140	Neo-angle corner unit, 16" x 24" x 16"	"	2	4		1,150	156		1,306	1,500
5200	Shower surround, 3 wall, polypropylene, 32" x 32"	1 Carp	4	2		615	71.50		686.50	795
5220	PVC, 32" x 32"		4	2		400	71.50		471.50	560
5240	Fiberglass		4	2		395	71.50		466.50	555
5250	2 wall, polypropylene, 32" x 32"		4	2		310	71.50		381.50	465
5270	PVC		4	2		375	71.50		446.50	530
5290	Fiberglass		4	2		375	71.50		446.50	535
5300	Tub doors, tempered glass & frame, obscure glass	1 Shee	8	1		222	39		261	310
5400	Clear glass		6	1.333		515	52		567	650
5600	Chrome plated, brass frame, obscure glass		8	1		291	39		330	385
5700	Clear glass		6	1.333		715	52		767	875
5900	Tub/shower enclosure, temp. glass, alum. frame, obscure glass		2	4		395	156		551	695
6200	Clear glass		1.50	5.333		825	208		1,033	1,250
6500	On chrome-plated brass frame, obscure glass		2	4		545	156		701	860
6600	Clear glass		1.50	5.333		1,175	208		1,383	1,625
6800	Tub surround, 3 wall, polypropylene	1 Carp	4	2		253	71.50		324.50	395
6900	PVC		4	2		370	71.50		441.50	525
7000	Fiberglass, obscure glass		4	2		390	71.50		461.50	545
7100	Clear glass		3	2.667		655	95		750	880

10 28 23 – Laundry Accessories

10 28 23.13 Built-In Ironing Boards		Crew	Daily Output	Labor-Hours	Unit	Material	Labor	Equipment	Total	Total Incl O&P
0010	**BUILT-IN IRONING BOARDS**									
0020	Including cabinet, board & light, 42"	1 Carp	2	4	Ea.	445	143		588	725

10 31 Manufactured Fireplaces

10 31 13 – Manufactured Fireplace Chimneys

10 31 13.10 Fireplace Chimneys		Crew	Daily Output	Labor-Hours	Unit	Material	Labor	Equipment	Total	Total Incl O&P
0010	**FIREPLACE CHIMNEYS**									
0500	Chimney dbl. wall, all stainless, over 8'-6", 7" diam., add to fireplace	1 Carp	33	.242	V.L.F.	88	8.65		96.65	111
0600	10" diameter, add to fireplace		32	.250		111	8.90		119.90	137
0700	12" diameter, add to fireplace		31	.258		173	9.20		182.20	205
0800	14" diameter, add to fireplace		30	.267		223	9.50		232.50	261
1000	Simulated brick chimney top, 4' high, 16" x 16"		10	.800	Ea.	455	28.50		483.50	545
1100	24" x 24"		7	1.143	"	560	40.50		600.50	685

10 31 13.20 Chimney Accessories		Crew	Daily Output	Labor-Hours	Unit	Material	Labor	Equipment	Total	Total Incl O&P
0010	**CHIMNEY ACCESSORIES**									
0020	Chimney screens, galv., 13" x 13" flue	1 Bric	8	1	Ea.	57	35.50		92.50	122
0050	24" x 24" flue		5	1.600		124	56.50		180.50	230
0200	Stainless steel, 13" x 13" flue		8	1		97.50	35.50		133	166
0250	20" x 20" flue		5	1.600		152	56.50		208.50	261
2400	Squirrel and bird screens, galvanized, 8" x 8" flue		16	.500		54.50	17.65		72.15	89.50
2450	13" x 13" flue		12	.667		66	23.50		89.50	112

10 31 Manufactured Fireplaces

10 31 16 – Manufactured Fireplace Forms

10 31 16.10 Fireplace Forms	Crew	Daily Output	Labor-Hours	Unit	Material	2019 Bare Costs Labor	Equipment	Total	Total Incl O&P
0010 **FIREPLACE FORMS**									
1800 Fireplace forms, no accessories, 32″ opening	1 Bric	3	2.667	Ea.	750	94		844	980
1900 36″ opening		2.50	3.200		955	113		1,068	1,250
2000 40″ opening		2	4		1,275	141		1,416	1,625
2100 78″ opening	↓	1.50	5.333	↓	1,850	188		2,038	2,350

10 31 23 – Prefabricated Fireplaces

10 31 23.10 Fireplace, Prefabricated	Crew	Daily Output	Labor-Hours	Unit	Material	Labor	Equipment	Total	Total Incl O&P
0010 **FIREPLACE, PREFABRICATED**, free standing or wall hung									
0100 With hood & screen, painted	1 Carp	1.30	6.154	Ea.	1,625	219		1,844	2,125
0150 Average		1	8		1,800	285		2,085	2,475
0200 Stainless steel		.90	8.889	↓	3,225	315		3,540	4,075
1500 Simulated logs, gas fired, 40,000 BTU, 2′ long, manual safety pilot		7	1.143	Set	540	40.50		580.50	665
1600 Adjustable flame remote pilot		6	1.333		1,275	47.50		1,322.50	1,475
1700 Electric, 1,500 BTU, 1′-6″ long, incandescent flame		7	1.143		262	40.50		302.50	355
1800 1,500 BTU, LED flame		6	1.333	↓	345	47.50		392.50	460
2000 Fireplace, built-in, 36″ hearth, radiant		1.30	6.154	Ea.	720	219		939	1,150
2100 Recirculating, small fan		1	8		910	285		1,195	1,475
2150 Large fan		.90	8.889		2,025	315		2,340	2,750
2200 42″ hearth, radiant		1.20	6.667		1,125	238		1,363	1,625
2300 Recirculating, small fan		.90	8.889		1,250	315		1,565	1,900
2350 Large fan		.80	10		1,450	355		1,805	2,175
2400 48″ hearth, radiant		1.10	7.273		2,375	259		2,634	3,025
2500 Recirculating, small fan		.80	10		2,900	355		3,255	3,800
2550 Large fan		.70	11.429		2,600	405		3,005	3,525
3000 See through, including doors		.80	10		2,350	355		2,705	3,200
3200 Corner (2 wall)	↓	1	8	↓	3,275	285		3,560	4,075

10 32 Fireplace Specialties

10 32 13 – Fireplace Dampers

10 32 13.10 Dampers	Crew	Daily Output	Labor-Hours	Unit	Material	Labor	Equipment	Total	Total Incl O&P
0010 **DAMPERS**									
0800 Damper, rotary control, steel, 30″ opening	1 Bric	6	1.333	Ea.	119	47		166	210
0850 Cast iron, 30″ opening		6	1.333		124	47		171	216
1200 Steel plate, poker control, 60″ opening		8	1		320	35.50		355.50	415
1250 84″ opening, special order		5	1.600		590	56.50		646.50	740
1400 "Universal" type, chain operated, 32″ x 20″ opening		8	1		251	35.50		286.50	335
1450 48″ x 24″ opening	↓	5	1.600	↓	375	56.50		431.50	505

10 32 23 – Fireplace Doors

10 32 23.10 Doors	Crew	Daily Output	Labor-Hours	Unit	Material	Labor	Equipment	Total	Total Incl O&P
0010 **DOORS**									
0400 Cleanout doors and frames, cast iron, 8″ x 8″	1 Bric	12	.667	Ea.	55	23.50		78.50	99.50
0450 12″ x 12″		10	.800		89	28		117	145
0500 18″ x 24″		8	1		150	35.50		185.50	224
0550 Cast iron frame, steel door, 24″ x 30″		5	1.600		315	56.50		371.50	440
1600 Dutch oven door and frame, cast iron, 12″ x 15″ opening		13	.615		131	21.50		152.50	181
1650 Copper plated, 12″ x 15″ opening	↓	13	.615	↓	257	21.50		278.50	320

10 35 Stoves

10 35 13 – Heating Stoves

10 35 13.10 Wood Burning Stoves		Crew	Daily Output	Labor-Hours	Unit	Material	2019 Bare Costs Labor	Equipment	Total	Total Incl O&P
0010	**WOOD BURNING STOVES**									
0015	Cast iron, less than 1,500 S.F.	2 Carp	1.30	12.308	Ea.	1,375	440		1,815	2,250
0020	1,500 to 2,000 S.F.		1	16		2,350	570		2,920	3,550
0030	greater than 2,000 S.F.		.80	20		2,800	715		3,515	4,250
0050	For gas log lighter, add					48			48	53

10 44 Fire Protection Specialties

10 44 16 – Fire Extinguishers

10 44 16.13 Portable Fire Extinguishers		Crew	Daily Output	Labor-Hours	Unit	Material	Labor	Equipment	Total	Total Incl O&P
0010	**PORTABLE FIRE EXTINGUISHERS**									
0140	CO_2, with hose and "H" horn, 10 lb.				Ea.	295			295	325
1000	Dry chemical, pressurized									
1040	Standard type, portable, painted, 2-1/2 lb.				Ea.	40.50			40.50	44.50
1080	10 lb.					86.50			86.50	95.50
1100	20 lb.					138			138	152
1120	30 lb.					435			435	480
2000	ABC all purpose type, portable, 2-1/2 lb.					23.50			23.50	26
2080	9-1/2 lb.					51.50			51.50	56.50

10 55 Postal Specialties

10 55 23 – Mail Boxes

10 55 23.10 Mail Boxes		Crew	Daily Output	Labor-Hours	Unit	Material	Labor	Equipment	Total	Total Incl O&P
0011	**MAIL BOXES**									
1900	Letter slot, residential	1 Carp	20	.400	Ea.	80.50	14.25		94.75	112
2400	Residential, galv. steel, small 20" x 7" x 9"	1 Clab	16	.500		203	13.75		216.75	246
2410	With galv. steel post, 54" long		6	1.333		250	36.50		286.50	335
2420	Large, 24" x 12" x 15"		16	.500		204	13.75		217.75	248
2430	With galv. steel post, 54" long		6	1.333		238	36.50		274.50	325
2440	Decorative, polyethylene, 22" x 10" x 10"		16	.500		60	13.75		73.75	88.50
2450	With alum. post, decorative, 54" long		6	1.333		200	36.50		236.50	281

10 56 Storage Assemblies

10 56 13 – Metal Storage Shelving

10 56 13.10 Shelving		Crew	Daily Output	Labor-Hours	Unit	Material	Labor	Equipment	Total	Total Incl O&P
0010	**SHELVING**									
0020	Metal, industrial, cross-braced, 3' W, 12" D	1 Sswk	175	.046	SF Shlf	8.05	1.82		9.87	12.05
0100	24" D		330	.024		6.35	.96		7.31	8.65
2200	Wide span, 1600 lb. capacity per shelf, 6' W, 24" D		380	.021		7.40	.84		8.24	9.60
2400	36" D		440	.018		6.30	.72		7.02	8.20
3000	Residential, vinyl covered wire, wardrobe, 12" D	1 Carp	195	.041	L.F.	15.30	1.46		16.76	19.25
3100	16" D		195	.041		11.65	1.46		13.11	15.25
3200	Standard, 6" D		195	.041		3.80	1.46		5.26	6.60
3300	9" D		195	.041		6.70	1.46		8.16	9.80
3400	12" D		195	.041		7.95	1.46		9.41	11.10
3500	16" D		195	.041		18.10	1.46		19.56	22.50
3600	20" D		195	.041		21	1.46		22.46	26
3700	Support bracket		80	.100	Ea.	7.30	3.57		10.87	13.95

10 57 Wardrobe and Closet Specialties

10 57 23 – Closet and Utility Shelving

10 57 23.19 Wood Closet and Utility Shelving		Crew	Daily Output	Labor-Hours	Unit	Material	2019 Bare Costs Labor	Equipment	Total	Total Incl O&P
0010	WOOD CLOSET AND UTILITY SHELVING									
0020	Pine, clear grade, no edge band, 1" x 8"	1 Carp	115	.070	L.F.	3.61	2.48		6.09	8.05
0100	1" x 10"		110	.073		4.49	2.59		7.08	9.20
0200	1" x 12"		105	.076		5.40	2.72		8.12	10.45
0450	1" x 18"		95	.084		8.10	3		11.10	13.90
0460	1" x 24"		85	.094		10.80	3.36		14.16	17.45
0600	Plywood, 3/4" thick with lumber edge, 12" wide		75	.107		1.92	3.80		5.72	8.40
0700	24" wide		70	.114		3.39	4.07		7.46	10.50
0900	Bookcase, clear grade pine, shelves 12" OC, 8" deep, per S.F. shelf		70	.114	S.F.	11.70	4.07		15.77	19.65
1000	12" deep shelves		65	.123	"	17.60	4.39		21.99	26.50
1200	Adjustable closet rod and shelf, 12" wide, 3' long		20	.400	Ea.	13.25	14.25		27.50	38
1300	8' long		15	.533	"	25.50	19		44.50	59.50
1500	Prefinished shelves with supports, stock, 8" wide		75	.107	L.F.	5.80	3.80		9.60	12.65
1600	10" wide		70	.114	"	5.50	4.07		9.57	12.80

10 73 Protective Covers

10 73 16 – Canopies

10 73 16.10 Canopies, Residential		Crew	Daily Output	Labor-Hours	Unit	Material	Labor	Equipment	Total	Total Incl O&P
0010	CANOPIES, RESIDENTIAL Prefabricated									
0500	Carport, free standing, baked enamel, alum., .032", 40 psf									
0520	16' x 8', 4 posts	2 Carp	3	5.333	Ea.	4,725	190		4,915	5,525
0600	20' x 10', 6 posts		2	8		5,000	285		5,285	5,975
0605	30' x 10', 8 posts		2	8		7,500	285		7,785	8,725
1000	Door canopies, extruded alum., .032", 42" projection, 4' wide	1 Carp	8	1		205	35.50		240.50	285
1020	6' wide	"	6	1.333		284	47.50		331.50	395
1040	8' wide	2 Carp	9	1.778		460	63.50		523.50	610
1060	10' wide		7	2.286		530	81.50		611.50	715
1080	12' wide		5	3.200		580	114		694	825
1200	54" projection, 4' wide	1 Carp	8	1		231	35.50		266.50	315
1220	6' wide	"	6	1.333		310	47.50		357.50	425
1240	8' wide	2 Carp	9	1.778		375	63.50		438.50	515
1260	10' wide		7	2.286		810	81.50		891.50	1,025
1280	12' wide		5	3.200		875	114		989	1,150
1300	Painted, add					20%				
1310	Bronze anodized, add					50%				
3000	Window awnings, aluminum, window 3' high, 4' wide	1 Carp	10	.800		160	28.50		188.50	223
3020	6' wide	"	8	1		193	35.50		228.50	271
3040	9' wide	2 Carp	9	1.778		495	63.50		558.50	650
3060	12' wide	"	5	3.200		510	114		624	750
3100	Window, 4' high, 4' wide	1 Carp	10	.800		290	28.50		318.50	365
3120	6' wide	"	8	1		350	35.50		385.50	445
3140	9' wide	2 Carp	9	1.778		655	63.50		718.50	825
3160	12' wide	"	5	3.200		730	114		844	990
3200	Window, 6' high, 4' wide	1 Carp	10	.800		350	28.50		378.50	430
3220	6' wide	"	8	1		390	35.50		425.50	490
3240	9' wide	2 Carp	9	1.778		945	63.50		1,008.50	1,150
3260	12' wide	"	5	3.200		1,450	114		1,564	1,775
3400	Roll-up aluminum, 2'-6" wide	1 Carp	14	.571		207	20.50		227.50	262
3420	3' wide		12	.667		212	24		236	274
3440	4' wide		10	.800		266	28.50		294.50	340
3460	6' wide		8	1		285	35.50		320.50	375

10 73 Protective Covers

10 73 16 – Canopies

10 73 16.10 Canopies, Residential		Crew	Daily Output	Labor-Hours	Unit	Material	Labor	Equipment	Total	Total Incl O&P
						2019 Bare Costs				
3480	9′ wide	2 Carp	9	1.778	Ea.	405	63.50		468.50	550
3500	12′ wide	"	5	3.200	↓	520	114		634	760
3600	Window awnings, canvas, 24″ drop, 3′ wide	1 Carp	30	.267	L.F.	59.50	9.50		69	81
3620	4′ wide		40	.200		45.50	7.15		52.65	62
3700	30″ drop, 3′ wide		30	.267		59.50	9.50		69	81
3720	4′ wide		40	.200		50	7.15		57.15	67
3740	5′ wide		45	.178		46	6.35		52.35	61
3760	6′ wide		48	.167		42.50	5.95		48.45	57
3780	8′ wide		48	.167		39.50	5.95		45.45	53.50
3800	10′ wide	↓	50	.160	↓	41	5.70		46.70	55

10 74 Manufactured Exterior Specialties

10 74 23 – Cupolas

10 74 23.10 Wood Cupolas

Line	Description	Crew	Daily Output	Labor-Hours	Unit	Material	Labor	Equipment	Total	Total Incl O&P
0010	**WOOD CUPOLAS**									
0020	Stock units, pine, painted, 18″ sq., 28″ high, alum. roof	1 Carp	4.10	1.951	Ea.	266	69.50		335.50	410
0100	Copper roof		3.80	2.105		292	75		367	445
0300	23″ square, 33″ high, aluminum roof		3.70	2.162		460	77		537	630
0400	Copper roof		3.30	2.424		615	86.50		701.50	820
0600	30″ square, 37″ high, aluminum roof		3.70	2.162		610	77		687	795
0700	Copper roof		3.30	2.424		740	86.50		826.50	960
0900	Hexagonal, 31″ wide, 46″ high, copper roof		4	2		980	71.50		1,051.50	1,200
1000	36″ wide, 50″ high, copper roof	↓	3.50	2.286		1,750	81.50		1,831.50	2,050
1200	For deluxe stock units, add to above					25%				
1400	For custom built units, add to above				↓	50%	50%			

10 74 33 – Weathervanes

10 74 33.10 Residential Weathervanes

Line	Description	Crew	Daily Output	Labor-Hours	Unit	Material	Labor	Equipment	Total	Total Incl O&P
0010	**RESIDENTIAL WEATHERVANES**									
0020	Residential types, 18″ to 24″	1 Carp	8	1	Ea.	136	35.50		171.50	209
0100	24″ to 48″	"	2	4	"	1,850	143		1,993	2,275

10 74 46 – Window Wells

10 74 46.10 Area Window Wells

Line	Description	Crew	Daily Output	Labor-Hours	Unit	Material	Labor	Equipment	Total	Total Incl O&P
0010	**AREA WINDOW WELLS**, Galvanized steel									
0020	20 ga., 3′-2″ wide, 1′ deep	1 Sswk	29	.276	Ea.	17.45	10.95		28.40	38
0100	2′ deep		23	.348		30	13.85		43.85	57.50
0300	16 ga., 3′-2″ wide, 1′ deep		29	.276		23.50	10.95		34.45	44.50
0400	3′ deep		23	.348		48	13.85		61.85	77
0600	Welded grating for above, 15 lb., painted		45	.178		95.50	7.05		102.55	117
0700	Galvanized		45	.178		125	7.05		132.05	149
0900	Translucent plastic cap for above	↓	60	.133	↓	21	5.30		26.30	32

10 75 Flagpoles

10 75 16 – Ground-Set Flagpoles

10 75 16.10 Flagpoles		Crew	Daily Output	Labor-Hours	Unit	Material	2019 Bare Costs Labor	Equipment	Total	Total Incl O&P
0010	**FLAGPOLES**, ground set									
0050	Not including base or foundation									
0100	Aluminum, tapered, ground set 20' high	K-1	2	8	Ea.	1,100	272	123	1,495	1,775
0200	25' high		1.70	9.412		1,150	320	145	1,615	1,950
0300	30' high		1.50	10.667		1,400	360	164	1,924	2,325
0500	40' high	↓	1.20	13.333	↓	3,075	455	205	3,735	4,350

Estimating Tips

General

- The items in this division are usually priced per square foot or each. Many of these items are purchased by the owner for installation by the contractor. Check the specifications for responsibilities and include time for receiving, storage, installation, and mechanical and electrical hookups in the appropriate divisions.
- Many items in Division 11 require some type of support system that is not usually furnished with the item. Examples of these systems include blocking for the attachment of casework and support angles for ceiling-hung projection screens. The required blocking or supports must be added to the estimate in the appropriate division.
- Some items in Division 11 may require assembly or electrical hookups. Verify the amount of assembly required or the need for a hard electrical connection and add the appropriate costs.

Reference Numbers

Reference numbers are shown at the beginning of some major classifications. These numbers refer to related items in the Reference Section. The reference information may be an estimating procedure, an alternate pricing method, or technical information.

11 30 Residential Equipment

11 30 13 – Residential Appliances

11 30 13.15 Cooking Equipment		Crew	Daily Output	Labor-Hours	Unit	Material	2019 Bare Costs Labor	Equipment	Total	Total Incl O&P
0010	**COOKING EQUIPMENT**									
0020	Cooking range, 30″ free standing, 1 oven, minimum	2 Clab	10	1.600	Ea.	470	44		514	595
0050	Maximum		4	4		2,325	110		2,435	2,750
0150	2 oven, minimum		10	1.600		1,025	44		1,069	1,200
0200	Maximum		10	1.600		3,400	44		3,444	3,800
0350	Built-in, 30″ wide, 1 oven, minimum	1 Elec	6	1.333		855	55		910	1,025
0400	Maximum	2 Carp	2	8		1,750	285		2,035	2,400
0500	2 oven, conventional, minimum		4	4		1,250	143		1,393	1,575
0550	1 conventional, 1 microwave, maximum		2	8		2,325	285		2,610	3,025
0700	Free standing, 1 oven, 21″ wide range, minimum	2 Clab	10	1.600		475	44		519	595
0750	21″ wide, maximum	″	4	4		670	110		780	920
0900	Countertop cooktops, 4 burner, standard, minimum	1 Elec	6	1.333		325	55		380	445
0950	Maximum		3	2.667		1,850	110		1,960	2,225
1050	As above, but with grill and griddle attachment, minimum		6	1.333		1,425	55		1,480	1,675
1100	Maximum		3	2.667		3,925	110		4,035	4,500
1250	Microwave oven, minimum		4	2		96	83		179	241
1300	Maximum		2	4		455	166		621	770
5380	Oven, built-in, standard		4	2		935	83		1,018	1,150
5390	Deluxe		2	4		3,150	166		3,316	3,725

11 30 13.16 Refrigeration Equipment		Crew	Daily Output	Labor-Hours	Unit	Material	Labor	Equipment	Total	Total Incl O&P
0010	**REFRIGERATION EQUIPMENT**									
2000	Deep freeze, 15 to 23 C.F., minimum	2 Clab	10	1.600	Ea.	640	44		684	780
2050	Maximum		5	3.200		800	88		888	1,025
2200	30 C.F., minimum		8	2		795	55		850	965
2250	Maximum		3	5.333		890	147		1,037	1,225
5200	Icemaker, automatic, 20 lbs./day	1 Plum	7	1.143		1,400	46		1,446	1,600
5350	51 lbs./day	″	2	4		1,400	162		1,562	1,825
5450	Refrigerator, no frost, 6 C.F.	2 Clab	15	1.067		380	29.50		409.50	465
5500	Refrigerator, no frost, 10 C.F. to 12 C.F., minimum		10	1.600		450	44		494	570
5600	Maximum		6	2.667		545	73.50		618.50	720
5750	14 C.F. to 16 C.F., minimum		9	1.778		565	49		614	700
5800	Maximum		5	3.200		945	88		1,033	1,200
5950	18 C.F. to 20 C.F., minimum		8	2		715	55		770	880
6000	Maximum		4	4		1,675	110		1,785	2,025
6150	21 C.F. to 29 C.F., minimum		7	2.286		1,100	63		1,163	1,300
6200	Maximum		3	5.333		2,450	147		2,597	2,950
6790	Energy-star qualified, 18 C.F., minimum G	2 Carp	4	4		515	143		658	800
6795	Maximum G		2	8		1,700	285		1,985	2,350
6797	21.7 C.F., minimum G		4	4		1,025	143		1,168	1,350
6799	Maximum G		4	4		1,750	143		1,893	2,175

11 30 13.17 Kitchen Cleaning Equipment		Crew	Daily Output	Labor-Hours	Unit	Material	Labor	Equipment	Total	Total Incl O&P
0010	**KITCHEN CLEANING EQUIPMENT**									
2750	Dishwasher, built-in, 2 cycles, minimum	L-1	4	2.500	Ea.	305	102		407	500
2800	Maximum		2	5		455	203		658	835
2950	4 or more cycles, minimum		4	2.500		420	102		522	630
2960	Average		4	2.500		560	102		662	780
3000	Maximum		2	5		1,900	203		2,103	2,425
3100	Energy-star qualified, minimum G		4	2.500		405	102		507	610
3110	Maximum G		2	5		1,775	203		1,978	2,275

11 30 Residential Equipment

11 30 13 – Residential Appliances

11 30 13.18 Waste Disposal Equipment		Crew	Daily Output	Labor-Hours	Unit	Material	2019 Bare Costs Labor	Equipment	Total	Total Incl O&P
0010	**WASTE DISPOSAL EQUIPMENT**									
1750	Compactor, residential size, 4 to 1 compaction, minimum	1 Carp	5	1.600	Ea.	710	57		767	875
1800	Maximum	"	3	2.667		1,150	95		1,245	1,425
3300	Garbage disposal, sink type, minimum	L-1	10	1		109	40.50		149.50	186
3350	Maximum	"	10	1		208	40.50		248.50	296

11 30 13.19 Kitchen Ventilation Equipment

		Crew	Daily Output	Labor-Hours	Unit	Material	Labor	Equipment	Total	Total Incl O&P
0010	**KITCHEN VENTILATION EQUIPMENT**									
4150	Hood for range, 2 speed, vented, 30" wide, minimum	L-3	5	2	Ea.	96	73.50		169.50	226
4200	Maximum		3	3.333		965	123		1,088	1,250
4300	42" wide, minimum		5	2		155	73.50		228.50	291
4330	Custom		5	2		1,625	73.50		1,698.50	1,925
4350	Maximum		3	3.333		1,975	123		2,098	2,375
4500	For ventless hood, 2 speed, add					17.80			17.80	19.60
4650	For vented 1 speed, deduct from maximum					64			64	70.50

11 30 13.24 Washers

		Crew	Daily Output	Labor-Hours	Unit	Material	Labor	Equipment	Total	Total Incl O&P
0010	**WASHERS**									
6650	Washing machine, automatic, minimum	1 Plum	3	2.667	Ea.	590	108		698	825
6700	Maximum		1	8		1,350	325		1,675	2,025
6750	Energy star qualified, front loading, minimum G		3	2.667		850	108		958	1,100
6760	Maximum G		1	8		1,825	325		2,150	2,550
6764	Top loading, minimum G		3	2.667		650	108		758	890
6766	Maximum G		3	2.667		1,000	108		1,108	1,275

11 30 13.25 Dryers

		Crew	Daily Output	Labor-Hours	Unit	Material	Labor	Equipment	Total	Total Incl O&P
0010	**DRYERS**									
6770	Electric, front loading, energy-star qualified, minimum G	L-2	3	5.333	Ea.	460	168		628	790
6780	Maximum G	"	2	8		1,500	252		1,752	2,075
7450	Vent kits for dryers	1 Carp	10	.800		46.50	28.50		75	98

11 30 15 – Miscellaneous Residential Appliances

11 30 15.13 Sump Pumps

		Crew	Daily Output	Labor-Hours	Unit	Material	Labor	Equipment	Total	Total Incl O&P
0010	**SUMP PUMPS**									
6400	Cellar drainer, pedestal, 1/3 HP, molded PVC base	1 Plum	3	2.667	Ea.	141	108		249	330
6450	Solid brass	"	2	4	"	239	162		401	530
6460	Sump pump, see also Section 22 14 29.16									

11 30 15.23 Water Heaters

		Crew	Daily Output	Labor-Hours	Unit	Material	Labor	Equipment	Total	Total Incl O&P
0010	**WATER HEATERS**									
6900	Electric, glass lined, 30 gallon, minimum	L-1	5	2	Ea.	895	81		976	1,125
6950	Maximum		3	3.333		1,250	135		1,385	1,600
7100	80 gallon, minimum		2	5		1,650	203		1,853	2,150
7150	Maximum		1	10		2,300	405		2,705	3,225
7180	Gas, glass lined, 30 gallon, minimum	2 Plum	5	3.200		1,600	129		1,729	1,950
7220	Maximum		3	5.333		2,225	215		2,440	2,800
7260	50 gallon, minimum		2.50	6.400		1,725	259		1,984	2,325
7300	Maximum		1.50	10.667		2,400	430		2,830	3,325
7310	Water heater, see also Section 22 33 30.13									

11 30 15.43 Air Quality

		Crew	Daily Output	Labor-Hours	Unit	Material	Labor	Equipment	Total	Total Incl O&P
0010	**AIR QUALITY**									
2450	Dehumidifier, portable, automatic, 15 pint	1 Elec	4	2	Ea.	208	83		291	365
2550	40 pint		3.75	2.133		240	88.50		328.50	410
3550	Heater, electric, built-in, 1250 watt, ceiling type, minimum		4	2		119	83		202	266
3600	Maximum		3	2.667		185	110		295	385

11 30 Residential Equipment

11 30 15 – Miscellaneous Residential Appliances

11 30 15.43 Air Quality		Crew	Daily Output	Labor-Hours	Unit	Material	2019 Bare Costs Labor	Equipment	Total	Total Incl O&P
3700	Wall type, minimum	1 Elec	4	2	Ea.	199	83		282	355
3750	Maximum		3	2.667		194	110		304	395
3900	1500 watt wall type, with blower		4	2		189	83		272	345
3950	3000 watt	↓	3	2.667		490	110		600	720
4850	Humidifier, portable, 8 gallons/day					164			164	180
5000	15 gallons/day				↓	198			198	218

11 30 33 – Retractable Stairs

11 30 33.10 Disappearing Stairway		Crew	Daily Output	Labor-Hours	Unit	Material	Labor	Equipment	Total	Total Incl O&P
0010	**DISAPPEARING STAIRWAY** No trim included									
0020	One piece, yellow pine, 8′-0″ ceiling	2 Carp	4	4	Ea.	241	143		384	500
0030	9′-0″ ceiling		4	4		276	143		419	540
0040	10′-0″ ceiling		3	5.333		238	190		428	575
0050	11′-0″ ceiling		3	5.333		315	190		505	660
0060	12′-0″ ceiling	↓	3	5.333		355	190		545	705
0100	Custom grade, pine, 8′-6″ ceiling, minimum	1 Carp	4	2		268	71.50		339.50	410
0150	Average		3.50	2.286		236	81.50		317.50	395
0200	Maximum		3	2.667		295	95		390	480
0500	Heavy duty, pivoted, from 7′-7″ to 12′-10″ floor to floor		3	2.667		1,375	95		1,470	1,650
0600	16′-0″ ceiling		2	4		1,550	143		1,693	1,925
0800	Economy folding, pine, 8′-6″ ceiling		4	2		176	71.50		247.50	310
0900	9′-6″ ceiling	↓	4	2		201	71.50		272.50	340
1100	Automatic electric, aluminum, floor to floor height, 8′ to 9′	2 Carp	1	16	↓	9,200	570		9,770	11,000

11 32 Unit Kitchens

11 32 13 – Metal Unit Kitchens

11 32 13.10 Commercial Unit Kitchens		Crew	Daily Output	Labor-Hours	Unit	Material	Labor	Equipment	Total	Total Incl O&P
0010	**COMMERCIAL UNIT KITCHENS**									
1500	Combination range, refrigerator and sink, 30″ wide, minimum	L-1	2	5	Ea.	1,400	203		1,603	1,875
1550	Maximum		1	10		1,150	405		1,555	1,950
1570	60″ wide, average		1.40	7.143		1,275	290		1,565	1,875
1590	72″ wide, average	↓	1.20	8.333	↓	1,800	340		2,140	2,550

11 41 Foodservice Storage Equipment

11 41 13 – Refrigerated Food Storage Cases

11 41 13.30 Wine Cellar		Crew	Daily Output	Labor-Hours	Unit	Material	Labor	Equipment	Total	Total Incl O&P
0010	**WINE CELLAR**, refrigerated, Redwood interior, carpeted, walk-in type									
0020	6′-8″ high, including racks									
0200	80″ W x 48″ D for 900 bottles	2 Carp	1.50	10.667	Ea.	4,225	380		4,605	5,250
0250	80″ W x 72″ D for 1300 bottles		1.33	12.030		5,525	430		5,955	6,775
0300	80″ W x 94″ D for 1900 bottles	↓	1.17	13.675	↓	6,350	490		6,840	7,800

11 81 Facility Maintenance Equipment

11 81 19 – Vacuum Cleaning Systems

11 81 19.10 Vacuum Cleaning		Crew	Daily Output	Labor-Hours	Unit	Material	Labor	Equipment	Total	Total Incl O&P
						2019 Bare Costs				
0010	**VACUUM CLEANING**									
0020	Central, 3 inlet, residential	1 Skwk	.90	8.889	Total	1,200	325		1,525	1,875
0400	5 inlet system, residential		.50	16		1,725	585		2,310	2,875
0600	7 inlet system, commercial		.40	20		2,200	730		2,930	3,650
0800	9 inlet system, residential	↓	.30	26.667		4,175	975		5,150	6,225
4010	Rule of thumb: First 1200 S.F., installed				↓				1,425	1,575
4020	For each additional S.F., add				S.F.				.26	.26

Division Notes

		CREW	DAILY OUTPUT	LABOR-HOURS	UNIT	BARE COSTS MAT.	LABOR	EQUIP.	TOTAL	TOTAL INCL O&P

Estimating Tips

General

- The items in this division are usually priced per square foot or each. Most of these items are purchased by the owner and installed by the contractor. Do not assume the items in Division 12 will be purchased and installed by the contractor. Check the specifications for responsibilities and include receiving, storage, installation, and mechanical and electrical hookups in the appropriate divisions.
- Some items in this division require some type of support system that is not usually furnished with the item. Examples of these systems include blocking for the attachment of casework and heavy drapery rods. The required blocking must be added to the estimate in the appropriate division.

Reference Numbers

Reference numbers are shown at the beginning of some major classifications. These numbers refer to related items in the Reference Section. The reference information may be an estimating procedure, an alternate pricing method, or technical information.

12 21 Window Blinds

12 21 13 – Horizontal Louver Blinds

12 21 13.13 Metal Horizontal Louver Blinds		Crew	Daily Output	Labor-Hours	Unit	2019 Bare Costs Material	Labor	Equipment	Total	Total Incl O&P
0010	METAL HORIZONTAL LOUVER BLINDS									
0020	Horizontal, 1" aluminum slats, solid color, stock	1 Carp	590	.014	S.F.	5.90	.48		6.38	7.30

12 21 13.33 Vinyl Horizontal Louver Blinds		Crew	Daily Output	Labor-Hours	Unit	Material	Labor	Equipment	Total	Total Incl O&P
0010	VINYL HORIZONTAL LOUVER BLINDS									
0015	1" composite, 48" wide, 48" high	1 Carp	30	.267	Ea.	32	9.50		41.50	50.50
0020	72" high		30	.267		39	9.50		48.50	58.50
0030	60" wide, 60" high		30	.267		46.50	9.50		56	67
0040	72" high		30	.267		46.50	9.50		56	67
0050	72" wide x 72" high		30	.267		56	9.50		65.50	77

12 22 Curtains and Drapes

12 22 16 – Drapery Track and Accessories

12 22 16.10 Drapery Hardware		Crew	Daily Output	Labor-Hours	Unit	Material	Labor	Equipment	Total	Total Incl O&P
0010	DRAPERY HARDWARE									
0030	Standard traverse, per foot, minimum	1 Carp	59	.136	L.F.	6.40	4.83		11.23	15.05
0100	Maximum		51	.157	"	16.10	5.60		21.70	27
0200	Decorative traverse, 28" to 48", minimum		22	.364	Ea.	26	12.95		38.95	50
0220	Maximum		21	.381		57.50	13.60		71.10	85.50
0300	48" to 84", minimum		20	.400		28	14.25		42.25	54.50
0320	Maximum		19	.421		73.50	15		88.50	106
0400	66" to 120", minimum		18	.444		38.50	15.85		54.35	68.50
0420	Maximum		17	.471		112	16.80		128.80	151
0500	84" to 156", minimum		16	.500		49.50	17.85		67.35	84
0520	Maximum		15	.533		153	19		172	200
0600	130" to 240", minimum		14	.571		72	20.50		92.50	113
0620	Maximum		13	.615		212	22		234	270
0700	Slide rings, each, minimum					1.34			1.34	1.47
0720	Maximum					2.18			2.18	2.40
4000	Traverse rods, adjustable, 28" to 48"	1 Carp	22	.364		32.50	12.95		45.45	57
4020	48" to 84"		20	.400		41	14.25		55.25	68.50
4040	66" to 120"		18	.444		49.50	15.85		65.35	80.50
4060	84" to 156"		16	.500		56.50	17.85		74.35	91.50
4080	100" to 180"		14	.571		66	20.50		86.50	106
4100	228" to 312"		13	.615		91.50	22		113.50	138
4500	Curtain rod, 28" to 48", single		22	.364		10.35	12.95		23.30	33
4510	Double		22	.364		17.65	12.95		30.60	41
4520	48" to 86", single		20	.400		17.75	14.25		32	43
4530	Double		20	.400		29.50	14.25		43.75	56
4540	66" to 120", single		18	.444		29.50	15.85		45.35	58.50
4550	Double		18	.444		46.50	15.85		62.35	77
4600	Valance, pinch pleated fabric, 12" deep, up to 54" long, minimum					42.50			42.50	46.50
4610	Maximum					105			105	116
4620	Up to 77" long, minimum					81.50			81.50	89.50
4630	Maximum					170			170	187
5000	Stationary rods, first 2'					8.65			8.65	9.50

12 23 Interior Shutters

12 23 10 – Wood Interior Shutters

12 23 10.10 Wood Interior Shutters		Crew	Daily Output	Labor-Hours	Unit	Material	2019 Bare Costs Labor	Equipment	Total	Total Incl O&P
0010	**WOOD INTERIOR SHUTTERS**, louvered									
0200	Two panel, 27" wide, 36" high	1 Carp	5	1.600	Set	165	57		222	276
0300	33" wide, 36" high		5	1.600		213	57		270	330
0500	47" wide, 36" high		5	1.600		286	57		343	410
1000	Four panel, 27" wide, 36" high		5	1.600		157	57		214	266
1100	33" wide, 36" high		5	1.600		201	57		258	315
1300	47" wide, 36" high		5	1.600		268	57		325	390
1400	Plantation shutters, 16" x 48"		5	1.600	Ea.	184	57		241	296
1450	16" x 96"		4	2		300	71.50		371.50	450
1460	36" x 96"		3	2.667		665	95		760	885

12 23 10.13 Wood Panels		Crew	Daily Output	Labor-Hours	Unit	Material	Labor	Equipment	Total	Total Incl O&P
0010	**WOOD PANELS**									
3000	Wood folding panels with movable louvers, 7" x 20" each	1 Carp	17	.471	Pr.	95	16.80		111.80	133
3300	8" x 28" each		17	.471		95	16.80		111.80	133
3450	9" x 36" each		17	.471		109	16.80		125.80	148
3600	10" x 40" each		17	.471		119	16.80		135.80	159
4000	Fixed louver type, stock units, 8" x 20" each		17	.471		97	16.80		113.80	135
4150	10" x 28" each		17	.471		82	16.80		98.80	118
4300	12" x 36" each		17	.471		97	16.80		113.80	135
4450	18" x 40" each		17	.471		139	16.80		155.80	180
5000	Insert panel type, stock, 7" x 20" each		17	.471		28	16.80		44.80	58
5150	8" x 28" each		17	.471		51	16.80		67.80	83.50
5300	9" x 36" each		17	.471		64.50	16.80		81.30	98.50
5450	10" x 40" each		17	.471		69.50	16.80		86.30	104
5600	Raised panel type, stock, 10" x 24" each		17	.471		118	16.80		134.80	158
5650	12" x 26" each		17	.471		118	16.80		134.80	158
5700	14" x 30" each		17	.471		130	16.80		146.80	171
5750	16" x 36" each		17	.471		144	16.80		160.80	186
6000	For custom built pine, add					22%				
6500	For custom built hardwood blinds, add					42%				

12 24 Window Shades

12 24 13 – Roller Window Shades

12 24 13.10 Shades		Crew	Daily Output	Labor-Hours	Unit	Material	Labor	Equipment	Total	Total Incl O&P
0010	**SHADES**									
0020	Basswood, roll-up, stain finish, 3/8" slats	1 Carp	300	.027	S.F.	20.50	.95		21.45	24.50
5011	Insulative shades G		125	.064		16.45	2.28		18.73	22
6011	Solar screening, fiberglass G		85	.094		7.85	3.36		11.21	14.20
8011	Interior insulative shutter									
8111	Stock unit, 15" x 60" G	1 Carp	17	.471	Pr.	17.70	16.80		34.50	47

12 32 Manufactured Wood Casework

12 32 23 – Hardwood Casework

12 32 23.10 Manufactured Wood Casework, Stock Units		Crew	Daily Output	Labor-Hours	Unit	Material	2019 Bare Costs Labor	Equipment	Total	Total Incl O&P
0010	**MANUFACTURED WOOD CASEWORK, STOCK UNITS**									
0700	Kitchen base cabinets, hardwood, not incl. counter tops,									
0710	24″ deep, 35″ high, prefinished									
0800	One top drawer, one door below, 12″ wide	2 Carp	24.80	.645	Ea.	315	23		338	390
0820	15″ wide		24	.667		330	24		354	400
0840	18″ wide		23.30	.687		350	24.50		374.50	425
0860	21″ wide		22.70	.705		360	25		385	435
0880	24″ wide		22.30	.717		435	25.50		460.50	525
1000	Four drawers, 12″ wide		24.80	.645		325	23		348	395
1020	15″ wide		24	.667		325	24		349	400
1040	18″ wide		23.30	.687		360	24.50		384.50	435
1060	24″ wide		22.30	.717		405	25.50		430.50	490
1200	Two top drawers, two doors below, 27″ wide		22	.727		445	26		471	535
1220	30″ wide		21.40	.748		505	26.50		531.50	600
1240	33″ wide		20.90	.766		525	27.50		552.50	625
1260	36″ wide		20.30	.788		550	28		578	645
1280	42″ wide		19.80	.808		580	29		609	685
1300	48″ wide		18.90	.847		620	30		650	730
1500	Range or sink base, two doors below, 30″ wide		21.40	.748		420	26.50		446.50	510
1520	33″ wide		20.90	.766		450	27.50		477.50	540
1540	36″ wide		20.30	.788		475	28		503	565
1560	42″ wide		19.80	.808		500	29		529	600
1580	48″ wide		18.90	.847		525	30		555	625
1800	For sink front units, deduct					182			182	201
2000	Corner base cabinets, 36″ wide, standard	2 Carp	18	.889		745	31.50		776.50	875
2100	Lazy Susan with revolving door	″	16.50	.970		1,000	34.50		1,034.50	1,150
4000	Kitchen wall cabinets, hardwood, 12″ deep with two doors									
4050	12″ high, 30″ wide	2 Carp	24.80	.645	Ea.	284	23		307	350
4400	15″ high, 30″ wide		24	.667		281	24		305	350
4420	33″ wide		23.30	.687		355	24.50		379.50	430
4440	36″ wide		22.70	.705		345	25		370	420
4450	42″ wide		22.70	.705		390	25		415	470
4700	24″ high, 30″ wide		23.30	.687		375	24.50		399.50	455
4720	36″ wide		22.70	.705		410	25		435	490
4740	42″ wide		22.30	.717		292	25.50		317.50	365
5000	30″ high, one door, 12″ wide		22	.727		271	26		297	340
5020	15″ wide		21.40	.748		284	26.50		310.50	360
5040	18″ wide		20.90	.766		310	27.50		337.50	390
5060	24″ wide		20.30	.788		365	28		393	445
5300	Two doors, 27″ wide		19.80	.808		405	29		434	495
5320	30″ wide		19.30	.829		425	29.50		454.50	515
5340	36″ wide		18.80	.851		480	30.50		510.50	575
5360	42″ wide		18.50	.865		525	31		556	625
5380	48″ wide		18.40	.870		590	31		621	700
6000	Corner wall, 30″ high, 24″ wide		18	.889		420	31.50		451.50	515
6050	30″ wide		17.20	.930		430	33		463	530
6100	36″ wide		16.50	.970		495	34.50		529.50	600
6500	Revolving Lazy Susan		15.20	1.053		132	37.50		169.50	207
7000	Broom cabinet, 84″ high, 24″ deep, 18″ wide		10	1.600		775	57		832	950
7500	Oven cabinets, 84″ high, 24″ deep, 27″ wide		8	2		1,175	71.50		1,246.50	1,425
7750	Valance board trim		396	.040	L.F.	18.80	1.44		20.24	23
7780	Toe kick trim	1 Carp	256	.031	″	3.55	1.11		4.66	5.75

12 32 23 – Hardwood Casework

12 32 23.10 Manufactured Wood Casework, Stock Units		Crew	Daily Output	Labor-Hours	Unit	Material	Labor	Equipment	Total	Total Incl O&P
						2019 Bare Costs				
7790	Base cabinet corner filler	1 Carp	16	.500	Ea.	48.50	17.85		66.35	83
7800	Cabinet filler, 3" x 24"		20	.400		19.75	14.25		34	45
7810	3" x 30"		20	.400		24.50	14.25		38.75	50.50
7820	3" x 42"		18	.444		34.50	15.85		50.35	64
7830	3" x 80"		16	.500		66	17.85		83.85	102
7850	Cabinet panel		50	.160	S.F.	10.80	5.70		16.50	21.50
9000	For deluxe models of all cabinets, add					40%				
9500	For custom built in place, add					25%	10%			
9558	Rule of thumb, kitchen cabinets not including									
9560	appliances & counter top, minimum	2 Carp	30	.533	L.F.	211	19		230	264
9600	Maximum	"	25	.640	"	445	23		468	530

12 32 23.30 Manufactured Wood Casework Vanities		Crew	Daily Output	Labor-Hours	Unit	Material	Labor	Equipment	Total	Total Incl O&P
0010	**MANUFACTURED WOOD CASEWORK VANITIES**									
8000	Vanity bases, 2 doors, 30" high, 21" deep, 24" wide	2 Carp	20	.800	Ea.	370	28.50		398.50	455
8050	30" wide		16	1		445	35.50		480.50	545
8100	36" wide		13.33	1.200		400	43		443	510
8150	48" wide		11.43	1.400		595	50		645	740
9000	For deluxe models of all vanities, add to above					40%				
9500	For custom built in place, add to above					25%	10%			

12 32 23.35 Manufactured Wood Casework Hardware		Crew	Daily Output	Labor-Hours	Unit	Material	Labor	Equipment	Total	Total Incl O&P
0010	**MANUFACTURED WOOD CASEWORK HARDWARE**									
1000	Catches, minimum	1 Carp	235	.034	Ea.	1.49	1.21		2.70	3.65
1020	Average		119.40	.067		4.28	2.39		6.67	8.65
1040	Maximum		80	.100		9	3.57		12.57	15.80
2000	Door/drawer pulls, handles									
2200	Handles and pulls, projecting, metal, minimum	1 Carp	48	.167	Ea.	5.30	5.95		11.25	15.65
2220	Average		42	.190		8.15	6.80		14.95	20
2240	Maximum		36	.222		10.95	7.90		18.85	25
2300	Wood, minimum		48	.167		5.45	5.95		11.40	15.80
2320	Average		42	.190		7.30	6.80		14.10	19.25
2340	Maximum		36	.222		10.10	7.90		18	24
2600	Flush, metal, minimum		48	.167		5.45	5.95		11.40	15.80
2620	Average		42	.190		7.40	6.80		14.20	19.35
2640	Maximum		36	.222		10.20	7.90		18.10	24.50
3000	Drawer tracks/glides, minimum		48	.167	Pr.	9.20	5.95		15.15	19.95
3020	Average		32	.250		16.60	8.90		25.50	33
3040	Maximum		24	.333		27	11.90		38.90	49
4000	Cabinet hinges, minimum		160	.050		3.16	1.78		4.94	6.45
4020	Average		95.24	.084		5.30	2.99		8.29	10.75
4040	Maximum		68	.118		13.85	4.19		18.04	22

12 34 Manufactured Plastic Casework

12 34 16 – Manufactured Solid-Plastic Casework

12 34 16.10 Outdoor Casework		Crew	Daily Output	Labor-Hours	Unit	Material	2019 Bare Costs Labor	Equipment	Total	Total Incl O&P
0010	**OUTDOOR CASEWORK**									
0020	Cabinet, base, sink/range, 36"	2 Carp	20.30	.788	Ea.	485	28		513	575
0100	Base, 36"		20.30	.788		4,650	28		4,678	5,175
0200	Filler strip, 1" x 30"		158	.101		35.50	3.61		39.11	45

12 36 Countertops

12 36 16 – Metal Countertops

12 36 16.10 Stainless Steel Countertops		Crew	Daily Output	Labor-Hours	Unit	Material	Labor	Equipment	Total	Total Incl O&P
0010	**STAINLESS STEEL COUNTERTOPS**									
3200	Stainless steel, custom	1 Carp	24	.333	S.F.	183	11.90		194.90	221

12 36 19 – Wood Countertops

12 36 19.10 Maple Countertops		Crew	Daily Output	Labor-Hours	Unit	Material	Labor	Equipment	Total	Total Incl O&P
0010	**MAPLE COUNTERTOPS**									
2900	Solid, laminated, 1-1/2" thick, no splash	1 Carp	28	.286	L.F.	91.50	10.20		101.70	118
3000	With square splash		28	.286	"	109	10.20		119.20	137
3400	Recessed cutting block with trim, 16" x 20" x 1"		8	1	Ea.	111	35.50		146.50	181
3410	Replace cutting block only		8	1	"	92	35.50		127.50	160

12 36 23 – Plastic Countertops

12 36 23.13 Plastic-Laminate-Clad Countertops		Crew	Daily Output	Labor-Hours	Unit	Material	Labor	Equipment	Total	Total Incl O&P
0010	**PLASTIC-LAMINATE-CLAD COUNTERTOPS**									
0020	Stock, 24" wide w/backsplash, minimum	1 Carp	30	.267	L.F.	17.50	9.50		27	35
0100	Maximum		25	.320		41	11.40		52.40	64
0300	Custom plastic, 7/8" thick, aluminum molding, no splash		30	.267		36	9.50		45.50	55
0400	Cove splash		30	.267		46	9.50		55.50	66
0600	1-1/4" thick, no splash		28	.286		39	10.20		49.20	59.50
0700	Square splash		28	.286		44	10.20		54.20	65.50
0900	Square edge, plastic face, 7/8" thick, no splash		30	.267		34.50	9.50		44	53.50
1000	With splash		30	.267		42	9.50		51.50	61.50
1200	For stainless channel edge, 7/8" thick, add					3.84			3.84	4.22
1300	1-1/4" thick, add					4.51			4.51	4.96
1500	For solid color suede finish, add					5.90			5.90	6.50
1700	For end splash, add				Ea.	22			22	24
1901	For cut outs, standard, add, minimum	1 Carp	32	.250			8.90		8.90	14.75
2000	Maximum		8	1		7.30	35.50		42.80	67
2010	Cut out in backsplash for elec. wall outlet		38	.211			7.50		7.50	12.40
2020	Cut out for sink		20	.400			14.25		14.25	23.50
2030	Cut out for stove top		18	.444			15.85		15.85	26
2100	Postformed, including backsplash and front edge		30	.267	L.F.	12.60	9.50		22.10	29.50
2110	Mitred, add		12	.667	Ea.		24		24	39.50
2200	Built-in place, 25" wide, plastic laminate		25	.320	L.F.	61.50	11.40		72.90	87

12 36 33 – Tile Countertops

12 36 33.10 Ceramic Tile Countertops		Crew	Daily Output	Labor-Hours	Unit	Material	Labor	Equipment	Total	Total Incl O&P
0010	**CERAMIC TILE COUNTERTOPS**									
2300	Ceramic tile mosaic	1 Carp	25	.320	L.F.	41.50	11.40		52.90	65

12 36 40 – Stone Countertops

12 36 40.10 Natural Stone Countertops		Crew	Daily Output	Labor-Hours	Unit	Material	Labor	Equipment	Total	Total Incl O&P
0010	**NATURAL STONE COUNTERTOPS**									
2500	Marble, stock, with splash, 1/2" thick, minimum	1 Bric	17	.471	L.F.	51.50	16.60		68.10	84.50
2700	3/4" thick, maximum		13	.615		130	21.50		151.50	179

12 36 Countertops

12 36 40 – Stone Countertops

12 36 40.10 Natural Stone Countertops		Crew	Daily Output	Labor-Hours	Unit	Material	2019 Bare Costs Labor	Equipment	Total	Total Incl O&P
2800	Granite, average, 1-1/4" thick, 24" wide, no splash	1 Bric	13.01	.615	L.F.	174	21.50		195.50	228

12 36 61 – Simulated Stone Countertops

12 36 61.16 Solid Surface Countertops

		Crew	Daily Output	Labor-Hours	Unit	Material	Labor	Equipment	Total	Total Incl O&P
0010	**SOLID SURFACE COUNTERTOPS**, Acrylic polymer									
2000	Pricing for order of 1-50 L.F.									
2100	25" wide, solid colors	2 Carp	20	.800	L.F.	72	28.50		100.50	127
2200	Patterned colors		20	.800		92	28.50		120.50	148
2300	Premium patterned colors		20	.800		122	28.50		150.50	182
2400	With silicone attached 4" backsplash, solid colors		19	.842		84.50	30		114.50	143
2500	Patterned colors		19	.842		107	30		137	168
2600	Premium patterned colors		19	.842		134	30		164	197
2700	With hard seam attached 4" backsplash, solid colors		15	1.067		84.50	38		122.50	156
2800	Patterned colors		15	1.067		107	38		145	181
2900	Premium patterned colors	↓	15	1.067	↓	134	38		172	210
3800	Sinks, pricing for order of 1-50 units									
3900	Single bowl, hard seamed, solid colors, 13" x 17"	1 Carp	2	4	Ea.	520	143		663	810
4000	10" x 15"		4.55	1.758		241	62.50		303.50	370
4100	Cutouts for sinks	↓	5.25	1.524	↓		54.50		54.50	90

12 36 61.17 Solid Surface Vanity Tops

		Crew	Daily Output	Labor-Hours	Unit	Material	Labor	Equipment	Total	Total Incl O&P
0010	**SOLID SURFACE VANITY TOPS**									
0015	Solid surface, center bowl, 17" x 19"	1 Carp	12	.667	Ea.	173	24		197	230
0020	19" x 25"		12	.667		191	24		215	250
0030	19" x 31"		12	.667		224	24		248	286
0040	19" x 37"		12	.667		261	24		285	325
0050	22" x 25"		10	.800		335	28.50		363.50	415
0060	22" x 31"		10	.800		390	28.50		418.50	475
0070	22" x 37"		10	.800		455	28.50		483.50	545
0080	22" x 43"		10	.800		525	28.50		553.50	620
0090	22" x 49"		10	.800		570	28.50		598.50	670
0110	22" x 55"		8	1		420	35.50		455.50	520
0120	22" x 61"		8	1		475	35.50		510.50	585
0220	Double bowl, 22" x 61"		8	1		535	35.50		570.50	650
0230	Double bowl, 22" x 73"	↓	8	1	↓	1,025	35.50		1,060.50	1,175
0240	For aggregate colors, add					35%				
0250	For faucets and fittings, see Section 22 41 39.10									

12 36 61.19 Quartz Agglomerate Countertops

		Crew	Daily Output	Labor-Hours	Unit	Material	Labor	Equipment	Total	Total Incl O&P
0010	**QUARTZ AGGLOMERATE COUNTERTOPS**									
0100	25" wide, 4" backsplash, color group A, minimum	2 Carp	15	1.067	L.F.	68.50	38		106.50	138
0110	Maximum		15	1.067		95.50	38		133.50	168
0120	Color group B, minimum		15	1.067		73.50	38		111.50	144
0130	Maximum		15	1.067		99.50	38		137.50	172
0140	Color group C, minimum		15	1.067		82	38		120	153
0150	Maximum		15	1.067		118	38		156	192
0160	Color group D, minimum		15	1.067		89	38		127	161
0170	Maximum	↓	15	1.067	↓	122	38		160	197

Division Notes

		CREW	DAILY OUTPUT	LABOR-HOURS	UNIT	BARE COSTS MAT.	LABOR	EQUIP.	TOTAL	TOTAL INCL O&P

Estimating Tips

General

- The items and systems in this division are usually estimated, purchased, supplied, and installed as a unit by one or more subcontractors. The estimator must ensure that all parties are operating from the same set of specifications and assumptions, and that all necessary items are estimated and will be provided. Many times the complex items and systems are covered, but the more common ones, such as excavation or a crane, are overlooked for the very reason that everyone assumes nobody could miss them. The estimator should be the central focus and be able to ensure that all systems are complete.
- It is important to consider factors such as site conditions, weather, shape and size of building, as well as labor availability as they may impact the overall cost of erecting special structures and systems included in this division.
- Another area where problems can develop in this division is at the interface between systems. The estimator must ensure, for instance, that anchor bolts, nuts, and washers are estimated and included for the air-supported structures and pre-engineered buildings to be bolted to their foundations. Utility supply is a common area where essential items or pieces of equipment can be missed or overlooked because each subcontractor may feel it is another's responsibility. The estimator should also be aware of certain items which may be supplied as part of a package but installed by others, and ensure that the installing contractor's estimate includes the cost of installation. Conversely, the estimator must also ensure that items are not costed by two different subcontractors, resulting in an inflated overall estimate.

13 30 00 Special Structures

- The foundations and floor slab, as well as rough mechanical and electrical, should be estimated, as this work is required for the assembly and erection of the structure. Generally, as noted in the data set, the pre-engineered building comes as a shell. Pricing is based on the size and structural design parameters stated in the reference section. Additional features, such as windows and doors with their related structural framing, must also be included by the estimator. Here again, the estimator must have a clear understanding of the scope of each portion of the work and all the necessary interfaces.

Reference Numbers

Reference numbers are shown at the beginning of some major classifications. These numbers refer to related items in the Reference Section. The reference information may be an estimating procedure, an alternate pricing method, or technical information.

Note: Not all subdivisions listed here necessarily appear. ■

13 11 Swimming Pools

13 11 13 – Below-Grade Swimming Pools

13 11 13.50 Swimming Pools		Crew	Daily Output	Labor-Hours	Unit	2019 Bare Costs Material	Labor	Equipment	Total	Total Incl O&P
0010	**SWIMMING POOLS** Residential in-ground, vinyl lined									
0020	Concrete sides, w/equip, sand bottom	B-52	300	.187	SF Surf	27	5.70	2	34.70	41.50
0100	Metal or polystyrene sides R131113-20	B-14	410	.117		22.50	3.42	.78	26.70	31.50
0200	Add for vermiculite bottom					1.73			1.73	1.90
0500	Gunite bottom and sides, white plaster finish									
0600	12′ x 30′ pool	B-52	145	.386	SF Surf	50.50	11.75	4.14	66.39	79.50
0720	16′ x 32′ pool		155	.361		45.50	11	3.88	60.38	72.50
0750	20′ x 40′ pool		250	.224		40.50	6.85	2.40	49.75	58.50
0810	Concrete bottom and sides, tile finish									
0820	12′ x 30′ pool	B-52	80	.700	SF Surf	51	21.50	7.50	80	99.50
0830	16′ x 32′ pool		95	.589		42	17.95	6.35	66.30	83
0840	20′ x 40′ pool		130	.431		33.50	13.15	4.62	51.27	63.50
1600	For water heating system, see Section 23 52 28.10									
1700	Filtration and deck equipment only, as % of total				Total				20%	20%
1800	Deck equipment, rule of thumb, 20′ x 40′ pool				SF Pool				1.18	1.30
3000	Painting pools, preparation + 3 coats, 20′ x 40′ pool, epoxy	2 Pord	.33	48.485	Total	1,750	1,450		3,200	4,300
3100	Rubber base paint, 18 gallons	"	.33	48.485	"	1,350	1,450		2,800	3,875

13 11 23 – On-Grade Swimming Pools

13 11 23.50 Swimming Pools		Crew	Daily Output	Labor-Hours	Unit	Material	Labor	Equipment	Total	Total Incl O&P
0010	**SWIMMING POOLS** Residential above ground, steel construction									
0100	Round, 15′ diam.	B-80A	3	8	Ea.	810	220	82	1,112	1,350
0120	18′ diam.		2.50	9.600		910	264	98.50	1,272.50	1,550
0140	21′ diam.		2	12		1,025	330	123	1,478	1,800
0160	24′ diam.		1.80	13.333		1,125	365	137	1,627	2,000
0180	27′ diam.		1.50	16		1,325	440	164	1,929	2,350
0200	30′ diam.		1	24		1,450	660	246	2,356	2,950
0220	Oval, 12′ x 24′		2.30	10.435		1,500	287	107	1,894	2,250
0240	15′ x 30′		1.80	13.333		2,075	365	137	2,577	3,025
0260	18′ x 33′		1	24		2,350	660	246	3,256	3,975

13 11 46 – Swimming Pool Accessories

13 11 46.50 Swimming Pool Equipment		Crew	Daily Output	Labor-Hours	Unit	Material	Labor	Equipment	Total	Total Incl O&P
0010	**SWIMMING POOL EQUIPMENT**									
0020	Diving stand, stainless steel, 3 meter	2 Carp	.40	40	Ea.	17,300	1,425		18,725	21,500
0300	1 meter		2.70	5.926		10,500	211		10,711	11,900
0600	Diving boards, 16′ long, aluminum		2.70	5.926		4,375	211		4,586	5,175
0700	Fiberglass		2.70	5.926		3,525	211		3,736	4,225
0800	14′ long, aluminum		2.70	5.926		4,125	211		4,336	4,900
0850	Fiberglass		2.70	5.926		3,500	211		3,711	4,200
1100	Bulkhead, movable, PVC, 8′-2″ wide	2 Clab	8	2		2,775	55		2,830	3,150
1120	7′-9″ wide		8	2		2,675	55		2,730	3,050
1140	7′-3″ wide		8	2		2,175	55		2,230	2,500
1160	6′-9″ wide		8	2		2,175	55		2,230	2,500
1200	Ladders, heavy duty, stainless steel, 2 tread	2 Carp	7	2.286		910	81.50		991.50	1,125
1500	4 tread	"	6	2.667		790	95		885	1,025
2100	Lights, underwater, 12 volt, with transformer, 300 watt	1 Elec	1	8		365	330		695	940
2200	110 volt, 500 watt, standard	"	1	8		320	330		650	890
3000	Pool covers, reinforced vinyl	3 Clab	1800	.013	S.F.	1.24	.37		1.61	1.97
3100	Vinyl, for winter, 400 S.F. max pool surface		3200	.008		.24	.21		.45	.60
3200	With water tubes, 400 S.F. max pool surface		3000	.008		.38	.22		.60	.78
3250	Sealed air bubble polyethylene solar blanket, 16 mils					.36			.36	.40
3300	Slides, tubular, fiberglass, aluminum handrails & ladder, 5′-0″, straight	2 Carp	1.60	10	Ea.	3,775	355		4,130	4,750
3320	8′-0″, curved	"	3	5.333	"	7,450	190		7,640	8,525

13 12 Fountains

13 12 13 – Exterior Fountains

13 12 13.10 Outdoor Fountains		Crew	Daily Output	Labor-Hours	Unit	Material	2019 Bare Costs Labor	Equipment	Total	Total Incl O&P
0010	OUTDOOR FOUNTAINS									
0100	Outdoor fountain, 48" high with bowl and figures	2 Clab	2	8	Ea.	460	220		680	870
0200	Commercial, concrete or cast stone, 40-60" H, simple		2	8		845	220		1,065	1,300
0220	Average		2	8		1,625	220		1,845	2,175
0240	Ornate		2	8		3,700	220		3,920	4,450
0260	Metal, 72" high		2	8		1,575	220		1,795	2,100
0280	90" high		2	8		2,125	220		2,345	2,725
0300	120" high		2	8		5,150	220		5,370	6,025
0320	Resin or fiberglass, 40-60" H, wall type		2	8		595	220		815	1,025
0340	Waterfall type		2	8		1,000	220		1,220	1,475

13 12 23 – Interior Fountains

13 12 23.10 Indoor Fountains		Crew	Daily Output	Labor-Hours	Unit	Material	Labor	Equipment	Total	Total Incl O&P
0010	INDOOR FOUNTAINS									
0100	Commercial, floor type, resin or fiberglass, lighted, cascade type	2 Clab	2	8	Ea.	365	220		585	770
0120	Tiered type		2	8		430	220		650	840
0140	Waterfall type		2	8		276	220		496	670

13 17 Tubs and Pools

13 17 13 – Hot Tubs

13 17 13.10 Redwood Hot Tub System		Crew	Daily Output	Labor-Hours	Unit	Material	Labor	Equipment	Total	Total Incl O&P
0010	REDWOOD HOT TUB SYSTEM									
7050	4' diameter x 4' deep	Q-1	1	16	Ea.	3,200	580		3,780	4,475
7150	6' diameter x 4' deep		.80	20		4,900	725		5,625	6,575
7200	8' diameter x 4' deep		.80	20		7,175	725		7,900	9,100

13 17 33 – Whirlpool Tubs

13 17 33.10 Whirlpool Bath		Crew	Daily Output	Labor-Hours	Unit	Material	Labor	Equipment	Total	Total Incl O&P
0010	WHIRLPOOL BATH									
6000	Whirlpool, bath with vented overflow, molded fiberglass									
6100	66" x 36" x 24"	Q-1	1	16	Ea.	935	580		1,515	1,975
6400	72" x 36" x 21"		1	16		1,225	580		1,805	2,300
6500	60" x 34" x 21"		1	16		1,250	580		1,830	2,325
6600	72" x 42" x 23"		1	16		1,300	580		1,880	2,375

13 24 Special Activity Rooms

13 24 16 – Saunas

13 24 16.50 Saunas and Heaters		Crew	Daily Output	Labor-Hours	Unit	Material	Labor	Equipment	Total	Total Incl O&P
0010	SAUNAS AND HEATERS									
0020	Prefabricated, incl. heater & controls, 7' high, 6' x 4', C/C	L-7	2.20	11.818	Ea.	5,250	365		5,615	6,375
0050	6' x 4', C/P		2	13		4,350	400		4,750	5,450
0400	6' x 5', C/C		2	13		5,425	400		5,825	6,625
0450	6' x 5', C/P		2	13		5,150	400		5,550	6,325
0600	6' x 6', C/C		1.80	14.444		7,100	445		7,545	8,550
0650	6' x 6', C/P		1.80	14.444		5,450	445		5,895	6,750
0800	6' x 9', C/C		1.60	16.250		7,850	505		8,355	9,475
0850	6' x 9', C/P		1.60	16.250		6,450	505		6,955	7,925
1000	8' x 12', C/C		1.10	23.636		10,700	730		11,430	13,000
1050	8' x 12', C/P		1.10	23.636		8,850	730		9,580	11,000
1200	8' x 8', C/C		1.40	18.571		8,600	575		9,175	10,400

13 24 Special Activity Rooms

13 24 16 – Saunas

13 24 16.50 Saunas and Heaters		Crew	Daily Output	Labor-Hours	Unit	Material	Labor	Equipment	Total	Total Incl O&P
						2019 Bare Costs				
1250	8′ x 8′, C/P	L-7	1.40	18.571	Ea.	7,675	575		8,250	9,375
1400	8′ x 10′, C/C		1.20	21.667		9,400	670		10,070	11,400
1450	8′ x 10′, C/P		1.20	21.667		8,325	670		8,995	10,300
1600	10′ x 12′, C/C		1	26		12,600	805		13,405	15,300
1650	10′ x 12′, C/P		1	26		13,600	805		14,405	16,300
1700	Door only, cedar, 2′ x 6′, w/ 1′ x 4′ tempered insulated glass window	2 Carp	3.40	4.706		625	168		793	965
1800	Prehung, incl. jambs, pulls & hardware	″	12	1.333		600	47.50		647.50	740
2500	Heaters only (incl. above), wall mounted, to 200 C.F.					960			960	1,050
2750	To 300 C.F.					1,050			1,050	1,150
3000	Floor standing, to 720 C.F., 10,000 watts, w/controls	1 Elec	3	2.667		2,475	110		2,585	2,900
3250	To 1,000 C.F., 16,000 watts	″	3	2.667		3,600	110		3,710	4,125

13 24 26 – Steam Baths

13 24 26.50 Steam Baths and Components

		Crew	Daily Output	Labor-Hours	Unit	Material	Labor	Equipment	Total	Total Incl O&P
0010	**STEAM BATHS AND COMPONENTS**									
0020	Heater, timer & head, single, to 140 C.F.	1 Plum	1.20	6.667	Ea.	2,300	269		2,569	2,975
0500	To 300 C.F.	″	1.10	7.273		2,625	294		2,919	3,375
2700	Conversion unit for residential tub, including door					3,900			3,900	4,275

13 34 Fabricated Engineered Structures

13 34 13 – Glazed Structures

13 34 13.13 Greenhouses

		Crew	Daily Output	Labor-Hours	Unit	Material	Labor	Equipment	Total	Total Incl O&P
0010	**GREENHOUSES**, Shell only, stock units, not incl. 2′ stub walls,									
0020	foundation, floors, heat or compartments									
0300	Residential type, free standing, 8′-6″ long x 7′-6″ wide	2 Carp	59	.271	SF Flr.	24	9.65		33.65	42
0400	10′-6″ wide		85	.188		44	6.70		50.70	59.50
0600	13′-6″ wide		108	.148		45.50	5.30		50.80	59
0700	17′-0″ wide		160	.100		44.50	3.57		48.07	55
0900	Lean-to type, 3′-10″ wide		34	.471		44	16.80		60.80	76
1000	6′-10″ wide		58	.276		28.50	9.85		38.35	48
1100	Wall mounted to existing window, 3′ x 3′	1 Carp	4	2	Ea.	2,125	71.50		2,196.50	2,450
1120	4′ x 5′	″	3	2.667	″	2,525	95		2,620	2,925
1200	Deluxe quality, free standing, 7′-6″ wide	2 Carp	55	.291	SF Flr.	83	10.35		93.35	109
1220	10′-6″ wide		81	.198		63	7.05		70.05	81
1240	13′-6″ wide		104	.154		55	5.50		60.50	70
1260	17′-0″ wide		150	.107		45.50	3.80		49.30	56.50
1400	Lean-to type, 3′-10″ wide		31	.516		104	18.40		122.40	146
1420	6′-10″ wide		55	.291		72.50	10.35		82.85	97
1440	8′-0″ wide		97	.165		64.50	5.90		70.40	80.50

13 34 13.19 Swimming Pool Enclosures

		Crew	Daily Output	Labor-Hours	Unit	Material	Labor	Equipment	Total	Total Incl O&P
0010	**SWIMMING POOL ENCLOSURES** Translucent, free standing									
0020	not including foundations, heat or light									
0200	Economy	2 Carp	200	.080	SF Hor.	50	2.85		52.85	59.50
0600	Deluxe	″	70	.229	″	92	8.15		100.15	114

13 34 63 – Natural Fiber Construction

13 34 63.50 Straw Bale Construction

		Crew	Daily Output	Labor-Hours	Unit	Material	Labor	Equipment	Total	Total Incl O&P
0010	**STRAW BALE CONSTRUCTION**									
2020	Straw bales in walls w/modified post and beam frame G	2 Carp	320	.050	S.F.	6.50	1.78		8.28	10.10

Estimating Tips

General

- Many products in Division 14 will require some type of support or blocking for installation not included with the item itself. Examples are supports for conveyors or tube systems, attachment points for lifts, and footings for hoists or cranes. Add these supports in the appropriate division.

14 10 00 Dumbwaiters
14 20 00 Elevators

- Dumbwaiters and elevators are estimated and purchased in a method similar to buying a car. The manufacturer has a base unit with standard features. Added to this base unit price will be whatever options the owner or specifications require. Increased load capacity, additional vertical travel, additional stops, higher speed, and cab finish options are items to be considered. When developing an estimate for dumbwaiters and elevators, remember that some items needed by the installers may have to be included as part of the general contract.

  Examples are:
  - ☐ shaftway
  - ☐ rail support brackets
  - ☐ machine room
  - ☐ electrical supply
  - ☐ sill angles
  - ☐ electrical connections
  - ☐ pits
  - ☐ roof penthouses
  - ☐ pit ladders

  Check the job specifications and drawings before pricing.
- Installation of elevators and handicapped lifts in historic structures can require significant additional costs. The associated structural requirements may involve cutting into and repairing finishes, moldings, flooring, etc. The estimator must account for these special conditions.

14 30 00 Escalators and Moving Walks

- Escalators and moving walks are specialty items installed by specialty contractors. There are numerous options associated with these items. For specific options, contact a manufacturer or contractor. In a method similar to estimating dumbwaiters and elevators, you should verify the extent of general contract work and add items as necessary.

14 40 00 Lifts
14 90 00 Other Conveying Equipment

- Products such as correspondence lifts, chutes, and pneumatic tube systems, as well as other items specified in this subdivision, may require trained installers. The general contractor might not have any choice as to who will perform the installation or when it will be performed. Long lead times are often required for these products, making early decisions in scheduling necessary.

Reference Numbers

Reference numbers are shown at the beginning of some major classifications. These numbers refer to related items in the Reference Section. The reference information may be an estimating procedure, an alternate pricing method, or technical information.

Note: Not all subdivisions listed here necessarily appear. ■

14 21 Electric Traction Elevators

14 21 33 – Electric Traction Residential Elevators

14 21 33.20 Residential Elevators		Crew	Daily Output	Labor-Hours	Unit	2019 Bare Costs Material	Labor	Equipment	Total	Total Incl O&P
0010	**RESIDENTIAL ELEVATORS**									
7000	Residential, cab type, 1 floor, 2 stop, economy model	2 Elev	.20	80	Ea.	10,800	4,425		15,225	19,000
7100	Custom model		.10	160		18,100	8,875		26,975	34,400
7200	2 floor, 3 stop, economy model		.12	133		16,000	7,400		23,400	29,600
7300	Custom model	↓	.06	267	↓	26,000	14,800		40,800	52,500

14 42 Wheelchair Lifts

14 42 13 – Inclined Wheelchair Lifts

14 42 13.10 Inclined Wheelchair Lifts and Stairclimbers		Crew	Daily Output	Labor-Hours	Unit	Material	Labor	Equipment	Total	Total Incl O&P
0010	**INCLINED WHEELCHAIR LIFTS AND STAIRCLIMBERS**									
7700	Stair climber (chair lift), single seat, minimum	2 Elev	1	16	Ea.	5,000	885		5,885	6,950
7800	Maximum	"	.20	80	"	6,825	4,425		11,250	14,800

Division 21 Fire Suppression

Estimating Tips

Pipe for fire protection and all uses is located in Subdivisions 21 11 13 and 22 11 13.

The labor adjustment factors listed in Subdivision 22 01 02.20 also apply to Division 21.

Many, but not all, areas in the U.S. require backflow protection in the fire system. Insurance underwriters may have specific requirements for the type of materials to be installed or design requirements based on the hazard to be protected. Local jurisdictions may have requirements not covered by code. It is advisable to be aware of any special conditions.

For your reference, the following is a list of the most applicable Fire Codes and Standards, which may be purchased from the NFPA, 1 Batterymarch Park, Quincy, MA 02169-7471.

- NFPA 1: Uniform Fire Code
- NFPA 10: Portable Fire Extinguishers
- NFPA 11: Low-, Medium-, and High-Expansion Foam
- NFPA 12: Carbon Dioxide Extinguishing Systems (Also companion 12A)
- NFPA 13: Installation of Sprinkler Systems (Also companion 13D, 13E, and 13R)
- NFPA 14: Installation of Standpipe and Hose Systems
- NFPA 15: Water Spray Fixed Systems for Fire Protection
- NFPA 16: Installation of Foam-Water Sprinkler and Foam-Water Spray Systems
- NFPA 17: Dry Chemical Extinguishing Systems (Also companion 17A)
- NFPA 18: Wetting Agents
- NFPA 20: Installation of Stationary Pumps for Fire Protection
- NFPA 22: Water Tanks for Private Fire Protection
- NFPA 24: Installation of Private Fire Service Mains and their Appurtenances
- NFPA 25: Inspection, Testing and Maintenance of Water-Based Fire Protection

Reference Numbers

Reference numbers are shown at the beginning of some major classifications. These numbers refer to related items in the Reference Section. The reference information may be an estimating procedure, an alternate pricing method, or technical information.

Note: "Powered in part by CINX™, based on liscensed proprietary information of Harrison Publishing House, Inc."

Note: Trade Service, in part, has been used as a reference source for some of the material prices used in Division 21.

21 05 23 – General-Duty Valves for Water-Based Fire-Suppression Piping

21 05 23.50 General-Duty Valves		Crew	Daily Output	Labor-Hours	Unit	2019 Bare Costs Material	Labor	Equipment	Total	Total Incl O&P
0010	**GENERAL-DUTY VALVES**, for water-based fire suppression									
6200	Valves and components									
6210	Wet alarm, includes									
6220	retard chamber, trim, gauges, alarm line strainer									
6260	3″ size	Q-12	3	5.333	Ea.	1,850	192		2,042	2,350
6280	4″ size	″	2	8		2,200	288		2,488	2,900
6300	6″ size	Q-13	4	8		2,000	288		2,288	2,675
6400	Dry alarm, includes									
6405	retard chamber, trim, gauges, alarm line strainer									
6410	1-1/2″ size	Q-12	3	5.333	Ea.	4,825	192		5,017	5,650
6420	2″ size		3	5.333		4,825	192		5,017	5,650
6430	3″ size		3	5.333		4,900	192		5,092	5,725
6440	4″ size		2	8		5,250	288		5,538	6,250
6450	6″ size	Q-13	3	10.667		6,050	385		6,435	7,275
6460	8″ size	″	3	10.667		8,875	385		9,260	10,400
6500	Check, swing, C.I. body, brass fittings, auto. ball drip									
6520	4″ size	Q-12	3	5.333	Ea.	400	192		592	755
6800	Check, wafer, butterfly type, C.I. body, bronze fittings									
6820	4″ size	Q-12	4	4	Ea.	1,300	144		1,444	1,650
8700	Floor control valve, includes trim and gauges, 2″ size		6	2.667		950	96		1,046	1,200
8710	2-1/2″ size		6	2.667		1,075	96		1,171	1,350
8720	3″ size		6	2.667		1,075	96		1,171	1,350
8730	4″ size		6	2.667		1,075	96		1,171	1,350
8740	6″ size		5	3.200		1,075	115		1,190	1,400
8800	Flow control valve, includes trim and gauges, 2″ size		2	8		4,925	288		5,213	5,900
8820	3″ size		1.50	10.667		6,050	385		6,435	7,275
8840	4″ size	Q-13	2.80	11.429		5,850	410		6,260	7,125
8860	6″ size	″	2	16		8,325	575		8,900	10,100
9200	Pressure operated relief valve, brass body	1 Spri	18	.444		650	17.75		667.75	745
9600	Waterflow indicator, vane type, with recycling retard and									
9610	two single pole retard switches, 2″ thru 6″ pipe size	1 Spri	8	1	Ea.	168	40		208	250

21 05 53 – Identification For Fire-Suppression Piping and Equipment

21 05 53.50 Identification		Crew	Daily Output	Labor-Hours	Unit	Material	Labor	Equipment	Total	Total Incl O&P
0010	**IDENTIFICATION**, for fire suppression piping and equipment									
3010	Plates and escutcheons for identification of fire dept. service/connections									
3100	Wall mount, round, aluminum									
3110	4″	1 Plum	96	.083	Ea.	24.50	3.37		27.87	32.50
3120	6″	″	96	.083	″	63.50	3.37		66.87	75.50
3200	Wall mount, round, cast brass									
3210	2-1/2″	1 Plum	70	.114	Ea.	58.50	4.62		63.12	72
3220	3″		70	.114		68.50	4.62		73.12	82.50
3230	4″		70	.114		117	4.62		121.62	137
3240	6″		70	.114		137	4.62		141.62	158
3250	For polished brass, add									
3260	For rough chrome, add									
3270	For polished chrome, add									
3300	Wall mount, square, cast brass									
3310	2-1/2″	1 Plum	70	.114	Ea.	156	4.62		160.62	180
3320	3″	″	70	.114	″	166	4.62		170.62	190
3330	For polished brass, add									
3340	For rough chrome, add									
3350	For polished chrome, add									

21 05 Common Work Results for Fire Suppression

21 05 53 – Identification For Fire-Suppression Piping and Equipment

21 05 53.50 Identification		Crew	Daily Output	Labor-Hours	Unit	Material	Labor	Equipment	Total	Total Incl O&P
						2019 Bare Costs				
3400	Wall mount, cast brass, multiple outlets									
3410	rect. 2 way	Q-1	5	3.200	Ea.	216	116		332	430
3420	rect. 3 way		4	4		505	145		650	800
3430	rect. 4 way		4	4		635	145		780	940
3440	square 4 way		4	4		630	145		775	930
3450	rect. 6 way		3	5.333		870	194		1,064	1,275
3460	For polished brass, add									
3470	For rough chrome, add									
3480	For polished chrome, add									
3500	Base mount, free standing fdc, cast brass									
3510	4"	1 Plum	60	.133	Ea.	117	5.40		122.40	138
3520	6"	"	60	.133	"	156	5.40		161.40	181
3530	For polished brass, add									
3540	For rough chrome, add									
3550	For polished chrome, add									

21 11 Facility Fire-Suppression Water-Service Piping

21 11 13 – Facility Water Distribution Piping

21 11 13.16 Pipe, Plastic		Crew	Daily Output	Labor-Hours	Unit	Material	Labor	Equipment	Total	Total Incl O&P
0010	**PIPE, PLASTIC**									
0020	CPVC, fire suppression (C-UL-S, FM, NFPA 13, 13D & 13R)									
0030	Socket joint, no couplings or hangers									
0100	SDR 13.5 (ASTM F442)									
0120	3/4" diameter	Q-12	420	.038	L.F.	1.69	1.37		3.06	4.11
0130	1" diameter		340	.047		2.61	1.69		4.30	5.65
0140	1-1/4" diameter		260	.062		4.16	2.21		6.37	8.20
0150	1-1/2" diameter		190	.084		5.75	3.03		8.78	11.30
0160	2" diameter		140	.114		9.15	4.11		13.26	16.80
0170	2-1/2" diameter		130	.123		16.90	4.42		21.32	26
0180	3" diameter		120	.133		25.50	4.79		30.29	36

21 11 13.18 Pipe Fittings, Plastic		Crew	Daily Output	Labor-Hours	Unit	Material	Labor	Equipment	Total	Total Incl O&P
0010	**PIPE FITTINGS, PLASTIC**									
0020	CPVC, fire suppression (C-UL-S, FM, NFPA 13, 13D & 13R)									
0030	Socket joint									
0100	90° elbow									
0120	3/4"	1 Plum	26	.308	Ea.	1.87	12.45		14.32	22.50
0130	1"		22.70	.352		4.12	14.25		18.37	28
0140	1-1/4"		20.20	.396		5.20	16		21.20	32.50
0150	1-1/2"		18.20	.440		7.40	17.75		25.15	37
0160	2"	Q-1	33.10	.483		9.20	17.55		26.75	39
0170	2-1/2"		24.20	.661		17.65	24		41.65	59
0180	3"		20.80	.769		24	28		52	72.50
0200	45° elbow									
0210	3/4"	1 Plum	26	.308	Ea.	2.58	12.45		15.03	23.50
0220	1"		22.70	.352		3.02	14.25		17.27	27
0230	1-1/4"		20.20	.396		4.37	16		20.37	31.50
0240	1-1/2"		18.20	.440		6.10	17.75		23.85	35.50
0250	2"	Q-1	33.10	.483		7.60	17.55		25.15	37.50
0260	2-1/2"		24.20	.661		13.65	24		37.65	54.50
0270	3"		20.80	.769		19.50	28		47.50	67.50
0300	Tee									

21 11 13 – Facility Water Distribution Piping

21 11 13.18 Pipe Fittings, Plastic		Crew	Daily Output	Labor-Hours	Unit	Material	Labor	Equipment	Total	Total Incl O&P
						2019 Bare Costs				
0310	3/4"	1 Plum	17.30	.462	Ea.	2.58	18.70		21.28	33.50
0320	1"		15.20	.526		5.10	21.50		26.60	40.50
0330	1-1/4"		13.50	.593		7.65	24		31.65	48
0340	1-1/2"	↓	12.10	.661		11.25	26.50		37.75	56.50
0350	2"	Q-1	20	.800		16.65	29		45.65	66
0360	2-1/2"		16.20	.988		27	36		63	89
0370	3"	↓	13.90	1.151	↓	41.50	42		83.50	115
0400	Tee, reducing x any size									
0420	1"	1 Plum	15.20	.526	Ea.	4.31	21.50		25.81	39.50
0430	1-1/4"		13.50	.593		7.90	24		31.90	48
0440	1-1/2"	↓	12.10	.661		9.60	26.50		36.10	54.50
0450	2"	Q-1	20	.800		19.30	29		48.30	68.50
0460	2-1/2"		16.20	.988		21	36		57	82
0470	3"	↓	13.90	1.151	↓	24.50	42		66.50	95
0500	Coupling									
0510	3/4"	1 Plum	26	.308	Ea.	1.80	12.45		14.25	22.50
0520	1"		22.70	.352		2.38	14.25		16.63	26
0530	1-1/4"		20.20	.396		3.47	16		19.47	30.50
0540	1-1/2"	↓	18.20	.440		4.95	17.75		22.70	34.50
0550	2"	Q-1	33.10	.483		6.70	17.55		24.25	36.50
0560	2-1/2"		24.20	.661		10.20	24		34.20	51
0570	3"	↓	20.80	.769	↓	13.30	28		41.30	60.50
0600	Coupling, reducing									
0610	1" x 3/4"	1 Plum	22.70	.352	Ea.	2.38	14.25		16.63	26
0620	1-1/4" x 1"		20.20	.396		3.61	16		19.61	30.50
0630	1-1/2" x 3/4"		18.20	.440		5.40	17.75		23.15	35
0640	1-1/2" x 1"		18.20	.440		5.20	17.75		22.95	35
0650	1-1/2" x 1-1/4"	↓	18.20	.440		4.95	17.75		22.70	34.50
0660	2" x 1"	Q-1	33.10	.483		6.95	17.55		24.50	36.50
0670	2" x 1-1/2"	"	33.10	.483	↓	6.70	17.55		24.25	36.50
0700	Cross									
0720	3/4"	1 Plum	13	.615	Ea.	4.05	25		29.05	45.50
0730	1"		11.30	.708		5.10	28.50		33.60	52.50
0740	1-1/4"		10.10	.792		7	32		39	60
0750	1-1/2"	↓	9.10	.879		9.50	35.50		45	69
0760	2"	Q-1	16.60	.964		15.80	35		50.80	75
0770	2-1/2"	"	12.10	1.322	↓	34.50	48		82.50	117
0800	Cap									
0820	3/4"	1 Plum	52	.154	Ea.	1.08	6.20		7.28	11.40
0830	1"		45	.178		1.55	7.20		8.75	13.50
0840	1-1/4"		40	.200		2.51	8.10		10.61	16
0850	1-1/2"	↓	36.40	.220		3.47	8.90		12.37	18.35
0860	2"	Q-1	66	.242		5.20	8.80		14	20
0870	2-1/2"		48.40	.331		7.50	12		19.50	28
0880	3"	↓	41.60	.385	↓	12.15	14		26.15	36.50
0900	Adapter, sprinkler head, female w/metal thd. insert (s x FNPT)									
0920	3/4" x 1/2"	1 Plum	52	.154	Ea.	4.93	6.20		11.13	15.60
0930	1" x 1/2"		45	.178		5.20	7.20		12.40	17.50
0940	1" x 3/4"	↓	45	.178	↓	8.20	7.20		15.40	21

21 11 16 – Facility Fire Hydrants

21 11 16.50 Fire Hydrants for Buildings		Crew	Daily Output	Labor-Hours	Unit	2019 Bare Costs Material	Labor	Equipment	Total	Total Incl O&P
0010	**FIRE HYDRANTS FOR BUILDINGS**									
3750	Hydrants, wall, w/caps, single, flush, polished brass									
3800	2-1/2" x 2-1/2"	Q-12	5	3.200	Ea.	260	115		375	475
3840	2-1/2" x 3"		5	3.200		520	115		635	760
3860	3" x 3"		4.80	3.333		415	120		535	655
3900	For polished chrome, add					20%				
3950	Double, flush, polished brass									
4000	2-1/2" x 2-1/2" x 4"	Q-12	5	3.200	Ea.	790	115		905	1,050
4040	2-1/2" x 2-1/2" x 6"		4.60	3.478		1,175	125		1,300	1,500
4080	3" x 3" x 4"		4.90	3.265		1,125	117		1,242	1,450
4120	3" x 3" x 6"		4.50	3.556		1,700	128		1,828	2,050
4200	For polished chrome, add					10%				
4350	Double, projecting, polished brass									
4400	2-1/2" x 2-1/2" x 4"	Q-12	5	3.200	Ea.	305	115		420	525
4450	2-1/2" x 2-1/2" x 6"	"	4.60	3.478	"	620	125		745	885
4460	Valve control, dbl. flush/projecting hydrant, cap &									
4470	chain, extension rod & cplg., escutcheon, polished brass	Q-12	8	2	Ea.	273	72		345	420
4480	Four-way square, flush, polished brass									
4540	2-1/2" (4) x 6"	Q-12	3.60	4.444	Ea.	3,700	160		3,860	4,350

21 11 19 – Fire-Department Connections

21 11 19.50 Connections for the Fire-Department		Crew	Daily Output	Labor-Hours	Unit	Material	Labor	Equipment	Total	Total Incl O&P
0010	**CONNECTIONS FOR THE FIRE-DEPARTMENT**									
4000	Storz type, with cap and chain									
6000	Roof manifold, horiz., brass, without valves & caps									
6040	2-1/2" x 2-1/2" x 4"	Q-12	4.80	3.333	Ea.	200	120		320	415
6060	2-1/2" x 2-1/2" x 6"		4.60	3.478		236	125		361	465
6080	2-1/2" x 2-1/2" x 2-1/2" x 4"		4.60	3.478		330	125		455	570
6090	2-1/2" x 2-1/2" x 2-1/2" x 6"		4.60	3.478		350	125		475	590
7000	Sprinkler line tester, cast brass					39			39	42.50
7140	Standpipe connections, wall, w/plugs & chains									
7160	Single, flush, brass, 2-1/2" x 2-1/2"	Q-12	5	3.200	Ea.	193	115		308	400
7180	2-1/2" x 3"	"	5	3.200	"	202	115		317	410
7240	For polished chrome, add					15%				
7280	Double, flush, polished brass									
7300	2-1/2" x 2-1/2" x 4"	Q-12	5	3.200	Ea.	820	115		935	1,100
7330	2-1/2" x 2-1/2" x 6"		4.60	3.478		910	125		1,035	1,200
7340	3" x 3" x 4"		4.90	3.265		1,050	117		1,167	1,350
7370	3" x 3" x 6"		4.50	3.556		1,275	128		1,403	1,600
7400	For polished chrome, add					15%				
7440	For sill cock combination, add				Ea.	97.50			97.50	107
7580	Double projecting, polished brass									
7600	2-1/2" x 2-1/2" x 4"	Q-12	5	3.200	Ea.	545	115		660	790
7630	2-1/2" x 2-1/2" x 6"	"	4.60	3.478	"	910	125		1,035	1,200
7680	For polished chrome, add					15%				
7900	Three way, flush, polished brass									
7920	2-1/2" (3) x 4"	Q-12	4.80	3.333	Ea.	1,950	120		2,070	2,350
7930	2-1/2" (3) x 6"	"	4.60	3.478		2,050	125		2,175	2,450
8000	For polished chrome, add					9%				
8020	Three way, projecting, polished brass									
8040	2-1/2" (3) x 4"	Q-12	4.80	3.333	Ea.	890	120		1,010	1,175
8070	2-1/2" (3) x 6"	"	4.60	3.478		1,775	125		1,900	2,150
8100	For polished chrome, add					12%				

21 11 Facility Fire-Suppression Water-Service Piping

21 11 19 – Fire-Department Connections

21 11 19.50 Connections for the Fire-Department		Crew	Daily Output	Labor-Hours	Unit	2019 Bare Costs Material	Labor	Equipment	Total	Total Incl O&P
8200	Four way, square, flush, polished brass,									
8240	2-1/2" (4) x 6"	Q-12	3.60	4.444	Ea.	1,475	160		1,635	1,900
8300	For polished chrome, add				"	10%				
8550	Wall, vertical, flush, cast brass									
8600	Two way, 2-1/2" x 2-1/2" x 4"	Q-12	5	3.200	Ea.	400	115		515	630
8660	Four way, 2-1/2" (4) x 6"		3.80	4.211		1,325	151		1,476	1,725
8680	Six way, 2-1/2" (6) x 6"		3.40	4.706		1,575	169		1,744	2,000
8700	For polished chrome, add					10%				
8800	Free standing siamese unit, polished brass, two way									
8820	2-1/2" x 2-1/2" x 4"	Q-12	2.50	6.400	Ea.	725	230		955	1,175
8850	2-1/2" x 2-1/2" x 6"		2	8		775	288		1,063	1,325
8860	3" x 3" x 4"		2.50	6.400		510	230		740	945
8890	3" x 3" x 6"		2	8		1,350	288		1,638	1,950
8940	For polished chrome, add					12%				
9100	Free standing siamese unit, polished brass, three way									
9120	2-1/2" x 2-1/2" x 2-1/2" x 6"	Q-12	2	8	Ea.	965	288		1,253	1,525
9160	For polished chrome, add				"	15%				

21 12 Fire-Suppression Standpipes

21 12 19 – Fire-Suppression Hose Racks

21 12 19.50 Fire Hose Racks

		Crew	Daily Output	Labor-Hours	Unit	Material	Labor	Equipment	Total	Total Incl O&P
0010	**FIRE HOSE RACKS**									
2600	Hose rack, swinging, for 1-1/2" diameter hose,									
2620	Enameled steel, 50' and 75' lengths of hose	Q-12	20	.800	Ea.	71	29		100	126
2640	100' and 125' lengths of hose		20	.800		94	29		123	152
2680	Chrome plated, 50' and 75' lengths of hose		20	.800		75.50	29		104.50	131
2700	100' and 125' lengths of hose		20	.800		144	29		173	207
2780	For hose rack nipple, 1-1/2" polished brass, add					35.50			35.50	39
2820	2-1/2" polished brass, add					54.50			54.50	60
2840	1-1/2" polished chrome, add					36			36	39.50
2860	2-1/2" polished chrome, add					71			71	78.50

21 12 23 – Fire-Suppression Hose Valves

21 12 23.70 Fire Hose Valves

		Crew	Daily Output	Labor-Hours	Unit	Material	Labor	Equipment	Total	Total Incl O&P
0010	**FIRE HOSE VALVES**									
0020	Angle, combination pressure adjust/restricting, rough brass									
0030	1-1/2"	1 Spri	12	.667	Ea.	98	26.50		124.50	152
0040	2-1/2"	"	7	1.143	"	213	45.50		258.50	310
0042	Nonpressure adjustable/restricting, rough brass									
0044	1-1/2"	1 Spri	12	.667	Ea.	49	26.50		75.50	98
0046	2-1/2"	"	7	1.143	"	128	45.50		173.50	216
0050	For polished brass, add					30%				
0060	For polished chrome, add					40%				
1000	Ball drip, automatic, rough brass, 1/2"	1 Spri	20	.400	Ea.	24	16		40	53
1010	3/4"	"	20	.400	"	24.50	16		40.50	53.50
1100	Ball, 175 lb., sprinkler system, FM/UL, threaded, bronze									
1120	Slow close									
1150	1" size	1 Spri	19	.421	Ea.	289	16.80		305.80	345
1160	1-1/4" size		15	.533		310	21.50		331.50	375
1170	1-1/2" size		13	.615		410	24.50		434.50	490
1180	2" size		11	.727		495	29		524	595

21 12 Fire-Suppression Standpipes

21 12 23 – Fire-Suppression Hose Valves

21 12 23.70 Fire Hose Valves		Crew	Daily Output	Labor-Hours	Unit	Material	2019 Bare Costs Labor	Equipment	Total	Total Incl O&P
1190	2-1/2" size	Q-12	15	1.067	Ea.	670	38.50		708.50	800
1230	For supervisory switch kit, all sizes									
1240	One circuit, add	1 Spri	48	.167	Ea.	179	6.65		185.65	208
1280	Quarter turn for trim									
1300	1/2" size	1 Spri	22	.364	Ea.	39	14.55		53.55	66.50
1310	3/4" size		20	.400		41.50	16		57.50	72.50
1320	1" size		19	.421		46	16.80		62.80	78.50
1330	1-1/4" size		15	.533		76	21.50		97.50	119
1340	1-1/2" size		13	.615		95	24.50		119.50	146
1350	2" size		11	.727		113	29		142	173
1400	Caps, polished brass with chain, 3/4"					60.50			60.50	67
1420	1"					76.50			76.50	84
1440	1-1/2"					21			21	23
1460	2-1/2"					32			32	35
1480	3"					39			39	43
1900	Escutcheon plate, for angle valves, polished brass, 1-1/2"					15.40			15.40	16.90
1920	2-1/2"					23			23	25.50
1940	3"					31.50			31.50	35
1980	For polished chrome, add					15%				
3000	Gate, hose, wheel handle, N.R.S., rough brass, 1-1/2"	1 Spri	12	.667		157	26.50		183.50	217
3040	2-1/2", 300 lb.	"	7	1.143		228	45.50		273.50	325
3080	For polished brass, add					40%				
3090	For polished chrome, add					50%				
5000	Pressure reducing rough brass, 1-1/2"	1 Spri	12	.667		335	26.50		361.50	410
5020	2-1/2"	"	7	1.143		395	45.50		440.50	510
5080	For polished brass, add					105%				
5090	For polished chrome, add					140%				

21 13 Fire-Suppression Sprinkler Systems

21 13 13 – Wet-Pipe Sprinkler Systems

21 13 13.50 Wet-Pipe Sprinkler System Components		Crew	Daily Output	Labor-Hours	Unit	Material	2019 Bare Costs Labor	Equipment	Total	Total Incl O&P
0010	**WET-PIPE SPRINKLER SYSTEM COMPONENTS**									
1100	Alarm, electric pressure switch (circuit closer)	1 Spri	26	.308	Ea.	111	12.30		123.30	142
1140	For explosion proof, max 20 psi, contacts close or open		26	.308		690	12.30		702.30	780
1220	Water motor, complete with gong		4	2		455	80		535	630
1900	Flexible sprinkler head connectors									
1910	Braided stainless steel hose with mounting bracket									
1920	1/2" and 3/4" outlet size									
1940	40" length	1 Spri	30	.267	Ea.	78	10.65		88.65	103
1960	60" length	"	22	.364	"	89.50	14.55		104.05	123
1982	May replace hard-pipe armovers									
1984	for wet and pre-action systems.									
2000	Release, emergency, manual, for hydraulic or pneumatic system	1 Spri	12	.667	Ea.	238	26.50		264.50	305
2060	Release, thermostatic, for hydraulic or pneumatic release line		20	.400		740	16		756	835
2200	Sprinkler cabinets, 6 head capacity		16	.500		85.50	20		105.50	127
2260	12 head capacity		16	.500		90.50	20		110.50	133
2340	Sprinkler head escutcheons, standard, brass tone, 1" size		40	.200		3.16	8		11.16	16.65
2360	Chrome, 1" size		40	.200		3.35	8		11.35	16.85
2400	Recessed type, bright brass		40	.200		11.20	8		19.20	25.50
2440	Chrome or white enamel		40	.200		3.74	8		11.74	17.25
2600	Sprinkler heads, not including supply piping									

21 13 Fire-Suppression Sprinkler Systems

21 13 13 – Wet-Pipe Sprinkler Systems

21 13 13.50 Wet-Pipe Sprinkler System Components		Crew	Daily Output	Labor-Hours	Unit	Material	Labor	Equipment	Total	Total Incl O&P
						2019 Bare Costs				
3700	Standard spray, pendent or upright, brass, 135°F to 286°F									
3720	1/2" NPT, 3/8" orifice	1 Spri	16	.500	Ea.	16.45	20		36.45	51
3730	1/2" NPT, 7/16" orifice		16	.500		16.35	20		36.35	51
3732	1/2" NPT, 7/16" orifice, chrome		16	.500		17.15	20		37.15	52
3740	1/2" NPT, 1/2" orifice		16	.500		10.60	20		30.60	44.50
3760	1/2" NPT, 17/32" orifice		16	.500		13.80	20		33.80	48
3780	3/4" NPT, 17/32" orifice		16	.500		12.75	20		32.75	47
3840	For chrome, add					3.74			3.74	4.11
4200	Sidewall, vertical brass, 135°F to 286°F									
4240	1/2" NPT, 1/2" orifice	1 Spri	16	.500	Ea.	27.50	20		47.50	63
4280	3/4" NPT, 17/32" orifice	"	16	.500		75	20		95	116
4360	For satin chrome, add					3.84			3.84	4.22
4500	Sidewall, horizontal, brass, 135°F to 286°F									
4520	1/2" NPT, 1/2" orifice	1 Spri	16	.500	Ea.	27.50	20		47.50	63.50
5600	Concealed, complete with cover plate									
5620	1/2" NPT, 1/2" orifice, 135°F to 212°F	1 Spri	9	.889	Ea.	25.50	35.50		61	86.50
6025	Residential sprinkler components (one and two family)									
6026	Water motor alarm with strainer	1 Spri	4	2	Ea.	455	80		535	630
6027	Fast response, glass bulb, 135°F to 155°F									
6028	1/2" NPT, pendent, brass	1 Spri	16	.500	Ea.	29.50	20		49.50	65.50
6029	1/2" NPT, sidewall, brass		16	.500		29.50	20		49.50	65.50
6030	1/2" NPT, pendent, brass, extended coverage		16	.500		25.50	20		45.50	61
6031	1/2" NPT, sidewall, brass, extended coverage		16	.500		25	20		45	60.50
6032	3/4" NPT sidewall, brass, extended coverage		16	.500		26	20		46	61.50
6033	For chrome, add					15%				
6034	For polyester/teflon coating, add					20%				
6100	Sprinkler head wrenches, standard head				Ea.	29			29	32
6120	Recessed head					40.50			40.50	44.50
6160	Tamper switch (valve supervisory switch)	1 Spri	16	.500		248	20		268	305
6165	Flow switch (valve supervisory switch)	"	16	.500		248	20		268	305

21 13 16 – Dry-Pipe Sprinkler Systems

21 13 16.50 Dry-Pipe Sprinkler System Components		Crew	Daily Output	Labor-Hours	Unit	Material	Labor	Equipment	Total	Total Incl O&P
0010	**DRY-PIPE SPRINKLER SYSTEM COMPONENTS**									
0600	Accelerator	1 Spri	8	1	Ea.	825	40		865	975
0800	Air compressor for dry pipe system, automatic, complete									
0820	30 gal. system capacity, 3/4 HP	1 Spri	1.30	6.154	Ea.	1,175	246		1,421	1,700
0860	30 gal. system capacity, 1 HP		1.30	6.154		1,475	246		1,721	2,025
0960	Air pressure maintenance control		24	.333		380	13.30		393.30	435
1600	Dehydrator package, incl. valves and nipples		12	.667		780	26.50		806.50	900
2600	Sprinkler heads, not including supply piping									
2640	Dry, pendent, 1/2" orifice, 3/4" or 1" NPT									
2660	1/2" to 6" length	1 Spri	14	.571	Ea.	135	23		158	186
2670	6-1/4" to 8" length		14	.571		140	23		163	192
2680	8-1/4" to 12" length		14	.571		146	23		169	199
2800	For each inch or fraction, add					3.90			3.90	4.29
6330	Valves and components									
6340	Alarm test/shut off valve, 1/2"	1 Spri	20	.400	Ea.	24	16		40	53
8000	Dry pipe air check valve, 3" size	Q-12	2	8		1,975	288		2,263	2,625
8200	Dry pipe valve, incl. trim and gauges, 3" size		2	8		2,825	288		3,113	3,600
8220	4" size		1	16		3,225	575		3,800	4,500
8240	6" size	Q-13	2	16		3,775	575		4,350	5,100
8280	For accelerator trim with gauges, add	1 Spri	8	1		264	40		304	355

Estimating Tips

22 10 00 Plumbing Piping and Pumps

This subdivision is primarily basic pipe and related materials. The pipe may be used by any of the mechanical disciplines, i.e., plumbing, fire protection, heating, and air conditioning.

Note: CPVC plastic piping approved for fire protection is located in 21 11 13.

- The labor adjustment factors listed in Subdivision 22 01 02.20 apply throughout Divisions 21, 22, and 23. CAUTION: the correct percentage may vary for the same items. For example, the percentage add for the basic pipe installation should be based on the maximum height that the installer must install for that particular section. If the pipe is to be located 14' above the floor but it is suspended on threaded rod from beams, the bottom flange of which is 18' high (4' rods), then the height is actually 18' and the add is 20%. The pipe cover, however, does not have to go above the 14' and so the add should be 10%.
- Most pipe is priced first as straight pipe with a joint (coupling, weld, etc.) every 10' and a hanger usually every 10'. There are exceptions with hanger spacing such as for cast iron pipe (5') and plastic pipe (3 per 10'). Following each type of pipe there are several lines listing sizes and the amount to be subtracted to delete couplings and hangers. This is for pipe that is to be buried or supported together on trapeze hangers. The reason that the couplings are deleted is that these runs are usually long, and frequently longer lengths of pipe are used. By deleting the couplings, the estimator is expected to look up and add back the correct reduced number of couplings.
- When preparing an estimate, it may be necessary to approximate the fittings. Fittings usually run between 25% and 50% of the cost of the pipe. The lower percentage is for simpler runs, and the higher number is for complex areas, such as mechanical rooms.
- For historic restoration projects, the systems must be as invisible as possible, and pathways must be sought for pipes, conduit, and ductwork. While installations in accessible spaces (such as basements and attics) are relatively straightforward to estimate, labor costs may be more difficult to determine when delivery systems must be concealed.

22 40 00 Plumbing Fixtures

- Plumbing fixture costs usually require two lines: the fixture itself and its "rough-in, supply, and waste."
- In the Assemblies Section (Plumbing D2010) for the desired fixture, the System Components Group at the center of the page shows the fixture on the first line. The rest of the list (fittings, pipe, tubing, etc.) will total up to what we refer to in the Unit Price section as "Rough-in, supply, waste, and vent." Note that for most fixtures we allow a nominal 5' of tubing to reach from the fixture to a main or riser.
- Remember that gas- and oil-fired units need venting.

Reference Numbers

Reference numbers are shown at the beginning of some major classifications. These numbers refer to related items in the Reference Section. The reference information may be an estimating procedure, an alternate pricing method, or technical information.

Note: Not all subdivisions listed here necessarily appear. ■

Note: "Powered in part by CINX™, based on liscensed proprietary information of Harrison Publishing House, Inc."

Note: Trade Service, in part, has been used as a reference source for some of the material prices used in Division 22.

22 05 Common Work Results for Plumbing

22 05 05 – Selective Demolition for Plumbing

22 05 05.10 Plumbing Demolition		Crew	Daily Output	Labor-Hours	Unit	Material	2019 Bare Costs Labor	Equipment	Total	Total Incl O&P
0010	**PLUMBING DEMOLITION**									
1020	Fixtures, including 10' piping									
1101	Bathtubs, cast iron	1 Clab	4	2	Ea.		55		55	91
1121	Fiberglass		6	1.333			36.50		36.50	60.50
1141	Steel		5	1.600			44		44	72.50
1200	Lavatory, wall hung	1 Plum	10	.800			32.50		32.50	53
1221	Counter top	1 Clab	16	.500			13.75		13.75	22.50
1301	Sink, single compartment		16	.500			13.75		13.75	22.50
1321	Double		10	.800			22		22	36.50
1400	Water closet, floor mounted	1 Plum	8	1			40.50		40.50	66.50
1421	Wall mounted	1 Clab	7	1.143			31.50		31.50	52
2001	Piping, metal, to 1-1/2" diameter		200	.040	L.F.		1.10		1.10	1.82
2051	2" thru 3-1/2" diameter		150	.053			1.47		1.47	2.42
2101	4" thru 6" diameter		100	.080			2.20		2.20	3.64
2160	Plastic pipe with fittings, up thru 1-1/2" diameter	1 Plum	250	.032			1.29		1.29	2.12
2162	2" thru 3" diameter	"	200	.040			1.62		1.62	2.65
2164	4" thru 6" diameter	Q-1	200	.080			2.91		2.91	4.77
2166	8" thru 14" diameter		150	.107			3.88		3.88	6.35
2168	16" diameter		100	.160			5.80		5.80	9.55
3000	Submersible sump pump	1 Plum	24	.333	Ea.		13.45		13.45	22
6000	Remove and reset fixtures, easy access		6	1.333			54		54	88.50
6100	Difficult access		4	2			81		81	133

22 05 23 – General-Duty Valves for Plumbing Piping

22 05 23.20 Valves, Bronze		Crew	Daily Output	Labor-Hours	Unit	Material	Labor	Equipment	Total	Total Incl O&P
0010	**VALVES, BRONZE**									
1750	Check, swing, class 150, regrinding disc, threaded									
1860	3/4"	1 Plum	20	.400	Ea.	91	16.15		107.15	127
1870	1"	"	19	.421	"	143	17		160	185
2850	Gate, N.R.S., soldered, 125 psi									
2940	3/4"	1 Plum	20	.400	Ea.	68	16.15		84.15	102
2950	1"	"	19	.421	"	76.50	17		93.50	113
5600	Relief, pressure & temperature, self-closing, ASME, threaded									
5640	3/4"	1 Plum	28	.286	Ea.	210	11.55		221.55	250
5650	1"		24	.333		335	13.45		348.45	390
5660	1-1/4"		20	.400		680	16.15		696.15	775
6400	Pressure, water, ASME, threaded									
6440	3/4"	1 Plum	28	.286	Ea.	144	11.55		155.55	177
6450	1"	"	24	.333	"	305	13.45		318.45	355
6900	Reducing, water pressure									
6920	300 psi to 25-75 psi, threaded or sweat									
6940	1/2"	1 Plum	24	.333	Ea.	460	13.45		473.45	525
6950	3/4"		20	.400		460	16.15		476.15	535
6960	1"		19	.421		715	17		732	815
8350	Tempering, water, sweat connections									
8400	1/2"	1 Plum	24	.333	Ea.	105	13.45		118.45	138
8440	3/4"	"	20	.400	"	144	16.15		160.15	186
8650	Threaded connections									
8700	1/2"	1 Plum	24	.333	Ea.	149	13.45		162.45	186
8740	3/4"	"	20	.400	"	1,025	16.15		1,041.15	1,150
8800	Water heater water & gas safety shut off									
8810	Protection against a leaking water heater									
8814	Shut off valve	1 Plum	16	.500	Ea.	193	20		213	245

22 05 23 – General-Duty Valves for Plumbing Piping

22 05 23.20 Valves, Bronze		Crew	Daily Output	Labor-Hours	Unit	Material	2019 Bare Costs Labor	Equipment	Total	Total Incl O&P
8818	Water heater dam	1 Plum	32	.250	Ea.	32	10.10		42.10	51.50
8822	Gas control wiring harness	↓	32	.250	↓	22	10.10		32.10	41
8830	Whole house flood safety shut off									
8834	Connections									
8838	3/4" NPT	1 Plum	12	.667	Ea.	950	27		977	1,100
8842	1" NPT		11	.727		975	29.50		1,004.50	1,125
8846	1-1/4" NPT	↓	10	.800	↓	1,000	32.50		1,032.50	1,150

22 05 29 – Hangers and Supports for Plumbing Piping and Equipment

22 05 29.10 Hangers & Supp. for Plumb'g/HVAC Pipe/Equip.		Crew	Daily Output	Labor-Hours	Unit	Material	Labor	Equipment	Total	Total Incl O&P
0010	**HANGERS AND SUPPORTS FOR PLUMB'G/HVAC PIPE/EQUIP.**									
8000	Pipe clamp, plastic, 1/2" CTS	1 Plum	80	.100	Ea.	.23	4.04		4.27	6.90
8010	3/4" CTS		73	.110		.24	4.43		4.67	7.50
8020	1" CTS		68	.118		.53	4.75		5.28	8.40
8080	Economy clamp, 1/4" CTS		175	.046		.05	1.85		1.90	3.09
8090	3/8" CTS		168	.048		.05	1.92		1.97	3.22
8100	1/2" CTS		160	.050		.05	2.02		2.07	3.38
8110	3/4" CTS		145	.055		.05	2.23		2.28	3.72
8200	Half clamp, 1/2" CTS		80	.100		.07	4.04		4.11	6.75
8210	3/4" CTS		73	.110		.10	4.43		4.53	7.35
8300	Suspension clamp, 1/2" CTS		80	.100		.24	4.04		4.28	6.90
8310	3/4" CTS		73	.110		.25	4.43		4.68	7.55
8320	1" CTS		68	.118		.53	4.75		5.28	8.40
8400	Insulator, 1/2" CTS		80	.100		.35	4.04		4.39	7.05
8410	3/4" CTS		73	.110		.35	4.43		4.78	7.65
8420	1" CTS		68	.118		.38	4.75		5.13	8.20
8500	J hook clamp with nail, 1/2" CTS		240	.033		.10	1.35		1.45	2.32
8501	3/4" CTS	↓	240	.033	↓	.10	1.35		1.45	2.32

22 05 48 – Vibration and Seismic Controls for Plumbing Piping and Equipment

22 05 48.10 Seismic Bracing Supports		Crew	Daily Output	Labor-Hours	Unit	Material	Labor	Equipment	Total	Total Incl O&P
0010	**SEISMIC BRACING SUPPORTS**									
0020	Clamps									
0030	C-clamp, for mounting on steel beam									
0040	3/8" threaded rod	1 Skwk	160	.050	Ea.	4.04	1.83		5.87	7.45
0050	1/2" threaded rod		160	.050		4.88	1.83		6.71	8.40
0060	5/8" threaded rod		160	.050		5.35	1.83		7.18	8.95
0070	3/4" threaded rod	↓	160	.050	↓	6.85	1.83		8.68	10.60
0100	Brackets									
0110	Beam side or wall malleable iron									
0120	3/8" threaded rod	1 Skwk	48	.167	Ea.	4.24	6.10		10.34	14.75
0130	1/2" threaded rod		48	.167		3.88	6.10		9.98	14.35
0140	5/8" threaded rod		48	.167		11.45	6.10		17.55	22.50
0150	3/4" threaded rod		48	.167		20.50	6.10		26.60	33
0160	7/8" threaded rod	↓	48	.167	↓	12.70	6.10		18.80	24
0170	For concrete installation, add						30%			
0180	Wall, welded steel									
0190	0 size 12" wide 18" deep	1 Skwk	34	.235	Ea.	180	8.60		188.60	212
0200	1 size 18" wide 24" deep		34	.235		216	8.60		224.60	252
0210	2 size 24" wide 30" deep	↓	34	.235	↓	285	8.60		293.60	330
0300	Rod, carbon steel									
0310	Continuous thread									
0320	1/4" thread	1 Skwk	144	.056	L.F.	2.30	2.03		4.33	5.90
0330	3/8" thread	↓	144	.056	↓	2.46	2.03		4.49	6.10

22 05 Common Work Results for Plumbing

22 05 48 – Vibration and Seismic Controls for Plumbing Piping and Equipment

22 05 48.10 Seismic Bracing Supports		Crew	Daily Output	Labor-Hours	Unit	2019 Bare Costs Material	Labor	Equipment	Total	Total Incl O&P
0340	1/2" thread	1 Skwk	144	.056	L.F.	3.87	2.03		5.90	7.65
0350	5/8" thread		144	.056		5.50	2.03		7.53	9.40
0360	3/4" thread		144	.056		9.65	2.03		11.68	14
0370	7/8" thread		144	.056		12.15	2.03		14.18	16.70
0380	For galvanized, add					30%				
0400	Channel, steel									
0410	3/4" x 1-1/2"	1 Skwk	80	.100	L.F.	2.63	3.65		6.28	8.95
0420	1-1/2" x 1-1/2"		70	.114		3.43	4.17		7.60	10.70
0430	1-7/8" x 1-1/2"		60	.133		17.60	4.87		22.47	27.50
0440	3" x 1-1/2"		50	.160		17.95	5.85		23.80	29.50
0450	Spring nuts									
0460	3/8"	1 Skwk	100	.080	Ea.	1.58	2.92		4.50	6.60
0470	1/2"	"	80	.100	"	1.66	3.65		5.31	7.90
0500	Welding, field									
0510	Cleaning and welding plates, bars, or rods									
0520	To existing beams, columns, or trusses									
0530	1" weld	1 Skwk	144	.056	Ea.	.23	2.03		2.26	3.62
0540	2" weld		72	.111		.46	4.06		4.52	7.25
0550	3" weld		54	.148		.69	5.40		6.09	9.75
0560	4" weld		36	.222		.91	8.10		9.01	14.50
0570	5" weld		30	.267		1.14	9.75		10.89	17.40
0580	6" weld		24	.333		1.37	12.15		13.52	21.50
0600	Vibration absorbers									
0610	Hangers, neoprene flex									
0620	10-120 lb. capacity	1 Skwk	8	1	Ea.	25.50	36.50		62	89
0630	75-550 lb. capacity		8	1		41	36.50		77.50	106
0640	250-1,100 lb. capacity		6	1.333		82.50	48.50		131	172
0650	1,000-4,000 lb. capacity		6	1.333		152	48.50		200.50	248

22 05 76 – Facility Drainage Piping Cleanouts

22 05 76.10 Cleanouts		Crew	Daily Output	Labor-Hours	Unit	Material	Labor	Equipment	Total	Total Incl O&P
0010	**CLEANOUTS**									
0060	Floor type									
0080	Round or square, scoriated nickel bronze top									
0100	2" pipe size	1 Plum	10	.800	Ea.	185	32.50		217.50	257
0120	3" pipe size		8	1		251	40.50		291.50	345
0140	4" pipe size		6	1.333		285	54		339	405

22 05 76.20 Cleanout Tees		Crew	Daily Output	Labor-Hours	Unit	Material	Labor	Equipment	Total	Total Incl O&P
0010	**CLEANOUT TEES**									
0100	Cast iron, B&S, with countersunk plug									
0220	3" pipe size	1 Plum	3.60	2.222	Ea.	154	90		244	315
0240	4" pipe size	"	3.30	2.424	"	240	98		338	425
0500	For round smooth access cover, same price									
4000	Plastic, tees and adapters. Add plugs									
4010	ABS, DWV									
4020	Cleanout tee, 1-1/2" pipe size	1 Plum	15	.533	Ea.	11.55	21.50		33.05	48.50

22 07 Plumbing Insulation

22 07 16 – Plumbing Equipment Insulation

22 07 16.10 Insulation for Plumbing Equipment		Crew	Daily Output	Labor-Hours	Unit	Material	Labor (2019 Bare Costs)	Equipment	Total	Total Incl O&P
0010	**INSULATION FOR PLUMBING EQUIPMENT**									
2900	Domestic water heater wrap kit									
2920	1-1/2" with vinyl jacket, 20 to 60 gal. G	1 Plum	8	1	Ea.	17.05	40.50		57.55	85.50

22 07 19 – Plumbing Piping Insulation

22 07 19.10 Piping Insulation		Crew	Daily Output	Labor-Hours	Unit	Material	Labor	Equipment	Total	Total Incl O&P
0010	**PIPING INSULATION**									
0230	Insulated protectors (ADA)									
0235	For exposed piping under sinks or lavatories									
0240	Vinyl coated foam, velcro tabs									
0245	P Trap, 1-1/4" or 1-1/2"	1 Plum	32	.250	Ea.	16.70	10.10		26.80	35
0260	Valve and supply cover									
0265	1/2", 3/8", and 7/16" pipe size	1 Plum	32	.250	Ea.	16.85	10.10		26.95	35
0285	1-1/4" pipe size	"	32	.250	"	13.20	10.10		23.30	31
0600	Pipe covering (price copper tube one size less than IPS)									
6600	Fiberglass, with all service jacket									
6840	1" wall, 1/2" iron pipe size G	Q-14	240	.067	L.F.	.88	2.20		3.08	4.68
6860	3/4" iron pipe size G		230	.070		.96	2.30		3.26	4.93
6870	1" iron pipe size G		220	.073		1.03	2.40		3.43	5.15
6900	2" iron pipe size G	↓	200	.080	↓	1.61	2.64		4.25	6.20
7879	Rubber tubing, flexible closed cell foam									
8100	1/2" wall, 1/4" iron pipe size G	1 Asbe	90	.089	L.F.	.83	3.26		4.09	6.40
8130	1/2" iron pipe size G		89	.090		.91	3.30		4.21	6.55
8140	3/4" iron pipe size G		89	.090		1.04	3.30		4.34	6.70
8150	1" iron pipe size G		88	.091		.84	3.34		4.18	6.50
8170	1-1/2" iron pipe size G		87	.092		1.70	3.37		5.07	7.55
8180	2" iron pipe size G		86	.093		2.06	3.41		5.47	8
8300	3/4" wall, 1/4" iron pipe size G		90	.089		.92	3.26		4.18	6.50
8330	1/2" iron pipe size G		89	.090		1.11	3.30		4.41	6.75
8340	3/4" iron pipe size G		89	.090		1.81	3.30		5.11	7.55
8350	1" iron pipe size G		88	.091		2.08	3.34		5.42	7.90
8380	2" iron pipe size G		86	.093		3.57	3.41		6.98	9.70
8444	1" wall, 1/2" iron pipe size G		86	.093		2.71	3.41		6.12	8.75
8445	3/4" iron pipe size G		84	.095		3.29	3.50		6.79	9.50
8446	1" iron pipe size G		84	.095		3.43	3.50		6.93	9.65
8447	1-1/4" iron pipe size G		82	.098		3.76	3.58		7.34	10.20
8448	1-1/2" iron pipe size G		82	.098		5.25	3.58		8.83	11.85
8449	2" iron pipe size G		80	.100		6.50	3.67		10.17	13.35
8450	2-1/2" iron pipe size G	↓	80	.100	↓	7.50	3.67		11.17	14.45
8456	Rubber insulation tape, 1/8" x 2" x 30' G				Ea.	21.50			21.50	23.50

22 11 Facility Water Distribution

22 11 13 – Facility Water Distribution Piping

22 11 13.23 Pipe/Tube, Copper		Crew	Daily Output	Labor-Hours	Unit	Material	Labor	Equipment	Total	Total Incl O&P
0010	**PIPE/TUBE, COPPER**, Solder joints									
1000	Type K tubing, couplings & clevis hanger assemblies 10' OC									
1180	3/4" diameter	1 Plum	74	.108	L.F.	8.55	4.37		12.92	16.55
1200	1" diameter	"	66	.121	"	12.70	4.90		17.60	22
2000	Type L tubing, couplings & clevis hanger assemblies 10' OC									
2140	1/2" diameter	1 Plum	81	.099	L.F.	3.38	3.99		7.37	10.25
2160	5/8" diameter	↓	79	.101	↓	5.55	4.09		9.64	12.85

22 11 Facility Water Distribution

22 11 13 – Facility Water Distribution Piping

22 11 13.23 Pipe/Tube, Copper		Crew	Daily Output	Labor-Hours	Unit	Material	Labor	Equipment	Total	Total Incl O&P
						2019 Bare Costs				
2180	3/4" diameter	1 Plum	76	.105	L.F.	4.39	4.25		8.64	11.80
2200	1" diameter		68	.118		7	4.75		11.75	15.55
2220	1-1/4" diameter		58	.138		11.65	5.55		17.20	22
3000	Type M tubing, couplings & clevis hanger assemblies 10' OC									
3140	1/2" diameter	1 Plum	84	.095	L.F.	3.64	3.85		7.49	10.30
3180	3/4" diameter		78	.103		5.20	4.14		9.34	12.50
3200	1" diameter		70	.114		8.40	4.62		13.02	16.85
3220	1-1/4" diameter		60	.133		11.10	5.40		16.50	21
3240	1-1/2" diameter		54	.148		14.20	6		20.20	25.50
3260	2" diameter		44	.182		21	7.35		28.35	35.50
4000	Type DWV tubing, couplings & clevis hanger assemblies 10' OC									
4100	1-1/4" diameter	1 Plum	60	.133	L.F.	12.45	5.40		17.85	22.50
4120	1-1/2" diameter		54	.148		12	6		18	23
4140	2" diameter		44	.182		18.55	7.35		25.90	32.50
4160	3" diameter	Q-1	58	.276		24.50	10.05		34.55	43
4180	4" diameter	"	40	.400		55.50	14.55		70.05	85

22 11 13.25 Pipe/Tube Fittings, Copper		Crew	Daily Output	Labor-Hours	Unit	Material	Labor	Equipment	Total	Total Incl O&P
0010	**PIPE/TUBE FITTINGS, COPPER**, Wrought unless otherwise noted									
0040	Solder joints, copper x copper									
0070	90° elbow, 1/4"	1 Plum	22	.364	Ea.	3.68	14.70		18.38	28
0100	1/2"		20	.400		1.20	16.15		17.35	28
0120	3/4"		19	.421		2.40	17		19.40	30.50
0250	45° elbow, 1/4"		22	.364		8.35	14.70		23.05	33
0280	1/2"		20	.400		3.02	16.15		19.17	30
0290	5/8"		19	.421		11.45	17		28.45	40.50
0300	3/4"		19	.421		4.15	17		21.15	32.50
0310	1"		16	.500		10.40	20		30.40	44.50
0320	1-1/4"		15	.533		18.60	21.50		40.10	56
0450	Tee, 1/4"		14	.571		8	23		31	47
0480	1/2"		13	.615		2.27	25		27.27	43.50
0490	5/8"		12	.667		18.90	27		45.90	65
0500	3/4"		12	.667		6.05	27		33.05	50.50
0510	1"		10	.800		16.20	32.50		48.70	71
0520	1-1/4"		9	.889		23.50	36		59.50	84.50
0612	Tee, reducing on the outlet, 1/4"		15	.533		15.60	21.50		37.10	52.50
0613	3/8"		15	.533		14.90	21.50		36.40	52
0614	1/2"		14	.571		13.95	23		36.95	53.50
0615	5/8"		13	.615		28	25		53	72
0616	3/4"		12	.667		9	27		36	54
0617	1"		11	.727		25	29.50		54.50	75.50
0618	1-1/4"		10	.800		30	32.50		62.50	86
0619	1-1/2"		9	.889		30.50	36		66.50	92.50
0620	2"		8	1		53	40.50		93.50	125
0621	2-1/2"	Q-1	9	1.778		129	64.50		193.50	248
0622	3"		8	2		149	72.50		221.50	283
0623	4"		6	2.667		287	97		384	475
0624	5"		5	3.200		1,775	116		1,891	2,150
0625	6"	Q-2	7	3.429		2,075	120		2,195	2,500
0626	8"	"	6	4		9,325	140		9,465	10,500
0630	Tee, reducing on the run, 1/4"	1 Plum	15	.533		22	21.50		43.50	60
0631	3/8"		15	.533		31	21.50		52.50	69.50
0632	1/2"		14	.571		19.20	23		42.20	59

22 11 13 – Facility Water Distribution Piping

22 11 13.25 Pipe/Tube Fittings, Copper		Crew	Daily Output	Labor-Hours	Unit	2019 Bare Costs Material	Labor	Equipment	Total	Total Incl O&P
0633	5/8"	1 Plum	13	.615	Ea.	28	25		53	72
0634	3/4"		12	.667		21	27		48	67
0635	1"		11	.727		24.50	29.50		54	75
0636	1-1/4"		10	.800		38.50	32.50		71	95.50
0637	1-1/2"		9	.889		68	36		104	134
0638	2"		8	1		78.50	40.50		119	153
0639	2-1/2"	Q-1	9	1.778		180	64.50		244.50	305
0640	3"		8	2		262	72.50		334.50	410
0641	4"		6	2.667		555	97		652	770
0642	5"		5	3.200		1,675	116		1,791	2,050
0643	6"	Q-2	7	3.429		2,550	120		2,670	3,000
0644	8"	"	6	4		9,575	140		9,715	10,700
0650	Coupling, 1/4"	1 Plum	24	.333		1.01	13.45		14.46	23
0680	1/2"		22	.364		.92	14.70		15.62	25
0690	5/8"		21	.381		3.94	15.40		19.34	30
0700	3/4"		21	.381		2.49	15.40		17.89	28
0710	1"		18	.444		4.94	17.95		22.89	35
0715	1-1/4"		17	.471		7	19		26	38.50
2000	DWV, solder joints, copper x copper									
2030	90° elbow, 1-1/4"	1 Plum	13	.615	Ea.	17.50	25		42.50	60.50
2050	1-1/2"		12	.667		23	27		50	69.50
2070	2"		10	.800		38	32.50		70.50	95
2090	3"	Q-1	10	1.600		85.50	58		143.50	190
2100	4"	"	9	1.778		460	64.50		524.50	615
2250	Tee, sanitary, 1-1/4"	1 Plum	9	.889		27	36		63	88.50
2270	1-1/2"		8	1		33.50	40.50		74	104
2290	2"		7	1.143		52.50	46		98.50	134
2310	3"	Q-1	7	2.286		200	83		283	355
2330	4"	"	6	2.667		485	97		582	695
2400	Coupling, 1-1/4"	1 Plum	14	.571		7.30	23		30.30	46
2420	1-1/2"		13	.615		9.05	25		34.05	51
2440	2"		11	.727		12.60	29.50		42.10	62
2460	3"	Q-1	11	1.455		29	53		82	119
2480	4"	"	10	1.600		64	58		122	166

22 11 13.44 Pipe, Steel		Crew	Daily Output	Labor-Hours	Unit	Material	Labor	Equipment	Total	Total Incl O&P
0010	**PIPE, STEEL**									
0050	Schedule 40, threaded, with couplings, and clevis hanger									
0060	assemblies sized for covering, 10' OC									
0540	Black, 1/4" diameter	1 Plum	66	.121	L.F.	5.80	4.90		10.70	14.40
0560	1/2" diameter		63	.127		3.91	5.15		9.06	12.70
0570	3/4" diameter		61	.131		4.25	5.30		9.55	13.40
0580	1" diameter		53	.151		4.42	6.10		10.52	14.85
0590	1-1/4" diameter	Q-1	89	.180		5.20	6.55		11.75	16.40
0600	1-1/2" diameter		80	.200		5.70	7.25		12.95	18.25
0610	2" diameter		64	.250		11.55	9.10		20.65	27.50

22 11 13.45 Pipe Fittings, Steel, Threaded		Crew	Daily Output	Labor-Hours	Unit	Material	Labor	Equipment	Total	Total Incl O&P
0010	**PIPE FITTINGS, STEEL, THREADED**									
5000	Malleable iron, 150 lb.									
5020	Black									
5040	90° elbow, straight									
5090	3/4"	1 Plum	14	.571	Ea.	4.46	23		27.46	43
5100	1"	"	13	.615		7.80	25		32.80	49.50

22 11 13.45 Pipe Fittings, Steel, Threaded

	22 11 13.45 Pipe Fittings, Steel, Threaded	Crew	Daily Output	Labor-Hours	Unit	Material	2019 Bare Costs Labor	Equipment	Total	Total Incl O&P
5120	1-1/2"	Q-1	20	.800	Ea.	17	29		46	66
5130	2"	"	18	.889		29.50	32.50		62	85.50
5450	Tee, straight									
5500	3/4"	1 Plum	9	.889	Ea.	7.35	36		43.35	67
5510	1"	"	8	1		12.55	40.50		53.05	80.50
5520	1-1/4"	Q-1	14	1.143		20.50	41.50		62	90.50
5530	1-1/2"		13	1.231		25.50	44.50		70	102
5540	2"		11	1.455		43	53		96	135
5650	Coupling									
5700	3/4"	1 Plum	18	.444	Ea.	6.20	17.95		24.15	36.50
5710	1"	"	15	.533		9.25	21.50		30.75	45.50
5720	1-1/4"	Q-1	26	.615		12.25	22.50		34.75	50
5730	1-1/2"		24	.667		16.20	24		40.20	58
5740	2"		21	.762		24	27.50		51.50	72

22 11 13.74 Pipe, Plastic

	22 11 13.74 Pipe, Plastic	Crew	Daily Output	Labor-Hours	Unit	Material	2019 Bare Costs Labor	Equipment	Total	Total Incl O&P
0010	**PIPE, PLASTIC**									
1800	PVC, couplings 10′ OC, clevis hanger assemblies, 3 per 10′									
1820	Schedule 40									
1860	1/2" diameter	1 Plum	54	.148	L.F.	4.76	6		10.76	15.05
1870	3/4" diameter		51	.157		5.30	6.35		11.65	16.20
1880	1" diameter		46	.174		9	7.05		16.05	21.50
1890	1-1/4" diameter		42	.190		9.80	7.70		17.50	23.50
1900	1-1/2" diameter		36	.222		9.85	9		18.85	25.50
1910	2" diameter	Q-1	59	.271		11.35	9.85		21.20	28.50
1920	2-1/2" diameter		56	.286		10.90	10.40		21.30	29
1930	3" diameter		53	.302		13.10	10.95		24.05	32.50
1940	4" diameter		48	.333		29.50	12.10		41.60	52
4100	DWV type, schedule 40, couplings 10′ OC, clevis hanger assy's, 3 per 10′									
4210	ABS, schedule 40, foam core type									
4212	Plain end black									
4214	1-1/2" diameter	1 Plum	39	.205	L.F.	8.25	8.30		16.55	22.50
4216	2" diameter	Q-1	62	.258		8.80	9.40		18.20	25
4218	3" diameter		56	.286		8.55	10.40		18.95	26.50
4220	4" diameter		51	.314		23.50	11.40		34.90	44.50
4222	6" diameter		42	.381		19.80	13.85		33.65	44.50
4240	To delete coupling & hangers, subtract									
4244	1-1/2" diam. to 6" diam.					43%	48%			
4400	PVC									
4410	1-1/4" diameter	1 Plum	42	.190	L.F.	8.80	7.70		16.50	22.50
4420	1-1/2" diameter	"	36	.222		8	9		17	23.50
4460	2" diameter	Q-1	59	.271		8.85	9.85		18.70	26
4470	3" diameter		53	.302		8.65	10.95		19.60	27.50
4480	4" diameter		48	.333		10.45	12.10		22.55	31.50
5300	CPVC, socket joint, couplings 10′ OC, clevis hanger assemblies, 3 per 10′									
5302	Schedule 40									
5304	1/2" diameter	1 Plum	54	.148	L.F.	5.60	6		11.60	15.95
5305	3/4" diameter		51	.157		6.60	6.35		12.95	17.65
5306	1" diameter		46	.174		10.85	7.05		17.90	23.50
5307	1-1/4" diameter		42	.190		12.25	7.70		19.95	26
5308	1-1/2" diameter		36	.222		12.15	9		21.15	28
5309	2" diameter	Q-1	59	.271		15.40	9.85		25.25	33
5360	CPVC, threaded, couplings 10′ OC, clevis hanger assemblies, 3 per 10′									

22 11 13.74 Pipe, Plastic		Crew	Daily Output	Labor-Hours	Unit	Material	Labor	Equipment	Total	Total Incl O&P
						2019 Bare Costs				
5380	Schedule 40									
5460	1/2" diameter	1 Plum	54	.148	L.F.	6.45	6		12.45	16.90
5470	3/4" diameter		51	.157		8.15	6.35		14.50	19.35
5480	1" diameter		46	.174		12.40	7.05		19.45	25
5490	1-1/4" diameter		42	.190		13.45	7.70		21.15	27.50
5500	1-1/2" diameter		36	.222		13.20	9		22.20	29.50
5510	2" diameter	Q-1	59	.271		16.60	9.85		26.45	34.50
6500	Residential installation, plastic pipe									
6510	Couplings 10' OC, strap hangers 3 per 10'									
6520	PVC, Schedule 40									
6530	1/2" diameter	1 Plum	138	.058	L.F.	1.06	2.34		3.40	5
6540	3/4" diameter		128	.063		1.21	2.53		3.74	5.50
6550	1" diameter		119	.067		1.79	2.72		4.51	6.40
6560	1-1/4" diameter		111	.072		2.23	2.91		5.14	7.25
6570	1-1/2" diameter		104	.077		2.37	3.11		5.48	7.70
6580	2" diameter	Q-1	197	.081		2.98	2.95		5.93	8.10
6590	2-1/2" diameter		162	.099		5.10	3.59		8.69	11.50
6600	4" diameter		123	.130		8.30	4.73		13.03	16.85
6700	PVC, DWV, Schedule 40									
6720	1-1/4" diameter	1 Plum	100	.080	L.F.	2.12	3.23		5.35	7.65
6730	1-1/2" diameter	"	94	.085		1.51	3.44		4.95	7.30
6740	2" diameter	Q-1	178	.090		2.10	3.27		5.37	7.65
6760	4" diameter	"	110	.145		6	5.30		11.30	15.30
7280	PEX, flexible, no couplings or hangers									
7282	Note: For labor costs add 25% to the couplings and fittings labor total.									
7285	For fittings see section 23 83 16.10 7000									
7300	Non-barrier type, hot/cold tubing rolls									
7310	1/4" diameter x 100'				L.F.	.52			.52	.57
7350	3/8" diameter x 100'					.55			.55	.61
7360	1/2" diameter x 100'					.64			.64	.70
7370	1/2" diameter x 500'					.64			.64	.70
7380	1/2" diameter x 1000'					.63			.63	.69
7400	3/4" diameter x 100'					1.01			1.01	1.11
7410	3/4" diameter x 500'					1.14			1.14	1.25
7420	3/4" diameter x 1000'					1.14			1.14	1.25
7460	1" diameter x 100'					1.96			1.96	2.16
7470	1" diameter x 300'					1.96			1.96	2.16
7480	1" diameter x 500'					1.96			1.96	2.16
7500	1-1/4" diameter x 100'					3.33			3.33	3.66
7510	1-1/4" diameter x 300'					3.33			3.33	3.66
7540	1-1/2" diameter x 100'					4.52			4.52	4.97
7550	1-1/2" diameter x 300'					4.50			4.50	4.95
7596	Most sizes available in red or blue									
7700	Non-barrier type, hot/cold tubing straight lengths									
7710	1/2" diameter x 20'				L.F.	.64			.64	.70
7750	3/4" diameter x 20'					1.14			1.14	1.25
7760	1" diameter x 20'					1.96			1.96	2.16
7770	1-1/4" diameter x 20'					3.44			3.44	3.78
7780	1-1/2" diameter x 20'					4.54			4.54	4.99
7790	2" diameter					8.85			8.85	9.75
7796	Most sizes available in red or blue									

22 11 13.76 Pipe Fittings, Plastic		Crew	Daily Output	Labor-Hours	Unit	2019 Bare Costs Material	Labor	Equipment	Total	Total Incl O&P
0010	**PIPE FITTINGS, PLASTIC**									
2700	PVC (white), schedule 40, socket joints									
2760	90° elbow, 1/2"	1 Plum	33.30	.240	Ea.	.51	9.70		10.21	16.50
2770	3/4"		28.60	.280		.59	11.30		11.89	19.20
2780	1"		25	.320		1.04	12.95		13.99	22
2790	1-1/4"		22.20	.360		1.82	14.55		16.37	26
2800	1-1/2"		20	.400		1.97	16.15		18.12	28.50
2810	2"	Q-1	36.40	.440		3.08	16		19.08	29.50
2820	2-1/2"		26.70	.599		9.90	22		31.90	47
2830	3"		22.90	.699		11.20	25.50		36.70	54
2840	4"		18.20	.879		20	32		52	74.50
3180	Tee, 1/2"	1 Plum	22.20	.360		.64	14.55		15.19	24.50
3190	3/4"		19	.421		.76	17		17.76	29
3200	1"		16.70	.479		1.38	19.35		20.73	33.50
3210	1-1/4"		14.80	.541		2.15	22		24.15	38.50
3220	1-1/2"		13.30	.602		2.61	24.50		27.11	43
3230	2"	Q-1	24.20	.661		3.80	24		27.80	43.50
3240	2-1/2"		17.80	.899		12.50	32.50		45	67.50
3250	3"		15.20	1.053		16.45	38.50		54.95	81
3260	4"		12.10	1.322		30	48		78	112
3380	Coupling, 1/2"	1 Plum	33.30	.240		.34	9.70		10.04	16.30
3390	3/4"		28.60	.280		.47	11.30		11.77	19.05
3400	1"		25	.320		.82	12.95		13.77	22
3410	1-1/4"		22.20	.360		1.13	14.55		15.68	25
3420	1-1/2"		20	.400		1.20	16.15		17.35	28
3430	2"	Q-1	36.40	.440		1.83	16		17.83	28
3440	2-1/2"		26.70	.599		4.06	22		26.06	40.50
3450	3"		22.90	.699		6.35	25.50		31.85	48.50
3460	4"		18.20	.879		9.20	32		41.20	62.50
4500	DWV, ABS, non pressure, socket joints									
4540	1/4 bend, 1-1/4"	1 Plum	20.20	.396	Ea.	4.73	16		20.73	31.50
4560	1-1/2"	"	18.20	.440		3.65	17.75		21.40	33
4570	2"	Q-1	33.10	.483		5.60	17.55		23.15	35
4650	1/8 bend, same as 1/4 bend									
4800	Tee, sanitary									
4820	1-1/4"	1 Plum	13.50	.593	Ea.	6.30	24		30.30	46.50
4830	1-1/2"	"	12.10	.661		5.40	26.50		31.90	50
4840	2"	Q-1	20	.800		8.30	29		37.30	56.50
5000	DWV, PVC, schedule 40, socket joints									
5040	1/4 bend, 1-1/4"	1 Plum	20.20	.396	Ea.	9.30	16		25.30	37
5060	1-1/2"	"	18.20	.440		2.66	17.75		20.41	32
5070	2"	Q-1	33.10	.483		4.19	17.55		21.74	33.50
5080	3"		20.80	.769		12.30	28		40.30	59.50
5090	4"		16.50	.970		24.50	35.50		60	85
5110	1/4 bend, long sweep, 1-1/2"	1 Plum	18.20	.440		6.15	17.75		23.90	36
5112	2"	Q-1	33.10	.483		6.85	17.55		24.40	36.50
5114	3"		20.80	.769		15.85	28		43.85	63.50
5116	4"		16.50	.970		30	35.50		65.50	91
5250	Tee, sanitary 1-1/4"	1 Plum	13.50	.593		10	24		34	50.50
5254	1-1/2"	"	12.10	.661		4.64	26.50		31.14	49
5255	2"	Q-1	20	.800		6.85	29		35.85	55
5256	3"		13.90	1.151		18	42		60	88.50

22 11 13 – Facility Water Distribution Piping

22 11 13.76 Pipe Fittings, Plastic		Crew	Daily Output	Labor-Hours	Unit	Material	2019 Bare Costs Labor	Equipment	Total	Total Incl O&P
5257	4"	Q-1	11	1.455	Ea.	33	53		86	124
5259	6"		6.70	2.388		133	87		220	288
5261	8"	Q-2	6.20	3.871		289	135		424	540
5264	2" x 1-1/2"	Q-1	22	.727		6.05	26.50		32.55	50
5266	3" x 1-1/2"		15.50	1.032		13.15	37.50		50.65	76
5268	4" x 3"		12.10	1.322		38.50	48		86.50	122
5271	6" x 4"		6.90	2.319		128	84.50		212.50	279
5314	Combination Y & 1/8 bend, 1-1/2"	1 Plum	12.10	.661		11.30	26.50		37.80	56.50
5315	2"	Q-1	20	.800		14.15	29		43.15	63
5317	3"		13.90	1.151		31	42		73	103
5318	4"		11	1.455		61.50	53		114.50	155
5324	Combination Y & 1/8 bend, reducing									
5325	2" x 2" x 1-1/2"	Q-1	22	.727	Ea.	15.90	26.50		42.40	61
5327	3" x 3" x 1-1/2"		15.50	1.032		28.50	37.50		66	92.50
5328	3" x 3" x 2"		15.30	1.046		21	38		59	86
5329	4" x 4" x 2"		12.20	1.311		32	47.50		79.50	113
5331	Wye, 1-1/4"	1 Plum	13.50	.593		12.80	24		36.80	53.50
5332	1-1/2"	"	12.10	.661		8.50	26.50		35	53.50
5333	2"	Q-1	20	.800		8.35	29		37.35	56.50
5334	3"		13.90	1.151		22.50	42		64.50	93
5335	4"		11	1.455		41	53		94	132
5336	6"		6.70	2.388		119	87		206	273
5337	8"	Q-2	6.20	3.871		211	135		346	455
5341	2" x 1-1/2"	Q-1	22	.727		10.25	26.50		36.75	55
5342	3" x 1-1/2"		15.50	1.032		15.15	37.50		52.65	78
5343	4" x 3"		12.10	1.322		33	48		81	116
5344	6" x 4"		6.90	2.319		90	84.50		174.50	237
5345	8" x 6"	Q-2	6.40	3.750		197	131		328	430
5347	Double wye, 1-1/2"	1 Plum	9.10	.879		19.15	35.50		54.65	79.50
5348	2"	Q-1	16.60	.964		21.50	35		56.50	81
5349	3"		10.40	1.538		44.50	56		100.50	141
5350	4"		8.25	1.939		90	70.50		160.50	215
5353	Double wye, reducing									
5354	2" x 2" x 1-1/2" x 1-1/2"	Q-1	16.80	.952	Ea.	19.50	34.50		54	78.50
5355	3" x 3" x 2" x 2"		10.60	1.509		33	55		88	127
5356	4" x 4" x 3" x 3"		8.45	1.893		71.50	69		140.50	192
5357	6" x 6" x 4" x 4"		7.25	2.207		250	80		330	405
5374	Coupling, 1-1/4"	1 Plum	20.20	.396		6.05	16		22.05	33
5376	1-1/2"	"	18.20	.440		1.24	17.75		18.99	30.50
5378	2"	Q-1	33.10	.483		1.71	17.55		19.26	31
5380	3"		20.80	.769		5.95	28		33.95	52.50
5390	4"		16.50	.970		10.15	35.50		45.65	69
5410	Reducer bushing, 2" x 1-1/4"		36.50	.438		3.50	15.95		19.45	30
5412	3" x 1-1/2"		27.30	.586		10.50	21.50		32	46.50
5414	4" x 2"		18.20	.879		18.25	32		50.25	72.50
5416	6" x 4"		11.10	1.441		47	52.50		99.50	138
5418	8" x 6"	Q-2	10.20	2.353		92	82.50		174.50	236
5500	CPVC, Schedule 80, threaded joints									
5540	90° elbow, 1/4"	1 Plum	32	.250	Ea.	12.95	10.10		23.05	31
5560	1/2"		30.30	.264		7.75	10.65		18.40	26
5570	3/4"		26	.308		11.25	12.45		23.70	33
5580	1"		22.70	.352		15.80	14.25		30.05	41
5590	1-1/4"		20.20	.396		30.50	16		46.50	60

22 11 Facility Water Distribution

22 11 13 – Facility Water Distribution Piping

22 11 13.76 Pipe Fittings, Plastic		Crew	Daily Output	Labor-Hours	Unit	Material	Labor	Equipment	Total	Total Incl O&P
						2019 Bare Costs				
5600	1-1/2"	1 Plum	18.20	.440	Ea.	32.50	17.75		50.25	65
5610	2"	Q-1	33.10	.483		44	17.55		61.55	77
5730	Coupling, 1/4"	1 Plum	32	.250		16.50	10.10		26.60	35
5732	1/2"		30.30	.264		13.95	10.65		24.60	33
5734	3/4"		26	.308		22.50	12.45		34.95	45
5736	1"		22.70	.352		25	14.25		39.25	51
5738	1-1/4"		20.20	.396		26.50	16		42.50	55.50
5740	1-1/2"		18.20	.440		28.50	17.75		46.25	60
5742	2"	Q-1	33.10	.483		33.50	17.55		51.05	66
5900	CPVC, Schedule 80, socket joints									
5904	90° elbow, 1/4"	1 Plum	32	.250	Ea.	12.35	10.10		22.45	30
5906	1/2"		30.30	.264		4.83	10.65		15.48	23
5908	3/4"		26	.308		6.15	12.45		18.60	27.50
5910	1"		22.70	.352		9.80	14.25		24.05	34.50
5912	1-1/4"		20.20	.396		21	16		37	50
5914	1-1/2"		18.20	.440		23.50	17.75		41.25	55
5916	2"	Q-1	33.10	.483		28.50	17.55		46.05	60.50
5930	45° elbow, 1/4"	1 Plum	32	.250		18.35	10.10		28.45	36.50
5932	1/2"		30.30	.264		5.90	10.65		16.55	24
5934	3/4"		26	.308		8.55	12.45		21	30
5936	1"		22.70	.352		13.60	14.25		27.85	38.50
5938	1-1/4"		20.20	.396		26.50	16		42.50	56
5940	1-1/2"		18.20	.440		27.50	17.75		45.25	59
5942	2"	Q-1	33.10	.483		30.50	17.55		48.05	62.50
5990	Coupling, 1/4"	1 Plum	32	.250		13.15	10.10		23.25	31
5992	1/2"		30.30	.264		5.10	10.65		15.75	23
5994	3/4"		26	.308		7.15	12.45		19.60	28.50
5996	1"		22.70	.352		9.60	14.25		23.85	34
5998	1-1/4"		20.20	.396		14.40	16		30.40	42.50
6000	1-1/2"		18.20	.440		18.10	17.75		35.85	49
6002	2"	Q-1	33.10	.483		21	17.55		38.55	52.50

22 11 19 – Domestic Water Piping Specialties

22 11 19.38 Water Supply Meters		Crew	Daily Output	Labor-Hours	Unit	Material	Labor	Equipment	Total	Total Incl O&P
0010	**WATER SUPPLY METERS**									
2000	Domestic/commercial, bronze									
2020	Threaded									
2060	5/8" diameter, to 20 GPM	1 Plum	16	.500	Ea.	51.50	20		71.50	90
2080	3/4" diameter, to 30 GPM		14	.571		94	23		117	141
2100	1" diameter, to 50 GPM		12	.667		143	27		170	201

22 11 19.42 Backflow Preventers		Crew	Daily Output	Labor-Hours	Unit	Material	Labor	Equipment	Total	Total Incl O&P
0010	**BACKFLOW PREVENTERS**, Includes valves									
0020	and four test cocks, corrosion resistant, automatic operation									
4000	Reduced pressure principle									
4100	Threaded, bronze, valves are ball									
4120	3/4" pipe size	1 Plum	16	.500	Ea.	465	20		485	550

22 11 19.50 Vacuum Breakers		Crew	Daily Output	Labor-Hours	Unit	Material	Labor	Equipment	Total	Total Incl O&P
0010	**VACUUM BREAKERS**									
0013	See also backflow preventers Section 22 11 19.42									
1000	Anti-siphon continuous pressure type									
1010	Max. 150 psi - 210°F									
1020	Bronze body									
1030	1/2" size	1 Stpi	24	.333	Ea.	183	13.85		196.85	224

22 11 Facility Water Distribution

22 11 19 – Domestic Water Piping Specialties

22 11 19.50 Vacuum Breakers		Crew	Daily Output	Labor-Hours	Unit	Material	2019 Bare Costs Labor	Equipment	Total	Total Incl O&P
1040	3/4" size	1 Stpi	20	.400	Ea.	183	16.60		199.60	229
1050	1" size		19	.421		187	17.50		204.50	235
1060	1-1/4" size		15	.533		370	22		392	440
1070	1-1/2" size		13	.615		455	25.50		480.50	540
1080	2" size		11	.727		465	30		495	565
1200	Max. 125 psi with atmospheric vent									
1210	Brass, in-line construction									
1220	1/4" size	1 Stpi	24	.333	Ea.	138	13.85		151.85	175
1230	3/8" size	"	24	.333		138	13.85		151.85	175
1260	For polished chrome finish, add					13%				
2000	Anti-siphon, non-continuous pressure type									
2010	Hot or cold water 125 psi - 210°F									
2020	Bronze body									
2030	1/4" size	1 Stpi	24	.333	Ea.	77.50	13.85		91.35	108
2040	3/8" size		24	.333		77.50	13.85		91.35	108
2050	1/2" size		24	.333		87	13.85		100.85	119
2060	3/4" size		20	.400		104	16.60		120.60	142
2070	1" size		19	.421		161	17.50		178.50	206
2080	1-1/4" size		15	.533		282	22		304	345
2090	1-1/2" size		13	.615		330	25.50		355.50	405
2100	2" size		11	.727		515	30		545	615
2110	2-1/2" size		8	1		1,475	41.50		1,516.50	1,700
2120	3" size		6	1.333		1,975	55.50		2,030.50	2,250
2150	For polished chrome finish, add					50%				

22 11 19.54 Water Hammer Arresters/Shock Absorbers		Crew	Daily Output	Labor-Hours	Unit	Material	Labor	Equipment	Total	Total Incl O&P
0010	**WATER HAMMER ARRESTERS/SHOCK ABSORBERS**									
0490	Copper									
0500	3/4" male IPS for 1 to 11 fixtures	1 Plum	12	.667	Ea.	29	27		56	76

22 13 Facility Sanitary Sewerage

22 13 16 – Sanitary Waste and Vent Piping

22 13 16.20 Pipe, Cast Iron		Crew	Daily Output	Labor-Hours	Unit	Material	Labor	Equipment	Total	Total Incl O&P
0010	**PIPE, CAST IRON**, Soil, on clevis hanger assemblies, 5' OC R221113-50									
0020	Single hub, service wt., lead & oakum joints 10' OC									
2120	2" diameter	Q-1	63	.254	L.F.	17.05	9.25		26.30	34
2140	3" diameter		60	.267		20	9.70		29.70	38
2160	4" diameter		55	.291		34	10.55		44.55	55
4000	No hub, couplings 10' OC									
4100	1-1/2" diameter	Q-1	71	.225	L.F.	16.20	8.20		24.40	31.50
4120	2" diameter		67	.239		17.30	8.70		26	33.50
4140	3" diameter		64	.250		19.70	9.10		28.80	36.50
4160	4" diameter		58	.276		33.50	10.05		43.55	53.50

22 13 16.30 Pipe Fittings, Cast Iron		Crew	Daily Output	Labor-Hours	Unit	Material	Labor	Equipment	Total	Total Incl O&P
0010	**PIPE FITTINGS, CAST IRON**, Soil									
0040	Hub and spigot, service weight, lead & oakum joints									
0080	1/4 bend, 2"	Q-1	16	1	Ea.	22	36.50		58.50	84
0120	3"		14	1.143		29.50	41.50		71	101
0140	4"		13	1.231		46.50	44.50		91	125
0340	1/8 bend, 2"		16	1		15.80	36.50		52.30	77
0350	3"		14	1.143		24.50	41.50		66	95

22 13 16 – Sanitary Waste and Vent Piping

22 13 16.30 Pipe Fittings, Cast Iron		Crew	Daily Output	Labor-Hours	Unit	Material	Labor	Equipment	Total	Total Incl O&P
						2019 Bare Costs				
0360	4"	Q-1	13	1.231	Ea.	36.50	44.50		81	114
0500	Sanitary tee, 2"		10	1.600		31	58		89	130
0540	3"		9	1.778		50	64.50		114.50	161
0620	4"		8	2		63.50	72.50		136	189
5990	No hub									
6000	Cplg. & labor required at joints not incl. in fitting									
6010	price. Add 1 coupling per joint for installed price									
6020	1/4 bend, 1-1/2"				Ea.	10.55			10.55	11.65
6060	2"					11.55			11.55	12.70
6080	3"					16.10			16.10	17.70
6120	4"					24			24	26
6184	1/4 bend, long sweep, 1-1/2"					27			27	29.50
6186	2"					25			25	27.50
6188	3"					30.50			30.50	33.50
6189	4"					48.50			48.50	53.50
6190	5"					94			94	103
6191	6"					107			107	118
6192	8"					297			297	325
6193	10"					610			610	670
6200	1/8 bend, 1-1/2"					8.90			8.90	9.80
6210	2"					9.95			9.95	10.95
6212	3"					13.30			13.30	14.65
6214	4"					17.45			17.45	19.15
6380	Sanitary tee, tapped, 1-1/2"					22.50			22.50	25
6382	2" x 1-1/2"					19.90			19.90	22
6384	2"					19.95			19.95	22
6386	3" x 2"					32			32	35
6388	3"					51			51	56.50
6390	4" x 1-1/2"					28.50			28.50	31
6392	4" x 2"					32.50			32.50	35.50
6393	4"					32.50			32.50	35.50
6394	6" x 1-1/2"					70.50			70.50	77.50
6396	6" x 2"					72			72	79
6459	Sanitary tee, 1-1/2"					14.85			14.85	16.30
6460	2"					15.90			15.90	17.50
6470	3"					19.60			19.60	21.50
6472	4"					37			37	41
8000	Coupling, standard (by CISPI Mfrs.)									
8020	1-1/2"	Q-1	48	.333	Ea.	16.45	12.10		28.55	38
8040	2"		44	.364		16.25	13.20		29.45	39.50
8080	3"		38	.421		19.70	15.30		35	46.50
8120	4"		33	.485		23	17.60		40.60	54

22 13 16.50 Shower Drains		Crew	Daily Output	Labor-Hours	Unit	Material	Labor	Equipment	Total	Total Incl O&P
0010	**SHOWER DRAINS**									
2780	Shower, with strainer, uniform diam. trap, bronze top									
2800	2" and 3" pipe size	Q-1	8	2	Ea.	355	72.50		427.50	510
2820	4" pipe size	"	7	2.286		405	83		488	580
2840	For galvanized body, add					198			198	218

22 13 16 – Sanitary Waste and Vent Piping

22 13 16.60 Traps		Crew	Daily Output	Labor-Hours	Unit	Material	Labor	Equipment	Total	Total Incl O&P
						2019 Bare Costs				
0010	TRAPS									
0030	Cast iron, service weight									
0050	Running P trap, without vent									
1100	2"	Q-1	16	1	Ea.	162	36.50		198.50	238
1150	4"	"	13	1.231		162	44.50		206.50	252
1160	6"	Q-2	17	1.412		775	49.50		824.50	935
3000	P trap, B&S, 2" pipe size	Q-1	16	1		41.50	36.50		78	105
3040	3" pipe size	"	14	1.143	↓	61.50	41.50		103	136
4700	Copper, drainage, drum trap									
4840	3" x 6" swivel, 1-1/2" pipe size	1 Plum	16	.500	Ea.	287	20		307	350
5100	P trap, standard pattern									
5200	1-1/4" pipe size	1 Plum	18	.444	Ea.	92.50	17.95		110.45	132
5240	1-1/2" pipe size		17	.471		103	19		122	144
5260	2" pipe size		15	.533		158	21.50		179.50	210
5280	3" pipe size	↓	11	.727	↓	495	29.50		524.50	595
6710	ABS DWV P trap, solvent weld joint									
6720	1-1/2" pipe size	1 Plum	18	.444	Ea.	11.40	17.95		29.35	42
6722	2" pipe size		17	.471		15	19		34	47.50
6724	3" pipe size		15	.533		59.50	21.50		81	101
6726	4" pipe size	↓	14	.571	↓	119	23		142	169
6732	PVC DWV P trap, solvent weld joint									
6733	1-1/2" pipe size	1 Plum	18	.444	Ea.	9.70	17.95		27.65	40
6734	2" pipe size		17	.471		11.85	19		30.85	44
6735	3" pipe size		15	.533		40	21.50		61.50	79.50
6736	4" pipe size		14	.571		91	23		114	138
6860	PVC DWV hub x hub, basin trap, 1-1/4" pipe size		18	.444		52	17.95		69.95	86.50
6870	Sink P trap, 1-1/2" pipe size		18	.444		14.90	17.95		32.85	46
6880	Tubular S trap, 1-1/2" pipe size	↓	17	.471	↓	25.50	19		44.50	59.50
6890	PVC sch. 40 DWV, drum trap									
6900	1-1/2" pipe size	1 Plum	16	.500	Ea.	37	20		57	73.50
6910	P trap, 1-1/2" pipe size		18	.444		9.20	17.95		27.15	39.50
6920	2" pipe size		17	.471		12.35	19		31.35	44.50
6930	3" pipe size		15	.533		42	21.50		63.50	81.50
6940	4" pipe size		14	.571		95	23		118	143
6950	P trap w/clean out, 1-1/2" pipe size		18	.444		15.60	17.95		33.55	46.50
6960	2" pipe size	↓	17	.471	↓	25.50	19		44.50	59

22 13 16.80 Vent Flashing and Caps		Crew	Daily Output	Labor-Hours	Unit	Material	Labor	Equipment	Total	Total Incl O&P
0010	VENT FLASHING AND CAPS									
0120	Vent caps									
0140	Cast iron									
0160	1-1/4" to 1-1/2" pipe	1 Plum	23	.348	Ea.	39.50	14.05		53.55	66.50
0170	2" to 2-1/8" pipe		22	.364		46	14.70		60.70	74.50
0180	2-1/2" to 3-5/8" pipe		21	.381		51.50	15.40		66.90	82
0190	4" to 4-1/8" pipe		19	.421		75	17		92	111
0200	5" to 6" pipe	↓	17	.471	↓	104	19		123	146
0300	PVC									
0320	1-1/4" to 1-1/2" pipe	1 Plum	24	.333	Ea.	13.75	13.45		27.20	37
0330	2" to 2-1/8" pipe	"	23	.348	"	15.55	14.05		29.60	40
0900	Vent flashing									
1350	Copper with neoprene ring									
1400	1-1/4" pipe	1 Plum	20	.400	Ea.	72.50	16.15		88.65	106
1430	1-1/2" pipe	↓	20	.400	↓	72.50	16.15		88.65	106

22 13 Facility Sanitary Sewerage

22 13 16 – Sanitary Waste and Vent Piping

22 13 16.80 Vent Flashing and Caps		Crew	Daily Output	Labor-Hours	Unit	2019 Bare Costs Material	Labor	Equipment	Total	Total Incl O&P
1440	2″ pipe	1 Plum	18	.444	Ea.	72.50	17.95		90.45	109
1450	3″ pipe		17	.471		87.50	19		106.50	128
1460	4″ pipe		16	.500		87.50	20		107.50	130
2980	Neoprene, one piece									
3000	1-1/4″ pipe	1 Plum	24	.333	Ea.	3.06	13.45		16.51	25.50
3030	1-1/2″ pipe		24	.333		3.06	13.45		16.51	25.50
3040	2″ pipe		23	.348		5.70	14.05		19.75	29.50
3050	3″ pipe		21	.381		6.75	15.40		22.15	33
3060	4″ pipe		20	.400		9.95	16.15		26.10	37.50

22 13 19 – Sanitary Waste Piping Specialties

22 13 19.13 Sanitary Drains		Crew	Daily Output	Labor-Hours	Unit	Material	Labor	Equipment	Total	Total Incl O&P
0010	**SANITARY DRAINS**									
2000	Floor, medium duty, CI, deep flange, 7″ diam. top									
2040	2″ and 3″ pipe size	Q-1	12	1.333	Ea.	244	48.50		292.50	350
2080	For galvanized body, add					116			116	127
2120	With polished bronze top					380			380	415

22 14 Facility Storm Drainage

22 14 26 – Facility Storm Drains

22 14 26.13 Roof Drains		Crew	Daily Output	Labor-Hours	Unit	Material	Labor	Equipment	Total	Total Incl O&P
0010	**ROOF DRAINS**									
3860	Roof, flat metal deck, CI body, 12″ CI dome									
3890	3″ pipe size	Q-1	14	1.143	Ea.	355	41.50		396.50	460

22 14 29 – Sump Pumps

22 14 29.16 Submersible Sump Pumps		Crew	Daily Output	Labor-Hours	Unit	Material	Labor	Equipment	Total	Total Incl O&P
0010	**SUBMERSIBLE SUMP PUMPS**									
7000	Sump pump, automatic									
7100	Plastic, 1-1/4″ discharge, 1/4 HP	1 Plum	6.40	1.250	Ea.	163	50.50		213.50	262
7500	Cast iron, 1-1/4″ discharge, 1/4 HP	″	6	1.333	″	215	54		269	325

22 31 Domestic Water Softeners

22 31 13 – Residential Domestic Water Softeners

22 31 13.10 Residential Water Softeners		Crew	Daily Output	Labor-Hours	Unit	Material	Labor	Equipment	Total	Total Incl O&P
0010	**RESIDENTIAL WATER SOFTENERS**									
7350	Water softener, automatic, to 30 grains per gallon	2 Plum	5	3.200	Ea.	375	129		504	620
7400	To 100 grains per gallon	″	4	4	″	855	162		1,017	1,200

22 33 Electric Domestic Water Heaters

22 33 30 – Residential, Electric Domestic Water Heaters

22 33 30.13 Residential, Small-Capacity Elec. Water Heaters		Crew	Daily Output	Labor-Hours	Unit	Material	2019 Bare Costs Labor	Equipment	Total	Total Incl O&P
0010	RESIDENTIAL, SMALL-CAPACITY ELECTRIC DOMESTIC WATER HEATERS									
1000	Residential, electric, glass lined tank, 5 yr., 10 gal., single element	1 Plum	2.30	3.478	Ea.	445	141		586	715
1060	30 gallon, double element		2.20	3.636		990	147		1,137	1,350
1080	40 gallon, double element		2	4		1,075	162		1,237	1,450
1100	52 gallon, double element		2	4		1,200	162		1,362	1,600
1120	66 gallon, double element		1.80	4.444		1,625	180		1,805	2,100
1140	80 gallon, double element		1.60	5		1,850	202		2,052	2,350

22 34 Fuel-Fired Domestic Water Heaters

22 34 13 – Instantaneous, Tankless, Gas Domestic Water Heaters

22 34 13.10 Instantaneous, Tankless, Gas Water Heaters

		Crew	Daily Output	Labor-Hours	Unit	Material	Labor	Equipment	Total	Total Incl O&P
0010	INSTANTANEOUS, TANKLESS, GAS WATER HEATERS									
9410	Natural gas/propane, 3.2 GPM G	1 Plum	2	4	Ea.	505	162		667	820
9420	6.4 GPM G		1.90	4.211		770	170		940	1,125
9430	8.4 GPM G		1.80	4.444		890	180		1,070	1,275
9440	9.5 GPM G		1.60	5		995	202		1,197	1,425

22 34 30 – Residential Gas Domestic Water Heaters

22 34 30.13 Residential, Atmos, Gas Domestic Wtr Heaters

		Crew	Daily Output	Labor-Hours	Unit	Material	Labor	Equipment	Total	Total Incl O&P
0010	RESIDENTIAL, ATMOSPHERIC, GAS DOMESTIC WATER HEATERS									
2000	Gas fired, foam lined tank, 10 yr., vent not incl.									
2040	30 gallon	1 Plum	2	4	Ea.	1,775	162		1,937	2,225
2100	75 gallon	"	1.50	5.333	"	2,775	215		2,990	3,400
3000	Tank leak safety, water & gas shut off see 22 05 23.20 8800									

22 34 46 – Oil-Fired Domestic Water Heaters

22 34 46.10 Residential Oil-Fired Water Heaters

		Crew	Daily Output	Labor-Hours	Unit	Material	Labor	Equipment	Total	Total Incl O&P
0010	RESIDENTIAL OIL-FIRED WATER HEATERS									
3000	Oil fired, glass lined tank, 5 yr., vent not included, 30 gallon	1 Plum	2	4	Ea.	1,275	162		1,437	1,675
3040	50 gallon	"	1.80	4.444	"	1,500	180		1,680	1,950

22 41 Residential Plumbing Fixtures

22 41 13 – Residential Water Closets, Urinals, and Bidets

22 41 13.13 Water Closets

		Crew	Daily Output	Labor-Hours	Unit	Material	Labor	Equipment	Total	Total Incl O&P
0010	WATER CLOSETS									
0150	Tank type, vitreous china, incl. seat, supply pipe w/stop, 1.6 gpf or noted									
0200	Wall hung									
0400	Two piece, close coupled	Q-1	5.30	3.019	Ea.	400	110		510	620
0960	For rough-in, supply, waste, vent and carrier	"	2.73	5.861	"	1,250	213		1,463	1,725
0999	Floor mounted									
1020	One piece, low profile	Q-1	5.30	3.019	Ea.	770	110		880	1,025
1100	Two piece, close coupled		5.30	3.019		203	110		313	405
1102	Economy		5.30	3.019		113	110		223	305
1110	Two piece, close coupled, dual flush		5.30	3.019		325	110		435	535
1140	Two piece, close coupled, 1.28 gpf, ADA G		5.30	3.019		325	110		435	540
1960	For color, add					30%				
1980	For rough-in, supply, waste and vent	Q-1	3.05	5.246	Ea.	365	191		556	720

22 41 Residential Plumbing Fixtures

22 41 16 – Residential Lavatories and Sinks

22 41 16.13 Lavatories		Crew	Daily Output	Labor-Hours	Unit	2019 Bare Costs Material	Labor	Equipment	Total	Total Incl O&P
0010	**LAVATORIES**, With trim, white unless noted otherwise									
0500	Vanity top, porcelain enamel on cast iron									
0600	20" x 18"	Q-1	6.40	2.500	Ea.	295	91		386	475
0640	33" x 19" oval		6.40	2.500		510	91		601	715
0720	19" round		6.40	2.500		395	91		486	585
0860	For color, add					25%				
1000	Cultured marble, 19" x 17", single bowl	Q-1	6.40	2.500	Ea.	121	91		212	282
1120	25" x 22", single bowl		6.40	2.500		161	91		252	325
1160	37" x 22", single bowl		6.40	2.500		187	91		278	355
1580	For color, same price									
1900	Stainless steel, self-rimming, 25" x 22", single bowl, ledge	Q-1	6.40	2.500	Ea.	310	91		401	490
1960	17" x 22", single bowl		6.40	2.500		300	91		391	480
2600	Steel, enameled, 20" x 17", single bowl		5.80	2.759		120	100		220	298
2900	Vitreous china, 20" x 16", single bowl		5.40	2.963		205	108		313	405
3200	22" x 13", single bowl		5.40	2.963		211	108		319	410
3580	Rough-in, supply, waste and vent for all above lavatories		2.30	6.957		249	253		502	690
4000	Wall hung									
4040	Porcelain enamel on cast iron, 16" x 14", single bowl	Q-1	8	2	Ea.	440	72.50		512.50	605
4180	20" x 18", single bowl	"	8	2	"	240	72.50		312.50	385
4580	For color, add					30%				
6000	Vitreous china, 18" x 15", single bowl with backsplash	Q-1	7	2.286	Ea.	153	83		236	305
6060	19" x 17", single bowl		7	2.286		116	83		199	264
6960	Rough-in, supply, waste and vent for above lavatories		1.66	9.639		485	350		835	1,100
7000	Pedestal type									
7600	Vitreous china, 27" x 21", white	Q-1	6.60	2.424	Ea.	665	88		753	880
7610	27" x 21", colored		6.60	2.424		850	88		938	1,075
7620	27" x 21", premium color		6.60	2.424		965	88		1,053	1,225
7660	26" x 20", white		6.60	2.424		645	88		733	855
7670	26" x 20", colored		6.60	2.424		825	88		913	1,050
7680	26" x 20", premium color		6.60	2.424		950	88		1,038	1,200
7700	24" x 20", white		6.60	2.424		465	88		553	655
7710	24" x 20", colored		6.60	2.424		590	88		678	795
7720	24" x 20", premium color		6.60	2.424		655	88		743	865
7760	21" x 18", white		6.60	2.424		245	88		333	415
7770	21" x 18", colored		6.60	2.424		271	88		359	445
7990	Rough-in, supply, waste and vent for pedestal lavatories		1.66	9.639		485	350		835	1,100

22 41 16.16 Sinks		Crew	Daily Output	Labor-Hours	Unit	Material	Labor	Equipment	Total	Total Incl O&P
0010	**SINKS**, With faucets and drain									
2000	Kitchen, counter top style, PE on CI, 24" x 21" single bowl	Q-1	5.60	2.857	Ea.	310	104		414	510
2100	31" x 22" single bowl		5.60	2.857		795	104		899	1,050
2200	32" x 21" double bowl		4.80	3.333		380	121		501	620
3000	Stainless steel, self rimming, 19" x 18" single bowl		5.60	2.857		615	104		719	845
3100	25" x 22" single bowl		5.60	2.857		680	104		784	920
3200	33" x 22" double bowl		4.80	3.333		985	121		1,106	1,275
3300	43" x 22" double bowl		4.80	3.333		1,150	121		1,271	1,450
4000	Steel, enameled, with ledge, 24" x 21" single bowl		5.60	2.857		525	104		629	750
4100	32" x 21" double bowl		4.80	3.333		545	121		666	800
4960	For color sinks except stainless steel, add					10%				
4980	For rough-in, supply, waste and vent, counter top sinks	Q-1	2.14	7.477		288	272		560	760
5000	Kitchen, raised deck, PE on CI									
5100	32" x 21", dual level, double bowl	Q-1	2.60	6.154	Ea.	445	224		669	855
5790	For rough-in, supply, waste & vent, sinks	"	1.85	8.649	"	288	315		603	830

22 41 Residential Plumbing Fixtures

22 41 19 – Residential Bathtubs

22 41 19.10 Baths		Crew	Daily Output	Labor-Hours	Unit	Material	Labor (2019 Bare Costs)	Equipment (2019 Bare Costs)	Total	Total Incl O&P
0010	**BATHS**									
0100	Tubs, recessed porcelain enamel on cast iron, with trim									
0180	48" x 42"	Q-1	4	4	Ea.	2,800	145		2,945	3,350
0220	72" x 36"	"	3	5.333	"	2,875	194		3,069	3,500
0300	Mat bottom									
0380	5' long	Q-1	4.40	3.636	Ea.	1,200	132		1,332	1,550
0480	Above floor drain, 5' long		4	4		845	145		990	1,175
0560	Corner 48" x 44"		4.40	3.636		2,825	132		2,957	3,325
2000	Enameled formed steel, 4'-6" long		5.80	2.759		505	100		605	720
4600	Module tub & showerwall surround, molded fiberglass									
4610	5' long x 34" wide x 76" high	Q-1	4	4	Ea.	770	145		915	1,075
9600	Rough-in, supply, waste and vent, for all above tubs, add	"	2.07	7.729	"	435	281		716	940

22 41 23 – Residential Showers

22 41 23.20 Showers

Line	Description	Crew	Daily Output	Labor-Hours	Unit	Material	Labor	Equipment	Total	Total Incl O&P
0010	**SHOWERS**									
1500	Stall, with drain only. Add for valve and door/curtain									
1520	32" square	Q-1	5	3.200	Ea.	1,175	116		1,291	1,500
1530	36" square		4.80	3.333		2,950	121		3,071	3,450
1540	Terrazzo receptor, 32" square		5	3.200		1,350	116		1,466	1,675
1560	36" square		4.80	3.333		1,475	121		1,596	1,825
1580	36" corner angle		4.80	3.333		1,750	121		1,871	2,125
3000	Fiberglass, one piece, with 3 walls, 32" x 32" square		5.50	2.909		345	106		451	555
3100	36" x 36" square		5.50	2.909		400	106		506	615
4200	Rough-in, supply, waste and vent for above showers		2.05	7.805		385	284		669	890

22 41 36 – Residential Laundry Trays

22 41 36.10 Laundry Sinks

Line	Description	Crew	Daily Output	Labor-Hours	Unit	Material	Labor	Equipment	Total	Total Incl O&P
0010	**LAUNDRY SINKS**, With trim									
0020	Porcelain enamel on cast iron, black iron frame									
0050	24" x 21", single compartment	Q-1	6	2.667	Ea.	595	97		692	815
0100	26" x 21", single compartment	"	6	2.667	"	625	97		722	845
3000	Plastic, on wall hanger or legs									
3020	18" x 23", single compartment	Q-1	6.50	2.462	Ea.	146	89.50		235.50	305
3100	20" x 24", single compartment		6.50	2.462		165	89.50		254.50	330
3200	36" x 23", double compartment		5.50	2.909		214	106		320	410
3300	40" x 24", double compartment		5.50	2.909		287	106		393	490
5000	Stainless steel, counter top, 22" x 17" single compartment		6	2.667		76.50	97		173.50	244
5200	33" x 22", double compartment		5	3.200		92	116		208	292
9600	Rough-in, supply, waste and vent, for all laundry sinks		2.14	7.477		288	272		560	760

22 41 39 – Residential Faucets, Supplies and Trim

22 41 39.10 Faucets and Fittings

Line	Description	Crew	Daily Output	Labor-Hours	Unit	Material	Labor	Equipment	Total	Total Incl O&P
0010	**FAUCETS AND FITTINGS**									
0150	Bath, faucets, diverter spout combination, sweat	1 Plum	8	1	Ea.	86.50	40.50		127	162
0200	For integral stops, IPS unions, add					111			111	122
0420	Bath, press-bal mix valve w/diverter, spout, shower head, arm/flange	1 Plum	8	1		176	40.50		216.50	260
0500	Drain, central lift, 1-1/2" IPS male		20	.400		49.50	16.15		65.65	81
0600	Trip lever, 1-1/2" IPS male		20	.400		60	16.15		76.15	92.50
1000	Kitchen sink faucets, top mount, cast spout		10	.800		84	32.50		116.50	145
1100	For spray, add		24	.333		17.15	13.45		30.60	41
1300	Single control lever handle									
1310	With pull out spray									

22 41 39 – Residential Faucets, Supplies and Trim

22 41 39.10 Faucets and Fittings		Crew	Daily Output	Labor-Hours	Unit	Material	2019 Bare Costs Labor	Equipment	Total	Total Incl O&P
1320	Polished chrome	1 Plum	10	.800	Ea.	197	32.50		229.50	270
2000	Laundry faucets, shelf type, IPS or copper unions		12	.667		61.50	27		88.50	112
2100	Lavatory faucet, centerset, without drain		10	.800		67	32.50		99.50	127
2120	With pop-up drain		6.66	1.201		50	48.50		98.50	135
2210	Porcelain cross handles and pop-up drain									
2220	Polished chrome	1 Plum	6.66	1.201	Ea.	209	48.50		257.50	310
2230	Polished brass	"	6.66	1.201	"	315	48.50		363.50	425
2260	Single lever handle and pop-up drain									
2280	Satin nickel	1 Plum	6.66	1.201	Ea.	275	48.50		323.50	380
2290	Polished chrome		6.66	1.201		197	48.50		245.50	296
2800	Self-closing, center set		10	.800		150	32.50		182.50	218
4000	Shower by-pass valve with union		18	.444		57.50	17.95		75.45	93
4200	Shower thermostatic mixing valve, concealed, with shower head trim kit		8	1		360	40.50		400.50	460
4220	Shower pressure balancing mixing valve									
4230	With shower head, arm, flange and diverter tub spout									
4240	Chrome	1 Plum	6.14	1.303	Ea.	420	52.50		472.50	545
4250	Satin nickel		6.14	1.303		555	52.50		607.50	695
4260	Polished graphite		6.14	1.303		555	52.50		607.50	695
5000	Sillcock, compact, brass, IPS or copper to hose		24	.333		10.60	13.45		24.05	33.50

22 41 39.70 Washer/Dryer Accessories		Crew	Daily Output	Labor-Hours	Unit	Material	Labor	Equipment	Total	Total Incl O&P
0010	**WASHER/DRYER ACCESSORIES**									
1020	Valves ball type single lever									
1030	1/2" diam., IPS	1 Plum	21	.381	Ea.	64	15.40		79.40	96
1040	1/2" diam., solder	"	21	.381	"	64	15.40		79.40	96
1050	Recessed box, 16 ga., two hose valves and drain									
1060	1/2" size, 1-1/2" drain	1 Plum	18	.444	Ea.	142	17.95		159.95	186
1070	1/2" size, 2" drain	"	17	.471	"	126	19		145	170
1080	With grounding electric receptacle									
1090	1/2" size, 1-1/2" drain	1 Plum	18	.444	Ea.	154	17.95		171.95	200
1100	1/2" size, 2" drain	"	17	.471	"	167	19		186	215
1110	With grounding and dryer receptacle									
1120	1/2" size, 1-1/2" drain	1 Plum	18	.444	Ea.	191	17.95		208.95	240
1130	1/2" size, 2" drain	"	17	.471	"	194	19		213	244
1140	Recessed box, 16 ga., ball valves with single lever and drain									
1150	1/2" size, 1-1/2" drain	1 Plum	19	.421	Ea.	267	17		284	320
1160	1/2" size, 2" drain	"	18	.444	"	230	17.95		247.95	283
1170	With grounding electric receptacle									
1180	1/2" size, 1-1/2" drain	1 Plum	19	.421	Ea.	267	17		284	320
1190	1/2" size, 2" drain	"	18	.444	"	257	17.95		274.95	315
1200	With grounding and dryer receptacles									
1210	1/2" size, 1-1/2" drain	1 Plum	19	.421	Ea.	252	17		269	305
1220	1/2" size, 2" drain	"	18	.444	"	281	17.95		298.95	340
1300	Recessed box, 20 ga., two hose valves and drain (economy type)									
1310	1/2" size, 1-1/2" drain	1 Plum	19	.421	Ea.	110	17		127	149
1320	1/2" size, 2" drain		18	.444		103	17.95		120.95	143
1330	Box with drain only		24	.333		64	13.45		77.45	92.50
1340	1/2" size, 1-1/2" ABS/PVC drain		19	.421		125	17		142	165
1350	1/2" size, 2" ABS/PVC drain		18	.444		128	17.95		145.95	171
1352	Box with drain and 15 A receptacle		24	.333		63.50	13.45		76.95	91.50
1360	1/2" size, 2" drain ABS/PVC, 15 A receptacle		24	.333		135	13.45		148.45	171
1400	Wall mounted									
1410	1/2" size, 1-1/2" plastic drain	1 Plum	19	.421	Ea.	31	17		48	62

22 41 Residential Plumbing Fixtures

22 41 39 – Residential Faucets, Supplies and Trim

22 41 39.70 Washer/Dryer Accessories		Crew	Daily Output	Labor-Hours	Unit	Material	Labor	Equipment	Total	Total Incl O&P
						2019 Bare Costs				
1420	1/2" size, 2" plastic drain	1 Plum	18	.444	Ea.	17.55	17.95		35.50	49
1500	Dryer vent kit									
1510	8' flex duct, clamps and outside hood	1 Plum	20	.400	Ea.	13.20	16.15		29.35	41
1980	Rough-in, supply, waste, and vent for washer boxes		3.46	2.310		325	93.50		418.50	515
9605	Washing machine valve assembly, hot & cold water supply, recessed		8	1		79.50	40.50		120	154
9610	Washing machine valve assembly, hot & cold water supply, mounted	↓	8	1	↓	64	40.50		104.50	137

22 42 Commercial Plumbing Fixtures

22 42 13 – Commercial Water Closets, Urinals, and Bidets

22 42 13.13 Water Closets

		Crew	Daily Output	Labor-Hours	Unit	Material	Labor	Equipment	Total	Total Incl O&P
0010	**WATER CLOSETS**									
3000	Bowl only, with flush valve, seat, 1.6 gpf unless noted									
3100	Wall hung	Q-1	5.80	2.759	Ea.	1,025	100		1,125	1,300
3200	For rough-in, supply, waste and vent, single WC		2.56	6.250		1,300	227		1,527	1,800
3300	Floor mounted		5.80	2.759		325	100		425	525
3370	For rough-in, supply, waste and vent, single WC	↓	2.84	5.634	↓	405	205		610	780
3390	Floor mounted children's size, 10-3/4" high									
3392	With automatic flush sensor, 1.6 gpf	Q-1	6.20	2.581	Ea.	665	94		759	890
3396	With automatic flush sensor, 1.28 gpf		6.20	2.581		620	94		714	835
3400	For rough-in, supply, waste and vent, single WC	↓	2.84	5.634	↓	405	205		610	780

22 42 16 – Commercial Lavatories and Sinks

22 42 16.13 Lavatories

		Crew	Daily Output	Labor-Hours	Unit	Material	Labor	Equipment	Total	Total Incl O&P
0010	**LAVATORIES**, With trim, white unless noted otherwise									
0020	Commercial lavatories same as residential. See Section 22 41 16									

22 42 16.40 Service Sinks

		Crew	Daily Output	Labor-Hours	Unit	Material	Labor	Equipment	Total	Total Incl O&P
0010	**SERVICE SINKS**									
6650	Service, floor, corner, PE on CI, 28" x 28"	Q-1	4.40	3.636	Ea.	1,050	132		1,182	1,375
6750	Vinyl coated rim guard, add					63.50			63.50	69.50
6755	Mop sink, molded stone, 22" x 18"	1 Plum	3.33	2.402		530	97		627	745
6760	Mop sink, molded stone, 24" x 36"		3.33	2.402		266	97		363	450
6770	Mop sink, molded stone, 24" x 36", w/rim 3 sides	↓	3.33	2.402		293	97		390	485
6790	For rough-in, supply, waste & vent, floor service sinks	Q-1	1.64	9.756	↓	995	355		1,350	1,675

22 42 39 – Commercial Faucets, Supplies, and Trim

22 42 39.10 Faucets and Fittings

		Crew	Daily Output	Labor-Hours	Unit	Material	Labor	Equipment	Total	Total Incl O&P
0010	**FAUCETS AND FITTINGS**									
2790	Faucets for lavatories									
2800	Self-closing, center set	1 Plum	10	.800	Ea.	150	32.50		182.50	218
2810	Automatic sensor and operator, with faucet head		6.15	1.301		485	52.50		537.50	620
3000	Service sink faucet, cast spout, pail hook, hose end	↓	14	.571	↓	76	23		99	122

22 42 39.30 Carriers and Supports

		Crew	Daily Output	Labor-Hours	Unit	Material	Labor	Equipment	Total	Total Incl O&P
0010	**CARRIERS AND SUPPORTS**, For plumbing fixtures									
0600	Plate type with studs, top back plate	1 Plum	7	1.143	Ea.	60	46		106	142
3000	Lavatory, concealed arm									
3050	Floor mounted, single									
3100	High back fixture	1 Plum	6	1.333	Ea.	630	54		684	785
3200	Flat slab fixture	"	6	1.333	"	545	54		599	690
8200	Water closet, residential									
8220	Vertical centerline, floor mount									
8240	Single, 3" caulk, 2" or 3" vent	1 Plum	6	1.333	Ea.	730	54		784	895

22 42 Commercial Plumbing Fixtures

22 42 39 – Commercial Faucets, Supplies, and Trim

22 42 39.30 Carriers and Supports		Crew	Daily Output	Labor-Hours	Unit	Material	Labor	Equipment	Total	Total Incl O&P
						2019 Bare Costs				
8260	4" caulk, 2" or 4" vent	1 Plum	6	1.333	Ea.	940	54		994	1,125

22 51 Swimming Pool Plumbing Systems

22 51 19 – Swimming Pool Water Treatment Equipment

22 51 19.50 Swimming Pool Filtration Equipment		Crew	Daily Output	Labor-Hours	Unit	Material	Labor	Equipment	Total	Total Incl O&P
0010	**SWIMMING POOL FILTRATION EQUIPMENT**									
0900	Filter system, sand or diatomite type, incl. pump, 6,000 gal./hr.	2 Plum	1.80	8.889	Total	2,200	360		2,560	3,025
1020	Add for chlorination system, 800 S.F. pool	"	3	5.333	Ea.	220	215		435	595

Estimating Tips

The labor adjustment factors listed in Subdivision 22 01 02.20 also apply to Division 23.

23 10 00 Facility Fuel Systems

- The prices in this subdivision for above- and below-ground storage tanks do not include foundations or hold-down slabs, unless noted. The estimator should refer to Divisions 3 and 31 for foundation system pricing. In addition to the foundations, required tank accessories, such as tank gauges, leak detection devices, and additional manholes and piping, must be added to the tank prices.

23 50 00 Central Heating Equipment

- When estimating the cost of an HVAC system, check to see who is responsible for providing and installing the temperature control system. It is possible to overlook controls, assuming that they would be included in the electrical estimate.
- When looking up a boiler, be careful on specified capacity. Some manufacturers rate their products on output while others use input.
- Include HVAC insulation for pipe, boiler, and duct (wrap and liner).
- Be careful when looking up mechanical items to get the correct pressure rating and connection type (thread, weld, flange).

23 70 00 Central HVAC Equipment

- Combination heating and cooling units are sized by the air conditioning requirements. (See Reference No. R236000-20 for the preliminary sizing guide.)
- A ton of air conditioning is nominally 400 CFM.
- Rectangular duct is taken off by the linear foot for each size, but its cost is usually estimated by the pound. Remember that SMACNA standards now base duct on internal pressure.
- Prefabricated duct is estimated and purchased like pipe: straight sections and fittings.
- Note that cranes or other lifting equipment are not included on any lines in Division 23. For example, if a crane is required to lift a heavy piece of pipe into place high above a gym floor, or to put a rooftop unit on the roof of a four-story building, etc., it must be added. Due to the potential for extreme variation—from nothing additional required to a major crane or helicopter—we feel that including a nominal amount for "lifting contingency" would be useless and detract from the accuracy of the estimate. When using equipment rental cost data from RSMeans, do not forget to include the cost of the operator(s).

Reference Numbers

Reference numbers are shown at the beginning of some major classifications. These numbers refer to related items in the Reference Section. The reference information may be an estimating procedure, an alternate pricing method, or technical information.

Note: Not all subdivisions listed here necessarily appear. ■

Note: "Powered in part by CINX™, based on liscensed proprietary information of Harrison Publishing House, Inc."

Note: Trade Service, in part, has been used as a reference source for some of the material prices used in Division 23.

23 05 05 – Selective Demolition for HVAC

23 05 05.10 HVAC Demolition		Crew	Daily Output	Labor-Hours	Unit	Material	2019 Bare Costs Labor	Equipment	Total	Total Incl O&P
0010	**HVAC DEMOLITION**									
0100	Air conditioner, split unit, 3 ton	Q-5	2	8	Ea.		299		299	490
0150	Package unit, 3 ton	Q-6	3	8	"		288		288	475
0260	Baseboard, hydronic fin tube, 1/2"	Q-5	117	.137	L.F.		5.10		5.10	8.40
0298	Boilers									
0300	Electric, up thru 148 kW	Q-19	2	12	Ea.		465		465	760
0310	150 thru 518 kW	"	1	24			930		930	1,525
0320	550 thru 2,000 kW	Q-21	.40	80			3,150		3,150	5,175
0330	2,070 kW and up	"	.30	107			4,200		4,200	6,900
0340	Gas and/or oil, up thru 150 MBH	Q-7	2.20	14.545			545		545	890
0350	160 thru 2,000 MBH		.80	40			1,500		1,500	2,450
0360	2,100 thru 4,500 MBH		.50	64			2,400		2,400	3,925
0370	4,600 thru 7,000 MBH		.30	107			4,000		4,000	6,550
0390	12,200 thru 25,000 MBH		.12	267			9,975		9,975	16,400
1000	Ductwork, 4" high, 8" wide	1 Clab	200	.040	L.F.		1.10		1.10	1.82
1100	6" high, 8" wide		165	.048			1.33		1.33	2.20
1200	10" high, 12" wide		125	.064			1.76		1.76	2.91
1300	12"-14" high, 16"-18" wide		85	.094			2.59		2.59	4.28
1500	30" high, 36" wide		56	.143			3.93		3.93	6.50
2200	Furnace, electric	Q-20	2	10	Ea.		365		365	600
2300	Gas or oil, under 120 MBH	Q-9	4	4			140		140	233
2340	Over 120 MBH	"	3	5.333			187		187	310
2800	Heat pump, package unit, 3 ton	Q-5	2.40	6.667			249		249	410
2840	Split unit, 3 ton		2	8			299		299	490
2950	Tank, steel, oil, 275 gal., above ground		10	1.600			60		60	98
2951	Tank, steel, oil, 550 gal., above ground		5	3.200			120		120	196
2952	Tank, steel, oil, 1000 gal., above ground		2.50	6.400			239		239	395
2960	Remove and reset		3	5.333			199		199	325
5090	Remove refrigerant from system	1 Stpi	40	.200	Lb.		8.30		8.30	13.65

23 07 HVAC Insulation

23 07 13 – Duct Insulation

23 07 13.10 Duct Thermal Insulation		Crew	Daily Output	Labor-Hours	Unit	Material	Labor	Equipment	Total	Total Incl O&P
0010	**DUCT THERMAL INSULATION**									
3000	Ductwork									
3020	Blanket type, fiberglass, flexible									
3030	Fire rated for grease and hazardous exhaust ducts									
3060	1-1/2" thick	Q-14	84	.190	S.F.	4.62	6.30		10.92	15.70
3090	Fire rated for plenums									
3100	1/2" x 24" x 25'	Q-14	1.94	8.247	Roll	172	272		444	650
3110	1/2" x 24" x 25'		98	.163	S.F.	3.43	5.40		8.83	12.85
3120	1/2" x 48" x 25'		1.04	15.385	Roll	340	510		850	1,225
3126	1/2" x 48" x 25'		104	.154	S.F.	3.39	5.10		8.49	12.30
3140	FSK vapor barrier wrap, .75 lb. density									
3160	1" thick G	Q-14	350	.046	S.F.	.23	1.51		1.74	2.79
3170	1-1/2" thick G	"	320	.050	"	.27	1.65		1.92	3.08
3210	Vinyl jacket, same as FSK									

23 09 Instrumentation and Control for HVAC

23 09 53 – Pneumatic and Electric Control System for HVAC

23 09 53.10 Control Components		Crew	Daily Output	Labor-Hours	Unit	Material	2019 Bare Costs Labor	Equipment	Total	Total Incl O&P
0010	**CONTROL COMPONENTS**									
5000	Thermostats									
5030	Manual	1 Stpi	8	1	Ea.	42	41.50		83.50	115
5040	1 set back, electric, timed G		8	1		75.50	41.50		117	151
5050	2 set back, electric, timed G		8	1		224	41.50		265.50	315

23 13 Facility Fuel-Storage Tanks

23 13 13 – Facility Underground Fuel-Oil, Storage Tanks

23 13 13.09 Single-Wall Steel Fuel-Oil Tanks

Line	Description	Crew	Daily Output	Labor-Hours	Unit	Material	Labor	Equipment	Total	Total Incl O&P
0010	**SINGLE-WALL STEEL FUEL-OIL TANKS**									
5000	Tanks, steel ugnd., sti-p3, not incl. hold-down bars									
5500	Excavation, pad, pumps and piping not included									
5510	Single wall, 500 gallon capacity, 7 ga. shell	Q-5	2.70	5.926	Ea.	1,850	222		2,072	2,400
5520	1,000 gallon capacity, 7 ga. shell	″	2.50	6.400		2,725	239		2,964	3,400
5530	2,000 gallon capacity, 1/4″ thick shell	Q-7	4.60	6.957		3,500	260		3,760	4,275
5535	2,500 gallon capacity, 7 ga. shell	Q-5	3	5.333		5,325	199		5,524	6,200
5610	25,000 gallon capacity, 3/8″ thick shell	Q-7	1.30	24.615		27,500	920		28,420	31,800
5630	40,000 gallon capacity, 3/8″ thick shell		.90	35.556		43,200	1,325		44,525	49,700
5640	50,000 gallon capacity, 3/8″ thick shell		.80	40		48,000	1,500		49,500	55,500

23 13 13.23 Glass-Fiber-Reinfcd-Plastic, Fuel-Oil, Storage

Line	Description	Crew	Daily Output	Labor-Hours	Unit	Material	Labor	Equipment	Total	Total Incl O&P
0010	**GLASS-FIBER-REINFCD-PLASTIC, UNDERGRND. FUEL-OIL, STORAGE**									
0210	Fiberglass, underground, single wall, UL listed, not including									
0220	manway or hold-down strap									
0230	1,000 gallon capacity	Q-5	2.46	6.504	Ea.	5,075	243		5,318	5,975
0240	2,000 gallon capacity	Q-7	4.57	7.002		7,475	262		7,737	8,650
0245	3,000 gallon capacity		3.90	8.205		8,250	305		8,555	9,575
0255	5,000 gallon capacity		3.20	10		10,600	375		10,975	12,300
0500	For manway, fittings and hold-downs, add					20%	15%			
2210	Fiberglass, underground, single wall, UL listed, including									
2220	hold-down straps, no manways									
2230	1,000 gallon capacity	Q-5	1.88	8.511	Ea.	5,575	320		5,895	6,650
2240	2,000 gallon capacity	Q-7	3.55	9.014	″	7,975	335		8,310	9,325

23 13 23 – Facility Aboveground Fuel-Oil, Storage Tanks

23 13 23.16 Steel

Line	Description	Crew	Daily Output	Labor-Hours	Unit	Material	Labor	Equipment	Total	Total Incl O&P
3001	**STEEL**, storage, above ground, including supports, coating									
3020	fittings, not including foundation, pumps or piping									
3040	Single wall, 275 gallon	Q-5	5	3.200	Ea.	520	120		640	765
3060	550 gallon	″	2.70	5.926		4,425	222		4,647	5,250
3080	1,000 gallon	Q-7	5	6.400		7,400	239		7,639	8,525
3320	Double wall, 500 gallon capacity	Q-5	2.40	6.667		1,850	249		2,099	2,450
3330	2,000 gallon capacity	Q-7	4.15	7.711		6,125	288		6,413	7,200
3340	4,000 gallon capacity		3.60	8.889		13,600	330		13,930	15,400
3350	6,000 gallon capacity		2.40	13.333		15,400	500		15,900	17,700
3360	8,000 gallon capacity		2	16		18,100	600		18,700	20,900
3370	10,000 gallon capacity		1.80	17.778		30,300	665		30,965	34,400
3380	15,000 gallon capacity		1.50	21.333		40,000	800		40,800	45,300
3390	20,000 gallon capacity		1.30	24.615		46,500	920		47,420	52,500
3400	25,000 gallon capacity		1.15	27.826		57,500	1,050		58,550	65,000
3410	30,000 gallon capacity		1	32		66,500	1,200		67,700	75,500

23 13 Facility Fuel-Storage Tanks

23 13 23 – Facility Aboveground Fuel-Oil, Storage Tanks

23 13 23.26 Horizontal, Conc., Abvgrd Fuel-Oil, Stor. Tanks		Crew	Daily Output	Labor-Hours	Unit	Material	2019 Bare Costs Labor	Equipment	Total	Total Incl O&P
0010	**HORIZONTAL, CONCRETE, ABOVEGROUND FUEL-OIL, STORAGE TANKS**									
0050	Concrete, storage, aboveground, including pad & pump									
0100	500 gallon	F-3	2	20	Ea.	10,100	655	236	10,991	12,400
0200	1,000 gallon	"	2	20	"	14,200	655	236	15,091	16,900

23 21 Hydronic Piping and Pumps

23 21 20 – Hydronic HVAC Piping Specialties

23 21 20.46 Expansion Tanks

		Crew	Daily Output	Labor-Hours	Unit	Material	Labor	Equipment	Total	Total Incl O&P
0010	**EXPANSION TANKS**									
1507	Underground fuel-oil storage tanks, see Section 23 13 13									
2000	Steel, liquid expansion, ASME, painted, 15 gallon capacity	Q-5	17	.941	Ea.	790	35		825	930
2040	30 gallon capacity		12	1.333		900	50		950	1,075
3000	Steel ASME expansion, rubber diaphragm, 19 gal. cap. accept.		12	1.333		2,775	50		2,825	3,150
3020	31 gallon capacity	↓	8	2	↓	3,275	75		3,350	3,725

23 21 23 – Hydronic Pumps

23 21 23.13 In-Line Centrifugal Hydronic Pumps

		Crew	Daily Output	Labor-Hours	Unit	Material	Labor	Equipment	Total	Total Incl O&P
0010	**IN-LINE CENTRIFUGAL HYDRONIC PUMPS**									
0600	Bronze, sweat connections, 1/40 HP, in line									
0640	3/4" size	Q-1	16	1	Ea.	288	36.50		324.50	375
1000	Flange connection, 3/4" to 1-1/2" size									
1040	1/12 HP	Q-1	6	2.667	Ea.	755	97		852	990
1060	1/8 HP		6	2.667		1,325	97		1,422	1,600
2101	Pumps, circulating, 3/4" to 1-1/2" size, 1/3 HP	↓	6	2.667	↓	990	97		1,087	1,250

23 23 Refrigerant Piping

23 23 16 – Refrigerant Piping Specialties

23 23 16.16 Refrigerant Line Sets

		Crew	Daily Output	Labor-Hours	Unit	Material	Labor	Equipment	Total	Total Incl O&P
0010	**REFRIGERANT LINE SETS**, Standard									
0100	Copper tube									
0110	1/2" insulation, both tubes									
0120	Combination 1/4" and 1/2" tubes									
0130	10' set	Q-5	42	.381	Ea.	48	14.25		62.25	76
0135	15' set		42	.381		69	14.25		83.25	99.50
0140	20' set		40	.400		74.50	14.95		89.45	107
0150	30' set		37	.432		105	16.15		121.15	143
0160	40' set		35	.457		132	17.10		149.10	173
0170	50' set		32	.500		163	18.70		181.70	211
0180	100' set	↓	22	.727	↓	355	27		382	435
0300	Combination 1/4" and 3/4" tubes									
0310	10' set	Q-5	40	.400	Ea.	53.50	14.95		68.45	83.50
0320	20' set		38	.421		94.50	15.75		110.25	130
0330	30' set		35	.457		142	17.10		159.10	184
0340	40' set		33	.485		185	18.15		203.15	234
0350	50' set		30	.533		233	19.95		252.95	289
0380	100' set	↓	20	.800	↓	530	30		560	635
0500	Combination 3/8" & 3/4" tubes									
0510	10' set	Q-5	28	.571	Ea.	61.50	21.50		83	103
0520	20' set	↓	36	.444	↓	95.50	16.60		112.10	133

23 23 Refrigerant Piping

23 23 16 – Refrigerant Piping Specialties

23 23 16.16 Refrigerant Line Sets		Crew	Daily Output	Labor-Hours	Unit	Material	2019 Bare Costs Labor	Equipment	Total	Total Incl O&P
0530	30′ set	Q-5	34	.471	Ea.	130	17.60		147.60	172
0540	40′ set		31	.516		171	19.30		190.30	220
0550	50′ set		28	.571		201	21.50		222.50	256
0580	100′ set		18	.889		595	33		628	710
0700	Combination 3/8″ & 1-1/8″ tubes									
0710	10′ set	Q-5	36	.444	Ea.	108	16.60		124.60	146
0720	20′ set		33	.485		180	18.15		198.15	228
0730	30′ set		31	.516		242	19.30		261.30	298
0740	40′ set		28	.571		365	21.50		386.50	435
0750	50′ set		26	.615		350	23		373	425
0900	Combination 1/2″ & 3/4″ tubes									
0910	10′ set	Q-5	37	.432	Ea.	65	16.15		81.15	98
0920	20′ set		35	.457		115	17.10		132.10	155
0930	30′ set		33	.485		173	18.15		191.15	221
0940	40′ set		30	.533		228	19.95		247.95	284
0950	50′ set		27	.593		287	22		309	350
0980	100′ set		17	.941		645	35		680	770
2100	Combination 1/2″ & 1-1/8″ tubes									
2110	10′ set	Q-5	35	.457	Ea.	111	17.10		128.10	150
2120	20′ set		31	.516		190	19.30		209.30	241
2130	30′ set		29	.552		285	20.50		305.50	350
2140	40′ set		25	.640		385	24		409	460
2150	50′ set		14	1.143		465	42.50		507.50	585
2300	For 1″ thick insulation add					30%	15%			
3000	Refrigerant line sets, min-split, flared									
3100	Combination 1/4″ & 3/8″ tubes									
3120	15′ set	Q-5	41	.390	Ea.	68.50	14.60		83.10	99.50
3140	25′ set		38.50	.416		97	15.55		112.55	133
3160	35′ set		37	.432		122	16.15		138.15	161
3180	50′ set		30	.533		166	19.95		185.95	216
3200	Combination 1/4″ & 1/2″ tubes									
3220	15′ set	Q-5	41	.390	Ea.	73	14.60		87.60	105
3240	25′ set		38.50	.416		99.50	15.55		115.05	135
3260	35′ set		37	.432		128	16.15		144.15	168
3280	50′ set		30	.533		173	19.95		192.95	223

23 31 HVAC Ducts and Casings

23 31 13 – Metal Ducts

23 31 13.13 Rectangular Metal Ducts		Crew	Daily Output	Labor-Hours	Unit	Material	Labor	Equipment	Total	Total Incl O&P
0010	**RECTANGULAR METAL DUCTS**									
0020	Fabricated rectangular, includes fittings, joints, supports,									
0021	allowance for flexible connections and field sketches.									
0030	Does not include "as-built dwgs." or insulation.									
0031	NOTE: Fabrication and installation are combined									
0040	as LABOR cost. Approx. 25% fittings assumed.									
0042	Fabrication/Inst. is to commercial quality standards									
0043	(SMACNA or equiv.) for structure, sealing, leak testing, etc.									
0100	Aluminum, alloy 3003-H14, under 100 lb.	Q-10	75	.320	Lb.	3.20	11.65		14.85	23
0110	100 to 500 lb.		80	.300		1.89	10.90		12.79	20
0120	500 to 1,000 lb.		95	.253		1.82	9.20		11.02	17.25
0140	1,000 to 2,000 lb.		120	.200		1.78	7.30		9.08	14.05

23 31 HVAC Ducts and Casings

23 31 13 – Metal Ducts

23 31 13.13 Rectangular Metal Ducts		Crew	Daily Output	Labor-Hours	Unit	2019 Bare Costs Material	Labor	Equipment	Total	Total Incl O&P
0500	Galvanized steel, under 200 lb.	Q-10	235	.102	Lb.	.57	3.72		4.29	6.80
0520	200 to 500 lb.	↓	245	.098	↓	.58	3.57		4.15	6.55
0540	500 to 1,000 lb.	↓	255	.094	↓	.56	3.43		3.99	6.30

23 31 16.22 Duct Board

		Crew	Daily Output	Labor-Hours	Unit	Material	Labor	Equipment	Total	Total Incl O&P
0010	DUCT BOARD									
3800	Rigid, resin bonded fibrous glass, FSK									
3810	Temperature, bacteria and fungi resistant									
3820	1″ thick	Q-14	150	.107	S.F.	1.63	3.52		5.15	7.75
3830	1-1/2″ thick	↓	130	.123	↓	2.15	4.07		6.22	9.20
3840	2″ thick	↓	120	.133	↓	3.71	4.40		8.11	11.50

23 33 Air Duct Accessories

23 33 13 – Dampers

23 33 13.13 Volume-Control Dampers

		Crew	Daily Output	Labor-Hours	Unit	Material	Labor	Equipment	Total	Total Incl O&P
0010	VOLUME-CONTROL DAMPERS									
6000	12″ x 12″	1 Shee	21	.381	Ea.	45	14.85		59.85	74
8000	Multi-blade dampers, parallel blade									
8100	8″ x 8″	1 Shee	24	.333	Ea.	116	13		129	150

23 33 13.16 Fire Dampers

		Crew	Daily Output	Labor-Hours	Unit	Material	Labor	Equipment	Total	Total Incl O&P
0010	FIRE DAMPERS									
3000	Fire damper, curtain type, 1-1/2 hr. rated, vertical, 6″ x 6″	1 Shee	24	.333	Ea.	34.50	13		47.50	59.50
3020	8″ x 6″	″	22	.364	″	34.50	14.20		48.70	61.50

23 33 46 – Flexible Ducts

23 33 46.10 Flexible Air Ducts

		Crew	Daily Output	Labor-Hours	Unit	Material	Labor	Equipment	Total	Total Incl O&P
0010	FLEXIBLE AIR DUCTS									
1300	Flexible, coated fiberglass fabric on corr. resist. metal helix									
1400	pressure to 12″ (WG) UL-181									
1500	Noninsulated, 3″ diameter	Q-9	400	.040	L.F.	1.21	1.40		2.61	3.66
1540	5″ diameter	↓	320	.050	↓	1.38	1.76		3.14	4.43
1560	6″ diameter	↓	280	.057	↓	1.61	2.01		3.62	5.10
1580	7″ diameter	↓	240	.067	↓	1.75	2.34		4.09	5.80
1900	Insulated, 1″ thick, PE jacket, 3″ diameter G	↓	380	.042	↓	2.66	1.48		4.14	5.40
1910	4″ diameter G	↓	340	.047	↓	2.83	1.65		4.48	5.85
1920	5″ diameter G	↓	300	.053	↓	3.01	1.87		4.88	6.40
1940	6″ diameter G	↓	260	.062	↓	3.14	2.16		5.30	7.05
1960	7″ diameter G	↓	220	.073	↓	3.59	2.55		6.14	8.20
1980	8″ diameter G	↓	180	.089	↓	3.96	3.12		7.08	9.55
2040	12″ diameter G	↓	100	.160	↓	5.55	5.60		11.15	15.40

23 33 53 – Duct Liners

23 33 53.10 Duct Liner Board

		Crew	Daily Output	Labor-Hours	Unit	Material	Labor	Equipment	Total	Total Incl O&P
0010	DUCT LINER BOARD									
3490	Board type, fiberglass liner, 3 lb. density									
3940	Board type, non-fibrous foam									
3950	Temperature, bacteria and fungi resistant									
3960	1″ thick G	Q-14	150	.107	S.F.	2.47	3.52		5.99	8.65
3970	1-1/2″ thick G	↓	130	.123	↓	4.23	4.07		8.30	11.50
3980	2″ thick G	↓	120	.133	↓	4.12	4.40		8.52	11.95

23 34 23 – HVAC Power Ventilators

23 34 23.10 HVAC Power Circulators and Ventilators		Crew	Daily Output	Labor-Hours	Unit	2019 Bare Costs Material	Labor	Equipment	Total	Total Incl O&P
0010	HVAC POWER CIRCULATORS AND VENTILATORS									
6650	Residential, bath exhaust, grille, back draft damper									
6660	50 CFM	Q-20	24	.833	Ea.	53	30.50		83.50	109
6670	110 CFM		22	.909		106	33		139	171
6900	Kitchen exhaust, grille, complete, 160 CFM		22	.909		107	33		140	173
6910	180 CFM		20	1		97	36.50		133.50	167
6920	270 CFM		18	1.111		235	40.50		275.50	325
6930	350 CFM	↓	16	1.250	↓	150	45.50		195.50	240
6940	Residential roof jacks and wall caps									
6944	Wall cap with back draft damper									
6946	3" & 4" diam. round duct	1 Shee	11	.727	Ea.	25	28.50		53.50	74.50
6948	6" diam. round duct	"	11	.727	"	77	28.50		105.50	132
6958	Roof jack with bird screen and back draft damper									
6960	3" & 4" diam. round duct	1 Shee	11	.727	Ea.	20.50	28.50		49	69.50
6962	3-1/4" x 10" rectangular duct	"	10	.800	"	37	31		68	93
8020	Attic, roof type									
8030	Aluminum dome, damper & curb									
8080	12" diameter, 1,000 CFM (gravity)	1 Elec	10	.800	Ea.	610	33		643	725
8090	16" diameter, 1,500 CFM (gravity)		9	.889		735	37		772	870
8100	20" diameter, 2,500 CFM (gravity)	↓	8	1	↓	900	41.50		941.50	1,050
8160	Plastic, ABS dome									
8180	1,050 CFM	1 Elec	14	.571	Ea.	179	23.50		202.50	236
8200	1,600 CFM	"	12	.667	"	268	27.50		295.50	340
8240	Attic, wall type, with shutter, one speed									
8250	12" diameter, 1,000 CFM	1 Elec	14	.571	Ea.	415	23.50		438.50	495
8260	14" diameter, 1,500 CFM		12	.667		450	27.50		477.50	535
8270	16" diameter, 2,000 CFM	↓	9	.889	↓	505	37		542	615
8290	Whole house, wall type, with shutter, one speed									
8300	30" diameter, 4,800 CFM	1 Elec	7	1.143	Ea.	1,075	47.50		1,122.50	1,275
8310	36" diameter, 7,000 CFM		6	1.333		1,175	55		1,230	1,400
8320	42" diameter, 10,000 CFM		5	1.600		1,325	66		1,391	1,550
8330	48" diameter, 16,000 CFM	↓	4	2		1,650	83		1,733	1,925
8340	For two speed, add				↓	99			99	109
8350	Whole house, lay-down type, with shutter, one speed									
8360	30" diameter, 4,500 CFM	1 Elec	8	1	Ea.	1,150	41.50		1,191.50	1,350
8370	36" diameter, 6,500 CFM		7	1.143		1,250	47.50		1,297.50	1,450
8380	42" diameter, 9,000 CFM		6	1.333		1,375	55		1,430	1,600
8390	48" diameter, 12,000 CFM	↓	5	1.600		1,550	66		1,616	1,800
8440	For two speed, add					74.50			74.50	82
8450	For 12 hour timer switch, add	1 Elec	32	.250	↓	74.50	10.35		84.85	99

23 37 Air Outlets and Inlets

23 37 13 – Diffusers, Registers, and Grilles

23 37 13.10 Diffusers		Crew	Daily Output	Labor-Hours	Unit	Material	2019 Bare Costs Labor	Equipment	Total	Total Incl O&P
0010	**DIFFUSERS**, Aluminum, opposed blade damper unless noted									
0100	Ceiling, linear, also for sidewall									
0120	2" wide	1 Shee	32	.250	L.F.	18.85	9.75		28.60	36.50
0160	4" wide		26	.308	"	25.50	12		37.50	48
0500	Perforated, 24" x 24" lay-in panel size, 6" x 6"		16	.500	Ea.	165	19.50		184.50	214
0520	8" x 8"		15	.533		173	21		194	225
0530	9" x 9"		14	.571		176	22.50		198.50	230
0590	16" x 16"		11	.727		205	28.50		233.50	273
1000	Rectangular, 1 to 4 way blow, 6" x 6"		16	.500		43	19.50		62.50	79.50
1010	8" x 8"		15	.533		61	21		82	102
1014	9" x 9"		15	.533		53.50	21		74.50	93
1016	10" x 10"		15	.533		83	21		104	126
1020	12" x 6"		15	.533		75	21		96	117
1040	12" x 9"		14	.571		79.50	22.50		102	125
1060	12" x 12"		12	.667		74.50	26		100.50	125
1070	14" x 6"		13	.615		81.50	24		105.50	130
1074	14" x 14"		12	.667		133	26		159	190
1150	18" x 18"		9	.889		117	34.50		151.50	187
1170	24" x 12"		10	.800		171	31		202	241
1180	24" x 24"		7	1.143		284	44.50		328.50	385
1500	Round, butterfly damper, steel, diffuser size, 6" diameter		18	.444		12.55	17.35		29.90	43
1520	8" diameter		16	.500		13.25	19.50		32.75	47
2000	T-bar mounting, 24" x 24" lay-in frame, 6" x 6"		16	.500		72.50	19.50		92	112
2020	8" x 8"		14	.571		72	22.50		94.50	117
2040	12" x 12"		12	.667		88.50	26		114.50	141
2060	16" x 16"		11	.727		115	28.50		143.50	173
2080	18" x 18"		10	.800		121	31		152	185
6000	For steel diffusers instead of aluminum, deduct					10%				

23 37 13.30 Grilles		Crew	Daily Output	Labor-Hours	Unit	Material	Labor	Equipment	Total	Total Incl O&P
0010	**GRILLES**									
0020	Aluminum, unless noted otherwise									
1000	Air return, steel, 6" x 6"	1 Shee	26	.308	Ea.	21	12		33	43
1020	10" x 6"		24	.333		21	13		34	44.50
1080	16" x 8"		22	.364		28.50	14.20		42.70	55
1100	12" x 12"		22	.364		28	14.20		42.20	54.50
1120	24" x 12"		18	.444		37.50	17.35		54.85	70.50
1180	16" x 16"		22	.364		36.50	14.20		50.70	63.50

23 37 13.60 Registers		Crew	Daily Output	Labor-Hours	Unit	Material	Labor	Equipment	Total	Total Incl O&P
0010	**REGISTERS**									
0980	Air supply									
3000	Baseboard, hand adj. damper, enameled steel									
3012	8" x 6"	1 Shee	26	.308	Ea.	7.45	12		19.45	28
3020	10" x 6"		24	.333		8.10	13		21.10	30.50
3040	12" x 5"		23	.348		6.85	13.55		20.40	30
3060	12" x 6"		23	.348		9.75	13.55		23.30	33.50
4000	Floor, toe operated damper, enameled steel									
4020	4" x 8"	1 Shee	32	.250	Ea.	13.60	9.75		23.35	31
4040	4" x 12"	"	26	.308	"	16	12		28	37.50
4300	Spiral pipe supply register									
4310	Steel, with air scoop									
4320	4" x 12", for 8" thru 13" diameter duct	1 Shee	25	.320	Ea.	81	12.50		93.50	110
4330	4" x 18", for 8" thru 13" diameter duct		18	.444		93.50	17.35		110.85	132

23 37 Air Outlets and Inlets

23 37 13 – Diffusers, Registers, and Grilles

23 37 13.60 Registers		Crew	Daily Output	Labor-Hours	Unit	Material	2019 Bare Costs Labor	Equipment	Total	Total Incl O&P
4340	6" x 12", for 14" thru 21" diameter duct	1 Shee	19	.421	Ea.	85.50	16.40		101.90	121
4350	6" x 16", for 14" thru 21" diameter duct		18	.444		97	17.35		114.35	136
4360	6" x 20", for 14" thru 21" diameter duct		17	.471		107	18.35		125.35	149
4370	6" x 24", for 14" thru 21" diameter duct		16	.500		129	19.50		148.50	175
4380	8" x 16", for 22" thru 31" diameter duct		19	.421		101	16.40		117.40	139
4390	8" x 18", for 22" thru 31" diameter duct		18	.444		109	17.35		126.35	149
4400	8" x 24", for 22" thru 31" diameter duct	↓	15	.533	↓	136	21		157	185

23 41 Particulate Air Filtration

23 41 13 – Panel Air Filters

23 41 13.10 Panel Type Air Filters

		Crew	Daily Output	Labor-Hours	Unit	Material	Labor	Equipment	Total	Total Incl O&P
0010	**PANEL TYPE AIR FILTERS**									
2950	Mechanical media filtration units									
3000	High efficiency type, with frame, non-supported G				MCFM	35			35	38.50
3100	Supported type G				"	54			54	59.50
5500	Throwaway glass or paper media type, 12" x 36" x 1"				Ea.	2.27			2.27	2.50

23 41 16 – Renewable-Media Air Filters

23 41 16.10 Disposable Media Air Filters

		Crew	Daily Output	Labor-Hours	Unit	Material	Labor	Equipment	Total	Total Incl O&P
0010	**DISPOSABLE MEDIA AIR FILTERS**									
5000	Renewable disposable roll				C.S.F.	5.25			5.25	5.75

23 41 19 – Washable Air Filters

23 41 19.10 Permanent Air Filters

		Crew	Daily Output	Labor-Hours	Unit	Material	Labor	Equipment	Total	Total Incl O&P
0010	**PERMANENT AIR FILTERS**									
4500	Permanent washable G				MCFM	25			25	27.50

23 41 23 – Extended Surface Filters

23 41 23.10 Expanded Surface Filters

		Crew	Daily Output	Labor-Hours	Unit	Material	Labor	Equipment	Total	Total Incl O&P
0010	**EXPANDED SURFACE FILTERS**									
4000	Medium efficiency, extended surface G				MCFM	6.15			6.15	6.75

23 42 Gas-Phase Air Filtration

23 42 13 – Activated-Carbon Air Filtration

23 42 13.10 Charcoal Type Air Filtration

		Crew	Daily Output	Labor-Hours	Unit	Material	Labor	Equipment	Total	Total Incl O&P
0010	**CHARCOAL TYPE AIR FILTRATION**									
0050	Activated charcoal type, full flow				MCFM	650			650	715
0060	Full flow, impregnated media 12" deep					225			225	248
0070	HEPA filter & frame for field erection					440			440	485
0080	HEPA filter-diffuser, ceiling install.				↓	350			350	385

23 43 Electronic Air Cleaners

23 43 13 – Washable Electronic Air Cleaners

23 43 13.10 Electronic Air Cleaners		Crew	Daily Output	Labor-Hours	Unit	Material	2019 Bare Costs Labor	Equipment	Total	Total Incl O&P
0010	**ELECTRONIC AIR CLEANERS**									
2000	Electronic air cleaner, duct mounted									
2150	1,000 CFM	1 Shee	4	2	Ea.	430	78		508	605
2200	1,200 CFM		3.80	2.105		505	82		587	690
2250	1,400 CFM		3.60	2.222		530	86.50		616.50	725

23 51 Breechings, Chimneys, and Stacks

23 51 23 – Gas Vents

23 51 23.10 Gas Chimney Vents

		Crew	Daily Output	Labor-Hours	Unit	Material	Labor	Equipment	Total	Total Incl O&P
0010	**GAS CHIMNEY VENTS**, Prefab metal, UL listed									
0020	Gas, double wall, galvanized steel									
0080	3″ diameter	Q-9	72	.222	V.L.F.	6.80	7.80		14.60	20.50
0100	4″ diameter	″	68	.235	″	8.35	8.25		16.60	23

23 52 Heating Boilers

23 52 13 – Electric Boilers

23 52 13.10 Electric Boilers, ASME

		Crew	Daily Output	Labor-Hours	Unit	Material	Labor	Equipment	Total	Total Incl O&P
0010	**ELECTRIC BOILERS, ASME**, Standard controls and trim									
1000	Steam, 6 KW, 20.5 MBH	Q-19	1.20	20	Ea.	4,300	775		5,075	6,000
1160	60 KW, 205 MBH		1	24		7,075	930		8,005	9,300
2000	Hot water, 7.5 KW, 25.6 MBH		1.30	18.462		5,200	715		5,915	6,900
2040	30 KW, 102 MBH		1.20	20		5,600	775		6,375	7,425
2060	45 KW, 164 MBH		1.20	20		5,700	775		6,475	7,550

23 52 23 – Cast-Iron Boilers

23 52 23.20 Gas-Fired Boilers

		Crew	Daily Output	Labor-Hours	Unit	Material	Labor	Equipment	Total	Total Incl O&P
0010	**GAS-FIRED BOILERS**, Natural or propane, standard controls, packaged									
1000	Cast iron, with insulated jacket									
3000	Hot water, gross output, 80 MBH	Q-7	1.46	21.918	Ea.	1,950	820		2,770	3,475
3020	100 MBH	″	1.35	23.704		2,275	885		3,160	3,950
7000	For tankless water heater, add					10%				
7050	For additional zone valves up to 312 MBH, add					197			197	217

23 52 23.30 Gas/Oil Fired Boilers

		Crew	Daily Output	Labor-Hours	Unit	Material	Labor	Equipment	Total	Total Incl O&P
0010	**GAS/OIL FIRED BOILERS**, Combination with burners and controls, packaged									
1000	Cast iron with insulated jacket									
2000	Steam, gross output, 720 MBH	Q-7	.43	74.074	Ea.	14,000	2,775		16,775	20,000
2900	Hot water, gross output									
2910	200 MBH	Q-6	.62	39.024	Ea.	10,400	1,400		11,800	13,700
2920	300 MBH		.49	49.080		10,400	1,775		12,175	14,300
2930	400 MBH		.41	57.971		12,200	2,100		14,300	16,800
2940	500 MBH		.36	67.039		13,100	2,425		15,525	18,400
3000	584 MBH	Q-7	.44	72.072		14,300	2,700		17,000	20,100

23 52 23.40 Oil-Fired Boilers

		Crew	Daily Output	Labor-Hours	Unit	Material	Labor	Equipment	Total	Total Incl O&P
0010	**OIL-FIRED BOILERS**, Standard controls, flame retention burner, packaged									
1000	Cast iron, with insulated flush jacket									
2000	Steam, gross output, 109 MBH	Q-7	1.20	26.667	Ea.	2,225	995		3,220	4,075
2060	207 MBH	″	.90	35.556	″	3,025	1,325		4,350	5,500
3000	Hot water, same price as steam									

23 52 Heating Boilers

23 52 26 – Steel Boilers

23 52 26.40 Oil-Fired Boilers		Crew	Daily Output	Labor-Hours	Unit	Material	Labor	Equipment	Total	Total Incl O&P
						2019 Bare Costs				
0010	**OIL-FIRED BOILERS**, Standard controls, flame retention burner									
5000	Steel, with insulated flush jacket									
7000	Hot water, gross output, 103 MBH	Q-6	1.60	15	Ea.	1,875	540		2,415	2,925
7020	122 MBH		1.45	16.506		2,025	595		2,620	3,200
7060	168 MBH		1.30	18.405		2,275	665		2,940	3,575
7080	225 MBH		1.22	19.704		3,525	710		4,235	5,050

23 52 28 – Swimming Pool Boilers

23 52 28.10 Swimming Pool Heaters		Crew	Daily Output	Labor-Hours	Unit	Material	Labor	Equipment	Total	Total Incl O&P
0010	**SWIMMING POOL HEATERS**, Not including wiring, external									
0020	piping, base or pad									
0160	Gas fired, input, 155 MBH	Q-6	1.50	16	Ea.	1,975	575		2,550	3,125
0200	199 MBH		1	24		2,125	865		2,990	3,775
0280	500 MBH		.40	60		8,625	2,150		10,775	13,100
2000	Electric, 12 KW, 4,800 gallon pool	Q-19	3	8		2,125	310		2,435	2,850
2020	15 KW, 7,200 gallon pool		2.80	8.571		2,150	330		2,480	2,900
2040	24 KW, 9,600 gallon pool		2.40	10		2,500	385		2,885	3,375
2100	57 KW, 24,000 gallon pool		1.20	20		3,700	775		4,475	5,325

23 52 88 – Burners

23 52 88.10 Replacement Type Burners		Crew	Daily Output	Labor-Hours	Unit	Material	Labor	Equipment	Total	Total Incl O&P
0010	**REPLACEMENT TYPE BURNERS**									
0990	Residential, conversion, gas fired, LP or natural									
1000	Gun type, atmospheric input 50 to 225 MBH	Q-1	2.50	6.400	Ea.	1,000	233		1,233	1,475
1020	100 to 400 MBH	"	2	8		1,675	291		1,966	2,325
1025	Burner, gas, 100 to 400 MBH G					1,675			1,675	1,850
1040	300 to 1,000 MBH	Q-1	1.70	9.412		5,050	340		5,390	6,100

23 54 Furnaces

23 54 13 – Electric-Resistance Furnaces

23 54 13.10 Electric Furnaces		Crew	Daily Output	Labor-Hours	Unit	Material	Labor	Equipment	Total	Total Incl O&P
0010	**ELECTRIC FURNACES**, Hot air, blowers, std. controls									
0011	not including gas, oil or flue piping									
1000	Electric, UL listed									
1070	10.2 MBH	Q-20	4.40	4.545	Ea.	350	165		515	660
1080	17.1 MBH		4.60	4.348		430	158		588	735
1100	34.1 MBH		4.40	4.545		540	165		705	865

23 54 16 – Fuel-Fired Furnaces

23 54 16.13 Gas-Fired Furnaces		Crew	Daily Output	Labor-Hours	Unit	Material	Labor	Equipment	Total	Total Incl O&P
0010	**GAS-FIRED FURNACES**									
3000	Gas, AGA certified, upflow, direct drive models									
3020	45 MBH input	Q-9	4	4	Ea.	670	140		810	970
3040	60 MBH input		3.80	4.211		695	148		843	1,000
3060	75 MBH input		3.60	4.444		755	156		911	1,100
3100	100 MBH input		3.20	5		785	176		961	1,150
3120	125 MBH input		3	5.333		815	187		1,002	1,200
3130	150 MBH input		2.80	5.714		1,825	201		2,026	2,325
3140	200 MBH input		2.60	6.154		3,400	216		3,616	4,100
4000	For starter plenum, add		16	1		98	35		133	166

23 54 Furnaces

23 54 16 – Fuel-Fired Furnaces

23 54 16.16 Oil-Fired Furnaces			Crew	Daily Output	Labor-Hours	Unit	2019 Bare Costs Material	Labor	Equipment	Total	Total Incl O&P
0010	**OIL-FIRED FURNACES**										
6000	Oil, UL listed, atomizing gun type burner										
6020	56 MBH output		Q-9	3.60	4.444	Ea.	3,125	156		3,281	3,675
6030	84 MBH output			3.50	4.571		2,975	160		3,135	3,525
6040	95 MBH output			3.40	4.706		2,900	165		3,065	3,450
6060	134 MBH output			3.20	5		2,900	176		3,076	3,500
6080	151 MBH output			3	5.333		3,225	187		3,412	3,850
6100	200 MBH input			2.60	6.154		3,775	216		3,991	4,500

23 54 16.21 Solid Fuel-Fired Furnaces

			Crew	Daily Output	Labor-Hours	Unit	Material	Labor	Equipment	Total	Total Incl O&P
0010	**SOLID FUEL-FIRED FURNACES**										
6020	Wood fired furnaces										
6030	Includes hot water coil, thermostat, and auto draft control										
6040	24″ long firebox	G	Q-9	4	4	Ea.	4,550	140		4,690	5,250
6050	30″ long firebox	G		3.60	4.444		5,300	156		5,456	6,075
6060	With fireplace glass doors	G		3.20	5		6,675	176		6,851	7,650
6200	Wood/oil fired furnaces, includes two thermostats										
6210	Includes hot water coil and auto draft control										
6240	24″ long firebox	G	Q-9	3.40	4.706	Ea.	5,600	165		5,765	6,425
6250	30″ long firebox	G		3	5.333		6,225	187		6,412	7,150
6260	With fireplace glass doors	G		2.80	5.714		7,625	201		7,826	8,700
6400	Wood/gas fired furnaces, includes two thermostats										
6410	Includes hot water coil and auto draft control										
6440	24″ long firebox	G	Q-9	2.80	5.714	Ea.	6,200	201		6,401	7,150
6450	30″ long firebox	G		2.40	6.667		6,950	234		7,184	8,050
6460	With fireplace glass doors	G		2	8		8,425	281		8,706	9,750
6600	Wood/oil/gas fired furnaces, optional accessories										
6610	Hot air plenum		Q-9	16	1	Ea.	112	35		147	181
6620	Safety heat dump			24	.667		97.50	23.50		121	146
6630	Auto air intake			18	.889		153	31		184	221
6640	Cold air return package			14	1.143		165	40		205	248
6650	Wood fork						53			53	58
6700	Wood fired outdoor furnace										
6740	24″ long firebox	G	Q-9	3.80	4.211	Ea.	4,100	148		4,248	4,775
6760	Wood fired outdoor furnace, optional accessories										
6770	Chimney section, stainless steel, 6″ ID x 3′ long	G	Q-9	36	.444	Ea.	194	15.60		209.60	239
6780	Chimney cap, stainless steel	G	″	40	.400	″	116	14.05		130.05	152
6800	Wood fired hot water furnace										
6820	Includes 200 gal. hot water storage, thermostat, and auto draft control										
6840	30″ long firebox	G	Q-9	2.10	7.619	Ea.	8,675	267		8,942	9,975
6850	Water to air heat exchanger										
6870	Includes mounting kit and blower relay										
6880	140 MBH, 18.75″ W x 18.75″ L		Q-9	7.50	2.133	Ea.	385	75		460	550
6890	200 MBH, 24″ W x 24″ L		″	7	2.286	″	550	80		630	740
6900	Water to water heat exchanger										
6940	100 MBH		Q-9	6.50	2.462	Ea.	390	86.50		476.50	570
6960	290 MBH		″	6	2.667	″	560	93.50		653.50	770
7000	Optional accessories										
7010	Large volume circulation pump (2 included)		Q-9	14	1.143	Ea.	253	40		293	345
7020	Air bleed fittings (package)			24	.667		53	23.50		76.50	97.50
7030	Domestic water preheater			6	2.667		229	93.50		322.50	405
7040	Smoke pipe kit			4	4		96	140		236	340

23 54 24.10 Furnace Components and Combinations		Crew	Daily Output	Labor-Hours	Unit	2019 Bare Costs Material	Labor	Equipment	Total	Total Incl O&P
0010	**FURNACE COMPONENTS AND COMBINATIONS**									
0080	Coils, A.C. evaporator, for gas or oil furnaces									
0090	Add-on, with holding charge									
0100	Upflow									
0120	1-1/2 ton cooling	Q-5	4	4	Ea.	217	150		367	485
0130	2 ton cooling		3.70	4.324		281	162		443	575
0140	3 ton cooling		3.30	4.848		430	181		611	765
0150	4 ton cooling		3	5.333		585	199		784	970
0160	5 ton cooling		2.70	5.926		630	222		852	1,050
0300	Downflow									
0330	2-1/2 ton cooling	Q-5	3	5.333	Ea.	335	199		534	695
0340	3-1/2 ton cooling		2.60	6.154		495	230		725	925
0350	5 ton cooling		2.20	7.273		520	272		792	1,025
0600	Horizontal									
0630	2 ton cooling	Q-5	3.90	4.103	Ea.	410	153		563	705
0640	3 ton cooling		3.50	4.571		445	171		616	770
0650	4 ton cooling		3.20	5		490	187		677	845
0660	5 ton cooling		2.90	5.517		490	206		696	880
2000	Cased evaporator coils for air handlers									
2100	1-1/2 ton cooling	Q-5	4.40	3.636	Ea.	315	136		451	575
2110	2 ton cooling		4.10	3.902		365	146		511	640
2120	2-1/2 ton cooling		3.90	4.103		385	153		538	675
2130	3 ton cooling		3.70	4.324		430	162		592	740
2140	3-1/2 ton cooling		3.50	4.571		540	171		711	870
2150	4 ton cooling		3.20	5		650	187		837	1,025
2160	5 ton cooling		2.90	5.517		640	206		846	1,050
3010	Air handler, modular									
3100	With cased evaporator cooling coil									
3120	1-1/2 ton cooling	Q-5	3.80	4.211	Ea.	985	157		1,142	1,325
3130	2 ton cooling		3.50	4.571		1,025	171		1,196	1,400
3140	2-1/2 ton cooling		3.30	4.848		1,100	181		1,281	1,500
3150	3 ton cooling		3.10	5.161		1,250	193		1,443	1,700
3160	3-1/2 ton cooling		2.90	5.517		1,500	206		1,706	2,000
3170	4 ton cooling		2.50	6.400		1,550	239		1,789	2,125
3180	5 ton cooling		2.10	7.619		1,850	285		2,135	2,525
3500	With no cooling coil									
3520	1-1/2 ton coil size	Q-5	12	1.333	Ea.	1,175	50		1,225	1,375
3530	2 ton coil size		10	1.600		1,275	60		1,335	1,500
3540	2-1/2 ton coil size		10	1.600		1,300	60		1,360	1,525
3554	3 ton coil size		9	1.778		1,275	66.50		1,341.50	1,525
3560	3-1/2 ton coil size		9	1.778		1,325	66.50		1,391.50	1,575
3570	4 ton coil size		8.50	1.882		1,400	70.50		1,470.50	1,675
3580	5 ton coil size		8	2		1,525	75		1,600	1,800
4000	With heater									
4120	5 kW, 17.1 MBH	Q-5	16	1	Ea.	470	37.50		507.50	580
4130	7.5 kW, 25.6 MBH		15.60	1.026		775	38.50		813.50	915
4140	10 kW, 34.2 MBH		15.20	1.053		1,225	39.50		1,264.50	1,425
4150	12.5 kW, 42.7 MBH		14.80	1.081		1,625	40.50		1,665.50	1,850
4160	15 kW, 51.2 MBH		14.40	1.111		1,775	41.50		1,816.50	2,025
4170	25 kW, 85.4 MBH		14	1.143		2,400	42.50		2,442.50	2,700
4180	30 kW, 102 MBH		13	1.231		2,775	46		2,821	3,125

23 62 Packaged Compressor and Condenser Units

23 62 13 – Packaged Air-Cooled Refrigerant Compressor and Condenser Units

23 62 13.10 Packaged Air-Cooled Refrig. Condensing Units		Crew	Daily Output	Labor-Hours	Unit	2019 Bare Costs Material	Labor	Equipment	Total	Total Incl O&P
0010	PACKAGED AIR-COOLED REFRIGERANT CONDENSING UNITS									
0020	Condensing unit									
0030	Air cooled, compressor, standard controls									
0050	1.5 ton	Q-5	2.50	6.400	Ea.	1,075	239		1,314	1,575
0100	2 ton		2.10	7.619		1,050	285		1,335	1,625
0200	2.5 ton		1.70	9.412		1,250	350		1,600	1,950
0300	3 ton		1.30	12.308		1,325	460		1,785	2,200
0350	3.5 ton		1.10	14.545		1,550	545		2,095	2,600
0400	4 ton		.90	17.778		1,775	665		2,440	3,050
0500	5 ton	↓	.60	26.667	↓	2,125	995		3,120	3,975

23 74 Packaged Outdoor HVAC Equipment

23 74 33 – Dedicated Outdoor-Air Units

23 74 33.10 Rooftop Air Conditioners		Crew	Daily Output	Labor-Hours	Unit	Material	Labor	Equipment	Total	Total Incl O&P
0010	**ROOFTOP AIR CONDITIONERS**, Standard controls, curb, economizer									
1000	Single zone, electric cool, gas heat									
1140	5 ton cooling, 112 MBH heating	Q-5	.56	28.521	Ea.	4,425	1,075		5,500	6,600
1150	7.5 ton cooling, 170 MBH heating		.50	32.258		5,825	1,200		7,025	8,400
1156	8.5 ton cooling, 170 MBH heating	↓	.46	34.783		7,100	1,300		8,400	9,925
1160	10 ton cooling, 200 MBH heating	Q-6	.67	35.982	↓	9,200	1,300		10,500	12,200

23 81 Decentralized Unitary HVAC Equipment

23 81 13 – Packaged Terminal Air-Conditioners

23 81 13.10 Packaged Cabinet Type Air-Conditioners		Crew	Daily Output	Labor-Hours	Unit	Material	Labor	Equipment	Total	Total Incl O&P
0010	**PACKAGED CABINET TYPE AIR-CONDITIONERS**, Cabinet, wall sleeve,									
0100	louver, electric heat, thermostat, manual changeover, 208 V									
0200	6,000 BTUH cooling, 8,800 BTU heat	Q-5	6	2.667	Ea.	705	99.50		804.50	940
0220	9,000 BTUH cooling, 13,900 BTU heat		5	3.200		970	120		1,090	1,275
0240	12,000 BTUH cooling, 13,900 BTU heat		4	4		1,475	150		1,625	1,875
0260	15,000 BTUH cooling, 13,900 BTU heat	↓	3	5.333	↓	1,450	199		1,649	1,925

23 81 19 – Self-Contained Air-Conditioners

23 81 19.10 Window Unit Air Conditioners		Crew	Daily Output	Labor-Hours	Unit	Material	Labor	Equipment	Total	Total Incl O&P
0010	**WINDOW UNIT AIR CONDITIONERS**									
4000	Portable/window, 15 amp, 125 V grounded receptacle required									
4060	5,000 BTUH	1 Carp	8	1	Ea.	305	35.50		340.50	395
4340	6,000 BTUH		8	1		252	35.50		287.50	335
4480	8,000 BTUH		6	1.333		355	47.50		402.50	470
4500	10,000 BTUH	↓	6	1.333		600	47.50		647.50	740
4520	12,000 BTUH	L-2	8	2	↓	1,950	63		2,013	2,250
4600	Window/thru-the-wall, 15 amp, 230 V grounded receptacle required									
4780	18,000 BTUH	L-2	6	2.667	Ea.	680	84		764	890
4940	25,000 BTUH		4	4		930	126		1,056	1,225
4960	29,000 BTUH	↓	4	4	↓	955	126		1,081	1,250

23 81 Decentralized Unitary HVAC Equipment

23 81 19 – Self-Contained Air-Conditioners

23 81 19.20 Self-Contained Single Package		Crew	Daily Output	Labor-Hours	Unit	Material	2019 Bare Costs Labor	Equipment	Total	Total Incl O&P
0010	**SELF-CONTAINED SINGLE PACKAGE**									
0100	Air cooled, for free blow or duct, not incl. remote condenser									
0110	Constant volume									
0200	3 ton cooling	Q-5	1	16	Ea.	3,825	600		4,425	5,175
0210	4 ton cooling	"	.80	20	"	4,175	750		4,925	5,800
1000	Water cooled for free blow or duct, not including tower									
1010	Constant volume									
1100	3 ton cooling	Q-6	1	24	Ea.	3,775	865		4,640	5,575
1300	For hot water or steam heat coils, add				"	12%	10%			

23 81 26 – Split-System Air-Conditioners

23 81 26.10 Split Ductless Systems		Crew	Daily Output	Labor-Hours	Unit	Material	Labor	Equipment	Total	Total Incl O&P
0010	**SPLIT DUCTLESS SYSTEMS**									
0100	Cooling only, single zone									
0110	Wall mount									
0120	3/4 ton cooling	Q-5	2	8	Ea.	1,100	299		1,399	1,725
0130	1 ton cooling		1.80	8.889		1,225	330		1,555	1,900
0140	1-1/2 ton cooling		1.60	10		1,975	375		2,350	2,800
0150	2 ton cooling	↓	1.40	11.429	↓	2,250	425		2,675	3,175
1000	Ceiling mount									
1020	2 ton cooling	Q-5	1.40	11.429	Ea.	2,000	425		2,425	2,900
1030	3 ton cooling	"	1.20	13.333	"	2,600	500		3,100	3,675
3000	Multizone									
3010	Wall mount									
3020	2 @ 3/4 ton cooling	Q-5	1.80	8.889	Ea.	3,475	330		3,805	4,375
5000	Cooling/Heating									
5010	Wall mount									
5110	1 ton cooling	Q-5	1.70	9.412	Ea.	1,300	350		1,650	2,025
5120	1-1/2 ton cooling	"	1.50	10.667	"	2,000	400		2,400	2,850
7000	Accessories for all split ductless systems									
7010	Add for ambient frost control	Q-5	8	2	Ea.	127	75		202	263
7020	Add for tube/wiring kit (line sets)									
7030	15' kit	Q-5	32	.500	Ea.	89	18.70		107.70	129
7036	25' kit		28	.571		138	21.50		159.50	186
7040	35' kit		24	.667		200	25		225	261
7050	50' kit	↓	20	.800	↓	203	30		233	272

23 81 43 – Air-Source Unitary Heat Pumps

23 81 43.10 Air-Source Heat Pumps		Crew	Daily Output	Labor-Hours	Unit	Material	Labor	Equipment	Total	Total Incl O&P
0010	**AIR-SOURCE HEAT PUMPS**, Not including interconnecting tubing									
1000	Air to air, split system, not including curbs, pads, fan coil and ductwork									
1012	Outside condensing unit only, for fan coil see Section 23 82 19.10									
1020	2 ton cooling, 8.5 MBH heat @ 0°F	Q-5	2	8	Ea.	1,775	299		2,074	2,450
1054	4 ton cooling, 24 MBH heat @ 0°F	"	.80	20	"	2,475	750		3,225	3,950
1500	Single package, not including curbs, pads, or plenums									
1520	2 ton cooling, 6.5 MBH heat @ 0°F	Q-5	1.50	10.667	Ea.	3,300	400		3,700	4,275
1580	4 ton cooling, 13 MBH heat @ 0°F	"	.96	16.667	"	4,375	625		5,000	5,850

23 81 Decentralized Unitary HVAC Equipment

23 81 46 – Water-Source Unitary Heat Pumps

23 81 46.10 Water Source Heat Pumps		Crew	Daily Output	Labor-Hours	Unit	2019 Bare Costs Material	Labor	Equipment	Total	Total Incl O&P
0010	**WATER SOURCE HEAT PUMPS**, Not incl. connecting tubing or water source									
2000	Water source to air, single package									
2100	1 ton cooling, 13 MBH heat @ 75°F	Q-5	2	8	Ea.	1,925	299		2,224	2,600
2200	4 ton cooling, 31 MBH heat @ 75°F	"	1.20	13.333	"	3,075	500		3,575	4,225

23 82 Convection Heating and Cooling Units

23 82 19 – Fan Coil Units

23 82 19.10 Fan Coil Air Conditioning		Crew	Daily Output	Labor-Hours	Unit	Material	Labor	Equipment	Total	Total Incl O&P
0010	**FAN COIL AIR CONDITIONING**									
0030	Fan coil AC, cabinet mounted, filters and controls									
0100	Chilled water, 1/2 ton cooling	Q-5	8	2	Ea.	585	75		660	770
0110	3/4 ton cooling	↓	7	2.286	↓	730	85.50		815.50	940
0120	1 ton cooling	↓	6	2.667	↓	855	99.50		954.50	1,100

23 82 29 – Radiators

23 82 29.10 Hydronic Heating		Crew	Daily Output	Labor-Hours	Unit	Material	Labor	Equipment	Total	Total Incl O&P
0010	**HYDRONIC HEATING**, Terminal units, not incl. main supply pipe									
1000	Radiation									
1100	Panel, baseboard, C.I., including supports, no covers	Q-5	46	.348	L.F.	46	13		59	72
3000	Radiators, cast iron									
3100	Free standing or wall hung, 6 tube, 25" high	Q-5	96	.167	Section	56.50	6.25		62.75	72.50
3200	4 tube, 19" high	"	96	.167	"	41.50	6.25		47.75	56
3250	Adj. brackets, 2 per wall radiator up to 30 sections	1 Stpi	32	.250	Ea.	65	10.40		75.40	88.50

23 82 36 – Finned-Tube Radiation Heaters

23 82 36.10 Finned Tube Radiation		Crew	Daily Output	Labor-Hours	Unit	Material	Labor	Equipment	Total	Total Incl O&P
0010	**FINNED TUBE RADIATION**, Terminal units, not incl. main supply pipe									
1310	Baseboard, pkgd, 1/2" copper tube, alum. fin, 7" high	Q-5	60	.267	L.F.	11.75	9.95		21.70	29.50
1320	3/4" copper tube, alum. fin, 7" high	↓	58	.276	↓	8.10	10.30		18.40	26
1340	1" copper tube, alum. fin, 8-7/8" high	↓	56	.286	↓	21.50	10.70		32.20	41
1360	1-1/4" copper tube, alum. fin, 8-7/8" high	↓	54	.296	↓	32	11.10		43.10	53
1500	Note: fin tube may also require corners, caps, etc.									

23 83 Radiant Heating Units

23 83 16 – Radiant-Heating Hydronic Piping

23 83 16.10 Radiant Floor Heating		Crew	Daily Output	Labor-Hours	Unit	Material	Labor	Equipment	Total	Total Incl O&P
0010	**RADIANT FLOOR HEATING**									
0100	Tubing, PEX (cross-linked polyethylene)									
0110	Oxygen barrier type for systems with ferrous materials									
0120	1/2"	Q-5	800	.020	L.F.	1.06	.75		1.81	2.40
0130	3/4"	↓	535	.030	↓	1.65	1.12		2.77	3.66
0140	1"	↓	400	.040	↓	2.32	1.50		3.82	5
0200	Non barrier type for ferrous free systems									
0210	1/2"	Q-5	800	.020	L.F.	.53	.75		1.28	1.81
0220	3/4"	↓	535	.030	↓	.96	1.12		2.08	2.90
0230	1"	↓	400	.040	↓	1.72	1.50		3.22	4.34
1000	Manifolds									
1110	Brass									
1120	With supply and return valves, flow meter, thermometer,									
1122	auto air vent and drain/fill valve.									

23 83 16.10 Radiant Floor Heating		Crew	Daily Output	Labor-Hours	Unit	2019 Bare Costs Material	Labor	Equipment	Total	Total Incl O&P
1130	1", 2 circuit	Q-5	14	1.143	Ea.	350	42.50		392.50	455
1140	1", 3 circuit		13.50	1.185		385	44.50		429.50	500
1150	1", 4 circuit		13	1.231		405	46		451	520
1154	1", 5 circuit		12.50	1.280		485	48		533	615
1158	1", 6 circuit		12	1.333		590	50		640	725
1162	1", 7 circuit		11.50	1.391		630	52		682	780
1166	1", 8 circuit		11	1.455		640	54.50		694.50	795
1172	1", 9 circuit		10.50	1.524		750	57		807	920
1174	1", 10 circuit		10	1.600		765	60		825	940
1178	1", 11 circuit		9.50	1.684		775	63		838	960
1182	1", 12 circuit		9	1.778		905	66.50		971.50	1,100
1610	Copper manifold header (cut to size)									
1620	1" header, 12 circuit 1/2" sweat outlets	Q-5	3.33	4.805	Ea.	110	180		290	415
1630	1-1/4" header, 12 circuit 1/2" sweat outlets		3.20	5		127	187		314	445
1640	1-1/4" header, 12 circuit 3/4" sweat outlets		3	5.333		129	199		328	465
1650	1-1/2" header, 12 circuit 3/4" sweat outlets		3.10	5.161		157	193		350	490
1660	2" header, 12 circuit 3/4" sweat outlets		2.90	5.517		235	206		441	600
3000	Valves									
3110	Thermostatic zone valve actuator with end switch	Q-5	40	.400	Ea.	49	14.95		63.95	78
3114	Thermostatic zone valve actuator	"	36	.444	"	91	16.60		107.60	128
3120	Motorized straight zone valve with operator complete									
3130	3/4"	Q-5	35	.457	Ea.	150	17.10		167.10	194
3140	1"		32	.500		163	18.70		181.70	211
3150	1-1/4"		29.60	.541		205	20		225	258
3500	4 way mixing valve, manual, brass									
3530	1"	Q-5	13.30	1.203	Ea.	218	45		263	315
3540	1-1/4"		11.40	1.404		237	52.50		289.50	345
3550	1-1/2"		11	1.455		292	54.50		346.50	410
3560	2"		10.60	1.509		430	56.50		486.50	565
3800	Mixing valve motor, 4 way for valves, 1" and 1-1/4"		34	.471		385	17.60		402.60	455
3810	Mixing valve motor, 4 way for valves, 1-1/2" and 2"		30	.533		410	19.95		429.95	485
5000	Radiant floor heating, zone control panel									
5120	4 zone actuator valve control, expandable	Q-5	20	.800	Ea.	167	30		197	233
5130	6 zone actuator valve control, expandable		18	.889		252	33		285	330
6070	Thermal track, straight panel for long continuous runs, 5.333 S.F.		40	.400		34	14.95		48.95	62
6080	Thermal track, utility panel, for direction reverse at run end, 5.333 S.F.		40	.400		34	14.95		48.95	62
6090	Combination panel, for direction reverse plus straight run, 5.333 S.F.		40	.400		34	14.95		48.95	62
7000	PEX tubing fittings									
7100	Compression type									
7116	Coupling									
7120	1/2" x 1/2"	1 Stpi	27	.296	Ea.	6.70	12.30		19	27.50
7124	3/4" x 3/4"	"	23	.348	"	15.05	14.45		29.50	40
7130	Adapter									
7132	1/2" x female sweat 1/2"	1 Stpi	27	.296	Ea.	4.35	12.30		16.65	25
7134	1/2" x female sweat 3/4"		26	.308		4.86	12.80		17.66	26.50
7136	5/8" x female sweat 3/4"		24	.333		6.95	13.85		20.80	30
7140	Elbow									
7142	1/2" x female sweat 1/2"	1 Stpi	27	.296	Ea.	7.85	12.30		20.15	28.50
7144	1/2" x female sweat 3/4"		26	.308		9.25	12.80		22.05	31
7146	5/8" x female sweat 3/4"		24	.333		10.40	13.85		24.25	34
7200	Insert type									
7206	PEX x male NPT									
7210	1/2" x 1/2"	1 Stpi	29	.276	Ea.	2.80	11.45		14.25	22

23 83 Radiant Heating Units

23 83 16 – Radiant-Heating Hydronic Piping

23 83 16.10 Radiant Floor Heating		Crew	Daily Output	Labor-Hours	Unit	Material	Labor	Equipment	Total	Total Incl O&P
						2019 Bare Costs				
7220	3/4" x 3/4"	1 Stpi	27	.296	Ea.	4.10	12.30		16.40	24.50
7230	1" x 1"	↓	26	.308	↓	6.85	12.80		19.65	28.50
7300	PEX coupling									
7310	1/2" x 1/2"	1 Stpi	30	.267	Ea.	.60	11.10		11.70	18.80
7320	3/4" x 3/4"		29	.276		.79	11.45		12.24	19.65
7330	1" x 1"	↓	28	.286	↓	1.37	11.85		13.22	21
7400	PEX stainless crimp ring									
7410	1/2" x 1/2"	1 Stpi	86	.093	Ea.	.52	3.87		4.39	6.90
7420	3/4" x 3/4"		84	.095		.71	3.96		4.67	7.30
7430	1" x 1"	↓	82	.098	↓	1.05	4.05		5.10	7.80

23 83 33 – Electric Radiant Heaters

23 83 33.10 Electric Heating		Crew	Daily Output	Labor-Hours	Unit	Material	Labor	Equipment	Total	Total Incl O&P
0010	**ELECTRIC HEATING**, not incl. conduit or feed wiring									
1100	Rule of thumb: Baseboard units, including control	1 Elec	4.40	1.818	kW	112	75.50		187.50	246
1300	Baseboard heaters, 2' long, 350 watt		8	1	Ea.	27.50	41.50		69	98
1400	3' long, 750 watt		8	1		32	41.50		73.50	103
1600	4' long, 1,000 watt		6.70	1.194		39	49.50		88.50	124
1800	5' long, 935 watt		5.70	1.404		46.50	58		104.50	146
2000	6' long, 1,500 watt		5	1.600		51.50	66		117.50	165
2400	8' long, 2,000 watt		4	2		65.50	83		148.50	207
2800	10' long, 1,875 watt	↓	3.30	2.424	↓	154	100		254	335
2950	Wall heaters with fan, 120 to 277 volt									
3170	1,000 watt	1 Elec	6	1.333	Ea.	94.50	55		149.50	194
3180	1,250 watt		5	1.600		102	66		168	220
3190	1,500 watt		4	2		102	83		185	247
3600	Thermostats, integral		16	.500		28	20.50		48.50	64.50
3800	Line voltage, 1 pole		8	1		17.15	41.50		58.65	86.50
5000	Radiant heating ceiling panels, 2' x 4', 500 watt		16	.500		395	20.50		415.50	470
5050	750 watt		16	.500		500	20.50		520.50	585
5300	Infrared quartz heaters, 120 volts, 1,000 watt		6.70	1.194		345	49.50		394.50	460
5350	1,500 watt		5	1.600		345	66		411	490
5400	240 volts, 1,500 watt		5	1.600		395	66		461	545
5450	2,000 watt		4	2		345	83		428	515
5500	3,000 watt	↓	3	2.667	↓	390	110		500	610

Estimating Tips

26 05 00 Common Work Results for Electrical

- Conduit should be taken off in three main categories—power distribution, branch power, and branch lighting—so the estimator can concentrate on systems and components, therefore making it easier to ensure all items have been accounted for.
- For cost modifications for elevated conduit installation, add the percentages to labor according to the height of installation and only to the quantities exceeding the different height levels, not to the total conduit quantities. Refer to subdivision 26 01 02.20 for labor adjustment factors.
- Remember that aluminum wiring of equal ampacity is larger in diameter than copper and may require larger conduit.
- If more than three wires at a time are being pulled, deduct percentages from the labor hours of that grouping of wires.
- When taking off grounding systems, identify separately the type and size of wire, and list each unique type of ground connection.
- The estimator should take the weights of materials into consideration when completing a takeoff. Topics to consider include: How will the materials be supported? What methods of support are available? How high will the support structure have to reach? Will the final support structure be able to withstand the total burden? Is the support material included or separate from the fixture, equipment, and material specified?
- Do not overlook the costs for equipment used in the installation. If scaffolding or highlifts are available in the field, contractors may use them in lieu of the proposed ladders and rolling staging.

26 20 00 Low-Voltage Electrical Transmission

- Supports and concrete pads may be shown on drawings for the larger equipment, or the support system may be only a piece of plywood for the back of a panelboard. In either case, they must be included in the costs.

26 40 00 Electrical and Cathodic Protection

- When taking off cathodic protection systems, identify the type and size of cable, and list each unique type of anode connection.

26 50 00 Lighting

- Fixtures should be taken off room by room using the fixture schedule, specifications, and the ceiling plan. For large concentrations of lighting fixtures in the same area, deduct the percentages from labor hours.

Reference Numbers

Reference numbers are shown at the beginning of some major classifications. These numbers refer to related items in the Reference Section. The reference information may be an estimating procedure, an alternate pricing method, or technical information.

Note: Not all subdivisions listed here necessarily appear. ■

Note: "Powered in part by CINX™, based on liscensed proprietary information of Harrison Publishing House, Inc."

Note: The following companies, in part, have been used as a reference source for some of the material prices used in Division 26:

Electriflex

Trade Service

26 05 05 – Selective Demolition for Electrical

26 05 05.10 Electrical Demolition		Crew	Daily Output	Labor-Hours	Unit	Material	2019 Bare Costs Labor	Equipment	Total	Total Incl O&P
0010	**ELECTRICAL DEMOLITION**									
0020	Electrical demolition, conduit to 10′ high, incl. fittings & hangers									
0100	Rigid galvanized steel, 1/2″ to 1″ diameter	1 Elec	242	.033	L.F.		1.37		1.37	2.23
0120	1-1/4″ to 2″	″	200	.040	″		1.66		1.66	2.70
0270	Armored cable (BX) avg. 50′ runs									
0280	#14, 2 wire	1 Elec	690	.012	L.F.		.48		.48	.78
0290	#14, 3 wire		571	.014			.58		.58	.95
0300	#12, 2 wire		605	.013			.55		.55	.89
0310	#12, 3 wire		514	.016			.64		.64	1.05
0320	#10, 2 wire		514	.016			.64		.64	1.05
0330	#10, 3 wire		425	.019			.78		.78	1.27
0340	#8, 3 wire		342	.023			.97		.97	1.58
0350	Non metallic sheathed cable (Romex)									
0360	#14, 2 wire	1 Elec	720	.011	L.F.		.46		.46	.75
0370	#14, 3 wire		657	.012			.50		.50	.82
0380	#12, 2 wire		629	.013			.53		.53	.86
0390	#10, 3 wire		450	.018			.74		.74	1.20
0400	Wiremold raceway, including fittings & hangers									
0420	No. 3000	1 Elec	250	.032	L.F.		1.32		1.32	2.16
0440	No. 4000		217	.037			1.53		1.53	2.49
0460	No. 6000		166	.048			2		2	3.26
0462	Plugmold with receptacle		114	.070			2.91		2.91	4.74
0465	Telephone/power pole		12	.667	Ea.		27.50		27.50	45
0470	Non-metallic, straight section		480	.017	L.F.		.69		.69	1.13
0500	Channels, steel, including fittings & hangers									
0520	3/4″ x 1-1/2″	1 Elec	308	.026	L.F.		1.08		1.08	1.75
0540	1-1/2″ x 1-1/2″		269	.030			1.23		1.23	2.01
0560	1-1/2″ x 1-7/8″		229	.035			1.45		1.45	2.36
1210	Panel boards, incl. removal of all breakers,									
1220	conduit terminations & wire connections									
1230	3 wire, 120/240 V, 100A, to 20 circuits	1 Elec	2.60	3.077	Ea.		127		127	208
1240	200 amps, to 42 circuits	2 Elec	2.60	6.154			255		255	415
1241	225 amps, to 42 circuits		2.40	6.667			276		276	450
1250	400 amps, to 42 circuits		2.20	7.273			300		300	490
1260	4 wire, 120/208 V, 125A, to 20 circuits	1 Elec	2.40	3.333			138		138	225
1270	200 amps, to 42 circuits	2 Elec	2.40	6.667			276		276	450
1720	Junction boxes, 4″ sq. & oct.	1 Elec	80	.100			4.14		4.14	6.75
1740	Handy box		107	.075			3.10		3.10	5.05
1760	Switch box		107	.075			3.10		3.10	5.05
1780	Receptacle & switch plates		257	.031			1.29		1.29	2.10
1800	Wire, THW-THWN-THHN, removed from									
1810	in place conduit, to 10′ high									
1830	#14	1 Elec	65	.123	C.L.F.		5.10		5.10	8.30
1840	#12		55	.145			6		6	9.85
1850	#10		45.50	.176			7.30		7.30	11.90
2000	Interior fluorescent fixtures, incl. supports									
2010	& whips, to 10′ high									
2100	Recessed drop-in 2′ x 2′, 2 lamp	2 Elec	35	.457	Ea.		18.95		18.95	31
2110	2′ x 2′, 4 lamp		30	.533			22		22	36
2140	2′ x 4′, 4 lamp		30	.533			22		22	36
2180	Surface mount, acrylic lens & hinged frame									
2220	2′ x 2′, 2 lamp	2 Elec	44	.364	Ea.		15.05		15.05	24.50
2260	2′ x 4′, 4 lamp	″	33	.485	″		20		20	33

26 05 Common Work Results for Electrical

26 05 05 – Selective Demolition for Electrical

26 05 05.10 Electrical Demolition

		Crew	Daily Output	Labor-Hours	Unit	2019 Bare Costs Material	Labor	Equipment	Total	Total Incl O&P
2300	Strip fixtures, surface mount									
2320	4′ long, 1 lamp	2 Elec	53	.302	Ea.		12.50		12.50	20.50
2380	8′ long, 2 lamp	"	40	.400	"		16.55		16.55	27
2460	Interior incandescent, surface, ceiling									
2470	or wall mount, to 12′ high									
2480	Metal cylinder type, 75 Watt	2 Elec	62	.258	Ea.		10.70		10.70	17.45
2600	Exterior fixtures, incandescent, wall mount									
2620	100 Watt	2 Elec	50	.320	Ea.		13.25		13.25	21.50
3000	Ceiling fan, tear out and remove	1 Elec	24	.333	"		13.80		13.80	22.50

26 05 19 – Low-Voltage Electrical Power Conductors and Cables

26 05 19.20 Armored Cable

		Crew	Daily Output	Labor-Hours	Unit	Material	Labor	Equipment	Total	Total Incl O&P
0010	**ARMORED CABLE**									
0051	600 volt, copper (BX), #14, 2 conductor, solid	1 Elec	240	.033	L.F.	.41	1.38		1.79	2.70
0101	3 conductor, solid		200	.040		.63	1.66		2.29	3.40
0151	#12, 2 conductor, solid		210	.038		.41	1.58		1.99	3.02
0201	3 conductor, solid		180	.044		.70	1.84		2.54	3.77
0251	#10, 2 conductor, solid		180	.044		.79	1.84		2.63	3.87
0301	3 conductor, solid		150	.053		1.08	2.21		3.29	4.79
0351	3 conductor, stranded		120	.067		2.40	2.76		5.16	7.15

26 05 19.55 Non-Metallic Sheathed Cable

		Crew	Daily Output	Labor-Hours	Unit	Material	Labor	Equipment	Total	Total Incl O&P
0010	**NON-METALLIC SHEATHED CABLE** 600 volt									
0100	Copper with ground wire (Romex)									
0151	#14, 2 wire	1 Elec	250	.032	L.F.	.18	1.32		1.50	2.36
0201	3 wire		230	.035		.25	1.44		1.69	2.63
0251	#12, 2 wire		220	.036		.24	1.51		1.75	2.73
0301	3 wire		200	.040		.40	1.66		2.06	3.14
0351	#10, 2 wire		200	.040		.41	1.66		2.07	3.15
0401	3 wire		140	.057		.62	2.37		2.99	4.54
0451	#8, 3 conductor		130	.062		1.16	2.55		3.71	5.45
0501	#6, 3 wire		120	.067		1.95	2.76		4.71	6.65
0550	SE type SER aluminum cable, 3 RHW and									
0601	1 bare neutral, 3 #8 & 1 #8	1 Elec	150	.053	L.F.	.56	2.21		2.77	4.21
0651	3 #6 & 1 #6	"	130	.062		.66	2.55		3.21	4.89
0701	3 #4 & 1 #6	2 Elec	220	.073		.70	3.01		3.71	5.70
0751	3 #2 & 1 #4		200	.080		1.22	3.31		4.53	6.75
0801	3 #1/0 & 1 #2		180	.089		1.79	3.68		5.47	7.95
0851	3 #2/0 & 1 #1		160	.100		2.08	4.14		6.22	9.05
0901	3 #4/0 & 1 #2/0		140	.114		2.90	4.73		7.63	10.90
2401	SEU service entrance cable, copper 2 conductors, #8 + #8 neutral	1 Elec	150	.053		1.03	2.21		3.24	4.73
2601	#6 + #8 neutral		130	.062		1.32	2.55		3.87	5.60
2801	#6 + #6 neutral		130	.062		1.54	2.55		4.09	5.85
3001	#4 + #6 neutral	2 Elec	220	.073		2.02	3.01		5.03	7.15
3201	#4 + #4 neutral		220	.073		2.23	3.01		5.24	7.35
3401	#3 + #5 neutral		210	.076		3.34	3.15		6.49	8.85
6500	Service entrance cap for copper SEU									
6600	100 amp	1 Elec	12	.667	Ea.	7.95	27.50		35.45	54
6700	150 amp		10	.800		10.70	33		43.70	66
6800	200 amp		8	1		17.25	41.50		58.75	86.50

26 05 19.90 Wire

		Crew	Daily Output	Labor-Hours	Unit	Material	Labor	Equipment	Total	Total Incl O&P
0010	**WIRE**, normal installation conditions in wireway, conduit, cable tray									
0021	600 volt, copper type THW, solid, #14	1 Elec	1300	.006	L.F.	.06	.25		.31	.48
0031	#12		1100	.007		.10	.30		.40	.60

26 05 19 – Low-Voltage Electrical Power Conductors and Cables

26 05 19.90 Wire		Crew	Daily Output	Labor-Hours	Unit	2019 Bare Costs Material	Labor	Equipment	Total	Total Incl O&P
0041	#10	1 Elec	1000	.008	L.F.	.16	.33		.49	.71
0050	Stranded, #14		13	.615	C.L.F.	7.60	25.50		33.10	50
0100	#12		11	.727		13.65	30		43.65	64
0120	#10		10	.800		21.50	33		54.50	77.50
0161	#6		650	.012	L.F.	.43	.51		.94	1.30
0181	#4	2 Elec	1060	.015		.79	.62		1.41	1.89
0201	#3		1000	.016		1.12	.66		1.78	2.31
0221	#2		900	.018		1	.74		1.74	2.30
0241	#1		800	.020		1.33	.83		2.16	2.82
0261	1/0		660	.024		2.12	1		3.12	3.97
0281	2/0		580	.028		2.01	1.14		3.15	4.08
0301	3/0		500	.032		2.30	1.32		3.62	4.69
0351	4/0		440	.036		4.26	1.51		5.77	7.15

26 05 26 – Grounding and Bonding for Electrical Systems

26 05 26.80 Grounding		Crew	Daily Output	Labor-Hours	Unit	Material	Labor	Equipment	Total	Total Incl O&P
0010	**GROUNDING**									
0030	Rod, copper clad, 8′ long, 1/2″ diameter	1 Elec	5.50	1.455	Ea.	19.75	60		79.75	120
0040	5/8″ diameter		5.50	1.455		22	60		82	123
0050	3/4″ diameter		5.30	1.509		35.50	62.50		98	141
0080	10′ long, 1/2″ diameter		4.80	1.667		22.50	69		91.50	138
0090	5/8″ diameter		4.60	1.739		23.50	72		95.50	143
0100	3/4″ diameter		4.40	1.818		49.50	75.50		125	177
0130	15′ long, 3/4″ diameter		4	2		49.50	83		132.50	190
0261	Wire, ground bare armored, #8-1 conductor		200	.040	L.F.	.69	1.66		2.35	3.46
0271	#6-1 conductor		180	.044	″	.84	1.84		2.68	3.92
0390	Bare copper wire, stranded, #8		11	.727	C.L.F.	43	30		73	96.50
0401	Bare copper, #6 wire		1000	.008	L.F.	.41	.33		.74	.99
0601	#2 stranded	2 Elec	1000	.016	″	.89	.66		1.55	2.06
1800	Water pipe ground clamps, heavy duty									
2000	Bronze, 1/2″ to 1″ diameter	1 Elec	8	1	Ea.	32	41.50		73.50	103
2100	1-1/4″ to 2″ diameter		8	1		29	41.50		70.50	99.50
2200	2-1/2″ to 3″ diameter		6	1.333		69.50	55		124.50	167

26 05 33 – Raceway and Boxes for Electrical Systems

26 05 33.13 Conduit		Crew	Daily Output	Labor-Hours	Unit	Material	Labor	Equipment	Total	Total Incl O&P
0010	**CONDUIT** To 10′ high, includes 2 terminations, 2 elbows,									
0020	11 beam clamps, and 11 couplings per 100 L.F.									
1750	Rigid galvanized steel, 1/2″ diameter	1 Elec	90	.089	L.F.	2.30	3.68		5.98	8.55
1770	3/4″ diameter		80	.100		4.74	4.14		8.88	11.95
1800	1″ diameter		65	.123		7.05	5.10		12.15	16.05
1830	1-1/4″ diameter		60	.133		4.82	5.50		10.32	14.30
1850	1-1/2″ diameter		55	.145		7.75	6		13.75	18.40
1870	2″ diameter		45	.178		9.95	7.35		17.30	23
5000	Electric metallic tubing (EMT), 1/2″ diameter		170	.047		.81	1.95		2.76	4.07
5020	3/4″ diameter		130	.062		1.18	2.55		3.73	5.45
5040	1″ diameter		115	.070		1.90	2.88		4.78	6.80
5060	1-1/4″ diameter		100	.080		3.03	3.31		6.34	8.75
5080	1-1/2″ diameter		90	.089		3.48	3.68		7.16	9.80
9100	PVC, schedule 40, 1/2″ diameter		190	.042		.93	1.74		2.67	3.86
9110	3/4″ diameter		145	.055		1.06	2.28		3.34	4.90
9120	1″ diameter		125	.064		1.58	2.65		4.23	6.05
9130	1-1/4″ diameter		110	.073		1.77	3.01		4.78	6.85
9140	1-1/2″ diameter		100	.080		2.08	3.31		5.39	7.70

26 05 33 – Raceway and Boxes for Electrical Systems

26 05 33.13 Conduit		Crew	Daily Output	Labor-Hours	Unit	Material	Labor	Equipment	Total	Total Incl O&P
						2019 Bare Costs				
9150	2" diameter	1 Elec	90	.089	L.F.	2.85	3.68		6.53	9.15
9995	Do not include labor when adding couplings to a fitting installation									
26 05 33.16 Boxes for Electrical Systems										
0010	**BOXES FOR ELECTRICAL SYSTEMS**									
0021	Pressed steel, octagon, 4"	1 Elec	18	.444	Ea.	3.46	18.40		21.86	34
0060	Covers, blank		64	.125		.87	5.20		6.07	9.40
0100	Extension rings		40	.200		4.46	8.30		12.76	18.40
0151	Square 4"		18	.444		5.20	18.40		23.60	35.50
0200	Extension rings		40	.200		4.59	8.30		12.89	18.55
0250	Covers, blank		64	.125		.81	5.20		6.01	9.35
0300	Plaster rings		64	.125		1.53	5.20		6.73	10.15
0651	Switchbox		24	.333		5.60	13.80		19.40	28.50
1100	Concrete, floor, 1 gang	↓	5.30	1.509	↓	123	62.50		185.50	238
26 05 33.17 Outlet Boxes, Plastic										
0010	**OUTLET BOXES, PLASTIC**									
0051	4" diameter, round, with 2 mounting nails	1 Elec	23	.348	Ea.	3.09	14.40		17.49	27
0101	Bar hanger mounted		23	.348		5.35	14.40		19.75	29.50
0201	Square with 2 mounting nails		23	.348		5.30	14.40		19.70	29.50
0300	Plaster ring		64	.125		2.02	5.20		7.22	10.65
0401	Switch box with 2 mounting nails, 1 gang		27	.296		3.10	12.25		15.35	23.50
0501	2 gang		23	.348		4.21	14.40		18.61	28
0601	3 gang	↓	18	.444	↓	5	18.40		23.40	35.50
26 05 33.18 Pull Boxes										
0010	**PULL BOXES**									
0100	Steel, pull box, NEMA 1, type SC, 6" W x 6" H x 4" D	1 Elec	8	1	Ea.	10	41.50		51.50	78.50
0200	8" W x 8" H x 4" D		8	1		12.45	41.50		53.95	81
0300	10" W x 12" H x 6" D	↓	5.30	1.509	↓	28.50	62.50		91	134
26 05 33.25 Conduit Fittings for Rigid Galvanized Steel										
0010	**CONDUIT FITTINGS FOR RIGID GALVANIZED STEEL**									
2280	LB, LR or LL fittings & covers, 1/2" diameter	1 Elec	16	.500	Ea.	6.95	20.50		27.45	41.50
2290	3/4" diameter		13	.615		8.20	25.50		33.70	50.50
2300	1" diameter		11	.727		12.80	30		42.80	63
2330	1-1/4" diameter		8	1		29.50	41.50		71	100
2350	1-1/2" diameter		6	1.333		23	55		78	115
2370	2" diameter		5	1.600		58.50	66		124.50	173
5280	Service entrance cap, 1/2" diameter		16	.500		4.90	20.50		25.40	39.50
5300	3/4" diameter		13	.615		5.85	25.50		31.35	48
5320	1" diameter		10	.800		4.79	33		37.79	59.50
5340	1-1/4" diameter		8	1		5.30	41.50		46.80	73.50
5360	1-1/2" diameter		6.50	1.231		10.45	51		61.45	94.50
5380	2" diameter	↓	5.50	1.455	↓	22	60		82	123
26 05 33.35 Flexible Metallic Conduit										
0010	**FLEXIBLE METALLIC CONDUIT**									
0050	Steel, 3/8" diameter	1 Elec	200	.040	L.F.	.37	1.66		2.03	3.11
0100	1/2" diameter		200	.040		.45	1.66		2.11	3.20
0200	3/4" diameter		160	.050		.57	2.07		2.64	4.01
0250	1" diameter		100	.080		1.30	3.31		4.61	6.85
0300	1-1/4" diameter		70	.114		1.37	4.73		6.10	9.20
0350	1-1/2" diameter		50	.160		2.93	6.60		9.53	14
0370	2" diameter	↓	40	.200	↓	3.62	8.30		11.92	17.50

26 05 39 – Underfloor Raceways for Electrical Systems

26 05 39.30 Conduit In Concrete Slab

	26 05 39.30 Conduit In Concrete Slab	Crew	Daily Output	Labor-Hours	Unit	2019 Bare Costs Material	Labor	Equipment	Total	Total Incl O&P
0010	**CONDUIT IN CONCRETE SLAB** Including terminations,									
0020	fittings and supports									
3230	PVC, schedule 40, 1/2" diameter	1 Elec	270	.030	L.F.	.54	1.23		1.77	2.59
3250	3/4" diameter		230	.035		.58	1.44		2.02	2.99
3270	1" diameter		200	.040		.77	1.66		2.43	3.55
3300	1-1/4" diameter		170	.047		1.04	1.95		2.99	4.33
3330	1-1/2" diameter		140	.057		1.27	2.37		3.64	5.25
3350	2" diameter		120	.067		1.58	2.76		4.34	6.25
4350	Rigid galvanized steel, 1/2" diameter		200	.040		2.02	1.66		3.68	4.93
4400	3/4" diameter		170	.047		4.46	1.95		6.41	8.10
4450	1" diameter		130	.062		6.90	2.55		9.45	11.75
4500	1-1/4" diameter		110	.073		4.23	3.01		7.24	9.55
4600	1-1/2" diameter		100	.080		7.25	3.31		10.56	13.40
4800	2" diameter		90	.089		8.95	3.68		12.63	15.85

26 05 39.40 Conduit In Trench

	26 05 39.40 Conduit In Trench	Crew	Daily Output	Labor-Hours	Unit	Material	Labor	Equipment	Total	Total Incl O&P
0010	**CONDUIT IN TRENCH** Includes terminations and fittings									
0020	Does not include excavation or backfill, see Section 31 23 16									
0200	Rigid galvanized steel, 2" diameter	1 Elec	150	.053	L.F.	8.60	2.21		10.81	13.05
0400	2-1/2" diameter	"	100	.080		11	3.31		14.31	17.50
0600	3" diameter	2 Elec	160	.100		12.85	4.14		16.99	21
0800	3-1/2" diameter	"	140	.114		16.75	4.73		21.48	26

26 05 80 – Wiring Connections

26 05 80.10 Motor Connections

	26 05 80.10 Motor Connections	Crew	Daily Output	Labor-Hours	Unit	Material	Labor	Equipment	Total	Total Incl O&P
0010	**MOTOR CONNECTIONS**									
0020	Flexible conduit and fittings, 115 volt, 1 phase, up to 1 HP motor	1 Elec	8	1	Ea.	5.70	41.50		47.20	74

26 05 90 – Residential Applications

26 05 90.10 Residential Wiring

	26 05 90.10 Residential Wiring	Crew	Daily Output	Labor-Hours	Unit	Material	Labor	Equipment	Total	Total Incl O&P
0010	**RESIDENTIAL WIRING**									
0020	20' avg. runs and #14/2 wiring incl. unless otherwise noted									
1000	Service & panel, includes 24' SE-AL cable, service eye, meter,									
1010	Socket, panel board, main bkr., ground rod, 15 or 20 amp									
1020	1-pole circuit breakers, and misc. hardware									
1100	100 amp, with 10 branch breakers	1 Elec	1.19	6.723	Ea.	325	278		603	815
1110	With PVC conduit and wire		.92	8.696		360	360		720	980
1120	With RGS conduit and wire		.73	10.959		575	455		1,030	1,375
1150	150 amp, with 14 branch breakers		1.03	7.767		845	320		1,165	1,450
1170	With PVC conduit and wire		.82	9.756		915	405		1,320	1,650
1180	With RGS conduit and wire		.67	11.940		1,225	495		1,720	2,150
1200	200 amp, with 18 branch breakers	2 Elec	1.80	8.889		1,000	370		1,370	1,700
1220	With PVC conduit and wire		1.46	10.959		1,075	455		1,530	1,925
1230	With RGS conduit and wire		1.24	12.903		1,450	535		1,985	2,475
1800	Lightning surge suppressor	1 Elec	32	.250		75	10.35		85.35	99.50
2000	Switch devices									
2100	Single pole, 15 amp, ivory, with a 1-gang box, cover plate,									
2110	Type NM (Romex) cable	1 Elec	17.10	.468	Ea.	15.50	19.35		34.85	48.50
2120	Type MC cable		14.30	.559		26	23		49	66.50
2130	EMT & wire		5.71	1.401		37.50	58		95.50	136
2150	3-way, #14/3, type NM cable		14.55	.550		9.80	23		32.80	48
2170	Type MC cable		12.31	.650		23.50	27		50.50	70
2180	EMT & wire		5	1.600		31.50	66		97.50	143
2200	4-way, #14/3, type NM cable		14.55	.550		18.50	23		41.50	57.50

26 05 90.10 Residential Wiring		Crew	Daily Output	Labor-Hours	Unit	Material	2019 Bare Costs Labor	Equipment	Total	Total Incl O&P
2220	Type MC cable	1 Elec	12.31	.650	Ea.	32.50	27		59.50	79.50
2230	EMT & wire		5	1.600		40.50	66		106.50	153
2250	S.P., 20 amp, #12/2, type NM cable		13.33	.600		11.30	25		36.30	53
2270	Type MC cable		11.43	.700		21	29		50	70.50
2280	EMT & wire		4.85	1.649		35.50	68.50		104	150
2290	S.P. rotary dimmer, 600 W, no wiring		17	.471		31.50	19.50		51	66.50
2300	S.P. rotary dimmer, 600 W, type NM cable		14.55	.550		35.50	23		58.50	76
2320	Type MC cable		12.31	.650		46	27		73	94.50
2330	EMT & wire		5	1.600		58.50	66		124.50	172
2350	3-way rotary dimmer, type NM cable		13.33	.600		23.50	25		48.50	66
2370	Type MC cable		11.43	.700		34	29		63	85
2380	EMT & wire	↓	4.85	1.649	↓	46.50	68.50		115	162
2400	Interval timer wall switch, 20 amp, 1-30 min., #12/2									
2410	Type NM cable	1 Elec	14.55	.550	Ea.	58.50	23		81.50	101
2420	Type MC cable		12.31	.650		64.50	27		91.50	115
2430	EMT & wire	↓	5	1.600	↓	82.50	66		148.50	199
2500	Decorator style									
2510	S.P., 15 amp, type NM cable	1 Elec	17.10	.468	Ea.	20.50	19.35		39.85	54.50
2520	Type MC cable		14.30	.559		31.50	23		54.50	72.50
2530	EMT & wire		5.71	1.401		42.50	58		100.50	142
2550	3-way, #14/3, type NM cable		14.55	.550		15	23		38	53.50
2570	Type MC cable		12.31	.650		29	27		56	76
2580	EMT & wire		5	1.600		37	66		103	149
2600	4-way, #14/3, type NM cable		14.55	.550		23.50	23		46.50	63
2620	Type MC cable		12.31	.650		37.50	27		64.50	85.50
2630	EMT & wire		5	1.600		45.50	66		111.50	158
2650	S.P., 20 amp, #12/2, type NM cable		13.33	.600		16.55	25		41.55	58.50
2670	Type MC cable		11.43	.700		26	29		55	76
2680	EMT & wire		4.85	1.649		41	68.50		109.50	156
2700	S.P., slide dimmer, type NM cable		17.10	.468		37	19.35		56.35	72.50
2720	Type MC cable		14.30	.559		47.50	23		70.50	90.50
2730	EMT & wire		5.71	1.401		60	58		118	161
2750	S.P., touch dimmer, type NM cable		17.10	.468		48.50	19.35		67.85	84.50
2770	Type MC cable		14.30	.559		59	23		82	103
2780	EMT & wire		5.71	1.401		71.50	58		129.50	173
2800	3-way touch dimmer, type NM cable		13.33	.600		49.50	25		74.50	95
2820	Type MC cable		11.43	.700		60.50	29		89.50	114
2830	EMT & wire	↓	4.85	1.649	↓	73	68.50		141.50	191
3000	Combination devices									
3100	S.P. switch/15 amp recpt., ivory, 1-gang box, plate									
3110	Type NM cable	1 Elec	11.43	.700	Ea.	23	29		52	72.50
3120	Type MC cable		10	.800		33.50	33		66.50	91
3130	EMT & wire		4.40	1.818		46	75.50		121.50	174
3150	S.P. switch/pilot light, type NM cable		11.43	.700		24	29		53	74
3170	Type MC cable		10	.800		35	33		68	92.50
3180	EMT & wire		4.43	1.806		47.50	75		122.50	174
3190	2-S.P. switches, 2-#14/2, no wiring		14	.571		13.25	23.50		36.75	53
3200	2-S.P. switches, 2-#14/2, type NM cables		10	.800		25	33		58	81.50
3220	Type MC cable		8.89	.900		40	37.50		77.50	105
3230	EMT & wire		4.10	1.951		48	81		129	185
3250	3-way switch/15 amp recpt., #14/3, type NM cable		10	.800		30.50	33		63.50	87.50
3270	Type MC cable		8.89	.900		44.50	37.50		82	110
3280	EMT & wire	↓	4.10	1.951	↓	52.50	81		133.50	190

26 05 90 – Residential Applications

26 05 90.10 Residential Wiring		Crew	Daily Output	Labor-Hours	Unit	Material	Labor	Equipment	Total	Total Incl O&P
						2019 Bare Costs				
3300	2-3 way switches, 2-#14/3, type NM cables	1 Elec	8.89	.900	Ea.	39	37.50		76.50	104
3320	Type MC cable		8	1		60.50	41.50		102	134
3330	EMT & wire		4	2		59.50	83		142.50	200
3350	S.P. switch/20 amp recpt., #12/2, type NM cable		10	.800		39	33		72	96.50
3370	Type MC cable		8.89	.900		44.50	37.50		82	110
3380	EMT & wire		4.10	1.951		63	81		144	202
3400	Decorator style									
3410	S.P. switch/15 amp recpt., type NM cable	1 Elec	11.43	.700	Ea.	28	29		57	78.50
3420	Type MC cable		10	.800		38.50	33		71.50	96.50
3430	EMT & wire		4.40	1.818		51	75.50		126.50	180
3450	S.P. switch/pilot light, type NM cable		11.43	.700		29.50	29		58.50	80
3470	Type MC cable		10	.800		40	33		73	98
3480	EMT & wire		4.40	1.818		52.50	75.50		128	181
3500	2-S.P. switches, 2-#14/2, type NM cables		10	.800		30	33		63	87
3520	Type MC cable		8.89	.900		45	37.50		82.50	111
3530	EMT & wire		4.10	1.951		53	81		134	191
3550	3-way/15 amp recpt., #14/3, type NM cable		10	.800		36	33		69	93.50
3570	Type MC cable		8.89	.900		49.50	37.50		87	116
3580	EMT & wire		4.10	1.951		57.50	81		138.50	196
3650	2-3 way switches, 2-#14/3, type NM cables		8.89	.900		44.50	37.50		82	110
3670	Type MC cable		8	1		66	41.50		107.50	140
3680	EMT & wire		4	2		64.50	83		147.50	206
3700	S.P. switch/20 amp recpt., #12/2, type NM cable		10	.800		44	33		77	103
3720	Type MC cable		8.89	.900		50	37.50		87.50	116
3730	EMT & wire		4.10	1.951		68.50	81		149.50	207
4000	Receptacle devices									
4010	Duplex outlet, 15 amp recpt., ivory, 1-gang box, plate									
4015	Type NM cable	1 Elec	14.55	.550	Ea.	8.70	23		31.70	46.50
4020	Type MC cable		12.31	.650		19.30	27		46.30	65
4030	EMT & wire		5.33	1.501		30.50	62		92.50	135
4050	With #12/2, type NM cable		12.31	.650		9.85	27		36.85	55
4070	Type MC cable		10.67	.750		19.40	31		50.40	72
4080	EMT & wire		4.71	1.699		34	70.50		104.50	153
4100	20 amp recpt., #12/2, type NM cable		12.31	.650		17.85	27		44.85	63.50
4120	Type MC cable		10.67	.750		27.50	31		58.50	80.50
4130	EMT & wire		4.71	1.699		42	70.50		112.50	162
4140	For GFI see Section 26 05 90.10 line 4300 below									
4150	Decorator style, 15 amp recpt., type NM cable	1 Elec	14.55	.550	Ea.	13.95	23		36.95	52.50
4170	Type MC cable		12.31	.650		24.50	27		51.50	71
4180	EMT & wire		5.33	1.501		36	62		98	141
4200	With #12/2, type NM cable		12.31	.650		15.10	27		42.10	60.50
4220	Type MC cable		10.67	.750		24.50	31		55.50	77.50
4230	EMT & wire		4.71	1.699		39.50	70.50		110	159
4250	20 amp recpt., #12/2, type NM cable		12.31	.650		23	27		50	69.50
4270	Type MC cable		10.67	.750		32.50	31		63.50	86.50
4280	EMT & wire		4.71	1.699		47.50	70.50		118	167
4300	GFI, 15 amp recpt., type NM cable		12.31	.650		21	27		48	67.50
4320	Type MC cable		10.67	.750		32	31		63	85.50
4330	EMT & wire		4.71	1.699		43	70.50		113.50	163
4350	GFI with #12/2, type NM cable		10.67	.750		22.50	31		53.50	75
4370	Type MC cable		9.20	.870		32	36		68	93.50
4380	EMT & wire		4.21	1.900		46.50	78.50		125	180
4400	20 amp recpt., #12/2, type NM cable		10.67	.750		49	31		80	104

26 05 90.10 Residential Wiring		Crew	Daily Output	Labor-Hours	Unit	Material	2019 Bare Costs Labor	Equipment	Total	Total Incl O&P
4420	Type MC cable	1 Elec	9.20	.870	Ea.	58.50	36		94.50	123
4430	EMT & wire		4.21	1.900		73	78.50		151.50	209
4500	Weather-proof cover for above receptacles, add	↓	32	.250	↓	2.06	10.35		12.41	19.15
4550	Air conditioner outlet, 20 amp-240 volt recpt.									
4560	30′ of #12/2, 2 pole circuit breaker									
4570	Type NM cable	1 Elec	10	.800	Ea.	60	33		93	120
4580	Type MC cable		9	.889		71	37		108	139
4590	EMT & wire		4	2		85	83		168	228
4600	Decorator style, type NM cable		10	.800		65	33		98	126
4620	Type MC cable		9	.889		76	37		113	144
4630	EMT & wire	↓	4	2	↓	89.50	83		172.50	234
4650	Dryer outlet, 30 amp-240 volt recpt., 20′ of #10/3									
4660	2 pole circuit breaker									
4670	Type NM cable	1 Elec	6.41	1.248	Ea.	54	51.50		105.50	144
4680	Type MC cable		5.71	1.401		63.50	58		121.50	164
4690	EMT & wire	↓	3.48	2.299	↓	76	95		171	239
4700	Range outlet, 50 amp-240 volt recpt., 30′ of #8/3									
4710	Type NM cable	1 Elec	4.21	1.900	Ea.	82.50	78.50		161	219
4720	Type MC cable		4	2		127	83		210	275
4730	EMT & wire		2.96	2.703		107	112		219	300
4750	Central vacuum outlet, type NM cable		6.40	1.250		53.50	52		105.50	144
4770	Type MC cable		5.71	1.401		67.50	58		125.50	169
4780	EMT & wire	↓	3.48	2.299	↓	88	95		183	252
4800	30 amp-110 volt locking recpt., #10/2 circ. bkr.									
4810	Type NM cable	1 Elec	6.20	1.290	Ea.	63	53.50		116.50	156
4820	Type MC cable		5.40	1.481		80	61.50		141.50	188
4830	EMT & wire	↓	3.20	2.500	↓	99	104		203	278
4900	Low voltage outlets									
4910	Telephone recpt., 20′ of 4/C phone wire	1 Elec	26	.308	Ea.	8.55	12.75		21.30	30.50
4920	TV recpt., 20′ of RG59U coax wire, F type connector	"	16	.500	"	17.10	20.50		37.60	53
4950	Door bell chime, transformer, 2 buttons, 60′ of bellwire									
4970	Economy model	1 Elec	11.50	.696	Ea.	55	29		84	108
4980	Custom model		11.50	.696		105	29		134	163
4990	Luxury model, 3 buttons	↓	9.50	.842	↓	188	35		223	264
6000	Lighting outlets									
6050	Wire only (for fixture), type NM cable	1 Elec	32	.250	Ea.	5.80	10.35		16.15	23.50
6070	Type MC cable		24	.333		11.65	13.80		25.45	35.50
6080	EMT & wire		10	.800		22	33		55	78.50
6100	Box (4″), and wire (for fixture), type NM cable		25	.320		14.85	13.25		28.10	38
6120	Type MC cable		20	.400		20.50	16.55		37.05	50
6130	EMT & wire	↓	11	.727	↓	31.50	30		61.50	83.50
6200	Fixtures (use with line 6050 or 6100 above)									
6210	Canopy style, economy grade	1 Elec	40	.200	Ea.	22.50	8.30		30.80	38.50
6220	Custom grade		40	.200		52.50	8.30		60.80	71
6250	Dining room chandelier, economy grade		19	.421		83	17.45		100.45	120
6260	Custom grade		19	.421		335	17.45		352.45	395
6270	Luxury grade		15	.533		1,250	22		1,272	1,400
6310	Kitchen fixture (fluorescent), economy grade		30	.267		71.50	11.05		82.55	96.50
6320	Custom grade		25	.320		149	13.25		162.25	186
6350	Outdoor, wall mounted, economy grade		30	.267		29.50	11.05		40.55	50.50
6360	Custom grade		30	.267		120	11.05		131.05	150
6370	Luxury grade		25	.320		243	13.25		256.25	289
6410	Outdoor PAR floodlights, 1 lamp, 150 watt	↓	20	.400	↓	26.50	16.55		43.05	56.50

26 05 90.10 Residential Wiring		Crew	Daily Output	Labor-Hours	Unit	2019 Bare Costs Material	Labor	Equipment	Total	Total Incl O&P
6420	2 lamp, 150 watt each	1 Elec	20	.400	Ea.	45	16.55		61.55	76.50
6425	Motion sensing, 2 lamp, 150 watt each		20	.400		110	16.55		126.55	148
6430	For infrared security sensor, add		32	.250		102	10.35		112.35	129
6450	Outdoor, quartz-halogen, 300 watt flood		20	.400		38.50	16.55		55.05	69
6600	Recessed downlight, round, pre-wired, 50 or 75 watt trim		30	.267		66	11.05		77.05	90.50
6610	With shower light trim		30	.267		90	11.05		101.05	117
6620	With wall washer trim		28	.286		92	11.85		103.85	120
6630	With eye-ball trim		28	.286		77	11.85		88.85	104
6700	Porcelain lamp holder		40	.200		3	8.30		11.30	16.80
6710	With pull switch		40	.200		11.30	8.30		19.60	26
6750	Fluorescent strip, 2-20 watt tube, wrap around diffuser, 24″		24	.333		48	13.80		61.80	75
6760	1-34 watt tube, 48″		24	.333		137	13.80		150.80	174
6770	2-34 watt tubes, 48″		20	.400		158	16.55		174.55	200
6800	Bathroom heat lamp, 1-250 watt		28	.286		31.50	11.85		43.35	54.50
6810	2-250 watt lamps		28	.286		76.50	11.85		88.35	104
6820	For timer switch, see Section 26 05 90.10 line 2400									
6900	Outdoor post lamp, incl. post, fixture, 35′ of #14/2									
6910	Type NM cable	1 Elec	3.50	2.286	Ea.	299	94.50		393.50	485
6920	Photo-eye, add		27	.296		28.50	12.25		40.75	51
6950	Clock dial time switch, 24 hr., w/enclosure, type NM cable		11.43	.700		78.50	29		107.50	134
6970	Type MC cable		11	.727		89	30		119	147
6980	EMT & wire		4.85	1.649		100	68.50		168.50	222
7000	Alarm systems									
7050	Smoke detectors, box, #14/3, type NM cable	1 Elec	14.55	.550	Ea.	34.50	23		57.50	75
7070	Type MC cable		12.31	.650		44.50	27		71.50	93
7080	EMT & wire		5	1.600		52.50	66		118.50	166
7090	For relay output to security system, add					11.15			11.15	12.25
8000	Residential equipment									
8050	Disposal hook-up, incl. switch, outlet box, 3′ of flex									
8060	20 amp-1 pole circ. bkr., and 25′ of #12/2									
8070	Type NM cable	1 Elec	10	.800	Ea.	28	33		61	85
8080	Type MC cable		8	1		38.50	41.50		80	110
8090	EMT & wire		5	1.600		56	66		122	170
8100	Trash compactor or dishwasher hook-up, incl. outlet box,									
8110	3′ of flex, 15 amp-1 pole circ. bkr., and 25′ of #14/2									
8130	Type MC cable	1 Elec	8	1	Ea.	27.50	41.50		69	97.50
8140	EMT & wire	″	5	1.600	″	42.50	66		108.50	155
8150	Hot water sink dispenser hook-up, use line 8100									
8200	Vent/exhaust fan hook-up, type NM cable	1 Elec	32	.250	Ea.	5.80	10.35		16.15	23.50
8220	Type MC cable		24	.333		11.65	13.80		25.45	35.50
8230	EMT & wire		10	.800		22	33		55	78.50
8250	Bathroom vent fan, 50 CFM (use with above hook-up)									
8260	Economy model	1 Elec	15	.533	Ea.	19.45	22		41.45	57.50
8270	Low noise model		15	.533		45.50	22		67.50	86
8280	Custom model		12	.667		120	27.50		147.50	177
8300	Bathroom or kitchen vent fan, 110 CFM									
8310	Economy model	1 Elec	15	.533	Ea.	67	22		89	110
8320	Low noise model	″	15	.533	″	93.50	22		115.50	139
8350	Paddle fan, variable speed (w/o lights)									
8360	Economy model (AC motor)	1 Elec	10	.800	Ea.	131	33		164	198
8362	With light kit		10	.800		171	33		204	242
8370	Custom model (AC motor)		10	.800		340	33		373	430
8372	With light kit		10	.800		380	33		413	475

26 05 Common Work Results for Electrical

26 05 90 – Residential Applications

26 05 90.10 Residential Wiring		Crew	Daily Output	Labor-Hours	Unit	Material	Labor	Equipment	Total	Total Incl O&P
						2019 Bare Costs				
8380	Luxury model (DC motor)	1 Elec	8	1	Ea.	330	41.50		371.50	430
8382	With light kit		8	1		370	41.50		411.50	475
8390	Remote speed switch for above, add	↓	12	.667	↓	34	27.50		61.50	82
8500	Whole house exhaust fan, ceiling mount, 36″, variable speed									
8510	Remote switch, incl. shutters, 20 amp-1 pole circ. bkr.									
8520	30′ of #12/2, type NM cable	1 Elec	4	2	Ea.	1,325	83		1,408	1,600
8530	Type MC cable		3.50	2.286		1,350	94.50		1,444.50	1,625
8540	EMT & wire	↓	3	2.667	↓	1,375	110		1,485	1,675
8600	Whirlpool tub hook-up, incl. timer switch, outlet box									
8610	3′ of flex, 20 amp-1 pole GFI circ. bkr.									
8620	30′ of #12/2, type NM cable	1 Elec	5	1.600	Ea.	128	66		194	248
8630	Type MC cable		4.20	1.905		135	79		214	277
8640	EMT & wire	↓	3.40	2.353	↓	150	97.50		247.50	325
8650	Hot water heater hook-up, incl. 1-2 pole circ. bkr., box;									
8660	3′ of flex, 20′ of #10/2, type NM cable	1 Elec	5	1.600	Ea.	28.50	66		94.50	139
8670	Type MC cable		4.20	1.905		41.50	79		120.50	175
8680	EMT & wire	↓	3.40	2.353	↓	49.50	97.50		147	214
9000	Heating/air conditioning									
9050	Furnace/boiler hook-up, incl. firestat, local on-off switch									
9060	Emergency switch, and 40′ of type NM cable	1 Elec	4	2	Ea.	53	83		136	193
9070	Type MC cable		3.50	2.286		68	94.50		162.50	229
9080	EMT & wire	↓	1.50	5.333	↓	92	221		313	460
9100	Air conditioner hook-up, incl. local 60 amp disc. switch									
9110	3′ sealtite, 40 amp, 2 pole circuit breaker									
9130	40′ of #8/2, type NM cable	1 Elec	3.50	2.286	Ea.	170	94.50		264.50	340
9140	Type MC cable		3	2.667		235	110		345	440
9150	EMT & wire	↓	1.30	6.154	↓	218	255		473	655
9200	Heat pump hook-up, 1-40 & 1-100 amp 2 pole circ. bkr.									
9210	Local disconnect switch, 3′ sealtite									
9220	40′ of #8/2 & 30′ of #3/2									
9230	Type NM cable	1 Elec	1.30	6.154	Ea.	505	255		760	970
9240	Type MC cable		1.08	7.407		555	305		860	1,100
9250	EMT & wire	↓	.94	8.511	↓	520	350		870	1,150
9500	Thermostat hook-up, using low voltage wire									
9520	Heating only, 25′ of #18-3	1 Elec	24	.333	Ea.	6.20	13.80		20	29.50
9530	Heating/cooling, 25′ of #18-4	″	20	.400	″	8.15	16.55		24.70	36

26 24 Switchboards and Panelboards

26 24 16 – Panelboards

26 24 16.10 Load Centers		Crew	Daily Output	Labor-Hours	Unit	Material	Labor	Equipment	Total	Total Incl O&P
0010	**LOAD CENTERS** (residential type)									
0100	3 wire, 120/240 V, 1 phase, including 1 pole plug-in breakers									
0200	100 amp main lugs, indoor, 8 circuits	1 Elec	1.40	5.714	Ea.	98	237		335	495
0300	12 circuits		1.20	6.667		123	276		399	585
0400	Rainproof, 8 circuits		1.40	5.714		126	237		363	525
0500	12 circuits	↓	1.20	6.667		151	276		427	615
0600	200 amp main lugs, indoor, 16 circuits	R-1A	1.80	8.889		203	330		533	765
0700	20 circuits		1.50	10.667		202	395		597	870
0800	24 circuits		1.30	12.308		253	460		713	1,025
1200	Rainproof, 16 circuits		1.80	8.889		245	330		575	810
1300	20 circuits	↓	1.50	10.667	↓	285	395		680	965

26 24 Switchboards and Panelboards

26 24 16 – Panelboards

26 24 16.10 Load Centers		Crew	Daily Output	Labor-Hours	Unit	Material	2019 Bare Costs Labor	Equipment	Total	Total Incl O&P
1400	24 circuits	R-1A	1.30	12.308	Ea.	335	460		795	1,125

26 24 16.20 Panelboard and Load Center Circuit Breakers

		Crew	Daily Output	Labor-Hours	Unit	Material	Labor	Equipment	Total	Total Incl O&P
0010	**PANELBOARD AND LOAD CENTER CIRCUIT BREAKERS**									
2000	Plug-in panel or load center, 120/240 volt, to 60 amp, 1 pole	1 Elec	12	.667	Ea.	6.15	27.50		33.65	52
2004	Circuit breaker, 120/240 volt, 20 A, 1 pole with NM cable		6.50	1.231		11	51		62	95
2006	30 A, 1 pole with NM cable		6.50	1.231		11	51		62	95
2010	2 pole		9	.889		25.50	37		62.50	88
2014	50 A, 2 pole with NM cable		5.50	1.455		30.50	60		90.50	132
2020	3 pole		7.50	1.067		84	44		128	165
2030	100 amp, 2 pole		6	1.333		103	55		158	203
2040	3 pole		4.50	1.778		115	73.50		188.50	247
2050	150-200 amp, 2 pole		3	2.667		257	110		367	465
2060	Plug-in tandem, 120/240 V, 2-15 A, 1 pole		11	.727		20.50	30		50.50	71.50
2070	1-15 A & 1-20 A		11	.727		17.30	30		47.30	68
2080	2-20 A		11	.727		20.50	30		50.50	71.50
2300	Ground fault, 240 volt, 30 amp, 1 pole		7	1.143		90.50	47.50		138	177
2310	2 pole	↓	6	1.333	↓	170	55		225	277

26 27 Low-Voltage Distribution Equipment

26 27 13 – Electricity Metering

26 27 13.10 Meter Centers and Sockets

		Crew	Daily Output	Labor-Hours	Unit	Material	Labor	Equipment	Total	Total Incl O&P
0010	**METER CENTERS AND SOCKETS**									
0100	Sockets, single position, 4 terminal, 100 amp	1 Elec	3.20	2.500	Ea.	45	104		149	219
0200	150 amp		2.30	3.478		63.50	144		207.50	305
0300	200 amp		1.90	4.211		98	174		272	390
0500	Double position, 4 terminal, 100 amp		2.80	2.857		215	118		333	430
0600	150 amp		2.10	3.810		287	158		445	570
0700	200 amp		1.70	4.706		555	195		750	930
1100	Meter centers and sockets, three phase, single pos, 7 terminal, 100 amp		2.80	2.857		118	118		236	320
1200	200 amp		2.10	3.810		228	158		386	510
1400	400 amp	↓	1.70	4.706	↓	840	195		1,035	1,250
2590	Basic meter device									
2600	1P 3W 120/240 V 4 jaw 125A sockets, 3 meter	2 Elec	1	16	Ea.	360	660		1,020	1,475
2610	4 meter		.90	17.778		445	735		1,180	1,700
2620	5 meter		.80	20		530	830		1,360	1,925
2630	6 meter		.60	26.667		555	1,100		1,655	2,425
2640	7 meter		.56	28.571		1,425	1,175		2,600	3,475
2660	10 meter	↓	.48	33.333	↓	1,925	1,375		3,300	4,375
2680	Rainproof 1P 3W 120/240 V 4 jaw 125A sockets									
2690	3 meter	2 Elec	1	16	Ea.	645	660		1,305	1,775
2710	6 meter		.60	26.667		1,100	1,100		2,200	3,025
2730	8 meter	↓	.52	30.769	↓	1,550	1,275		2,825	3,775
2750	1P 3W 120/240 V 4 jaw sockets									
2760	with 125A circuit breaker, 3 meter	2 Elec	1	16	Ea.	1,200	660		1,860	2,400
2780	5 meter		.80	20		1,900	830		2,730	3,450
2800	7 meter		.56	28.571		2,725	1,175		3,900	4,925
2820	10 meter	↓	.48	33.333	↓	3,800	1,375		5,175	6,450
2830	Rainproof 1P 3W 120/240 V 4 jaw sockets									
2840	with 125A circuit breaker, 3 meter	2 Elec	1	16	Ea.	1,200	660		1,860	2,400
2870	6 meter		.60	26.667		2,225	1,100		3,325	4,250
2890	8 meter	↓	.52	30.769	↓	3,050	1,275		4,325	5,425

26 27 Low-Voltage Distribution Equipment

26 27 13 – Electricity Metering

26 27 13.10 Meter Centers and Sockets		Crew	Daily Output	Labor-Hours	Unit	Material	2019 Bare Costs Labor	Equipment	Total	Total Incl O&P
3250	1P 3W 120/240 V 4 jaw sockets									
3260	with 200A circuit breaker, 3 meter	2 Elec	1	16	Ea.	1,800	660		2,460	3,075
3290	6 meter		.60	26.667		3,625	1,100		4,725	5,775
3310	8 meter		.56	28.571		4,900	1,175		6,075	7,300
3330	Rainproof 1P 3W 120/240 V 4 jaw sockets									
3350	with 200A circuit breaker, 3 meter	2 Elec	1	16	Ea.	1,800	660		2,460	3,075
3380	6 meter		.60	26.667		3,625	1,100		4,725	5,775
3400	8 meter		.52	30.769		4,900	1,275		6,175	7,450

26 27 23 – Indoor Service Poles

26 27 23.40 Surface Raceway		Crew	Daily Output	Labor-Hours	Unit	Material	Labor	Equipment	Total	Total Incl O&P
0010	**SURFACE RACEWAY**									
0090	Metal, straight section									
0100	No. 500	1 Elec	100	.080	L.F.	1.27	3.31		4.58	6.80
0110	No. 700		100	.080		1.43	3.31		4.74	6.95
0200	No. 1000		90	.089		1.46	3.68		5.14	7.60
0400	No. 1500, small pancake		90	.089		2.20	3.68		5.88	8.40
0600	No. 2000, base & cover, blank		90	.089		2.24	3.68		5.92	8.45
0800	No. 3000, base & cover, blank		75	.107		4.38	4.42		8.80	12
2400	Fittings, elbows, No. 500		40	.200	Ea.	2.29	8.30		10.59	16
2800	Elbow cover, No. 2000		40	.200		3.67	8.30		11.97	17.55
2880	Tee, No. 500		42	.190		4.23	7.90		12.13	17.50
2900	No. 2000		27	.296		12.20	12.25		24.45	33.50
3000	Switch box, No. 500		16	.500		14.75	20.50		35.25	50
3400	Telephone outlet, No. 1500		16	.500		17.90	20.50		38.40	53.50
3600	Junction box, No. 1500		16	.500		11.30	20.50		31.80	46.50
3800	Plugmold wired sections, No. 2000									
4000	1 circuit, 6 outlets, 3′ long	1 Elec	8	1	Ea.	38.50	41.50		80	110
4100	2 circuits, 8 outlets, 6′ long	"	5.30	1.509	"	61.50	62.50		124	170

26 27 26 – Wiring Devices

26 27 26.10 Low Voltage Switching		Crew	Daily Output	Labor-Hours	Unit	Material	Labor	Equipment	Total	Total Incl O&P
0010	**LOW VOLTAGE SWITCHING**									
3600	Relays, 120 V or 277 V standard	1 Elec	12	.667	Ea.	46	27.50		73.50	95.50
3800	Flush switch, standard		40	.200		11.60	8.30		19.90	26.50
4000	Interchangeable		40	.200		15.20	8.30		23.50	30
4100	Surface switch, standard		40	.200		8.25	8.30		16.55	22.50
4200	Transformer 115 V to 25 V		12	.667		129	27.50		156.50	187
4400	Master control, 12 circuit, manual		4	2		129	83		212	277
4500	25 circuit, motorized		4	2		143	83		226	292
4600	Rectifier, silicon		12	.667		46	27.50		73.50	95.50
4800	Switchplates, 1 gang, 1, 2 or 3 switch, plastic		80	.100		5.05	4.14		9.19	12.30
5000	Stainless steel		80	.100		11.40	4.14		15.54	19.30
5400	2 gang, 3 switch, stainless steel		53	.151		23	6.25		29.25	35.50
5500	4 switch, plastic		53	.151		10.50	6.25		16.75	22
5600	2 gang, 4 switch, stainless steel		53	.151		22	6.25		28.25	34.50
5700	6 switch, stainless steel		53	.151		43.50	6.25		49.75	58
5800	3 gang, 9 switch, stainless steel		32	.250		66	10.35		76.35	89.50

26 27 26.20 Wiring Devices Elements		Crew	Daily Output	Labor-Hours	Unit	Material	Labor	Equipment	Total	Total Incl O&P
0010	**WIRING DEVICES ELEMENTS**									
0200	Toggle switch, quiet type, single pole, 15 amp	1 Elec	40	.200	Ea.	.47	8.30		8.77	14
0600	3 way, 15 amp		23	.348		1.90	14.40		16.30	25.50
0900	4 way, 15 amp		15	.533		10.90	22		32.90	48

26 27 Low-Voltage Distribution Equipment

26 27 26 – Wiring Devices

26 27 26.20 Wiring Devices Elements		Crew	Daily Output	Labor-Hours	Unit	Material	Labor	Equipment	Total	Total Incl O&P
						2019 Bare Costs				
1650	Dimmer switch, 120 volt, incandescent, 600 watt, 1 pole G	1 Elec	16	.500	Ea.	23	20.50		43.50	59
2460	Receptacle, duplex, 120 volt, grounded, 15 amp		40	.200		1.51	8.30		9.81	15.15
2470	20 amp		27	.296		9.55	12.25		21.80	30.50
2490	Dryer, 30 amp		15	.533		4.39	22		26.39	41
2500	Range, 50 amp		11	.727		10.80	30		40.80	61
2600	Wall plates, stainless steel, 1 gang		80	.100		2.58	4.14		6.72	9.60
2800	2 gang		53	.151		4.37	6.25		10.62	15
3200	Lampholder, keyless		26	.308		17.45	12.75		30.20	40
3400	Pullchain with receptacle	↓	22	.364	↓	21	15.05		36.05	47.50

26 27 73 – Door Chimes

26 27 73.10 Doorbell System		Crew	Daily Output	Labor-Hours	Unit	Material	Labor	Equipment	Total	Total Incl O&P
0010	**DOORBELL SYSTEM**, incl. transformer, button & signal									
0100	6″ bell	1 Elec	4	2	Ea.	141	83		224	290
0200	Buzzer		4	2		111	83		194	258
1000	Door chimes, 2 notes		16	.500		31	20.50		51.50	68.50
1020	with ambient light		12	.667		123	27.50		150.50	180
1100	Tube type, 3 tube system		12	.667		231	27.50		258.50	299
1180	4 tube system		10	.800		495	33		528	600
1900	For transformer & button, add		5	1.600		16.25	66		82.25	126
3000	For push button only	↓	24	.333	↓	.95	13.80		14.75	23.50

26 28 Low-Voltage Circuit Protective Devices

26 28 16 – Enclosed Switches and Circuit Breakers

26 28 16.10 Circuit Breakers		Crew	Daily Output	Labor-Hours	Unit	Material	Labor	Equipment	Total	Total Incl O&P
0010	**CIRCUIT BREAKERS** (in enclosure)									
0100	Enclosed (NEMA 1), 600 volt, 3 pole, 30 amp	1 Elec	3.20	2.500	Ea.	515	104		619	735
0200	60 amp		2.80	2.857		620	118		738	880
0400	100 amp	↓	2.30	3.478	↓	710	144		854	1,025

26 28 16.20 Safety Switches		Crew	Daily Output	Labor-Hours	Unit	Material	Labor	Equipment	Total	Total Incl O&P
0010	**SAFETY SWITCHES**									
0100	General duty 240 volt, 3 pole NEMA 1, fusible, 30 amp	1 Elec	3.20	2.500	Ea.	85.50	104		189.50	263
0200	60 amp		2.30	3.478		145	144		289	395
0300	100 amp		1.90	4.211		246	174		420	555
0400	200 amp	↓	1.30	6.154		535	255		790	1,000
0500	400 amp	2 Elec	1.80	8.889		1,400	370		1,770	2,125
9010	Disc. switch, 600 volt 3 pole fusible, 30 amp, to 10 HP motor	1 Elec	3.20	2.500		345	104		449	550
9050	60 amp, to 30 HP motor		2.30	3.478		795	144		939	1,100
9070	100 amp, to 60 HP motor	↓	1.90	4.211	↓	795	174		969	1,150

26 28 16.40 Time Switches		Crew	Daily Output	Labor-Hours	Unit	Material	Labor	Equipment	Total	Total Incl O&P
0010	**TIME SWITCHES**									
0100	Single pole, single throw, 24 hour dial	1 Elec	4	2	Ea.	130	83		213	278
0200	24 hour dial with reserve power		3.60	2.222		590	92		682	800
0300	Astronomic dial		3.60	2.222		235	92		327	410
0400	Astronomic dial with reserve power		3.30	2.424		895	100		995	1,150
0500	7 day calendar dial		3.30	2.424		191	100		291	375
0600	7 day calendar dial with reserve power		3.20	2.500		222	104		326	415
0700	Photo cell 2,000 watt	↓	8	1	↓	26	41.50		67.50	96

26 32 Packaged Generator Assemblies

26 32 13 – Engine Generators

26 32 13.16 Gas-Engine-Driven Generator Sets		Crew	Daily Output	Labor-Hours	Unit	Material	2019 Bare Costs Labor	Equipment	Total	Total Incl O&P
0010	**GAS-ENGINE-DRIVEN GENERATOR SETS**									
0020	Gas or gasoline operated, includes battery,									
0050	charger, & muffler									
0200	7.5 kW	R-3	.83	24.096	Ea.	8,700	985	156	9,841	11,300
0300	11.5 kW		.71	28.169		12,300	1,150	183	13,633	15,700
0400	20 kW		.63	31.746		14,500	1,300	206	16,006	18,400
0500	35 kW		.55	36.364		17,300	1,475	236	19,011	21,700

26 33 Battery Equipment

26 33 43 – Battery Chargers

26 33 43.55 Electric Vehicle Charging		Crew	Daily Output	Labor-Hours	Unit	Material	Labor	Equipment	Total	Total Incl O&P
0010	**ELECTRIC VEHICLE CHARGING**									
0020	Level 2, wall mounted									
2100	Light duty, hard wired G	1 Elec	20.48	.391	Ea.	660	16.15		676.15	750
2110	plug in G	"	30.72	.260	"	660	10.80		670.80	745

26 36 Transfer Switches

26 36 23 – Automatic Transfer Switches

26 36 23.10 Automatic Transfer Switch Devices		Crew	Daily Output	Labor-Hours	Unit	Material	Labor	Equipment	Total	Total Incl O&P
0010	**AUTOMATIC TRANSFER SWITCH DEVICES**									
0100	Switches, enclosed 480 volt, 3 pole, 30 amp	1 Elec	2.30	3.478	Ea.	3,375	144		3,519	3,950
0200	60 amp	"	1.90	4.211	"	3,375	174		3,549	4,000

26 41 Facility Lightning Protection

26 41 13 – Lightning Protection for Structures

26 41 13.13 Lightning Protection for Buildings		Crew	Daily Output	Labor-Hours	Unit	Material	Labor	Equipment	Total	Total Incl O&P
0010	**LIGHTNING PROTECTION FOR BUILDINGS**									
0200	Air terminals & base, copper									
0400	3/8" diameter x 10" (to 75' high)	1 Elec	8	1	Ea.	24	41.50		65.50	94
1000	Aluminum, 1/2" diameter x 12" (to 75' high)		8	1		19.50	41.50		61	89
1020	1/2" diameter x 24"		7.30	1.096		21.50	45.50		67	97.50
1040	1/2" diameter x 60"		6.70	1.194		28	49.50		77.50	112
2000	Cable, copper, 220 lb. per thousand ft. (to 75' high)		320	.025	L.F.	3.05	1.04		4.09	5.05
2500	Aluminum, 101 lb. per thousand ft. (to 75' high)		280	.029	"	.95	1.18		2.13	2.98
3000	Arrester, 175 volt AC to ground		8	1	Ea.	146	41.50		187.50	229

26 51 13 – Interior Lighting Fixtures, Lamps, and Ballasts

26 51 13.50 Interior Lighting Fixtures		Crew	Daily Output	Labor-Hours	Unit	2019 Bare Costs Material	Labor	Equipment	Total	Total Incl O&P
0010	**INTERIOR LIGHTING FIXTURES** Including lamps, mounting									
0030	hardware and connections									
0100	Fluorescent, C.W. lamps, troffer, recess mounted in grid, RS									
0130	Grid ceiling mount									
0200	Acrylic lens, 1′ W x 4′ L, two 40 watt	1 Elec	5.70	1.404	Ea.	51.50	58		109.50	152
0300	2′ W x 2′ L, two U40 watt		5.70	1.404		55	58		113	156
0600	2′ W x 4′ L, four 40 watt	↓	4.70	1.702	↓	62.50	70.50		133	184
1000	Surface mounted, RS									
1030	Acrylic lens with hinged & latched door frame									
1100	1′ W x 4′ L, two 40 watt	1 Elec	7	1.143	Ea.	66	47.50		113.50	150
1200	2′ W x 2′ L, two U40 watt		7	1.143		71	47.50		118.50	155
1500	2′ W x 4′ L, four 40 watt		5.30	1.509		85	62.50		147.50	196
1501	2′ W x 4′ L, six 40 watt T8	↓	5.20	1.538	↓	85	63.50		148.50	198
2100	Strip fixture									
2200	4′ long, one 40 watt, RS	1 Elec	8.50	.941	Ea.	31	39		70	97.50
2300	4′ long, two 40 watt, RS	"	8	1		43.50	41.50		85	116
2600	8′ long, one 75 watt, SL	2 Elec	13.40	1.194		55	49.50		104.50	142
2700	8′ long, two 75 watt, SL	"	12.40	1.290	↓	66.50	53.50		120	160
4450	Incandescent, high hat can, round alzak reflector, prewired									
4470	100 watt	1 Elec	8	1	Ea.	68	41.50		109.50	142
4480	150 watt		8	1		104	41.50		145.50	183
4500	300 watt	↓	6.70	1.194	↓	241	49.50		290.50	345
5200	Ceiling, surface mounted, opal glass drum									
5300	8″, one 60 watt lamp	1 Elec	10	.800	Ea.	66.50	33		99.50	127
5400	10″, two 60 watt lamps		8	1		74	41.50		115.50	149
5500	12″, four 60 watt lamps		6.70	1.194		104	49.50		153.50	196
6900	Mirror light, fluorescent, RS, acrylic enclosure, two 40 watt		8	1		113	41.50		154.50	192
6910	One 40 watt		8	1		97	41.50		138.50	175
6920	One 20 watt	↓	12	.667	↓	83.50	27.50		111	137

26 51 13.55 Interior LED Fixtures		Crew	Daily Output	Labor-Hours	Unit	Material	Labor	Equipment	Total	Total Incl O&P
0010	**INTERIOR LED FIXTURES** Incl. lamps and mounting hardware									
0100	Downlight, recess mounted, 7.5″ diameter, 25 watt G	1 Elec	8	1	Ea.	335	41.50		376.50	440
0120	10″ diameter, 36 watt G		8	1		360	41.50		401.50	465
0160	cylinder, 10 watts G		8	1		104	41.50		145.50	182
0180	20 watts G		8	1		129	41.50		170.50	210
1000	Troffer, recess mounted, 2′ x 4′, 3,200 lumens G		5.30	1.509		138	62.50		200.50	253
1010	4,800 lumens G		5	1.600		178	66		244	305
1020	6,400 lumens G		4.70	1.702		190	70.50		260.50	325
1100	Troffer retrofit lamp, 38 watt G		21	.381		69.50	15.75		85.25	102
1110	60 watt G		20	.400		142	16.55		158.55	183
1120	100 watt G		18	.444		206	18.40		224.40	257
1200	Troffer, volumetric recess mounted, 2′ x 2′ G		5.70	1.404		320	58		378	445
2000	Strip, surface mounted, one light bar 4′ long, 3,500 K G		8.50	.941		310	39		349	405
2010	5,000 K G		8	1		276	41.50		317.50	375
2020	Two light bar 4′ long, 5,000 K G		7	1.143		430	47.50		477.50	550
3000	Linear, suspended mounted, one light bar 4′ long, 37 watt G	↓	6.70	1.194		168	49.50		217.50	265
3010	One light bar 8′ long, 74 watt G	2 Elec	12.20	1.311		310	54.50		364.50	430
3020	Two light bar 4′ long, 74 watt G	1 Elec	5.70	1.404		335	58		393	460
3030	Two light bar 8′ long, 148 watt G	2 Elec	8.80	1.818		380	75.50		455.50	540
4000	High bay, surface mounted, round, 150 watts G		5.41	2.959		425	122		547	670
4010	2 bars, 164 watts G		5.41	2.959		385	122		507	625
4020	3 bars, 246 watts G	↓	5.01	3.197	↓	520	132		652	785

26 51 Interior Lighting

26 51 13 – Interior Lighting Fixtures, Lamps, and Ballasts

26 51 13.55 Interior LED Fixtures			Crew	Daily Output	Labor-Hours	Unit	2019 Bare Costs Material	Labor	Equipment	Total	Total Incl O&P
4030	4 bars, 328 watts	G	2 Elec	4.60	3.478	Ea.	750	144		894	1,050
4040	5 bars, 410 watts	G	3 Elec	4.20	5.716		840	237		1,077	1,300
4050	6 bars, 492 watts	G		3.80	6.324		940	262		1,202	1,450
4060	7 bars, 574 watts	G		3.39	7.075		1,025	293		1,318	1,600
4070	8 bars, 656 watts	G		2.99	8.029		1,050	330		1,380	1,700
5000	Track, lighthead, 6 watt	G	1 Elec	32	.250		55.50	10.35		65.85	78
5010	9 watt	G	"	32	.250		63	10.35		73.35	86.50
6000	Garage, surface mounted, 103 watts	G	2 Elec	6.50	2.462		990	102		1,092	1,275
6100	pendent mounted, 80 watts	G		6.50	2.462		720	102		822	955
6200	95 watts	G		6.50	2.462		855	102		957	1,100
6300	125 watts	G		6.50	2.462		905	102		1,007	1,150

26 51 13.70 Residential Fixtures		Crew	Daily Output	Labor-Hours	Unit	Material	Labor	Equipment	Total	Total Incl O&P
0010	**RESIDENTIAL FIXTURES**									
0400	Fluorescent, interior, surface, circline, 32 watt & 40 watt	1 Elec	20	.400	Ea.	165	16.55		181.55	209
0500	2′ x 2′, two U-tube 32 watt T8		8	1		130	41.50		171.50	210
0700	Shallow under cabinet, two 20 watt		16	.500		70.50	20.50		91	112
0900	Wall mounted, 4′ L, two 32 watt T8, with baffle		10	.800		164	33		197	234
2000	Incandescent, exterior lantern, wall mounted, 60 watt		16	.500		61	20.50		81.50	101
2100	Post light, 150 W, with 7′ post		4	2		268	83		351	430
2500	Lamp holder, weatherproof with 150 W PAR		16	.500		35	20.50		55.50	72.50
2550	With reflector and guard		12	.667		69.50	27.50		97	121
2600	Interior pendent, globe with shade, 150 W		20	.400		204	16.55		220.55	252

26 55 Special Purpose Lighting

26 55 59 – Display Lighting

26 55 59.10 Track Lighting		Crew	Daily Output	Labor-Hours	Unit	Material	Labor	Equipment	Total	Total Incl O&P
0010	**TRACK LIGHTING**									
0100	8′ section	2 Elec	10.60	1.509	Ea.	71	62.50		133.50	180
0300	3 circuits, 4′ section	1 Elec	6.70	1.194		129	49.50		178.50	223
0400	8′ section	2 Elec	10.60	1.509		133	62.50		195.50	248
0500	12′ section	"	8.80	1.818		160	75.50		235.50	299
1000	Feed kit, surface mounting	1 Elec	16	.500		14.95	20.50		35.45	50.50
1100	End cover		24	.333		8.25	13.80		22.05	31.50
1200	Feed kit, stem mounting, 1 circuit		16	.500		51	20.50		71.50	90
1300	3 circuit		16	.500		51	20.50		71.50	90
2000	Electrical joiner, for continuous runs, 1 circuit		32	.250		39.50	10.35		49.85	60.50
2100	3 circuit		32	.250		83	10.35		93.35	108
2200	Fixtures, spotlight, 75 W PAR halogen		16	.500		51.50	20.50		72	91
2210	50 W MR16 halogen		16	.500		173	20.50		193.50	224
3000	Wall washer, 250 W tungsten halogen		16	.500		133	20.50		153.50	180
3100	Low voltage, 25/50 W, 1 circuit		16	.500		134	20.50		154.50	181
3120	3 circuit		16	.500		197	20.50		217.50	251

26 56 Exterior Lighting

26 56 13 – Lighting Poles and Standards

26 56 13.10 Lighting Poles		Crew	Daily Output	Labor-Hours	Unit	2019 Bare Costs Material	Labor	Equipment	Total	Total Incl O&P
0010	**LIGHTING POLES**									
6420	Wood pole, 4-1/2" x 5-1/8", 8' high	1 Elec	6	1.333	Ea.	385	55		440	515
6440	12' high		5.70	1.404		560	58		618	710
6460	20' high		4	2		795	83		878	1,000

26 56 23 – Area Lighting

26 56 23.10 Exterior Fixtures		Crew	Daily Output	Labor-Hours	Unit	Material	Labor	Equipment	Total	Total Incl O&P
0010	**EXTERIOR FIXTURES** With lamps									
0400	Quartz, 500 watt	1 Elec	5.30	1.509	Ea.	61.50	62.50		124	170
1100	Wall pack, low pressure sodium, 35 watt		4	2		202	83		285	360
1150	55 watt		4	2		240	83		323	400

26 56 26 – Landscape Lighting

26 56 26.20 Landscape Fixtures		Crew	Daily Output	Labor-Hours	Unit	Material	Labor	Equipment	Total	Total Incl O&P
0010	**LANDSCAPE FIXTURES**									
7380	Landscape recessed uplight, incl. housing, ballast, transformer									
7390	& reflector, not incl. conduit, wire, trench									
7420	Incandescent, 250 watt	1 Elec	5	1.600	Ea.	635	66		701	810
7440	Quartz, 250 watt	"	5	1.600	"	605	66		671	775

26 56 26.50 Landscape LED Fixtures		Crew	Daily Output	Labor-Hours	Unit	Material	Labor	Equipment	Total	Total Incl O&P
0010	**LANDSCAPE LED FIXTURES**									
0100	12 volt alum bullet hooded-BLK	1 Elec	5	1.600	Ea.	97	66		163	215
0200	12 volt alum bullet hooded-BRZ		5	1.600		97	66		163	215
0300	12 volt alum bullet hooded-GRN		5	1.600		97	66		163	215
1000	12 volt alum large bullet hooded-BLK		5	1.600		71.50	66		137.50	187
1100	12 volt alum large bullet hooded-BRZ		5	1.600		71.50	66		137.50	187
1200	12 volt alum large bullet hooded-GRN		5	1.600		71.50	66		137.50	187
2000	12 volt large bullet landscape light fixture		5	1.600		71.50	66		137.50	187
2100	12 volt alum light large bullet		5	1.600		71.50	66		137.50	187
2200	12 volt alum bullet light		5	1.600		71.50	66		137.50	187

26 56 33 – Walkway Lighting

26 56 33.10 Walkway Luminaire		Crew	Daily Output	Labor-Hours	Unit	Material	Labor	Equipment	Total	Total Incl O&P
0010	**WALKWAY LUMINAIRE**									
6500	Bollard light, lamp & ballast, 42" high with polycarbonate lens									
7200	Incandescent, 150 watt	1 Elec	3	2.667	Ea.	715	110		825	965

26 61 Lighting Systems and Accessories

26 61 23 – Lamps Applications

26 61 23.10 Lamps		Crew	Daily Output	Labor-Hours	Unit	Material	Labor	Equipment	Total	Total Incl O&P
0010	**LAMPS**									
0081	Fluorescent, rapid start, cool white, 2' long, 20 watt	1 Elec	100	.080	Ea.	3.75	3.31		7.06	9.55
0101	4' long, 40 watt		90	.089		2.48	3.68		6.16	8.75
1351	High pressure sodium, 70 watt		30	.267		15.85	11.05		26.90	35.50
1371	150 watt		30	.267		16.20	11.05		27.25	36

Estimating Tips

27 20 00 Data Communications
27 30 00 Voice Communications
27 40 00 Audio-Video Communications

- When estimating material costs for special systems, it is always prudent to obtain manufacturers' quotations for equipment prices and special installation requirements that may affect the total cost.
- For cost modifications for elevated tray installation, add the percentages to labor according to the height of the installation and only to the quantities exceeding the different height levels, not to the total tray quantities. Refer to subdivision 26 01 02.20 for labor adjustment factors.
- Do not overlook the costs for equipment used in the installation. If scissor lifts and boom lifts are available in the field, contractors may use them in lieu of the proposed ladders and rolling staging.

Reference Numbers

Reference numbers are shown at the beginning of some major classifications. These numbers refer to related items in the Reference Section. The reference information may be an estimating procedure, an alternate pricing method, or technical information.

Note: Not all subdivisions listed here necessarily appear. ■

Note: Trade Service, in part, has been used as a reference source for some of the material prices used in Division 27.

27 13 Communications Backbone Cabling

27 13 23 – Communications Optical Fiber Backbone Cabling

	27 13 23.13 Communications Optical Fiber	Crew	Daily Output	Labor-Hours	Unit	2019 Bare Costs Material	Labor	Equipment	Total	Total Incl O&P
0010	**COMMUNICATIONS OPTICAL FIBER**									
0040	Specialized tools & techniques cause installation costs to vary.									
0070	Fiber optic, cable, bulk simplex, single mode	1 Elec	8	1	C.L.F.	23	41.50		64.50	92.50
0080	Multi mode		8	1		29.50	41.50		71	100
0090	4 strand, single mode		7.34	1.090		37.50	45		82.50	115
0095	Multi mode		7.34	1.090		50	45		95	129
0100	12 strand, single mode		6.67	1.199		75.50	49.50		125	165
0105	Multi mode		6.67	1.199		99.50	49.50		149	190
0150	Jumper				Ea.	36			36	39.50
0200	Pigtail					34			34	37.50
0300	Connector	1 Elec	24	.333		25.50	13.80		39.30	50.50
0350	Finger splice		32	.250		37.50	10.35		47.85	58.50
0400	Transceiver (low cost bi-directional)		8	1		440	41.50		481.50	555
0450	Rack housing, 4 rack spaces, 12 panels (144 fibers)		2	4		610	166		776	940
1000	Cable, 62.5 microns, direct burial, 4 fiber	R-15	1200	.040	L.F.	.90	1.62	.24	2.76	3.90
1020	Indoor, 2 fiber	R-19	1000	.020		.44	.83		1.27	1.83
1040	Outdoor, aerial/duct	"	1670	.012		.65	.50		1.15	1.53
1060	50 microns, direct burial, 8 fiber	R-22	4000	.009		1.27	.35		1.62	1.98
1080	12 fiber		4000	.009		2.05	.35		2.40	2.84
1100	Indoor, 12 fiber		759	.049		1.99	1.86		3.85	5.25
1120	Connectors, 62.5 micron cable, transmission	R-19	40	.500	Ea.	14.15	21		35.15	49.50
1140	Cable splice		40	.500		14.90	21		35.90	50.50
1160	125 micron cable, transmission		16	1.250		14.90	52		66.90	101
1180	Receiver, 1.2 mile range		20	1		261	41.50		302.50	355
1200	1.9 mile range		20	1		221	41.50		262.50	310
1220	6.2 mile range		5	4		278	166		444	575
1240	Transmitter, 1.2 mile range		20	1		287	41.50		328.50	385
1260	1.9 mile range		20	1		279	41.50		320.50	375
1280	6.2 mile range		5	4		415	166		581	725
1300	Modem, 1.2 mile range		5	4		172	166		338	460
1320	6.2 mile range		5	4		315	166		481	615
1340	1.9 mile range, 12 channel		5	4		2,000	166		2,166	2,475
1360	Repeater, 1.2 mile range		10	2		370	83		453	540
1380	1.9 mile range		10	2		475	83		558	655
1400	6.2 mile range		5	4		910	166		1,076	1,275
1420	1.2 mile range, digital		5	4		435	166		601	750

27 41 Audio-Video Systems

27 41 33 – Master Antenna Television Systems

	27 41 33.10 TV Systems	Crew	Daily Output	Labor-Hours	Unit	Material	Labor	Equipment	Total	Total Incl O&P
0010	**TV SYSTEMS,** not including rough-in wires, cables & conduits									
0100	Master TV antenna system									
0200	VHF reception & distribution, 12 outlets	1 Elec	6	1.333	Outlet	118	55		173	220
0800	VHF & UHF reception & distribution, 12 outlets		6	1.333	"	251	55		306	365
5000	Antenna, small		6	1.333	Ea.	55.50	55		110.50	151
5100	Large		4	2		233	83		316	390
5110	Rotor unit		8	1		64	41.50		105.50	138
5120	Single booster		8	1		31	41.50		72.50	102
5130	Antenna pole, 10′		3.20	2.500		19.70	104		123.70	191
6100	Satellite TV system	2 Elec	1	16		2,175	660		2,835	3,450
6110	Dish, mesh, 10′ diam.	"	2.40	6.667		1,725	276		2,001	2,350

27 41 33 – Master Antenna Television Systems

27 41 33.10 TV Systems		Crew	Daily Output	Labor-Hours	Unit	2019 Bare Costs Material	Labor	Equipment	Total	Total Incl O&P
6111	Two way RF/IF tapeoff	1 Elec	36	.222	Ea.	3.59	9.20		12.79	18.95
6112	Two way RF/IF splitter		24	.333		13.80	13.80		27.60	37.50
6113	Line amplifier		24	.333		13.15	13.80		26.95	37
6114	Line splitters		36	.222		4.59	9.20		13.79	20
6115	Line multi switches		8	1		660	41.50		701.50	795
6120	Motor unit		2.40	3.333		263	138		401	515
7000	Home theater, widescreen, 42", high definition, TV		10.24	.781		400	32.50		432.50	495
7050	Flat wall mount bracket		10.24	.781		68.50	32.50		101	129
7100	7 channel home theater receiver		10.24	.781		385	32.50		417.50	480
7200	Home theater speakers		10.24	.781	Set	216	32.50		248.50	290
7300	Home theater programmable remote		10.24	.781	Ea.	250	32.50		282.50	330
8000	Main video splitter		4	2		3,275	83		3,358	3,725
8010	Video distribution units		4	2		141	83		224	290

Division Notes

		CREW	DAILY OUTPUT	LABOR-HOURS	UNIT	BARE COSTS MAT.	LABOR	EQUIP.	TOTAL	TOTAL INCL O&P

Estimating Tips

- When estimating material costs for electronic safety and security systems, it is always prudent to obtain manufacturers' quotations for equipment prices and special installation requirements that may affect the total cost.
- Fire alarm systems consist of control panels, annunciator panels, batteries with rack, charger, and fire alarm actuating and indicating devices. Some fire alarm systems include speakers, telephone lines, door closer controls, and other components. Be careful not to overlook the costs related to installation for these items. Also be aware of costs for integrated automation instrumentation and terminal devices, control equipment, control wiring, and programming. Insurance underwriters may have specific requirements for the type of materials to be installed or design requirements based on the hazard to be protected. Local jurisdictions may have requirements not covered by code. It is advisable to be aware of any special conditions.
- Security equipment includes items such as CCTV, access control, and other detection and identification systems to perform alert and alarm functions. Be sure to consider the costs related to installation for this security equipment, such as for integrated automation instrumentation and terminal devices, control equipment, control wiring, and programming.

Reference Numbers

Reference numbers are shown at the beginning of some major classifications. These numbers refer to related items in the Reference Section. The reference information may be an estimating procedure, an alternate pricing method, or technical information.

Note: Trade Service, in part, has been used as a reference source for some of the material prices used in Division 28.

28 31 Intrusion Detection

28 31 16 – Intrusion Detection Systems Infrastructure

28 31 16.50 Intrusion Detection		Crew	Daily Output	Labor-Hours	Unit	2019 Bare Costs Material	Labor	Equipment	Total	Total Incl O&P
0010	**INTRUSION DETECTION**, not including wires & conduits									
0100	Burglar alarm, battery operated, mechanical trigger	1 Elec	4	2	Ea.	281	83		364	445
0200	Electrical trigger		4	2		335	83		418	505
0400	For outside key control, add		8	1		87	41.50		128.50	163
0600	For remote signaling circuitry, add		8	1		144	41.50		185.50	227
0800	Card reader, flush type, standard		2.70	2.963		815	123		938	1,100
1000	Multi-code		2.70	2.963		1,225	123		1,348	1,525
1200	Door switches, hinge switch		5.30	1.509		64.50	62.50		127	173
1400	Magnetic switch		5.30	1.509		105	62.50		167.50	218
2800	Ultrasonic motion detector, 12 V		2.30	3.478		200	144		344	455
3000	Infrared photoelectric detector		4	2		135	83		218	283
3200	Passive infrared detector		4	2		238	83		321	395
3420	Switchmats, 30" x 5'		5.30	1.509		105	62.50		167.50	218
3440	30" x 25'		4	2		206	83		289	360
3460	Police connect panel		4	2		285	83		368	450
3480	Telephone dialer		5.30	1.509		390	62.50		452.50	530
3500	Alarm bell		4	2		106	83		189	251
3520	Siren	↓	4	2	↓	148	83		231	298

28 46 Fire Detection and Alarm

28 46 11 – Fire Sensors and Detectors

28 46 11.21 Carbon-Monoxide Detection Sensors		Crew	Daily Output	Labor-Hours	Unit	Material	Labor	Equipment	Total	Total Incl O&P
0010	**CARBON-MONOXIDE DETECTION SENSORS**									
8400	Smoke and carbon monoxide alarm battery operated photoelectric low profile	1 Elec	24	.333	Ea.	59	13.80		72.80	87.50
8410	low profile photoelectric battery powered		24	.333		36.50	13.80		50.30	62.50
8420	photoelectric low profile sealed lithium		24	.333		66.50	13.80		80.30	96
8430	Photoelectric low profile sealed lithium smoke and CO with voice combo		24	.333		41.50	13.80		55.30	68
8500	Carbon monoxide sensor, wall mount 1Mod 1 relay output smoke & heat		24	.333		505	13.80		518.80	580
8700	Carbon monoxide detector, battery operated, wall mounted		16	.500		46	20.50		66.50	84.50
8710	Hardwired, wall and ceiling mounted		8	1		90	41.50		131.50	167
8720	Duct mounted	↓	8	1	↓	350	41.50		391.50	455

28 46 11.27 Other Sensors		Crew	Daily Output	Labor-Hours	Unit	Material	Labor	Equipment	Total	Total Incl O&P
0010	**OTHER SENSORS**									
5200	Smoke detector, ceiling type	1 Elec	6.20	1.290	Ea.	123	53.50		176.50	222
5240	Smoke detector, addressable type		6	1.333		227	55		282	340
5420	Duct addressable type		3.20	2.500		525	104		629	745
8300	Smoke alarm with integrated strobe light 120 V, 16DB 60 fpm flash rate		16	.500		112	20.50		132.50	157
8310	Photoelectric smoke detector with strobe 120 V, 90 DB ceiling mount		12	.667		205	27.50		232.50	270
8320	120 V, 90 DB wall mount	↓	12	.667	↓	189	27.50		216.50	252

28 46 21 – Fire Alarm

28 46 21.50 Alarm Panels and Devices		Crew	Daily Output	Labor-Hours	Unit	Material	Labor	Equipment	Total	Total Incl O&P
0010	**ALARM PANELS AND DEVICES**, not including wires & conduits									
5600	Strobe and horn	1 Elec	5.30	1.509	Ea.	140	62.50		202.50	256
5610	Strobe and horn (ADA type)		5.30	1.509		168	62.50		230.50	287
5620	Visual alarm (ADA type)		6.70	1.194		117	49.50		166.50	210
5800	Fire alarm horn		6.70	1.194		55.50	49.50		105	142
6600	Drill switch		8	1		380	41.50		421.50	490
6800	Master box		2.70	2.963		6,575	123		6,698	7,425
7000	Break glass station		8	1		58	41.50		99.50	131
7800	Remote annunciator, 8 zone lamp	↓	1.80	4.444	↓	213	184		397	535

28 46 21.50 Alarm Panels and Devices		Crew	Daily Output	Labor-Hours	Unit	2019 Bare Costs Material	Labor	Equipment	Total	Total Incl O&P
8000	12 zone lamp	2 Elec	2.60	6.154	Ea.	415	255		670	875
8200	16 zone lamp	"	2.20	7.273	↓	375	300		675	900

Division Notes

		CREW	DAILY OUTPUT	LABOR-HOURS	UNIT	BARE COSTS				TOTAL INCL O&P
						MAT.	LABOR	EQUIP.	TOTAL	

Estimating Tips

31 05 00 Common Work Results for Earthwork

- Estimating the actual cost of performing earthwork requires careful consideration of the variables involved. This includes items such as type of soil, whether water will be encountered, dewatering, whether banks need bracing, disposal of excavated earth, and length of haul to fill or spoil sites, etc. If the project has large quantities of cut or fill, consider raising or lowering the site to reduce costs, while paying close attention to the effect on site drainage and utilities.
- If the project has large quantities of fill, creating a borrow pit on the site can significantly lower the costs.
- It is very important to consider what time of year the project is scheduled for completion. Bad weather can create large cost overruns from dewatering, site repair, and lost productivity from cold weather.

Reference Numbers

Reference numbers are shown at the beginning of some major classifications. These numbers refer to related items in the Reference Section. The reference information may be an estimating procedure, an alternate pricing method, or technical information.

Note: Not all subdivisions listed here necessarily appear. ■

31 05 Common Work Results for Earthwork

31 05 13 – Soils for Earthwork

31 05 13.10 Borrow		Crew	Daily Output	Labor-Hours	Unit	Material	Labor (2019 Bare Costs)	Equipment (2019 Bare Costs)	Total (2019 Bare Costs)	Total Incl O&P
0010	BORROW									
0020	Spread, 200 HP dozer, no compaction, 2 mile RT haul									
0200	Common borrow	B-15	600	.047	C.Y.	12.60	1.56	4.04	18.20	21

31 05 16 – Aggregates for Earthwork

31 05 16.10 Borrow		Crew	Daily Output	Labor-Hours	Unit	Material	Labor (2019 Bare Costs)	Equipment (2019 Bare Costs)	Total (2019 Bare Costs)	Total Incl O&P
0010	BORROW									
0020	Spread, with 200 HP dozer, no compaction, 2 mile RT haul									
0100	Bank run gravel	B-15	600	.047	L.C.Y.	18.65	1.56	4.04	24.25	27.50
0300	Crushed stone (1.40 tons per C.Y.), 1-1/2"		600	.047		27.50	1.56	4.04	33.10	37
0320	3/4"		600	.047		27.50	1.56	4.04	33.10	37
0340	1/2"		600	.047		24.50	1.56	4.04	30.10	34
0360	3/8"		600	.047		16.95	1.56	4.04	22.55	25.50
0400	Sand, washed, concrete		600	.047		36	1.56	4.04	41.60	47
0500	Dead or bank sand		600	.047		18.10	1.56	4.04	23.70	27

31 11 Clearing and Grubbing

31 11 10 – Clearing and Grubbing Land

31 11 10.10 Clear and Grub Site		Crew	Daily Output	Labor-Hours	Unit	Material	Labor (2019 Bare Costs)	Equipment (2019 Bare Costs)	Total (2019 Bare Costs)	Total Incl O&P
0010	CLEAR AND GRUB SITE									
0020	Cut & chip light trees to 6" diam.	B-7	1	48	Acre		1,425	1,675	3,100	4,175
0150	Grub stumps and remove	B-30	2	12			415	1,025	1,440	1,800
0200	Cut & chip medium trees to 12" diam.	B-7	.70	68.571			2,025	2,400	4,425	5,975
0250	Grub stumps and remove	B-30	1	24			825	2,050	2,875	3,600
0300	Cut & chip heavy trees to 24" diam.	B-7	.30	160			4,725	5,625	10,350	14,000
0350	Grub stumps and remove	B-30	.50	48			1,650	4,075	5,725	7,225
0400	If burning is allowed, deduct cut & chip								40%	40%

31 13 Selective Tree and Shrub Removal and Trimming

31 13 13 – Selective Tree and Shrub Removal

31 13 13.10 Selective Clearing		Crew	Daily Output	Labor-Hours	Unit	Material	Labor (2019 Bare Costs)	Equipment (2019 Bare Costs)	Total (2019 Bare Costs)	Total Incl O&P
0010	SELECTIVE CLEARING									
0020	Clearing brush with brush saw	A-1C	.25	32	Acre		880	110	990	1,575
0100	By hand	1 Clab	.12	66.667			1,825		1,825	3,025
0300	With dozer, ball and chain, light clearing	B-11A	2	8			260	645	905	1,150
0400	Medium clearing	"	1.50	10.667			345	860	1,205	1,525

31 14 Earth Stripping and Stockpiling

31 14 13 – Soil Stripping and Stockpiling

31 14 13.23 Topsoil Stripping and Stockpiling		Crew	Daily Output	Labor-Hours	Unit	Material	Labor (2019 Bare Costs)	Equipment (2019 Bare Costs)	Total (2019 Bare Costs)	Total Incl O&P
0010	TOPSOIL STRIPPING AND STOCKPILING									
1400	Loam or topsoil, remove and stockpile on site									
1420	6" deep, 200' haul	B-10B	865	.009	C.Y.		.35	1.49	1.84	2.21
1430	300' haul		520	.015			.58	2.48	3.06	3.68
1440	500' haul		225	.036			1.33	5.75	7.08	8.50
1450	Alternate method: 6" deep, 200' haul		5090	.002	S.Y.		.06	.25	.31	.38
1460	500' haul		1325	.006	"		.23	.97	1.20	1.44
1500	Loam or topsoil, remove/stockpile on site									

31 14 Earth Stripping and Stockpiling

31 14 13 – Soil Stripping and Stockpiling

31 14 13.23 Topsoil Stripping and Stockpiling		Crew	Daily Output	Labor-Hours	Unit	Material	2019 Bare Costs Labor	Equipment	Total	Total Incl O&P
1510	By hand, 6" deep, 50' haul, less than 100 S.Y.	B-1	100	.240	S.Y.		6.75		6.75	11.15
1520	By skid steer, 6" deep, 100' haul, 101-500 S.Y.	B-62	500	.048			1.45	.35	1.80	2.77
1530	100' haul, 501-900 S.Y.	"	900	.027			.81	.19	1	1.54
1540	200' haul, 901-1,100 S.Y.	B-63	1000	.040			1.10	.17	1.27	2.01
1550	By dozer, 200' haul, 1,101-4,000 S.Y.	B-10B	4000	.002	↓		.07	.32	.39	.47

31 22 Grading

31 22 16 – Fine Grading

31 22 16.10 Finish Grading		Crew	Daily Output	Labor-Hours	Unit	Material	Labor	Equipment	Total	Total Incl O&P
0010	**FINISH GRADING**									
0012	Finish grading area to be paved with grader, small area	B-11L	400	.040	S.Y.		1.30	1.64	2.94	3.95
0100	Large area		2000	.008			.26	.33	.59	.79
0200	Grade subgrade for base course, roadways	↓	3500	.005			.15	.19	.34	.45
1020	For large parking lots	B-32C	5000	.010			.32	.44	.76	1
1050	For small irregular areas	"	2000	.024			.79	1.09	1.88	2.50
1100	Fine grade for slab on grade, machine	B-11L	1040	.015			.50	.63	1.13	1.52
1150	Hand grading	B-18	700	.034			.97	.06	1.03	1.66
1200	Fine grade granular base for sidewalks and bikeways	B-62	1200	.020	↓		.60	.15	.75	1.16
2550	Hand grade select gravel	2 Clab	60	.267	C.S.F.		7.35		7.35	12.10
3000	Hand grade select gravel, including compaction, 4" deep	B-18	555	.043	S.Y.		1.22	.07	1.29	2.09
3100	6" deep		400	.060			1.69	.10	1.79	2.90
3120	8" deep	↓	300	.080			2.25	.14	2.39	3.87
3300	Finishing grading slopes, gentle	B-11L	8900	.002			.06	.07	.13	.18
3310	Steep slopes	"	7100	.002	↓		.07	.09	.16	.22

31 23 Excavation and Fill

31 23 16 – Excavation

31 23 16.13 Excavating, Trench		Crew	Daily Output	Labor-Hours	Unit	Material	Labor	Equipment	Total	Total Incl O&P
0010	**EXCAVATING, TRENCH**									
0011	Or continuous footing									
0050	1' to 4' deep, 3/8 C.Y. excavator	B-11C	150	.107	B.C.Y.		3.46	2.13	5.59	8.05
0060	1/2 C.Y. excavator	B-11M	200	.080			2.60	1.94	4.54	6.40
0090	4' to 6' deep, 1/2 C.Y. excavator	"	200	.080			2.60	1.94	4.54	6.40
0100	5/8 C.Y. excavator	B-12Q	250	.064			2.09	2.35	4.44	6.05
0300	1/2 C.Y. excavator, truck mounted	B-12J	200	.080			2.62	4.31	6.93	9.05
1352	4' to 6' deep, 1/2 C.Y. excavator w/trench box	B-13H	188	.085			2.78	5	7.78	10.10
1354	5/8 C.Y. excavator	"	235	.068			2.23	4	6.23	8.05
1400	By hand with pick and shovel 2' to 6' deep, light soil	1 Clab	8	1			27.50		27.50	45.50
1500	Heavy soil	"	4	2	↓		55		55	91
5020	Loam & sandy clay with no sheeting or dewatering included									
5050	1' to 4' deep, 3/8 C.Y. tractor loader/backhoe	B-11C	162	.099	B.C.Y.		3.21	1.98	5.19	7.45
5060	1/2 C.Y. excavator	B-11M	216	.074			2.41	1.80	4.21	5.95
5080	4' to 6' deep, 1/2 C.Y. excavator	"	216	.074			2.41	1.80	4.21	5.95
5090	5/8 C.Y. excavator	B-12Q	276	.058			1.90	2.13	4.03	5.45
5130	1/2 C.Y. excavator, truck mounted	B-12J	216	.074			2.42	3.99	6.41	8.40
5352	4' to 6' deep, 1/2 C.Y. excavator w/trench box	B-13H	205	.078			2.55	4.59	7.14	9.25
5354	5/8 C.Y. excavator	"	257	.062	↓		2.04	3.66	5.70	7.40
6020	Sand & gravel with no sheeting or dewatering included									
6050	1' to 4' deep, 3/8 C.Y. excavator	B-11C	165	.097	B.C.Y.		3.15	1.94	5.09	7.35

31 23 Excavation and Fill

31 23 16 – Excavation

31 23 16.13 Excavating, Trench		Crew	Daily Output	Labor-Hours	Unit	Material	2019 Bare Costs Labor	Equipment	Total	Total Incl O&P
6060	1/2 C.Y. excavator	B-11M	220	.073	B.C.Y.		2.36	1.76	4.12	5.85
6080	4′ to 6′ deep, 1/2 C.Y. excavator	"	220	.073			2.36	1.76	4.12	5.85
6090	5/8 C.Y. excavator	B-12Q	275	.058			1.90	2.14	4.04	5.50
6130	1/2 C.Y. excavator, truck mounted	B-12J	220	.073			2.38	3.92	6.30	8.25
6352	4′ to 6′ deep, 1/2 C.Y. excavator w/trench box	B-13H	209	.077			2.50	4.50	7	9.05
6354	5/8 C.Y. excavator	"	261	.061			2	3.60	5.60	7.25
7020	Dense hard clay with no sheeting or dewatering included									
7050	1′ to 4′ deep, 3/8 C.Y. excavator	B-11C	132	.121	B.C.Y.		3.94	2.43	6.37	9.15
7060	1/2 C.Y. excavator	B-11M	176	.091			2.95	2.20	5.15	7.30
7080	4′ to 6′ deep, 1/2 C.Y. excavator	"	176	.091			2.95	2.20	5.15	7.30
7090	5/8 C.Y. excavator	B-12Q	220	.073			2.38	2.67	5.05	6.85
7130	1/2 C.Y. excavator, truck mounted	B-12J	176	.091			2.97	4.90	7.87	10.30

31 23 16.14 Excavating, Utility Trench

		Crew	Daily Output	Labor-Hours	Unit	Material	Labor	Equipment	Total	Total Incl O&P
0010	**EXCAVATING, UTILITY TRENCH**									
0011	Common earth									
0050	Trenching with chain trencher, 12 HP, operator walking									
0100	4″ wide trench, 12″ deep	B-53	800	.010	L.F.		.28	.08	.36	.54
1000	Backfill by hand including compaction, add									
1050	4″ wide trench, 12″ deep	A-1G	800	.010	L.F.		.28	.07	.35	.52

31 23 16.16 Structural Excavation for Minor Structures

		Crew	Daily Output	Labor-Hours	Unit	Material	Labor	Equipment	Total	Total Incl O&P
0010	**STRUCTURAL EXCAVATION FOR MINOR STRUCTURES**									
0015	Hand, pits to 6′ deep, sandy soil	1 Clab	8	1	B.C.Y.		27.50		27.50	45.50
0100	Heavy soil or clay		4	2			55		55	91
1100	Hand loading trucks from stock pile, sandy soil		12	.667			18.35		18.35	30.50
1300	Heavy soil or clay		8	1			27.50		27.50	45.50
1500	For wet or muck hand excavation, add to above								50%	50%

31 23 16.42 Excavating, Bulk Bank Measure

		Crew	Daily Output	Labor-Hours	Unit	Material	Labor	Equipment	Total	Total Incl O&P
0010	**EXCAVATING, BULK BANK MEASURE**									
0011	Common earth piled									
0020	For loading onto trucks, add								15%	15%
0200	Excavator, hydraulic, crawler mtd., 1 C.Y. cap. = 100 C.Y./hr.	B-12A	800	.020	B.C.Y.		.65	.94	1.59	2.11
0310	Wheel mounted, 1/2 C.Y. cap. = 40 C.Y./hr.	B-12E	320	.050			1.64	1.39	3.03	4.22
1200	Front end loader, track mtd., 1-1/2 C.Y. cap. = 70 C.Y./hr.	B-10N	560	.014			.54	1.06	1.60	2.04
1500	Wheel mounted, 3/4 C.Y. cap. = 45 C.Y./hr.	B-10R	360	.022			.83	.83	1.66	2.28
5000	Excavating, bulk bank measure, sandy clay & loam piled									
5020	For loading onto trucks, add								15%	15%
5100	Excavator, hydraulic, crawler mtd., 1 C.Y. cap. = 120 C.Y./hr.	B-12A	960	.017	B.C.Y.		.55	.78	1.33	1.76
5610	Wheel mounted, 1/2 C.Y. cap. = 44 C.Y./hr.	B-12E	352	.045	"		1.49	1.26	2.75	3.84
8000	For hauling excavated material, see Section 31 23 23.20									

31 23 23 – Fill

31 23 23.13 Backfill

		Crew	Daily Output	Labor-Hours	Unit	Material	Labor	Equipment	Total	Total Incl O&P
0010	**BACKFILL**									
0015	By hand, no compaction, light soil	1 Clab	14	.571	L.C.Y.		15.70		15.70	26
0100	Heavy soil		11	.727	"		20		20	33
0300	Compaction in 6″ layers, hand tamp, add to above		20.60	.388	E.C.Y.		10.70		10.70	17.65
0400	Roller compaction operator walking, add	B-10A	100	.080			3	1.82	4.82	6.90
0500	Air tamp, add	B-9D	190	.211			5.85	1.42	7.27	11.25
0600	Vibrating plate, add	A-1D	60	.133			3.67	.52	4.19	6.60
0800	Compaction in 12″ layers, hand tamp, add to above	1 Clab	34	.235			6.45		6.45	10.70
1300	Dozer backfilling, bulk, up to 300′ haul, no compaction	B-10B	1200	.007	L.C.Y.		.25	1.08	1.33	1.59
1400	Air tamped, add	B-11B	80	.200	E.C.Y.		6.30	3.59	9.89	14.35

31 23 23.16 Fill By Borrow and Utility Bedding

	31 23 23.16 Fill By Borrow and Utility Bedding	Crew	Daily Output	Labor-Hours	Unit	Material	2019 Bare Costs Labor	Equipment	Total	Total Incl O&P
0010	**FILL BY BORROW AND UTILITY BEDDING**									
0049	Utility bedding, for pipe & conduit, not incl. compaction									
0050	Crushed or screened bank run gravel	B-6	150	.160	L.C.Y.	21	4.84	2.13	27.97	34
0100	Crushed stone 3/4" to 1/2"		150	.160		27.50	4.84	2.13	34.47	40.50
0200	Sand, dead or bank		150	.160		18.10	4.84	2.13	25.07	30
0500	Compacting bedding in trench	A-1D	90	.089	E.C.Y.		2.44	.35	2.79	4.42
0600	If material source exceeds 2 miles, add for extra mileage.									
0610	See Section 31 23 23.20 for hauling mileage add.									

31 23 23.17 General Fill

	31 23 23.17 General Fill	Crew	Daily Output	Labor-Hours	Unit	Material	Labor	Equipment	Total	Total Incl O&P
0010	**GENERAL FILL**									
0011	Spread dumped material, no compaction									
0020	By dozer	B-10B	1000	.008	L.C.Y.		.30	1.29	1.59	1.91
0100	By hand	1 Clab	12	.667	"		18.35		18.35	30.50
0500	Gravel fill, compacted, under floor slabs, 4" deep	B-37	10000	.005	S.F.	.40	.14	.02	.56	.69
0600	6" deep		8600	.006		.61	.16	.02	.79	.96
0700	9" deep		7200	.007		1.01	.19	.02	1.22	1.45
0800	12" deep		6000	.008		1.41	.23	.03	1.67	1.98
1000	Alternate pricing method, 4" deep		120	.400	E.C.Y.	30.50	11.70	1.27	43.47	54
1100	6" deep		160	.300		30.50	8.75	.95	40.20	49
1200	9" deep		200	.240		30.50	7	.76	38.26	46
1300	12" deep		220	.218		30.50	6.35	.69	37.54	45

31 23 23.20 Hauling

	31 23 23.20 Hauling	Crew	Daily Output	Labor-Hours	Unit	Material	Labor	Equipment	Total	Total Incl O&P
0010	**HAULING**									
0011	Excavated or borrow, loose cubic yards									
0012	no loading equipment, including hauling, waiting, loading/dumping									
0013	time per cycle (wait, load, travel, unload or dump & return)									
0014	8 C.Y. truck, 15 MPH avg., cycle 0.5 miles, 10 min. wait/ld./uld.	B-34A	320	.025	L.C.Y.		.82	1.06	1.88	2.52
0016	cycle 1 mile		272	.029			.97	1.25	2.22	2.96
0018	cycle 2 miles		208	.038			1.27	1.63	2.90	3.88
0020	cycle 4 miles		144	.056			1.83	2.36	4.19	5.60
0022	cycle 6 miles		112	.071			2.35	3.03	5.38	7.20
0024	cycle 8 miles		88	.091			2.99	3.86	6.85	9.15
0026	20 MPH avg., cycle 0.5 mile		336	.024			.78	1.01	1.79	2.40
0028	cycle 1 mile		296	.027			.89	1.15	2.04	2.72
0030	cycle 2 miles		240	.033			1.10	1.41	2.51	3.36
0032	cycle 4 miles		176	.045			1.50	1.93	3.43	4.58
0034	cycle 6 miles		136	.059			1.94	2.50	4.44	5.95
0036	cycle 8 miles		112	.071			2.35	3.03	5.38	7.20
0044	25 MPH avg., cycle 4 miles		192	.042			1.37	1.77	3.14	4.20
0046	cycle 6 miles		160	.050			1.65	2.12	3.77	5.05
0048	cycle 8 miles		128	.063			2.06	2.65	4.71	6.30
0050	30 MPH avg., cycle 4 miles		216	.037			1.22	1.57	2.79	3.73
0052	cycle 6 miles		176	.045			1.50	1.93	3.43	4.58
0054	cycle 8 miles		144	.056			1.83	2.36	4.19	5.60
0114	15 MPH avg., cycle 0.5 mile, 15 min. wait/ld./uld.		224	.036			1.17	1.52	2.69	3.60
0116	cycle 1 mile		200	.040			1.32	1.70	3.02	4.03
0118	cycle 2 miles		168	.048			1.57	2.02	3.59	4.80
0120	cycle 4 miles		120	.067			2.19	2.83	5.02	6.70
0122	cycle 6 miles		96	.083			2.74	3.54	6.28	8.40
0124	cycle 8 miles		80	.100			3.29	4.25	7.54	10.05
0126	20 MPH avg., cycle 0.5 mile		232	.034			1.13	1.46	2.59	3.48
0128	cycle 1 mile		208	.038			1.27	1.63	2.90	3.88

31 23 23.20 Hauling		Crew	Daily Output	Labor-Hours	Unit	Material	2019 Bare Costs Labor	Equipment	Total	Total Incl O&P
0130	cycle 2 miles	B-34A	184	.043	L.C.Y.		1.43	1.85	3.28	4.38
0132	cycle 4 miles		144	.056			1.83	2.36	4.19	5.60
0134	cycle 6 miles		112	.071			2.35	3.03	5.38	7.20
0136	cycle 8 miles		96	.083			2.74	3.54	6.28	8.40
0144	25 MPH avg., cycle 4 miles		152	.053			1.73	2.23	3.96	5.30
0146	cycle 6 miles		128	.063			2.06	2.65	4.71	6.30
0148	cycle 8 miles		112	.071			2.35	3.03	5.38	7.20
0150	30 MPH avg., cycle 4 miles		168	.048			1.57	2.02	3.59	4.80
0152	cycle 6 miles		144	.056			1.83	2.36	4.19	5.60
0154	cycle 8 miles		120	.067			2.19	2.83	5.02	6.70
0214	15 MPH avg., cycle 0.5 mile, 20 min. wait/ld./uld.		176	.045			1.50	1.93	3.43	4.58
0216	cycle 1 mile		160	.050			1.65	2.12	3.77	5.05
0218	cycle 2 miles		136	.059			1.94	2.50	4.44	5.95
0220	cycle 4 miles		104	.077			2.53	3.27	5.80	7.75
0222	cycle 6 miles		88	.091			2.99	3.86	6.85	9.15
0224	cycle 8 miles		72	.111			3.66	4.72	8.38	11.20
0226	20 MPH avg., cycle 0.5 mile		176	.045			1.50	1.93	3.43	4.58
0228	cycle 1 mile		168	.048			1.57	2.02	3.59	4.80
0230	cycle 2 miles		144	.056			1.83	2.36	4.19	5.60
0232	cycle 4 miles		120	.067			2.19	2.83	5.02	6.70
0234	cycle 6 miles		96	.083			2.74	3.54	6.28	8.40
0236	cycle 8 miles		88	.091			2.99	3.86	6.85	9.15
0244	25 MPH avg., cycle 4 miles		128	.063			2.06	2.65	4.71	6.30
0246	cycle 6 miles		112	.071			2.35	3.03	5.38	7.20
0248	cycle 8 miles		96	.083			2.74	3.54	6.28	8.40
0250	30 MPH avg., cycle 4 miles		136	.059			1.94	2.50	4.44	5.95
0252	cycle 6 miles		120	.067			2.19	2.83	5.02	6.70
0254	cycle 8 miles		104	.077			2.53	3.27	5.80	7.75
0314	15 MPH avg., cycle 0.5 mile, 25 min. wait/ld./uld.		144	.056			1.83	2.36	4.19	5.60
0316	cycle 1 mile		128	.063			2.06	2.65	4.71	6.30
0318	cycle 2 miles		112	.071			2.35	3.03	5.38	7.20
0320	cycle 4 miles		96	.083			2.74	3.54	6.28	8.40
0322	cycle 6 miles		80	.100			3.29	4.25	7.54	10.05
0324	cycle 8 miles		64	.125			4.11	5.30	9.41	12.60
0326	20 MPH avg., cycle 0.5 mile		144	.056			1.83	2.36	4.19	5.60
0328	cycle 1 mile		136	.059			1.94	2.50	4.44	5.95
0330	cycle 2 miles		120	.067			2.19	2.83	5.02	6.70
0332	cycle 4 miles		104	.077			2.53	3.27	5.80	7.75
0334	cycle 6 miles		88	.091			2.99	3.86	6.85	9.15
0336	cycle 8 miles		80	.100			3.29	4.25	7.54	10.05
0344	25 MPH avg., cycle 4 miles		112	.071			2.35	3.03	5.38	7.20
0346	cycle 6 miles		96	.083			2.74	3.54	6.28	8.40
0348	cycle 8 miles		88	.091			2.99	3.86	6.85	9.15
0350	30 MPH avg., cycle 4 miles		112	.071			2.35	3.03	5.38	7.20
0352	cycle 6 miles		104	.077			2.53	3.27	5.80	7.75
0354	cycle 8 miles		96	.083			2.74	3.54	6.28	8.40
0414	15 MPH avg., cycle 0.5 mile, 30 min. wait/ld./uld.		120	.067			2.19	2.83	5.02	6.70
0416	cycle 1 mile		112	.071			2.35	3.03	5.38	7.20
0418	cycle 2 miles		96	.083			2.74	3.54	6.28	8.40
0420	cycle 4 miles		80	.100			3.29	4.25	7.54	10.05
0422	cycle 6 miles		72	.111			3.66	4.72	8.38	11.20
0424	cycle 8 miles		64	.125			4.11	5.30	9.41	12.60
0426	20 MPH avg., cycle 0.5 mile		120	.067			2.19	2.83	5.02	6.70

31 23 23.20 Hauling		Crew	Daily Output	Labor-Hours	Unit	Material	2019 Bare Costs Labor	Equipment	Total	Total Incl O&P
0428	cycle 1 mile	B-34A	112	.071	L.C.Y.		2.35	3.03	5.38	7.20
0430	cycle 2 miles		104	.077			2.53	3.27	5.80	7.75
0432	cycle 4 miles		88	.091			2.99	3.86	6.85	9.15
0434	cycle 6 miles		80	.100			3.29	4.25	7.54	10.05
0436	cycle 8 miles		72	.111			3.66	4.72	8.38	11.20
0444	25 MPH avg., cycle 4 miles		96	.083			2.74	3.54	6.28	8.40
0446	cycle 6 miles		88	.091			2.99	3.86	6.85	9.15
0448	cycle 8 miles		80	.100			3.29	4.25	7.54	10.05
0450	30 MPH avg., cycle 4 miles		96	.083			2.74	3.54	6.28	8.40
0452	cycle 6 miles		88	.091			2.99	3.86	6.85	9.15
0454	cycle 8 miles		80	.100			3.29	4.25	7.54	10.05
0514	15 MPH avg., cycle 0.5 mile, 35 min. wait/ld./uld.		104	.077			2.53	3.27	5.80	7.75
0516	cycle 1 mile		96	.083			2.74	3.54	6.28	8.40
0518	cycle 2 miles		88	.091			2.99	3.86	6.85	9.15
0520	cycle 4 miles		72	.111			3.66	4.72	8.38	11.20
0522	cycle 6 miles		64	.125			4.11	5.30	9.41	12.60
0524	cycle 8 miles		56	.143			4.70	6.05	10.75	14.40
0526	20 MPH avg., cycle 0.5 mile		104	.077			2.53	3.27	5.80	7.75
0528	cycle 1 mile		96	.083			2.74	3.54	6.28	8.40
0530	cycle 2 miles		96	.083			2.74	3.54	6.28	8.40
0532	cycle 4 miles		80	.100			3.29	4.25	7.54	10.05
0534	cycle 6 miles		72	.111			3.66	4.72	8.38	11.20
0536	cycle 8 miles		64	.125			4.11	5.30	9.41	12.60
0544	25 MPH avg., cycle 4 miles		88	.091			2.99	3.86	6.85	9.15
0546	cycle 6 miles		80	.100			3.29	4.25	7.54	10.05
0548	cycle 8 miles		72	.111			3.66	4.72	8.38	11.20
0550	30 MPH avg., cycle 4 miles		88	.091			2.99	3.86	6.85	9.15
0552	cycle 6 miles		80	.100			3.29	4.25	7.54	10.05
0554	cycle 8 miles		72	.111			3.66	4.72	8.38	11.20
1014	12 C.Y. truck, cycle 0.5 mile, 15 MPH avg., 15 min. wait/ld./uld.	B-34B	336	.024			.78	1.69	2.47	3.15
1016	cycle 1 mile		300	.027			.88	1.89	2.77	3.52
1018	cycle 2 miles		252	.032			1.04	2.25	3.29	4.20
1020	cycle 4 miles		180	.044			1.46	3.15	4.61	5.85
1022	cycle 6 miles		144	.056			1.83	3.94	5.77	7.35
1024	cycle 8 miles		120	.067			2.19	4.73	6.92	8.80
1025	cycle 10 miles		96	.083			2.74	5.90	8.64	11
1026	20 MPH avg., cycle 0.5 mile		348	.023			.76	1.63	2.39	3.03
1028	cycle 1 mile		312	.026			.84	1.82	2.66	3.39
1030	cycle 2 miles		276	.029			.95	2.05	3	3.83
1032	cycle 4 miles		216	.037			1.22	2.63	3.85	4.89
1034	cycle 6 miles		168	.048			1.57	3.38	4.95	6.30
1036	cycle 8 miles		144	.056			1.83	3.94	5.77	7.35
1038	cycle 10 miles		120	.067			2.19	4.73	6.92	8.80
1040	25 MPH avg., cycle 4 miles		228	.035			1.15	2.49	3.64	4.64
1042	cycle 6 miles		192	.042			1.37	2.95	4.32	5.50
1044	cycle 8 miles		168	.048			1.57	3.38	4.95	6.30
1046	cycle 10 miles		144	.056			1.83	3.94	5.77	7.35
1050	30 MPH avg., cycle 4 miles		252	.032			1.04	2.25	3.29	4.20
1052	cycle 6 miles		216	.037			1.22	2.63	3.85	4.89
1054	cycle 8 miles		180	.044			1.46	3.15	4.61	5.85
1056	cycle 10 miles		156	.051			1.69	3.63	5.32	6.75
1060	35 MPH avg., cycle 4 miles		264	.030			1	2.15	3.15	4
1062	cycle 6 miles		228	.035			1.15	2.49	3.64	4.64

31 23 23.20 Hauling		Crew	Daily Output	Labor-Hours	Unit	Material	2019 Bare Costs Labor	Equipment	Total	Total Incl O&P
1064	cycle 8 miles	B-34B	204	.039	L.C.Y.		1.29	2.78	4.07	5.20
1066	cycle 10 miles		180	.044			1.46	3.15	4.61	5.85
1068	cycle 20 miles		120	.067			2.19	4.73	6.92	8.80
1069	cycle 30 miles		84	.095			3.13	6.75	9.88	12.60
1070	cycle 40 miles		72	.111			3.66	7.90	11.56	14.65
1072	40 MPH avg., cycle 6 miles		240	.033			1.10	2.36	3.46	4.40
1074	cycle 8 miles		216	.037			1.22	2.63	3.85	4.89
1076	cycle 10 miles		192	.042			1.37	2.95	4.32	5.50
1078	cycle 20 miles		120	.067			2.19	4.73	6.92	8.80
1080	cycle 30 miles		96	.083			2.74	5.90	8.64	11
1082	cycle 40 miles		72	.111			3.66	7.90	11.56	14.65
1084	cycle 50 miles		60	.133			4.39	9.45	13.84	17.60
1094	45 MPH avg., cycle 8 miles		216	.037			1.22	2.63	3.85	4.89
1096	cycle 10 miles		204	.039			1.29	2.78	4.07	5.20
1098	cycle 20 miles		132	.061			1.99	4.30	6.29	8
1100	cycle 30 miles		108	.074			2.44	5.25	7.69	9.80
1102	cycle 40 miles		84	.095			3.13	6.75	9.88	12.60
1104	cycle 50 miles		72	.111			3.66	7.90	11.56	14.65
1106	50 MPH avg., cycle 10 miles		216	.037			1.22	2.63	3.85	4.89
1108	cycle 20 miles		144	.056			1.83	3.94	5.77	7.35
1110	cycle 30 miles		108	.074			2.44	5.25	7.69	9.80
1112	cycle 40 miles		84	.095			3.13	6.75	9.88	12.60
1114	cycle 50 miles		72	.111			3.66	7.90	11.56	14.65
1214	15 MPH avg., cycle 0.5 mile, 20 min. wait/ld./uld.		264	.030			1	2.15	3.15	4
1216	cycle 1 mile		240	.033			1.10	2.36	3.46	4.40
1218	cycle 2 miles		204	.039			1.29	2.78	4.07	5.20
1220	cycle 4 miles		156	.051			1.69	3.63	5.32	6.75
1222	cycle 6 miles		132	.061			1.99	4.30	6.29	8
1224	cycle 8 miles		108	.074			2.44	5.25	7.69	9.80
1225	cycle 10 miles		96	.083			2.74	5.90	8.64	11
1226	20 MPH avg., cycle 0.5 mile		264	.030			1	2.15	3.15	4
1228	cycle 1 mile		252	.032			1.04	2.25	3.29	4.20
1230	cycle 2 miles		216	.037			1.22	2.63	3.85	4.89
1232	cycle 4 miles		180	.044			1.46	3.15	4.61	5.85
1234	cycle 6 miles		144	.056			1.83	3.94	5.77	7.35
1236	cycle 8 miles		132	.061			1.99	4.30	6.29	8
1238	cycle 10 miles		108	.074			2.44	5.25	7.69	9.80
1240	25 MPH avg., cycle 4 miles		192	.042			1.37	2.95	4.32	5.50
1242	cycle 6 miles		168	.048			1.57	3.38	4.95	6.30
1244	cycle 8 miles		144	.056			1.83	3.94	5.77	7.35
1246	cycle 10 miles		132	.061			1.99	4.30	6.29	8
1250	30 MPH avg., cycle 4 miles		204	.039			1.29	2.78	4.07	5.20
1252	cycle 6 miles		180	.044			1.46	3.15	4.61	5.85
1254	cycle 8 miles		156	.051			1.69	3.63	5.32	6.75
1256	cycle 10 miles		144	.056			1.83	3.94	5.77	7.35
1260	35 MPH avg., cycle 4 miles		216	.037			1.22	2.63	3.85	4.89
1262	cycle 6 miles		192	.042			1.37	2.95	4.32	5.50
1264	cycle 8 miles		168	.048			1.57	3.38	4.95	6.30
1266	cycle 10 miles		156	.051			1.69	3.63	5.32	6.75
1268	cycle 20 miles		108	.074			2.44	5.25	7.69	9.80
1269	cycle 30 miles		72	.111			3.66	7.90	11.56	14.65
1270	cycle 40 miles		60	.133			4.39	9.45	13.84	17.60
1272	40 MPH avg., cycle 6 miles	↓	192	.042	↓		1.37	2.95	4.32	5.50

31 23 23.20 Hauling		Crew	Daily Output	Labor-Hours	Unit	Material	2019 Bare Costs Labor	Equipment	Total	Total Incl O&P
1274	cycle 8 miles	B-34B	180	.044	L.C.Y.		1.46	3.15	4.61	5.85
1276	cycle 10 miles		156	.051			1.69	3.63	5.32	6.75
1278	cycle 20 miles		108	.074			2.44	5.25	7.69	9.80
1280	cycle 30 miles		84	.095			3.13	6.75	9.88	12.60
1282	cycle 40 miles		72	.111			3.66	7.90	11.56	14.65
1284	cycle 50 miles		60	.133			4.39	9.45	13.84	17.60
1294	45 MPH avg., cycle 8 miles		180	.044			1.46	3.15	4.61	5.85
1296	cycle 10 miles		168	.048			1.57	3.38	4.95	6.30
1298	cycle 20 miles		120	.067			2.19	4.73	6.92	8.80
1300	cycle 30 miles		96	.083			2.74	5.90	8.64	11
1302	cycle 40 miles		72	.111			3.66	7.90	11.56	14.65
1304	cycle 50 miles		60	.133			4.39	9.45	13.84	17.60
1306	50 MPH avg., cycle 10 miles		180	.044			1.46	3.15	4.61	5.85
1308	cycle 20 miles		132	.061			1.99	4.30	6.29	8
1310	cycle 30 miles		96	.083			2.74	5.90	8.64	11
1312	cycle 40 miles		84	.095			3.13	6.75	9.88	12.60
1314	cycle 50 miles		72	.111			3.66	7.90	11.56	14.65
1414	15 MPH avg., cycle 0.5 mile, 25 min. wait/ld./uld.		204	.039			1.29	2.78	4.07	5.20
1416	cycle 1 mile		192	.042			1.37	2.95	4.32	5.50
1418	cycle 2 miles		168	.048			1.57	3.38	4.95	6.30
1420	cycle 4 miles		132	.061			1.99	4.30	6.29	8
1422	cycle 6 miles		120	.067			2.19	4.73	6.92	8.80
1424	cycle 8 miles		96	.083			2.74	5.90	8.64	11
1425	cycle 10 miles		84	.095			3.13	6.75	9.88	12.60
1426	20 MPH avg., cycle 0.5 mile		216	.037			1.22	2.63	3.85	4.89
1428	cycle 1 mile		204	.039			1.29	2.78	4.07	5.20
1430	cycle 2 miles		180	.044			1.46	3.15	4.61	5.85
1432	cycle 4 miles		156	.051			1.69	3.63	5.32	6.75
1434	cycle 6 miles		132	.061			1.99	4.30	6.29	8
1436	cycle 8 miles		120	.067			2.19	4.73	6.92	8.80
1438	cycle 10 miles		96	.083			2.74	5.90	8.64	11
1440	25 MPH avg., cycle 4 miles		168	.048			1.57	3.38	4.95	6.30
1442	cycle 6 miles		144	.056			1.83	3.94	5.77	7.35
1444	cycle 8 miles		132	.061			1.99	4.30	6.29	8
1446	cycle 10 miles		108	.074			2.44	5.25	7.69	9.80
1450	30 MPH avg., cycle 4 miles		168	.048			1.57	3.38	4.95	6.30
1452	cycle 6 miles		156	.051			1.69	3.63	5.32	6.75
1454	cycle 8 miles		132	.061			1.99	4.30	6.29	8
1456	cycle 10 miles		120	.067			2.19	4.73	6.92	8.80
1460	35 MPH avg., cycle 4 miles		180	.044			1.46	3.15	4.61	5.85
1462	cycle 6 miles		156	.051			1.69	3.63	5.32	6.75
1464	cycle 8 miles		144	.056			1.83	3.94	5.77	7.35
1466	cycle 10 miles		132	.061			1.99	4.30	6.29	8
1468	cycle 20 miles		96	.083			2.74	5.90	8.64	11
1469	cycle 30 miles		72	.111			3.66	7.90	11.56	14.65
1470	cycle 40 miles		60	.133			4.39	9.45	13.84	17.60
1472	40 MPH avg., cycle 6 miles		168	.048			1.57	3.38	4.95	6.30
1474	cycle 8 miles		156	.051			1.69	3.63	5.32	6.75
1476	cycle 10 miles		144	.056			1.83	3.94	5.77	7.35
1478	cycle 20 miles		96	.083			2.74	5.90	8.64	11
1480	cycle 30 miles		84	.095			3.13	6.75	9.88	12.60
1482	cycle 40 miles		60	.133			4.39	9.45	13.84	17.60
1484	cycle 50 miles		60	.133			4.39	9.45	13.84	17.60

31 23 23.20 Hauling		Crew	Daily Output	Labor-Hours	Unit	Material	2019 Bare Costs Labor	Equipment	Total	Total Incl O&P
1494	45 MPH avg., cycle 8 miles	B-34B	156	.051	L.C.Y.		1.69	3.63	5.32	6.75
1496	cycle 10 miles		144	.056			1.83	3.94	5.77	7.35
1498	cycle 20 miles		108	.074			2.44	5.25	7.69	9.80
1500	cycle 30 miles		84	.095			3.13	6.75	9.88	12.60
1502	cycle 40 miles		72	.111			3.66	7.90	11.56	14.65
1504	cycle 50 miles		60	.133			4.39	9.45	13.84	17.60
1506	50 MPH avg., cycle 10 miles		156	.051			1.69	3.63	5.32	6.75
1508	cycle 20 miles		120	.067			2.19	4.73	6.92	8.80
1510	cycle 30 miles		96	.083			2.74	5.90	8.64	11
1512	cycle 40 miles		72	.111			3.66	7.90	11.56	14.65
1514	cycle 50 miles		60	.133			4.39	9.45	13.84	17.60
1614	15 MPH avg., cycle 0.5 mile, 30 min. wait/ld./uld.		180	.044			1.46	3.15	4.61	5.85
1616	cycle 1 mile		168	.048			1.57	3.38	4.95	6.30
1618	cycle 2 miles		144	.056			1.83	3.94	5.77	7.35
1620	cycle 4 miles		120	.067			2.19	4.73	6.92	8.80
1622	cycle 6 miles		108	.074			2.44	5.25	7.69	9.80
1624	cycle 8 miles		84	.095			3.13	6.75	9.88	12.60
1625	cycle 10 miles		84	.095			3.13	6.75	9.88	12.60
1626	20 MPH avg., cycle 0.5 mile		180	.044			1.46	3.15	4.61	5.85
1628	cycle 1 mile		168	.048			1.57	3.38	4.95	6.30
1630	cycle 2 miles		156	.051			1.69	3.63	5.32	6.75
1632	cycle 4 miles		132	.061			1.99	4.30	6.29	8
1634	cycle 6 miles		120	.067			2.19	4.73	6.92	8.80
1636	cycle 8 miles		108	.074			2.44	5.25	7.69	9.80
1638	cycle 10 miles		96	.083			2.74	5.90	8.64	11
1640	25 MPH avg., cycle 4 miles		144	.056			1.83	3.94	5.77	7.35
1642	cycle 6 miles		132	.061			1.99	4.30	6.29	8
1644	cycle 8 miles		108	.074			2.44	5.25	7.69	9.80
1646	cycle 10 miles		108	.074			2.44	5.25	7.69	9.80
1650	30 MPH avg., cycle 4 miles		144	.056			1.83	3.94	5.77	7.35
1652	cycle 6 miles		132	.061			1.99	4.30	6.29	8
1654	cycle 8 miles		120	.067			2.19	4.73	6.92	8.80
1656	cycle 10 miles		108	.074			2.44	5.25	7.69	9.80
1660	35 MPH avg., cycle 4 miles		156	.051			1.69	3.63	5.32	6.75
1662	cycle 6 miles		144	.056			1.83	3.94	5.77	7.35
1664	cycle 8 miles		132	.061			1.99	4.30	6.29	8
1666	cycle 10 miles		120	.067			2.19	4.73	6.92	8.80
1668	cycle 20 miles		84	.095			3.13	6.75	9.88	12.60
1669	cycle 30 miles		72	.111			3.66	7.90	11.56	14.65
1670	cycle 40 miles		60	.133			4.39	9.45	13.84	17.60
1672	40 MPH avg., cycle 6 miles		144	.056			1.83	3.94	5.77	7.35
1674	cycle 8 miles		132	.061			1.99	4.30	6.29	8
1676	cycle 10 miles		120	.067			2.19	4.73	6.92	8.80
1678	cycle 20 miles		96	.083			2.74	5.90	8.64	11
1680	cycle 30 miles		72	.111			3.66	7.90	11.56	14.65
1682	cycle 40 miles		60	.133			4.39	9.45	13.84	17.60
1684	cycle 50 miles		48	.167			5.50	11.80	17.30	22
1694	45 MPH avg., cycle 8 miles		144	.056			1.83	3.94	5.77	7.35
1696	cycle 10 miles		132	.061			1.99	4.30	6.29	8
1698	cycle 20 miles		96	.083			2.74	5.90	8.64	11
1700	cycle 30 miles		84	.095			3.13	6.75	9.88	12.60
1702	cycle 40 miles		60	.133			4.39	9.45	13.84	17.60
1704	cycle 50 miles	↓	60	.133	↓		4.39	9.45	13.84	17.60

31 23 23.20 Hauling		Crew	Daily Output	Labor-Hours	Unit	Material	2019 Bare Costs Labor	Equipment	Total	Total Incl O&P
1706	50 MPH avg., cycle 10 miles	B-34B	132	.061	L.C.Y.		1.99	4.30	6.29	8
1708	cycle 20 miles		108	.074			2.44	5.25	7.69	9.80
1710	cycle 30 miles		84	.095			3.13	6.75	9.88	12.60
1712	cycle 40 miles		72	.111			3.66	7.90	11.56	14.65
1714	cycle 50 miles		60	.133			4.39	9.45	13.84	17.60
2000	Hauling, 8 C.Y. truck, small project cost per hour	B-34A	8	1	Hr.		33	42.50	75.50	101
2100	12 C.Y. truck	B-34B	8	1			33	71	104	132
2150	16.5 C.Y. truck	B-34C	8	1			33	77.50	110.50	140
2175	18 C.Y. 8 wheel truck	B-34I	8	1			33	88	121	151
2200	20 C.Y. truck	B-34D	8	1			33	79.50	112.50	142
9014	18 C.Y. truck, 8 wheels,15 min. wait/ld./uld.,15 MPH, cycle 0.5 mi.	B-34I	504	.016	L.C.Y.		.52	1.40	1.92	2.40
9016	cycle 1 mile		450	.018			.59	1.57	2.16	2.68
9018	cycle 2 miles		378	.021			.70	1.86	2.56	3.19
9020	cycle 4 miles		270	.030			.97	2.61	3.58	4.47
9022	cycle 6 miles		216	.037			1.22	3.26	4.48	5.60
9024	cycle 8 miles		180	.044			1.46	3.92	5.38	6.70
9025	cycle 10 miles		144	.056			1.83	4.90	6.73	8.40
9026	20 MPH avg., cycle 0.5 mile		522	.015			.50	1.35	1.85	2.32
9028	cycle 1 mile		468	.017			.56	1.51	2.07	2.58
9030	cycle 2 miles		414	.019			.64	1.70	2.34	2.92
9032	cycle 4 miles		324	.025			.81	2.18	2.99	3.73
9034	cycle 6 miles		252	.032			1.04	2.80	3.84	4.80
9036	cycle 8 miles		216	.037			1.22	3.26	4.48	5.60
9038	cycle 10 miles		180	.044			1.46	3.92	5.38	6.70
9040	25 MPH avg., cycle 4 miles		342	.023			.77	2.06	2.83	3.54
9042	cycle 6 miles		288	.028			.91	2.45	3.36	4.19
9044	cycle 8 miles		252	.032			1.04	2.80	3.84	4.80
9046	cycle 10 miles		216	.037			1.22	3.26	4.48	5.60
9050	30 MPH avg., cycle 4 miles		378	.021			.70	1.86	2.56	3.19
9052	cycle 6 miles		324	.025			.81	2.18	2.99	3.73
9054	cycle 8 miles		270	.030			.97	2.61	3.58	4.47
9056	cycle 10 miles		234	.034			1.12	3.01	4.13	5.15
9060	35 MPH avg., cycle 4 miles		396	.020			.66	1.78	2.44	3.05
9062	cycle 6 miles		342	.023			.77	2.06	2.83	3.54
9064	cycle 8 miles		288	.028			.91	2.45	3.36	4.19
9066	cycle 10 miles		270	.030			.97	2.61	3.58	4.47
9068	cycle 20 miles		162	.049			1.62	4.35	5.97	7.45
9070	cycle 30 miles		126	.063			2.09	5.60	7.69	9.60
9072	cycle 40 miles		90	.089			2.92	7.85	10.77	13.40
9074	40 MPH avg., cycle 6 miles		360	.022			.73	1.96	2.69	3.35
9076	cycle 8 miles		324	.025			.81	2.18	2.99	3.73
9078	cycle 10 miles		288	.028			.91	2.45	3.36	4.19
9080	cycle 20 miles		180	.044			1.46	3.92	5.38	6.70
9082	cycle 30 miles		144	.056			1.83	4.90	6.73	8.40
9084	cycle 40 miles		108	.074			2.44	6.55	8.99	11.20
9086	cycle 50 miles		90	.089			2.92	7.85	10.77	13.40
9094	45 MPH avg., cycle 8 miles		324	.025			.81	2.18	2.99	3.73
9096	cycle 10 miles		306	.026			.86	2.30	3.16	3.94
9098	cycle 20 miles		198	.040			1.33	3.56	4.89	6.10
9100	cycle 30 miles		144	.056			1.83	4.90	6.73	8.40
9102	cycle 40 miles		126	.063			2.09	5.60	7.69	9.60
9104	cycle 50 miles		108	.074			2.44	6.55	8.99	11.20
9106	50 MPH avg., cycle 10 miles		324	.025			.81	2.18	2.99	3.73

31 23 23.20 Hauling		Crew	Daily Output	Labor-Hours	Unit	Material	2019 Bare Costs Labor	Equipment	Total	Total Incl O&P
9108	cycle 20 miles	B-34I	216	.037	L.C.Y.		1.22	3.26	4.48	5.60
9110	cycle 30 miles		162	.049			1.62	4.35	5.97	7.45
9112	cycle 40 miles		126	.063			2.09	5.60	7.69	9.60
9114	cycle 50 miles		108	.074			2.44	6.55	8.99	11.20
9214	20 min. wait/ld./uld.,15 MPH, cycle 0.5 mi.		396	.020			.66	1.78	2.44	3.05
9216	cycle 1 mile		360	.022			.73	1.96	2.69	3.35
9218	cycle 2 miles		306	.026			.86	2.30	3.16	3.94
9220	cycle 4 miles		234	.034			1.12	3.01	4.13	5.15
9222	cycle 6 miles		198	.040			1.33	3.56	4.89	6.10
9224	cycle 8 miles		162	.049			1.62	4.35	5.97	7.45
9225	cycle 10 miles		144	.056			1.83	4.90	6.73	8.40
9226	20 MPH avg., cycle 0.5 mile		396	.020			.66	1.78	2.44	3.05
9228	cycle 1 mile		378	.021			.70	1.86	2.56	3.19
9230	cycle 2 miles		324	.025			.81	2.18	2.99	3.73
9232	cycle 4 miles		270	.030			.97	2.61	3.58	4.47
9234	cycle 6 miles		216	.037			1.22	3.26	4.48	5.60
9236	cycle 8 miles		198	.040			1.33	3.56	4.89	6.10
9238	cycle 10 miles		162	.049			1.62	4.35	5.97	7.45
9240	25 MPH avg., cycle 4 miles		288	.028			.91	2.45	3.36	4.19
9242	cycle 6 miles		252	.032			1.04	2.80	3.84	4.80
9244	cycle 8 miles		216	.037			1.22	3.26	4.48	5.60
9246	cycle 10 miles		198	.040			1.33	3.56	4.89	6.10
9250	30 MPH avg., cycle 4 miles		306	.026			.86	2.30	3.16	3.94
9252	cycle 6 miles		270	.030			.97	2.61	3.58	4.47
9254	cycle 8 miles		234	.034			1.12	3.01	4.13	5.15
9256	cycle 10 miles		216	.037			1.22	3.26	4.48	5.60
9260	35 MPH avg., cycle 4 miles		324	.025			.81	2.18	2.99	3.73
9262	cycle 6 miles		288	.028			.91	2.45	3.36	4.19
9264	cycle 8 miles		252	.032			1.04	2.80	3.84	4.80
9266	cycle 10 miles		234	.034			1.12	3.01	4.13	5.15
9268	cycle 20 miles		162	.049			1.62	4.35	5.97	7.45
9270	cycle 30 miles		108	.074			2.44	6.55	8.99	11.20
9272	cycle 40 miles		90	.089			2.92	7.85	10.77	13.40
9274	40 MPH avg., cycle 6 miles		288	.028			.91	2.45	3.36	4.19
9276	cycle 8 miles		270	.030			.97	2.61	3.58	4.47
9278	cycle 10 miles		234	.034			1.12	3.01	4.13	5.15
9280	cycle 20 miles		162	.049			1.62	4.35	5.97	7.45
9282	cycle 30 miles		126	.063			2.09	5.60	7.69	9.60
9284	cycle 40 miles		108	.074			2.44	6.55	8.99	11.20
9286	cycle 50 miles		90	.089			2.92	7.85	10.77	13.40
9294	45 MPH avg., cycle 8 miles		270	.030			.97	2.61	3.58	4.47
9296	cycle 10 miles		252	.032			1.04	2.80	3.84	4.80
9298	cycle 20 miles		180	.044			1.46	3.92	5.38	6.70
9300	cycle 30 miles		144	.056			1.83	4.90	6.73	8.40
9302	cycle 40 miles		108	.074			2.44	6.55	8.99	11.20
9304	cycle 50 miles		90	.089			2.92	7.85	10.77	13.40
9306	50 MPH avg., cycle 10 miles		270	.030			.97	2.61	3.58	4.47
9308	cycle 20 miles		198	.040			1.33	3.56	4.89	6.10
9310	cycle 30 miles		144	.056			1.83	4.90	6.73	8.40
9312	cycle 40 miles		126	.063			2.09	5.60	7.69	9.60
9314	cycle 50 miles		108	.074			2.44	6.55	8.99	11.20
9414	25 min. wait/ld./uld.,15 MPH, cycle 0.5 mi.		306	.026			.86	2.30	3.16	3.94
9416	cycle 1 mile		288	.028			.91	2.45	3.36	4.19

31 23 23.20 Hauling		Crew	Daily Output	Labor-Hours	Unit	Material	2019 Bare Costs Labor	Equipment	Total	Total Incl O&P
9418	cycle 2 miles	B-34I	252	.032	L.C.Y.		1.04	2.80	3.84	4.80
9420	cycle 4 miles		198	.040			1.33	3.56	4.89	6.10
9422	cycle 6 miles		180	.044			1.46	3.92	5.38	6.70
9424	cycle 8 miles		144	.056			1.83	4.90	6.73	8.40
9425	cycle 10 miles		126	.063			2.09	5.60	7.69	9.60
9426	20 MPH avg., cycle 0.5 mile		324	.025			.81	2.18	2.99	3.73
9428	cycle 1 mile		306	.026			.86	2.30	3.16	3.94
9430	cycle 2 miles		270	.030			.97	2.61	3.58	4.47
9432	cycle 4 miles		234	.034			1.12	3.01	4.13	5.15
9434	cycle 6 miles		198	.040			1.33	3.56	4.89	6.10
9436	cycle 8 miles		180	.044			1.46	3.92	5.38	6.70
9438	cycle 10 miles		144	.056			1.83	4.90	6.73	8.40
9440	25 MPH avg., cycle 4 miles		252	.032			1.04	2.80	3.84	4.80
9442	cycle 6 miles		216	.037			1.22	3.26	4.48	5.60
9444	cycle 8 miles		198	.040			1.33	3.56	4.89	6.10
9446	cycle 10 miles		180	.044			1.46	3.92	5.38	6.70
9450	30 MPH avg., cycle 4 miles		252	.032			1.04	2.80	3.84	4.80
9452	cycle 6 miles		234	.034			1.12	3.01	4.13	5.15
9454	cycle 8 miles		198	.040			1.33	3.56	4.89	6.10
9456	cycle 10 miles		180	.044			1.46	3.92	5.38	6.70
9460	35 MPH avg., cycle 4 miles		270	.030			.97	2.61	3.58	4.47
9462	cycle 6 miles		234	.034			1.12	3.01	4.13	5.15
9464	cycle 8 miles		216	.037			1.22	3.26	4.48	5.60
9466	cycle 10 miles		198	.040			1.33	3.56	4.89	6.10
9468	cycle 20 miles		144	.056			1.83	4.90	6.73	8.40
9470	cycle 30 miles		108	.074			2.44	6.55	8.99	11.20
9472	cycle 40 miles		90	.089			2.92	7.85	10.77	13.40
9474	40 MPH avg., cycle 6 miles		252	.032			1.04	2.80	3.84	4.80
9476	cycle 8 miles		234	.034			1.12	3.01	4.13	5.15
9478	cycle 10 miles		216	.037			1.22	3.26	4.48	5.60
9480	cycle 20 miles		144	.056			1.83	4.90	6.73	8.40
9482	cycle 30 miles		126	.063			2.09	5.60	7.69	9.60
9484	cycle 40 miles		90	.089			2.92	7.85	10.77	13.40
9486	cycle 50 miles		90	.089			2.92	7.85	10.77	13.40
9494	45 MPH avg., cycle 8 miles		234	.034			1.12	3.01	4.13	5.15
9496	cycle 10 miles		216	.037			1.22	3.26	4.48	5.60
9498	cycle 20 miles		162	.049			1.62	4.35	5.97	7.45
9500	cycle 30 miles		126	.063			2.09	5.60	7.69	9.60
9502	cycle 40 miles		108	.074			2.44	6.55	8.99	11.20
9504	cycle 50 miles		90	.089			2.92	7.85	10.77	13.40
9506	50 MPH avg., cycle 10 miles		234	.034			1.12	3.01	4.13	5.15
9508	cycle 20 miles		180	.044			1.46	3.92	5.38	6.70
9510	cycle 30 miles		144	.056			1.83	4.90	6.73	8.40
9512	cycle 40 miles		108	.074			2.44	6.55	8.99	11.20
9514	cycle 50 miles		90	.089			2.92	7.85	10.77	13.40
9614	30 min. wait/ld./uld.,15 MPH, cycle 0.5 mi.		270	.030			.97	2.61	3.58	4.47
9616	cycle 1 mile		252	.032			1.04	2.80	3.84	4.80
9618	cycle 2 miles		216	.037			1.22	3.26	4.48	5.60
9620	cycle 4 miles		180	.044			1.46	3.92	5.38	6.70
9622	cycle 6 miles		162	.049			1.62	4.35	5.97	7.45
9624	cycle 8 miles		126	.063			2.09	5.60	7.69	9.60
9625	cycle 10 miles		126	.063			2.09	5.60	7.69	9.60
9626	20 MPH avg., cycle 0.5 mile	↓	270	.030	↓		.97	2.61	3.58	4.47

31 23 23.20 Hauling		Crew	Daily Output	Labor-Hours	Unit	Material	2019 Bare Costs Labor	Equipment	Total	Total Incl O&P
9628	cycle 1 mile	B-34I	252	.032	L.C.Y.		1.04	2.80	3.84	4.80
9630	cycle 2 miles		234	.034			1.12	3.01	4.13	5.15
9632	cycle 4 miles		198	.040			1.33	3.56	4.89	6.10
9634	cycle 6 miles		180	.044			1.46	3.92	5.38	6.70
9636	cycle 8 miles		162	.049			1.62	4.35	5.97	7.45
9638	cycle 10 miles		144	.056			1.83	4.90	6.73	8.40
9640	25 MPH avg., cycle 4 miles		216	.037			1.22	3.26	4.48	5.60
9642	cycle 6 miles		198	.040			1.33	3.56	4.89	6.10
9644	cycle 8 miles		180	.044			1.46	3.92	5.38	6.70
9646	cycle 10 miles		162	.049			1.62	4.35	5.97	7.45
9650	30 MPH avg., cycle 4 miles		216	.037			1.22	3.26	4.48	5.60
9652	cycle 6 miles		198	.040			1.33	3.56	4.89	6.10
9654	cycle 8 miles		180	.044			1.46	3.92	5.38	6.70
9656	cycle 10 miles		162	.049			1.62	4.35	5.97	7.45
9660	35 MPH avg., cycle 4 miles		234	.034			1.12	3.01	4.13	5.15
9662	cycle 6 miles		216	.037			1.22	3.26	4.48	5.60
9664	cycle 8 miles		198	.040			1.33	3.56	4.89	6.10
9666	cycle 10 miles		180	.044			1.46	3.92	5.38	6.70
9668	cycle 20 miles		126	.063			2.09	5.60	7.69	9.60
9670	cycle 30 miles		108	.074			2.44	6.55	8.99	11.20
9672	cycle 40 miles		90	.089			2.92	7.85	10.77	13.40
9674	40 MPH avg., cycle 6 miles		216	.037			1.22	3.26	4.48	5.60
9676	cycle 8 miles		198	.040			1.33	3.56	4.89	6.10
9678	cycle 10 miles		180	.044			1.46	3.92	5.38	6.70
9680	cycle 20 miles		144	.056			1.83	4.90	6.73	8.40
9682	cycle 30 miles		108	.074			2.44	6.55	8.99	11.20
9684	cycle 40 miles		90	.089			2.92	7.85	10.77	13.40
9686	cycle 50 miles		72	.111			3.66	9.80	13.46	16.75
9694	45 MPH avg., cycle 8 miles		216	.037			1.22	3.26	4.48	5.60
9696	cycle 10 miles		198	.040			1.33	3.56	4.89	6.10
9698	cycle 20 miles		144	.056			1.83	4.90	6.73	8.40
9700	cycle 30 miles		126	.063			2.09	5.60	7.69	9.60
9702	cycle 40 miles		108	.074			2.44	6.55	8.99	11.20
9704	cycle 50 miles		90	.089			2.92	7.85	10.77	13.40
9706	50 MPH avg., cycle 10 miles		198	.040			1.33	3.56	4.89	6.10
9708	cycle 20 miles		162	.049			1.62	4.35	5.97	7.45
9710	cycle 30 miles		126	.063			2.09	5.60	7.69	9.60
9712	cycle 40 miles		108	.074			2.44	6.55	8.99	11.20
9714	cycle 50 miles	▼	90	.089	▼		2.92	7.85	10.77	13.40

31 23 23.24 Compaction, Structural		Crew	Daily Output	Labor-Hours	Unit	Material	Labor	Equipment	Total	Total Incl O&P
0010	**COMPACTION, STRUCTURAL**									
0020	Steel wheel tandem roller, 5 tons	B-10E	8	1	Hr.		37.50	19.05	56.55	82.50
0050	Air tamp, 6" to 8" lifts, common fill	B-9	250	.160	E.C.Y.		4.46	.94	5.40	8.45
0060	Select fill	"	300	.133			3.72	.78	4.50	7
0600	Vibratory plate, 8" lifts, common fill	A-1D	200	.040			1.10	.16	1.26	1.99
0700	Select fill	"	216	.037	▼		1.02	.14	1.16	1.84

31 25 Erosion and Sedimentation Controls

31 25 14 – Stabilization Measures for Erosion and Sedimentation Control

31 25 14.16 Rolled Erosion Control Mats and Blankets		Crew	Daily Output	Labor-Hours	Unit	2019 Bare Costs Material	Labor	Equipment	Total	Total Incl O&P
0010	**ROLLED EROSION CONTROL MATS AND BLANKETS**									
0020	Jute mesh, 100 S.Y. per roll, 4′ wide, stapled G	B-80A	2400	.010	S.Y.	.99	.28	.10	1.37	1.65
0100	Plastic netting, stapled, 2″ x 1″ mesh, 20 mil G	B-1	2500	.010		.24	.27		.51	.71
0120	Revegetation mat, webbed G	2 Clab	1000	.016		3.10	.44		3.54	4.14
0200	Polypropylene mesh, stapled, 6.5 oz./S.Y. G	B-1	2500	.010		1.73	.27		2	2.35
0300	Tobacco netting, or jute mesh #2, stapled G	″	2500	.010	↓	.27	.27		.54	.75
0600	Straw in polymeric netting, biodegradable log	A-2	1000	.024	L.F.	6.30	.70	.20	7.20	8.30
1000	Silt fence, install and maintain, remove G	B-62	1300	.018	″	.49	.56	.13	1.18	1.61
1100	Allow 10% per month for maintenance; 6-month max life									

31 31 Soil Treatment

31 31 16 – Termite Control

31 31 16.13 Chemical Termite Control		Crew	Daily Output	Labor-Hours	Unit	Material	Labor	Equipment	Total	Total Incl O&P
0010	**CHEMICAL TERMITE CONTROL**									
0020	Slab and walls, residential	1 Skwk	1200	.007	SF Flr.	.33	.24		.57	.76
0030	SS mesh, no chemicals, avg 1,400 S.F. home, min G		1000	.008	″	.33	.29		.62	.85
0400	Insecticides for termite control, minimum		14.20	.563	Gal.	71	20.50		91.50	112
0500	Maximum	↓	11	.727	″	121	26.50		147.50	177

Division Notes

	CREW	DAILY OUTPUT	LABOR-HOURS	UNIT	BARE COSTS MAT.	LABOR	EQUIP.	TOTAL	TOTAL INCL O&P

Division 32 Exterior Improvements

Estimating Tips

32 01 00 Operations and Maintenance of Exterior Improvements

- Recycling of asphalt pavement is becoming very popular and is an alternative to removal and replacement. It can be a good value engineering proposal if removed pavement can be recycled, either at the project site or at another site that is reasonably close to the project site. Sections on repair of flexible and rigid pavement are included.

32 10 00 Bases, Ballasts, and Paving

- When estimating paving, keep in mind the project schedule. Also note that prices for asphalt and concrete are generally higher in the cold seasons. Lines for pavement markings, including tactile warning systems and fence lines, are included.

32 90 00 Planting

- The timing of planting and guarantee specifications often dictate the costs for establishing tree and shrub growth and a stand of grass or ground cover. Establish the work performance schedule to coincide with the local planting season. Maintenance and growth guarantees can add 20–100% to the total landscaping cost and can be contractually cumbersome. The cost to replace trees and shrubs can be as high as 5% of the total cost, depending on the planting zone, soil conditions, and time of year.

Reference Numbers

Reference numbers are shown at the beginning of some major classifications. These numbers refer to related items in the Reference Section. The reference information may be an estimating procedure, an alternate pricing method, or technical information.

Note: Not all subdivisions listed here necessarily appear. ■

32 01 Operation and Maintenance of Exterior Improvements

32 01 13 – Flexible Paving Surface Treatment

32 01 13.66 Fog Seal		Crew	Daily Output	Labor-Hours	Unit	Material	Labor	Equipment	Total	Total Incl O&P
						2019 Bare Costs				
0010	**FOG SEAL**									
0012	Sealcoating, 2 coat coal tar pitch emulsion over 10,000 S.Y.	B-45	5000	.003	S.Y.	.85	.10	.16	1.11	1.28
0030	1,000 to 10,000 S.Y.	″	3000	.005		.85	.16	.27	1.28	1.51
0100	Under 1,000 S.Y.	B-1	1050	.023		.85	.64		1.49	2
0300	Petroleum resistant, over 10,000 S.Y.	B-45	5000	.003		1.42	.10	.16	1.68	1.90
0320	1,000 to 10,000 S.Y.	″	3000	.005		1.42	.16	.27	1.85	2.13
0400	Under 1,000 S.Y.	B-1	1050	.023		1.42	.64		2.06	2.62

32 06 Schedules for Exterior Improvements

32 06 10 – Schedules for Bases, Ballasts, and Paving

32 06 10.10 Sidewalks, Driveways and Patios

		Crew	Daily Output	Labor-Hours	Unit	Material	Labor	Equipment	Total	Total Incl O&P
0010	**SIDEWALKS, DRIVEWAYS AND PATIOS** No base									
0021	Asphaltic concrete, 2″ thick	B-37	6480	.007	S.F.	.75	.22	.02	.99	1.22
0101	2-1/2″ thick	″	5950	.008	″	.96	.24	.03	1.23	1.47
0300	Concrete, 3,000 psi, CIP, 6 x 6 - W1.4 x W1.4 mesh,									
0310	broomed finish, no base, 4″ thick	B-24	600	.040	S.F.	2.22	1.32		3.54	4.60
0350	5″ thick		545	.044		2.75	1.45		4.20	5.40
0400	6″ thick		510	.047		3.20	1.55		4.75	6.05
0450	For bank run gravel base, 4″ thick, add	B-18	2500	.010		.43	.27	.02	.72	.95
0520	8″ thick, add	″	1600	.015		.88	.42	.03	1.33	1.69
1000	Crushed stone, 1″ thick, white marble	2 Clab	1700	.009		.49	.26		.75	.97
1050	Bluestone	″	1700	.009		.20	.26		.46	.65
1700	Redwood, prefabricated, 4′ x 4′ sections	2 Carp	316	.051		4.84	1.81		6.65	8.30
1750	Redwood planks, 1″ thick, on sleepers	″	240	.067		4.84	2.38		7.22	9.25
2250	Stone dust, 4″ thick	B-62	900	.027	S.Y.	4.90	.81	.19	5.90	6.95

32 06 10.20 Steps

		Crew	Daily Output	Labor-Hours	Unit	Material	Labor	Equipment	Total	Total Incl O&P
0010	**STEPS**									
0011	Incl. excav., borrow & concrete base as required									
0100	Brick steps	B-24	35	.686	LF Riser	17.20	22.50		39.70	56
0200	Railroad ties	2 Clab	25	.640		3.59	17.60		21.19	33
0300	Bluestone treads, 12″ x 2″ or 12″ x 1-1/2″	B-24	30	.800		43	26.50		69.50	90.50
0600	Precast concrete, see Section 03 41 23.50									
4025	Steel edge strips, incl. stakes, 1/4″ x 5″	B-1	390	.062	L.F.	4.88	1.73		6.61	8.20
4050	Edging, landscape timber or railroad ties, 6″ x 8″	2 Carp	170	.094	″	2.32	3.36		5.68	8.10

32 11 Base Courses

32 11 23 – Aggregate Base Courses

32 11 23.23 Base Course Drainage Layers

		Crew	Daily Output	Labor-Hours	Unit	Material	Labor	Equipment	Total	Total Incl O&P
0010	**BASE COURSE DRAINAGE LAYERS**									
0011	For roadways and large areas									
0051	3/4″ stone compacted to 3″ deep	B-36	36000	.001	S.F.	.30	.04	.04	.38	.44
0101	6″ deep		35100	.001		.59	.04	.04	.67	.76
0201	9″ deep		25875	.002		.86	.05	.06	.97	1.10
0305	12″ deep		21150	.002		2.40	.06	.07	2.53	2.82
0306	Crushed 1-1/2″ stone base, compacted to 4″ deep		47000	.001		.09	.03	.03	.15	.18
0307	6″ deep		35100	.001		1.20	.04	.04	1.28	1.43
0308	8″ deep		27000	.001		1.60	.05	.06	1.71	1.90
0309	12″ deep		16200	.002		2.40	.08	.10	2.58	2.88
0350	Bank run gravel, spread and compacted									

32 11 Base Courses

32 11 23 – Aggregate Base Courses

32 11 23.23 Base Course Drainage Layers		Crew	Daily Output	Labor-Hours	Unit	2019 Bare Costs Material	Labor	Equipment	Total	Total Incl O&P
0371	6″ deep	B-32	54000	.001	S.F.	.40	.02	.04	.46	.51
0391	9″ deep		39600	.001		.59	.03	.06	.68	.76
0401	12″ deep		32400	.001		.81	.03	.07	.91	1.02
6900	For small and irregular areas, add						50%	50%		

32 11 26 – Asphaltic Base Courses

32 11 26.19 Bituminous-Stabilized Base Courses

Line	Description	Crew	Daily Output	Labor-Hours	Unit	Material	Labor	Equipment	Total	Total Incl O&P
0010	BITUMINOUS-STABILIZED BASE COURSES									
0020	And large paved areas									
0700	Liquid application to gravel base, asphalt emulsion	B-45	6000	.003	Gal.	4.56	.08	.14	4.78	5.30
0800	Prime and seal, cut back asphalt		6000	.003	″	5.40	.08	.14	5.62	6.20
1000	Macadam penetration crushed stone, 2 gal./S.Y., 4″ thick		6000	.003	S.Y.	9.10	.08	.14	9.32	10.35
1100	6″ thick, 3 gal./S.Y.		4000	.004		13.70	.12	.20	14.02	15.45
1200	8″ thick, 4 gal./S.Y.		3000	.005		18.25	.16	.27	18.68	20.50
8900	For small and irregular areas, add						50%	50%		

32 12 Flexible Paving

32 12 16 – Asphalt Paving

32 12 16.14 Paving Asphaltic Concrete

Line	Description	Crew	Daily Output	Labor-Hours	Unit	Material	Labor	Equipment	Total	Total Incl O&P
0010	PAVING ASPHALTIC CONCRETE									
0020	6″ stone base, 2″ binder course, 1″ topping	B-25C	9000	.005	S.F.	2.20	.17	.26	2.63	2.97
0025	2″ binder course, 2″ topping		9000	.005		2.65	.17	.26	3.08	3.47
0030	3″ binder course, 2″ topping		9000	.005		3.05	.17	.26	3.48	3.90
0035	4″ binder course, 2″ topping		9000	.005		3.43	.17	.26	3.86	4.32
0040	1-1/2″ binder course, 1″ topping		9000	.005		2.01	.17	.26	2.44	2.76
0042	3″ binder course, 1″ topping		9000	.005		2.59	.17	.26	3.02	3.40
0045	3″ binder course, 3″ topping		9000	.005		3.50	.17	.26	3.93	4.40
0050	4″ binder course, 3″ topping		9000	.005		3.89	.17	.26	4.32	4.83
0055	4″ binder course, 4″ topping		9000	.005		4.34	.17	.26	4.77	5.30
0300	Binder course, 1-1/2″ thick		35000	.001		.58	.04	.07	.69	.78
0400	2″ thick		25000	.002		.75	.06	.09	.90	1.03
0500	3″ thick		15000	.003		1.17	.10	.15	1.42	1.61
0600	4″ thick		10800	.004		1.53	.14	.21	1.88	2.14
0800	Sand finish course, 3/4″ thick		41000	.001		.31	.04	.06	.41	.46
0900	1″ thick		34000	.001		.39	.04	.07	.50	.57
1000	Fill pot holes, hot mix, 2″ thick	B-16	4200	.008		.81	.22	.14	1.17	1.41
1100	4″ thick		3500	.009		1.19	.27	.16	1.62	1.92
1120	6″ thick		3100	.010		1.59	.30	.18	2.07	2.45
1140	Cold patch, 2″ thick	B-51	3000	.016		.90	.46	.07	1.43	1.82
1160	4″ thick		2700	.018		1.72	.51	.07	2.30	2.81
1180	6″ thick		1900	.025		2.68	.72	.10	3.50	4.25

32 13 Rigid Paving

32 13 13 – Concrete Paving

32 13 13.25 Concrete Pavement, Highways		Crew	Daily Output	Labor-Hours	Unit	Material	2019 Bare Costs Labor	Equipment	Total	Total Incl O&P
0010	**CONCRETE PAVEMENT, HIGHWAYS**									
0015	Including joints, finishing and curing									
0021	Fixed form, 12′ pass, unreinforced, 6″ thick	B-26	18000	.005	S.F.	3.80	.15	.17	4.12	4.62
0101	8″ thick	"	13500	.007		5.10	.20	.23	5.53	6.25
0701	Finishing, broom finish small areas	2 Cefi	1215	.013	↓		.47		.47	.76

32 14 Unit Paving

32 14 13 – Precast Concrete Unit Paving

32 14 13.18 Precast Concrete Plantable Pavers

		Crew	Daily Output	Labor-Hours	Unit	Material	Labor	Equipment	Total	Total Incl O&P
0010	**PRECAST CONCRETE PLANTABLE PAVERS** (50% grass)									
0300	3/4″ crushed stone base for plantable pavers, 6″ depth	B-62	1000	.024	S.Y.	4.15	.73	.17	5.05	5.95
0400	8″ depth		900	.027		5.55	.81	.19	6.55	7.65
0500	10″ depth		800	.030		6.90	.91	.22	8.03	9.35
0600	12″ depth	↓	700	.034	↓	8.30	1.04	.25	9.59	11.10
0700	Hydro seeding plantable pavers	B-81A	20	.800	M.S.F.	11.45	24	18.50	53.95	72.50
0800	Apply fertilizer and seed to plantable pavers	1 Clab	8	1	"	44	27.50		71.50	94

32 14 16 – Brick Unit Paving

32 14 16.10 Brick Paving

		Crew	Daily Output	Labor-Hours	Unit	Material	Labor	Equipment	Total	Total Incl O&P
0010	**BRICK PAVING**									
0012	4″ x 8″ x 1-1/2″, without joints (4.5 bricks/S.F.)	D-1	110	.145	S.F.	2.55	4.71		7.26	10.65
0100	Grouted, 3/8″ joint (3.9 bricks/S.F.)		90	.178		2.07	5.75		7.82	11.90
0200	4″ x 8″ x 2-1/4″, without joints (4.5 bricks/S.F.)		110	.145		2.39	4.71		7.10	10.50
0300	Grouted, 3/8″ joint (3.9 bricks/S.F.)		90	.178		2.07	5.75		7.82	11.90
0455	Pervious brick paving, 4″ x 8″ x 3-1/4″, without joints (4.5 bricks/S.F.)	↓	110	.145		3.64	4.71		8.35	11.85
0500	Bedding, asphalt, 3/4″ thick	B-25	5130	.017		.64	.52	.52	1.68	2.13
0540	Course washed sand bed, 1″ thick	B-18	5000	.005		.36	.14	.01	.51	.63
0580	Mortar, 1″ thick	D-1	300	.053		.66	1.73		2.39	3.60
0620	2″ thick		200	.080		1.32	2.59		3.91	5.75
1500	Brick on 1″ thick sand bed laid flat, 4.5/S.F.		100	.160		2.91	5.20		8.11	11.85
2000	Brick pavers, laid on edge, 7.2/S.F.	↓	70	.229	↓	4.44	7.40		11.84	17.25

32 14 23 – Asphalt Unit Paving

32 14 23.10 Asphalt Blocks

		Crew	Daily Output	Labor-Hours	Unit	Material	Labor	Equipment	Total	Total Incl O&P
0010	**ASPHALT BLOCKS**									
0020	Rectangular, 6″ x 12″ x 1-1/4″, w/bed & neopr. adhesive	D-1	135	.119	S.F.	9.40	3.84		13.24	16.75
0100	3″ thick		130	.123		13.15	3.99		17.14	21
0300	Hexagonal tile, 8″ wide, 1-1/4″ thick		135	.119		9.40	3.84		13.24	16.75
0400	2″ thick		130	.123		13.15	3.99		17.14	21
0500	Square, 8″ x 8″, 1-1/4″ thick		135	.119		9.40	3.84		13.24	16.75
0600	2″ thick	↓	130	.123	↓	13.15	3.99		17.14	21

32 14 40 – Stone Paving

32 14 40.10 Stone Pavers

		Crew	Daily Output	Labor-Hours	Unit	Material	Labor	Equipment	Total	Total Incl O&P
0010	**STONE PAVERS**									
1100	Flagging, bluestone, irregular, 1″ thick,	D-1	81	.198	S.F.	10.50	6.40		16.90	22
1150	Snapped random rectangular, 1″ thick		92	.174		15.95	5.65		21.60	27
1200	1-1/2″ thick		85	.188		19.15	6.10		25.25	31
1250	2″ thick		83	.193		22.50	6.25		28.75	35
1300	Slate, natural cleft, irregular, 3/4″ thick		92	.174		9.65	5.65		15.30	20
1310	1″ thick		85	.188		11.25	6.10		17.35	22.50
1351	Random rectangular, gauged, 1/2″ thick	↓	105	.152	↓	21	4.93		25.93	31.50

32 14 Unit Paving

32 14 40 – Stone Paving

32 14 40.10 Stone Pavers		Crew	Daily Output	Labor-Hours	Unit	2019 Bare Costs Material	Labor	Equipment	Total	Total Incl O&P
1400	Random rectangular, butt joint, gauged, 1/4″ thick	D-1	150	.107	S.F.	22.50	3.45		25.95	30.50
1450	For sand rubbed finish, add					9.60			9.60	10.55
1500	For interior setting, add								25%	25%
1550	Granite blocks, 3-1/2″ x 3-1/2″ x 3-1/2″	D-1	92	.174	S.F.	20.50	5.65		26.15	32

32 16 Curbs, Gutters, Sidewalks, and Driveways

32 16 13 – Curbs and Gutters

32 16 13.13 Cast-in-Place Concrete Curbs and Gutters

		Crew	Daily Output	Labor-Hours	Unit	Material	Labor	Equipment	Total	Total Incl O&P
0010	**CAST-IN-PLACE CONCRETE CURBS AND GUTTERS**									
0290	Forms only, no concrete									
0300	Concrete, wood forms, 6″ x 18″, straight	C-2	500	.096	L.F.	3	3.06		6.06	8.35
0400	6″ x 18″, radius	″	200	.240		3.13	7.65		10.78	16.15
0404	Concrete, wood forms, 6″ x 18″, straight & concrete	C-2A	500	.096		6.45	3.32		9.77	12.55
0406	6″ x 18″, radius	″	200	.240		6.60	8.30		14.90	21

32 16 13.23 Precast Concrete Curbs and Gutters

		Crew	Daily Output	Labor-Hours	Unit	Material	Labor	Equipment	Total	Total Incl O&P
0010	**PRECAST CONCRETE CURBS AND GUTTERS**									
0550	Precast, 6″ x 18″, straight	B-29	700	.069	L.F.	9.35	2.03	1.23	12.61	14.95
0600	6″ x 18″, radius	″	325	.148	″	10.40	4.37	2.65	17.42	21.50

32 16 13.33 Asphalt Curbs

		Crew	Daily Output	Labor-Hours	Unit	Material	Labor	Equipment	Total	Total Incl O&P
0010	**ASPHALT CURBS**									
0012	Curbs, asphaltic, machine formed, 8″ wide, 6″ high, 40 L.F./ton	B-27	1000	.032	L.F.	1.64	.90	.33	2.87	3.64
0100	8″ wide, 8″ high, 30 L.F./ton		900	.036		2.19	1	.36	3.55	4.46
0150	Asphaltic berm, 12″ W, 3″ to 6″ H, 35 L.F./ton, before pavement		700	.046		.04	1.28	.46	1.78	2.68
0200	12″ W, 1-1/2″ to 4″ H, 60 L.F./ton, laid with pavement	B-2	1050	.038		.02	1.06		1.08	1.79

32 16 13.43 Stone Curbs

		Crew	Daily Output	Labor-Hours	Unit	Material	Labor	Equipment	Total	Total Incl O&P
0010	**STONE CURBS**									
1000	Granite, split face, straight, 5″ x 16″	D-13	275	.175	L.F.	14.95	5.95	1.35	22.25	28
1100	6″ x 18″	″	250	.192		19.65	6.55	1.49	27.69	34
1300	Radius curbing, 6″ x 18″, over 10′ radius	B-29	260	.185		24	5.45	3.32	32.77	39
1400	Corners, 2′ radius	″	80	.600	Ea.	81	17.75	10.80	109.55	130
1600	Edging, 4-1/2″ x 12″, straight	D-13	300	.160	L.F.	7.50	5.45	1.24	14.19	18.65
1800	Curb inlets (guttermouth) straight	B-29	41	1.171	Ea.	180	34.50	21	235.50	278
2000	Indian granite (Belgian block)									
2100	Jumbo, 10-1/2″ x 7-1/2″ x 4″, grey	D-1	150	.107	L.F.	9.80	3.45		13.25	16.55
2150	Pink		150	.107		8.50	3.45		11.95	15.10
2200	Regular, 9″ x 4-1/2″ x 4-1/2″, grey		160	.100		4.38	3.24		7.62	10.20
2250	Pink		160	.100		5.85	3.24		9.09	11.80
2300	Cubes, 4″ x 4″ x 4″, grey		175	.091		3.62	2.96		6.58	8.90
2350	Pink		175	.091		3.79	2.96		6.75	9.10
2400	6″ x 6″ x 6″, pink		155	.103		12.90	3.34		16.24	19.75
2500	Alternate pricing method for Indian granite									
2550	Jumbo, 10-1/2″ x 7-1/2″ x 4″ (30 lb.), grey				Ton	560			560	615
2600	Pink					495			495	545
2650	Regular, 9″ x 4-1/2″ x 4-1/2″ (20 lb.), grey					310			310	340
2700	Pink					410			410	450
2750	Cubes, 4″ x 4″ x 4″ (5 lb.), grey					440			440	480
2800	Pink					485			485	535
2850	6″ x 6″ x 6″ (25 lb.), pink					505			505	555
2900	For pallets, add					22			22	24

32 31 13.15 Chain Link Fence		Crew	Daily Output	Labor-Hours	Unit	2019 Bare Costs Material	Labor	Equipment	Total	Total Incl O&P
0010	**CHAIN LINK FENCE**									
0020	1-5/8" post 10' OC, 1-3/8" top rail, 2" corner post galv. stl., 3' high	B-1	185	.130	L.F.	9.95	3.65		13.60	17
0050	4' high		170	.141		10.60	3.98		14.58	18.20
0100	6' high		115	.209		12.30	5.90		18.20	23
0150	Add for gate 3' wide, 1-3/8" frame 3' high		12	2	Ea.	97.50	56.50		154	200
0170	4' high		10	2.400		121	67.50		188.50	245
0190	6' high		10	2.400		143	67.50		210.50	270
0200	Add for gate 4' wide, 1-3/8" frame 3' high		9	2.667		117	75		192	253
0220	4' high		9	2.667		127	75		202	264
0240	6' high		8	3		150	84.50		234.50	305
0350	Aluminized steel, 9 ga. wire, 3' high		185	.130	L.F.	9.05	3.65		12.70	16
0380	4' high		170	.141		9.55	3.98		13.53	17.05
0400	6' high		115	.209		12.65	5.90		18.55	23.50
0450	Add for gate 3' wide, 1-3/8" frame 3' high		12	2	Ea.	155	56.50		211.50	263
0470	4' high		10	2.400		168	67.50		235.50	297
0490	6' high		10	2.400		187	67.50		254.50	315
0500	Add for gate 4' wide, 1-3/8" frame 3' high		10	2.400		162	67.50		229.50	291
0520	4' high		9	2.667		151	75		226	290
0540	6' high		8	3		201	84.50		285.50	360
0620	Vinyl covered 9 ga. wire, 3' high		185	.130	L.F.	8.65	3.65		12.30	15.60
0640	4' high		170	.141		7.35	3.98		11.33	14.65
0660	6' high		115	.209		9.90	5.90		15.80	20.50
0720	Add for gate 3' wide, 1-3/8" frame 3' high		12	2	Ea.	110	56.50		166.50	214
0740	4' high		10	2.400		129	67.50		196.50	254
0760	6' high		10	2.400		155	67.50		222.50	282
0780	Add for gate 4' wide, 1-3/8" frame 3' high		10	2.400		120	67.50		187.50	244
0800	4' high		9	2.667		124	75		199	261
0820	6' high		8	3		172	84.50		256.50	330
0860	Tennis courts, 11 ga. wire, 2-1/2" post 10' OC, 1-5/8" top rail									
0900	2-1/2" corner post, 10' high	B-1	95	.253	L.F.	8.85	7.10		15.95	21.50
0920	12' high		80	.300	"	8.70	8.45		17.15	23.50
1000	Add for gate 3' wide, 1-5/8" frame 10' high		10	2.400	Ea.	231	67.50		298.50	365
1040	Aluminized, 11 ga. wire 10' high		95	.253	L.F.	11.70	7.10		18.80	24.50
1100	12' high		80	.300	"	12.70	8.45		21.15	28
1140	Add for gate 3' wide, 1-5/8" frame, 10' high		10	2.400	Ea.	177	67.50		244.50	305
1250	Vinyl covered 11 ga. wire, 10' high		95	.253	L.F.	10.15	7.10		17.25	23
1300	12' high		80	.300	"	12.15	8.45		20.60	27.50
1400	Add for gate 3' wide, 1-3/8" frame, 10' high		10	2.400	Ea.	335	67.50		402.50	480

32 31 13.80 Residential Chain Link Gate		Crew	Daily Output	Labor-Hours	Unit	Material	Labor	Equipment	Total	Total Incl O&P
0010	**RESIDENTIAL CHAIN LINK GATE**									
0110	Residential 4' gate, single incl. hardware and concrete	B-80C	10	2.400	Ea.	125	70	20.50	215.50	276
0120	5'		10	2.400		135	70	20.50	225.50	287
0130	6'		10	2.400		146	70	20.50	236.50	299
0510	Residential 4' gate, double incl. hardware and concrete		10	2.400		223	70	20.50	313.50	385
0520	5'		10	2.400		239	70	20.50	329.50	400
0530	6'		10	2.400		278	70	20.50	368.50	445

32 31 13.82 Internal Chain Link Gate		Crew	Daily Output	Labor-Hours	Unit	Material	Labor	Equipment	Total	Total Incl O&P
0010	**INTERNAL CHAIN LINK GATE**									
0110	Internal 6' gate, single incl. post flange, hardware and concrete	B-80C	10	2.400	Ea.	273	70	20.50	363.50	440
0120	8'		10	2.400		310	70	20.50	400.50	485
0130	10'		10	2.400		430	70	20.50	520.50	610
0510	Internal 6' gate, double incl. post flange, hardware and concrete		10	2.400		500	70	20.50	590.50	690

32 31 Fences and Gates

32 31 13 – Chain Link Fences and Gates

32 31 13.82 Internal Chain Link Gate		Crew	Daily Output	Labor-Hours	Unit	2019 Bare Costs Material	Labor	Equipment	Total	Total Incl O&P
0520	8′	B-80C	10	2.400	Ea.	575	70	20.50	665.50	770
0530	10′		10	2.400		725	70	20.50	815.50	935

32 31 13.84 Industrial Chain Link Gate

Code	Description	Crew	Daily Output	Labor-Hours	Unit	Material	Labor	Equipment	Total	Total Incl O&P
0010	**INDUSTRIAL CHAIN LINK GATE**									
0110	Industrial 8′ gate, single incl. hardware and concrete	B-80C	10	2.400	Ea.	465	70	20.50	555.50	655
0120	10′		10	2.400		530	70	20.50	620.50	725
0510	Industrial 8′ gate, double incl. hardware and concrete		10	2.400		730	70	20.50	820.50	940
0520	10′		10	2.400		830	70	20.50	920.50	1,050

32 31 13.88 Chain Link Transom

Code	Description	Crew	Daily Output	Labor-Hours	Unit	Material	Labor	Equipment	Total	Total Incl O&P
0010	**CHAIN LINK TRANSOM**									
0110	Add for, single transom, 3′ wide, incl. components & hardware	B-80C	10	2.400	Ea.	116	70	20.50	206.50	266
0120	Add for, double transom, 6′ wide, incl. components & hardware	″	10	2.400	″	124	70	20.50	214.50	275

32 31 23 – Plastic Fences and Gates

32 31 23.10 Fence, Vinyl

Code	Description	Crew	Daily Output	Labor-Hours	Unit	Material	Labor	Equipment	Total	Total Incl O&P
0010	**FENCE, VINYL**									
0011	White, steel reinforced, stainless steel fasteners									
0020	Picket, 4″ x 4″ posts @ 6′-0″ OC, 3′ high	B-1	140	.171	L.F.	24.50	4.83		29.33	35
0030	4′ high		130	.185		26.50	5.20		31.70	38
0040	5′ high		120	.200		29	5.65		34.65	41.50
0100	Board (semi-privacy), 5″ x 5″ posts @ 7′-6″ OC, 5′ high		130	.185		27	5.20		32.20	38
0120	6′ high		125	.192		29.50	5.40		34.90	41.50
0200	Basket weave, 5″ x 5″ posts @ 7′-6″ OC, 5′ high		160	.150		26.50	4.23		30.73	36
0220	6′ high		150	.160		29.50	4.51		34.01	40
0300	Privacy, 5″ x 5″ posts @ 7′-6″ OC, 5′ high		130	.185		26	5.20		31.20	37
0320	6′ high		150	.160		29.50	4.51		34.01	40
0350	Gate, 5′ high		9	2.667	Ea.	320	75		395	480
0360	6′ high		9	2.667		365	75		440	530
0400	For posts set in concrete, add		25	.960		9.60	27		36.60	55
0500	Post and rail fence, 2 rail		150	.160	L.F.	6.10	4.51		10.61	14.15
0510	3 rail		150	.160		7.90	4.51		12.41	16.15
0515	4 rail		150	.160		10.70	4.51		15.21	19.25

32 31 26 – Wire Fences and Gates

32 31 26.10 Fences, Misc. Metal

Code	Description	Crew	Daily Output	Labor-Hours	Unit	Material	Labor	Equipment	Total	Total Incl O&P
0010	**FENCES, MISC. METAL**									
0012	Chicken wire, posts @ 4′, 1″ mesh, 4′ high	B-80C	410	.059	L.F.	3.48	1.70	.49	5.67	7.20
0100	2″ mesh, 6′ high		350	.069		3.99	2	.58	6.57	8.30
0200	Galv. steel, 12 ga., 2″ x 4″ mesh, posts 5′ OC, 3′ high		300	.080		2.76	2.33	.68	5.77	7.60
0300	5′ high		300	.080		3.33	2.33	.68	6.34	8.25
0400	14 ga., 1″ x 2″ mesh, 3′ high		300	.080		3.37	2.33	.68	6.38	8.30
0500	5′ high		300	.080		4.51	2.33	.68	7.52	9.55
1000	Kennel fencing, 1-1/2″ mesh, 6′ long, 3′-6″ wide, 6′-2″ high	2 Clab	4	4	Ea.	500	110		610	730
1050	12′ long		4	4		705	110		815	955
1200	Top covers, 1-1/2″ mesh, 6′ long		15	1.067		136	29.50		165.50	199
1250	12′ long		12	1.333		188	36.50		224.50	268

32 31 29 – Wood Fences and Gates

32 31 29.10 Fence, Wood		Crew	Daily Output	Labor-Hours	Unit	2019 Bare Costs Material	Labor	Equipment	Total	Total Incl O&P
0010	**FENCE, WOOD**									
0011	Basket weave, 3/8" x 4" boards, 2" x 4"									
0020	stringers on spreaders, 4" x 4" posts									
0050	No. 1 cedar, 6' high	B-80C	160	.150	L.F.	26	4.37	1.27	31.64	37
0070	Treated pine, 6' high		150	.160		37.50	4.66	1.35	43.51	50.50
0090	Vertical weave, 6' high		145	.166		23	4.82	1.40	29.22	35
0200	Board fence, 1" x 4" boards, 2" x 4" rails, 4" x 4" post									
0220	Preservative treated, 2 rail, 3' high	B-80C	145	.166	L.F.	10.40	4.82	1.40	16.62	21
0240	4' high		135	.178		12.15	5.15	1.50	18.80	23.50
0260	3 rail, 5' high		130	.185		12.70	5.35	1.56	19.61	24.50
0300	6' high		125	.192		15.75	5.60	1.62	22.97	28.50
0320	No. 2 grade western cedar, 2 rail, 3' high		145	.166		12.55	4.82	1.40	18.77	23.50
0340	4' high		135	.178		12.10	5.15	1.50	18.75	23.50
0360	3 rail, 5' high		130	.185		14.10	5.35	1.56	21.01	26
0400	6' high		125	.192		14.95	5.60	1.62	22.17	27.50
0420	No. 1 grade cedar, 2 rail, 3' high		145	.166		13.65	4.82	1.40	19.87	24.50
0440	4' high		135	.178		15	5.15	1.50	21.65	26.50
0460	3 rail, 5' high		130	.185		17.95	5.35	1.56	24.86	30.50
0500	6' high		125	.192		21.50	5.60	1.62	28.72	34.50
0540	Shadow box, 1" x 6" board, 2" x 4" rail, 4" x 4" post									
0560	Pine, pressure treated, 3 rail, 6' high	B-80C	150	.160	L.F.	25	4.66	1.35	31.01	36.50
0600	Gate, 3'-6" wide		8	3	Ea.	135	87.50	25.50	248	320
0620	No. 1 cedar, 3 rail, 4' high		130	.185	L.F.	19.10	5.35	1.56	26.01	31.50
0640	6' high		125	.192		26.50	5.60	1.62	33.72	40
0860	Open rail fence, split rails, 2 rail, 3' high, no. 1 cedar		160	.150		9.65	4.37	1.27	15.29	19.25
0870	No. 2 cedar		160	.150		8.40	4.37	1.27	14.04	17.85
0880	3 rail, 4' high, no. 1 cedar		150	.160		12.20	4.66	1.35	18.21	22.50
0890	No. 2 cedar		150	.160		8.10	4.66	1.35	14.11	18.10
0920	Rustic rails, 2 rail, 3' high, no. 1 cedar		160	.150		14	4.37	1.27	19.64	24
0930	No. 2 cedar		160	.150		11.60	4.37	1.27	17.24	21.50
0940	3 rail, 4' high		150	.160		12.50	4.66	1.35	18.51	23
0950	No. 2 cedar		150	.160		8.45	4.66	1.35	14.46	18.50
0960	Picket fence, gothic, pressure treated pine									
1000	2 rail, 3' high	B-80C	140	.171	L.F.	7.90	4.99	1.45	14.34	18.55
1020	3 rail, 4' high		130	.185	"	9.95	5.35	1.56	16.86	21.50
1040	Gate, 3'-6" wide		9	2.667	Ea.	76	77.50	22.50	176	237
1060	No. 2 cedar, 2 rail, 3' high		140	.171	L.F.	9.20	4.99	1.45	15.64	19.95
1100	3 rail, 4' high		130	.185	"	9.35	5.35	1.56	16.26	21
1120	Gate, 3'-6" wide		9	2.667	Ea.	81.50	77.50	22.50	181.50	243
1140	No. 1 cedar, 2 rail, 3' high		140	.171	L.F.	13.90	4.99	1.45	20.34	25
1160	3 rail, 4' high		130	.185		18.45	5.35	1.56	25.36	31
1200	Rustic picket, molded pine, 2 rail, 3' high		140	.171		8.70	4.99	1.45	15.14	19.40
1220	No. 1 cedar, 2 rail, 3' high		140	.171		10.60	4.99	1.45	17.04	21.50
1240	Stockade fence, no. 1 cedar, 3-1/4" rails, 6' high		160	.150		13.50	4.37	1.27	19.14	23.50
1260	8' high		155	.155		18.70	4.51	1.31	24.52	29.50
1300	No. 2 cedar, treated wood rails, 6' high		160	.150		13.85	4.37	1.27	19.49	24
1320	Gate, 3'-6" wide		8	3	Ea.	89.50	87.50	25.50	202.50	271
1360	Treated pine, treated rails, 6' high		160	.150	L.F.	13.95	4.37	1.27	19.59	24
1400	8' high		150	.160	"	20.50	4.66	1.35	26.51	31.50

32 31 Fences and Gates

32 31 29 – Wood Fences and Gates

32 31 29.20 Fence, Wood Rail		Crew	Daily Output	Labor-Hours	Unit	2019 Bare Costs Material	Labor	Equipment	Total	Total Incl O&P
0010	**FENCE, WOOD RAIL**									
0012	Picket, No. 2 cedar, Gothic, 2 rail, 3′ high	B-1	160	.150	L.F.	8.40	4.23		12.63	16.25
0050	Gate, 3′-6″ wide	B-80C	9	2.667	Ea.	79	77.50	22.50	179	240
0400	3 rail, 4′ high		150	.160	L.F.	9.45	4.66	1.35	15.46	19.60
0500	Gate, 3′-6″ wide		9	2.667	Ea.	96	77.50	22.50	196	259
1200	Stockade, No. 2 cedar, treated wood rails, 6′ high		160	.150	L.F.	9.40	4.37	1.27	15.04	18.95
1250	Gate, 3′ wide		9	2.667	Ea.	98.50	77.50	22.50	198.50	262
1300	No. 1 cedar, 3-1/4″ cedar rails, 6′ high		160	.150	L.F.	20	4.37	1.27	25.64	30.50
1500	Gate, 3′ wide		9	2.667	Ea.	230	77.50	22.50	330	405
2700	Prefabricated redwood or cedar, 4′ high		160	.150	L.F.	16.30	4.37	1.27	21.94	26.50
2800	6′ high		150	.160		24	4.66	1.35	30.01	35
3300	Board, shadow box, 1″ x 6″, treated pine, 6′ high		160	.150		12.85	4.37	1.27	18.49	22.50
3400	No. 1 cedar, 6′ high		150	.160		25.50	4.66	1.35	31.51	37
3900	Basket weave, No. 1 cedar, 6′ high		160	.150		35	4.37	1.27	40.64	47
4200	Gate, 3′-6″ wide	B-1	9	2.667	Ea.	184	75		259	325
5000	Fence rail, redwood, 2″ x 4″, merch. grade, 8′	″	2400	.010	L.F.	2.58	.28		2.86	3.31

32 32 Retaining Walls

32 32 13 – Cast-in-Place Concrete Retaining Walls

32 32 13.10 Retaining Walls, Cast Concrete		Crew	Daily Output	Labor-Hours	Unit	Material	Labor	Equipment	Total	Total Incl O&P
0010	**RETAINING WALLS, CAST CONCRETE**									
1800	Concrete gravity wall with vertical face including excavation & backfill									
1850	No reinforcing									
1900	6′ high, level embankment	C-17C	36	2.306	L.F.	85	85	15.50	185.50	252
2000	33° slope embankment	″	32	2.594	″	105	96	17.45	218.45	294
2800	Reinforced concrete cantilever, incl. excavation, backfill & reinf.									
2900	6′ high, 33° slope embankment	C-17C	35	2.371	L.F.	82.50	87.50	15.95	185.95	253

32 32 23 – Segmental Retaining Walls

32 32 23.13 Segmental Conc. Unit Masonry Retaining Walls		Crew	Daily Output	Labor-Hours	Unit	Material	Labor	Equipment	Total	Total Incl O&P
0010	**SEGMENTAL CONC. UNIT MASONRY RETAINING WALLS**									
7100	Segmental retaining wall system, incl. pins and void fill									
7120	base and backfill not included									
7140	Large unit, 8″ high x 18″ wide x 20″ deep, 3 plane split	B-62	300	.080	S.F.	13.25	2.42	.58	16.25	19.20
7150	Straight split		300	.080		13.20	2.42	.58	16.20	19.15
7160	Medium, lt. wt., 8″ high x 18″ wide x 12″ deep, 3 plane split		400	.060		6.75	1.81	.44	9	10.85
7170	Straight split		400	.060		9.90	1.81	.44	12.15	14.30
7180	Small unit, 4″ x 18″ x 10″ deep, 3 plane split		400	.060		15.60	1.81	.44	17.85	20.50
7190	Straight split		400	.060		11.35	1.81	.44	13.60	15.95
7200	Cap unit, 3 plane split		300	.080		13.90	2.42	.58	16.90	19.90
7210	Cap unit, straight split		300	.080		13.90	2.42	.58	16.90	19.90
7250	Geo-grid soil reinforcement 4′ x 50′	2 Clab	22500	.001		.74	.02		.76	.84
7255	Geo-grid soil reinforcement 6′ x 150′	″	22500	.001		.58	.02		.60	.67

32 32 26 – Metal Crib Retaining Walls

32 32 26.10 Metal Bin Retaining Walls		Crew	Daily Output	Labor-Hours	Unit	Material	Labor	Equipment	Total	Total Incl O&P
0010	**METAL BIN RETAINING WALLS**									
0011	Aluminized steel bin, excavation									
0020	and backfill not included, 10′ wide									
0100	4′ high, 5.5′ deep	B-13	650	.074	S.F.	29	2.18	.90	32.08	36.50
0200	8′ high, 5.5′ deep		615	.078		33.50	2.31	.95	36.76	41.50

32 32 Retaining Walls

32 32 26 – Metal Crib Retaining Walls

32 32 26.10 Metal Bin Retaining Walls		Crew	Daily Output	Labor-Hours	Unit	Material	Labor	Equipment	Total	Total Incl O&P
						2019 Bare Costs				
0300	10′ high, 7.7′ deep	B-13	580	.083	S.F.	37	2.45	1	40.45	46
0400	12′ high, 7.7′ deep		530	.091		40	2.68	1.10	43.78	49.50
0500	16′ high, 7.7′ deep		515	.093		42.50	2.76	1.13	46.39	52.50

32 32 29 – Timber Retaining Walls

32 32 29.10 Landscape Timber Retaining Walls

		Crew	Daily Output	Labor-Hours	Unit	Material	Labor	Equipment	Total	Total Incl O&P
0010	**LANDSCAPE TIMBER RETAINING WALLS**									
0100	Treated timbers, 6″ x 6″	1 Clab	265	.030	L.F.	2.54	.83		3.37	4.16
0110	6″ x 8′	″	200	.040	″	5.75	1.10		6.85	8.15
0120	Drilling holes in timbers for fastening, 1/2″	1 Carp	450	.018	Inch		.63		.63	1.05
0130	5/8″	″	450	.018	″		.63		.63	1.05
0140	Reinforcing rods for fastening, 1/2″	1 Clab	312	.026	L.F.	.38	.71		1.09	1.59
0150	5/8″	″	312	.026	″	.59	.71		1.30	1.82
0160	Reinforcing fabric	2 Clab	2500	.006	S.Y.	2.17	.18		2.35	2.68
0170	Gravel backfill		28	.571	C.Y.	16.95	15.70		32.65	44.50
0180	Perforated pipe, 4″ diameter with silt sock		1200	.013	L.F.	1.01	.37		1.38	1.72
0190	Galvanized 60d common nails	1 Clab	625	.013	Ea.	.20	.35		.55	.80
0200	20d common nails	″	3800	.002	″	.04	.06		.10	.15

32 32 53 – Stone Retaining Walls

32 32 53.10 Retaining Walls, Stone

		Crew	Daily Output	Labor-Hours	Unit	Material	Labor	Equipment	Total	Total Incl O&P
0010	**RETAINING WALLS, STONE**									
0015	Including excavation, concrete footing and									
0020	stone 3′ below grade. Price is exposed face area.									
0200	Decorative random stone, to 6′ high, 1′-6″ thick, dry set	D-1	35	.457	S.F.	73.50	14.80		88.30	106
0300	Mortar set		40	.400		76	12.95		88.95	105
0500	Cut stone, to 6′ high, 1′-6″ thick, dry set		35	.457		76	14.80		90.80	109
0600	Mortar set		40	.400		77	12.95		89.95	107
0800	Random stone, 6′ to 10′ high, 2′ thick, dry set		45	.356		79	11.50		90.50	106
0900	Mortar set		50	.320		82	10.35		92.35	107
1100	Cut stone, 6′ to 10′ high, 2′ thick, dry set		45	.356		79.50	11.50		91	107
1200	Mortar set		50	.320		82.50	10.35		92.85	108

32 33 Site Furnishings

32 33 33 – Site Manufactured Planters

32 33 33.10 Planters

		Crew	Daily Output	Labor-Hours	Unit	Material	Labor	Equipment	Total	Total Incl O&P
0010	**PLANTERS**									
0012	Concrete, sandblasted, precast, 48″ diameter, 24″ high	2 Clab	15	1.067	Ea.	650	29.50		679.50	770
0300	Fiberglass, circular, 36″ diameter, 24″ high		15	1.067		720	29.50		749.50	840
1200	Wood, square, 48″ side, 24″ high		15	1.067		1,600	29.50		1,629.50	1,825
1300	Circular, 48″ diameter, 30″ high		10	1.600		1,125	44		1,169	1,300
1600	Planter/bench, 72″		5	3.200		3,800	88		3,888	4,325

32 33 43 – Site Seating and Tables

32 33 43.13 Site Seating

		Crew	Daily Output	Labor-Hours	Unit	Material	Labor	Equipment	Total	Total Incl O&P
0010	**SITE SEATING**									
0012	Seating, benches, park, precast conc., w/backs, wood rails, 4′ long	2 Clab	5	3.200	Ea.	625	88		713	830
0100	8′ long		4	4		1,050	110		1,160	1,325
0500	Steel barstock pedestals w/backs, 2″ x 3″ wood rails, 4′ long		10	1.600		1,350	44		1,394	1,550
0510	8′ long		7	2.286		1,700	63		1,763	1,975
0800	Cast iron pedestals, back & arms, wood slats, 4′ long		8	2		435	55		490	570
0820	8′ long		5	3.200		1,050	88		1,138	1,300

32 33 Site Furnishings

32 33 43 – Site Seating and Tables

32 33 43.13 Site Seating		Crew	Daily Output	Labor-Hours	Unit	Material	2019 Bare Costs Labor	Equipment	Total	Total Incl O&P
1700	Steel frame, fir seat, 10′ long	2 Clab	10	1.600	Ea.	400	44		444	515

32 84 Planting Irrigation

32 84 23 – Underground Sprinklers

32 84 23.10 Sprinkler Irrigation System

		Crew	Daily Output	Labor-Hours	Unit	Material	Labor	Equipment	Total	Total Incl O&P
0010	**SPRINKLER IRRIGATION SYSTEM**									
0011	For lawns									
0800	Residential system, custom, 1″ supply	B-20	2000	.012	S.F.	.26	.34		.60	.85
0900	1-1/2″ supply	″	1800	.013	″	.43	.38		.81	1.09

32 91 Planting Preparation

32 91 13 – Soil Preparation

32 91 13.16 Mulching

		Crew	Daily Output	Labor-Hours	Unit	Material	Labor	Equipment	Total	Total Incl O&P
0010	**MULCHING**									
0100	Aged barks, 3″ deep, hand spread	1 Clab	100	.080	S.Y.	3.55	2.20		5.75	7.55
0150	Skid steer loader	B-63	13.50	2.963	M.S.F.	395	81.50	12.90	489.40	585
0160	Skid steer loader	″	1500	.027	S.Y.	3.55	.73	.12	4.40	5.25
0200	Hay, 1″ deep, hand spread	1 Clab	475	.017	″	.46	.46		.92	1.28
0250	Power mulcher, small	B-64	180	.089	M.S.F.	51	2.66	1.86	55.52	62.50
0350	Large	B-65	530	.030	″	51	.90	.89	52.79	58.50
0400	Humus peat, 1″ deep, hand spread	1 Clab	700	.011	S.Y.	3.02	.31		3.33	3.84
0450	Push spreader	″	2500	.003	″	3.02	.09		3.11	3.47
0550	Tractor spreader	B-66	700	.011	M.S.F.	335	.41	.31	335.72	370
0600	Oat straw, 1″ deep, hand spread	1 Clab	475	.017	S.Y.	.57	.46		1.03	1.40
0650	Power mulcher, small	B-64	180	.089	M.S.F.	63.50	2.66	1.86	68.02	76
0700	Large	B-65	530	.030	″	63.50	.90	.89	65.29	72
0750	Add for asphaltic emulsion	B-45	1770	.009	Gal.	5.85	.27	.46	6.58	7.35
0800	Peat moss, 1″ deep, hand spread	1 Clab	900	.009	S.Y.	4.88	.24		5.12	5.75
0850	Push spreader	″	2500	.003	″	4.88	.09		4.97	5.50
0950	Tractor spreader	B-66	700	.011	M.S.F.	540	.41	.31	540.72	595
1000	Polyethylene film, 6 mil	2 Clab	2000	.008	S.Y.	.54	.22		.76	.95
1100	Redwood nuggets, 3″ deep, hand spread	1 Clab	150	.053	″	3.16	1.47		4.63	5.90
1150	Skid steer loader	B-63	13.50	2.963	M.S.F.	350	81.50	12.90	444.40	535
1200	Stone mulch, hand spread, ceramic chips, economy	1 Clab	125	.064	S.Y.	7.05	1.76		8.81	10.65
1250	Deluxe	″	95	.084	″	10.70	2.32		13.02	15.65
1300	Granite chips	B-1	10	2.400	C.Y.	74	67.50		141.50	194
1400	Marble chips		10	2.400		217	67.50		284.50	350
1600	Pea gravel		28	.857		115	24		139	166
1700	Quartz		10	2.400		192	67.50		259.50	325
1800	Tar paper, 15 lb. felt	1 Clab	800	.010	S.Y.	.49	.28		.77	.98
1900	Wood chips, 2″ deep, hand spread	″	220	.036	″	1.63	1		2.63	3.44
1950	Skid steer loader	B-63	20.30	1.970	M.S.F.	181	54	8.60	243.60	298

32 91 13.26 Planting Beds

		Crew	Daily Output	Labor-Hours	Unit	Material	Labor	Equipment	Total	Total Incl O&P
0010	**PLANTING BEDS**									
0100	Backfill planting pit, by hand, on site topsoil	2 Clab	18	.889	C.Y.		24.50		24.50	40.50
0200	Prepared planting mix, by hand	″	24	.667			18.35		18.35	30.50
0300	Skid steer loader, on site topsoil	B-62	340	.071			2.13	.51	2.64	4.08
0400	Prepared planting mix	″	410	.059			1.77	.43	2.20	3.39
1000	Excavate planting pit, by hand, sandy soil	2 Clab	16	1			27.50		27.50	45.50

32 91 Planting Preparation

32 91 13 – Soil Preparation

32 91 13.26 Planting Beds		Crew	Daily Output	Labor-Hours	Unit	Material	2019 Bare Costs Labor	Equipment	Total	Total Incl O&P
1100	Heavy soil or clay	2 Clab	8	2	C.Y.		55		55	91
1200	1/2 C.Y. backhoe, sandy soil	B-11C	150	.107			3.46	2.13	5.59	8.05
1300	Heavy soil or clay	"	115	.139			4.52	2.78	7.30	10.50
2000	Mix planting soil, incl. loam, manure, peat, by hand	2 Clab	60	.267		46.50	7.35		53.85	63
2100	Skid steer loader	B-62	150	.160		46.50	4.84	1.16	52.50	60
3000	Pile sod, skid steer loader	"	2800	.009	S.Y.		.26	.06	.32	.50
3100	By hand	2 Clab	400	.040			1.10		1.10	1.82
4000	Remove sod, F.E. loader	B-10S	2000	.004			.15	.18	.33	.44
4100	Sod cutter	B-12K	3200	.005			.16	.37	.53	.68
4200	By hand	2 Clab	240	.067			1.83		1.83	3.03

32 91 19 – Landscape Grading

32 91 19.13 Topsoil Placement and Grading		Crew	Daily Output	Labor-Hours	Unit	Material	Labor	Equipment	Total	Total Incl O&P
0010	**TOPSOIL PLACEMENT AND GRADING**									
0300	Fine grade, base course for paving, see Section 32 11 23.23									
0701	Furnish and place, truck dumped, unscreened, 4" deep	B-10S	12000	.001	S.F.	.44	.03	.03	.50	.55
0801	6" deep	"	7400	.001	"	.50	.04	.05	.59	.67
0900	Fine grading and seeding, incl. lime, fertilizer & seed,									
1001	With equipment	B-14	9000	.005	S.F.	.07	.16	.04	.27	.38

32 92 Turf and Grasses

32 92 19 – Seeding

32 92 19.13 Mechanical Seeding		Crew	Daily Output	Labor-Hours	Unit	Material	Labor	Equipment	Total	Total Incl O&P
0010	**MECHANICAL SEEDING**									
0020	Mechanical seeding, 215 lb./acre	B-66	1.50	5.333	Acre	575	190	146	911	1,100
0100	44 lb./M.S.Y.	"	2500	.003	S.Y.	.20	.11	.09	.40	.51
0101	44 lb./M.S.Y.	1 Clab	13950	.001	S.F.	.02	.02		.04	.05
0300	Fine grading and seeding incl. lime, fertilizer & seed,									
0310	with equipment	B-14	1000	.048	S.Y.	.46	1.40	.32	2.18	3.17
0400	Fertilizer hand push spreader, 35 lb./M.S.F.	1 Clab	200	.040	M.S.F.	9.90	1.10		11	12.70
0600	Limestone hand push spreader, 50 lb./M.S.F.		180	.044		4.87	1.22		6.09	7.35
0800	Grass seed hand push spreader, 4.5 lb./M.S.F.		180	.044		22	1.22		23.22	26.50
1000	Hydro or air seeding for large areas, incl. seed and fertilizer	B-81	8900	.002	S.Y.	.52	.05	.07	.64	.73
1100	With wood fiber mulch added	"	8900	.002	"	2.03	.05	.07	2.15	2.39
1300	Seed only, over 100 lb., field seed, minimum				Lb.	1.92			1.92	2.11
1400	Maximum					1.83			1.83	2.01
1500	Lawn seed, minimum					1.45			1.45	1.60
1600	Maximum					2.55			2.55	2.81
1800	Aerial operations, seeding only, field seed	B-58	50	.480	Acre	625	14.50	64.50	704	780
1900	Lawn seed		50	.480		470	14.50	64.50	549	615
2100	Seed and liquid fertilizer, field seed		50	.480		700	14.50	64.50	779	865
2200	Lawn seed		50	.480		545	14.50	64.50	624	695

32 92 23 – Sodding

32 92 23.10 Sodding Systems		Crew	Daily Output	Labor-Hours	Unit	Material	Labor	Equipment	Total	Total Incl O&P
0010	**SODDING SYSTEMS**									
0020	Sodding, 1" deep, bluegrass sod, on level ground, over 8 M.S.F.	B-63	22	1.818	M.S.F.	198	50	7.95	255.95	310
0200	4 M.S.F.		17	2.353		249	64.50	10.25	323.75	390
0300	1,000 S.F.		13.50	2.963		325	81.50	12.90	419.40	505
0500	Sloped ground, over 8 M.S.F.		6	6.667		198	183	29	410	555
0600	4 M.S.F.		5	8		249	220	35	504	680
0700	1,000 S.F.		4	10		325	275	43.50	643.50	860

32 92 Turf and Grasses

32 92 23 – Sodding

32 92 23.10 Sodding Systems		Crew	Daily Output	Labor-Hours	Unit	2019 Bare Costs Material	Labor	Equipment	Total	Total Incl O&P
1000	Bent grass sod, on level ground, over 6 M.S.F.	B-63	20	2	M.S.F.	262	55	8.70	325.70	390
1100	3 M.S.F.		18	2.222		276	61	9.70	346.70	415
1200	Sodding 1,000 S.F. or less		14	2.857		305	78.50	12.45	395.95	480
1500	Sloped ground, over 6 M.S.F.		15	2.667		262	73.50	11.65	347.15	420
1600	3 M.S.F.		13.50	2.963		276	81.50	12.90	370.40	455
1700	1,000 S.F.		12	3.333		305	91.50	14.55	411.05	505

32 93 Plants

32 93 13 – Ground Covers

32 93 13.10 Ground Cover Plants

		Crew	Daily Output	Labor-Hours	Unit	Material	Labor	Equipment	Total	Total Incl O&P
0010	**GROUND COVER PLANTS**									
0012	Plants, pachysandra, in prepared beds	B-1	15	1.600	C	84	45		129	167
0200	Vinca minor, 1 yr., bare root, in prepared beds		12	2	"	72	56.50		128.50	172
0600	Stone chips, in 50 lb. bags, Georgia marble		520	.046	Bag	4.79	1.30		6.09	7.40
0700	Onyx gemstone		260	.092		16.25	2.60		18.85	22
0800	Quartz		260	.092		16.85	2.60		19.45	23
0900	Pea gravel, truckload lots		28	.857	Ton	15.75	24		39.75	57.50

32 93 33 – Shrubs

32 93 33.10 Shrubs and Trees

		Crew	Daily Output	Labor-Hours	Unit	Material	Labor	Equipment	Total	Total Incl O&P
0010	**SHRUBS AND TREES**									
0011	Evergreen, in prepared beds, B&B									
0100	Arborvitae pyramidal, 4′-5′	B-17	30	1.067	Ea.	106	33	22	161	195
0150	Globe, 12″-15″	B-1	96	.250		26.50	7.05		33.55	41
0300	Cedar, blue, 8′-10′	B-17	18	1.778		243	55	36.50	334.50	400
0500	Hemlock, Canadian, 2-1/2′-3′	B-1	36	.667		33	18.80		51.80	67.50
0550	Holly, Savannah, 8′-10′ H		9.68	2.479		310	70		380	455
0600	Juniper, andorra, 18″-24″		80	.300		52.50	8.45		60.95	72
0620	Wiltoni, 15″-18″		80	.300		27	8.45		35.45	44
0640	Skyrocket, 4-1/2′-5′	B-17	55	.582		113	17.95	12	142.95	167
0660	Blue pfitzer, 2′-2-1/2′	B-1	44	.545		40	15.35		55.35	69.50
0680	Ketleerie, 2-1/2′-3′		50	.480		57.50	13.50		71	86
0700	Pine, black, 2-1/2′-3′		50	.480		64	13.50		77.50	93
0720	Mugo, 18″-24″		60	.400		56.50	11.25		67.75	81
0740	White, 4′-5′	B-17	75	.427		54.50	13.20	8.80	76.50	91
0800	Spruce, blue, 18″-24″	B-1	60	.400		70.50	11.25		81.75	96
0840	Norway, 4′-5′	B-17	75	.427		89	13.20	8.80	111	129
0900	Yew, denisforma, 12″-15″	B-1	60	.400		37.50	11.25		48.75	60
1000	Capitata, 18″-24″		30	.800		35	22.50		57.50	75.50
1100	Hicksi, 2′-2-1/2′		30	.800		112	22.50		134.50	160

32 93 33.20 Shrubs

		Crew	Daily Output	Labor-Hours	Unit	Material	Labor	Equipment	Total	Total Incl O&P
0010	**SHRUBS**									
0011	Broadleaf Evergreen, planted in prepared beds									
0100	Andromeda, 15″-18″, cont	B-1	96	.250	Ea.	34.50	7.05		41.55	49.50
0200	Azalea, 15″-18″, cont		96	.250		31	7.05		38.05	45.50
0300	Barberry, 9″-12″, cont		130	.185		19.25	5.20		24.45	29.50
0400	Boxwood, 15″-18″, B&B		96	.250		46	7.05		53.05	62.50
0500	Euonymus, emerald gaiety, 12″-15″, cont		115	.209		25.50	5.90		31.40	37.50
0600	Holly, 15″-18″, B&B		96	.250		41	7.05		48.05	56.50
0900	Mount laurel, 18″-24″, B&B		80	.300		75.50	8.45		83.95	97
1000	Paxistema, 9″-12″ H		130	.185		22.50	5.20		27.70	33

32 93 Plants

32 93 33 – Shrubs

32 93 33.20 Shrubs		Crew	Daily Output	Labor-Hours	Unit	2019 Bare Costs Material	Labor	Equipment	Total	Total Incl O&P
1100	Rhododendron, 18″-24″, cont	B-1	48	.500	Ea.	39.50	14.10		53.60	67
1200	Rosemary, 1 gal. cont		600	.040		19.05	1.13		20.18	23
2000	Deciduous, planted in prepared beds, amelanchier, 2′-3′, B&B		57	.421		126	11.85		137.85	158
2100	Azalea, 15″-18″, B&B		96	.250		31	7.05		38.05	45.50
2300	Bayberry, 2′-3′, B&B		57	.421		31.50	11.85		43.35	54
2600	Cotoneaster, 15″-18″, B&B	↓	80	.300		28	8.45		36.45	44.50
2800	Dogwood, 3′-4′, B&B	B-17	40	.800		34	24.50	16.50	75	96
2900	Euonymus, alatus compacta, 15″-18″, cont	B-1	80	.300		28	8.45		36.45	44.50
3200	Forsythia, 2′-3′, cont	″	60	.400		18.75	11.25		30	39
3300	Hibiscus, 3′-4′, B&B	B-17	75	.427		43.50	13.20	8.80	65.50	78.50
3400	Honeysuckle, 3′-4′, B&B	B-1	60	.400		28.50	11.25		39.75	50
3500	Hydrangea, 2′-3′, B&B	″	57	.421		32	11.85		43.85	55
3600	Lilac, 3′-4′, B&B	B-17	40	.800		19.30	24.50	16.50	60.30	79.50
3900	Privet, bare root, 18″-24″	B-1	80	.300		15.95	8.45		24.40	31.50
4100	Quince, 2′-3′, B&B	″	57	.421		30	11.85		41.85	52.50
4200	Russian olive, 3′-4′, B&B	B-17	75	.427		25	13.20	8.80	47	58.50
4400	Spirea, 3′-4′, B&B	B-1	70	.343		21	9.65		30.65	39
4500	Viburnum, 3′-4′, B&B	B-17	40	.800	↓	27	24.50	16.50	68	88

32 93 43 – Trees

32 93 43.20 Trees			Crew	Daily Output	Labor-Hours	Unit	Material	Labor	Equipment	Total	Total Incl O&P
0010	**TREES**										
0011	Deciduous, in prep. beds, balled & burlapped (B&B)										
0100	Ash, 2″ caliper	G	B-17	8	4	Ea.	208	124	82.50	414.50	525
0200	Beech, 5′-6′	G		50	.640		221	19.75	13.20	253.95	290
0300	Birch, 6′-8′, 3 stems	G		20	1.600		169	49.50	33	251.50	305
0500	Crabapple, 6′-8′	G		20	1.600		144	49.50	33	226.50	276
0600	Dogwood, 4′-5′	G		40	.800		147	24.50	16.50	188	221
0700	Eastern redbud, 4′-5′	G		40	.800		150	24.50	16.50	191	224
0800	Elm, 8′-10′	G		20	1.600		345	49.50	33	427.50	495
0900	Ginkgo, 6′-7′	G		24	1.333		155	41	27.50	223.50	269
1000	Hawthorn, 8′-10′, 1″ caliper	G		20	1.600		162	49.50	33	244.50	296
1100	Honeylocust, 10′-12′, 1-1/2″ caliper	G		10	3.200		214	99	66	379	470
1300	Larch, 8′	G		32	1		131	31	20.50	182.50	219
1400	Linden, 8′-10′, 1″ caliper	G		20	1.600		148	49.50	33	230.50	281
1500	Magnolia, 4′-5′	G		20	1.600		118	49.50	33	200.50	248
1600	Maple, red, 8′-10′, 1-1/2″ caliper	G		10	3.200		204	99	66	369	460
1700	Mountain ash, 8′-10′, 1″ caliper	G		16	2		190	62	41	293	355
1800	Oak, 2-1/2″-3″ caliper	G		6	5.333		330	165	110	605	750
2100	Planetree, 9′-11′, 1-1/4″ caliper	G		10	3.200		272	99	66	437	535
2200	Plum, 6′-8′, 1″ caliper	G		20	1.600		82	49.50	33	164.50	208
2300	Poplar, 9′-11′, 1-1/4″ caliper	G		10	3.200		102	99	66	267	350
2500	Sumac, 2′-3′	G		75	.427		45.50	13.20	8.80	67.50	81
2700	Tulip, 5′-6′	G		40	.800		48	24.50	16.50	89	111
2800	Willow, 6′-8′, 1″ caliper	G	↓	20	1.600	↓	99.50	49.50	33	182	228

32 94 Planting Accessories

32 94 13 – Landscape Edging

32 94 13.20 Edging		Crew	Daily Output	Labor-Hours	Unit	Material	2019 Bare Costs Labor	Equipment	Total	Total Incl O&P
0010	**EDGING**									
0050	Aluminum alloy, including stakes, 1/8″ x 4″, mill finish	B-1	390	.062	L.F.	2.17	1.73		3.90	5.25
0051	Black paint		390	.062		2.52	1.73		4.25	5.65
0052	Black anodized		390	.062		2.91	1.73		4.64	6.05
0100	Brick, set horizontally, 1-1/2 bricks per L.F.	D-1	370	.043		1.42	1.40		2.82	3.89
0150	Set vertically, 3 bricks per L.F.	″	135	.119		3.60	3.84		7.44	10.35
0200	Corrugated aluminum, roll, 4″ wide	1 Carp	650	.012		2.21	.44		2.65	3.16
0250	6″ wide	″	550	.015		2.76	.52		3.28	3.90
0600	Railroad ties, 6″ x 8″	2 Carp	170	.094		2.32	3.36		5.68	8.10
0650	7″ x 9″		136	.118		2.58	4.19		6.77	9.80
0750	Redwood 2″ x 4″		330	.048		2.30	1.73		4.03	5.40
0800	Steel edge strips, incl. stakes, 1/4″ x 5″	B-1	390	.062		4.88	1.73		6.61	8.20
0850	3/16″ x 4″	″	390	.062		3.86	1.73		5.59	7.10

32 94 50 – Tree Guying

32 94 50.10 Tree Guying Systems		Crew	Daily Output	Labor-Hours	Unit	Material	Labor	Equipment	Total	Total Incl O&P
0010	**TREE GUYING SYSTEMS**									
0015	Tree guying including stakes, guy wire and wrap									
0100	Less than 3″ caliper, 2 stakes	2 Clab	35	.457	Ea.	13.30	12.55		25.85	35.50
0200	3″ to 4″ caliper, 3 stakes	″	21	.762	″	19.65	21		40.65	56
1000	Including arrowhead anchor, cable, turnbuckles and wrap									
1100	Less than 3″ caliper, 3″ anchors	2 Clab	20	.800	Ea.	21	22		43	59.50
1200	3″ to 6″ caliper, 4″ anchors		15	1.067		31.50	29.50		61	83
1300	6″ caliper, 6″ anchors		12	1.333		23.50	36.50		60	86.50
1400	8″ caliper, 8″ anchors		9	1.778		115	49		164	208

32 96 Transplanting

32 96 23 – Plant and Bulb Transplanting

32 96 23.23 Planting		Crew	Daily Output	Labor-Hours	Unit	Material	Labor	Equipment	Total	Total Incl O&P
0010	**PLANTING**									
0012	Moving shrubs on site, 12″ ball	B-62	28	.857	Ea.		26	6.25	32.25	49.50
0100	24″ ball	″	22	1.091	″		33	7.95	40.95	63.50

32 96 23.43 Moving Trees		Crew	Daily Output	Labor-Hours	Unit	Material	Labor	Equipment	Total	Total Incl O&P
0010	**MOVING TREES**, On site									
0300	Moving trees on site, 36″ ball	B-6	3.75	6.400	Ea.		193	85.50	278.50	415
0400	60″ ball	″	1	24	″		725	320	1,045	1,550

Division Notes

		CREW	DAILY OUTPUT	LABOR-HOURS	UNIT	BARE COSTS MAT.	BARE COSTS LABOR	BARE COSTS EQUIP.	BARE COSTS TOTAL	TOTAL INCL O&P

Estimating Tips

33 10 00 Water Utilities
33 30 00 Sanitary Sewerage Utilities
33 40 00 Storm Drainage Utilities

- Never assume that the water, sewer, and drainage lines will go in at the early stages of the project. Consider the site access needs before dividing the site in half with open trenches, loose pipe, and machinery obstructions. Always inspect the site to establish that the site drawings are complete. Check off all existing utilities on your drawings as you locate them. Be especially careful with underground utilities because appurtenances are sometimes buried during regrading or repaving operations. If you find any discrepancies, mark up the site plan for further research. Differing site conditions can be very costly if discovered later in the project.
- See also Section 33 01 00 for restoration of pipe where removal/ replacement may be undesirable. Use of new types of piping materials can reduce the overall project cost. Owners/design engineers should consider the installing contractor as a valuable source of current information on utility products and local conditions that could lead to significant cost savings.

Reference Numbers

Reference numbers are shown at the beginning of some major classifications. These numbers refer to related items in the Reference Section. The reference information may be an estimating procedure, an alternate pricing method, or technical information.

Note: Not all subdivisions listed here necessarily appear. ■

Note: "Powered in part by CINX™, based on liscensed proprietary information of Harrison Publishing House, Inc."

Note: Trade Service, in part, has been used as a reference source for some of the material prices used in Division 33

33 05 Common Work Results for Utilities

33 05 07 – Trenchless Installation of Utility Piping

33 05 07.36 Microtunneling		Crew	Daily Output	Labor-Hours	Unit	Material	2019 Bare Costs Labor	Equipment	Total	Total Incl O&P
0010	**MICROTUNNELING**									
0011	Not including excavation, backfill, shoring,									
0020	or dewatering, average 50′/day, slurry method									
0100	24″ to 48″ outside diameter, minimum				L.F.				965	965
0110	Adverse conditions, add				″				500	500
1000	Rent microtunneling machine, average monthly lease				Month				97,500	107,000
1010	Operating technician				Day				630	690
1100	Mobilization and demobilization, minimum				Job				41,200	45,900
1110	Maximum				″				445,500	490,500

33 05 61 – Concrete Manholes

33 05 61.10 Storm Drainage Manholes, Frames and Covers		Crew	Daily Output	Labor-Hours	Unit	Material	Labor	Equipment	Total	Total Incl O&P
0010	**STORM DRAINAGE MANHOLES, FRAMES & COVERS**									
0020	Excludes footing, excavation, backfill (See line items for frame & cover)									
0050	Brick, 4′ inside diameter, 4′ deep	D-1	1	16	Ea.	590	520		1,110	1,525
1110	Precast, 4′ ID, 4′ deep	B-22	4.10	7.317	″	865	220	47.50	1,132.50	1,375

33 05 63 – Concrete Vaults and Chambers

33 05 63.13 Precast Concrete Utility Structures		Crew	Daily Output	Labor-Hours	Unit	Material	Labor	Equipment	Total	Total Incl O&P
0010	**PRECAST CONCRETE UTILITY STRUCTURES**, 6″ thick									
0050	5′ x 10′ x 6′ high, ID	B-13	2	24	Ea.	1,800	710	291	2,801	3,475
0350	Hand hole, precast concrete, 1-1/2″ thick									
0400	1′-0″ x 2′-0″ x 1′-9″, ID, light duty	B-1	4	6	Ea.	500	169		669	830
0450	4′-6″ x 3′-2″ x 2′-0″, OD, heavy duty	B-6	3	8	″	1,550	242	107	1,899	2,225

33 11 Groundwater Sources

33 11 13 – Potable Water Supply Wells

33 11 13.10 Wells and Accessories		Crew	Daily Output	Labor-Hours	Unit	Material	Labor	Equipment	Total	Total Incl O&P
0010	**WELLS & ACCESSORIES**									
0011	Domestic									
0100	Drilled, 4″ to 6″ diameter	B-23	120	.333	L.F.		9.30	22.50	31.80	40
1500	Pumps, installed in wells to 100′ deep, 4″ submersible									
1520	3/4 HP	Q-1	2.66	6.015	Ea.	860	219		1,079	1,300
1600	1 HP	″	2.29	6.987	″	925	254		1,179	1,450

33 14 Water Utility Transmission and Distribution

33 14 13 – Public Water Utility Distribution Piping

33 14 13.15 Water Supply, Ductile Iron Pipe		Crew	Daily Output	Labor-Hours	Unit	Material	Labor	Equipment	Total	Total Incl O&P
0010	**WATER SUPPLY, DUCTILE IRON PIPE**									
0020	Not including excavation or backfill									
2000	Pipe, class 50 water piping, 18′ lengths									
2020	Mechanical joint, 4″ diameter	B-21A	200	.200	L.F.	36	6.70	1.86	44.56	52.50
2040	6″ diameter		160	.250		43	8.40	2.33	53.73	64
3000	Push-on joint, 4″ diameter		400	.100		20.50	3.35	.93	24.78	29
3020	6″ diameter		333.33	.120		23	4.02	1.12	28.14	33
8000	Piping, fittings, mechanical joint, AWWA C110									
8006	90° bend, 4″ diameter	B-20A	16	2	Ea.	175	65		240	300
8020	6″ diameter		12.80	2.500		256	81		337	415
8200	Wye or tee, 4″ diameter		10.67	2.999		375	97.50		472.50	575
8220	6″ diameter		8.53	3.751		570	122		692	825

33 14 Water Utility Transmission and Distribution

33 14 13 – Public Water Utility Distribution Piping

33 14 13.15 Water Supply, Ductile Iron Pipe		Crew	Daily Output	Labor-Hours	Unit	Material	2019 Bare Costs Labor	Equipment	Total	Total Incl O&P
8398	45° bend, 4" diameter	B-20A	16	2	Ea.	214	65		279	345
8400	6" diameter		12.80	2.500		271	81		352	430
8405	8" diameter		10.67	2.999		390	97.50		487.50	590
8450	Decreaser, 6" x 4" diameter		14.22	2.250		234	73		307	375
8460	8" x 6" diameter		11.64	2.749		360	89		449	540
8550	Piping, butterfly valves, cast iron									
8560	4" diameter	B-20	6	4	Ea.	440	113		553	670
8700	Joint restraint, ductile iron mechanical joints									
8710	4" diameter	B-20A	32	1	Ea.	32	32.50		64.50	89
8720	6" diameter		25.60	1.250		38.50	40.50		79	109
8730	8" diameter		21.33	1.500		57	48.50		105.50	143
8740	10" diameter		18.28	1.751		88.50	57		145.50	191
8750	12" diameter		16.84	1.900		117	61.50		178.50	230
8760	14" diameter		16	2		145	65		210	267
8770	16" diameter		11.64	2.749		180	89		269	345
8780	18" diameter		11.03	2.901		254	94		348	435
8785	20" diameter		9.14	3.501		315	114		429	530
8790	24" diameter		7.53	4.250		430	138		568	695
9600	Steel sleeve with tap, 4" diameter	B-20	3	8		425	225		650	840
9620	6" diameter	"	2	12		495	340		835	1,100

33 14 13.25 Water Supply, Polyvinyl Chloride Pipe		Crew	Daily Output	Labor-Hours	Unit	Material	Labor	Equipment	Total	Total Incl O&P
0010	**WATER SUPPLY, POLYVINYL CHLORIDE PIPE**									
2100	PVC pipe, Class 150, 1-1/2" diameter	Q-1A	750	.013	L.F.	.52	.54		1.06	1.46
2120	2" diameter		686	.015		.88	.59		1.47	1.95
2140	2-1/2" diameter		500	.020		1.19	.82		2.01	2.65
2160	3" diameter	B-20	430	.056		1.62	1.57		3.19	4.38
8700	PVC pipe, joint restraint									
8710	4" diameter	B-20A	32	1	Ea.	46	32.50		78.50	104
8720	6" diameter		25.60	1.250		56.50	40.50		97	129
8730	8" diameter		21.33	1.500		83.50	48.50		132	172
8740	10" diameter		18.28	1.751		149	57		206	258
8750	12" diameter		16.84	1.900		158	61.50		219.50	274
8760	14" diameter		16	2		223	65		288	350
8770	16" diameter		11.64	2.749		300	89		389	480
8780	18" diameter		11.03	2.901		370	94		464	560
8785	20" diameter		9.14	3.501		465	114		579	695
8790	24" diameter		7.53	4.250		540	138		678	815

33 14 17 – Site Water Utility Service Laterals

33 14 17.15 Tapping, Crosses and Sleeves		Crew	Daily Output	Labor-Hours	Unit	Material	Labor	Equipment	Total	Total Incl O&P
0010	**TAPPING, CROSSES AND SLEEVES**									
4000	Drill and tap pressurized main (labor only)									
4100	6" main, 1" to 2" service	Q-1	3	5.333	Ea.		194		194	320
4150	8" main, 1" to 2" service	"	2.75	5.818	"		211		211	345
4500	Tap and insert gate valve									
4600	8" main, 4" branch	B-21	3.20	8.750	Ea.		259	40.50	299.50	470
4650	6" branch		2.70	10.370			305	48	353	560
4700	10" main, 4" branch		2.70	10.370			305	48	353	560
4750	6" branch		2.35	11.915			350	55	405	640
4800	12" main, 6" branch		2.35	11.915			350	55	405	640

33 31 Sanitary Sewerage Piping

33 31 11 – Public Sanitary Sewerage Gravity Piping

33 31 11.15 Sewage Collection, Concrete Pipe		Crew	Daily Output	Labor-Hours	Unit	Material	2019 Bare Costs Labor	Equipment	Total	Total Incl O&P
0010	SEWAGE COLLECTION, CONCRETE PIPE									
0020	See Section 33 42 11.60 for sewage/drainage collection, concrete pipe									

33 31 11.25 Sewage Collection, Polyvinyl Chloride Pipe		Crew	Daily Output	Labor-Hours	Unit	Material	Labor	Equipment	Total	Total Incl O&P
0010	SEWAGE COLLECTION, POLYVINYL CHLORIDE PIPE									
0020	Not including excavation or backfill									
2000	20′ lengths, SDR 35, B&S, 4″ diameter	B-20	375	.064	L.F.	1.67	1.80		3.47	4.82
2040	6″ diameter		350	.069		3.67	1.93		5.60	7.25
2080	13′ lengths, SDR 35, B&S, 8″ diameter	↓	335	.072		6.60	2.02		8.62	10.60
2120	10″ diameter	B-21	330	.085		11.40	2.51	.39	14.30	17.10
4000	Piping, DWV PVC, no exc./bkfill., 10′ L, Sch 40, 4″ diameter	B-20	375	.064		3.80	1.80		5.60	7.15
4010	6″ diameter		350	.069		8.30	1.93		10.23	12.30
4020	8″ diameter		335	.072	↓	13.05	2.02		15.07	17.75
4030	Fittings, 1/4 bend DWV PVC, 4″ diameter		19	1.263	Ea.	24.50	35.50		60	86
4040	6″ diameter		12	2		112	56.50		168.50	216
4050	8″ diameter		11	2.182		112	61.50		173.50	225
4060	1/8 bend DWV PVC, 4″ diameter		19	1.263		20	35.50		55.50	81
4070	6″ diameter		12	2		75	56.50		131.50	176
4080	8″ diameter		11	2.182		121	61.50		182.50	235
4090	Tee DWV PVC, 4″ diameter		12	2		33	56.50		89.50	130
4100	6″ diameter		10	2.400		133	67.50		200.50	258
4110	8″ diameter	↓	9	2.667	↓	289	75		364	445

33 34 Onsite Wastewater Disposal

33 34 13 – Septic Tanks

33 34 13.13 Concrete Septic Tanks		Crew	Daily Output	Labor-Hours	Unit	Material	Labor	Equipment	Total	Total Incl O&P
0010	CONCRETE SEPTIC TANKS									
0011	Not including excavation or piping									
0015	Septic tanks, precast, 1,000 gallon	B-21	8	3.500	Ea.	1,050	103	16.20	1,169.20	1,350
0060	1,500 gallon		7	4		1,575	118	18.50	1,711.50	1,950
0100	2,000 gallon	↓	5	5.600		2,275	166	26	2,467	2,800
0900	Concrete riser 24″ x 12″ with standard lid	1 Clab	6	1.333		87	36.50		123.50	157
0905	24″ x 12″ with heavy duty lid		6	1.333		110	36.50		146.50	182
0910	24″ x 8″ with standard lid		6	1.333		81.50	36.50		118	151
0915	24″ x 8″ with heavy duty lid		6	1.333		105	36.50		141.50	177
0917	24″ x 12″ extension		12	.667		59	18.35		77.35	95
0920	24″ x 8″ extension		12	.667		53.50	18.35		71.85	89.50
0950	HDPE riser 20″ x 12″ with standard lid		8	1		138	27.50		165.50	198
0951	HDPE 20″ x 12″ with heavy duty lid		8	1		152	27.50		179.50	213
0955	HDPE 24″ x 12″ with standard lid		8	1		141	27.50		168.50	202
0956	HDPE 24″ x 12″ with heavy duty lid		8	1		162	27.50		189.50	224
0960	HDPE 20″ x 6″ extension		48	.167		31	4.58		35.58	41.50
0962	HDPE 20″ x 12″ extension		48	.167		56	4.58		60.58	69
0965	HDPE 24″ x 6″ extension		48	.167		36	4.58		40.58	47
0967	HDPE 24″ x 12″ extension	↓	48	.167		58	4.58		62.58	71
1150	Leaching field chambers, 13′ x 3′-7″ x 1′-4″, standard	B-13	16	3		505	88.50	36.50	630	745
1420	Leaching pit, precast concrete, 6′ diameter, 3′ deep	B-21	4.70	5.957	↓	840	176	27.50	1,043.50	1,250

33 34 13.33 Polyethylene Septic Tanks		Crew	Daily Output	Labor-Hours	Unit	Material	Labor	Equipment	Total	Total Incl O&P
0010	POLYETHYLENE SEPTIC TANKS									
0015	High density polyethylene, 1,000 gallon	B-21	8	3.500	Ea.	1,300	103	16.20	1,419.20	1,625
0020	1,250 gallon	↓	8	3.500	↓	1,300	103	16.20	1,419.20	1,625

33 34 Onsite Wastewater Disposal

33 34 13 – Septic Tanks

33 34 13.33 Polyethylene Septic Tanks		Crew	Daily Output	Labor-Hours	Unit	2019 Bare Costs Material	Labor	Equipment	Total	Total Incl O&P
0025	1,500 gallon	B-21	7	4	Ea.	1,375	118	18.50	1,511.50	1,750

33 34 16 – Septic Tank Effluent Filters

33 34 16.13 Septic Tank Gravity Effluent Filters		Crew	Daily Output	Labor-Hours	Unit	Material	Labor	Equipment	Total	Total Incl O&P
0010	SEPTIC TANK GRAVITY EFFLUENT FILTERS									
3000	Effluent filter, 4″ diameter	1 Skwk	8	1	Ea.	42	36.50		78.50	107
3020	6″ diameter	″	7	1.143	″	54.50	41.50		96	130

33 34 51 – Drainage Field Systems

33 34 51.10 Drainage Field Excavation and Fill		Crew	Daily Output	Labor-Hours	Unit	Material	Labor	Equipment	Total	Total Incl O&P
0010	DRAINAGE FIELD EXCAVATION AND FILL									
2200	Septic tank & drainage field excavation with 3/4 C.Y. backhoe	B-12F	145	.110	C.Y.		3.61	4.70	8.31	11.10
2400	4′ trench for disposal field, 3/4 C.Y. backhoe	″	335	.048	L.F.		1.56	2.03	3.59	4.81
2600	Gravel fill, run of bank	B-6	150	.160	C.Y.	16.95	4.84	2.13	23.92	29
2800	Crushed stone, 3/4″	″	150	.160	″	38	4.84	2.13	44.97	52.50

33 34 51.13 Utility Septic Tank Tile Drainage Field		Crew	Daily Output	Labor-Hours	Unit	Material	Labor	Equipment	Total	Total Incl O&P
0010	UTILITY SEPTIC TANK TILE DRAINAGE FIELD									
0015	Distribution box, concrete, 5 outlets	2 Clab	20	.800	Ea.	93	22		115	139
0020	7 outlets		16	1		94.50	27.50		122	150
0025	9 outlets		8	2		555	55		610	700
0115	Distribution boxes, HDPE, 5 outlets		20	.800		77.50	22		99.50	122
0117	6 outlets		15	1.067		80.50	29.50		110	137
0118	7 outlets		15	1.067		74.50	29.50		104	131
0120	8 outlets		10	1.600		81.50	44		125.50	162
0240	Distribution boxes, outlet flow leveler	1 Clab	50	.160		2.47	4.40		6.87	9.95
0300	Precast concrete, galley, 4′ x 4′ x 4′	B-21	16	1.750		244	51.50	8.10	303.60	360
0350	HDPE infiltration chamber 12″ H x 15″ W	2 Clab	300	.053	L.F.	6.95	1.47		8.42	10.05
0351	12″ H x 15″ W end cap	1 Clab	32	.250	Ea.	18.80	6.90		25.70	32
0355	chamber 12″ H x 22″ W	2 Clab	300	.053	L.F.	6.75	1.47		8.22	9.85
0356	12″ H x 22″ W end cap	1 Clab	32	.250	Ea.	16.15	6.90		23.05	29
0360	chamber 13″ H x 34″ W	2 Clab	300	.053	L.F.	14.25	1.47		15.72	18.10
0361	13″ H x 34″ W end cap	1 Clab	32	.250	Ea.	50.50	6.90		57.40	67
0365	chamber 16″ H x 34″ W	2 Clab	300	.053	L.F.	19.15	1.47		20.62	23.50
0366	16″ H x 34″ W end cap	1 Clab	32	.250	Ea.	15.75	6.90		22.65	28.50
0370	chamber 8″ H x 16″ W	2 Clab	300	.053	L.F.	9.85	1.47		11.32	13.25
0371	8″ H x 16″ W end cap	1 Clab	32	.250	Ea.	11.05	6.90		17.95	23.50

33 41 Subdrainage

33 41 16 – Subdrainage Piping

33 41 16.25 Piping, Subdrainage, Corrugated Metal		Crew	Daily Output	Labor-Hours	Unit	Material	Labor	Equipment	Total	Total Incl O&P
0010	PIPING, SUBDRAINAGE, CORRUGATED METAL									
0021	Not including excavation and backfill									
2010	Aluminum, perforated									
2020	6″ diameter, 18 ga.	B-20	380	.063	L.F.	6.70	1.78		8.48	10.35
2200	8″ diameter, 16 ga.	″	370	.065		8.45	1.83		10.28	12.25
2220	10″ diameter, 16 ga.	B-21	360	.078		10.55	2.30	.36	13.21	15.80
3000	Uncoated galvanized, perforated									
3020	6″ diameter, 18 ga.	B-20	380	.063	L.F.	6.45	1.78		8.23	10
3200	8″ diameter, 16 ga.	″	370	.065		7.95	1.83		9.78	11.70
3220	10″ diameter, 16 ga.	B-21	360	.078		8.40	2.30	.36	11.06	13.45
3240	12″ diameter, 16 ga.	″	285	.098		9.35	2.90	.45	12.70	15.60
4000	Steel, perforated, asphalt coated									

33 41 Subdrainage

33 41 16 – Subdrainage Piping

33 41 16.25 Piping, Subdrainage, Corrugated Metal		Crew	Daily Output	Labor-Hours	Unit	2019 Bare Costs Material	Labor	Equipment	Total	Total Incl O&P
4020	6" diameter, 18 ga.	B-20	380	.063	L.F.	6.60	1.78		8.38	10.25
4030	8" diameter, 18 ga.	"	370	.065		8.35	1.83		10.18	12.20
4040	10" diameter, 16 ga.	B-21	360	.078		10.50	2.30	.36	13.16	15.75
4050	12" diameter, 16 ga.		285	.098		11.80	2.90	.45	15.15	18.25
4060	18" diameter, 16 ga.	↓	205	.137	↓	18.60	4.04	.63	23.27	28

33 42 Stormwater Conveyance

33 42 11 – Stormwater Gravity Piping

33 42 11.60 Sewage/Drainage Collection, Concrete Pipe

		Crew	Daily Output	Labor-Hours	Unit	Material	Labor	Equipment	Total	Total Incl O&P
0010	**SEWAGE/DRAINAGE COLLECTION, CONCRETE PIPE**									
0020	Not including excavation or backfill									
1020	8" diameter	B-14	224	.214	L.F.	8.45	6.25	1.43	16.13	21
1030	10" diameter	"	216	.222	"	9.35	6.50	1.48	17.33	22.50
3780	Concrete slotted pipe, class 4 mortar joint									
3800	12" diameter	B-21	168	.167	L.F.	31	4.93	.77	36.70	43.50
3840	18" diameter	"	152	.184	"	35.50	5.45	.85	41.80	49.50
3900	Concrete slotted pipe, Class 4 O-ring joint									
3940	12" diameter	B-21	168	.167	L.F.	28	4.93	.77	33.70	40
3960	18" diameter	"	152	.184	"	32	5.45	.85	38.30	45

33 42 33 – Stormwater Curbside Drains and Inlets

33 42 33.13 Catch Basins

		Crew	Daily Output	Labor-Hours	Unit	Material	Labor	Equipment	Total	Total Incl O&P
0010	**CATCH BASINS**									
0011	Not including footing & excavation									
1600	Frames & grates, C.I., 24" square, 500 lb.	B-6	7.80	3.077	Ea.	355	93	41	489	590

33 52 Hydrocarbon Transmission and Distribution

33 52 16 – Gas Hydrocarbon Piping

33 52 16.13 Steel Natural Gas Piping

		Crew	Daily Output	Labor-Hours	Unit	Material	Labor	Equipment	Total	Total Incl O&P
0010	**STEEL NATURAL GAS PIPING**									
0020	Not including excavation or backfill, tar coated and wrapped									
4000	Pipe schedule 40, plain end									
4040	1" diameter	Q-4	300	.107	L.F.	5.20	4.09	.19	9.48	12.60
4080	2" diameter	"	280	.114	"	8.15	4.39	.20	12.74	16.35

33 52 16.20 Piping, Gas Service and Distribution, P.E.

		Crew	Daily Output	Labor-Hours	Unit	Material	Labor	Equipment	Total	Total Incl O&P
0010	**PIPING, GAS SERVICE AND DISTRIBUTION, POLYETHYLENE**									
0020	Not including excavation or backfill									
1000	60 psi coils, compression coupling @ 100', 1/2" diameter, SDR 11	B-20A	608	.053	L.F.	.48	1.71		2.19	3.34
1010	1" diameter, SDR 11		544	.059		1.13	1.91		3.04	4.38
1040	1-1/4" diameter, SDR 11		544	.059		1.64	1.91		3.55	4.94
1100	2" diameter, SDR 11		488	.066		2.54	2.13		4.67	6.30
1160	3" diameter, SDR 11	↓	408	.078		5.60	2.54		8.14	10.35
1500	60 psi 40' joints with coupling, 3" diameter, SDR 11	B-21A	408	.098		6.90	3.29	.91	11.10	13.95
1540	4" diameter, SDR 11		352	.114		11.70	3.81	1.06	16.57	20.50
1600	6" diameter, SDR 11		328	.122		31.50	4.09	1.13	36.72	42.50
1640	8" diameter, SDR 11	↓	272	.147	↓	47	4.93	1.37	53.30	61

Estimating Tips

- When estimating costs for the installation of electrical power generation equipment, factors to review include access to the job site, access and setting up at the installation site, required connections, uncrating pads, anchors, leveling, final assembly of the components, and temporary protection from physical damage, such as environmental exposure.
- Be aware of the costs of equipment supports, concrete pads, and vibration isolators. Cross-reference them against other trades' specifications. Also, review site and structural drawings for items that must be included in the estimates.
- It is important to include items that are not documented in the plans and specifications but must be priced. These items include, but are not limited to, testing, dust protection, roof penetration, core drilling concrete floors and walls, patching, cleanup, and final adjustments. Add a contingency or allowance for utility company fees for power hookups, if needed.
- The project size and scope of electrical power generation equipment will have a significant impact on cost. The intent of RSMeans cost data is to provide a benchmark cost so that owners, engineers, and electrical contractors will have a comfortable number with which to start a project. Additionally, there are many websites available to use for research and to obtain a vendor's quote to finalize costs.

Reference Numbers

Reference numbers are shown at the beginning of some major classifications. These numbers refer to related items in the Reference Section. The reference information may be an estimating procedure, an alternate pricing method, or technical information.

Note: Trade Service, in part, has been used as a reference source for some of the material prices used in Division 48.

48 15 13.50 Wind Turbines and Components			Crew	Daily Output	Labor-Hours	Unit	Material	Labor (2019 Bare Costs)	Equipment (2019 Bare Costs)	Total (2019 Bare Costs)	Total Incl O&P
0010	**WIND TURBINES & COMPONENTS**										
0500	Complete system, grid connected										
1000	20 kW, 31′ diam., incl. labor & material	G				System				49,900	49,900
2000	2.4 kW, 12′ diam., incl. labor & material	G				System				18,000	18,000
2900	Component system										
3200	1,000 W, 9′ diam.	G	1 Elec	2.05	3.902	Ea.	1,700	162		1,862	2,150
3400	Mounting hardware										
3500	30′ guyed tower kit	G	2 Clab	5.12	3.125	Ea.	405	86		491	585
3505	3′ galvanized helical earth screw	G	1 Clab	8	1		54.50	27.50		82	105
3510	Attic mount kit	G	1 Rofc	2.56	3.125		216	96		312	410
3520	Roof mount kit	G	1 Clab	3.41	2.346	↓	264	64.50		328.50	395
8900	Equipment										
9100	DC to AC inverter for, 48 V, 4,000 W	G	1 Elec	2	4	Ea.	2,300	166		2,466	2,800

Reference Section

All the reference information is in one section, making it easy to find what you need to know and easy to use the data set on a daily basis. This section is visually identified by a vertical black bar on the page edges.

In this Reference Section, we've included Equipment Rental Costs, a listing of rental and operating costs; Crew Listings, a full listing of all crews, equipment, and their costs; Location Factors for adjusting costs to the region you are in; Reference Tables, where you will find explanations, estimating information and procedures, and technical data; an explanation of all the Abbreviations in the data set; and sample Estimating Forms.

Table of Contents

Estimating Tips

- This section contains the average costs to rent and operate hundreds of pieces of construction equipment. This is useful information when one is estimating the time and material requirements of any particular operation in order to establish a unit or total cost. Bare equipment costs shown on a unit cost line include, not only rental, but also operating costs for equipment under normal use.

Rental Costs

- Equipment rental rates are obtained from the following industry sources throughout North America: contractors, suppliers, dealers, manufacturers, and distributors.
- Rental rates vary throughout the country, with larger cities generally having lower rates. Lease plans for new equipment are available for periods in excess of six months, with a percentage of payments applying toward purchase.
- Monthly rental rates vary from 2% to 5% of the purchase price of the equipment depending on the anticipated life of the equipment and its wearing parts.
- Weekly rental rates are about 1/3 of the monthly rates, and daily rental rates are about 1/3 of the weekly rate.
- Rental rates can also be treated as reimbursement costs for contractor-owned equipment. Owned equipment costs include depreciation, loan payments, interest, taxes, insurance, storage, and major repairs.

Operating Costs

- The operating costs include parts and labor for routine servicing, such as the repair and replacement of pumps, filters, and worn lines. Normal operating expendables, such as fuel, lubricants, tires, and electricity (where applicable), are also included.
- Extraordinary operating expendables with highly variable wear patterns, such as diamond bits and blades, are excluded. These costs can be found as material costs in the Unit Price section.
- The hourly operating costs listed do not include the operator's wages.

Equipment Cost/Day

- Any power equipment required by a crew is shown in the Crew Listings with a daily cost.
- This daily cost of equipment needed by a crew includes both the rental cost and the operating cost and is based on dividing the weekly rental rate by 5 (the number of working days in the week), then adding the hourly operating cost multiplied by 8 (the number of hours in a day). This "Equipment Cost/Day" is shown in the far right column of the Equipment Rental section.
- If equipment is needed for only one or two days, it is best to develop your own cost by including components for daily rent and hourly operating costs. This is important when the listed Crew for a task does not contain the equipment needed, such as a crane for lifting mechanical heating/cooling equipment up onto a roof.
- If the quantity of work is less than the crew's Daily Output shown for a Unit Price line item that includes a bare unit equipment cost, the recommendation is to estimate one day's rental cost and operating cost for equipment shown in the Crew Listing for that line item.
- Please note, in some cases the equipment description in the crew is followed by a time period in parenthesis. For example: (daily) or (monthly). In these cases the equipment cost/day is calculated by adding the rental cost per time period to the hourly operating cost multiplied by 8.

Mobilization, Demobilization Costs

- The cost to move construction equipment from an equipment yard or rental company to the job site and back again is not included in equipment rental costs listed in the Reference Section. It is also not included in the bare equipment cost of any Unit Price line item or in any equipment costs shown in the Crew Listings.
- Mobilization (to the site) and demobilization (from the site) costs can be found in the Unit Price section.
- If a piece of equipment is already at the job site, it is not appropriate to utilize mobilization or demobilization costs again in an estimate. ■

01 54 33 | Equipment Rental

		01 54 33 Equipment Rental	UNIT	HOURLY OPER. COST	RENT PER DAY	RENT PER WEEK	RENT PER MONTH	EQUIPMENT COST/DAY	
10	0010	**CONCRETE EQUIPMENT RENTAL** without operators R015433-10							10
	0200	Bucket, concrete lightweight, 1/2 C.Y.	Ea.	.88	24.50	74	222	21.80	
	0300	1 C.Y.		.98	28.50	85.50	257	24.95	
	0400	1-1/2 C.Y.		1.24	38.50	115	345	32.90	
	0500	2 C.Y.		1.34	47	141	425	38.90	
	0580	8 C.Y.		6.49	265	795	2,375	210.90	
	0600	Cart, concrete, self-propelled, operator walking, 10 C.F.		2.86	58.50	176	530	58.10	
	0700	Operator riding, 18 C.F.		4.82	98.50	296	890	97.75	
	0800	Conveyer for concrete, portable, gas, 16″ wide, 26′ long		10.64	130	390	1,175	163.10	
	0900	46′ long		11.02	155	465	1,400	181.15	
	1000	56′ long		11.18	162	485	1,450	186.45	
	1100	Core drill, electric, 2-1/2 H.P., 1″ to 8″ bit diameter		1.57	56.50	170	510	46.55	
	1150	11 H.P., 8″ to 18″ cores		5.40	115	345	1,025	112.20	
	1200	Finisher, concrete floor, gas, riding trowel, 96″ wide		9.66	148	445	1,325	166.30	
	1300	Gas, walk-behind, 3 blade, 36″ trowel		2.04	24.50	73.50	221	31	
	1400	4 blade, 48″ trowel		3.07	28	84.50	254	41.50	
	1500	Float, hand-operated (Bull float), 48″ wide		.08	13.85	41.50	125	8.95	
	1570	Curb builder, 14 H.P., gas, single screw		14.04	285	855	2,575	283.30	
	1590	Double screw		15.04	340	1,025	3,075	325.35	
	1600	Floor grinder, concrete and terrazzo, electric, 22″ path		3.04	190	570	1,700	138.30	
	1700	Edger, concrete, electric, 7″ path		1.18	56.50	170	510	43.45	
	1750	Vacuum pick-up system for floor grinders, wet/dry		1.62	98.50	296	890	72.15	
	1800	Mixer, powered, mortar and concrete, gas, 6 C.F., 18 H.P.		7.42	127	380	1,150	135.35	
	1900	10 C.F., 25 H.P.		9.00	150	450	1,350	162	
	2000	16 C.F.		9.36	173	520	1,550	178.85	
	2100	Concrete, stationary, tilt drum, 2 C.Y.		7.23	243	730	2,200	203.85	
	2120	Pump, concrete, truck mounted, 4″ line, 80′ boom		29.88	1,000	3,025	9,075	844.05	
	2140	5″ line, 110′ boom		37.46	1,375	4,150	12,500	1,130	
	2160	Mud jack, 50 C.F. per hr.		6.45	123	370	1,100	125.60	
	2180	225 C.F. per hr.		8.55	145	435	1,300	155.40	
	2190	Shotcrete pump rig, 12 C.Y./hr.		13.97	218	655	1,975	242.75	
	2200	35 C.Y./hr.		15.80	242	725	2,175	271.40	
	2600	Saw, concrete, manual, gas, 18 H.P.		5.54	47.50	143	430	72.90	
	2650	Self-propelled, gas, 30 H.P.		7.90	78.50	235	705	110.15	
	2675	V-groove crack chaser, manual, gas, 6 H.P.		1.64	19	57	171	24.50	
	2700	Vibrators, concrete, electric, 60 cycle, 2 H.P.		.47	9.15	27.50	82.50	9.25	
	2800	3 H.P.		.56	11.65	35	105	11.50	
	2900	Gas engine, 5 H.P.		1.54	16.35	49	147	22.15	
	3000	8 H.P.		2.08	16.50	49.50	149	26.55	
	3050	Vibrating screed, gas engine, 8 H.P.		2.81	95	285	855	79.45	
	3120	Concrete transit mixer, 6 x 4, 250 H.P., 8 C.Y., rear discharge		50.72	615	1,850	5,550	775.80	
	3200	Front discharge		58.88	740	2,225	6,675	916.05	
	3300	6 x 6, 285 H.P., 12 C.Y., rear discharge		58.15	710	2,125	6,375	890.20	
	3400	Front discharge	↓	60.59	750	2,250	6,750	934.70	
20	0010	**EARTHWORK EQUIPMENT RENTAL** without operators R015433-10							20
	0040	Aggregate spreader, push type, 8′ to 12′ wide	Ea.	2.60	27.50	82.50	248	37.30	
	0045	Tailgate type, 8′ wide		2.55	33	99	297	40.15	
	0055	Earth auger, truck mounted, for fence & sign posts, utility poles		13.85	460	1,375	4,125	385.85	
	0060	For borings and monitoring wells		42.65	690	2,075	6,225	756.20	
	0070	Portable, trailer mounted		2.30	33	99	297	38.20	
	0075	Truck mounted, for caissons, water wells		85.39	2,900	8,725	26,200	2,428	
	0080	Horizontal boring machine, 12″ to 36″ diameter, 45 H.P.		22.77	188	565	1,700	295.15	
	0090	12″ to 48″ diameter, 65 H.P.		31.25	330	985	2,950	447	
	0095	Auger, for fence posts, gas engine, hand held		.45	6.35	19	57	7.40	
	0100	Excavator, diesel hydraulic, crawler mounted, 1/2 C.Y. cap.		21.72	450	1,350	4,050	443.75	
	0120	5/8 C.Y. capacity		29.04	590	1,775	5,325	587.30	
	0140	3/4 C.Y. capacity		32.66	700	2,100	6,300	681.25	
	0150	1 C.Y. capacity	↓	41.21	700	2,100	6,300	749.70	

01 54 33 | Equipment Rental

			UNIT	HOURLY OPER. COST	RENT PER DAY	RENT PER WEEK	RENT PER MONTH	EQUIPMENT COST/DAY
20	0200	1-1/2 C.Y. capacity	Ea.	48.59	865	2,600	7,800	908.70
	0300	2 C.Y. capacity		56.58	1,050	3,125	9,375	1,078
	0320	2-1/2 C.Y. capacity		82.64	1,300	3,900	11,700	1,441
	0325	3-1/2 C.Y. capacity		120.12	2,150	6,475	19,400	2,256
	0330	4-1/2 C.Y. capacity		151.63	2,750	8,275	24,800	2,868
	0335	6 C.Y. capacity		192.39	3,400	10,200	30,600	3,579
	0340	7 C.Y. capacity		175.20	3,225	9,650	29,000	3,332
	0342	Excavator attachments, bucket thumbs		3.40	250	750	2,250	177.20
	0345	Grapples		3.14	215	645	1,925	154.10
	0346	Hydraulic hammer for boom mounting, 4,000 ft. lb.		13.48	375	1,125	3,375	332.85
	0347	5,000 ft. lb.		15.95	465	1,400	4,200	407.60
	0348	8,000 ft. lb.		23.54	675	2,025	6,075	593.35
	0349	12,000 ft. lb.		25.72	800	2,400	7,200	685.75
	0350	Gradall type, truck mounted, 3 ton @ 15′ radius, 5/8 C.Y.		43.44	860	2,575	7,725	862.55
	0370	1 C.Y. capacity		59.40	1,175	3,550	10,700	1,185
	0400	Backhoe-loader, 40 to 45 H.P., 5/8 C.Y. capacity		11.90	207	620	1,850	219.20
	0450	45 H.P. to 60 H.P., 3/4 C.Y. capacity		18.02	293	880	2,650	320.20
	0460	80 H.P., 1-1/4 C.Y. capacity		20.36	375	1,125	3,375	387.85
	0470	112 H.P., 1-1/2 C.Y. capacity		32.99	590	1,775	5,325	618.90
	0482	Backhoe-loader attachment, compactor, 20,000 lb.		6.44	150	450	1,350	141.50
	0485	Hydraulic hammer, 750 ft. lb.		3.68	103	310	930	91.45
	0486	Hydraulic hammer, 1,200 ft. lb.		6.55	198	595	1,775	171.40
	0500	Brush chipper, gas engine, 6″ cutter head, 35 H.P.		9.17	110	330	990	139.35
	0550	Diesel engine, 12″ cutter head, 130 H.P.		23.67	335	1,000	3,000	389.40
	0600	15″ cutter head, 165 H.P.		26.59	400	1,200	3,600	452.70
	0750	Bucket, clamshell, general purpose, 3/8 C.Y.		1.40	41	123	370	35.80
	0800	1/2 C.Y.		1.52	50.50	152	455	42.55
	0850	3/4 C.Y.		1.64	57.50	173	520	47.75
	0900	1 C.Y.		1.70	62.50	188	565	51.20
	0950	1-1/2 C.Y.		2.79	87	261	785	74.50
	1000	2 C.Y.		2.92	95	285	855	80.40
	1010	Bucket, dragline, medium duty, 1/2 C.Y.		.82	24.50	74	222	21.35
	1020	3/4 C.Y.		.78	26	78	234	21.85
	1030	1 C.Y.		.80	27	81.50	245	22.70
	1040	1-1/2 C.Y.		1.26	41.50	125	375	35.10
	1050	2 C.Y.		1.29	45	135	405	37.35
	1070	3 C.Y.		2.08	66	198	595	56.25
	1200	Compactor, manually guided 2-drum vibratory smooth roller, 7.5 H.P.		7.22	207	620	1,850	181.75
	1250	Rammer/tamper, gas, 8″		2.21	46.50	140	420	45.65
	1260	15″		2.63	53.50	160	480	53.05
	1300	Vibratory plate, gas, 18″ plate, 3,000 lb. blow		2.13	23.50	70.50	212	31.10
	1350	21″ plate, 5,000 lb. blow		2.62	32.50	98	294	40.55
	1370	Curb builder/extruder, 14 H.P., gas, single screw		14.03	283	850	2,550	282.25
	1390	Double screw		15.04	340	1,025	3,075	325.30
	1500	Disc harrow attachment, for tractor		.47	79.50	239	715	51.60
	1810	Feller buncher, shearing & accumulating trees, 100 H.P.		39.20	815	2,450	7,350	803.55
	1860	Grader, self-propelled, 25,000 lb.		33.35	760	2,275	6,825	721.80
	1910	30,000 lb.		32.85	660	1,975	5,925	657.85
	1920	40,000 lb.		51.89	1,275	3,825	11,500	1,180
	1930	55,000 lb.		66.93	1,650	4,925	14,800	1,520
	1950	Hammer, pavement breaker, self-propelled, diesel, 1,000 to 1,250 lb.		28.40	475	1,425	4,275	512.20
	2000	1,300 to 1,500 lb.		42.80	975	2,925	8,775	927.40
	2050	Pile driving hammer, steam or air, 4,150 ft. lb. @ 225 bpm		12.15	565	1,700	5,100	437.20
	2100	8,750 ft. lb. @ 145 bpm		14.35	810	2,425	7,275	599.75
	2150	15,000 ft. lb. @ 60 bpm		14.68	840	2,525	7,575	622.40
	2200	24,450 ft. lb. @ 111 bpm		15.69	935	2,800	8,400	685.50
	2250	Leads, 60′ high for pile driving hammers up to 20,000 ft. lb.		3.67	85.50	256	770	80.55
	2300	90′ high for hammers over 20,000 ft. lb.		5.45	152	455	1,375	134.60

01 54 33 | Equipment Rental

			UNIT	HOURLY OPER. COST	RENT PER DAY	RENT PER WEEK	RENT PER MONTH	EQUIPMENT COST/DAY	
20	2350	Diesel type hammer, 22,400 ft. lb.	Ea.	17.82	475	1,425	4,275	427.55	20
	2400	41,300 ft. lb.		25.68	600	1,800	5,400	565.45	
	2450	141,000 ft. lb.		41.33	950	2,850	8,550	900.65	
	2500	Vib. elec. hammer/extractor, 200 kW diesel generator, 34 H.P.		41.37	690	2,075	6,225	745.95	
	2550	80 H.P.		73.03	1,000	3,000	9,000	1,184	
	2600	150 H.P.		135.14	1,925	5,775	17,300	2,236	
	2800	Log chipper, up to 22″ diameter, 600 H.P.		46.26	665	2,000	6,000	770.10	
	2850	Logger, for skidding & stacking logs, 150 H.P.		43.53	835	2,500	7,500	848.25	
	2860	Mulcher, diesel powered, trailer mounted		18.04	220	660	1,975	276.35	
	2900	Rake, spring tooth, with tractor		14.72	360	1,075	3,225	332.75	
	3000	Roller, vibratory, tandem, smooth drum, 20 H.P.		7.81	150	450	1,350	152.45	
	3050	35 H.P.		10.13	252	755	2,275	232.05	
	3100	Towed type vibratory compactor, smooth drum, 50 H.P.		25.27	365	1,100	3,300	422.15	
	3150	Sheepsfoot, 50 H.P.		25.64	375	1,125	3,375	430.15	
	3170	Landfill compactor, 220 H.P.		70.01	1,575	4,750	14,300	1,510	
	3200	Pneumatic tire roller, 80 H.P.		12.92	390	1,175	3,525	338.35	
	3250	120 H.P.		19.39	640	1,925	5,775	540.10	
	3300	Sheepsfoot vibratory roller, 240 H.P.		62.21	1,375	4,125	12,400	1,323	
	3320	340 H.P.		83.82	2,075	6,250	18,800	1,921	
	3350	Smooth drum vibratory roller, 75 H.P.		23.34	635	1,900	5,700	566.70	
	3400	125 H.P.		27.61	715	2,150	6,450	650.90	
	3410	Rotary mower, brush, 60″, with tractor		18.79	350	1,050	3,150	360.30	
	3420	Rototiller, walk-behind, gas, 5 H.P.		2.14	89	267	800	70.50	
	3422	8 H.P.		2.81	103	310	930	84.50	
	3440	Scrapers, towed type, 7 C.Y. capacity		6.44	123	370	1,100	125.50	
	3450	10 C.Y. capacity		7.20	165	495	1,475	156.65	
	3500	15 C.Y. capacity		7.40	190	570	1,700	173.20	
	3525	Self-propelled, single engine, 14 C.Y. capacity		133.29	2,575	7,725	23,200	2,611	
	3550	Dual engine, 21 C.Y. capacity		141.37	2,350	7,050	21,200	2,541	
	3600	31 C.Y. capacity		187.85	3,500	10,500	31,500	3,603	
	3640	44 C.Y. capacity		232.68	4,500	13,500	40,500	4,561	
	3650	Elevating type, single engine, 11 C.Y. capacity		61.87	1,125	3,350	10,100	1,165	
	3700	22 C.Y. capacity		114.62	2,325	6,950	20,900	2,307	
	3710	Screening plant, 110 H.P. w/5′ x 10′ screen		21.13	385	1,150	3,450	399.05	
	3720	5′ x 16′ screen		26.68	490	1,475	4,425	508.40	
	3850	Shovel, crawler-mounted, front-loading, 7 C.Y. capacity		218.65	3,800	11,400	34,200	4,029	
	3855	12 C.Y. capacity		336.90	5,275	15,800	47,400	5,855	
	3860	Shovel/backhoe bucket, 1/2 C.Y.		2.69	70.50	212	635	63.95	
	3870	3/4 C.Y.		2.67	79.50	238	715	68.90	
	3880	1 C.Y.		2.76	88	264	790	74.85	
	3890	1-1/2 C.Y.		2.95	103	310	930	85.65	
	3910	3 C.Y.		3.44	140	420	1,250	111.50	
	3950	Stump chipper, 18″ deep, 30 H.P.		6.90	212	635	1,900	182.20	
	4110	Dozer, crawler, torque converter, diesel 80 H.P.		25.25	450	1,350	4,050	472	
	4150	105 H.P.		34.34	560	1,675	5,025	609.70	
	4200	140 H.P.		41.27	840	2,525	7,575	835.20	
	4260	200 H.P.		63.16	1,300	3,925	11,800	1,290	
	4310	300 H.P.		80.73	1,975	5,925	17,800	1,831	
	4360	410 H.P.		106.74	2,350	7,075	21,200	2,269	
	4370	500 H.P.		133.38	2,900	8,725	26,200	2,812	
	4380	700 H.P.		230.17	5,300	15,900	47,700	5,021	
	4400	Loader, crawler, torque conv., diesel, 1-1/2 C.Y., 80 H.P.		29.54	590	1,775	5,325	591.30	
	4450	1-1/2 to 1-3/4 C.Y., 95 H.P.		30.27	675	2,025	6,075	647.15	
	4510	1-3/4 to 2-1/4 C.Y., 130 H.P.		47.76	1,025	3,100	9,300	1,002	
	4530	2-1/2 to 3-1/4 C.Y., 190 H.P.		57.79	1,225	3,700	11,100	1,202	
	4560	3-1/2 to 5 C.Y., 275 H.P.		71.41	1,475	4,450	13,400	1,461	
	4610	Front end loader, 4WD, articulated frame, diesel, 1 to 1-1/4 C.Y., 70 H.P.		16.62	273	820	2,450	297	
	4620	1-1/2 to 1-3/4 C.Y., 95 H.P.	↓	20.00	320	965	2,900	352.95	

01 54 33 | Equipment Rental

		01 54 33 Equipment Rental	UNIT	HOURLY OPER. COST	RENT PER DAY	RENT PER WEEK	RENT PER MONTH	EQUIPMENT COST/DAY
20	4650	1-3/4 to 2 C.Y., 130 H.P.	Ea.	21.07	385	1,150	3,450	398.55
	4710	2-1/2 to 3-1/2 C.Y., 145 H.P.		29.53	490	1,475	4,425	531.20
	4730	3 to 4-1/2 C.Y., 185 H.P.		32.09	515	1,550	4,650	566.70
	4760	5-1/4 to 5-3/4 C.Y., 270 H.P.		53.19	915	2,750	8,250	975.55
	4810	7 to 9 C.Y., 475 H.P.		91.17	1,750	5,275	15,800	1,784
	4870	9 to 11 C.Y., 620 H.P.		131.92	2,625	7,850	23,600	2,625
	4880	Skid-steer loader, wheeled, 10 C.F., 30 H.P. gas		9.57	163	490	1,475	174.55
	4890	1 C.Y., 78 H.P., diesel		18.44	410	1,225	3,675	392.50
	4892	Skid-steer attachment, auger		.75	137	410	1,225	87.95
	4893	Backhoe		.74	118	355	1,075	76.90
	4894	Broom		.71	123	370	1,100	79.65
	4895	Forks		.15	26.50	80	240	17.25
	4896	Grapple		.72	107	320	960	69.80
	4897	Concrete hammer		1.05	177	530	1,600	114.40
	4898	Tree spade		.60	100	300	900	64.80
	4899	Trencher		.65	108	325	975	70.20
	4900	Trencher, chain, boom type, gas, operator walking, 12 H.P.		4.18	50	150	450	63.40
	4910	Operator riding, 40 H.P.		16.69	360	1,075	3,225	348.50
	5000	Wheel type, diesel, 4′ deep, 12″ wide		68.71	935	2,800	8,400	1,110
	5100	6′ deep, 20″ wide		87.58	2,150	6,475	19,400	1,996
	5150	Chain type, diesel, 5′ deep, 8″ wide		16.30	350	1,050	3,150	340.40
	5200	Diesel, 8′ deep, 16″ wide		89.66	1,875	5,600	16,800	1,837
	5202	Rock trencher, wheel type, 6″ wide x 18″ deep		47.12	700	2,100	6,300	797
	5206	Chain type, 18″ wide x 7′ deep		104.70	3,075	9,200	27,600	2,678
	5210	Tree spade, self-propelled		13.69	390	1,175	3,525	344.55
	5250	Truck, dump, 2-axle, 12 ton, 8 C.Y. payload, 220 H.P.		23.95	247	740	2,225	339.60
	5300	Three axle dump, 16 ton, 12 C.Y. payload, 400 H.P.		44.63	350	1,050	3,150	567.05
	5310	Four axle dump, 25 ton, 18 C.Y. payload, 450 H.P.		50.00	510	1,525	4,575	705
	5350	Dump trailer only, rear dump, 16-1/2 C.Y.		5.74	147	440	1,325	133.95
	5400	20 C.Y.		6.20	165	495	1,475	148.60
	5450	Flatbed, single axle, 1-1/2 ton rating		19.05	71.50	214	640	195.20
	5500	3 ton rating		23.12	102	305	915	245.95
	5550	Off highway rear dump, 25 ton capacity		62.86	1,425	4,250	12,800	1,353
	5600	35 ton capacity		67.10	1,550	4,675	14,000	1,472
	5610	50 ton capacity		84.12	1,775	5,325	16,000	1,738
	5620	65 ton capacity		89.83	1,925	5,800	17,400	1,879
	5630	100 ton capacity		121.61	2,850	8,550	25,700	2,683
	6000	Vibratory plow, 25 H.P., walking	↓	6.79	60.50	181	545	90.55
40	0010	**GENERAL EQUIPMENT RENTAL** without operators (R015433 -10)						
	0020	Aerial lift, scissor type, to 20′ high, 1200 lb. capacity, electric	Ea.	3.49	50.50	151	455	58.15
	0030	To 30′ high, 1,200 lb. capacity		3.78	67.50	202	605	70.70
	0040	Over 30′ high, 1,500 lb. capacity		5.15	122	365	1,100	114.20
	0070	Articulating boom, to 45′ high, 500 lb. capacity, diesel (R015433 -15)		9.95	275	825	2,475	244.60
	0075	To 60′ high, 500 lb. capacity		13.70	510	1,525	4,575	414.60
	0080	To 80′ high, 500 lb. capacity		16.10	560	1,675	5,025	463.80
	0085	To 125′ high, 500 lb. capacity		18.40	775	2,325	6,975	612.20
	0100	Telescoping boom to 40′ high, 500 lb. capacity, diesel		11.27	320	965	2,900	283.15
	0105	To 45′ high, 500 lb. capacity		12.54	315	945	2,825	289.35
	0110	To 60′ high, 500 lb. capacity		16.41	535	1,600	4,800	451.25
	0115	To 80′ high, 500 lb. capacity		21.33	835	2,500	7,500	670.65
	0120	To 100′ high, 500 lb. capacity		28.80	860	2,575	7,725	745.40
	0125	To 120′ high, 500 lb. capacity		29.25	840	2,525	7,575	739
	0195	Air compressor, portable, 6.5 CFM, electric		.91	12.65	38	114	14.85
	0196	Gasoline		.65	18	54	162	16.05
	0200	Towed type, gas engine, 60 CFM		9.46	51.50	155	465	106.70
	0300	160 CFM		10.51	53	159	475	115.85
	0400	Diesel engine, rotary screw, 250 CFM		12.12	118	355	1,075	167.95
	0500	365 CFM	↓	16.05	142	425	1,275	213.35

01 54 33 | Equipment Rental

		01 54 33 Equipment Rental	UNIT	HOURLY OPER. COST	RENT PER DAY	RENT PER WEEK	RENT PER MONTH	EQUIPMENT COST/DAY	
40	0550	450 CFM	Ea.	20.01	177	530	1,600	266.05	40
	0600	600 CFM		34.20	240	720	2,150	417.60	
	0700	750 CFM		34.72	252	755	2,275	428.80	
	0930	Air tools, breaker, pavement, 60 lb.		.57	10.35	31	93	10.75	
	0940	80 lb.		.57	10.65	32	96	10.90	
	0950	Drills, hand (jackhammer), 65 lb.		.67	17.35	52	156	15.80	
	0960	Track or wagon, swing boom, 4" drifter		54.83	940	2,825	8,475	1,004	
	0970	5" drifter		63.49	1,100	3,325	9,975	1,173	
	0975	Track mounted quarry drill, 6" diameter drill		102.22	1,625	4,875	14,600	1,793	
	0980	Dust control per drill		1.04	25	75.50	227	23.45	
	0990	Hammer, chipping, 12 lb.		.60	27	81	243	21	
	1000	Hose, air with couplings, 50' long, 3/4" diameter		.07	10	30	90	6.55	
	1100	1" diameter		.08	13	39	117	8.40	
	1200	1-1/2" diameter		.22	35	105	315	22.75	
	1300	2" diameter		.24	40	120	360	25.90	
	1400	2-1/2" diameter		.36	56.50	170	510	36.85	
	1410	3" diameter		.42	37.50	112	335	25.75	
	1450	Drill, steel, 7/8" x 2'		.08	12.50	37.50	113	8.20	
	1460	7/8" x 6'		.12	19	57	171	12.30	
	1520	Moil points		.03	4.73	14.20	42.50	3.05	
	1525	Pneumatic nailer w/accessories		.48	31	92.50	278	22.35	
	1530	Sheeting driver for 60 lb. breaker		.04	7.50	22.50	67.50	4.85	
	1540	For 90 lb. breaker		.13	10.15	30.50	91.50	7.15	
	1550	Spade, 25 lb.		.50	7.15	21.50	64.50	8.30	
	1560	Tamper, single, 35 lb.		.59	39.50	119	355	28.55	
	1570	Triple, 140 lb.		.89	59.50	179	535	42.90	
	1580	Wrenches, impact, air powered, up to 3/4" bolt		.43	12.85	38.50	116	11.10	
	1590	Up to 1-1/4" bolt		.58	23	69.50	209	18.50	
	1600	Barricades, barrels, reflectorized, 1 to 99 barrels		.03	5.50	16.55	49.50	3.55	
	1610	100 to 200 barrels		.02	4.27	12.80	38.50	2.75	
	1620	Barrels with flashers, 1 to 99 barrels		.03	6.20	18.55	55.50	4	
	1630	100 to 200 barrels		.03	4.95	14.85	44.50	3.20	
	1640	Barrels with steady burn type C lights		.05	8.15	24.50	73.50	5.30	
	1650	Illuminated board, trailer mounted, with generator		3.29	135	405	1,225	107.30	
	1670	Portable barricade, stock, with flashers, 1 to 6 units		.03	6.15	18.50	55.50	4	
	1680	25 to 50 units		.03	5.75	17.25	52	3.70	
	1685	Butt fusion machine, wheeled, 1.5 HP electric, 2" - 8" diameter pipe		2.64	203	610	1,825	143.10	
	1690	Tracked, 20 HP diesel, 4" - 12" diameter pipe		11.27	525	1,575	4,725	405.15	
	1695	83 HP diesel, 8" - 24" diameter pipe		51.47	2,525	7,575	22,700	1,927	
	1700	Carts, brick, gas engine, 1,000 lb. capacity		2.95	69.50	208	625	65.20	
	1800	1,500 lb., 7-1/2' lift		2.92	65	195	585	62.40	
	1822	Dehumidifier, medium, 6 lb./hr., 150 CFM		1.19	74	222	665	53.95	
	1824	Large, 18 lb./hr., 600 CFM		2.20	140	420	1,250	101.60	
	1830	Distributor, asphalt, trailer mounted, 2,000 gal., 38 H.P. diesel		11.03	340	1,025	3,075	293.20	
	1840	3,000 gal., 38 H.P. diesel		12.90	365	1,100	3,300	323.25	
	1850	Drill, rotary hammer, electric		1.12	27	80.50	242	25.05	
	1860	Carbide bit, 1-1/2" diameter, add to electric rotary hammer		.03	5.10	15.30	46	3.30	
	1865	Rotary, crawler, 250 H.P.		136.18	2,225	6,650	20,000	2,419	
	1870	Emulsion sprayer, 65 gal., 5 H.P. gas engine		2.78	103	310	930	84.20	
	1880	200 gal., 5 H.P. engine		7.25	173	520	1,550	161.95	
	1900	Floor auto-scrubbing machine, walk-behind, 28" path		5.64	365	1,100	3,300	265.15	
	1930	Floodlight, mercury vapor, or quartz, on tripod, 1,000 watt		.46	22	66.50	200	16.95	
	1940	2,000 watt		.59	27.50	82	246	21.15	
	1950	Floodlights, trailer mounted with generator, one - 300 watt light		3.56	76	228	685	74.05	
	1960	two 1000 watt lights		4.50	84.50	254	760	86.80	
	2000	four 300 watt lights		4.26	96.50	290	870	92.05	
	2005	Foam spray rig, incl. box trailer, compressor, generator, proportioner		25.54	515	1,550	4,650	514.30	
	2015	Forklift, pneumatic tire, rough terr., straight mast, 5,000 lb., 12' lift, gas		18.65	212	635	1,900	276.20	

01 54 33 | Equipment Rental

		01 54 33 Equipment Rental	UNIT	HOURLY OPER. COST	RENT PER DAY	RENT PER WEEK	RENT PER MONTH	EQUIPMENT COST/DAY	
40	2025	8,000 lb, 12′ lift	Ea.	22.75	350	1,050	3,150	392	40
	2030	5,000 lb., 12′ lift, diesel		15.45	237	710	2,125	265.60	
	2035	8,000 lb, 12′ lift, diesel		16.75	268	805	2,425	295	
	2045	All terrain, telescoping boom, diesel, 5,000 lb., 10′ reach, 19′ lift		17.25	410	1,225	3,675	383	
	2055	6,600 lb., 29′ reach, 42′ lift		21.10	400	1,200	3,600	408.80	
	2065	10,000 lb., 31′ reach, 45′ lift		23.10	465	1,400	4,200	464.75	
	2070	Cushion tire, smooth floor, gas, 5,000 lb. capacity		8.25	76.50	230	690	112	
	2075	8,000 lb. capacity		11.37	89	267	800	144.30	
	2085	Diesel, 5,000 lb. capacity		7.75	83.50	250	750	112	
	2090	12,000 lb. capacity		12.05	130	390	1,175	174.40	
	2095	20,000 lb. capacity		17.25	150	450	1,350	228.05	
	2100	Generator, electric, gas engine, 1.5 kW to 3 kW		2.58	10.15	30.50	91.50	26.70	
	2200	5 kW		3.22	12.65	38	114	33.35	
	2300	10 kW		5.93	33.50	101	305	67.65	
	2400	25 kW		7.41	84.50	254	760	110.05	
	2500	Diesel engine, 20 kW		9.21	74	222	665	118.05	
	2600	50 kW		15.95	96.50	289	865	185.35	
	2700	100 kW		28.60	133	400	1,200	308.80	
	2800	250 kW		54.35	255	765	2,300	587.85	
	2850	Hammer, hydraulic, for mounting on boom, to 500 ft. lb.		2.90	90.50	271	815	77.40	
	2860	1,000 ft. lb.		4.61	135	405	1,225	117.85	
	2900	Heaters, space, oil or electric, 50 MBH		1.46	8	24	72	16.50	
	3000	100 MBH		2.72	11.85	35.50	107	28.85	
	3100	300 MBH		7.92	40	120	360	87.40	
	3150	500 MBH		13.16	45	135	405	132.25	
	3200	Hose, water, suction with coupling, 20′ long, 2″ diameter		.02	3.13	9.40	28	2.05	
	3210	3″ diameter		.03	4.10	12.30	37	2.70	
	3220	4″ diameter		.03	4.80	14.40	43	3.10	
	3230	6″ diameter		.11	17	51	153	11.10	
	3240	8″ diameter		.27	45	135	405	29.15	
	3250	Discharge hose with coupling, 50′ long, 2″ diameter		.01	1.33	4	12	.90	
	3260	3″ diameter		.01	2.22	6.65	19.95	1.40	
	3270	4″ diameter		.02	3.47	10.40	31	2.25	
	3280	6″ diameter		.06	8.65	26	78	5.70	
	3290	8″ diameter		.24	40	120	360	25.90	
	3295	Insulation blower		.83	6	18.05	54	10.25	
	3300	Ladders, extension type, 16′ to 36′ long		.18	31.50	95	285	20.45	
	3400	40′ to 60′ long		.64	112	335	1,000	72.10	
	3405	Lance for cutting concrete		2.21	64	192	575	56.05	
	3407	Lawn mower, rotary, 22″, 5 H.P.		1.06	25	75	225	23.45	
	3408	48″ self-propelled		2.90	127	380	1,150	99.20	
	3410	Level, electronic, automatic, with tripod and leveling rod		1.05	61.50	185	555	45.40	
	3430	Laser type, for pipe and sewer line and grade		2.18	153	460	1,375	109.40	
	3440	Rotating beam for interior control		.90	62	186	560	44.40	
	3460	Builder's optical transit, with tripod and rod		.10	17	51	153	11	
	3500	Light towers, towable, with diesel generator, 2,000 watt		4.27	98	294	880	92.95	
	3600	4,000 watt		4.51	102	305	915	97.10	
	3700	Mixer, powered, plaster and mortar, 6 C.F., 7 H.P.		2.06	21	62.50	188	28.95	
	3800	10 C.F., 9 H.P.		2.24	33.50	101	305	38.15	
	3850	Nailer, pneumatic		.48	32.50	97	291	23.25	
	3900	Paint sprayers complete, 8 CFM		.85	59.50	179	535	42.60	
	4000	17 CFM		1.60	107	320	960	76.80	
	4020	Pavers, bituminous, rubber tires, 8′ wide, 50 H.P., diesel		32.02	550	1,650	4,950	586.20	
	4030	10′ wide, 150 H.P.		95.91	1,875	5,650	17,000	1,897	
	4050	Crawler, 8′ wide, 100 H.P., diesel		87.85	2,000	5,975	17,900	1,898	
	4060	10′ wide, 150 H.P.		104.28	2,275	6,825	20,500	2,199	
	4070	Concrete paver, 12′ to 24′ wide, 250 H.P.		87.88	1,625	4,900	14,700	1,683	
	4080	Placer-spreader-trimmer, 24′ wide, 300 H.P.	↓	117.87	2,475	7,425	22,300	2,428	

01 54 33 | Equipment Rental

			UNIT	HOURLY OPER. COST	RENT PER DAY	RENT PER WEEK	RENT PER MONTH	EQUIPMENT COST/DAY	
40	4100	Pump, centrifugal gas pump, 1-1/2" diam., 65 GPM	Ea.	3.93	52.50	157	470	62.90	40
	4200	2" diameter, 130 GPM		4.99	64.50	194	580	78.75	
	4300	3" diameter, 250 GPM		5.13	63.50	191	575	79.25	
	4400	6" diameter, 1,500 GPM		22.31	197	590	1,775	296.45	
	4500	Submersible electric pump, 1-1/4" diameter, 55 GPM		.40	17.65	53	159	13.80	
	4600	1-1/2" diameter, 83 GPM		.45	20.50	61	183	15.75	
	4700	2" diameter, 120 GPM		1.65	25	75.50	227	28.30	
	4800	3" diameter, 300 GPM		3.04	45	135	405	51.35	
	4900	4" diameter, 560 GPM		14.80	163	490	1,475	216.35	
	5000	6" diameter, 1,590 GPM		22.14	200	600	1,800	297.15	
	5100	Diaphragm pump, gas, single, 1-1/2" diameter		1.13	57.50	172	515	43.45	
	5200	2" diameter		3.99	70.50	212	635	74.30	
	5300	3" diameter		4.06	74.50	224	670	77.30	
	5400	Double, 4" diameter		6.05	155	465	1,400	141.40	
	5450	Pressure washer, 5 GPM, 3,000 psi		3.88	53.50	160	480	63.05	
	5460	7 GPM, 3,000 psi		4.95	63	189	565	77.40	
	5500	Trash pump, self-priming, gas, 2" diameter		3.83	23.50	71	213	44.80	
	5600	Diesel, 4" diameter		6.70	94.50	284	850	110.40	
	5650	Diesel, 6" diameter		16.90	165	495	1,475	234.20	
	5655	Grout pump		18.75	272	815	2,450	313	
	5700	Salamanders, L.P. gas fired, 100,000 BTU		2.89	14.15	42.50	128	31.65	
	5705	50,000 BTU		1.67	11.65	35	105	20.35	
	5720	Sandblaster, portable, open top, 3 C.F. capacity		.60	26.50	79.50	239	20.70	
	5730	6 C.F. capacity		1.01	39.50	118	355	31.65	
	5740	Accessories for above		.14	23	69	207	14.90	
	5750	Sander, floor		.77	19.35	58	174	17.75	
	5760	Edger		.52	15.85	47.50	143	13.65	
	5800	Saw, chain, gas engine, 18" long		1.76	22.50	67.50	203	27.55	
	5900	Hydraulic powered, 36" long		.78	67.50	202	605	46.65	
	5950	60" long		.78	69.50	208	625	47.85	
	6000	Masonry, table mounted, 14" diameter, 5 H.P.		1.32	81	243	730	59.20	
	6050	Portable cut-off, 8 H.P.		1.82	32.50	97.50	293	34	
	6100	Circular, hand held, electric, 7-1/4" diameter		.23	4.93	14.80	44.50	4.80	
	6200	12" diameter		.24	7.85	23.50	70.50	6.60	
	6250	Wall saw, w/hydraulic power, 10 H.P.		3.30	33.50	101	305	46.60	
	6275	Shot blaster, walk-behind, 20" wide		4.75	272	815	2,450	201	
	6280	Sidewalk broom, walk-behind		2.25	80	240	720	65.95	
	6300	Steam cleaner, 100 gallons per hour		3.35	80	240	720	74.80	
	6310	200 gallons per hour		4.35	96.50	290	870	92.75	
	6340	Tar kettle/pot, 400 gallons		16.53	76.50	230	690	178.20	
	6350	Torch, cutting, acetylene-oxygen, 150' hose, excludes gases		.45	14.85	44.50	134	12.50	
	6360	Hourly operating cost includes tips and gas		20.98				167.85	
	6410	Toilet, portable chemical		.13	22.50	67	201	14.45	
	6420	Recycle flush type		.16	27.50	83	249	17.90	
	6430	Toilet, fresh water flush, garden hose,		.19	33	99	297	21.35	
	6440	Hoisted, non-flush, for high rise		.16	27	81	243	17.45	
	6465	Tractor, farm with attachment		17.42	300	900	2,700	319.35	
	6480	Trailers, platform, flush deck, 2 axle, 3 ton capacity		1.69	21.50	65	195	26.55	
	6500	25 ton capacity		6.25	138	415	1,250	133	
	6600	40 ton capacity		8.06	197	590	1,775	182.50	
	6700	3 axle, 50 ton capacity		8.75	218	655	1,975	200.95	
	6800	75 ton capacity		11.11	290	870	2,600	262.90	
	6810	Trailer mounted cable reel for high voltage line work		5.90	282	845	2,525	216.20	
	6820	Trailer mounted cable tensioning rig		11.71	560	1,675	5,025	428.65	
	6830	Cable pulling rig		73.99	3,125	9,375	28,100	2,467	
	6850	Portable cable/wire puller, 8,000 lb. max. pulling capacity		3.71	205	615	1,850	152.70	
	6900	Water tank trailer, engine driven discharge, 5,000 gallons		7.18	153	460	1,375	149.45	
	6925	10,000 gallons	↓	9.78	208	625	1,875	203.25	

01 54 33 | Equipment Rental

			UNIT	HOURLY OPER. COST	RENT PER DAY	RENT PER WEEK	RENT PER MONTH	EQUIPMENT COST/DAY	
40	6950	Water truck, off highway, 6,000 gallons	Ea.	71.97	810	2,425	7,275	1,061	40
	7010	Tram car for high voltage line work, powered, 2 conductor		6.90	152	455	1,375	146.20	
	7020	Transit (builder's level) with tripod		.10	17	51	153	11	
	7030	Trench box, 3,000 lb., 6′ x 8′		.56	93.50	281	845	60.70	
	7040	7,200 lb., 6′ x 20′		.72	120	360	1,075	77.75	
	7050	8,000 lb., 8′ x 16′		1.08	180	540	1,625	116.65	
	7060	9,500 lb., 8′ x 20′		1.21	225	675	2,025	144.65	
	7065	11,000 lb., 8′ x 24′		1.27	212	635	1,900	137.15	
	7070	12,000 lb., 10′ x 20′		1.49	255	765	2,300	164.95	
	7100	Truck, pickup, 3/4 ton, 2 wheel drive		9.26	59.50	179	535	109.90	
	7200	4 wheel drive		9.51	76	228	685	121.70	
	7250	Crew carrier, 9 passenger		12.70	88	264	790	154.35	
	7290	Flat bed truck, 20,000 lb. GVW		15.31	128	385	1,150	199.45	
	7300	Tractor, 4 x 2, 220 H.P.		22.32	208	625	1,875	303.55	
	7410	330 H.P.		32.43	285	855	2,575	430.40	
	7500	6 x 4, 380 H.P.		36.20	330	990	2,975	487.60	
	7600	450 H.P.		44.36	400	1,200	3,600	594.90	
	7610	Tractor, with A frame, boom and winch, 225 H.P.		24.81	283	850	2,550	368.50	
	7620	Vacuum truck, hazardous material, 2,500 gallons		12.83	300	900	2,700	282.65	
	7625	5,000 gallons		13.06	425	1,275	3,825	359.50	
	7650	Vacuum, HEPA, 16 gallon, wet/dry		.85	17.15	51.50	155	17.10	
	7655	55 gallon, wet/dry		.78	26.50	80	240	22.25	
	7660	Water tank, portable		.73	155	465	1,400	98.85	
	7690	Sewer/catch basin vacuum, 14 C.Y., 1,500 gallons		17.37	640	1,925	5,775	523.95	
	7700	Welder, electric, 200 amp		3.83	16	48	144	40.20	
	7800	300 amp		5.56	19.35	58	174	56.10	
	7900	Gas engine, 200 amp		8.97	23	69	207	85.60	
	8000	300 amp		10.16	24	72	216	95.65	
	8100	Wheelbarrow, any size		.06	10.35	31	93	6.70	
	8200	Wrecking ball, 4,000 lb.	↓	2.51	72	216	650	63.25	
50	0010	**HIGHWAY EQUIPMENT RENTAL** without operators R015433-10							50
	0050	Asphalt batch plant, portable drum mixer, 100 ton/hr.	Ea.	88.67	1,500	4,475	13,400	1,604	
	0060	200 ton/hr.		102.29	1,600	4,775	14,300	1,773	
	0070	300 ton/hr.		120.22	1,875	5,600	16,800	2,082	
	0100	Backhoe attachment, long stick, up to 185 H.P., 10-1/2′ long		.37	24.50	74	222	17.75	
	0140	Up to 250 H.P., 12′ long		.41	27.50	83	249	19.90	
	0180	Over 250 H.P., 15′ long		.57	37.50	113	340	27.15	
	0200	Special dipper arm, up to 100 H.P., 32′ long		1.16	77	231	695	55.50	
	0240	Over 100 H.P., 33′ long		1.45	96.50	290	870	69.60	
	0280	Catch basin/sewer cleaning truck, 3 ton, 9 C.Y., 1,000 gal.		35.50	410	1,225	3,675	528.95	
	0300	Concrete batch plant, portable, electric, 200 C.Y./hr.		24.26	540	1,625	4,875	519.05	
	0520	Grader/dozer attachment, ripper/scarifier, rear mounted, up to 135 H.P.		3.16	61.50	184	550	62.10	
	0540	Up to 180 H.P.		4.15	92.50	278	835	88.80	
	0580	Up to 250 H.P.		5.87	148	445	1,325	135.95	
	0700	Pvmt. removal bucket, for hyd. excavator, up to 90 H.P.		2.17	56.50	169	505	51.10	
	0740	Up to 200 H.P.		2.31	72	216	650	61.70	
	0780	Over 200 H.P.		2.53	88.50	265	795	73.20	
	0900	Aggregate spreader, self-propelled, 187 H.P.		50.75	715	2,150	6,450	836	
	1000	Chemical spreader, 3 C.Y.		3.18	46.50	139	415	53.20	
	1900	Hammermill, traveling, 250 H.P.		67.55	2,250	6,750	20,300	1,890	
	2000	Horizontal borer, 3″ diameter, 13 H.P. gas driven		5.43	57.50	172	515	77.85	
	2150	Horizontal directional drill, 20,000 lb. thrust, 78 H.P. diesel		27.66	685	2,050	6,150	631.30	
	2160	30,000 lb. thrust, 115 H.P.		34.00	1,050	3,125	9,375	897	
	2170	50,000 lb. thrust, 170 H.P.		48.74	1,325	3,975	11,900	1,185	
	2190	Mud trailer for HDD, 1,500 gallons, 175 H.P., gas		25.58	157	470	1,400	298.65	
	2200	Hydromulcher, diesel, 3,000 gallon, for truck mounting		17.48	255	765	2,300	292.90	
	2300	Gas, 600 gallon		7.52	107	320	960	124.15	
	2400	Joint & crack cleaner, walk behind, 25 H.P.	↓	3.17	52.50	158	475	56.95	

01 54 33 | Equipment Rental

			UNIT	HOURLY OPER. COST	RENT PER DAY	RENT PER WEEK	RENT PER MONTH	EQUIPMENT COST/DAY	
50	2500	Filler, trailer mounted, 400 gallons, 20 H.P.	Ea.	8.37	220	660	1,975	198.95	50
	3000	Paint striper, self-propelled, 40 gallon, 22 H.P.		6.78	163	490	1,475	152.25	
	3100	120 gallon, 120 H.P.		19.28	410	1,225	3,675	399.25	
	3200	Post drivers, 6″ I-Beam frame, for truck mounting		12.45	390	1,175	3,525	334.60	
	3400	Road sweeper, self-propelled, 8′ wide, 90 H.P.		36.01	690	2,075	6,225	703.10	
	3450	Road sweeper, vacuum assisted, 4 C.Y., 220 gallons		58.45	650	1,950	5,850	857.60	
	4000	Road mixer, self-propelled, 130 H.P.		46.37	800	2,400	7,200	851	
	4100	310 H.P.		75.23	2,100	6,275	18,800	1,857	
	4220	Cold mix paver, incl. pug mill and bitumen tank, 165 H.P.		95.25	2,250	6,725	20,200	2,107	
	4240	Pavement brush, towed		3.44	96.50	290	870	85.50	
	4250	Paver, asphalt, wheel or crawler, 130 H.P., diesel		94.52	2,200	6,600	19,800	2,076	
	4300	Paver, road widener, gas, 1′ to 6′, 67 H.P.		46.80	940	2,825	8,475	939.40	
	4400	Diesel, 2′ to 14′, 88 H.P.		56.55	1,125	3,350	10,100	1,122	
	4600	Slipform pavers, curb and gutter, 2 track, 75 H.P.		58.01	1,225	3,650	11,000	1,194	
	4700	4 track, 165 H.P.		35.79	815	2,450	7,350	776.35	
	4800	Median barrier, 215 H.P.		58.61	1,300	3,900	11,700	1,249	
	4901	Trailer, low bed, 75 ton capacity		10.74	273	820	2,450	249.90	
	5000	Road planer, walk behind, 10″ cutting width, 10 H.P.		2.46	33.50	100	300	39.70	
	5100	Self-propelled, 12″ cutting width, 64 H.P.		8.28	117	350	1,050	136.25	
	5120	Traffic line remover, metal ball blaster, truck mounted, 115 H.P.		46.70	785	2,350	7,050	843.60	
	5140	Grinder, truck mounted, 115 H.P.		51.04	810	2,425	7,275	893.35	
	5160	Walk-behind, 11 H.P.		3.57	54.50	164	490	61.35	
	5200	Pavement profiler, 4′ to 6′ wide, 450 H.P.		217.23	3,425	10,300	30,900	3,798	
	5300	8′ to 10′ wide, 750 H.P.		332.58	4,525	13,600	40,800	5,381	
	5400	Roadway plate, steel, 1″ x 8′ x 20′		.09	14.65	44	132	9.50	
	5600	Stabilizer, self-propelled, 150 H.P.		41.26	640	1,925	5,775	715.05	
	5700	310 H.P.		76.41	1,700	5,075	15,200	1,626	
	5800	Striper, truck mounted, 120 gallon paint, 460 H.P.		48.89	490	1,475	4,425	686.10	
	5900	Thermal paint heating kettle, 115 gallons		7.73	26.50	80	240	77.85	
	6000	Tar kettle, 330 gallon, trailer mounted		12.31	60	180	540	134.50	
	7000	Tunnel locomotive, diesel, 8 to 12 ton		29.85	600	1,800	5,400	598.75	
	7005	Electric, 10 ton		29.33	685	2,050	6,150	644.70	
	7010	Muck cars, 1/2 C.Y. capacity		2.30	26	77.50	233	33.95	
	7020	1 C.Y. capacity		2.52	33.50	101	305	40.35	
	7030	2 C.Y. capacity		2.66	37.50	113	340	43.90	
	7040	Side dump, 2 C.Y. capacity		2.88	46.50	140	420	51.05	
	7050	3 C.Y. capacity		3.86	51.50	154	460	61.70	
	7060	5 C.Y. capacity		5.64	66.50	199	595	84.90	
	7100	Ventilating blower for tunnel, 7-1/2 H.P.		2.14	51	153	460	47.75	
	7110	10 H.P.		2.43	53.50	160	480	51.40	
	7120	20 H.P.		3.55	69.50	208	625	70	
	7140	40 H.P.		6.16	91.50	275	825	104.25	
	7160	60 H.P.		8.71	98.50	295	885	128.70	
	7175	75 H.P.		10.40	153	460	1,375	175.20	
	7180	200 H.P.		20.85	300	905	2,725	347.75	
	7800	Windrow loader, elevating		54.10	1,350	4,025	12,100	1,238	
60	0010	**LIFTING AND HOISTING EQUIPMENT RENTAL** without operators R015433-10							60
	0150	Crane, flatbed mounted, 3 ton capacity	Ea.	14.45	195	585	1,750	232.65	
	0200	Crane, climbing, 106′ jib, 6,000 lb. capacity, 410 fpm R312316-45		39.84	1,775	5,350	16,100	1,389	
	0300	101′ jib, 10,250 lb. capacity, 270 fpm		46.57	2,275	6,800	20,400	1,733	
	0500	Tower, static, 130′ high, 106′ jib, 6,200 lb. capacity at 400 fpm		45.29	2,075	6,200	18,600	1,602	
	0520	Mini crawler spider crane, up to 24″ wide, 1,990 lb. lifting capacity		12.54	535	1,600	4,800	420.30	
	0525	Up to 30″ wide, 6,450 lb. lifting capacity		14.57	635	1,900	5,700	496.55	
	0530	Up to 52″ wide, 6,680 lb. lifting capacity		23.17	775	2,325	6,975	650.40	
	0535	Up to 55″ wide, 8,920 lb. lifting capacity		25.87	860	2,575	7,725	722	
	0540	Up to 66″ wide, 13,350 lb. lifting capacity		35.03	1,325	4,000	12,000	1,080	
	0600	Crawler mounted, lattice boom, 1/2 C.Y., 15 tons at 12′ radius		37.07	950	2,850	8,550	866.60	
	0700	3/4 C.Y., 20 tons at 12′ radius		50.57	1,200	3,625	10,900	1,130	

01 54 33 | Equipment Rental

		Description	Unit	Hourly Oper. Cost	Rent Per Day	Rent Per Week	Rent Per Month	Equipment Cost/Day	
60	0800	1 C.Y., 25 tons at 12′ radius	Ea.	67.62	1,400	4,225	12,700	1,386	60
	0900	1-1/2 C.Y., 40 tons at 12′ radius		66.51	1,425	4,300	12,900	1,392	
	1000	2 C.Y., 50 tons at 12′ radius		89.03	2,200	6,600	19,800	2,032	
	1100	3 C.Y., 75 tons at 12′ radius		75.49	1,875	5,650	17,000	1,734	
	1200	100 ton capacity, 60′ boom		86.17	1,975	5,950	17,900	1,879	
	1300	165 ton capacity, 60′ boom		106.42	2,275	6,850	20,600	2,221	
	1400	200 ton capacity, 70′ boom		138.63	3,175	9,550	28,700	3,019	
	1500	350 ton capacity, 80′ boom		182.75	4,075	12,200	36,600	3,902	
	1600	Truck mounted, lattice boom, 6 x 4, 20 tons at 10′ radius		39.88	915	2,750	8,250	869	
	1700	25 tons at 10′ radius		42.86	1,400	4,175	12,500	1,178	
	1800	8 x 4, 30 tons at 10′ radius		45.67	1,475	4,425	13,300	1,250	
	1900	40 tons at 12′ radius		48.70	1,550	4,625	13,900	1,315	
	2000	60 tons at 15′ radius		53.85	1,650	4,950	14,900	1,421	
	2050	82 tons at 15′ radius		59.60	1,775	5,350	16,100	1,547	
	2100	90 tons at 15′ radius		66.59	1,950	5,825	17,500	1,698	
	2200	115 tons at 15′ radius		75.12	2,175	6,525	19,600	1,906	
	2300	150 tons at 18′ radius		81.33	2,275	6,850	20,600	2,021	
	2350	165 tons at 18′ radius		87.30	2,425	7,275	21,800	2,153	
	2400	Truck mounted, hydraulic, 12 ton capacity		29.59	390	1,175	3,525	471.75	
	2500	25 ton capacity		36.46	485	1,450	4,350	581.70	
	2550	33 ton capacity		50.82	900	2,700	8,100	946.60	
	2560	40 ton capacity		49.62	900	2,700	8,100	937	
	2600	55 ton capacity		53.94	915	2,750	8,250	981.50	
	2700	80 ton capacity		75.94	1,475	4,400	13,200	1,487	
	2720	100 ton capacity		75.18	1,550	4,675	14,000	1,536	
	2740	120 ton capacity		103.12	1,825	5,500	16,500	1,925	
	2760	150 ton capacity		110.25	2,050	6,125	18,400	2,107	
	2800	Self-propelled, 4 x 4, with telescoping boom, 5 ton		15.18	230	690	2,075	259.45	
	2900	12-1/2 ton capacity		21.48	335	1,000	3,000	371.85	
	3000	15 ton capacity		34.52	535	1,600	4,800	596.20	
	3050	20 ton capacity		24.09	660	1,975	5,925	587.75	
	3100	25 ton capacity		36.80	615	1,850	5,550	664.40	
	3150	40 ton capacity		45.03	650	1,950	5,850	750.25	
	3200	Derricks, guy, 20 ton capacity, 60′ boom, 75′ mast		22.81	435	1,300	3,900	442.50	
	3300	100′ boom, 115′ mast		36.15	750	2,250	6,750	739.20	
	3400	Stiffleg, 20 ton capacity, 70′ boom, 37′ mast		25.48	565	1,700	5,100	543.85	
	3500	100′ boom, 47′ mast		39.43	910	2,725	8,175	860.50	
	3550	Helicopter, small, lift to 1,250 lb. maximum, w/pilot		99.44	3,525	10,600	31,800	2,916	
	3600	Hoists, chain type, overhead, manual, 3/4 ton		.15	.33	1	3	1.35	
	3900	10 ton		.79	6	18	54	9.90	
	4000	Hoist and tower, 5,000 lb. cap., portable electric, 40′ high		5.14	252	755	2,275	192.10	
	4100	For each added 10′ section, add		.12	19.65	59	177	12.75	
	4200	Hoist and single tubular tower, 5,000 lb. electric, 100′ high		6.98	350	1,050	3,150	265.80	
	4300	For each added 6′-6″ section, add		.21	34.50	103	310	22.25	
	4400	Hoist and double tubular tower, 5,000 lb., 100′ high		7.59	385	1,150	3,450	290.75	
	4500	For each added 6′-6″ section, add		.23	37.50	113	340	24.40	
	4550	Hoist and tower, mast type, 6,000 lb., 100′ high		8.26	400	1,200	3,600	306.10	
	4570	For each added 10′ section, add		.13	23	69	207	14.85	
	4600	Hoist and tower, personnel, electric, 2,000 lb., 100′ @ 125 fpm		17.55	1,075	3,200	9,600	780.40	
	4700	3,000 lb., 100′ @ 200 fpm		20.08	1,200	3,625	10,900	885.65	
	4800	3,000 lb., 150′ @ 300 fpm		22.28	1,350	4,075	12,200	993.30	
	4900	4,000 lb., 100′ @ 300 fpm		23.05	1,375	4,150	12,500	1,014	
	5000	6,000 lb., 100′ @ 275 fpm	▼	24.77	1,450	4,350	13,100	1,068	
	5100	For added heights up to 500′, add	L.F.	.01	1.67	5	15	1.10	
	5200	Jacks, hydraulic, 20 ton	Ea.	.05	2	6	18	1.60	
	5500	100 ton		.40	12	36	108	10.40	
	6100	Jacks, hydraulic, climbing w/50′ jackrods, control console, 30 ton cap.		2.17	145	435	1,300	104.40	
	6150	For each added 10′ jackrod section, add	▼	.05	3.33	10	30	2.40	

01 54 33 | Equipment Rental

			UNIT	HOURLY OPER. COST	RENT PER DAY	RENT PER WEEK	RENT PER MONTH	EQUIPMENT COST/DAY	
60	6300	50 ton capacity	Ea.	3.49	232	695	2,075	166.95	60
	6350	For each added 10′ jackrod section, add		.06	4	12	36	2.90	
	6500	125 ton capacity		9.13	610	1,825	5,475	438.05	
	6550	For each added 10′ jackrod section, add		.62	41	123	370	29.55	
	6600	Cable jack, 10 ton capacity with 200′ cable		1.82	122	365	1,100	87.60	
	6650	For each added 50′ of cable, add	↓	.22	14.65	44	132	10.55	
70	0010	**WELLPOINT EQUIPMENT RENTAL** without operators R015433-10							70
	0020	Based on 2 months rental							
	0100	Combination jetting & wellpoint pump, 60 H.P. diesel	Ea.	15.72	360	1,075	3,225	340.75	
	0200	High pressure gas jet pump, 200 H.P., 300 psi	"	33.93	305	920	2,750	455.45	
	0300	Discharge pipe, 8″ diameter	L.F.	.01	.58	1.74	5.20	.40	
	0350	12″ diameter		.01	.83	2.50	7.50	.60	
	0400	Header pipe, flows up to 150 GPM, 4″ diameter		.01	.53	1.59	4.77	.40	
	0500	400 GPM, 6″ diameter		.01	.62	1.86	5.60	.45	
	0600	800 GPM, 8″ diameter		.01	.83	2.50	7.50	.60	
	0700	1,500 GPM, 10″ diameter		.01	.88	2.64	7.90	.65	
	0800	2,500 GPM, 12″ diameter		.03	1.70	5.10	15.30	1.20	
	0900	4,500 GPM, 16″ diameter		.03	2.18	6.55	19.65	1.55	
	0950	For quick coupling aluminum and plastic pipe, add	↓	.03	2.27	6.80	20.50	1.65	
	1100	Wellpoint, 25′ long, with fittings & riser pipe, 1-1/2″ or 2″ diameter	Ea.	.07	4.52	13.55	40.50	3.25	
	1200	Wellpoint pump, diesel powered, 4″ suction, 20 H.P.		7.02	207	620	1,850	180.20	
	1300	6″ suction, 30 H.P.		9.42	257	770	2,300	229.35	
	1400	8″ suction, 40 H.P.		12.76	350	1,050	3,150	312.10	
	1500	10″ suction, 75 H.P.		18.83	410	1,225	3,675	395.65	
	1600	12″ suction, 100 H.P.		27.32	660	1,975	5,925	613.55	
	1700	12″ suction, 175 H.P.	↓	39.09	725	2,175	6,525	747.75	
80	0010	**MARINE EQUIPMENT RENTAL** without operators R015433-10							80
	0200	Barge, 400 ton, 30′ wide x 90′ long	Ea.	17.69	1,150	3,475	10,400	836.50	
	0240	800 ton, 45′ wide x 90′ long		22.21	1,425	4,275	12,800	1,033	
	2000	Tugboat, diesel, 100 H.P.		29.66	230	690	2,075	375.25	
	2040	250 H.P.		57.58	415	1,250	3,750	710.70	
	2080	380 H.P.		125.36	1,250	3,750	11,300	1,753	
	3000	Small work boat, gas, 16-foot, 50 H.P.		11.38	46.50	139	415	118.85	
	4000	Large, diesel, 48-foot, 200 H.P.	↓	74.90	1,325	3,975	11,900	1,394	

Crew No.	Bare Costs		Incl. Subs O&P		Cost Per Labor-Hour	
Crew A-1	**Hr.**	**Daily**	**Hr.**	**Daily**	**Bare Costs**	**Incl. O&P**
1 Building Laborer	$27.50	$220.00	$45.45	$363.60	$27.50	$45.45
1 Concrete Saw, Gas Manual		72.90		80.19	9.11	10.02
8 L.H., Daily Totals		$292.90		$443.79	$36.61	$55.47
Crew A-1A	**Hr.**	**Daily**	**Hr.**	**Daily**	**Bare Costs**	**Incl. O&P**
1 Skilled Worker	$36.50	$292.00	$60.65	$485.20	$36.50	$60.65
1 Shot Blaster, 20″		201.00		221.10	25.13	27.64
8 L.H., Daily Totals		$493.00		$706.30	$61.63	$88.29
Crew A-1B	**Hr.**	**Daily**	**Hr.**	**Daily**	**Bare Costs**	**Incl. O&P**
1 Building Laborer	$27.50	$220.00	$45.45	$363.60	$27.50	$45.45
1 Concrete Saw		110.15		121.17	13.77	15.15
8 L.H., Daily Totals		$330.15		$484.76	$41.27	$60.60
Crew A-1C	**Hr.**	**Daily**	**Hr.**	**Daily**	**Bare Costs**	**Incl. O&P**
1 Building Laborer	$27.50	$220.00	$45.45	$363.60	$27.50	$45.45
1 Chain Saw, Gas, 18″		27.55		30.31	3.44	3.79
8 L.H., Daily Totals		$247.55		$393.90	$30.94	$49.24
Crew A-1D	**Hr.**	**Daily**	**Hr.**	**Daily**	**Bare Costs**	**Incl. O&P**
1 Building Laborer	$27.50	$220.00	$45.45	$363.60	$27.50	$45.45
1 Vibrating Plate, Gas, 18″		31.10		34.21	3.89	4.28
8 L.H., Daily Totals		$251.10		$397.81	$31.39	$49.73
Crew A-1E	**Hr.**	**Daily**	**Hr.**	**Daily**	**Bare Costs**	**Incl. O&P**
1 Building Laborer	$27.50	$220.00	$45.45	$363.60	$27.50	$45.45
1 Vibrating Plate, Gas, 21″		40.55		44.60	5.07	5.58
8 L.H., Daily Totals		$260.55		$408.20	$32.57	$51.03
Crew A-1F	**Hr.**	**Daily**	**Hr.**	**Daily**	**Bare Costs**	**Incl. O&P**
1 Building Laborer	$27.50	$220.00	$45.45	$363.60	$27.50	$45.45
1 Rammer/Tamper, Gas, 8″		45.65		50.22	5.71	6.28
8 L.H., Daily Totals		$265.65		$413.82	$33.21	$51.73
Crew A-1G	**Hr.**	**Daily**	**Hr.**	**Daily**	**Bare Costs**	**Incl. O&P**
1 Building Laborer	$27.50	$220.00	$45.45	$363.60	$27.50	$45.45
1 Rammer/Tamper, Gas, 15″		53.05		58.35	6.63	7.29
8 L.H., Daily Totals		$273.05		$421.95	$34.13	$52.74
Crew A-1H	**Hr.**	**Daily**	**Hr.**	**Daily**	**Bare Costs**	**Incl. O&P**
1 Building Laborer	$27.50	$220.00	$45.45	$363.60	$27.50	$45.45
1 Exterior Steam Cleaner		74.80		82.28	9.35	10.29
8 L.H., Daily Totals		$294.80		$445.88	$36.85	$55.73
Crew A-1J	**Hr.**	**Daily**	**Hr.**	**Daily**	**Bare Costs**	**Incl. O&P**
1 Building Laborer	$27.50	$220.00	$45.45	$363.60	$27.50	$45.45
1 Cultivator, Walk-Behind, 5 H.P.		70.50		77.55	8.81	9.69
8 L.H., Daily Totals		$290.50		$441.15	$36.31	$55.14
Crew A-1K	**Hr.**	**Daily**	**Hr.**	**Daily**	**Bare Costs**	**Incl. O&P**
1 Building Laborer	$27.50	$220.00	$45.45	$363.60	$27.50	$45.45
1 Cultivator, Walk-Behind, 8 H.P.		84.50		92.95	10.56	11.62
8 L.H., Daily Totals		$304.50		$456.55	$38.06	$57.07
Crew A-1M	**Hr.**	**Daily**	**Hr.**	**Daily**	**Bare Costs**	**Incl. O&P**
1 Building Laborer	$27.50	$220.00	$45.45	$363.60	$27.50	$45.45
1 Snow Blower, Walk-Behind		65.95		72.55	8.24	9.07
8 L.H., Daily Totals		$285.95		$436.14	$35.74	$54.52

Crew No.	Bare Costs		Incl. Subs O&P		Cost Per Labor-Hour	
Crew A-2	**Hr.**	**Daily**	**Hr.**	**Daily**	**Bare Costs**	**Incl. O&P**
2 Laborers	$27.50	$440.00	$45.45	$727.20	$29.10	$48.02
1 Truck Driver (light)	32.30	258.40	53.15	425.20		
1 Flatbed Truck, Gas, 1.5 Ton		195.20		214.72	8.13	8.95
24 L.H., Daily Totals		$893.60		$1367.12	$37.23	$56.96
Crew A-2A	**Hr.**	**Daily**	**Hr.**	**Daily**	**Bare Costs**	**Incl. O&P**
2 Laborers	$27.50	$440.00	$45.45	$727.20	$29.10	$48.02
1 Truck Driver (light)	32.30	258.40	53.15	425.20		
1 Flatbed Truck, Gas, 1.5 Ton		195.20		214.72		
1 Concrete Saw		110.15		121.17	12.72	14.00
24 L.H., Daily Totals		$1003.75		$1488.29	$41.82	$62.01
Crew A-2B	**Hr.**	**Daily**	**Hr.**	**Daily**	**Bare Costs**	**Incl. O&P**
1 Truck Driver (light)	$32.30	$258.40	$53.15	$425.20	$32.30	$53.15
1 Flatbed Truck, Gas, 1.5 Ton		195.20		214.72	24.40	26.84
8 L.H., Daily Totals		$453.60		$639.92	$56.70	$79.99
Crew A-3A	**Hr.**	**Daily**	**Hr.**	**Daily**	**Bare Costs**	**Incl. O&P**
1 Equip. Oper. (light)	$35.65	$285.20	$58.50	$468.00	$35.65	$58.50
1 Pickup Truck, 4x4, 3/4 Ton		121.70		133.87	15.21	16.73
8 L.H., Daily Totals		$406.90		$601.87	$50.86	$75.23
Crew A-3B	**Hr.**	**Daily**	**Hr.**	**Daily**	**Bare Costs**	**Incl. O&P**
1 Equip. Oper. (medium)	$37.45	$299.60	$61.45	$491.60	$35.17	$57.77
1 Truck Driver (heavy)	32.90	263.20	54.10	432.80		
1 Dump Truck, 12 C.Y., 400 H.P.		567.05		623.76		
1 F.E. Loader, W.M., 2.5 C.Y.		531.20		584.32	68.64	75.50
16 L.H., Daily Totals		$1661.05		$2132.47	$103.82	$133.28
Crew A-3C	**Hr.**	**Daily**	**Hr.**	**Daily**	**Bare Costs**	**Incl. O&P**
1 Equip. Oper. (light)	$35.65	$285.20	$58.50	$468.00	$35.65	$58.50
1 Loader, Skid Steer, 78 H.P.		392.50		431.75	49.06	53.97
8 L.H., Daily Totals		$677.70		$899.75	$84.71	$112.47
Crew A-3D	**Hr.**	**Daily**	**Hr.**	**Daily**	**Bare Costs**	**Incl. O&P**
1 Truck Driver (light)	$32.30	$258.40	$53.15	$425.20	$32.30	$53.15
1 Pickup Truck, 4x4, 3/4 Ton		121.70		133.87		
1 Flatbed Trailer, 25 Ton		133.00		146.30	31.84	35.02
8 L.H., Daily Totals		$513.10		$705.37	$64.14	$88.17
Crew A-3E	**Hr.**	**Daily**	**Hr.**	**Daily**	**Bare Costs**	**Incl. O&P**
1 Equip. Oper. (crane)	$37.90	$303.20	$62.20	$497.60	$35.40	$58.15
1 Truck Driver (heavy)	32.90	263.20	54.10	432.80		
1 Pickup Truck, 4x4, 3/4 Ton		121.70		133.87	7.61	8.37
16 L.H., Daily Totals		$688.10		$1064.27	$43.01	$66.52
Crew A-3F	**Hr.**	**Daily**	**Hr.**	**Daily**	**Bare Costs**	**Incl. O&P**
1 Equip. Oper. (crane)	$37.90	$303.20	$62.20	$497.60	$35.40	$58.15
1 Truck Driver (heavy)	32.90	263.20	54.10	432.80		
1 Pickup Truck, 4x4, 3/4 Ton		121.70		133.87		
1 Truck Tractor, 6x4, 380 H.P.		487.60		536.36		
1 Lowbed Trailer, 75 Ton		249.90		274.89	53.70	59.07
16 L.H., Daily Totals		$1425.60		$1875.52	$89.10	$117.22

Crews - Residential

Crew No.	Bare Costs		Incl. Subs O&P		Cost Per Labor-Hour	
Crew A-3G	**Hr.**	**Daily**	**Hr.**	**Daily**	**Bare Costs**	**Incl. O&P**
1 Equip. Oper. (crane)	$37.90	$303.20	$62.20	$497.60	$35.40	$58.15
1 Truck Driver (heavy)	32.90	263.20	54.10	432.80		
1 Pickup Truck, 4x4, 3/4 Ton		121.70		133.87		
1 Truck Tractor, 6x4, 450 H.P.		594.90		654.39		
1 Lowbed Trailer, 75 Ton		249.90		274.89	60.41	66.45
16 L.H., Daily Totals		$1532.90		$1993.55	$95.81	$124.60
Crew A-3H	**Hr.**	**Daily**	**Hr.**	**Daily**	**Bare Costs**	**Incl. O&P**
1 Equip. Oper. (crane)	$37.90	$303.20	$62.20	$497.60	$37.90	$62.20
1 Hyd. Crane, 12 Ton (Daily)		708.90		779.79	88.61	97.47
8 L.H., Daily Totals		$1012.10		$1277.39	$126.51	$159.67
Crew A-3I	**Hr.**	**Daily**	**Hr.**	**Daily**	**Bare Costs**	**Incl. O&P**
1 Equip. Oper. (crane)	$37.90	$303.20	$62.20	$497.60	$37.90	$62.20
1 Hyd. Crane, 25 Ton (Daily)		785.85		864.43	98.23	108.05
8 L.H., Daily Totals		$1089.05		$1362.04	$136.13	$170.25
Crew A-3J	**Hr.**	**Daily**	**Hr.**	**Daily**	**Bare Costs**	**Incl. O&P**
1 Equip. Oper. (crane)	$37.90	$303.20	$62.20	$497.60	$37.90	$62.20
1 Hyd. Crane, 40 Ton (Daily)		1280.00		1408.00	160.00	176.00
8 L.H., Daily Totals		$1583.20		$1905.60	$197.90	$238.20
Crew A-3K	**Hr.**	**Daily**	**Hr.**	**Daily**	**Bare Costs**	**Incl. O&P**
1 Equip. Oper. (crane)	$37.90	$303.20	$62.20	$497.60	$35.42	$58.13
1 Equip. Oper. (oiler)	32.95	263.60	54.05	432.40		
1 Hyd. Crane, 55 Ton (Daily)		1299.00		1428.90		
1 P/U Truck, 3/4 Ton (Daily)		140.45		154.50	89.97	98.96
16 L.H., Daily Totals		$2006.25		$2513.40	$125.39	$157.09
Crew A-3L	**Hr.**	**Daily**	**Hr.**	**Daily**	**Bare Costs**	**Incl. O&P**
1 Equip. Oper. (crane)	$37.90	$303.20	$62.20	$497.60	$35.42	$58.13
1 Equip. Oper. (oiler)	32.95	263.60	54.05	432.40		
1 Hyd. Crane, 80 Ton (Daily)		2056.00		2261.60		
1 P/U Truck, 3/4 Ton (Daily)		140.45		154.50	137.28	151.01
16 L.H., Daily Totals		$2763.25		$3346.09	$172.70	$209.13
Crew A-3M	**Hr.**	**Daily**	**Hr.**	**Daily**	**Bare Costs**	**Incl. O&P**
1 Equip. Oper. (crane)	$37.90	$303.20	$62.20	$497.60	$35.42	$58.13
1 Equip. Oper. (oiler)	32.95	263.60	54.05	432.40		
1 Hyd. Crane, 100 Ton (Daily)		2179.00		2396.90		
1 P/U Truck, 3/4 Ton (Daily)		140.45		154.50	144.97	159.46
16 L.H., Daily Totals		$2886.25		$3481.40	$180.39	$217.59
Crew A-3N	**Hr.**	**Daily**	**Hr.**	**Daily**	**Bare Costs**	**Incl. O&P**
1 Equip. Oper. (crane)	$37.90	$303.20	$62.20	$497.60	$37.90	$62.20
1 Tower Crane (monthly)		1181.00		1299.10	147.63	162.39
8 L.H., Daily Totals		$1484.20		$1796.70	$185.53	$224.59
Crew A-3P	**Hr.**	**Daily**	**Hr.**	**Daily**	**Bare Costs**	**Incl. O&P**
1 Equip. Oper. (light)	$35.65	$285.20	$58.50	$468.00	$35.65	$58.50
1 A.T. Forklift, 31' reach, 45' lift		464.75		511.23	58.09	63.90
8 L.H., Daily Totals		$749.95		$979.23	$93.74	$122.40
Crew A-3Q	**Hr.**	**Daily**	**Hr.**	**Daily**	**Bare Costs**	**Incl. O&P**
1 Equip. Oper. (light)	$35.65	$285.20	$58.50	$468.00	$35.65	$58.50
1 Pickup Truck, 4x4, 3/4 Ton		121.70		133.87		
1 Flatbed Trailer, 3 Ton		26.55		29.20	18.53	20.38
8 L.H., Daily Totals		$433.45		$631.08	$54.18	$78.88

Crew No.	Bare Costs		Incl. Subs O&P		Cost Per Labor-Hour	
Crew A-3R	**Hr.**	**Daily**	**Hr.**	**Daily**	**Bare Costs**	**Incl. O&P**
1 Equip. Oper. (light)	$35.65	$285.20	$58.50	$468.00	$35.65	$58.50
1 Forklift, Smooth Floor, 8,000 Lb.		144.30		158.73	18.04	19.84
8 L.H., Daily Totals		$429.50		$626.73	$53.69	$78.34
Crew A-4	**Hr.**	**Daily**	**Hr.**	**Daily**	**Bare Costs**	**Incl. O&P**
2 Carpenters	$35.65	$570.40	$58.90	$942.40	$33.72	$55.57
1 Painter, Ordinary	29.85	238.80	48.90	391.20		
24 L.H., Daily Totals		$809.20		$1333.60	$33.72	$55.57
Crew A-5	**Hr.**	**Daily**	**Hr.**	**Daily**	**Bare Costs**	**Incl. O&P**
2 Laborers	$27.50	$440.00	$45.45	$727.20	$28.03	$46.31
.25 Truck Driver (light)	32.30	64.60	53.15	106.30		
.25 Flatbed Truck, Gas, 1.5 Ton		48.80		53.68	2.71	2.98
18 L.H., Daily Totals		$553.40		$887.18	$30.74	$49.29
Crew A-6	**Hr.**	**Daily**	**Hr.**	**Daily**	**Bare Costs**	**Incl. O&P**
1 Instrument Man	$36.50	$292.00	$60.65	$485.20	$35.63	$58.92
1 Rodman/Chainman	34.75	278.00	57.20	457.60		
1 Level, Electronic		45.40		49.94	2.84	3.12
16 L.H., Daily Totals		$615.40		$992.74	$38.46	$62.05
Crew A-7	**Hr.**	**Daily**	**Hr.**	**Daily**	**Bare Costs**	**Incl. O&P**
1 Chief of Party	$42.30	$338.40	$69.65	$557.20	$37.85	$62.50
1 Instrument Man	36.50	292.00	60.65	485.20		
1 Rodman/Chainman	34.75	278.00	57.20	457.60		
1 Level, Electronic		45.40		49.94	1.89	2.08
24 L.H., Daily Totals		$953.80		$1549.94	$39.74	$64.58
Crew A-8	**Hr.**	**Daily**	**Hr.**	**Daily**	**Bare Costs**	**Incl. O&P**
1 Chief of Party	$42.30	$338.40	$69.65	$557.20	$37.08	$61.17
1 Instrument Man	36.50	292.00	60.65	485.20		
2 Rodmen/Chainmen	34.75	556.00	57.20	915.20		
1 Level, Electronic		45.40		49.94	1.42	1.56
32 L.H., Daily Totals		$1231.80		$2007.54	$38.49	$62.74
Crew A-9	**Hr.**	**Daily**	**Hr.**	**Daily**	**Bare Costs**	**Incl. O&P**
1 Asbestos Foreman	$37.20	$297.60	$62.65	$501.20	$36.76	$61.91
7 Asbestos Workers	36.70	2055.20	61.80	3460.80		
64 L.H., Daily Totals		$2352.80		$3962.00	$36.76	$61.91
Crew A-10A	**Hr.**	**Daily**	**Hr.**	**Daily**	**Bare Costs**	**Incl. O&P**
1 Asbestos Foreman	$37.20	$297.60	$62.65	$501.20	$36.87	$62.08
2 Asbestos Workers	36.70	587.20	61.80	988.80		
24 L.H., Daily Totals		$884.80		$1490.00	$36.87	$62.08
Crew A-10B	**Hr.**	**Daily**	**Hr.**	**Daily**	**Bare Costs**	**Incl. O&P**
1 Asbestos Foreman	$37.20	$297.60	$62.65	$501.20	$36.83	$62.01
3 Asbestos Workers	36.70	880.80	61.80	1483.20		
32 L.H., Daily Totals		$1178.40		$1984.40	$36.83	$62.01
Crew A-10C	**Hr.**	**Daily**	**Hr.**	**Daily**	**Bare Costs**	**Incl. O&P**
3 Asbestos Workers	$36.70	$880.80	$61.80	$1483.20	$36.70	$61.80
1 Flatbed Truck, Gas, 1.5 Ton		195.20		214.72	8.13	8.95
24 L.H., Daily Totals		$1076.00		$1697.92	$44.83	$70.75

Crew No.	Bare Costs		Incl. Subs O&P		Cost Per Labor-Hour	

Crew A-10D	Hr.	Daily	Hr.	Daily	Bare Costs	Incl. O&P
2 Asbestos Workers	$36.70	$587.20	$61.80	$988.80	$36.06	$59.96
1 Equip. Oper. (crane)	37.90	303.20	62.20	497.60		
1 Equip. Oper. (oiler)	32.95	263.60	54.05	432.40		
1 Hydraulic Crane, 33 Ton		946.60		1041.26	29.58	32.54
32 L.H., Daily Totals		$2100.60		$2960.06	$65.64	$92.50

Crew A-11	Hr.	Daily	Hr.	Daily	Bare Costs	Incl. O&P
1 Asbestos Foreman	$37.20	$297.60	$62.65	$501.20	$36.76	$61.91
7 Asbestos Workers	36.70	2055.20	61.80	3460.80		
2 Chip. Hammers, 12 Lb., Elec.		42.00		46.20	0.66	0.72
64 L.H., Daily Totals		$2394.80		$4008.20	$37.42	$62.63

Crew A-12	Hr.	Daily	Hr.	Daily	Bare Costs	Incl. O&P
1 Asbestos Foreman	$37.20	$297.60	$62.65	$501.20	$36.76	$61.91
7 Asbestos Workers	36.70	2055.20	61.80	3460.80		
1 Trk-Mtd Vac, 14 CY, 1500 Gal.		523.95		576.35		
1 Flatbed Truck, 20,000 GVW		199.45		219.40	11.30	12.43
64 L.H., Daily Totals		$3076.20		$4757.74	$48.07	$74.34

Crew A-13	Hr.	Daily	Hr.	Daily	Bare Costs	Incl. O&P
1 Equip. Oper. (light)	$35.65	$285.20	$58.50	$468.00	$35.65	$58.50
1 Trk-Mtd Vac, 14 CY, 1500 Gal.		523.95		576.35		
1 Flatbed Truck, 20,000 GVW		199.45		219.40	90.42	99.47
8 L.H., Daily Totals		$1008.60		$1263.74	$126.08	$157.97

Crew B-1	Hr.	Daily	Hr.	Daily	Bare Costs	Incl. O&P
1 Labor Foreman (outside)	$29.50	$236.00	$48.75	$390.00	$28.17	$46.55
2 Laborers	27.50	440.00	45.45	727.20		
24 L.H., Daily Totals		$676.00		$1117.20	$28.17	$46.55

Crew B-1A	Hr.	Daily	Hr.	Daily	Bare Costs	Incl. O&P
1 Labor Foreman (outside)	$29.50	$236.00	$48.75	$390.00	$28.17	$46.55
2 Laborers	27.50	440.00	45.45	727.20		
2 Cutting Torches		25.00		27.50		
2 Sets of Gases		335.70		369.27	15.03	16.53
24 L.H., Daily Totals		$1036.70		$1513.97	$43.20	$63.08

Crew B-1B	Hr.	Daily	Hr.	Daily	Bare Costs	Incl. O&P
1 Labor Foreman (outside)	$29.50	$236.00	$48.75	$390.00	$30.60	$50.46
2 Laborers	27.50	440.00	45.45	727.20		
1 Equip. Oper. (crane)	37.90	303.20	62.20	497.60		
2 Cutting Torches		25.00		27.50		
2 Sets of Gases		335.70		369.27		
1 Hyd. Crane, 12 Ton		471.75		518.92	26.01	28.62
32 L.H., Daily Totals		$1811.65		$2530.49	$56.61	$79.08

Crew B-1C	Hr.	Daily	Hr.	Daily	Bare Costs	Incl. O&P
1 Labor Foreman (outside)	$29.50	$236.00	$48.75	$390.00	$28.17	$46.55
2 Laborers	27.50	440.00	45.45	727.20		
1 Telescoping Boom Lift, to 60'		451.25		496.38	18.80	20.68
24 L.H., Daily Totals		$1127.25		$1613.58	$46.97	$67.23

Crew B-1D	Hr.	Daily	Hr.	Daily	Bare Costs	Incl. O&P
2 Laborers	$27.50	$440.00	$45.45	$727.20	$27.50	$45.45
1 Small Work Boat, Gas, 50 H.P.		118.85		130.74		
1 Pressure Washer, 7 GPM		77.40		85.14	12.27	13.49
16 L.H., Daily Totals		$636.25		$943.08	$39.77	$58.94

Crew B-1E	Hr.	Daily	Hr.	Daily	Bare Costs	Incl. O&P
1 Labor Foreman (outside)	$29.50	$236.00	$48.75	$390.00	$28.00	$46.27
3 Laborers	27.50	660.00	45.45	1090.80		
1 Work Boat, Diesel, 200 H.P.		1394.00		1533.40		
2 Pressure Washers, 7 GPM		154.80		170.28	48.40	53.24
32 L.H., Daily Totals		$2444.80		$3184.48	$76.40	$99.52

Crew B-1F	Hr.	Daily	Hr.	Daily	Bare Costs	Incl. O&P
2 Skilled Workers	$36.50	$584.00	$60.65	$970.40	$33.50	$55.58
1 Laborer	27.50	220.00	45.45	363.60		
1 Small Work Boat, Gas, 50 H.P.		118.85		130.74		
1 Pressure Washer, 7 GPM		77.40		85.14	8.18	8.99
24 L.H., Daily Totals		$1000.25		$1549.88	$41.68	$64.58

Crew B-1G	Hr.	Daily	Hr.	Daily	Bare Costs	Incl. O&P
2 Laborers	$27.50	$440.00	$45.45	$727.20	$27.50	$45.45
1 Small Work Boat, Gas, 50 H.P.		118.85		130.74	7.43	8.17
16 L.H., Daily Totals		$558.85		$857.93	$34.93	$53.62

Crew B-1H	Hr.	Daily	Hr.	Daily	Bare Costs	Incl. O&P
2 Skilled Workers	$36.50	$584.00	$60.65	$970.40	$33.50	$55.58
1 Laborer	27.50	220.00	45.45	363.60		
1 Small Work Boat, Gas, 50 H.P.		118.85		130.74	4.95	5.45
24 L.H., Daily Totals		$922.85		$1464.73	$38.45	$61.03

Crew B-1J	Hr.	Daily	Hr.	Daily	Bare Costs	Incl. O&P
1 Labor Foreman (inside)	$28.00	$224.00	$46.30	$370.40	$27.75	$45.88
1 Laborer	27.50	220.00	45.45	363.60		
16 L.H., Daily Totals		$444.00		$734.00	$27.75	$45.88

Crew B-1K	Hr.	Daily	Hr.	Daily	Bare Costs	Incl. O&P
1 Carpenter Foreman (inside)	$36.15	$289.20	$59.75	$478.00	$35.90	$59.33
1 Carpenter	35.65	285.20	58.90	471.20		
16 L.H., Daily Totals		$574.40		$949.20	$35.90	$59.33

Crew B-2	Hr.	Daily	Hr.	Daily	Bare Costs	Incl. O&P
1 Labor Foreman (outside)	$29.50	$236.00	$48.75	$390.00	$27.90	$46.11
4 Laborers	27.50	880.00	45.45	1454.40		
40 L.H., Daily Totals		$1116.00		$1844.40	$27.90	$46.11

Crew B-2A	Hr.	Daily	Hr.	Daily	Bare Costs	Incl. O&P
1 Labor Foreman (outside)	$29.50	$236.00	$48.75	$390.00	$28.17	$46.55
2 Laborers	27.50	440.00	45.45	727.20		
1 Telescoping Boom Lift, to 60'		451.25		496.38	18.80	20.68
24 L.H., Daily Totals		$1127.25		$1613.58	$46.97	$67.23

Crew B-3	Hr.	Daily	Hr.	Daily	Bare Costs	Incl. O&P
1 Labor Foreman (outside)	$29.50	$236.00	$48.75	$390.00	$31.29	$51.55
2 Laborers	27.50	440.00	45.45	727.20		
1 Equip. Oper. (medium)	37.45	299.60	61.45	491.60		
2 Truck Drivers (heavy)	32.90	526.40	54.10	865.60		
1 Crawler Loader, 3 C.Y.		1202.00		1322.20		
2 Dump Trucks, 12 C.Y., 400 H.P.		1134.10		1247.51	48.67	53.54
48 L.H., Daily Totals		$3838.10		$5044.11	$79.96	$105.09

Crew B-3A	Hr.	Daily	Hr.	Daily	Bare Costs	Incl. O&P
4 Laborers	$27.50	$880.00	$45.45	$1454.40	$29.49	$48.65
1 Equip. Oper. (medium)	37.45	299.60	61.45	491.60		
1 Hyd. Excavator, 1.5 C.Y.		908.70		999.57	22.72	24.99
40 L.H., Daily Totals		$2088.30		$2945.57	$52.21	$73.64

Crew No.	Bare Costs		Incl. Subs O&P		Cost Per Labor-Hour	
Crew B-3B	**Hr.**	**Daily**	**Hr.**	**Daily**	**Bare Costs**	**Incl. O&P**
2 Laborers	$27.50	$440.00	$45.45	$727.20	$31.34	$51.61
1 Equip. Oper. (medium)	37.45	299.60	61.45	491.60		
1 Truck Driver (heavy)	32.90	263.20	54.10	432.80		
1 Backhoe Loader, 80 H.P.		387.85		426.63		
1 Dump Truck, 12 C.Y., 400 H.P.		567.05		623.76	29.84	32.82
32 L.H., Daily Totals		$1957.70		$2701.99	$61.18	$84.44
Crew B-3C	**Hr.**	**Daily**	**Hr.**	**Daily**	**Bare Costs**	**Incl. O&P**
3 Laborers	$27.50	$660.00	$45.45	$1090.80	$29.99	$49.45
1 Equip. Oper. (medium)	37.45	299.60	61.45	491.60		
1 Crawler Loader, 4 C.Y.		1461.00		1607.10	45.66	50.22
32 L.H., Daily Totals		$2420.60		$3189.50	$75.64	$99.67
Crew B-4	**Hr.**	**Daily**	**Hr.**	**Daily**	**Bare Costs**	**Incl. O&P**
1 Labor Foreman (outside)	$29.50	$236.00	$48.75	$390.00	$28.73	$47.44
4 Laborers	27.50	880.00	45.45	1454.40		
1 Truck Driver (heavy)	32.90	263.20	54.10	432.80		
1 Truck Tractor, 220 H.P.		303.55		333.90		
1 Flatbed Trailer, 40 Ton		182.50		200.75	10.13	11.14
48 L.H., Daily Totals		$1865.25		$2811.86	$38.86	$58.58
Crew B-5	**Hr.**	**Daily**	**Hr.**	**Daily**	**Bare Costs**	**Incl. O&P**
1 Labor Foreman (outside)	$29.50	$236.00	$48.75	$390.00	$29.89	$49.31
3 Laborers	27.50	660.00	45.45	1090.80		
1 Equip. Oper. (medium)	37.45	299.60	61.45	491.60		
1 Air Compressor, 250 cfm		167.95		184.75		
2 Breakers, Pavement, 60 lb.		21.50		23.65		
2 -50' Air Hoses, 1.5"		45.50		50.05		
1 Crawler Loader, 3 C.Y.		1202.00		1322.20	35.92	39.52
40 L.H., Daily Totals		$2632.55		$3553.05	$65.81	$88.83
Crew B-5A	**Hr.**	**Daily**	**Hr.**	**Daily**	**Bare Costs**	**Incl. O&P**
1 Labor Foreman (outside)	$29.50	$236.00	$48.75	$390.00	$30.90	$50.92
6 Laborers	27.50	1320.00	45.45	2181.60		
2 Equip. Oper. (medium)	37.45	599.20	61.45	983.20		
1 Equip. Oper. (light)	35.65	285.20	58.50	468.00		
2 Truck Drivers (heavy)	32.90	526.40	54.10	865.60		
1 Air Compressor, 365 cfm		213.35		234.69		
2 Breakers, Pavement, 60 lb.		21.50		23.65		
8 -50' Air Hoses, 1"		67.20		73.92		
2 Dump Trucks, 8 C.Y., 220 H.P.		679.20		747.12	10.22	11.24
96 L.H., Daily Totals		$3948.05		$5967.77	$41.13	$62.16
Crew B-5B	**Hr.**	**Daily**	**Hr.**	**Daily**	**Bare Costs**	**Incl. O&P**
1 Powderman	$36.50	$292.00	$60.65	$485.20	$35.02	$57.64
2 Equip. Oper. (medium)	37.45	599.20	61.45	983.20		
3 Truck Drivers (heavy)	32.90	789.60	54.10	1298.40		
1 F.E. Loader, W.M., 2.5 C.Y.		531.20		584.32		
3 Dump Trucks, 12 C.Y., 400 H.P.		1701.15		1871.27		
1 Air Compressor, 365 cfm		213.35		234.69	50.95	56.05
48 L.H., Daily Totals		$4126.50		$5457.07	$85.97	$113.69

Crew No.	Bare Costs		Incl. Subs O&P		Cost Per Labor-Hour	
Crew B-5C	**Hr.**	**Daily**	**Hr.**	**Daily**	**Bare Costs**	**Incl. O&P**
3 Laborers	$27.50	$660.00	$45.45	$1090.80	$32.08	$52.78
1 Equip. Oper. (medium)	37.45	299.60	61.45	491.60		
2 Truck Drivers (heavy)	32.90	526.40	54.10	865.60		
1 Equip. Oper. (crane)	37.90	303.20	62.20	497.60		
1 Equip. Oper. (oiler)	32.95	263.60	54.05	432.40		
2 Dump Trucks, 12 C.Y., 400 H.P.		1134.10		1247.51		
1 Crawler Loader, 4 C.Y.		1461.00		1607.10		
1 S.P. Crane, 4x4, 25 Ton		664.40		730.84	50.93	56.02
64 L.H., Daily Totals		$5312.30		$6963.45	$83.00	$108.80
Crew B-5D	**Hr.**	**Daily**	**Hr.**	**Daily**	**Bare Costs**	**Incl. O&P**
1 Labor Foreman (outside)	$29.50	$236.00	$48.75	$390.00	$30.39	$50.11
3 Laborers	27.50	660.00	45.45	1090.80		
1 Equip. Oper. (medium)	37.45	299.60	61.45	491.60		
1 Truck Driver (heavy)	32.90	263.20	54.10	432.80		
1 Air Compressor, 250 cfm		167.95		184.75		
2 Breakers, Pavement, 60 lb.		21.50		23.65		
2 -50' Air Hoses, 1.5"		45.50		50.05		
1 Crawler Loader, 3 C.Y.		1202.00		1322.20		
1 Dump Truck, 12 C.Y., 400 H.P.		567.05		623.76	41.75	45.92
48 L.H., Daily Totals		$3462.80		$4609.60	$72.14	$96.03
Crew B-6	**Hr.**	**Daily**	**Hr.**	**Daily**	**Bare Costs**	**Incl. O&P**
2 Laborers	$27.50	$440.00	$45.45	$727.20	$30.22	$49.80
1 Equip. Oper. (light)	35.65	285.20	58.50	468.00		
1 Backhoe Loader, 48 H.P.		320.20		352.22	13.34	14.68
24 L.H., Daily Totals		$1045.40		$1547.42	$43.56	$64.48
Crew B-6A	**Hr.**	**Daily**	**Hr.**	**Daily**	**Bare Costs**	**Incl. O&P**
.5 Labor Foreman (outside)	$29.50	$118.00	$48.75	$195.00	$31.88	$52.51
1 Laborer	27.50	220.00	45.45	363.60		
1 Equip. Oper. (medium)	37.45	299.60	61.45	491.60		
1 Vacuum Truck, 5000 Gal.		359.50		395.45	17.98	19.77
20 L.H., Daily Totals		$997.10		$1445.65	$49.85	$72.28
Crew B-6B	**Hr.**	**Daily**	**Hr.**	**Daily**	**Bare Costs**	**Incl. O&P**
2 Labor Foremen (outside)	$29.50	$472.00	$48.75	$780.00	$28.17	$46.55
4 Laborers	27.50	880.00	45.45	1454.40		
1 S.P. Crane, 4x4, 5 Ton		259.45		285.39		
1 Flatbed Truck, Gas, 1.5 Ton		195.20		214.72		
1 Butt Fusion Mach., 4"-12" diam.		405.15		445.67	17.91	19.70
48 L.H., Daily Totals		$2211.80		$3180.18	$46.08	$66.25
Crew B-6C	**Hr.**	**Daily**	**Hr.**	**Daily**	**Bare Costs**	**Incl. O&P**
2 Labor Foremen (outside)	$29.50	$472.00	$48.75	$780.00	$28.17	$46.55
4 Laborers	27.50	880.00	45.45	1454.40		
1 S.P. Crane, 4x4, 12 Ton		371.85		409.04		
1 Flatbed Truck, Gas, 3 Ton		245.95		270.55		
1 Butt Fusion Mach., 8"-24" diam.		1927.00		2119.70	53.02	58.32
48 L.H., Daily Totals		$3896.80		$5033.68	$81.18	$104.87
Crew B-7	**Hr.**	**Daily**	**Hr.**	**Daily**	**Bare Costs**	**Incl. O&P**
1 Labor Foreman (outside)	$29.50	$236.00	$48.75	$390.00	$29.49	$48.67
4 Laborers	27.50	880.00	45.45	1454.40		
1 Equip. Oper. (medium)	37.45	299.60	61.45	491.60		
1 Brush Chipper, 12", 130 H.P.		389.40		428.34		
1 Crawler Loader, 3 C.Y.		1202.00		1322.20		
2 Chain Saws, Gas, 36" Long		93.30		102.63	35.10	38.61
48 L.H., Daily Totals		$3100.30		$4189.17	$64.59	$87.27

Crew No.	Bare Costs		Incl. Subs O&P		Cost Per Labor-Hour	
Crew B-7A	**Hr.**	**Daily**	**Hr.**	**Daily**	**Bare Costs**	**Incl. O&P**
2 Laborers	$27.50	$440.00	$45.45	$727.20	$30.22	$49.80
1 Equip. Oper. (light)	35.65	285.20	58.50	468.00		
1 Rake w/Tractor		332.75		366.02		
2 Chain Saws, Gas, 18"		55.10		60.61	16.16	17.78
24 L.H., Daily Totals		$1113.05		$1621.84	$46.38	$67.58
Crew B-7B	**Hr.**	**Daily**	**Hr.**	**Daily**	**Bare Costs**	**Incl. O&P**
1 Labor Foreman (outside)	$29.50	$236.00	$48.75	$390.00	$29.98	$49.44
4 Laborers	27.50	880.00	45.45	1454.40		
1 Equip. Oper. (medium)	37.45	299.60	61.45	491.60		
1 Truck Driver (heavy)	32.90	263.20	54.10	432.80		
1 Brush Chipper, 12", 130 H.P.		389.40		428.34		
1 Crawler Loader, 3 C.Y.		1202.00		1322.20		
2 Chain Saws, Gas, 36" Long		93.30		102.63		
1 Dump Truck, 8 C.Y., 220 H.P.		339.60		373.56	36.15	39.76
56 L.H., Daily Totals		$3703.10		$4995.53	$66.13	$89.21
Crew B-7C	**Hr.**	**Daily**	**Hr.**	**Daily**	**Bare Costs**	**Incl. O&P**
1 Labor Foreman (outside)	$29.50	$236.00	$48.75	$390.00	$29.98	$49.44
4 Laborers	27.50	880.00	45.45	1454.40		
1 Equip. Oper. (medium)	37.45	299.60	61.45	491.60		
1 Truck Driver (heavy)	32.90	263.20	54.10	432.80		
1 Brush Chipper, 12", 130 H.P.		389.40		428.34		
1 Crawler Loader, 3 C.Y.		1202.00		1322.20		
2 Chain Saws, Gas, 36" Long		93.30		102.63		
1 Dump Truck, 12 C.Y., 400 H.P.		567.05		623.76	40.21	44.23
56 L.H., Daily Totals		$3930.55		$5245.73	$70.19	$93.67
Crew B-8	**Hr.**	**Daily**	**Hr.**	**Daily**	**Bare Costs**	**Incl. O&P**
1 Labor Foreman (outside)	$29.50	$236.00	$48.75	$390.00	$32.17	$52.96
2 Laborers	27.50	440.00	45.45	727.20		
2 Equip. Oper. (medium)	37.45	599.20	61.45	983.20		
2 Truck Drivers (heavy)	32.90	526.40	54.10	865.60		
1 Hyd. Crane, 25 Ton		581.70		639.87		
1 Crawler Loader, 3 C.Y.		1202.00		1322.20		
2 Dump Trucks, 12 C.Y., 400 H.P.		1134.10		1247.51	52.10	57.31
56 L.H., Daily Totals		$4719.40		$6175.58	$84.28	$110.28
Crew B-9	**Hr.**	**Daily**	**Hr.**	**Daily**	**Bare Costs**	**Incl. O&P**
1 Labor Foreman (outside)	$29.50	$236.00	$48.75	$390.00	$27.90	$46.11
4 Laborers	27.50	880.00	45.45	1454.40		
1 Air Compressor, 250 cfm		167.95		184.75		
2 Breakers, Pavement, 60 lb.		21.50		23.65		
2 -50' Air Hoses, 1.5"		45.50		50.05	5.87	6.46
40 L.H., Daily Totals		$1350.95		$2102.84	$33.77	$52.57
Crew B-9A	**Hr.**	**Daily**	**Hr.**	**Daily**	**Bare Costs**	**Incl. O&P**
2 Laborers	$27.50	$440.00	$45.45	$727.20	$29.30	$48.33
1 Truck Driver (heavy)	32.90	263.20	54.10	432.80		
1 Water Tank Trailer, 5000 Gal.		149.45		164.40		
1 Truck Tractor, 220 H.P.		303.55		333.90		
2 -50' Discharge Hoses, 3"		2.80		3.08	18.99	20.89
24 L.H., Daily Totals		$1159.00		$1661.38	$48.29	$69.22

Crew No.	Bare Costs		Incl. Subs O&P		Cost Per Labor-Hour	
Crew B-9B	**Hr.**	**Daily**	**Hr.**	**Daily**	**Bare Costs**	**Incl. O&P**
2 Laborers	$27.50	$440.00	$45.45	$727.20	$29.30	$48.33
1 Truck Driver (heavy)	32.90	263.20	54.10	432.80		
2 -50' Discharge Hoses, 3"		2.80		3.08		
1 Water Tank Trailer, 5000 Gal.		149.45		164.40		
1 Truck Tractor, 220 H.P.		303.55		333.90		
1 Pressure Washer		63.05		69.36	21.62	23.78
24 L.H., Daily Totals		$1222.05		$1730.73	$50.92	$72.11
Crew B-9D	**Hr.**	**Daily**	**Hr.**	**Daily**	**Bare Costs**	**Incl. O&P**
1 Labor Foreman (outside)	$29.50	$236.00	$48.75	$390.00	$27.90	$46.11
4 Common Laborers	27.50	880.00	45.45	1454.40		
1 Air Compressor, 250 cfm		167.95		184.75		
2 -50' Air Hoses, 1.5"		45.50		50.05		
2 Air Powered Tampers		57.10		62.81	6.76	7.44
40 L.H., Daily Totals		$1386.55		$2142.01	$34.66	$53.55
Crew B-9E	**Hr.**	**Daily**	**Hr.**	**Daily**	**Bare Costs**	**Incl. O&P**
1 Cement Finisher	$35.70	$285.60	$57.70	$461.60	$31.60	$51.58
1 Laborer	27.50	220.00	45.45	363.60		
1 Chip. Hammers, 12 Lb., Elec.		21.00		23.10	1.31	1.44
16 L.H., Daily Totals		$526.60		$848.30	$32.91	$53.02
Crew B-10	**Hr.**	**Daily**	**Hr.**	**Daily**	**Bare Costs**	**Incl. O&P**
1 Equip. Oper. (medium)	37.45	299.60	61.45	491.60	37.45	61.45
8 L.H., Daily Totals		$299.60		$491.60	$37.45	$61.45
Crew B-10A	**Hr.**	**Daily**	**Hr.**	**Daily**	**Bare Costs**	**Incl. O&P**
1 Equip. Oper. (medium)	$37.45	$299.60	$61.45	$491.60	$37.45	$61.45
1 Roller, 2-Drum, W.B., 7.5 H.P.		181.75		199.93	22.72	24.99
8 L.H., Daily Totals		$481.35		$691.52	$60.17	$86.44
Crew B-10B	**Hr.**	**Daily**	**Hr.**	**Daily**	**Bare Costs**	**Incl. O&P**
1 Equip. Oper. (medium)	$37.45	$299.60	$61.45	$491.60	$37.45	$61.45
1 Dozer, 200 H.P.		1290.00		1419.00	161.25	177.38
8 L.H., Daily Totals		$1589.60		$1910.60	$198.70	$238.82
Crew B-10C	**Hr.**	**Daily**	**Hr.**	**Daily**	**Bare Costs**	**Incl. O&P**
1 Equip. Oper. (medium)	$37.45	$299.60	$61.45	$491.60	$37.45	$61.45
1 Dozer, 200 H.P.		1290.00		1419.00		
1 Vibratory Roller, Towed, 23 Ton		422.15		464.37	214.02	235.42
8 L.H., Daily Totals		$2011.75		$2374.97	$251.47	$296.87
Crew B-10D	**Hr.**	**Daily**	**Hr.**	**Daily**	**Bare Costs**	**Incl. O&P**
1 Equip. Oper. (medium)	$37.45	$299.60	$61.45	$491.60	$37.45	$61.45
1 Dozer, 200 H.P.		1290.00		1419.00		
1 Sheepsft. Roller, Towed		430.15		473.17	215.02	236.52
8 L.H., Daily Totals		$2019.75		$2383.76	$252.47	$297.97
Crew B-10E	**Hr.**	**Daily**	**Hr.**	**Daily**	**Bare Costs**	**Incl. O&P**
1 Equip. Oper. (medium)	$37.45	$299.60	$61.45	$491.60	$37.45	$61.45
1 Tandem Roller, 5 Ton		152.45		167.69	19.06	20.96
8 L.H., Daily Totals		$452.05		$659.29	$56.51	$82.41
Crew B-10F	**Hr.**	**Daily**	**Hr.**	**Daily**	**Bare Costs**	**Incl. O&P**
1 Equip. Oper. (medium)	$37.45	$299.60	$61.45	$491.60	$37.45	$61.45
1 Tandem Roller, 10 Ton		232.05		255.26	29.01	31.91
8 L.H., Daily Totals		$531.65		$746.86	$66.46	$93.36

Crew B-10G	Bare Costs Hr.	Bare Costs Daily	Incl. Subs O&P Hr.	Incl. Subs O&P Daily	Cost Per Labor-Hour Bare Costs	Cost Per Labor-Hour Incl. O&P
1 Equip. Oper. (medium)	$37.45	$299.60	$61.45	$491.60	$37.45	$61.45
1 Sheepsfoot Roller, 240 H.P.		1323.00		1455.30	165.38	181.91
8 L.H., Daily Totals		$1622.60		$1946.90	$202.82	$243.36

Crew B-10H	Hr.	Daily	Hr.	Daily	Bare Costs	Incl. O&P
1 Equip. Oper. (medium)	$37.45	$299.60	$61.45	$491.60	$37.45	$61.45
1 Diaphragm Water Pump, 2"		74.30		81.73		
1 -20' Suction Hose, 2"		2.05		2.25		
2 -50' Discharge Hoses, 2"		1.80		1.98	9.77	10.75
8 L.H., Daily Totals		$377.75		$577.57	$47.22	$72.20

Crew B-10I	Hr.	Daily	Hr.	Daily	Bare Costs	Incl. O&P
1 Equip. Oper. (medium)	$37.45	$299.60	$61.45	$491.60	$37.45	$61.45
1 Diaphragm Water Pump, 4"		141.40		155.54		
1 -20' Suction Hose, 4"		3.10		3.41		
2 -50' Discharge Hoses, 4"		4.50		4.95	18.63	20.49
8 L.H., Daily Totals		$448.60		$655.50	$56.08	$81.94

Crew B-10J	Hr.	Daily	Hr.	Daily	Bare Costs	Incl. O&P
1 Equip. Oper. (medium)	$37.45	$299.60	$61.45	$491.60	$37.45	$61.45
1 Centrifugal Water Pump, 3"		79.25		87.17		
1 -20' Suction Hose, 3"		2.70		2.97		
2 -50' Discharge Hoses, 3"		2.80		3.08	10.59	11.65
8 L.H., Daily Totals		$384.35		$584.83	$48.04	$73.10

Crew B-10K	Hr.	Daily	Hr.	Daily	Bare Costs	Incl. O&P
1 Equip. Oper. (medium)	$37.45	$299.60	$61.45	$491.60	$37.45	$61.45
1 Centr. Water Pump, 6"		296.45		326.10		
1 -20' Suction Hose, 6"		11.10		12.21		
2 -50' Discharge Hoses, 6"		11.40		12.54	39.87	43.86
8 L.H., Daily Totals		$618.55		$842.45	$77.32	$105.31

Crew B-10L	Hr.	Daily	Hr.	Daily	Bare Costs	Incl. O&P
1 Equip. Oper. (medium)	$37.45	$299.60	$61.45	$491.60	$37.45	$61.45
1 Dozer, 80 H.P.		472.00		519.20	59.00	64.90
8 L.H., Daily Totals		$771.60		$1010.80	$96.45	$126.35

Crew B-10M	Hr.	Daily	Hr.	Daily	Bare Costs	Incl. O&P
1 Equip. Oper. (medium)	$37.45	$299.60	$61.45	$491.60	$37.45	$61.45
1 Dozer, 300 H.P.		1831.00		2014.10	228.88	251.76
8 L.H., Daily Totals		$2130.60		$2505.70	$266.32	$313.21

Crew B-10N	Hr.	Daily	Hr.	Daily	Bare Costs	Incl. O&P
1 Equip. Oper. (medium)	$37.45	$299.60	$61.45	$491.60	$37.45	$61.45
1 F.E. Loader, T.M., 1.5 C.Y.		591.30		650.43	73.91	81.30
8 L.H., Daily Totals		$890.90		$1142.03	$111.36	$142.75

Crew B-10O	Hr.	Daily	Hr.	Daily	Bare Costs	Incl. O&P
1 Equip. Oper. (medium)	$37.45	$299.60	$61.45	$491.60	$37.45	$61.45
1 F.E. Loader, T.M., 2.25 C.Y.		1002.00		1102.20	125.25	137.78
8 L.H., Daily Totals		$1301.60		$1593.80	$162.70	$199.22

Crew B-10P	Hr.	Daily	Hr.	Daily	Bare Costs	Incl. O&P
1 Equip. Oper. (medium)	$37.45	$299.60	$61.45	$491.60	$37.45	$61.45
1 Crawler Loader, 3 C.Y.		1202.00		1322.20	150.25	165.28
8 L.H., Daily Totals		$1501.60		$1813.80	$187.70	$226.72

Crew B-10Q	Hr.	Daily	Hr.	Daily	Bare Costs	Incl. O&P
1 Equip. Oper. (medium)	$37.45	$299.60	$61.45	$491.60	$37.45	$61.45
1 Crawler Loader, 4 C.Y.		1461.00		1607.10	182.63	200.89
8 L.H., Daily Totals		$1760.60		$2098.70	$220.07	$262.34

Crew B-10R	Hr.	Daily	Hr.	Daily	Bare Costs	Incl. O&P
1 Equip. Oper. (medium)	$37.45	$299.60	$61.45	$491.60	$37.45	$61.45
1 F.E. Loader, W.M., 1 C.Y.		297.00		326.70	37.13	40.84
8 L.H., Daily Totals		$596.60		$818.30	$74.58	$102.29

Crew B-10S	Hr.	Daily	Hr.	Daily	Bare Costs	Incl. O&P
1 Equip. Oper. (medium)	$37.45	$299.60	$61.45	$491.60	$37.45	$61.45
1 F.E. Loader, W.M., 1.5 C.Y.		352.95		388.25	44.12	48.53
8 L.H., Daily Totals		$652.55		$879.85	$81.57	$109.98

Crew B-10T	Hr.	Daily	Hr.	Daily	Bare Costs	Incl. O&P
1 Equip. Oper. (medium)	$37.45	$299.60	$61.45	$491.60	$37.45	$61.45
1 F.E. Loader, W.M., 2.5 C.Y.		531.20		584.32	66.40	73.04
8 L.H., Daily Totals		$830.80		$1075.92	$103.85	$134.49

Crew B-10U	Hr.	Daily	Hr.	Daily	Bare Costs	Incl. O&P
1 Equip. Oper. (medium)	$37.45	$299.60	$61.45	$491.60	$37.45	$61.45
1 F.E. Loader, W.M., 5.5 C.Y.		975.55		1073.11	121.94	134.14
8 L.H., Daily Totals		$1275.15		$1564.70	$159.39	$195.59

Crew B-10V	Hr.	Daily	Hr.	Daily	Bare Costs	Incl. O&P
1 Equip. Oper. (medium)	$37.45	$299.60	$61.45	$491.60	$37.45	$61.45
1 Dozer, 700 H.P.		5021.00		5523.10	627.63	690.39
8 L.H., Daily Totals		$5320.60		$6014.70	$665.08	$751.84

Crew B-10W	Hr.	Daily	Hr.	Daily	Bare Costs	Incl. O&P
1 Equip. Oper. (medium)	$37.45	$299.60	$61.45	$491.60	$37.45	$61.45
1 Dozer, 105 H.P.		609.70		670.67	76.21	83.83
8 L.H., Daily Totals		$909.30		$1162.27	$113.66	$145.28

Crew B-10X	Hr.	Daily	Hr.	Daily	Bare Costs	Incl. O&P
1 Equip. Oper. (medium)	$37.45	$299.60	$61.45	$491.60	$37.45	$61.45
1 Dozer, 410 H.P.		2269.00		2495.90	283.63	311.99
8 L.H., Daily Totals		$2568.60		$2987.50	$321.07	$373.44

Crew B-10Y	Hr.	Daily	Hr.	Daily	Bare Costs	Incl. O&P
1 Equip. Oper. (medium)	$37.45	$299.60	$61.45	$491.60	$37.45	$61.45
1 Vibr. Roller, Towed, 12 Ton		566.70		623.37	70.84	77.92
8 L.H., Daily Totals		$866.30		$1114.97	$108.29	$139.37

Crew B-11A	Hr.	Daily	Hr.	Daily	Bare Costs	Incl. O&P
1 Equipment Oper. (med.)	$37.45	$299.60	$61.45	$491.60	$32.48	$53.45
1 Laborer	27.50	220.00	45.45	363.60		
1 Dozer, 200 H.P.		1290.00		1419.00	80.63	88.69
16 L.H., Daily Totals		$1809.60		$2274.20	$113.10	$142.14

Crew B-11B	Hr.	Daily	Hr.	Daily	Bare Costs	Incl. O&P
1 Equipment Oper. (light)	$35.65	$285.20	$58.50	$468.00	$31.57	$51.98
1 Laborer	27.50	220.00	45.45	363.60		
1 Air Powered Tamper		28.55		31.41		
1 Air Compressor, 365 cfm		213.35		234.69		
2 -50' Air Hoses, 1.5"		45.50		50.05	17.96	19.76
16 L.H., Daily Totals		$792.60		$1147.74	$49.54	$71.73

Crew No.	Bare Costs		Incl. Subs O&P		Cost Per Labor-Hour	
Crew B-11C	**Hr.**	**Daily**	**Hr.**	**Daily**	**Bare Costs**	**Incl. O&P**
1 Equipment Oper. (med.)	$37.45	$299.60	$61.45	$491.60	$32.48	$53.45
1 Laborer	27.50	220.00	45.45	363.60		
1 Backhoe Loader, 48 H.P.		320.20		352.22	20.01	22.01
16 L.H., Daily Totals		$839.80		$1207.42	$52.49	$75.46
Crew B-11K	**Hr.**	**Daily**	**Hr.**	**Daily**	**Bare Costs**	**Incl. O&P**
1 Equipment Oper. (med.)	$37.45	$299.60	$61.45	$491.60	$32.48	$53.45
1 Laborer	27.50	220.00	45.45	363.60		
1 Trencher, Chain Type, 8' D		1837.00		2020.70	114.81	126.29
16 L.H., Daily Totals		$2356.60		$2875.90	$147.29	$179.74
Crew B-11L	**Hr.**	**Daily**	**Hr.**	**Daily**	**Bare Costs**	**Incl. O&P**
1 Equipment Oper. (med.)	$37.45	$299.60	$61.45	$491.60	$32.48	$53.45
1 Laborer	27.50	220.00	45.45	363.60		
1 Grader, 30,000 Lbs.		657.85		723.63	41.12	45.23
16 L.H., Daily Totals		$1177.45		$1578.84	$73.59	$98.68
Crew B-11M	**Hr.**	**Daily**	**Hr.**	**Daily**	**Bare Costs**	**Incl. O&P**
1 Equipment Oper. (med.)	$37.45	$299.60	$61.45	$491.60	$32.48	$53.45
1 Laborer	27.50	220.00	45.45	363.60		
1 Backhoe Loader, 80 H.P.		387.85		426.63	24.24	26.66
16 L.H., Daily Totals		$907.45		$1281.84	$56.72	$80.11
Crew B-11W	**Hr.**	**Daily**	**Hr.**	**Daily**	**Bare Costs**	**Incl. O&P**
1 Equipment Operator (med.)	$37.45	$299.60	$61.45	$491.60	$32.83	$53.99
1 Common Laborer	27.50	220.00	45.45	363.60		
10 Truck Drivers (heavy)	32.90	2632.00	54.10	4328.00		
1 Dozer, 200 H.P.		1290.00		1419.00		
1 Vibratory Roller, Towed, 23 Ton		422.15		464.37		
10 Dump Trucks, 8 C.Y., 220 H.P.		3396.00		3735.60	53.21	58.53
96 L.H., Daily Totals		$8259.75		$10802.17	$86.04	$112.52
Crew B-11Y	**Hr.**	**Daily**	**Hr.**	**Daily**	**Bare Costs**	**Incl. O&P**
1 Labor Foreman (outside)	$29.50	$236.00	$48.75	$390.00	$31.04	$51.15
5 Common Laborers	27.50	1100.00	45.45	1818.00		
3 Equipment Operators (med.)	37.45	898.80	61.45	1474.80		
1 Dozer, 80 H.P.		472.00		519.20		
2 Rollers, 2-Drum, W.B., 7.5 H.P.		363.50		399.85		
4 Vibrating Plates, Gas, 21"		162.20		178.42	13.86	15.24
72 L.H., Daily Totals		$3232.50		$4780.27	$44.90	$66.39
Crew B-12A	**Hr.**	**Daily**	**Hr.**	**Daily**	**Bare Costs**	**Incl. O&P**
1 Equip. Oper. (crane)	$37.90	$303.20	$62.20	$497.60	$32.70	$53.83
1 Laborer	27.50	220.00	45.45	363.60		
1 Hyd. Excavator, 1 C.Y.		749.70		824.67	46.86	51.54
16 L.H., Daily Totals		$1272.90		$1685.87	$79.56	$105.37
Crew B-12B	**Hr.**	**Daily**	**Hr.**	**Daily**	**Bare Costs**	**Incl. O&P**
1 Equip. Oper. (crane)	$37.90	$303.20	$62.20	$497.60	$32.70	$53.83
1 Laborer	27.50	220.00	45.45	363.60		
1 Hyd. Excavator, 1.5 C.Y.		908.70		999.57	56.79	62.47
16 L.H., Daily Totals		$1431.90		$1860.77	$89.49	$116.30
Crew B-12C	**Hr.**	**Daily**	**Hr.**	**Daily**	**Bare Costs**	**Incl. O&P**
1 Equip. Oper. (crane)	$37.90	$303.20	$62.20	$497.60	$32.70	$53.83
1 Laborer	27.50	220.00	45.45	363.60		
1 Hyd. Excavator, 2 C.Y.		1078.00		1185.80	67.38	74.11
16 L.H., Daily Totals		$1601.20		$2047.00	$100.08	$127.94

Crew No.	Bare Costs		Incl. Subs O&P		Cost Per Labor-Hour	
Crew B-12D	**Hr.**	**Daily**	**Hr.**	**Daily**	**Bare Costs**	**Incl. O&P**
1 Equip. Oper. (crane)	$37.90	$303.20	$62.20	$497.60	$32.70	$53.83
1 Laborer	27.50	220.00	45.45	363.60		
1 Hyd. Excavator, 3.5 C.Y.		2256.00		2481.60	141.00	155.10
16 L.H., Daily Totals		$2779.20		$3342.80	$173.70	$208.93
Crew B-12E	**Hr.**	**Daily**	**Hr.**	**Daily**	**Bare Costs**	**Incl. O&P**
1 Equip. Oper. (crane)	$37.90	$303.20	$62.20	$497.60	$32.70	$53.83
1 Laborer	27.50	220.00	45.45	363.60		
1 Hyd. Excavator, .5 C.Y.		443.75		488.13	27.73	30.51
16 L.H., Daily Totals		$966.95		$1349.33	$60.43	$84.33
Crew B-12F	**Hr.**	**Daily**	**Hr.**	**Daily**	**Bare Costs**	**Incl. O&P**
1 Equip. Oper. (crane)	$37.90	$303.20	$62.20	$497.60	$32.70	$53.83
1 Laborer	27.50	220.00	45.45	363.60		
1 Hyd. Excavator, .75 C.Y.		681.25		749.38	42.58	46.84
16 L.H., Daily Totals		$1204.45		$1610.58	$75.28	$100.66
Crew B-12G	**Hr.**	**Daily**	**Hr.**	**Daily**	**Bare Costs**	**Incl. O&P**
1 Equip. Oper. (crane)	$37.90	$303.20	$62.20	$497.60	$32.70	$53.83
1 Laborer	27.50	220.00	45.45	363.60		
1 Crawler Crane, 15 Ton		866.60		953.26		
1 Clamshell Bucket, .5 C.Y.		42.55		46.81	56.82	62.50
16 L.H., Daily Totals		$1432.35		$1861.27	$89.52	$116.33
Crew B-12H	**Hr.**	**Daily**	**Hr.**	**Daily**	**Bare Costs**	**Incl. O&P**
1 Equip. Oper. (crane)	$37.90	$303.20	$62.20	$497.60	$32.70	$53.83
1 Laborer	27.50	220.00	45.45	363.60		
1 Crawler Crane, 25 Ton		1386.00		1524.60		
1 Clamshell Bucket, 1 C.Y.		51.20		56.32	89.83	98.81
16 L.H., Daily Totals		$1960.40		$2442.12	$122.53	$152.63
Crew B-12I	**Hr.**	**Daily**	**Hr.**	**Daily**	**Bare Costs**	**Incl. O&P**
1 Equip. Oper. (crane)	$37.90	$303.20	$62.20	$497.60	$32.70	$53.83
1 Laborer	27.50	220.00	45.45	363.60		
1 Crawler Crane, 20 Ton		1130.00		1243.00		
1 Dragline Bucket, .75 C.Y.		21.85		24.04	71.99	79.19
16 L.H., Daily Totals		$1675.05		$2128.24	$104.69	$133.01
Crew B-12J	**Hr.**	**Daily**	**Hr.**	**Daily**	**Bare Costs**	**Incl. O&P**
1 Equip. Oper. (crane)	$37.90	$303.20	$62.20	$497.60	$32.70	$53.83
1 Laborer	27.50	220.00	45.45	363.60		
1 Gradall, 5/8 C.Y.		862.55		948.80	53.91	59.30
16 L.H., Daily Totals		$1385.75		$1810.01	$86.61	$113.13
Crew B-12K	**Hr.**	**Daily**	**Hr.**	**Daily**	**Bare Costs**	**Incl. O&P**
1 Equip. Oper. (crane)	$37.90	$303.20	$62.20	$497.60	$32.70	$53.83
1 Laborer	27.50	220.00	45.45	363.60		
1 Gradall, 3 Ton, 1 C.Y.		1185.00		1303.50	74.06	81.47
16 L.H., Daily Totals		$1708.20		$2164.70	$106.76	$135.29
Crew B-12L	**Hr.**	**Daily**	**Hr.**	**Daily**	**Bare Costs**	**Incl. O&P**
1 Equip. Oper. (crane)	$37.90	$303.20	$62.20	$497.60	$32.70	$53.83
1 Laborer	27.50	220.00	45.45	363.60		
1 Crawler Crane, 15 Ton		866.60		953.26		
1 F.E. Attachment, .5 C.Y.		63.95		70.34	58.16	63.98
16 L.H., Daily Totals		$1453.75		$1884.81	$90.86	$117.80

Crew No.	Bare Costs		Incl. Subs O&P		Cost Per Labor-Hour	
Crew B-12M	**Hr.**	**Daily**	**Hr.**	**Daily**	**Bare Costs**	**Incl. O&P**
1 Equip. Oper. (crane)	$37.90	$303.20	$62.20	$497.60	$32.70	$53.83
1 Laborer	27.50	220.00	45.45	363.60		
1 Crawler Crane, 20 Ton		1130.00		1243.00		
1 F.E. Attachment, .75 C.Y.		68.90		75.79	74.93	82.42
16 L.H., Daily Totals		$1722.10		$2179.99	$107.63	$136.25
Crew B-12N	**Hr.**	**Daily**	**Hr.**	**Daily**	**Bare Costs**	**Incl. O&P**
1 Equip. Oper. (crane)	$37.90	$303.20	$62.20	$497.60	$32.70	$53.83
1 Laborer	27.50	220.00	45.45	363.60		
1 Crawler Crane, 25 Ton		1386.00		1524.60		
1 F.E. Attachment, 1 C.Y.		74.85		82.33	91.30	100.43
16 L.H., Daily Totals		$1984.05		$2468.14	$124.00	$154.26
Crew B-12O	**Hr.**	**Daily**	**Hr.**	**Daily**	**Bare Costs**	**Incl. O&P**
1 Equip. Oper. (crane)	$37.90	$303.20	$62.20	$497.60	$32.70	$53.83
1 Laborer	27.50	220.00	45.45	363.60		
1 Crawler Crane, 40 Ton		1392.00		1531.20		
1 F.E. Attachment, 1.5 C.Y.		85.65		94.22	92.35	101.59
16 L.H., Daily Totals		$2000.85		$2486.61	$125.05	$155.41
Crew B-12P	**Hr.**	**Daily**	**Hr.**	**Daily**	**Bare Costs**	**Incl. O&P**
1 Equip. Oper. (crane)	$37.90	$303.20	$62.20	$497.60	$32.70	$53.83
1 Laborer	27.50	220.00	45.45	363.60		
1 Crawler Crane, 40 Ton		1392.00		1531.20		
1 Dragline Bucket, 1.5 C.Y.		35.10		38.61	89.19	98.11
16 L.H., Daily Totals		$1950.30		$2431.01	$121.89	$151.94
Crew B-12Q	**Hr.**	**Daily**	**Hr.**	**Daily**	**Bare Costs**	**Incl. O&P**
1 Equip. Oper. (crane)	$37.90	$303.20	$62.20	$497.60	$32.70	$53.83
1 Laborer	27.50	220.00	45.45	363.60		
1 Hyd. Excavator, 5/8 C.Y.		587.30		646.03	36.71	40.38
16 L.H., Daily Totals		$1110.50		$1507.23	$69.41	$94.20
Crew B-12S	**Hr.**	**Daily**	**Hr.**	**Daily**	**Bare Costs**	**Incl. O&P**
1 Equip. Oper. (crane)	$37.90	$303.20	$62.20	$497.60	$32.70	$53.83
1 Laborer	27.50	220.00	45.45	363.60		
1 Hyd. Excavator, 2.5 C.Y.		1441.00		1585.10	90.06	99.07
16 L.H., Daily Totals		$1964.20		$2446.30	$122.76	$152.89
Crew B-12T	**Hr.**	**Daily**	**Hr.**	**Daily**	**Bare Costs**	**Incl. O&P**
1 Equip. Oper. (crane)	$37.90	$303.20	$62.20	$497.60	$32.70	$53.83
1 Laborer	27.50	220.00	45.45	363.60		
1 Crawler Crane, 75 Ton		1734.00		1907.40		
1 F.E. Attachment, 3 C.Y.		111.50		122.65	115.34	126.88
16 L.H., Daily Totals		$2368.70		$2891.25	$148.04	$180.70
Crew B-12V	**Hr.**	**Daily**	**Hr.**	**Daily**	**Bare Costs**	**Incl. O&P**
1 Equip. Oper. (crane)	$37.90	$303.20	$62.20	$497.60	$32.70	$53.83
1 Laborer	27.50	220.00	45.45	363.60		
1 Crawler Crane, 75 Ton		1734.00		1907.40		
1 Dragline Bucket, 3 C.Y.		56.25		61.88	111.89	123.08
16 L.H., Daily Totals		$2313.45		$2830.47	$144.59	$176.90
Crew B-12Y	**Hr.**	**Daily**	**Hr.**	**Daily**	**Bare Costs**	**Incl. O&P**
1 Equip. Oper. (crane)	$37.90	$303.20	$62.20	$497.60	$30.97	$51.03
2 Laborers	27.50	440.00	45.45	727.20		
1 Hyd. Excavator, 3.5 C.Y.		2256.00		2481.60	94.00	103.40
24 L.H., Daily Totals		$2999.20		$3706.40	$124.97	$154.43

Crew No.	Bare Costs		Incl. Subs O&P		Cost Per Labor-Hour	
Crew B-12Z	**Hr.**	**Daily**	**Hr.**	**Daily**	**Bare Costs**	**Incl. O&P**
1 Equip. Oper. (crane)	$37.90	$303.20	$62.20	$497.60	$30.97	$51.03
2 Laborers	27.50	440.00	45.45	727.20		
1 Hyd. Excavator, 2.5 C.Y.		1441.00		1585.10	60.04	66.05
24 L.H., Daily Totals		$2184.20		$2809.90	$91.01	$117.08
Crew B-13	**Hr.**	**Daily**	**Hr.**	**Daily**	**Bare Costs**	**Incl. O&P**
1 Labor Foreman (outside)	$29.50	$236.00	$48.75	$390.00	$29.57	$48.79
4 Laborers	27.50	880.00	45.45	1454.40		
1 Equip. Oper. (crane)	37.90	303.20	62.20	497.60		
1 Hyd. Crane, 25 Ton		581.70		639.87	12.12	13.33
48 L.H., Daily Totals		$2000.90		$2981.87	$41.69	$62.12
Crew B-13A	**Hr.**	**Daily**	**Hr.**	**Daily**	**Bare Costs**	**Incl. O&P**
1 Labor Foreman (outside)	$29.50	$236.00	$48.75	$390.00	$32.17	$52.96
2 Laborers	27.50	440.00	45.45	727.20		
2 Equipment Operators (med.)	37.45	599.20	61.45	983.20		
2 Truck Drivers (heavy)	32.90	526.40	54.10	865.60		
1 Crawler Crane, 75 Ton		1734.00		1907.40		
1 Crawler Loader, 4 C.Y.		1461.00		1607.10		
2 Dump Trucks, 8 C.Y., 220 H.P.		679.20		747.12	69.18	76.10
56 L.H., Daily Totals		$5675.80		$7227.62	$101.35	$129.06
Crew B-13B	**Hr.**	**Daily**	**Hr.**	**Daily**	**Bare Costs**	**Incl. O&P**
1 Labor Foreman (outside)	$29.50	$236.00	$48.75	$390.00	$30.05	$49.54
4 Laborers	27.50	880.00	45.45	1454.40		
1 Equip. Oper. (crane)	37.90	303.20	62.20	497.60		
1 Equip. Oper. (oiler)	32.95	263.60	54.05	432.40		
1 Hyd. Crane, 55 Ton		981.50		1079.65	17.53	19.28
56 L.H., Daily Totals		$2664.30		$3854.05	$47.58	$68.82
Crew B-13C	**Hr.**	**Daily**	**Hr.**	**Daily**	**Bare Costs**	**Incl. O&P**
1 Labor Foreman (outside)	$29.50	$236.00	$48.75	$390.00	$30.05	$49.54
4 Laborers	27.50	880.00	45.45	1454.40		
1 Equip. Oper. (crane)	37.90	303.20	62.20	497.60		
1 Equip. Oper. (oiler)	32.95	263.60	54.05	432.40		
1 Crawler Crane, 100 Ton		1879.00		2066.90	33.55	36.91
56 L.H., Daily Totals		$3561.80		$4841.30	$63.60	$86.45
Crew B-13D	**Hr.**	**Daily**	**Hr.**	**Daily**	**Bare Costs**	**Incl. O&P**
1 Laborer	$27.50	$220.00	$45.45	$363.60	$32.70	$53.83
1 Equip. Oper. (crane)	37.90	303.20	62.20	497.60		
1 Hyd. Excavator, 1 C.Y.		749.70		824.67		
1 Trench Box		77.75		85.53	51.72	56.89
16 L.H., Daily Totals		$1350.65		$1771.40	$84.42	$110.71
Crew B-13E	**Hr.**	**Daily**	**Hr.**	**Daily**	**Bare Costs**	**Incl. O&P**
1 Laborer	$27.50	$220.00	$45.45	$363.60	$32.70	$53.83
1 Equip. Oper. (crane)	37.90	303.20	62.20	497.60		
1 Hyd. Excavator, 1.5 C.Y.		908.70		999.57		
1 Trench Box		77.75		85.53	61.65	67.82
16 L.H., Daily Totals		$1509.65		$1946.30	$94.35	$121.64
Crew B-13F	**Hr.**	**Daily**	**Hr.**	**Daily**	**Bare Costs**	**Incl. O&P**
1 Laborer	$27.50	$220.00	$45.45	$363.60	$32.70	$53.83
1 Equip. Oper. (crane)	37.90	303.20	62.20	497.60		
1 Hyd. Excavator, 3.5 C.Y.		2256.00		2481.60		
1 Trench Box		77.75		85.53	145.86	160.45
16 L.H., Daily Totals		$2856.95		$3428.32	$178.56	$214.27

Crew B-13G	Bare Costs Hr.	Bare Costs Daily	Incl. Subs O&P Hr.	Incl. Subs O&P Daily	Cost Per Labor-Hour Bare Costs	Cost Per Labor-Hour Incl. O&P
1 Laborer	$27.50	$220.00	$45.45	$363.60	$32.70	$53.83
1 Equip. Oper. (crane)	37.90	303.20	62.20	497.60		
1 Hyd. Excavator, .75 C.Y.		681.25		749.38		
1 Trench Box		77.75		85.53	47.44	52.18
16 L.H., Daily Totals		$1282.20		$1696.10	$80.14	$106.01

Crew B-13H	Hr.	Daily	Hr.	Daily	Bare Costs	Incl. O&P
1 Laborer	$27.50	$220.00	$45.45	$363.60	$32.70	$53.83
1 Equip. Oper. (crane)	37.90	303.20	62.20	497.60		
1 Gradall, 5/8 C.Y.		862.55		948.80		
1 Trench Box		77.75		85.53	58.77	64.65
16 L.H., Daily Totals		$1463.50		$1895.53	$91.47	$118.47

Crew B-13I	Hr.	Daily	Hr.	Daily	Bare Costs	Incl. O&P
1 Laborer	$27.50	$220.00	$45.45	$363.60	$32.70	$53.83
1 Equip. Oper. (crane)	37.90	303.20	62.20	497.60		
1 Gradall, 3 Ton, 1 C.Y.		1185.00		1303.50		
1 Trench Box		77.75		85.53	78.92	86.81
16 L.H., Daily Totals		$1785.95		$2250.22	$111.62	$140.64

Crew B-13J	Hr.	Daily	Hr.	Daily	Bare Costs	Incl. O&P
1 Laborer	$27.50	$220.00	$45.45	$363.60	$32.70	$53.83
1 Equip. Oper. (crane)	37.90	303.20	62.20	497.60		
1 Hyd. Excavator, 2.5 C.Y.		1441.00		1585.10		
1 Trench Box		77.75		85.53	94.92	104.41
16 L.H., Daily Totals		$2041.95		$2531.82	$127.62	$158.24

Crew B-13K	Hr.	Daily	Hr.	Daily	Bare Costs	Incl. O&P
2 Equip. Opers. (crane)	$37.90	$606.40	$62.20	$995.20	$37.90	$62.20
1 Hyd. Excavator, .75 C.Y.		681.25		749.38		
1 Hyd. Hammer, 4000 ft-lb		332.85		366.13		
1 Hyd. Excavator, .75 C.Y.		681.25		749.38	105.96	116.56
16 L.H., Daily Totals		$2301.75		$2860.09	$143.86	$178.76

Crew B-13L	Hr.	Daily	Hr.	Daily	Bare Costs	Incl. O&P
2 Equip. Opers. (crane)	$37.90	$606.40	$62.20	$995.20	$37.90	$62.20
1 Hyd. Excavator, 1.5 C.Y.		908.70		999.57		
1 Hyd. Hammer, 5000 ft-lb		407.60		448.36		
1 Hyd. Excavator, .75 C.Y.		681.25		749.38	124.85	137.33
16 L.H., Daily Totals		$2603.95		$3192.51	$162.75	$199.53

Crew B-13M	Hr.	Daily	Hr.	Daily	Bare Costs	Incl. O&P
2 Equip. Opers. (crane)	$37.90	$606.40	$62.20	$995.20	$37.90	$62.20
1 Hyd. Excavator, 2.5 C.Y.		1441.00		1585.10		
1 Hyd. Hammer, 8000 ft-lb		593.35		652.68		
1 Hyd. Excavator, 1.5 C.Y.		908.70		999.57	183.94	202.33
16 L.H., Daily Totals		$3549.45		$4232.56	$221.84	$264.53

Crew B-13N	Hr.	Daily	Hr.	Daily	Bare Costs	Incl. O&P
2 Equip. Opers. (crane)	$37.90	$606.40	$62.20	$995.20	$37.90	$62.20
1 Hyd. Excavator, 3.5 C.Y.		2256.00		2481.60		
1 Hyd. Hammer, 12,000 ft-lb		685.75		754.33		
1 Hyd. Excavator, 1.5 C.Y.		908.70		999.57	240.65	264.72
16 L.H., Daily Totals		$4456.85		$5230.69	$278.55	$326.92

Crew B-14	Bare Costs Hr.	Bare Costs Daily	Incl. Subs O&P Hr.	Incl. Subs O&P Daily	Cost Per Labor-Hour Bare Costs	Cost Per Labor-Hour Incl. O&P
1 Labor Foreman (outside)	$29.50	$236.00	$48.75	$390.00	$29.19	$48.17
4 Laborers	27.50	880.00	45.45	1454.40		
1 Equip. Oper. (light)	35.65	285.20	58.50	468.00		
1 Backhoe Loader, 48 H.P.		320.20		352.22	6.67	7.34
48 L.H., Daily Totals		$1721.40		$2664.62	$35.86	$55.51

Crew B-14A	Hr.	Daily	Hr.	Daily	Bare Costs	Incl. O&P
1 Equip. Oper. (crane)	$37.90	$303.20	$62.20	$497.60	$34.43	$56.62
.5 Laborer	27.50	110.00	45.45	181.80		
1 Hyd. Excavator, 4.5 C.Y.		2868.00		3154.80	239.00	262.90
12 L.H., Daily Totals		$3281.20		$3834.20	$273.43	$319.52

Crew B-14B	Hr.	Daily	Hr.	Daily	Bare Costs	Incl. O&P
1 Equip. Oper. (crane)	$37.90	$303.20	$62.20	$497.60	$34.43	$56.62
.5 Laborer	27.50	110.00	45.45	181.80		
1 Hyd. Excavator, 6 C.Y.		3579.00		3936.90	298.25	328.07
12 L.H., Daily Totals		$3992.20		$4616.30	$332.68	$384.69

Crew B-14C	Hr.	Daily	Hr.	Daily	Bare Costs	Incl. O&P
1 Equip. Oper. (crane)	$37.90	$303.20	$62.20	$497.60	$34.43	$56.62
.5 Laborer	27.50	110.00	45.45	181.80		
1 Hyd. Excavator, 7 C.Y.		3332.00		3665.20	277.67	305.43
12 L.H., Daily Totals		$3745.20		$4344.60	$312.10	$362.05

Crew B-14F	Hr.	Daily	Hr.	Daily	Bare Costs	Incl. O&P
1 Equip. Oper. (crane)	$37.90	$303.20	$62.20	$497.60	$34.43	$56.62
.5 Laborer	27.50	110.00	45.45	181.80		
1 Hyd. Shovel, 7 C.Y.		4029.00		4431.90	335.75	369.32
12 L.H., Daily Totals		$4442.20		$5111.30	$370.18	$425.94

Crew B-14G	Hr.	Daily	Hr.	Daily	Bare Costs	Incl. O&P
1 Equip. Oper. (crane)	$37.90	$303.20	$62.20	$497.60	$34.43	$56.62
.5 Laborer	27.50	110.00	45.45	181.80		
1 Hyd. Shovel, 12 C.Y.		5855.00		6440.50	487.92	536.71
12 L.H., Daily Totals		$6268.20		$7119.90	$522.35	$593.33

Crew B-14J	Hr.	Daily	Hr.	Daily	Bare Costs	Incl. O&P
1 Equip. Oper. (medium)	$37.45	$299.60	$61.45	$491.60	$34.13	$56.12
.5 Laborer	27.50	110.00	45.45	181.80		
1 F.E. Loader, 8 C.Y.		1784.00		1962.40	148.67	163.53
12 L.H., Daily Totals		$2193.60		$2635.80	$182.80	$219.65

Crew B-14K	Hr.	Daily	Hr.	Daily	Bare Costs	Incl. O&P
1 Equip. Oper. (medium)	$37.45	$299.60	$61.45	$491.60	$34.13	$56.12
.5 Laborer	27.50	110.00	45.45	181.80		
1 F.E. Loader, 10 C.Y.		2625.00		2887.50	218.75	240.63
12 L.H., Daily Totals		$3034.60		$3560.90	$252.88	$296.74

Crew B-15	Hr.	Daily	Hr.	Daily	Bare Costs	Incl. O&P
1 Equipment Oper. (med.)	$37.45	$299.60	$61.45	$491.60	$33.43	$54.96
.5 Laborer	27.50	110.00	45.45	181.80		
2 Truck Drivers (heavy)	32.90	526.40	54.10	865.60		
2 Dump Trucks, 12 C.Y., 400 H.P.		1134.10		1247.51		
1 Dozer, 200 H.P.		1290.00		1419.00	86.58	95.23
28 L.H., Daily Totals		$3360.10		$4205.51	$120.00	$150.20

Crew No.	Bare Costs		Incl. Subs O&P		Cost Per Labor-Hour	
Crew B-16	**Hr.**	**Daily**	**Hr.**	**Daily**	**Bare Costs**	**Incl. O&P**
1 Labor Foreman (outside)	$29.50	$236.00	$48.75	$390.00	$29.35	$48.44
2 Laborers	27.50	440.00	45.45	727.20		
1 Truck Driver (heavy)	32.90	263.20	54.10	432.80		
1 Dump Truck, 12 C.Y., 400 H.P.		567.05		623.76	17.72	19.49
32 L.H., Daily Totals		$1506.25		$2173.76	$47.07	$67.93
Crew B-17	**Hr.**	**Daily**	**Hr.**	**Daily**	**Bare Costs**	**Incl. O&P**
2 Laborers	$27.50	$440.00	$45.45	$727.20	$30.89	$50.88
1 Equip. Oper. (light)	35.65	285.20	58.50	468.00		
1 Truck Driver (heavy)	32.90	263.20	54.10	432.80		
1 Backhoe Loader, 48 H.P.		320.20		352.22		
1 Dump Truck, 8 C.Y., 220 H.P.		339.60		373.56	20.62	22.68
32 L.H., Daily Totals		$1648.20		$2353.78	$51.51	$73.56
Crew B-17A	**Hr.**	**Daily**	**Hr.**	**Daily**	**Bare Costs**	**Incl. O&P**
2 Labor Foremen (outside)	$29.50	$472.00	$48.75	$780.00	$29.90	$49.48
6 Laborers	27.50	1320.00	45.45	2181.60		
1 Skilled Worker Foreman (out)	38.50	308.00	63.95	511.60		
1 Skilled Worker	36.50	292.00	60.65	485.20		
80 L.H., Daily Totals		$2392.00		$3958.40	$29.90	$49.48
Crew B-17B	**Hr.**	**Daily**	**Hr.**	**Daily**	**Bare Costs**	**Incl. O&P**
2 Laborers	$27.50	$440.00	$45.45	$727.20	$30.89	$50.88
1 Equip. Oper. (light)	35.65	285.20	58.50	468.00		
1 Truck Driver (heavy)	32.90	263.20	54.10	432.80		
1 Backhoe Loader, 48 H.P.		320.20		352.22		
1 Dump Truck, 12 C.Y., 400 H.P.		567.05		623.76	27.73	30.50
32 L.H., Daily Totals		$1875.65		$2603.97	$58.61	$81.37
Crew B-18	**Hr.**	**Daily**	**Hr.**	**Daily**	**Bare Costs**	**Incl. O&P**
1 Labor Foreman (outside)	$29.50	$236.00	$48.75	$390.00	$28.17	$46.55
2 Laborers	27.50	440.00	45.45	727.20		
1 Vibrating Plate, Gas, 21"		40.55		44.60	1.69	1.86
24 L.H., Daily Totals		$716.55		$1161.81	$29.86	$48.41
Crew B-19	**Hr.**	**Daily**	**Hr.**	**Daily**	**Bare Costs**	**Incl. O&P**
1 Pile Driver Foreman (outside)	$38.00	$304.00	$64.75	$518.00	$35.34	$59.69
4 Pile Drivers	36.00	1152.00	61.35	1963.20		
1 Equip. Oper. (crane)	37.90	303.20	62.20	497.60		
1 Building Laborer	27.50	220.00	45.45	363.60		
1 Crawler Crane, 40 Ton		1392.00		1531.20		
1 Lead, 90' High		134.60		148.06		
1 Hammer, Diesel, 22k ft-lb		427.55		470.31	34.90	38.39
56 L.H., Daily Totals		$3933.35		$5491.97	$70.24	$98.07
Crew B-19A	**Hr.**	**Daily**	**Hr.**	**Daily**	**Bare Costs**	**Incl. O&P**
1 Pile Driver Foreman (outside)	$38.00	$304.00	$64.75	$518.00	$35.34	$59.69
4 Pile Drivers	36.00	1152.00	61.35	1963.20		
1 Equip. Oper. (crane)	37.90	303.20	62.20	497.60		
1 Common Laborer	27.50	220.00	45.45	363.60		
1 Crawler Crane, 75 Ton		1734.00		1907.40		
1 Lead, 90' High		134.60		148.06		
1 Hammer, Diesel, 41k ft-lb		565.45		622.00	43.47	47.81
56 L.H., Daily Totals		$4413.25		$6019.85	$78.81	$107.50

Crew No.	Bare Costs		Incl. Subs O&P		Cost Per Labor-Hour	
Crew B-19B	**Hr.**	**Daily**	**Hr.**	**Daily**	**Bare Costs**	**Incl. O&P**
1 Pile Driver Foreman (outside)	$38.00	$304.00	$64.75	$518.00	$35.34	$59.69
4 Pile Drivers	36.00	1152.00	61.35	1963.20		
1 Equip. Oper. (crane)	37.90	303.20	62.20	497.60		
1 Common Laborer	27.50	220.00	45.45	363.60		
1 Crawler Crane, 40 Ton		1392.00		1531.20		
1 Lead, 90' High		134.60		148.06		
1 Hammer, Diesel, 22k ft-lb		427.55		470.31		
1 Barge, 400 Ton		836.50		920.15	49.83	54.82
56 L.H., Daily Totals		$4769.85		$6412.11	$85.18	$114.50
Crew B-19C	**Hr.**	**Daily**	**Hr.**	**Daily**	**Bare Costs**	**Incl. O&P**
1 Pile Driver Foreman (outside)	$38.00	$304.00	$64.75	$518.00	$35.34	$59.69
4 Pile Drivers	36.00	1152.00	61.35	1963.20		
1 Equip. Oper. (crane)	37.90	303.20	62.20	497.60		
1 Common Laborer	27.50	220.00	45.45	363.60		
1 Crawler Crane, 75 Ton		1734.00		1907.40		
1 Lead, 90' High		134.60		148.06		
1 Hammer, Diesel, 41k ft-lb		565.45		622.00		
1 Barge, 400 Ton		836.50		920.15	58.40	64.24
56 L.H., Daily Totals		$5249.75		$6940.01	$93.75	$123.93
Crew B-20	**Hr.**	**Daily**	**Hr.**	**Daily**	**Bare Costs**	**Incl. O&P**
1 Labor Foreman (outside)	$29.50	$236.00	$48.75	$390.00	$28.17	$46.55
2 Laborers	27.50	440.00	45.45	727.20		
24 L.H., Daily Totals		$676.00		$1117.20	$28.17	$46.55
Crew B-20A	**Hr.**	**Daily**	**Hr.**	**Daily**	**Bare Costs**	**Incl. O&P**
1 Labor Foreman (outside)	$29.50	$236.00	$48.75	$390.00	$32.42	$53.38
1 Laborer	27.50	220.00	45.45	363.60		
1 Plumber	40.40	323.20	66.30	530.40		
1 Plumber Apprentice	32.30	258.40	53.00	424.00		
32 L.H., Daily Totals		$1037.60		$1708.00	$32.42	$53.38
Crew B-21	**Hr.**	**Daily**	**Hr.**	**Daily**	**Bare Costs**	**Incl. O&P**
1 Labor Foreman (outside)	$29.50	$236.00	$48.75	$390.00	$29.56	$48.79
2 Laborers	27.50	440.00	45.45	727.20		
.5 Equip. Oper. (crane)	37.90	151.60	62.20	248.80		
.5 S.P. Crane, 4x4, 5 Ton		129.72		142.70	4.63	5.10
28 L.H., Daily Totals		$957.33		$1508.70	$34.19	$53.88
Crew B-21A	**Hr.**	**Daily**	**Hr.**	**Daily**	**Bare Costs**	**Incl. O&P**
1 Labor Foreman (outside)	$29.50	$236.00	$48.75	$390.00	$33.52	$55.14
1 Laborer	27.50	220.00	45.45	363.60		
1 Plumber	40.40	323.20	66.30	530.40		
1 Plumber Apprentice	32.30	258.40	53.00	424.00		
1 Equip. Oper. (crane)	37.90	303.20	62.20	497.60		
1 S.P. Crane, 4x4, 12 Ton		371.85		409.04	9.30	10.23
40 L.H., Daily Totals		$1712.65		$2614.64	$42.82	$65.37
Crew B-21B	**Hr.**	**Daily**	**Hr.**	**Daily**	**Bare Costs**	**Incl. O&P**
1 Labor Foreman (outside)	$29.50	$236.00	$48.75	$390.00	$29.98	$49.46
3 Laborers	27.50	660.00	45.45	1090.80		
1 Equip. Oper. (crane)	37.90	303.20	62.20	497.60		
1 Hyd. Crane, 12 Ton		471.75		518.92	11.79	12.97
40 L.H., Daily Totals		$1670.95		$2497.32	$41.77	$62.43

Crew No.	Bare Costs		Incl. Subs O&P		Cost Per Labor-Hour	
Crew B-21C	**Hr.**	**Daily**	**Hr.**	**Daily**	**Bare Costs**	**Incl. O&P**
1 Labor Foreman (outside)	$29.50	$236.00	$48.75	$390.00	$30.05	$49.54
4 Laborers	27.50	880.00	45.45	1454.40		
1 Equip. Oper. (crane)	37.90	303.20	62.20	497.60		
1 Equip. Oper. (oiler)	32.95	263.60	54.05	432.40		
2 Cutting Torches		25.00		27.50		
2 Sets of Gases		335.70		369.27		
1 Lattice Boom Crane, 90 Ton		1698.00		1867.80	36.76	40.44
56 L.H., Daily Totals		$3741.50		$5038.97	$66.81	$89.98

Crew B-22	Hr.	Daily	Hr.	Daily	Bare Costs	Incl. O&P
1 Labor Foreman (outside)	$29.50	$236.00	$48.75	$390.00	$30.11	$49.68
2 Laborers	27.50	440.00	45.45	727.20		
.75 Equip. Oper. (crane)	37.90	227.40	62.20	373.20		
.75 S.P. Crane, 4x4, 5 Ton		194.59		214.05	6.49	7.13
30 L.H., Daily Totals		$1097.99		$1704.45	$36.60	$56.81

Crew B-22A	Hr.	Daily	Hr.	Daily	Bare Costs	Incl. O&P
1 Labor Foreman (outside)	$29.50	$236.00	$48.75	$390.00	$31.78	$52.50
1 Skilled Worker	36.50	292.00	60.65	485.20		
2 Laborers	27.50	440.00	45.45	727.20		
1 Equipment Operator, Crane	37.90	303.20	62.20	497.60		
1 S.P. Crane, 4x4, 5 Ton		259.45		285.39		
1 Butt Fusion Mach., 4"-12" diam.		405.15		445.67	16.61	18.28
40 L.H., Daily Totals		$1935.80		$2831.06	$48.40	$70.78

Crew B-22B	Hr.	Daily	Hr.	Daily	Bare Costs	Incl. O&P
1 Labor Foreman (outside)	$29.50	$236.00	$48.75	$390.00	$31.78	$52.50
1 Skilled Worker	36.50	292.00	60.65	485.20		
2 Laborers	27.50	440.00	45.45	727.20		
1 Equip. Oper. (crane)	37.90	303.20	62.20	497.60		
1 S.P. Crane, 4x4, 5 Ton		259.45		285.39		
1 Butt Fusion Mach., 8"-24" diam.		1927.00		2119.70	54.66	60.13
40 L.H., Daily Totals		$3457.65		$4505.10	$86.44	$112.63

Crew B-22C	Hr.	Daily	Hr.	Daily	Bare Costs	Incl. O&P
1 Skilled Worker	$36.50	$292.00	$60.65	$485.20	$32.00	$53.05
1 Laborer	27.50	220.00	45.45	363.60		
1 Butt Fusion Mach., 2"-8" diam.		143.10		157.41	8.94	9.84
16 L.H., Daily Totals		$655.10		$1006.21	$40.94	$62.89

Crew B-23	Hr.	Daily	Hr.	Daily	Bare Costs	Incl. O&P
1 Labor Foreman (outside)	$29.50	$236.00	$48.75	$390.00	$27.90	$46.11
4 Laborers	27.50	880.00	45.45	1454.40		
1 Drill Rig, Truck-Mounted		2428.00		2670.80		
1 Flatbed Truck, Gas, 3 Ton		245.95		270.55	66.85	73.53
40 L.H., Daily Totals		$3789.95		$4785.74	$94.75	$119.64

Crew B-23A	Hr.	Daily	Hr.	Daily	Bare Costs	Incl. O&P
1 Labor Foreman (outside)	$29.50	$236.00	$48.75	$390.00	$31.48	$51.88
1 Laborer	27.50	220.00	45.45	363.60		
1 Equip. Oper. (medium)	37.45	299.60	61.45	491.60		
1 Drill Rig, Truck-Mounted		2428.00		2670.80		
1 Pickup Truck, 3/4 Ton		109.90		120.89	105.75	116.32
24 L.H., Daily Totals		$3293.50		$4036.89	$137.23	$168.20

Crew B-23B	Hr.	Daily	Hr.	Daily	Bare Costs	Incl. O&P
1 Labor Foreman (outside)	$29.50	$236.00	$48.75	$390.00	$31.48	$51.88
1 Laborer	27.50	220.00	45.45	363.60		
1 Equip. Oper. (medium)	37.45	299.60	61.45	491.60		
1 Drill Rig, Truck-Mounted		2428.00		2670.80		
1 Pickup Truck, 3/4 Ton		109.90		120.89		
1 Centr. Water Pump, 6"		296.45		326.10	118.10	129.91
24 L.H., Daily Totals		$3589.95		$4362.98	$149.58	$181.79

Crew B-24	Hr.	Daily	Hr.	Daily	Bare Costs	Incl. O&P
1 Cement Finisher	$35.70	$285.60	$57.70	$461.60	$32.95	$54.02
1 Laborer	27.50	220.00	45.45	363.60		
1 Carpenter	35.65	285.20	58.90	471.20		
24 L.H., Daily Totals		$790.80		$1296.40	$32.95	$54.02

Crew B-25	Hr.	Daily	Hr.	Daily	Bare Costs	Incl. O&P
1 Labor Foreman (outside)	$29.50	$236.00	$48.75	$390.00	$30.40	$50.11
7 Laborers	27.50	1540.00	45.45	2545.20		
3 Equip. Oper. (medium)	37.45	898.80	61.45	1474.80		
1 Asphalt Paver, 130 H.P.		2076.00		2283.60		
1 Tandem Roller, 10 Ton		232.05		255.26		
1 Roller, Pneum. Whl., 12 Ton		338.35		372.19	30.07	33.08
88 L.H., Daily Totals		$5321.20		$7321.04	$60.47	$83.19

Crew B-25B	Hr.	Daily	Hr.	Daily	Bare Costs	Incl. O&P
1 Labor Foreman (outside)	$29.50	$236.00	$48.75	$390.00	$30.98	$51.06
7 Laborers	27.50	1540.00	45.45	2545.20		
4 Equip. Oper. (medium)	37.45	1198.40	61.45	1966.40		
1 Asphalt Paver, 130 H.P.		2076.00		2283.60		
2 Tandem Rollers, 10 Ton		464.10		510.51		
1 Roller, Pneum. Whl., 12 Ton		338.35		372.19	29.98	32.98
96 L.H., Daily Totals		$5852.85		$8067.90	$60.97	$84.04

Crew B-25C	Hr.	Daily	Hr.	Daily	Bare Costs	Incl. O&P
1 Labor Foreman (outside)	$29.50	$236.00	$48.75	$390.00	$31.15	$51.33
3 Laborers	27.50	660.00	45.45	1090.80		
2 Equip. Oper. (medium)	37.45	599.20	61.45	983.20		
1 Asphalt Paver, 130 H.P.		2076.00		2283.60		
1 Tandem Roller, 10 Ton		232.05		255.26	48.08	52.89
48 L.H., Daily Totals		$3803.25		$5002.85	$79.23	$104.23

Crew B-25D	Hr.	Daily	Hr.	Daily	Bare Costs	Incl. O&P
1 Labor Foreman (outside)	$29.50	$236.00	$48.75	$390.00	$31.31	$51.59
3 Laborers	27.50	660.00	45.45	1090.80		
2.125 Equip. Oper. (medium)	37.45	636.65	61.45	1044.65		
.125 Truck Driver (heavy)	32.90	32.90	54.10	54.10		
.125 Truck Tractor, 6x4, 380 H.P.		60.95		67.05		
.125 Dist. Tanker, 3000 Gallon		40.41		44.45		
1 Asphalt Paver, 130 H.P.		2076.00		2283.60		
1 Tandem Roller, 10 Ton		232.05		255.26	48.19	53.01
50 L.H., Daily Totals		$3974.96		$5229.90	$79.50	$104.60

Crew B-25E	Hr.	Daily	Hr.	Daily	Bare Costs	Incl. O&P
1 Labor Foreman (outside)	$29.50	$236.00	$48.75	$390.00	$31.46	$51.83
3 Laborers	27.50	660.00	45.45	1090.80		
2.250 Equip. Oper. (medium)	37.45	674.10	61.45	1106.10		
.25 Truck Driver (heavy)	32.90	65.80	54.10	108.20		
.25 Truck Tractor, 6x4, 380 H.P.		121.90		134.09		
.25 Dist. Tanker, 3000 Gallon		80.81		88.89		
1 Asphalt Paver, 130 H.P.		2076.00		2283.60		
1 Tandem Roller, 10 Ton		232.05		255.26	48.28	53.11
52 L.H., Daily Totals		$4146.66		$5456.94	$79.74	$104.94

Crew B-26	Bare Costs Hr.	Bare Costs Daily	Incl. Subs O&P Hr.	Incl. Subs O&P Daily	Cost Per Labor-Hour Bare Costs	Cost Per Labor-Hour Incl. O&P
1 Labor Foreman (outside)	$29.50	$236.00	$48.75	$390.00	$31.28	$51.50
6 Laborers	27.50	1320.00	45.45	2181.60		
2 Equip. Oper. (medium)	37.45	599.20	61.45	983.20		
1 Rodman (reinf.)	38.95	311.60	64.40	515.20		
1 Cement Finisher	35.70	285.60	57.70	461.60		
1 Grader, 30,000 Lbs.		657.85		723.63		
1 Paving Mach. & Equip.		2428.00		2670.80	35.07	38.57
88 L.H., Daily Totals		$5838.25		$7926.03	$66.34	$90.07

Crew B-26A	Hr.	Daily	Hr.	Daily	Bare Costs	Incl. O&P
1 Labor Foreman (outside)	$29.50	$236.00	$48.75	$390.00	$31.28	$51.50
6 Laborers	27.50	1320.00	45.45	2181.60		
2 Equip. Oper. (medium)	37.45	599.20	61.45	983.20		
1 Rodman (reinf.)	38.95	311.60	64.40	515.20		
1 Cement Finisher	35.70	285.60	57.70	461.60		
1 Grader, 30,000 Lbs.		657.85		723.63		
1 Paving Mach. & Equip.		2428.00		2670.80		
1 Concrete Saw		110.15		121.17	36.32	39.95
88 L.H., Daily Totals		$5948.40		$8047.20	$67.60	$91.45

Crew B-26B	Hr.	Daily	Hr.	Daily	Bare Costs	Incl. O&P
1 Labor Foreman (outside)	$29.50	$236.00	$48.75	$390.00	$31.79	$52.33
6 Laborers	27.50	1320.00	45.45	2181.60		
3 Equip. Oper. (medium)	37.45	898.80	61.45	1474.80		
1 Rodman (reinf.)	38.95	311.60	64.40	515.20		
1 Cement Finisher	35.70	285.60	57.70	461.60		
1 Grader, 30,000 Lbs.		657.85		723.63		
1 Paving Mach. & Equip.		2428.00		2670.80		
1 Concrete Pump, 110' Boom		1130.00		1243.00	43.92	48.31
96 L.H., Daily Totals		$7267.85		$9660.64	$75.71	$100.63

Crew B-26C	Hr.	Daily	Hr.	Daily	Bare Costs	Incl. O&P
1 Labor Foreman (outside)	$29.50	$236.00	$48.75	$390.00	$30.66	$50.50
6 Laborers	27.50	1320.00	45.45	2181.60		
1 Equip. Oper. (medium)	37.45	299.60	61.45	491.60		
1 Rodman (reinf.)	38.95	311.60	64.40	515.20		
1 Cement Finisher	35.70	285.60	57.70	461.60		
1 Paving Mach. & Equip.		2428.00		2670.80		
1 Concrete Saw		110.15		121.17	31.73	34.90
80 L.H., Daily Totals		$4990.95		$6831.97	$62.39	$85.40

Crew B-27	Hr.	Daily	Hr.	Daily	Bare Costs	Incl. O&P
1 Labor Foreman (outside)	$29.50	$236.00	$48.75	$390.00	$28.00	$46.27
3 Laborers	27.50	660.00	45.45	1090.80		
1 Berm Machine		325.30		357.83	10.17	11.18
32 L.H., Daily Totals		$1221.30		$1838.63	$38.17	$57.46

Crew B-28	Hr.	Daily	Hr.	Daily	Bare Costs	Incl. O&P
2 Carpenters	$35.65	$570.40	$58.90	$942.40	$32.93	$54.42
1 Laborer	27.50	220.00	45.45	363.60		
24 L.H., Daily Totals		$790.40		$1306.00	$32.93	$54.42

Crew B-29	Hr.	Daily	Hr.	Daily	Bare Costs	Incl. O&P
1 Labor Foreman (outside)	$29.50	$236.00	$48.75	$390.00	$29.57	$48.79
4 Laborers	27.50	880.00	45.45	1454.40		
1 Equip. Oper. (crane)	37.90	303.20	62.20	497.60		
1 Gradall, 5/8 C.Y.		862.55		948.80	17.97	19.77
48 L.H., Daily Totals		$2281.75		$3290.80	$47.54	$68.56

Crew B-30	Bare Costs Hr.	Bare Costs Daily	Incl. Subs O&P Hr.	Incl. Subs O&P Daily	Cost Per Labor-Hour Bare Costs	Cost Per Labor-Hour Incl. O&P
1 Equip. Oper. (medium)	$37.45	$299.60	$61.45	$491.60	$34.42	$56.55
2 Truck Drivers (heavy)	32.90	526.40	54.10	865.60		
1 Hyd. Excavator, 1.5 C.Y.		908.70		999.57		
2 Dump Trucks, 12 C.Y., 400 H.P.		1134.10		1247.51	85.12	93.63
24 L.H., Daily Totals		$2868.80		$3604.28	$119.53	$150.18

Crew B-31	Hr.	Daily	Hr.	Daily	Bare Costs	Incl. O&P
1 Labor Foreman (outside)	$29.50	$236.00	$48.75	$390.00	$27.90	$46.11
4 Laborers	27.50	880.00	45.45	1454.40		
1 Air Compressor, 250 cfm		167.95		184.75		
1 Sheeting Driver		7.15		7.87		
2 -50' Air Hoses, 1.5"		45.50		50.05	5.51	6.07
40 L.H., Daily Totals		$1336.60		$2087.06	$33.41	$52.18

Crew B-32	Hr.	Daily	Hr.	Daily	Bare Costs	Incl. O&P
1 Laborer	$27.50	$220.00	$45.45	$363.60	$34.96	$57.45
3 Equip. Oper. (medium)	37.45	898.80	61.45	1474.80		
1 Grader, 30,000 Lbs.		657.85		723.63		
1 Tandem Roller, 10 Ton		232.05		255.26		
1 Dozer, 200 H.P.		1290.00		1419.00	68.12	74.93
32 L.H., Daily Totals		$3298.70		$4236.29	$103.08	$132.38

Crew B-32A	Hr.	Daily	Hr.	Daily	Bare Costs	Incl. O&P
1 Laborer	$27.50	$220.00	$45.45	$363.60	$34.13	$56.12
2 Equip. Oper. (medium)	37.45	599.20	61.45	983.20		
1 Grader, 30,000 Lbs.		657.85		723.63		
1 Roller, Vibratory, 25 Ton		650.90		715.99	54.53	59.98
24 L.H., Daily Totals		$2127.95		$2786.43	$88.66	$116.10

Crew B-32B	Hr.	Daily	Hr.	Daily	Bare Costs	Incl. O&P
1 Laborer	$27.50	$220.00	$45.45	$363.60	$34.13	$56.12
2 Equip. Oper. (medium)	37.45	599.20	61.45	983.20		
1 Dozer, 200 H.P.		1290.00		1419.00		
1 Roller, Vibratory, 25 Ton		650.90		715.99	80.87	88.96
24 L.H., Daily Totals		$2760.10		$3481.79	$115.00	$145.07

Crew B-32C	Hr.	Daily	Hr.	Daily	Bare Costs	Incl. O&P
1 Labor Foreman (outside)	$29.50	$236.00	$48.75	$390.00	$32.81	$54.00
2 Laborers	27.50	440.00	45.45	727.20		
3 Equip. Oper. (medium)	37.45	898.80	61.45	1474.80		
1 Grader, 30,000 Lbs.		657.85		723.63		
1 Tandem Roller, 10 Ton		232.05		255.26		
1 Dozer, 200 H.P.		1290.00		1419.00	45.41	49.96
48 L.H., Daily Totals		$3754.70		$4989.89	$78.22	$103.96

Crew B-33A	Hr.	Daily	Hr.	Daily	Bare Costs	Incl. O&P
1 Equip. Oper. (medium)	$37.45	$299.60	$61.45	$491.60	$37.45	$61.45
.25 Equip. Oper. (medium)	37.45	74.90	61.45	122.90		
1 Scraper, Towed, 7 C.Y.		125.50		138.05		
1.250 Dozers, 300 H.P.		2288.75		2517.63	241.43	265.57
10 L.H., Daily Totals		$2788.75		$3270.18	$278.88	$327.02

Crew B-33B	Hr.	Daily	Hr.	Daily	Bare Costs	Incl. O&P
1 Equip. Oper. (medium)	$37.45	$299.60	$61.45	$491.60	$37.45	$61.45
.25 Equip. Oper. (medium)	37.45	74.90	61.45	122.90		
1 Scraper, Towed, 10 C.Y.		156.65		172.32		
1.250 Dozers, 300 H.P.		2288.75		2517.63	244.54	268.99
10 L.H., Daily Totals		$2819.90		$3304.44	$281.99	$330.44

Crew No.	Bare Costs		Incl. Subs O&P		Cost Per Labor-Hour	
Crew B-33C	**Hr.**	**Daily**	**Hr.**	**Daily**	**Bare Costs**	**Incl. O&P**
1 Equip. Oper. (medium)	$37.45	$299.60	$61.45	$491.60	$37.45	$61.45
.25 Equip. Oper. (medium)	37.45	74.90	61.45	122.90		
1 Scraper, Towed, 15 C.Y.		173.20		190.52		
1.250 Dozers, 300 H.P.		2288.75		2517.63	246.19	270.81
10 L.H., Daily Totals		$2836.45		$3322.65	$283.64	$332.26
Crew B-33D	**Hr.**	**Daily**	**Hr.**	**Daily**	**Bare Costs**	**Incl. O&P**
1 Equip. Oper. (medium)	$37.45	$299.60	$61.45	$491.60	$37.45	$61.45
.25 Equip. Oper. (medium)	37.45	74.90	61.45	122.90		
1 S.P. Scraper, 14 C.Y.		2611.00		2872.10		
.25 Dozer, 300 H.P.		457.75		503.52	306.88	337.56
10 L.H., Daily Totals		$3443.25		$3990.13	$344.32	$399.01
Crew B-33E	**Hr.**	**Daily**	**Hr.**	**Daily**	**Bare Costs**	**Incl. O&P**
1 Equip. Oper. (medium)	$37.45	$299.60	$61.45	$491.60	$37.45	$61.45
.25 Equip. Oper. (medium)	37.45	74.90	61.45	122.90		
1 S.P. Scraper, 21 C.Y.		2541.00		2795.10		
.25 Dozer, 300 H.P.		457.75		503.52	299.88	329.86
10 L.H., Daily Totals		$3373.25		$3913.13	$337.32	$391.31
Crew B-33F	**Hr.**	**Daily**	**Hr.**	**Daily**	**Bare Costs**	**Incl. O&P**
1 Equip. Oper. (medium)	$37.45	$299.60	$61.45	$491.60	$37.45	$61.45
.25 Equip. Oper. (medium)	37.45	74.90	61.45	122.90		
1 Elev. Scraper, 11 C.Y.		1165.00		1281.50		
.25 Dozer, 300 H.P.		457.75		503.52	162.28	178.50
10 L.H., Daily Totals		$1997.25		$2399.53	$199.72	$239.95
Crew B-33G	**Hr.**	**Daily**	**Hr.**	**Daily**	**Bare Costs**	**Incl. O&P**
1 Equip. Oper. (medium)	$37.45	$299.60	$61.45	$491.60	$37.45	$61.45
.25 Equip. Oper. (medium)	37.45	74.90	61.45	122.90		
1 Elev. Scraper, 22 C.Y.		2307.00		2537.70		
.25 Dozer, 300 H.P.		457.75		503.52	276.48	304.12
10 L.H., Daily Totals		$3139.25		$3655.72	$313.93	$365.57
Crew B-33K	**Hr.**	**Daily**	**Hr.**	**Daily**	**Bare Costs**	**Incl. O&P**
1 Equipment Operator (med.)	$37.45	$299.60	$61.45	$491.60	$34.61	$56.88
.25 Equipment Operator (med.)	37.45	74.90	61.45	122.90		
.5 Laborer	27.50	110.00	45.45	181.80		
1 S.P. Scraper, 31 C.Y.		3603.00		3963.30		
.25 Dozer, 410 H.P.		567.25		623.98	297.88	327.66
14 L.H., Daily Totals		$4654.75		$5383.57	$332.48	$384.54
Crew B-34A	**Hr.**	**Daily**	**Hr.**	**Daily**	**Bare Costs**	**Incl. O&P**
1 Truck Driver (heavy)	$32.90	$263.20	$54.10	$432.80	$32.90	$54.10
1 Dump Truck, 8 C.Y., 220 H.P.		339.60		373.56	42.45	46.70
8 L.H., Daily Totals		$602.80		$806.36	$75.35	$100.80
Crew B-34B	**Hr.**	**Daily**	**Hr.**	**Daily**	**Bare Costs**	**Incl. O&P**
1 Truck Driver (heavy)	$32.90	$263.20	$54.10	$432.80	$32.90	$54.10
1 Dump Truck, 12 C.Y., 400 H.P.		567.05		623.76	70.88	77.97
8 L.H., Daily Totals		$830.25		$1056.56	$103.78	$132.07
Crew B-34C	**Hr.**	**Daily**	**Hr.**	**Daily**	**Bare Costs**	**Incl. O&P**
1 Truck Driver (heavy)	$32.90	$263.20	$54.10	$432.80	$32.90	$54.10
1 Truck Tractor, 6x4, 380 H.P.		487.60		536.36		
1 Dump Trailer, 16.5 C.Y.		133.95		147.35	77.69	85.46
8 L.H., Daily Totals		$884.75		$1116.51	$110.59	$139.56

Crew No.	Bare Costs		Incl. Subs O&P		Cost Per Labor-Hour	
Crew B-34D	**Hr.**	**Daily**	**Hr.**	**Daily**	**Bare Costs**	**Incl. O&P**
1 Truck Driver (heavy)	$32.90	$263.20	$54.10	$432.80	$32.90	$54.10
1 Truck Tractor, 6x4, 380 H.P.		487.60		536.36		
1 Dump Trailer, 20 C.Y.		148.60		163.46	79.53	87.48
8 L.H., Daily Totals		$899.40		$1132.62	$112.43	$141.58
Crew B-34E	**Hr.**	**Daily**	**Hr.**	**Daily**	**Bare Costs**	**Incl. O&P**
1 Truck Driver (heavy)	$32.90	$263.20	$54.10	$432.80	$32.90	$54.10
1 Dump Truck, Off Hwy., 25 Ton		1353.00		1488.30	169.13	186.04
8 L.H., Daily Totals		$1616.20		$1921.10	$202.03	$240.14
Crew B-34F	**Hr.**	**Daily**	**Hr.**	**Daily**	**Bare Costs**	**Incl. O&P**
1 Truck Driver (heavy)	$32.90	$263.20	$54.10	$432.80	$32.90	$54.10
1 Dump Truck, Off Hwy., 35 Ton		1472.00		1619.20	184.00	202.40
8 L.H., Daily Totals		$1735.20		$2052.00	$216.90	$256.50
Crew B-34G	**Hr.**	**Daily**	**Hr.**	**Daily**	**Bare Costs**	**Incl. O&P**
1 Truck Driver (heavy)	$32.90	$263.20	$54.10	$432.80	$32.90	$54.10
1 Dump Truck, Off Hwy., 50 Ton		1738.00		1911.80	217.25	238.97
8 L.H., Daily Totals		$2001.20		$2344.60	$250.15	$293.07
Crew B-34H	**Hr.**	**Daily**	**Hr.**	**Daily**	**Bare Costs**	**Incl. O&P**
1 Truck Driver (heavy)	$32.90	$263.20	$54.10	$432.80	$32.90	$54.10
1 Dump Truck, Off Hwy., 65 Ton		1879.00		2066.90	234.88	258.36
8 L.H., Daily Totals		$2142.20		$2499.70	$267.77	$312.46
Crew B-34I	**Hr.**	**Daily**	**Hr.**	**Daily**	**Bare Costs**	**Incl. O&P**
1 Truck Driver (heavy)	$32.90	$263.20	$54.10	$432.80	$32.90	$54.10
1 Dump Truck, 18 C.Y., 450 H.P.		705.00		775.50	88.13	96.94
8 L.H., Daily Totals		$968.20		$1208.30	$121.03	$151.04
Crew B-34J	**Hr.**	**Daily**	**Hr.**	**Daily**	**Bare Costs**	**Incl. O&P**
1 Truck Driver (heavy)	$32.90	$263.20	$54.10	$432.80	$32.90	$54.10
1 Dump Truck, Off Hwy., 100 Ton		2683.00		2951.30	335.38	368.91
8 L.H., Daily Totals		$2946.20		$3384.10	$368.27	$423.01
Crew B-34K	**Hr.**	**Daily**	**Hr.**	**Daily**	**Bare Costs**	**Incl. O&P**
1 Truck Driver (heavy)	$32.90	$263.20	$54.10	$432.80	$32.90	$54.10
1 Truck Tractor, 6x4, 450 H.P.		594.90		654.39		
1 Lowbed Trailer, 75 Ton		249.90		274.89	105.60	116.16
8 L.H., Daily Totals		$1108.00		$1362.08	$138.50	$170.26
Crew B-34L	**Hr.**	**Daily**	**Hr.**	**Daily**	**Bare Costs**	**Incl. O&P**
1 Equip. Oper. (light)	$35.65	$285.20	$58.50	$468.00	$35.65	$58.50
1 Flatbed Truck, Gas, 1.5 Ton		195.20		214.72	24.40	26.84
8 L.H., Daily Totals		$480.40		$682.72	$60.05	$85.34
Crew B-34M	**Hr.**	**Daily**	**Hr.**	**Daily**	**Bare Costs**	**Incl. O&P**
1 Equip. Oper. (light)	$35.65	$285.20	$58.50	$468.00	$35.65	$58.50
1 Flatbed Truck, Gas, 3 Ton		245.95		270.55	30.74	33.82
8 L.H., Daily Totals		$531.15		$738.54	$66.39	$92.32
Crew B-34N	**Hr.**	**Daily**	**Hr.**	**Daily**	**Bare Costs**	**Incl. O&P**
1 Truck Driver (heavy)	$32.90	$263.20	$54.10	$432.80	$35.17	$57.77
1 Equip. Oper. (medium)	37.45	299.60	61.45	491.60		
1 Truck Tractor, 6x4, 380 H.P.		487.60		536.36		
1 Flatbed Trailer, 40 Ton		182.50		200.75	41.88	46.07
16 L.H., Daily Totals		$1232.90		$1661.51	$77.06	$103.84

Crew B-34P	Bare Costs Hr.	Bare Costs Daily	Incl. Subs O&P Hr.	Incl. Subs O&P Daily	Cost Per Labor-Hour Bare Costs	Cost Per Labor-Hour Incl. O&P
1 Pipe Fitter	$41.55	$332.40	$68.15	$545.20	$37.10	$60.92
1 Truck Driver (light)	32.30	258.40	53.15	425.20		
1 Equip. Oper. (medium)	37.45	299.60	61.45	491.60		
1 Flatbed Truck, Gas, 3 Ton		245.95		270.55		
1 Backhoe Loader, 48 H.P.		320.20		352.22	23.59	25.95
24 L.H., Daily Totals		$1456.55		$2084.76	$60.69	$86.87

Crew B-34Q	Hr.	Daily	Hr.	Daily	Bare Costs	Incl. O&P
1 Pipe Fitter	$41.55	$332.40	$68.15	$545.20	$37.25	$61.17
1 Truck Driver (light)	32.30	258.40	53.15	425.20		
1 Equip. Oper. (crane)	37.90	303.20	62.20	497.60		
1 Flatbed Trailer, 25 Ton		133.00		146.30		
1 Dump Truck, 8 C.Y., 220 H.P.		339.60		373.56		
1 Hyd. Crane, 25 Ton		581.70		639.87	43.93	48.32
24 L.H., Daily Totals		$1948.30		$2627.73	$81.18	$109.49

Crew B-34R	Hr.	Daily	Hr.	Daily	Bare Costs	Incl. O&P
1 Pipe Fitter	$41.55	$332.40	$68.15	$545.20	$37.25	$61.17
1 Truck Driver (light)	32.30	258.40	53.15	425.20		
1 Equip. Oper. (crane)	37.90	303.20	62.20	497.60		
1 Flatbed Trailer, 25 Ton		133.00		146.30		
1 Dump Truck, 8 C.Y., 220 H.P.		339.60		373.56		
1 Hyd. Crane, 25 Ton		581.70		639.87		
1 Hyd. Excavator, 1 C.Y.		749.70		824.67	75.17	82.68
24 L.H., Daily Totals		$2698.00		$3452.40	$112.42	$143.85

Crew B-34S	Hr.	Daily	Hr.	Daily	Bare Costs	Incl. O&P
2 Pipe Fitters	$41.55	$664.80	$68.15	$1090.40	$38.48	$63.15
1 Truck Driver (heavy)	32.90	263.20	54.10	432.80		
1 Equip. Oper. (crane)	37.90	303.20	62.20	497.60		
1 Flatbed Trailer, 40 Ton		182.50		200.75		
1 Truck Tractor, 6x4, 380 H.P.		487.60		536.36		
1 Hyd. Crane, 80 Ton		1487.00		1635.70		
1 Hyd. Excavator, 2 C.Y.		1078.00		1185.80	101.10	111.21
32 L.H., Daily Totals		$4466.30		$5579.41	$139.57	$174.36

Crew B-34T	Hr.	Daily	Hr.	Daily	Bare Costs	Incl. O&P
2 Pipe Fitters	$41.55	$664.80	$68.15	$1090.40	$38.48	$63.15
1 Truck Driver (heavy)	32.90	263.20	54.10	432.80		
1 Equip. Oper. (crane)	37.90	303.20	62.20	497.60		
1 Flatbed Trailer, 40 Ton		182.50		200.75		
1 Truck Tractor, 6x4, 380 H.P.		487.60		536.36		
1 Hyd. Crane, 80 Ton		1487.00		1635.70	67.41	74.15
32 L.H., Daily Totals		$3388.30		$4393.61	$105.88	$137.30

Crew B-34U	Hr.	Daily	Hr.	Daily	Bare Costs	Incl. O&P
1 Truck Driver (heavy)	$32.90	$263.20	$54.10	$432.80	$34.27	$56.30
1 Equip. Oper. (light)	35.65	285.20	58.50	468.00		
1 Truck Tractor, 220 H.P.		303.55		333.90		
1 Flatbed Trailer, 25 Ton		133.00		146.30	27.28	30.01
16 L.H., Daily Totals		$984.95		$1381.01	$61.56	$86.31

Crew B-34V	Hr.	Daily	Hr.	Daily	Bare Costs	Incl. O&P
1 Truck Driver (heavy)	$32.90	$263.20	$54.10	$432.80	$35.48	$58.27
1 Equip. Oper. (crane)	37.90	303.20	62.20	497.60		
1 Equip. Oper. (light)	35.65	285.20	58.50	468.00		
1 Truck Tractor, 6x4, 450 H.P.		594.90		654.39		
1 Equipment Trailer, 50 Ton		200.95		221.04		
1 Pickup Truck, 4x4, 3/4 Ton		121.70		133.87	38.23	42.05
24 L.H., Daily Totals		$1769.15		$2407.70	$73.71	$100.32

Crew B-34W	Hr.	Daily	Hr.	Daily	Bare Costs	Incl. O&P
5 Truck Drivers (heavy)	$32.90	$1316.00	$54.10	$2164.00	$33.98	$55.86
2 Equip. Opers. (crane)	37.90	606.40	62.20	995.20		
1 Equip. Oper. (mechanic)	38.00	304.00	62.35	498.80		
1 Laborer	27.50	220.00	45.45	363.60		
4 Truck Tractors, 6x4, 380 H.P.		1950.40		2145.44		
2 Equipment Trailers, 50 Ton		401.90		442.09		
2 Flatbed Trailers, 40 Ton		365.00		401.50		
1 Pickup Truck, 4x4, 3/4 Ton		121.70		133.87		
1 S.P. Crane, 4x4, 20 Ton		587.75		646.52	47.59	52.35
72 L.H., Daily Totals		$5873.15		$7791.02	$81.57	$108.21

Crew B-35	Hr.	Daily	Hr.	Daily	Bare Costs	Incl. O&P
1 Labor Foreman (outside)	$29.50	$236.00	$48.75	$390.00	$34.36	$56.67
1 Skilled Worker	36.50	292.00	60.65	485.20		
1 Welder (plumber)	40.40	323.20	66.30	530.40		
1 Laborer	27.50	220.00	45.45	363.60		
1 Equip. Oper. (crane)	37.90	303.20	62.20	497.60		
1 Welder, Electric, 300 amp		56.10		61.71		
1 Hyd. Excavator, .75 C.Y.		681.25		749.38	18.43	20.28
40 L.H., Daily Totals		$2111.75		$3077.89	$52.79	$76.95

Crew B-35A	Hr.	Daily	Hr.	Daily	Bare Costs	Incl. O&P
1 Labor Foreman (outside)	$29.50	$236.00	$48.75	$390.00	$33.18	$54.69
2 Laborers	27.50	440.00	45.45	727.20		
1 Skilled Worker	36.50	292.00	60.65	485.20		
1 Welder (plumber)	40.40	323.20	66.30	530.40		
1 Equip. Oper. (crane)	37.90	303.20	62.20	497.60		
1 Equip. Oper. (oiler)	32.95	263.60	54.05	432.40		
1 Welder, Gas Engine, 300 amp		95.65		105.22		
1 Crawler Crane, 75 Ton		1734.00		1907.40	32.67	35.94
56 L.H., Daily Totals		$3687.65		$5075.42	$65.85	$90.63

Crew B-36	Hr.	Daily	Hr.	Daily	Bare Costs	Incl. O&P
1 Labor Foreman (outside)	$29.50	$236.00	$48.75	$390.00	$31.88	$52.51
2 Laborers	27.50	440.00	45.45	727.20		
2 Equip. Oper. (medium)	37.45	599.20	61.45	983.20		
1 Dozer, 200 H.P.		1290.00		1419.00		
1 Aggregate Spreader		37.30		41.03		
1 Tandem Roller, 10 Ton		232.05		255.26	38.98	42.88
40 L.H., Daily Totals		$2834.55		$3815.68	$70.86	$95.39

Crew B-36A	Hr.	Daily	Hr.	Daily	Bare Costs	Incl. O&P
1 Labor Foreman (outside)	$29.50	$236.00	$48.75	$390.00	$33.47	$55.06
2 Laborers	27.50	440.00	45.45	727.20		
4 Equip. Oper. (medium)	37.45	1198.40	61.45	1966.40		
1 Dozer, 200 H.P.		1290.00		1419.00		
1 Aggregate Spreader		37.30		41.03		
1 Tandem Roller, 10 Ton		232.05		255.26		
1 Roller, Pneum. Whl., 12 Ton		338.35		372.19	33.89	37.28
56 L.H., Daily Totals		$3772.10		$5171.07	$67.36	$92.34

Crew B-36B	Hr.	Daily	Hr.	Daily	Bare Costs	Incl. O&P
1 Labor Foreman (outside)	$29.50	$236.00	$48.75	$390.00	$33.40	$54.94
2 Laborers	27.50	440.00	45.45	727.20		
4 Equip. Oper. (medium)	37.45	1198.40	61.45	1966.40		
1 Truck Driver (heavy)	32.90	263.20	54.10	432.80		
1 Grader, 30,000 Lbs.		657.85		723.63		
1 F.E. Loader, Crl, 1.5 C.Y.		647.15		711.87		
1 Dozer, 300 H.P.		1831.00		2014.10		
1 Roller, Vibratory, 25 Ton		650.90		715.99		
1 Truck Tractor, 6x4, 450 H.P.		594.90		654.39		
1 Water Tank Trailer, 5000 Gal.		149.45		164.40	70.80	77.88
64 L.H., Daily Totals		$6668.85		$8500.77	$104.20	$132.82

Crew B-36C	Hr.	Daily	Hr.	Daily	Bare Costs	Incl. O&P
1 Labor Foreman (outside)	$29.50	$236.00	$48.75	$390.00	$34.95	$57.44
3 Equip. Oper. (medium)	37.45	898.80	61.45	1474.80		
1 Truck Driver (heavy)	32.90	263.20	54.10	432.80		
1 Grader, 30,000 Lbs.		657.85		723.63		
1 Dozer, 300 H.P.		1831.00		2014.10		
1 Roller, Vibratory, 25 Ton		650.90		715.99		
1 Truck Tractor, 6x4, 450 H.P.		594.90		654.39		
1 Water Tank Trailer, 5000 Gal.		149.45		164.40	97.10	106.81
40 L.H., Daily Totals		$5282.10		$6570.11	$132.05	$164.25

Crew B-36E	Hr.	Daily	Hr.	Daily	Bare Costs	Incl. O&P
1 Labor Foreman (outside)	$29.50	$236.00	$48.75	$390.00	$35.37	$58.11
4 Equip. Oper. (medium)	37.45	1198.40	61.45	1966.40		
1 Truck Driver (heavy)	32.90	263.20	54.10	432.80		
1 Grader, 30,000 Lbs.		657.85		723.63		
1 Dozer, 300 H.P.		1831.00		2014.10		
1 Roller, Vibratory, 25 Ton		650.90		715.99		
1 Truck Tractor, 6x4, 380 H.P.		487.60		536.36		
1 Dist. Tanker, 3000 Gallon		323.25		355.57	82.30	90.53
48 L.H., Daily Totals		$5648.20		$7134.86	$117.67	$148.64

Crew B-37	Hr.	Daily	Hr.	Daily	Bare Costs	Incl. O&P
1 Labor Foreman (outside)	$29.50	$236.00	$48.75	$390.00	$29.19	$48.17
4 Laborers	27.50	880.00	45.45	1454.40		
1 Equip. Oper. (light)	35.65	285.20	58.50	468.00		
1 Tandem Roller, 5 Ton		152.45		167.69	3.18	3.49
48 L.H., Daily Totals		$1553.65		$2480.09	$32.37	$51.67

Crew B-37A	Hr.	Daily	Hr.	Daily	Bare Costs	Incl. O&P
2 Laborers	$27.50	$440.00	$45.45	$727.20	$29.10	$48.02
1 Truck Driver (light)	32.30	258.40	53.15	425.20		
1 Flatbed Truck, Gas, 1.5 Ton		195.20		214.72		
1 Tar Kettle, T.M.		134.50		147.95	13.74	15.11
24 L.H., Daily Totals		$1028.10		$1515.07	$42.84	$63.13

Crew B-37B	Hr.	Daily	Hr.	Daily	Bare Costs	Incl. O&P
3 Laborers	$27.50	$660.00	$45.45	$1090.80	$28.70	$47.38
1 Truck Driver (light)	32.30	258.40	53.15	425.20		
1 Flatbed Truck, Gas, 1.5 Ton		195.20		214.72		
1 Tar Kettle, T.M.		134.50		147.95	10.30	11.33
32 L.H., Daily Totals		$1248.10		$1878.67	$39.00	$58.71

Crew B-37C	Hr.	Daily	Hr.	Daily	Bare Costs	Incl. O&P
2 Laborers	$27.50	$440.00	$45.45	$727.20	$29.90	$49.30
2 Truck Drivers (light)	32.30	516.80	53.15	850.40		
2 Flatbed Trucks, Gas, 1.5 Ton		390.40		429.44		
1 Tar Kettle, T.M.		134.50		147.95	16.40	18.04
32 L.H., Daily Totals		$1481.70		$2154.99	$46.30	$67.34

Crew B-37D	Hr.	Daily	Hr.	Daily	Bare Costs	Incl. O&P
1 Laborer	$27.50	$220.00	$45.45	$363.60	$29.90	$49.30
1 Truck Driver (light)	32.30	258.40	53.15	425.20		
1 Pickup Truck, 3/4 Ton		109.90		120.89	6.87	7.56
16 L.H., Daily Totals		$588.30		$909.69	$36.77	$56.86

Crew B-37E	Hr.	Daily	Hr.	Daily	Bare Costs	Incl. O&P
3 Laborers	$27.50	$660.00	$45.45	$1090.80	$31.46	$51.80
1 Equip. Oper. (light)	35.65	285.20	58.50	468.00		
1 Equip. Oper. (medium)	37.45	299.60	61.45	491.60		
2 Truck Drivers (light)	32.30	516.80	53.15	850.40		
4 Barrels w/ Flasher		16.00		17.60		
1 Concrete Saw		110.15		121.17		
1 Rotary Hammer Drill		25.05		27.56		
1 Hammer Drill Bit		3.30		3.63		
1 Loader, Skid Steer, 30 H.P.		174.55		192.01		
1 Conc. Hammer Attach.		114.40		125.84		
1 Vibrating Plate, Gas, 18"		31.10		34.21		
2 Flatbed Trucks, Gas, 1.5 Ton		390.40		429.44	15.45	16.99
56 L.H., Daily Totals		$2626.55		$3852.24	$46.90	$68.79

Crew B-37F	Hr.	Daily	Hr.	Daily	Bare Costs	Incl. O&P
3 Laborers	$27.50	$660.00	$45.45	$1090.80	$28.70	$47.38
1 Truck Driver (light)	32.30	258.40	53.15	425.20		
4 Barrels w/ Flasher		16.00		17.60		
1 Concrete Mixer, 10 C.F.		162.00		178.20		
1 Air Compressor, 60 cfm		106.70		117.37		
1 -50' Air Hose, 3/4"		6.55		7.21		
1 Spade (Chipper)		8.30		9.13		
1 Flatbed Truck, Gas, 1.5 Ton		195.20		214.72	15.46	17.01
32 L.H., Daily Totals		$1413.15		$2060.22	$44.16	$64.38

Crew B-37G	Hr.	Daily	Hr.	Daily	Bare Costs	Incl. O&P
1 Labor Foreman (outside)	$29.50	$236.00	$48.75	$390.00	$29.19	$48.17
4 Laborers	27.50	880.00	45.45	1454.40		
1 Equip. Oper. (light)	35.65	285.20	58.50	468.00		
1 Berm Machine		325.30		357.83		
1 Tandem Roller, 5 Ton		152.45		167.69	9.95	10.95
48 L.H., Daily Totals		$1878.95		$2837.93	$39.14	$59.12

Crew B-37H	Hr.	Daily	Hr.	Daily	Bare Costs	Incl. O&P
1 Labor Foreman (outside)	$29.50	$236.00	$48.75	$390.00	$29.19	$48.17
4 Laborers	27.50	880.00	45.45	1454.40		
1 Equip. Oper. (light)	35.65	285.20	58.50	468.00		
1 Tandem Roller, 5 Ton		152.45		167.69		
1 Flatbed Truck, Gas, 1.5 Ton		195.20		214.72		
1 Tar Kettle, T.M.		134.50		147.95	10.04	11.05
48 L.H., Daily Totals		$1883.35		$2842.76	$39.24	$59.22

Crew No.	Bare Costs		Incl. Subs O&P		Cost Per Labor-Hour	
Crew B-37I	**Hr.**	**Daily**	**Hr.**	**Daily**	**Bare Costs**	**Incl. O&P**
3 Laborers	$27.50	$660.00	$45.45	$1090.80	$31.46	$51.80
1 Equip. Oper. (light)	35.65	285.20	58.50	468.00		
1 Equip. Oper. (medium)	37.45	299.60	61.45	491.60		
2 Truck Drivers (light)	32.30	516.80	53.15	850.40		
4 Barrels w/ Flasher		16.00		17.60		
1 Concrete Saw		110.15		121.17		
1 Rotary Hammer Drill		25.05		27.56		
1 Hammer Drill Bit		3.30		3.63		
1 Air Compressor, 60 cfm		106.70		117.37		
1 -50' Air Hose, 3/4"		6.55		7.21		
1 Spade (Chipper)		8.30		9.13		
1 Loader, Skid Steer, 30 H.P.		174.55		192.01		
1 Conc. Hammer Attach.		114.40		125.84		
1 Concrete Mixer, 10 C.F.		162.00		178.20		
1 Vibrating Plate, Gas, 18"		31.10		34.21		
2 Flatbed Trucks, Gas, 1.5 Ton		390.40		429.44	20.51	22.56
56 L.H., Daily Totals		$2910.10		$4164.15	$51.97	$74.36
Crew B-37J	**Hr.**	**Daily**	**Hr.**	**Daily**	**Bare Costs**	**Incl. O&P**
1 Labor Foreman (outside)	$29.50	$236.00	$48.75	$390.00	$29.19	$48.17
4 Laborers	27.50	880.00	45.45	1454.40		
1 Equip. Oper. (light)	35.65	285.20	58.50	468.00		
1 Air Compressor, 60 cfm		106.70		117.37		
1 -50' Air Hose, 3/4"		6.55		7.21		
2 Concrete Mixers, 10 C.F.		324.00		356.40		
2 Flatbed Trucks, Gas, 1.5 Ton		390.40		429.44		
1 Shot Blaster, 20"		201.00		221.10	21.43	23.57
48 L.H., Daily Totals		$2429.85		$3443.92	$50.62	$71.75
Crew B-37K	**Hr.**	**Daily**	**Hr.**	**Daily**	**Bare Costs**	**Incl. O&P**
1 Labor Foreman (outside)	$29.50	$236.00	$48.75	$390.00	$29.19	$48.17
4 Laborers	27.50	880.00	45.45	1454.40		
1 Equip. Oper. (light)	35.65	285.20	58.50	468.00		
1 Air Compressor, 60 cfm		106.70		117.37		
1 -50' Air Hose, 3/4"		6.55		7.21		
2 Flatbed Trucks, Gas, 1.5 Ton		390.40		429.44		
1 Shot Blaster, 20"		201.00		221.10	14.68	16.15
48 L.H., Daily Totals		$2105.85		$3087.51	$43.87	$64.32
Crew B-38	**Hr.**	**Daily**	**Hr.**	**Daily**	**Bare Costs**	**Incl. O&P**
2 Laborers	$27.50	$440.00	$45.45	$727.20	$30.22	$49.80
1 Equip. Oper. (light)	35.65	285.20	58.50	468.00		
1 Backhoe Loader, 48 H.P.		320.20		352.22		
1 Hyd. Hammer (1200 lb.)		171.40		188.54		
1 F.E. Loader, W.M., 4 C.Y.		566.70		623.37		
1 Pvmt. Rem. Bucket		61.70		67.87	46.67	51.33
24 L.H., Daily Totals		$1845.20		$2427.20	$76.88	$101.13
Crew B-39	**Hr.**	**Daily**	**Hr.**	**Daily**	**Bare Costs**	**Incl. O&P**
1 Labor Foreman (outside)	$29.50	$236.00	$48.75	$390.00	$27.83	$46.00
5 Laborers	27.50	1100.00	45.45	1818.00		
1 Air Compressor, 250 cfm		167.95		184.75		
2 Breakers, Pavement, 60 lb.		21.50		23.65		
2 -50' Air Hoses, 1.5"		45.50		50.05	4.89	5.38
48 L.H., Daily Totals		$1570.95		$2466.45	$32.73	$51.38

Crew No.	Bare Costs		Incl. Subs O&P		Cost Per Labor-Hour	
Crew B-40	**Hr.**	**Daily**	**Hr.**	**Daily**	**Bare Costs**	**Incl. O&P**
1 Pile Driver Foreman (outside)	$38.00	$304.00	$64.75	$518.00	$35.34	$59.69
4 Pile Drivers	36.00	1152.00	61.35	1963.20		
1 Building Laborer	27.50	220.00	45.45	363.60		
1 Equip. Oper. (crane)	37.90	303.20	62.20	497.60		
1 Crawler Crane, 40 Ton		1392.00		1531.20		
1 Vibratory Hammer & Gen.		2236.00		2459.60	64.79	71.26
56 L.H., Daily Totals		$5607.20		$7333.20	$100.13	$130.95
Crew B-40B	**Hr.**	**Daily**	**Hr.**	**Daily**	**Bare Costs**	**Incl. O&P**
1 Labor Foreman (outside)	$29.50	$236.00	$48.75	$390.00	$30.48	$50.23
3 Laborers	27.50	660.00	45.45	1090.80		
1 Equip. Oper. (crane)	37.90	303.20	62.20	497.60		
1 Equip. Oper. (oiler)	32.95	263.60	54.05	432.40		
1 Lattice Boom Crane, 40 Ton		1315.00		1446.50	27.40	30.14
48 L.H., Daily Totals		$2777.80		$3857.30	$57.87	$80.36
Crew B-41	**Hr.**	**Daily**	**Hr.**	**Daily**	**Bare Costs**	**Incl. O&P**
1 Labor Foreman (outside)	$29.50	$236.00	$48.75	$390.00	$28.58	$47.20
4 Laborers	27.50	880.00	45.45	1454.40		
.25 Equip. Oper. (crane)	37.90	75.80	62.20	124.40		
.25 Equip. Oper. (oiler)	32.95	65.90	54.05	108.10		
.25 Crawler Crane, 40 Ton		348.00		382.80	7.91	8.70
44 L.H., Daily Totals		$1605.70		$2459.70	$36.49	$55.90
Crew B-42	**Hr.**	**Daily**	**Hr.**	**Daily**	**Bare Costs**	**Incl. O&P**
1 Labor Foreman (outside)	$29.50	$236.00	$48.75	$390.00	$31.11	$51.29
4 Laborers	27.50	880.00	45.45	1454.40		
1 Equip. Oper. (crane)	37.90	303.20	62.20	497.60		
1 Welder	40.40	323.20	66.30	530.40		
1 Hyd. Crane, 25 Ton		581.70		639.87		
1 Welder, Gas Engine, 300 amp		95.65		105.22		
1 Horz. Boring Csg. Mch.		447.00		491.70	20.08	22.09
56 L.H., Daily Totals		$2866.75		$4109.19	$51.19	$73.38
Crew B-43	**Hr.**	**Daily**	**Hr.**	**Daily**	**Bare Costs**	**Incl. O&P**
1 Labor Foreman (outside)	$29.50	$236.00	$48.75	$390.00	$27.90	$46.11
4 Laborers	27.50	880.00	45.45	1454.40		
1 Drill Rig, Truck-Mounted		2428.00		2670.80	60.70	66.77
40 L.H., Daily Totals		$3544.00		$4515.20	$88.60	$112.88
Crew B-44	**Hr.**	**Daily**	**Hr.**	**Daily**	**Bare Costs**	**Incl. O&P**
1 Pile Driver Foreman (outside)	$38.00	$304.00	$64.75	$518.00	$34.36	$57.91
4 Pile Drivers	36.00	1152.00	61.35	1963.20		
1 Equip. Oper. (crane)	37.90	303.20	62.20	497.60		
2 Laborers	27.50	440.00	45.45	727.20		
1 Crawler Crane, 40 Ton		1392.00		1531.20		
1 Lead, 60' High		80.55		88.61		
1 Hammer, Diesel, 15K ft.-lbs.		622.40		684.64	32.73	36.01
64 L.H., Daily Totals		$4294.15		$6010.44	$67.10	$93.91
Crew B-45	**Hr.**	**Daily**	**Hr.**	**Daily**	**Bare Costs**	**Incl. O&P**
1 Building Laborer	$27.50	$220.00	$45.45	$363.60	$30.20	$49.77
1 Truck Driver (heavy)	32.90	263.20	54.10	432.80		
1 Dist. Tanker, 3000 Gallon		323.25		355.57		
1 Truck Tractor, 6x4, 380 H.P.		487.60		536.36	50.68	55.75
16 L.H., Daily Totals		$1294.05		$1688.34	$80.88	$105.52

Crew B-46	Hr.	Daily	Hr.	Daily	Bare Costs	Incl. O&P
1 Pile Driver Foreman (outside)	$38.00	$304.00	$64.75	$518.00	$32.08	$53.97
2 Pile Drivers	36.00	576.00	61.35	981.60		
3 Laborers	27.50	660.00	45.45	1090.80		
1 Chain Saw, Gas, 36" Long		46.65		51.31	0.97	1.07
48 L.H., Daily Totals		$1586.65		$2641.72	$33.06	$55.04

Crew B-47	Hr.	Daily	Hr.	Daily	Bare Costs	Incl. O&P
1 Blast Foreman (outside)	$29.50	$236.00	$48.75	$390.00	$28.50	$47.10
1 Driller	27.50	220.00	45.45	363.60		
1 Air Track Drill, 4"		1004.00		1104.40		
1 Air Compressor, 600 cfm		417.60		459.36		
2 -50' Air Hoses, 3"		51.50		56.65	92.07	101.28
16 L.H., Daily Totals		$1929.10		$2374.01	$120.57	$148.38

Crew B-47A	Hr.	Daily	Hr.	Daily	Bare Costs	Incl. O&P
1 Drilling Foreman (outside)	$29.50	$236.00	$48.75	$390.00	$33.45	$55.00
1 Equip. Oper. (heavy)	37.90	303.20	62.20	497.60		
1 Equip. Oper. (oiler)	32.95	263.60	54.05	432.40		
1 Air Track Drill, 5"		1173.00		1290.30	48.88	53.76
24 L.H., Daily Totals		$1975.80		$2610.30	$82.33	$108.76

Crew B-47C	Hr.	Daily	Hr.	Daily	Bare Costs	Incl. O&P
1 Laborer	$27.50	$220.00	$45.45	$363.60	$31.57	$51.98
1 Equip. Oper. (light)	35.65	285.20	58.50	468.00		
1 Air Compressor, 750 cfm		428.80		471.68		
2 -50' Air Hoses, 3"		51.50		56.65		
1 Air Track Drill, 4"		1004.00		1104.40	92.77	102.05
16 L.H., Daily Totals		$1989.50		$2464.33	$124.34	$154.02

Crew B-47E	Hr.	Daily	Hr.	Daily	Bare Costs	Incl. O&P
1 Labor Foreman (outside)	$29.50	$236.00	$48.75	$390.00	$28.00	$46.27
3 Laborers	27.50	660.00	45.45	1090.80		
1 Flatbed Truck, Gas, 3 Ton		245.95		270.55	7.69	8.45
32 L.H., Daily Totals		$1141.95		$1751.35	$35.69	$54.73

Crew B-47G	Hr.	Daily	Hr.	Daily	Bare Costs	Incl. O&P
1 Labor Foreman (outside)	$29.50	$236.00	$48.75	$390.00	$28.17	$46.55
2 Laborers	27.50	440.00	45.45	727.20		
1 Air Track Drill, 4"		1004.00		1104.40		
1 Air Compressor, 600 cfm		417.60		459.36		
2 -50' Air Hoses, 3"		51.50		56.65		
1 Gunite Pump Rig		313.00		344.30	74.42	81.86
24 L.H., Daily Totals		$2462.10		$3081.91	$102.59	$128.41

Crew B-47H	Hr.	Daily	Hr.	Daily	Bare Costs	Incl. O&P
1 Skilled Worker Foreman (out)	$38.50	$308.00	$63.95	$511.60	$37.00	$61.48
3 Skilled Workers	36.50	876.00	60.65	1455.60		
1 Flatbed Truck, Gas, 3 Ton		245.95		270.55	7.69	8.45
32 L.H., Daily Totals		$1429.95		$2237.74	$44.69	$69.93

Crew B-48	Hr.	Daily	Hr.	Daily	Bare Costs	Incl. O&P
1 Labor Foreman (outside)	$29.50	$236.00	$48.75	$390.00	$29.57	$48.79
4 Laborers	27.50	880.00	45.45	1454.40		
1 Equip. Oper. (crane)	37.90	303.20	62.20	497.60		
1 Centr. Water Pump, 6"		296.45		326.10		
1 -20' Suction Hose, 6"		11.10		12.21		
1 -50' Discharge Hose, 6"		5.70		6.27		
1 Drill Rig, Truck-Mounted		2428.00		2670.80	57.11	62.82
48 L.H., Daily Totals		$4160.45		$5357.38	$86.68	$111.61

Crew B-49	Hr.	Daily	Hr.	Daily	Bare Costs	Incl. O&P
1 Labor Foreman (outside)	$29.50	$236.00	$48.75	$390.00	$30.77	$51.21
5 Laborers	27.50	1100.00	45.45	1818.00		
1 Equip. Oper. (crane)	37.90	303.20	62.20	497.60		
2 Pile Drivers	36.00	576.00	61.35	981.60		
1 Hyd. Crane, 25 Ton		581.70		639.87		
1 Centr. Water Pump, 6"		296.45		326.10		
1 -20' Suction Hose, 6"		11.10		12.21		
1 -50' Discharge Hose, 6"		5.70		6.27		
1 Drill Rig, Truck-Mounted		2428.00		2670.80	46.15	50.77
72 L.H., Daily Totals		$5538.15		$7342.44	$76.92	$101.98

Crew B-50	Hr.	Daily	Hr.	Daily	Bare Costs	Incl. O&P
1 Pile Driver Foreman (outside)	$38.00	$304.00	$64.75	$518.00	$33.03	$55.56
6 Pile Drivers	36.00	1728.00	61.35	2944.80		
1 Equip. Oper. (crane)	37.90	303.20	62.20	497.60		
5 Laborers	27.50	1100.00	45.45	1818.00		
1 Crawler Crane, 40 Ton		1392.00		1531.20		
1 Lead, 60' High		80.55		88.61		
1 Hammer, Diesel, 15K ft.-lbs.		622.40		684.64		
1 Air Compressor, 600 cfm		417.60		459.36		
2 -50' Air Hoses, 3"		51.50		56.65		
1 Chain Saw, Gas, 36" Long		46.65		51.31	25.10	27.61
104 L.H., Daily Totals		$6045.90		$8650.17	$58.13	$83.17

Crew B-51	Hr.	Daily	Hr.	Daily	Bare Costs	Incl. O&P
1 Labor Foreman (outside)	$29.50	$236.00	$48.75	$390.00	$28.63	$47.28
4 Laborers	27.50	880.00	45.45	1454.40		
1 Truck Driver (light)	32.30	258.40	53.15	425.20		
1 Flatbed Truck, Gas, 1.5 Ton		195.20		214.72	4.07	4.47
48 L.H., Daily Totals		$1569.60		$2484.32	$32.70	$51.76

Crew B-52	Hr.	Daily	Hr.	Daily	Bare Costs	Incl. O&P
1 Labor Foreman (outside)	$29.50	$236.00	$48.75	$390.00	$30.48	$50.34
1 Carpenter	35.65	285.20	58.90	471.20		
4 Laborers	27.50	880.00	45.45	1454.40		
.5 Rodman (reinf.)	38.95	155.80	64.40	257.60		
.5 Equip. Oper. (medium)	37.45	149.80	61.45	245.80		
.5 Crawler Loader, 3 C.Y.		601.00		661.10	10.73	11.81
56 L.H., Daily Totals		$2307.80		$3480.10	$41.21	$62.14

Crew B-53	Hr.	Daily	Hr.	Daily	Bare Costs	Incl. O&P
1 Building Laborer	$27.50	$220.00	$45.45	$363.60	$27.50	$45.45
1 Trencher, Chain, 12 H.P.		63.40		69.74	7.92	8.72
8 L.H., Daily Totals		$283.40		$433.34	$35.42	$54.17

Crew B-54	Hr.	Daily	Hr.	Daily	Bare Costs	Incl. O&P
1 Equip. Oper. (light)	$35.65	$285.20	$58.50	$468.00	$35.65	$58.50
1 Trencher, Chain, 40 H.P.		348.50		383.35	43.56	47.92
8 L.H., Daily Totals		$633.70		$851.35	$79.21	$106.42

Crew B-54A	Hr.	Daily	Hr.	Daily	Bare Costs	Incl. O&P
.17 Labor Foreman (outside)	$29.50	$40.12	$48.75	$66.30	$36.29	$59.60
1 Equipment Operator (med.)	37.45	299.60	61.45	491.60		
1 Wheel Trencher, 67 H.P.		1110.00		1221.00	118.59	130.45
9.36 L.H., Daily Totals		$1449.72		$1778.90	$154.88	$190.05

Crew B-54B	Hr.	Daily	Hr.	Daily	Bare Costs	Incl. O&P
.25 Labor Foreman (outside)	$29.50	$59.00	$48.75	$97.50	$35.86	$58.91
1 Equipment Operator (med.)	37.45	299.60	61.45	491.60		
1 Wheel Trencher, 150 H.P.		1996.00		2195.60	199.60	219.56
10 L.H., Daily Totals		$2354.60		$2784.70	$235.46	$278.47

Crew No.	Bare Costs		Incl. Subs O&P		Cost Per Labor-Hour	
Crew B-54D	**Hr.**	**Daily**	**Hr.**	**Daily**	**Bare Costs**	**Incl. O&P**
1 Laborer	$27.50	$220.00	$45.45	$363.60	$32.48	$53.45
1 Equipment Operator (med.)	37.45	299.60	61.45	491.60		
1 Rock Trencher, 6" Width		797.00		876.70	49.81	54.79
16 L.H., Daily Totals		$1316.60		$1731.90	$82.29	$108.24
Crew B-54E	**Hr.**	**Daily**	**Hr.**	**Daily**	**Bare Costs**	**Incl. O&P**
1 Laborer	$27.50	$220.00	$45.45	$363.60	$32.48	$53.45
1 Equipment Operator (med.)	37.45	299.60	61.45	491.60		
1 Rock Trencher, 18" Width		2678.00		2945.80	167.38	184.11
16 L.H., Daily Totals		$3197.60		$3801.00	$199.85	$237.56
Crew B-55	**Hr.**	**Daily**	**Hr.**	**Daily**	**Bare Costs**	**Incl. O&P**
1 Laborer	$27.50	$220.00	$45.45	$363.60	$29.90	$49.30
1 Truck Driver (light)	32.30	258.40	53.15	425.20		
1 Truck-Mounted Earth Auger		756.20		831.82		
1 Flatbed Truck, Gas, 3 Ton		245.95		270.55	62.63	68.90
16 L.H., Daily Totals		$1480.55		$1891.17	$92.53	$118.20
Crew B-56	**Hr.**	**Daily**	**Hr.**	**Daily**	**Bare Costs**	**Incl. O&P**
2 Laborers	$27.50	$440.00	$45.45	$727.20	$27.50	$45.45
1 Air Track Drill, 4"		1004.00		1104.40		
1 Air Compressor, 600 cfm		417.60		459.36		
1 -50' Air Hose, 3"		25.75		28.32	90.46	99.51
16 L.H., Daily Totals		$1887.35		$2319.28	$117.96	$144.96
Crew B-57	**Hr.**	**Daily**	**Hr.**	**Daily**	**Bare Costs**	**Incl. O&P**
1 Labor Foreman (outside)	$29.50	$236.00	$48.75	$390.00	$29.98	$49.46
3 Laborers	27.50	660.00	45.45	1090.80		
1 Equip. Oper. (crane)	37.90	303.20	62.20	497.60		
1 Crawler Crane, 25 Ton		1386.00		1524.60		
1 Clamshell Bucket, 1 C.Y.		51.20		56.32		
1 Centr. Water Pump, 6"		296.45		326.10		
1 -20' Suction Hose, 6"		11.10		12.21		
20 -50' Discharge Hoses, 6"		114.00		125.40	46.47	51.12
40 L.H., Daily Totals		$3057.95		$4023.03	$76.45	$100.58
Crew B-58	**Hr.**	**Daily**	**Hr.**	**Daily**	**Bare Costs**	**Incl. O&P**
2 Laborers	$27.50	$440.00	$45.45	$727.20	$30.22	$49.80
1 Equip. Oper. (light)	35.65	285.20	58.50	468.00		
1 Backhoe Loader, 48 H.P.		320.20		352.22		
1 Small Helicopter, w/ Pilot		2916.00		3207.60	134.84	148.33
24 L.H., Daily Totals		$3961.40		$4755.02	$165.06	$198.13
Crew B-59	**Hr.**	**Daily**	**Hr.**	**Daily**	**Bare Costs**	**Incl. O&P**
1 Truck Driver (heavy)	$32.90	$263.20	$54.10	$432.80	$32.90	$54.10
1 Truck Tractor, 220 H.P.		303.55		333.90		
1 Water Tank Trailer, 5000 Gal.		149.45		164.40	56.63	62.29
8 L.H., Daily Totals		$716.20		$931.10	$89.53	$116.39
Crew B-60	**Hr.**	**Daily**	**Hr.**	**Daily**	**Bare Costs**	**Incl. O&P**
1 Labor Foreman (outside)	$29.50	$236.00	$48.75	$390.00	$30.93	$50.97
3 Laborers	27.50	660.00	45.45	1090.80		
1 Equip. Oper. (crane)	37.90	303.20	62.20	497.60		
1 Equip. Oper. (light)	35.65	285.20	58.50	468.00		
1 Crawler Crane, 40 Ton		1392.00		1531.20		
1 Lead, 60' High		80.55		88.61		
1 Hammer, Diesel, 15K ft.-lbs.		622.40		684.64		
1 Backhoe Loader, 48 H.P.		320.20		352.22	50.32	55.35
48 L.H., Daily Totals		$3899.55		$5103.06	$81.24	$106.31

Crew No.	Bare Costs		Incl. Subs O&P		Cost Per Labor-Hour	
Crew B-61	**Hr.**	**Daily**	**Hr.**	**Daily**	**Bare Costs**	**Incl. O&P**
1 Labor Foreman (outside)	$29.50	$236.00	$48.75	$390.00	$27.90	$46.11
4 Laborers	27.50	880.00	45.45	1454.40		
1 Cement Mixer, 2 C.Y.		203.85		224.24		
1 Air Compressor, 160 cfm		115.85		127.44	7.99	8.79
40 L.H., Daily Totals		$1435.70		$2196.07	$35.89	$54.90
Crew B-62	**Hr.**	**Daily**	**Hr.**	**Daily**	**Bare Costs**	**Incl. O&P**
2 Laborers	$27.50	$440.00	$45.45	$727.20	$30.22	$49.80
1 Equip. Oper. (light)	35.65	285.20	58.50	468.00		
1 Loader, Skid Steer, 30 H.P.		174.55		192.01	7.27	8.00
24 L.H., Daily Totals		$899.75		$1387.20	$37.49	$57.80
Crew B-62A	**Hr.**	**Daily**	**Hr.**	**Daily**	**Bare Costs**	**Incl. O&P**
2 Laborers	$27.50	$440.00	$45.45	$727.20	$30.22	$49.80
1 Equip. Oper. (light)	35.65	285.20	58.50	468.00		
1 Loader, Skid Steer, 30 H.P.		174.55		192.01		
1 Trencher Attachment		70.20		77.22	10.20	11.22
24 L.H., Daily Totals		$969.95		$1464.43	$40.41	$61.02
Crew B-63	**Hr.**	**Daily**	**Hr.**	**Daily**	**Bare Costs**	**Incl. O&P**
5 Laborers	$27.50	$1100.00	$45.45	$1818.00	$27.50	$45.45
1 Loader, Skid Steer, 30 H.P.		174.55		192.01	4.36	4.80
40 L.H., Daily Totals		$1274.55		$2010.01	$31.86	$50.25
Crew B-63B	**Hr.**	**Daily**	**Hr.**	**Daily**	**Bare Costs**	**Incl. O&P**
1 Labor Foreman (inside)	$28.00	$224.00	$46.30	$370.40	$29.66	$48.92
2 Laborers	27.50	440.00	45.45	727.20		
1 Equip. Oper. (light)	35.65	285.20	58.50	468.00		
1 Loader, Skid Steer, 78 H.P.		392.50		431.75	12.27	13.49
32 L.H., Daily Totals		$1341.70		$1997.35	$41.93	$62.42
Crew B-64	**Hr.**	**Daily**	**Hr.**	**Daily**	**Bare Costs**	**Incl. O&P**
1 Laborer	$27.50	$220.00	$45.45	$363.60	$29.90	$49.30
1 Truck Driver (light)	32.30	258.40	53.15	425.20		
1 Power Mulcher (small)		139.35		153.29		
1 Flatbed Truck, Gas, 1.5 Ton		195.20		214.72	20.91	23.00
16 L.H., Daily Totals		$812.95		$1156.81	$50.81	$72.30
Crew B-65	**Hr.**	**Daily**	**Hr.**	**Daily**	**Bare Costs**	**Incl. O&P**
1 Laborer	$27.50	$220.00	$45.45	$363.60	$29.90	$49.30
1 Truck Driver (light)	32.30	258.40	53.15	425.20		
1 Power Mulcher (Large)		276.35		303.99		
1 Flatbed Truck, Gas, 1.5 Ton		195.20		214.72	29.47	32.42
16 L.H., Daily Totals		$949.95		$1307.51	$59.37	$81.72
Crew B-66	**Hr.**	**Daily**	**Hr.**	**Daily**	**Bare Costs**	**Incl. O&P**
1 Equip. Oper. (light)	$35.65	$285.20	$58.50	$468.00	$35.65	$58.50
1 Loader-Backhoe, 40 H.P.		219.20		241.12	27.40	30.14
8 L.H., Daily Totals		$504.40		$709.12	$63.05	$88.64
Crew B-67	**Hr.**	**Daily**	**Hr.**	**Daily**	**Bare Costs**	**Incl. O&P**
1 Millwright	$37.90	$303.20	$60.40	$483.20	$36.77	$59.45
1 Equip. Oper. (light)	35.65	285.20	58.50	468.00		
1 R.T. Forklift, 5,000 Lb., diesel		265.60		292.16	16.60	18.26
16 L.H., Daily Totals		$854.00		$1243.36	$53.38	$77.71
Crew B-67B	**Hr.**	**Daily**	**Hr.**	**Daily**	**Bare Costs**	**Incl. O&P**
1 Millwright Foreman (inside)	$38.40	$307.20	$61.20	$489.60	$38.15	$60.80
1 Millwright	37.90	303.20	60.40	483.20		
16 L.H., Daily Totals		$610.40		$972.80	$38.15	$60.80

Crew No.	Bare Costs		Incl. Subs O&P		Cost Per Labor-Hour	
Crew B-68	**Hr.**	**Daily**	**Hr.**	**Daily**	**Bare Costs**	**Incl. O&P**
2 Millwrights	$37.90	$606.40	$60.40	$966.40	$37.15	$59.77
1 Equip. Oper. (light)	35.65	285.20	58.50	468.00		
1 R.T. Forklift, 5,000 Lb., diesel		265.60		292.16	11.07	12.17
24 L.H., Daily Totals		$1157.20		$1726.56	$48.22	$71.94
Crew B-68A	**Hr.**	**Daily**	**Hr.**	**Daily**	**Bare Costs**	**Incl. O&P**
1 Millwright Foreman (inside)	$38.40	$307.20	$61.20	$489.60	$38.07	$60.67
2 Millwrights	37.90	606.40	60.40	966.40		
1 Forklift, Smooth Floor, 8,000 Lb.		144.30		158.73	6.01	6.61
24 L.H., Daily Totals		$1057.90		$1614.73	$44.08	$67.28
Crew B-68B	**Hr.**	**Daily**	**Hr.**	**Daily**	**Bare Costs**	**Incl. O&P**
1 Millwright Foreman (inside)	$38.40	$307.20	$61.20	$489.60	$39.69	$64.24
2 Millwrights	37.90	606.40	60.40	966.40		
2 Electricians	41.40	662.40	67.55	1080.80		
2 Plumbers	40.40	646.40	66.30	1060.80		
1 R.T. Forklift, 5,000 Lb., gas		276.20		303.82	4.93	5.43
56 L.H., Daily Totals		$2498.60		$3901.42	$44.62	$69.67
Crew B-68C	**Hr.**	**Daily**	**Hr.**	**Daily**	**Bare Costs**	**Incl. O&P**
1 Millwright Foreman (inside)	$38.40	$307.20	$61.20	$489.60	$39.52	$63.86
1 Millwright	37.90	303.20	60.40	483.20		
1 Electrician	41.40	331.20	67.55	540.40		
1 Plumber	40.40	323.20	66.30	530.40		
1 R.T. Forklift, 5,000 Lb., gas		276.20		303.82	8.63	9.49
32 L.H., Daily Totals		$1541.00		$2347.42	$48.16	$73.36
Crew B-68D	**Hr.**	**Daily**	**Hr.**	**Daily**	**Bare Costs**	**Incl. O&P**
1 Labor Foreman (inside)	$28.00	$224.00	$46.30	$370.40	$30.38	$50.08
1 Laborer	27.50	220.00	45.45	363.60		
1 Equip. Oper. (light)	35.65	285.20	58.50	468.00		
1 R.T. Forklift, 5,000 Lb., gas		276.20		303.82	11.51	12.66
24 L.H., Daily Totals		$1005.40		$1505.82	$41.89	$62.74
Crew B-68E	**Hr.**	**Daily**	**Hr.**	**Daily**	**Bare Costs**	**Incl. O&P**
1 Struc. Steel Foreman (inside)	$40.25	$322.00	$69.85	$558.80	$39.85	$69.13
3 Struc. Steel Workers	39.75	954.00	68.95	1654.80		
1 Welder	39.75	318.00	68.95	551.60		
1 Forklift, Smooth Floor, 8,000 Lb.		144.30		158.73	3.61	3.97
40 L.H., Daily Totals		$1738.30		$2923.93	$43.46	$73.10
Crew B-68F	**Hr.**	**Daily**	**Hr.**	**Daily**	**Bare Costs**	**Incl. O&P**
1 Skilled Worker Foreman (out)	$38.50	$308.00	$63.95	$511.60	$37.17	$61.75
2 Skilled Workers	36.50	584.00	60.65	970.40		
1 R.T. Forklift, 5,000 Lb., gas		276.20		303.82	11.51	12.66
24 L.H., Daily Totals		$1168.20		$1785.82	$48.67	$74.41
Crew B-68G	**Hr.**	**Daily**	**Hr.**	**Daily**	**Bare Costs**	**Incl. O&P**
2 Structural Steel Workers	$39.75	$636.00	$68.95	$1103.20	$39.75	$68.95
1 R.T. Forklift, 5,000 Lb., gas		276.20		303.82	17.26	18.99
16 L.H., Daily Totals		$912.20		$1407.02	$57.01	$87.94
Crew B-69	**Hr.**	**Daily**	**Hr.**	**Daily**	**Bare Costs**	**Incl. O&P**
1 Labor Foreman (outside)	$29.50	$236.00	$48.75	$390.00	$30.48	$50.23
3 Laborers	27.50	660.00	45.45	1090.80		
1 Equip. Oper. (crane)	37.90	303.20	62.20	497.60		
1 Equip. Oper. (oiler)	32.95	263.60	54.05	432.40		
1 Hyd. Crane, 80 Ton		1487.00		1635.70	30.98	34.08
48 L.H., Daily Totals		$2949.80		$4046.50	$61.45	$84.30

Crew No.	Bare Costs		Incl. Subs O&P		Cost Per Labor-Hour	
Crew B-69A	**Hr.**	**Daily**	**Hr.**	**Daily**	**Bare Costs**	**Incl. O&P**
1 Labor Foreman (outside)	$29.50	$236.00	$48.75	$390.00	$30.86	$50.71
3 Laborers	27.50	660.00	45.45	1090.80		
1 Equip. Oper. (medium)	37.45	299.60	61.45	491.60		
1 Concrete Finisher	35.70	285.60	57.70	461.60		
1 Curb/Gutter Paver, 2-Track		1194.00		1313.40	24.88	27.36
48 L.H., Daily Totals		$2675.20		$3747.40	$55.73	$78.07
Crew B-69B	**Hr.**	**Daily**	**Hr.**	**Daily**	**Bare Costs**	**Incl. O&P**
1 Labor Foreman (outside)	$29.50	$236.00	$48.75	$390.00	$30.86	$50.71
3 Laborers	27.50	660.00	45.45	1090.80		
1 Equip. Oper. (medium)	37.45	299.60	61.45	491.60		
1 Cement Finisher	35.70	285.60	57.70	461.60		
1 Curb/Gutter Paver, 4-Track		776.35		853.99	16.17	17.79
48 L.H., Daily Totals		$2257.55		$3287.99	$47.03	$68.50
Crew B-70	**Hr.**	**Daily**	**Hr.**	**Daily**	**Bare Costs**	**Incl. O&P**
1 Labor Foreman (outside)	$29.50	$236.00	$48.75	$390.00	$32.05	$52.78
3 Laborers	27.50	660.00	45.45	1090.80		
3 Equip. Oper. (medium)	37.45	898.80	61.45	1474.80		
1 Grader, 30,000 Lbs.		657.85		723.63		
1 Ripper, Beam & 1 Shank		88.80		97.68		
1 Road Sweeper, S.P., 8' wide		703.10		773.41		
1 F.E. Loader, W.M., 1.5 C.Y.		352.95		388.25	32.19	35.41
56 L.H., Daily Totals		$3597.50		$4938.57	$64.24	$88.19
Crew B-71	**Hr.**	**Daily**	**Hr.**	**Daily**	**Bare Costs**	**Incl. O&P**
1 Labor Foreman (outside)	$29.50	$236.00	$48.75	$390.00	$32.05	$52.78
3 Laborers	27.50	660.00	45.45	1090.80		
3 Equip. Oper. (medium)	37.45	898.80	61.45	1474.80		
1 Pvmt. Profiler, 750 H.P.		5381.00		5919.10		
1 Road Sweeper, S.P., 8' wide		703.10		773.41		
1 F.E. Loader, W.M., 1.5 C.Y.		352.95		388.25	114.95	126.44
56 L.H., Daily Totals		$8231.85		$10036.36	$147.00	$179.22
Crew B-72	**Hr.**	**Daily**	**Hr.**	**Daily**	**Bare Costs**	**Incl. O&P**
1 Labor Foreman (outside)	$29.50	$236.00	$48.75	$390.00	$32.73	$53.86
3 Laborers	27.50	660.00	45.45	1090.80		
4 Equip. Oper. (medium)	37.45	1198.40	61.45	1966.40		
1 Pvmt. Profiler, 750 H.P.		5381.00		5919.10		
1 Hammermill, 250 H.P.		1890.00		2079.00		
1 Windrow Loader		1238.00		1361.80		
1 Mix Paver, 165 H.P.		2107.00		2317.70		
1 Roller, Pneum. Whl., 12 Ton		338.35		372.19	171.16	188.28
64 L.H., Daily Totals		$13048.75		$15496.99	$203.89	$242.14
Crew B-73	**Hr.**	**Daily**	**Hr.**	**Daily**	**Bare Costs**	**Incl. O&P**
1 Labor Foreman (outside)	$29.50	$236.00	$48.75	$390.00	$33.97	$55.86
2 Laborers	27.50	440.00	45.45	727.20		
5 Equip. Oper. (medium)	37.45	1498.00	61.45	2458.00		
1 Road Mixer, 310 H.P.		1857.00		2042.70		
1 Tandem Roller, 10 Ton		232.05		255.26		
1 Hammermill, 250 H.P.		1890.00		2079.00		
1 Grader, 30,000 Lbs.		657.85		723.63		
.5 F.E. Loader, W.M., 1.5 C.Y.		176.47		194.12		
.5 Truck Tractor, 220 H.P.		151.78		166.95		
.5 Water Tank Trailer, 5000 Gal.		74.72		82.20	78.75	86.62
64 L.H., Daily Totals		$7213.88		$9119.06	$112.72	$142.49

Crew No.	Bare Costs		Incl. Subs O&P		Cost Per Labor-Hour	
Crew B-74	**Hr.**	**Daily**	**Hr.**	**Daily**	**Bare Costs**	**Incl. O&P**
1 Labor Foreman (outside)	$29.50	$236.00	$48.75	$390.00	$34.08	$56.02
1 Laborer	27.50	220.00	45.45	363.60		
4 Equip. Oper. (medium)	37.45	1198.40	61.45	1966.40		
2 Truck Drivers (heavy)	32.90	526.40	54.10	865.60		
1 Grader, 30,000 Lbs.		657.85		723.63		
1 Ripper, Beam & 1 Shank		88.80		97.68		
2 Stabilizers, 310 H.P.		3252.00		3577.20		
1 Flatbed Truck, Gas, 3 Ton		245.95		270.55		
1 Chem. Spreader, Towed		53.20		58.52		
1 Roller, Vibratory, 25 Ton		650.90		715.99		
1 Water Tank Trailer, 5000 Gal.		149.45		164.40		
1 Truck Tractor, 220 H.P.		303.55		333.90	84.40	92.84
64 L.H., Daily Totals		$7582.50		$9527.47	$118.48	$148.87
Crew B-75	**Hr.**	**Daily**	**Hr.**	**Daily**	**Bare Costs**	**Incl. O&P**
1 Labor Foreman (outside)	$29.50	$236.00	$48.75	$390.00	$34.24	$56.30
1 Laborer	27.50	220.00	45.45	363.60		
4 Equip. Oper. (medium)	37.45	1198.40	61.45	1966.40		
1 Truck Driver (heavy)	32.90	263.20	54.10	432.80		
1 Grader, 30,000 Lbs.		657.85		723.63		
1 Ripper, Beam & 1 Shank		88.80		97.68		
2 Stabilizers, 310 H.P.		3252.00		3577.20		
1 Dist. Tanker, 3000 Gallon		323.25		355.57		
1 Truck Tractor, 6x4, 380 H.P.		487.60		536.36		
1 Roller, Vibratory, 25 Ton		650.90		715.99	97.51	107.26
56 L.H., Daily Totals		$7378.00		$9159.24	$131.75	$163.56
Crew B-76	**Hr.**	**Daily**	**Hr.**	**Daily**	**Bare Costs**	**Incl. O&P**
1 Dock Builder Foreman (outside)	$38.00	$304.00	$64.75	$518.00	$36.31	$61.11
5 Dock Builders	36.00	1440.00	61.35	2454.00		
2 Equip. Oper. (crane)	37.90	606.40	62.20	995.20		
1 Equip. Oper. (oiler)	32.95	263.60	54.05	432.40		
1 Crawler Crane, 50 Ton		2032.00		2235.20		
1 Barge, 400 Ton		836.50		920.15		
1 Hammer, Diesel, 15K ft.-lbs.		622.40		684.64		
1 Lead, 60' High		80.55		88.61		
1 Air Compressor, 600 cfm		417.60		459.36		
2 -50' Air Hoses, 3"		51.50		56.65	56.12	61.73
72 L.H., Daily Totals		$6654.55		$8844.20	$92.42	$122.84
Crew B-76A	**Hr.**	**Daily**	**Hr.**	**Daily**	**Bare Costs**	**Incl. O&P**
1 Labor Foreman (outside)	$29.50	$236.00	$48.75	$390.00	$29.73	$49.03
5 Laborers	27.50	1100.00	45.45	1818.00		
1 Equip. Oper. (crane)	37.90	303.20	62.20	497.60		
1 Equip. Oper. (oiler)	32.95	263.60	54.05	432.40		
1 Crawler Crane, 50 Ton		2032.00		2235.20		
1 Barge, 400 Ton		836.50		920.15	44.82	49.30
64 L.H., Daily Totals		$4771.30		$6293.35	$74.55	$98.33
Crew B-77	**Hr.**	**Daily**	**Hr.**	**Daily**	**Bare Costs**	**Incl. O&P**
1 Labor Foreman (outside)	$29.50	$236.00	$48.75	$390.00	$28.86	$47.65
3 Laborers	27.50	660.00	45.45	1090.80		
1 Truck Driver (light)	32.30	258.40	53.15	425.20		
1 Crack Cleaner, 25 H.P.		56.95		62.65		
1 Crack Filler, Trailer Mtd.		198.95		218.85		
1 Flatbed Truck, Gas, 3 Ton		245.95		270.55	12.55	13.80
40 L.H., Daily Totals		$1656.25		$2458.03	$41.41	$61.45

Crew No.	Bare Costs		Incl. Subs O&P		Cost Per Labor-Hour	
Crew B-78	**Hr.**	**Daily**	**Hr.**	**Daily**	**Bare Costs**	**Incl. O&P**
1 Labor Foreman (outside)	$29.50	$236.00	$48.75	$390.00	$27.90	$46.11
4 Laborers	27.50	880.00	45.45	1454.40		
1 Paint Striper, S.P., 40 Gallon		152.25		167.47		
1 Flatbed Truck, Gas, 3 Ton		245.95		270.55		
1 Pickup Truck, 3/4 Ton		109.90		120.89	12.70	13.97
40 L.H., Daily Totals		$1624.10		$2403.31	$40.60	$60.08
Crew B-78B	**Hr.**	**Daily**	**Hr.**	**Daily**	**Bare Costs**	**Incl. O&P**
2 Laborers	$27.50	$440.00	$45.45	$727.20	$28.41	$46.90
.25 Equip. Oper. (light)	35.65	71.30	58.50	117.00		
1 Pickup Truck, 3/4 Ton		109.90		120.89		
1 Line Rem.,11 H.P.,Walk Behind		61.35		67.48		
.25 Road Sweeper, S.P., 8' wide		175.78		193.35	19.28	21.21
18 L.H., Daily Totals		$858.33		$1225.93	$47.68	$68.11
Crew B-78C	**Hr.**	**Daily**	**Hr.**	**Daily**	**Bare Costs**	**Incl. O&P**
1 Labor Foreman (outside)	$29.50	$236.00	$48.75	$390.00	$28.63	$47.28
4 Laborers	27.50	880.00	45.45	1454.40		
1 Truck Driver (light)	32.30	258.40	53.15	425.20		
1 Paint Striper, T.M., 120 Gal.		686.10		754.71		
1 Flatbed Truck, Gas, 3 Ton		245.95		270.55		
1 Pickup Truck, 3/4 Ton		109.90		120.89	21.71	23.88
48 L.H., Daily Totals		$2416.35		$3415.74	$50.34	$71.16
Crew B-78D	**Hr.**	**Daily**	**Hr.**	**Daily**	**Bare Costs**	**Incl. O&P**
2 Labor Foremen (outside)	$29.50	$472.00	$48.75	$780.00	$28.38	$46.88
7 Laborers	27.50	1540.00	45.45	2545.20		
1 Truck Driver (light)	32.30	258.40	53.15	425.20		
1 Paint Striper, T.M., 120 Gal.		686.10		754.71		
1 Flatbed Truck, Gas, 3 Ton		245.95		270.55		
3 Pickup Trucks, 3/4 Ton		329.70		362.67		
1 Air Compressor, 60 cfm		106.70		117.37		
1 -50' Air Hose, 3/4"		6.55		7.21		
1 Breaker, Pavement, 60 lb.		10.75		11.82	17.32	19.05
80 L.H., Daily Totals		$3656.15		$5274.73	$45.70	$65.93
Crew B-78E	**Hr.**	**Daily**	**Hr.**	**Daily**	**Bare Costs**	**Incl. O&P**
2 Labor Foremen (outside)	$29.50	$472.00	$48.75	$780.00	$28.23	$46.64
9 Laborers	27.50	1980.00	45.45	3272.40		
1 Truck Driver (light)	32.30	258.40	53.15	425.20		
1 Paint Striper, T.M., 120 Gal.		686.10		754.71		
1 Flatbed Truck, Gas, 3 Ton		245.95		270.55		
4 Pickup Trucks, 3/4 Ton		439.60		483.56		
2 Air Compressors, 60 cfm		213.40		234.74		
2 -50' Air Hoses, 3/4"		13.10		14.41		
2 Breakers, Pavement, 60 lb.		21.50		23.65	16.87	18.56
96 L.H., Daily Totals		$4330.05		$6259.22	$45.10	$65.20
Crew B-78F	**Hr.**	**Daily**	**Hr.**	**Daily**	**Bare Costs**	**Incl. O&P**
2 Labor Foremen (outside)	$29.50	$472.00	$48.75	$780.00	$28.13	$46.47
11 Laborers	27.50	2420.00	45.45	3999.60		
1 Truck Driver (light)	32.30	258.40	53.15	425.20		
1 Paint Striper, T.M., 120 Gal.		686.10		754.71		
1 Flatbed Truck, Gas, 3 Ton		245.95		270.55		
7 Pickup Trucks, 3/4 Ton		769.30		846.23		
3 Air Compressors, 60 cfm		320.10		352.11		
3 -50' Air Hoses, 3/4"		19.65		21.61		
3 Breakers, Pavement, 60 lb.		32.25		35.48	18.51	20.36
112 L.H., Daily Totals		$5223.75		$7485.48	$46.64	$66.83

Crew No.	Bare Costs		Incl. Subs O&P		Cost Per Labor-Hour	
Crew B-79	**Hr.**	**Daily**	**Hr.**	**Daily**	**Bare Costs**	**Incl. O&P**
1 Labor Foreman (outside)	$29.50	$236.00	$48.75	$390.00	$28.86	$47.65
3 Laborers	27.50	660.00	45.45	1090.80		
1 Truck Driver (light)	32.30	258.40	53.15	425.20		
1 Paint Striper, T.M., 120 Gal.		686.10		754.71		
1 Heating Kettle, 115 Gallon		77.85		85.64		
1 Flatbed Truck, Gas, 3 Ton		245.95		270.55		
2 Pickup Trucks, 3/4 Ton		219.80		241.78	30.74	33.82
40 L.H., Daily Totals		$2384.10		$3258.67	$59.60	$81.47

Crew No.	Bare Costs		Incl. Subs O&P		Cost Per Labor-Hour	
Crew B-79B	**Hr.**	**Daily**	**Hr.**	**Daily**	**Bare Costs**	**Incl. O&P**
1 Laborer	$27.50	$220.00	$45.45	$363.60	$27.50	$45.45
1 Set of Gases		167.85		184.63	20.98	23.08
8 L.H., Daily Totals		$387.85		$548.24	$48.48	$68.53

Crew No.	Bare Costs		Incl. Subs O&P		Cost Per Labor-Hour	
Crew B-79C	**Hr.**	**Daily**	**Hr.**	**Daily**	**Bare Costs**	**Incl. O&P**
1 Labor Foreman (outside)	$29.50	$236.00	$48.75	$390.00	$28.47	$47.02
5 Laborers	27.50	1100.00	45.45	1818.00		
1 Truck Driver (light)	32.30	258.40	53.15	425.20		
1 Paint Striper, T.M., 120 Gal.		686.10		754.71		
1 Heating Kettle, 115 Gallon		77.85		85.64		
1 Flatbed Truck, Gas, 3 Ton		245.95		270.55		
3 Pickup Trucks, 3/4 Ton		329.70		362.67		
1 Air Compressor, 60 cfm		106.70		117.37		
1 -50' Air Hose, 3/4"		6.55		7.21		
1 Breaker, Pavement, 60 lb.		10.75		11.82	26.14	28.75
56 L.H., Daily Totals		$3058.00		$4243.16	$54.61	$75.77

Crew No.	Bare Costs		Incl. Subs O&P		Cost Per Labor-Hour	
Crew B-79D	**Hr.**	**Daily**	**Hr.**	**Daily**	**Bare Costs**	**Incl. O&P**
2 Labor Foremen (outside)	$29.50	$472.00	$48.75	$780.00	$28.60	$47.24
5 Laborers	27.50	1100.00	45.45	1818.00		
1 Truck Driver (light)	32.30	258.40	53.15	425.20		
1 Paint Striper, T.M., 120 Gal.		686.10		754.71		
1 Heating Kettle, 115 Gallon		77.85		85.64		
1 Flatbed Truck, Gas, 3 Ton		245.95		270.55		
4 Pickup Trucks, 3/4 Ton		439.60		483.56		
1 Air Compressor, 60 cfm		106.70		117.37		
1 -50' Air Hose, 3/4"		6.55		7.21		
1 Breaker, Pavement, 60 lb.		10.75		11.82	24.59	27.04
64 L.H., Daily Totals		$3403.90		$4754.05	$53.19	$74.28

Crew No.	Bare Costs		Incl. Subs O&P		Cost Per Labor-Hour	
Crew B-79E	**Hr.**	**Daily**	**Hr.**	**Daily**	**Bare Costs**	**Incl. O&P**
2 Labor Foremen (outside)	$29.50	$472.00	$48.75	$780.00	$28.38	$46.88
7 Laborers	27.50	1540.00	45.45	2545.20		
1 Truck Driver (light)	32.30	258.40	53.15	425.20		
1 Paint Striper, T.M., 120 Gal.		686.10		754.71		
1 Heating Kettle, 115 Gallon		77.85		85.64		
1 Flatbed Truck, Gas, 3 Ton		245.95		270.55		
5 Pickup Trucks, 3/4 Ton		549.50		604.45		
2 Air Compressors, 60 cfm		213.40		234.74		
2 -50' Air Hoses, 3/4"		13.10		14.41		
2 Breakers, Pavement, 60 lb.		21.50		23.65	22.59	24.85
80 L.H., Daily Totals		$4077.80		$5738.54	$50.97	$71.73

Crew No.	Bare Costs		Incl. Subs O&P		Cost Per Labor-Hour	
Crew B-80	**Hr.**	**Daily**	**Hr.**	**Daily**	**Bare Costs**	**Incl. O&P**
1 Labor Foreman (outside)	$29.50	$236.00	$48.75	$390.00	$28.17	$46.55
2 Laborers	27.50	440.00	45.45	727.20		
1 Flatbed Truck, Gas, 3 Ton		245.95		270.55		
1 Earth Auger, Truck-Mtd.		385.85		424.44	26.32	28.96
24 L.H., Daily Totals		$1307.80		$1812.18	$54.49	$75.51

Crew No.	Bare Costs		Incl. Subs O&P		Cost Per Labor-Hour	
Crew B-80A	**Hr.**	**Daily**	**Hr.**	**Daily**	**Bare Costs**	**Incl. O&P**
3 Laborers	$27.50	$660.00	$45.45	$1090.80	$27.50	$45.45
1 Flatbed Truck, Gas, 3 Ton		245.95		270.55	10.25	11.27
24 L.H., Daily Totals		$905.95		$1361.35	$37.75	$56.72

Crew No.	Bare Costs		Incl. Subs O&P		Cost Per Labor-Hour	
Crew B-80B	**Hr.**	**Daily**	**Hr.**	**Daily**	**Bare Costs**	**Incl. O&P**
3 Laborers	$27.50	$660.00	$45.45	$1090.80	$29.54	$48.71
1 Equip. Oper. (light)	35.65	285.20	58.50	468.00		
1 Crane, Flatbed Mounted, 3 Ton		232.65		255.91	7.27	8.00
32 L.H., Daily Totals		$1177.85		$1814.71	$36.81	$56.71

Crew No.	Bare Costs		Incl. Subs O&P		Cost Per Labor-Hour	
Crew B-80C	**Hr.**	**Daily**	**Hr.**	**Daily**	**Bare Costs**	**Incl. O&P**
2 Laborers	$27.50	$440.00	$45.45	$727.20	$29.10	$48.02
1 Truck Driver (light)	32.30	258.40	53.15	425.20		
1 Flatbed Truck, Gas, 1.5 Ton		195.20		214.72		
1 Manual Fence Post Auger, Gas		7.40		8.14	8.44	9.29
24 L.H., Daily Totals		$901.00		$1375.26	$37.54	$57.30

Crew No.	Bare Costs		Incl. Subs O&P		Cost Per Labor-Hour	
Crew B-81	**Hr.**	**Daily**	**Hr.**	**Daily**	**Bare Costs**	**Incl. O&P**
1 Laborer	$27.50	$220.00	$45.45	$363.60	$30.20	$49.77
1 Truck Driver (heavy)	32.90	263.20	54.10	432.80		
1 Hydromulcher, T.M., 3000 Gal.		292.90		322.19		
1 Truck Tractor, 220 H.P.		303.55		333.90	37.28	41.01
16 L.H., Daily Totals		$1079.65		$1452.49	$67.48	$90.78

Crew No.	Bare Costs		Incl. Subs O&P		Cost Per Labor-Hour	
Crew B-81A	**Hr.**	**Daily**	**Hr.**	**Daily**	**Bare Costs**	**Incl. O&P**
1 Laborer	$27.50	$220.00	$45.45	$363.60	$29.90	$49.30
1 Truck Driver (light)	32.30	258.40	53.15	425.20		
1 Hydromulcher, T.M., 600 Gal.		124.15		136.57		
1 Flatbed Truck, Gas, 3 Ton		245.95		270.55	23.13	25.44
16 L.H., Daily Totals		$848.50		$1195.91	$53.03	$74.74

Crew No.	Bare Costs		Incl. Subs O&P		Cost Per Labor-Hour	
Crew B-82	**Hr.**	**Daily**	**Hr.**	**Daily**	**Bare Costs**	**Incl. O&P**
1 Laborer	$27.50	$220.00	$45.45	$363.60	$31.57	$51.98
1 Equip. Oper. (light)	35.65	285.20	58.50	468.00		
1 Horiz. Borer, 6 H.P.		77.85		85.64	4.87	5.35
16 L.H., Daily Totals		$583.05		$917.24	$36.44	$57.33

Crew No.	Bare Costs		Incl. Subs O&P		Cost Per Labor-Hour	
Crew B-82A	**Hr.**	**Daily**	**Hr.**	**Daily**	**Bare Costs**	**Incl. O&P**
2 Laborers	$27.50	$440.00	$45.45	$727.20	$31.57	$51.98
2 Equip. Opers. (light)	35.65	570.40	58.50	936.00		
2 Dump Trucks, 8 C.Y., 220 H.P.		679.20		747.12		
1 Flatbed Trailer, 25 Ton		133.00		146.30		
1 Horiz. Dir. Drill, 20k lb. Thrust		631.30		694.43		
1 Mud Trailer for HDD, 1500 Gal.		298.65		328.51		
1 Pickup Truck, 4x4, 3/4 Ton		121.70		133.87		
1 Flatbed Trailer, 3 Ton		26.55		29.20		
1 Loader, Skid Steer, 78 H.P.		392.50		431.75	71.34	78.47
32 L.H., Daily Totals		$3293.30		$4174.39	$102.92	$130.45

Crew No.	Bare Costs		Incl. Subs O&P		Cost Per Labor-Hour	
Crew B-82B	**Hr.**	**Daily**	**Hr.**	**Daily**	**Bare Costs**	**Incl. O&P**
2 Laborers	$27.50	$440.00	$45.45	$727.20	$31.57	$51.98
2 Equip. Opers. (light)	35.65	570.40	58.50	936.00		
2 Dump Trucks, 8 C.Y., 220 H.P.		679.20		747.12		
1 Flatbed Trailer, 25 Ton		133.00		146.30		
1 Horiz. Dir. Drill, 30k lb. Thrust		897.00		986.70		
1 Mud Trailer for HDD, 1500 Gal.		298.65		328.51		
1 Pickup Truck, 4x4, 3/4 Ton		121.70		133.87		
1 Flatbed Trailer, 3 Ton		26.55		29.20		
1 Loader, Skid Steer, 78 H.P.		392.50		431.75	79.64	87.61
32 L.H., Daily Totals		$3559.00		$4466.66	$111.22	$139.58

Crew B-82C	Bare Costs Hr.	Bare Costs Daily	Incl. Subs O&P Hr.	Incl. Subs O&P Daily	Bare Costs	Incl. O&P
2 Laborers	$27.50	$440.00	$45.45	$727.20	$31.57	$51.98
2 Equip. Opers. (light)	35.65	570.40	58.50	936.00		
2 Dump Trucks, 8 C.Y., 220 H.P.		679.20		747.12		
1 Flatbed Trailer, 25 Ton		133.00		146.30		
1 Horiz. Dir. Drill, 50k lb. Thrust		1185.00		1303.50		
1 Mud Trailer for HDD, 1500 Gal.		298.65		328.51		
1 Pickup Truck, 4x4, 3/4 Ton		121.70		133.87		
1 Flatbed Trailer, 3 Ton		26.55		29.20		
1 Loader, Skid Steer, 78 H.P.		392.50		431.75	88.64	97.51
32 L.H., Daily Totals		$3847.00		$4783.46	$120.22	$149.48

Crew B-82D	Hr.	Daily	Hr.	Daily	Bare Costs	Incl. O&P
1 Equip. Oper. (light)	$35.65	$285.20	$58.50	$468.00	$35.65	$58.50
1 Mud Trailer for HDD, 1500 Gal.		298.65		328.51	37.33	41.06
8 L.H., Daily Totals		$583.85		$796.51	$72.98	$99.56

Crew B-83	Hr.	Daily	Hr.	Daily	Bare Costs	Incl. O&P
1 Tugboat Captain	$37.45	$299.60	$61.45	$491.60	$32.48	$53.45
1 Tugboat Hand	27.50	220.00	45.45	363.60		
1 Tugboat, 250 H.P.		710.70		781.77	44.42	48.86
16 L.H., Daily Totals		$1230.30		$1636.97	$76.89	$102.31

Crew B-84	Hr.	Daily	Hr.	Daily	Bare Costs	Incl. O&P
1 Equip. Oper. (medium)	$37.45	$299.60	$61.45	$491.60	$37.45	$61.45
1 Rotary Mower/Tractor		360.30		396.33	45.04	49.54
8 L.H., Daily Totals		$659.90		$887.93	$82.49	$110.99

Crew B-85	Hr.	Daily	Hr.	Daily	Bare Costs	Incl. O&P
3 Laborers	$27.50	$660.00	$45.45	$1090.80	$30.57	$50.38
1 Equip. Oper. (medium)	37.45	299.60	61.45	491.60		
1 Truck Driver (heavy)	32.90	263.20	54.10	432.80		
1 Telescoping Boom Lift, to 80'		670.65		737.72		
1 Brush Chipper, 12", 130 H.P.		389.40		428.34		
1 Pruning Saw, Rotary		6.60		7.26	26.67	29.33
40 L.H., Daily Totals		$2289.45		$3188.51	$57.24	$79.71

Crew B-86	Hr.	Daily	Hr.	Daily	Bare Costs	Incl. O&P
1 Equip. Oper. (medium)	$37.45	$299.60	$61.45	$491.60	$37.45	$61.45
1 Stump Chipper, S.P.		182.20		200.42	22.77	25.05
8 L.H., Daily Totals		$481.80		$692.02	$60.23	$86.50

Crew B-86A	Hr.	Daily	Hr.	Daily	Bare Costs	Incl. O&P
1 Equip. Oper. (medium)	$37.45	$299.60	$61.45	$491.60	$37.45	$61.45
1 Grader, 30,000 Lbs.		657.85		723.63	82.23	90.45
8 L.H., Daily Totals		$957.45		$1215.23	$119.68	$151.90

Crew B-86B	Hr.	Daily	Hr.	Daily	Bare Costs	Incl. O&P
1 Equip. Oper. (medium)	$37.45	$299.60	$61.45	$491.60	$37.45	$61.45
1 Dozer, 200 H.P.		1290.00		1419.00	161.25	177.38
8 L.H., Daily Totals		$1589.60		$1910.60	$198.70	$238.82

Crew B-87	Hr.	Daily	Hr.	Daily	Bare Costs	Incl. O&P
1 Laborer	$27.50	$220.00	$45.45	$363.60	$35.46	$58.25
4 Equip. Oper. (medium)	37.45	1198.40	61.45	1966.40		
2 Feller Bunchers, 100 H.P.		1607.10		1767.81		
1 Log Chipper, 22" Tree		770.10		847.11		
1 Dozer, 105 H.P.		609.70		670.67		
1 Chain Saw, Gas, 36" Long		46.65		51.31	75.84	83.42
40 L.H., Daily Totals		$4451.95		$5666.90	$111.30	$141.67

Crew B-88	Bare Costs Hr.	Bare Costs Daily	Incl. Subs O&P Hr.	Incl. Subs O&P Daily	Bare Costs	Incl. O&P
1 Laborer	$27.50	$220.00	$45.45	$363.60	$36.03	$59.16
6 Equip. Oper. (medium)	37.45	1797.60	61.45	2949.60		
2 Feller Bunchers, 100 H.P.		1607.10		1767.81		
1 Log Chipper, 22" Tree		770.10		847.11		
2 Log Skidders, 50 H.P.		1696.50		1866.15		
1 Dozer, 105 H.P.		609.70		670.67		
1 Chain Saw, Gas, 36" Long		46.65		51.31	84.47	92.91
56 L.H., Daily Totals		$6747.65		$8516.25	$120.49	$152.08

Crew B-89	Hr.	Daily	Hr.	Daily	Bare Costs	Incl. O&P
1 Skilled Worker	$36.50	$292.00	$60.65	$485.20	$32.00	$53.05
1 Building Laborer	27.50	220.00	45.45	363.60		
1 Flatbed Truck, Gas, 3 Ton		245.95		270.55		
1 Concrete Saw		110.15		121.17		
1 Water Tank, 65 Gal.		98.85		108.74	28.43	31.28
16 L.H., Daily Totals		$966.95		$1349.24	$60.43	$84.33

Crew B-89A	Hr.	Daily	Hr.	Daily	Bare Costs	Incl. O&P
1 Skilled Worker	$36.50	$292.00	$60.65	$485.20	$32.00	$53.05
1 Laborer	27.50	220.00	45.45	363.60		
1 Core Drill (Large)		112.20		123.42	7.01	7.71
16 L.H., Daily Totals		$624.20		$972.22	$39.01	$60.76

Crew B-89B	Hr.	Daily	Hr.	Daily	Bare Costs	Incl. O&P
1 Equip. Oper. (light)	$35.65	$285.20	$58.50	$468.00	$33.98	$55.83
1 Truck Driver (light)	32.30	258.40	53.15	425.20		
1 Wall Saw, Hydraulic, 10 H.P.		46.60		51.26		
1 Generator, Diesel, 100 kW		308.80		339.68		
1 Water Tank, 65 Gal.		98.85		108.74		
1 Flatbed Truck, Gas, 3 Ton		245.95		270.55	43.76	48.14
16 L.H., Daily Totals		$1243.80		$1663.42	$77.74	$103.96

Crew B-90	Hr.	Daily	Hr.	Daily	Bare Costs	Incl. O&P
1 Labor Foreman (outside)	$29.50	$236.00	$48.75	$390.00	$31.14	$51.29
3 Laborers	27.50	660.00	45.45	1090.80		
2 Equip. Oper. (light)	35.65	570.40	58.50	936.00		
2 Truck Drivers (heavy)	32.90	526.40	54.10	865.60		
1 Road Mixer, 310 H.P.		1857.00		2042.70		
1 Dist. Truck, 2000 Gal.		293.20		322.52	33.60	36.96
64 L.H., Daily Totals		$4143.00		$5647.62	$64.73	$88.24

Crew B-90A	Hr.	Daily	Hr.	Daily	Bare Costs	Incl. O&P
1 Labor Foreman (outside)	$29.50	$236.00	$48.75	$390.00	$33.47	$55.06
2 Laborers	27.50	440.00	45.45	727.20		
4 Equip. Oper. (medium)	37.45	1198.40	61.45	1966.40		
2 Graders, 30,000 Lbs.		1315.70		1447.27		
1 Tandem Roller, 10 Ton		232.05		255.26		
1 Roller, Pneum. Whl., 12 Ton		338.35		372.19	33.68	37.05
56 L.H., Daily Totals		$3760.50		$5158.31	$67.15	$92.11

Crew B-90B	Hr.	Daily	Hr.	Daily	Bare Costs	Incl. O&P
1 Labor Foreman (outside)	$29.50	$236.00	$48.75	$390.00	$32.81	$54.00
2 Laborers	27.50	440.00	45.45	727.20		
3 Equip. Oper. (medium)	37.45	898.80	61.45	1474.80		
1 Roller, Pneum. Whl., 12 Ton		338.35		372.19		
1 Road Mixer, 310 H.P.		1857.00		2042.70	45.74	50.31
48 L.H., Daily Totals		$3770.15		$5006.89	$78.54	$104.31

Crew No.	Bare Costs		Incl. Subs O&P		Cost Per Labor-Hour	
Crew B-90C	**Hr.**	**Daily**	**Hr.**	**Daily**	**Bare Costs**	**Incl. O&P**
1 Labor Foreman (outside)	$29.50	$236.00	$48.75	$390.00	$31.87	$52.47
4 Laborers	27.50	880.00	45.45	1454.40		
3 Equip. Oper. (medium)	37.45	898.80	61.45	1474.80		
3 Truck Drivers (heavy)	32.90	789.60	54.10	1298.40		
3 Road Mixers, 310 H.P.		5571.00		6128.10	63.31	69.64
88 L.H., Daily Totals		$8375.40		$10745.70	$95.17	$122.11
Crew B-90D	**Hr.**	**Daily**	**Hr.**	**Daily**	**Bare Costs**	**Incl. O&P**
1 Labor Foreman (outside)	$29.50	$236.00	$48.75	$390.00	$31.20	$51.39
6 Laborers	27.50	1320.00	45.45	2181.60		
3 Equip. Oper. (medium)	37.45	898.80	61.45	1474.80		
3 Truck Drivers (heavy)	32.90	789.60	54.10	1298.40		
3 Road Mixers, 310 H.P.		5571.00		6128.10	53.57	58.92
104 L.H., Daily Totals		$8815.40		$11472.90	$84.76	$110.32
Crew B-90E	**Hr.**	**Daily**	**Hr.**	**Daily**	**Bare Costs**	**Incl. O&P**
1 Labor Foreman (outside)	$29.50	$236.00	$48.75	$390.00	$31.64	$52.11
4 Laborers	27.50	880.00	45.45	1454.40		
3 Equip. Oper. (medium)	37.45	898.80	61.45	1474.80		
1 Truck Driver (heavy)	32.90	263.20	54.10	432.80		
1 Road Mixer, 310 H.P.		1857.00		2042.70	25.79	28.37
72 L.H., Daily Totals		$4135.00		$5794.70	$57.43	$80.48
Crew B-91	**Hr.**	**Daily**	**Hr.**	**Daily**	**Bare Costs**	**Incl. O&P**
1 Labor Foreman (outside)	$29.50	$236.00	$48.75	$390.00	$33.40	$54.94
2 Laborers	27.50	440.00	45.45	727.20		
4 Equip. Oper. (medium)	37.45	1198.40	61.45	1966.40		
1 Truck Driver (heavy)	32.90	263.20	54.10	432.80		
1 Dist. Tanker, 3000 Gallon		323.25		355.57		
1 Truck Tractor, 6x4, 380 H.P.		487.60		536.36		
1 Aggreg. Spreader, S.P.		836.00		919.60		
1 Roller, Pneum. Whl., 12 Ton		338.35		372.19		
1 Tandem Roller, 10 Ton		232.05		255.26	34.64	38.11
64 L.H., Daily Totals		$4354.85		$5955.38	$68.04	$93.05
Crew B-91B	**Hr.**	**Daily**	**Hr.**	**Daily**	**Bare Costs**	**Incl. O&P**
1 Laborer	$27.50	$220.00	$45.45	$363.60	$32.48	$53.45
1 Equipment Oper. (med.)	37.45	299.60	61.45	491.60		
1 Road Sweeper, Vac. Assist.		857.60		943.36	53.60	58.96
16 L.H., Daily Totals		$1377.20		$1798.56	$86.08	$112.41
Crew B-91C	**Hr.**	**Daily**	**Hr.**	**Daily**	**Bare Costs**	**Incl. O&P**
1 Laborer	$27.50	$220.00	$45.45	$363.60	$29.90	$49.30
1 Truck Driver (light)	32.30	258.40	53.15	425.20		
1 Catch Basin Cleaning Truck		528.95		581.85	33.06	36.37
16 L.H., Daily Totals		$1007.35		$1370.65	$62.96	$85.67
Crew B-91D	**Hr.**	**Daily**	**Hr.**	**Daily**	**Bare Costs**	**Incl. O&P**
1 Labor Foreman (outside)	$29.50	$236.00	$48.75	$390.00	$32.31	$53.19
5 Laborers	27.50	1100.00	45.45	1818.00		
5 Equip. Oper. (medium)	37.45	1498.00	61.45	2458.00		
2 Truck Drivers (heavy)	32.90	526.40	54.10	865.60		
1 Aggreg. Spreader, S.P.		836.00		919.60		
2 Truck Tractors, 6x4, 380 H.P.		975.20		1072.72		
2 Dist. Tankers, 3000 Gallon		646.50		711.15		
2 Pavement Brushes, Towed		171.00		188.10		
2 Rollers Pneum. Whl., 12 Ton		676.70		744.37	31.78	34.96
104 L.H., Daily Totals		$6665.80		$9167.54	$64.09	$88.15

Crew No.	Bare Costs		Incl. Subs O&P		Cost Per Labor-Hour	
Crew B-92	**Hr.**	**Daily**	**Hr.**	**Daily**	**Bare Costs**	**Incl. O&P**
1 Labor Foreman (outside)	$29.50	$236.00	$48.75	$390.00	$28.00	$46.27
3 Laborers	27.50	660.00	45.45	1090.80		
1 Crack Cleaner, 25 H.P.		56.95		62.65		
1 Air Compressor, 60 cfm		106.70		117.37		
1 Tar Kettle, T.M.		134.50		147.95		
1 Flatbed Truck, Gas, 3 Ton		245.95		270.55	17.00	18.70
32 L.H., Daily Totals		$1440.10		$2079.31	$45.00	$64.98
Crew B-93	**Hr.**	**Daily**	**Hr.**	**Daily**	**Bare Costs**	**Incl. O&P**
1 Equip. Oper. (medium)	$37.45	$299.60	$61.45	$491.60	$37.45	$61.45
1 Feller Buncher, 100 H.P.		803.55		883.90	100.44	110.49
8 L.H., Daily Totals		$1103.15		$1375.51	$137.89	$171.94
Crew B-94A	**Hr.**	**Daily**	**Hr.**	**Daily**	**Bare Costs**	**Incl. O&P**
1 Laborer	$27.50	$220.00	$45.45	$363.60	$27.50	$45.45
1 Diaphragm Water Pump, 2"		74.30		81.73		
1 -20' Suction Hose, 2"		2.05		2.25		
2 -50' Discharge Hoses, 2"		1.80		1.98	9.77	10.75
8 L.H., Daily Totals		$298.15		$449.57	$37.27	$56.20
Crew B-94B	**Hr.**	**Daily**	**Hr.**	**Daily**	**Bare Costs**	**Incl. O&P**
1 Laborer	$27.50	$220.00	$45.45	$363.60	$27.50	$45.45
1 Diaphragm Water Pump, 4"		141.40		155.54		
1 -20' Suction Hose, 4"		3.10		3.41		
2 -50' Discharge Hoses, 4"		4.50		4.95	18.63	20.49
8 L.H., Daily Totals		$369.00		$527.50	$46.13	$65.94
Crew B-94C	**Hr.**	**Daily**	**Hr.**	**Daily**	**Bare Costs**	**Incl. O&P**
1 Laborer	$27.50	$220.00	$45.45	$363.60	$27.50	$45.45
1 Centrifugal Water Pump, 3"		79.25		87.17		
1 -20' Suction Hose, 3"		2.70		2.97		
2 -50' Discharge Hoses, 3"		2.80		3.08	10.59	11.65
8 L.H., Daily Totals		$304.75		$456.82	$38.09	$57.10
Crew B-94D	**Hr.**	**Daily**	**Hr.**	**Daily**	**Bare Costs**	**Incl. O&P**
1 Laborer	$27.50	$220.00	$45.45	$363.60	$27.50	$45.45
1 Centr. Water Pump, 6"		296.45		326.10		
1 -20' Suction Hose, 6"		11.10		12.21		
2 -50' Discharge Hoses, 6"		11.40		12.54	39.87	43.86
8 L.H., Daily Totals		$538.95		$714.45	$67.37	$89.31
Crew C-1	**Hr.**	**Daily**	**Hr.**	**Daily**	**Bare Costs**	**Incl. O&P**
2 Carpenters	$35.65	$570.40	$58.90	$942.40	$31.52	$52.25
1 Carpenter Helper	27.30	218.40	45.75	366.00		
1 Laborer	27.50	220.00	45.45	363.60		
32 L.H., Daily Totals		$1008.80		$1672.00	$31.52	$52.25
Crew C-2	**Hr.**	**Daily**	**Hr.**	**Daily**	**Bare Costs**	**Incl. O&P**
1 Carpenter Foreman (outside)	$37.65	$301.20	$62.20	$497.60	$31.84	$52.83
2 Carpenters	35.65	570.40	58.90	942.40		
2 Carpenter Helpers	27.30	436.80	45.75	732.00		
1 Laborer	27.50	220.00	45.45	363.60		
48 L.H., Daily Totals		$1528.40		$2535.60	$31.84	$52.83
Crew C-2A	**Hr.**	**Daily**	**Hr.**	**Daily**	**Bare Costs**	**Incl. O&P**
1 Carpenter Foreman (outside)	$37.65	$301.20	$62.20	$497.60	$34.63	$57.01
3 Carpenters	35.65	855.60	58.90	1413.60		
1 Cement Finisher	35.70	285.60	57.70	461.60		
1 Laborer	27.50	220.00	45.45	363.60		
48 L.H., Daily Totals		$1662.40		$2736.40	$34.63	$57.01

Crew C-3	Bare Costs Hr.	Bare Costs Daily	Incl. Subs O&P Hr.	Incl. Subs O&P Daily	Cost Per Labor-Hour Bare Costs	Cost Per Labor-Hour Incl. O&P
1 Rodman Foreman (outside)	$40.95	$327.60	$67.70	$541.60	$34.49	$56.97
3 Rodmen (reinf.)	38.95	934.80	64.40	1545.60		
1 Equip. Oper. (light)	35.65	285.20	58.50	468.00		
3 Laborers	27.50	660.00	45.45	1090.80		
3 Stressing Equipment		31.20		34.32		
.5 Grouting Equipment		77.70		85.47	1.70	1.87
64 L.H., Daily Totals		$2316.50		$3765.79	$36.20	$58.84

Crew C-4	Hr.	Daily	Hr.	Daily	Bare Costs	Incl. O&P
1 Rodman Foreman (outside)	$40.95	$327.60	$67.70	$541.60	$36.59	$60.49
2 Rodmen (reinf.)	38.95	623.20	64.40	1030.40		
1 Building Laborer	27.50	220.00	45.45	363.60		
3 Stressing Equipment		31.20		34.32	0.97	1.07
32 L.H., Daily Totals		$1202.00		$1969.92	$37.56	$61.56

Crew C-4A	Hr.	Daily	Hr.	Daily	Bare Costs	Incl. O&P
2 Rodmen (reinf.)	$38.95	$623.20	$64.40	$1030.40	$38.95	$64.40
4 Stressing Equipment		41.60		45.76	2.60	2.86
16 L.H., Daily Totals		$664.80		$1076.16	$41.55	$67.26

Crew C-5	Hr.	Daily	Hr.	Daily	Bare Costs	Incl. O&P
1 Rodman Foreman (outside)	$40.95	$327.60	$67.70	$541.60	$35.29	$58.27
2 Rodmen (reinf.)	38.95	623.20	64.40	1030.40		
1 Equip. Oper. (crane)	37.90	303.20	62.20	497.60		
2 Building Laborers	27.50	440.00	45.45	727.20		
1 Hyd. Crane, 25 Ton		581.70		639.87	12.12	13.33
48 L.H., Daily Totals		$2275.70		$3436.67	$47.41	$71.60

Crew C-6	Hr.	Daily	Hr.	Daily	Bare Costs	Incl. O&P
1 Labor Foreman (outside)	$29.50	$236.00	$48.75	$390.00	$29.20	$48.04
4 Laborers	27.50	880.00	45.45	1454.40		
1 Cement Finisher	35.70	285.60	57.70	461.60		
2 Gas Engine Vibrators		53.10		58.41	1.11	1.22
48 L.H., Daily Totals		$1454.70		$2364.41	$30.31	$49.26

Crew C-6A	Hr.	Daily	Hr.	Daily	Bare Costs	Incl. O&P
2 Cement Finishers	$35.70	$571.20	$57.70	$923.20	$35.70	$57.70
1 Concrete Vibrator, Elec, 2 HP		9.25		10.18	0.58	0.64
16 L.H., Daily Totals		$580.45		$933.38	$36.28	$58.34

Crew B-89C	Hr.	Daily	Hr.	Daily	Bare Costs	Incl. O&P
1 Cement Finisher	$35.70	$285.60	$57.70	$461.60	$35.70	$57.70
1 Masonry cut-off saw, gas		34.00		37.40	4.25	4.67
8 L.H., Daily Totals		$319.60		$499.00	$39.95	$62.38

Crew C-7	Hr.	Daily	Hr.	Daily	Bare Costs	Incl. O&P
1 Labor Foreman (outside)	$29.50	$236.00	$48.75	$390.00	$30.34	$49.91
5 Laborers	27.50	1100.00	45.45	1818.00		
1 Cement Finisher	35.70	285.60	57.70	461.60		
1 Equip. Oper. (medium)	37.45	299.60	61.45	491.60		
1 Equip. Oper. (oiler)	32.95	263.60	54.05	432.40		
2 Gas Engine Vibrators		53.10		58.41		
1 Concrete Bucket, 1 C.Y.		24.95		27.45		
1 Hyd. Crane, 55 Ton		981.50		1079.65	14.72	16.19
72 L.H., Daily Totals		$3244.35		$4759.10	$45.06	$66.10

Crew C-8	Bare Costs Hr.	Bare Costs Daily	Incl. Subs O&P Hr.	Incl. Subs O&P Daily	Cost Per Labor-Hour Bare Costs	Cost Per Labor-Hour Incl. O&P
1 Labor Foreman (outside)	$29.50	$236.00	$48.75	$390.00	$31.55	$51.71
3 Laborers	27.50	660.00	45.45	1090.80		
2 Cement Finishers	35.70	571.20	57.70	923.20		
1 Equip. Oper. (medium)	37.45	299.60	61.45	491.60		
1 Concrete Pump (Small)		844.05		928.46	15.07	16.58
56 L.H., Daily Totals		$2610.85		$3824.05	$46.62	$68.29

Crew C-8A	Hr.	Daily	Hr.	Daily	Bare Costs	Incl. O&P
1 Labor Foreman (outside)	$29.50	$236.00	$48.75	$390.00	$30.57	$50.08
3 Laborers	27.50	660.00	45.45	1090.80		
2 Cement Finishers	35.70	571.20	57.70	923.20		
48 L.H., Daily Totals		$1467.20		$2404.00	$30.57	$50.08

Crew C-8B	Hr.	Daily	Hr.	Daily	Bare Costs	Incl. O&P
1 Labor Foreman (outside)	$29.50	$236.00	$48.75	$390.00	$29.89	$49.31
3 Laborers	27.50	660.00	45.45	1090.80		
1 Equip. Oper. (medium)	37.45	299.60	61.45	491.60		
1 Vibrating Power Screed		79.45		87.39		
1 Roller, Vibratory, 25 Ton		650.90		715.99		
1 Dozer, 200 H.P.		1290.00		1419.00	50.51	55.56
40 L.H., Daily Totals		$3215.95		$4194.78	$80.40	$104.87

Crew C-8C	Hr.	Daily	Hr.	Daily	Bare Costs	Incl. O&P
1 Labor Foreman (outside)	$29.50	$236.00	$48.75	$390.00	$30.86	$50.71
3 Laborers	27.50	660.00	45.45	1090.80		
1 Cement Finisher	35.70	285.60	57.70	461.60		
1 Equip. Oper. (medium)	37.45	299.60	61.45	491.60		
1 Shotcrete Rig, 12 C.Y./hr		242.75		267.02		
1 Air Compressor, 160 cfm		115.85		127.44		
4 -50' Air Hoses, 1"		33.60		36.96		
4 -50' Air Hoses, 2"		103.60		113.96	10.33	11.36
48 L.H., Daily Totals		$1977.00		$2979.38	$41.19	$62.07

Crew C-8D	Hr.	Daily	Hr.	Daily	Bare Costs	Incl. O&P
1 Labor Foreman (outside)	$29.50	$236.00	$48.75	$390.00	$32.09	$52.60
1 Laborer	27.50	220.00	45.45	363.60		
1 Cement Finisher	35.70	285.60	57.70	461.60		
1 Equipment Oper. (light)	35.65	285.20	58.50	468.00		
1 Air Compressor, 250 cfm		167.95		184.75		
2 -50' Air Hoses, 1"		16.80		18.48	5.77	6.35
32 L.H., Daily Totals		$1211.55		$1886.43	$37.86	$58.95

Crew C-8E	Hr.	Daily	Hr.	Daily	Bare Costs	Incl. O&P
1 Labor Foreman (outside)	$29.50	$236.00	$48.75	$390.00	$30.56	$50.22
3 Laborers	27.50	660.00	45.45	1090.80		
1 Cement Finisher	35.70	285.60	57.70	461.60		
1 Equipment Oper. (light)	35.65	285.20	58.50	468.00		
1 Shotcrete Rig, 35 C.Y./hr		271.40		298.54		
1 Air Compressor, 250 cfm		167.95		184.75		
4 -50' Air Hoses, 1"		33.60		36.96		
4 -50' Air Hoses, 2"		103.60		113.96	12.01	13.21
48 L.H., Daily Totals		$2043.35		$3044.61	$42.57	$63.43

Crew No.	Bare Costs		Incl. Subs O&P		Cost Per Labor-Hour	
Crew C-9	**Hr.**	**Daily**	**Hr.**	**Daily**	**Bare Costs**	**Incl. O&P**
1 Cement Finisher	$35.70	$285.60	$57.70	$461.60	$31.59	$51.77
2 Laborers	27.50	440.00	45.45	727.20		
1 Equipment Oper. (light)	35.65	285.20	58.50	468.00		
1 Grout Pump, 50 C.F./hr.		125.60		138.16		
1 Air Compressor, 160 cfm		115.85		127.44		
2 -50' Air Hoses, 1"		16.80		18.48		
2 -50' Air Hoses, 2"		51.80		56.98	9.69	10.66
32 L.H., Daily Totals		$1320.85		$1997.86	$41.28	$62.43
Crew C-10	**Hr.**	**Daily**	**Hr.**	**Daily**	**Bare Costs**	**Incl. O&P**
1 Laborer	$27.50	$220.00	$45.45	$363.60	$32.97	$53.62
2 Cement Finishers	35.70	571.20	57.70	923.20		
24 L.H., Daily Totals		$791.20		$1286.80	$32.97	$53.62
Crew C-10B	**Hr.**	**Daily**	**Hr.**	**Daily**	**Bare Costs**	**Incl. O&P**
3 Laborers	$27.50	$660.00	$45.45	$1090.80	$30.78	$50.35
2 Cement Finishers	35.70	571.20	57.70	923.20		
1 Concrete Mixer, 10 C.F.		162.00		178.20		
2 Trowels, 48" Walk-Behind		83.00		91.30	6.13	6.74
40 L.H., Daily Totals		$1476.20		$2283.50	$36.91	$57.09
Crew C-10C	**Hr.**	**Daily**	**Hr.**	**Daily**	**Bare Costs**	**Incl. O&P**
1 Laborer	$27.50	$220.00	$45.45	$363.60	$32.97	$53.62
2 Cement Finishers	35.70	571.20	57.70	923.20		
1 Trowel, 48" Walk-Behind		41.50		45.65	1.73	1.90
24 L.H., Daily Totals		$832.70		$1332.45	$34.70	$55.52
Crew C-10D	**Hr.**	**Daily**	**Hr.**	**Daily**	**Bare Costs**	**Incl. O&P**
1 Laborer	$27.50	$220.00	$45.45	$363.60	$32.97	$53.62
2 Cement Finishers	35.70	571.20	57.70	923.20		
1 Vibrating Power Screed		79.45		87.39		
1 Trowel, 48" Walk-Behind		41.50		45.65	5.04	5.54
24 L.H., Daily Totals		$912.15		$1419.85	$38.01	$59.16
Crew C-10E	**Hr.**	**Daily**	**Hr.**	**Daily**	**Bare Costs**	**Incl. O&P**
1 Laborer	$27.50	$220.00	$45.45	$363.60	$32.97	$53.62
2 Cement Finishers	35.70	571.20	57.70	923.20		
1 Vibrating Power Screed		79.45		87.39		
1 Cement Trowel, 96" Ride-On		166.30		182.93	10.24	11.26
24 L.H., Daily Totals		$1036.95		$1557.13	$43.21	$64.88
Crew C-10F	**Hr.**	**Daily**	**Hr.**	**Daily**	**Bare Costs**	**Incl. O&P**
1 Laborer	$27.50	$220.00	$45.45	$363.60	$32.97	$53.62
2 Cement Finishers	35.70	571.20	57.70	923.20		
1 Telescoping Boom Lift, to 60'		451.25		496.38	18.80	20.68
24 L.H., Daily Totals		$1242.45		$1783.18	$51.77	$74.30
Crew C-11	**Hr.**	**Daily**	**Hr.**	**Daily**	**Bare Costs**	**Incl. O&P**
1 Skilled Worker Foreman	$38.50	$308.00	$63.95	$511.60	$36.99	$61.34
5 Skilled Workers	36.50	1460.00	60.65	2426.00		
1 Equip. Oper. (crane)	37.90	303.20	62.20	497.60		
1 Lattice Boom Crane, 150 Ton		2021.00		2223.10	36.09	39.70
56 L.H., Daily Totals		$4092.20		$5658.30	$73.08	$101.04

Crew No.	Bare Costs		Incl. Subs O&P		Cost Per Labor-Hour	
Crew C-12	**Hr.**	**Daily**	**Hr.**	**Daily**	**Bare Costs**	**Incl. O&P**
1 Carpenter Foreman (outside)	$37.65	$301.20	$62.20	$497.60	$35.00	$57.76
3 Carpenters	35.65	855.60	58.90	1413.60		
1 Laborer	27.50	220.00	45.45	363.60		
1 Equip. Oper. (crane)	37.90	303.20	62.20	497.60		
1 Hyd. Crane, 12 Ton		471.75		518.92	9.83	10.81
48 L.H., Daily Totals		$2151.75		$3291.32	$44.83	$68.57
Crew C-13	**Hr.**	**Daily**	**Hr.**	**Daily**	**Bare Costs**	**Incl. O&P**
2 Struc. Steel Workers	$39.75	$636.00	$68.95	$1103.20	$38.38	$65.60
1 Carpenter	35.65	285.20	58.90	471.20		
1 Welder, Gas Engine, 300 amp		95.65		105.22	3.99	4.38
24 L.H., Daily Totals		$1016.85		$1679.62	$42.37	$69.98
Crew C-14	**Hr.**	**Daily**	**Hr.**	**Daily**	**Bare Costs**	**Incl. O&P**
1 Carpenter Foreman (outside)	$37.65	$301.20	$62.20	$497.60	$32.41	$53.54
3 Carpenters	35.65	855.60	58.90	1413.60		
2 Carpenter Helpers	27.30	436.80	45.75	732.00		
4 Laborers	27.50	880.00	45.45	1454.40		
2 Rodmen (reinf.)	38.95	623.20	64.40	1030.40		
2 Rodman Helpers	27.30	436.80	45.75	732.00		
2 Cement Finishers	35.70	571.20	57.70	923.20		
1 Equip. Oper. (crane)	37.90	303.20	62.20	497.60		
1 Hyd. Crane, 80 Ton		1487.00		1635.70	10.93	12.03
136 L.H., Daily Totals		$5895.00		$8916.50	$43.35	$65.56
Crew C-14A	**Hr.**	**Daily**	**Hr.**	**Daily**	**Bare Costs**	**Incl. O&P**
1 Carpenter Foreman (outside)	$37.65	$301.20	$62.20	$497.60	$35.68	$58.89
16 Carpenters	35.65	4563.20	58.90	7539.20		
4 Rodmen (reinf.)	38.95	1246.40	64.40	2060.80		
2 Laborers	27.50	440.00	45.45	727.20		
1 Cement Finisher	35.70	285.60	57.70	461.60		
1 Equip. Oper. (medium)	37.45	299.60	61.45	491.60		
1 Gas Engine Vibrator		26.55		29.20		
1 Concrete Pump (Small)		844.05		928.46	4.35	4.79
200 L.H., Daily Totals		$8006.60		$12735.66	$40.03	$63.68
Crew C-14B	**Hr.**	**Daily**	**Hr.**	**Daily**	**Bare Costs**	**Incl. O&P**
1 Carpenter Foreman (outside)	$37.65	$301.20	$62.20	$497.60	$35.68	$58.84
16 Carpenters	35.65	4563.20	58.90	7539.20		
4 Rodmen (reinf.)	38.95	1246.40	64.40	2060.80		
2 Laborers	27.50	440.00	45.45	727.20		
2 Cement Finishers	35.70	571.20	57.70	923.20		
1 Equip. Oper. (medium)	37.45	299.60	61.45	491.60		
1 Gas Engine Vibrator		26.55		29.20		
1 Concrete Pump (Small)		844.05		928.46	4.19	4.60
208 L.H., Daily Totals		$8292.20		$13197.26	$39.87	$63.45
Crew C-14C	**Hr.**	**Daily**	**Hr.**	**Daily**	**Bare Costs**	**Incl. O&P**
1 Carpenter Foreman (outside)	$37.65	$301.20	$62.20	$497.60	$33.94	$55.99
6 Carpenters	35.65	1711.20	58.90	2827.20		
2 Rodmen (reinf.)	38.95	623.20	64.40	1030.40		
4 Laborers	27.50	880.00	45.45	1454.40		
1 Cement Finisher	35.70	285.60	57.70	461.60		
1 Gas Engine Vibrator		26.55		29.20	0.24	0.26
112 L.H., Daily Totals		$3827.75		$6300.40	$34.18	$56.25

Crew No.	Bare Costs		Incl. Subs O&P		Cost Per Labor-Hour	
Crew C-14D	**Hr.**	**Daily**	**Hr.**	**Daily**	**Bare Costs**	**Incl. O&P**
1 Carpenter Foreman (outside)	$37.65	$301.20	$62.20	$497.60	$35.42	$58.45
18 Carpenters	35.65	5133.60	58.90	8481.60		
2 Rodmen (reinf.)	38.95	623.20	64.40	1030.40		
2 Laborers	27.50	440.00	45.45	727.20		
1 Cement Finisher	35.70	285.60	57.70	461.60		
1 Equip. Oper. (medium)	37.45	299.60	61.45	491.60		
1 Gas Engine Vibrator		26.55		29.20		
1 Concrete Pump (Small)		844.05		928.46	4.35	4.79
200 L.H., Daily Totals		$7953.80		$12647.66	$39.77	$63.24
Crew C-14E	**Hr.**	**Daily**	**Hr.**	**Daily**	**Bare Costs**	**Incl. O&P**
1 Carpenter Foreman (outside)	$37.65	$301.20	$62.20	$497.60	$34.81	$57.42
2 Carpenters	35.65	570.40	58.90	942.40		
4 Rodmen (reinf.)	38.95	1246.40	64.40	2060.80		
3 Laborers	27.50	660.00	45.45	1090.80		
1 Cement Finisher	35.70	285.60	57.70	461.60		
1 Gas Engine Vibrator		26.55		29.20	0.30	0.33
88 L.H., Daily Totals		$3090.15		$5082.40	$35.12	$57.75
Crew C-14F	**Hr.**	**Daily**	**Hr.**	**Daily**	**Bare Costs**	**Incl. O&P**
1 Labor Foreman (outside)	$29.50	$236.00	$48.75	$390.00	$33.19	$53.98
2 Laborers	27.50	440.00	45.45	727.20		
6 Cement Finishers	35.70	1713.60	57.70	2769.60		
1 Gas Engine Vibrator		26.55		29.20	0.37	0.41
72 L.H., Daily Totals		$2416.15		$3916.01	$33.56	$54.39
Crew C-14G	**Hr.**	**Daily**	**Hr.**	**Daily**	**Bare Costs**	**Incl. O&P**
1 Labor Foreman (outside)	$29.50	$236.00	$48.75	$390.00	$32.47	$52.92
2 Laborers	27.50	440.00	45.45	727.20		
4 Cement Finishers	35.70	1142.40	57.70	1846.40		
1 Gas Engine Vibrator		26.55		29.20	0.47	0.52
56 L.H., Daily Totals		$1844.95		$2992.80	$32.95	$53.44
Crew C-14H	**Hr.**	**Daily**	**Hr.**	**Daily**	**Bare Costs**	**Incl. O&P**
1 Carpenter Foreman (outside)	$37.65	$301.20	$62.20	$497.60	$35.18	$57.92
2 Carpenters	35.65	570.40	58.90	942.40		
1 Rodman (reinf.)	38.95	311.60	64.40	515.20		
1 Laborer	27.50	220.00	45.45	363.60		
1 Cement Finisher	35.70	285.60	57.70	461.60		
1 Gas Engine Vibrator		26.55		29.20	0.55	0.61
48 L.H., Daily Totals		$1715.35		$2809.61	$35.74	$58.53
Crew C-14L	**Hr.**	**Daily**	**Hr.**	**Daily**	**Bare Costs**	**Incl. O&P**
1 Carpenter Foreman (outside)	$37.65	$301.20	$62.20	$497.60	$33.10	$54.59
6 Carpenters	35.65	1711.20	58.90	2827.20		
4 Laborers	27.50	880.00	45.45	1454.40		
1 Cement Finisher	35.70	285.60	57.70	461.60		
1 Gas Engine Vibrator		26.55		29.20	0.28	0.30
96 L.H., Daily Totals		$3204.55		$5270.01	$33.38	$54.90
Crew C-14M	**Hr.**	**Daily**	**Hr.**	**Daily**	**Bare Costs**	**Incl. O&P**
1 Carpenter Foreman (outside)	$37.65	$301.20	$62.20	$497.60	$34.51	$56.81
2 Carpenters	35.65	570.40	58.90	942.40		
1 Rodman (reinf.)	38.95	311.60	64.40	515.20		
2 Laborers	27.50	440.00	45.45	727.20		
1 Cement Finisher	35.70	285.60	57.70	461.60		
1 Equip. Oper. (medium)	37.45	299.60	61.45	491.60		
1 Gas Engine Vibrator		26.55		29.20		
1 Concrete Pump (Small)		844.05		928.46	13.60	14.96
64 L.H., Daily Totals		$3079.00		$4593.26	$48.11	$71.77

Crew No.	Bare Costs		Incl. Subs O&P		Cost Per Labor-Hour	
Crew C-15	**Hr.**	**Daily**	**Hr.**	**Daily**	**Bare Costs**	**Incl. O&P**
1 Carpenter Foreman (outside)	$37.65	$301.20	$62.20	$497.60	$33.53	$55.13
2 Carpenters	35.65	570.40	58.90	942.40		
3 Laborers	27.50	660.00	45.45	1090.80		
2 Cement Finishers	35.70	571.20	57.70	923.20		
1 Rodman (reinf.)	38.95	311.60	64.40	515.20		
72 L.H., Daily Totals		$2414.40		$3969.20	$33.53	$55.13
Crew C-16	**Hr.**	**Daily**	**Hr.**	**Daily**	**Bare Costs**	**Incl. O&P**
1 Labor Foreman (outside)	$29.50	$236.00	$48.75	$390.00	$31.55	$51.71
3 Laborers	27.50	660.00	45.45	1090.80		
2 Cement Finishers	35.70	571.20	57.70	923.20		
1 Equip. Oper. (medium)	37.45	299.60	61.45	491.60		
1 Gunite Pump Rig		313.00		344.30		
2 -50' Air Hoses, 3/4"		13.10		14.41		
2 -50' Air Hoses, 2"		51.80		56.98	6.75	7.42
56 L.H., Daily Totals		$2144.70		$3311.29	$38.30	$59.13
Crew C-16A	**Hr.**	**Daily**	**Hr.**	**Daily**	**Bare Costs**	**Incl. O&P**
1 Laborer	$27.50	$220.00	$45.45	$363.60	$34.09	$55.58
2 Cement Finishers	35.70	571.20	57.70	923.20		
1 Equip. Oper. (medium)	37.45	299.60	61.45	491.60		
1 Gunite Pump Rig		313.00		344.30		
2 -50' Air Hoses, 3/4"		13.10		14.41		
2 -50' Air Hoses, 2"		51.80		56.98		
1 Telescoping Boom Lift, to 60'		451.25		496.38	25.91	28.50
32 L.H., Daily Totals		$1919.95		$2690.47	$60.00	$84.08
Crew C-17	**Hr.**	**Daily**	**Hr.**	**Daily**	**Bare Costs**	**Incl. O&P**
2 Skilled Worker Foremen (out)	$38.50	$616.00	$63.95	$1023.20	$36.90	$61.31
8 Skilled Workers	36.50	2336.00	60.65	3881.60		
80 L.H., Daily Totals		$2952.00		$4904.80	$36.90	$61.31
Crew C-17A	**Hr.**	**Daily**	**Hr.**	**Daily**	**Bare Costs**	**Incl. O&P**
2 Skilled Worker Foremen (out)	$38.50	$616.00	$63.95	$1023.20	$36.91	$61.32
8 Skilled Workers	36.50	2336.00	60.65	3881.60		
.125 Equip. Oper. (crane)	37.90	37.90	62.20	62.20		
.125 Hyd. Crane, 80 Ton		185.88		204.46	2.29	2.52
81 L.H., Daily Totals		$3175.78		$5171.46	$39.21	$63.85
Crew C-17B	**Hr.**	**Daily**	**Hr.**	**Daily**	**Bare Costs**	**Incl. O&P**
2 Skilled Worker Foremen (out)	$38.50	$616.00	$63.95	$1023.20	$36.92	$61.33
8 Skilled Workers	36.50	2336.00	60.65	3881.60		
.25 Equip. Oper. (crane)	37.90	75.80	62.20	124.40		
.25 Hyd. Crane, 80 Ton		371.75		408.93		
.25 Trowel, 48" Walk-Behind		10.38		11.41	4.66	5.13
82 L.H., Daily Totals		$3409.93		$5449.54	$41.58	$66.46
Crew C-17C	**Hr.**	**Daily**	**Hr.**	**Daily**	**Bare Costs**	**Incl. O&P**
2 Skilled Worker Foremen (out)	$38.50	$616.00	$63.95	$1023.20	$36.94	$61.34
8 Skilled Workers	36.50	2336.00	60.65	3881.60		
.375 Equip. Oper. (crane)	37.90	113.70	62.20	186.60		
.375 Hyd. Crane, 80 Ton		557.63		613.39	6.72	7.39
83 L.H., Daily Totals		$3623.32		$5704.79	$43.65	$68.73
Crew C-17D	**Hr.**	**Daily**	**Hr.**	**Daily**	**Bare Costs**	**Incl. O&P**
2 Skilled Worker Foremen (out)	$38.50	$616.00	$63.95	$1023.20	$36.95	$61.35
8 Skilled Workers	36.50	2336.00	60.65	3881.60		
.5 Equip. Oper. (crane)	37.90	151.60	62.20	248.80		
.5 Hyd. Crane, 80 Ton		743.50		817.85	8.85	9.74
84 L.H., Daily Totals		$3847.10		$5971.45	$45.80	$71.09

Crew No.	Bare Costs		Incl. Subs O&P		Cost Per Labor-Hour	
Crew C-17E	**Hr.**	**Daily**	**Hr.**	**Daily**	**Bare Costs**	**Incl. O&P**
2 Skilled Worker Foremen (out)	$38.50	$616.00	$63.95	$1023.20	$36.90	$61.31
8 Skilled Workers	36.50	2336.00	60.65	3881.60		
1 Hyd. Jack with Rods		104.40		114.84	1.30	1.44
80 L.H., Daily Totals		$3056.40		$5019.64	$38.20	$62.75

Crew No.	Bare Costs		Incl. Subs O&P		Cost Per Labor-Hour	
Crew C-18	**Hr.**	**Daily**	**Hr.**	**Daily**	**Bare Costs**	**Incl. O&P**
.125 Labor Foreman (outside)	$29.50	$29.50	$48.75	$48.75	$27.72	$45.82
1 Laborer	27.50	220.00	45.45	363.60		
1 Concrete Cart, 10 C.F.		58.10		63.91	6.46	7.10
9 L.H., Daily Totals		$307.60		$476.26	$34.18	$52.92

Crew No.	Bare Costs		Incl. Subs O&P		Cost Per Labor-Hour	
Crew C-19	**Hr.**	**Daily**	**Hr.**	**Daily**	**Bare Costs**	**Incl. O&P**
.125 Labor Foreman (outside)	$29.50	$29.50	$48.75	$48.75	$27.72	$45.82
1 Laborer	27.50	220.00	45.45	363.60		
1 Concrete Cart, 18 C.F.		97.75		107.53	10.86	11.95
9 L.H., Daily Totals		$347.25		$519.88	$38.58	$57.76

Crew No.	Bare Costs		Incl. Subs O&P		Cost Per Labor-Hour	
Crew C-20	**Hr.**	**Daily**	**Hr.**	**Daily**	**Bare Costs**	**Incl. O&P**
1 Labor Foreman (outside)	$29.50	$236.00	$48.75	$390.00	$30.02	$49.39
5 Laborers	27.50	1100.00	45.45	1818.00		
1 Cement Finisher	35.70	285.60	57.70	461.60		
1 Equip. Oper. (medium)	37.45	299.60	61.45	491.60		
2 Gas Engine Vibrators		53.10		58.41		
1 Concrete Pump (Small)		844.05		928.46	14.02	15.42
64 L.H., Daily Totals		$2818.35		$4148.06	$44.04	$64.81

Crew No.	Bare Costs		Incl. Subs O&P		Cost Per Labor-Hour	
Crew C-21	**Hr.**	**Daily**	**Hr.**	**Daily**	**Bare Costs**	**Incl. O&P**
1 Labor Foreman (outside)	$29.50	$236.00	$48.75	$390.00	$30.02	$49.39
5 Laborers	27.50	1100.00	45.45	1818.00		
1 Cement Finisher	35.70	285.60	57.70	461.60		
1 Equip. Oper. (medium)	37.45	299.60	61.45	491.60		
2 Gas Engine Vibrators		53.10		58.41		
1 Concrete Conveyer		186.45		205.10	3.74	4.12
64 L.H., Daily Totals		$2160.75		$3424.70	$33.76	$53.51

Crew No.	Bare Costs		Incl. Subs O&P		Cost Per Labor-Hour	
Crew C-22	**Hr.**	**Daily**	**Hr.**	**Daily**	**Bare Costs**	**Incl. O&P**
1 Rodman Foreman (outside)	$40.95	$327.60	$67.70	$541.60	$39.16	$64.73
4 Rodmen (reinf.)	38.95	1246.40	64.40	2060.80		
.125 Equip. Oper. (crane)	37.90	37.90	62.20	62.20		
.125 Equip. Oper. (oiler)	32.95	32.95	54.05	54.05		
.125 Hyd. Crane, 25 Ton		72.71		79.98	1.73	1.90
42 L.H., Daily Totals		$1717.56		$2798.63	$40.89	$66.63

Crew No.	Bare Costs		Incl. Subs O&P		Cost Per Labor-Hour	
Crew C-23	**Hr.**	**Daily**	**Hr.**	**Daily**	**Bare Costs**	**Incl. O&P**
2 Skilled Worker Foremen (out)	$38.50	$616.00	$63.95	$1023.20	$36.69	$60.81
6 Skilled Workers	36.50	1752.00	60.65	2911.20		
1 Equip. Oper. (crane)	37.90	303.20	62.20	497.60		
1 Equip. Oper. (oiler)	32.95	263.60	54.05	432.40		
1 Lattice Boom Crane, 90 Ton		1698.00		1867.80	21.23	23.35
80 L.H., Daily Totals		$4632.80		$6732.20	$57.91	$84.15

Crew No.	Bare Costs		Incl. Subs O&P		Cost Per Labor-Hour	
Crew C-24	**Hr.**	**Daily**	**Hr.**	**Daily**	**Bare Costs**	**Incl. O&P**
2 Skilled Worker Foremen (out)	$38.50	$616.00	$63.95	$1023.20	$36.69	$60.81
6 Skilled Workers	36.50	1752.00	60.65	2911.20		
1 Equip. Oper. (crane)	37.90	303.20	62.20	497.60		
1 Equip. Oper. (oiler)	32.95	263.60	54.05	432.40		
1 Lattice Boom Crane, 150 Ton		2021.00		2223.10	25.26	27.79
80 L.H., Daily Totals		$4955.80		$7087.50	$61.95	$88.59

Crew No.	Bare Costs		Incl. Subs O&P		Cost Per Labor-Hour	
Crew C-25	**Hr.**	**Daily**	**Hr.**	**Daily**	**Bare Costs**	**Incl. O&P**
2 Rodmen (reinf.)	$38.95	$623.20	$64.40	$1030.40	$30.98	$52.98
2 Rodmen Helpers	23.00	368.00	41.55	664.80		
32 L.H., Daily Totals		$991.20		$1695.20	$30.98	$52.98

Crew No.	Bare Costs		Incl. Subs O&P		Cost Per Labor-Hour	
Crew C-27	**Hr.**	**Daily**	**Hr.**	**Daily**	**Bare Costs**	**Incl. O&P**
2 Cement Finishers	$35.70	$571.20	$57.70	$923.20	$35.70	$57.70
1 Concrete Saw		110.15		121.17	6.88	7.57
16 L.H., Daily Totals		$681.35		$1044.37	$42.58	$65.27

Crew No.	Bare Costs		Incl. Subs O&P		Cost Per Labor-Hour	
Crew C-28	**Hr.**	**Daily**	**Hr.**	**Daily**	**Bare Costs**	**Incl. O&P**
1 Cement Finisher	$35.70	$285.60	$57.70	$461.60	$35.70	$57.70
1 Portable Air Compressor, Gas		16.05		17.66	2.01	2.21
8 L.H., Daily Totals		$301.65		$479.26	$37.71	$59.91

Crew No.	Bare Costs		Incl. Subs O&P		Cost Per Labor-Hour	
Crew C-29	**Hr.**	**Daily**	**Hr.**	**Daily**	**Bare Costs**	**Incl. O&P**
1 Laborer	$27.50	$220.00	$45.45	$363.60	$27.50	$45.45
1 Pressure Washer		63.05		69.36	7.88	8.67
8 L.H., Daily Totals		$283.05		$432.95	$35.38	$54.12

Crew No.	Bare Costs		Incl. Subs O&P		Cost Per Labor-Hour	
Crew C-30	**Hr.**	**Daily**	**Hr.**	**Daily**	**Bare Costs**	**Incl. O&P**
1 Laborer	$27.50	$220.00	$45.45	$363.60	$27.50	$45.45
1 Concrete Mixer, 10 C.F.		162.00		178.20	20.25	22.27
8 L.H., Daily Totals		$382.00		$541.80	$47.75	$67.72

Crew No.	Bare Costs		Incl. Subs O&P		Cost Per Labor-Hour	
Crew C-31	**Hr.**	**Daily**	**Hr.**	**Daily**	**Bare Costs**	**Incl. O&P**
1 Cement Finisher	$35.70	$285.60	$57.70	$461.60	$35.70	$57.70
1 Grout Pump		313.00		344.30	39.13	43.04
8 L.H., Daily Totals		$598.60		$805.90	$74.83	$100.74

Crew No.	Bare Costs		Incl. Subs O&P		Cost Per Labor-Hour	
Crew C-32	**Hr.**	**Daily**	**Hr.**	**Daily**	**Bare Costs**	**Incl. O&P**
1 Cement Finisher	$35.70	$285.60	$57.70	$461.60	$31.60	$51.58
1 Laborer	27.50	220.00	45.45	363.60		
1 Crack Chaser Saw, Gas, 6 H.P.		24.50		26.95		
1 Vacuum Pick-Up System		72.15		79.36	6.04	6.64
16 L.H., Daily Totals		$602.25		$931.51	$37.64	$58.22

Crew No.	Bare Costs		Incl. Subs O&P		Cost Per Labor-Hour	
Crew D-1	**Hr.**	**Daily**	**Hr.**	**Daily**	**Bare Costs**	**Incl. O&P**
1 Bricklayer	$35.25	$282.00	$58.75	$470.00	$32.38	$53.98
1 Bricklayer Helper	29.50	236.00	49.20	393.60		
16 L.H., Daily Totals		$518.00		$863.60	$32.38	$53.98

Crew No.	Bare Costs		Incl. Subs O&P		Cost Per Labor-Hour	
Crew D-2	**Hr.**	**Daily**	**Hr.**	**Daily**	**Bare Costs**	**Incl. O&P**
3 Bricklayers	$35.25	$846.00	$58.75	$1410.00	$32.95	$54.93
2 Bricklayer Helpers	29.50	472.00	49.20	787.20		
40 L.H., Daily Totals		$1318.00		$2197.20	$32.95	$54.93

Crew No.	Bare Costs		Incl. Subs O&P		Cost Per Labor-Hour	
Crew D-3	**Hr.**	**Daily**	**Hr.**	**Daily**	**Bare Costs**	**Incl. O&P**
3 Bricklayers	$35.25	$846.00	$58.75	$1410.00	$33.08	$55.12
2 Bricklayer Helpers	29.50	472.00	49.20	787.20		
.25 Carpenter	35.65	71.30	58.90	117.80		
42 L.H., Daily Totals		$1389.30		$2315.00	$33.08	$55.12

Crew No.	Bare Costs		Incl. Subs O&P		Cost Per Labor-Hour	
Crew D-4	**Hr.**	**Daily**	**Hr.**	**Daily**	**Bare Costs**	**Incl. O&P**
1 Bricklayer	$35.25	$282.00	$58.75	$470.00	$30.25	$50.36
3 Bricklayer Helpers	29.50	708.00	49.20	1180.80		
1 Building Laborer	27.50	220.00	45.45	363.60		
1 Grout Pump, 50 C.F./hr.		125.60		138.16	3.14	3.45
40 L.H., Daily Totals		$1335.60		$2152.56	$33.39	$53.81

Crew No.	Bare Costs		Incl. Subs O&P		Cost Per Labor-Hour	
Crew D-5	**Hr.**	**Daily**	**Hr.**	**Daily**	**Bare Costs**	**Incl. O&P**
1 Block Mason Helper	29.50	236.00	49.20	393.60	29.50	49.20
8 L.H., Daily Totals		$236.00		$393.60	$29.50	$49.20
Crew D-6	**Hr.**	**Daily**	**Hr.**	**Daily**	**Bare Costs**	**Incl. O&P**
3 Bricklayers	$35.25	$846.00	$58.75	$1410.00	$32.38	$53.98
3 Bricklayer Helpers	29.50	708.00	49.20	1180.80		
48 L.H., Daily Totals		$1554.00		$2590.80	$32.38	$53.98
Crew D-7	**Hr.**	**Daily**	**Hr.**	**Daily**	**Bare Costs**	**Incl. O&P**
1 Tile Layer	$34.45	$275.60	$55.60	$444.80	$30.45	$49.15
1 Tile Layer Helper	26.45	211.60	42.70	341.60		
16 L.H., Daily Totals		$487.20		$786.40	$30.45	$49.15
Crew D-8	**Hr.**	**Daily**	**Hr.**	**Daily**	**Bare Costs**	**Incl. O&P**
3 Bricklayers	$35.25	$846.00	$58.75	$1410.00	$32.95	$54.93
2 Bricklayer Helpers	29.50	472.00	49.20	787.20		
40 L.H., Daily Totals		$1318.00		$2197.20	$32.95	$54.93
Crew D-9	**Hr.**	**Daily**	**Hr.**	**Daily**	**Bare Costs**	**Incl. O&P**
3 Bricklayers	$35.25	$846.00	$58.75	$1410.00	$32.38	$53.98
3 Bricklayer Helpers	29.50	708.00	49.20	1180.80		
48 L.H., Daily Totals		$1554.00		$2590.80	$32.38	$53.98
Crew D-10	**Hr.**	**Daily**	**Hr.**	**Daily**	**Bare Costs**	**Incl. O&P**
1 Bricklayer Foreman (outside)	$37.25	$298.00	$62.10	$496.80	$34.98	$58.06
1 Bricklayer	35.25	282.00	58.75	470.00		
1 Bricklayer Helper	29.50	236.00	49.20	393.60		
1 Equip. Oper. (crane)	37.90	303.20	62.20	497.60		
1 S.P. Crane, 4x4, 12 Ton		371.85		409.04	11.62	12.78
32 L.H., Daily Totals		$1491.05		$2267.03	$46.60	$70.84
Crew D-11	**Hr.**	**Daily**	**Hr.**	**Daily**	**Bare Costs**	**Incl. O&P**
2 Bricklayers	$35.25	$564.00	$58.75	$940.00	$33.33	$55.57
1 Bricklayer Helper	29.50	236.00	49.20	393.60		
24 L.H., Daily Totals		$800.00		$1333.60	$33.33	$55.57
Crew D-12	**Hr.**	**Daily**	**Hr.**	**Daily**	**Bare Costs**	**Incl. O&P**
2 Bricklayers	$35.25	$564.00	$58.75	$940.00	$32.38	$53.98
2 Bricklayer Helpers	29.50	472.00	49.20	787.20		
32 L.H., Daily Totals		$1036.00		$1727.20	$32.38	$53.98
Crew D-13	**Hr.**	**Daily**	**Hr.**	**Daily**	**Bare Costs**	**Incl. O&P**
1 Bricklayer Foreman (outside)	$37.25	$298.00	$62.10	$496.80	$34.11	$56.70
2 Bricklayers	35.25	564.00	58.75	940.00		
2 Bricklayer Helpers	29.50	472.00	49.20	787.20		
1 Equip. Oper. (crane)	37.90	303.20	62.20	497.60		
1 S.P. Crane, 4x4, 12 Ton		371.85		409.04	7.75	8.52
48 L.H., Daily Totals		$2009.05		$3130.64	$41.86	$65.22
Crew D-14	**Hr.**	**Daily**	**Hr.**	**Daily**	**Bare Costs**	**Incl. O&P**
3 Bricklayers	$35.25	$846.00	$58.75	$1410.00	$33.81	$56.36
1 Bricklayer Helper	29.50	236.00	49.20	393.60		
32 L.H., Daily Totals		$1082.00		$1803.60	$33.81	$56.36
Crew E-1	**Hr.**	**Daily**	**Hr.**	**Daily**	**Bare Costs**	**Incl. O&P**
2 Struc. Steel Workers	$39.75	$636.00	$68.95	$1103.20	$39.75	$68.95
1 Welder, Gas Engine, 300 amp		95.65		105.22	5.98	6.58
16 L.H., Daily Totals		$731.65		$1208.42	$45.73	$75.53

Crew No.	Bare Costs		Incl. Subs O&P		Cost Per Labor-Hour	
Crew E-2	**Hr.**	**Daily**	**Hr.**	**Daily**	**Bare Costs**	**Incl. O&P**
1 Struc. Steel Foreman (outside)	$41.75	$334.00	$72.45	$579.60	$39.77	$68.41
4 Struc. Steel Workers	39.75	1272.00	68.95	2206.40		
1 Equip. Oper. (crane)	37.90	303.20	62.20	497.60		
1 Lattice Boom Crane, 90 Ton		1698.00		1867.80	35.38	38.91
48 L.H., Daily Totals		$3607.20		$5151.40	$75.15	$107.32
Crew E-3	**Hr.**	**Daily**	**Hr.**	**Daily**	**Bare Costs**	**Incl. O&P**
1 Struc. Steel Foreman (outside)	$41.75	$334.00	$72.45	$579.60	$40.42	$70.12
2 Struc. Steel Workers	39.75	636.00	68.95	1103.20		
1 Welder, Gas Engine, 300 amp		95.65		105.22	3.99	4.38
24 L.H., Daily Totals		$1065.65		$1788.02	$44.40	$74.50
Crew E-3A	**Hr.**	**Daily**	**Hr.**	**Daily**	**Bare Costs**	**Incl. O&P**
1 Struc. Steel Foreman (outside)	$41.75	$334.00	$72.45	$579.60	$40.42	$70.12
2 Struc. Steel Workers	39.75	636.00	68.95	1103.20		
1 Welder, Gas Engine, 300 amp		95.65		105.22		
1 Telescoping Boom Lift, to 40'		283.15		311.46	15.78	17.36
24 L.H., Daily Totals		$1348.80		$2099.48	$56.20	$87.48
Crew E-4	**Hr.**	**Daily**	**Hr.**	**Daily**	**Bare Costs**	**Incl. O&P**
1 Struc. Steel Foreman (outside)	$41.75	$334.00	$72.45	$579.60	$40.25	$69.83
3 Struc. Steel Workers	39.75	954.00	68.95	1654.80		
1 Welder, Gas Engine, 300 amp		95.65		105.22	2.99	3.29
32 L.H., Daily Totals		$1383.65		$2339.61	$43.24	$73.11
Crew E-5	**Hr.**	**Daily**	**Hr.**	**Daily**	**Bare Costs**	**Incl. O&P**
1 Struc. Steel Foreman (outside)	$41.75	$334.00	$72.45	$579.60	$39.77	$68.59
7 Struc. Steel Workers	39.75	2226.00	68.95	3861.20		
1 Equip. Oper. (crane)	37.90	303.20	62.20	497.60		
1 Lattice Boom Crane, 90 Ton		1698.00		1867.80		
1 Welder, Gas Engine, 300 amp		95.65		105.22	24.91	27.40
72 L.H., Daily Totals		$4656.85		$6911.42	$64.68	$95.99
Crew E-6	**Hr.**	**Daily**	**Hr.**	**Daily**	**Bare Costs**	**Incl. O&P**
1 Struc. Steel Foreman (outside)	$41.75	$334.00	$72.45	$579.60	$39.49	$68.04
12 Struc. Steel Workers	39.75	3816.00	68.95	6619.20		
1 Equip. Oper. (crane)	37.90	303.20	62.20	497.60		
1 Equip. Oper. (light)	35.65	285.20	58.50	468.00		
1 Lattice Boom Crane, 90 Ton		1698.00		1867.80		
1 Welder, Gas Engine, 300 amp		95.65		105.22		
1 Air Compressor, 160 cfm		115.85		127.44		
2 Impact Wrenches		37.00		40.70	16.22	17.84
120 L.H., Daily Totals		$6684.90		$10305.55	$55.71	$85.88
Crew E-7	**Hr.**	**Daily**	**Hr.**	**Daily**	**Bare Costs**	**Incl. O&P**
1 Struc. Steel Foreman (outside)	$41.75	$334.00	$72.45	$579.60	$39.77	$68.59
7 Struc. Steel Workers	39.75	2226.00	68.95	3861.20		
1 Equip. Oper. (crane)	37.90	303.20	62.20	497.60		
1 Lattice Boom Crane, 90 Ton		1698.00		1867.80		
2 Welder, Gas Engine, 300 amp		191.30		210.43	26.24	28.86
72 L.H., Daily Totals		$4752.50		$7016.63	$66.01	$97.45
Crew E-8	**Hr.**	**Daily**	**Hr.**	**Daily**	**Bare Costs**	**Incl. O&P**
1 Struc. Steel Foreman (outside)	$41.75	$334.00	$72.45	$579.60	$39.76	$68.65
9 Struc. Steel Workers	39.75	2862.00	68.95	4964.40		
1 Equip. Oper. (crane)	37.90	303.20	62.20	497.60		
1 Lattice Boom Crane, 90 Ton		1698.00		1867.80		
4 Welder, Gas Engine, 300 amp		382.60		420.86	23.64	26.01
88 L.H., Daily Totals		$5579.80		$8330.26	$63.41	$94.66

Crew No.	Bare Costs		Incl. Subs O&P		Cost Per Labor-Hour	
Crew E-9	**Hr.**	**Daily**	**Hr.**	**Daily**	**Bare Costs**	**Incl. O&P**
2 Struc. Steel Foremen (outside)	$41.75	$668.00	$72.45	$1159.20	$39.33	$67.60
5 Struc. Steel Workers	39.75	1590.00	68.95	2758.00		
1 Welder Foreman (outside)	41.75	334.00	72.45	579.60		
5 Welders	39.75	1590.00	68.95	2758.00		
1 Equip. Oper. (crane)	37.90	303.20	62.20	497.60		
1 Equip. Oper. (oiler)	32.95	263.60	54.05	432.40		
1 Equip. Oper. (light)	35.65	285.20	58.50	468.00		
1 Lattice Boom Crane, 90 Ton		1698.00		1867.80		
5 Welder, Gas Engine, 300 amp		478.25		526.08	17.00	18.70
128 L.H., Daily Totals		$7210.25		$11046.67	$56.33	$86.30

Crew E-10	Hr.	Daily	Hr.	Daily	Bare Costs	Incl. O&P
1 Struc. Steel Foreman (outside)	$41.75	$334.00	$72.45	$579.60	$40.42	$70.12
2 Struc. Steel Workers	39.75	636.00	68.95	1103.20		
1 Welder, Gas Engine, 300 amp		95.65		105.22		
1 Flatbed Truck, Gas, 3 Ton		245.95		270.55	14.23	15.66
24 L.H., Daily Totals		$1311.60		$2058.56	$54.65	$85.77

Crew E-11	Hr.	Daily	Hr.	Daily	Bare Costs	Incl. O&P
2 Painters, Struc. Steel	$30.35	$485.60	$53.90	$862.40	$30.96	$52.94
1 Building Laborer	27.50	220.00	45.45	363.60		
1 Equip. Oper. (light)	35.65	285.20	58.50	468.00		
1 Air Compressor, 250 cfm		167.95		184.75		
1 Sandblaster, Portable, 3 C.F.		20.70		22.77		
1 Set Sand Blasting Accessories		14.90		16.39	6.36	7.00
32 L.H., Daily Totals		$1194.35		$1917.91	$37.32	$59.93

Crew E-11A	Hr.	Daily	Hr.	Daily	Bare Costs	Incl. O&P
2 Painters, Struc. Steel	$30.35	$485.60	$53.90	$862.40	$30.96	$52.94
1 Building Laborer	27.50	220.00	45.45	363.60		
1 Equip. Oper. (light)	35.65	285.20	58.50	468.00		
1 Air Compressor, 250 cfm		167.95		184.75		
1 Sandblaster, Portable, 3 C.F.		20.70		22.77		
1 Set Sand Blasting Accessories		14.90		16.39		
1 Telescoping Boom Lift, to 60'		451.25		496.38	20.46	22.51
32 L.H., Daily Totals		$1645.60		$2414.28	$51.42	$75.45

Crew E-11B	Hr.	Daily	Hr.	Daily	Bare Costs	Incl. O&P
2 Painters, Struc. Steel	$30.35	$485.60	$53.90	$862.40	$29.40	$51.08
1 Building Laborer	27.50	220.00	45.45	363.60		
2 Paint Sprayer, 8 C.F.M.		85.20		93.72		
1 Telescoping Boom Lift, to 60'		451.25		496.38	22.35	24.59
24 L.H., Daily Totals		$1242.05		$1816.10	$51.75	$75.67

Crew E-12	Hr.	Daily	Hr.	Daily	Bare Costs	Incl. O&P
1 Welder Foreman (outside)	$41.75	$334.00	$72.45	$579.60	$38.70	$65.47
1 Equip. Oper. (light)	35.65	285.20	58.50	468.00		
1 Welder, Gas Engine, 300 amp		95.65		105.22	5.98	6.58
16 L.H., Daily Totals		$714.85		$1152.82	$44.68	$72.05

Crew E-13	Hr.	Daily	Hr.	Daily	Bare Costs	Incl. O&P
1 Welder Foreman (outside)	$41.75	$334.00	$72.45	$579.60	$39.72	$67.80
.5 Equip. Oper. (light)	35.65	142.60	58.50	234.00		
1 Welder, Gas Engine, 300 amp		95.65		105.22	7.97	8.77
12 L.H., Daily Totals		$572.25		$918.82	$47.69	$76.57

Crew E-14	Hr.	Daily	Hr.	Daily	Bare Costs	Incl. O&P
1 Struc. Steel Worker	$39.75	$318.00	$68.95	$551.60	$39.75	$68.95
1 Welder, Gas Engine, 300 amp		95.65		105.22	11.96	13.15
8 L.H., Daily Totals		$413.65		$656.82	$51.71	$82.10

Crew No.	Bare Costs		Incl. Subs O&P		Cost Per Labor-Hour	
Crew E-16	**Hr.**	**Daily**	**Hr.**	**Daily**	**Bare Costs**	**Incl. O&P**
1 Welder Foreman (outside)	$41.75	$334.00	$72.45	$579.60	$40.75	$70.70
1 Welder	39.75	318.00	68.95	551.60		
1 Welder, Gas Engine, 300 amp		95.65		105.22	5.98	6.58
16 L.H., Daily Totals		$747.65		$1236.42	$46.73	$77.28

Crew E-17	Hr.	Daily	Hr.	Daily	Bare Costs	Incl. O&P
1 Struc. Steel Foreman (outside)	$41.75	$334.00	$72.45	$579.60	$40.75	$70.70
1 Structural Steel Worker	39.75	318.00	68.95	551.60		
16 L.H., Daily Totals		$652.00		$1131.20	$40.75	$70.70

Crew E-18	Hr.	Daily	Hr.	Daily	Bare Costs	Incl. O&P
1 Struc. Steel Foreman (outside)	$41.75	$334.00	$72.45	$579.60	$39.69	$68.15
3 Structural Steel Workers	39.75	954.00	68.95	1654.80		
1 Equipment Operator (med.)	37.45	299.60	61.45	491.60		
1 Lattice Boom Crane, 20 Ton		869.00		955.90	21.73	23.90
40 L.H., Daily Totals		$2456.60		$3681.90	$61.41	$92.05

Crew E-19	Hr.	Daily	Hr.	Daily	Bare Costs	Incl. O&P
1 Struc. Steel Foreman (outside)	$41.75	$334.00	$72.45	$579.60	$39.05	$66.63
1 Structural Steel Worker	39.75	318.00	68.95	551.60		
1 Equip. Oper. (light)	35.65	285.20	58.50	468.00		
1 Lattice Boom Crane, 20 Ton		869.00		955.90	36.21	39.83
24 L.H., Daily Totals		$1806.20		$2555.10	$75.26	$106.46

Crew E-20	Hr.	Daily	Hr.	Daily	Bare Costs	Incl. O&P
1 Struc. Steel Foreman (outside)	$41.75	$334.00	$72.45	$579.60	$38.92	$66.68
5 Structural Steel Workers	39.75	1590.00	68.95	2758.00		
1 Equip. Oper. (crane)	37.90	303.20	62.20	497.60		
1 Equip. Oper. (oiler)	32.95	263.60	54.05	432.40		
1 Lattice Boom Crane, 40 Ton		1315.00		1446.50	20.55	22.60
64 L.H., Daily Totals		$3805.80		$5714.10	$59.47	$89.28

Crew E-22	Hr.	Daily	Hr.	Daily	Bare Costs	Incl. O&P
1 Skilled Worker Foreman (out)	$38.50	$308.00	$63.95	$511.60	$37.17	$61.75
2 Skilled Workers	36.50	584.00	60.65	970.40		
24 L.H., Daily Totals		$892.00		$1482.00	$37.17	$61.75

Crew E-24	Hr.	Daily	Hr.	Daily	Bare Costs	Incl. O&P
3 Structural Steel Workers	$39.75	$954.00	$68.95	$1654.80	$39.17	$67.08
1 Equipment Operator (med.)	37.45	299.60	61.45	491.60		
1 Hyd. Crane, 25 Ton		581.70		639.87	18.18	20.00
32 L.H., Daily Totals		$1835.30		$2786.27	$57.35	$87.07

Crew E-25	Hr.	Daily	Hr.	Daily	Bare Costs	Incl. O&P
1 Welder	$39.75	$318.00	$68.95	$551.60	$39.75	$68.95
1 Cutting Torch		12.50		13.75	1.56	1.72
8 L.H., Daily Totals		$330.50		$565.35	$41.31	$70.67

Crew E-26	Hr.	Daily	Hr.	Daily	Bare Costs	Incl. O&P
1 Struc. Steel Foreman (outside)	$41.75	$334.00	$72.45	$579.60	$40.49	$69.66
1 Struc. Steel Worker	39.75	318.00	68.95	551.60		
1 Welder	39.75	318.00	68.95	551.60		
.25 Electrician	41.40	82.80	67.55	135.10		
.25 Plumber	40.40	80.80	66.30	132.60		
1 Welder, Gas Engine, 300 amp		95.65		105.22	3.42	3.76
28 L.H., Daily Totals		$1229.25		$2055.72	$43.90	$73.42

Crew No.	Bare Costs		Incl. Subs O&P		Cost Per Labor-Hour	
Crew E-27	**Hr.**	**Daily**	**Hr.**	**Daily**	**Bare Costs**	**Incl. O&P**
1 Struc. Steel Foreman (outside)	$41.75	$334.00	$72.45	$579.60	$39.77	$68.54
6 Struc. Steel Workers	39.75	1908.00	68.95	3309.60		
1 Equip. Oper. (crane)	37.90	303.20	62.20	497.60		
1 Hyd. Crane, 12 Ton		471.75		518.92		
1 Hyd. Crane, 80 Ton		1487.00		1635.70	30.61	33.67
64 L.H., Daily Totals		$4503.95		$6541.43	$70.37	$102.21
Crew F-3	**Hr.**	**Daily**	**Hr.**	**Daily**	**Bare Costs**	**Incl. O&P**
2 Carpenters	$35.65	$570.40	$58.90	$942.40	$32.76	$54.30
2 Carpenter Helpers	27.30	436.80	45.75	732.00		
1 Equip. Oper. (crane)	37.90	303.20	62.20	497.60		
1 Hyd. Crane, 12 Ton		471.75		518.92	11.79	12.97
40 L.H., Daily Totals		$1782.15		$2690.93	$44.55	$67.27
Crew F-4	**Hr.**	**Daily**	**Hr.**	**Daily**	**Bare Costs**	**Incl. O&P**
2 Carpenters	$35.65	$570.40	$58.90	$942.40	$32.76	$54.30
2 Carpenter Helpers	27.30	436.80	45.75	732.00		
1 Equip. Oper. (crane)	37.90	303.20	62.20	497.60		
1 Hyd. Crane, 55 Ton		981.50		1079.65	24.54	26.99
40 L.H., Daily Totals		$2291.90		$3251.65	$57.30	$81.29
Crew F-5	**Hr.**	**Daily**	**Hr.**	**Daily**	**Bare Costs**	**Incl. O&P**
2 Carpenters	$35.65	$570.40	$58.90	$942.40	$31.48	$52.33
2 Carpenter Helpers	27.30	436.80	45.75	732.00		
32 L.H., Daily Totals		$1007.20		$1674.40	$31.48	$52.33
Crew F-6	**Hr.**	**Daily**	**Hr.**	**Daily**	**Bare Costs**	**Incl. O&P**
2 Carpenters	$35.65	$570.40	$58.90	$942.40	$32.84	$54.18
2 Building Laborers	27.50	440.00	45.45	727.20		
1 Equip. Oper. (crane)	37.90	303.20	62.20	497.60		
1 Hyd. Crane, 12 Ton		471.75		518.92	11.79	12.97
40 L.H., Daily Totals		$1785.35		$2686.13	$44.63	$67.15
Crew F-7	**Hr.**	**Daily**	**Hr.**	**Daily**	**Bare Costs**	**Incl. O&P**
2 Carpenters	$35.65	$570.40	$58.90	$942.40	$31.57	$52.17
2 Building Laborers	27.50	440.00	45.45	727.20		
32 L.H., Daily Totals		$1010.40		$1669.60	$31.57	$52.17
Crew G-1	**Hr.**	**Daily**	**Hr.**	**Daily**	**Bare Costs**	**Incl. O&P**
1 Roofer Foreman (outside)	$32.65	$261.20	$59.00	$472.00	$28.75	$51.93
4 Roofers Composition	30.65	980.80	55.35	1771.20		
2 Roofer Helpers	23.00	368.00	41.55	664.80		
1 Application Equipment		181.15		199.26		
1 Tar Kettle/Pot		178.20		196.02		
1 Crew Truck		154.35		169.79	9.17	10.09
56 L.H., Daily Totals		$2123.70		$3473.07	$37.92	$62.02
Crew G-2	**Hr.**	**Daily**	**Hr.**	**Daily**	**Bare Costs**	**Incl. O&P**
1 Plasterer	$34.85	$278.80	$57.00	$456.00	$30.80	$50.53
1 Plasterer Helper	30.05	240.40	49.15	393.20		
1 Building Laborer	27.50	220.00	45.45	363.60		
1 Grout Pump, 50 C.F./hr.		125.60		138.16	5.23	5.76
24 L.H., Daily Totals		$864.80		$1350.96	$36.03	$56.29

Crew No.	Bare Costs		Incl. Subs O&P		Cost Per Labor-Hour	
Crew G-2A	**Hr.**	**Daily**	**Hr.**	**Daily**	**Bare Costs**	**Incl. O&P**
1 Roofer Composition	$30.65	$245.20	$55.35	$442.80	$27.05	$47.45
1 Roofer Helper	23.00	184.00	41.55	332.40		
1 Building Laborer	27.50	220.00	45.45	363.60		
1 Foam Spray Rig, Trailer-Mtd.		514.30		565.73		
1 Pickup Truck, 3/4 Ton		109.90		120.89	26.01	28.61
24 L.H., Daily Totals		$1273.40		$1825.42	$53.06	$76.06
Crew G-3	**Hr.**	**Daily**	**Hr.**	**Daily**	**Bare Costs**	**Incl. O&P**
2 Sheet Metal Workers	$39.00	$624.00	$64.70	$1035.20	$33.25	$55.08
2 Building Laborers	27.50	440.00	45.45	727.20		
32 L.H., Daily Totals		$1064.00		$1762.40	$33.25	$55.08
Crew G-4	**Hr.**	**Daily**	**Hr.**	**Daily**	**Bare Costs**	**Incl. O&P**
1 Labor Foreman (outside)	$29.50	$236.00	$48.75	$390.00	$28.17	$46.55
2 Building Laborers	27.50	440.00	45.45	727.20		
1 Flatbed Truck, Gas, 1.5 Ton		195.20		214.72		
1 Air Compressor, 160 cfm		115.85		127.44	12.96	14.26
24 L.H., Daily Totals		$987.05		$1459.36	$41.13	$60.81
Crew G-5	**Hr.**	**Daily**	**Hr.**	**Daily**	**Bare Costs**	**Incl. O&P**
1 Roofer Foreman (outside)	$32.65	$261.20	$59.00	$472.00	$27.99	$50.56
2 Roofers Composition	30.65	490.40	55.35	885.60		
2 Roofer Helpers	23.00	368.00	41.55	664.80		
1 Application Equipment		181.15		199.26	4.53	4.98
40 L.H., Daily Totals		$1300.75		$2221.67	$32.52	$55.54
Crew G-6A	**Hr.**	**Daily**	**Hr.**	**Daily**	**Bare Costs**	**Incl. O&P**
2 Roofers Composition	$30.65	$490.40	$55.35	$885.60	$30.65	$55.35
1 Small Compressor, Electric		14.85		16.34		
2 Pneumatic Nailers		44.70		49.17	3.72	4.09
16 L.H., Daily Totals		$549.95		$951.11	$34.37	$59.44
Crew G-7	**Hr.**	**Daily**	**Hr.**	**Daily**	**Bare Costs**	**Incl. O&P**
1 Carpenter	$35.65	$285.20	$58.90	$471.20	$35.65	$58.90
1 Small Compressor, Electric		14.85		16.34		
1 Pneumatic Nailer		22.35		24.59	4.65	5.12
8 L.H., Daily Totals		$322.40		$512.12	$40.30	$64.02
Crew H-1	**Hr.**	**Daily**	**Hr.**	**Daily**	**Bare Costs**	**Incl. O&P**
2 Glaziers	$34.75	$556.00	$57.20	$915.20	$37.25	$63.08
2 Struc. Steel Workers	39.75	636.00	68.95	1103.20		
32 L.H., Daily Totals		$1192.00		$2018.40	$37.25	$63.08
Crew H-2	**Hr.**	**Daily**	**Hr.**	**Daily**	**Bare Costs**	**Incl. O&P**
2 Glaziers	$34.75	$556.00	$57.20	$915.20	$32.33	$53.28
1 Building Laborer	27.50	220.00	45.45	363.60		
24 L.H., Daily Totals		$776.00		$1278.80	$32.33	$53.28
Crew H-3	**Hr.**	**Daily**	**Hr.**	**Daily**	**Bare Costs**	**Incl. O&P**
1 Glazier	$34.75	$278.00	$57.20	$457.60	$31.02	$51.48
1 Helper	27.30	218.40	45.75	366.00		
16 L.H., Daily Totals		$496.40		$823.60	$31.02	$51.48
Crew H-4	**Hr.**	**Daily**	**Hr.**	**Daily**	**Bare Costs**	**Incl. O&P**
1 Carpenter	$35.65	$285.20	$58.90	$471.20	$33.46	$55.37
1 Carpenter Helper	27.30	218.40	45.75	366.00		
.5 Electrician	41.40	165.60	67.55	270.20		
20 L.H., Daily Totals		$669.20		$1107.40	$33.46	$55.37

Crews - Residential

Crew No.	Bare Costs		Incl. Subs O&P		Cost Per Labor-Hour	
Crew J-1	**Hr.**	**Daily**	**Hr.**	**Daily**	**Bare Costs**	**Incl. O&P**
3 Plasterers	$34.85	$836.40	$57.00	$1368.00	$32.93	$53.86
2 Plasterer Helpers	30.05	480.80	49.15	786.40		
1 Mixing Machine, 6 C.F.		135.35		148.88	3.38	3.72
40 L.H., Daily Totals		$1452.55		$2303.28	$36.31	$57.58
Crew J-2	**Hr.**	**Daily**	**Hr.**	**Daily**	**Bare Costs**	**Incl. O&P**
3 Plasterers	$34.85	$836.40	$57.00	$1368.00	$33.30	$54.31
2 Plasterer Helpers	30.05	480.80	49.15	786.40		
1 Lather	35.15	281.20	56.55	452.40		
1 Mixing Machine, 6 C.F.		135.35		148.88	2.82	3.10
48 L.H., Daily Totals		$1733.75		$2755.68	$36.12	$57.41
Crew J-3	**Hr.**	**Daily**	**Hr.**	**Daily**	**Bare Costs**	**Incl. O&P**
1 Terrazzo Worker	$33.95	$271.60	$54.80	$438.40	$31.70	$51.17
1 Terrazzo Helper	29.45	235.60	47.55	380.40		
1 Floor Grinder, 22" Path		138.30		152.13		
1 Terrazzo Mixer		178.85		196.74	19.82	21.80
16 L.H., Daily Totals		$824.35		$1167.67	$51.52	$72.98
Crew J-4	**Hr.**	**Daily**	**Hr.**	**Daily**	**Bare Costs**	**Incl. O&P**
2 Cement Finishers	$35.70	$571.20	$57.70	$923.20	$32.97	$53.62
1 Laborer	27.50	220.00	45.45	363.60		
1 Floor Grinder, 22" Path		138.30		152.13		
1 Floor Edger, 7" Path		43.45		47.80		
1 Vacuum Pick-Up System		72.15		79.36	10.58	11.64
24 L.H., Daily Totals		$1045.10		$1566.09	$43.55	$65.25
Crew J-4A	**Hr.**	**Daily**	**Hr.**	**Daily**	**Bare Costs**	**Incl. O&P**
2 Cement Finishers	$35.70	$571.20	$57.70	$923.20	$31.60	$51.58
2 Laborers	27.50	440.00	45.45	727.20		
1 Floor Grinder, 22" Path		138.30		152.13		
1 Floor Edger, 7" Path		43.45		47.80		
1 Vacuum Pick-Up System		72.15		79.36		
1 Floor Auto Scrubber		265.15		291.67	16.22	17.84
32 L.H., Daily Totals		$1530.25		$2221.36	$47.82	$69.42
Crew J-4B	**Hr.**	**Daily**	**Hr.**	**Daily**	**Bare Costs**	**Incl. O&P**
1 Laborer	$27.50	$220.00	$45.45	$363.60	$27.50	$45.45
1 Floor Auto Scrubber		265.15		291.67	33.14	36.46
8 L.H., Daily Totals		$485.15		$655.26	$60.64	$81.91
Crew J-6	**Hr.**	**Daily**	**Hr.**	**Daily**	**Bare Costs**	**Incl. O&P**
2 Painters	$29.85	$477.60	$48.90	$782.40	$30.71	$50.44
1 Building Laborer	27.50	220.00	45.45	363.60		
1 Equip. Oper. (light)	35.65	285.20	58.50	468.00		
1 Air Compressor, 250 cfm		167.95		184.75		
1 Sandblaster, Portable, 3 C.F.		20.70		22.77		
1 Set Sand Blasting Accessories		14.90		16.39	6.36	7.00
32 L.H., Daily Totals		$1186.35		$1837.91	$37.07	$57.43
Crew J-7	**Hr.**	**Daily**	**Hr.**	**Daily**	**Bare Costs**	**Incl. O&P**
2 Painters	$29.85	$477.60	$48.90	$782.40	$29.85	$48.90
1 Floor Belt Sander		17.75		19.52		
1 Floor Sanding Edger		13.65		15.02	1.96	2.16
16 L.H., Daily Totals		$509.00		$816.94	$31.81	$51.06

Crew No.	Bare Costs		Incl. Subs O&P		Cost Per Labor-Hour	
Crew K-1	**Hr.**	**Daily**	**Hr.**	**Daily**	**Bare Costs**	**Incl. O&P**
1 Carpenter	$35.65	$285.20	$58.90	$471.20	$33.98	$56.02
1 Truck Driver (light)	32.30	258.40	53.15	425.20		
1 Flatbed Truck, Gas, 3 Ton		245.95		270.55	15.37	16.91
16 L.H., Daily Totals		$789.55		$1166.94	$49.35	$72.93
Crew K-2	**Hr.**	**Daily**	**Hr.**	**Daily**	**Bare Costs**	**Incl. O&P**
1 Struc. Steel Foreman (outside)	$41.75	$334.00	$72.45	$579.60	$37.93	$64.85
1 Struc. Steel Worker	39.75	318.00	68.95	551.60		
1 Truck Driver (light)	32.30	258.40	53.15	425.20		
1 Flatbed Truck, Gas, 3 Ton		245.95		270.55	10.25	11.27
24 L.H., Daily Totals		$1156.35		$1826.94	$48.18	$76.12
Crew L-1	**Hr.**	**Daily**	**Hr.**	**Daily**	**Bare Costs**	**Incl. O&P**
.25 Electrician	$41.40	$82.80	$67.55	$135.10	$40.60	$66.55
1 Plumber	40.40	323.20	66.30	530.40		
10 L.H., Daily Totals		$406.00		$665.50	$40.60	$66.55
Crew L-2	**Hr.**	**Daily**	**Hr.**	**Daily**	**Bare Costs**	**Incl. O&P**
1 Carpenter	$35.65	$285.20	$58.90	$471.20	$31.48	$52.33
1 Carpenter Helper	27.30	218.40	45.75	366.00		
16 L.H., Daily Totals		$503.60		$837.20	$31.48	$52.33
Crew L-3	**Hr.**	**Daily**	**Hr.**	**Daily**	**Bare Costs**	**Incl. O&P**
1 Carpenter	$35.65	$285.20	$58.90	$471.20	$36.80	$60.63
.25 Electrician	41.40	82.80	67.55	135.10		
10 L.H., Daily Totals		$368.00		$606.30	$36.80	$60.63
Crew L-3A	**Hr.**	**Daily**	**Hr.**	**Daily**	**Bare Costs**	**Incl. O&P**
1 Carpenter Foreman (outside)	$37.65	$301.20	$62.20	$497.60	$38.10	$63.03
.5 Sheet Metal Worker	39.00	156.00	64.70	258.80		
12 L.H., Daily Totals		$457.20		$756.40	$38.10	$63.03
Crew L-4	**Hr.**	**Daily**	**Hr.**	**Daily**	**Bare Costs**	**Incl. O&P**
1 Skilled Worker	$36.50	$292.00	$60.65	$485.20	$31.90	$53.20
1 Helper	27.30	218.40	45.75	366.00		
16 L.H., Daily Totals		$510.40		$851.20	$31.90	$53.20
Crew L-5	**Hr.**	**Daily**	**Hr.**	**Daily**	**Bare Costs**	**Incl. O&P**
1 Struc. Steel Foreman (outside)	$41.75	$334.00	$72.45	$579.60	$39.77	$68.49
5 Struc. Steel Workers	39.75	1590.00	68.95	2758.00		
1 Equip. Oper. (crane)	37.90	303.20	62.20	497.60		
1 Hyd. Crane, 25 Ton		581.70		639.87	10.39	11.43
56 L.H., Daily Totals		$2808.90		$4475.07	$50.16	$79.91
Crew L-5A	**Hr.**	**Daily**	**Hr.**	**Daily**	**Bare Costs**	**Incl. O&P**
1 Struc. Steel Foreman (outside)	$41.75	$334.00	$72.45	$579.60	$39.79	$68.14
2 Structural Steel Workers	39.75	636.00	68.95	1103.20		
1 Equip. Oper. (crane)	37.90	303.20	62.20	497.60		
1 S.P. Crane, 4x4, 25 Ton		664.40		730.84	20.76	22.84
32 L.H., Daily Totals		$1937.60		$2911.24	$60.55	$90.98
Crew L-5B	**Hr.**	**Daily**	**Hr.**	**Daily**	**Bare Costs**	**Incl. O&P**
1 Struc. Steel Foreman (outside)	$41.75	$334.00	$72.45	$579.60	$39.17	$65.49
2 Structural Steel Workers	39.75	636.00	68.95	1103.20		
2 Electricians	41.40	662.40	67.55	1080.80		
2 Steamfitters/Pipefitters	41.55	664.80	68.15	1090.40		
1 Equip. Oper. (crane)	37.90	303.20	62.20	497.60		
1 Common Laborer	27.50	220.00	45.45	363.60		
1 Hyd. Crane, 80 Ton		1487.00		1635.70	20.65	22.72
72 L.H., Daily Totals		$4307.40		$6350.90	$59.83	$88.21

Crew No.	Bare Costs		Incl. Subs O&P		Cost Per Labor-Hour	
Crew L-6	**Hr.**	**Daily**	**Hr.**	**Daily**	**Bare Costs**	**Incl. O&P**
1 Plumber	$40.40	$323.20	$66.30	$530.40	$40.73	$66.72
.5 Electrician	41.40	165.60	67.55	270.20		
12 L.H., Daily Totals		$488.80		$800.60	$40.73	$66.72
Crew L-7	**Hr.**	**Daily**	**Hr.**	**Daily**	**Bare Costs**	**Incl. O&P**
1 Carpenter	$35.65	$285.20	$58.90	$471.20	$30.95	$51.47
2 Carpenter Helpers	27.30	436.80	45.75	732.00		
.25 Electrician	41.40	82.80	67.55	135.10		
26 L.H., Daily Totals		$804.80		$1338.30	$30.95	$51.47
Crew L-8	**Hr.**	**Daily**	**Hr.**	**Daily**	**Bare Costs**	**Incl. O&P**
1 Carpenter	$35.65	$285.20	$58.90	$471.20	$33.26	$55.12
1 Carpenter Helper	27.30	218.40	45.75	366.00		
.5 Plumber	40.40	161.60	66.30	265.20		
20 L.H., Daily Totals		$665.20		$1102.40	$33.26	$55.12
Crew L-9	**Hr.**	**Daily**	**Hr.**	**Daily**	**Bare Costs**	**Incl. O&P**
1 Skilled Worker Foreman	$38.50	$308.00	$63.95	$511.60	$33.40	$55.53
1 Skilled Worker	36.50	292.00	60.65	485.20		
2 Helpers	27.30	436.80	45.75	732.00		
.5 Electrician	41.40	165.60	67.55	270.20		
36 L.H., Daily Totals		$1202.40		$1999.00	$33.40	$55.53
Crew L-10	**Hr.**	**Daily**	**Hr.**	**Daily**	**Bare Costs**	**Incl. O&P**
1 Struc. Steel Foreman (outside)	$41.75	$334.00	$72.45	$579.60	$39.80	$67.87
1 Structural Steel Worker	39.75	318.00	68.95	551.60		
1 Equip. Oper. (crane)	37.90	303.20	62.20	497.60		
1 Hyd. Crane, 12 Ton		471.75		518.92	19.66	21.62
24 L.H., Daily Totals		$1426.95		$2147.72	$59.46	$89.49
Crew L-11	**Hr.**	**Daily**	**Hr.**	**Daily**	**Bare Costs**	**Incl. O&P**
2 Wreckers	$27.90	$446.40	$47.10	$753.60	$32.34	$53.73
1 Equip. Oper. (crane)	37.90	303.20	62.20	497.60		
1 Equip. Oper. (light)	35.65	285.20	58.50	468.00		
1 Hyd. Excavator, 2.5 C.Y.		1441.00		1585.10		
1 Loader, Skid Steer, 78 H.P.		392.50		431.75	57.30	63.03
32 L.H., Daily Totals		$2868.30		$3736.05	$89.63	$116.75
Crew M-1	**Hr.**	**Daily**	**Hr.**	**Daily**	**Bare Costs**	**Incl. O&P**
3 Elevator Constructors	$55.45	$1330.80	$90.20	$2164.80	$52.67	$85.69
1 Elevator Apprentice	44.35	354.80	72.15	577.20		
5 Hand Tools		49.50		54.45	1.55	1.70
32 L.H., Daily Totals		$1735.10		$2796.45	$54.22	$87.39
Crew M-3	**Hr.**	**Daily**	**Hr.**	**Daily**	**Bare Costs**	**Incl. O&P**
1 Electrician Foreman (outside)	$43.40	$347.20	$70.85	$566.80	$42.37	$69.18
1 Common Laborer	27.50	220.00	45.45	363.60		
.25 Equipment Operator (med.)	37.45	74.90	61.45	122.90		
1 Elevator Constructor	55.45	443.60	90.20	721.60		
1 Elevator Apprentice	44.35	354.80	72.15	577.20		
.25 S.P. Crane, 4x4, 20 Ton		146.94		161.63	4.32	4.75
34 L.H., Daily Totals		$1587.44		$2513.73	$46.69	$73.93

Crew No.	Bare Costs		Incl. Subs O&P		Cost Per Labor-Hour	
Crew M-4	**Hr.**	**Daily**	**Hr.**	**Daily**	**Bare Costs**	**Incl. O&P**
1 Electrician Foreman (outside)	$43.40	$347.20	$70.85	$566.80	$41.87	$68.38
1 Common Laborer	27.50	220.00	45.45	363.60		
.25 Equipment Operator, Crane	37.90	75.80	62.20	124.40		
.25 Equip. Oper. (oiler)	32.95	65.90	54.05	108.10		
1 Elevator Constructor	55.45	443.60	90.20	721.60		
1 Elevator Apprentice	44.35	354.80	72.15	577.20		
.25 S.P. Crane, 4x4, 40 Ton		187.56		206.32	5.21	5.73
36 L.H., Daily Totals		$1694.86		$2668.02	$47.08	$74.11
Crew Q-1	**Hr.**	**Daily**	**Hr.**	**Daily**	**Bare Costs**	**Incl. O&P**
1 Plumber	$40.40	$323.20	$66.30	$530.40	$36.35	$59.65
1 Plumber Apprentice	32.30	258.40	53.00	424.00		
16 L.H., Daily Totals		$581.60		$954.40	$36.35	$59.65
Crew Q-1A	**Hr.**	**Daily**	**Hr.**	**Daily**	**Bare Costs**	**Incl. O&P**
.25 Plumber Foreman (outside)	$42.40	$84.80	$69.55	$139.10	$40.80	$66.95
1 Plumber	40.40	323.20	66.30	530.40		
10 L.H., Daily Totals		$408.00		$669.50	$40.80	$66.95
Crew Q-1C	**Hr.**	**Daily**	**Hr.**	**Daily**	**Bare Costs**	**Incl. O&P**
1 Plumber	$40.40	$323.20	$66.30	$530.40	$36.72	$60.25
1 Plumber Apprentice	32.30	258.40	53.00	424.00		
1 Equip. Oper. (medium)	37.45	299.60	61.45	491.60		
1 Trencher, Chain Type, 8' D		1837.00		2020.70	76.54	84.20
24 L.H., Daily Totals		$2718.20		$3466.70	$113.26	$144.45
Crew Q-2	**Hr.**	**Daily**	**Hr.**	**Daily**	**Bare Costs**	**Incl. O&P**
1 Plumber	$40.40	$323.20	$66.30	$530.40	$35.00	$57.43
2 Plumber Apprentices	32.30	516.80	53.00	848.00		
24 L.H., Daily Totals		$840.00		$1378.40	$35.00	$57.43
Crew Q-3	**Hr.**	**Daily**	**Hr.**	**Daily**	**Bare Costs**	**Incl. O&P**
2 Plumbers	$40.40	$646.40	$66.30	$1060.80	$36.35	$59.65
2 Plumber Apprentices	32.30	516.80	53.00	848.00		
32 L.H., Daily Totals		$1163.20		$1908.80	$36.35	$59.65
Crew Q-4	**Hr.**	**Daily**	**Hr.**	**Daily**	**Bare Costs**	**Incl. O&P**
2 Plumbers	$40.40	$646.40	$66.30	$1060.80	$38.38	$62.98
1 Welder (plumber)	40.40	323.20	66.30	530.40		
1 Plumber Apprentice	32.30	258.40	53.00	424.00		
1 Welder, Electric, 300 amp		56.10		61.71	1.75	1.93
32 L.H., Daily Totals		$1284.10		$2076.91	$40.13	$64.90
Crew Q-5	**Hr.**	**Daily**	**Hr.**	**Daily**	**Bare Costs**	**Incl. O&P**
1 Steamfitter	$41.55	$332.40	$68.15	$545.20	$37.40	$61.35
1 Steamfitter Apprentice	33.25	266.00	54.55	436.40		
16 L.H., Daily Totals		$598.40		$981.60	$37.40	$61.35
Crew Q-6	**Hr.**	**Daily**	**Hr.**	**Daily**	**Bare Costs**	**Incl. O&P**
1 Steamfitter	$41.55	$332.40	$68.15	$545.20	$36.02	$59.08
2 Steamfitter Apprentices	33.25	532.00	54.55	872.80		
24 L.H., Daily Totals		$864.40		$1418.00	$36.02	$59.08
Crew Q-7	**Hr.**	**Daily**	**Hr.**	**Daily**	**Bare Costs**	**Incl. O&P**
2 Steamfitters	$41.55	$664.80	$68.15	$1090.40	$37.40	$61.35
2 Steamfitter Apprentices	33.25	532.00	54.55	872.80		
32 L.H., Daily Totals		$1196.80		$1963.20	$37.40	$61.35

Crew Q-8	Bare Costs Hr.	Bare Costs Daily	Incl. Subs O&P Hr.	Incl. Subs O&P Daily	Cost Per Labor-Hour Bare Costs	Cost Per Labor-Hour Incl. O&P
2 Steamfitters	$41.55	$664.80	$68.15	$1090.40	$39.48	$64.75
1 Welder (steamfitter)	41.55	332.40	68.15	545.20		
1 Steamfitter Apprentice	33.25	266.00	54.55	436.40		
1 Welder, Electric, 300 amp		56.10		61.71	1.75	1.93
32 L.H., Daily Totals		$1319.30		$2133.71	$41.23	$66.68

Crew Q-9	Hr.	Daily	Hr.	Daily	Bare Costs	Incl. O&P
1 Sheet Metal Worker	$39.00	$312.00	$64.70	$517.60	$35.10	$58.23
1 Sheet Metal Apprentice	31.20	249.60	51.75	414.00		
16 L.H., Daily Totals		$561.60		$931.60	$35.10	$58.23

Crew Q-10	Hr.	Daily	Hr.	Daily	Bare Costs	Incl. O&P
2 Sheet Metal Workers	$39.00	$624.00	$64.70	$1035.20	$36.40	$60.38
1 Sheet Metal Apprentice	31.20	249.60	51.75	414.00		
24 L.H., Daily Totals		$873.60		$1449.20	$36.40	$60.38

Crew Q-11	Hr.	Daily	Hr.	Daily	Bare Costs	Incl. O&P
2 Sheet Metal Workers	$39.00	$624.00	$64.70	$1035.20	$35.10	$58.23
2 Sheet Metal Apprentices	31.20	499.20	51.75	828.00		
32 L.H., Daily Totals		$1123.20		$1863.20	$35.10	$58.23

Crew Q-12	Hr.	Daily	Hr.	Daily	Bare Costs	Incl. O&P
1 Sprinkler Installer	$39.95	$319.60	$65.70	$525.60	$35.95	$59.13
1 Sprinkler Apprentice	31.95	255.60	52.55	420.40		
16 L.H., Daily Totals		$575.20		$946.00	$35.95	$59.13

Crew Q-13	Hr.	Daily	Hr.	Daily	Bare Costs	Incl. O&P
2 Sprinkler Installers	$39.95	$639.20	$65.70	$1051.20	$35.95	$59.13
2 Sprinkler Apprentices	31.95	511.20	52.55	840.80		
32 L.H., Daily Totals		$1150.40		$1892.00	$35.95	$59.13

Crew Q-14	Hr.	Daily	Hr.	Daily	Bare Costs	Incl. O&P
1 Asbestos Worker	$36.70	$293.60	$61.80	$494.40	$33.02	$55.60
1 Asbestos Apprentice	29.35	234.80	49.40	395.20		
16 L.H., Daily Totals		$528.40		$889.60	$33.02	$55.60

Crew Q-15	Hr.	Daily	Hr.	Daily	Bare Costs	Incl. O&P
1 Plumber	$40.40	$323.20	$66.30	$530.40	$36.35	$59.65
1 Plumber Apprentice	32.30	258.40	53.00	424.00		
1 Welder, Electric, 300 amp		56.10		61.71	3.51	3.86
16 L.H., Daily Totals		$637.70		$1016.11	$39.86	$63.51

Crew Q-16	Hr.	Daily	Hr.	Daily	Bare Costs	Incl. O&P
2 Plumbers	$40.40	$646.40	$66.30	$1060.80	$37.70	$61.87
1 Plumber Apprentice	32.30	258.40	53.00	424.00		
1 Welder, Electric, 300 amp		56.10		61.71	2.34	2.57
24 L.H., Daily Totals		$960.90		$1546.51	$40.04	$64.44

Crew Q-17	Hr.	Daily	Hr.	Daily	Bare Costs	Incl. O&P
1 Steamfitter	$41.55	$332.40	$68.15	$545.20	$37.40	$61.35
1 Steamfitter Apprentice	33.25	266.00	54.55	436.40		
1 Welder, Electric, 300 amp		56.10		61.71	3.51	3.86
16 L.H., Daily Totals		$654.50		$1043.31	$40.91	$65.21

Crew Q-17A	Hr.	Daily	Hr.	Daily	Bare Costs	Incl. O&P
1 Steamfitter	$41.55	$332.40	$68.15	$545.20	$37.57	$61.63
1 Steamfitter Apprentice	33.25	266.00	54.55	436.40		
1 Equip. Oper. (crane)	37.90	303.20	62.20	497.60		
1 Hyd. Crane, 12 Ton		471.75		518.92		
1 Welder, Electric, 300 amp		56.10		61.71	21.99	24.19
24 L.H., Daily Totals		$1429.45		$2059.84	$59.56	$85.83

Crew Q-18	Hr.	Daily	Hr.	Daily	Bare Costs	Incl. O&P
2 Steamfitters	$41.55	$664.80	$68.15	$1090.40	$38.78	$63.62
1 Steamfitter Apprentice	33.25	266.00	54.55	436.40		
1 Welder, Electric, 300 amp		56.10		61.71	2.34	2.57
24 L.H., Daily Totals		$986.90		$1588.51	$41.12	$66.19

Crew Q-19	Hr.	Daily	Hr.	Daily	Bare Costs	Incl. O&P
1 Steamfitter	$41.55	$332.40	$68.15	$545.20	$38.73	$63.42
1 Steamfitter Apprentice	33.25	266.00	54.55	436.40		
1 Electrician	41.40	331.20	67.55	540.40		
24 L.H., Daily Totals		$929.60		$1522.00	$38.73	$63.42

Crew Q-20	Hr.	Daily	Hr.	Daily	Bare Costs	Incl. O&P
1 Sheet Metal Worker	$39.00	$312.00	$64.70	$517.60	$36.36	$60.09
1 Sheet Metal Apprentice	31.20	249.60	51.75	414.00		
.5 Electrician	41.40	165.60	67.55	270.20		
20 L.H., Daily Totals		$727.20		$1201.80	$36.36	$60.09

Crew Q-21	Hr.	Daily	Hr.	Daily	Bare Costs	Incl. O&P
2 Steamfitters	$41.55	$664.80	$68.15	$1090.40	$39.44	$64.60
1 Steamfitter Apprentice	33.25	266.00	54.55	436.40		
1 Electrician	41.40	331.20	67.55	540.40		
32 L.H., Daily Totals		$1262.00		$2067.20	$39.44	$64.60

Crew Q-22	Hr.	Daily	Hr.	Daily	Bare Costs	Incl. O&P
1 Plumber	$40.40	$323.20	$66.30	$530.40	$36.35	$59.65
1 Plumber Apprentice	32.30	258.40	53.00	424.00		
1 Hyd. Crane, 12 Ton		471.75		518.92	29.48	32.43
16 L.H., Daily Totals		$1053.35		$1473.33	$65.83	$92.08

Crew Q-22A	Hr.	Daily	Hr.	Daily	Bare Costs	Incl. O&P
1 Plumber	$40.40	$323.20	$66.30	$530.40	$34.52	$56.74
1 Plumber Apprentice	32.30	258.40	53.00	424.00		
1 Laborer	27.50	220.00	45.45	363.60		
1 Equip. Oper. (crane)	37.90	303.20	62.20	497.60		
1 Hyd. Crane, 12 Ton		471.75		518.92	14.74	16.22
32 L.H., Daily Totals		$1576.55		$2334.53	$49.27	$72.95

Crew Q-23	Hr.	Daily	Hr.	Daily	Bare Costs	Incl. O&P
1 Plumber Foreman (outside)	$42.40	$339.20	$69.55	$556.40	$40.08	$65.77
1 Plumber	40.40	323.20	66.30	530.40		
1 Equip. Oper. (medium)	37.45	299.60	61.45	491.60		
1 Lattice Boom Crane, 20 Ton		869.00		955.90	36.21	39.83
24 L.H., Daily Totals		$1831.00		$2534.30	$76.29	$105.60

Crew R-1	Hr.	Daily	Hr.	Daily	Bare Costs	Incl. O&P
1 Electrician Foreman	$41.90	$335.20	$68.40	$547.20	$38.72	$63.17
3 Electricians	41.40	993.60	67.55	1621.20		
2 Electrician Apprentices	33.10	529.60	54.00	864.00		
48 L.H., Daily Totals		$1858.40		$3032.40	$38.72	$63.17

Crew No.	Bare Costs		Incl. Subs O&P		Cost Per Labor-Hour	
Crew R-1A	**Hr.**	**Daily**	**Hr.**	**Daily**	**Bare Costs**	**Incl. O&P**
1 Electrician	$41.40	$331.20	$67.55	$540.40	$37.25	$60.77
1 Electrician Apprentice	33.10	264.80	54.00	432.00		
16 L.H., Daily Totals		$596.00		$972.40	$37.25	$60.77
Crew R-1B	**Hr.**	**Daily**	**Hr.**	**Daily**	**Bare Costs**	**Incl. O&P**
1 Electrician	$41.40	$331.20	$67.55	$540.40	$35.87	$58.52
2 Electrician Apprentices	33.10	529.60	54.00	864.00		
24 L.H., Daily Totals		$860.80		$1404.40	$35.87	$58.52
Crew R-1C	**Hr.**	**Daily**	**Hr.**	**Daily**	**Bare Costs**	**Incl. O&P**
2 Electricians	$41.40	$662.40	$67.55	$1080.80	$37.25	$60.77
2 Electrician Apprentices	33.10	529.60	54.00	864.00		
1 Portable Cable Puller, 8000 lb.		152.70		167.97	4.77	5.25
32 L.H., Daily Totals		$1344.70		$2112.77	$42.02	$66.02
Crew R-2	**Hr.**	**Daily**	**Hr.**	**Daily**	**Bare Costs**	**Incl. O&P**
1 Electrician Foreman	$41.90	$335.20	$68.40	$547.20	$38.60	$63.04
3 Electricians	41.40	993.60	67.55	1621.20		
2 Electrician Apprentices	33.10	529.60	54.00	864.00		
1 Equip. Oper. (crane)	37.90	303.20	62.20	497.60		
1 S.P. Crane, 4x4, 5 Ton		259.45		285.39	4.63	5.10
56 L.H., Daily Totals		$2421.05		$3815.40	$43.23	$68.13
Crew R-3	**Hr.**	**Daily**	**Hr.**	**Daily**	**Bare Costs**	**Incl. O&P**
1 Electrician Foreman	$41.90	$335.20	$68.40	$547.20	$40.90	$66.82
1 Electrician	41.40	331.20	67.55	540.40		
.5 Equip. Oper. (crane)	37.90	151.60	62.20	248.80		
.5 S.P. Crane, 4x4, 5 Ton		129.72		142.70	6.49	7.13
20 L.H., Daily Totals		$947.73		$1479.10	$47.39	$73.95
Crew R-4	**Hr.**	**Daily**	**Hr.**	**Daily**	**Bare Costs**	**Incl. O&P**
1 Struc. Steel Foreman (outside)	$41.75	$334.00	$72.45	$579.60	$40.48	$69.37
3 Struc. Steel Workers	39.75	954.00	68.95	1654.80		
1 Electrician	41.40	331.20	67.55	540.40		
1 Welder, Gas Engine, 300 amp		95.65		105.22	2.39	2.63
40 L.H., Daily Totals		$1714.85		$2880.01	$42.87	$72.00
Crew R-5	**Hr.**	**Daily**	**Hr.**	**Daily**	**Bare Costs**	**Incl. O&P**
1 Electrician Foreman	$41.90	$335.20	$68.40	$547.20	$36.32	$59.70
4 Electrician Linemen	41.40	1324.80	67.55	2161.60		
2 Electrician Operators	41.40	662.40	67.55	1080.80		
4 Electrician Groundmen	27.30	873.60	45.75	1464.00		
1 Crew Truck		154.35		169.79		
1 Flatbed Truck, 20,000 GVW		199.45		219.40		
1 Pickup Truck, 3/4 Ton		109.90		120.89		
.2 Hyd. Crane, 55 Ton		196.30		215.93		
.2 Hyd. Crane, 12 Ton		94.35		103.79		
.2 Earth Auger, Truck-Mtd.		77.17		84.89		
1 Tractor w/Winch		368.50		405.35	13.64	15.00
88 L.H., Daily Totals		$4396.02		$6573.62	$49.95	$74.70

Crew No.	Bare Costs		Incl. Subs O&P		Cost Per Labor-Hour	
Crew R-6	**Hr.**	**Daily**	**Hr.**	**Daily**	**Bare Costs**	**Incl. O&P**
1 Electrician Foreman	$41.90	$335.20	$68.40	$547.20	$36.32	$59.70
4 Electrician Linemen	41.40	1324.80	67.55	2161.60		
2 Electrician Operators	41.40	662.40	67.55	1080.80		
4 Electrician Groundmen	27.30	873.60	45.75	1464.00		
1 Crew Truck		154.35		169.79		
1 Flatbed Truck, 20,000 GVW		199.45		219.40		
1 Pickup Truck, 3/4 Ton		109.90		120.89		
.2 Hyd. Crane, 55 Ton		196.30		215.93		
.2 Hyd. Crane, 12 Ton		94.35		103.79		
.2 Earth Auger, Truck-Mtd.		77.17		84.89		
1 Tractor w/Winch		368.50		405.35		
3 Cable Trailers		648.60		713.46		
.5 Tensioning Rig		214.32		235.76		
.5 Cable Pulling Rig		1233.50		1356.85	37.46	41.21
88 L.H., Daily Totals		$6492.44		$8879.69	$73.78	$100.91
Crew R-7	**Hr.**	**Daily**	**Hr.**	**Daily**	**Bare Costs**	**Incl. O&P**
1 Electrician Foreman	$41.90	$335.20	$68.40	$547.20	$29.73	$49.52
5 Electrician Groundmen	27.30	1092.00	45.75	1830.00		
1 Crew Truck		154.35		169.79	3.22	3.54
48 L.H., Daily Totals		$1581.55		$2546.99	$32.95	$53.06
Crew R-8	**Hr.**	**Daily**	**Hr.**	**Daily**	**Bare Costs**	**Incl. O&P**
1 Electrician Foreman	$41.90	$335.20	$68.40	$547.20	$36.78	$60.42
3 Electrician Linemen	41.40	993.60	67.55	1621.20		
2 Electrician Groundmen	27.30	436.80	45.75	732.00		
1 Pickup Truck, 3/4 Ton		109.90		120.89		
1 Crew Truck		154.35		169.79	5.51	6.06
48 L.H., Daily Totals		$2029.85		$3191.07	$42.29	$66.48
Crew R-9	**Hr.**	**Daily**	**Hr.**	**Daily**	**Bare Costs**	**Incl. O&P**
1 Electrician Foreman	$41.90	$335.20	$68.40	$547.20	$34.41	$56.76
1 Electrician Lineman	41.40	331.20	67.55	540.40		
2 Electrician Operators	41.40	662.40	67.55	1080.80		
4 Electrician Groundmen	27.30	873.60	45.75	1464.00		
1 Pickup Truck, 3/4 Ton		109.90		120.89		
1 Crew Truck		154.35		169.79	4.13	4.54
64 L.H., Daily Totals		$2466.65		$3923.07	$38.54	$61.30
Crew R-10	**Hr.**	**Daily**	**Hr.**	**Daily**	**Bare Costs**	**Incl. O&P**
1 Electrician Foreman	$41.90	$335.20	$68.40	$547.20	$39.13	$64.06
4 Electrician Linemen	41.40	1324.80	67.55	2161.60		
1 Electrician Groundman	27.30	218.40	45.75	366.00		
1 Crew Truck		154.35		169.79		
3 Tram Cars		438.60		482.46	12.35	13.59
48 L.H., Daily Totals		$2471.35		$3727.05	$51.49	$77.65
Crew R-11	**Hr.**	**Daily**	**Hr.**	**Daily**	**Bare Costs**	**Incl. O&P**
1 Electrician Foreman	$41.90	$335.20	$68.40	$547.20	$38.99	$63.75
4 Electricians	41.40	1324.80	67.55	2161.60		
1 Equip. Oper. (crane)	37.90	303.20	62.20	497.60		
1 Common Laborer	27.50	220.00	45.45	363.60		
1 Crew Truck		154.35		169.79		
1 Hyd. Crane, 12 Ton		471.75		518.92	11.18	12.30
56 L.H., Daily Totals		$2809.30		$4258.71	$50.17	$76.05

Crew No.	Bare Costs		Incl. Subs O&P		Cost Per Labor-Hour	
Crew R-12	**Hr.**	**Daily**	**Hr.**	**Daily**	**Bare Costs**	**Incl. O&P**
1 Carpenter Foreman (inside)	$36.15	$289.20	$59.75	$478.00	$33.27	$55.23
4 Carpenters	35.65	1140.80	58.90	1884.80		
4 Common Laborers	27.50	880.00	45.45	1454.40		
1 Equip. Oper. (medium)	37.45	299.60	61.45	491.60		
1 Steel Worker	39.75	318.00	68.95	551.60		
1 Dozer, 200 H.P.		1290.00		1419.00		
1 Pickup Truck, 3/4 Ton		109.90		120.89	15.91	17.50
88 L.H., Daily Totals		$4327.50		$6400.29	$49.18	$72.73
Crew R-13	**Hr.**	**Daily**	**Hr.**	**Daily**	**Bare Costs**	**Incl. O&P**
1 Electrician Foreman	$41.90	$335.20	$68.40	$547.20	$39.72	$64.89
3 Electricians	41.40	993.60	67.55	1621.20		
.25 Equip. Oper. (crane)	37.90	75.80	62.20	124.40		
1 Equipment Oiler	32.95	263.60	54.05	432.40		
.25 Hydraulic Crane, 33 Ton		236.65		260.32	5.63	6.20
42 L.H., Daily Totals		$1904.85		$2985.51	$45.35	$71.08
Crew R-15	**Hr.**	**Daily**	**Hr.**	**Daily**	**Bare Costs**	**Incl. O&P**
1 Electrician Foreman	$41.90	$335.20	$68.40	$547.20	$40.52	$66.18
4 Electricians	41.40	1324.80	67.55	2161.60		
1 Equipment Oper. (light)	35.65	285.20	58.50	468.00		
1 Telescoping Boom Lift, to 40'		283.15		311.46	5.90	6.49
48 L.H., Daily Totals		$2228.35		$3488.26	$46.42	$72.67
Crew R-15A	**Hr.**	**Daily**	**Hr.**	**Daily**	**Bare Costs**	**Incl. O&P**
1 Electrician Foreman	$41.90	$335.20	$68.40	$547.20	$35.89	$58.82
2 Electricians	41.40	662.40	67.55	1080.80		
2 Common Laborers	27.50	440.00	45.45	727.20		
1 Equip. Oper. (light)	35.65	285.20	58.50	468.00		
1 Telescoping Boom Lift, to 40'		283.15		311.46	5.90	6.49
48 L.H., Daily Totals		$2005.95		$3134.67	$41.79	$65.31
Crew R-18	**Hr.**	**Daily**	**Hr.**	**Daily**	**Bare Costs**	**Incl. O&P**
.25 Electrician Foreman	$41.90	$83.80	$68.40	$136.80	$36.33	$59.28
1 Electrician	41.40	331.20	67.55	540.40		
2 Electrician Apprentices	33.10	529.60	54.00	864.00		
26 L.H., Daily Totals		$944.60		$1541.20	$36.33	$59.28
Crew R-19	**Hr.**	**Daily**	**Hr.**	**Daily**	**Bare Costs**	**Incl. O&P**
.5 Electrician Foreman	$41.90	$167.60	$68.40	$273.60	$41.50	$67.72
2 Electricians	41.40	662.40	67.55	1080.80		
20 L.H., Daily Totals		$830.00		$1354.40	$41.50	$67.72
Crew R-21	**Hr.**	**Daily**	**Hr.**	**Daily**	**Bare Costs**	**Incl. O&P**
1 Electrician Foreman	$41.90	$335.20	$68.40	$547.20	$41.43	$67.61
3 Electricians	41.40	993.60	67.55	1621.20		
.1 Equip. Oper. (medium)	37.45	29.96	61.45	49.16		
.1 S.P. Crane, 4x4, 25 Ton		66.44		73.08	2.03	2.23
32.8 L.H., Daily Totals		$1425.20		$2290.64	$43.45	$69.84
Crew R-22	**Hr.**	**Daily**	**Hr.**	**Daily**	**Bare Costs**	**Incl. O&P**
.66 Electrician Foreman	$41.90	$221.23	$68.40	$361.15	$37.91	$61.85
2 Electricians	41.40	662.40	67.55	1080.80		
2 Electrician Apprentices	33.10	529.60	54.00	864.00		
37.28 L.H., Daily Totals		$1413.23		$2305.95	$37.91	$61.85

Crew No.	Bare Costs		Incl. Subs O&P		Cost Per Labor-Hour	
Crew R-30	**Hr.**	**Daily**	**Hr.**	**Daily**	**Bare Costs**	**Incl. O&P**
.25 Electrician Foreman (outside)	$43.40	$86.80	$70.85	$141.70	$33.00	$54.20
1 Electrician	41.40	331.20	67.55	540.40		
2 Laborers (Semi-Skilled)	27.50	440.00	45.45	727.20		
26 L.H., Daily Totals		$858.00		$1409.30	$33.00	$54.20

Location Factors - Residential

Costs shown in *Residential Costs with RSMeans data* are based on national averages for materials and installation. To adjust these costs to a specific location, simply multiply the base cost by the factor for that city. The data is arranged alphabetically by state and postal zip code numbers. For a city not listed, use the factor for a nearby city with similar economic characteristics.

STATE	CITY	Residential
ALABAMA		
350-352	Birmingham	.84
354	Tuscaloosa	.84
355	Jasper	.83
356	Decatur	.83
357-358	Huntsville	.84
359	Gadsden	.81
360-361	Montgomery	.82
362	Anniston	.80
363	Dothan	.84
364	Evergreen	.79
365-366	Mobile	.82
367	Selma	.81
368	Phenix City	.82
369	Butler	.80
ALASKA		
995-996	Anchorage	1.21
997	Fairbanks	1.24
998	Juneau	1.22
999	Ketchikan	1.24
ARIZONA		
850,853	Phoenix	.87
851,852	Mesa/Tempe	.85
855	Globe	.84
856-857	Tucson	.85
859	Show Low	.86
860	Flagstaff	.88
863	Prescott	.87
864	Kingman	.85
865	Chambers	.87
ARKANSAS		
716	Pine Bluff	.78
717	Camden	.75
718	Texarkana	.78
719	Hot Springs	.74
720-722	Little Rock	.81
723	West Memphis	.78
724	Jonesboro	.77
725	Batesville	.75
726	Harrison	.76
727	Fayetteville	.74
728	Russellville	.75
729	Fort Smith	.82
CALIFORNIA		
900-902	Los Angeles	1.14
903-905	Inglewood	1.13
906-908	Long Beach	1.12
910-912	Pasadena	1.12
913-916	Van Nuys	1.15
917-918	Alhambra	1.16
919-921	San Diego	1.10
922	Palm Springs	1.12
923-924	San Bernardino	1.13
925	Riverside	1.14
926-927	Santa Ana	1.14
928	Anaheim	1.14
930	Oxnard	1.13
931	Santa Barbara	1.13
932-933	Bakersfield	1.12
934	San Luis Obispo	1.15
935	Mojave	1.13
936-938	Fresno	1.17
939	Salinas	1.20
940-941	San Francisco	1.32
942,956-958	Sacramento	1.19
943	Palo Alto	1.27
944	San Mateo	1.30
945	Vallejo	1.23
946	Oakland	1.29
947	Berkeley	1.32
948	Richmond	1.31
949	San Rafael	1.30
950	Santa Cruz	1.24
951	San Jose	1.30
952	Stockton	1.20
953	Modesto	1.18

STATE	CITY	Residential
CALIFORNIA (CONT'D)		
954	Santa Rosa	1.27
955	Eureka	1.24
959	Marysville	1.20
960	Redding	1.24
961	Susanville	1.24
COLORADO		
800-802	Denver	.89
803	Boulder	.91
804	Golden	.87
805	Fort Collins	.89
806	Greeley	.88
807	Fort Morgan	.88
808-809	Colorado Springs	.86
810	Pueblo	.86
811	Alamosa	.84
812	Salida	.83
813	Durango	.87
814	Montrose	.83
815	Grand Junction	.92
816	Glenwood Springs	.82
CONNECTICUT		
060	New Britain	1.09
061	Hartford	1.08
062	Willimantic	1.09
063	New London	1.09
064	Meriden	1.09
065	New Haven	1.09
066	Bridgeport	1.10
067	Waterbury	1.09
068	Norwalk	1.10
069	Stamford	1.11
D.C.		
200-205	Washington	.92
DELAWARE		
197	Newark	1.01
198	Wilmington	1.00
199	Dover	1.01
FLORIDA		
320,322	Jacksonville	.81
321	Daytona Beach	.83
323	Tallahassee	.81
324	Panama City	.81
325	Pensacola	.83
326,344	Gainesville	.80
327-328,347	Orlando	.82
329	Melbourne	.82
330-332,340	Miami	.81
333	Fort Lauderdale	.82
334,349	West Palm Beach	.82
335-336,346	Tampa	.81
337	St. Petersburg	.80
338	Lakeland	.78
339,341	Fort Myers	.80
342	Sarasota	.82
GEORGIA		
300-303,399	Atlanta	.88
304	Statesboro	.79
305	Gainesville	.82
306	Athens	.81
307	Dalton	.84
308-309	Augusta	.87
310-312	Macon	.83
313-314	Savannah	.84
315	Waycross	.80
316	Valdosta	.74
317,398	Albany	.83
318-319	Columbus	.82
HAWAII		
967	Hilo	1.18
968	Honolulu	1.22

Location Factors - Residential

STATE	CITY	Residential
STATES & POSS.		
969	Guam	.95
IDAHO		
832	Pocatello	.89
833	Twin Falls	.89
834	Idaho Falls	.87
835	Lewiston	.98
836-837	Boise	.90
838	Coeur d'Alene	.99
ILLINOIS		
600-603	North Suburban	1.23
604	Joliet	1.23
605	South Suburban	1.23
606-608	Chicago	1.25
609	Kankakee	1.16
610-611	Rockford	1.12
612	Rock Island	.97
613	La Salle	1.13
614	Galesburg	1.03
615-616	Peoria	1.06
617	Bloomington	1.03
618-619	Champaign	1.04
620-622	East St. Louis	1.01
623	Quincy	1.03
624	Effingham	1.04
625	Decatur	1.03
626-627	Springfield	1.04
628	Centralia	1.03
629	Carbondale	1.00
INDIANA		
460	Anderson	.90
461-462	Indianapolis	.93
463-464	Gary	1.03
465-466	South Bend	.90
467-468	Fort Wayne	.89
469	Kokomo	.91
470	Lawrenceburg	.86
471	New Albany	.87
472	Columbus	.91
473	Muncie	.90
474	Bloomington	.93
475	Washington	.91
476-477	Evansville	.90
478	Terre Haute	.90
479	Lafayette	.91
IOWA		
500-503,509	Des Moines	.92
504	Mason City	.85
505	Fort Dodge	.84
506-507	Waterloo	.87
508	Creston	.88
510-511	Sioux City	.89
512	Sibley	.76
513	Spencer	.78
514	Carroll	.88
515	Council Bluffs	.87
516	Shenandoah	.88
520	Dubuque	.88
521	Decorah	.87
522-524	Cedar Rapids	.94
525	Ottumwa	.87
526	Burlington	.94
527-528	Davenport	.99
KANSAS		
660-662	Kansas City	.99
664-666	Topeka	.87
667	Fort Scott	.91
668	Emporia	.84
669	Belleville	.82
670-672	Wichita	.81
673	Independence	.90
674	Salina	.80
675	Hutchinson	.80
676	Hays	.83
677	Colby	.86
678	Dodge City	.83
679	Liberal	.82
KENTUCKY		
400-402	Louisville	.86
403-405	Lexington	.87

STATE	CITY	Residential
KENTUCKY (CONT'D)		
406	Frankfort	.90
407-409	Corbin	.83
410	Covington	.85
411-412	Ashland	.88
413-414	Campton	.87
415-416	Pikeville	.90
417-418	Hazard	.86
420	Paducah	.88
421-422	Bowling Green	.87
423	Owensboro	.87
424	Henderson	.88
425-426	Somerset	.83
427	Elizabethtown	.83
LOUISIANA		
700-701	New Orleans	.86
703	Thibodaux	.82
704	Hammond	.78
705	Lafayette	.86
706	Lake Charles	.86
707-708	Baton Rouge	.85
710-711	Shreveport	.82
712	Monroe	.80
713-714	Alexandria	.81
MAINE		
039	Kittery	.87
040-041	Portland	.92
042	Lewiston	.91
043	Augusta	.90
044	Bangor	.90
045	Bath	.88
046	Machias	.87
047	Houlton	.88
048	Rockland	.89
049	Waterville	.88
MARYLAND		
206	Waldorf	.92
207-208	College Park	.89
209	Silver Spring	.90
210-212	Baltimore	.93
214	Annapolis	.91
215	Cumberland	.91
216	Easton	.87
217	Hagerstown	.90
218	Salisbury	.85
219	Elkton	.94
MASSACHUSETTS		
010-011	Springfield	1.07
012	Pittsfield	1.03
013	Greenfield	1.05
014	Fitchburg	1.08
015-016	Worcester	1.10
017	Framingham	1.12
018	Lowell	1.12
019	Lawrence	1.13
020-022, 024	Boston	1.18
023	Brockton	1.10
025	Buzzards Bay	1.08
026	Hyannis	1.07
027	New Bedford	1.10
MICHIGAN		
480,483	Royal Oak	.98
481	Ann Arbor	.99
482	Detroit	1.02
484-485	Flint	.93
486	Saginaw	.89
487	Bay City	.90
488-489	Lansing	.90
490	Battle Creek	.88
491	Kalamazoo	.87
492	Jackson	.90
493,495	Grand Rapids	.88
494	Muskegon	.86
496	Traverse City	.82
497	Gaylord	.86
498-499	Iron Mountain	.86
MINNESOTA		
550-551	Saint Paul	1.07
553-555	Minneapolis	1.08
556-558	Duluth	.99

Location Factors - Residential

STATE	CITY	Residential
MINNESOTA (CONT'D)		
559	Rochester	.99
560	Mankato	.96
561	Windom	.91
562	Willmar	.94
563	St. Cloud	1.02
564	Brainerd	.93
565	Detroit Lakes	.90
566	Bemidji	.91
567	Thief River Falls	.90
MISSISSIPPI		
386	Clarksdale	.73
387	Greenville	.81
388	Tupelo	.74
389	Greenwood	.77
390-392	Jackson	.83
393	Meridian	.79
394	Laurel	.78
395	Biloxi	.83
396	McComb	.74
397	Columbus	.75
MISSOURI		
630-631	St. Louis	1.03
633	Bowling Green	.98
634	Hannibal	.95
635	Kirksville	.91
636	Flat River	.98
637	Cape Girardeau	.91
638	Sikeston	.89
639	Poplar Bluff	.89
640-641	Kansas City	1.02
644-645	St. Joseph	.97
646	Chillicothe	.96
647	Harrisonville	.99
648	Joplin	.91
650-651	Jefferson City	.92
652	Columbia	.89
653	Sedalia	.90
654-655	Rolla	.95
656-658	Springfield	.88
MONTANA		
590-591	Billings	.88
592	Wolf Point	.87
593	Miles City	.89
594	Great Falls	.88
595	Havre	.84
596	Helena	.88
597	Butte	.86
598	Missoula	.85
599	Kalispell	.86
NEBRASKA		
680-681	Omaha	.90
683-685	Lincoln	.91
686	Columbus	.89
687	Norfolk	.89
688	Grand Island	.87
689	Hastings	.89
690	McCook	.83
691	North Platte	.88
692	Valentine	.83
693	Alliance	.83
NEVADA		
889-891	Las Vegas	1.02
893	Ely	1.02
894-895	Reno	.90
897	Carson City	.90
898	Elko	.99
NEW HAMPSHIRE		
030	Nashua	.98
031	Manchester	.97
032-033	Concord	.97
034	Keene	.93
035	Littleton	.93
036	Charleston	.92
037	Claremont	.92
038	Portsmouth	.95

STATE	CITY	Residential
NEW JERSEY		
070-071	Newark	1.21
072	Elizabeth	1.22
073	Jersey City	1.19
074-075	Paterson	1.20
076	Hackensack	1.19
077	Long Branch	1.16
078	Dover	1.19
079	Summit	1.20
080,083	Vineland	1.16
081	Camden	1.17
082,084	Atlantic City	1.20
085-086	Trenton	1.18
087	Point Pleasant	1.17
088-089	New Brunswick	1.23
NEW MEXICO		
870-872	Albuquerque	.85
873	Gallup	.85
874	Farmington	.85
875	Santa Fe	.85
877	Las Vegas	.85
878	Socorro	.85
879	Truth/Consequences	.84
880	Las Cruces	.83
881	Clovis	.84
882	Roswell	.84
883	Carrizozo	.84
884	Tucumcari	.85
NEW YORK		
100-102	New York	1.39
103	Staten Island	1.35
104	Bronx	1.36
105	Mount Vernon	1.18
106	White Plains	1.24
107	Yonkers	1.24
108	New Rochelle	1.21
109	Suffern	1.18
110	Queens	1.36
111	Long Island City	1.39
112	Brooklyn	1.40
113	Flushing	1.38
114	Jamaica	1.37
115,117,118	Hicksville	1.25
116	Far Rockaway	1.37
119	Riverhead	1.26
120-122	Albany	1.02
123	Schenectady	1.03
124	Kingston	1.14
125-126	Poughkeepsie	1.22
127	Monticello	1.15
128	Glens Falls	.97
129	Plattsburgh	.99
130-132	Syracuse	.98
133-135	Utica	.96
136	Watertown	.94
137-139	Binghamton	.99
140-142	Buffalo	1.09
143	Niagara Falls	1.04
144-146	Rochester	1.02
147	Jamestown	.95
148-149	Elmira	.97
NORTH CAROLINA		
270,272-274	Greensboro	.93
271	Winston-Salem	.93
275-276	Raleigh	.93
277	Durham	.93
278	Rocky Mount	.91
279	Elizabeth City	.90
280	Gastonia	.95
281-282	Charlotte	.94
283	Fayetteville	.93
284	Wilmington	.92
285	Kinston	.91
286	Hickory	.92
287-288	Asheville	.93
289	Murphy	.94
NORTH DAKOTA		
580-581	Fargo	.89
582	Grand Forks	.88
583	Devils Lake	.91
584	Jamestown	.89
585	Bismarck	.89

Location Factors - Residential

STATE	CITY	Residential
NORTH DAKOTA (CONT'D)		
586	Dickinson	.88
587	Minot	.87
588	Williston	.88
OHIO		
430-432	Columbus	.91
433	Marion	.92
434-436	Toledo	.96
437-438	Zanesville	.90
439	Steubenville	.93
440	Lorain	.95
441	Cleveland	.97
442-443	Akron	.97
444-445	Youngstown	.94
446-447	Canton	.93
448-449	Mansfield	.91
450	Hamilton	.92
451-452	Cincinnati	.91
453-454	Dayton	.92
455	Springfield	.92
456	Chillicothe	.94
457	Athens	.90
458	Lima	.90
OKLAHOMA		
730-731	Oklahoma City	.83
734	Ardmore	.81
735	Lawton	.82
736	Clinton	.81
737	Enid	.82
738	Woodward	.77
739	Guymon	.81
740-741	Tulsa	.82
743	Miami	.80
744	Muskogee	.80
745	McAlester	.74
746	Ponca City	.79
747	Durant	.78
748	Shawnee	.79
749	Poteau	.79
OREGON		
970-972	Portland	1.00
973	Salem	.99
974	Eugene	.99
975	Medford	.98
976	Klamath Falls	.98
977	Bend	1.01
978	Pendleton	.99
979	Vale	.96
PENNSYLVANIA		
150-152	Pittsburgh	1.00
153	Washington	.96
154	Uniontown	.94
155	Bedford	.91
156	Greensburg	.94
157	Indiana	.95
158	Dubois	.92
159	Johnstown	.90
160	Butler	.93
161	New Castle	.93
162	Kittanning	.92
163	Oil City	.93
164-165	Erie	.94
166	Altoona	.89
167	Bradford	.95
168	State College	.92
169	Wellsboro	.92
170-171	Harrisburg	.95
172	Chambersburg	.89
173-174	York	.91
175-176	Lancaster	.93
177	Williamsport	.90
178	Sunbury	.92
179	Pottsville	.92
180	Lehigh Valley	1.02
181	Allentown	1.03
182	Hazleton	.92
183	Stroudsburg	.95
184-185	Scranton	.96
186-187	Wilkes-Barre	.92
188	Montrose	.92
189	Doylestown	1.08

STATE	CITY	Residential
PENNSYLVANIA (CONT'D)		
190-191	Philadelphia	1.17
193	Westchester	1.09
194	Norristown	1.08
195-196	Reading	.96
PUERTO RICO		
009	San Juan	.77
RHODE ISLAND		
028	Newport	1.07
029	Providence	1.07
SOUTH CAROLINA		
290-292	Columbia	.95
293	Spartanburg	.94
294	Charleston	.95
295	Florence	.91
296	Greenville	.93
297	Rock Hill	.93
298	Aiken	.94
299	Beaufort	.84
SOUTH DAKOTA		
570-571	Sioux Falls	.90
572	Watertown	.85
573	Mitchell	.79
574	Aberdeen	.88
575	Pierre	.84
576	Mobridge	.78
577	Rapid City	.87
TENNESSEE		
370-372	Nashville	.84
373-374	Chattanooga	.84
375,380-381	Memphis	.85
376	Johnson City	.77
377-379	Knoxville	.81
382	McKenzie	.72
383	Jackson	.75
384	Columbia	.78
385	Cookeville	.70
TEXAS		
750	McKinney	.80
751	Waxahachie	.80
752-753	Dallas	.83
754	Greenville	.80
755	Texarkana	.81
756	Longview	.79
757	Tyler	.81
758	Palestine	.77
759	Lufkin	.79
760-761	Fort Worth	.82
762	Denton	.85
763	Wichita Falls	.84
764	Eastland	.83
765	Temple	.81
766-767	Waco	.84
768	Brownwood	.81
769	San Angelo	.79
770-772	Houston	.83
773	Huntsville	.80
774	Wharton	.80
775	Galveston	.82
776-777	Beaumont	.85
778	Bryan	.78
779	Victoria	.81
780	Laredo	.80
781-782	San Antonio	.81
783-784	Corpus Christi	.83
785	McAllen	.83
786-787	Austin	.79
788	Del Rio	.80
789	Giddings	.80
790-791	Amarillo	.79
792	Childress	.81
793-794	Lubbock	.81
795-796	Abilene	.82
797	Midland	.84
798-799,885	El Paso	.80
UTAH		
840-841	Salt Lake City	.85
842,844	Ogden	.83
843	Logan	.83

Location Factors - Residential

STATE	CITY	Residential
UTAH (CONT'D)		
845	Price	.84
846-847	Provo	.84
VERMONT		
050	White River Jct.	.94
051	Bellows Falls	.99
052	Bennington	.98
053	Brattleboro	.99
054	Burlington	.93
056	Montpelier	.97
057	Rutland	.92
058	St. Johnsbury	.95
059	Guildhall	.95
VIRGINIA		
220-221	Fairfax	1.01
222	Arlington	1.02
223	Alexandria	1.03
224-225	Fredericksburg	.99
226	Winchester	1.00
227	Culpeper	1.02
228	Harrisonburg	.85
229	Charlottesville	.88
230-232	Richmond	1.02
233-235	Norfolk	.96
236	Newport News	.95
237	Portsmouth	.86
238	Petersburg	.95
239	Farmville	.86
240-241	Roanoke	1.00
242	Bristol	.90
243	Pulaski	.85
244	Staunton	.86
245	Lynchburg	.98
246	Grundy	.82
WASHINGTON		
980-981,987	Seattle	1.05
982	Everett	1.05
983-984	Tacoma	1.02
985	Olympia	1.00
986	Vancouver	.97
988	Wenatchee	.95
989	Yakima	.99
990-992	Spokane	.97
993	Richland	.97
994	Clarkston	.93
WEST VIRGINIA		
247-248	Bluefield	.92
249	Lewisburg	.91
250-253	Charleston	.95
254	Martinsburg	.89
255-257	Huntington	.96
258-259	Beckley	.92
260	Wheeling	.92
261	Parkersburg	.91
262	Buckhannon	.91
263-264	Clarksburg	.91
265	Morgantown	.92
266	Gassaway	.92
267	Romney	.91
268	Petersburg	.92
WISCONSIN		
530,532	Milwaukee	1.05
531	Kenosha	1.06
534	Racine	1.04
535	Beloit	1.00
537	Madison	1.00
538	Lancaster	.99
539	Portage	.97
540	New Richmond	.97
541-543	Green Bay	1.02
544	Wausau	.97
545	Rhinelander	.95
546	La Crosse	.95
547	Eau Claire	.98
548	Superior	.96
549	Oshkosh	.96
WYOMING		
820	Cheyenne	.85
821	Yellowstone Nat. Pk.	.83
822	Wheatland	.79

STATE	CITY	Residential
WYOMING (CONT'D)		
823	Rawlins	.86
824	Worland	.83
825	Riverton	.82
826	Casper	.82
827	Newcastle	.84
828	Sheridan	.85
829-831	Rock Springs	.88
CANADIAN FACTORS (reflect Canadian currency)		
ALBERTA		
	Calgary	1.07
	Edmonton	1.07
	Fort McMurray	1.09
	Lethbridge	1.07
	Lloydminster	1.02
	Medicine Hat	1.02
	Red Deer	1.02
BRITISH COLUMBIA		
	Kamloops	.99
	Prince George	.98
	Vancouver	1.01
	Victoria	1.00
MANITOBA		
	Brandon	1.06
	Portage la Prairie	.97
	Winnipeg	.95
NEW BRUNSWICK		
	Bathurst	.89
	Dalhousie	.90
	Fredericton	.94
	Moncton	.90
	Newcastle	.89
	Saint John	.99
NEWFOUNDLAND		
	Corner Brook	1.05
	St. John's	1.04
NORTHWEST TERRITORIES		
	Yellowknife	1.11
NOVA SCOTIA		
	Bridgewater	.92
	Dartmouth	1.01
	Halifax	1.00
	New Glasgow	1.00
	Sydney	.99
	Truro	.92
	Yarmouth	1.00
ONTARIO		
	Barrie	1.10
	Brantford	1.09
	Cornwall	1.08
	Hamilton	1.05
	Kingston	1.08
	Kitchener	1.03
	London	1.06
	North Bay	1.15
	Oshawa	1.05
	Ottawa	1.05
	Owen Sound	1.09
	Peterborough	1.06
	Sarnia	1.09
	Sault Ste. Marie	1.03
	St. Catharines	1.04
	Sudbury	1.01
	Thunder Bay	1.06
	Timmins	1.05
	Toronto	1.07
	Windsor	1.05
PRINCE EDWARD ISLAND		
	Charlottetown	.90
	Summerside	.94
QUEBEC		
	Cap-de-la-Madeleine	1.06
	Charlesbourg	1.06
	Chicoutimi	1.09
	Gatineau	1.06
	Granby	1.06

STATE	CITY	Residential
QUEBEC (CONT'D)		
	Hull	1.06
	Joliette	1.06
	Laval	1.06
	Montreal	1.05
	Quebec City	1.05
	Rimouski	1.10
	Rouyn-Noranda	1.06
	Saint-Hyacinthe	1.06
	Sherbrooke	1.06
	Sorel	1.06
	Saint-Jerome	1.06
	Trois-Rivieres	1.17
SASKATCHEWAN		
	Moose Jaw	.89
	Prince Albert	.88
	Regina	1.07
	Saskatoon	1.03
YUKON		
	Whitehorse	1.03

R011105-05 Tips for Accurate Estimating

1. Use pre-printed or columnar forms for orderly sequence of dimensions and locations and for recording telephone quotations.
2. Use only the front side of each paper or form except for certain pre-printed summary forms.
3. Be consistent in listing dimensions: For example, length x width x height. This helps in rechecking to ensure that, the total length of partitions is appropriate for the building area.
4. Use printed (rather than measured) dimensions where given.
5. Add up multiple printed dimensions for a single entry where possible.
6. Measure all other dimensions carefully.
7. Use each set of dimensions to calculate multiple related quantities.
8. Convert foot and inch measurements to decimal feet when listing. Memorize decimal equivalents to .01 parts of a foot (1/8″ equals approximately .01′).
9. Do not "round off" quantities until the final summary.
10. Mark drawings with different colors as items are taken off.
11. Keep similar items together, different items separate.
12. Identify location and drawing numbers to aid in future checking for completeness.
13. Measure or list everything on the drawings or mentioned in the specifications.
14. It may be necessary to list items not called for to make the job complete.
15. Be alert for: Notes on plans such as N.T.S. (not to scale); changes in scale throughout the drawings; reduced size drawings; discrepancies between the specifications and the drawings.
16. Develop a consistent pattern of performing an estimate. For example:
    a. Start the quantity takeoff at the lower floor and move to the next higher floor.
    b. Proceed from the main section of the building to the wings.
    c. Proceed from south to north or vice versa, clockwise or counterclockwise.
    d. Take off floor plan quantities first, elevations next, then detail drawings.
17. List all gross dimensions that can be either used again for different quantities, or used as a rough check of other quantities for verification (exterior perimeter, gross floor area, individual floor areas, etc.).
18. Utilize design symmetry or repetition (repetitive floors, repetitive wings, symmetrical design around a center line, similar room layouts, etc.). Note: Extreme caution is needed here so as not to omit or duplicate an area.
19. Do not convert units until the final total is obtained. For instance, when estimating concrete work, keep all units to the nearest cubic foot, then summarize and convert to cubic yards.
20. When figuring alternatives, it is best to total all items involved in the basic system, then total all items involved in the alternates. Therefore you work with positive numbers in all cases. When adds and deducts are used, it is often confusing whether to add or subtract a portion of an item; especially on a complicated or involved alternate.

R011105-50 Metric Conversion Factors

Description: This table is primarily for converting customary U.S. units in the left hand column to SI metric units in the right hand column. In addition, conversion factors for some commonly encountered Canadian and non-SI metric units are included.

	If You Know		Multiply By		To Find
Length	Inches	x	25.4[a]	=	Millimeters
	Feet	x	0.3048[a]	=	Meters
	Yards	x	0.9144[a]	=	Meters
	Miles (statute)	x	1.609	=	Kilometers
Area	Square inches	x	645.2	=	Square millimeters
	Square feet	x	0.0929	=	Square meters
	Square yards	x	0.8361	=	Square meters
Volume (Capacity)	Cubic inches	x	16,387	=	Cubic millimeters
	Cubic feet	x	0.02832	=	Cubic meters
	Cubic yards	x	0.7646	=	Cubic meters
	Gallons (U.S. liquids)[b]	x	0.003785	=	Cubic meters[c]
	Gallons (Canadian liquid)[b]	x	0.004546	=	Cubic meters[c]
	Ounces (U.S. liquid)[b]	x	29.57	=	Milliliters[c, d]
	Quarts (U.S. liquid)[b]	x	0.9464	=	Liters[c, d]
	Gallons (U.S. liquid)[b]	x	3.785	=	Liters[c, d]
Force	Kilograms force[d]	x	9.807	=	Newtons
	Pounds force	x	4.448	=	Newtons
	Pounds force	x	0.4536	=	Kilograms force[d]
	Kips	x	4448	=	Newtons
	Kips	x	453.6	=	Kilograms force[d]
Pressure, Stress, Strength (Force per unit area)	Kilograms force per square centimeter[d]	x	0.09807	=	Megapascals
	Pounds force per square inch (psi)	x	0.006895	=	Megapascals
	Kips per square inch	x	6.895	=	Megapascals
	Pounds force per square inch (psi)	x	0.07031	=	Kilograms force per square centimeter[d]
	Pounds force per square foot	x	47.88	=	Pascals
	Pounds force per square foot	x	4.882	=	Kilograms force per square meter[d]
Flow	Cubic feet per minute	x	0.4719	=	Liters per second
	Gallons per minute	x	0.0631	=	Liters per second
	Gallons per hour	x	1.05	=	Milliliters per second
Bending Moment Or Torque	Inch-pounds force	x	0.01152	=	Meter-kilograms force[d]
	Inch-pounds force	x	0.1130	=	Newton-meters
	Foot-pounds force	x	0.1383	=	Meter-kilograms force[d]
	Foot-pounds force	x	1.356	=	Newton-meters
	Meter-kilograms force[d]	x	9.807	=	Newton-meters
Mass	Ounces (avoirdupois)	x	28.35	=	Grams
	Pounds (avoirdupois)	x	0.4536	=	Kilograms
	Tons (metric)	x	1000	=	Kilograms
	Tons, short (2000 pounds)	x	907.2	=	Kilograms
	Tons, short (2000 pounds)	x	0.9072	=	Megagrams[e]
Mass per Unit Volume	Pounds mass per cubic foot	x	16.02	=	Kilograms per cubic meter
	Pounds mass per cubic yard	x	0.5933	=	Kilograms per cubic meter
	Pounds mass per gallon (U.S. liquid)[b]	x	119.8	=	Kilograms per cubic meter
	Pounds mass per gallon (Canadian liquid)[b]	x	99.78	=	Kilograms per cubic meter
Temperature	Degrees Fahrenheit		(F-32)/1.8	=	Degrees Celsius
	Degrees Fahrenheit		(F+459.67)/1.8	=	Degrees Kelvin
	Degrees Celsius		C+273.15	=	Degrees Kelvin

[a]The factor given is exact
[b]One U.S. gallon = 0.8327 Canadian gallon
[c]1 liter = 1000 milliliters = 1000 cubic centimeters
1 cubic decimeter = 0.001 cubic meter
[d]Metric but not SI unit
[e]Called "tonne" in England and "metric ton" in other metric countries

R011105-60 Weights and Measures

Measures of Length
1 Mile = 1760 Yards = 5280 Feet
1 Yard = 3 Feet = 36 inches
1 Foot = 12 Inches
1 Mil = 0.001 Inch
1 Fathom = 2 Yards = 6 Feet
1 Rod = 5.5 Yards = 16.5 Feet
1 Hand = 4 Inches
1 Span = 9 Inches
1 Micro-inch = One Millionth Inch or 0.000001 Inch
1 Micron = One Millionth Meter + 0.00003937 Inch

Surveyor's Measure
1 Mile = 8 Furlongs = 80 Chains
1 Furlong = 10 Chains = 220 Yards
1 Chain = 4 Rods = 22 Yards = 66 Feet = 100 Links
1 Link = 7.92 Inches

Square Measure
1 Square Mile = 640 Acres = 6400 Square Chains
1 Acre = 10 Square Chains = 4840 Square Yards = 43,560 Sq. Ft.
1 Square Chain = 16 Square Rods = 484 Square Yards = 4356 Sq. Ft.
1 Square Rod = 30.25 Square Yards = 272.25 Square Feet = 625 Square Lines
1 Square Yard = 9 Square Feet
1 Square Foot = 144 Square Inches
An Acre equals a Square 208.7 Feet per Side

Cubic Measure
1 Cubic Yard = 27 Cubic Feet
1 Cubic Foot = 1728 Cubic Inches
1 Cord of Wood = 4 x 4 x 8 Feet = 128 Cubic Feet
1 Perch of Masonry = 16½ x 1½ x 1 Foot = 24.75 Cubic Feet

Avoirdupois or Commercial Weight
1 Gross or Long Ton = 2240 Pounds
1 Net or Short Ton = 2000 Pounds
1 Pound = 16 Ounces = 7000 Grains
1 Ounce = 16 Drachms = 437.5 Grains
1 Stone = 14 Pounds

Power
1 British Thermal Unit per Hour = 0.2931 Watts
1 Ton (Refrigeration) = 3.517 Kilowatts
1 Horsepower (Boiler) = 9.81 Kilowatts
1 Horsepower (550 ft-lb/s) = 0.746 Kilowatts

Shipping Measure
For Measuring Internal Capacity of a Vessel:
1 Register Ton = 100 Cubic Feet

For Measurement of Cargo:
Approximately 40 Cubic Feet of Merchandise is considered a Shipping Ton, unless that bulk would weigh more than 2000 Pounds, in which case Freight Charge may be based upon weight.

40 Cubic Feet = 32.143 U.S. Bushels = 31.16 Imp. Bushels

Liquid Measure
1 Imperial Gallon = 1.2009 U.S. Gallon = 277.42 Cu. In.
1 Cubic Foot = 7.48 U.S. Gallons

R011110-10 Architectural Fees

Tabulated below are typical percentage fees by project size, for good professional architectural service. Fees may vary from those listed depending upon degree of design difficulty and economic conditions in any particular area.

Rates can be interpolated horizontally and vertically. Various portions of the same project requiring different rates should be adjusted proportionately. For alterations, add 50% to the fee for the first $500,000 of project cost and add 25% to the fee for project cost over $500,000.

Architectural fees tabulated below include Structural, Mechanical and Electrical Engineering Fees. They do not include the fees for special consultants such as kitchen planning, security, acoustical, interior design, etc.

Civil Engineering fees are included in the Architectural fee for project sites requiring minimal design such as city sites. However, separate Civil Engineering fees must be added when utility connections require design, drainage calculations are needed, stepped foundations are required, or provisions are required to protect adjacent wetlands.

	Total Project Size in Thousands of Dollars						
Building Types	100	250	500	1,000	5,000	10,000	50,000
Factories, garages, warehouses, repetitive housing	9.0%	8.0%	7.0%	6.2%	5.3%	4.9%	4.5%
Apartments, banks, schools, libraries, offices, municipal buildings	12.2	12.3	9.2	8.0	7.0	6.6	6.2
Churches, hospitals, homes, laboratories, museums, research	15.0	13.6	12.7	11.9	9.5	8.8	8.0
Memorials, monumental work, decorative furnishings	—	16.0	14.5	13.1	10.0	9.0	8.3

R012909-80 Sales Tax by State

State sales tax on materials is tabulated below (5 states have no sales tax). Many states allow local jurisdictions, such as a county or city, to levy additional sales tax.

Some projects may be sales tax exempt, particularly those constructed with public funds.

State	Tax (%)	State	Tax (%)	State	Tax (%)	State	Tax (%)
Alabama	4	Illinois	6.25	Montana	0	Rhode Island	7
Alaska	0	Indiana	7	Nebraska	5.5	South Carolina	6
Arizona	5.6	Iowa	6	Nevada	6.85	South Dakota	4.5
Arkansas	6.5	Kansas	6.5	New Hampshire	0	Tennessee	7
California	7.25	Kentucky	6	New Jersey	6.625	Texas	6.25
Colorado	2.9	Louisiana	4.45	New Mexico	5.125	Utah	5.95
Connecticut	6.35	Maine	5.5	New York	4	Vermont	6
Delaware	0	Maryland	6	North Carolina	4.75	Virginia	5.3
District of Columbia	5.75	Massachusetts	6.25	North Dakota	5	Washington	6.5
Florida	6	Michigan	6	Ohio	5.75	West Virginia	6
Georgia	4	Minnesota	6.875	Oklahoma	4.5	Wisconsin	5
Hawaii	4	Mississippi	7	Oregon	0	Wyoming	4
Idaho	6	Missouri	4.225	Pennsylvania	6	Average	5.10%

Sales Tax by Province (Canada)

GST - a value-added tax, which the government imposes on most goods and services provided in or imported into Canada. PST - a retail sales tax, which five of the provinces impose on the prices of most goods and some services. QST - a value-added tax, similar to the federal GST, which Quebec imposes. HST - Three provinces have combined their retail sales taxes with the federal GST into one harmonized tax.

Province	PST (%)	QST (%)	GST(%)	HST(%)
Alberta	0	0	5	0
British Columbia	7	0	5	0
Manitoba	8	0	5	0
New Brunswick	0	0	0	15
Newfoundland	0	0	0	15
Northwest Territories	0	0	5	0
Nova Scotia	0	0	0	15
Ontario	0	0	0	13
Prince Edward Island	0	0	0	15
Quebec	0	9.975	5	0
Saskatchewan	6	0	5	0
Yukon	0	0	5	0

R012909-85 Unemployment Taxes and Social Security Taxes

State unemployment tax rates vary not only from state to state, but also with the experience rating of the contractor. The federal unemployment tax rate is 6.0% of the first $7,000 of wages. This is reduced by a credit of up to 5.4% for timely payment to the state. The minimum federal unemployment tax is 0.6% after all credits.

Social security (FICA) for 2019 is estimated at time of publication to be 7.65% of wages up to $128,400.

R012909-86 Unemployment Tax by State

Information is from the U.S. Department of Labor, state unemployment tax rates.

State	Tax (%)	State	Tax (%)	State	Tax (%)	State	Tax (%)
Alabama	6.74	Illinois	7.75	Montana	6.12	Rhode Island	9.79
Alaska	5.4	Indiana	7.474	Nebraska	5.4	South Carolina	5.46
Arizona	8.91	Iowa	8	Nevada	5.4	South Dakota	9.5
Arkansas	6.0	Kansas	7.6	New Hampshire	7.5	Tennessee	10.0
California	6.2	Kentucky	10.0	New Jersey	5.8	Texas	7.5
Colorado	8.9	Louisiana	6.2	New Mexico	5.4	Utah	7.2
Connecticut	6.8	Maine	5.4	New York	8.5	Vermont	8.4
Delaware	8.0	Maryland	7.50	North Carolina	5.76	Virginia	6.27
District of Columbia	7	Massachusetts	11.13	North Dakota	10.72	Washington	5.7
Florida	5.4	Michigan	10.3	Ohio	8.7	West Virginia	7.5
Georgia	5.4	Minnesota	9.0	Oklahoma	5.5	Wisconsin	12.0
Hawaii	5.6	Mississippi	5.4	Oregon	5.4	Wyoming	8.8
Idaho	5.4	Missouri	9.75	Pennsylvania	10.89	Median	7.47%

General Requirements R0131 Project Management & Coordination

R013113-40 Builder's Risk Insurance

Builder's risk insurance is insurance on a building during construction. Premiums are paid by the owner or the contractor. Blasting, collapse and underground insurance would raise total insurance costs.

R013113-50 General Contractor's Overhead

There are two distinct types of overhead on a construction project: Project overhead and main office overhead. Project overhead includes those costs at a construction site not directly associated with the installation of construction materials. Examples of project overhead costs include the following:

1. Superintendent
2. Construction office and storage trailers
3. Temporary sanitary facilities
4. Temporary utilities
5. Security fencing
6. Photographs
7. Cleanup
8. Performance and payment bonds

The above project overhead items are also referred to as general requirements and therefore are estimated in Division 1. Division 1 is the first division listed in the CSI MasterFormat but it is usually the last division estimated. The sum of the costs in Divisions 1 through 49 is referred to as the sum of the direct costs.

All construction projects also include indirect costs. The primary components of indirect costs are the contractor's main office overhead and profit. The amount of the main office overhead expense varies depending on the following:

1. Owner's compensation
2. Project managers' and estimators' wages
3. Clerical support wages
4. Office rent and utilities
5. Corporate legal and accounting costs
6. Advertising
7. Automobile expenses
8. Association dues
9. Travel and entertainment expenses

These costs are usually calculated as a percentage of annual sales volume. This percentage can range from 35% for a small contractor doing less than $500,000 to 5% for a large contractor with sales in excess of $100 million.

R013113-55 Installing Contractor's Overhead

Installing contractors (subcontractors) also incur costs for general requirements and main office overhead.

Included within the total incl. overhead and profit costs is a percent mark-up for overhead that includes:

1. Compensation and benefits for office staff and project managers
2. Office rent, utilities, business equipment, and maintenance
3. Corporate legal and accounting costs
4. Advertising
5. Vehicle expenses (for office staff and project managers)
6. Association dues
7. Travel, entertainment
8. Insurance
9. Small tools and equipment

R013113-60 Workers' Compensation Insurance Rates by Trade

The table below tabulates the national averages for workers' compensation insurance rates by trade and type of building. The average "Insurance Rate" is multiplied by the "% of Building Cost" for each trade. This produces the "Workers' Compensation" cost by % of total labor cost, to be added for each trade by building type to determine the weighted average workers' compensation rate for the building types analyzed.

Trade	Insurance Rate (% Labor Cost)		% of Building Cost			Workers' Compensation		
	Range	Average	Office Bldgs.	Schools & Apts.	Mfg.	Office Bldgs.	Schools & Apts.	Mfg.
Excavation, Grading, etc.	2.3 % to 19.2%	8.5%	4.8%	4.9%	4.5%	0.41%	0.42%	0.38%
Piles & Foundations	3.9 to 27.3	13.4	7.1	5.2	8.7	0.95	0.70	1.17
Concrete	3.5 to 25.6	11.8	5.0	14.8	3.7	0.59	1.75	0.44
Masonry	3.6 to 50.4	13.8	6.9	7.5	1.9	0.95	1.04	0.26
Structural Steel	4.9 to 43.8	21.2	10.7	3.9	17.6	2.27	0.83	3.73
Miscellaneous & Ornamental Metals	2.8 to 27.0	10.6	2.8	4.0	3.6	0.30	0.42	0.38
Carpentry & Millwork	3.7 to 31.9	13.0	3.7	4.0	0.5	0.48	0.52	0.07
Metal or Composition Siding	5.1 to 113.7	19.0	2.3	0.3	4.3	0.44	0.06	0.82
Roofing	5.7 to 103.7	29.0	2.3	2.6	3.1	0.67	0.75	0.90
Doors & Hardware	3.1 to 31.9	11.0	0.9	1.4	0.4	0.10	0.15	0.04
Sash & Glazing	3.9 to 24.3	12.1	3.5	4.0	1.0	0.42	0.48	0.12
Lath & Plaster	2.7 to 34.4	10.7	3.3	6.9	0.8	0.35	0.74	0.09
Tile, Marble & Floors	2.2 to 21.5	8.7	2.6	3.0	0.5	0.23	0.26	0.04
Acoustical Ceilings	1.9 to 29.7	8.5	2.4	0.2	0.3	0.20	0.02	0.03
Painting	3.7 to 36.7	11.2	1.5	1.6	1.6	0.17	0.18	0.18
Interior Partitions	3.7 to 31.9	13.0	3.9	4.3	4.4	0.51	0.56	0.57
Miscellaneous Items	2.1 to 97.7	11.2	5.2	3.7	9.7	0.58	0.42	1.09
Elevators	1.4 to 11.4	4.7	2.1	1.1	2.2	0.10	0.05	0.10
Sprinklers	2.0 to 16.4	6.7	0.5	—	2.0	0.03	—	0.13
Plumbing	1.5 to 14.6	6.3	4.9	7.2	5.2	0.31	0.45	0.33
Heat., Vent., Air Conditioning	3.0 to 17.0	8.3	13.5	11.0	12.9	1.12	0.91	1.07
Electrical	1.9 to 11.3	5.2	10.1	8.4	11.1	0.53	0.44	0.58
Total	1.4 % to 113.7%	—	100.0%	100.0%	100.0%	11.71%	11.15%	12.52%
		Overall Weighted Average 11.79%						

Workers' Compensation Insurance Rates by States

The table below lists the weighted average Workers' Compensation base rate for each state with a factor comparing this with the national average of 11.8%.

State	Weighted Average	Factor	State	Weighted Average	Factor	State	Weighted Average	Factor
Alabama	13.9%	129	Kentucky	11.3%	105	North Dakota	7.4%	69
Alaska	11.0	102	Louisiana	20.0	185	Ohio	6.2	57
Arizona	9.1	84	Maine	9.8	91	Oklahoma	8.7	81
Arkansas	5.7	53	Maryland	11.2	104	Oregon	9.0	83
California	23.6	219	Massachusetts	9.1	84	Pennsylvania	24.9	231
Colorado	8.1	75	Michigan	8.0	74	Rhode Island	10.6	98
Connecticut	17.0	157	Minnesota	16.2	150	South Carolina	20.3	188
Delaware	10.7	99	Mississippi	11.4	106	South Dakota	10.3	95
District of Columbia	9.1	84	Missouri	14.0	130	Tennessee	7.9	73
Florida	11.3	105	Montana	7.8	72	Texas	6.2	57
Georgia	32.9	305	Nebraska	13.4	124	Utah	7.4	69
Hawaii	9.2	85	Nevada	8.9	82	Vermont	10.8	100
Idaho	9.2	85	New Hampshire	11.3	105	Virginia	7.4	69
Illinois	20.1	186	New Jersey	15.4	143	Washington	8.9	82
Indiana	3.7	34	New Mexico	13.7	127	West Virginia	4.7	44
Iowa	12.1	112	New York	19.1	177	Wisconsin	13.3	123
Kansas	6.6	61	North Carolina	17.1	158	Wyoming	6.3	58
			Weighted Average for U.S. is	11.8% of payroll = 100%				

The weighted average skilled worker rate for 35 trades is 11.8%. For bidding purposes, apply the full value of Workers' Compensation directly to total labor costs, or if labor is 38%, materials 42% and overhead and profit 20% of total cost, carry 38/80 x 11.8% = 6.0% of cost (before overhead and profit) into overhead. Rates vary not only from state to state but also with the experience rating of the contractor.

Rates are the most current available at the time of publication.

R015423-10 Steel Tubular Scaffolding

On new construction, tubular scaffolding is efficient up to 60′ high or five stories. Above this it is usually better to use a hung scaffolding if construction permits. Swing scaffolding operations may interfere with tenants. In this case, the tubular is more practical at all heights.

In repairing or cleaning the front of an existing building the cost of tubular scaffolding per S.F. of building front increases as the height increases above the first tier. The first tier cost is relatively high due to leveling and alignment.

The minimum efficient crew for erecting and dismantling is three workers. They can set up and remove 18 frame sections per day up to 5 stories high. For 6 to 12 stories high, a crew of four is most efficient. Use two or more on top and two on the bottom for handing up or hoisting. They can also set up and remove 18 frame sections per day. At 7' horizontal spacing, this will run about 800 S.F. per day of erecting and dismantling. Time for placing and removing planks must be added to the above. A crew of three can place and remove 72 planks per day up to 5 stories. For over 5 stories, a crew of four can place and remove 80 planks per day.

The table below shows the number of pieces required to erect tubular steel scaffolding for 1000 S.F. of building frontage. This area is made up of a scaffolding system that is 12 frames (11 bays) long by 2 frames high.

For jobs under twenty-five frames, add 50% to rental cost. Rental rates will be lower for jobs over three months duration. Large quantities for long periods can reduce rental rates by 20%.

Description of Component	Number of Pieces for 1000 S.F. of Building Front	Unit
5′ Wide Standard Frame, 6′-4″ High	24	Ea.
Leveling Jack & Plate	24	↓
Cross Brace	44	↓
Side Arm Bracket, 21″	12	↓
Guardrail Post	12	↓
Guardrail, 7′ section	22	↓
Stairway Section	2	↓
Stairway Starter Bar	1	↓
Stairway Inside Handrail	2	↓
Stairway Outside Handrail	2	↓
Walk-Thru Frame Guardrail	2	↓

Scaffolding is often used as falsework over 15′ high during construction of cast-in-place concrete beams and slabs. Two foot wide scaffolding is generally used for heavy beam construction. The span between frames depends upon the load to be carried with a maximum span of 5′.

Heavy duty shoring frames with a capacity of 10,000#/leg can be spaced up to 10′ O.C. depending upon form support design and loading.

Scaffolding used as horizontal shoring requires less than half the material required with conventional shoring.

On new construction, erection is done by carpenters.

Rolling towers supporting horizontal shores can reduce labor and speed the job. For maintenance work, catwalks with spans up to 70′ can be supported by the rolling towers.

R015423-20 Pump Staging

Pump staging is generally not available for rent. The table below shows the number of pieces required to erect pump staging for 2400 S.F. of building frontage. This area is made up of a pump jack system that is 3 poles (2 bays) wide by 2 poles high.

Item	Number of Pieces for 2400 S.F. of Building Front	Unit
Aluminum pole section, 24′ long	6	Ea.
Aluminum splice joint, 6′ long	3	↓
Aluminum foldable brace	3	↓
Aluminum pump jack	3	↓
Aluminum support for workbench/back safety rail	3	↓
Aluminum scaffold plank/workbench, 14″ wide x 24′ long	4	↓
Safety net, 22′ long	2	↓
Aluminum plank end safety rail	2	↓

The cost in place for this 2400 S.F. will depend on how many uses are realized during the life of the equipment.

R015436-50 Mobilization

Costs to move rented construction equipment to a job site from an equipment dealer's or contractor's yard (mobilization) or off the job site (demobilization) are not included in the rental or operating rates, nor in the equipment cost on a unit price line or in a crew listing. These costs can be found consolidated in the Mobilization section of the data and elsewhere in particular site work sections. If a piece of equipment is already on the job site, it is not appropriate to include mob/demob costs in a new estimate that requires use of that equipment. The following table identifies approximate sizes of rented construction equipment that would be hauled on a towed trailer. Because this listing is not all-encompassing, the user can infer as to what size trailer might be required for a piece of equipment not listed.

3-ton Trailer	20-ton Trailer	40-ton Trailer	50-ton Trailer
20 H.P. Excavator	110 H.P. Excavator	200 H.P. Excavator	270 H.P. Excavator
50 H.P. Skid Steer	165 H.P. Dozer	300 H.P. Dozer	Small Crawler Crane
35 H.P. Roller	150 H.P. Roller	400 H.P. Scraper	500 H.P. Scraper
40 H.P. Trencher	Backhoe	450 H.P. Art. Dump Truck	500 H.P. Art. Dump Truck

R024119-10 Demolition Defined

Whole Building Demolition - Demolition of the whole building with no concern for any particular building element, component, or material type being demolished. This type of demolition is accomplished with large pieces of construction equipment that break up the structure, load it into trucks and haul it to a disposal site, but disposal or dump fees are not included. Demolition of below-grade foundation elements, such as footings, foundation walls, grade beams, slabs on grade, etc., is not included. Certain mechanical equipment containing flammable liquids or ozone-depleting refrigerants, electric lighting elements, communication equipment components, and other building elements may contain hazardous waste, and must be removed, either selectively or carefully, as hazardous waste before the building can be demolished.

Foundation Demolition - Demolition of below-grade foundation footings, foundation walls, grade beams, and slabs on grade. This type of demolition is accomplished by hand or pneumatic hand tools, and does not include saw cutting, or handling, loading, hauling, or disposal of the debris.

Gutting - Removal of building interior finishes and electrical/mechanical systems down to the load-bearing and sub-floor elements of the rough building frame, with no concern for any particular building element, component, or material type being demolished. This type of demolition is accomplished by hand or pneumatic hand tools, and includes loading into trucks, but not hauling, disposal or dump fees, scaffolding, or shoring. Certain mechanical equipment containing flammable liquids or ozone-depleting refrigerants, electric lighting elements, communication equipment components, and other building elements may contain hazardous waste, and must be removed, either selectively or carefully, as hazardous waste, before the building is gutted.

Selective Demolition - Demolition of a selected building element, component, or finish, with some concern for surrounding or adjacent elements, components, or finishes (see the first Subdivision (s) at the beginning of appropriate Divisions). This type of demolition is accomplished by hand or pneumatic hand tools, and does not include handling, loading, storing, hauling, or disposal of the debris, scaffolding, or shoring. "Gutting" methods may be used in order to save time, but damage that is caused to surrounding or adjacent elements, components, or finishes may have to be repaired at a later time.

Careful Removal - Removal of a piece of service equipment, building element or component, or material type, with great concern for both the removed item and surrounding or adjacent elements, components or finishes. The purpose of careful removal may be to protect the removed item for later re-use, preserve a higher salvage value of the removed item, or replace an item while taking care to protect surrounding or adjacent elements, components, connections, or finishes from cosmetic and/or structural damage. An approximation of the time required to perform this type of removal is 1/3 to 1/2 the time it would take to install a new item of like kind. This type of removal is accomplished by hand or pneumatic hand tools, and does not include loading, hauling, or storing the removed item, scaffolding, shoring, or lifting equipment.

Cutout Demolition - Demolition of a small quantity of floor, wall, roof, or other assembly, with concern for the appearance and structural integrity of the surrounding materials. This type of demolition is accomplished by hand or pneumatic hand tools, and does not include saw cutting, handling, loading, hauling, or disposal of debris, scaffolding, or shoring.

Rubbish Handling - Work activities that involve handling, loading or hauling of debris. Generally, the cost of rubbish handling must be added to the cost of all types of demolition, with the exception of whole building demolition.

Minor Site Demolition - Demolition of site elements outside the footprint of a building. This type of demolition is accomplished by hand or pneumatic hand tools, or with larger pieces of construction equipment, and may include loading a removed item onto a truck (check the Crew for equipment used). It does not include saw cutting, hauling or disposal of debris, and, sometimes, handling or loading.

R024119-20 Dumpsters

Dumpster rental costs on construction sites are presented in two ways.

The cost per week rental includes the delivery of the dumpster; its pulling or emptying once per week, and its final removal. The assumption is made that the dumpster contractor could choose to empty a dumpster by simply bringing in an empty unit and removing the full one. These costs also include the disposal of the materials in the dumpster.

The Alternate Pricing can be used when actual planned conditions are not approximated by the weekly numbers. For example, these lines can be used when a dumpster is needed for 4 weeks and will need to be emptied 2 or 3 times per week. Conversely the Alternate Pricing lines can be used when a dumpster will be rented for several weeks or months but needs to be emptied only a few times over this period.

R040130-10 Cleaning Face Brick

On smooth brick a person can clean 70 S.F. an hour; on rough brick 50 S.F. per hour. Use one gallon muriatic acid to 20 gallons of water for 1000 S.F. Do not use acid solution until wall is at least seven days old, but a mild soap solution may be used after two days.

Time has been allowed for clean-up in brick prices.

R040513-10 Cement Mortar (material only)

Type N - 1:1:6 mix by volume. Use everywhere above grade except as noted below. - 1:3 mix using conventional masonry cement which saves handling two separate bagged materials.

Type M - 1:1/4:3 mix by volume, or 1 part cement, 1/4 (10% by wt.) lime, 3 parts sand. Use for heavy loads and where earthquakes or hurricanes may occur. Also for reinforced brick, sewers, manholes and everywhere below grade.

Mix Proportions by Volume and Compressive Strength of Mortar

Where Used	Allowable Proportions by Volume					Compressive Strength @ 28 days
	Mortar Type	Portland Cement	Masonry Cement	Hydrated Lime	Masonry Sand	
Plain Masonry	M	1	1	—	6	2500 psi
		1	—	1/4	3	
	S	1/2	1	—	4	1800 psi
		1	—	1/4 to 1/2	4	
	N	—	1	—	3	750 psi
		1	—	1/2 to 1-1/4	6	
	O	—	1	—	3	350 psi
		1	—	1-1/4 to 2-1/2	9	
	K	1	—	2-1/2 to 4	12	75 psi
Reinforced Masonry	PM	1	1	—	6	2500 psi
	PL	1	—	1/4 to 1/2	4	2500 psi

Note: The total aggregate should be between 2.25 to 3 times the sum of the cement and lime used.

The labor cost to mix the mortar is included in the productivity and labor cost of unit price lines in unit cost sections for brickwork, blockwork and stonework.

The material cost of mixed mortar is included in the material cost of those same unit price lines and includes the cost of renting and operating a 10 C.F. mixer at the rate of 200 C.F. per day.

There are two types of mortar color used. One type is the inert additive type with about 100 lbs. per M brick as the typical quantity required. These colors are also available in smaller-batch-sized bags (1 lb. to 15 lb.) which can be placed directly into the mixer without measuring. The other type is premixed and replaces the masonry cement. Dark green color has the highest cost.

R040519-50 Masonry Reinforcing

Horizontal joint reinforcing helps prevent wall cracks where wall movement may occur and in many locations is required by code. Horizontal joint reinforcing is generally not considered to be structural reinforcing and an unreinforced wall may still contain joint reinforcing.

Reinforcing strips come in 10′ and 12′ lengths and in truss and ladder shapes, with and without drips. Field labor runs between 2.7 to 5.3 hours per 1000 L.F. for wall thicknesses up to 12″.

The wire meets ASTM A82 for cold drawn steel wire and the typical size is 9 ga. sides and ties with 3/16″ diameter also available. Typical finish is mill galvanized with zinc coating at .10 oz. per S.F. Class I (.40 oz. per S.F.) and Class III (.80 oz. per S.F.) are also available, as is hot dipped galvanizing at 1.50 oz. per S.F.

R042110-10 Economy in Bricklaying

Have adequate supervision. Be sure bricklayers are always supplied with materials so there is no waiting. Place experienced bricklayers at corners and openings.

Use only screened sand for mortar. Otherwise, labor time will be wasted picking out pebbles. Use seamless metal tubs for mortar as they do not leak or catch the trowel. Locate stack and mortar for easy wheeling.

Have brick delivered for stacking. This makes for faster handling, reduces chipping and breakage, and requires less storage space. Many dealers will deliver select common in 2′ x 3′ x 4′ pallets or face brick packaged. This affords quick handling with a crane or forklift and easy tonging in units of ten, which reduces waste.

Use wider bricks for one wythe wall construction. Keep scaffolding away from the wall to allow mortar to fall clear and not stain the wall.

On large jobs develop specialized crews for each type of masonry unit.

Consider designing for prefabricated panel construction on high rise projects.

Avoid excessive corners or openings. Each opening adds about 50% to the labor cost for area of opening.

Bolting stone panels and using window frames as stops reduce labor costs and speed up erection.

R042110-20 Common and Face Brick

Common building brick manufactured according to ASTM C62 and facing brick manufactured according to ASTM C216 are the two standard bricks available for general building use.

Building brick is made in three grades: SW, where high resistance to damage caused by cyclic freezing is required; MW, where moderate resistance to cyclic freezing is needed; and NW, where little resistance to cyclic freezing is needed. Facing brick is made in only the two grades SW and MW. Additionally, facing brick is available in three types: FBS, for general use; FBX, for general use where a higher degree of precision and lower permissible variation in size than FBS are needed; and FBA, for general use to produce characteristic architectural effects resulting from non-uniformity in size and texture of the units.

In figuring the material cost of brickwork, an allowance of 25% mortar waste and 3% brick breakage was included. If bricks are delivered palletized with 280 to 300 per pallet, or packaged, allow only 1-1/2% for breakage. Packaged or palletized delivery is practical when a job is big enough to have a crane or other equipment available to handle a package of brick. This is so on all industrial work but not always true on small commercial buildings.

The use of buff and gray face is increasing, and there is a continuing trend to the Norman, Roman, Jumbo and SCR brick.

Common red clay brick for backup is not used that often. Concrete block is the most usual backup material with occasional use of sand lime or cement brick. Building brick is commonly used in solid walls for strength and as a fire stop.

Brick panels built on the ground and then crane erected to the upper floors have proven to be economical. This allows the work to be done under cover and without scaffolding.

R042110-50 Brick, Block & Mortar Quantities

Running Bond						For Other Bonds Standard Size Add to S.F. Quantities in Table to Left		
Number of Brick per S.F. of Wall - Single Wythe with 3/8″ Joints				C.F. of Mortar per M Bricks, Waste Included				
Type Brick	Nominal Size (incl. mortar) L H W	Modular Coursing	Number of Brick per S.F.	3/8″ Joint	1/2″ Joint	Bond Type	Description	Factor
Standard	8 x 2-2/3 x 4	3C=8″	6.75	10.3	12.9	Common	full header every fifth course	+20%
Economy	8 x 4 x 4	1C=4″	4.50	11.4	14.6		full header every sixth course	+16.7%
Engineer	8 x 3-1/5 x 4	5C=16″	5.63	10.6	13.6	English	full header every second course	+50%
Fire	9 x 2-1/2 x 4-1/2	2C=5″	6.40	550 # Fireclay	—	Flemish	alternate headers every course	+33.3%
Jumbo	12 x 4 x 6 or 8	1C=4″	3.00	23.8	30.8		every sixth course	+5.6%
Norman	12 x 2-2/3 x 4	3C=8″	4.50	14.0	17.9	Header = W x H exposed		+100%
Norwegian	12 x 3-1/5 x 4	5C=16″	3.75	14.6	18.6	Rowlock = H x W exposed		+100%
Roman	12 x 2 x 4	2C=4″	6.00	13.4	17.0	Rowlock stretcher = L x W exposed		+33.3%
SCR	12 x 2-2/3 x 6	3C=8″	4.50	21.8	28.0	Soldier = H x L exposed		—
Utility	12 x 4 x 4	1C=4″	3.00	15.4	19.6	Sailor = W x L exposed		-33.3%

Concrete Blocks Nominal Size	Approximate Weight per S.F.		Blocks per 100 S.F.	Mortar per M block, waste included	
	Standard	Lightweight		Partitions	Back up
2″ x 8″ x 16″	20 PSF	15 PSF	113	27 C.F.	36 C.F.
4″	30	20		41	51
6″	42	30		56	66
8″	55	38		72	82
10″	70	47		87	97
12″	85	55		102	112

Brick & Mortar Quantities
©Brick Industry Association. 2009 Feb. Technical Notes on Brick Construction 10:
Dimensioning and Estimating Brick Masonry. Reston (VA): BIA. Table 1 Modular Brick Sizes and Table 4 Quantity Estimates for Brick Masonry.

R042210-20 Concrete Block

The material cost of special block such as corner, jamb and head block can be figured at the same price as ordinary block of equal size. Labor on specials is about the same as equal-sized regular block.

Bond beams and 16″ high lintel blocks are more expensive than regular units of equal size. Lintel blocks are 8″ long and either 8″ or 16″ high.

Use of a motorized mortar spreader box will speed construction of continuous walls.

Hollow non-load-bearing units are made according to ASTM C129 and hollow load-bearing units according to ASTM C90.

R053100-10 Decking Descriptions

General - All Deck Products

A steel deck is made by cold forming structural grade sheet steel into a repeating pattern of parallel ribs. The strength and stiffness of the panels are the result of the ribs and the material properties of the steel. Deck lengths can be varied to suit job conditions, but because of shipping considerations, are usually less than 40 feet. Standard deck width varies with the product used but full sheets are usually 12″, 18″, 24″, 30″, or 36″. The deck is typically furnished in a standard width with the ends cut square. Any cutting for width, such as at openings or for angular fit, is done at the job site.

The deck is typically attached to the building frame with arc puddle welds, self-drilling screws, or powder or pneumatically driven pins. Sheet to sheet fastening is done with screws, button punching (crimping), or welds.

Composite Floor Deck

After installation and adequate fastening, a floor deck serves several purposes. It (a) acts as a working platform, (b) stabilizes the frame, (c) serves as a concrete form for the slab, and (d) reinforces the slab to carry the design loads applied during the life of the building. Composite decks are distinguished by the presence of shear connector devices as part of the deck. These devices are designed to mechanically lock the concrete and deck together so that the concrete and the deck work together to carry subsequent floor loads. These shear connector devices can be rolled-in embossments, lugs, holes, or wires welded to the panels. The deck profile can also be used to interlock concrete and steel.

Composite deck finishes are either galvanized (zinc coated) or phosphatized/painted. Galvanized deck has a zinc coating on both the top and bottom surfaces. The phosphatized/painted deck has a bare (phosphatized) top surface that will come into contact with the concrete. This bare top surface can be expected to develop rust before the concrete is placed. The bottom side of the deck has a primer coat of paint.

A composite floor deck is normally installed so the panel ends do not overlap on the supporting beams. Shear lugs or panel profile shapes often prevent a tight metal to metal fit if the panel ends overlap; the air gap caused by overlapping will prevent proper fusion with the structural steel supports when the panel end laps are shear stud welded.

Adequate end bearing of the deck must be obtained as shown on the drawings. If bearing is actually less in the field than shown on the drawings, further investigation is required.

Roof Deck

A roof deck is not designed to act compositely with other materials. A roof deck acts alone in transferring horizontal and vertical loads into the building frame. Roof deck rib openings are usually narrower than floor deck rib openings. This provides adequate support of the rigid thermal insulation board.

A roof deck is typically installed to endlap approximately 2″ over supports. However, it can be butted (or lapped more than 2″) to solve field fit problems. Since designers frequently use the installed deck system as part of the horizontal bracing system (the deck as a diaphragm), any fastening substitution or change should be approved by the designer. Continuous perimeter support of the deck is necessary to limit edge deflection in the finished roof and may be required for diaphragm shear transfer.

Standard roof deck finishes are galvanized or primer painted. The standard factory applied paint for roof decks is a primer paint and is not intended to weather for extended periods of time. Field painting or touching up of abrasions and deterioration of the primer coat or other protective finishes is the responsibility of the contractor.

Cellular Deck

A cellular deck is made by attaching a bottom steel sheet to a roof deck or composite floor deck panel. A cellular deck can be used in the same manner as a floor deck. Electrical, telephone, and data wires are easily run through the chase created between the deck panel and the bottom sheet.

When used as part of the electrical distribution system, the cellular deck must be installed so that the ribs line up and create a smooth cell transition at abutting ends. The joint that occurs at butting cell ends must be taped or otherwise sealed to prevent wet concrete from seeping into the cell. Cell interiors must be free of welding burrs, or other sharp intrusions, to prevent damage to wires.

When used as a roof deck, the bottom flat plate is usually left exposed to view. Care must be maintained during erection to keep good alignment and prevent damage.

A cellular deck is sometimes used with the flat plate on the top side to provide a flat working surface. Installation of the deck for this purpose requires special methods for attachment to the frame because the flat plate, now on the top, can prevent direct access to the deck material that is bearing on the structural steel. It may be advisable to treat the flat top surface to prevent slipping.

A cellular deck is always furnished galvanized or painted over galvanized.

Form Deck

A form deck can be any floor or roof deck product used as a concrete form. Connections to the frame are by the same methods used to anchor floor and roof decks. Welding washers are recommended when welding a deck that is less than 20 gauge thickness.

A form deck is furnished galvanized, prime painted, or uncoated. A galvanized deck must be used for those roof deck systems where a form deck is used to carry a lightweight insulating concrete fill.

Wood, Plastics & Comp. — R0611 Wood Framing

R061110-30 Lumber Product Material Prices

The price of forest products fluctuates widely from location to location and from season to season depending upon economic conditions. The bare material prices in the unit cost sections of the data set show the National Average material prices in effect Jan. 1 of this data year. It must be noted that lumber prices in general may change significantly during the year.

Availability of certain items depends upon geographic location and must be checked prior to firm-price bidding.

Wood, Plastics & Comp. — R0616 Sheathing

R061636-20 Plywood

There are two types of plywood used in construction: interior, which is moisture-resistant but not waterproofed, and exterior, which is waterproofed.

The grade of the exterior surface of the plywood sheets is designated by the first letter: A, for smooth surface with patches allowed; B, for solid surface with patches and plugs allowed; C, which may be surface plugged or may have knot holes up to 1″ wide; and D, which is used only for interior type plywood and may have knot holes up to 2-1/2″ wide. "Structural Grade" is specifically designed for engineered applications such as box beams. All CC & DD grades have roof and floor spans marked on them.

Underlayment-grade plywood runs from 1/4″ to 1-1/4″ thick. Thicknesses 5/8″ and over have optional tongue and groove joints which eliminate the need for blocking the edges. Underlayment 19/32″ and over may be referred to as Sturd-i-Floor.

The price of plywood can fluctuate widely due to geographic and economic conditions.

Typical uses for various plywood grades are as follows:

AA-AD Interior — cupboards, shelving, paneling, furniture

BB Plyform — concrete form plywood

CDX — wall and roof sheathing

Structural — box beams, girders, stressed skin panels

AA-AC Exterior — fences, signs, siding, soffits, etc.

Underlayment — base for resilient floor coverings

Overlaid HDO — high density for concrete forms & highway signs

Overlaid MDO — medium density for painting, siding, soffits & signs

303 Siding — exterior siding, textured, striated, embossed, etc.

Thermal & Moist. Protec. — R0731 Shingles & Shakes

R073126-20 Roof Slate

16″, 18″ and 20″ are standard lengths, and slate usually comes in random widths. For standard 3/16″ thickness use 1-1/2″ copper nails. Allow for 3% breakage.

Thermal & Moist. Protec. — R0752 Modified Bituminous Membrane Roofing

R075213-30 Modified Bitumen Roofing

The cost of modified bitumen roofing is highly dependent on the type of installation that is planned. Installation is based on the type of modifier used in the bitumen. The two most popular modifiers are atactic polypropylene (APP) and styrene butadiene styrene (SBS). The modifiers are added to heated bitumen during the manufacturing process to change its characteristics. A polyethylene, polyester or fiberglass reinforcing sheet is then sandwiched between layers of this bitumen. When completed, the result is a pre-assembled, built-up roof that has increased elasticity and weatherability. Some manufacturers include a surfacing material such as ceramic or mineral granules, metal particles or sand.

The preferred method of adhering SBS-modified bitumen roofing to the substrate is with hot-mopped asphalt (much the same as built-up roofing). This installation method requires a tar kettle/pot to heat the asphalt, as well as the labor, tools and equipment necessary to distribute and spread the hot asphalt.

The alternative method for applying APP and SBS modified bitumen is as follows. A skilled installer uses a torch to melt a small pool of bitumen off the membrane. This pool must form across the entire roll for proper adhesion. The installer must unroll the roofing at a pace slow enough to melt the bitumen, but fast enough to prevent damage to the rest of the membrane.

Modified bitumen roofing provides the advantages of both built-up and single-ply roofing. Labor costs are reduced over those of built-up roofing because only a single ply is necessary. The elasticity of single-ply roofing is attained with the reinforcing sheet and polymer modifiers. Modifieds have some self-healing characteristics and because of their multi-layer construction, they offer the reliability and safety of built-up roofing.

R081313-20 Steel Door Selection Guide

Standard steel doors are classified into four levels, as recommended by the Steel Door Institute in the chart below. Each of the four levels offers a range of construction models and designs to meet architectural requirements for preference and appearance, including full flush, seamless, and stile & rail. Recommended minimum gauge requirements are also included.

For complete standard steel door construction specifications and available sizes, refer to the Steel Door Institute Technical Data Series, ANSI A250.8-98 (SDI-100), and ANSI A250.4-94 Test Procedure and Acceptance Criteria for Physical Endurance of Steel Door and Hardware Reinforcements.

Level		Model	Construction	For Full Flush or Seamless		
				Min. Gauge	Thickness (in)	Thickness (mm)
I	Standard Duty	1 2	Full Flush Seamless	20	0.032	0.8
II	Heavy Duty	1 2	Full Flush Seamless	18	0.042	1.0
III	Extra Heavy Duty	1 2 3	Full Flush Seamless *Stile & Rail	16	0.053	1.3
IV	Maximum Duty	1 2	Full Flush Seamless	14	0.067	1.6

*Stiles & rails are 16 gauge; flush panels, when specified, are 18 gauge.

Openings R0852 Wood Windows

R085216-10 Window Estimates

To ensure a complete window estimate, be sure to include the material and labor costs for each window, as well as the material and labor costs for an interior wood trim set.

Openings R0853 Plastic Windows

R085313-20 Replacement Windows

Replacement windows are typically measured per United Inch.

United Inches are calculated by rounding the width and height of the window opening up to the nearest inch, then adding the two figures.

The labor cost for replacement windows includes removal of sash, existing sash balance or weights, parting bead where necessary and installation of new window.

Debris hauling and dump fees are not included.

R087110-10 Hardware Finishes

This table describes hardware finishes used throughout the industry. It also shows the base metal and the respective symbols in the three predominate systems of identification. Many of these are used in pricing descriptions in Division Eight.

US"	BMHA*	CDN^	Base	Description
US P	600	CP	Steel	Primed for Painting
US 1B	601	C1B	Steel	Bright Black Japanned
US 2C	602	C2C	Steel	Zinc Plated
US 2G	603	C2G	Steel	Zinc Plated
US 3	605	C3	Brass	Bright Brass, Clear Coated
US 4	606	C4	Brass	Satin Brass, Clear Coated
US 5	609	C5	Brass	Satin Brass, Blackened, Satin Relieved, Clear Coated
US 7	610	C7	Brass	Satin Brass, Blackened, Bright Relieved, Clear Coated
US 9	611	C9	Bronze	Bright Bronze, Clear Coated
US 10	612	C10	Bronze	Satin Bronze, Clear Coated
US 10A	641	C10A	Steel	Antiqued Bronze, Oiled and Lacquered
US 10B	613	C10B	Bronze	Antiqued Bronze, Oiled
US 11	616	C11	Bronze	Satin Bronze, Blackened, Satin Relieved, Clear Coated
US 14	618	C14	Brass/Bronze	Bright Nickel Plated, Clear Coated
US 15	619	C15	Brass/Bronze	Satin Nickel, Clear Coated
US 15A	620	C15A	Brass/Bronze	Satin Nickel Plated, Blackened, Satin Relieved, Clear Coated
US 17A	621	C17A	Brass/Bronze	Nickel Plated, Blackened, Relieved, Clear Coated
US 19	622	C19	Brass/Bronze	Flat Black Coated
US 20	623	C20	Brass/Bronze	Statuary Bronze, Light
US 20A	624	C20A	Brass/Bronze	Statuary Bronze, Dark
US 26	625	C26	Brass/Bronze	Bright Chromium
US 26D	626	C26D	Brass/Bronze	Satin Chromium
US 20	627	C27	Aluminum	Satin Aluminum Clear
US 28	628	C28	Aluminum	Anodized Dull Aluminum
US 32	629	C32	Stainless Steel	Bright Stainless Steel
US 32D	630	C32D	Stainless Steel	Stainless Steel
US 3	632	C3	Steel	Bright Brass Plated, Clear Coated
US 4	633	C4	Steel	Satin Brass, Clear Coated
US 7	636	C7	Steel	Satin Brass Plated, Blackened, Bright Relieved, Clear Coated
US 9	637	C9	Steel	Bright Bronze Plated, Clear Coated
US 5	638	C5	Steel	Satin Brass Plated, Blackened, Bright Relieved, Clear Coated
US 10	639	C10	Steel	Satin Bronze Plated, Clear Coated
US 10B	640	C10B	Steel	Antique Bronze, Oiled
US 10A	641	C10A	Steel	Antiqued Bronze, Oiled and Lacquered
US 11	643	C11	Steel	Satin Bronze Plated, Blackened, Bright Relieved, Clear Coated
US 14	645	C14	Steel	Bright Nickel Plated, Clear Coated
US 15	646	C15	Steel	Satin Nickel Plated, Clear Coated
US 15A	647	C15A	Steel	Nickel Plated, Blackened, Bright Relieved, Clear Coated
US 17A	648	C17A	Steel	Nickel Plated, Blackened, Relieved, Clear Coated
US 20	649	C20	Steel	Statuary Bronze, Light
US 20A	650	C20A	Steel	Statuary Bronze, Dark
US 26	651	C26	Steel	Bright Chromium Plated
US 26D	652	C26D	Steel	Satin Chromium Plated

* - BMHA Builders Hardware Manufacturing Association
" - US Equivalent
^ - Canadian Equivalent
Japanning is imitating Asian lacquer work

R092000-50 Lath, Plaster and Gypsum Board

Gypsum board lath is available in 3/8″ thick x 16″ wide x 4′ long sheets as a base material for multi-layer plaster applications. It is also available as a base for either multi-layer or veneer plaster applications in 1/2″ and 5/8″ thick–4′ wide x 8′, 10′ or 12′ long sheets. Fasteners are screws or blued ring shank nails for wood framing and screws for metal framing.

Metal lath is available in diamond mesh patterns with flat or self-furring profiles. Paper backing is available for applications where excessive plaster waste needs to be avoided. A slotted mesh ribbed lath should be used in areas where the span between structural supports is greater than normal. Most metal lath comes in 27″ x 96″ sheets. Diamond mesh weighs 1.75, 2.5 or 3.4 pounds per square yard, slotted mesh lath weighs 2.75 or 3.4 pounds per square yard. Metal lath can be nailed, screwed or tied in place.

Many **accessories** are available. Corner beads, flat reinforcing strips, casing beads, control and expansion joints, furring brackets and channels are some examples. Note that accessories are not included in plaster or stucco line items.

Plaster is defined as a material or combination of materials that when mixed with a suitable amount of water, forms a plastic mass or paste. When applied to a surface, the paste adheres to it and subsequently hardens, preserving in a rigid state the form or texture imposed during the period of elasticity.

Gypsum plaster is made from ground calcined gypsum. It is mixed with aggregates and water for use as a base coat plaster.

Vermiculite plaster is a fire-retardant plaster covering used on steel beams, concrete slabs and other heavy construction materials. Vermiculite is a group name for certain clay minerals, hydrous silicates or aluminum, magnesium and iron that have been expanded by heat.

Perlite plaster is a plaster using perlite as an aggregate instead of sand. Perlite is a volcanic glass that has been expanded by heat.

Gauging plaster is a mix of gypsum plaster and lime putty that when applied produces a quick drying finish coat.

Veneer plaster is a one or two component gypsum plaster used as a thin finish coat over special gypsum board.

Keenes cement is a white cementitious material manufactured from gypsum that has been burned at a high temperature and ground to a fine powder. Alum is added to accelerate the set. The resulting plaster is hard and strong and accepts and maintains a high polish, hence it is used as a finishing plaster.

Stucco is a Portland cement based plaster used primarily as an exterior finish.

Plaster is used on both interior and exterior surfaces. Generally it is applied in multiple-coat systems. A three-coat system uses the terms scratch, brown and finish to identify each coat. A two-coat system uses base and finish to describe each coat. Each type of plaster and application system has attributes that are chosen by the designer to best fit the intended use.

Gypsum Plaster Quantities for 100 S.Y.	2 Coat, 5/8″ Thick		3 Coat, 3/4″ Thick		
	Base	Finish	Scratch	Brown	Finish
	1:3 Mix	2:1 Mix	1:2 Mix	1:3 Mix	2:1 Mix
Gypsum plaster	1,300 lb.		1,350 lb.	650 lb.	
Sand	1.75 C.Y.		1.85 C.Y.	1.35 C.Y.	
Finish hydrated lime		340 lb.			340 lb.
Gauging plaster		170 lb.			170 lb.

Vermiculite or Perlite Plaster Quantities for 100 S.Y.	2 Coat, 5/8″ Thick		3 Coat, 3/4″ Thick		
	Base	Finish	Scratch	Brown	Finish
Gypsum plaster	1,250 lb.		1,450 lb.	800 lb.	
Vermiculite or perlite	7.8 bags		8.0 bags	3.3 bags	
Finish hydrated lime		340 lb.			340 lb.
Gauging plaster		170 lb.			170 lb.

Stucco-Three-Coat System Quantities for 100 S.Y.	On Wood Frame	On Masonry
Portland cement	29 bags	21 bags
Sand	2.6 C.Y.	2.0 C.Y.
Hydrated lime	180 lb.	120 lb.

R092910-10 Levels of Gypsum Board Finish

In the past, contract documents often used phrases such as "industry standard" and "workmanlike finish" to specify the expected quality of gypsum board wall and ceiling installations. The vagueness of these descriptions led to unacceptable work and disputes.

In order to resolve this problem, four major trade associations concerned with the manufacture, erection, finish and decoration of gypsum board wall and ceiling systems developed an industry-wide *Recommended Levels of Gypsum Board Finish.*

The finish of gypsum board walls and ceilings for specific final decoration is dependent on a number of factors. A primary consideration is the location of the surface and the degree of decorative treatment desired. Painted and unpainted surfaces in warehouses and other areas where appearance is normally not critical may simply require the taping of wallboard joints and 'spotting' of fastener heads. Blemish-free, smooth, monolithic surfaces often intended for painted and decorated walls and ceilings in habitated structures, ranging from single-family dwellings through monumental buildings, require additional finishing prior to the application of the final decoration.

Other factors to be considered in determining the level of finish of the gypsum board surface are (1) the type of angle of surface illumination (both natural and artificial lighting), and (2) the paint and method of application or the type and finish of wallcovering specified as the final decoration. Critical lighting conditions, gloss paints, and thin wall coverings require a higher level of gypsum board finish than heavily textured surfaces which are subsequently painted or surfaces which are to be decorated with heavy grade wall coverings.

The following descriptions were developed by the Association of the Wall and Ceiling Industries-International (AWCI), Ceiling & Interior Systems Construction Association (CISCA), Gypsum Association (GA), and Painting and Decorating Contractors of America (PDCA) as a guide.

Level 0: Used in temporary construction or wherever the final decoration has not been determined. Unfinished. No taping, finishing or corner beads are required. Also could be used where non-predecorated panels will be used in demountable-type partitions that are to be painted as a final finish. .

Level 1: Frequently used in plenum areas above ceilings, in attics, in areas where the assembly would generally be concealed, or in building service corridors and other areas not normally open to public view. Some degree of sound and smoke control is provided; in some geographic areas, this level is referred to as "fire-taping," although this level of finish does not typically meet fire-resistant assembly requirements. Where a fire resistance rating is required for the gypsum board assembly, details of construction should be in accordance with reports of fire tests of assemblies that have met the requirements of the fire rating acceptable.

All joints and interior angles shall have tape embedded in joint compound. Accessories are optional at specifier discretion in corridors and other areas with pedestrian traffic. Tape and fastener heads need not be covered with joint compound. Surface shall be free of excess joint compound. Tool marks and ridges are acceptable.

Level 2: It may be specified for standard gypsum board surfaces in garages, warehouse storage, or other similar areas where surface appearance is not of primary importance.

All joints and interior angles shall have tape embedded in joint compound and shall be immediately wiped with a joint knife or trowel, leaving a thin coating of joint compound over all joints and interior angles. Fastener heads and accessories shall be covered with a coat of joint compound. Surface shall be free of excess joint compound. Tool marks and ridges are acceptable.

Level 3: Typically used in areas receiving heavy texture (spray or hand applied) finishes before final painting, or where commercial-grade (heavy duty) wall coverings are to be applied as the final decoration. This level of finish should not be used where smooth painted surfaces or where lighter weight wall coverings are specified. The prepared surface shall be coated with a drywall primer prior to the application of final finishes.

All joints and interior angles shall have tape embedded in joint compound and shall be immediately wiped with a joint knife or trowel, leaving a thin coating of joint compound over all joints and interior angles. One additional coat of joint compound shall be applied over all joints and interior angles. Fastener heads and accessories shall be covered with two separate coats of joint compound. All joint compounds shall be smooth and free of tool marks and ridges. The prepared surface shall be covered with a drywall primer prior to the application of the final decoration.

Level 4: This level should be used where residential grade (light duty) wall coverings, flat paints, or light textures are to be applied. The prepared surface shall be coated with a drywall primer prior to the application of final finishes. Release agents for wall coverings are specifically formulated to minimize damage if coverings are subsequently removed.

The weight, texture, and sheen level of the wall covering material selected should be taken into consideration when specifying wall coverings over this level of drywall treatment. Joints and fasteners must be sufficiently concealed if the wall covering material is lightweight, contains limited pattern, has a glossy finish, or has any combination of these features. In critical lighting areas, flat paints applied over light textures tend to reduce joint photographing. Gloss, semi-gloss, and enamel paints are not recommended over this level of finish.

All joints and interior angles shall have tape embedded in joint compound and shall be immediately wiped with a joint knife or trowel, leaving a thin coating of joint compound over all joints and interior angles. In addition, two separate coats of joint compound shall be applied over all flat joints and one separate coat of joint compound applied over interior angles. Fastener heads and accessories shall be covered with three separate coats of joint compound. All joint compounds shall be smooth and free of tool marks and ridges. The prepared surface shall be covered with a drywall primer like Sheetrock first coat prior to the application of the final decoration.

Level 5: The highest quality finish is the most effective method to provide a uniform surface and minimize the possibility of joint photographing and of fasteners showing through the final decoration. This level of finish is required where gloss, semi-gloss, or enamel is specified; when flat joints are specified over an untextured surface; or where critical lighting conditions occur. The prepared surface shall be coated with a drywall primer prior to the application of the final decoration.

All joints and interior angles shall have tape embedded in joint compound and be immediately wiped with a joint knife or trowel, leaving a thin coating of joint compound over all joints and interior angles. Two separate coats of joint compound shall be applied over all flat joints and one separate coat of joint compound applied over interior angles. Fastener heads and accessories shall be covered with three separate coats of joint compound.

A thin skim coat of joint compound shall be trowel applied to the entire surface. Excess compound is immediately troweled off, leaving a film or skim coating of compound completely covering the paper. As an alternative to a skim coat, a material manufactured especially for this purpose may be applied such as Sheetrock Tuff-Hide primer surfacer. The surface must be smooth and free of tool marks and ridges. The prepared surface shall be covered with a drywall primer prior to the application of the final decoration.

R097223-10 Wall Covering

The table below lists the quantities required for 100 S.F. of wall covering.

Description	Medium-Priced Paper	Expensive Paper
Paper	1.6 dbl. rolls	1.6 dbl. rolls
Wall sizing	0.25 gallon	0.25 gallon
Vinyl wall paste	0.6 gallon	0.6 gallon
Apply sizing	0.3 hour	0.3 hour
Apply paper	1.2 hours	1.5 hours

Most wallpapers now come in double rolls only.
To remove old paper, allow 1.3 hours per 100 S.F.

R099100-10 Painting Estimating Techniques

Proper estimating methodology is needed to obtain an accurate painting estimate. There is no known reliable shortcut or square foot method. The following steps should be followed:

- List all surfaces to be painted, with an accurate quantity (area) of each. Items having similar surface condition, finish, application method and accessibility may be grouped together.
- List all the tasks required for each surface to be painted, including surface preparation, masking, and protection of adjacent surfaces. Surface preparation may include minor repairs, washing, sanding and puttying.
- Select the proper Means line for each task. Review and consider all adjustments to labor and materials for type of paint and location of work. Apply the height adjustment carefully. For instance, when applying the adjustment for work over 8' high to a wall that is 12' high, apply the adjustment only to the area between 8' and 12' high, and not to the entire wall.

When applying more than one percent (%) adjustment, apply each to the base cost of the data, rather than applying one percentage adjustment on top of the other.

When estimating the cost of painting walls and ceilings remember to add the brushwork for all cut-ins at inside corners and around windows and doors as a LF measure. One linear foot of cut-in with a brush equals one square foot of painting.

All items for spray painting include the labor for roll-back.

Deduct for openings greater than 100 SF or openings that extend from floor to ceiling and are greater than 5' wide. Do not deduct small openings.

The cost of brushes, rollers, ladders and spray equipment are considered part of a painting contractor's overhead, and should not be added to the estimate. The cost of rented equipment such as scaffolding and swing staging should be added to the estimate.

R099100-20 Painting

Item	Coat	One Gallon Covers			In 8 Hours a Laborer Covers			Labor-Hours per 100 S.F.		
		Brush	Roller	Spray	Brush	Roller	Spray	Brush	Roller	Spray
Paint wood siding	prime	250 S.F.	225 S.F.	290 S.F.	1150 S.F.	1300 S.F.	2275 S.F.	.695	.615	.351
	others	270	250	290	1300	1625	2600	.615	.492	.307
Paint exterior trim	prime	400	—	—	650	—	—	1.230	—	—
	1st	475	—	—	800	—	—	1.000	—	—
	2nd	520	—	—	975	—	—	.820	—	—
Paint shingle siding	prime	270	255	300	650	975	1950	1.230	.820	.410
	others	360	340	380	800	1150	2275	1.000	.695	.351
Stain shingle siding	1st	180	170	200	750	1125	2250	1.068	.711	.355
	2nd	270	250	290	900	1325	2600	.888	.603	.307
Paint brick masonry	prime	180	135	160	750	800	1800	1.066	1.000	.444
	1st	270	225	290	815	975	2275	.981	.820	.351
	2nd	340	305	360	815	1150	2925	.981	.695	.273
Paint interior plaster or drywall	prime	400	380	495	1150	2000	3250	.695	.400	.246
	others	450	425	495	1300	2300	4000	.615	.347	.200
Paint interior doors and windows	prime	400	—	—	650	—	—	1.230	—	—
	1st	425	—	—	800	—	—	1.000	—	—
	2nd	450	—	—	975	—	—	.820	—	—

R131113-20 Swimming Pools

Pool prices given per square foot of surface area include pool structure, filter and chlorination equipment, pumps, related piping, ladders/steps, maintenance kit, skimmer and vacuum system. Decks and electrical service to equipment are not included.

Residential in-ground pool construction can be divided into two categories: vinyl lined and gunite. Vinyl lined pool walls are constructed of different materials including wood, concrete, plastic or metal. The bottom is often graded with sand over which the vinyl liner is installed. Vermiculite or soil cement bottoms may be substituted for an added cost.

Gunite pool construction is used both in residential and municipal installations. These structures are steel reinforced for strength and finished with a white cement limestone plaster.

Municipal pools will have a higher cost because plumbing codes require more expensive materials, chlorination equipment and higher filtration rates.

Municipal pools greater than 1,800 S.F. require gutter systems to control waves. This gutter may be formed into the concrete wall. Often a vinyl/stainless steel gutter or gutter/wall system is specified, which will raise the pool cost.

Competition pools usually require tile bottoms and sides with contrasting lane striping, which will also raise the pool cost.

R221113-50 Pipe Material Considerations

1. Malleable fittings should be used for gas service.
2. Malleable fittings are used where there are stresses/strains due to expansion and vibration.
3. Cast fittings may be broken as an aid to disassembling heating lines frozen by long use, temperature and minerals.
4. A cast iron pipe is extensively used for underground and submerged service.
5. Type M (light wall) copper tubing is available in hard temper only and is used for nonpressure and less severe applications than K and L.
6. Type L (medium wall) copper tubing, available hard or soft for interior service.
7. Type K (heavy wall) copper tubing, available in hard or soft temper for use where conditions are severe. For underground and interior service.
8. Hard drawn tubing requires fewer hangers or supports but should not be bent. Silver brazed fittings are recommended, but soft solder is normally used.
9. Type DMV (very light wall) copper tubing designed for drainage, waste and vent plus other non-critical pressure services.

Domestic/Imported Pipe and Fittings Costs

The prices shown in this publication for steel/cast iron pipe and steel, cast iron, and malleable iron fittings are based on domestic production sold at the normal trade discounts. The above listed items of foreign manufacture may be available at prices 1/3 to 1/2 of those shown. Some imported items after minor machining or finishing operations are being sold as domestic to further complicate the system.

Caution: Most pipe prices in this data set also include a coupling and pipe hangers which for the larger sizes can add significantly to the per foot cost and should be taken into account when comparing "book cost" with the quoted supplier's cost.

Exterior Improvements R3292 Turf & Grasses

R329219-50 Seeding

The type of grass is determined by light, shade and moisture content of soil plus intended use. Fertilizer should be disked 4″ before seeding. For steep slopes disk five tons of mulch and lay two tons of hay or straw on surface per acre after seeding. Surface mulch can be staked, lightly disked or tar emulsion sprayed. Material for mulch can be wood chips, peat moss, partially rotted hay or straw, wood fibers and sprayed emulsions. Hemp seed blankets with fertilizer are also available. For spring seeding, watering is necessary. Late fall seeding may have to be reseeded in the spring. Hydraulic seeding, power mulching, and aerial seeding can be used on large areas.

R331113-80 Piping Designations

There are several systems currently in use to describe pipe and fittings. The following paragraphs will help to identify and clarify classifications of piping systems used for water distribution.

Piping may be classified by schedule. Piping schedules include 5S, 10S, 10, 20, 30, Standard, 40, 60, Extra Strong, 80, 100, 120, 140, 160 and Double Extra Strong. These schedules are dependent upon the pipe wall thickness. The wall thickness of a particular schedule may vary with pipe size.

Ductile iron pipe for water distribution is classified by Pressure Classes such as Class 150, 200, 250, 300 and 350. These classes are actually the rated water working pressure of the pipe in pounds per square inch (psi). The pipe in these pressure classes is designed to withstand the rated water working pressure plus a surge allowance of 100 psi.

The American Water Works Association (AWWA) provides standards for various types of **plastic pipe.** C-900 is the specification for polyvinyl chloride (PVC) piping used for water distribution in sizes ranging from 4″ through 12″. C-901 is the specification for polyethylene (PE) pressure pipe, tubing and fittings used for water distribution in sizes ranging from 1/2″ through 3″. C-905 is the specification for PVC piping sizes 14″ and greater.

PVC pressure-rated pipe is identified using the standard dimensional ratio (SDR) method. This method is defined by the American Society for Testing and Materials (ASTM) Standard D 2241. This pipe is available in SDR numbers 64, 41, 32.5, 26, 21, 17, and 13.5. A pipe with an SDR of 64 will have the thinnest wall while a pipe with an SDR of 13.5 will have the thickest wall. When the pressure rating (PR) of a pipe is given in psi, it is based on a line supplying water at 73 degrees F.

The National Sanitation Foundation (NSF) seal of approval is applied to products that can be used with potable water. These products have been tested to ANSI/NSF Standard 14.

Valves and strainers are classified by American National Standards Institute (ANSI) Classes. These Classes are 125, 150, 200, 250, 300, 400, 600, 900, 1500 and 2500. Within each class there is an operating pressure range dependent upon temperature. Design parameters should be compared to the appropriate material dependent, pressure-temperature rating chart for accurate valve selection.

Abbreviations

Abbreviation	Meaning
A	Area Square Feet; Ampere
AAFES	Army and Air Force Exchange Service
ABS	Acrylonitrile Butadiene Stryrene; Asbestos Bonded Steel
A.C., AC	Alternating Current; Air-Conditioning; Asbestos Cement; Plywood Grade A & C
ACI	American Concrete Institute
ACR	Air Conditioning Refrigeration
ADA	Americans with Disabilities Act
AD	Plywood, Grade A & D
Addit.	Additional
Adh.	Adhesive
Adj.	Adjustable
af	Audio-frequency
AFFF	Aqueous Film Forming Foam
AFUE	Annual Fuel Utilization Efficiency
AGA	American Gas Association
Agg.	Aggregate
A.H., Ah	Ampere Hours
A hr.	Ampere-hour
A.H.U., AHU	Air Handling Unit
A.I.A.	American Institute of Architects
AIC	Ampere Interrupting Capacity
Allow.	Allowance
alt., alt	Alternate
Alum.	Aluminum
a.m.	Ante Meridiem
Amp.	Ampere
Anod.	Anodized
ANSI	American National Standards Institute
APA	American Plywood Association
Approx.	Approximate
Apt.	Apartment
Asb.	Asbestos
A.S.B.C.	American Standard Building Code
Asbe.	Asbestos Worker
ASCE	American Society of Civil Engineers
A.S.H.R.A.E.	American Society of Heating, Refrig. & AC Engineers
ASME	American Society of Mechanical Engineers
ASTM	American Society for Testing and Materials
Attchmt.	Attachment
Avg., Ave.	Average
AWG	American Wire Gauge
AWWA	American Water Works Assoc.
Bbl.	Barrel
B&B, BB	Grade B and Better; Balled & Burlapped
B&S	Bell and Spigot
B.&W.	Black and White
b.c.c.	Body-centered Cubic
B.C.Y.	Bank Cubic Yards
BE	Bevel End
B.F.	Board Feet
Bg. cem.	Bag of Cement
BHP	Boiler Horsepower; Brake Horsepower
B.I.	Black Iron
bidir.	bidirectional
Bit., Bitum.	Bituminous
Bit., Conc.	Bituminous Concrete
Bk.	Backed
Bkrs.	Breakers
Bldg., bldg	Building
Blk.	Block
Bm.	Beam
Boil.	Boilermaker
bpm	Blows per Minute
BR	Bedroom
Brg., brng.	Bearing
Brhe.	Bricklayer Helper
Bric.	Bricklayer
Brk., brk	Brick
brkt	Bracket
Brs.	Brass
Brz.	Bronze
Bsn.	Basin
Btr.	Better
BTU	British Thermal Unit
BTUH	BTU per Hour
Bu.	Bushels
BUR	Built-up Roofing
BX	Interlocked Armored Cable
°C	Degree Centigrade
c	Conductivity, Copper Sweat
C	Hundred; Centigrade
C/C	Center to Center, Cedar on Cedar
C-C	Center to Center
Cab	Cabinet
Cair.	Air Tool Laborer
Cal.	Caliper
Calc	Calculated
Cap.	Capacity
Carp.	Carpenter
C.B.	Circuit Breaker
C.C.A.	Chromate Copper Arsenate
C.C.F.	Hundred Cubic Feet
cd	Candela
cd/sf	Candela per Square Foot
CD	Grade of Plywood Face & Back
CDX	Plywood, Grade C & D, exterior glue
Cefi.	Cement Finisher
Cem.	Cement
CF	Hundred Feet
C.F.	Cubic Feet
CFM	Cubic Feet per Minute
CFRP	Carbon Fiber Reinforced Plastic
c.g.	Center of Gravity
CHW	Chilled Water; Commercial Hot Water
C.I., CI	Cast Iron
C.I.P., CIP	Cast in Place
Circ.	Circuit
C.L.	Carload Lot
CL	Chain Link
Clab.	Common Laborer
Clam	Common Maintenance Laborer
C.L.F.	Hundred Linear Feet
CLF	Current Limiting Fuse
CLP	Cross Linked Polyethylene
cm	Centimeter
CMP	Corr. Metal Pipe
CMU	Concrete Masonry Unit
CN	Change Notice
Col.	Column
CO_2	Carbon Dioxide
Comb.	Combination
comm.	Commercial, Communication
Compr.	Compressor
Conc.	Concrete
Cont., cont	Continuous; Continued, Container
Corkbd.	Cork Board
Corr.	Corrugated
Cos	Cosine
Cot	Cotangent
Cov.	Cover
C/P	Cedar on Paneling
CPA	Control Point Adjustment
Cplg.	Coupling
CPM	Critical Path Method
CPVC	Chlorinated Polyvinyl Chloride
C.Pr.	Hundred Pair
CRC	Cold Rolled Channel
Creos.	Creosote
Crpt.	Carpet & Linoleum Layer
CRT	Cathode-ray Tube
CS	Carbon Steel, Constant Shear Bar Joist
Csc	Cosecant
C.S.F.	Hundred Square Feet
CSI	Construction Specifications Institute
CT	Current Transformer
CTS	Copper Tube Size
Cu	Copper, Cubic
Cu. Ft.	Cubic Foot
cw	Continuous Wave
C.W.	Cool White; Cold Water
Cwt.	100 Pounds
C.W.X.	Cool White Deluxe
C.Y.	Cubic Yard (27 cubic feet)
C.Y./Hr.	Cubic Yard per Hour
Cyl.	Cylinder
d	Penny (nail size)
D	Deep; Depth; Discharge
Dis., Disch.	Discharge
Db	Decibel
Dbl.	Double
DC	Direct Current
DDC	Direct Digital Control
Demob.	Demobilization
d.f.t.	Dry Film Thickness
d.f.u.	Drainage Fixture Units
D.H.	Double Hung
DHW	Domestic Hot Water
DI	Ductile Iron
Diag.	Diagonal
Diam., Dia	Diameter
Distrib.	Distribution
Div.	Division
Dk.	Deck
D.L.	Dead Load; Diesel
DLH	Deep Long Span Bar Joist
dlx	Deluxe
Do.	Ditto
DOP	Dioctyl Phthalate Penetration Test (Air Filters)
Dp., dp	Depth
D.P.S.T.	Double Pole, Single Throw
Dr.	Drive
DR	Dimension Ratio
Drink.	Drinking
D.S.	Double Strength
D.S.A.	Double Strength A Grade
D.S.B.	Double Strength B Grade
Dty.	Duty
DWV	Drain Waste Vent
DX	Deluxe White, Direct Expansion
dyn	Dyne
e	Eccentricity
E	Equipment Only; East; Emissivity
Ea.	Each
EB	Encased Burial
Econ.	Economy
E.C.Y	Embankment Cubic Yards
EDP	Electronic Data Processing
EIFS	Exterior Insulation Finish System
E.D.R.	Equiv. Direct Radiation
Eq.	Equation
EL	Elevation
Elec.	Electrician; Electrical
Elev.	Elevator; Elevating
EMT	Electrical Metallic Conduit; Thin Wall Conduit
Eng.	Engine, Engineered
EPDM	Ethylene Propylene Diene Monomer
EPS	Expanded Polystyrene
Eqhv.	Equip. Oper., Heavy
Eqlt.	Equip. Oper., Light
Eqmd.	Equip. Oper., Medium
Eqmm.	Equip. Oper., Master Mechanic
Eqol.	Equip. Oper., Oilers
Equip.	Equipment
ERW	Electric Resistance Welded

Abbreviations

Abbreviation	Meaning
E.S.	Energy Saver
Est.	Estimated
esu	Electrostatic Units
E.W.	Each Way
EWT	Entering Water Temperature
Excav.	Excavation
excl	Excluding
Exp., exp	Expansion, Exposure
Ext., ext	Exterior; Extension
Extru.	Extrusion
f.	Fiber Stress
F	Fahrenheit; Female; Fill
Fab., fab	Fabricated; Fabric
FBGS	Fiberglass
F.C.	Footcandles
f.c.c.	Face-centered Cubic
f'c.	Compressive Stress in Concrete; Extreme Compressive Stress
F.E.	Front End
FEP	Fluorinated Ethylene Propylene (Teflon)
F.G.	Flat Grain
F.H.A.	Federal Housing Administration
Fig.	Figure
Fin.	Finished
FIPS	Female Iron Pipe Size
Fixt.	Fixture
FJP	Finger jointed and primed
Fl. Oz.	Fluid Ounces
Flr.	Floor
Flrs.	Floors
FM	Frequency Modulation; Factory Mutual
Fmg.	Framing
FM/UL	Factory Mutual/Underwriters Labs
Fdn.	Foundation
FNPT	Female National Pipe Thread
Fori.	Foreman, Inside
Foro.	Foreman, Outside
Fount.	Fountain
fpm	Feet per Minute
FPT	Female Pipe Thread
Fr	Frame
F.R.	Fire Rating
FRK	Foil Reinforced Kraft
FSK	Foil/Scrim/Kraft
FRP	Fiberglass Reinforced Plastic
FS	Forged Steel
FSC	Cast Body; Cast Switch Box
Ft., ft	Foot; Feet
Ftng.	Fitting
Ftg.	Footing
Ft lb.	Foot Pound
Furn.	Furniture
FVNR	Full Voltage Non-Reversing
FVR	Full Voltage Reversing
FXM	Female by Male
Fy.	Minimum Yield Stress of Steel
g	Gram
G	Gauss
Ga.	Gauge
Gal., gal.	Gallon
Galv., galv	Galvanized
GC/MS	Gas Chromatograph/Mass Spectrometer
Gen.	General
GFI	Ground Fault Interrupter
GFRC	Glass Fiber Reinforced Concrete
Glaz.	Glazier
GPD	Gallons per Day
gpf	Gallon per Flush
GPH	Gallons per Hour
gpm, GPM	Gallons per Minute
GR	Grade
Gran.	Granular
Grnd.	Ground
GVW	Gross Vehicle Weight
GWB	Gypsum Wall Board
H	High Henry
HC	High Capacity
H.D., HD	Heavy Duty; High Density
H.D.O.	High Density Overlaid
HDPE	High Density Polyethylene Plastic
Hdr.	Header
Hdwe.	Hardware
H.I.D., HID	High Intensity Discharge
Help.	Helper Average
HEPA	High Efficiency Particulate Air Filter
Hg	Mercury
HIC	High Interrupting Capacity
HM	Hollow Metal
HMWPE	High Molecular Weight Polyethylene
HO	High Output
Horiz.	Horizontal
H.P., HP	Horsepower; High Pressure
H.P.F.	High Power Factor
Hr.	Hour
Hrs./Day	Hours per Day
HSC	High Short Circuit
Ht.	Height
Htg.	Heating
Htrs.	Heaters
HVAC	Heating, Ventilation & Air-Conditioning
Hvy.	Heavy
HW	Hot Water
Hyd.; Hydr.	Hydraulic
Hz	Hertz (cycles)
I.	Moment of Inertia
IBC	International Building Code
I.C.	Interrupting Capacity
ID	Inside Diameter
I.D.	Inside Dimension; Identification
I.F.	Inside Frosted
I.M.C.	Intermediate Metal Conduit
In.	Inch
Incan.	Incandescent
Incl.	Included; Including
Int.	Interior
Inst.	Installation
Insul., insul	Insulation/Insulated
I.P.	Iron Pipe
I.P.S., IPS	Iron Pipe Size
IPT	Iron Pipe Threaded
I.W.	Indirect Waste
J	Joule
J.I.C.	Joint Industrial Council
K	Thousand; Thousand Pounds; Heavy Wall Copper Tubing, Kelvin
K.A.H.	Thousand Amp. Hours
kcmil	Thousand Circular Mils
KD	Knock Down
K.D.A.T.	Kiln Dried After Treatment
kg	Kilogram
kG	Kilogauss
kgf	Kilogram Force
kHz	Kilohertz
Kip	1000 Pounds
KJ	Kilojoule
K.L.	Effective Length Factor
K.L.F.	Kips per Linear Foot
Km	Kilometer
KO	Knock Out
K.S.F.	Kips per Square Foot
K.S.I.	Kips per Square Inch
kV	Kilovolt
kVA	Kilovolt Ampere
kVAR	Kilovar (Reactance)
KW	Kilowatt
KWh	Kilowatt-hour
L	Labor Only; Length; Long; Medium Wall Copper Tubing
Lab.	Labor
lat	Latitude
Lath.	Lather
Lav.	Lavatory
lb.; #	Pound
L.B., LB	Load Bearing; L Conduit Body
L. & E.	Labor & Equipment
lb./hr.	Pounds per Hour
lb./L.F.	Pounds per Linear Foot
lbf/sq.in.	Pound-force per Square Inch
L.C.L.	Less than Carload Lot
L.C.Y.	Loose Cubic Yard
Ld.	Load
LE	Lead Equivalent
LED	Light Emitting Diode
L.F.	Linear Foot
L.F. Hdr	Linear Feet of Header
L.F. Nose	Linear Foot of Stair Nosing
L.F. Rsr	Linear Foot of Stair Riser
Lg.	Long; Length; Large
L & H	Light and Heat
LH	Long Span Bar Joist
L.H.	Labor Hours
L.L., LL	Live Load
L.L.D.	Lamp Lumen Depreciation
lm	Lumen
lm/sf	Lumen per Square Foot
lm/W	Lumen per Watt
LOA	Length Over All
log	Logarithm
L-O-L	Lateralolet
long.	Longitude
L.P., LP	Liquefied Petroleum; Low Pressure
L.P.F.	Low Power Factor
LR	Long Radius
L.S.	Lump Sum
Lt.	Light
Lt. Ga.	Light Gauge
L.T.L.	Less than Truckload Lot
Lt. Wt.	Lightweight
L.V.	Low Voltage
M	Thousand; Material; Male; Light Wall Copper Tubing
M^2CA	Meters Squared Contact Area
m/hr.; M.H.	Man-hour
mA	Milliampere
Mach.	Machine
Mag. Str.	Magnetic Starter
Maint.	Maintenance
Marb.	Marble Setter
Mat; Mat'l.	Material
Max.	Maximum
MBF	Thousand Board Feet
MBH	Thousand BTU's per hr.
MC	Metal Clad Cable
MCC	Motor Control Center
M.C.F.	Thousand Cubic Feet
MCFM	Thousand Cubic Feet per Minute
M.C.M.	Thousand Circular Mils
MCP	Motor Circuit Protector
MD	Medium Duty
MDF	Medium-density fibreboard
M.D.O.	Medium Density Overlaid
Med.	Medium
MF	Thousand Feet
M.F.B.M.	Thousand Feet Board Measure
Mfg.	Manufacturing
Mfrs.	Manufacturers
mg	Milligram
MGD	Million Gallons per Day
MGPH	Million Gallons per Hour
MH, M.H.	Manhole; Metal Halide; Man-Hour
MHz	Megahertz
Mi.	Mile
MI	Malleable Iron; Mineral Insulated
MIPS	Male Iron Pipe Size
mj	Mechanical Joint
m	Meter
mm	Millimeter
Mill.	Millwright
Min., min.	Minimum, Minute

Abbreviations

Misc.	Miscellaneous	PCM	Phase Contrast Microscopy	SBS	Styrene Butadiere Styrene
ml	Milliliter, Mainline	PDCA	Painting and Decorating Contractors of America	SC	Screw Cover
M.L.F.	Thousand Linear Feet			SCFM	Standard Cubic Feet per Minute
Mo.	Month	P.E., PE	Professional Engineer; Porcelain Enamel; Polyethylene; Plain End	Scaf.	Scaffold
Mobil.	Mobilization			Sch., Sched.	Schedule
Mog.	Mogul Base			S.C.R.	Modular Brick
MPH	Miles per Hour	P.E.C.I.	Porcelain Enamel on Cast Iron	S.D.	Sound Deadening
MPT	Male Pipe Thread	Perf.	Perforated	SDR	Standard Dimension Ratio
MRGWB	Moisture Resistant Gypsum Wallboard	PEX	Cross Linked Polyethylene	S.E.	Surfaced Edge
		Ph.	Phase	Sel.	Select
MRT	Mile Round Trip	P.I.	Pressure Injected	SER, SEU	Service Entrance Cable
ms	Millisecond	Pile.	Pile Driver	S.F.	Square Foot
M.S.F.	Thousand Square Feet	Pkg.	Package	S.F.C.A.	Square Foot Contact Area
Mstz.	Mosaic & Terrazzo Worker	Pl.	Plate	S.F. Flr.	Square Foot of Floor
M.S.Y.	Thousand Square Yards	Plah.	Plasterer Helper	S.F.G.	Square Foot of Ground
Mtd., mtd., mtd	Mounted	Plas.	Plasterer	S.F. Hor.	Square Foot Horizontal
Mthe.	Mosaic & Terrazzo Helper	plf	Pounds Per Linear Foot	SFR	Square Feet of Radiation
Mtng.	Mounting	Pluh.	Plumber Helper	S.F. Shlf.	Square Foot of Shelf
Mult.	Multi; Multiply	Plum.	Plumber	S4S	Surface 4 Sides
MUTCD	Manual on Uniform Traffic Control Devices	Ply.	Plywood	Shee.	Sheet Metal Worker
		p.m.	Post Meridiem	Sin.	Sine
M.V.A.	Million Volt Amperes	Pntd.	Painted	Skwk.	Skilled Worker
M.V.A.R.	Million Volt Amperes Reactance	Pord.	Painter, Ordinary	SL	Saran Lined
MV	Megavolt	pp	Pages	S.L.	Slimline
MW	Megawatt	PP, PPL	Polypropylene	Sldr.	Solder
MXM	Male by Male	P.P.M.	Parts per Million	SLH	Super Long Span Bar Joist
MYD	Thousand Yards	Pr.	Pair	S.N.	Solid Neutral
N	Natural; North	P.E.S.B.	Pre-engineered Steel Building	SO	Stranded with oil resistant inside insulation
nA	Nanoampere	Prefab.	Prefabricated		
NA	Not Available; Not Applicable	Prefin.	Prefinished	S-O-L	Socketolet
N.B.C.	National Building Code	Prop.	Propelled	sp	Standpipe
NC	Normally Closed	PSF, psf	Pounds per Square Foot	S.P.	Static Pressure; Single Pole; Self-Propelled
NEMA	National Electrical Manufacturers Assoc.	PSI, psi	Pounds per Square Inch		
		PSIG	Pounds per Square Inch Gauge	Spri.	Sprinkler Installer
NEHB	Bolted Circuit Breaker to 600V.	PSP	Plastic Sewer Pipe	spwg	Static Pressure Water Gauge
NFPA	National Fire Protection Association	Pspr.	Painter, Spray	S.P.D.T.	Single Pole, Double Throw
NLB	Non-Load-Bearing	Psst.	Painter, Structural Steel	SPF	Spruce Pine Fir; Sprayed Polyurethane Foam
NM	Non-Metallic Cable	P.T.	Potential Transformer		
nm	Nanometer	P. & T.	Pressure & Temperature	S.P.S.T.	Single Pole, Single Throw
No.	Number	Ptd.	Painted	SPT	Standard Pipe Thread
NO	Normally Open	Ptns.	Partitions	Sq.	Square; 100 Square Feet
N.O.C.	Not Otherwise Classified	Pu	Ultimate Load	Sq. Hd.	Square Head
Nose.	Nosing	PVC	Polyvinyl Chloride	Sq. In.	Square Inch
NPT	National Pipe Thread	Pvmt.	Pavement	S.S.	Single Strength; Stainless Steel
NQOD	Combination Plug-on/Bolt on Circuit Breaker to 240V.	PRV	Pressure Relief Valve	S.S.B.	Single Strength B Grade
		Pwr.	Power	sst, ss	Stainless Steel
N.R.C., NRC	Noise Reduction Coefficient/ Nuclear Regulator Commission	Q	Quantity Heat Flow	Sswk.	Structural Steel Worker
		Qt.	Quart	Sswl.	Structural Steel Welder
N.R.S.	Non Rising Stem	Quan., Qty.	Quantity	St.; Stl.	Steel
ns	Nanosecond	Q.C.	Quick Coupling	STC	Sound Transmission Coefficient
NTP	Notice to Proceed	r	Radius of Gyration	Std.	Standard
nW	Nanowatt	R	Resistance	Stg.	Staging
OB	Opposing Blade	R.C.P.	Reinforced Concrete Pipe	STK	Select Tight Knot
OC	On Center	Rect.	Rectangle	STP	Standard Temperature & Pressure
OD	Outside Diameter	recpt.	Receptacle	Stpi.	Steamfitter, Pipefitter
O.D.	Outside Dimension	Reg.	Regular	Str.	Strength; Starter; Straight
ODS	Overhead Distribution System	Reinf.	Reinforced	Strd.	Stranded
O.G.	Ogee	Req'd.	Required	Struct.	Structural
O.H.	Overhead	Res.	Resistant	Sty.	Story
O&P	Overhead and Profit	Resi.	Residential	Subj.	Subject
Oper.	Operator	RF	Radio Frequency	Subs.	Subcontractors
Opng.	Opening	RFID	Radio-frequency Identification	Surf.	Surface
Orna.	Ornamental	Rgh.	Rough	Sw.	Switch
OSB	Oriented Strand Board	RGS	Rigid Galvanized Steel	Swbd.	Switchboard
OS&Y	Outside Screw and Yoke	RHW	Rubber, Heat & Water Resistant; Residential Hot Water	S.Y.	Square Yard
OSHA	Occupational Safety and Health Act			Syn.	Synthetic
		rms	Root Mean Square	S.Y.P.	Southern Yellow Pine
Ovhd.	Overhead	Rnd.	Round	Sys.	System
OWG	Oil, Water or Gas	Rodm.	Rodman	t.	Thickness
Oz.	Ounce	Rofc.	Roofer, Composition	T	Temperature; Ton
P.	Pole; Applied Load; Projection	Rofp.	Roofer, Precast	Tan	Tangent
p.	Page	Rohe.	Roofer Helpers (Composition)	T.C.	Terra Cotta
Pape.	Paperhanger	Rots.	Roofer, Tile & Slate	T & C	Threaded and Coupled
P.A.P.R.	Powered Air Purifying Respirator	R.O.W.	Right of Way	T.D.	Temperature Difference
PAR	Parabolic Reflector	RPM	Revolutions per Minute	TDD	Telecommunications Device for the Deaf
P.B., PB	Push Button	R.S.	Rapid Start		
Pc., Pcs.	Piece, Pieces	Rsr	Riser	T.E.M.	Transmission Electron Microscopy
P.C.	Portland Cement; Power Connector	RT	Round Trip	temp	Temperature, Tempered, Temporary
P.C.F.	Pounds per Cubic Foot	S.	Suction; Single Entrance; South	TFFN	Nylon Jacketed Wire

Abbreviations

Abbreviation	Meaning
TFE	Tetrafluoroethylene (Teflon)
T. & G.	Tongue & Groove; Tar & Gravel
Th., Thk.	Thick
Thn.	Thin
Thrded	Threaded
Tilf.	Tile Layer, Floor
Tilh.	Tile Layer, Helper
THHN	Nylon Jacketed Wire
THW.	Insulated Strand Wire
THWN	Nylon Jacketed Wire
T.L., TL	Truckload
T.M.	Track Mounted
Tot.	Total
T-O-L	Threadolet
tmpd	Tempered
TPO	Thermoplastic Polyolefin
T.S.	Trigger Start
Tr.	Trade
Transf.	Transformer
Trhv.	Truck Driver, Heavy
Trlr	Trailer
Trlt.	Truck Driver, Light
TTY	Teletypewriter
TV	Television
T.W.	Thermoplastic Water Resistant Wire
UCI	Uniform Construction Index
UF	Underground Feeder
UGND	Underground Feeder
UHF	Ultra High Frequency
U.I.	United Inch
U.L., UL	Underwriters Laboratory
Uld.	Unloading
Unfin.	Unfinished
UPS	Uninterruptible Power Supply
URD	Underground Residential Distribution
US	United States
USGBC	U.S. Green Building Council
USP	United States Primed
UTMCD	Uniform Traffic Manual For Control Devices
UTP	Unshielded Twisted Pair
V	Volt
VA	Volt Amperes
VAT	Vinyl Asbestos Tile
V.C.T.	Vinyl Composition Tile
VAV	Variable Air Volume
VC	Veneer Core
VDC	Volts Direct Current
Vent.	Ventilation
Vert.	Vertical
V.F.	Vinyl Faced
V.G.	Vertical Grain
VHF	Very High Frequency
VHO	Very High Output
Vib.	Vibrating
VLF	Vertical Linear Foot
VOC	Volatile Organic Compound
Vol.	Volume
VRP	Vinyl Reinforced Polyester
W	Wire; Watt; Wide; West
w/	With
W.C., WC	Water Column; Water Closet
W.F.	Wide Flange
W.G.	Water Gauge
Wldg.	Welding
W. Mile	Wire Mile
W-O-L	Weldolet
W.R.	Water Resistant
Wrck.	Wrecker
WSFU	Water Supply Fixture Unit
W.S.P.	Water, Steam, Petroleum
WT., Wt.	Weight
WWF	Welded Wire Fabric
XFER	Transfer
XFMR	Transformer
XHD	Extra Heavy Duty
XHHW	Cross-Linked Polyethylene Wire
XLPE	Insulation
XLP	Cross-linked Polyethylene
Xport	Transport
Y	Wye
yd	Yard
yr	Year
Δ	Delta
%	Percent
~	Approximately
Ø	Phase; diameter
@	At
#	Pound; Number
<	Less Than
>	Greater Than
Z	Zone

RESIDENTIAL COST ESTIMATE

OWNER'S NAME: APPRAISER:

RESIDENCE ADDRESS: PROJECT:

CITY, STATE, ZIP CODE: DATE:

CLASS OF CONSTRUCTION
- ☐ ECONOMY
- ☐ AVERAGE
- ☐ CUSTOM
- ☐ LUXURY

RESIDENCE TYPE
- ☐ 1 STORY
- ☐ 1-1/2 STORY
- ☐ 2 STORY
- ☐ 2-1/2 STORY
- ☐ 3 STORY
- ☐ BI-LEVEL
- ☐ TRI-LEVEL

CONFIGURATION
- ☐ DETACHED
- ☐ TOWN/ROW HOUSE
- ☐ SEMI-DETACHED

OCCUPANCY
- ☐ ONE STORY
- ☐ TWO FAMILY
- ☐ THREE FAMILY
- ☐ OTHER ____________

EXTERIOR WALL SYSTEM
- ☐ WOOD SIDING—WOOD FRAME
- ☐ BRICK VENEER—WOOD FRAME
- ☐ STUCCO ON WOOD FRAME
- ☐ PAINTED CONCRETE BLOCK
- ☐ SOLID MASONRY (AVERAGE & CUSTOM)
- ☐ STONE VENEER—WOOD FRAME
- ☐ SOLID BRICK (LUXURY)
- ☐ SOLID STONE (LUXURY)

*LIVING AREA (Main Building)	
First Level	____________ S.F.
Second Level	____________ S.F.
Third Level	____________ S.F.
Total	____________ S.F.

*LIVING AREA (Wing or Ell) ( )	
First Level	____________ S.F.
Second Level	____________ S.F.
Third Level	____________ S.F.
Total	____________ S.F.

*LIVING AREA (Wing or Ell) ( )	
First Level	____________ S.F.
Second Level	____________ S.F.
Third Level	____________ S.F.
Total	____________ S.F.

*Basement Area is not part of living area.

MAIN BUILDING	COSTS PER S.F. LIVING AREA
Cost per Square Foot of Living Area, from Page ________	$
Basement Addition: ________ % Finished, ________ % Unfinished	+
Roof Cover Adjustment: ________ Type, Page ________ (Add or Deduct)	( )
Central Air Conditioning: ☐ Separate Ducts ☐ Heating Ducts, Page ________	+
Heating System Adjustment: ________ Type, Page ________ (Add or Deduct)	( )
Main Building: Adjusted Cost per S.F. of Living Area	$

MAIN BUILDING TOTAL COST	$ /S.F.	x	S.F.	x		=	$
	Cost per S.F. Living Area		Living Area		Town/Row House Multiplier (Use 1 for Detached)		TOTAL COST

WING OR ELL ( ) ________ STORY	COSTS PER S.F. LIVING AREA
Cost per Square Foot of Living Area, from Page ________	$
Basement Addition: ________ % Finished, ________ % Unfinished	+
Roof Cover Adjustment: ________ Type, Page ________ (Add or Deduct)	( )
Central Air Conditioning: ☐ Separate Ducts ☐ Heating Ducts, Page ________	+
Heating System Adjustment: ________ Type, Page ________ (Add or Deduct)	( )
Wing or Ell ( ): Adjusted Cost per S.F. of Living Area	$

WING OR ELL ( ) TOTAL COST	$ /S.F.	x	S.F.	=	$
	Cost per S.F. Living Area		Living Area		TOTAL COST

WING OR ELL ( ) ________ STORY	COSTS PER S.F. LIVING AREA
Cost per Square Foot of Living Area, from Page ________	$
Basement Addition: ________ % Finished, ________ % Unfinished	+
Roof Cover Adjustment: ________ Type, Page ________ (Add or Deduct)	( )
Central Air Conditioning: ☐ Separate Ducts ☐ Heating Ducts, Page ________	+
Heating System Adjustment: ________ Type, Page ________ (Add or Deduct)	( )
Wing or Ell ( ): Adjusted Cost per S.F. of Living Area	$

WING OR ELL ( ) TOTAL COST	$ /S.F.	x	S.F.	=	$
	Cost per S.F. Living Area		Living Area		TOTAL COST

TOTAL THIS PAGE ____________

Page 1 of 2

RESIDENTIAL COST ESTIMATE

Total Page 1			$
	QUANTITY	UNIT COST	
Additional Bathrooms: ________ Full, ________ Half			
Finished Attic: ________ Ft. x ________ Ft.	S.F.		+
Breezeway: ☐ Open ☐ Enclosed ________ Ft. x ________ Ft.	S.F.		+
Covered Porch: ☐ Open ☐ Enclosed ________ Ft. x ________ Ft.	S.F.		+
Fireplace: ☐ Interior Chimney ☐ Exterior Chimney ☐ No. of Flues ☐ Additional Fireplaces			+
Appliances:			+
Kitchen Cabinets Adjustment: (+/–)			
☐ Garage ☐ Carport: ________ Car(s) Description ________________ (+/–)			
Miscellaneous:			+

ADJUSTED TOTAL BUILDING COST $

REPLACEMENT COST	
ADJUSTED TOTAL BUILDING COST	$ ________
Site Improvements	
(A) Paving & Sidewalks	$ ________
(B) Landscaping	$ ________
(C) Fences	$ ________
(D) Swimming Pool	$ ________
(E) Miscellaneous	$ ________
TOTAL	$ ________
Location Factor	x ________
Location Replacement Cost	$ ________
Depreciation	–$ ________
LOCAL DEPRECIATED COST	$ ________

INSURANCE COST	
ADJUSTED TOTAL BUILDING COST	$ ________
Insurance Exclusions	
(A) Footings, Site Work, Underground Piping	–$ ________
(B) Architects' Fees	–$ ________
Total Building Cost Less Exclusion	$ ________
Location Factor	x ________
LOCAL INSURABLE REPLACEMENT COST	$ ________

SKETCH AND ADDITIONAL CALCULATIONS

Page 2 of 2

C

D

E

F

G

H

I

N

O

P

Q

R

S

T

U

V

W

X

Y

Z

Cost Data Selection Guide

The following table provides definitive information on the content of each cost data publication. The number of lines of data provided in each unit price or assemblies division, as well as the number of crews, is listed for each data set. The presence of other elements such as reference tables, square foot models, equipment rental costs, historical cost indexes, and city cost indexes, is also indicated. You can use the table to help select the RSMeans data set that has the quantity and type of information you most need in your work.

Unit Cost Divisions	Building Construction	Mechanical	Electrical	Commercial Renovation	Square Foot	Site Work Landsc.	Green Building	Interior	Concrete Masonry	Open Shop	Heavy Construction	Light Commercial	Facilities Construction	Plumbing	Residential
1	584	406	427	531	0	516	200	326	467	583	521	273	1056	416	178
2	779	278	86	735	0	995	207	397	218	778	737	479	1222	285	274
3	1744	340	230	1265	0	1536	1041	354	2273	1744	1929	537	2027	316	444
4	961	21	0	921	0	726	180	615	1159	929	616	534	1176	0	448
5	1889	158	155	1093	0	852	1787	1106	729	1889	1025	979	1906	204	746
6	2453	18	18	2111	0	110	589	1528	281	2449	123	2141	2125	22	2661
7	1596	215	128	1634	0	580	763	532	523	1593	26	1329	1697	227	1049
8	2140	80	3	2733	0	255	1140	1813	105	2142	0	2328	2966	0	1552
9	2107	86	45	1931	0	309	455	2193	412	2048	15	1756	2356	54	1521
10	1089	17	10	685	0	232	32	899	136	1089	34	589	1180	237	224
11	1097	201	166	541	0	135	56	925	29	1064	0	231	1117	164	110
12	548	0	2	298	0	219	147	1551	14	515	0	273	1574	23	217
13	744	149	158	253	0	366	125	254	78	720	267	109	760	115	104
14	273	36	0	223	0	0	0	257	0	273	0	12	293	16	6
21	130	0	41	37	0	0	0	296	0	130	0	121	668	688	259
22	1165	7559	160	1226	0	1573	1063	849	20	1154	1682	875	7506	9416	719
23	1198	7001	581	940	0	157	901	789	38	1181	110	890	5240	1918	485
25	0	0	14	14	0	0	0	0	0	0	0	0	0	0	0
26	1512	491	10455	1293	0	811	644	1159	55	1438	600	1360	10236	399	636
27	94	0	447	101	0	0	0	71	0	94	39	67	388	0	56
28	143	79	223	124	0	0	28	97	0	127	0	70	209	57	41
31	1511	733	610	807	0	3266	289	7	1218	1456	3282	605	1570	660	614
32	838	49	8	905	0	4475	355	406	315	809	1891	440	1752	142	487
33	1248	1080	534	252	0	3040	38	0	239	525	3090	128	1707	2089	154
34	107	0	47	4	0	190	0	0	31	62	221	0	136	0	0
35	18	0	0	0	0	327	0	0	0	18	442	0	84	0	0
41	62	0	0	33	0	8	0	22	0	61	31	0	68	14	0
44	75	79	0	0	0	0	0	0	0	0	0	0	75	75	0
46	23	16	0	0	0	274	261	0	0	23	264	0	33	33	0
48	8	0	36	2	0	0	21	0	0	8	15	8	21	0	8
Totals	26136	19092	14584	20692	0	20952	10322	16446	8340	24902	16960	16134	51148	17570	12993

Assem Div	Building Construction	Mechanical	Elec-trical	Commercial Renovation	Square Foot	Site Work Landscape	Assemblies	Green Building	Interior	Concrete Masonry	Heavy Construction	Light Commercial	Facilities Construction	Plumbing	Asm Div	Residential
A		15	0	188	164	577	598	0	0	536	571	154	24	0	1	378
B		0	0	848	2554	0	5661	56	329	1976	368	2094	174	0	2	211
C		0	0	647	954	0	1334	0	1641	146	0	844	251	0	3	588
D		1057	941	712	1858	72	2538	330	824	0	0	1345	1104	1088	4	851
E		0	0	86	261	0	301	0	5	0	0	258	5	0	5	391
F		0	0	0	114	0	143	0	0	0	0	114	0	0	6	357
G		527	447	318	312	3378	792	0	0	535	1349	205	293	677	7	307
															8	760
															9	80
															10	0
															11	0
															12	0
Totals		1599	1388	2799	6217	4027	11367	386	2799	3193	2288	5014	1851	1765		3923

Reference Section	Building Construction Costs	Mechanical	Electrical	Commercial Renovation	Square Foot	Site Work Landscape	Assem.	Green Building	Interior	Concrete Masonry	Open Shop	Heavy Construction	Light Commercial	Facilities Construction	Plumbing	Resi.
Reference Tables	yes	yes	yes	yes	no	yes	yes	yes	yes	yes	yes	yes	yes	yes	yes	yes
Models					111			25					50			28
Crews	582	582	582	561		582		582	582	582	560	582	560	561	582	560
Equipment Rental Costs	yes	yes	yes	yes		yes		yes	yes	yes	yes	yes	yes	yes	yes	yes
Historical Cost Indexes	yes	yes	yes	yes	yes	yes	yes	yes	yes	yes	yes	yes	yes	yes	yes	no
City Cost Indexes	yes	yes	yes	yes	yes	yes	yes	yes	yes	yes	yes	yes	yes	yes	yes	yes

RSMeans Data Online

Cloud-based access to North America's leading construction cost database.

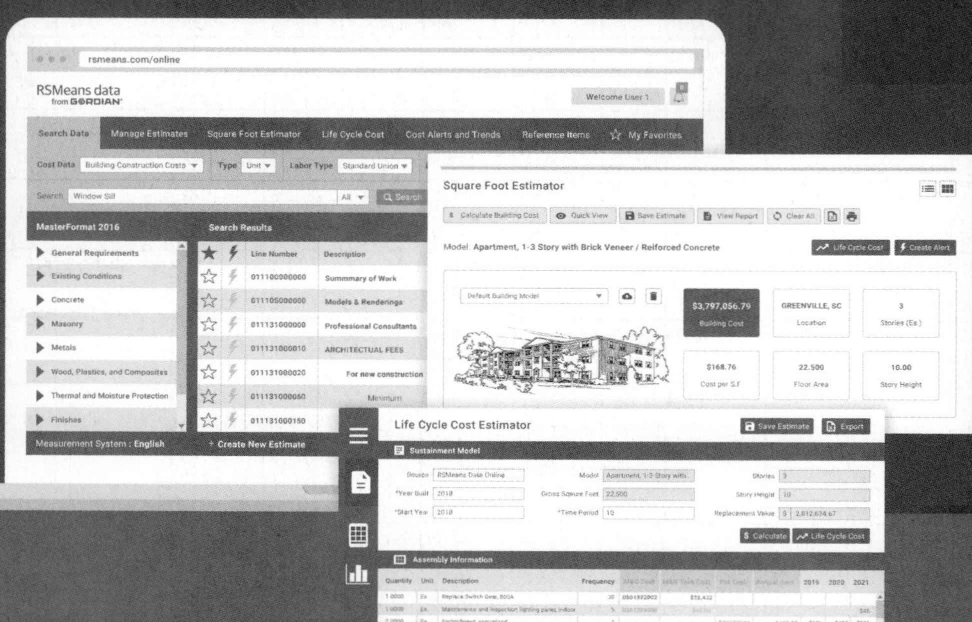

Plan your next budget with accurate prices and insight.

Estimate in great detail or at a conceptual level.

Validate unfamiliar costs or scopes of work.

Maintain existing buildings with schedules and costs.

RSMeans data from GORDIAN®

Learn more at rsmeans.com/online

RSMeans Data Online **Core**

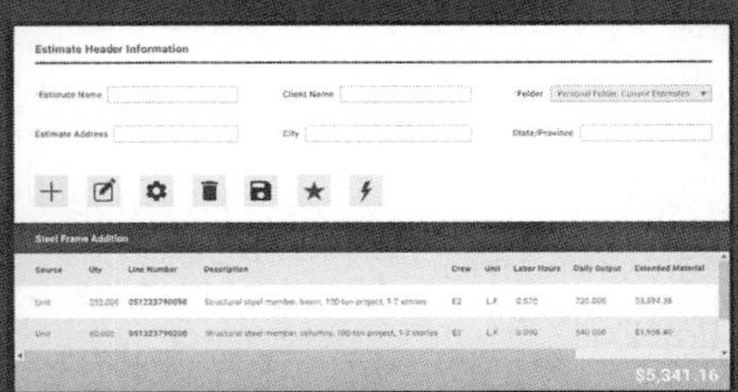

rsmeans.com/core

The Core tier of RSMeans Data Online provides reliable construction cost data along with the tools necessary to quickly access costs at the material or task level. Users can create unit line estimates with RSMeans data from Gordian and share them with ease from the web-based application.

Key Features:

- Search
- Estimate
- Share

Data Available:

16 datasets and packages curated by project type

RSMeans Data Online **Complete**

The Complete tier of RSMeans Data Online is designed to provide the latest in construction costs with comprehensive tools for projects of varying scopes. Unit, assembly or square foot price data is available 24/7 from the web-based application. Harness the power of RSMeans data from Gordian with expanded features and powerful tools like the square foot model estimator and cost trends analysis.

Key Features:

- Alerts
- Square Foot Estimator
- Trends

Data Available:

20 datasets and packages curated by project type

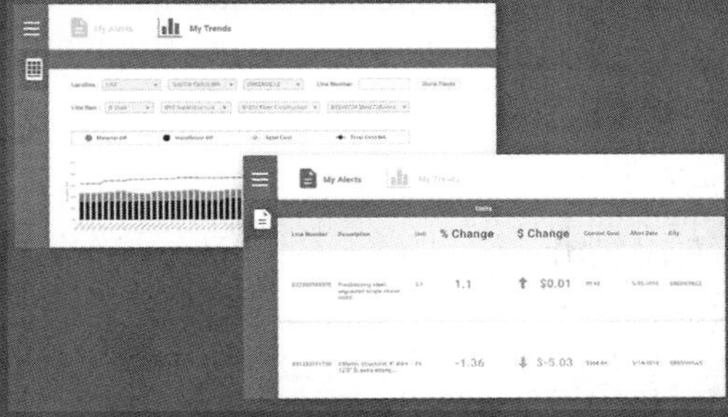

rsmeans.com/complete

RSMeans Data Online **Complete Plus**

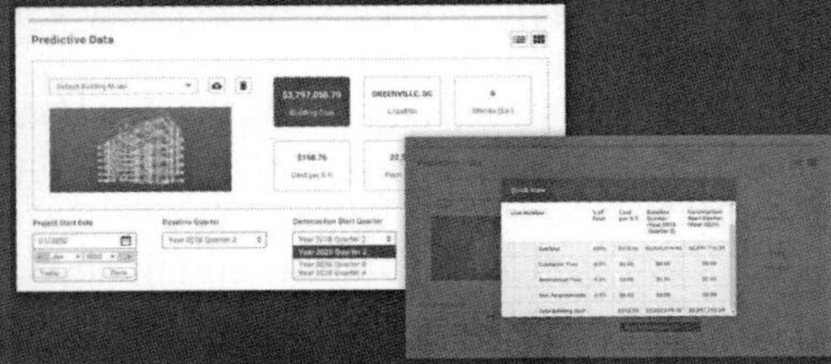

rsmeans.com/completeplus

The Complete Plus tier of RSMeans Data Online was developed to take project planning and estimating to the next level with predictive cost data. Users gain access to an exclusive set of tools and features and the most comprehensive database in the industry. Leverage everything RSMeans Data Online has to offer with the all-inclusive Complete Plus tier.

Key Features:

- Predictive Cost Data
- Life Cycle Costing
- Full History

Data Available:

Full library and renovation models

2019 Seminar Schedule

877-620-6245

Note: call for exact dates, locations, and details as some cities are subject to change.

Location	Dates
Seattle, WA	January and August
Dallas/Ft. Worth, TX	January
Austin, TX	February
Jacksonville, FL	February
Anchorage, AK	March and September
Las Vegas, NV	March
Washington, DC	April and September
Charleston, SC	April
Toronto	May
Denver, CO	May
San Francisco, CA	June
Bethesda, MD	June
Dallas, TX	September
Raleigh, NC	October
Baltimore, MD	November
Orlando, FL	November
San Diego, CA	December
San Antonio, TX	December

Gordian also offers a suite of online RSMeans data self-paced offerings.
Check our website RSMeans.com/products/training.aspx for more information.

Facilities Construction Estimating

In this two-day course, professionals working in facilities management can get help with their daily challenges to establish budgets for all phases of a project.

Some of what you'll learn:

- Determining the full scope of a project
- Identifying the scope of risks and opportunities
- Creative solutions to estimating issues
- Organizing estimates for presentation and discussion
- Special techniques for repair/remodel and maintenance projects
- Negotiating project change orders

Who should attend: facility managers, engineers, contractors, facility tradespeople, planners, and project managers.

Mechanical & Electrical Estimating

This two-day course teaches attendees how to prepare more accurate and complete mechanical/electrical estimates, avoid the pitfalls of omission and double-counting, and understand the composition and rationale within the RSMeans mechanical/electrical database.

Some of what you'll learn:

- The unique way mechanical and electrical systems are interrelated
- M&E estimates—conceptual, planning, budgeting, and bidding stages
- Order of magnitude, square foot, assemblies, and unit price estimating
- Comparative cost analysis of equipment and design alternatives

Who should attend: architects, engineers, facilities managers, mechanical and electrical contractors, and others who need a highly reliable method for developing, understanding, and evaluating mechanical and electrical contracts.

Construction Cost Estimating: Concepts and Practice

This one or two day introductory course to improve estimating skills and effectiveness starts with the details of interpreting bid documents and ends with the summary of the estimate and bid submission.

Some of what you'll learn:

- Using the plans and specifications to create estimates
- The takeoff process—deriving all tasks with correct quantities
- Developing pricing using various sources; how subcontractor pricing fits in
- Summarizing the estimate to arrive at the final number
- Formulas for area and cubic measure, adding waste and adjusting productivity to specific projects
- Evaluating subcontractors' proposals and prices
- Adding insurance and bonds
- Understanding how labor costs are calculated
- Submitting bids and proposals

Who should attend: project managers, architects, engineers, owners' representatives, contractors, and anyone who's responsible for budgeting or estimating construction projects.

Assessing Scope of Work for Facilities Construction Estimating

This two-day practical training program addresses the vital importance of understanding the scope of projects in order to produce accurate cost estimates for facility repair and remodeling.

Some of what you'll learn:

- Discussions of site visits, plans/specs, record drawings of facilities, and site-specific lists
- Review of CSI divisions, including means, methods, materials, and the challenges of scoping each topic
- Exercises in scope identification and scope writing for accurate estimating of projects
- Hands-on exercises that require scope, take-off, and pricing

Who should attend: corporate and government estimators, planners, facility managers, and others who need to produce accurate project estimates.

Maintenance & Repair Estimating for Facilities

This two-day course teaches attendees how to plan, budget, and estimate the cost of ongoing and preventive maintenance and repair for existing buildings and grounds.

Some of what you'll learn:

- The most financially favorable maintenance, repair, and replacement scheduling and estimating
- Auditing and value engineering facilities
- Preventive planning and facilities upgrading
- Determining both in-house and contract-out service costs
- Annual, asset-protecting M&R plan

Who should attend: facility managers, maintenance supervisors, buildings and grounds superintendents, plant managers, planners, estimators, and others involved in facilities planning and budgeting.

Practical Project Management for Construction Professionals

In this two-day course, acquire the essential knowledge and develop the skills to effectively and efficiently execute the day-to-day responsibilities of the construction project manager.

Some of what you'll learn:

- General conditions of the construction contract
- Contract modifications: change orders and construction change directives
- Negotiations with subcontractors and vendors
- Effective writing: notification and communications
- Dispute resolution: claims and liens

Who should attend: architects, engineers, owners' representatives, and project managers.

Life Cycle Cost Estimating for Facility Asset Managers

Life Cycle Cost Estimating will take the attendee through choosing the correct RSMeans database to use and then correctly applying RSMeans data to their specific life cycle application. Conceptual estimating through RSMeans new building models, conceptual estimating of major existing building projects through RSMeans renovation models, pricing specific renovation elements, estimating repair, replacement and preventive maintenance costs today and forward up to 30 years will be covered.

Some of what you'll learn:

- Cost implications of managing assets
- Planning projects and initial & life cycle costs
- How to use RSMeans data online

Who should attend: facilities owners and managers and anyone involved in the financial side of the decision making process in the planning, design, procurement, and operation of facility real assets.

Please bring a laptop with ability to access the internet.

Building Systems and the Construction Process

This one-day course was written to assist novices and those outside the industry in obtaining a solid understanding of the construction process - from both a building systems and construction administration approach.

Some of what you'll learn:

- Various systems used and how components come together to create a building
- Start with foundation and end with the physical systems of the structure such as HVAC and Electrical
- Focus on the process from start of design through project closeout

This training session requires you to bring a laptop computer to class.

Who should attend: building professionals or novices to help make the crossover to the construction industry; suited for anyone responsible for providing high level oversight on construction projects.

Training for our Online Estimating Solution

Construction estimating is vital to the decision-making process at each state of every project. Our online solution works the way you do. It's systematic, flexible and intuitive. In this one-day class you will see how you can estimate any phase of any project faster and better.

Some of what you'll learn:

- Customizing our online estimating solution
- Making the most of RSMeans "Circle Reference" numbers
- How to integrate your cost data
- Generating reports, exporting estimates to MS Excel, sharing, collaborating and more

Also offered as a self-paced or on-site training program!

Training for our CD Estimating Solution

This one-day course helps users become more familiar with the functionality of the CD. Each menu, icon, screen, and function found in the program is explained in depth. Time is devoted to hands-on estimating exercises.

Some of what you'll learn:

- Searching the database using all navigation methods
- Exporting RSMeans data to your preferred spreadsheet format
- Viewing crews, assembly components, and much more
- Automatically regionalizing the database

This training session requires you to bring a laptop computer to class.

When you register for this course you will receive an outline for your laptop requirements.

Also offered as a self-paced or on-site training program!

Site Work Estimating with RSMeans data

This one-day program focuses directly on site work costs. Accurately scoping, quantifying, and pricing site preparation, underground utility work, and improvements to exterior site elements are often the most difficult estimating tasks on any project. Some of what you'll learn:

- Evaluation of site work and understanding site scope including: site clearing, grading, excavation, disposal and trucking of materials, backfill and compaction, underground utilities, paving, sidewalks, and seeding & planting.
- Unit price site work estimates—Correct use of RSMeans site work cost data to develop a cost estimate.
- Using and modifying assemblies—Save valuable time when estimating site work activities using custom assemblies.

Who should attend: Engineers, contractors, estimators, project managers, owner's representatives, and others who are concerned with the proper preparation and/or evaluation of site work estimates.

Please bring a laptop with ability to access the internet.

Facilities Estimating Using the CD

This two-day class combines hands-on skill-building with best estimating practices and real-life problems. You will learn key concepts, tips, pointers, and guidelines to save time and avoid cost oversights and errors.

Some of what you'll learn:

- Estimating process concepts
- Customizing and adapting RSMeans cost data
- Establishing scope of work to account for all known variables
- Budget estimating: when, why, and how
- Site visits: what to look for and what you can't afford to overlook
- How to estimate repair and remodeling variables

This training session requires you to bring a laptop computer to class.

Who should attend: facility managers, architects, engineers, contractors, facility tradespeople, planners, project managers, and anyone involved with JOC, SABRE, or IDIQ.

Registration Information

Register early to save up to $100!!!

Register 45+ days before date of a class and save $50 off each class. This savings cannot be combined with any other promotional or discounting of the regular price of classes!

How to register

By Phone

Register by phone at 877-620-6245

Online

Register online at
RSMeans.com/products/seminars.aspx

Note: Purchase Orders or Credits Cards are required to register.

Two-day seminar registration fee - $1,200.

One-Day Construction Cost Estimating or Building Systems and the Construction Process - $765.

Government pricing

All federal government employees save off the regular seminar price. Other promotional discounts cannot be combined with the government discount. Call 781-422-5115 for government pricing.

CANCELLATION POLICY:

If you are unable to attend a seminar, substitutions may be made at any time before the session starts by notifying the seminar registrar at 1-781-422-5115 or your sales representative.

If you cancel twenty-one (21) days or more prior to the seminar, there will be no penalty and your registration fees will be refunded. These cancellations must be received by the seminar registrar or your sales representative and will be confirmed to be eligible for cancellation.

If you cancel fewer than twenty-one (21) days prior to the seminar, you will forfeit the registration fee.

In the unfortunate event of an RSMeans cancellation, RSMeans will work with you to reschedule your attendance in the same seminar at a later date or will fully refund your registration fee. RSMeans cannot be responsible for any non-refundable travel expenses incurred by you or another as a result of your registration, attendance at, or cancellation of an RSMeans seminar.

Any on-demand training modules are not eligible for cancellation, substitution, transfer, return or refund.

AACE approved courses

Many seminars described and offered here have been approved for 14 hours (1.4 recertification credits) of credit by the AACE International Certification Board toward meeting the continuing education requirements for recertification as a Certified Cost Engineer/Certified Cost Consultant.

AIA Continuing Education

We are registered with the AIA Continuing Education System (AIA/CES) and are committed to developing quality learning activities in accordance with the CES criteria. Many seminars meet the AIA/CES criteria for Quality Level 2. AIA members may receive 14 learning units (LUs) for each two-day RSMeans course.

Daily course schedule

The first day of each seminar session begins at 8:30 a.m. and ends at 4:30 p.m. The second day begins at 8:00 a.m. and ends at 4:00 p.m. Participants are urged to bring a hand-held calculator since many actual problems will be worked out in each session.

Continental breakfast

Your registration includes the cost of a continental breakfast and a morning and afternoon refreshment break. These informal segments allow you to discuss topics of mutual interest with other seminar attendees. (You are free to make your own lunch and dinner arrangements.)

Hotel/transportation arrangements

We arrange to hold a block of rooms at most host hotels. To take advantage of special group rates when making your reservation, be sure to mention that you are attending the RSMeans Institute data seminar. You are, of course, free to stay at the lodging place of your choice. (Hotel reservations and transportation arrangements should be made directly by seminar attendees.)

Important

Class sizes are limited, so please register as soon as possible.

Note: Pricing subject to change.